Building Energy and Environment

Building Energy and Environment

Editors

Shi-Jie Cao
Wei Feng

MDPI

Basel • Beijing • Wuhan • Barcelona • Belgrade • Novi Sad • Cluj • Manchester

Editors
Shi-Jie Cao
Southeast University
Nanjing, China

Wei Feng
Chinese Academy of Sciences
Shenzhen, China

Editorial Office
MDPI
St. Alban-Anlage 66
4052 Basel, Switzerland

This is a reprint of articles from the Topic published online in the open access journals *Energies* (ISSN 1996-1073), *Buildings* (ISSN 2075-5309), and *Designs* (ISSN 2411-9660) (available at: https://www.mdpi.com/topics/building).

For citation purposes, cite each article independently as indicated on the article page online and as indicated below:

Lastname, A.A.; Lastname, B.B. Article Title. *Journal Name* **Year**, *Volume Number*, Page Range.

ISBN 978-3-0365-9624-2 (Hbk)
ISBN 978-3-0365-9625-9 (PDF)
doi.org/10.3390/books978-3-0365-9625-9

Contents

energies

MDPI

Article

eSCIFI: An Energy Saving Mechanism for WLANs Based on Machine Learning

Guilherme Henrique Apostolo [1,2,*], Flavia Bernardini [1], Luiz C. Schara Magalhães [2,3] and Débora C. Muchaluat-Saade [1,2]

[1] Institute of Computing, Universidade Federal Fluminense, Niterói 24210-240, Brazil; fcbernardini@ic.uff.br (F.B.); debora@midiacom.uff.br (D.C.M.-S.)
[2] MídiaCom Lab, Universidade Federal Fluminense, Niterói 24210-240, Brazil; schara@telecom.ufff.br
[3] Telecommunications Engineering Department, Universidade Federal Fluminense, Niterói 24210-240, Brazil
* Correspondence: guilherme_apostolo@midiacom.uff.br

Abstract: As wireless local area networks grow in size to provide access to users, power consumption becomes an important issue. Power savings in a large-scale Wi-Fi network, with low impact to user service, is undoubtedly desired. In this work, we propose and evaluate the eSCIFI energy saving mechanism for Wireless Local Area Networks (WLANs). eSCIFI is an energy saving mechanism that uses machine learning algorithms as occupancy demand estimators. The eSCIFI mechanism is designed to cope with a broader range of WLANs, which includes Wi-Fi networks such as the Fluminense Federal University (UFF) SCIFI network. The eSCIFI can cope with WLANs that cannot acquire data in a real time manner and/or possess a limited CPU power. The eSCIFI design also includes two clustering algorithms, named cSCIFI and cSCIFI+, that help to guarantee the network's coverage. eSCIFI uses those network clusters and machine learning predictions as input features to an energy state decision algorithm that then decides which Access Points (AP) can be switched off during the day. To evaluate eSCIFI performance, we conducted several trace-driven simulations comparing the eSCIFI mechanism using both clustering algorithms with other energy saving mechanisms found in the literature using the UFF SCIFI network traces. The results showed that eSCIFI mechanism using the cSCIFI+ clustering algorithm achieves the best performance and that it can save up to 64.32% of the UFF SCIFI network energy without affecting the user coverage.

Keywords: WLAN energy saving mechanism; machine learning; RoD strategy mechanisms; smart buildings; Wi-Fi networks

Citation: Apostolo, G.H.; Bernardini, F.; Magalhães, L.C.S.; Muchaluat-Saade, D.C. eSCIFI: An Energy Saving Mechanism for WLANs Based on Machine Learning. *Energies* **2022**, *15*, 462. https://doi.org/10.3390/en15020462

Academic Editors: Shi-Jie Cao and Wei Feng

Received: 13 December 2021
Accepted: 6 January 2022
Published: 10 January 2022

1. Introduction

The presence of Wireless Local Area Networks (WLANs) on shopping centers, convention centers, and commercial and university buildings has been increasing daily [1]. To cope with the increasing demand, new wireless Access Points (*APs*) are added to the network infrastructure in order to supply user demand with good Internet connection [2]. However, the deployment of new *APs* not only rises the network infrastructure maintenance cost, but also its operation costs [3]. These higher operation costs are mostly caused by energy consumption [4,5].

The energy consumption of large-scale wireless networks has raised concerns among researchers [3,4,6–9]. There are several studies in the literature proposing Resource on Demand (RoDs) strategies to improve the energy efficiency of those networks [7,10–14]. Wi-Fi RoD strategy management systems, or simply RoD strategy mechanisms, implement algorithms and policies that decide which *APs* should be switched off to save energy and which *APs* must stay switched on to cope with the traffic demands [1].

Some mechanisms use real time data acquisition or sophisticated RoD strategies that require great processing requirements to create their energy saving solutions [4,15,16].

However, there are wireless network scenarios where the Central Processing Unit (CPU) power is limited. This restriction makes the use of real-time occupancy's measurement and prediction impractical. Nevertheless, even those network scenarios could benefit from RoD strategy mechanisms and few to no adjustments would be required.

The Fluminense Federal University (UFF) SCIFI wireless network is a large-scale network developed by UFF, initially financed by RNP (Brazilian National Research and Education Network) [17]. The SCIFI network is a low-cost solution for large-scale wireless networks. Its implementation is open source and allows the control and management of those networks. The SCIFI network has two main components: The SCIFI smart controller; and the running APs, operating under the open source OpenWRT firmware [18].

It is possible to apply machine learning predictions to estimate APs occupancy demand in the UFF SCIFI wireless network scenario [19]. The key idea behind that is to use machine learning models predicted occupancy demand to control the power state of the APs during the day. The responsible for switching off APs according to the estimated demand for each time slot is the Wi-Fi network controller. The Wi-fi controller can do that by using RoD strategy mechanisms based on machine learning estimated demand. Machine Learning models are responsible for future occupancy demand estimations that our RoD mechanism bases its decisions on. Consequently, we need to conduct an analysis on the performance of machine learning models for our scenario.Some work use the Wi-Fi infrastructure to gather information about the Wi-Fi network occupancy history and use different classification and regression machine learning models to predict network usage [3,20–24]. However, to the best of our knowledge, none of them have investigated the combination of regression and classification predictions to improve the demand estimation accuracy or the combination of two RoD strategy algorithms to ensure client's association and the network minimum coverage for Wi-Fi networks.

This work proposes eSCIFI, an energy saving mechanism for WLAN. eSCIFI uses machine learning models to predict the wireless network future demand, therefore it can work in wireless networks where the controller's CPU power does not allow real time data acquisition to estimate this demand. eSCIFI uses two RoD strategy algorithms to ensure clienst' association and the network minimum coverage: the AP clustering algorithm and the double threshold algorithm. The eSCIFI mechanism can determine which AP should be active or turned off during certain moments of the day in order to cope with the actual network demand and also save energy.

The main contributions of this work are:

- An energy saving mechanism named eSCIFI for WLAN, which can work in scenarios where real data acquisition is not possible;
- Analysis of how eSCIFI can cope with the network demand while saving energy and the comparison of its results with other RoD strategy mechanisms found in the literature. When compared to related work, eSCIFI achieved the highest energy saving factor (64.32%) without affecting the user access coverage.

The remainder of this work is organized as follows. Section 2 presents the related work to WLAN energy saving mechanisms. Section 3 describes the eSCIFI energy saving mechanism solution proposed in our work. Section 4 covers the evaluation of the proposed eSCIFI energy saving mechanism. Finally, Section 5 concludes this work, pointing out some enhancements and applications that might be explored in future work.

2. Related Work

Based on the work of Budzisz et al. [25], Jardosh et al. [2] and Lorincz et al [26], we developed an extended taxonomy that helped us to compare distinct RoD strategy mechanisms for WLANs. Our taxonomy consists of seven non-overlapping categories, corresponding to the main characteristics of related work: (1) network type, (2) WLAN application scenario, (3) control scheme, (4) operation strategy, (5) metrics, (6) algorithm type, and (7) evaluation method.

Most related work have developed RoD strategy mechanisms for Wi-Fi (IEEE 802.11) networks. However, there are great contributions in the literature that developed RoD strategy mechanisms for mesh [27] and cellular networks [8,24,28,29]. Those wireless network types have distinct characteristics, but the strategies and algorithms used on their RoD strategy mechanisms are interchangeable and sometimes even overlapping. It is important to notice that an RoD strategy mechanism developed and tested for a specific wireless network can be used in other wireless network types. Therefore, the network type category does not mean any sort of limitation to the RoD strategy mechanism applicability, but only describes the type of network used as the motivation and experimental scenario.

Most of the RoD strategy mechanisms were developed for application scenarios where they depend on homogeneous WLANs to operate, such as [2,7,11,12,14]. In those cases, the RoD strategy mechanism is implemented to fully cope with the WLAN technology without depending on any other wireless networks that might work in that area to help to implement their energy saving strategies. However, there are some RoD strategy mechanisms that were designed to operate in heterogeneous WLAN scenarios such as [9,30,31]. In the heterogeneous WLAN application scenarios, the WLAN can rely on other wireless technologies such as Bluetooth or in a separate wake-up radio transceiver to detect user activity while the WLAN infrastructure is turned off. The RoD strategy mechanism developed for heterogeneous WLAN application scenarios can usually achieve higher energy saving rates without affecting their user Quality of Service (QoS), since there is always a supportive wireless network to detect new users instantly. However homogeneous networks are less complex in terms of deployment, control and management, due to their independent WLAN nature.

The control scheme category expresses how the RoD strategy mechanism implements its energy saving strategy. The control scheme can be centralized or distributed. RoD strategy mechanisms with centralized control scheme uses a central controller to supervise the network and send the commands to APs. Centralized control schemes are more common for large wireless networks since most of them already have a central controller and their APs usually are not powerful enough to implement the algorithms and calculations needed. However, the centralized control scheme can be subdivided into two categories depending whether the central controller is designed for a Software Defined Network (SDN) or not.

SDNs separate the control and data plane by introducing a centralized controller that is responsible for resolving flows forwarding policies and assigning them to the switches' forwarding tables [10]. Some related work [10,14–16] developed energy saving mechanisms for SDN-based networks with a centralized SDN controller. The use of SDN controllers allows those energy saving mechanisms to use some collected network information such as network topology and traffic usage easily. However, not every large scale WLAN controller is based on the SDN paradigm and therefore can not count on all its advantages.

There are some proposed energy saving mechanisms in our related work that do not consider the controller to be SDN-based [3,8,27,32]. Those energy saving mechanisms also work with a centralized control scheme, but with non-SDN controllers which make them a feasible solution to WLANs where not all SDN advantages are present. On the other hand, in a distributed network, the WLAN elements are all responsible for controlling their energy state and deciding whether they can be turned off or not. However, it is important to highlight that a distributed control scheme does not necessarily mean that each WLAN AP works independently of the other. In [33,34], the Wi-Fi APs implement an energy saving strategy without a central controller, but they use out-of-band communication between them to decide which APs can be turned off.

RoD strategy mechanisms can be classified into two operation strategies: demand-driven or schedule-driven. Demand-driven strategies collect real-time information from the WLAN resources to estimate user demand [2]. The advantage of these strategies is that they can generate an energy saving in the WLAN while satisfying the user demand. However demand-driven strategies have a higher CPU power cost due to the overhead of assessing user demands continuously [6]. Demand-driven strategies are more suitable in scenarios where the user demand may unpredictably vary over time such as in stadiums [7]. On the other hand, schedule-driven strategies use predefined schedules to produce its energy saving. These schedules can be obtained with machine learning models trained with WLAN historical usage data [3,12,29] or can be based on the administrator's experience [14]. The advantage of using schedule-driven strategies is their low CPU power requirements. Schedule-driven strategies are only suitable for scenarios where user demand is predictable, such as university networks [3,12,14].

The RoD strategy mechanisms can be divided into 4 metrics subsets according to the metrics they use to minimize the energy consumption. The most common and most intuitive metrics are the traffic metrics subset. The traffic metrics subset comprises any network traffic related metric such as number of associated users [3,12,32], throughput [8] or more sophisticated ones such as channel utilization [2]. Traffic metrics are often used and measured in a network, and therefore they are easily accessible, but they might not be enough to guarantee the QoS or coverage alone. Coverage metrics are used to ensure that the whole radio area network [2,6,14] and users [7,15,16] will be covered. Coverage implies that the RoD energy saving strategies will guarantee that all users can connect to at least one active radio. QoS metrics are often used in studies that try to minimize the impact on the user's service [13,28], but they also imply smaller savings or more complex algorithms to work. Energy metrics [4,26] consider the reducing energy quantitative for the analysis of switching on/off strategies. A clear implication is that the user's traffic or QoS constraints can not be met. One important thing to highlight is that every metric alone has its advantages and weaknesses, therefore most related work uses a combination of metrics to guarantee the user's demand will be met.

RoD strategy mechanisms can also be divided by the type of algorithm used for making the energy status decision for the WLAN resources based on the available metrics. Heuristic algorithms can rapidly determine a solution within reasonable time using reasonable resources [26]. As the name suggests, heuristic algorithms are based on heuristics solutions that are easier to implement and usually based on thresholds [6,12,29] or other metrics combination rules [30]. Heuristic algorithms are usually most suitable for WLAN scenarios where the CPU power and/or computational time required are low. On the other hand, optimization algorithms are based on different mathematical problems and solvers that guarantee the best possible solution to a specified problem [4]. Optimization algorithms require more time and resources to provide their solution and therefore are suitable for WLAN scenarios where the CPU power and/or computational time required are high. Related work show that optimization algorithms achieve better results when compared to heuristic ones [15,16], however, Lorincz et al. [26] concluded in their work that "heuristic algorithms can be valuable alternatives offering good solution in reasonable amount of time".

Lastly related work can be divided according to the experimental test made to evaluate their RoD strategy mechanism performance. Simulation tests are those that make use of simulation software such as Matlab [35], Scenargie [36] or NS-3 [14] to recreate their WLAN scenarios and evaluate performance. Trace-driven tests are those that use network traces to reproduce a real network scenario comparing how their network would respond to the changes in that scenario using distinct energy saving mechanism [3,12,24,29]. Testbed experiments are those where a real WLAN infrastructure is used, but a limited set of users and their behavior are simulated [1,7,37]. There are related work that refer to their tests as real network scenario tests [2,7], however, they do not analyze the real infrastructure in a regular usage scenario with undefined users or behaviors and therefore we classified them as testbed.

Table 1 compares related work to our proposed eSCIFI mechanism. It is important to highlight that eSCIFI can be used in a wider range of WLAN networks than most of the mechanisms presented in related work that have a centralized control scheme, since eSCIFI can cope with non-SDN-based wireless networks, does not have a high-CPU-power controller and cannot collect data in real time. Those characteristics make any energy saving mechanism that presents optimization algorithms or demand-driven strategies unpractical. However, it is worth to mention that, in WLAN scenarios that present those characteristics, eSCIFI can work normally, but it might not be the best practical solution since it might not take advantage of those characteristics.

The eSCIFI characteristics make it a feasible solution for our motivation and evaluation scenario once it allows the development of an energy saving mechanism that can cope with the UFF SCIFI pure Wi-Fi network characteristics. eSCIFI presents a centralized controlling scheme, a schedule-driven operation strategy based on machine learning, using heuristic algorithms, traffic and coverage metrics.

Therefore, we can summarize the eSCIFI key contributions and advantages as:

- Introduces a centralized non-SDN-based control scheme;
- Proposes lightweight heuristic algorithms based on machine learning;
- Does not require real-time data acquisition;
- Does not require high-CPU-power controllers.

Table 1. RoD strategy mechanism related work comparison.

Related Work	Network Type	WLAN Scenario	Control Scheme	Operating Strategy	Metrics	Algorithm Type	Evaluation Method
Capone et al. [27]	Mesh Networks	Homogeneous	Centralized	Demand-driven	Coverage and QoS	Optimization	Simulation
Chen et al. [10]	IEEE 802.11	Homogeneous	Centralized	Demand-driven	QoS	Optimization	Testbed
Chin et al. [11]	IEEE 802.11	Homogeneous	Centralized	Demand-driven	Traffic and Coverage	Optimization	Simulation
Dalmasso et al. [8]	Cellular Networks	Homogeneous	Centralized	Demand-driven	Traffic	Heuristic	Trace-driven
Debele et al. [1]	IEEE 802.11	Homogeneous	Centralized	Demand-driven	Traffic	Heuristic	Testbed
Donevski et al. [24]	Cellular Networks	Homogeneous	Centralized	Schedule-driven	Traffic	Heuristic	Trace-driven
Fang et al. [12]	IEEE 802.11	Homogeneous	Centralized	Schedule-driven	Traffic	Heuristic	Trace-driven
Ganji et al. [35]	IEEE 802.11	Homogeneous	Centralized	Demand-driven	Coverage	Optimization	Simulation
Gomez et al. [7]	IEEE 802.11	Homogeneous	Centralized	Demand-driven	Traffic and QoS	Heuristic	Testbed
Jardosh et al. [6]	IEEE 802.11	Homogeneous	Centralized	Demand-driven	Traffic and Coverage	Heuristic	Simulation
Jardosh et al. [2]	IEEE 802.11	Homogeneous	Centralized	Demand-driven	Traffic and Coverage	Heuristic	Testbed
Kumazoe et al. [33,34]	IEEE 802.11	Homogeneous	Distributed	Demand-driven	Traffic and QoS	Heuristic	Simulation
Lee et al. [13]	IEEE 802.11	Homogeneous	Centralized	Demand-driven	Coverage and QoS	Optimization	Simulation and Testbed
Liu et al. [28]	Cellular Networks	Homogeneous	Centralized	Demand-driven	QoS	Optimization	Simulation
Lorincz et al. [4]	IEEE 802.11	Homogeneous	Centralized	Demand-driven	Energy and Traffic	Optimization	Simulation
Lorincz et al. [26]	IEEE 802.11	Homogeneous	Centralized	Demand-driven	Energy and Traffic	Heuristic	Simulation
Lyu et al. [3]	IEEE 802.11	Homogeneous	Centralized	Schedule-driven	Traffic	Heuristic	Trace-driven
Marsan et al. [32]	IEEE 802.11	Homogeneous	Centralized	Schedule-driven	Traffic	Heuristic	Trace-driven
Nagareda et al. [36]	IEEE 802.11	Homogeneous	Centralized	Demand-driven	Traffic and Coverage	Heuristic	Simulation
Rossi et al. [9]	IEEE 802.11	Heterogeneous	Distributed	Demand-driven	Traffic and Coverage	Heuristic	Simulation and Testbed
Silva et al. [14]	IEEE 802.11	Homogeneous	Centralized	Schedule-driven	Traffic and Coverage	Heuristic	Simulation
Tanaka at al. [30]	IEEE 802.11	Heterogeneous	Centralized	Demand-driven	Traffic	Heuristic	Simulation
Vallero et al. [29]	Cellular Networks	Homogeneous	Centralized	Schedule-driven	Traffic	Heuristic	Trace-driven
Wu et al. [38]	IEEE 802.11	Homogeneous	Centralized	Demand-driven	Traffic	Optimization	Simulation
Xu et al. [15,16]	IEEE 802.11	Homogeneous	Centralized	Demand-driven	Traffic	Optimization	Simulation
Yaodong Zhang et al. [37]	IEEE 802.11	Heterogeneous	Centralized	Demand-driven	Traffic	Optimization	Testbed
Yomo et al. [31]	IEEE 802.11	Heterogeneous	Distributed	Demand-driven	Traffic	Heuristic	Simulation
eSCIFI	IEEE 802.11	Homogeneous	Centralized	Schedule-driven	Traffic and Coverage	Heuristic	Trace-driven

3. Proposed eSCIFI Mechanism

In our previous work [39], we created a dataset using real user data collected from a subset of APs of the UFF SCIFI network located in a specific building of the engineering campus (the H building). That dataset provides the occupancy estimations for the H building during a period of 6 months (from April 2018 to September 2018). These features makes our dataset one of the biggest, most recent [3,12,32], and the only publicly available to the best of our knowledge. From the occupancy analysis [39], it was possible to observe that most network APs at the H building are switched on despite being idle. That active idleness causes an unnecessary waste of energy. Therefore, an energy saving WLAN mechanism based on RoD strategies, or simply RoD strategy mechanisms, that effectively controls WLAN resources can help to prevent those energy waste while coping with the user demand.

This work proposes the eSCIFI energy saving mechanism for WLANs. eSCIFI uses machine learning prediction models and other RoD strategies to create an energy saving mechanism. eSCIFI can also work with non-SDN large wireless networks and/or large wireless networks where real-time data acquisition is not possible. Those possibilities make the eSCIFI a feasible solution for a greater number of wireless networks in use, especially university networks, such as the UFF SCIFI network, which was used for evaluating our proposal.

3.1. eScifi Mechanism Overview

Figure 1 shows eSCIFI main architectural components and its major steps, which are (i) the unified methodology; (ii) the hybrid model; (iii) heuristic algorithm.

The first step, shown in the left upper part of the figure, is to use our unified methodology to create the datasets and select the best regression and classification model configuration parameters. Later on, in the hybrid model, we combine the best trained regression and classification models selected in our unified methodology to give the future *AP* occupancy estimation. That occupancy estimation is used by our heuristic algorithm to define which *APs* should be turned on or off.

In the heuristic mechanism, we first extract the *AP* statistics from the dataset. Later on, the heuristic network clusters formation uses the *AP* neighborhood list and the *AP* statistics to create the network clusters that can guarantee a minimum network coverage. Finally, the energy state decision algorithm uses the defined network clusters and the *AP* occupancy estimation to decide which *APs* should be switched on/off to cope with the user demand. At the end of this process, our heuristic mechanism provides an energy scheduling of all *APs* in the network for an entire day that can guarantee a minimum coverage to the network while coping with the user demand. That way the eSCIFI mechanism needs to run only once a day to generate the energy scheduling of all *APs* in the network. Therefore, the eSCIFI mechanism can run at any moment of low activity in the network such as late night hours after midnight in our case. This scheme guarantees that eSCIFI can run at any network controller without burdening its processing capacity.

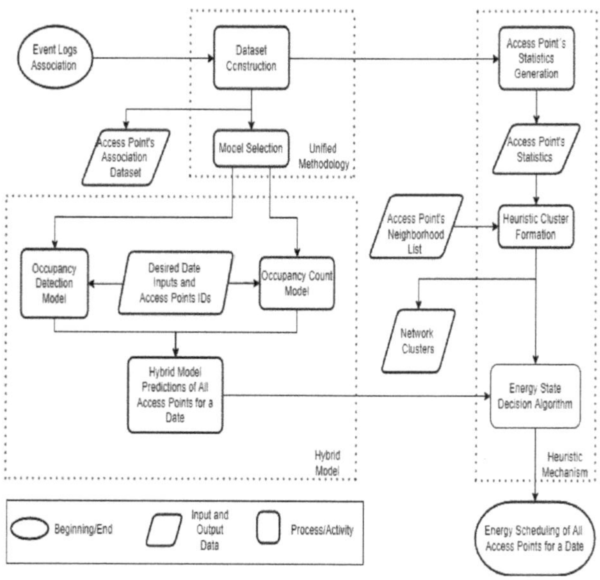

Figure 1. eSCIFI architecture.

3.2. Unified Methodology and Model Selection

The unified methodology proposed in [39] explains how the occupancy count (the amount of devices connected to an *AP*) and occupancy detection (if the *AP* is occupied or not) datasets were created. In summary, we have processed *AP* event logs to filter information about the association status between mobile stations and *APs*. Each day was divided into 144 time slots (10 min each), and for each time slot the number of associated devices was processed. This was computed for all the *APs* involved. The datasets show occupancy count and detection of 28 *APs* in a classroom building at UFF's Engineering Campus over a period of 6 months, from April to September 2018. Those datasets were crucial to extract the *AP*'s statistics that were necessary for the network clusters formation. They were also crucial to the model selection process.

The model selection process in our unified methodology compares several model configuration and hyperparameters in order to determine the best classification and regression models for our evaluation scenario. The evaluation involved the use of multiple classification (for occupancy detection) and regression (for occupancy count) models using a variety of configurations and algorithms. The following algorithms were used as classification models: Decision Tree (DT); K-NN; Random Forest (RF); and MultiLayer Perceptron neural network. As regression models, we used DT, K-NN, RF, XG optimized gradient boosting, support vector machine (SVM), stochastic gradient descent (SGD) algorithms and MultiLayer Perceptron neural network.

To evaluate these models, we applied a train/test split on our data where the dataset's association data from April to August were used for training, and the dataset's association data for the month of September were used for testing. We used several metrics to evaluate our classification (such as Accuracy and F-1 score) and regression (such as root mean square error and mean average percentage error) models performance. From the results shown in [39], it was possible to select the best classification and regression model for our scenario.

Results showed that the best classification and regression models were a single-label regressor and classifier trained using the decision tree algorithm in a collective manner where only one classifier and regressor were trained to predict the occupancy of all *APs* based on the previous data of all *APs*. The output feature is the occupancy estimation for the specific time slot. The models used the following attributes: Month, Day, Day of the Week, Holiday, Access Point Id, Hour, and Minute.

Therefore only one classifier and one regressor is needed for our scenario. Those classification and regression models used the decision tree machine learning algorithm and three input features (*AP* identification, day of the week and holiday). *AP* identification (APid) carries the access point identification number, day of the week indicates the respective week day and holiday indicates if the day is a day with lectures or not. Those are the machine learning classification and regression models selected and they will be used on the hybrid model to provide future usage predictions for the H building UFF SCIFI Wi-Fi network in our evaluation.

3.3. Hybrid Model

The results in [39] showed that even the best regression model has significant Root Mean Squared Percentage Error for a specific time slot tj ($RMSPE_{tj}$) values during night and morning time slots, but the $RMSPE_{tj}$ values for time slots after midday decrease. On the other hand, the best classification model has relatively higher accuracy for a specific time slot t_j (A_{t_j}) values for night and morning time slots than for the rest of the day. Therefore we propose a hybrid model. The hybrid model combines the accuracy results given by the classification models with the regression results given by the regression models in order to create a better occupancy count estimation. Considering CMR as the classification results matrix that shows the occupancy detection estimations provided by the classifier for the APs and RMR as the regression results matrix that shows the occupancy count estimations provided by the regressor, we can define that the hybrid model estimation HMR is the Hadamard product result between both CMR and RMR matrices. Equation (1) shows the Hadamard product that produces the hybrid model results matrix that is used as the demand estimation by our mechanism.

$$HMR = CMR \circ RMR \tag{1}$$

Figure 2 shows how the hybrid model demand prediction results are closer to the real demand than the regression model demand predictions for the month of September 2018. In fact, Figure 2 shows that the hybrid model results can reduce the over demand prediction that happened on the weekends (September 1, 2, 8, 9, 15, 16, 22, 23, 29, 30) and on the Brazil's Independence day public holiday (September 7).

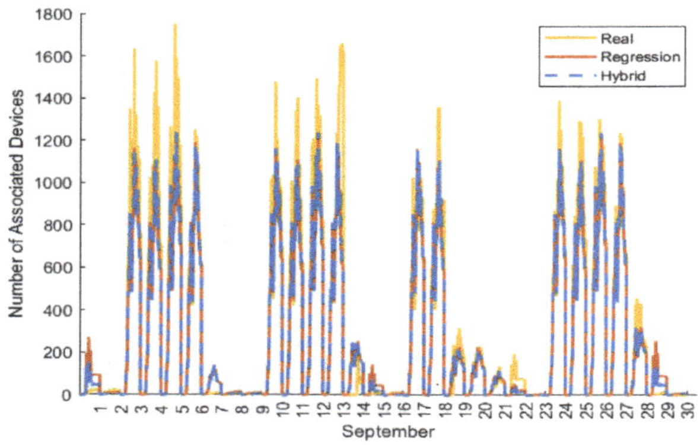

Figure 2. Hybrid model results compared with the real demand and the demand given by the regression results for the whole month of September.

The Hybrid model created only uses the APid, day of the week and holiday attributes as input features. Consequently, there are only 14 possible demand estimations for a specific AP (one for each regular day of the week and one for each holiday on these days). Therefore, we decided to compare the results of our hybrid model with a mean estimator. The occupancy count prediction provided by the mean estimator for a specific set of input features (APid, day of the week and holiday) is the average occupancy count of that specific set of input features in the association history. We compared the results of this mean estimator with the results of our hybrid model. Table 2 shows that the hybrid model had better Root Mean Squared Error ($\overline{RMSE}$), overall Root Mean Percentage Error ($\overline{RMSPE}$) and overall Mean Absolute Percentage Error ($\overline{MAPE}$) results when compared to the mean estimator model. Those better results shown in Table 2 can be explained by the fact that the hybrid model has reduced the error predictions that happened on weekends and on public holidays when compared to the mean estimation results. Those reduced demands on weekend and on public holidays were more significant than the errors caused in night time slots by the hybrid model results, and therefore the overall $\overline{RMSE}$, $\overline{RMSPE}$ and $\overline{MAPE}$ results were better.

It is important to highlight that the difference between the results shown in Table 2 are not significant enough to prove that the hybrid model is a better regression prediction model than the pure regression model or mean estimator for all scenarios. The mean estimator results in our case scenario are very close to those achieved by the hybrid model. However, those results achieved by the mean estimator for our case scenario were only possible due the H building occupancy characteristics. The H building has only classrooms, so its occupation mainly occurs through lectures and exam applications. The lecture's schedule did not change drastically throughout the entire dataset which makes the occupancy behavior periodical and well behaved in our case. This behavior might not be common for other buildings in the university that have other room types, such as professor's offices or laboratories, or even other scenarios such as parks or malls. On other scenarios similar to ours, the mean estimator can be a viable option due to its simplicity. The use of the mean estimator does not impose any change to the eSCIFI operation. However, we decided to use the hybrid model since it has shown better results in our case scenario, specifically on weekends and holidays.

Table 2. Mean Estimator and Hybrid models performance results.

Metrics	Mean	Hybrid
$\overline{RMSE}$	8.4161	8.3996
$\overline{RMSPE}$	0.2977	0.2968
$\overline{MAPE}$	0.4189	0.4096

3.4. Heuristic Mechanism

The heuristic mechanism is responsible for providing the SCIFI APs energy state (on or off) schedule for a date. It is important to highlight that we only control the APs wireless interface energy state due to UFF SCIFI existing infrastructure that only allows us to control its energy state. However, in WLANs where the APs are connected to Power over Ethernet (PoE) switches, eSCIFI could normally control the energy state of the AP and not only its wireless interface.

Our heuristic mechanism has two main components: the heuristic cluster formation algorithm and the energy state decision algorithm. The clustering algorithm creates the AP clusters based on their neighborhood in order to guarantee the network coverage area to the clients. The energy state decision algorithm provides the energy state of all APs for a specific time slot and date based on the machine learning occupancy predictions and clusters. In the following sections, we detail the heuristic cluster formation algorithm and the energy state decision algorithm and its challenges.

3.4.1. Heuristic Cluster Formation: cSCIFI and cSCIFI+

Jardsoh et al. [2] proposed a clustering algorithm called green clustering. The idea behind the green clustering algorithm is to create clusters of APs that are in proximity of each other. Several APs in a large wireless network have overlapping coverage areas in order to cope with higher user demand. Those APs are in a spatially neighboring condition that allows one of them to provide coverage to the users of all APs in its vicinity. Therefore it is possible to create clusters of neighboring APs where any user within the cluster coverage is able to connect to the network as long as at least one AP in the cluster is turned on. We propose two heuristic cluster formation algorithms, cSCIFI (cluster SCIFI) and cSCIFI+ (cluster SCIFI+). Those clustering algorithms are based on the green clustering algorithm of Jardsoh et al. [2]. However, we introduced some basic changes to improve the cSCIFI and cSCIFI+ clustering formation process.

Our clustering algorithms need two input features to work: the neighborhood list and the AP statistics. To create a neighborhood list, we need to define the vicinity criteria. Only APs that are considered neighbors can belong to the same cluster. Jardsoh et al. [2,6] have used the spatial distance between APs and the median number of beacon messages and the median signal strength of the beacons as vicinity criteria. In our cSCIFI and cSCIFI+ algorithms, we are going to use the APs' signal quality scan to define our vicinity criterion. The SCIFI network periodically runs a signal quality scan that informs the different signal quality values received from the other APs that a certain AP has scanned. The signal quality is a measurement that goes from 0 to 100 and takes into consideration the Received Signal Strength Indication (RSSI) and other network parameters. We considered APs with a measured signal quality above 50 to be neighbors. Therefore, the neighbors of an AP are: (i) the APs on the same side of the building and floor; (ii) the APs that are in rooms directly above and below (e.g., neighbors to the AP in room 303 are the APs in rooms 403 and 203). With the established vicinity criteria, we can determine which APs are neighbors and create a neighborhood set list for each AP.

We describe our cluster formation algorithms as follows. Consider V_i as the neighborhood set of AP_i, C as our cluster set, and C_i as the cluster formed starting from AP_i. In our cSCIFI clustering algorithm, we first start by selecting AP_s with the biggest neighborhood set, forming a new cluster C_s and adding AP_s to its newly formed cluster C_s . When AP_s is added to its cluster C_s, the cSCIFI algorithm also removes AP_s from all other AP neighborhood sets and update their number of neighbors. Then, the algorithm steps through all the APs in AP_s neighborhood set V_s and adds AP_h that has the biggest neighborhood set, as long as every new AP_h added to C_s is in the neighborhood set of all other APs already included in C_s. We call this the neighboring condition. As long as AP_s has APs on its neighborhood set V_s that satisfy the neighboring condition, those APs are added to cluster C_s and removed from the other AP neighborhood sets.

When there are no more APs in the AP_s neighborhood set or there are no more APs that satisfy the neighboring condition, the algorithm moves to the next AP with the biggest neighbor set and continues the cluster formation until there are no more APs left and the cluster set C is finished.

Algorithm 1 shows the cSCIFI cluster function code, where we can see that every AP will only be in one cluster and that every AP is on the vicinity of all other APs inside its cluster. The neighboring condition (line 5) allows any user in the cluster coverage area to connect to any of the powered-on APs, since they are all each other's neighbors.

Algorithm 1 *cSCIFI*

```
1: function Create_Cluster_cSCIFI (Cluster_Head, Cluster_head_list_of_neighbors):
2:    Cluster_auxiliary_list = [Cluster_Head]
3:    sort APs in Cluster_head_list_of_neighbors according with the number of
neighbors in their neighborhood list
4:    for AP in Cluster_head_list_of_neighbors:
5:       if AP in neighborhood list of all Cluster_auxiliary_list elements:
6:          add AP to Cluster_auxiliary_list
7:          remove AP from neighborhood list of all APs
8:    return Cluster_auxiliary_list
```

cSCIFI+ is simpler and more aggressive than cSCIFI. The cSCIFI+ clustering algorithm works like the cSCIFI, but now the APs added to a certain cluster C_i do not need to cope with the neighboring condition. In cSCIFI+, all neighbors in the AP_A neighborhood AP set V_A are added to cluster C_A.

cSCIFI+ guarantees that the size of cluster set C will be the smallest possible. However, users from a switched-off AP in the cluster can only connect to AP_A that initiated that cluster. Considering the clusters formed with cSCIFI, users from any AP can connect to other APs in that cluster, which might balance the load between the switched-on APs. Algorithm 2 shows the cSCIFI+ cluster function code, where we can see that only the AP that initiated the cluster formation can assure connection to all users from switched-off APs, which may sometimes cause congestion.

Algorithm 2 *cSCIFI+*

```
function Create_Cluster_cSCIFI+ (Cluster_Head, Cluster_head_list_of_neighbors):
   Cluster_auxiliary_list = [Cluster_Head]
   for AP in Cluster_head_list_of_neighbors:
      add AP to Cluster_auxiliary_list
      remove AP from neighborhood list of all APs
   return Cluster_auxiliary_list
```

The cSCIFI and cSCIFI+ greedy algorithms alone cannot guarantee that the best cluster set is formed in cases where there is a tie between APs. A solution would be creating all cluster possibilities, choosing each one of the tied APs as the first choice. After creating all possible sets C, we would select the one that has the minimum number of clusters. Those multiple cluster sets creation can cause a exponential growth in the execution time. Trying to minimize those problems, we simplified the cSCIFI and cSCIFI+ selection in cases of ties. The cSCIFI and cSCIFI+ will only create multiple cluster sets when there are ties between

APs that will be selected to initiate a cluster formation. This selection criteria will guarantee that only different clusters initiation will be taken into relevance and not all possible cluster internal formations, which will minimize the possible solution set.

In the cSCIFI algorithm, we also added another selection criterion for cases where there are ties between APs to be added to cluster C_i where an AP_i has already initiated it. In those cases, AP_j with the highest number of neighbors in the AP_i neighborhood set V_i is selected. APs with the same number of neighbors in their sets can generate different clusters, since some of their neighbors might not be in AP_i neighborhood set V_i. Therefore, in cases of ties, it is the best option to select AP_j that has the biggest number of matching neighbors to the APs in the neighborhood set V_i. This change in the internal cluster formation process guarantees that the next APs to be added will be the ones that will contribute to a bigger cluster size.

Those characteristics cited previously minimizes the execution time and guarantee that a possible cluster set C will be selected independent of their appearances on the cluster neighborhood list. This is an important advantage to our clustering algorithms when compared to the green clustering algorithm proposed by Jardosh et al. [2], since we do not need to worry about the AP order of appearance in the neighborhood list construction process.

The last characteristic of our clustering algorithms is the cluster head election. The cluster head is the AP that will be always switched on and will be responsible for guaranteeing the cluster coverage area. In cSCIFI+, the cluster head will always be the one that initiated the cluster formation. This AP is the only AP that can be the cluster head, since this is the only AP that has a guaranteed neighboring condition to all other APs in the cluster. On the other hand, the election of the clusters head in the cSCIFI algorithm can be more sophisticated since all APs in the clusters obey the neighboring condition. In clusters formed with cSCIFI, the cluster head will change throughout the day using the statistics of the APs. In those clusters, the average association of each AP is calculated for the night (0 a.m.–7 a.m.), morning (7 a.m.–1 p.m.) and afternoon/evening (1 p.m.–11 p.m. 59′) periods. The AP with the highest night average association will be selected to be the cluster head for the night period and so on.

3.4.2. Energy State Decision Algorithm

The energy state decision algorithm is responsible for providing the energy scheduling of all APs for a date. It runs once a day and uses the traffic demand estimated by the hybrid machine learning model (the user association number in our SCIFI network scenario) to calculate the cluster demand for specific moments of a day and then decide which APs in a cluster can be switched off. The energy state decision algorithm is the last step on the eSCIFI energy saving mechanism, and it is responsible for actively deciding which APs will be switched on or off and to provide the energy scheduling to the SCIFI controller. The SCIFI controller, based on this energy scheduling, will then control the AP wireless interface switching on and off for the specified periods.

Our energy state decision algorithm uses the RoD policy proposed in the work of Dalmasso et al. [8]. However, our energy state decision algorithm works using machine learning occupancy estimations instead of real traffic data and therefore presents some modification in the RoD policy design. This RoD policy has two main components: the time window and the double threshold criteria. The time window defines how long it will take before the algorithm reconfigures the AP's energy state. The time window size tw informs on which frequency the network will be reconfigured and also the demand estimation resolution. A small time window will allow the energy state decision algorithm to perceive short bursts in the traffic demand variations. On the other hand, a large time window will only perceive the average traffic where instant or momentarily bursts in the traffic demand will fade. At first, a smaller time window size seems always the best choice, however a smaller time window size means more rounds of energy state decisions will have to be made by the algorithm and that the controller will have to reconfigure the network more frequently. Related work [3,8,12,14,24] state that small time window sizes are not necessary. In fact, depending on the network traffic profile, those changes in the traffic demand can take hours to happen. Therefore, the selection of the time window size is a parameter that needs to be decided based on the network scenario. In Section 4.1, we will deeply discuss the selection of the time window size.

The main concept behind our energy saving strategy is moving the traffic demand from switched off APs to the cluster head AP or other switched on APs in the cluster that can handle them. In the work of Damalso et al. [8], the APs in a cluster can be switched off based on the actual traffic demand (real time traffic data) at the beginning of each time window. However, the eSCIFI mechanism uses machine learning models to estimate demand. Therefore, in our energy decision algorithm, the decisions made for each time window will take into consideration the demand estimated for its whole duration and not just the demand at the beginning of the time window.

All APs in the network have the same maximum user threshold T_{max} for a time window. This maximum user threshold T_{max} defines how much traffic (or how much associations in our case) the APs can handle for the duration of the time window. The cluster head of every cluster will always be switched on guaranteeing a traffic capacity of T_{max} for the cluster. In our energy state decision algorithm, the double threshold criteria defines which APs in a cluster can be switched off based on the traffic demand estimated by the machine learning hybrid model for the assessed time window. However, this energy state decision algorithm varies depending whether cSCIFI or cSCIFI+ algorithms are used.

In the cSCIFI algorithm, all APs in a cluster are neighbors between themselves. Therefore, in the cSCIFI case, the double threshold criteria defines that APs with estimated traffic demand below a minimum threshold T_{min} for the whole time window are switched off as long as the available traffic capacity provided by all APs that are switched on can handle their estimated traffic. Considering DM_i as the traffic demand of AP_i for a time window, d as the number of switched on APs, o as the number of switched off APs, $sum_{a=1}^{d} DM_a$ as the traffic demand of all d switched on APs and $sum_{a=1}^{o} DM_a$ as the traffic demand of all o switched off APs, we can define how our energy state decision algorithm decides if an AP_i will be switched off based on the double threshold criteria if the cSCIFI algorithm is used. Equation (2) shows the double threshold criteria, where the first criterion defines if the traffic demand is too low for the AP_i to be switched on and the second criterion defines if the cluster switched-on APs can handle the AP_i traffic. If there are more d switched-on APs in the cluster, the cluster maximum traffic capacity CCA increases to $(d+1)T_{max}$ because cSCIFI guarantees that all APs inside a cluster can provide connection to any mobile station trying to connect to any AP in the cluster.

$$\begin{cases} DM_i < T_{min} \\ CCA - (\sum_{a=1}^{d} DM_a + \sum_{b=1}^{o} DM_b) \geqslant DM_i, \quad \text{where } CCA = (d+1)T_{max} \end{cases} \tag{2}$$

On the other hand, in cSCIFI+, all APs in a cluster are neighbors only of the cluster head. Therefore, APs with estimated traffic demand below a minimum threshold T_{min} for the whole time window are switched off as long as the available traffic capacity provided by the cluster head can handle their estimated traffic. Equation (3) shows the double threshold criteria if cSCIFI+ is used, where the first criterion defines if the traffic demand is too low for AP_i to be switched on and the second criterion defines if the cluster head can handle the AP_i traffic. In cSCIFI+, the cluster maximum traffic capacity CCA is fixed to T_{max} because cSCIFI+ guarantees that only the cluster head can provide connection to any mobile station trying to connect to any AP in the cluster, except the AP itself.

$$\begin{cases} DM_i < T_{min} \\ CCA - (\sum_{a=1}^{d} DM_a + \sum_{b=1}^{o} DM_b) \geqslant DM_i, \quad \text{where } CCA = T_{max} \end{cases} \tag{3}$$

eSCIFI has several parameters that must be configured and that may depend on the network usage profile, such as the selection of the time window size and T_{min} value. In the next section, we evaluate how the different components in the eSCIFI architecture affect the mechanism energy saving capacity and the network coverage to its users. We also compare eSCIFI to other related work about energy saving mechanisms that are applicable in our evaluation scenario.

4. eSCIFI Evaluation

To evaluate how eSCIFI impacts on the network performance, we performed trace-driven simulations using the real association trace data collected from the UFF SCIFI network. Our trace-driven tests use UFF SCIFI's association traces to reproduce a real network scenario. The idea is to compare how the network would respond to the changes in that scenario using distinct energy saving mechanisms. A trace-driven test does not require using a network simulator as NS-3 for example. It allows estimating the metrics by simply inputting the real association traces to the eSCIFI mechanism and then evaluating if eSCIFI can cope with user demand while saving energy. To perform our simulations, we are going to use the association data collected for one week in September 2018 from the H building at UFF. The week used in our collected data is formed by a weekend (1 and 2 September 2018) and 5 weekdays from Monday to Friday (24–28 September 2018). The weekends used are apart from the weekday dates because there were not complete association history traces for the weekend before or after those weekdays. This might have happened for several reasons such as energy outages or network failures for example. However, the weekend (1 and 2 September 2018) contains the association data for all APs in the H building, and therefore will be used to represent Saturday and Sunday in our trace-driven simulations. We are also going to use the Brazil's Independence day public holiday (September 7) to compare and evaluate how eSCIFI impacts the network on holidays.

The work of [40] presents a mathematical formula, indicated in Equation (4), that allows us to determine the energy saving factor ESF achieved with the AP wireless network interface shut down during periods of time. The formula gives the saved energy percentage when shutting down the AP wireless network interface compared to the total energy that would be consumed if its interface works the whole time.

Terms P_{ext_on} and P_{ext_off} of Equation (4) represent the measured power values in Watts, in the AP external power source, when the wireless network interface is powered on and off, respectively. Terms t_{on} and t_{total} represent the amount of time the AP stayed with its wireless interface switched on and the total analysis elapsed time, respectively. Equation (4) provides the percentage of energy that could have been saved by switching off the wireless interfaces of the AP during the idle time slots. This formula can be easily extended to also provide the network's energy saving factor. To do so, the terms t_{on} and t_{total} must change in order to represent the sum of time that all APs on the network and the total time multiplied by the number of APs in the network, respectively.

$$ESF = \frac{P_{ext\_on} - P_{ext\_off}}{P_{ext\_on}}(1 - \frac{t_{on}}{t_{total}}) \quad (4)$$

From Equation (4), it is possible to notice that ESF reaches its maximum power saving factor value, ESF_{max}, when $t_{on} = 0$. This condition represents the scenario where the wireless interface of all APs in the network are switched off during the whole time. However, it is also possible to notice that, depending on the scenario and switching off scheme, ESF_{max} can assume several values. Therefore, the normalized energy saving factor, $\overline{ESF}$ given by Equation (5), can better indicate the performance of the mechanism in different scenarios. The normalized energy saving factor $\overline{ESF}$ is limited between 0% and 100% and represents the percentage of the maximum energy saving factor that could be saved.

$$\overline{ESF}(\%) = \frac{ESF(\%)}{ESF_{max}(\%)} \quad (5)$$

The work of [12] defines the coverage ratio loss CR formula, indicated in Equation (6). The coverage ratio loss is the number of uncovered clients U_l by the energy saving mechanism over the total clients in the network U within a certain period of time. The coverage ratio loss gives the percentage of clients that could not successfully access the network under the evaluated period.

$$CR(\%) = (\frac{U_l}{U} \times 100) \quad (6)$$

The analysis in this section will evaluate the normalized energy saving factor (Equation (5)) and the coverage ratio loss (Equation (6)) to compare how the eSCIFI mechanism impacts on the network performance. To calculate the coverage ratio loss, we must know the parameter T_{max} that indicates the maximum number of association an AP might support in a time slot. We defined $T_{max} = 300$, which is roughly the maximum number of APs associated in a time slot registered plus 10%. In our experimental scenario, only the wireless network interface will be switched off. Table 3 shows the consumed power measured for the AP model present in the UFF SCIFI network when the wireless interface is switched on and off (P_{ext_on} and P_{ext_off}). Table 3 also shows what would be the maximum power saving factor, ESF_{max}, which represents the power saving factor percentage if the wireless interface of all APs were switched off the whole time. Therefore, in our evaluation scenario the maximum energy saving factor percentage that could be reached by switching off the wireless interface of the entire network during the whole evaluation period is 23,93%. That information is required by the normalized energy saving factor calculations.

Table 3. AP's consumed power and maximum power saving factor percentage.

P_{ext_on} (W)	P_{ext_off} (W)	ESF_{max} (%)
1.111	0.845	23.93

We are going to evaluate how several components from the eSCIFI architecture impact on the network performance. eSCFI using the cSCIFI and the cSCIFI+ clustering algorithms will also be compared with other mechanisms proposed in the literature. The eSCIFI energy saving mechanisms will be compared with SEAR, ACE and ECMA mechanisms proposed by Jardosh et al. [2], Fang et al. [12] and Silva et al.[14], respectively. The SEAR mechanism uses the green clustering algorithms and a single threshold where only the T_{min} parameter is used as the RoD strategy. In the SEAR mechanism, the network APs are grouped into clusters, the cluster head is always switched on and the other APs in the clusters remain switched off as long as their traffic demand is lower than T_{min}. The ACE mechanism uses an inactivity time window based on machine learning occupancy detection results as its RoD strategy and does not have a coverage guarantee. APs that remain unused by a whole time window size are switched off the whole time window duration period. The ECMA mechanism uses the SEAR mechanism for night hours (between 0 a.m.–6:59 a.m.) and keeps the whole network switched on the rest of the day. The Baseline mechanism where all the APs in the network remain switched on between 7 a.m.–11:59 a.m. and switched off between 0 a.m.–6:59 a.m. is also used for comparison.

We will evaluate how the time window size and the minimum threshold value affect the network performance. After that, we will also compare the eSCIFI mechanism performance on regular weekday with its performance on a public holiday. Our last analysis will compare the SEAR green clustering, eSCIFI with cSCIFI and cSCIFI+ clustering algorithms performances using different neighborhood lists. Our trace-driven simulation scripts were developed in Python.

4.1. Time Window Size Analysis

In Section 3.4.2, we have seen that the time window tw defines how long it takes before reconfiguring the network APs energy state. A bigger time window is desired since it will minimize the number of times the controller will need to change the APs working status, which will minimize the controller tasks over a day. On the other side, a bigger time window may not notice small traffic demand bursts, which may lead to network coverage losses during these bursts due to unnoticed behaviors. Therefore, we need to evaluate how the eSCIFI time window size may affect the network energy saving and coverage loss. We tested the eSCIFI mechanisms using 5 different time window values (10 min, 30 min, 1 h, 1:30 h and 2 h). Those time window values were selected based on our time slots size and correspond to 1, 3, 6, 9 and 12 time slots, respectively. Those time windows were selected based on the lecture duration time at UFF, which usually takes 2 h. The real and predicted association values for time windows bigger than one time slot (10 min) is the sum of the devices connected during the corresponding amount of time slots. We evaluated eSCIFI using the cSCIFI and cSCIFI+ clustering algorithms and $T_{min} = 72$ with different time windows to evaluate the normalized energy saving factor and coverage loss. Those fixed parameters were used because they delivered the best normalized energy saving factor percentage and coverage ratio loss to all possible time windows. We will also evaluate the time window size effect in the SEAR, ACE, ECMA and Baseline mechanisms.

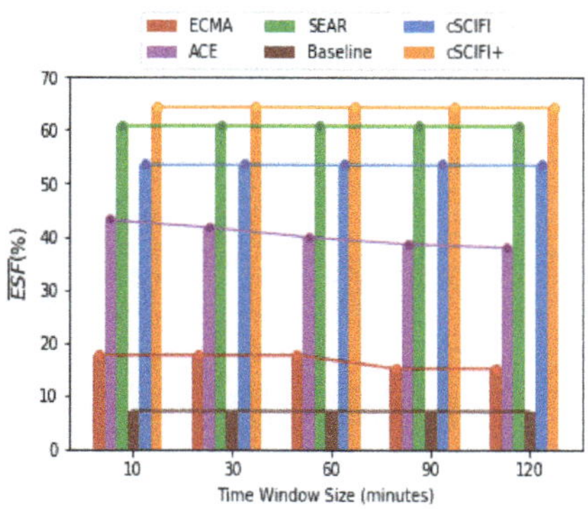

Figure 3. Normalized energy saving factor for different time window sizes.

As we can see in Figure 3, the selected time window sizes have not affected the normalized energy saving factor $\overline{ESF}$ for SEAR and eSCIFI using the cSCIFI and the CSCIFI+ clustering algorithms. Only ECMA and ACE had their normalized energy saving factor negatively affected by the time window size. The baseline estimator does not depend on the time window (its scheduling presents fixed switching on/off periods), and therefore we can see that its normalized energy saving factor does not change. This result means that for our evaluation scenario it is possible to use a 2-h time window resolution without affecting the normalized energy saving factor for our eSCIFI mechanism. This would allow the eSCIFI mechanism to compute less energy state changes in the APs and consequently less tasks to be execute by the wireless network controller.

Figure 4 shows how the different time window sizes affects the coverage ratio. As we can see only Baseline and ACE presented coverage ratio losses in this evaluation scenario. The baseline estimator has a fixed coverage loss that does not depend on the time window size. The Baseline loss occurs due to unattended users in the night hours where all APs are switched off. However, the ACE mechanism shows a small decrease in the coverage loss as the time window grows. That result was expected because ACE uses the time window size as an inactivity criteria to switch off APs, and therefore a large time window would require a longer period of inactivity, which would be harder to achieve and consequently would lower the chances of mistakenly switching off APs.

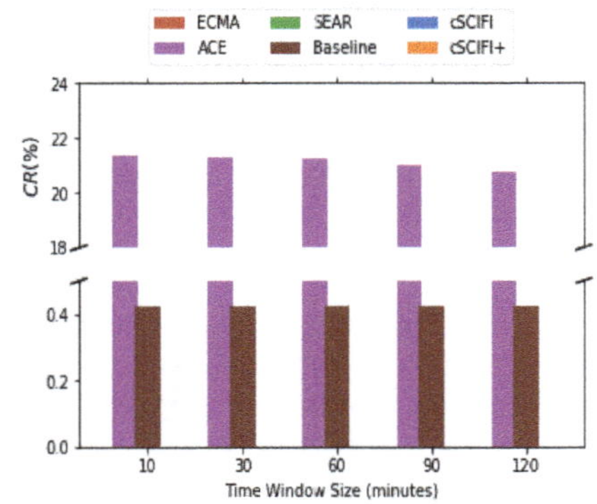

Figure 4. Coverage loss for different time window sizes.

4.2. Minimum Threshold Analysis

The last parameter on our eSCIFI mechanism that needs to be evaluated is the T_{min} value selection. T_{min} defines the minimum number of associations that an AP must have during the time window duration to be switched on. If the number of associations is below T_{min}, the AP will be evaluated to be switched off by the energy state decision algorithm. In this section, we evaluate how the T_{min} value affects the normalized energy saving factor and the coverage ratio. To do so, we varied the value assumed by T_{min} during one time slot including all multiples of 9 ranging from 9 to 90. Therefore, the T_{min} value will be proportional to the time window size used. Therefore, if the time window has a size w of time slots, the T_{min} values assumed will be $w \times T_{min}$. We fixed the time window size to 12 time slots (2 h or 120 min).

Figure 5 shows the normalized energy saving factor achieved by eSCIFI using the cSCFI and cSCIFI+ clustering algorithms, SEAR, ACE and ECMA. As it can be seen in Figure 5, SEAR and eSCIFI using cSCIFI+ got the best energy saving percentages on our evaluation scenario. eSCIFI using cSCIFI had a smaller energy saving percentage because it has a different cluster set that is bigger than the ones formed by the SEAR eSCIFI using cSCIFI+. From Figure 5, we can also see that the normalized energy factor $\overline{ESF}$ grows as T_{min} grows until it reaches $T_{min} = 54$, after that, the energy factor stays the same for all mechanisms. This result was expected and it is the same result achieved by Dalmasso et al. [8]. This asymptotic characteristic in the normalized energy saving factor curve happens because, for values of T_{min} higher than 54, the cluster maximum capacity CAA threshold is reached requiring those same APs to be turned on anyway.

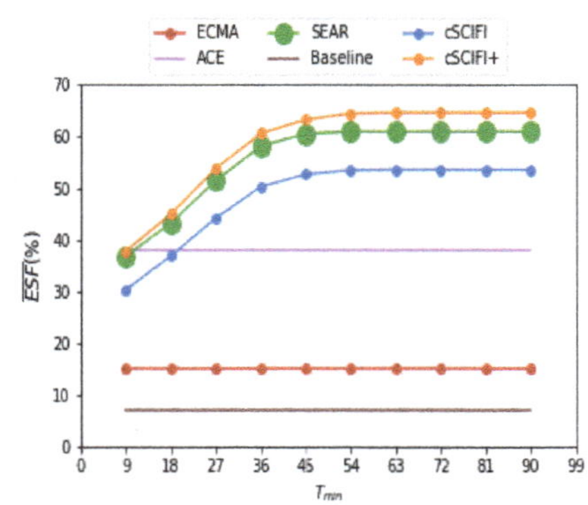

Figure 5. Normalized Energy saving factor for different T_{min} values.

Higher T_{min} values mean that APs will require a higher number of associations in a time window to be switched on according to the first criteria, which means it will be harder for them to be switched on. However, those APs will have their demand transferred to the cluster head (or other switched on AP in the case where cSCIFI has been used). That will mean that the cluster maximum capacity CAA threshold will be reached sooner and the APs will have to be turned on anyway. Therefore the normalized energy saving factor is limited and there is a T_{min} value that reaches it. Increasing T_{min} after its optimum value will not change the normalized energy saving factor. The possible explanations behind that may be that after $T_{min} = 54$, SEAR already reaches the minimum required APs to guarantee coverage (only the cluster heads may be switched on) or the switched on APs after that value present traffic demands much higher than the maximum value of $T_{min} = 90$. Figure 5 shows that ECMA has a steady normalized energy saving factor that does not depend on the T_{min} value. This might happen because ECMA applies the SEAR mechanism in night hours (between 0 a.m–6:59 a.m) and keeps the whole network switched on the rest of the day. The network has very little traffic demands in night hours, therefore few APs are required to be turned on or will have enough traffic to trigger the cluster maximum capacity threshold. That way, ECMA already reaches its highest energy saving factor with a $T_{min} = 9$. ACE and Baseline do not present a T_{min} parameter for energy state decision and therefore their normalized energy saving factors do not change.

We also evaluate how different T_{min} values affect the coverage ratio. As we can see in Figure 6, eSCIFI using both clustering algorithms (cSCIFI and cSCIFI+), SEAR and ECMA strategy had no coverage ratio loss at all for any value of T_{min}. This results showed that none of the mechanisms had overpassed the maximum cluster capacity at any moment. Only Baseline and ACE present coverage losses. However, as we have already mentioned previously, those mechanisms do not change their energy state decisions based on a minimum threshold T_{min} parameter. Therefore, their coverage ratio results are the same showed in Figure 4, where the time window size is $tw = 120$ min.

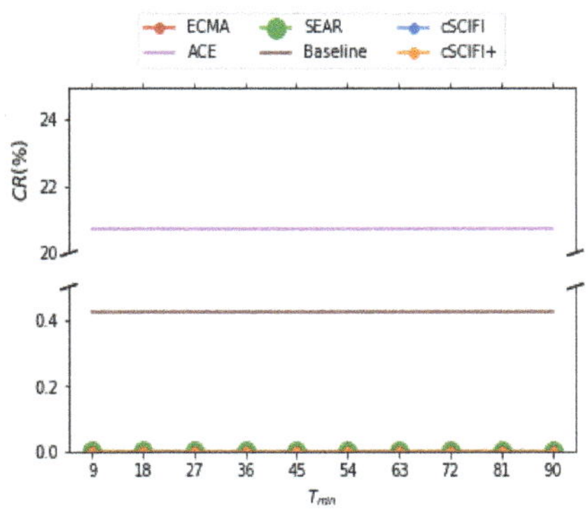

Figure 6. Coverage ratio loss for different T_{min} values.

4.3. Weekday Versus Holiday Analysis

The eSCIFI uses machine learning prediction models to estimate traffic demands. In our scenario, the hybrid model uses a holiday input feature that distinguishes normal weekdays from public holidays and university student holidays. The hybrid model uses this feature to differentiate the network demand variation that happens between regular day and holidays. Here, we will evaluate if eSCIFI using the hybrid model can better cope with the holiday demand than SEAR, ACE and ECMA. We compare Brazil's Independence Day public holiday (Friday, September 7) and the Friday used in our regular week. We compared the mechanism using the parameters that gave the best normalized energy saving factor and smallest coverage ratio loss ($T_{min} = 54$ and $tw = 120$ min).

Figure 7 shows the normalized energy saving factor achieved by the different mechanisms. As we can see, eSCIFI using both clustering algorithms and SEAR kept the normalized energy saving factor stable. eSCIFI using cSCIFI+ has the biggest normalized energy saving factor for holiday and weekdays. Baseline and ECMA also remain with their normalized energy saving factor unchanged. For Baseline, this happens because the decision is only based on time schedules and not on traffic demand estimations and therefore is unaffected. For ECMA, this result happened because the traffic demand for holiday or weekday remains unchanged, which did not change the SEAR *APs* switching on/off schedule during night hours. Only ACE had a reduction on the normalized energy saving factor in our holiday evaluation.

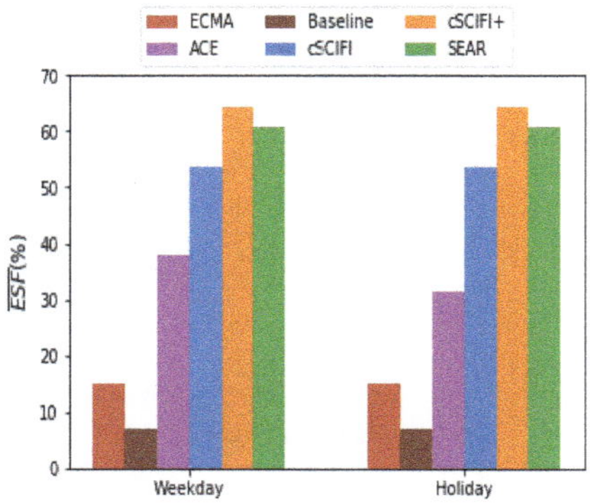

Figure 7. Normalized Energy saving factor comparison between a holiday and a weekday.

As we have seen in Figure 2, the demand on that holiday (Friday, September 7) was much smaller than the demand presented for the regular weekday (Friday, September 28). Figure 2 also shows that the holiday demand predicted by the hybrid model is much bigger than the real one, differently from the regular Friday where the hybrid model prediction was very close to the real traffic. Those results would first suggest that the normalized energy saving factor achieved by eSCIFI and SEAR for the public holiday should have been smaller as it happened with ACE. However, Figure 2 shows that the hybrid model wrong estimations have not even reached 200 associated devices for the whole network in any moment of the day on September 7. Figure 2 also shows that the regular Friday has not even reached 500 associated devices for the whole network on September 28. Those association values are very low considering that $T_{max} = 300$. Therefore, we can presume that, for the evaluated regular and holiday Friday, the network is working with the minimum set of *APs* switched on (only the cluster heads) and that is the reason why SEAR and eSCIFI using both clustering algorithms have their normalized energy saving factor unchanged. In fact, we analyzed how the real data would affect the normalized energy saving factor results in that analysis for the mechanisms and it showed that it would not have changed much (less than 3.3% for all mechanisms) in the results.

Figure 8 shows the mechanisms' coverage ratio loss for the regular weekday and for the holiday. Only ACE and Baseline present some energy loss since they are the only mechanisms that do not have a coverage guarantee. The Baseline coverage loss remains the same, which shows that the traffic demand for night hours (0 a.m–6:59 a.m) on both our holiday or weekday remains unchanged. The smaller coverage ratio loss for our holiday when compared to our weekday on the ACE mechanism case can be explained by the smaller traffic demand estimated for the whole day. Another explanation for the ACE mechanism reduced coverage ratio can be on the fact that the ACE mechanism has a smaller normalized energy saving factor on our holiday, which means it has a smaller number of *APs* switched off or that they are switched off for a short period of time.

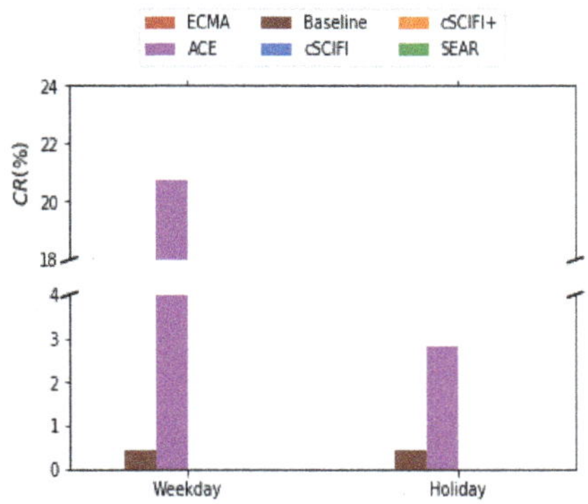

Figure 8. Coverage ratio loss comparison between a holiday and a weekday.

The results shown in this section cannot give us a precise conclusion on whether or not our algorithm could cope with the holiday demand without sacrificing the normalized energy saving factor. A large number of holidays in distinct weekdays and with distinct demand estimations would be necessary to understand it better. However, results indicated that the mechanism performance on holidays is not related to any change on its function, but it is in fact intimately related to the correct traffic estimations given by the hybrid model when compared to the real traffic data.

4.4. SEAR vs. eSCIFI Clustering Algorithms

As we have seen in Section 4.1, SEAR had a better normalized energy saving factor result than eSCIFI using the cSCIFI clustering algorithm. However, the clustering algorithm developed by Jardosh et al. [2] does not have the same optimizations criterion we have implemented on our both algorithms. Therefore, the work of Jardosh et al. [2] is susceptible to the order of appearance of APs in the neighborhood list of other APs. This order affects which will be the next APs selected by the Jardosh et al. [2] green clustering algorithm to fill the cluster in case of ties between the number of neighbors. The order of appearance of APs in the neighborhood list impacts its result since it does not have any tie breaker rule in the selection of the next AP to be put in the cluster in case of a tie in the number of neighbors between APs. cSCIFI and cSCIFI+ do not have this disadvantage, and therefore we can guarantee that the clusters formed will not depend on the order of appearance. Here, we will compare the normalized energy saving factor achieved by the SEAR and eSCIFI mechanism using both clustering algorithm using 3 different orders of appearance of APs in the neighborhood lists of the APs. The two first neighborhood lists present cases where the APs position inside the neighborhood lists are randomized and the third represent the neighborhood list we have used for all tests we have done before for the SEAR mechanism. We will compare the SEAR and eSCIFI mechanism using both clustering algorithms with the parameters that gave the best normalized energy saving factor and no coverage ratio loss ($T_{min} = 54$ and $tw = 120$).

As we can see in Figure 9, SEAR energy saving result is heavily affected by the order of appearance of APs in the neighborhood list, while eSCIFI using cSCIFI and cSCIFI+ are not affected at all. This result shows that the changes we have implemented on cSCIFI and cSCIFI+ have turned our algorithm unaffected by the order of appearance of AP in the neighborhood list. This is a clear advantage since it will not require an optimization on the neighborhood list formation process that in a huge network topology might be unpractical to be done.

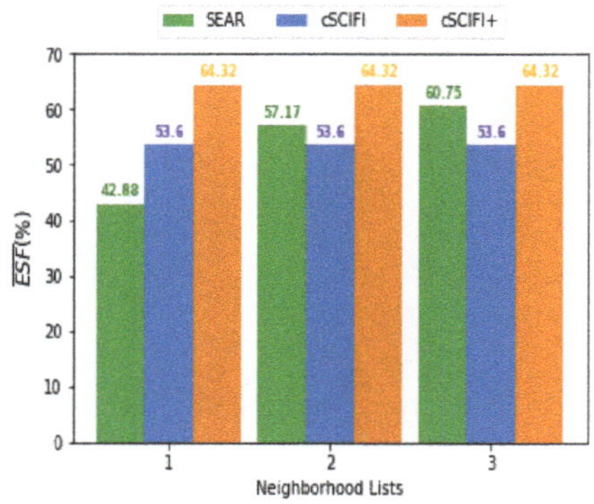

Figure 9. Normalized Energy saving factor comparison between SEAR and eSCIFI using both clustering algorithm with different neighborhood lists.

4.5. Best Results Comparison

Table 4 shows the best energy saving factor and coverage ratio loss results that can be achieved by the distinct mechanisms for our trace-driven test. Each mechanism uses distinct algorithms with a set of parameters as we have seen in the previous sections. As seen in Table 4, eSCIFI using cSCIFI+ achieved the highest energy saving factor among all mechanisms (64.32%) while presenting 0% coverage ratio loss.

ECMA, SEAR and eSCIFI using cSCIFI also achieved 0% coverage ratio loss. However, those mechanisms achieved lower energy saving factors when compared to the eSCIFI mechanism using cSCIFI+. SEAR got similar results to cSCIFI+ and better results than eSCIFI using cSCIFI. However, as explained previously and shown in Figure 9, SEAR energy saving result is affected by the neighborhood list ordering, while eSCIFI proposals using cSCIFI or cSCIFI+ are not affected at all. Therefore, our results show that eSCIFI using the cSCIFI+ algorithm achieved the best energy saving and coverage ratio loss results for our scenario.

Table 4. Mechanism's best results comparison.

Metrics	Baseline	ECMA	ACE	SEAR	cSCIFI	cSCIFI+
Coverage Ratio Loss (CR %)	0.42	**0**	20.73	**0**	**0**	**0**
Energy Saving Factor ($\overline{ESF}$)	6.98	**17.72**	43.27	**60.75**	**53.60**	**64.32**

5. Conclusions

We presented the eSCIFI energy saving mechanism and its main architecture. eSCIFI uses traffic demand estimations given by machine learning models to manage the energy state of *APs* and it was designed to cope with a broader variety of wireless networks, specially those that cannot collect traffic data in a real time manner and/or have a limited CPU power.

We evaluated the normalized energy saving factor and the coverage ratio loss of our proposed mechanism. We also reproduced and compared eSCIFI results to the ones achieved by ACE, ECMA and SEAR. Those results showed that, for the UFF SCIFI network scenario, eSCIFI produced the best results. The best energy saving mechanism was the eSCIFI using the cSCIFI+ mechanism that can save up to 64.32% of the total energy consumed in a week without affecting the network coverage and user's association capacity.

eSCIFI has not been tested and implemented in real network scenarios yet. As future work, we plan to implement eSCIFI on the UFF SCIFI controller and do some future experiments using the real network infrastructure. The practical usage will give us some real insights about how to properly tune eSCIFI parameters according to a real implementation.

We also plan to use more sophisticated metrics as the average throughput and delay to evaluate the network performance and user coverage on those real network tests. We hope that those tests and new features will allow us to fully understand the eSCIFI possibilities and overcome its limitations.

Author Contributions: Conceptualization, G.H.A. and D.C.M.-S.; methodology, G.H.A., F.B. and D.C.M.-S.; validation, G.H.A., F.B., D.C.M.-S. and L.C.S.M.; formal analysis, G.H.A.; investigation, G.H.A.; resources, G.H.A. and L.C.S.M.; data curation, G.H.A., F.B. and L.C.S.M.; writing—original draft preparation, G.H.A. and D.C.M.-S.; writing—review and editing, G.H.A., F.B., D.C.M.-S. and L.C.S.M.; supervision, F.B., D.C.M.-S. and L.C.S.M.; project administration, D.C.M.-S.; funding acquisition, D.C.M.-S. All authors have read and agreed to the published version of the manuscript.

Funding: This research was partially funded by the Research Foundation of the State of Rio de Janeiro (FAPERJ), the Research Foundation of the State of São Paulo (FAPESP), the Coordination for the Improvement of Higher Education Personnel (CAPES), CAPES PRINT, and the Brazilian National Council for Scientific and Technological Development (CNPq).

Institutional Review Board Statement: Not applicable.

Informed Consent Statement: Not applicable.

Data Availability Statement: The UFF SCIFI network association datasets (for classification and regression) are available at https://github.com/midiacom/UFF-SCIFI-Datasets, accessed on 3 December 2021.

Acknowledgments: We thank FAPERJ, FAPESP, CAPES, CAPES PRINT, and CNPq for their financial support.

Conflicts of Interest: The authors declare no conflict of interest.

References

1. Debele, F.G.; Li, N.; Meo, M.; Ricca, M.; Zhang, Y. Experimenting resourceondemand strategies for green WLANs. *ACM SIGMETRICS Perform. Eval. Rev.* **2014**, *42*, 61–66. [CrossRef]
2. Jardosh, A.P.; Papagiannaki, K.; Belding, E.M.; Almeroth, K.C.; Iannaccone, G.; Vinnakota, B. Green WLANs: on-demand WLAN infrastructures. *Mob. Netw. Appl.* **2009**, *14*, 798–814. [CrossRef]
3. Lyu, F.; Fang, L.; Xue, G.; Xue, H.; Li, M. Large-scale full WiFi coverage: Deployment and management strategy based on user spatio-temporal association analytics. *IEEE Internet Things J.* **2019**, *6*, 9386–9398. [CrossRef]
4. Lorincz, J.; Capone, A.; Bogarelli, M. Energy savings in wireless access networks through optimized network management. In Proceedings of the IEEE 5th International Symposium on Wireless Pervasive Computing, Modena, Italy, 5–7 May 2010; pp. 449–454.
5. Cui, Y.; Ma, X.; Wang, H.; Stojmenovic, I.; Liu, J. A survey of energy efficient wireless transmission and modeling in mobile cloud computing. *Mob. Netw. Appl.* **2013**, *18*, 148–155. [CrossRef]
6. Jardosh, A.P.; Iannaccone, G.; Papagiannaki, K.; Vinnakota, B. Towards an energy-star WLAN infrastructure. In Proceedings of the Eighth IEEE Workshop on Mobile Computing Systems and Applications, Tucson, AZ, USA, 8–9 March 2007; pp. 85–90.
7. Gomez, K.; Sengul, C.; Bayer, N.; Riggio, R.; Rasheed, T.; Miorandi, D. MORFEO: Saving energy in wireless access infrastructures. In Proceedings of the 2013 IEEE 14th International Symposium on A World of Wireless, Mobile and Multimedia Networks (WoWMoM), Madrid, Spain, 4–7 June 2013; pp. 1–6.
8. Dalmasso, M.; Meo, M.; Renga, D. Radio resource management for improving energy self-sufficiency of green mobile networks. *ACM SIGMETRICS Perform. Eval. Rev.* **2016**, *44*, 82–87. [CrossRef]
9. Rossi, C.; Casetti, C.; Chiasserini, C.F.; Borgiattino, C. Cooperative energy-efficient management of federated WiFi networks. *IEEE Trans. Mob. Comput.* **2015**, *14*, 2201–2215. [CrossRef]
10. Chen, Y.J.; Shen, Y.H.; Wang, L.C. Achieving energy saving with QoS guarantee for WLAN using SDN. In Proceedings of the 2016 IEEE International Conference on Communications (ICC), Kuala Lumpur, Malaysia, 23–27 May 2016; pp. 1–7.
11. Chin, K.W. A green scheduler for enterprise WLANs. In Proceedings of the 2011 IEEE Australasian Telecommunication Networks and Applications Conference (ATNAC), Melbourne, VIC, Australia, 9–11 November 2011; pp. 1–3.
12. Fang, L.; Xue, G.; Lyu, F.; Sheng, H.; Zou, F.; Li, M. Intelligent Large-Scale *AP* Control with Remarkable Energy Saving in Campus WiFi System. In Proceedings of the 2018 IEEE 24th International Conference on Parallel and Distributed Systems (ICPADS), Singapore, 11–13 December 2018; pp. 69–76.
13. Lee, K.; Kim, Y.; Kim, S.; Shin, J.; Shin, S.; Chong, S. Just-in-time WLANs: On-demand interference-managed WLAN infrastructures. In Proceedings of the IEEE INFOCOM 2016-the 35th Annual IEEE International Conference on Computer Communications, San Francisco, CA, USA, 10–14 April 2016; pp. 1–9.

14. Silva P.; Almeida, N.T.; Campos, R. Energy consumption management for dense Wi-Fi networks. In Proceedings of the 2019 Wireless Days (WD), Manchester, UK, 24–26 April 2019; pp. 1–8.
15. Xu, C.; Han, Z.; Zhao, G.; Yu, S. A sleeping and offloading optimization scheme for energy-efficient WLANs. *IEEE Commun. Lett.* **2016**, *21*, 877–880. [CrossRef]
16. Xu, C.; Wang, J.; Zhu, Z.; Niyato, D. Energy-Efficient WLANs with resource and re-association scheduling optimization. *IEEE Trans. Netw. Serv. Manag.* **2019**, *16*, 563–577. [CrossRef]
17. Magalhães, L.C.S.; Balbi, H.D.; Corrêa, C.; Valle, R.D.T.d.; Stanton, M. Scifi—A software-based controller for efficient wireless networks. In *UbuntuNet-Connect 2013*; UbuntuNet Alliance: Kigali, Rwanda, 2013.
18. Fainelli, F. The OpenWrt embedded development framework. In Proceedings of the Free and Open Source Software Developers European Meeting, Brussels, Belgium, 23–24 February 2008; p. 106.
19. Apostolo, G.H.; Bernardini, F.; Magalhães, L.C.S.; Muchaluat-Saade, D.C. An Experimental Analysis for Detecting Wi-Fi Network Associations Using Multi-label Learning. In Proceedings of the IEEE 2020 International Conference on Systems, Signals and Image Processing (IWSSIP), Niterói, Brazil, 1–3 July 2020; pp. 423–428.
20. Wang, W.; Chen, J.; Hong, T.; Zhu, N. Occupancy prediction through Markov based feedback recurrent neural network (M-FRNN) algorithm with WiFi probe technology. *Build. Environ.* **2018**, *138*, 160–170. [CrossRef]
21. Zou, H.; Jiang, H.; Yang, J.; Xie, L.; Spanos, C. Non-intrusive occupancy sensing in commercial buildings. *Energy Build.* **2017**, *154*, 633–643. [CrossRef]
22. Ghai, S.K.; Thanayankizil, L.V.; Seetharam, D.P.; Chakraborty, D. Occupancy detection in commercial buildings using opportunistic context sources. In Proceedings of the 2012 IEEE International Conference on Pervasive Computing and Communications Workshops, Lugano, Switzerland, 19–23 March 2012; pp. 463–466.
23. Balaji, B.; Xu, J.; Nwokafor, A.; Gupta, R.; Agarwal, Y. Sentinel: occupancy based HVAC actuation using existing WiFi infrastructure within commercial buildings. In Proceedings of the 11th ACM Conference on Embedded Networked Sensor Systems, Roma, Italy, 11–15 November 2013; pp. 1–14.
24. Donevski, I.; Vallero, G.; Marsan, M.A. Neural networks for cellular base station switching. In Proceedings of the IEEE INFOCOM 2019-IEEE Conference on Computer Communications Workshops (INFOCOM WKSHPS), Paris, France, 29 April–2 May 2019; pp. 738–743.
25. Budzisz, Ł.; Ganji, F.; Rizzo, G.; Marsan, M.A.; Meo, M.; Zhang, Y.; Koutitas, G.; Tassiulas, L.; Lambert, S.; Lannoo, B.; others. Dynamic resource provisioning for energy efficiency in wireless access networks: A survey and an outlook. *IEEE Commun. Surv. Tutor.* **2014**, *16*, 2259–2285. [CrossRef]
26. Lorincz, J.; Bogarelli, M.; Capone, A.; Begušić, D. Heuristic approach for optimized energy savings in wireless access networks. In *SoftCOM 2010, Proceedings of the 18th International Conference on Software, Telecommunications and Computer Networks, Split-Bol, Croatia, 23–25 September 2010*; IEEE: New York, NY, USA, 2010; pp. 60–65.
27. Capone, A.; Malandra, F.; Sansò, B. Energy savings in wireless mesh networks in a time-variable context. *Mob. Netw. Appl.* **2012**, *17*, 298–311. [CrossRef]
28. Liu, L.; Xu, S.; Hu, J.; Cui, L.; Min, G. Balancing of the quality-of-service, energy and revenue of base stations in wireless networks via tullock contests. In Proceedings of the International Symposium on Quality of Service, Phoenix, AZ, USA, 24–25 June 2019; pp. 1–8.
29. Vallero, G.; Renga, D.; Meo, M.; Marsan, M.A. Greener RAN operation through machine learning. *IEEE Trans. Netw. Serv. Manag.* **2019**, *16*, 896–908. [CrossRef]
30. Tanaka, T.; Abe, K.; Aust, S.; Ito, T.; Yomo, H.; Sakata, S. Automatic and cooperative sleep control strategies for power-saving in radio-on-demand WLANs. In Proceedings of the 2013 IEEE Green Technologies Conference (GreenTech), Denver, CO, USA, 4–5 April 2013; pp. 293–300.
31. Yomo, H.; Kondo, Y.; Namba, K.; Tang, S.; Kimura, T.; Ito, T. Wake-up ID and protocol design for radio-on-demand wireless LAN. In Proceedings of the 2012 IEEE 23rd International Symposium on Personal, Indoor and Mobile Radio Communications-(PIMRC), Sydney, Australia , 9–12 September 2012; pp. 419–424.
32. Marsan, M.A.; Chiaraviglio, L.; Ciullo, D.; Meo, M. A simple analytical model for the energy-efficient activation of access points in dense WLANs. In Proceedings of the 1st International Conference on Energy-Efficient Computing and Networking, Athens, Greece, 14–15 October 2010; pp. 159–168.
33. Kumazoe, K.; Nobayashi, D.; Fukuda, Y.; Ikenaga, T.; Abe, K. Multiple station aggregation procedure for radio-on-demand WLANs. In Proceedings of the IEEE 2012 Seventh International Conference on Broadband, Wireless Computing, Communication and Applications, Victoria, BC, Canada, 12–14 November 2012; pp. 156–161.
34. Kumazoe, K.; Nobayashi, D.; Fukuda, Y.; Ikenaga, T. Station aggregation scheme considering channel interference for radio on demand networks. In Proceedings of the 2013 IEEE Pacific Rim Conference on Communications, Computers and Signal Processing (PACRIM), Victoria, BC, Canada, 27–29 August 2013; pp. 265–270.
35. Ganji, F.; Budzisz, Ł.; Wolisz, A. Assessment of the power saving potential in dense enterprise WLANs. In Proceedings of the IEEE 24th International Symposium on Personal, Indoor, and Mobile Radio Communications, London, UK, 8–11 September 2013.
36. Nagareda, R.; Hasegawa, A.; Shibata, T.; Obana, S. A proposal of power saving scheme for wireless access networks with access point sharing. In Proceedings of the 2012 IEEE International Conference on Computing, Networking and Communications (ICNC), Maui, HI, USA, 30 January–2 February 2012; pp. 1128–1132.

37. Zhang, Y.; Jiang, C.; Wang, J.; Han, Z.; Yuan, J.; Cao, J. Green Wi-Fi implementation and management in dense autonomous environments for Smart Cities. *IEEE Trans. Ind. Inform.* **2017**, *14*, 1552–1563. [CrossRef]
38. Wu, W.; Luo, J.; Dong, K.; Yang, M.; Ling, Z. Energy-efficient user association with congestion avoidance and migration constraint in green WLANs. *Wirel. Commun. Mob. Comput.* **2018**, *2018*, 9596141. [CrossRef]
39. Apostolo, G.H.; Bernardini, F.; Magalhães, L.C.S.; Muchaluat-Saade, D.C. A Unified Methodology to Predict Wi-Fi Network Usage in Smart Buildings. *IEEE Access* **2021**, *9*, 11455–11469. [CrossRef]
40. Haratcherev, I.; Fiorito, M.; Balageas, C. Low-power sleep mode and out-of-band wake-up for indoor access points. In Proceedings of the 2009 IEEE Globecom Workshops, Honolulu, HI, USA, 30 November–4 December 2009; pp. 1–6.

buildings

MDPI

Article

The Impact of Climate Change on a University Campus' Energy Use: Use of Machine Learning and Building Characteristics

Haekyung Im [1,*], Ravi S. Srinivasan [1], Daniel Maxwell [2], Ruth L. Steiner [3] and Sayar Karmakar [4]

1 UrbSys (Urban Building Energy, Sensing, Controls, Big Data Analysis, and Visualization) Laboratory, M.E. Rinker, Sr. School of Construction Management, University of Florida, Gainesville, FL 32603, USA; sravi@ufl.edu
2 1 Tigert Hall, University of Florida, Gainesville, FL 32611, USA; danielmaxwell@ufl.edu
3 Department of Urban and Regional Planning, University of Florida, 431 Architecture Building, Gainesville, FL 32611, USA; rsteiner@ufl.edu
4 Department of Statistics, University of Florida, Gainesville, FL 32611, USA; sayarkarmakar@ufl.edu
* Correspondence: haekyung.rosa@ufl.edu

Abstract: Global warming is expected to increase 1.5 °C between 2030 and 2052. This may lead to an increase in building energy consumption. With the changing climate, university campuses need to prepare to mitigate risks with building energy forecasting models. Although many scholars have developed buildings energy models (BEMs), only a few have focused on the interpretation of the meaning of BEM, including climate change and its impacts. Additionally, despite several review papers on BEMs, there is no comprehensive guideline indicating which variables are appropriate to use to explain building energy consumption. This study developed building energy prediction models by using statistical analysis: multivariate regression models, multiple linear regression (MLR) models, and relative importance analysis. The outputs are electricity (ELC) and steam (STM) consumption. The independent variables used as inputs are building characteristics, temporal variables, and meteorological variables. Results showed that categorizing the campus buildings by building type is critical, and the equipment power density is the most important factor for ELC consumption, while the heating degree is the most critical factor for STM consumption. The laboratory building type is the most STM-consumed building type, so it needs to be monitored closely. The prediction models give an insight into which building factors remain essential and applicable to campus building policy and campus action plans. Increasing STM is to raise awareness of the severity of climate change through future weather scenarios.

Keywords: building energy modelling; regression analysis; machine learning; climate change; university campus; energy consumption prediction

Citation: Im, H.; Srinivasan, R.S.; Maxwell, D.; Steiner, R.L.; Karmakar, S. The Impact of Climate Change on a University Campus' Energy Use: Use of Machine Learning and Building Characteristics. *Buildings* **2022**, *12*, 108. https://doi.org/10.3390/buildings12020108

Academic Editors: Shi-Jie Cao and Wei Feng

Received: 13 December 2021
Accepted: 19 January 2022
Published: 23 January 2022

1. Introduction

Even though accelerating issues on climate change are arousing people's awareness, many are still uninformed about how climate change affects building energy consumption. Building operations are responsible for 28% of total emissions, while embodied carbon, which is from building materials and construction, is responsible for an additional 11% annually [1]. This is nearly 40% of CO_2 emissions coming from the building and building construction sector, which is responsible for over one-third of global energy consumption [2]. Global warming—a product of climate change—is also expected to significantly increase building energy use for cooling. Climate change phenomena are not uniform across the globe, so energy consumption direction or protocol can differ by climate zone. Therefore, this study focuses on the energy use of a Philadelphia (PA)-based university's campus under climate zone 4A, comprised of a humid subtropical climate [3].

College and university campuses use an average of 18.9 kilowatt-hours (kWh) of electricity (ELC) and 17 cubic feet (ft^2) of natural gas per square foot annually [4]. Universities have consumed a large portion of energy use and have updated their sustainability plan to reduce greenhouse gas emissions and energy consumption. Universities serve as pioneers for green societies and make immediate decisions and policies to address climate change, compared to other national-level organizations. Since university campuses include various building types, such as libraries, offices, laboratories, hospitals, and housing, these installations remain uniquely positioned to analyze the energy consumption of "mixed-use" group buildings. Essentially, a college or university campus serves as an independent district. Thus, typical university campuses can offer informative representations of urban energy consumption.

To minimize energy consumption, building energy consumption patterns need to be monitored and analyzed. By doing so, universities identify ways to effectively manage energy use and plan building renovation with an understanding about the relationship between building characteristics and energy consumption. In addition, energy consumption prediction can be beneficial for universities to forecast reliable energy budgets and identify opportunities of energy conservation [5].

This study investigates global climate change's influence on building energy consumption, the building's behavior, and other energy-related factors on campus. In this case study, ELC and steam (STM) were considered. ELC was provided by Penn Power, and the STM distribution system consisted of underground pipelines. Building characteristics, meteorological variables, and temporal variables were used to develop BEMs with a bottom-up approach for ELC and STM consumption. Among various bottom-up approaches, a multivariate regression and multiple linear regression (MLR) were chosen.

Thanks to rapid technology development, electronic appliance efficiency is improving, and construction materials are evolving for better insulation. In addition, smart homes and intelligent cities are becoming more popular, and the number of electronic devices people own is rising fast. To eliminate complex interventions, the impact of climate change on the energy consumption of existing buildings is analyzed purely by excluding technological advances. Furthermore, occupancy behavior was not considered in the focus on weather features and building characteristics. Consequently, operation features such as set point temperature, fan schedules, HVAC schedule (heating and cooling), lighting schedule, class schedule, building-operation hours, and building use schedule (occupancy schedule) were excluded.

2. Literature Review

To determine the method for this research, the statistical analysis used in BEMs were examined. Additionally, research on the BEMs of university campuses were analyzed to narrow the target. Then, the status of climate change and how global warming will affect building energy consumption were investigated. Finally, based on previous literature reviews, variable selections from other research allowed us to decide how to structure the data frame. Additionally, via literature reviews, this study intends to investigate factors that affect energy consumption the through interpretation of statistical models, with the following research questions (RQ):

RQ1. What are the appropriate building- and weather-related variables to be included in the predictive model?

RQ2. What is the relationship between input variables and energy consumption (output) by energy types considered by relative importance?

RQ3. How will each energy type change with the effects of global warming?

2.1. BEMs Focused on Statistical Analysis

Several review articles have summarized and defined Urban Building Energy Modeling (UBEM) [6,7]. The prevailing UBEM has two opposite modeling approaches: top-down or bottom-up. The top-down approach works at an aggregated level, typically aimed at fitting historical timelines based on national energy consumption while the bottom-up approach is built up from data as disaggregated components.

Since this study is focused on the bottom-up approach, this paper specifically discusses regression modeling approaches. Baker and Rylatt used clustering, simple regression, and MLR [8]. Kavousian et al. used stepwise selection to choose predictors in a MLR model because all input variables are numeric, not categorical [9]. Hsu used a Bayesian multilevel regression model to analyze the value of different measurements for predicting energy use and found that benchmarking data alone explains energy use as well as benchmarking and auditing data together [10]. Zeng et al. considered the multivariate regression model to be the best method in the case study with simplified inputs related to building energy [11]. Walter and Sohn also developed a multivariate regression model to predict Energy Unit Intensity (EUI) by using numerical predictors and categorical indicator variables [12]. For example, numerical predictors were operating hours, occupant density, etc., and categorical indicator variables were climate zone, heating system type, etc. This model measured the contribution of building characteristics and systems-to-energy use based on the cross-validation (CV) approach.

Most previous research in BEMs has failed to interpret the BEM, used insufficient samples, or missed some important variables that are equipment power density (EPD) or meteorological variables. Most research has focused on maximizing performance accuracy of machine learning (ML) models and finding the best model by comparing different ML techniques. This research has shown the improvement of the ML models' accuracy. According to Kikumoto et al., as a result of increasing temperature, energy simulation showed a 15% increase in the heat load of two residential buildings [13]. A small sample size for application made it difficult to generalize future energy consumption. However, protocol or interpretations of the final ML models were frequently missing, so this study aims to interpret the regression models with a larger sample size.

2.2. BEMs Focused on University Campuses

Owing to the trend of growing energy consumption, Hong et al. investigated the energy waste in universities and suggested optimization options of campus buildings in South Korea [14]. The amount of ELC had risen about 19.7% over three years because the equipment used for heat had increased. Hong et al. also found that higher monthly average temperatures led to more A/C use on campus [15]. They developed several energy saving scenarios to suggest possible solutions. Chung and Rhee observed that equipment loads and occupancy schedules of university buildings for education and research are difficult to control [16]. However, they found that universities have a high potential to reduce energy losses caused by unnecessary energy consumption, low thermal performance, and airtightness. Because of the many random users rather than the operators themselves in the university buildings, retrofitting the existing buildings into low-energy buildings is crucial [14].

Guan et al. analyzed the ELC, heating, and water usage of campus buildings in Norway, which has a similar climate with climate zone category 5A [17]. They said that UBEM plays a critical role in learning about the efficient energy planning of future urban energy systems and smart systems. Several Chinese universities and colleges have initiated sustainable education and developed incentive policies to encourage students and faculty members to save energy [15]. Universities in the U.S. have taken actions on energy policies and movement for green campus development.

2.3. Climate Change

Recently, more studies have aimed at the future impact on energy consumption owing to climate change. Examples include energy use over long-term climate change for use in life cycle assessment applications with nine typical Florida residential houses [18]. Additionally, Fathi et al. and Fathi and Srinivasan expanded the sample size and targeted a university campus instead of residential buildings, wherein Principal Component Analysis (PCA) and Autoregressive Integrated Moving Average (ARIMA) techniques were selected for energy prediction with climate change [19,20]. Another example is that Godoy-Shimizu et al. estimated urban-scale energy consumption using physics-based models for future weather [21].

Fumo and Rafe Biswas (2015) predicted total ELC using only outdoor dry-bulb temperature and added the global horizontal radiation to produce MLR models without considering building characteristics [22]. Higher resolution for the time interval, which was hourly data, leads to models with a lower quality than regression models using daily data. All three regression models (simple linear, simple quadratic, and MLR) with daily resolution had better R-square values but lower accuracies based on root mean square error (RMSE) than with hourly resolution.

Mohammadiziazi and Bilec used RF to analyze office buildings' EUI due to climate change [23]. They found that energy consumption will increase between 8.9% and 63.1% compared to the 2012 baseline for different geographic regions between 2030 and 2080. Campagna and Fiorito found that 65% of studies on climate change impacts on building energy consumption focused on climate zone C [24]. Therefore, studying climate zone A in this study can contribute to mitigating biased sample selection.

3. Methods

In this study, the hourly energy consumption data were obtained, which consist of ELC and STM. To answer the research questions, the BEMs' development process follows six-steps, namely: (1) set data frame with processing, (2) conduct descriptive statistical analysis, (3) select variables including relative importance analysis, (4) develop multivariate regression models, (5) develop multiple linear regression (MLR) models, (6) estimate energy use of a university campus with future weather data, and (7) analyze university campus' action plans (Figure 1).

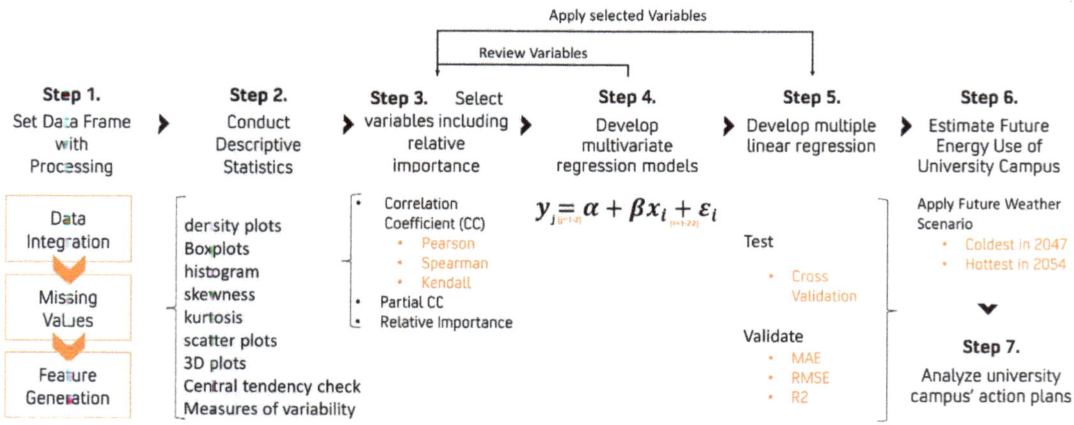

Figure 1. Conceptual framework.

Step 1 is setting data frame with processing. Four separate datasets (energy consumption variables, building variables, meteorological variables, and temporal variables) were merged to form one complete dataset. After collecting and cleaning all the data, the merged data was divided into a training and a testing dataset. The training dataset is from 1 July 2015 to 30 June 2016. For validation, a 10-fold cross-validation was conducted. The testing dataset is from 1 July 2016 to 14 August 2016.

Step 2 is conducting a descriptive statistical analysis. To observe the distribution, density plots, boxplots, histogram, skewness, and kurtosis were used. To understand the data and to detect outliers, scatter plots, three-dimensional (3D) plots, relative importance, correlation coefficient (CC), partial CC, and CC matrix plots are used. Additionally, central tendency (mean, mode, and median), measures of variability (range and interquartile range), variance, and standard deviation were measured.

Step 3 is selecting variables, including importance analysis. The multivariate regression method was used to determine independent variables because all buildings share the same independent variables. Based on other literatures and experts in machine learnings and statistics, different variables for multivariate regression models were used and compared to each other based on R^2 and RMSE. Four sets of multivariate regression models were developed, consisting of two linear regression models for ELC and STM. For reliable predictions, a large amount of historical data is required. Thus, the hourly data of 18 buildings with 22 input (independent) variables and 2 (dependent) variables were each observed and used to meet the requirement to develop the multivariate regression models.

The variables were analyzed based on CC, partial CC, and relative importance. To check correlations between variables, three CC measuring methods were used: (1) Pearson CC with the linear dependence between two numeric variables, (2) Spearman for polynomial relationship, and (3) Kendall between categorical input variables and numeric output variables. Kendall's tau and Spearman's rho were used to estimate a rank-based measure of association. To examine the categorical variables, we referred to relative importance as well. For relative importance, rank bootstrap confidence intervals were obtained by using the percentile method. Bootstraps were replicated 100 times in order to calculate confidence intervals. Metrics for relative importance are normalized to sum to 100%. This was used for all numeric variables and meteorological variables alone to examine the influence of variables on energy consumption. Lastly, partial CCs were measured to select meaningful variables.

Step 4 is developing multivariate regression models. From MLR development, variables were selected to finalize the multivariate regression model. As shown in Equation (1), independent numeric variables were standardized with a mean (μ) of 0 and standard deviation (σ) of 1 (unit variance):

$$X' = (X - \mu)/\sigma \quad (1)$$

Step 5 is developing MLR Models BEMs. As a feature selection method, stepwise selection methods were used. The performance was measured through Mean Absolute Error (MAE), RMSE, and adjusted R square (R^2). MAE is a key performance indicator (KPI) for measuring forecast accuracy. These measures are used to check the regression model's accuracy to validate and test and can be calculated using Equations (2)–(4):

$$\text{MAE} = \frac{1}{n}\sum_{i=1}^{n}\left|y_{predict,i} - y_{data,i}\right| \quad (2)$$

$$\text{RMSE} = \sqrt{\frac{\sum_{i=1}^{n}\left(y_{predict,i} - y_{data,i}\right)^2}{n}} \quad (3)$$

$$R^2 = 1 - \frac{\sum_{i=1}^{n}\left(y_{predict,i} - y_{data,i}\right)^2}{\sum_{i=1}^{n}\left(y_{data,i} - y_{data,i}\right)^2} \quad (4)$$

According to Hoff and Perez, MAE is commonly accepted as a measure of dispersion because of its lesser sensitivity to distant outliers and lesser subjection to interpretation when expressed in relative (percentage) terms [25]. A 10-fold CV was used to validate BEMs and, to test the models, regression models' accuracies were checked with a validation dataset. Based on MLR models, significant variables for each energy consumption type are revealed.

The sixth step is to estimate the energy use of university campuses with future weather data. As the last step, to estimate the operational energy consumption under long-term climate change, the average values from the hottest scenario representing 2054 were used as inputs into the MLR models. Likewise, the average values of building characteristics were used to predict energy consumption in 2054. Lastly, Step 7 is to analyze university campus' action plans to learn lessons explained in Section 5.6. Specifically, in my research, energy action plans of 18 universities were analyzed to show how a leading energy group deals with climate change.

4. Variable Selection

From Step 1 to Step 3, potential independent variables were scrutinized to build BEMs to explain energy consumptions. This section demonstrates details about the variable selection process, which includes a literature review and the analysis of results.

It should be noted that Kikumoto et al. did not include the building characteristics in predicting heat loads [13]. Whereas Im et al. concentrated on interpreting the regression models to explain the relationships among building characteristics, energy consumption, and weather [26,27], Im et al. (2019) developed polynomial regression models on chilled water (CHW) and ELC consumption of campus buildings by comparing BEMs with hourly and daily data [24]. Additionally, Im et al. (2020) attempted to predict CHW with lasso regression models using future weather data [25]. However, both studies did not use solar radiation data, which is considered a significant factor in BEM [28]. In conclusion, this study focused on variable selection and the interpretation of BEMs with larger sample sizes, as well as additional building characteristics and weather data, including solar radiation. Considering ELC and STM allows us to grasp campus energy consumption comprehensively.

Out of 194 buildings, 48 buildings had one year of two energy consumption information: ELC and STM. Out of 48 buildings, 18 buildings' data were used after excluding buildings missing building characteristics data to develop building energy modeling (BEM) by adopting multivariate regression models with a 14.5-month period. Developing multivariate regression models is to achieve a comprehensive understanding of energy use depending on the energy type by comparing the variables' impact. The sample size could be larger when final regression models were developed separately, like 43 buildings for ELC and 32 buildings for STM (Table 1). From these separate data frames, building types with less than three buildings were eliminated because of the small size in the number of buildings. As a result, all food and health-related buildings were eliminated. Eventually, 39 buildings for ELC and 30 buildings for STM were secured for multiple regression model analysis with the same time frame.

Table 1. Building number by energy types and building types.

Building Type	ELC	STM
Education	7	4
Laboratory	8	7
Lodge	6	7
Office	10	6
Public Assembly	8	6
Food	3	1
Health	1	1
Sub-total number of buildings	43	32
Total number of buildings	39	30

4.1. Energy Consumption Variables

Energy consumption variables were comprised of electricity consumption (ELC, kilo-British Thermal Unit (kBTU)/Gross Square Feet (GSF)) and steam consumption (STM, kBTU/GSF). Gross square feet (GSF) is highly related to several variables that have a value per unit square feet, such as U-values, EPD, and LPD. Therefore, GSF was included as a denominator of the dependent variable by dividing energy consumption data in kBTU with GSF. Equipment, lighting, and plug loads are three main categories contributing to the energy consumptions in the buildings. Equipment, such as heating, ventilation, and air conditioning (HVAC) systems and water heaters, is the other main internal load, which contributes to STM in this case study. Lighting and plug loads contribute to ELC. Plug loads are growing as electronics become more pervasive with the ever-accelerating progress of technology, even in construction, which used to be known as a conservative field.

4.2. Building Variables

Building variables comprised of building thermo-physical properties and other power densities. Building variables included the U-value of Wall (U-Wall, Btu/h °F.ft^2), U-value of Windows (U-Windows, Btu/h °F.ft^2), U-value of Roof (U-Roof, Btu/h °F.ft^2), Window-Wall Ratio (WWR), building height (feet), construction year (year), Building Age (year), Renovation Age (year), equipment power density (EPD, W/ft^2), and Lighting Power Density (LPD, W/ft^2).

Based on the regression and the plot, the beta coefficient (β: slope of the regression models) shows that ELC consumption and Building Age are statistically, negatively related (Figure 2). The initial plot with Building Age could lead to the misinterpretation of energy consumption, which indicates that buildings consume less energy as time goes by, regardless of the deterioration of the building. Thus, it is important to use construction year instead of building age. As shown in Figure 3, ELC consumption is higher in newer buildings than in older buildings. This may be because buildings are evolving with advanced technology and more outlets. Newer buildings have more opportunity (more outlets or electronic appliances) of energy use compared to older buildings.

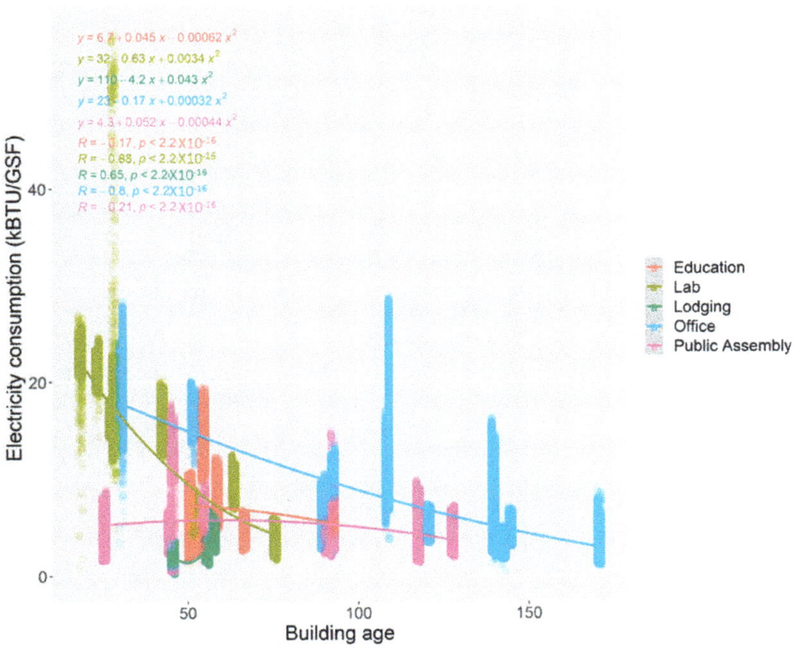

Figure 2. Scatter plot of electricity consumption and Building Age.

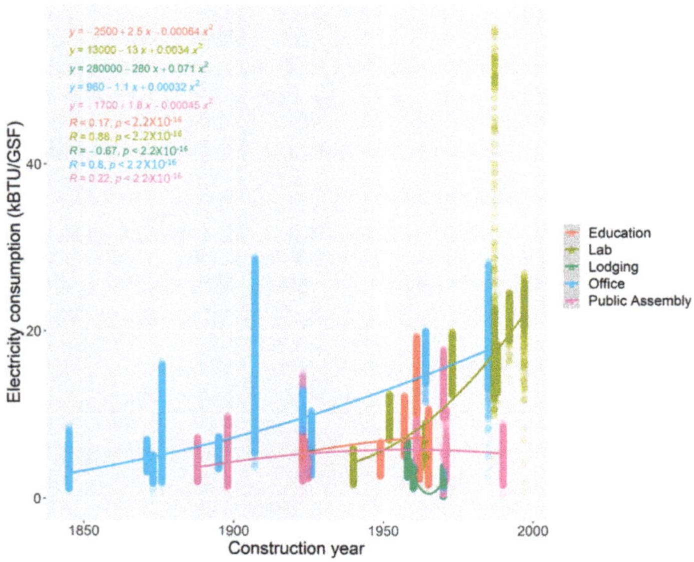

Figure 3. Scatter plot of electricity consumption and construction year.

Gao et al. selected four variables (wall area, building height, building orientation, and glazing area) through the feature reduction process to improve the performance of models out of eight independent variables (relative compactness, surface area, wall area, roof area, building height, orientation, glazing area, and glazing area distribution of a residential building) [29]. Compared to the study by Gao et al., in this study, window wall ratios (WWR) were used, which is a combination of wall area and glazing. Building orientation was not used.

Godoy-Shimizu (2018) mentioned that several studies considered building height to estimate the building energy performance [18]. Additionally, a survey of the literature showed no previous studies used building height and the number of floors at the same time. Even though Capozzoli et al. considered both building height and the number of floors to predict heating energy consumption in schools, they a had too-low Pearson CC to include them in the MLR model [30]. Therefore, they excluded both variables in the MLR model because of the low CC.

Im et al. (2019) used the number of floors, the building height, and their interaction term in previous research and found that they are significant considering the *p*-value [21]. However, the number of floors was removed from the regression model due to high multicollinearity (Pearson CC: 0.9). The building height improved the regression models better than the number of floors. Using both building height and the number of floors might be worthy to research in commercial building because of its variety of space and height. Lastly, the year of building renovation was never used for building energy prediction. Furthermore, Construction Year was used to be the input of the BEM.

4.3. Meteorological Variables

Meteorological variables included solar radiation, outdoor air temperature, heating degree (HD), cooling degree (CD), relative humidity, pressure, and wind speed. The future meteorological data was obtained through the open source from the nonprofit energy weather research organization the Slipstream Group [25]. They developed the weather scenarios as a representative location for each ASHRAE climate zone with their proprietary algorithm. The mentioned algorithm uses raw climate data for future weather from the NARCCAP (North American Regional Climate Change Assessment Program). Future weather scenarios for Pennsylvania (PA) were unavailable; however, those scenarios for Baltimore, Maryland (MD) were available, which is in the same climate zone 4 and moist (A) category as PA. Predicted Maryland weather data has three future weather scenarios—coldest (2047), average, and hottest (2054). The hottest weather scenario of MD was chosen to consider the biggest impact on global warming.

The preliminary studies by Im et al. used two independent meteorological variables: (1) outdoor air temperature and (2) relative humidity, representing years 2015, 2016, and 2054 [26,27]. Common factors were selected among 15 historical meteorological variables and 11 future meteorological variables. There were four climate variables in common: temperature, humidity, pressure, and wind speed. Fathi and Srinivasan (2019) used temperature, solar radiation, and humidity as meteorological factors [20]. Daut et al. (2012) also revealed a strong linear relationship between solar radiation and surface temperature [26]. Therefore, solar radiation was added to the four climate variables as a final list in this study. HD and CD were generated from temperature with the setpoint standard 18.33 °C (65 °F). Because of the high multicollinearity, temperature and HD/CD cannot be used together.

4.4. Temporal Variables

Temporal variables included the number of the week throughout the year (week's mumber), type of day (weekday, Saturday, or Sunday), numeric hour, categorical hour (0–23), hour type (working, evening, or night), business day type (business day or non-business day), and season type (spring, summer, fall, or winter).

According to Wang and Srinivasan, some researchers utilized occupancy indicators such as the time of day and day type [31]. For example, Dong et al. used the time of the day as a categorical variable to predict building energy consumption [28]. Dong et al. and Kotchen used the month of the year as a categorical variable [32,33]. Based on literature review and an expert's opinion in Information Systems and Operations Management, in this study, temporal variables used as occupancy indicators are type of day, numeric hour, categorical hour, hour type, and business day type. These temporal variables allow for observation of the occupancy condition and pattern by remedying the missing occupancy information. Boiron et al. used the month, day of the week, and hour to develop a regression model [34]. This can be effective for observing behavior patterns at residential buildings, but it does not reflect change over time despite slight improvements in R^2 of campus data. Additionally, there is less of an opportunity to change human behavior in the campus setting than to change the technology and machinery being used by the buildings. Season type was used instead of the month of the year in this study to consider the campus's characteristics. As a result, data were grouped into four seasons based on the academic calendar.

Both ELC and STM are consumed by occupants, so the presence of the users makes the difference in energy consumption. Therefore, holiday and non-holiday were separated as categorical variables. In addition, electricity consumption is for other uses rather than cooling and heating such as appliances, electronics, and lab equipment.

5. Results

5.1. Process of Variable Selection (Multivariate Regression Model Development)

Multivariate regression models were developed with 177,408 observations to estimate the ELC and STM consumption (Table 2). All numeric predictors were normalized to compare the β coefficient regardless of the unit of the variables. These models were developed with four sets of different variables. Variables in Test 1 consist of 15 numeric variables and one nominal variable. Variables in Test 2 consist of 15 numeric variables, a nominal variable, and one ordinal variable. Variables in Test 3 consist of 15 numeric variables and one nominal variable. Variables in Test 4 consist of 18 numeric variables including three interaction terms, two nominal variables, and two ordinal variables. R^2 and RMSE in Table 2 are averaged values of two models for ELC and STM. Detailed information with β coefficients is shown in Appendix A.

Table 2. Different Variables Sets Input in Multivariate regression models.

Input Variables	Data Type	Test 1	Test 2	Test 3	Test 4
$X1_{U\text{-}Wall}$	Numeric	◉	◉	◉	◉
$X2_{U\text{-}Window}$	Numeric	◉	◉	◉	◉
$X3_{U\text{-}Roof}$	Numeric	◉	◉	◉	◉
$X4_{WWR}$	Numeric	◉	◉	◉	◉
$X5_{Bldg\cdot Height}$	Numeric	◉	◉	◉	◉
$X6\text{-}1_{Const\cdot Year}$	Numeric				◉
$X6\text{-}2_{Bldg\cdot Age}$	Numeric	◉	◉	◉	
$X6\text{-}3_{Renov\cdot}$	Numeric	◉	◉	◉	◉
$X7_{Date}$	Numeric	◉			
$X8_{EPD}$	Numeric	◉	◉	◉	◉
$X9_{LPD}$	Numeric	◉	◉	◉	◉
$X10_{Solar}$	Numeric	◉	◉	◉	◉
$X11\text{-}1_{Temp\cdot}$	Numeric	◉	◉	◉	
$X11\text{-}2_{HD}$	Numeric				◉
$X11\text{-}3_{CD}$	Numeric				◉
$X12_{Humid\cdot}$	Numeric	◉	◉	◉	◉
$X13_{Pressure}$	Numeric	◉	◉	◉	◉
$X14_{WindSpeed}$	Numeric	◉	◉	◉	◉
$X15_{Weeknum}$	Numeric		◉	◉	
$X16_{Type\ of\ day}$	Nominal	◉	◉	◉	
$X17\text{-}0_{Numeric\ Hour}$	Numeric			◉	
$X17_{Categorical\ Hour\ 1\text{–}23}$	Ordinal		◉		
$X18_{Hour\ Type\ 1\text{–}2}$	Ordinal				◉
$X19_{Business\ day\ Type}$	Nominal				◉
$X20_{Bldg\ Type\ 1\text{–}4}$	Nominal				◉
$X21_{Season\ Type\ 1\text{–}3}$	Ordinal				◉
X22: X2 × X4	Numeric				◉
X23: X6-1 × X6-3	Numeric				◉
X24: X1 × X4	Numeric				◉
Average RMSE		2.78	2.71	2.6	2.12
Average R^2		0.66	0.66	0.68	0.81

There was an overlap between time variables and weather information, which reflects seasonal changes. Meteorological variables' change, construction year, and Renovation Age are reflected by time. Therefore, year, month of the year, and day were excluded for regression analysis. Adding Categorical Hour and Week's Number to Test 2, instead of only the Date as in relative importance with bootstrap confidence interval were analyzed with a one-year train dataset.

Test 1 improved the ELC and STM models' accuracy based on RMSE. Week's number was used in Test 2 and Test 3 to observe the chronological order. Because of the continuous cyclical feature of the hour, the numeric hour was used in Test 3 and transformed with sine and cosine. Then, in both cases (ELC and STM), the models' accuracy and R^2 improved. As a final step, new variables were added based on expert feedback in statistics, and some variables were eliminated. On top of three interaction terms (X22: $X1_{U\text{-}Window} \times X4_{WWR}$, X23: $X6\text{-}1_{Construction\ Year} \times X6\text{-}3_{Renov\cdot}$, X24: $X1U\text{-}_{Wall} \times X4\ _{WWR}$), four new variables are construction year instead of building age, hour type, building types, and business day type. Business day type was used to consider holidays instead of the type of day. The temperature was used initially in Test 1 through Test 3, but the R^2 and RMSE showed improvement in terms of the statistical model's accuracy and explanation of energy consumption when HD and CD were considered instead of temperature. The last change in variables enhances all two models significantly. Inaccuracy of ELC MLR were increased based on higher RMSE, but R^2 was improved.

Based on the result of Test 2, among temporal variables, the hour variable was analyzed as a categorical variable. The MLR model for ELC, 6 am–10 pm, showed a significant *p*-

value while 11 pm–5 am was insignificant, which was the nighttime, when people were asleep. For STM, the days of the week and 4 am–8 pm were significant. The rest of the hour variables were not. Descriptive data analysis is performed to capture the behavior of energy consumption over time. The hour variable is insignificant for STM, but a predictable pattern was observed in ELC (continuous increase from 3 am to 2 pm and continuous decrease during the rest of the time). Therefore, the hour variable was required by adopting sine and cosine as a periodical factor.

In conclusion, ELC usage had the longest hour timeframe that showed a meaningful difference with other times. STM was the next. According to the 2D plots, only ELC had an hourly consumption pattern as a cycle of the day, particularly owing to occupancy behavior rather than weather. Therefore, the hour variable for the ELC model, rather than STM, needed to be considered based on the findings in the multivariate regression model and descriptive analysis. Additionally, week number were insignificant for ELC. Thus, this study used multivariate analysis to compare the impact of variables for each type of energy consumption in order to grasp campus energy consumption comprehensively.

A lower U-value is better insulation for energy consumption. This applies to WWR because a high window rate compared to the wall can increase the chance of infiltration, leading to more cooling in the summer or heating in the winter season. Therefore, positive relationships are expected between the U-value of wall, the U-value of window, and the U-value of roof and energy consumption as well as between WWR and energy consumption. Final multivariate regression models were developed with 177,408 observations and 19 predictors with 3 interaction terms, consisting of 3 MLR models for ELC and STM (Table 3).

Table 3. Building number by energy types and building types.

Building Type	ELC	STM
$X1_{U\text{-}Wall}$	0.084	−0.131
$X2_{U\text{-}Window}$	−0.137	−0.001
$X3_{U\text{-}Roof}$	0.358 **	0.103
$X4_{WWR}$	−0.279 *	−0.086
$X5_{Bldg\cdot Height}$	0.056	0.130
$X6\text{-}1_{Const\cdot Year}$	−0.267 *	−0.115
$X6\text{-}3_{Renov\cdot Age}$	−0.257 *	0.139
$X8_{EPD}$	0.458 **	0.157
$X9_{LPD}$	0.070	0.070
$X10_{Solar}$	0.124	0.123
$X11\text{-}2_{HD}$	0.008	0.392 *
$X11\text{-}3_{CD}$	0.0315	−0.059
$X12_{Humid\cdot}$	−0.034	−0.033
$X13_{pressure}$	−0.007	0.021
$X14_{WindSpeed}$	0.042	0.045

Notes: Correlation is significant at different levels (2-tailed) as follows: ** 0.001 level and * 0.01 level.

5.2. Comparison of Multivariate Regression Model by Energy Type

Among building variables, there were three outstanding β coefficients in the multivariate regression model (Table 4). ELC had a high β coefficient with EPD (1.58), compared with STM (1.24). STM had a positive β coefficient with a U-Window (0.61), opposite to ELC (−0.81). STM had a higher beta with building height (0.751), compared with meteorological ELC (0.349).

Table 4. Multivariate regression model (18 buildings) without standardization.

Building Type (Unit: kBtu/GSF)	ELC	STM
Intercept	588.71	120.81
$X1_{U\text{-}Wall}$	53.48	35.20
$X2_{U\text{-}Window}$	22.05	−12.53
$X3_{U\text{-}Roof}$	−10.53	−3.27
$X4_{WWR}$	160.66	24.36
$X5_{Height}$	0.15	−0.01
$X6\text{-}1_{Construction\ Year}$	−0.30	−0.07
$X6\text{-}3_{Renov.}$	−15.33	−0.41
$X8_{EPD}$	1.00	0.79
$X9_{LPD}$	−3.71	2.80
$X10_{Solar}$	−0.01	0.00
$X11\text{-}2_{HD}$	−0.43	0.23
$X11\text{-}3_{CD}$	1.29	−0.06
$X12_{Humid}$	0.03	0.00
$X13_{Pressure}$	−0.04	0.01
$X14_{WindSpeed}$	−0.13	0.06
$X15_{bldg\ Type\_Lab}$	7.98	10.87
$X15_{bldg\ Type\_office}$	−2.83	−2.08
$X15_{bldg\ Type\_public}$	7.99	4.18
$X18\text{-}1_{Hour\ Type\_night\ hour}$	−0.51	−0.11
$X18\text{-}2_{Hour\ Type\_working\ hour}$	−1.52	0.39
$X19_{Business\ Day\ Type\_business\ day}$	−0.55	−0.04
$X21\text{-}1_{Season\ Type\_spring}$	−1.20	−0.17
$X21\text{-}2_{Season\ Type\_summer}$	1.77	−0.16
$X21\text{-}3_{Season\ Type\_winter}$	−0.31	0.04
X22: X2 × X4	−155.71	−98.53
X23: X6-1 × X6-3	−145.48	7.73
X24: X1 × X4	0.01	0.00
Adjusted R^2 (%)	94.71	66.17

Among variables, solar radiation and wind speed were insignificant for ELC, considering the 0.05 level of *p*-values. Additionally, the week (Monday to Friday) was significant, but differentiating Saturday and Sunday was meaningless. Therefore, for ELC, categorizing the day of the week into business day or non-business day is better than the type of week. Both ELC and STM have negative β coefficients with temperature, which are −0.049 and −1.729, respectively.

5.3. Result of MLR

Cross-validation was conducted for the validation, and Table 5 shows the MLR models' results. The MLR model could explain 76.74% of the ELC consumption and 56.59% of STM based on R^2. The R^2 of ELC was the highest among the three energy consumption types for validation. Additionally, MAE, RMSE, and R^2 were measured to check the regression model's accuracy to test as shown in Table 6. ELC and STM had similar values in MAE (0.00 and 0.01) and RMSE (2.62 and 1.62). The R^2s of model accuracy were ELC (88.28) and STM (74.47), respectively.

Table 5. Multiple linear regression.

Building Type	ELC	STM
Intercept	7.33	4.37
$X1_{U\text{-}Wall}$	1.39	−0.09
$X2_{U\text{-}Window}$	−0.81	0.61
$X3_{U\text{-}Roof}$	1.31	0.99
$X4_{WWR}$	2.18	1.34
$X5_{Height}$	0.31	0.52
$X6\text{-}1_{Construction\ Year}$	3.33	2.37
$X6\text{-}3_{Renov\cdot Age}$	29.04	60.51
$X8_{EPD}$	1.58	1.24
$X9_{LPD}$	−0.51	-
$X10_{Solar}$	−0.04	0.08
$X11_{HD}$	0.03	1.55
$X11_{CD}$	0.46	−0.20
$X12_{Humid\cdot}$	0.02	0.03
$X13_{Pressure}$	−0.05	0.05
$X14_{WindSpeed}$	−0.03	0.11
$X18\text{-}1_{Hour\ Type\_night}$	−0.17	−0.08
$X18\text{-}2_{Hour\ Type\_working}$	0.18	0.18
$X19_{Bizday\_business\ day}$	−0.24	−0.02
$X20\text{-}1_{bldg\ Type\_Office}$	1.78	0.90
$X20\text{-}2_{bldg\ Type\_Lab}$	1.07	−0.47
$X20\text{-}3_{bldg\ Type\_Lodge}$	−1.25	−0.15
$X20\text{-}4_{bldg\ Type\_Public}$	0.63	0.16
$X21\text{-}1_{Season\ Type\_Spring}$	−0.01	0.03
$X21\text{-}2_{Season\ Type\_Summer}$	−0.57	−0.02
$X21\text{-}3_{Season\ Type\_Winter}$	−0.70	−0.01
X22: X2 × X4	−1.17	−1.07
X23: X6-1 × X6-3	−29.54	−61.06
X24: X1 × X4	−2.20	−0.81
Adjusted R^2 (%)	76.74	56.59

Table 6. BEMs' accuracy test result of MLR.

	ELC	STM
Length	39,072	29,568
X Mean	7.33	2.54
Y Mean	7.33	2.54
MAE	0.00	0.01
RMSE	2.62	1.62
Adjusted R^2 (%)	88.28	74.47

5.4. Relative Importance Analytics

Relative importances with bootstrap confidence intervals (%) were analyzed with a one-year train dataset (177,408 observations), and metrics were normalized to sum to 100% (Figure 4).

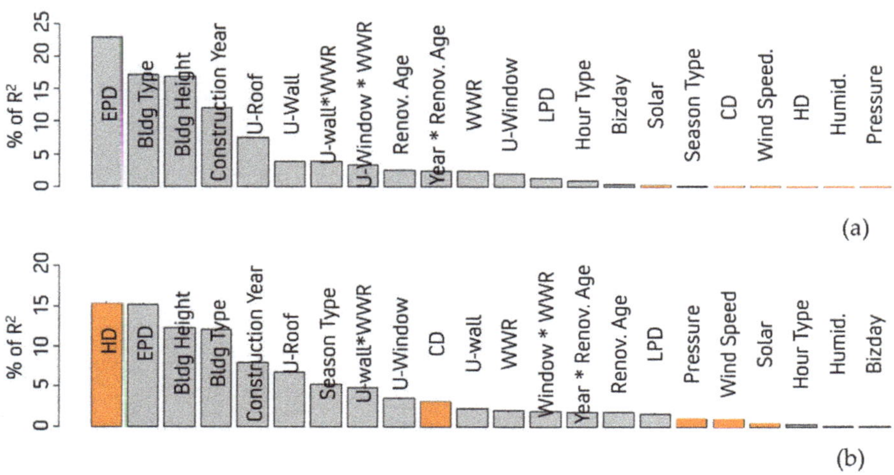

Figure 4. Relative importance of all independent variables with 95% bootstrap confidence intervals: (**a**) electricity (ELC) and (**b**) steam (STM).

Meteorological variables explained 0.06% (relative importance: 0.36%) for ELC and 4.36% (relative importance: 20.51%) for STM as revealed from R^2 of multivariate regression models. The relative contribution of solar radiation for ELC is 0.19% despite the low relative contribution rate as a most-related variable among weather variables. The relative contribution of temperature for ELC is 0.12%, which consists of CD: 0.11% and HD: 0.01%. The relationship between outdoor temperature and heating/cooling is obvious, but the relative importance analysis confirms the contribution of temperature to energy consumption. Based on relative importance analysis, HD is the most crucial predictor to predict STM. CD is the next crucial variable among meteorological variables, but other variables come prior to CD. When only weather variables were considered, relative contributions with 95% confidence intervals for HD are between 82.64% and 85.16% for STM. Relative contributions with 95% confidence intervals for CD are between 82.64% and 85.16% for STM. With other all variables, relative contributions for HD are 15.26% for STM. Relative contributions for CD are 3.07% for STM. Universities need to be aware of climate change because temperature-related variables are the top variable for STM. According to CD's rank for STM, only HD was a significantly important variable for STM consumption, and a temperature below 65 °F did not affect the heating significantly.

Proportions of variance explained by the models were the same as adjusted R^2 of multivariate regression models. To answer RQ1, the top four variables that accounted for more than 50% of the relative importance were as follows: EPD, building type, building height, and construction year for ELC; HD, EPD, building height, and building type for STM. building height are common for both energy type. EPD, construction year and building type are common for ELC and STM. Therefore, universities should choose energy efficient EPD and decide on reasonable building height in the design phase to save ELC and STM consumption. Additionally, building energy should be analyzed by building type because the laboratory building type showed totally different patterns compared to other building types.

To be specific, energy consumption patterns by building type are analyzed as follows: through descriptive statistical analysis, the rapid increase of ELC consumption in the laboratory and the office was observed. One of them may be the increasing demand of computing work in laboratory and office buildings, because this trend is not present in lodging, education, and public assembly. Laboratories are the most energy consuming building type. The laboratory building type has a larger variance in STM consumption compared to other building types, and the laboratory requires much higher STM consumption, as a regression model shows (Figure 5). This finding aligns with the findings by Ferguson et al., saying that priority should be given to buildings with high energy demands, such as research in university settings [35]. Therefore, the university needs to track the cause of energy consumption and replace it with energy efficient equipment (better EPD) or improve the insulation of building components for laboratory buildings. According to the regression model with temperature and STM consumption, laboratory buildings had the steepest slope, and office buildings had the second steepest slope. The steep slope means that more STM is consumed when the temperature is higher. Evaluating laboratory hoods is suggested to universities to improve energy efficiency, user operations, or to arrange for removing unneeded equipment [35]. Washington University St. Louis implemented low-flow fume hoods with hood occupancy controls, which led to a 40% reduction in energy use [36]. Purchasing Energy Star equipment for its offices and laboratories can be challenging for individual faculty, so universities may promote it financially or provide an endorsement of the products.

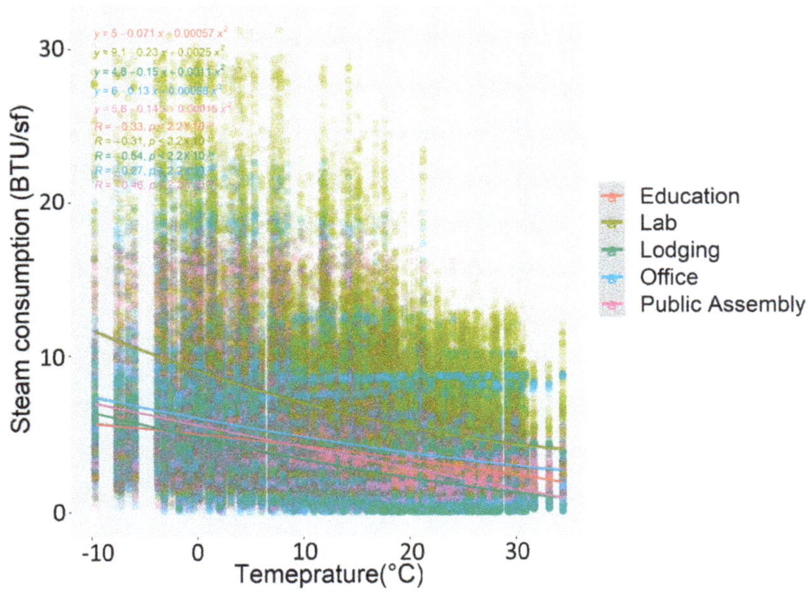

Figure 5. Scatter plot of STM consumption by temperature.

The education building type consumes STM with the smallest variance compared to other building types with the least outliers. The small variance means that estimating STM of the lodge building type is easier than other building types because it is clustered, which causes less error in prediction models.

5.5. Result of Correlation Coefficient (CC) and the Partial CC

Both CC and the partial CC were considered for analysis. A partial CC was run to determine the relationship between an individual variable and each energy consumption while controlling for the rest of numeric variables. For ELC, EPD has CC (0.505) and partial CC (0.458). U-Roof has CC (0.438) and partial CC (0.358), respectively. The relationships between EPD and ELC and between U-Roof and ELC show a direct relationship with little difference when comparing the CC and the partial CC.

For STM, the CC and partial CC of building height (0.521; 0.130), construction year (−0.429; −0.115), and EPD (0.546; 0.157) indicate that the rest of the variables had a very large influence in controlling for the relationship between these three variables and STM. In other words, these three variables are related with other variables closely, so direct relation with STM is relatively weak. There was a moderate, positive CC between HD and STM. All the results of CC, partial CC, and relative importance indicated that temperature is a significant variable for STM and has a direct relation with STM.

Bruan et al. (2014) found that temperature has the highest influence on building energy consumption with R^2 being equal to 0.92 for ELC and R^2 being equal to 0.85 for gas consumption [37]. Unlike Bruan et al.'s study, temperature was insignificant for ELC, while solar radiation was the most critical weather factor for ELC among meteorological variables. The end use needs to be compared for the appropriate comparison. Based on analyzing the relative importance of MLR models, the most effective key variables to answer for RQ1 about the appropriate building- and weather-related variables to be included in the predictive model are as follows: EPD, building type, building height, and construction year for ELC and HD, EPD, building height, and building type for STM.

Regression diagnostics plots were created to check the linearity, normality of residuals, homogeneity or residuals variance, and independence of residuals error terms. The null hypothesis of the studentized Breusch–Pagan test (BPtest) was that the residuals have constant variance. So, a *p*-value less than 0.05 would mean that the homoscedasticity assumption would have to be rejected. Both final models passed the BPtest. Furthermore, an increase in the STM consumption was observed because it is influenced by temperature (Figures 4–6).

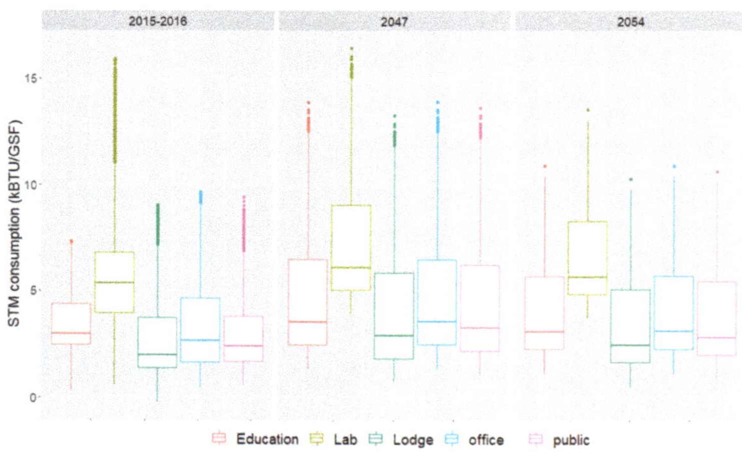

Figure 6. Boxplots for STM prediction by building type in 2047 with coldest scenario and 2054 with hottest scenario.

As observed in β coefficient analysis, the weather variable does not influence ELC consumption when STM and CHW are used as another energy source. Based on Figure 6, STM consumption will increase with the humid subtropical climate in both the hottest and coldest scenarios. Climate change will contribute to increasing STM for heating as well as countering the effect of global warming, which can easily be neglected.

5.6. Retro-Commissioning of University Campuses

Universities tend to respond to the challenge of global climate change more effectively as educational institutions compared to other building types. Other building types try to meet the requirements of building energy standards passively and inexpensively when energy policies change. Therefore, checking how universities respond to climate change can be meaningful to glimpse the trajectory for the overall building sector.

When prioritizing and implementing the deferred maintenance program, building operating efficiency and optimization should be fully considered [38]. The monitoring results lead to selecting energy-intensive buildings, which consume more campus energy than other buildings [39]. Based on this selection of the buildings, one should conduct energy audits and retro-commission them. The potential to save energy can be different from the amount of energy consumption. Thus, during the selecting process and retro-commissioning, potential factors that can bring significant change in energy saving need to be examined.

Most energy saving-related information was a prediction, and not actual energy saving results in the energy action plans. These plans tend to be unevaluated, especially when they do not achieve their goal. Universities need to pay attention to actual results by their action plans to use as a reference to evolve energy saving strategies.

Virginia Tech found that 35 percent of all buildings (50 buildings) on campus accounted for over 70 percent of overall university energy costs. Thus, they chose the top ten buildings to focus on energy saving [40]. Boston University reduced its energy consumption by 4% through its energy saving plan after eight years from 2006 while growing in size by 14% [41]. University of Pennsylvania (UPenn) also upgraded lighting in 45 buildings [42]. At UPenn, carbon dioxide equivalent has reduced 32% from STM and 9% from ELC since 2014. Nonetheless, CHW increased 17%, but emitted the lowest carbon dioxide among these three energy types [42].

To save energy, many universities have established energy action plans. Commonly observed action items from 18 universities are shown in Table 7: (1) replacing lights with energy-efficient lights; (2) installing occupancy sensor; (3) equipment reinforcement; (4) upgrading windows; (5) implementing renewable energy system (e.g., solar panels); (6) renovating roofs as green roofs or insulating roofs; and (7) managing building automation (monitoring system).

Table 7. Summary of universities' energy efficiency strategies.

University	The number of LEED Certified Buildings	Energy Efficient Lights	Motion Sensor	Equipment	Window	Geothermal System	Solar Panel	Roof
University of California Irvine	30	Yes					Yes	
University of Wisconsin at Oshkosh	6				Yes	Yes	Yes	Yes
Columbia University	7	Yes	Yes			Yes		Yes
Stanford University	4					Yes	Yes	Yes
Washington University St. Louis	21	Yes		Yes			Yes	Yes
UPenn	40	Yes						
New York University	11		Yes	Yes				
Vanderbilt University	21	Yes			Yes			
Quinnipiac University	11	Yes			Yes			
Colby College	17						Yes	
American University	4 #							
Cornell University	26	Yes			Yes			
Bentley University	5 *						Yes	
Pomona College	10				Yes		Yes	Yes
Iowa State University	22			Yes		Yes		
University of Colorado–Boulder	28						Yes	
Harvard University	134							
University of North Carolina at Chapel Hill	4 #							Yes
Average (Rank)	22	39%	11%	17%	28%	22%	44%	33%

means that LEED certifications for new construction are required; * means Energy Star certified buildings.

Table 7 showed that installing solar panels, photovoltaic panels, or solar thermal panels was the most approachable practice, and replacing lights followed next. Improving exterior building shell envelopes was also common practice. This included renovating the roof with a green roof or adding insulation (3rd place) and replacing windows (4th place).

Boston University reduced energy consumption by 2.4 million kWh per year with 8000 LED replacements [41]. Washington University St. Louis upgraded lights with low wattage bulbs, which saved an average of 376,394 kWh per year in energy. Its overall lighting plan has saved more than 20.6 million kWh in total.

Equipment reinforcement includes equipment installation and replacement. Monitoring occupancy behavior is crucial for energy analysis, and installing occupancy sensors can advance operation management. A/C is handled by a centralized chilled water plant system, allowing greater efficiencies [43]. The chiller plant has free cooling capacity that enables universities to reduce energy cost by using the cool ambient air to cool the chilled water. An ice storage system is used for cooling during peak electrical demands. Another method is converting the chiller plant to variable primary flow [39]. It is also beneficial to use economizer ventilation for heat removal [35].

Renovating windows and roofs is one of the most practical renovating methods. Universities can replace single-paned windows with double-paned or triple-paned, thermally efficient glass [35]. Adding insulation during roof replacements or renovating roofs as green roofs is another practical renovation solution.

Implementing a renewable energy system to campus buildings is another energy saving method [41]. The University of Wisconsin at Oshkosh installed solar or photovoltaic panels, which produce 3 million BTUs per day. They provide 70% of the hot water needs for that building with the rate at a total of 47.1 kW, or enough to power four homes. UPenn analyzed its buildings and building systems and determined nine buildings for HVAC and systems replacement.

Leveraging campus building automation systems (BAS) achieves an optimal balance of occupant comfort and energy efficiency through effective building automation and control. A BAS is an integrating component for supply and exhaust fans, pumps, and air handling units, together with components such as flow control valves, air dampers, mixing boxes, instrumentation, thermostats, and humidity control. As a comprehensive monitoring and optimizing system, temperature, pressure, humidity, air balance, and flow rates (both air and water) are controlled to perform effective building occupant safety and comfort and efficient building operation [44]. Additionally, evaluating the feasibility of additional metering during each building renovation project can be implemented [40]. Increasing metering to monitor electricity, steam, and chilled water across campus is beneficial to supervising and managing electrical consumption [35,39,43]. For example, Power Monitoring Expert (PME) software can be used to predict and achieve optimum efficiency [45]. Through BA management, universities can prioritize and identify ways to optimize energy efficiency.

Monitoring temperature control and energy usage throughout the campus saved more than $5 million per year in electricity costs. For example, UPenn used a steam trap testing program over five years and reduced lost steam costs by over 1.2 million dollars [43]. Overall, closely coordinating with facilities services to request and accomplish the prompt repair of building mechanical, electrical, and plumbing systems can bring a difference in energy saving.

6. Discussion

One of the main findings of this research is that there are various perspectives, such as numeric and categorical, to use the same data by forming the data differently. Variables can be in multiple formats depending on research questions, model accuracy, or data availability. For example, when hourly data with date is used, it can be used as (1) timestamp (numeric time with the same interval), (2) categorical variable: periods (categorized by year, month, day, hour, season), type of day, or day of week (3) numeric variable: transformed time considered a periodical feature or time of the day. Another example is that building material that can be used as categorical, binomial, or numeric data. The material itself can be used as categorical data itself, and binomial data can be used when the research focuses on the existence of the material. The other way is quantifying building characteristics, such as the thickness of the wall, density of wall, the thermal conductivity of wall, R-value of the wall, and U-value of the wall. However, since these building characteristics share common elements in their calculation, multicollinearity issues should be handled properly.

Except for the season type for STM, temporal variables are insignificant to estimate energy consumption. Among meteorological variables, the temperature is the solely key variable and expected to increase as evident in STM increase; temperature increase is because of global warming. The developed BEMs showed that the EPD is the most important factor for ELC and the second most important factor for STM consumption. The prediction models give an insight of which factors remain essential and applicable to campus building policy to prevent wasting energy in buildings because of climate change. Therefore, universities need to focus on making action plans accordingly by energy type. For ELC, in order of importance, the vital building variables are EPD, building height, construction year, and U-value of roof. For STM, the vital building variable does not exist.

To predict the future energy use of university campus buildings, derived MLR models were used to estimate energy consumption in 2047 and 2054. In the campus setting, ELC is mainly used for lighting and other appliances, which are barely influenced by the weather. There is significant impact of global warming on building-energy use for heating in the university campus according to prediction result of STM consumption. According to Fan et al. (2014), some independent weather variables, such as the maximum dry-bulb temperature of the prediction day, were used regardless of the BEM methodology [30]. Fan et al. removed these weather variables from all seven different ML methods, although some researchers found that the relative humidity and wind speed are significant [22,23]. Based on CC, partial CC, and relative important inspection, this study's findings on meteorological variables were aligned with Fan et al.'s findings. On one hand, temperature was the only meteorological predictor, showing significant relevant results with energy consumption for cooling and heating. Meteorological variables except for solar radiation are insignificant to measure ELC. This is because "electricity is used for many more end-uses other than space heating and cooling [46]". In other words, people use light or other appliances regardless of the weather. On the other hand, temperature above 65 °F is an important variable for STM. It is because STM is used for heating, and CHW is used for cooling in the university campus, unlike residential buildings.

The more variables that are added, the better the explanation that can be made. Therefore, a polynomial regression model with reduced variables was not used, as it had an overfitting problem with variables having a higher power than two. However, a polynomial regression method has merit because it uses only selected variables without the intervention of other minor influential variables.

The developed BEMs showed that the EPD is the second most important factor for STM consumption. However, a low partial CC between EPD and STM was observed, indicating that there is a high indirect impact by other variables related with EPD. Additionally, EPDs of individual buildings are not related to the central controlled boiler for STM. More investigation is necessary for the better explanation of multivariate regression and MLR models.

Universities have saved energy consumption through the renovation of buildings, but building-related variables were not critical based on the developed MLR. This could be because mean WWR was used in this study instead of eastern WWR, western WWR, southern WWR, and northern WWR. Likewise, other building characteristics are averaged values. Additionally, occupancy features and operating variables were inaccessible. This includes building use schedule, heat gain through lights and people, and number of occupants [47]. For more accurate prediction of energy consumption, more variables need to be evaluated.

7. Conclusions

To fight global warming, we need to be aware of the importance of energy saving. In building construction, there are a variety of strategies to reduce building energy consumption. This study used historical data to raise the awareness of people who are linked to this university on the impact of climate change and how to deal with it from a building aspect. The findings can help universities to reduce energy consumption and cost. To answer RQ2, the relationship between input variables and energy consumption by energy types was examined, considering descriptive statistics, relative importance analysis, and BEMs.

This study examined the crucial variables for ELC and STM consumption via statistical analysis methods. In order to explain relationships between variables and energy consumption as well as interpretations of the MLR models, this study analyzed various descriptive statistics. The developed multivariate regression models with different variable sets were analyzed, and stepwise feature selection was used to derive MLR models. For ELC, both multivariate regression models and MLR models suggest that the meteorological variables are insignificant, unlike for STM. The most important four variables out of 22 variables are EPD, building type, building height, and construction year for ELC. These are HD, EPD, building height, and building type for STM.

Among meteorological variables, only the temperature was the outstanding weather variable to estimate STM. EPD was the critical variable for ELC, so universities should focus on EPD when they devise campus energy action plans. Building height was another crucial factor contributing to ELC, so building energy must be considered in the design phase to decide the appropriate building height. The U-value of the roof was critical for ELC. Upgrading windows is regarded as an easy and widespread implementation to boost building energy efficiency. However, the roof needs to be considered as well in order to create an energy-efficient building.

This study observed significant increases in STM consumption in the future (2047 and 2054) in humid subtropical climate zones. The future energy use of university campus buildings needs to be closely monitored as temperature increases due to global warming. Therefore, continuous efforts are required to combat climate change through retro-commissioning by observing and implementing energy retrofit projects based on previous audits. There are many ways that universities can save energy by adopting an energy policy for the campus. For example, they can upgrade chiller programming and investigate the effects of scheduling building fans, increasing set-points, and reducing areas of dehumidification to reduce ELC use. For STM, continuing a steam maintenance program such as replacing steam traps and repairing steam and condensation leaks is the key to energy saving.

One of the limitations of this work was that the sample data was insufficient to categorize by building type. According to Zhai and Helman, a campus-wide analysis of energy use and building characteristics differentiates by building type [31]. That research classified the University of Michigan buildings into laboratories, clinics, service buildings, campus buildings, and residential buildings. Additionally, Amber et al. categorized the campus buildings into administrative buildings or academic buildings [4]. This seems too general to categorize building types, but having two building types can provide a larger sample size for each category. For further research, the sample size should be enlarged to categorize the buildings by building types for a more in-depth analysis. Lastly, more detailed information is required to investigate building renovations because aesthetic improvements or interior renovations do not affect energy consumption.

Author Contributions: Conceptualization, H.I. and R.S.S.; methodology, H.I., R.S.S., and S.K.; software, H.I.; validation, H.I.; formal analysis, H.I.; investigation, H.I.; resources, R.S.S.; data curation, H.I., R.S.S., D.M.; writing—original draft preparation, H.I.; writing—review and editing, H.I., R.S.S., R.L.S.; visualization, H.I.; supervision, R.S.S.; funding acquisition, S.K. All authors have read and agreed to the published version of the manuscript.

Funding: The fifth author's research is partially funded by NSF-DMS 2124222.

Data Availability Statement: Owing to privacy concerns and confidentiality, we are not able to disclose the university campus name and specific geographic location (address). Nevertheless, we were fortunate to receive high-fidelity university campus building data to conduct this study. If you have further questions, please contact Dr. Ravi Srinivasan at sravi@ufl.edu.

Conflicts of Interest: The authors declare no conflict of interest.

Appendix A. Multiple Linear Regression Models

Table A1. Multivariate Regression Models.

Test Number	1	2	3	4	1	2	3	4
Energy Type	**ELC**	**ELC**	**ELC**	**ELC**	**STM**	**STM**	**STM**	**STM**
Intercept	1.49	7.35	7.01	7.33	−5.67	3.99	3.01	4.37
$X1_{U\text{-}Wall}$	7.08	0.31	1.18	1.39	−0.97	−0.47	−1.71	−0.09
$X2_{U\text{-}Window}$	−1.68	−0.51	−1.61	−0.81	−3.18	−0.78	−2.44	0.61
$X3_{U\text{-}Roof}$	9.05	1.27	4.78	1.31	2.39	0.32	1.20	0.99
$X4_{WWR}$	−13.18	−1.01	−3.72	2.18	−3.88	−0.28	−1.04	1.34
$X5_{Height}$	0.02	0.35	1.34	0.31	0.02	0.75	2.95	0.52
$X6\text{-}1_{construction\ year}$				3.33				2.37
$X6\text{-}2_{Bldg\cdot Age}$	−0.04	−1.30	−4.23		−0.01	−0.51	−1.64	
$X6\text{-}3_{Renov\cdot}$	−0.07	−1.09	−4.69	29.04	0.04	0.53	2.37	60.51
$X7_{Date}$	0.00				0.00			
$X8_{EPD}$	1.28	3.35	11.18	1.58	0.42	0.97	3.21	1.24
$X9_{LPD}$	2.22	0.27	1.20	−0.51	1.05	0.25	1.11	-
$X10_{Solar}$	0.00	0.03	−0.02	−0.04	0.00	0.08	−0.34	0.08
$X11\text{-}1_{Temp\cdot}$	0.00	−0.05	−0.05		−0.19	−1.73	−1.73	
$X11\text{-}2_{HD}$				0.03				1.55
$X11\text{-}3_{CD}$				0.46				−0.20
$X12_{Humid\cdot}$	−0.01	−0.03	−0.05	0.02	0.00	−0.11	−0.19	0.03
$X13_{pressure}$	-	−0.05	−0.30	−0.05	0.01	0.03	0.23	0.05
$X14_{WindSpeed}$	0.06	0.01	0.11	−0.03	0.06	0.07	0.61	0.11
$X15_{Weeknum}$		0.00	0.00			0.00	−0.03	
$X16\text{-}1_{Type\ of\ Day\_Weekday}$	0.75	0.72	0.74		0.09	0.03	0.07	
$X16\text{-}2_{Type\ of\ Day\_Sunday}$	−0.02	−0.02	−0.01		0.04	0.01	0.03	
$X17_{Numeric\ Hour}$			−0.04				−0.08	
$X17\text{-}1_{Hour}$		−0.06				−0.01		
$X17\text{-}2_{Hour}$		−0.06				−0.03		
$X17\text{-}3_{Hour}$		−0.11				0.03		
$X17\text{-}4_{Hour}$		−0.09				0.15		
$X17\text{-}5_{Hour}$		−0.07				0.27		
$X17\text{-}6_{Hour}$		0.19				0.51		
$X17\text{-}7_{Hour}$		0.49				0.77		
$X17\text{-}8_{Hour}$		0.84				1.01		
$X17\text{-}9_{Hour}$		1.26				1.10		
$X17\text{-}10_{Hour}$		1.60				1.07		
$X17\text{-}11_{Hour}$		1.72				1.10		
$X17\text{-}12_{Hour}$		1.73				0.98		
$X17\text{-}13_{Hour}$		1.79				0.96		
$X17\text{-}14_{Hour}$		1.79				0.94		
$X17\text{-}15_{Hour}$		1.75				0.80		
$X17\text{-}16_{Hour}$		1.66				0.66		
$X17\text{-}17_{Hour}$		1.42				0.58		
$X17\text{-}18_{Hour}$		1.13				0.46		
$X17\text{-}19_{Hour}$		0.89				0.37		
$X17\text{-}20_{Hour}$		0.58				0.21		
$X17\text{-}21_{Hour}$		0.43				0.14		
$X17\text{-}22_{Hour}$		0.28				0.06		
$X17\text{-}23_{Hour}$		0.11				0.03		
$X18\text{-}1_{Hour\ Type\_night}$				−0.17				−0.08
$X18\text{-}2_{Hour\ Type\_working}$				0.18				0.18
$X19_{Business\ day\ Type}$				−0.24				−0.02
$X20\text{-}1_{bldg\ Type\_office}$				1.78				0.90
$X20\text{-}2_{bldg\ Type\_Lab}$				1.07				−0.47
$X20\text{-}3_{bldg\ Type\_Lodge}$				−1.25				−0.15
$X20\text{-}4_{bldg\ Type\_public}$				0.63				0.16
$X21\text{-}1_{Season\ Type\_spring}$				−1.20				−0.17
$X21\text{-}2_{Season\ Type\_summer}$				1.77				−0.16
$X21\text{-}3_{Season\ Type\_winter}$				−0.31				0.04
X22: X2 × X4				−1.17				−1.07
X23: X6-1 × X6-3				−29.54				−61.06
X24: X1 × X4				−2.20				−0.81
RMSE	2.82	2.77	2.57	2.62	2.74	2.65	2.63	1.62
R^2	0.76	0.76	0.80	0.88	0.56	0.55	0.56	0.74

References

1. *Why The Building Sector?—Architecture 2030*; Architecture 2030: Santa Fe, NM, USA.
2. International Energy Agency Buildings. A Source of Enormous Untapped Efficiency Potential. Available online: https://www.iea.org/topics/buildings (accessed on 27 September 2020).
3. International Code Council 2012 IECC—International Energy Conservation Code. Available online: https://basc.pnnl.gov/images/iecc-climate-zone-map (accessed on 27 September 2020).
4. Business Energy Advisor Colleges and Universities. Available online: https://esource.bizenergyadvisor.com/article/colleges-and-universities (accessed on 14 September 2020).
5. Amber, K.P.; Aslam, M.W.; Mahmood, A.; Kousar, A.; Younis, M.Y.; Akbar, B.; Chaudhary, G.Q.; Hussain, S.K. Energy Consumption Forecasting for University Sector Buildings. *Energies* **2017**, *10*, 1579. [CrossRef]
6. Reinhart, C.F.; Cerezo Davila, C. Urban Building Energy Modeling—A Review of a Nascent Field. *Build. Environ.* **2016**, *97*, 196–202. [CrossRef]
7. Li, W.; Zhou, Y.; Cetin, K.; Eom, J.; Wang, Y.; Chen, G.; Zhang, X. Modeling Urban Building Energy Use: A Review of Modeling Approaches and Procedures. *Energy* **2017**, *141*, 2445–2457. [CrossRef]
8. Baker K.J.; Rylatt, R.M. Improving the Prediction of UK Domestic Energy-Demand Using Annual Consumption-Data. *Appl. Energy* **2008**, *85*, 475–482. [CrossRef]
9. Kavousian, A.; Rajagopal, R.; Fischer, M. Determinants of Residential Electricity Consumption: Using Smart Meter Data to Examine the Effect of Climate, Building Characteristics, Appliance Stock, and Occupants' Behavior. *Energy* **2013**, *55*, 184–194. [CrossRef]
10. Hsu, D. How Much Information Disclosure of Building Energy Performance Is Necessary? *Energy Policy* **2014**, *64*, 263–272. [CrossRef]
11. Zeng, A.; Liu, S.; Yu, Y. Comparative Study of Data Driven Methods in Building Electricity Use Prediction. *Energy Build.* **2019**, *194*, 289–300. [CrossRef]
12. Walter, T.; Sohn, M.D. A Regression-Based Approach to Estimating Retrofit Savings Using the Building Performance Database. *Appl. Energy* **2016**, *179*, 996–1005. [CrossRef]
13. Kikumoto, H.; Ooka, R.; Arima, Y.; Yamanaka, T. Study on the Future Weather Data Considering the Global and Local Climate Change for Building Energy Simulation. *Sustain. Cities Soc.* **2015**, *14*, 404–413. [CrossRef]
14. Hong, W.; Kim, J.; Lee, C.; Jeon, G. Energy Consumption and the Power Saving Potential of a University in Korea: Using a Field Survey. *J. Asian Archit. Build. Eng.* **2011**, *10*, 445–452. [CrossRef]
15. Hong, T.; Chen, Y.; Luo, X.; Luo, N.; Lee, S.H. Ten Questions on Urban Building Energy Modeling. *Build. Environ.* **2020**, *168*, 106508. [CrossRef]
16. Chung, M.H.; Rhee, E.K. Potential Opportunities for Energy Conservation in Existing Buildings on University Campus: A Field Survey in Korea. *Energy Build.* **2014**, *78*, 176–182. [CrossRef]
17. Guan, J.; Nord, N.; Chen, S. Energy Planning of University Campus Building Complex: Energy Usage and Coincidental Analysis of Individual Buildings with a Case Study. *Energy Build.* **2016**, *124*, 99–111. [CrossRef]
18. Kim, D.-S. Forecasting Environmental Impact Assessment of Residential Buildings in Florida under Future Climate Change. Ph.D. Thesis, University Of Florida, Gainesville, FL, USA, 2018; p. 133.
19. Fathi, S.; Srinivasan, R.; Ries, R. *Campus Energy Use Prediction (CEUP) Using Artificial Intelligence (AI) to Study Climate Change Impacts*; IBPSA: Rome, Italy, 2019; pp. 3594–3601.
20. Fathi, S.; Srinivasan, R. Climate Change Impacts on Campus Buildings Energy Use: An AI-Based Scenario Analysis. In Proceedings of the 1st ACM International Workshop on Urban Building Energy Sensing, Controls, Big Data Analysis, and Visualization; Association for Computing Machinery: New York, NY, USA, 2019; pp. 112–119.
21. Godoy-Shimizu, D.; Steadman, P.; Hamilton, I.; Donn, M.; Evans, S.; Moreno, G.; Shayesteh, H. Energy Use and Height in Office Buildings. *Build. Res. Inf.* **2018**, *46*, 845–863. [CrossRef]
22. Fumo, N.; Rafe Biswas, M.A. Regression Analysis for Prediction of Residential Energy Consumption. *Renew. Sustain. Energy Rev.* **2015**, *47*, 332–343. [CrossRef]
23. Mohammadiziazi, R.; Bilec, M.M. Application of Machine Learning for Predicting Building Energy Use at Different Temporal and Spatial Resolution under Climate Change in USA. *Buildings* **2020**, *10*, 139. [CrossRef]
24. Campagna, L.M.; Fiorito, F. On the Impact of Climate Change on Building Energy Consumptions: A Meta-Analysis. *Energies* **2022**, *15*, 354. [CrossRef]
25. Hoff, T.E.; Perez, R. Modeling PV Fleet Output Variability. *Sol. Energy* **2012**, *86*, 2177–2189. [CrossRef]
26. Im, H.; Srinivasan, R.; Fathi, S. Building Energy Use Prediction Owing to Climate Change: A Case Study of a University Campus. In Proceedings of the 1st ACM International Workshop on Urban Building Energy Sensing, Controls, Big Data Analysis, and Visualization; Association for Computing Machinery: New York, NY, USA, 2019; pp. 43–50.
27. Im, H.; Srinivasan, R.; Jia, M. Forecasting Chilled Water Consumption under Climate Change: Regression Analysis of University Campus Buildings. In *Construction Research Congress 2020: Computer Applications*; American Society of Civil Engineers: Reston, VA, USA, 2020; pp. 896–904. [CrossRef]

28. Fathi, S.; Srinivasan, R.S. Analysis of Energy Performance of University Campus Buildings Using Statistical and Energy Modeling Approaches. In Proceedings of the 2015 Winter Simulation Conference (WSC), Huntington Beach, CA, USA, 6–9 December 2015; pp. 3356–3366.
29. Gao, W.; Alsarraf, J.; Moayedi, H.; Shahsavar, A.; Nguyen, H. Comprehensive Preference Learning and Feature Validity for Designing Energy-Efficient Residential Buildings Using Machine Learning Paradigms. *Appl. Soft Comput.* **2019**, *84*, 105748. [CrossRef]
30. Capozzoli, A.; Grassi, D.; Causone, F. Estimation Models of Heating Energy Consumption in Schools for Local Authorities Planning. *Energy Build.* **2015**, *105*, 302–313. [CrossRef]
31. Wang, Z.; Srinivasan, R.S. A Review of Artificial Intelligence Based Building Energy Use Prediction: Contrasting the Capabilities of Single and Ensemble Prediction Models. *Renew. Sustain. Energy Rev.* **2017**, *75*, 796–808. [CrossRef]
32. Dong, Z.; Liu, J.; Liu, B.; Li, K.; Li, X. Hourly Energy Consumption Prediction of an Office Building Based on Ensemble Learning and Energy Consumption Pattern Classification. *Energy Build.* **2021**, *241*, 110929. [CrossRef]
33. Kotchen, M.J. Longer-Run Evidence on Whether Building Energy Codes Reduce Residential Energy Consumption. *J. Assoc. Environ. Resour. Econ.* **2017**, *4*, 135–153. [CrossRef]
34. Boiron, A.; Lo, S.; Marot, A. *Predicting Future Energy Consumption*; CS229 Project Report; Stanford University: Stanford, CA, USA, 2012; p. 5.
35. Ferguson, T.; Leivano, R.; Mageau, M.; Bates, T.; Brown, T.; Axler, R.; Stark, S.; Sawyer, J.; Bushnell, T.; Novek, K.; et al. *University of Minnesota Duluth Energy Action Plan—Version 2.0*; UMD Sustainability: College Park, MD, USA, 2011.
36. Green Labs | Sustainability. Available online: https://sustainability.wustl.edu/get-involved/staff-faculty/green-labs-2/ (accessed on 24 November 2020).
37. Braun, M.R.; Altan, H.; Beck, S.B.M. Using Regression Analysis to Predict the Future Energy Consumption of a Supermarket in the UK. *Appl. Energy* **2014**, *130*, 305–313. [CrossRef]
38. Initiative 10—Building Energy Efficiency Upgrades and Optimization—Capital Investment. Available online: https://utilities.tamu.edu/energy-action-plan-2015/initiative-10-building-energy-efficiency-capital-upgrades/ (accessed on 20 November 2020).
39. Energy Action Plan. 2020. Available online: https://utilities.tamu.edu/energy-action-plan-2020/ (accessed on 20 November 2020).
40. Briehl, A.; Owczarski, M. $3 Million Investment Provides Boost to University Energy Efficiency Efforts. Available online: https://www.vtnews.vt.edu/content/vtnews_vt_edu/en/articles/2020/02/ops-energyactionplan.html (accessed on 24 November 2020).
41. 25 of the Most Energy Efficient Colleges. Available online: https://www.electricchoice.com/blog/25-of-the-most-energy-efficient-colleges/ (accessed on 23 November 2020).
42. Conserving Energy. Available online: https://www.sustainability.upenn.edu/initiatives/conserving-energy (accessed on 20 November 2020).
43. Energy Management. Available online: https://www.facilities.upenn.edu/sustainability/energy-management (accessed on 20 November 2020).
44. Initiative 3—Comprehensive Building Automation Management. Available online: https://utilities.tamu.edu/energy-action-plan-2020/initiative-3-comprehensive-building-automation-management/ (accessed on 20 November 2020).
45. Initiative 4—Comprehensive Utility Metering, Data Management, Billing, and Reporting. Available online: https://utilities.tamu.edu/energy-action-plan-2020/initiative-4-comprehensive-utility-metering-data-management-billing-and-reporting/ (accessed on 20 November 2020).
46. Comstock, O.; Jarzomski, K. Consumption and Saturation Trends of Residential Miscellaneous End-Use Loads. In Proceedings of the 2012 ACEEE Summer Study on Energy Efficiency in Buildings, Pacific Grove, CA, USA, 12–17 August 2012; pp. 9-63–9-74.
47. Amasyali, K.; El-Gohary, N.M. A Review of Data-Driven Building Energy Consumption Prediction Studies. *Renew. Sustain. Energy Rev.* **2018**, *81*, 1192–1205. [CrossRef]

energies
MDPI

Article

Assessment of Waterfront Office Redevelopment Plan on Optimal Building Arrangements with Rooftop Photovoltaics: A Case Study for Shinagawa, Tokyo

Younghun Choi [1], Takuro Kobashi [1,*], Yoshiki Yamagata [1] and Akito Murayama [2]

[1] Center for Global Environment Research, National Institute for Environment Studies, Tsukuba 305-8056, Japan; zlzlaga@gmail.com (Y.C.); yamagata@sdm.keio.ac.jp (Y.Y.)
[2] Department of Urban Engineering, School of Engineering, The University of Tokyo, Tokyo 113-8656, Japan; murayama@up.t.u-tokyo.ac.jp
* Correspondence: kobashi.takuro@nies.go.jp

Abstract: Designing waterfront redevelopment generally focuses on attractiveness, leisure, and beauty, resulting in various types of building and block shapes with limited considerations on environmental aspects. However, increasing climate change impacts necessitate these buildings to be sustainable, resilient, and zero CO_2 emissions. By producing five scenarios (plus existing buildings) with constant floor areas, we investigated how buildings and district forms with building integrated photovoltaics (BIPV) affect energy consumption and production, self-sufficiency, CO_2 emission, and energy costs in the context of waterfront redevelopment in Tokyo. From estimated hourly electricity demands of the buildings, techno-economic analyses were conducted for rooftop PV systems for 2018 and 2030 with declining costs of rooftop PV systems. We found that environmental building designs with rooftop PV system are increasingly economical in Tokyo with CO_2 emission reduction of 2–9% that depends on rooftop sizes. Payback periods drop from 14 years in 2018 to 6 years in 2030. Toward net-zero CO_2 emissions by 2050, immediate actions are necessary to install rooftop PVs on existing and new buildings with energy efficiency improvements by construction industry and building owners. To facilitate such actions, national and local governments need to adopt appropriate policies.

Keywords: building; electricity demand; photovoltaics; techno-economic analysis; urban decarbonization; CO_2 emission

Citation: Choi, Y.; Kobashi, T.; Yamagata, Y.; Murayama, A. Assessment of Waterfront Office Redevelopment Plan on Optimal Building Arrangements with Rooftop Photovoltaics: A Case Study for Shinagawa, Tokyo. *Energies* **2022**, *15*, 883. https://doi.org/10.3390/en15030883

Academic Editor: Wilfried van Sark

Received: 3 January 2022
Accepted: 24 January 2022
Published: 26 January 2022

1. Introduction

About 75% of global power consumption and 60–70% of greenhouse gas emissions originate from cities [1,2]. However, as the center of economic competitiveness and innovation, cities are also the sources of solutions [2,3]. Smart city is one of the necessary ingredients to urban sustainability contributing on recent urban challenges such as rapid expansion of urban population and decarbonization. Increasing digitization, development of Information and Communication Technology (ICT) and artificial intelligence (AI) is expected to play substantial roles on the development of decentralized urban power systems. In addition, declining costs of PV systems and EVs with increasingly tighter regulations are rapidly introducing these technologies into urban energy systems, which are integrated by the smart city technologies as distributed energy resources (DER) [4]. Studies indicated that rooftop PVs plus EVs as batteries can play substantial roles on urban decarbonization supplying up to 95% of affordable CO_2-free electricity to urban dwellers in nine Japanese cities known as the SolarEV City concept [5,6].

The Government of Japan announced that Japan aims to reach net-zero emission by 2050. Therefore, it is critical that all urban planning processes are to be assessed for future zero-emission. As Japan is constituted in four main islands with long coastlines, waterfront redevelopments are one of the higher priority policy options for many local governments to increase life quality of citizens and to attract tourists. As a consequence, many redevelopments of river or coastal sides are taking place [7–9] with benefits to improve economic values, environmental conditions, transport and social services, economic investment opportunities on currently degraded areas. At the time of rapid energy transition toward net-zero CO_2 emission, this waterfront redevelopment planning must also integrate energy efficient building, block design, renewable energies such as tidal power, hydroelectric power, and solar power for their energy demands.

Energy demands for office buildings are created for various services such as lighting, space cooling and heating, office appliances, elevator, etc. Space heating and cooling demands (e.g., about 28% of the total office building energy demand in Japan [10]) are controlled by various factors such as building wall materials, efficiency of heating, ventilation, and air conditioning (HVAC) system, building shapes, and influence of shades by neighboring buildings. Therefore, to achieve efficient building energy systems, it is necessary to conduct energy assessments in the early planning phase of redevelopment with proper tools and methods [11–14]. In addition, retrofitting existing buildings needs to be considered for a rapid reduction of CO_2 emission to reach net-zero emission by 2050 [15,16]. Expected rapid developments of PV technologies for the coming decades in terms of costs, efficiency, weights, and design, will provide unprecedented opportunities for these measures to be effective and beneficial to building owners.

Urban building energy modeling (UBEM) with three-dimensional (3D) representation are rapidly developing, and more and more applied to assess sustainable urban building forms [13,17–19]. For example, "Rhinoceros 3D" is a computer 3D graphics for computer-aided design (CAD), and its plug-in Grasshopper is a visual programming environment. "Grasshopper" hosts various energy modeling tools such as Ladybugs and Honeybee, which further connect with a well-known building energy modeling tool such as "EnergyPlus". The analysis can be made for a building or building blocks in various resolution in time and space. As often hourly building energy demand is not publicly available owing to privacy, etc., these models are important for the assessments on the viability of variable renewable energies (VREs) such as BIPV in urban environments.

The tools have been utilized for various applications. Natanian et al. [12] analyzed various nearly zero energy building and district types between courtyard, scatter, slab, high-rise, and courtyard in the hot/dry climate of Mediterranean. They found the courtyard typology performs to be the best option in terms of energy balance, but with less optimal performance in daylight utilization. Then, also Natanian et al. [20] introduced an energy and environmental quality evaluation workflow. Zhang et al. [21] compared energy demand and solar potentials of different block types in the hot and humid climate of Singapore. They found solar energy harvesting amount can increase up to 200% depending on block types with other variables constant except morphology. Chang et al. [14] investigated relationship between design parameters and urban performance parameters such as energy demands, solar harvesting potential, and sky view factor for university campus design in Shenzhen, China. They applied statistical approaches and identified optimal building coverage ratio and sky view factor.

Actual implementation of renewable energy projects such as BIPVs depends on financial merits in comparison to existing energy systems such as grid electricity [22]. Techno-economic analyses can assess if a renewable energy project is viable considering the costs of technologies, discount rate, project period, degradation, electricity tariff, insolation changes, etc. [23]. As the cost of PV systems is expected to drop further [24], the viability of PV projects also improves significantly in the coming decades, increasing potentials of rooftops PVs [25]. Many studies have been conducted to test viability of rooftop PV systems coupled with battery and EV as battery for households [26–28]. For example, Lang et al. investigated residential and commercial buildings for viability using techno-economic analyses for Germany, Switzerland, and Austria [29]. They found that the rooftop PVs are already attractive to many buildings without subsidies. However, few studies have investigated impacts of building and block design for waterfront office redevelopment on energy demand, rooftop PV generation, CO_2 emissions, considering declining costs of PV systems from 2018 to 2030.

In this study, we conducted environmental and energy analyses for waterfront office building redevelopment in Shinagawa, Tokyo, Japan as a test site (Figure 1). We produced five scenarios with different building and district forms (scenario 1–5) in comparison to existing buildings (scenario 0), which include nine buildings on average. Energy demands of buildings and rooftop PV generation in an hourly resolution are estimated for all the scenarios using "Rhinoceros 3D" and its plug-ins Grasshopper, etc., considering energy balance between in- and out-side of buildings with a weather file for 2018 as an input. Total floor areas of buildings and site area in all the scenarios are set constant for comparison purposes. Then, techno-economic analyses were conducted using System Advisor Model (SAM) [30] to assess the viability of rooftop PV systems for 2018 and 2030 to evaluate impacts of increasingly cheaper rooftop PV systems. Finally, environmental and energy indicators such as CO_2 emission, self-sufficiency, self-consumption, and energy sufficiency were evaluated for each scenario as well as financial indicators (net present values (NPV), payback periods, and levelized costs of electricity (LCOE)).

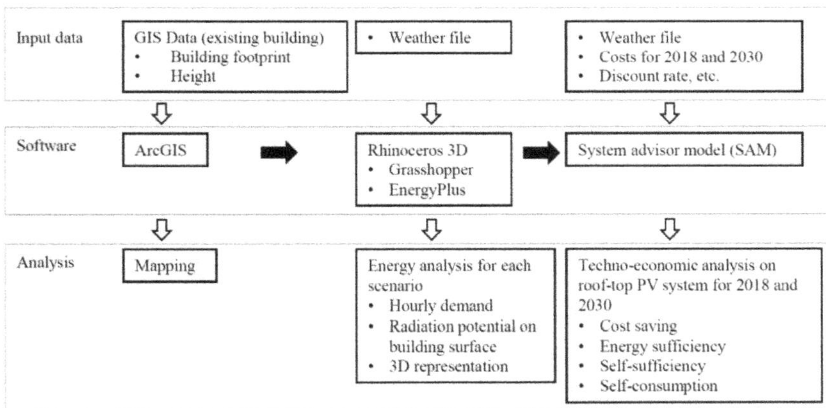

Figure 1. Evaluation workflow for waterfront building environmental energy analyses.

In the following Section 2, methodologies of the analyses were presented using Rhinoceros 3D, Grasshopper, and SAM. In the Section 3, estimated hourly energy demands for scenarios were presented, and various indicators were calculated and compared between scenarios. The implications of the results were discussed in the Section 4. Finally, we summarize and conclude our findings in the Section 5.

2. Materials and Methods

The test site, Shinagawa area (35.6° N, 139.7° W) is located near Shinagawa Railway Station in Minato Wards, Tokyo, Japan. The Shinagawa Station is one of the busiest railway stations in Japan with annually 380,000 users. Land use of Shinagawa is divided by the Shinagawa Station. West side of the station is mainly for residential-oriented mixed-use area, and east side is office/industry-oriented mixed-use area where the test site is located. The test site (Figure 2) is near harbor along Tokyo Bay with canals going through the middle of the district. Currently, this waterfront area is not actively utilized as a recreation area considering their potentials. Shinagawa experiences maximum daily average temperature of 30 °C in summer and minimum temperature of 0 °C (Figure 3) with snow fall only occurring a few times a year. Coastal regions of Japan along Pacific Ocean including Shinagawa have generally fine weather in winter, as reflected in high quality PV generation but with shorter daytime (Figure 3).

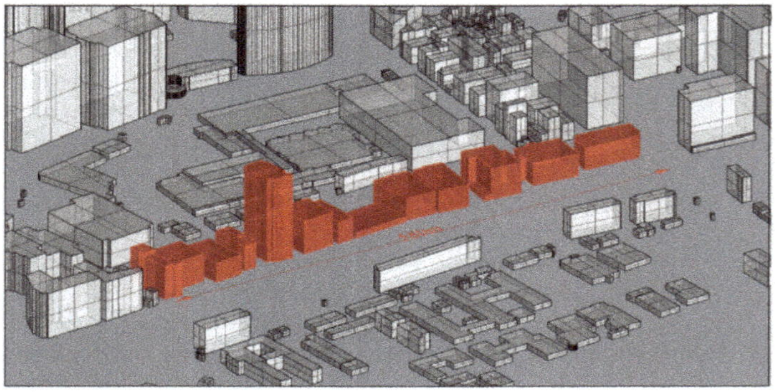

Figure 2. "Existing buildings" with surrounding buildings in Shinagawa area, Tokyo. Red colored buildings were analyzed as exiting buildings (scenario zero). It is noted that there is a canal in front of the existing buildings.

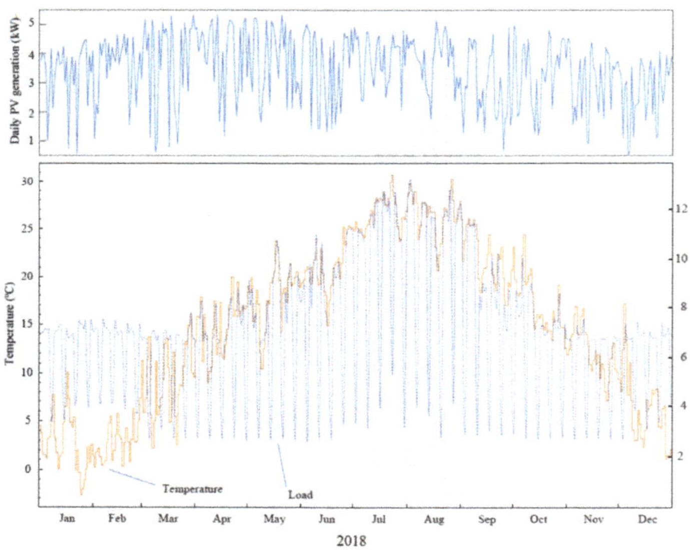

Figure 3. PV generation, daily outside temperature, and estimated demands of existing buildings for the test site in Shinagawa for 2018. Blue bands represent hot days with heightened cooling demand.

Environmental energy analyses were conducted as in a workflow chart in Figure 1. "ArcGIS", a Geographic Information System (GIS) program, was utilized to create GIS database and mapping [31]. Footprint and height data of buildings were obtained from publicly available dataset [32], and 3D polygons of buildings were produced by extruding foot-print areas with the corresponding heights of buildings (Figure 2). Then, the data was saved as a shapefile by "ArcGIS". "Rhinoceros 3D (version 6)" is a 3-dimensional computer aided design (3D CAD) software developed by Robert McNeel & Associates (Seattle, WA, USA). "Rhinoceros 3D" and its plugin, "Grasshopper", provide various analysis tools (e.g., Ladybug and Honeybee for energy analyses) for building designers, allowing them to work with independently developed software such as "EnergyPlus", "Radiance", and "Daysim" [33]. "EnergyPlus" is a well-known program for whole building energy analyses developed by US Department of Energy (DOE) [34]. We used these programs to estimate hourly energy demand and PV generation potentials for buildings [14], considering building usage patterns, materials of walls, windows and rooftops, weather, and urban context such as shades of nearby buildings. All the specification is available in https://doi.org/10.17632/wfpkdc6rd7.1 (accessed on 5 January 2022) for Grasshopper. Accuracy of EnergyPlus has been tested and validated during its on-going development [35].

The shapefile was loaded into "Rhinceros 3D" (Figure 1). "Grasshopper" provides a platform to build sequences of energy analyses. A weather file (epw) is necessary to estimate hourly load and PV electricity generation. We used "SIREN" to produce a weather file for Shinagawa, Tokyo for 2018 [36]. "EnergyPlus", integrated within Ladybug, analyzes building hourly energy demands (heating, cooling, lighting, and appliances) considering the influences of shading from neighboring building. PV electricity generation was also calculated on the surface of buildings (kWh·m^{-2}) in hourly resolution (Figure 3), also considering shading (Figure 4). Annual radiation amounts on the surface of the buildings were calculated for each mesh with an average area of 8 m^2 (Figure 4) Maximum annual radiation amount was calculated as 1383 kWh·m^{-2} (Figure 4). Above-ground-window/wall ratio for north, west, south, and east faced walls were set as 0.4, 0.35, 0.2, 0.15, respectively (Figure 5). Floor heights were set to 3 m. Space heating and cooling demands were converted to electricity demand by coefficient of performance (COP) with the values of 2.27 and 2.51, respectively [37]. Owing to the rapid development of the building energy analyses tools, the analyses between the program are smoothly linked, and results can be readily projected as 3D building representation in Rhinoceros 3D (Figure 5). Grasshopper files for energy and radiation analyses with weather files are available as https://doi.org/10.17632/wfpkdc6rd7.1 (accessed on 5 January 2022).

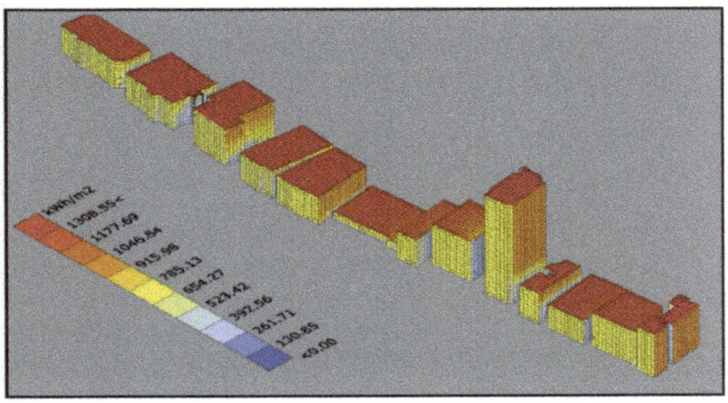

Figure 4. Annual radiation amount (kWh·m^{-2}) on the surfaces of the existing buildings calculated by Grasshopper. The direction of increasing radiation in the colorbar represents north.

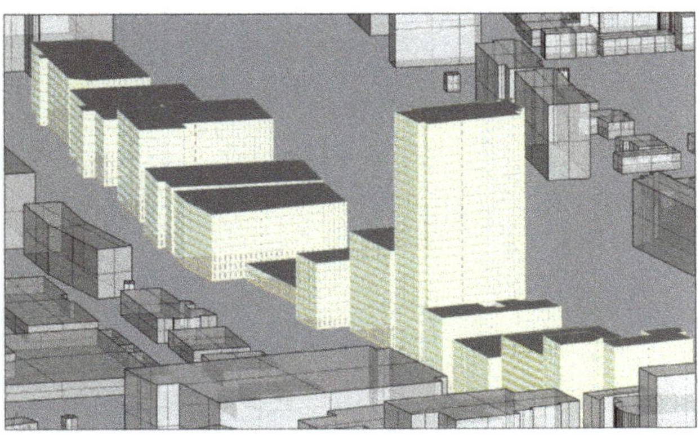

Figure 5. Existing buildings (yellow) with windows for the analyses.

Techno-economic analyses evaluate the viability of renewable energy projects such as rooftop PV systems [25,28], comparing with existing energy systems. The analyses consider project periods, discount rates, costs of PV systems, degradation, various energy losses, tariffs, etc. [23]. To investigate the impacts of declining PV system costs, we conducted techno-economic analyses on rooftop PV systems on large office buildings for 2018 and 2030 (Table 1). The methods generally follow those of our earlier studies [5,6]. System Advisor Model (SAM) was used for our analyses. The software is publicly available and developed by National Renewable Energy Laboratory (NREL) of the U.S. DOE [30]. PV degradation rate is set to annually 0.5% and soiling on the surface of PV reduce generation by 5% [30]. Hourly energy demands of buildings were estimated using the aforementioned "Rhinoceros 3D". A weather file for 2018 was applied to SAM analyses. We used a project period of 25 years with a discount rate of 3% for the rooftop PV systems. Currency exchange rate of 110 yen/$ was used. Other parameters used for the analyses are listed in Table 1. An electricity tariff price for high-voltage users was utilized for the analysis, which is cheaper than that for low voltage users (households, etc.) in Japan. SAM files with a weather file for the analyses were made available as https://doi.org/10.17632/wfpkdc6rd7.1 (accessed on 5 January 2022).

Table 1. Parameters used for techno-economic analyses [5]. Small-scale PV system costs are for 2018 (2030), respectively. Maintenance costs for PV system include inverter replacements.

Items	2018 (2030)
Small-scale PV system cost ($ kW^{-1})	2.15 (0.88)
PV system maintenance cost ($ kW^{-1}·yr^{-1})	31.4
Electricity to buy ($ kWh^{-1})	0.15
Electricity to sell ($ kWh^{-1})	0.08
PV tilt angle (degree)	30
Grid emission factor (kgCO_2·kWh^{-1})	0.455

We used net present values (*NPVs*) as a primary financial indicator and identified optimal PV capacity for each scenario using a function of SAM, "Parametrics". *NPV* of a PV project is a sum of discounted annual net saving over the project period including all the costs incurred (e.g., capital, and annual maintenance costs) [30,38].

Therefore, *NPV* is defined as:

$$NPV(p,t) = \sum_{n=1}^{N} \frac{Cash\ Flow(p,n,t)}{(1+R_d)^n} - System\ Cost(p,t) \quad (1)$$

where,

p = PV capacity (kW)
t = Project first year (yr)
N = Project period (yr)
R_d = Discount rate
And

$$Cash\ Flow\ (p, n, t) = Electricity\ Cost_{Base}\ (n, t) - Electricity\ Cost_{System}\ (p, n, t) \quad (2)$$

Electricity Cost$_{base}$ and *Electricity Cost*$_{System}$ are the costs of purchased grid electricity without and with PV systems, respectively. *System Cost* is the initial investment cost of PV systems [6].

Simple payback period (hereafter, payback period) is the time to recover the project cost of an investment, and can be expressed as the duration (e.g., years) from the initial investment to the time when the following condition is satisfied [38].

$$\sum_{n=1}^{t} \Delta I_n \leq \sum_{n=1}^{t} \Delta S_n \quad (3)$$

where ΔI is non discounted incremental investment costs ($) and ΔS is non discounted sum value of the annual cash flows net annual costs ($). t represents the time when the condition is satisfied for the first time, payback period.

Levelized cost of electricity ($LCOE$) is a measure of the average net present cost of PV electricity generation for its lifetime. It is useful to compare various sources of energy. $LCOE$ ($ kWh^{-1}) can be expressed as a following equation [30]:

$$LCOE = \frac{-C_0 - \sum_{n=1}^{N} \frac{C_n}{(1+R_d)^n}}{\sum_{n=1}^{N} \frac{Q_n}{(1+R_d)^n}} \quad (4)$$

where C_0 is the initial investment cost ($), C_n is the annual project costs ($) in year n, and Q_n is electricity (kWh) generated by the PV system in year n.

Analyzed results are also evaluated with following five environmental and energy indicators [6]. 1. Energy Sufficiency (ES) is how total PV generation can be compared to total annual demand. 2. Self-Sufficiency (SS) is how much PV generation can supply to local building demand considering hourly demand-supply balance. 3. Self-Consumption (SC) is how much PV generation can be consumed locally. 4. Cost Saving (CS) is how much energy costs can be saved by installing PV systems including capital and maintenance costs. 5. CO_2 emission reduction by PV systems is calculated by comparing CO_2 emission from gird electricity consumption before and after the system installation. All the indicators are expressed in percentage. Equations of the indicators can be expressed as:

$$\text{ES} = \text{Total annual PV generation (kWh)/total annual demand (kWh)} \times 100\ (\%) \quad (5)$$

$$\text{SS} = \text{Total PV electricity amount locally consumed/total annual demand (kWh)} \times 100\ (\%) \quad (6)$$

$$\text{SC} = \text{Total PV electricity amount locally consumed/total annual PV generation (kWh)} \times 100\ (\%) \quad (7)$$

$$\text{Cost saving} = (\text{NPV}/25)/(\text{grid electricity cost})_{\text{base}} \times 100\ (\%) \quad (8)$$

$$CO_2 \text{ emission reduction} = \{1 - (CO_2 \text{ emission from grid electricity consumption})_{\text{system}}/(CO_2 \text{ emission from grid electricity consumption})_{\text{base}}\} \times 100\ (\%) \quad (9)$$

where CO_2 emission from grid electricity consumption = total imported grid electricity (kWh) × emission factor ($kgCO_2 \cdot kWh^{-1}$). Subscripts, "system" and "base" indicate building energy systems "with PV system" and "without PV system", respectively.

To evaluate various building shapes in a block for energy demand and PV generation, we produced five scenarios (scenario 1–5) in comparison to existing buildings (scenario 0) (Figure 6 and Table 2). Five scenarios are characterized by "Low-rise", "High-rise", "Center corridor", "Courtyard", and "Korean style". We set building widths in a range from 15 m to 50 m following general building shapes in the area. The "Low-rise" buildings (scenario 1) give pedestrian continuity along the front street. Therefore, they have advantages for small shops. "High-rise" buildings (scenario 2) are more independent to other buildings, which tends to foster unique identity to represent one company or residential buildings. Buildings with the "center corridor" (scenario 3) have a common open space between buildings. The open spaces around the buildings offer places for many community activities to workers, shops, and offices. "Courtyard" style (scenario 4) has similar shape as Scenario 3, but it is a typical courtyard type in Europe. Korean style (scenario 5) was adopted from building arrangements from Cheongye river, Seoul, South Korea, which is a well-known redeveloped waterfront area.

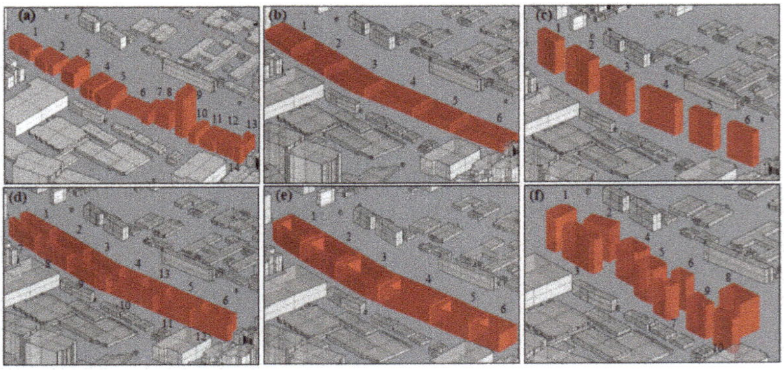

Figure 6. Existing buildings and five scenarios. (**a**) Scenario 0: existing buildings, (**b**) Scenario 1, (**c**) Scenario 2, (**d**) Scenario 3, (**e**) Scenario 4, and (**f**) Scenario 5. All the scenarios have the same FAR. Numbers next to buildings are building identification numbers.

Table 2. Characteristics of district scenarios. For the calculation of "surface area/volume", "above-ground surface area" was used. Numbers in parentheses were calculated by standard deviations divided by averages. See also Figure 5 for 3D building representation of the scenarios. Bldg. is an abbreviation for building.

Scenario Number	0	1	2	3	4	5	Average	Standard Deviation
Character	**Existing**	**Low-Rise**	**High-Rise**	**Center Corridor**	**Courtyard**	**Korean Style**		
Number of bldgs.	14	6	6	13	6	10	9	4 (40%)
Average bldg. height	27	18	66	33	33	54	39	18 (46%)
Average number of floors	9	6	22	11	11	18	13	6 (46%)
Floor area (m^2)	185,000	185,000	185,000	185,000	185,000	185,000	185,000	0 (0%)
FAR (%)	400	400	400	400	400	400	400	0 (0%)
BCR (%)	44	67	18	36	36	22	37	17 (46%)
Total bldg. volume (m^3)	555,000	555,000	555,000	555,000	555,000	555,000	555,000	0 (0%)
Total surface area (m^2)	106,000	94,000	85,000	121,000	128,000	94,000	105,000	17,000 (16%)
Above-ground surface area (m^2)	86,000	64,000	77,000	104,000	111,000	84,000	87,000	17,000 (20%)
Surface area/volume (m^{-1})	0.15	0.11	0.14	0.19	0.20	0.15	0.16	0.03 (20%)
Total rooftop area (m^2)	20,200	30,800	8400	16,800	16,800	10,300	17,200	8000 (46%)
Total rooftop PV capacity (kW)	2890	4400	1200	2400	2400	1470	2460	1140 (46%)

In the following energy modeling, it is assumed that all the buildings in the scenarios are set to be used as "office" for EnergyPlus. The analyses were conducted in an hourly resolution with weather information in 2018 for Shinagawa (Figure 3). To compare various building morphology in comparison to existing buildings, floor area ratio (*FAR*) is set as a control variable. *FAR* is used to regulate building volumes and thus number of people in the districts or cities, which is inherently relate to necessary sizes of public services and goods such as water-sewer, road services, sun light availability, openness, and noises in the cities [39]. *FAR* is defined as:

$$FAR\ (\%) = total\ floor\ area/site\ area \times 100 \quad (10)$$

As the site area (46,250 m^2) is common for all the scenarios, the total floor areas of buildings in the scenarios are also constant (Table 2). This results in total building volumes to be the same for all the scenarios (Table 2). Another important building indicator to regulate building forms, "building coverage ratio (*BCR*)" is defined as:

$$BCR\ (\%) = building\ area/site\ area \times 100 \quad (11)$$

Along with the *FAR*, *BCR* controls the shape and heights of buildings as well as occupied area by buildings in the site [40]. It is known that controlling *BCR* is an important policy measure to prevent spreading of fire. *BCR* varies between the scenarios from 18% to 67% (Table 2). Total surface areas above ground varies between the scenarios by 20%. Surface area to volume ratio of buildings, which is an important indicator for energy balance of buildings, also varies by 20% among the scenarios (Table 2). Total rooftop areas vary by 46% between the scenarios (Table 2). 70% of the total rooftop area is considered to be available for PV installation as a PV panel with 20% efficiency needs areas of about 5 m^2 plus an additional 2 m^2 for management or shaded areas, etc. Thus, 7 $m^2 \cdot kW^{-1}$ is used as a coefficient to calculate maximum rooftop PV capacity for each scenario (Table 2).

3. Results

3.1. Building Energy Demands

Estimated energy demands for buildings include interior lighting, interior electric equipment, space heating, and space cooling with typical office use activity in an hourly resolution. Interior equipment consumes the largest amount of electricity by 68% of the total, and lighting is 12% (Table 3). In addition, space heating and cooling are 2% and 18%, respectively. Demands for lighting and interior equipment are constant among the scenarios (Table 3) as they are generally functions of floor area. Space heating, which shows the largest variability (13% of average) between the scenarios, has significant correlation with surface area/volume ratio, explaining 97% of variance (Figure 7). Space cooling, although much smaller variability (0.8% of average), has significant correction with building height or number of floors (Table 3, Figure 7). Little variability of space cooling among the scenarios indicate that floor space or volume of building (set constant in this study) is the most important factor, and surface area-volume ratio has little impacts on space cooling. Unit floor electricity consumptions ($kWh \cdot m^{-2}$) are generally consistent with available observed data of 290 $kWh \cdot m^{-2}$ for office building electricity consumption for 2019–2020 [41]. Slightly older data of 2014 shows a larger average value of 389 $kWh \cdot m^{-2}$ for office buildings in Kanto area [42].

Table 3. Calculated building energy demands for the scenarios. Parentheses at average section represent percentages of demand components to the total demand. Parentheses at standard deviation section represent percentages of standard deviation divided by averages.

Scenario Number	0	1	2	3	4	5	Average	Standard Deviation
Name	Existing	Low-Rise	High-Rise	Center-Corridor	Courtyard	Korean Style		
Lighting (GWh)	7.7	7.7	7.7	7.7	7.7	7.7	7.7 (12%)	0 (0%)
Interior Equipment (GWh)	42.2	42.1	42.2	42.2	42.2	42.2	42.2 (68%)	0 (0%)
Space heating (GWh)	1.2	0.9	1.1	1.3	1.4	1.2	1.2 (2%)	0.2 (13%)
Space cooling (GWh)	11.3	11.2	11.5	11.3	11.3	11.4	11.3 (18%)	0.1 (0.8%)
Total consumption electricity (GWh)	62.4	62.0	62.5	62.6	62.6	62.5	62.4	0.2 (0.3%)
Unit floor electricity consumption ($kWh \cdot m^{-1}$)	337	336	338	338	338	338	337	1.0 (0.3%)

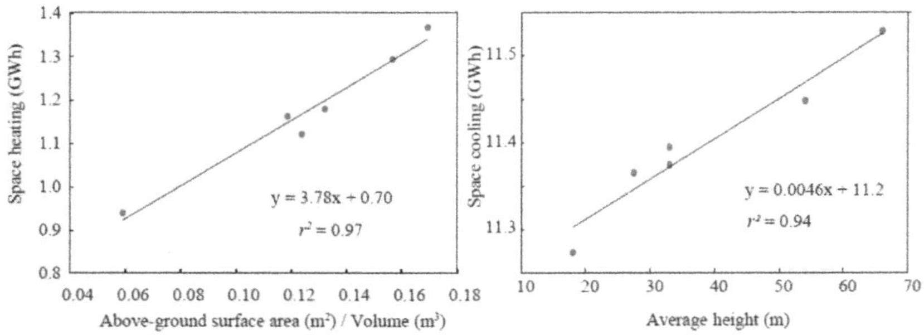

Figure 7. Influences of surface area/volume ratios and building heights on space heating (**left**) and cooling (**right**), respectively.

Space cooling is the second largest demand and shows clear relationship with outside temperatures from April to November (Figure 3). On the other hand, demand for space heating is so small that its correlation with out-side temperature in winter is not clear (Figure 3). Total energy demands show clear weekly cycles (national holidays are not considered) (Figure 3). Lighting and interior equipment do not have seasonal changes. Estimated hourly total demands for all the scenarios are highly correlated ($r > 0.99$) with that of the existing buildings (scenarios 0) as high demand components (i.e., interior equipment, lighting, and space cooling) do not show differences between the scenarios.

To investigate the impacts of shading, we conducted a test "with shades" and "without shades" for the building # 8 in the existing buildings (scenario 0; Figure 6). The shades were produced by the building # 9, which is the tallest building in all the buildings considered and located south of the building # 8 (Figure 6). The analysis was conducted in the same way for the scenario analyses for one year in an hourly resolution. Results show that the influence of shades are negligible in terms of the total annual energy demand (Table 4). However, the demand for space heating increases by shades by 5.7%, but the demand for space cooling reduces by 0.5% (Table 4). As absolute numbers are similar between demands for space heating and cooling but with opposite signs, they cancel each other with little change in the total (Table 4). Therefore, we conclude that shading has little influence on the total building energy demands in the settings we considered.

Table 4. Influences of shades on building energy demand of building # 8 of scenario 0. Parentheses for "shaded" and "no shade" are ratios to the total. Parentheses for "difference" are ratios with "no shade". "Difference" is calculated as "Shaded"–"No shade".

	Shaded (%)	No Shade (%)	Difference (%)
Space heating (MWh)	104.3 (1.8)	98.7 (1.7)	5.6 (5.7%)
Space cooling (MWh)	1065.4 (18.2)	1071.2 (18.3)	−5.8 (−0.5%)
Interior lighting (MWh)	725.7 (12.4)	725.7 (12.4)	0 (0.0%)
Interior equipment (MWh)	3955.3 (67.6)	3955.3 (67.6)	0 (0.0%)
Total (MWh)	5850.7 (100.0)	5850.9 (100.0)	−0.2 (0.0%)

3.2. Technoeconomic Analysis with Rooftop PV Systems

Building integrated PV (BIPV) are increasingly important for urban decarbonization when costs of PV systems are declining and land areas for PV are limited such as for Japan. However, BIPV potential for buildings in urban environment are affected by building forms and relationships between neighboring buildings for available sunlight (Figure 4). Analyses on existing buildings show that 81% of the rooftop area can receive more than 90% of the maximum solar radiation (Table 5). On the other hand, southern faced façade with no shades receives only 60–70% of the maximum solar radiation. As the cost of installation of rooftop PV systems are lower than that for facades [43], the first priority should be given to the rooftop, but newly developing PV materials with light weight such as perovskite PV could allow future application on façade more economic and easier than at present. All other scenarios have constant heights for all the buildings such that the rooftops of these buildings receive the maximum solar radiation.

Table 5. Radiation analysis on the existing buildings (scenario 0; Figure 4). Top and bottom represent ranges of annual radiation received on meshes. "% of max radiation" indicates percentage of the maximum annual radiation (1383 kWh·m^{-2}). Rooftop and façade areas are percentages of the total areas. For example, 81% of the rooftop area received more than 90% of the maximum annual radiation.

Top (kWh·m^{-2})	Bottom (kWh·m^{-2})	% of Max Radiation	Rooftop Area (%)	Façade Area (%)
1383	1244	90	81	0
1383	1106	80	91	0
1383	968	70	96	0
1383	830	60	98	10
1383	691	50	100	19
1383	553	40	100	51
1383	415	30	100	56
1383	277	20	100	60
1383	138	10	100	78
1383	0	0	100	100

As total rooftop areas are variable between the scenarios, amounts of PV generation is highly variable (Table 2). "Low-rise" buildings (scenario 1) have the largest rooftop area and PV capacity (4.4 MW) installed. Annual PV generation is 5.53 GWh, which supplies 8.9% of demand (Table 5). As the PV generation is small compared with the building demands, all PV generated electricity is consumed on-site (100% self-consumption) for all the scenarios (Table 6). Thus, self-sufficiency and energy sufficiency are equal. In addition, as CO_2 emission reduction is equal to the amount of grid electricity replaced by PV electricity, the values for CO_2 emission reduction are the same with self-sufficiency and energy-sufficiency (Table 6). The situation is highly different when it is compared with residential houses where rooftop PV generation often exceed household demands [25].

Table 6. Energy indicators for 2018 and 2030. Results are the same for both years as optimal PV capacities are the same for both years. 70% of the rooftop area is used for the maximum PV installation.

2018 and 2030	S0	S1	S2	S3	S4	S5
Rooftop PV generation (GWh·yr^{-1})	3.63	5.53	1.51	3.02	3.02	1.85
Self-sufficiency (%)	5.8	8.9	2.4	4.8	4.8	3.0
Self-consumption (%)	100	100	100	100	100	100
Energy sufficiency (%)	5.8	8.9	2.4	4.8	4.8	3.0
CO_2 emission reduction	5.8	8.9	2.4	4.8	4.8	3.0

If PV are installed on the façade receiving greater than 50% (691 kWh·m^{-2}) of the maximum solar insolation for the existing buildings, the corresponding façade area is 12,425 m^2 (19% of the total façade area including windows) and the façade PV is expected to generate annually 2.1 GWh of electricity, which is about the same amount with the annual rooftop PV generation, emphasizing the need to utilize the façade to further increase on-site PV generation.

Rapidly declining cost of PV system creates increasing opportunities for affordable decarbonization of buildings with BIPV. For 2018, LCOE of 0.13 \$·kWh^{-1} is slightly lower than high-voltage electricity price of 0.15 \$·kWh^{-1}, indicating that rooftop PV is already economic in 2018. As self-consumption is 100% for all the scenarios, simple payback period are the same values for all the scenarios. Payback period is 14 years in 2018, which is still longer than generally considered promising investment opportunities of less than

10 years [44]. Cost saving is less than 1%, which indicates difficulty to motivate building owners to invest on on-site PV systems in 2018. By 2030, economic situations of rooftop PV for these buildings improve significantly. LCOE becomes 0.07 $\$\cdot kWh^{-1}$, which is half of the tariff price. Payback period of 6 years is less than half of that of 2018. However, cost saving for energy expense is only 1–3% owing smaller PV generation in comparison to the demand. NPVs in 2030 become nearly four times larger than that in 2018. The maximum NPV reaches $7.0 million for the low-rise buildings (scenario 1), which is 3.7 times larger than that for the high-rise building (Table 7).

Table 7. Financial indicators of rooftop PV systems for 2018 and 2030.

2018	S0	S1	S2	S3	S4	S5
LCOE ($\$\cdot kWh^{-1}$)	0.13	0.13	0.13	0.13	0.13	0.13
NPV (million $)	1.21	1.85	0.50	1.00	1.00	0.62
Payback period (yr)	14	14	14	14	14	14
Cost saving (%)	0.5	0.8	0.2	0.4	0.4	0.3
2030	S0	S1	S2	S3	S4	S5
LCOE ($\$\cdot kWh^{-1}$)	0.07	0.07	0.07	0.07	0.07	0.07
NPV (million $)	4.60	7.00	1.91	3.82	3.82	2.33
Payback period (yr)	6	6	6	6	6	6
Cost saving (%)	2.0	3.%	0.8	1.6	1.6	1.0

4. Discussion

In fall 2020, Japanese government declared a goal to reach carbon neutral by 2050, which substantially changed social atmosphere toward carbon neutrality. Primary sources of carbon-free energy will be renewable energies, in particular, "solar power" [45]. As available lands for PV installation are limited in Japan and for the sake of saving natural lands [46], it is critical to maximize the rooftop uses for PV generation in a physically maximum extent. Although feed-in-tariffs (FITs) helped successfully expand PV in Japan [47], rooftop PVs of large buildings have received little attention as rooftop PV generation is rather small compared to large building demands. As our analyses indicated, declining cost of PV systems for the next decades will create large economic benefits to install PV on the rooftops of large buildings. In addition, total rooftop areas of large buildings in cities are not negligible and difficult to be replaced on the ground. Therefore, it is important to place adequate policy measures to facilitate expansion of rooftop PV systems on large buildings.

A few relevant policies can be recommended in this regard. First, current building form regulations mainly through FAR and BCR could be upgraded to promote the installation of rooftop PV on new buildings. One measure is to introduce a performance-based regulation under the designated FAR where energy efficiency is set as one of the core performances mandated to buildings. Another measure is to craft a form-based regulation designating wall setbacks, maximum height, minimum "Roof Area Ratio", etc. to facilitate building forms that allow effective rooftop PV installation. Second, in Japan where population and economy are shrinking in many cities, it is possible to reduce the designated FAR in existing urban area, which will prevent the construction of high-rise buildings that have less opportunities to install rooftop PV. However, this measure should be carefully discussed in relation to the compact city-plus-network concept that promotes high density urban area to reduce energy consumption in transport sector.

Toward carbon neutrality, critical measure for buildings are energy efficiency improvements including electrification [48]. Lighting, electric equipment, space cooling and heating have large potentials to reduce energy consumption but providing the same or better services by switching to more efficient and smart apparatus [49]. Passive solutions also play important roles particularly during building design phases, which are, for example, building materials for thermal insulation, building forms, window-to-wall ratio leading to reduced energy consumption. However, building design should be considered not solely from the energy efficiency requirement but also from other perspectives including the livability inside the building and the quality of environment outside building [12]. Buildings also need to prepare for adaptation to increasing climate change by enhancing resilience, etc., which often overlaps with mitigation measures such as developing decentralized energy systems [50].

In a city scale, rooftop PVs coupled with electric vehicles (EV) are possibly highly effective tools to decarbonize urban energy systems [5,6]. EVs are important means to decarbonize transport sector by replacing internal combustion of fossil fuels with electricity, but also play roles as energy storage for VRE such as rooftop PVs [25] or wind power [51]. It has been shown that rooftop PVs combined with EVs in a city scale can supply up to 95% of electricity to cities in Japan [6]. In the case of the special districts of Tokyo including Shinagawa, the PV plus EV systems can supply 53% of the annual demands. As large buildings of the central urban area such as this study consumes all the on-site PV generation, EVs or batteries have no roles to play as energy storage. However, Tokyo has various types of districts with independent houses and small buildings with smaller demands. Rooftop PVs of these buildings produce excess electricity, which supplies part of the demands of the large buildings by coupling with EVs. To fully understand the supply and demand balance of urban power systems, it is necessary to analyze disaggregated rooftop PV generation coupled with EVs within cities considering grid constrains.

5. Conclusions

This study established a workflow of assessing waterfront office building redevelopment plans with rooftop photovoltaics (PV), different building shapes and arrangements. We produced five scenarios in comparison to existing buildings with the same floor area ratio (FAR) and total floor areas. Demands for space heating are found to strongly correlate with surface area to volume ratio, although space heating demand for buildings are small in comparison to the total demand of buildings for the cases we considered. Shades by neighboring buildings affect space heating and cooling demands in opposite signs and cancel each other. Therefore, shades have little influence on the total energy demand for the buildings. Rooftop PV is already economic in 2018, and by 2030 it improves substantially with payback periods reaching 6 years. However, the rooftop PV contribution on the total demands of buildings or CO_2 emission reduction is small (2–9%), as rooftop areas are limited on the large buildings. It is noted that places such as Japan where lands are limited for PV installation, should implement mandatory regulation of installing rooftop PVs on buildings as such investments should not be issues for building owners. In addition, it is recommended to upgrade building form regulation to promote the installation of rooftop PV on new buildings.

Author Contributions: Conceptualization, Y.C. and T.K.; methodology, Y.C. and T.K.; investigation, Y.C. and T.K., writing—original draft, Y.C., T.K. and A.M.; validation, T.K.; formal analysis, T.K.; investigation, T.K.; software, T.K.; writing—review and editing, T.K.; visualization, T.K.; supervision, T.K.; funding acquisition, Y.Y. All authors have read and agreed to the published version of the manuscript.

Funding: This research received no external funding.

Data Availability Statement: Data to this article can be found online at: https://doi.org/10.17632/wfpkdc6rd7.1 (accessed on 5 January 2022).

Acknowledgments: We appreciate Soowon Chang at Purdue University for sharing Grasshopper files for the building energy analyses.

Conflicts of Interest: The authors declare no conflict of interest.

References

1. United Nations Human Settlements Programme. *Cities and Climate Change: Global Report on Human Settlements, 2011*; Routledge: London, UK, 2011.
2. Dodman, D. Blaming cities for climate change? An analysis of urban greenhouse gas emissions inventories. *Environ. Urban.* **2009**, *21*, 185–201. [CrossRef]
3. Nam, T.; Pardo, T.A. Smart city as urban innovation: Focusing on management, policy, and context. In Proceedings of the 5th International Conference on Theory and Practice of Electronic Governance, Tallin, Estonia, 26–28 September 2011; pp. 185–194. [CrossRef]
4. Kobashi, T.; Yamagata, Y.; Yoshida, T.; Chang, S.; Mochizuki, Y.; Ahl, A.; Aleksejeva, J. Chapter 9—Smart city and ICT infrastructure with vehicle to X applications toward urban decarbonization. In *Urban Systems Design*; Yamagata, Y., Yang, P.P.J., Eds.; Elsevier: Amsterdam, The Netherlands, 2020; pp. 289–333. ISBN 978-0-12-816055-8.
5. Kobashi, T.; Yoshida, T.; Yamagata, Y.; Naito, K.; Pfenninger, S.; Say, K.; Takeda, Y.; Ahl, A.; Yarime, M.; Hara, K. On the potential of "Photovoltaics + Electric vehicles" for deep decarbonization of Kyoto's power systems: Techno-economic-social considerations. *Appl. Energy* **2020**, *275*, 115419. [CrossRef]
6. Kobashi, T.; Jittrapirom, P.; Yoshida, T.; Hirano, Y.; Yamagata, Y. SolarEV City concept: Building the next urban power and mobility systems. *Environ. Res. Lett.* **2021**, *16*, 024042. [CrossRef]
7. Chen, Y. Financialising urban redevelopment: Transforming Shanghai's waterfront. *Land Use Policy* **2020**, *112*, 105126. [CrossRef]
8. Papatheochari, T.; Coccossis, H. Development of a waterfront regeneration tool to support local decision making in the context of integrated coastal zone management. *Ocean Coast. Manag.* **2019**, *169*, 284–295. [CrossRef]
9. Keyvanfar, A.; Shafaghat, A.; Mohamad, S.; Abdullahi, M.M.; Ahmad, H.; Derus, N.H.M.; Khorami, M. A sustainable historicwaterfront revitalization decision support tool for attracting tourists. *Sustainability* **2018**, *10*, 215. [CrossRef]
10. ECCJ. Features of Office Building Energy Consumption. Available online: https://www.eccj.or.jp/office_bldg/01.html (accessed on 13 August 2021).
11. Naboni, E.; Natanian, J.; Brizzi, G.; Florio, P.; Chokhachian, A.; Galanos, T.; Rastogi, P. A digital workflow to quantify regenerative urban design in the context of a changing climate. *Renew. Sustain. Energy Rev.* **2019**, *113*, 109255. [CrossRef]
12. Natanian, J.; Aleksandrowicz, O.; Auer, T. A parametric approach to optimizing urban form, energy balance and environmental quality: The case of Mediterranean districts. *Appl. Energy* **2019**, *254*, 113637. [CrossRef]
13. Han, T.; Huang, Q.; Zhang, A.; Zhang, Q. Simulation-based decision support tools in the early design stages of a green building-A review. *Sustainability* **2018**, *10*, 3696. [CrossRef]
14. Chang, S.; Saha, N.; Castro-Lacouture, D.; Yang, P.P.J. Multivariate relationships between campus design parameters and energy performance using reinforcement learning and parametric modeling. *Appl. Energy* **2019**, *249*, 253–264. [CrossRef]
15. Chang, S.; Castro-Lacouture, D.; Yamagata, Y. Decision support for retrofitting building envelopes using multi-objective optimization under uncertainties. *J. Build. Eng.* **2020**, *32*, 101413. [CrossRef]
16. Chen, Y.; Hong, T.; Piette, M.A. Automatic generation and simulation of urban building energy models based on city datasets for city-scale building retrofit analysis. *Appl. Energy* **2017**, *205*, 323–335. [CrossRef]
17. Li, W.; Zhou, Y.; Cetin, K.; Eom, J.; Wang, Y.; Chen, G.; Zhang, X. Modeling urban building energy use: A review of modeling approaches and procedures. *Energy* **2017**, *141*, 2445–2457. [CrossRef]
18. Johari, F.; Peronato, G.; Sadeghian, P.; Zhao, X.; Widén, J. Urban building energy modeling: State of the art and future prospects. *Renew. Sustain. Energy Rev.* **2020**, *128*, 109902. [CrossRef]
19. Ferrando, M.; Causone, F.; Hong, T.; Chen, Y. Urban building energy modeling (UBEM) tools: A state-of-the-art review of bottom-up physics-based approaches. *Sustain. Cities Soc.* **2020**, *62*, 102408. [CrossRef]
20. Natanian, J.; Auer, T. Beyond nearly zero energy urban design: A holistic microclimatic energy and environmental quality evaluation workflow. *Sustain. Cities Soc.* **2020**, *56*, 102094. [CrossRef]
21. Zhang, J.; Xu, L.; Shabunko, V.; Tay, S.E.R.; Sun, H.; Lau, S.S.Y.; Reindl, T. Impact of urban block typology on building solar potential and energy use efficiency in tropical high-density city. *Appl. Energy* **2019**, *240*, 513–533. [CrossRef]
22. Li, Y.; Liu, C. Techno-economic analysis for constructing solar photovoltaic projects on building envelopes. *Build. Environ.* **2018**, *127*, 37–46. [CrossRef]
23. Hoppmann, J.; Volland, J.; Schmidt, T.S.; Hoffmann, V.H. The economic viability of battery storage for residential solar photovoltaic systems—A review and a simulation model. *Renew. Sustain. Energy Rev.* **2014**, *39*, 1101–1118. [CrossRef]
24. BNEF. *New Energy Outlook 2018*; BNEF: London, UK, 2018.
25. Kobashi, T.; Say, K.; Wang, J.; Yarime, M.; Wang, D.; Yoshida, T.; Yamagata, Y. Techno-economic assessment of photovoltaics plus electric vehicles towards household-sector decarbonization in Kyoto and Shenzhen by the year 2030. *J. Clean. Prod.* **2020**, *253*, 119933. [CrossRef]
26. Wilkinson, S.; John, M.; Morrison, G.M. Rooftop PV and the renewable energy transition; a review of driving forces and analytical frameworks. *Sustainability* **2021**, *13*, 5613. [CrossRef]
27. Zhou, Y.; Cao, S.; Hensen, J.L.M. An energy paradigm transition framework from negative towards positive district energy sharing networks—Battery cycling aging, advanced battery management strategies, flexible vehicles-to-buildings interactions, uncertainty and sensitivity analysis. *Appl. Energy* **2021**, *288*, 116606. [CrossRef]

28. Say, K.; Schill, W.P.; John, M. Degrees of displacement: The impact of household PV battery prosumage on utility generation and storage. *Appl. Energy* **2020**, *276*, 115466. [CrossRef]
29. Lang, T.; Ammann, D.; Girod, B. Profitability in absence of subsidies: A techno-economic analysis of rooftop photovoltaic self-consumption in residential and commercial buildings. *Renew. Energy* **2016**, *87*, 77–87. [CrossRef]
30. Blair, N.; Diorio, N.; Freeman, J.; Gilman, P.; Janzou, S.; Neises, T.W.; Wagner, M.J. *System Advisor Model (SAM) General Description*; National Renewable Energy Laboratory: Golden, CO, USA, 2018.
31. Booth, B.; Mitchell, A. *Getting Started with ArcGIS GIS by ESRI*; ESRI: Redlands, CA, USA, 2001; p. 260.
32. Geospatial Information Authority of Japan. Fundamental Geospatial Data. Available online: https://fgd.gsi.go.jp/download/menu.php (accessed on 5 March 2020).
33. Roudsari, M.S.; Pak, M. Ladybug: A parametric environmental plugin for grasshopper to help designers create an environmentally-conscious design. In Proceedings of the BS2013: 13TH Conference of International Building Performance Simulation Association, Chambery, France, 26–28 August 2013; pp. 3128–3135.
34. Crawley, D.B.; Lawrie, L.K.; Winkelmann, F.C.; Buhl, W.F.; Huang, Y.J.; Pedersen, C.O.; Strand, R.K.; Liesen, R.J.; Fisher, D.E.; Witte, M.J.; et al. EnergyPlus: Creating a new-generation building energy simulation program. *Energy Build.* **2001**, *33*, 319–331. [CrossRef]
35. EnergyPlus. Testing and Validation. Available online: https://energyplus.net/testing (accessed on 3 June 2021).
36. King, A. SIREN: SEN's interactive renewable energy network tool. In *Transition towards 100% Renewable Energy*; Sayigh, A., Ed.; Springer: Cham, Switzerland, 2018; p. 536, ISBN 978-3-319-69844-1.
37. An, H.J.; Yoon, J.H.; An, Y.S.; Heo, E. Heating and cooling performance of office buildings with a-Si BIPV windows considering operating conditions in temperate climates: The case of Korea. *Sustainability* **2018**, *10*, 4856. [CrossRef]
38. Short, W.; Packey, D.J.; Holt, T. A Manual for the Economic Evaluation of Energy Efficiency and Renewable Energy Technologies. Available online: https://www.nrel.gov/docs/legosti/old/5173.pdf (accessed on 13 August 2020).
39. Joshi, K.K.; Kono, T. Optimization of floor area ratio regulation in a growing city. *Reg. Sci. Urban Econ.* **2009**, *39*, 502–511. [CrossRef]
40. Usui, H. Variation in Building Heights under Zoning Regulations of Building Coverage Ratio and Floor Area Ratio: Theoretical. In Proceedings of the ISUF 2020 Virtual Conference Proceedings, Virtual, 23 February 2020.
41. BEMA. 43rd Report on Building Energy Consumption (2019–2020). Available online: http://www.bema.or.jp/_src/11317/digest43.pdf?v=1627864413127 (accessed on 15 August 2021).
42. Arakawa, S.; Kuboi, D.; Yoshizawa, A.; Naraoka, S. Actual condition survey on energy consumption in office buildings for energy saving. *Techinical Pap. Annu. Meet. Soc. Heat. Air-Cond. Sanit. Eng.* **2016**, *8*, 73–76.
43. SUPSI. *Building Integrated Photovoltaics: A Practical Handbook for Solar Buildings' Stakeholders*; Status Report; SUPSI—Swiss BIPV Competence Centre: Manno, Switzerland, 2020.
44. Li, H.X.; Zhang, Y.; Li, Y.; Huang, J.; Costin, G.; Zhang, P. Exploring payback-year based feed-in tariff mechanisms in Australia. *Energy Policy* **2021**, *150*, 112133. [CrossRef]
45. METI. Summary for Basic Energy Plan of Japan (Draft). Available online: https://www.enecho.meti.go.jp/committee/council/basic_policy_subcommittee/2021/046/046_004.pdf (accessed on 16 August 2021).
46. Kim, J.Y.; Koide, D.; Ishihama, F.; Kadoya, T.; Nishihiro, J. Current site planning of medium to large solar power systems accelerates the loss of the remaining semi-natural and agricultural habitats. *Sci. Total Environ.* **2021**, *779*, 146475. [CrossRef]
47. Cozzi, L.; Gould, T.; Bouckart, S.; Crow, D.; Kim, T.-Y.; McGlade, C.; Olejarnik, P.; Wanner, B.; Wetzel, D. *World Energy Outlook 2020*; OECD Publishing: Paris, France, 2020; Volume 2050, pp. 1–461.
48. Belussi, L.; Barozzi, B.; Bellazzi, A.; Danza, L.; Devitofrancesco, A.; Fanciulli, C.; Ghellere, M.; Guazzi, G.; Meroni, I.; Salamone, F.; et al. A review of performance of zero energy buildings and energy efficiency solutions. *J. Build. Eng.* **2019**, *25*, 100772. [CrossRef]
49. Lovins, A. *Reinventing Fire: Bold Business Solutions for the New Energy Era*; Chelsea Green Publishing: Hartford, VT, USA, 2011.
50. Economidou, M.; Todeschi, V.; Bertoldi, P.; D'Agostino, D.; Zangheri, P.; Castellazzi, L. Review of 50 years of EU energy efficiency policies for buildings. *Energy Build.* **2020**, *225*, 110322. [CrossRef]
51. Lund, H.; Kempton, W. Integration of renewable energy into the transport and electricity sectors through V2G. *Energy Policy* **2008**, *36*, 3578–3587. [CrossRef]

energies

Article

A Forecasting Method for Macro-Control Policy of Heating Energy Consumption and Carbon Emissions Based on Building Area and Energy Intensity: A Case Study of Northern China

Yu Zheng [1,*], Wenlong Yin [2], Wenjie Zhang [2], Jinhan Liang [3], Kangyong Liu [2] and Kuan Wang [4]

1 School of Mechanical, Electronic and Control Engineering, Beijing Jiaotong University, Beijing 100044, China
2 School of Energy and Power Engineering, Nanjing University of Science and Technology, Nanjing 210094, China; 13251611050@163.com (W.Y.); zhangwenjie001@139.com (W.Z.); conan1016@live.com (K.L.)
3 Norendar Internation Ltd., Shijiazhaung 050030, China; liang289922@126.com
4 China Railway Construction Group Co., Ltd., Beijing 100043, China; wangkuan@ztjs.cn
* Correspondence: 19116054@bjtu.edu.cn

Abstract: In response to extreme climate change, China has set a goal of reaching emission peak by 2030 and achieving carbon neutrality by 2050. Energy conservation and emission reduction of building heating in northern China are key to achieving this goal. Based on building area prediction and energy consumption intensity, this paper establishes a model for calculating heating energy consumption and carbon emissions in northern residential buildings at the macro level, which provides a basis for formulating policies related to heating and emission reduction in northern buildings. Based on the research method of scenario projection, combining the carbon emission subsets and future heating energy mix projections in northern China, the heating energy consumption in northern China decreases to 175, 149 and 135 Mtce in 2050 under the baseline, medium control and strict control scenarios, respectively. The heating energy consumption in the northern region should be controlled at least under the medium control scheme. Under this scenario, building heating carbon emissions in the north could be reduced to 450 $MtCO_2$ by 2050, or 280 $MtCO_2$ if more stringent abatement technologies are adopted. In order to achieve this goal, a combination of energy-saving technologies must be used. The use of biomass and solar technologies should be emphasized in rural heating, while envelope renovation makes the greatest contribution to heating energy savings, with envelope renovation contributing up to 92 Mtce to urban heating.

Keywords: heating; northern China; energy saving; carbon emission; model forecasting

Citation: Zheng, Y.; Yin, W.; Zhang, W.; Liang, J.; Liu, K.; Wang, K. A Forecasting Method for Macro-Control Policy of Heating Energy Consumption and Carbon Emissions Based on Building Area and Energy Intensity: A Case Study of Northern China. *Energies* **2022**, *15*, 1153. https://doi.org/10.3390/en15031153

Academic Editors: Shi-Jie Cao, Wei Feng and Vincenzo Costanzo

Received: 29 December 2021
Accepted: 2 February 2022
Published: 4 February 2022

1. Introduction

In response to extreme climate change, in 2018 the Intergovernmental Panel on Climate Change released a report showing that the world must limit global warming to 1.5 °C [1]. Countries can only achieve this goal if they achieve zero carbon emissions by the mid-century. To this end, China announced at the 75th session of the United Nations General Assembly that it would take stronger policy measures to peak CO_2 emissions by 2030 and achieve carbon neutrality by 2060 [2].

The 2030 and 2060 carbon emission targets provide a concrete timeline for the transformation of China's energy mix. In view of the current situation in China, China's energy reform and building energy efficiency work must start from the national situation and find a way to meet its own characteristics. The northern region of China is located to the north of the Qinling–Huaihe line, covering two climatic regions of severe cold and cold regions. Due to the low temperature in winter, the heating in northern towns is mainly concentrated heating. Fifteen provinces and municipalities with a large coverage rate of central heating in northern China are Beijing, Tianjin, Hebei, Shanxi, Inner Mongolia, Heilongjiang, Jilin,

Liaoning, Shandong, Henan, Shaanxi, Gansu, Qinghai, Ningxia and Xinjiang. In the northern part of China, due to the winter heating demand, the adjustment of its heating energy use structure and carbon emission reduction is the key to achieving the national carbon emission target. The statistics and prediction of heating energy consumption and the carbon emission data of northern buildings are important for a comprehensive understanding of the current situation of building energy consumption in northern China, discovering the problematic aspects of building energy use, controlling building carbon emissions and guiding the development of building energy conservation work, and the research results can provide strong data support for this.

The current domestic and international energy consumption calculation methods can be divided into two categories: one is the building energy consumption obtained by statistical analysis using survey data, and the other is the building energy consumption obtained by using calculation models based on macroscopic data [3]. Microscopic data statistics usually adopt the method of sampling survey to conduct statistics on various energy consumption data of various buildings, and the United States, Japan, Denmark and other developed countries have set up special building energy consumption statistics institutions to investigate building energy consumption. The U.S. Energy Information Administration (EIA) divides buildings into two categories, residential buildings and commercial buildings, and conducts surveys and statistics every four years, and then processes the data through regression analysis, engineering models and adjusted estimates to decompose energy consumption into different energy-using terminals [4]. The U.S. housing metering energy model uses the National Energy Modeling System (NEMS) to analyze and calculate building energy [5]. The model includes both a top-down macroeconomic model and a bottom-up energy supply and demand model.

The econometric model LEAP [6], jointly developed by the Stockholm Environment Association and Boston University, USA, can be used as a scenario-based energy-environment modeling tool for energy demand analysis and its corresponding environmental impact analysis and cost-benefit analysis. Based on the LEAP model, Liu, J.L. et al. [7] developed a building sector energy system model (PECE-Building) by setting three scenarios—baseline scenario (BAU), autonomous contribution scenario (NDC) and enhanced low carbon (ELC)—and analyzed the energy demand and CO_2 emission trends of the building sector under different development paths from 2013 to 2050, concluding that heating is the most important area of emission reduction in northern cities and towns. Li Xinyi [8] from Chongqing University used cluster analysis and statistical data-based definitions to select residential building prototypes to construct community- and city-scale residential building area energy consumption models around community-scale and urban-scale residential buildings, while using machine learning methods to construct a residential building energy demand prediction tool that took into account the behavioral characteristics of people and future meteorological corrections. Xueling Liu et al. [9] considered climate and population changes, proposed an evaluation method for per capita energy consumption per unit area and simulated and calculated the cooling and heating energy consumption of various types of buildings in the Tianjin area for the next 30 years according to the different building functions. Sascha Leiber et al. [10] proposed a novel visual appearance only based heating energy prediction method for single-family houses using powerful image analysis and computer vision techniques that can be used in widely distributed rural buildings in the north of China. The Tsinghua Building Energy Efficiency Research Center has established the China Building Energy Consumption Model (CBEM) that is based on energy intensity and macro-validated by statistical data [11], yielding a heating energy consumption of 201 Mtce for northern cities and towns in China in 2017, accounting for 21% of building energy consumption.

However, some of these models are from a macro perspective, mainly aimed at fitting the historical data of national energy consumption and carbon emissions, to facilitate the study of the relationship between the energy sector and the overall economy. In addition, the economic development in northern China is not balanced, and the rural area is large.

Some models do not separate the rural areas but focus on urban areas. It is impossible to calculate the heating energy consumption and carbon emissions in rural areas more accurately. At the same time, for different building types, accurate division is also very important.

2. Method

Based on the research on macro building energy consumption at home and abroad, this paper calculates the living and public building area by analyzing factors such as population, per capita living area and economic indicators. Starting from the terminal energy consumption, based on the characteristics of China's energy structure system and building energy consumption, combined with the design and actual operation characteristics of urban central heating and rural heating, this paper establishes a macro prediction calculation model of heating energy consumption of northern civil buildings based on the energy consumption intensity of per capita building area and unit building area. Combined with carbon emissions and future heating energy structure prediction in northern China, the model calculates the total amount of heating carbon emissions in the future, which provides data support for controlling building carbon emissions and guiding building energy efficiency. In the calculation process, considering the heating difference between urban and rural areas, the northern civil buildings are divided into urban buildings and rural residential buildings from the area. Urban buildings include urban residential buildings and public buildings. The total heating energy consumption of urban buildings, including residential buildings and public buildings, and rural residential buildings are calculated respectively, which realizes the attention to urban areas, rural areas and different types of buildings.

The calculation model constructed in this paper is based on the calculation method of building area and unit energy intensity, and after deriving the heating energy consumption, the corresponding carbon emission intensity is then calculated according to the energy structure and carbon emission coefficient. Firstly, the northern civil buildings are divided into urban buildings and rural residential buildings, among which urban buildings are further divided into residential buildings and public buildings. In calculating the building area, a more accurate calculation forecast is made for each type of building area by considering the influence of factors such as housing demand, economic and social development, land resource constraints and urbanization rate. Similarly, when calculating energy consumption intensity, it is necessary to combine factors such as heating methods, heat source efficiency, losses, etc., while taking into account changes in energy policies and changes in heating energy structures, in order to arrive at relatively accurate results. After calculating the floor area and energy consumption intensity, the total heating energy consumption in northern regions can be easily obtained; finally, the heating carbon emissions are calculated based on the energy structure and carbon emission coefficient. The calculation method is shown in Figure 1.

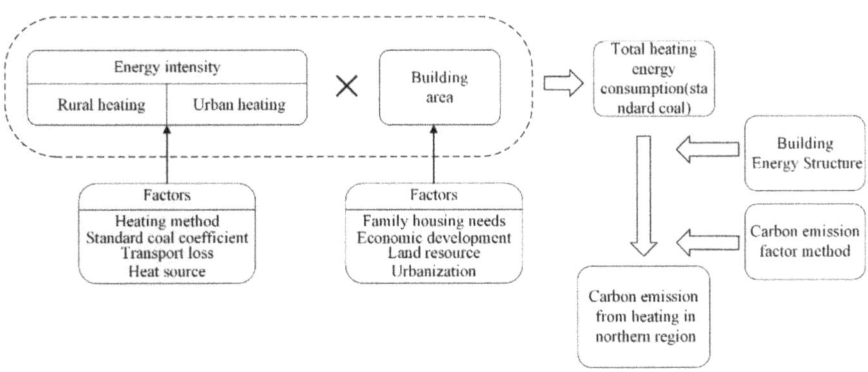

Figure 1. Heating energy consumption and carbon emission research method.

The research flow of the article is shown in Figure 2, and the conclusion is in Section 7.

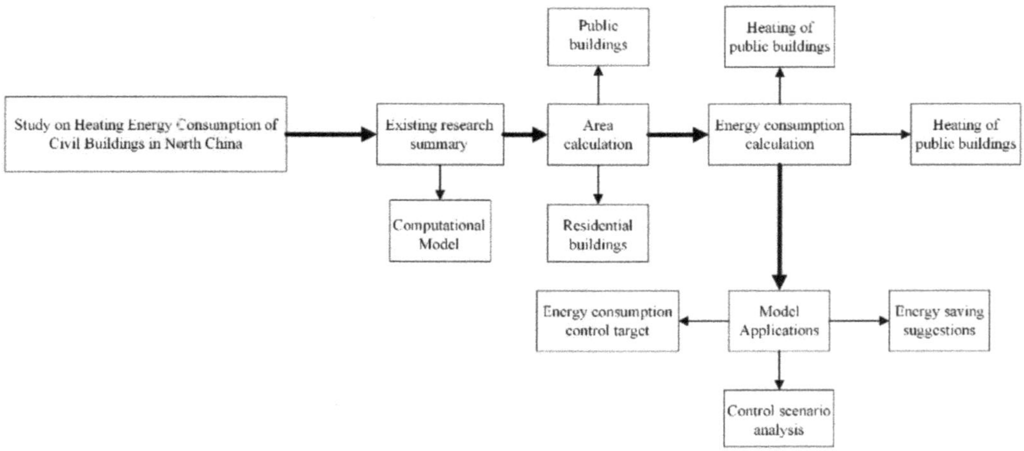

Figure 2. Article Research Flow Chart.

3. Forecasting of Heating Area in Northern China

Building area data are the working basis of building energy efficiency. Whether it is to calculate the total amount of building energy consumption, analyze energy-saving potential or formulate energy-saving goals, it must be based on building area data. However, because China's construction field only gathers statistics on the amount of new construction over the years and does not count the corresponding amount of demolition management, the current various types of building stock cannot be given directly by statistical data. Therefore, we can only calculate the floor area by other means and cannot give precise data. The current functional characteristics of civil buildings in China are divided into two categories for statistical analysis: residential buildings and public buildings.

3.1. Residential Buildings

3.1.1. Residential Area Status

Residential buildings can be divided into urban and rural residential buildings according to their geographical distribution and living habits. Different organizations have given different values of residential area per capita in China by sampling research or estimation methods. As shown in Figure 3, the data released by the National Bureau of Statistics [12,13] are the result of a sample survey of residential building area, based on the per capita residential area converted from the registered population. The Building Energy Efficiency Association [14] and the Building Energy Conservation Research Center of Tsinghua University [15] use the relevant data in the Statistical Yearbook to calculate the residential building area by using the 'fuzzy neural network' and 'existing + new construction–demolition' methods and then calculate the corresponding per capita residential building area according to the number of urban and rural residents.

There are three main reasons for the differences in the data of each agency: one is the data source—the per capita area data of each agency are mostly the result of sample surveys or different methods based on historical data; the second is the statistical caliber of the population—the number of permanent residents and the registered population will lead to large differences in the per capita area; the third is about the scope of the residential area—the three concepts of building area, usable area and floor area are quite different. The residential area discussed in this article is the building area, which is analyzed and calculated according to the per capita residential area index of the permanent population.

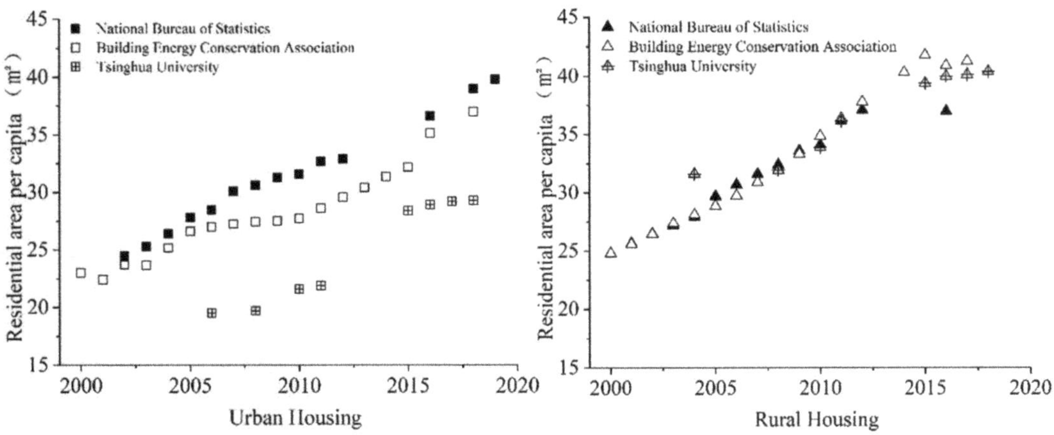

Figure 3. Comparison of calculated values of per capita residential area by different institutions.

With reference to the research data on residential building area by various institutions, using the number of permanent residents and the urbanization rate in the China Statistical Yearbook [16] over the years, the existing building area of urban and rural residential buildings can be calculated, as shown in Figure 4.

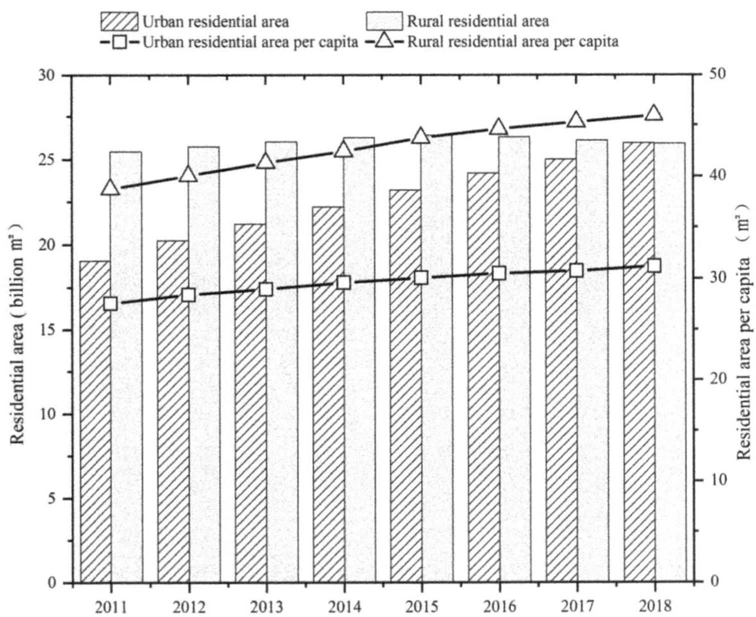

Figure 4. Urban and rural residential construction area and per capita residential area over the years.

3.1.2. Per Capita Residential Space Needs and Constraints

(1) GDP per capita. Foreign experience in housing development shows that the demand for housing area per capita gradually increases with urbanization. The global housing floor area per capita in each region is related to the degree of development, which is about 54 m^2 in OECD member countries (about 45 m^2 on average in Europe and Central Asia), about 30 m^2 in Asia (except the Middle East) and about 20 m^2 in sub-

Saharan Africa, and databases of agencies such as Eurostat, the International Energy Agency and Japan's Ministry of Land, Infrastructure, Transport and Tourism show that the GDP per capita (PPP) in medium-sized developed countries is $20,000 to $40,000 when the housing floor area per capita is 32~45 m^2.

(2) Resident income. With the development of the social economy, the increase in residents' income gradually increases the demand for housing. The China General Social Survey (CGSS) [17] conducted a questionnaire survey on the annual income and residential area of residents in China. The sample covered all provinces and cities in China except Hong Kong, Macao, Taiwan and Tibet. The sample selected in this paper is housing property owners with an annual income between ¥0.5 and ¥250,000 yuan in urban areas, and, after excluding missing values and outliers, the valid sample data are 1761 groups to analyze the demand for residential area per capita in urban areas in China. The relationship between the annual income of residents and housing floor area is shown in Figure 5. From the figure, it can be seen that the overall housing floor area increases with the increase in annual income. When the annual income of residents reaches ¥180,000 (corresponding to a GDP per capita of about $40,000), the household housing floor area is around 130 m^2.

(3) Household size. The number of households has a significant positive effect on the housing demand of Chinese residents, and the explanatory power of the number of households on housing demand is significantly better than that of the total population indicator [18]. The survey results of the National Health Planning Commission's China Family Development Report show that the average size of urban households was 2.84 persons in 2015 [19], and two-person and three-person households accounted for 21.9% and 31.7% of the total number of surveyed households, respectively. Accordingly, it is deduced that when the income of residents grows to a certain extent, the expected value of future urban residential area per capita to meet the needs of families should be around 45 m^2.

(4) Land resource constraints. According to China's territorial spatial planning, China's urban spatial index will be at 116,700 square kilometers in 2030 [20]. At present, the proportion of residential land in China accounts for about 31% of the urban construction land area [21], of which the ratio of residential to residential land is about 95%. The proportion of residential land in China is too low, which is generally around 50% internationally, assuming that the proportion of residential land reaches 35% in 2030 and the proportion of residential remains unchanged. Research studies show that the average volume ratio of urban residential land is about 0.995 when the per capita energy consumption of residents is the lowest. Under the land resource constraint, a reasonable residential area per capita can be calculated by using the area and proportion of urban residential land and the population size, combined with the housing floor area ratio. The specific measurement method is shown in Equation (1). From this, it can be estimated that under the constraints of land resources and urbanization development, the housing floor area per capita should be controlled between 36 and 39 m^2 in 2020–2030.

$$a = A \cdot \rho_{re} \cdot \rho_{re,1} \cdot \theta \div P \tag{1}$$

where, a is the per capita residential area in town, m^2/person; A is the spatial control index of town, billion m^2; ρ_{re} is the proportion of residential land; $\rho_{re,1}$ is the proportion of residential to residential land; θ is the volume ratio of residential land in town; and P is the number of resident population in town, billion.

Compared with urban residential area, the rural residential area constraint is more relaxed. Data from the China Statistical Yearbook show that the rural residential area per capita increased from 26.5 to 37.1 m^2/person between 2002 and 2012, an increase of 40.0%, with an average annual increase of 3.4%; the rural residential area per capita converted by the Standard Determination Institute [22] using the cumulative method of area over the years increased from 24.6 to 40.9 m^2/person between 2002 and 2010, an increase of 66.3%, with an average annual growth rate of 5.2%. With the development of urbanization and urban–rural integration, the growth rate of rural residential area per capita is expected to decrease in the future.

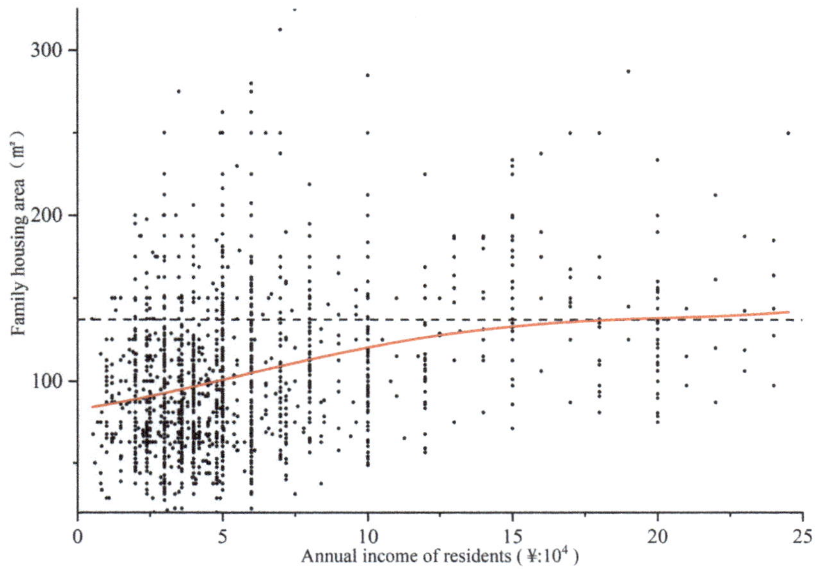

Figure 5. Residential annual income and housing construction area scatter plot.

3.2. Public Buildings

Because of the economic development of rural areas, there are fewer public buildings, so they are neglected, and all public buildings referred to in this paper refer to urban buildings. There are many types of public buildings, and the energy consumption of different types of public buildings varies greatly, so classifying public buildings is beneficial for the subsequent energy consumption research. Based on the consideration of building energy consumption and building quantity, combined with the data of various types of public buildings indexes in the China Statistical Yearbook, this paper will focus on the area calculation of medical buildings, office buildings, educational buildings, commercial buildings, cultural buildings and transportation buildings. Based on the current situation research and the existing data collected, this paper analyzes and refines the calculation methods of various types of public building area and the indexes that need to be predicted, as shown in Figure 6.

For all kinds of infrastructure and commercial public building area calculations, this paper refers to the model proposed by Gong et al. [23] to calculate and predict; the calculation method is shown in Figure 6. Other categories include public buildings such as gardens, communication buildings and religious buildings. The results of the census of large public buildings in Beijing [24] show that the public building area of 11 categories in 2005 was about 34 million m^2. Using the data in the Statistical Yearbook [25], the estimated public building area in Beijing in the same year was 46.8 million m^2, so the other categories of public buildings in Beijing that were not counted were about 25% of the total. In this paper, we will make reference to this ratio to estimate other types of public buildings.

3.3. Civil Building Area Forecast

With a population of only 1.4 billion in 2020 and a relatively low fertility rate, China has entered the era of negative population growth [26]. Famous population forecasters at home and abroad tend to forecast the peak year of China's population to be shifted forward, around 2023–2029. According to the results of a study by Professors Xianling

Zhang and Zhenwu Zhai on negative population growth in China [27], if the total fertility rate is maintained at 1.6% in the future, negative population growth might occur in 2027. In this scenario, the total population of China will not exceed 1.41 billion in 2030.

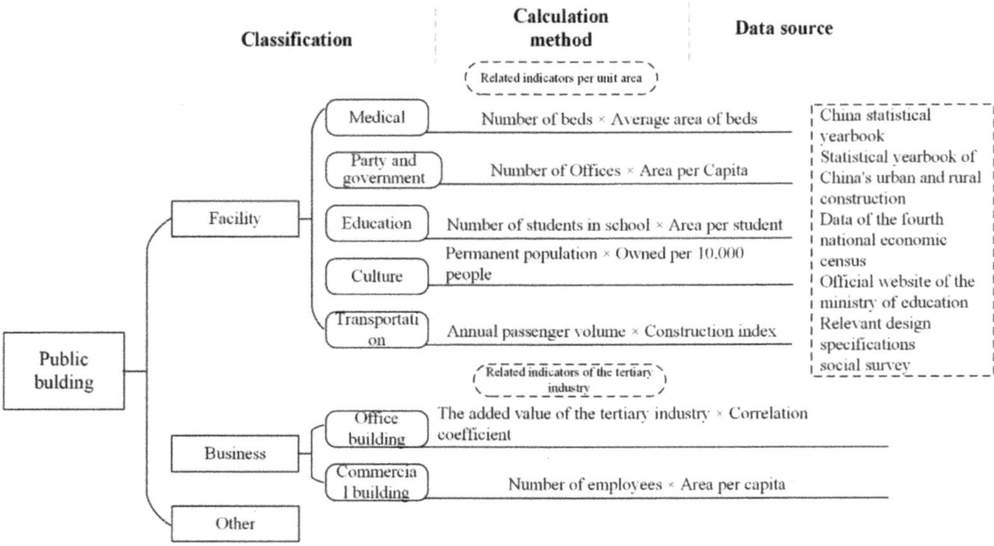

Figure 6. Classification of public buildings in China and calculation methods.

International experience shows that high economic growth is accompanied by rapid urbanization, and the urbanization process gradually slows down after the economic growth rate steps down. The 'saturation value' of the urbanization rate varies significantly from country to country. According to the research results of the Development Research Center of the State Council, the urbanization rate is about 65~75% when the per capita GDP is US$10,000~20,000. From 2014 to 2020, China's urbanization rate will keep an average annual growth rate of about 1 percentage point, and the urbanization rate of the resident population will exceed 60% by 2020 [28].

The China Statistical Yearbook shows that in 2019 the urbanization rate of the permanent population in the country reached 60.6%. Assuming that China's urbanization rate will continue to grow by 1% per year in the next 10 years, it will reach 70% in 2030 and a plateau of 75% in 2040 (referring to the experience of international urbanization rates, combined with the age structure of China's population, rural arable land and food, the constraints of production, rural labor and other aspects, we assume that the peak urbanization rate in China is 75%). The future development trend of China's population and urbanization rate is shown in Figure 7. The total population will reach a peak of 1.412 billion in 2027, and the urbanization rate will enter a plateau of 75% in 2040. At this time, the urban population will reach a peak of 1.03 billion, and the rural population will continue to decrease.

For urban dwellings, taking into account household housing demand, economic and social development and land resource constraints, this paper assumes that the future per capita dwelling area in China's urban areas will be between 32 and 42 m^2. For rural dwellings, considering further urbanization development in the future, the growth rate of rural dwellings per capita will be reduced, and the rural dwelling area per capita is expected to reach a peak of 55 m^2/person in 2035. In summary, the prediction of residential area per capita in China in this paper is shown in Figure 7.

Based on historical data, with reference to the development trend of developed countries and combined with China's national conditions, this paper makes predictions for each type of public building area index in China in the future, as shown in Table 1.

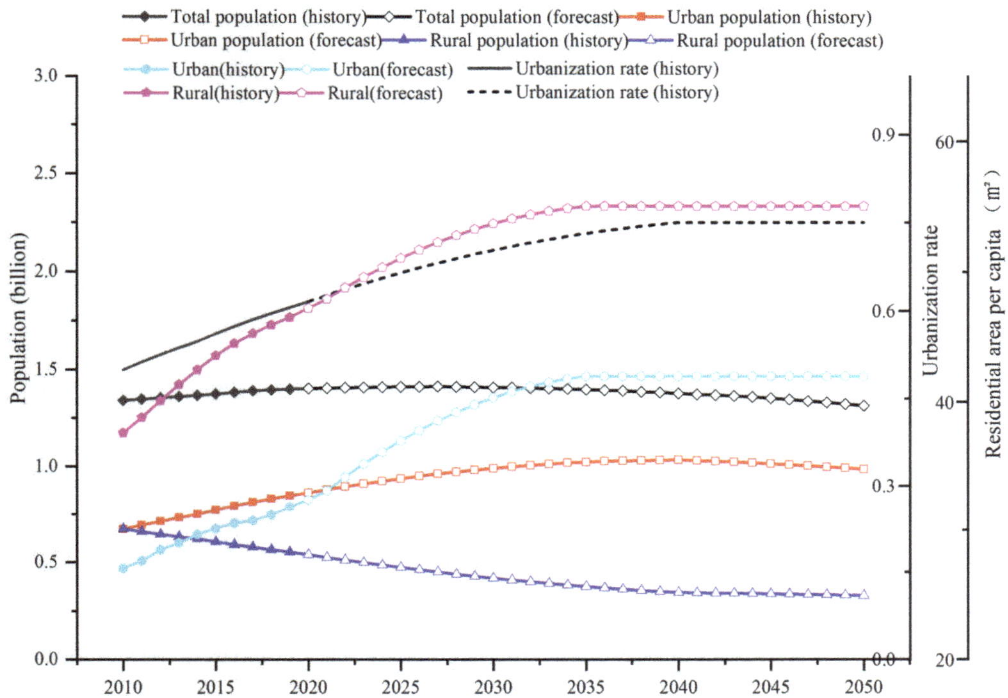

Figure 7. Forecast of future urbanization rate, corresponding population and residential area per capita in China.

Table 1. Forecast of future indicators for various types of public buildings in China.

Building Type	Indicator	2020	2030	2040	2050
Medical Building	Number of beds (beds per 1000 people)	6.0	8.5	9.0	7.5
	Average bed area (m^2)	107	137	146	151
Party and Government Building	Number of employees (million)	91.13	91.52	89.57	85.37
	Area per capita (m^2)	33.0	37.6	39.9	40.0
Office Building	Value added of tertiary industry (trillion)	60	132	227	332
Education Building	Number of students (million)	308.44	348.01	367.05	364.54
	Average student area (m^2/person)	13.5	14.3	14.8	15.0
Business—Accommodation	Number of employees (million)	5.17	8.23	8.86	8.91
	Area per capita (m^2/person)	32	62	70	75
Business—Restaurant	Number of employees (million)	6.31	7.01	7.04	6.81
	Area per capita (m^2/person)	35	56	62	65
Business—Wholesale	Number of employees (million)	21.13	23.47	24.54	25.32
	Area per capita (m^2/person)	36	39	40	40
Business—Retail	Number of employees (million)	21.03	23.43	24.36	23.79
	Area per capita (m^2/person)	65	73	78	80
Cultural Building	Owned per 10,000 people (m^2)	480	593	648	801
Transportation—Passenger Terminal	Annual passenger volume (billion)	17.1	15	13.5	12.8
Transportation—Terminal	Annual passenger volume (billion)	0.63	0.75	0.85	0.92

According to the forecasts of population, urbanization rate and various civil building related index parameters, the forecast results of the total scale of urban residential, rural residential and public buildings in China in the future are shown in Table 2.

Table 2. Various types of construction area in China in the future (billion m^2).

Building Type	2020	2030	2040	2050
Urban residence	27.6	39.9	43.4	41.4
Rural residence	25.3	22.4	18.9	18.1
Public building	15.2	19.7	21.8	21.9
Total area	68.1	82.1	84.2	81.3

Based on the historical population data of each province in the China Statistical Yearbook, combined with the forecast of the future population of each province in the report of the calibration institute, the proportion of the population of each province to the total population can be obtained. The civil building area in China is split according to the proportion of population, and the various types of civil building area in the northern region are shown in Table 3.

Table 3. Future projections of various types of floor space in northern China (billion m^2).

Building Type	2020	2030	2040	2050
Urban residence	11.6	16.7	18.0	16.9
Rural residence	10.6	9.4	7.8	7.4
Public building	6.4	8.2	9.0	8.9
Total area	28.6	34.3	34.8	33.2

3.4. Heating Area in Northern Towns

The building area in urban areas is the sum of residential area and public building area in towns and cities, and the total number of public buildings in rural areas is smaller and ignored for the time being. Heating in northern towns includes the centralized heating area and a part of the decentralized heating area that is not covered by the centralized heating pipeline network. According to the table of centralized heating in cities in various regions in the China Statistical Yearbook, it was calculated that the centralized heating rate in northern urban areas of China was about 65% in 2018. However, the non-operating centralized heating data are not counted in the Statistical Yearbook, and, referring to the status of the centralized heating area in the special planning of heating in each region [29,30], the comparison reveals that the planned heating area is about 1.2 times the area in the Statistical Yearbook. This leads to a correction of the centralized heating area in northern towns and cities, which yielded a centralized heating rate of about 85% and a centralized heating area of 13.6 billion m^2 in 2018. According to the current heating situation and heating plan in northern China, it is assumed that the centralized heating rate in northern urban areas will reach 90% in 2035 and 95% in 2050. The prediction of heating area in northern urban areas of China in the future in this paper is shown in Table 4.

Table 4. Future heating area forecast in northern urban areas of China (billion m^2).

	2020	2030	2040	2050
Heating area in northern towns	18.0	24.9	27.0	25.8
Central heating area	15.4	22.1	24.8	24.5

4. Northern Civil Building Heating Energy Consumption

4.1. Energy Consumption in Northern Towns

The heat consumed by building heating is included in the building heating energy consumption according to the actual heat source and the type of energy consumed by the transmission and distribution system, as well as the electricity, gas or standard coal converted according to the heat provided and the actual energy efficiency of the system [31]. The energy conversion in China is mainly based on the conversion of low-level calories of various types of energy into the standard coal equivalent, and, according to the appendix of the China Energy Statistical Yearbook, the conversion coefficients of various types of energy into standard coal are shown in Table 5. The coal consumption method for electricity supply refers to the amount of standard coal consumed by thermal power plants for every 1 kWh of electricity supplied to the outside, which will be used as the standard coal equivalent coefficient for electricity conversion in this paper.

Table 5. Standard coal coefficients of various energy conversions.

Energy Type	Raw Coal	Liquefied Petroleum Gas	Natural Gas	Thermal Power (Standard Coal Equivalent)	Electricity (Coal Consumption for Power Supply)
Conversion factor to standard coal	0.7143 kgce/kg	1.7143 kgce/kg	1.3 kgce/m^3	0.03412 kgce/MJ	0.31 kgce/kWh

4.1.1. Heat Source

The heat source of heating in northern cities and towns in China presents a pattern of combined heat and power, with coal-fired boilers as the leading source, and a combination of multiple energy sources and multiple heat sources, mainly divided into centralized heating and decentralized heating. At present, centralized heating can be divided into cogeneration, regional boilers, heat pumps, etc. According to the heating method, decentralized heating is mainly gas wall-hung furnaces, household coal-fired furnaces and direct electric heating. Clean heating uses clean energy such as natural gas, electricity, geothermal, biomass, solar energy, industrial waste heat, clean coal (ultra-low emissions) and nuclear energy to achieve low emissions and low energy consumption through high-efficiency energy-use systems. The heating energy used in northern areas of China is mainly coal-fired, and, at the end of 2016, the coal-fired heating area accounted for about 83% of the total heating area, while natural gas, electricity, geothermal energy, biomass, solar energy and industrial waste heat together accounted for about 17%, with renewable energy reaching 4% [32]. The current heating methods in northern cities and towns are mainly cogeneration, boilers, wall-hung furnaces and auxiliary electric-driven heat pumps. The efficiency of heat sources for different heating methods in northern China was calculated using the available energy apportionment method [32], and the calculation results and the proportion of heat sources used are shown in Table 6.

Table 6. Efficiency and proportion of heat sources for different heating methods in 2018.

Heating Method	Centralized Heating					Decentralized Heating		Renewable Energy
	Coal-Fired Cogeneration	Gas-Fired Cogeneration	Coal-Fired Boiler	Gas Boiler	Centralized Heat Pump	Gas Wall-Mounted Furnace	Electric	
Heat source efficiency (kgce/GJ)	21	23	40	35	28	32	84	-
Proportion (%)	45	3	30	2.5	1	10	4.5	4

4.1.2. Energy Consumption for Transmission and Distribution

(1) Transmission and distribution consumption. The electricity consumption of the heating system pipe network pumps during the heating period includes the electricity consumption of the circulating pumps at the heat source, the circulating pumps at the heat station and the circulating pumps on the heat consumer side of the heat pump heating. Research shows that the current electricity consumption of the secondary network of heat stations in northern areas is between 1 and 4 kWh/m^2, which has a large potential for energy saving [33]. Unreasonable water pump selection, inlet and outlet pressure losses and excessive secondary network flow are the causes of high transmission and distribution power consumption.

(2) Heat loss of pipe network. The delivery efficiency of China's heating pipe network is between 70% and 80%, while the delivery efficiency of foreign countries reaches 92–95% [34]. The reason for such a large gap is that the heat loss of the pipe network abroad is only insulation loss, while China has insulation loss, imbalance loss and leakage loss, of which imbalance loss is more serious. The heat loss of the primary network

of urban heating is about 10%, while the secondary network has serious running and leaking phenomena due to the early construction age and insufficient maintenance, and the average heat loss is about 20% [35]. According to the Technical Specification for Energy Conservation and Renovation of Heating Systems, the transmission efficiency of the primary heating network shall not be less than 95%, and the transmission efficiency of the secondary heating network shall not be less than 92% [36]. The clean heating plan also puts forward energy-saving renovation requirements for heating pipe networks, with 84,000 km of new heating pipe networks in northern areas and 50,000 km of heating pipe network renovation completed—16,000 km of primary networks and 34,000 km of secondary networks renovated.

4.1.3. Building Heating Consumption

JGJ 26-2010 gives the specific calculation formula of the heat consumption index of buildings, which is very detailed, but the amount of data is too much. In order to facilitate the calculation, this study simplifies the calculation method of the heat consumption index of buildings in this standard, see Equation (2). Multiplying by the heating hours results in the building heat consumption, see Equation (3).

$$q_H = \left(K + C_p \cdot \rho \cdot N \cdot I \cdot \varepsilon\right)\left(t_n - t_e\right) - q_{I.H} \quad (2)$$

$$Q_H = q_H \cdot T_H \quad (3)$$

where q_H is the heat consumption index of the building, W/m^2; K is the integrated heat transfer coefficient of the building envelope (obtained by converting the average heat transfer coefficient of windows, walls, doors and roofs), W/(m^2-K); C_p is the specific heat capacity of air, taken as 0.28 Wh/(kg-K); ρ is the density of air (taken as the value under the temperature t_e), kg/m^3; N is the number of air changes, taken as 0.5 1/h; I is the building floor height, taken as 3 m; ε is the ventilation volume correction coefficient, taken as 0.6; t_n is the calculated indoor temperature, taken as 18 °C; t_e is the average outdoor temperature during the heating period (°C); $q_{I.H}$ is the unit building area and unit time—the internal heat gain of the building is taken as 3.8 W/m^2; Q_H is the heat consumption of the building, kWh/m^2; and T_H is the complete heating time in a heating period, h.

The northern region of China includes two climatic zones of severe cold and cold areas, where the outdoor temperature, heating hours and thermal performance of the envelope differ greatly. According to JGJ 26-2010, the average outdoor temperature during the heating period in the cold region (Inner Mongolia, Liaoning, Jilin, Heilongjiang, Qinghai and Xinjiang) is −5.6 °C and the average heating duration is 174.1 days; the average outdoor temperature during the heating period in the cold region (Beijing, Tianjin, Hebei, Shanxi, Shandong, Henan, Shaanxi, Gansu and Ningxia) is −0.76 °C and the average heating duration is 121.7 days [37]. The area ratio of the two climatic zones, harsh and cold, is estimated to be 2.8:7.2 by population ratio. Due to the geographical location of the severe cold regions, the heat transfer coefficient of their building envelopes is lower than that of the cold regions by 0.1–0.3 W/(m^2-K) [38]. According to the current rate of building energy renovation, the proportion of energy-efficient buildings to urban civil construction area exceeded 50% in 2016 [32], and the proportion of energy-efficient buildings would have been about 60% in 2018. The comprehensive relevant data above, according to Equations (2) and (3), can achieve the current building heat consumption in the cold and severe cold areas of northern towns in China, as shown in Table 7.

Table 7. Heat consumption of buildings in severe cold and cold areas of northern towns in China, 2018.

Climate Zone	Building Type	Average Heat Transfer Coefficient (W/(m^2·K))	Building Heat Consumption (kWh/m^2)	Average Heat Consumption (kWh/m^2)
Severe cold area	Non-Energy-Efficient Building	1.1	125.4	78.3
	Energy-Efficient Building	0.75	89.5	
Cold area	Non-Energy-Efficient Building	1.3	78.0	
	Energy-Efficient Building	0.95	58.1	

4.1.4. Calculation Method of Heating Energy Consumption in Northern Towns

The heating energy consumption (excluding renewable) in northern towns is calculated as shown in Equations (4)–(6).

$$e_c = \left(\frac{Q_H}{\eta_1 \cdot \eta_2}\right) \times \sum_i (p_{c,i} \cdot k_{c,i}) \tag{4}$$

$$e_d = Q_H \times \sum_j \left(p_{d,j} \cdot k_{d,j}\right) \tag{5}$$

$$E_H = (e_c \cdot A_c + e_d \cdot A_d) \times (1 - \alpha_{re}) \tag{6}$$

where e_c is the energy intensity of central heating, kgce/(m^2-a); η_1 is the primary network efficiency; η_2 is the secondary network efficiency; Q_L is the transmission and distribution consumption loss, kWh/m^2; p is the heat source efficiency of different heating methods, kgce/GJ; k is the proportion of different heating methods used; i is the different central heating methods (coal-fired cogeneration, gas-fired cogeneration, coal-fired boiler, gas boiler, centralized heat pump); j is the different decentralized heating methods (gas wall-hung stove, electric heating); e_d is the energy intensity of decentralized heating, kgce/(m^2-a); E_H is the total energy consumption of heating in northern towns, billion kgce; A_c is the centralized heating area, billion m^2; A_d is the decentralized heating area, billion m^2; and α_{re} is the proportion of renewable energy for heating. When the national average coal consumption for electricity supply is 310 kgce/kWh, the above formula can be used to calculate that the current energy consumption intensity of centralized heating in China is 11.0 kgce/(m^2-a), the energy consumption intensity of decentralized heating is 13.3 kgce/(m^2-a), the average energy consumption intensity of urban heating is 11.4 kgce/(m^2-a) and the total energy consumption of urban heating in northern areas is 180 Mtce.

4.2. Northern Rural Heating Energy Consumption

The type of energy in China's rural areas is mainly coal-fired and biomass (straw, fuel wood, biogas), and, with the promotion of clean energy, the proportion of electricity, natural gas and other energy sources is gradually increasing. Most of the rural areas in the north are decentralized for heating, with a large number of earth beds, stoves and earth heaters, and energy use is mainly coal and biomass. The cost of clean energy is high, and only a few rural houses use natural gas, electric energy and renewable energy for heating. The thermal performance of the envelope structure is poor, and only 20% of rural heating buildings have taken energy-saving measures. At present, the ratio of coal to biomass energy use in rural winter heating energy consumption in northern areas is about 3:1, and the average heating energy intensity is about 5 kgce/(m^2-a) except for biomass [39,40].

The calculation of energy consumption for heating rural dwellings is shown below:

$$E_r = e_r \times A_r \tag{7}$$

where E_r is the total energy consumption of rural residential heating, billion kgce; e_r is the energy consumption intensity of heating, kgce/(m^2-a); and A_r is the area of rural residential houses in northern areas, billion m^2. By analyzing the various rural energy uses, using Equation (7), we can calculate the residential heating energy intensity of 5.3 kgce/(m^2-a) and the total energy consumption of 57.54 Mtce in 2018 in rural areas in northern China.

5. Scenario-Based Analysis for Predicting Building Heating Energy Consumption in Northern Regions

5.1. Scenario Construction

Using the 2018 energy intensity data as a benchmark, the future energy intensity for heating buildings in the north is set under three scenarios: (1) baseline control, i.e., development according to the current control measures; (2) medium control, i.e., adopting slightly stricter control measures than the benchmark scenario; (3) strict control, i.e., stricter measures are taken to control energy consumption intensity.

The energy intensity settings for the three future scenarios are shown in Table 8, and the settings are described below.

(1) Prediction of energy consumption intensity of heating in northern towns. China is currently continuing to promote heating reform in northern cities and towns, and energy-saving efforts are being promoted in three areas: heat sources, heat networks and heat users. In terms of heating heat source, the proportion of centralized heating will be further increased, expanding from cities to urban and rural areas. Considering the requirements of the clean heating policy, centralized heating will be dominated by high-efficiency cogeneration, and small- and medium-sized coal-fired boilers will be gradually replaced. Decentralized heating will be dominated by gas wall-hung boilers, and electric energy will be mainly used for all kinds of electric-driven heat pump heating. In addition, the proportion of renewable energy for heating will continue to increase, and the demand for primary energy such as coal and natural gas will decline. For the transmission and distribution network, with the energy-saving transformation of old primary and secondary pipeline networks and heat exchange stations, the efficiency of the pipeline network will gradually improve, and the heat loss and electricity consumption of transmission and distribution will be reduced. For building terminals, energy-saving renovation of existing buildings will be accelerated, and for new buildings 65% or even 75% energy-saving standards will be strictly enforced to reduce the heat demand for heating at the user end. Referring to the requirements of the Energy Consumption Standard for Civil Buildings (GB/T 51161-2016) for the design value of heating energy intensity, the constrained value of centralized heating energy intensity is 6.8 kgce/(m^2-a) and the guide value is 3.6 kgce/(m^2-a) in Shijiazhuang City, Hebei Province, for example. In this paper, it is assumed that the heating energy intensity will reach the standard constraint level in 2030 under the baseline control and will decrease to 5.3 kgce/(m^2-a) in 2050; under the medium control, it is controlled at about 4.3 kgce/(m^2-a) in 2050; and under the strict control, it approaches the standard guideline level and decreases to 3.9 kgce/(m^2-a) in 2050.
(2) Prediction of energy consumption intensity of rural residential heating. With the continuous development of the rural economy and the improvement of the urbanization level, the demand for rural heating is increasing, and, when urbanization reaches a high level, the standard of living in rural areas should not be much different from that in cities and towns, but, due to the larger residential area per capita in rural areas, the building energy efficiency is backward and contains energy consumption for split heating, thus the intensity of energy consumption for heating in rural buildings will gradually rise. If our government can strengthen the regulation and guidance of rural energy use in the future, focus on promoting the use of renewable energy sources such as biomass, biogas and the development of photovoltaic and solar thermal technologies to replace some of the energy consumption increased by the improvement in the living standards of rural residents, the energy intensity of rural residential heating is expected to be controlled. In this paper, we assume that under the baseline control, the energy intensity of rural residential heating will increase rapidly, increasing by 19% to 6.3 kgce/(m^2-a) in 2030 and 7.1 kgce/(m^2-a) in 2050; under the medium control scenario, the energy intensity will increase slowly, reaching 6.6 kgce/(m^2-a) in 2050; and under the strict control, the energy intensity of rural houses will increase slowly by 19% by 2030 and will be controlled at about 6.1 kgce/(m^2-a) in 2050.

5.2. Prediction Results of Building Heating Energy Consumption under Different Scenarios

(1) Baseline control scenario. Although the intensity of heating energy consumption in rural dwellings is increasing, due to the decreasing residential area, heating energy consumption in rural dwellings will peak at about 59.92 Mtce in 2026 and is expected to be controlled to 52.3 Mtce in 2050. Heating energy consumption in northern towns is decreasing year by year with the implementation of heating energy conservation

policies and may be reduced to 122 Mtce in 2050. The total heating energy consumption in the north will peak in 2025 at about 237 Mtce.

(2) Medium control scenario. The energy consumption for heating in northern towns will be about 153 Mtce in 2030 and may decrease to 100 Mtce in 2050, a 45% reduction from 2018. Heating energy consumption in rural dwellings will peak at 58.57 Mtce in 2022 and then decline continuously under various energy efficiency measures to 56.59 Mtce in 2030 and 48.39 Mtce in 2050. The total heating energy consumption in the whole northern region will decrease to 210 Mtce in 2030 and to about 148 Mtce in 2050.

(3) Strict control scenario. The energy consumption for future heating in northern areas under strict control scenarios is decreasing. Heating energy consumption in northern cities and towns may reach 141 Mtce in 2030 and be as low as 0.90 Mtce in 2050, with significant energy savings. Rural heating energy consumption will drop to 54.76 Mtce in 2030 and 44.92 Mtce in 2050.

Table 8. Future energy intensity scenarios for building heating in northern areas of China (kgce/(m^2-a)).

	2018		2030			2050	
	Current	Baseline	Medium	Strict	Baseline	Medium	Strict
Heating of northern towns	11.4	7.3 −35%	6.6 −42%	6.0 −47%	5.3 −54%	4.3 −62%	3.9 −66%
Heating of rural residences	5.3	6.3 19%	6.0 13%	5.8 9%	7.1 34%	6.6 25%	6.1 15%

The China Building Energy Conservation Association calculated the urban building energy consumption of each province (city) in China, in which the urban building energy consumption of 15 provinces (cities) in the northern region accounted for 41% of the total building energy consumption in the country [41]. Tsinghua University's research on national rural energy use showed that the rural residential energy consumption in 15 provinces (cities) in the northern region accounted for 62% of national rural energy consumption [15]. A comprehensive estimate shows that building energy consumption in northern China accounts for about 55% of the total building energy consumption in the country. In addition, heating energy has always been an important part of building energy consumption in the north. Figure 8 gives the heating energy consumption of residential buildings in northern regions under different scenarios. With low carbon requirements, building heating energy consumption in northern regions should be kept under at least a medium control scheme if we want to achieve peak carbon as early as possible.

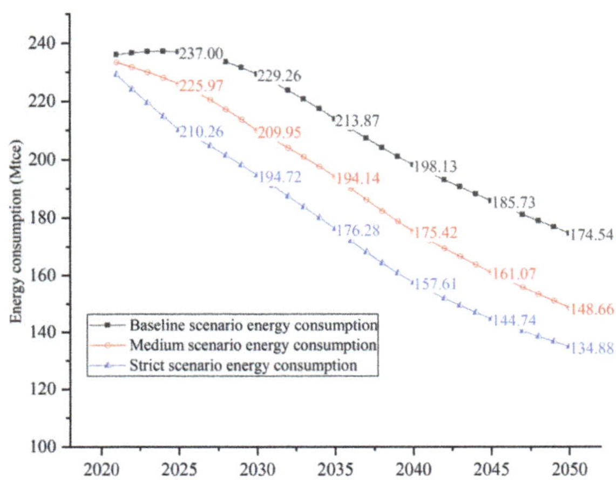

Figure 8. Total heating energy consumption in northern buildings under different scenarios.

5.3. Carbon Emissions from Building Heating in Northern Regions

(1) Carbon emission factors. The most widely used carbon emission calculation method internationally is the carbon emission factor method proposed by IPCC, whose basic idea is to use the activity data of various emission sources and the product of emission factors as the carbon emission estimation value according to the carbon emission inventory list. The emission factors are more regional, and the carbon emission calculation standards for buildings in China are shown in Table 9 [42]. The electricity consumed for building energy consumption mainly comes from thermal power generation and renewable energy generation, and renewable energy generation is mainly clean energy such as wind, water, photovoltaic and nuclear energy, which can be regarded as zero carbon emission, so the electricity carbon emission factor can be obtained by dividing the carbon emissions from thermal power generation by the total national electricity generation. According to the data published by the State Grid, thermal power generation accounted for 70% of the total power generation in China in 2018, and the national electricity carbon emission factor was estimated to be 0.5 $kgCO_2/kWh$ with reference to the electricity carbon emission factor of each region. According to the National Energy Board's forecast of China's medium- and long-term power generation capacity [43], the share of renewable energy generation will reach 40% in 2030 and about 50% in 2050, then the power carbon emission factor may drop to 0.52 $kgCO_2/kWh$ in 2030 and about 0.38 $kgCO_2/kWh$ in 2050.

(2) Forecast of the energy structure. In terms of heating, for a long time to come, the main energy source for heating in northern towns will still be coal, and rural areas should continue to develop renewable energy sources such as photovoltaic, wind power and biomass to reduce the proportion of coal and other energy consumption in residential heating. In order to achieve the goal of carbon neutrality, coal should be gradually phased out from the heating energy system in the future and the proportion of renewable energy and clean energy should be increased. With reference to the proportion of energy use in the current energy balance sheet, the energy structure of urban and rural heating under the medium control scenario is projected as shown in Figure 9, and a new energy structure is projected after strengthening carbon emission control measures and increasing the utilization of new and clean energy sources such as rural biomass, nuclear energy and urban industrial waste heat under the medium control scenario.

(3) Prediction of carbon emissions from building heating under the medium control scenario. Figure 10 shows the future trend of building heating carbon emissions in the northern region of China under the medium control scenario. Under the medium control scenario, the carbon emissions from building heating in the northern region can be reduced to 450 $MtCO_2$ in 2050, and to 280 $MtCO_2$ with the adoption of more stringent energy-saving and emission reduction technologies.

Table 9. Carbon emission factors of various energy sources in China in 2018.

Energy Type	Coal	Liquefied Petroleum Gas	Natural Gas	Biomass	Electricity
Carbon emission factor	94.44 $kgCO_2/GJ$	61.81 $kgCO_2/GJ$	55.54 $kgCO_2/GJ$	76.5 $kgCO_2/GJ$	0.58 $kgCO_2/kWh$

As can be seen from the Figure 10, building heating carbon emissions in the northern region are dominated by urban heating, accounting for about 75% of the total building heating carbon emissions in the north. In the medium scenario, urban heating carbon emissions reduce from 579 $MtCO_2$ in 2021 to 376 $MtCO_2$ in 2050, a reduction of 203 $MtCO_2$, accounting for 78% of the total emission reductions. In the medium control scenario to strengthen the reduction of carbon emissions, urban heating carbon emissions reduce from 572 $MtCO_2$ in 2021 to 209.8 $MtCO_2$ in 2050, a reduction of 363 $MtCO_2$. Therefore, urban building energy-saving and emission reduction should be the main goals of northern heating.

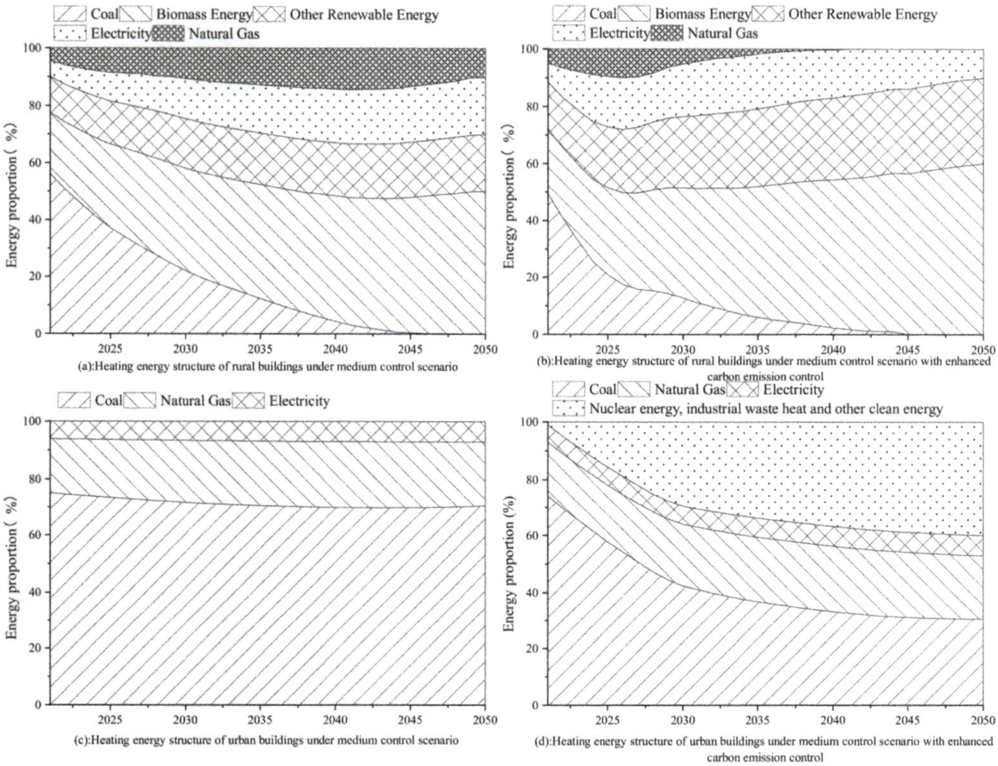

Figure 9. Heating energy structure forecast for northern region.

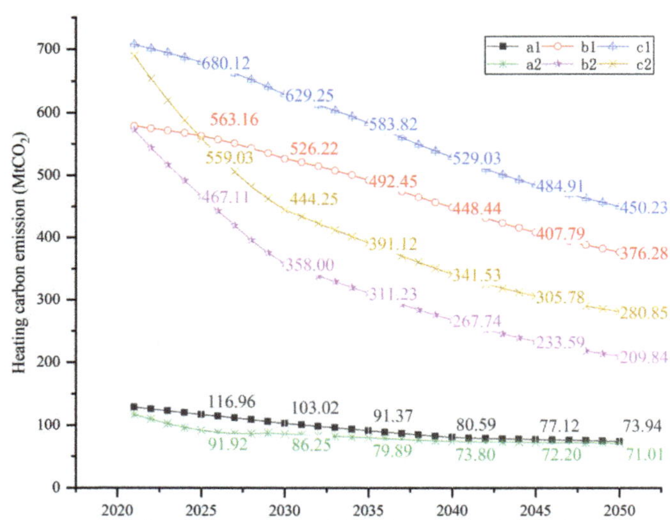

a:Rural buildings heating carbon emissions. b:Urban buildings heating carbon emissions. c:Northern buildings heating carbon emissions. 1: Carbon emissions under a medium control scenario. 2: Carbon emissions under a medium control scenario with enhanced c-arbon emission control.

Figure 10. Future projection of heat carbon emissions from building heating in northern China.

6. Analysis of the Implementation Path of Heating Energy Saving in Northern Buildings

Based on the existing northern clean heating policies, the energy intensity of urban heating in northern regions under the three scenarios set in this paper has a decreasing trend. In order to facilitate the analysis of energy savings from heating energy efficiency technologies, this part sets a non-energy-efficient heating energy scenario for heating energy consumption in northern towns, i.e., at the current heating energy intensity, the heating energy consumption in future heating of northern towns without any additional measures is compared with the heating energy consumption in a medium control scenario as shown in Figure 11. The implementation of 65% energy-saving standards for new civil buildings began in 2010, before which the stock of existing buildings in northern towns was about 10.5 billion m^2. Although the total energy consumption of existing buildings and the proportion they account for is gradually decreasing due to their demolition or collapse year by year, due to their huge stock, without energy-saving renovation and operation and maintenance, the heating energy consumption in northern cities and towns will still reach 235 Mtce in 2050. Thus, the energy-saving potential of building heating energy-saving renovation and operation and maintenance is huge.

(1) Energy-saving renovation of envelope structure. The building envelope is the part that has the greatest impact on heating energy consumption. At present, the comprehensive heat transfer coefficient of the non-energy-saving building envelope in China is around 1.1~1.3 W/(m^2-K), while the comprehensive heat transfer coefficient of the building envelope that reaches the 65% energy-saving standard is about 0.8~1.0 W/(m^2-K). Currently, some northern regions have started to implement 75% energy efficiency standards, and the integrated heat transfer coefficient of new buildings has been reduced to 0.4~0.6 W/(m^2-K) [44,45]. As of 2018, about 60% of urban buildings in northern China had completed an energy-saving renovation, and assuming that all northern urban areas under the medium control scenario complete an energy-saving renovation of existing building envelopes in 2030 and the integrated heat transfer coefficient reaches 0.55–0.75 W/(m^2-K), heating energy consumption will drop to 193 Mtce in 2030. With the update of energy-saving standards, the heat transfer coefficient of the building envelope will be further reduced in the future, and, if the energy efficiency level of 75% is fully achieved in 2050, the heating energy consumption may reach 144 Mtce, which is 92 Mtce lower than the non-energy-saving scenario.

(2) Heat source restructuring. From the perspective of heating the heat source structure, heating in northern cities and towns in China is still mainly coal-fired, mainly through coal-fired cogeneration and coal-fired boilers for centralized heating, accounting for about 45% and 30% of heating methods, respectively. According to the requirements of clean heating planning, the scope of centralized heating should be expanded in the future, the proportion of gas heating should be increased, the use of loose coal should be reduced and heat sources such as electric heat pumps, industrial waste heat and renewable energy should be developed for heating. The future adjustment of the heat source structure in northern areas is shown in Figure 12; energy consumption for heating will reach 233 Mtce in 2030, decreasing to 224 Mtce in 2050, a reduction of 0.12 Mtce over the non-energy-efficient scenario.

(3) Heating source efficiency. At present, the heating capacity of cogeneration units has not been fully developed, and the waste heat of spent steam has not been fully utilized; most of the units have a heat-to-electricity ratio below 1.5. While the actual operation can be up to about three times, the heat-to-electricity ratio still needs to be improved. Compared with cogeneration, the boiler efficiency is obviously low. The thermal efficiency of large coal-fired boilers is around 70%, and the efficiency of gas-fired boilers is above 85%, so energy-saving renovation of boiler rooms can effectively improve the heating efficiency of boilers. With the development of science and technology, the efficiency of various heat pumps for heating is also improving. The ground source heat pump COP is usually between 3 and 5, while the heat pump

efficiency can be increased to about 8 through the inverter centrifugal technology. Assuming that the heating coal consumption of various heating methods can reach the level of Table 10 in the future through new energy-saving technologies for heating, the heating energy consumption can be reduced to 213 Mtce in 2050, and the energy saving is about 0.23 billion tce compared with the non-energy-saving scenario.

(4) Energy-saving renovation of transmission and distribution pipeline network. At present, the heat loss of China's urban centralized heating pipe network is more serious, and GB/T 51161 limits the heat loss rate of the pipe network and the index of transmission and distribution consumption, as shown in Table 11. If the heat loss rate of the pipe network in 2050 under the medium control scenario can be controlled within the district heating constraint value, the pipe network efficiency will reach 95% and the transmission and distribution consumption will drop from 3 to 2 kWh/(m^2-a), then heating energy consumption can reach 219 Mtce in 2030 and be controlled to 192 Mtce in 2050, which is a 44 Mtce reduction compared to non-energy-saving scenarios.

(5) Heat metering. Heat metering buildings set temperature control valves indoors. When the indoor temperature is higher than the temperature set by the temperature control valve, the heating system temperature control valve will reduce the opening or close the valve to provide less heat to the room and vice versa to increase the heat supply. However, since its introduction, the heat metering policy has not been popularized due to uneven heat supply, lax monitoring and unreasonable charges. However, heat metering has an important role for heating energy saving. Actual cases show that the annual heat consumption of a centrally heated building has decreased from 87 to 63 kWh/(m^2-a) since the implementation of heat metering retrofit, which is a 28% reduction [46]. This article assumes that under the medium scenario full heat metering charges will be reached in 2050, the annual heat consumption of urban heating buildings in northern regions will be reduced from 74 to 53 kWh/(m^2-a), and the heating energy consumption will be about 170 Mtce. Compared with the non-energy-saving scenario, the energy saving will be about 65 Mtce.

Table 10. Future coal consumption for heating from various heating sources.

	2018	2030	2050
Coal-fired cogeneration	21	20	19
Gas cogeneration	23	21	20
Coal-fired boiler	39	37	35
Gas boiler	35	33	32
Centralized heat pump	28	26	25
Gas wall-mounted furnace	32	31	30
Electricity	84	79	75

Table 11. Heat loss of pipe network and limit of transmission and distribution consumption index.

		Constraint Value	Guide Value
Index of heat loss rate of pipe network (%)	District central heating	5.0	3.0
	Centralized heating in community	2.0	1.0
Water pump power consumption index of pipe network (kWh/(m^2-a))	Heating for 4 months	1.7	1.0
	Heating for 5 months	2.1	1.3
	Heating for 6 months	2.5	1.5
	Heating for 7 months	2.9	1.8

The heating energy consumption and energy savings in 2050 after adopting the above energy-saving technologies, compared to the non-energy-saving scenario, are shown in Table 12. However, different energy-saving technologies are interrelated; for example, the improvement of the maintenance structure will at the same time enhance the effect of heat metering and the change in the heat source structure will also affect the heat source efficiency. Therefore, the energy consumption and energy saving in the table are the results of a single action relative to the non-energy-saving scenarios, and the baseline, medium and strict control scenarios in the previous section are a comprehensive consideration of various technology development policy changes, which are reflected in the energy consumption intensity.

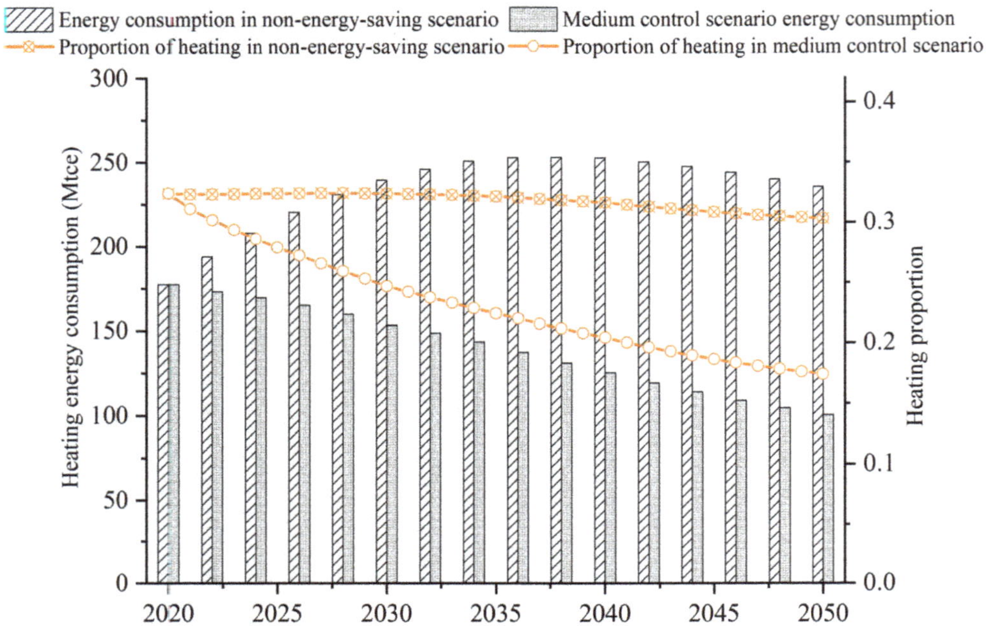

Figure 11. Comparison of heating energy consumption in the non-energy-saving scenario and the medium control scenario.

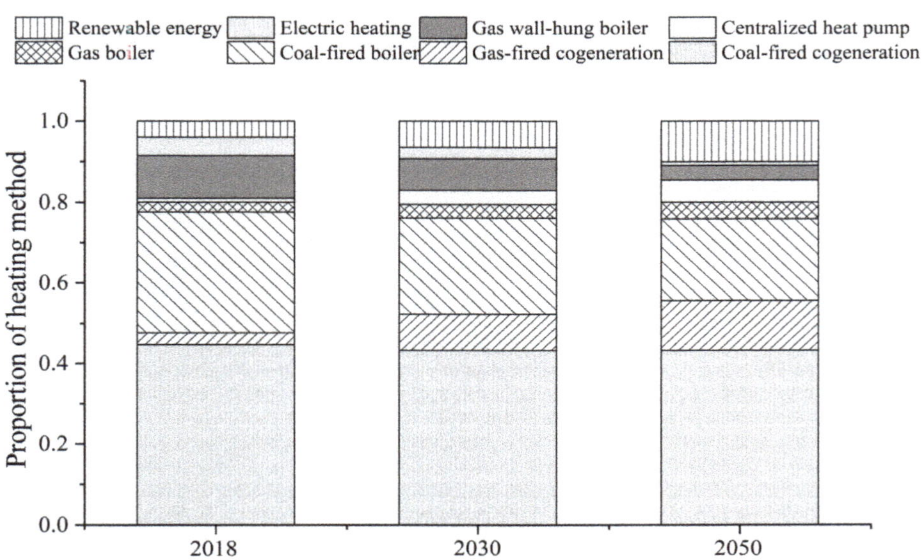

Figure 12. Comparison of heating heat source structures.

Table 12. Heating energy consumption and energy savings with energy efficiency measures compared to non-energy-efficient scenario (Mtce).

Energy-Saving Heating Technology	Energy Consumption in 2050	Energy Saving
Renovation of envelope structure (75% energy saving)	144	92
Heat source restructuring	224	12
Heating source efficiency improvement	213	23
Renovation of pipeline	192	44
Heat metering	170	65

Figure 13 compares the energy savings of various energy-efficient technologies for heating in northern towns and cities and clearly shows the difference in energy consumption between the non-energy-efficient scenario and the medium control scenario. It can be seen that building envelope renovation contributes the most to heating energy saving, and building energy saving should focus on this. Heat source structure adjustment has the least effect on heating energy consumption, but, from the perspective of carbon emissions, improving the heat source structure can effectively reduce carbon emissions and promote carbon peaking. The analysis of energy savings of each type of technology in the figure is the result of its single action. In order to achieve the target of urban heating energy consumption under the medium control scenario, all types of energy-saving technologies should also be used in combination with China's national conditions.

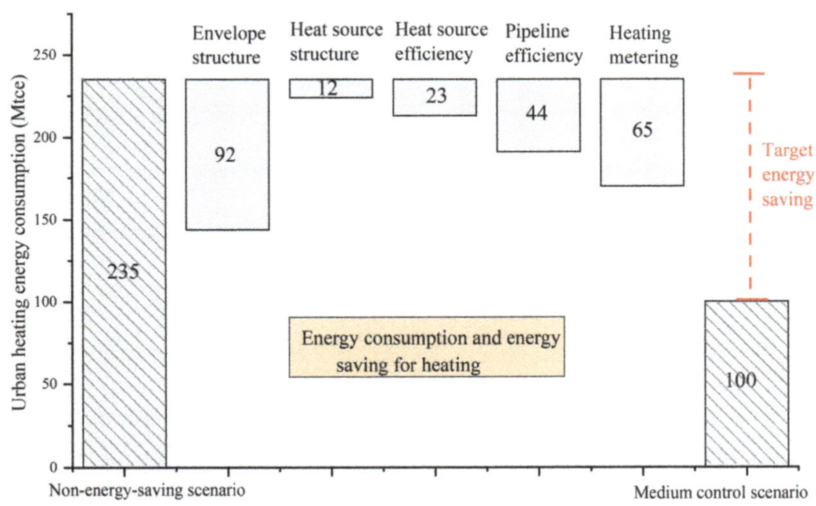

Figure 13. Comparison of energy savings of heating energy-saving technologies in northern towns.

According to the above analysis of various building energy efficiency technologies, combined with China's building energy efficiency policy measures, in order to achieve the energy consumption target of the medium control scenario, the development path of civil building heating energy efficiency in northern China is shown in Figure 14.

For the vast rural areas in northern China, the different climatic conditions, geography and economic development should perhaps prompt us to shift our focus from simply reducing the energy intensity of rural heating as a way to reduce carbon emissions to reducing the energy intensity of heating and developing the use of clean energy for heating to reduce carbon emissions. Biomass and solar energy are widely distributed and accessible in rural areas with very different natural and energy conditions. Improving the use of biomass and thus reducing carbon emissions has already been described. Solar energy for rural heating can be divided into active and passive types, according to whether mechanical power is required to drive the elements [47]. However, due to the low energy flow density, instability and mismatch between the heating characteristics of solar energy and the building load, solar energy is often used in combination with multiple energy sources for complementary heating to enhance the contribution of solar energy in renewable energy and system energy efficiency [48].

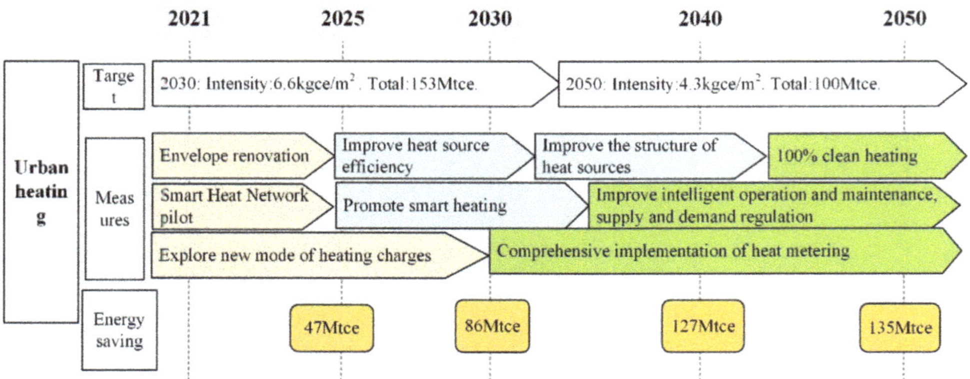

Figure 14. Development path of civil building energy efficiency in northern towns.

An example is the common coupled solar–air source heat pump system. Chengyang Jiang [49] et al. proposed a ribbed collector for DXSASHPWH, where the collector can absorb heat in the air and also in the solar radiation, and the average COP of the system can reach 6, which is much higher than the conventional-type system. Shan Ming [10] et al. selected a demonstration household located in the rural Pinggu District, Beijing; the heating system was a solar hot water collector system plus 4 kW flow inverter low-temperature air source heat pump hot water machine, the end unit used geothermal radiant heating, with the whole heating season solar hot water circulation pump power consumption of about 151 kWh, a low-temperature air source heat pump hot water system power consumption of 6104 kWh and carbon emissions of 2790 kg, much lower than the carbon emissions of traditional heating methods [50]. Meysam, Huide Fu [51,52] et al. designed a PV/T system based on heat pipes by combining cylindrical heat pipes with photovoltaic modules and combined it with a heat pump to design a solar heat pump system based on cylindrical heat pipes, where the average COP of the heat pump system could reach 4.87 and the photovoltaic efficiency was above 11%. The green electricity generated by the PV system can also be used for heat pump heating, thus reducing the carbon emissions from heating. This is promising in provinces such as Tibet, Inner Mongolia and Qinghai, where solar energy resources are abundant. In addition, a solar–geothermal source heat pump coupling system, solar–water source heat pump coupling system, etc. can achieve the same energy-saving and emission reduction effect, which needs to be determined according to the actual local climate, geography and economic situation. Due to the instability of solar energy and the various conditions in different regions, such systems are more complex, and the specific form is difficult to determine, so no further detailed calculations will be made, but rather they will be grouped with other green energy sources under the category of rural renewable energy.

7. Conclusions

By analyzing the historical data of civil buildings and relevant energy consumption models, this paper classifies and predicts the civil building area in the north by considering various factors, calculates and predicts the building energy consumption of different types of buildings, and establishes a macro-level calculation model of heating energy consumption of civil buildings in the north to obtain detailed data of heating energy consumption of buildings in the north of China. By analyzing the year-to-year changes in energy consumption and future development trends, the control target of building heating energy consumption in northern China under the carbon emission constraint is determined. Through the analysis of building energy-saving technologies, the key points of heating energy saving and emission reduction are clarified, providing a relevant basis for the formulation of heating energy consumption and carbon emission control policies in northern China. The main conclusions are as follows:

(1) Population, urbanization rate and the economic development level are the main influencing factors of building area. As urbanization accelerates, urban residential and public building area will increase, reaching a peak around 2040, while rural residential building area will continue to decline. Among public buildings, medical, education, office and commercial buildings account for 73% of the total public buildings, and area control should focus on these buildings.

(2) Under the total energy and carbon emission constraints, total energy consumption in the northern region should be controlled at least below the medium control scenario in the future. Under this target, China's total population will peak at 1.412 billion in 2027; the urbanization rate will enter a plateau after reaching 75% in 2040; the residential building area in the northern region will peak at 35.2 billion m^2 in 2035; the heating energy consumption in northern towns will decrease, falling to 6.6 kgce/(m^2-a) in 2030 and to 4.3 kgce/(m^2-a) in 2050; and rural residential heating energy consumption will increase with economic development, growing to 6.0 kgce/(m^2-a) in 2030 and possibly to 6.6 kgce/(m^2-a) in 2050.

(3) To achieve the medium control target, it is necessary to make comprehensive use of various energy-saving technologies. In rural areas, attention should be paid to the effective use of biomass energy, local use of solar, geothermal and other green renewable energy. In terms of urban heating energy-saving technology, the transformation of the building envelope contributes the most to heating energy saving, reaching 92 Mtce. Building energy saving should focus on this, supplemented by technical means such as pipe network and heat source transformation to control heating energy consumption. At the same time, the development and utilization of clean heating technology should be strengthened to replace coal consumption in urban heating.

Author Contributions: Conceptualization, Y.Z. and W.Z.; methodology, Y.Z. and W.Z.; formal analysis, W.Y. and J.L.; investigation, W.Y. and J.L.; data curation, K.W. and K.L.; writing—original draft preparation, W.Y. and J.L.; writing—review and editing, Y.Z. and W.Z.; visualization, K.W. and K.L. All authors have read and agreed to the published version of the manuscript.

Funding: This research was funded by the National Natural Science Foundation of China, grant number 51908287; the China National Key R&D Program, grant number 2018YFC0704400; and the National Natural Science Foundation of Jiangsu Province, grant number BK20180484.

Institutional Review Board Statement: Not applicable.

Informed Consent Statement: Not applicable.

Data Availability Statement: The data presented in this study are available on request from the corresponding author. The data are not publicly available due to privacy.

Conflicts of Interest: The authors declare no conflict of interest.

References

1. Andrew, N. *The IPCC's Report on Global Warming of 1.5 °C Spells out the Urgency of Action*; International Institute for Environment and Development: London, UK, 2018.
2. Jinping, X. Address to the general debate of the 75th session of the United Nations General Assembly. *Gaz. StateCouncil People Repub. China* **2020**, *28*, 5–7.
3. Beibei, Q. Research on China's Building Energy Consumption Statistical Method. Master's Thesis, Chongqing University, Chongqing, China, 2014.

4. EIA. Residential Energy Consumption Survey Household Questionnaire. 2009. Available online: https://www.iea.org/reports/armenia-energy-profile (accessed on 5 October 2021).
5. U.S. DOE Energy Information Administration. *Model Documentation Report: Residential Sector Demand Module of the National Energy Modeling System*; DOE/EIA-M067(97); U.S. DOE Energy Information Administration: Washington, DC, USA, 2005.
6. Clark, V.; Heaps, C. LEAP: Long-range Energy Alternatives Planning System. Available online: http://www.energycommunity.org (accessed on 23 November 2021).
7. Liu, J.; Xiang, Q.; Wang, K.; Zou, J.; Kong, Y. Mid- to long-term low carbon development pathways of China's building sector. *Resour. Sci.* **2019**, *41*, 509–520.
8. Xinyi, L. A Study About Residential Building Stock Energy Modelling. Ph.D. Thesis, Chongqing University, Chongqing, China, 2018.
9. Xueling, L.; Weijuan, F.; Yuanming, W. Simulation and prediction of building energy consumption based on climate and population change. In Proceedings of the 2021 Seminar on Construction and Efficient Operation of Heat Supply Projects, Chengdu, China, 21 April 2021.
10. Despotovic, M.; Koch, D.; Leiber, S.; Doeller, M.; Sakeena, M.; Zeppelzauer, M. Prediction and analysis of heating energy demand for detached houses by computer vision. *Energy Build.* **2019**, *193*, 29–35. [CrossRef]
11. Xiu, Y. Study of China's Building Energy Efficiency Based on Energy Data. Ph.D. Thesis, Tsinghua University, Beijing, China, 2009.
12. National Bureau of Statistics of People's Republic of China. *China Statistical Yearbook-2013*; China Statistics Press: Beijing, China, 2013.
13. Menghui, W. Significant achievements have been made in the development of housing and urban-rural construction during the "13th Five-Year Plan" period. *China Eng. Consult.* **2020**, *12*, 16–18.
14. China Association of Building Energy Efficiency-Energy Consumption Statistics Professional Committee. *China Energy Consumption Research Report 2020*; China Association of Building Energy Efficiency: Beijing, China, 2020.
15. Research Center for Building Energy Efficiency, Tsinghua University. *2020 Annual Report on China Building Energy Efficiency*; China Construction Industry Press: Beijing, China, 2020; pp. 2–5.
16. National Bureau of Statistics. China Statistical Yearbook. Available online: http://www.stats.gov.cn/tjsj/ndsj/ (accessed on 5 October 2021).
17. China Survey and Data Center, National People's University of China. China General Social Survey (CGSS). Available online: http://cgss.ruc.edu.cn/ (accessed on 5 October 2021).
18. Sun, W.K. Changes in the number of households and housing demand in China. *Soc. Sci. J.* **2020**, *6*, 160–166.
19. National Health and Family Planning Commission. *China Family Development Report 2015*; China Population Press: Beijing, China, 2015.
20. State Council of the People's Republic of China. *Notice on the Issuance of the National Land Planning Outline (2016–2030)*; State Council of the People's Republic of China: Beijing, China, 2017; pp. 35–64.
21. Ministry of Housing and Urban-Rural Development of People's Republic of China. *China Urban and Rural Construction Statistical Yearbook*; China Plan Press: Beijing, China, 2018.
22. Institute of Standardization, Ministry of Housing and Urban-Rural Development. *Energy Consumption Control Strategy of Civil Buildings*; China Construction Industry Press: Beijing, China, 2016.
23. Tongdan, G.; Wenjie, Z.; Jinhan, L.; Changqing, L.; Kai, M. Forecast and analysis of the total amount of civil buildings in China in the future based on population driven. *Sustainability* **2021**, *13*, 14051. [CrossRef]
24. Wang, Y.; Wei, Q.; Xue, Z.; Jiang, Y. Large-scale public building energy conservation consultation (3)-Investigation and analysis investigation and analysis of energy consumption in large public buildings. *Constr. Sci. Technol.* **2007**, *2*, 17–19.
25. Ministry of Housing and Urban-Rural Development of People's Republic of China. *China Urban Construction Statistical Yearbook*; China Construction Industry Press: Beijing, China, 2006.
26. Wang, Y.; Qiao, J. Trends and social determinants of adolescent marriage and fertility in China. *Lancet Glob. Health* **2020**, *8*, e873–e874. [CrossRef]
27. Xianling, Z.; Zhenwu, Z.; Tao, T. Trends and patterns of negative population growth in China. *Popul. Res.* **2020**, *44*, 3–20.
28. National New Urbanization Plan (2014–2020). *Rural Work Newslett.* **2014**, *6*, 32–48.
29. Tianjin Development and Reform Commission: "Thirteenth Five-Year Plan" for Tianjin Heating Development. Available online: http://fzgg.tj.gov.cn/zwgk_47325/zcfg_47338/zcwjx/fgwj/202012/t20201219_5069573.html (accessed on 26 November 2021).
30. Department of Housing and Urban-Rural Development. Hebei: "Thirteenth Five-Year Plan" for Urban Heating in Hebei Province. Available online: http://zfcxjst.hebei.gov.cn/cszsdw/hbsrqrlglzx/tfwj/201705/t20170526_228033.html (accessed on 20 November 2021).
31. Ministry of Housing and Urban-Rural Development of People's Republic of China. *Standard for Energy Consumption of Building*; GBT 51161-2016; China Construction Industry Press: Beijing, China, 2016.
32. National Development and Reform Commission of the People's Republic of China: Clean Heating Plan in Winter in Northern area (2017–2021). Available online: http://www.gov.cn/xinwen/2017-12/20/content_5248855.htm (accessed on 23 November 2021).
33. Research Center for Building Energy Efficiency, Tsinghua University. *2019 Annual Report on China Building Energy Efficiency*; China Construction Industry Press: Beijing, China, 2019.

34. Li, Y.; Zhang, X.; Zhou, T. A heat transfer calculation method and primary verification for high-tempture fluid transmission and distribution network from EAST. *Fusion Engineering and Design* **2020**, *151*, 111403. [CrossRef]
35. Xu, Q.; Wang, K.; Zou, Z.; Zhong, L.; Akkurt, N.; Feng, J.; Xiong, Y.; Han, J.; Wang, J.; Du, Y. A new type of two-supply, one-return, triple pipe-structured heat loss model based on a low temperature district heating system. *Energy* **2021**, *218*, 119569. [CrossRef]
36. Beijing Urban Construction Science Technology Promoting Association. *Technical Code for Retrofitting Heating System on Energy Efficiency*; GB/T 50893-2013; China Construction Industry Press: Beijing, China, 2014.
37. Ministry of Housing and Urban-Rural Development of People's Republic of China. *Design Standard for Energy Efficiency of Residential Buildings in Severe Cold and Cold Zones*; JGJ 26–2010; China Construction Industry Press: Beijing, China, 2010.
38. Ministry of Housing and Urban-Rural Development of People's Republic of China. *Design Standard for Energy Efficiency of Residential Buildings in Severe Cold and Cold Zones*; JGJ 26–2018; China Construction Industry Press: Beijing, China, 2018.
39. Zhao, N.; Li, B.; Li, H.; Li, G.; Wu, R.; Hong, Q.; Mperejekumana, P.; Liu, S.; Zhou, Y.; Ahmad, R.; et al. The potential co-benefits for health, economy and climate by substituting raw coal with waste cooking oil as a winter heating fuel in rural households of Northern China. *Environ. Res.* **2021**, *194*, 110683. [CrossRef] [PubMed]
40. Research Center for Building Energy Efficiency, Tsinghua University. *2016 Annual Report on China Building Energy Efficiency*; China Construction Industry Press: Beijing, China, 2016.
41. China Association of Building Energy Efficiency-Energy Consumption Statistics Professional Committee: China Building Energy Consumption Study Report 2018. Available online: https://www.cabee.org/site/content/22960.html (accessed on 22 November 2021).
42. Ministry of Housing and Urban-Rural Development of People's Republic of China. *Standard for Building Carbon Emission Calculation*; GBT 51366-2019; China Construction Industry Press: Beijing, China, 2019.
43. National Energy Administration: China's Mid- to Long-Term Power Generation Capacity and Power Demand Development Forecast. Available online: http://www.nea.gov.cn/2013-02/20/c_132180424_2.htm (accessed on 22 November 2021).
44. Department of Housing and Urban-Rural Development. Hebei. *Design Standard for Energy Efficiency of Residential Buildings in Hebei Province (Energy Saving 75%)*; DB13(J)185-2015; China Construction Industry Press: Hebei, China, 2015.
45. Department of Housing and Urban-Rural Development. Jilin. *Design Standard for Energy Efficiency of Residential Buildings in Jilin Province (Energy Saving 75%)*; DB22/T 5034-2019; China Construction Industry Press: Jilin, China, 2019.
46. Chunhua, S. Study on Heat Using Mode and Heat Consumption Characteristics in Buildings Adopting Heat Metering Systems. Ph.D. Thesis, Chongqing University, Chongqing, China, 2012.
47. Zhang, X.; Congcong, Z.; Wenjie, X.; Haidong, Z. Utilization of solar energy in green buildings. *Sci. Technol. Inf.* **2019**, *4*, 37–38.
48. Shanshan, X.; Deying, L. Development and application of solar-multi-energy complementary heating technology. In Proceedings of the 2021 Seminar on Construction and Efficient Operation of Heat Supply Projects, Chengdu, China, 21 April 2021.
49. Chengyang, J.; Yanjun, D. Direct expansion solar assisted heat pump using novel fin- tube collector/evaporator. *CIESC J.* **2016**, *67*, 318–325.
50. Ming, S.; Yanqing, L.; Rongjiang, M.; Mengsi, D.; Xingli, D.; Xudong, Y.; Yongjie, Z.; Jiandong, Y. Comparison of economic and emission performances of different technologies from coal to clean energy in northern rural China. *Environ. Sustain. Dev.* **2020**, *45*, 43–49.
51. Long, H.; Tai, T.T.; Chow, J.J. Building—integrated heat pipe photovoltaic/thermal system for use in Hong Kong. *Sol. Energy* **2017**, *155*, 1084–1091. [CrossRef]
52. Dehui, F. Numerical and Experimental Study on a Heat Pipe Photovoltaic/Thermal System. Ph.D. Thesis, University of Science and Technology of China, Hefei, China, 2012.

energies

MDPI

Article

Smart Evaluation Index of Roof SHS Suitability

Juan Zhao [1,*], Yifei Bai [1], Junmei Gao [1], Tianwei Qiang [1] and Pei Liang [2,*]

1 School of Urban Planning and Municipal Engineering, Xi'an Polytechnic University, Xi'an 710048, China; 201421044@stu.xpu.edu.cn (Y.B.); dun2022@126.com (J.G.); 20060902@xpu.edu.cn (T.Q.)
2 College of Optical and Electronic Technology, China Jiliang University, Hangzhou 310018, China
* Correspondence: juanzhao@xpu.edu.cn (J.Z.); plianghust@cjlu.edu.cn (P.L.); Tel.: +86-137-2060-6092 (J.Z.); +86-571-8687-5622 or 86-150-5818-1796 (P.L.)

Abstract: The instability of solar energy and its resource distribution characteristics make it difficult to judge its suitability in practical engineering applications, which hinders its promotion and application. In order to better promote the effective use of solar energy and promote the solar heating system, it is necessary to put forward a simple method of judging the suitability of the solar heating system for engineering application. This study puts forward "F, Q" as the basis for judging the suitability of solar heating systems built on the roof. Two types of public buildings, office buildings and three-star hotels, are taken as the research objects. DeST software is used to change the heating area of the building by superimposing floors to simulate the heat load of the building when the heating area changes. A dynamic simulation coupling model of solar heating system is established in the TRNSYS software to analyze the operating status of the system under all working conditions. The functional relationship between "F, Q" and solar energy guarantee rate is established, and the solar energy contribution rate is divided into three regions of $F < 30\%$, $30\% \leq F \leq 50\%$, and $F > 50\%$. The evaluation standard of the building suitability of the solar energy heating system is established according to the scope of "F, Q" in different regions (An office building for, e.g., if the contribution rate of solar heating system is required to be greater than 50%, the "F" of these four areas should be greater than 0.11388, 0.15543, 0.10572, and 0.04511.), and the effectiveness of "F" is verified through actual cases verified by other scholars in the research. The method proposed in this paper is helpful to judge the suitability of solar heating systems in different regions and different types of conventional buildings, so as to better promote solar heating systems.

Keywords: solar heating; building suitability; dynamic simulation; solar contribution rate; fitting regression

Citation: Zhao, J.; Bai, Y.; Gao, J.; Qiang, T.; Liang, P. Smart Evaluation Index of Roof SHS Suitability. *Energies* **2022**, *15*, 1164. https://doi.org/10.3390/en15031164

Academic Editors: Shi-Jie Cao, Carlo Roselli and Antonio Rosato

Received: 21 December 2021
Accepted: 31 January 2022
Published: 4 February 2022

1. Introduction

With the population growth and economic development, the global energy consumption is rapidly growing in recent decades. As one of the most energy-consuming areas, building energy consumption accounts for about 28% of China's energy consumption [1,2], nearly 40% of which is related to heating, ventilation, and air-conditioning systems [3]. At present, the energy structure of is still dominated by fossil energy in China. The fossil energy not only brings great progress to industrial development, but also brings more and more serious environmental and climate problems [4]. In the context of low carbon, energy has become an issue of global consensus, how to ensure energy security and achieve the goal of "carbon peak, carbon neutralization" is the main problem the world faces. Energy saving and emission reduction in the field of buildings is the key to achieve the goal of carbon peak and carbon neutralization. Therefore, more and more attention has been paid to how to develop and utilize renewable energy in the field of architecture. Renewable energy includes solar energy, wind energy, geothermal energy, etc., among which solar energy is recognized as an important renewable energy because of its uniform distribution and cleanness. As a well-known form of solar thermal application, solar water-heating system accounts for 80% of the world's solar thermal utilization [5]. Martinopoulos et al. [6]

showed that although solar heating system is a mature technology, its market penetration and public acceptance are relatively low. Obviously, the instability of solar energy and the characteristics of its resource distribution make it difficult to judge its suitability in practical engineering applications, which hinder its promotion and application. Therefore, many scholars have studied and improved the solar energy collecting device and the heat storage device of the solar energy heating system [7–9].

In terms of suitability evaluation, Wang et al. [10] established the suitability assessment model of solar thermal system for residential buildings based on energy efficiency, economic, social, and environmental factors, according to the multi-attribute decision analysis method and GIS technology, and conducted a comprehensive evaluation of the suitability of solar thermal system for 31 provincial capital cities in China. Gassawd et al. [11] analyzed the regional, technical, and economic performance of solar hot-water systems throughout the United States. Hansen et al. [12] evaluated the potential of solar photothermal utilization and its impact on the energy systems of Germany, Austria, Italy, and Denmark by analyzing the existing energy system. Zhou et al. [13] studied the applicability of daily total solar radiation calculation models for different climate regions in China, and determined the most suitable model form for each climate region. Based on this model, a general calculation model of the daily total solar radiation applicable to different climatic regions in China was established.

In addition, there are many studies that evaluate the suitability of a particular solar water-heating system. Chen et al. [14] proposed a comprehensive evaluation method for regional adaptability of solar hot-water systems in college bathrooms, and evaluated the regional adaptability of 33 provincial capitals in China. Raisul et al. [15] analyzed the performance characteristics of different types of solar hot-water systems, and analyzed their economy. Stefano et al. [16] studied the economic benefits and main performance indicators for evaluating solar heating systems, and proposed an economic comparison based on deterministic systems for optimization. The results show that with the decrease in heating dates and the increase in solar energy resources, the benefits generated by solar heating have been significantly improved.

Comprehensive analysis of existing studies shows that most of the current studies on the suitability of solar heating systems focus on regional suitability analysis and the analysis and evaluation of the suitability of a specific solar hot water system [17–20]. In practical engineering, the judgment of the suitability of solar heating systems often involves multiple factors, such as local meteorological conditions, building types, heating area, and system characteristics, which brings certain difficulties to engineering applications. In order to solve the above problems, a simple index F for evaluating suitability is proposed, and the research method is as follows: take two types of public buildings, an office building and a three-star hotel, as the research object; maintain the same heating area; change the floor number to change the heating area; analyze the solar energy contribution rate of the building when the heating area changes; and evaluate the building suitability.

2. Proposal of Simple Evaluation Index

Two types of public buildings, an office building and a three-star hotel, as the research object, in order to analyze the change rule of solar energy contribution rate when the heating area changes. Modeling in the DeST software, the simulation building is gradually superimposed from one layer to ten layers by stacking the floor, and the heating area also increases regularly. In the process of overlaying the floor, the area of the top floor of the building remains unchanged, and the area of the solar collector laid on the top floor of the building remains unchanged. This paper analyzes the solar energy contribution rate of the building when its heating area changes, and evaluates the SHS suitability.

2.1. Ratio of Solar Collecting Heat to Building Heating Load

According to the Technical Standards for Solar Heating Engineering [21], the total area of solar collectors of a direct solar heating system with short-term heat storage shall be calculated according to Formula (1):

$$A_c = \frac{86400 Q_J f}{J_T \eta_{cd}(1 - \eta_L)} \quad (1)$$

In the formula:

A_c—short-term direct thermal storage system collector total area, m^2;

Q_J—Design load of solar collector system, W;

J_T—Average daily solar irradiation in December on the lighting surface of the local collector, $[J/(m^2 \cdot d)]$;

f Solar energy guarantee rate, %;

η_{cd}—Average collector efficiency based on total area, %;

η_L—Heat loss rate of pipelines and heat storage devices, %.

According to the above formula, the contribution rate of solar system is proportional to the heat collected by solar collector and inversely proportional to the building heat load. Additionally, "Solar Heating Engineering Technical Standards" determines the richness of solar energy resources by solar energy contribution rate, as shown in Table 1 [21].

Table 1. Range of recommended values for solar guarantee rate of solar heating systems in different areas.

Resource Regionalization	Solar Guarantee Rate of Short-Term Heat Storage System	Solar Guarantee Rate of Seasonal Heat Storage System
I Resource-richer area	≥50%	≥60%
II Resource-rich area	30~50%	40~60%
III Resource-general area	10~30%	20~40%
IV Resource-poor area	5~10%	10~20%

Based on the above, the ratio of heat collection to building heat load can be used as the basis to judge the building suitability of solar heating system. The collector heat load ratio "Q" is proposed as (2).

$$Q = \frac{S_j}{H_j} \quad (2)$$

In the formula:

Q—the ratio of the total heat collected by the Q-solar collector to the cumulative heat load of the building;

S_j—Total heat collected by solar collector in heating season, kW·h;

H_j—Building heat load, namely cumulative heat load of building in heating season, kW·h.

When the "Q" increases or decreases, the change also presents certain correlation and regularity with the change of solar energy contribution rate of the building. Therefore, "Q" can be used as a variable to analyze and evaluate the building suitability of solar heating system by analyzing the relationship between "Q" and solar energy contribution rate.

2.2. Ratio of Solar Collector Effective Area to Building Heating Area

Since the calculation of "Q" requires the heat collected by the solar collector and the building heat load, the calculation of these two parameters is more complicated when achieving the accurate results. For most conventional buildings, the building heat load will change regularly when the heating area is changed without considering the effect of the shape coefficient. When the solar collector area changes, it will affect the heat collected by the solar system. Both of the changes will ultimately change the contribution of the overall solar heating system.

Based on the above, the index of "F" is proposed,

$$F = \frac{A_j}{S_t} \quad (3)$$

In the formula:

F—the ratio of solar collector effective area to building heating area;

A_j—total effective area of solar collector on top floor of building, m^2;

S_t—building heating area, m^2.

The building suitability of SHS could be evaluated by the relationship between "F" and solar contribution rate.

3. Research Methods and Purposes

In this paper, two index of "F" and "Q" are proposed to evaluate the suitability of solar heating system in different regions and different types of conventional buildings. The specific research process is as follows:

(1) Building model is established in DeST software, and two types of public buildings, an office building and a three-star hotel, are taken as the research objects. The specific parameter settings are shown in Table 2.

Table 2. Indoor parameters.

Indoor Parameter	Office Building	Three-Star Hotel
Date of heating	From 15 November to 15 March of the following year, a total of 120 days	
Heating time	07:00~20:00	24 h continuous heating
Personnel density	$0.1/m^2$	$0.07/m^2$
Lighting power	$18\ W/m^2$	$15\ W/m^2$
Equipment power	$13/m^2$	$13\ W/m^2$
Indoor temperature	25 °C	26 °C

Changing the heating area by overlaying it from one floor to ten floors, ten groups of incremental data are obtained. Four cities with different abundance of solar energy resources (Hanzhong, Xi'an, Xining, and Lhasa) are selected to simulate the heat load of the two types of buildings by DeST software, and the hourly heat load, hourly unit area heat load index, and the maximum heat load in heating season under different conditions are obtained.

(2) According to the requirements of this research, the corresponding solar heating system is designed, as shown in Figure 1a, which is mainly composed of solar collector array, plate heat exchanger, heat pump, heat exchange pump, heat storage water tank, auxiliary heat source, heat pump, and indoor heating terminal. The solar energy heating system converts solar energy into heat energy by solar collector, and stores solar radiation energy by regenerative water tank, when the regenerative water tank reaches the heating temperature, the indoor heat shall be supplied, and when the regenerative temperature of the regenerative water tank does not meet the heating requirements, the supplementary heat shall be supplied by the auxiliary heat source; when the temperature of the water tank is less than the temperature of the backwater, the heating backwater shall be directly reheated by the auxiliary heat source, and the regenerative water tank shall not participate in the heating process.

According to the designed solar heating system, the dynamic simulation coupling model is established in TRNSYS software, as shown in Figure 1b. The operation state of the solar heating system is analyzed when the heating area changes, and the parameters characterizing the performance of the system, such as fluid temperature, flow rate, and auxiliary heat source heating, are obtained.

(3) By processing and calculating the data obtained in the first and second steps, ten groups of "F, Q" and their corresponding solar contribution rates are obtained, respectively; the functional relationship between "F, Q" and solar energy contribution rate is established by fitting regression.

The mathematical models used in the regression analysis of this study are mainly as follows:

(1) Logarithmic regression formula model:

$$y = a\ln(x) + b \tag{4}$$

(2) Unitary linear regression formula model:

$$y = ax + b \tag{5}$$

According to the Technical Standard of Solar Heating Engineering, the contribution rate of solar energy is divided into three regions, and the range of "F, Q" is determined by the established function relationship [21].

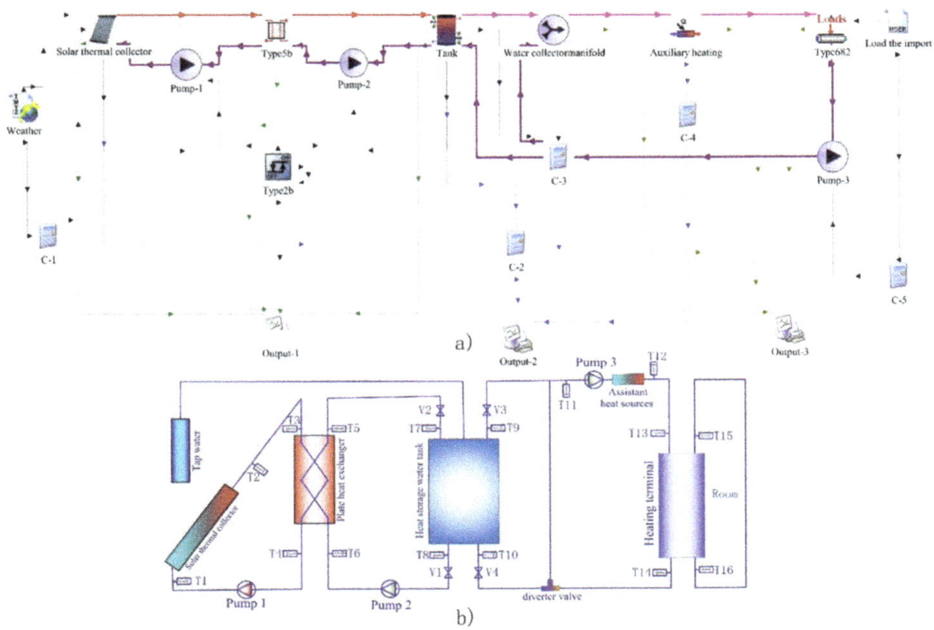

Figure 1. (**a**) TRNSYS dynamic simulation flow chart; (**b**) Structure diagram of solar heating system.

4. System Operation Status and Suitability

4.1. Solar Energy Contribution Rate in Xi'an

Taking Xi'an as an example, an architectural model is established in DeST software, and two types of public buildings, an office building and a three-star hotel, are the research objects. According to the preliminary design conditions, the heating end is selected as the hot water floor radiation, and the inlet water temperature is 40 °C. The layout of the solar collectors should consider the placement of the roof area, and, finally, 120 collectors were arranged. The effective heat collection area is 1.92 m^2, and every 4 blocks are connected in series. According to the "Technical Specification for Solar Heating Engineering", the hot water storage tank is selected as a 100 L/square meter heat collector, and the total hot water storage tank is 25 m^3. The operation status of the same solar heating system in two different types of buildings is analyzed [21].

The area of each floor remains the same, and when the heating area gradually increases from one floor to ten floors by superimposing floors, the "F" keeps shrinking and keeps approaching zero.

Figure 2a,b show the variation trend of building thermal load, auxiliary heating, and solar heat collection when "F" is taken as an independent variable for two types of buildings, and office building and a three-star hotel. It can be seen that the three kinds of heat are gradually increasing, but with the decrease in "F", the building heat load, and the auxiliary heat increase rapidly, while the change of solar heat collection is small. This is mainly because the volume of the building increases in the process of stacking floors but the area of the top floor does not change, that is, the total area of solar collectors laid on the top floor remains the same, so the growth of solar heat collection is limited and the change is small.

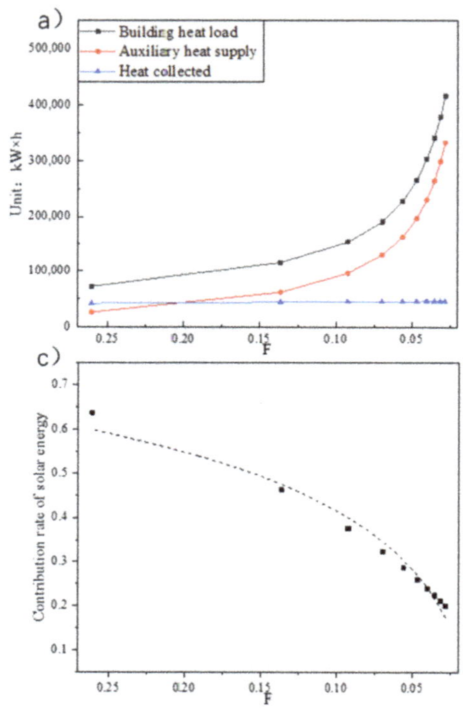

Figure 2. When F is the independent variable: the analysis of the three kinds of heat of office building and three-star hotel is shown in graph (**a**) and graph (**b**); the analysis of solar energy guarantee rate is shown in graph (**c**) and graph (**d**).

Figure 2c,d show the changing trend of the solar energy contribution rate of two types of buildings, an office building and a three-star hotel, with the decrease in "F". The mathematical model is established by fitting regression to the data of ten coordinate points in the figure, and the corresponding regression equation and R^2 (coefficient of determination) are obtained.

Among them, the solar energy contribution rate of an office building when "F" is the independent variable, the change trend is determined by regression analysis as shown in Figure 2c, and the corresponding regression equation and R^2 are as follows:

$$y = 0.1914\ln(x) + 0.8563$$

$$R^2 = 0.9746$$

The solar energy guarantee rate of three-star hotel with "F" as the independent variable, through regression analysis to determine the trend as shown in Figure 2d, the corresponding regression equation and R^2 are as follows:

$$y = 0.1562\ln(x) + 0.6344$$

$$R^2 = 0.9716$$

Figure 3a,b show the change trend of the solar energy contribution rate of the two types of buildings as "Q" decreases. By fitting regression to the data of ten coordinate points in the figure, the corresponding linear regression equation and R^2 are established and obtained.

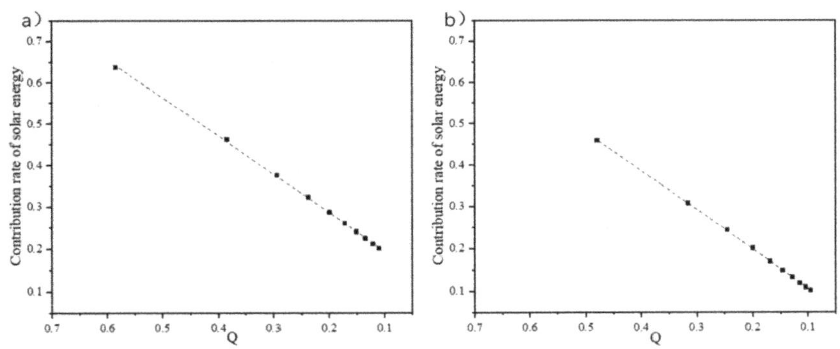

Figure 3. When Q is the independent variable, the analysis of solar guarantee rate of office building and three-star hotel is shown in graph (**a**) and graph (**b**).

Among them, the solar energy contribution rate of an office building when "Q" is used as the independent variable, the change trend is determined by regression analysis as shown in Figure 3a, and the corresponding regression equation and R^2 are as follows:

$$y = 0.9247x + 0.1012R^2 = 0.9995$$

$$R^2 = 0.9995$$

The solar energy contribution rate of a three-star hotel with "Q" as the independent variable, through regression analysis to determine the trend as shown in Figure 3b, the corresponding regression equation and R^2 are as follows:

$$y = 0.9349x + 0.0117$$

$$R^2 = 0.9999$$

From Figures 2 and 3, it can be seen that with the decrease in "F/Q", the solar energy contribution rate of the two buildings shows a downward trend. However, the difference is that the mathematical model established when fitting regression to the change of solar energy contribution rate with "F" as the independent variable is the logarithmic regression equation; the mathematical model established with "Q" as the independent variable is the unary linear regression equation. That is, under the condition of "F" and "Q" as independent variables, the solar energy contribution rate presents two different trends. As the calculation of solar energy contribution rate is more complicated, the calculation of "F/Q" is relatively simple. Therefore, after establishing the functional relationship between "F, Q" and solar energy contribution rate, the range of solar energy contribution rate can be determined by calculating "F" or "Q", so as to make a simple evaluation of the building suitability of solar heating system.

4.2. Analysis of Building Suitability of Solar Heating System

Based on the above conclusions, this study selected four cities: Hanzhong, Xi'an, Xining, and Lhasa, and evaluated and judged the suitability of solar heating systems in the local buildings using "F" and "Q" methods, and analyzed and compared the pros and cons of the two methods.

Figure 4a,b, respectively show the analysis results of the solar energy contribution rate of office buildings and three-star hotels in the four regions when "F" is used as the independent variable. According to the "Technical Specification for Solar Heating Engineering", the contribution rate of solar energy can be divided into three regions [21],

(1) The area where the value of solar energy contribution rate is less than 0.3;
(2) The area where the value of solar contribution rate is between 0.3 and 0.5;
(3) It is the area where the value of solar energy contribution rate is greater than 0.5.

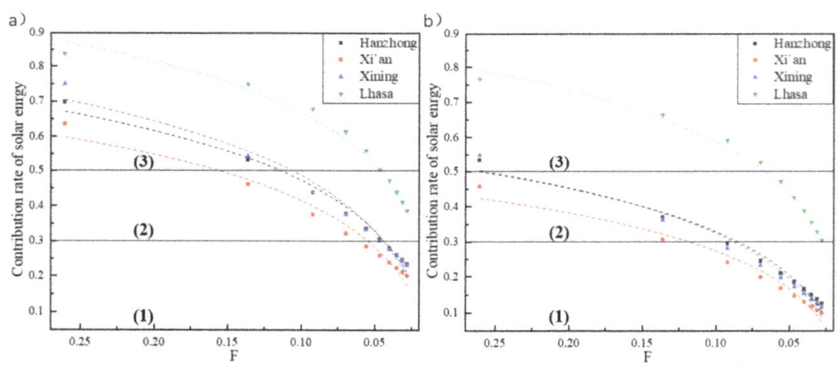

Figure 4. When F is the independent variable, in the four regions, the solar guarantee rate of office buildings and three-star hotels is shown in graph (**a**) and graph (**b**).

The greater the value of the solar energy contribution rate, the better the suitability of the building for the solar heating system.

Regression analysis was performed on the solar energy contribution rate of the four regions when "F" was used as the independent variable, and the fitted curve is shown in Figure 4. The established mathematical model, and the regression equation and R^2 obtained are shown in Table A1 in Appendix A.

When the solar energy contribution rate is divided into the three regions of "(1) (2) (3)", the value of "F" can be inversely deduced through the established functional relationship to determine the range of "F" in the three regions. According to the mathematical model established in Table A1 in the Appendix A, the "F" range of each region is derived, and the results are shown in Table A2 in the Appendix A.

Table A2 divides the "F" of the two types of buildings in the four regions, that is, the ratio of the area of the collector to the heating area. Take the office building in Xi'an as an example, when $F < 0.05467$, it can be judged that the solar energy contribution rate of the building is less than 30%; when $0.05467 \leq F \leq 0.15543$, it can be judged that the solar energy contribution rate of the building is within the range of 30~50%; when $F > 0.15543$, it can be judged that the solar energy contribution rate of the building is greater than 50%. Therefore, in actual engineering, a preliminary judgment can be made on the suitability of the building's solar heating system by calculating the "F" of the building.

When "Q" is used as the independent variable, and the relationship between "Q" and solar energy contribution rate is analyzed, the suitability of the building for solar heating system can also be judged accordingly.

Figures 5 and 6, respectively show the analysis results of the solar energy contribution rate of office building and three-star hotels in the four regions when "Q" is used as the independent variable. Divide the solar energy contribution rate into the same three regions as above, and perform regression analysis on the solar energy contribution rate of the four regions when "Q" is used as the independent variable. The fitted curves are shown in Figures 5 and 6. Additionally, the mathematical model was established and the regression equation and R^2 were obtained as shown in Table A3 in Appendix A.

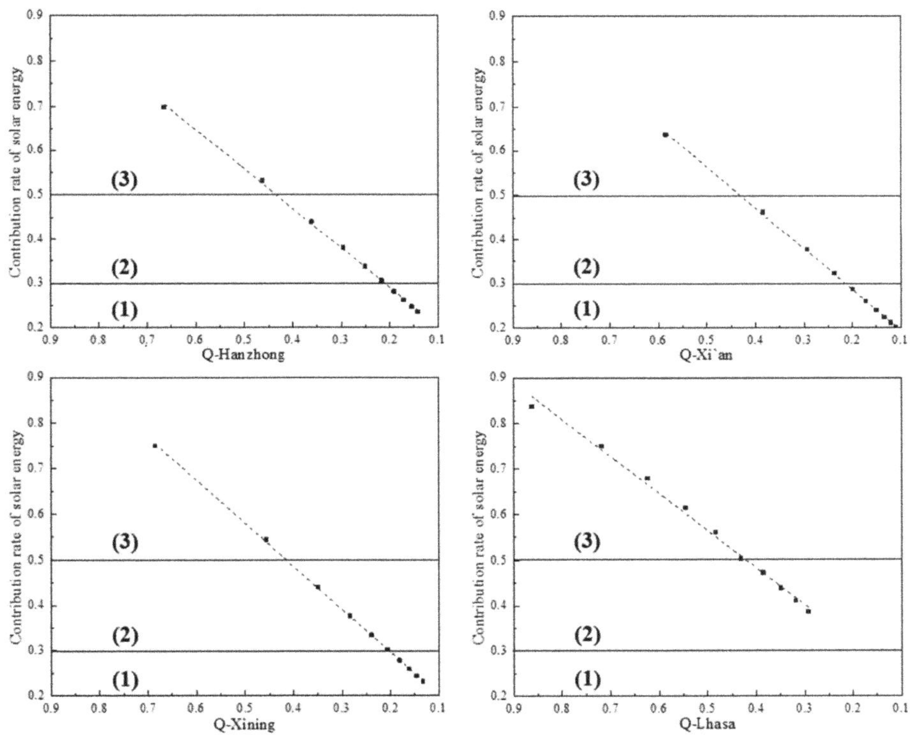

Figure 5. When Q is the independent variable, the solar energy contribution rate of office building in the four regions.

From the above, the contribution rate of solar energy is divided into three areas "(1)(2)(3)", and the value of "Q" can be deduced through the established functional relationship to determine "Q" in the three areas. Range from the mathematical model established in Table A3, the range of "Q" in each area is deduced, and the results are shown in Table A4 in the Appendix A.

Table A4 divides the "Q" of the two types of buildings in the four regions, that is, the "ratio of solar heat collection to building heat load". Take the office building in Xi'an as an example. When $Q < 0.21499$, it can be judged that the solar energy contribution rate of the building is less than 30%; when $0.21499 \leq Q \leq 0.43128$, it can be judged that the solar energy contribution rate of the building is within the range of 30–50%; When $Q > 0.43128$, it can be judged that the solar energy contribution rate of the building is greater than 50%. Therefore, in actual engineering, a preliminary judgment can be made on the suitability of the building's solar heating system by calculating the "Q" of the building.

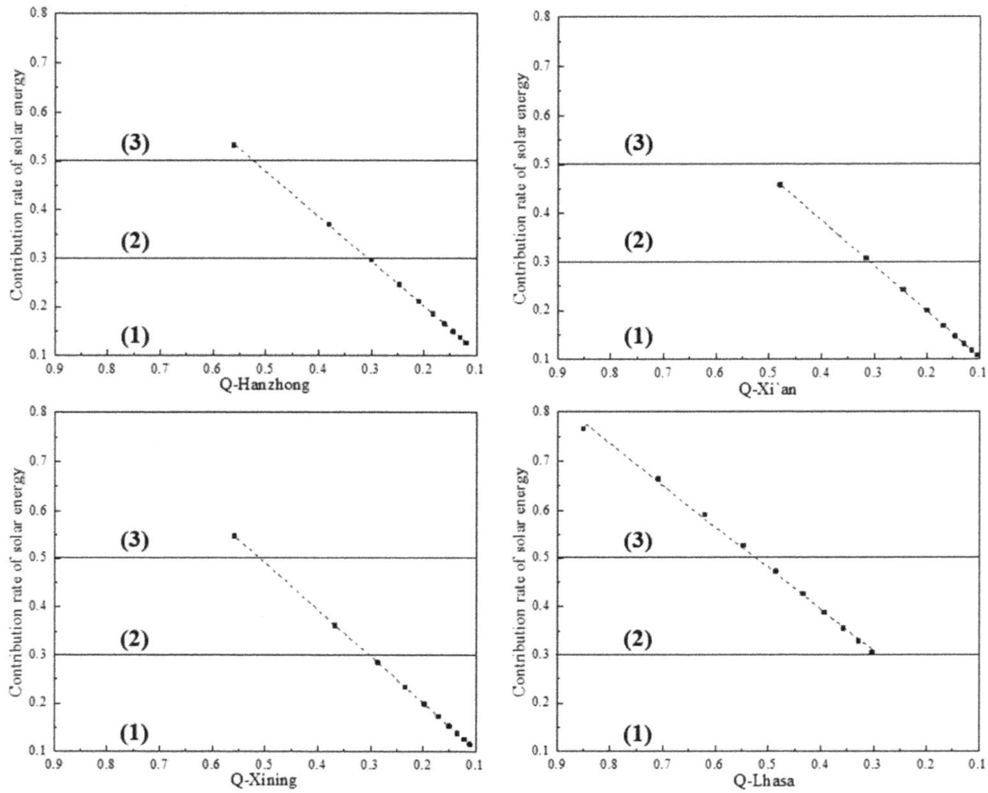

Figure 6. When Q is the independent variable, the solar energy contribution rate of three-star hotels in the four regions.

4.3. Case Validation

In this study, the effectiveness of the proposed method is verified by referring to the case of solar heating system in the research results of other scholars, and according to the functional relationship established in the proposed method.

According to Li Yue, Long T, and other scholars' research on solar heating systems, the research objects are all office buildings in the Lhasa area [22–24]. As can be seen from the above, when "F" is used as the basis for judging office buildings in Lhasa, the established regression equation is as follows:

$$y = 0.2127\ln(x) + 1.1591$$

$$R^2 = 0.9838$$

The results obtained according to the regression equation established in this study and the solar contribution rates obtained by other scholars in the case studies verified by actual measurements are shown in Table 3:

Table 3. Model Validation.

Parameter	Case 1	Case 2	Case 3
Solar collector area	580.7m^2	198.47 m^2	–
Heating area	4750 m^2	7500 m^2	–
"F"	0.12225	0.02646	0.49
Contribution rate of Solar energy (In the case of other scholars)	70%	42.79%	99.8%
Contribution rate of Solar energy (Obtained in this study)	71.2%	38.65%	100.7%

It can be seen that there are some errors between the solar contribution rate calculated by the method based on "F" proposed in this study and the research results verified by other scholars through actual measurement, but they are basically consistent. It can be seen that the evaluation criteria based on the method proposed in this study have certain reference value.

The deviations may be caused by the following problems: (1) The main work of this study is simulation calculation. Due to the limitations of simulation software, there may be deviations from the experimental results in reality; (2) The indoor parameters of the building model used in this study are set according to the specifications, but in actual cases, the interior design parameters of the same type and different buildings, such as personnel, equipment, and lighting, may not be exactly the same, which will affect the calculation of the building thermal load; (3) There may be some differences between the solar heating system used in the actual case and the solar heating system designed in this study.

4.4. Discussion

Through the above research results, the pros and cons of the two methods "F" and "Q" for evaluating the suitability of solar heating systems are compared, and the following conclusions are obtained:

(1) Compared with "F", the calculation of "Q" is more complicated. This is mainly because "F" is the ratio of the area of the solar collector to the heating area, and "Q" is the ratio of the solar heat collection to the heat load of the building. The calculation of the area of the solar collector and the heating area is relatively simple, while the calculation of the solar heat collection and the heat load of the building in the actual project is more cumbersome;

(2) By comparing the R^2 value (coefficient of determination) of the regression equations in Tables A1 and A3, it can be found that the regression equation fitted is better when analyzing the contribution rate of solar energy with "Q" as the independent variable. That is, compared with "F", the regression equation with "Q" as the independent variable is more accurate in judging the solar energy contribution rate;

(3) Compared with "Q", "F" is more limited. This is mainly because "F" is used as an independent variable to judge the suitability of solar heating systems, at present, it can only be temporarily applied to conventional buildings with similar body shape coefficients. The difference between the shape coefficients of buildings with the same volume will affect the judgment of the building's heat load, which will increase the error of the solar energy contribution rate and make the accuracy of the regression equation obtained with "F" as the independent variable worse.

5. Conclusions

In this paper, the building model is established by DeST software, and the dynamic simulation model of solar heating system is established by TRNSYS software. In the process of analyzing the architectural suitability of solar heating systems in the four cities of Hanzhong, Xi'an, Xining, and Lhasa based on simulation data, the following research results have been mainly achieved:

(1) The indexes of "F/Q" are proposed as a basis for judging the suitability of solar heating systems;
(2) In the DeST simulation software, the heating area is changed by superimposing the floors, and the simulated building is superimposed from one to ten to make the "F/Q" change regularly, and the TRNSYS software is used to analyze and obtain the corresponding solar energy contribution rate. Establish the functional relationship between "F/Q" and solar energy contribution rate by fitting regression, and the results are shown in Tables A1 and A3;
(3) The solar energy contribution rate is divided into three regions of $F < 30\%$, $30\% \leq F \leq 50\%$, and $F > 50\%$. The value of "F,Q" is inversely deduced through the functional relationship between "F/Q" and solar energy contribution rate, and the range of "F/Q" in the three divided regions is determined. The results are shown in Tables A2 and A4. (Take an office building, for example, when the contribution rate of solar heating system is less than 30%, "F" of Hanzhong, Xi 'an, Xining and Lhasa is less than 0.00919, 0.01166, 0.00932, and 0.00361, respectively. If the contribution rate of solar heating system is required to be greater than 50%, the "F" of these four areas should be greater than 0.11388, 0.15543, 0.10572, and 0.04511. Additionally, for a three-star hotel, the "Q" should be, respectively, greater than 0.52302, 0.52230, 0.51001, and 0.52379 when it is required to be greater than 50%).
(4) The validity of the "F" proposed in this study is verified by comparison with the actual cases verified by other scholars in their studies. It can be seen that the method proposed in this study has certain reference value and provides a new idea for the suitability evaluation of solar heating system.

This paper puts forward "F" and "Q" as the simple method for judging the contribution rate of roof solar heating system built. It is helpful to better analyze and evaluate the suitability of solar heating system in different types of conventional buildings in actual projects, so as to better promote solar heating system and promote its application in winter heating.

Author Contributions: Conceptualization, J.Z. and P.L.; methodology, J.Z.; software, Y.B.; validation, P.L., T.Q. and J.G.; formal analysis, Y.B.; resources, J.Z.; data curation, Y.B.; writing—original draft preparation, J.Z. and Y.B.; writing—review and editing, J.G.; visualization, T.Q.; supervision, P.L.; project administration, P.L.; funding acquisition, Xi'an Polytechnic University. All authors have read and agreed to the published version of the manuscript.

Funding: This research was funded by [Xi'an Polytechnic University], grant number [107020491].

Data Availability Statement: Not applicable.

Conflicts of Interest: The authors declare no conflict of interest.

Appendix A

Table A1. The mathematical model established when F is the independent variable.

Building Type	City	Regression Equation y = a*ln(x) +b	Standard Error	R^2 (Decision Coefficient)
Office building	Hanzhong	y = 0.2074ln(x) + 0.9506	a ± 0.00841 b ± 0.02412	R^2 = 0.987
	Xi'an	y = 0.1914ln(x) + 0.8563	a ± 0.01093 b ± 0.03138	R^2 = 0.9746
	Xining	y = 0.228ln(x) + 1.0123	a ± 0.01291 b ± 0.03707	R^2 = 0.975
	Lhasa	y = 0.2127ln(x) + 1.1591	a ± 0.00964 b ± 0.02766	R^2 = 0.9838

Table A1. *Cont.*

Building Type	City	Regression Equation y = a*ln(x) +b	Standard Error	R^2 (Decision Coefficient)
Three-star hotel	Hanzhong	y = 0.1781ln(x) + 0.7391	a ± 0.00918 b ± 0.02634	$R^2 = 0.9792$
	Xi'an	y = 0.1562ln(x) + 0.6344	a ± 0.00945 b ± 0.02712	$R^2 = 0.9716$
	Xining	y = 0.1872ln(x) + 0.7559	a ± 0.012 b ± 0.03443	$R^2 = 0.9682$
	Lhasa	y = 0.2147ln(x) + 1.0818	a ± 0.00729 b ± 0.02091	$R^2 = 0.9909$

Table A2. Results of regional division using F for two types of buildings.

Building Type	City	(1)	(2)	(3)
Office building	Hanzhong	F < 0.00919	0.04342 ≤ F ≤ 0.11388	F > 0.11388
	Xi'an	F < 0.01166	0.05467 ≤ F ≤ 0.15543	F > 0.15543
	Xining	F < 0.00932	0.04398 ≤ F ≤ 0.10572	F > 0.10572
	Lhasa	F < 0.00361	0.01761 ≤ F ≤ 0.04511	F > 0.04511
Three-star hotel	Hanzhong	F < 0.01842	0.08497 ≤ F ≤ 0.26119	F > 0.26119
	Xi'an	F < 0.02574	0.11756 ≤ F ≤ 0.42298	F > 0.42298
	Xining	F < 0.01897	0.08757 ≤ F ≤ 0.25487	F > 0.25487
	Lhasa	F < 0.00546	0.02622 ≤ F ≤ 0.06655	F > 0.06655

Table A3. The mathematical model established when Q is the independent variable.

Building Type	City	Regression Equation y = a*x + b	Standard Error	R^2 (Decision Coefficient)
Office building	Hanzhong	y = 0.8942x + 0.1115	a ± 0.00321 b ± 0.0097	$R^2 = 0.9991$
	Xi'an	y = 0.9247x + 0.1012	a ± 0.00195 b ± 0.00702	$R^2 = 0.9995$
	Xining	y = 0.9459x + 0.1073	a ± 0.00169 b ± 0.00516	$R^2 = 0.9998$
	Lhasa	y = 0.8131x + 0.1583	a ± 0.01145 b ± 0.02156	$R^2 = 0.9944$
Three-star hotel	Hanzhong	y = 0.9233x + 0.0171	a ± 0.00143 b ± 0.00478	$R^2 = 0.9999$
	Xi'an	y = 0.9349x + 0.0117	a ± 0.00068 b ± 0.00296	$R^2 = 0.9999$
	Xining	y = 0.9639x + 0.0084	a ± 0.00036 b ± 0.00136	$R^2 = 1$
	Lhasa	y = 0.8534x + 0.053	a ± 0.00917 b ± 0.01623	$R^2 = 0.998$

Table A4. Results of regional division of two types of buildings using Q.

Building Type	City	(1)	(2)	(3)
Office building	Hanzhong	Q < 0.21080	0.21080 ≤ Q ≤ 0.43447	Q > 0.43447
	Xi'an	Q < 0.21499	0.21499 ≤ Q ≤ 0.43128	Q > 0.43128
	Xining	Q < 0.20372	0.20372 ≤ Q ≤ 0.41516	Q > 0.41516
	Lhasa	Q < 0.17427	0.17427 ≤ Q ≤ 0.42024	Q > 0.42024
Three-star hotel	Hanzhong	Q < 0.30640	0.30640 ≤ Q ≤ 0.52302	Q > 0.52302
	Xi'an	Q < 0.30838	0.30838 ≤ Q ≤ 0.52230	Q > 0.52230
	Xining	Q < 0.30252	0.30252 ≤ Q ≤ 0.51001	Q > 0.51001
	Lhasa	Q < 0.28943	0.28943 ≤ Q ≤ 0.52379	Q > 0.52379

References

1. Ma, H.T.; Du, N.; Yu, S.J.; Lu, W.Q.; Zhang, Z.Y.; Deng, N.; Li, C. Analysis of typical public building energy consumption in northern China. *Energy Build.* **2016**, *136*, 139–150. [CrossRef]
2. Masoso, O.T.; Grobler, L.J. The dark side of occupants' behaviour on building energy use. *Energy Build.* **2010**, *42*, 173–177. [CrossRef]
3. Yu, H.; Niu, J.L.; Chung, T.M. Study on performance of energy-efficient retrofitting measures on commercial building external walls in cooling-dominant cities. *Appl. Energy* **2013**, *103*, 97–108.
4. Ge, T.S.; Wang, R.Z.; Xu, Z.Y.; Pan, Q.W.; Du, S.; Chen, X.M.; Ma, T.; Wu, X.N.; Sun, X.L.; Chen, J.F. Solar heating and cooling: Present and future development. *Renew. Energy* **2017**, *126*, 1126–1140. [CrossRef]
5. Wang, Z.Y.; Yang, W.S.; Qiu, F.; Zhang, X.M.; Zhao, X.D. Solar water heating: From theory, application, marketing and research. *Renew. Sustain. Energy Rev.* **2015**, *41*, 68–84. [CrossRef]
6. Martinopoulos, G.; Tsalikis, G. Active solar heating systems for energy efficient buildings in Greece: A technical economic and environmental evaluation. *Energy Build.* **2014**, *68*, 130–137. [CrossRef]
7. Yang, X.H.; Guo, J.F.; Yang, B.; Cheng, H.N.; Wei, P.; He, Y.-L. Design of non-uniformly distributed annular fins for a shell-and-tube thermal energy storage unit. *Appl. Energy* **2020**, *279*, 115772. [CrossRef]
8. Guo, J.F.; Liu, Z.; Du, Z.; Yu, J.B.; Yang, X.H.; Yan, J.Y. Effect of fin-metal foam structure on thermal energy storage: An experimental study. *Renew. Energy* **2021**, *172*, 57–70. [CrossRef]
9. Guo, J.F.; Liu, Z.; Yang, B.; Yang, X.H.; Yan, J.Y. Melting assessment on the angled fin design for a novel latent heat thermal energy storage tube. *Renew. Energy* **2022**, *183*, 406–422. [CrossRef]
10. Wang, L.Z.; Tan, H.W.; Zhuang, Z.; Lei, Y. Regional adaptability evaluation of solar thermal system based on multiple attribute decision making method. *Acta Energ. Sol. Sin.* **2014**, *35*, 649–655.
11. Cassard, H.; Denholm, P.; Ong, S. Technical and economic performance of residential solar water heating in the United States. *Renew. Sustain. Energy Rev.* **2011**, *15*, 3789–3800. [CrossRef]
12. Hansen, K.; Mathiesen, B.V. Comprehensive assessment of the role and potential for solar thermal in future energy systems. *Sol. Energy.* **2018**, *169*, 144–152. [CrossRef]
13. Zhou, Y.; Liu, Y.F.; Wang, D.J.; Liu, X.J. Applicability analysis of daily total solar radiation calculation model and optimization of general calculation model in Different climatic regions of China. *Acta Energ. Sol. Sin.* **2021**, *217*, 1–7.
14. Chen, S.Q.; Pan, Y.Y.; Tan, H.W. Comprehensive evaluation of regional adaptability of solar hot water system in university bathroom. *J. Zhejiang Univ. Eng. Sci.* **2017**, *51*, 8.
15. Islam, M.R.; Sumathy, K.; Khan, S.U. Solar water heating systems and their market trends. *Renew. Sustain. Energy Rev.* **2013**, *17*, 1–25. [CrossRef]
16. Poppi, S.; Sommerfeldt, N.; Bales, C.; Madani, H.; Lundqvist, P. Techno-economic review of solar heat pump systems for residential heating applications. *Renew. Sustain. Energy Rev.* **2018**, *81*, 22–32. [CrossRef]
17. Ding, Y.; Liu, X.; Li, B.Z. Application suitability analysis of solar hot water in chongqing. *Acta Energ. Sol. Sin.* **2014**, *35*, 299–305.
18. Fan, M.; Liang, H.B.; You, S.J.; Zhang, H.; Yin, B.Q.; Wu, X.T. Applicability analysis of the solar heating system with parabolic trough solar collectors in different regions of China. *Appl. Energy* **2018**, *221*, 100–111. [CrossRef]
19. Thygesen, R. An Analysis of Different Solar-Assisted Heating Systems and Their Effect on the Energy Performance of Multifamily Buildings—A Swedish Case. *Energies* **2017**, *10*, 88. [CrossRef]
20. Yang, W.B.; Zhang, H.; Liang, X.F. Experimental performance evaluation and parametric study of a solar-ground source heat pump system operated in heating modes. *Energies* **2018**, *149*, 173–189. [CrossRef]
21. Ministry of Housing and Urban-Rural Development of the People's Republic of China. *Technical Code for Solar Heating System*; GB 50495-2019; China Architecture & Building Press: Beijing, China, 2019.
22. Li, Y.; Fan, R.; Xiao, H.H. Research on Optimization of Solar Heating System in Tibet, China. *Build. Sci.* **2019**, *35*, 21–28.
23. Long, T.H.; Qiao, Z.Y.; Wang, M.L.; Li, Y.C.; Lu, J.; Li, W.Y.; Zheng, L.; Zeng, L.Y.; Huang, S. Performance analysis and optimization of a solar-air source heat pump heating system in Tibet, China. *Energy Build.* **2020**, *220*, 110084. [CrossRef]
24. Cheng, J. *Application Study of Solar Energy and Water Source Heat Pump Combined Heating in Xizang*; Southwest Jiaotong University: Chengdu, China, 2005.

energies

Article

Assessment of the Possibility of Using Various Types of Renewable Energy Sources Installations in Single-Family Buildings as Part of Saving Final Energy Consumption in Polish Conditions

Piotr Wróblewski [1,2,*] and Mariusz Niekurzak [3,*]

[1] Faculty of Engineering, University of Technology and Economics H. Chodkowska in Warsaw, Jutrzenki 135, 02-231 Warsaw, Poland
[2] Faculty of Mechatronics, Armament and Aerospace, Military University of Technology, ul. Gen. Sylwestra Kaliskiego 2, 00-908 Warsaw, Poland
[3] Faculty of Management, AGH University of Science and Technology, 30-067 Krakow, Poland
* Correspondence: piotr.wroblewski@wat.edu.pl (P.W.); mniekurz@zarz.agh.edu.pl (M.N.)

Abstract: The aim of this work is to develop a model of heat supply to buildings with almost zero energy consumption, indicating the significant importance of heat losses and gains in heating installations. The prepared model is to indicate the need for changes in the structure and topology of heating installations, resulting from the changing heat demand of buildings. The need to create a new model is heightened by changes that relate to tightening legal regulations related to energy consumption and demand, which must meet the standards of buildings in Poland from 2021. The article presents the assumptions and results of analyses of the use of energy installations in residential buildings that use renewable energy sources to balance energy consumption in various areas of its use. To achieve this goal, calculations were made using simulations of the impact of the use of installations using renewable energy sources on the energy performance of a building with different quality of partitions and improvement of energy efficiency in accordance with the Polish standard PN-EN 12831. The test results allow to choose the most advantageous, from the point of view of economic profitability, option of replacing installations in residential buildings, and they also allow to determine the possibilities of meeting national obligations in the field of final energy reduction and increasing the share of renewable energy sources in meeting its demand in accordance with the EU obligations imposed on Poland. Thermomodernization of buildings in the temperate climate zone allows for a reduction of 38% of energy demand over the entire life cycle of a building and a reduction of CO_2 emissions by 99%.

Keywords: energy-efficient construction; renewable energy; thermal modernization; external partitions

Citation: Wróblewski, P.; Niekurzak, M. Assessment of the Possibility of Using Various Types of Renewable Energy Sources Installations in Single-Family Buildings as Part of Saving Final Energy Consumption in Polish Conditions. *Energies* **2022**, *15*, 1329. https://doi.org/10.3390/en15041329

Academic Editor: Shi-Jie Cao

Received: 17 January 2022
Accepted: 9 February 2022
Published: 11 February 2022

1. Introduction

The electrification of heating systems in construction is becoming a necessity and an obligation under Polish law. This is the best way to reduce greenhouse gas emissions. One of the basic challenges of the modern world is to reduce energy consumption in residential construction. This is due to economic and ecological reasons. The higher the consumption, the higher the emission of pollutants into the atmosphere. Reducing consumption also contributes to increasing the world's energy security. In countries with cool and temperate climates, construction is one of the largest energy consumers. A certain amount of energy is used to produce materials and build buildings, but much more energy is used to heat them. The value of the energy used for heating over the life of an average building is about twice the cost of constructing it. Building warm buildings is in the interest of all countries, as experience so far has shown that reducing energy consumption to space heating is one of the cheapest ways to reduce its deficit. Therefore, innovative design and construction solutions are constantly being developed to increase the thermal insulation requirements of buildings, which will contribute to reducing energy consumption for their heating. In recent years, the concept of energy-saving and passive construction has gained in popularity. One of the goals of building a building is to reduce the operating costs of a building, which can be achieved thanks to the proper design of the building. The development of the construction sector requires the introduction of low-energy solutions to the market. Currently, buildings erected in traditional technology consume about 40% of the total energy consumption. The trend of low-energy construction in the European Union, regarding the energy performance of buildings, assumes that buildings should be nearly zero-energy buildings [1]. The prerequisite for the creation of the above-mentioned directive was the need for actions aimed at a significant reduction in energy consumption, reducing the Union's energy dependence and the need to reduce energy consumption from non-renewable sources aimed at reducing greenhouse gas emissions by at least 20%. The current construction law in Poland regulates the energy requirements for newly constructed buildings through the selection of appropriate building materials and compliance with construction requirements. Thanks to these regulations, the construction industry meets both legal, energy, and environmental requirements. While research on the reduction of CO_2 emissions in transport has been widely conducted, and researchers propose a number of remedial solutions [2–8], buildings built before 2010 that do not meet energy efficiency requirements and that must undergo transformation remain a problem.

With this in mind, the article presents various variants of thermal modernization of these buildings, including research on the quality parameters of external partitions and the selection of an appropriate source of renewable energy heating installations. The results presented in the paper will allow future investors to make an informed decision in selecting the appropriate thermomodernization variant to meet the legal requirements for their building. The presented variants take into account various degrees of the achieved energy efficiency and aspects of the economic assessment of the incurred expenditure.

2. Review of the Literature

The building's energy needs depend on the scope of its equipment with heating, ventilation and air-conditioning systems, hot water, and internal lighting. The preparation of the energy balance for the steady states allows to present the dependencies of the flow of individual energy levels in all separate aspects of its consumption in the building. The annual unit demand for non-renewable primary energy determines the total efficiency of the building and includes, apart from the final energy demand, additional inputs of non-renewable primary energy for supplying each used energy carrier to the building boundary [9]. The demand for utility energy contains important information about the energy quality of the building. It gives an image of the state of thermal insulation of external partitions, the degree of elimination of thermal bridges and their impact on heat losses and determines the type of ventilation used. From the amount of usable energy demand, it can additionally be deduced how the body of the building was designed and the benefits from solar radiation through transparent partitions that were used. The most important parameter for the user of the building is the final energy demand [10,11]. Activities rationalizing the demand for final energy in a building result from the optimization of the utility energy demand in a given field and the structure of its coverage by the installed energy systems. The annual amount of saved energy resources from non-renewable primary sources in a given area of its demand in a building is proportional to the amount of final energy used from renewable energy sources and supplied through an installation based on these resources, and to the difference in inputs of non-renewable primary energy resources for producing and providing a given type of carrier for that equivalent final energy that is consumed in a conventional or alternative installation [12,13]. Building energy quality, determined by the parameters thermal insulation of building partitions and the quality of energy systems, determined by the partial efficiency of carrier production, accumulation, transport, and regulation, has a significant impact on the size of devices and installations using renewable energy sources. The analysis of the profitability of using various renewable energy installations in single-family housing is presented in the works [14,15].

Greenhouse gas emissions in the construction sector are mainly related to the production of heat for heating and domestic hot water preparation based on fossil fuels, mainly hard coal. This applies to both system heat sources (boiler rooms, combined heat, and power plants) and local building heat sources. The key challenge in the process of achieving climate neutrality of the construction industry in Poland is so there is a shift away from coal in local heat sources in the first place. This process should be carried out by the end of 2030 at the latest. At the same time, the process of decarbonizing the heating sector and the electric power sector should be carried out, as electrification of heating will play a key role in the process of achieving climate neutrality. Therefore, the key challenge is to conduct an economically effective thermal modernization of the existing resources and to meet the energy needs of modernized renewable energy sources buildings. Achieving climate neutrality in Polish construction requires the government to link in practice the country's energy policy with the policy of increasing energy efficiency and the use of renewable energy sources in construction, including ambitious goals for improving energy efficiency in documents such as the National Energy and Climate Plan and Poland's Energy Policy until 2040. Therefore, both NECP, PEP2040, and similar documents require modification by the government so that they take into account the goals for zero-emission construction. Adopting a policy of maximum use of the renewable energy potential in construction, e.g., by the obligation to install a PV or solar heating installation in each building using an individual heat source, will also strengthen the achievement of the climate neutrality goal in 2050. It is necessary to combine the strategy of the "Clean Air" program with the energy efficiency policy and other related initiatives (including, in particular, those related to the support of renewable energy sources). It requires coordination of various support programs (renewable energy, thermal modernization, smog elimination, etc.) in order to effectively use funds for zero-emission construction. First of all, innovative technologies and solutions as well as additional activities, unprofitable from the investor's point of view, but leading to the creation of low-emission buildings, should be subsidized.

In an energy-efficient building, the proper design and construction of all external partitions has a fundamental influence on obtaining the proper thermal comfort of rooms [16,17]. The energy efficiency of a house is determined, among other things, by the heat transfer coefficient of each of its elements, such as the floor on the ground, walls, roof, windows, and doors. The lower the value of the heat transfer coefficient, the better the insulation properties the partition exhibits. According to the currently binding construction law, the heat transfer coefficient (U) cannot exceed 0.3 $W/(m^2 \cdot K)$ in the case of floors on the ground and ceilings above unheated cellars; 0.2 $W/(m^2 \cdot K)$ for external walls; 0.15 $W/(m^2 \cdot K)$ for roofs, flat roofs, and ceilings under unheated attics. The U value of the entire partition depends on the lambda thermal conductivity coefficient (λ) and the thickness of all its layers. The lower the value of this coefficient, the better the thermal insulation of the external partition by specific thickness. In construction, thermal insulation materials should be used that have λ within 0.03–0.05 $W/(m \cdot K)$ [18–20]. Walls are one of the most important elements of a home. They are exposed to unfavorable weather conditions such as rain, snow, wind, frost, and UV rays [21]. Therefore, they must be durable and, at the same time, provide good thermal insulation, tightness and soundproofing, be non-absorbent, and resistant to frost and color fading caused by the action of the sun [22–25]. Single-layer walls in an energy-saving house are rarely planned, although there are technologies that allow them to achieve the required thermal parameters, there are often construction problems that result in the lack of proper airtightness of partitions and the formation of thermal bridges. Three-layer walls (Figure 1) are the best type of partitions that ensure durability, tightness, thermal accumulation, and very good acoustic insulation of the house [26,27].

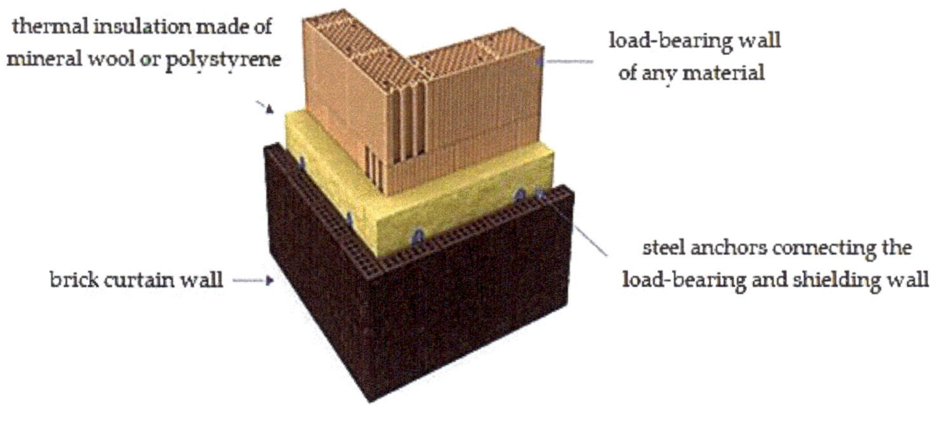

Figure 1. Construction of a three-layer wall. Source: own study.

The calculation of the energy performance of a building is only accurate if the tightness of the building is measured [28,29]. It can be measured by various methods: theoretically [30,31], empirically [32–36], or by single-component models [37]. Relander et al. tested the tightness of connections between the basement wall and the wall with a wooden structure. They presented the results of their work in [38]. Hallik et al. conducted research on air leaks through various wooden joints filled with three different polyurethane foams [39]. Kalamees et al. measured air leakage levels for eight joints that were used for various sealing solutions in a prefabricated timber envelope structure [40]. Pinto et al. measured the air permeability through a set of window frames [41]. Langmans et al. investigated the influence of climatic conditions on the tightness of typical tape connections [42]. Research on air permeability in various types of structures carried out by the authors [43–45] allows to obtain information on how air passes through the various parts of the envelope. Usually 20–25% of the heat supplied to the building is lost through windows and external doors. In order to maintain an appropriate level of annual energy consumption in buildings and to enable the economically viable use of renewable energy sources, the building, depending on the construction technology, must meet the criteria presented in Table 1.

Table 1. Criteria and standards for residential buildings—a list of parameters.

Requirements	Standard Building	Low-Energy Building	Building Passive
The maximum heat transfer coefficient of the building partition, $W/m^2{\cdot}K$:			
- external wall	0.30	0.18	0.15
- roof / flat roof	0.25	0.20	0.10
- windows	1.80/1.70	1.70	0.80
Energy demand for heating purposes, W/m^2	≤100	25–35	≤10
Annual energy demand for heating purposes, $kWh/(m^2{\cdot}a)$	≤100	30–40	≤15
Building tightness n50 (h^{-1})	≤3.0		0.2–0.6

Source: own study.

The energy balance is an important element in the design of low-energy and passive buildings. The purpose of this construction industry is to reduce the demand for heat from the central heating system or to eliminate it completely. The energy balance shows that losses in a traditional house are very large. In Polish climatic conditions, the largest component of the total energy consumption is the heating demand of buildings, therefore an energy-saving object is a building with low energy demand for heating purposes [46]. Table 2 presents the energy classification of buildings. In order to make a preliminary profitability analysis of the choice of renewable energy sources technology, it is therefore necessary to estimate the amount of energy necessary for a given facility to cover, among others, heat losses generated by external walls, roof, ceiling, floors on the ground or ceiling above an unheated basement, windows, and external doors. However, it is also necessary to take into account such systems as: lighting, hot water preparation utility, ventilation, air conditioning, etc.

Table 2. Energy classification of buildings according to Association for Sustainable Development.

Energy Class	Energy Rating	Indicator EA, [kWh/m^2/Year]
A$^+$	Passive	≥20
A	Low energy	20–45
B	Energy saving	45–80
C	Moderately energy efficient	80–100
D	Moderately energy-consuming	100–150
E	Energy-consuming	150–250
F	Highly energy-consuming	≤250

Source: own study.

Thanks to the use of renewable energy sources, including solar panels, solar collectors, heat pumps, or the use of three-layer glazing in windows with a low heat transfer coefficient caused the intense development of construction with zero energy consumption in the world [47]. Research on low-energy buildings due to the ongoing climate change in the world and the desire to reduce the sources of emissions are still of interest to many scientists. Not only the tropical and hot climate is conducive to the development of low-emission and zero-emission construction technologies, but also more and more research works are taking up the topic of using this technology also in moderate climates. For example, Szejnwald et al. [48] stated during their research that thermal modernization of buildings is one of the best available techniques in economic terms, enabling the reduction of a significant amount of energy demand for the examined buildings. Researchers Yi et al. [49] in their research found that the impact of climate and the change of seasons has a significant impact on the energy demand of buildings. In turn, the results of the research carried out by Szalay and Zold [50] prove that the appropriate location of buildings in the area in terms of the shape of the structure has a significant impact on obtaining low heat transfer coefficients through the walls of buildings. Srinivasan et al. [51] presented the results on how energy produced from renewable sources in hot climates reduces the energy demand in buildings throughout their lifetime. Research by Robert and Kummert [52] using the "Morphing" method has shown how global warming has a significant impact on zero-energy buildings. Many researchers, including Zhou et al. [53], in their works indicate the advantages related to the selection of appropriate HVAC systems and related technologies for energy buildings. In turn, Pikas et al. [54] studied the energy costs of buildings energetic.

Despite these earlier works, while the issues related to new low-energy construction have been legally regulated and widely described in scientific studies, buildings built before 2010 remain a problem, which should be thermomodernized, preferably in such a way that they can use renewable energy sources to the maximum. Problems with the selection of an appropriate renewable heat source in Poland, its profitability and technical conditions of the building are still the subject of research and analysis. There is no work on the principles of selecting the appropriate renewable energy sources technology as part of the thermal modernization of buildings located in the temperate climate zone. In Poland, there is a small number of buildings that meet the concepts of low- and zero-energy buildings and zero-emission buildings. In a building where the energy demand for domestic hot water preparation constitutes 60% of the energy balance, it becomes necessary to look for solutions in innovative heat sources, taking into account renewable energy technologies.

Therefore, the aim of this study is to evaluate and analyze the costs and energy consumption of thermomodernization of zero-energy, low-carbon residential buildings in temperate zones. This work, through the analyses and results obtained, allows investors to choose the most advantageous option in terms of economic profitability of the solution for the replacement of installations for residential buildings located in the temperate climate zone. The results showed that the thermal modernization of the building requires additional expenses estimated at 38% of the initial cost of the building. This will result in annual savings of €550 and an approximately 99% reduction in carbon dioxide emissions.

3. Materials and Methods

The article aims to show the methodology of selecting heat supplies to residential buildings located in the temperate climate zone, taking into account the economic application of various renewable energy technologies. The prepared model is to indicate a different alternative to the choice of renewable energy related to the mandatory thermal modernization of buildings, taking into account the construction parameters of buildings and their topology. The need to create a new model is intensified by changes that relate to tightening legal regulations related to energy consumption and demand, which must meet the standards of buildings in Poland from 2021. In order to achieve the goal, the article presents the assumptions and results of analyses of the use of energy installations in residential buildings that use renewable energy sources to balance energy consumption in various areas of its use, taking into account the location of the construction industry in moderate climate conditions. The impact of the use of the installation has been calculated using renewable energy sources for the energy performance of a building with different quality of partitions and improving energy efficiency. The research used weather data, simulations to determine energy consumption, and economic profitability of the investment.

3.1. Location and Climatic Data

This study was carried out in the south of Poland in Kraków, located in the temperate climate zone. The city has a very high social potential, which favors the development of renewable energy technologies. However, due to climatic conditions and low winter sun content, the country has less favorable conditions for the development of zero-energy targets in low-energy construction. However, Poland is undergoing important reforms to achieve zero energy and low carbon emissions by 2050. This study is intended to support this approach. According to the Köppen classification, Kraków is located in the oceanic climate zone (Cfb) with some features of a humid subtropical climate (Cfa). According to Wincenty Okołowicz's classification, the city lies in the warm temperate climate zone. In the reference period 1991–2019, the annual mean temperature was 9.9 °C, the annual mean maximum temperature was 14.4 °C and the yearly mean minimum temperature was 5.4 °C. The highest recorded temperature is 38.8 °C and the lowest −26.8 °C. The hottest month is July with Wed. tmax 26 °C, avg. tavg 20.4 °C, and avg. tmin 14.9 °C, and the coldest January with an Wed. tmax 2.2 °C, avg. tavg −0.7 °C, and avg. tmin −3.6 °C. Wed tavg is above 10 °C for 7 months from April to October and above 0 °C for 11 months (only January has a slightly negative tavg). Maximum temperatures exceeding 20 °C may occur from mid-February to mid-December (the earliest was recorded on 23 February, at the latest on December 5), and temperatures exceeding 10 °C are common throughout year. A 24-h frost may occur from December to February (sporadically in November and March), but usually it occurs only a dozen or so days a year and is within the limits of moderate frost (e.g., in 2019 the lowest tmax was −5 °C). The meridional circulation, intensified by the fen effect (mountain wind), plays an important role, raising the temperature in winter to a dozen or so degrees above zero. The annual rainfall is 670 mm. The highest amounts of rainfall occur in June and May, and the lowest in December. The amount of rainfall in summer decreases with the change of climate, and it increases in spring and autumn. Snowfall is less and less frequent. Usually there are 30–40 days with snowfall, but only a dozen or so days with accumulation above 1 cm. Due to the small number of days with 24-h frost, snow usually falls at a mild plus temperature (0–2 °C) and melts quickly. The insolation in the latest available data ranged from 203 h in July to 32 h in December (ref. 1961–1990), but it has increased significantly recently. It is related to the approaching border of a humid subtropical climate and more and more often reaching the city's regions of sunny southern European highlands. Krakow's climate is one of the warmest in Poland, especially when it comes to maximum temperatures. In terms of average monthly temperatures, however, it loses in the winter to western Poland, which results from lower minimum temperatures than in western Poland (western Poland, closer to the ocean, has more clouds, so at night the temperature does not drop so much).

3.2. Case Study: Building Energy Parameters

Building characteristics:	
Type and purpose	Residential building, single-family house
Standard	According to the regulations of 2002–2009
Number of stories	1
The height of the story, m	2.7
Heated area, m^2	120
Roof	gable
Total area of the facade, m^2	104
Insulated facade surface, m^2	104
Total area of windows, m^2	34
Cubature, m^3	247.07
Location	Kraków, Poland
Insolation, kWh/m^2	1080
Airing, m/s	17
Climate zone, °C	I–7.7
Heating system:	
Building heating system	Solid fuel boiler
Type	Old type (soot)
Heat distributor	Heater
Efficiency of the device, %	60
Efficiency of distribution and regulation, %	62
The temperature in the house, °C	20
Fuel	Coal
Calorific value, MJ/kg	26
Price 1 t, EUR	197.8
Cost 1 kWh, EUR	0.12
Max. power of the heating device, kW	13.2
Efficiency, %	60
Annual energy demand, kWh	2900
Building insulation parameters-roof:	
Material, $\lambda = 0.042$	Mineral wool
Thickness, cm	30
U-factor, $W/m^2{\cdot}K$	0.13
Insulating parameters of the building–walls:	
U-factor of a wall without additional insulation	1.25
Material, $\lambda = 0.045$	standard Styrofoam
Thickness, cm	15
U-factor, $W/m^2{\cdot}K$	0.24
Exterior doors, U	1.3
Insulating parameters of the building–floors:	
Material, $\lambda = 0.045$	Styrofoam XPS
Thickness, cm	10
U-factor, $W/m^2{\cdot}K$	0.19
Foundations	Traditional insulated

Insulating parameters of the building–ventilation:	
Tightness	Airtight without recuperation
Trial n50	3
Ventilation	Gravitational
Patency, %	100

3.3. Simulation Tools

The single-family house used for the research was designed for the needs of the Engineering Project based on the study concerning the energy house (Figure 2). The object was spatially modeled in the Termocadia program by Intersoft, and then thermal calculations were carried out in the compatible ArCADia TERMOCAD 7.1 program in accordance with the PN-EN 12831 standard. This software is commonly used in simulation, optimization, and modeling and served as the basis for thousands of research works. For calculation purposes, in accordance with the requirements of energy construction, the building's partitions were selected: windows, doors, flat roof, internal partitions, floor on the ground, and three variants of external walls were analyzed in detail. Subsequently, the renewable energy source devices necessary to determine the primary and final energy demand of the facility were selected, and the energy characteristics of the building were made.

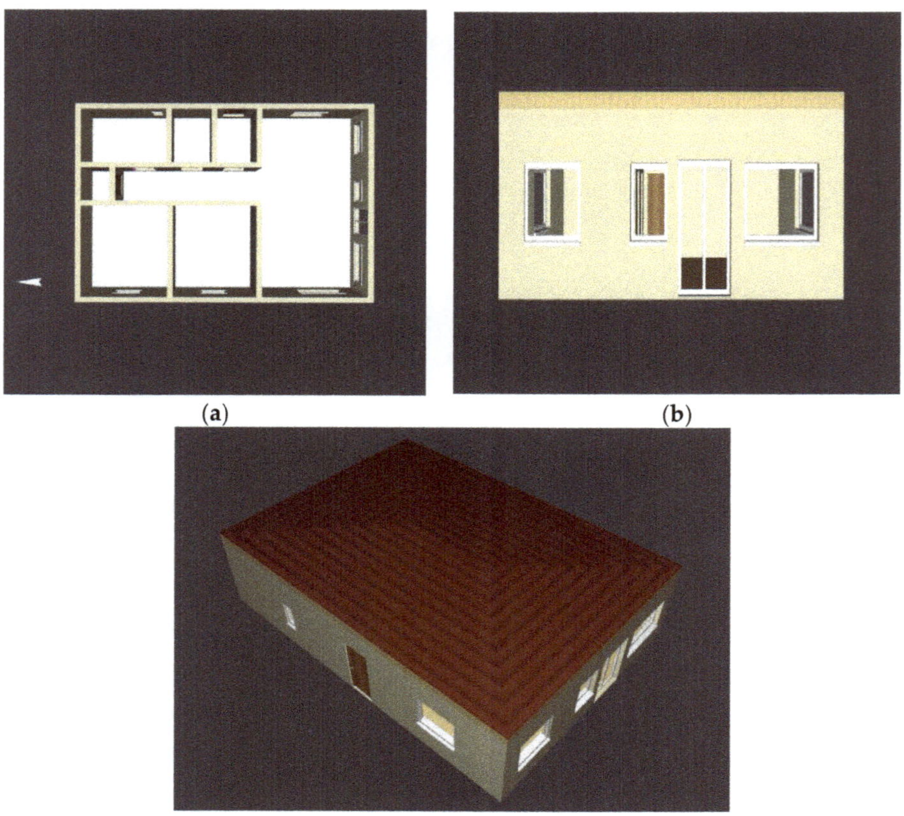

(c)

Figure 2. Analyzed building. (**a**) Building plan in 3D; (**b**) southern facade of the building; (**c**) 3D top view of the object. Source: own study.

3.4. Requirements for the Heat Transfer Coefficient U_c

The method of calculating the U_c heat transfer coefficient for external partitions in the tested facility was made in accordance with the standard PN-EN 12831-1:2017 [54]:

- Heat transfer coefficient U_c:

$$U_c = \frac{1}{R_T}, \left[\frac{W}{m^2 \cdot K}\right] \tag{1}$$

where:

R_T—resistance to heat transfer of the partition, $\left[\frac{m^2 \cdot K}{W}\right]$.

- Heat transfer coefficient resistance (thermal insulation) R_T:

$$R_T = R_{si} + \sum R_i + R_{se}, \left[\frac{m^2 \cdot K}{W}\right] \tag{2}$$

where:

R_{si}—resistance to heat transfer on the inner surface of the partition, $\left[\frac{m^2 \cdot K}{W}\right]$;

R_i—design thermal resistance for the i-th layer of the partition, $\left[\frac{m^2 \cdot K}{W}\right]$;

R_{se}—resistance to heat transfer on the outer surface of the partition, $\left[\frac{m^2 \cdot K}{W}\right]$.

For internal partitions, the resistance R_{se} assumes the values R_{si}. Computational thermal resistance for i-th layer of the partition R_i:

$$R_i = \frac{d_i}{\lambda_i}, \left[\frac{m^2 \cdot K}{W}\right] \tag{3}$$

where:

d_i—thickness of the i-th layer of the partition, [m];

λ_i—heat conduction coefficient, $\left[\frac{W}{m \cdot K}\right]$.

3.5. The Use of Renewable Energy Sources in Buildings and Their Impact on Energy and Ecological Efficiency

In this part of the research, the impact of the use of selected renewable energy sources (i.e., heat pump, photovoltaic installation, induction boiler, mini wind farm) on the energy performance of the building was analyzed by making variant calculations of the demand for non-renewable primary energy for various options of equipping the building with heating installations using resources. In this part of the research, the impact of the use of selected renewable energy sources. The value of the EP index of the annual demand for non-renewable primary energy for a building is influenced by [55]:

- The amount of energy demand in individual areas of use;
- The amount of auxiliary energy demand;
- Efficiency of systems in individual areas of use;
- Values of the coefficients of input of non-renewable primary energy in and for the delivery of an energy carrier or a given type of energy to the building.

The use of renewable energy sources to supply buildings with heat allows, in most cases, to significantly reduce their EP indicators of the demand for non-renewable primary energy. For the analyzed building, the following variants of the application of renewable energy sources for energy supply for central heating were analyzed, and domestic hot water preparation was analyzed as well. In the calculations of energy characteristics, the values of heat transfer coefficients for external partitions and separating heated and unheated spaces were assumed, equal to the maximum values in force in the standards. The calculations were carried out for various values of heat transfer coefficients for external partitions, applicable in a given period of their construction.

4. Results and Discussion

Currently, dynamic simulation methods are gaining more and more popularity, supplementing or replacing the static calculations used so far. This paper presents the results of dynamic simulations adapted to typical single-family houses. The data analysis technique was based on the simulation methods of an example residential building located in a temperate climate zone. The analyses were based on simulation studies in which actual data and parameters of buildings built in Poland in the years 1950–2021 were applied. Information has been collected on the basis of data contained in the provisions of the construction law, technical standards, and energy certificates of buildings. Due to the changing climatic conditions, energy requirements, and EU regulations, buildings built in Poland in 1950–2021 are responsible for 38% of greenhouse gas emissions. Therefore, they should be thoroughly modernized to meet the applicable EU regulations and directives.

Simulation studies on the basis of an example building with a standard usable area became the authors' motivation to analyze the technologies used in construction, taking into account building partition coefficients since 1950, in terms of the possibility of using renewable energy sources in them. The simulation tests were supported by a computational methodology with the use of simulations of the impact of the use of installations using renewable energy sources on the energy performance of a building with different quality of partitions and the improvement of energy efficiency in accordance with the PN-EN 12831 standard. Correct planning and design of installations based on the use of renewable sources is of key importance to ensure their correct and effective operation [56]. The calculations of various variants of the thermomodernization project in the field of building partitions consisted of introducing an additional layer of external insulation of the building walls, additional insulation of the roof, and replacement of window and door joinery in accordance with the requirements of the standards. During the simulation tests, all possible cases were given, taking into account real data, the parameters of which were used for the final thermal calculations. For each of them, appropriate materials were selected, which enabled the calculation and achievement of the correct value of the heat transfer coefficient Uc for all three variants of possible partitions used in single-family housing in the years 1950–2021. In the research, due to the location of Poland in the zone of the temperate climate, the most common sources of renewable energy technologies to be used for this climate zone, which will make it possible to achieve the required final energy demand, were analyzed. These studies give a picture of the state of thermal insulation of external partitions, the degree of elimination of thermal bridges and their impact on heat losses, and the use of the best renewable energy source for the analyzed building in technical and economic terms. The analysis of the obtained results allows to conclude that the use of dynamic simulations at the design stage of renewable energy installations can significantly help in the selection of the appropriate technology for residential buildings and the elimination of many potential operational problems.

4.1. Building Energy Parameters–Calculations

The aim of this part of the work was to analyze and compare the installation costs of individual solutions using renewable energy sources along with their impact on the environment. Then, those that were both cheap and relatively easily available were selected.

- Usable energy demand during the year [56]:

$$E_U = Q_{P,H} + Q_{P,W}, \left[\frac{\text{kWh}}{\text{year}}\right] \tag{4}$$

where:

$Q_{P,H}$—annual demand for primary energy by the heating and ventilation system;
$Q_{P,W}$—the annual primary energy requirement of the hot water system.

Four people live in the building, and if each of them consumes 35 dm^3 of water at a temperature of 45 °C per day = 140 dm^3 per day for four people. We assume that there are no residents at home for 14 days during the year (trips, holidays). So, they use hot water 351 days a year.

$$4 \times 35\ \mathrm{dm}^3 = 140 \left[\frac{\mathrm{dm}^3}{\mathrm{days}}\right] \tag{5}$$

Heat the water to 45 °C. The water delivered to the house is at a temperature of 10 °C. Thus, the difference is 35 °C.

$$45\ °\mathrm{C} - 10\ °\mathrm{C} = 35\ °\mathrm{C} \tag{6}$$

- Energy necessary to heat 1 dm^3 of water by 1 °C = $4.2\left[\frac{\mathrm{kJ}}{\mathrm{kg \cdot K}}\right]$:

$$\begin{gathered} 140\left[\frac{\mathrm{kg}}{\mathrm{days}}\right] \times 35\ \mathrm{K} \times 4.2\left[\frac{\mathrm{kJ}}{\mathrm{kg \cdot K}}\right] = 20,580\left[\frac{\mathrm{kJ}}{\mathrm{days}}\right] \\ 20,580\left[\frac{\mathrm{kJ}}{\mathrm{days}}\right] \div 3600 = 5.72\left[\frac{\mathrm{kWh}}{\mathrm{days}}\right] \\ 5.72\left[\frac{\mathrm{kWh}}{\mathrm{day}}\right] \times 351\ \mathrm{days} = 2007.72\left[\frac{\mathrm{kWh}}{\mathrm{year}}\right] \end{gathered} \tag{7}$$

Annual primary energy demand for the heating and ventilation system $Q_{P,H}$. The house has an area of 120 m^2. Its heat energy requirement is 150 $\left[\frac{\mathrm{kWh}}{\mathrm{m}^2 \cdot \mathrm{year}}\right]$ (an older, insulated house).

$$120\ \mathrm{m}^2 \times 150\left[\frac{\mathrm{kWh}}{\mathrm{m}^2 \cdot \mathrm{year}}\right] = 18,000\left[\frac{\mathrm{kWh}}{\mathrm{year}}\right] \tag{8}$$

Thus, the demand for primary energy during the year is:

$$E_U = 18,000\left[\frac{\mathrm{kWh}}{\mathrm{year}}\right] + 2007.72\left[\frac{\mathrm{kWh}}{\mathrm{year}}\right] = 20,007.72\left[\frac{\mathrm{kWh}}{\mathrm{year}}\right] \tag{9}$$

$$EK = \frac{EU}{\eta}, \left[\frac{\mathrm{kWh}}{\mathrm{year}}\right] \tag{10}$$

where:

EU—usable energy;
η—efficiency of the heating system.

Thus, the final energy demand is:

$$EK = \frac{20,007.72}{60\%}\left[\frac{\mathrm{kWh}}{\mathrm{year}}\right] = 33,346.20\left[\frac{\mathrm{kWh}}{\mathrm{year}}\right] \tag{11}$$

In order to determine the value of the primary energy, the amount of final energy should be multiplied by a factor appropriate for the given energy carrier. For hard coal, the coefficient is 1.1.

$$EP = 1.1 \times 33,346.20\left[\frac{\mathrm{kWh}}{\mathrm{year}}\right] = 36,680.82\left[\frac{\mathrm{kWh}}{\mathrm{year}}\right] \tag{12}$$

A coal-fired boiler with a capacity of B = 13.2 kW and a calorific value Qi = 26 MJ/kg has the following chemical composition: C = 65%; H = 4%; S = 1%; O = 1%; ash = 12%; total humidity = 8%.

$$\begin{gathered} \mathrm{E_{CO_2}} = \mathrm{V_{CO_2}} \times \mathrm{G_{pal}} \times \rho_{\mathrm{CO_2}} \\ \left[\frac{\mathrm{kg \cdot CO_2}}{\mathrm{h}}\right] = \left[\frac{\mathrm{m^3 \cdot CO_2}}{\mathrm{kg}}\right] \times \left[\frac{\mathrm{kg\ fuel}}{\mathrm{m^3 \cdot CO_2}}\right] \times \left[\frac{\mathrm{m^3 \cdot CO_2}}{\mathrm{kg}}\right] \\ \frac{65\%\ \mathrm{C}}{100\%\ \mathrm{w \cdot k}} = 0.65\left[\frac{\mathrm{kg\ C}}{\mathrm{kg\ fuel}}\right] \\ \mathrm{C} + \mathrm{O_2} \rightarrow \mathrm{CO_2}\ 1\,\mathrm{kmol\ C} + 1\,\mathrm{kmol\,O_2} \rightarrow 1\,\mathrm{kmol\,CO_2} \\ 12\,\mathrm{kg\,C} + 22.4\mathrm{m^3 O_2} \rightarrow 22.4\mathrm{m^3 CO_2} \end{gathered} \tag{13}$$

$$
\begin{gathered}
0.65\left[\frac{kgC}{kg\ fuel}\right] \\
V_{CO_2} = \frac{0.65\times 22.4}{12} = 1.21\cdot\frac{\frac{kgC}{kg\ fuel}\cdot m^3CO_2}{kgC} = \left[\frac{m^3\cdot CO_2}{kg}\right] \\
\text{Fuel stream}:\ G_{pal} = \frac{B}{Qi} = \frac{13.2}{26{,}000}\cdot\frac{\frac{kgJ}{s}}{\frac{kgJ}{kg\ fuel}} = \frac{kg\ fuel}{s} = \left[\frac{3600s}{h}\right] \\
G_{fuel} = \frac{13.2}{26{,}000}\times 3600 = 1.83\left[\frac{kg}{h}\right] \\
\rho_{CO_2} = \left[\frac{m\cdot CO_2}{V}\right] = \frac{12+2\times 16}{22.4} = \frac{44}{22.4} = 1.96\left[\frac{kgCO_2}{m^3\cdot CO_2}\right] \\
E_{CO_2} = 1.21\times 1.83\times 1.96 = 4.34\left[\frac{kgCO_2}{h}\right] \\
4.34\left[\frac{kg\cdot CO_2}{h}\right]\times 24\ h\times 270\ \text{days heating period} = 28{,}123.20\left[\frac{kg\cdot CO_2}{year}\right]
\end{gathered} \tag{14}
$$

During the 270-day heating period (1 September–31 May), the boiler will release to the atmosphere 28,123.20 kg CO_2.

$$
E_k = 33{,}346.20\left[\frac{kWh}{year}\right] = \frac{33{,}346.20}{120\ m^2}\left[\frac{kWh}{m^2\cdot year}\right] = 277.89\left[\frac{kWh}{m^2\cdot year}\right] \tag{15}
$$

The final energy of E_k is 277.89 kWh/m^2 per year. This value exceeds the maximum E_A index given in the reference range (Table 2). Therefore, the analyzed building is highly energy-consuming and has an F energy class of over 250 kWh/m^2. The planned energy purchase is: 33,346.20 [kWh/year], while the cost of energy purchase is 33,346.20 [kWh/year] 0.12 [EUR/ kWh] = 4001.55 [EUR/year].

4.2. Requirements for the Heat Transfer Coefficient U_c

The coefficients R_{se}, R_{si}, and $U_{c,max}$ were calculated based on the standard PN-EN 12831-1:2017 and presented in the Tables 3 and 4.

Table 3. Values of the coefficients R_{se} and R_{si}.

Heat Flux Direction	Resistance to Heat Transfer on the Outer Surface R_{se}, $\left[\frac{m^2\cdot K}{W}\right]$	Resistance to Heat Transfer on the Inner Surface R_{si}, $\left[\frac{m^2\cdot K}{W}\right]$
Up	0.04	0.10
Down	0.04	0.17
Horizontal	0.04	0.13

Source: own study.

Table 4. Maximum values of the permeation coefficient warm $U_{c,max}$.

Partition	from 1 January 2017 r.	from 1 January 2021 r.
Exterior walls	0.23	0.20
Interior walls	1.00	1.00
Roofs, flat roofs, and ceilings under unheated attics or over crossings	0.18	0.15
Floors on the ground	0.30	0.30
Ceilings over heated underground spaces and inter-story ceilings	1.00	1.00
Windows (except roof windows), balcony doors, and unopened transparent surfaces	1.1	0.9
Doors in external partitions or partitions between heated and unheated rooms	1.5	1.3

Source: own study.

Figure 3 shows the measured dependencies of air permeability on the pressure difference with the use of different variants of insulation used in construction.

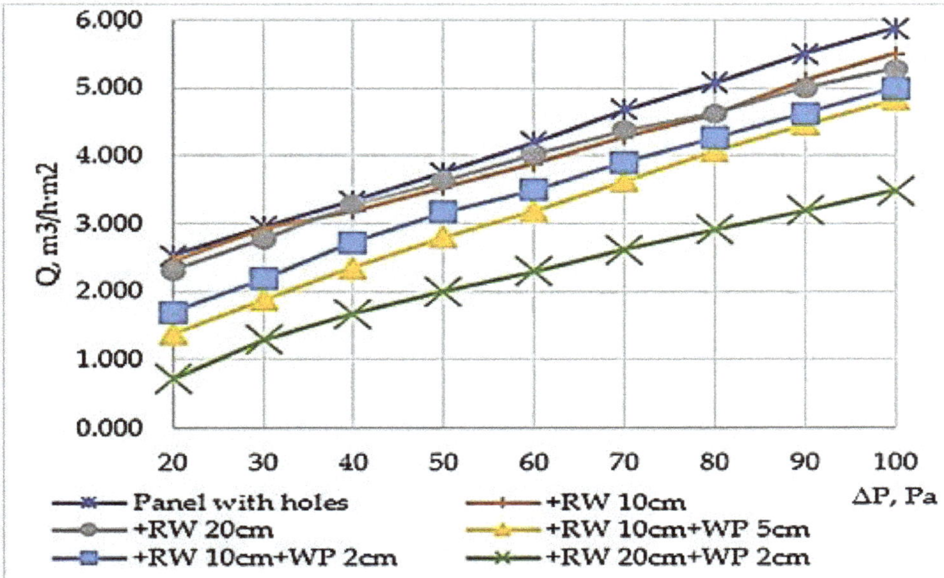

Figure 3. The dependence of air permeability on the pressure difference when using different variants of insulation used in construction. Source: own study based on [28].

As can be seen from Figure 3, the thicker rock wool thermal insulation layer used without wind protection boards had a greater impact on the reduction of air permeability. A thicker rock wool thermal insulation layer and a lower air permeability of wind protection layer shows the lowest air permeability results for the whole structure. For the analyzed building, the values of thermomodernization parameters and material characteristics of all partitions were calculated, which will be used in the further part of the research to select the appropriate renewable energy source technology.

Ceiling: the design uses a flat roof, i.e., the ceiling above the last floor of the building, which also fulfills the role of a roof with the material characteristics presented in Table 5.

Table 5. Characteristics of the partition-flat roof.

Material	The Thickness of the Partition Layer d, [m]	Thermal Conductivity λ, [W/m·K]	Layer Resistance R, $\left[\frac{m^2 \cdot K}{W}\right]$
Corrugated sheet	0.024	58.000	0.000
Plates URSA XPS-DRAIN 140 mm	0.280	0.040	7.000
Reinforced concrete with 1% steel	0.150	2.300	0.065
Plaster or cement-lime finish	0.000	0.820	0.000

Source: own study.

Calculation results of the coefficient values:

Insulation thickness d, m	0.45
Total resistance R_T, $\frac{m^2 \cdot K}{W}$	7.21
Total permeation coefficient U_c, $\frac{m^2 \cdot K}{W}$	0.14

Ground on the ground: the design uses a multi-layer ground floor with the material characteristics presented in Table 6.

Table 6. Characteristics of the partition-floor on the ground.

Material	The Thickness of the Partition Layer d, [m]	Thermal Conductivity λ, [W/m·K]	Layer Resistance R, $\left[\frac{m^2 \cdot K}{W}\right]$
Sand	0.400	2.000	0.200
Concrete of medium density 1800	0.200	1.150	0.174
Styrofoam 40	0.160	0.040	4.000
Steel reinforced concrete 1%	0.200	2.300	0.087
Styrofoam 40	0.070	0.040	1.750
Concrete of medium density 2000	0.050	1.350	0.037

Source: own study.

On the ground floor:
- Insulation thickness d, m 1.08
- Total resistance R_T, $\frac{m^2 \cdot K}{W}$ 6.42
- Total permeation coefficient U_c, $\frac{m^2 \cdot K}{W}$ 0.16

Exterior doors:
- Total resistance R_T, $\frac{m^2 \cdot K}{W}$ 6.42
- Total permeation coefficient U_c, $\frac{m^2 \cdot K}{W}$ 0.16

Interior doors:
- Total resistance R_T, $\frac{m^2 \cdot K}{W}$ 0.83
- Total permeation coefficient U_c, $\frac{m^2 \cdot K}{W}$ 1.20

Windows:
- Permeability coefficient b, % 0.14
- Total resistance R_T, $\frac{m^2 \cdot K}{W}$ 0.83
- Total permeation coefficient U_c, $\frac{m^2 \cdot K}{W}$ 1.20

External walls: the material characteristics of the walls are shown in Table 7.

Table 7. Characteristics of the partition-external walls.

Material	The Thickness of the Partition layer d, [m]	Thermal Conductivity λ, [W/m·K]	Layer Resistance R, $\left[\frac{m^2 \cdot K}{W}\right]$
POROTHERM 25P + ordinary mortar	0.250	0.300	0.833

Source: own study.

Calculation results of the coefficient values:
- Insulation thickness d, m 1.08
- Total resistance R_T, $\frac{m^2 \cdot K}{W}$ 1.09
- Total permeation coefficient U_c, $\frac{m^2 \cdot K}{W}$ 0.91

4.3. The Use of Renewable Energy Sources in Buildings and Their Impact on Energy and Ecological Efficiency

Reducing energy consumption and the use of renewable energy in the residential building sector is a priority action in the Member States of the European Union. These activities allow for the effective and sustainable use of the potential of fossil fuels and for the reduction of gaseous and dust emissions resulting from the process of energy combustion of fuel. The constantly growing prices of energy carriers also force the necessity to improve the energy performance of a given building, in particular a newly erected building. The aim of this part of the research is to compare the energy needs of a single-family residential building with a specific cubature and heating area. The building's energy consumption depends not only on the thermal insulation of building partitions, but also on the efficiency of the building's technical systems. Therefore, in the calculations, several variants of heating systems were introduced, including those supported by energy from renewable sources. For those under consideration in cases, the total costs of energy supplied to the building were estimated, that is, the cost of heating, domestic hot water, and electricity.

(a) Air Source Heat Pump

One of the solutions adopted for the research analysis is the use of the HEWALEX PCCO MONO 11 kW heat pump with a 3-kW heater with the technical parameters shown in Table 8. The data applies to the A7/W35 heating mode (temperature of the lower source 7 °C; temperature of the upper source 35 °C).

Table 8. Heating system air-to-water heat pump.

Heating System	
Source of heat	Heat pump
Type	Air-water
The efficiency of the device, %	490
Efficiency of distribution and regulation, %	98
The temperature in the house, °C	20
Fuel	Electricity
Cost 1 kWh, EUR	0.14
Maximum power of the heating device, kW	11.6
Water heating	Heat pump to CWU
Heating mode	A7/W35
Tray	Modern tray

Source: own study.

$$EK = \frac{EU}{\eta} = \frac{20{,}007.72}{490\%}\left[\frac{\text{kWh}}{\text{year}}\right] = 4083.21\left[\frac{\text{kWh}}{\text{year}}\right]$$
$$EP = 3 \times 4083.21\left[\frac{\text{kWh}}{\text{year}}\right] = 12{,}249.62\left[\frac{\text{kWh}}{\text{year}}\right]$$

The calculated rates of CO_2 emissions in [kg/MWh] for end consumers of electricity are presented in Table 9.

Table 9. Heating system air-to-water heat pump.

Indicators	Measuring Unit, [kg/MWh]
Carbon dioxide (CO_2)	719
Sulfur oxide (SOx/SO_2)	0.511
Nitric oxide (NOx/NO_2)	0.576
Carbon monoxide (CO)	0.233
Total dust	0.029

Source: own study.

The CO_2 emission is 0.719 kg CO_2 /kWh

$$Emission\ CO_2 = 0.719\left[\frac{\text{kg } CO_2}{\text{kWh}}\right] \times 4083.21\left[\frac{\text{kWh}}{\text{year}}\right] = 2935.83\left[\frac{\text{kg } CO_2}{\text{year}}\right]$$
$$E_k = 4083.21\left[\frac{\text{kWh}}{\text{year}}\right] = \frac{4083.21}{120\text{ m}^2}\left[\frac{\text{kWh}}{\text{m}^2\text{ year}}\right] = 34.02\left[\frac{\text{kWh}}{\text{m}^2\text{ year}}\right]$$

The energy parameters of the building after the application of the air-to-water heat pump are presented in Table 10.

Purchase of energy to central heating and domestic hot water are: 4083.21 [kWh/year]. Energy purchase cost up to central heating and domestic hot water: 4083.21 [kWh/year]·0.14 [EUR/kWh] = 571.65 [EUR/year]. Purchase of energy to power household appliances and lighting the building: 2900 [kWh/year]. The cost of purchasing energy to power household appliances and lighting the building: 2900 [kWh/year] 0.14 [EUR/kWh] = 406 [EUR/year]. It is necessary to purchase energy for central heating, domestic hot water, and to power household appliances and building lighting: 4083.21 [kWh/year] +2900 [kWh/year] = 6983.21 [kWh/year]. Energy purchase cost up to central heating, domestic hot water, and to power household appliances and building lighting: 571.65 [EUR/year] + 406 [EUR/year] = 977.65 [EUR/year].

Table 10. Energy parameters of the building after using the air-to-water heat pump.

Building Type	Single-Family House
Heated area, m^2	120
Energy performance index	Low energy A
The amount of the annual demand for usable energy, kWh/year	20,007.72
The volume of the annual demand for final energy, kWh/year	4083.21
Annual primary energy demand, kWh/year	12,249.62
Value of the issue CO_2, kg CO_2/year	2935.83
Power, kW	11.6

Source: own study.

(b) Ground Source Heat Pump

Similar calculations were made for the applicability of the ground heat pump for the analyzed building. The results are presented in Table 11.

Table 11. Energy parameters of the building after using the brine-water heat pump.

Building Type	Single-Family House
Heated area, m^2	120
Energy performance index	Low energy A
The amount of the annual demand for usable energy, kWh/year	20,007.72
The volume of the annual demand for final energy, kWh/year	4446.16
Annual primary energy demand, kWh/year	13,338.48
Value of the issue CO_2, kg CO_2/year	3196.79
Power, kW	12.5

Source: own study.

Purchase of energy to central heating, domestic hot water: 4446.16 [kWh/year]. Energy purchase cost up to central heating: 4446.16 [kWh/year]·0.14 [EUR/kWh] = 622.47 [EUR/year]. Purchase of energy to power household appliances and lighting the building: 2900 [kWh/year]. The cost of purchasing energy to power household appliances and lighting the building: 2900 [kWh/year]·0.14 [EUR/kWh] = 406 [EUR/year]. Purchase of energy to central heating, domestic hot water, and to power household appliances and building lighting: 4446.16 [kWh/year] + 2900 [kWh/year] = 7346.16 [kWh/year]. Energy purchase cost up to central heating, domestic hot water, and to power household appliances and building lighting: 622.47 [EUR/year] + 406 [EUR/year] = 1028.47 [EUR/year].

(c) Air Source Heat Pump and Solar Panels

A HEWALEX PCCO MONO 11kW heat pump with a 3-kW heater was used. Data refer to A7/W35 heating mode (heat source temperature 7 °C, heat source temperature 35 °C).

The annual electricity consumption of the heat pump is $E_k = 4083.21\left[\frac{\text{kWh}}{\text{year}}\right]$. It is assumed that by consuming 1000 kWh annually, a photovoltaic installation with a total capacity is needed 1 kWp.

$$Power\ of\ the\ photovoltaic\ installation = \frac{E_k\left[\frac{\text{kWh}}{\text{year}}\right] \times 1.2\ \text{kWp}}{1000\ \text{kW}}\ [\text{kWp}] \quad (16)$$

In addition, devices that use electricity in the house (washing machine, dishwasher, etc.) and lighting should be supplied with electricity obtained from the photovoltaic installation. During the year, electricity consumption for these purposes is 2900 kWh/year.

$$Power\ of\ the\ photovoltaic\ installation = \frac{\left(4083.21\left[\frac{\text{kWh}}{\text{year}}\right] + 2900\left[\frac{\text{kWh}}{\text{year}}\right]\right) \times 1.2\ \text{kWp}}{1000\ \text{kW}} = 8.4[\text{kWp}]$$

Additionally, we install monocrystalline photovoltaic panels on the roof (SUNTECH STP370S-B60/WNH-370 Wp). The power of one module is 0.33 kWp. How many panels need to be installed to be produced is calculated below 5 kWp.

$$\frac{8.4\ [\text{kWp}]}{0.33\ [\text{kWp}]} = 25.5\ \text{pieces} \approx 26\ \text{pieces}$$

The efficiency of the photovoltaic installation is 20.3%. Dimensions of one module are $175 \times 1039 \times 35\ \text{mm}^3$. The area required for the installation is 47.3 m^2. The CO_2 emission of the photovoltaic installation is 0.

$$E_k = \left(4083.21\left[\frac{\text{kWh}}{\text{year}}\right] + 2900\left[\frac{\text{kWh}}{\text{year}}\right]\right) = 6983.21\left[\frac{\text{kWh}}{\text{year}}\right] = \frac{6983.21}{120\ \text{m}^2}\left[\frac{\text{kWh}}{\text{m}^2\text{year}}\right] = 58.2\left[\frac{\text{kWh}}{\text{m}^2\text{year}}\right]$$

The energy parameters of the building after the use of an air source heat pump an photovoltaic panels are shown in Table 12.

Table 12. Building energy parameters after using an air heat pump and photovoltaic panels.

Building Type	Single-Family House
Heated area, m^2	120
Energy performance index	Energy-saving B
The amount of the annual demand for usable energy, kWh/year	20,007.72
The volume of the annual demand for final energy, kWh/year	0
Annual primary energy demand, kWh/year	0
Value of the issue CO_2, kg CO_2/year	2935.83
Power, kW	8.4

Source: own study.

(d) Induction Boiler and Solar Panels

$$\begin{aligned} Boiler\ power &= \text{Surface}\ [\text{m}^2] \cdot \text{The height of the rooms}\ [\text{m}] \cdot 60\ \text{W} \\ Boiler\ power &= 120\ [\text{m}^2] \times 2.7\ [\text{m}] \cdot 60\ [\text{W}] = 19{,}400\ [\text{W}] = 19.44\ [\text{kW}] \end{aligned} \quad (17)$$

The technical parameters of the heating boiler are presented in Table 13.

Table 13. Heating system: induction boiler.

Heating System	
Source of heat	Induction boiler
The efficiency of the device	0.986
Temperature in the house, °C	20
Fuel	Electricity
Cost 1 kWh, EUR	0.14
Heating device power, kW	21
Water heating	Electricity
Adjustable outlet temperature, °C	>80
Permissible outlet temperature, °C	90

Source: own study.

$$Power\ of\ the\ photovoltaic\ installation = \frac{E\left[\frac{\text{kWh}}{\text{year}}\right] \times 1.2\ \text{kWp}}{1000\ \text{kW}}\ [\text{kWp}] \quad (18)$$

We supply electricity obtained from the photovoltaic installation to devices that use electricity at home (washing machine, dishwasher, etc.) and lighting. During the year, electricity consumption for these purposes is 2900 kWh/year. Electricity consumption is 20,007 kWh/year.

$$Power\ of\ the\ photovoltaic\ installation = \frac{\left(21,000\left[\frac{\text{kWh}}{\text{year}}\right] + 2900\left[\frac{\text{kWh}}{\text{year}}\right]\right) \times 1.2\ \text{kWp}}{1000\ \text{kW}} = 28.7\ [\text{kWp}]$$

The power of one module is 0.33 kWp. How many panels need to be installed to be produced is calculated below 5 kWp.

$$\frac{28.7\ [\text{kWp}]}{0.33\ [\text{kWp}]} = 87\ \text{pieces}$$

To supply household appliances, lighting, and the energy demand of an induction boiler with electricity, you need a photovoltaic installation with a capacity of 28.7 kWp.

The efficiency of the photovoltaic installation is 20.3%. Dimensions of one module are 1756 × 1039 × 35 mm. The area required for the installation is 158.7 m^2. The roof area of the building is 172 m^2, and there are no roof windows, so the solar panels will fit on the roof. CO_2 emission = 0. The energy parameters of the building with the use of an induction boiler are shown in Table 14.

Purchase of energy: 0 [kWh/year]

Energy purchase cost: 0 [kWh/year] × 0.14 [EUR/kWh] = 0 [EUR/year]

Table 14. Building energy parameters after using an induction boiler.

Building Type	Single-Family House
Heated area, m^2	120
Energy performance index	Passive A$^+$
The amount of the annual demand for usable energy, kWh/rok	20,007.72
The volume of the annual demand for final energy, kWh/rok	0
Annual primary energy demand, kWh/rok	0
Value of the issue CO_2, kg CO_2/year	0
Computational power demand, kW	21,000

Source: own study.

(e) Use of an Air Source Heat Pump and a Vertical swing wind turbine

Due to the location of the analyzed building in an area with good ventilation, the use of an air heat pump and a vertical line wind turbine were further analyzed. The technical parameters of the turbine are presented in Table 15. In order to supply the building with electricity needed for central heating and domestic hot water, the heat pump requires 4083.21 kWh.

P_{el}—rated power of the wind farm: 2 kW,
h_n—height of the turbine mast: 9 m,
V_0—wind speed at height h_0 = 30 m: 7 m/s.

Table 15. Technical parameters of the turbine FLTXNY FH-2000.

Technical Parameters of the Wind Turbine	
Rated power, kW	2
Rated voltage, V	24/48
Take-off wind speed, m/s	2.5
Rated wind speed, m/s	12
Maximum wind speed, m/s	40
Rotor diameter, m	2
Number of rotor blades, pcs.	3
Propeller material	Aluminum

Source: own study.

If the turbine is to be installed at a height other than that at which the wind speed is known, the wind speed should be determined for the installation height of the turbine.

$$V_n = V_0\left(\frac{h_n}{h_0}\right)^{\alpha} = 7\frac{\text{m}}{\text{s}}\left(\frac{9\text{ m}}{30\text{ m}}\right)^{0.165} = 5.7\left[\frac{\text{m}}{\text{s}}\right] \quad (19)$$

Reading from the standards, the power plant utilization factor for the rated wind speed adopted for the determination of the turbine power is equal to 7 m/s, and the wind speed at the height of 9 m is equal to 5.7 m/s; the factor amounts to 36%.

$$E_{el} = 36\% \times P_{el} \times t = 36\% \times 2\text{ kW} \times 8760\text{ h} = 6307.20\left[\frac{\text{kWh}}{\text{year}}\right] \quad (20)$$

To supply the building with the electricity needed for central heating and domestic hot water, the heat pump requires 4083.21 kWh. The wind farm will produce 6307.20 kWh per year.

$$6307.20\left[\frac{\text{kWh}}{\text{year}}\right] - 4083.21\left[\frac{\text{kWh}}{\text{year}}\right] = 2224\left[\frac{\text{kWh}}{\text{year}}\right]$$

There is an overproduction of 2224 kWh/year. It can be used to supply electricity to domestic appliances and building lighting 2900 kWh/year.

$$2224\left[\frac{\text{kWh}}{\text{year}}\right] - 2900\left[\frac{\text{kWh}}{\text{year}}\right] = -676\left[\frac{\text{kWh}}{\text{year}}\right]$$

After the building is supplied with electricity for purposes, central heating, domestic hot water, lighting, and electrical demand are lacking 676 kWh annually. The cost of purchasing the missing electricity is EUR 94.75 per year.

$$676\left[\frac{\text{kWh}}{\text{year}}\right] \times 0.14\text{ EUR} = 94.75\left[\frac{\text{EUR}}{\text{year}}\right]$$

(f) The use of an Induction Boiler and a Wind Turbine with a Vertical Rotation Line

In order to supply the building with electricity needed for central heating and domestic hot water, the induction boiler requires 21,000 kWh. The technical parameters of the turbine are shown in Table 16.

P_{el}—rated power of the wind farm: 6 kW,
h_n—turbine rotor height: 12 m,
V_0—wind speed at height h_0 = 30 m: 7 m/s.

The technical parameters of the Spine + ARC 1500 turbine are shown in Table 16.

Table 16. Technical parameters of the Spine + ARC 1500 turbine.

Technical Parameters of the Wind Turbine	
Rated power, kW	6
Rated voltage, V	48
Take-off wind speed, m/s	2
Rated wind speed, m/s	11
Maximum wind speed, m/s	40
Rotor diameter, m	3
Number of rotor blades, pcs	3
Propeller material	Fiberglass reinforced with polyester resin
Weight, kg	156

Source: own study.

If the turbine is to be installed at a height other than that at which the wind speed is known, the wind speed should be determined for the installation height of the turbine.

$$V_n = V_0\left(\frac{h_n}{h_0}\right)^{\alpha} = 7\frac{\text{m}}{\text{s}}\left(\frac{12\ \text{m}}{30\ \text{m}}\right)^{0.165} = 6\left[\frac{\text{m}}{\text{s}}\right]$$

Reading from the standards, the power plant utilization factor for the rated wind speed adopted for the determination of the turbine power is equal to 7 m/s, and the wind speed at the height of 12 m is equal to 6 m/s; the factor amounts to 39%.

$$E_{el} = 39\% \times P_{el} \times t = 39\% \times 6\ \text{kW} \times 8760\ \text{h} = 20,498.40\left[\frac{\text{kWh}}{\text{year}}\right]$$

To supply the building with electricity needed for central heating and domestic hot water, the heat pump needs 20,007.72 kWh. The wind farm will produce 20,498.40 kWh annually.

$$20,498.40\left[\frac{\text{kWh}}{\text{year}}\right] \times 20,007.72\left[\frac{\text{kWh}}{\text{year}}\right] = 490.70\left[\frac{\text{kWh}}{\text{year}}\right]$$

There is an overproduction of 490.70 kWh/year. We can use it to supply electricity to domestic appliances and building lighting 2900 kWh/year.

$$490.70\left[\frac{\text{kWh}}{\text{year}}\right] - 2900\left[\frac{\text{kWh}}{\text{year}}\right] = -2409.32\left[\frac{\text{kWh}}{\text{year}}\right]$$

After supplying the building with electricity for central heating, domestic hot water, lighting, and electrical appliances, it is necessary to buy an additional 2409.32 kWh per year. The purchase cost is:

$$2409.32\left[\frac{\text{kWh}}{\text{year}}\right] \times 0.14\ \text{EUR} = 337.31\ \text{EUR}$$

Table 17 shows the total purchase costs of the analyzed solutions. On the other hand, Figure 3 presents a summary of the quantity and costs of purchasing the missing electricity during the year.

Table 17. Purchase costs of the applied solutions.

No.	Device Name	The Cost of Purchasing the Device and Assembly, EUR	The Cost of Purchasing Additional Elements, EUR	Total Cost of Purchase and Assembly, EUR	Amount of Missing Electricity to Meet the Demand, [kWh/Year]	The Cost of Purchasing the Missing Energy, EUR	Sum of Costs in the First year, EUR/Year
1.	Air-to-water heat pump	6477.77	323.89	6801.65	1508.02	980.30	7781.96
2.	Ground source heat pump	6261.84	1079.63	7341.47	1586.40	1031.04	8372.51
3.	Air-to-water heat pump + Solar panels	11,517.47	1943.33	13,385.22	-	-	13,385.22
4.	Induction boiler + Monocrystal line photovoltaic panels	18,209.00	2159.26	20,368.25	-	-	20,368.25
5.	Air-to-water heat pump + A wind turbine Ista Breeze i-2000	8097.21	388.67	8485.87	373.12	242.48	8728.36
6.	Air-to-water heat pump + FLTX FH-2000 wind turbine	10,364.43	388.67	10,753.09	145.97	95.01	10,848.10
7.	Induction boiler + A wind turbine Windspot 7.5 kW	11,767.94	453.44	12,221.38	Overproduction 870.18	-	12,221.38
8.	Induction boiler + A wind turbine Spine + ARC 1500 6 kW	14,251.08	431.85	14,682.94	520.38	338.14	15,021.07

Source: own study.

Figure 4 shows the amount of electricity that was insufficient to meet the needs of the analyzed building and the costs that must be incurred for this purpose. Ensuring the independence of the house from the power grid means that the selected solution will cover the household's demand for electricity necessary to power household appliances, lighting rooms, central heating, and hot water. The largest amount of electricity must be additionally purchased in the case of using a ground heat pump (7346.16 kWh/year; 1028.47 EUR/year). Similar values were observed for the installation of the air heat pump (6983.21 kWh/year; 977.65 EUR/year). The use of an air heat pump in the building in conjunction with a photovoltaic installation and an induction boiler in conjunction with a photovoltaic installation ensured that the total electricity demand was met and there was no need to purchase additional energy. There was also no overproduction of energy. Using a heat pump air-water in combination with a wind turbine with a vertical or horizontal axis of rotation, one should take into account the necessity to pay EUR 87–260 per year for the missing amount of electricity. A 2.2-kW horizontal swing turbine generates 20% less energy compared to a 2-kW vertical rotation turbine. This translates into almost EUR 152 higher costs of purchasing the missing energy. However, when using an induction boiler in combination with a wind turbine with a horizontal axis of rotation, after ensuring the self-sufficiency of the building from electricity, there was an overproduction of 4023.30 kWh/year. Additional energy can be used to power other devices that, for example, will appear on the farm after some time and have not been included in the demand. Then, you can save EUR 565.72 per year at the electricity price of EUR 0.14/kWh. Then, the research compared the total expenses that must be incurred in the first 12 months from the installation and purchase of a new one energy sources. These costs are shown in Figure 5.

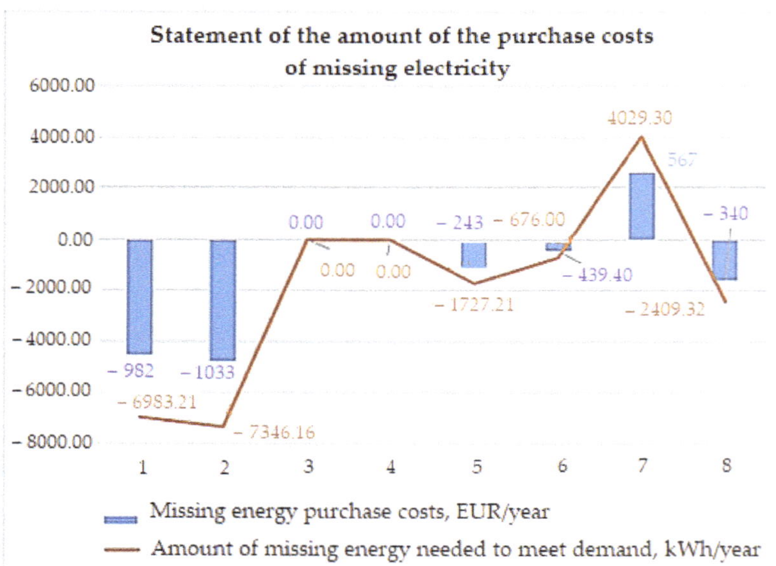

Figure 4. Summary of the quantity and costs of purchasing the missing electricity during the year: (1) 11.6 kW air-to-water heat pump; (2) ground source heat pump 12.5 kW; (3) 11.6 kW air-to-water heat pump + 8.4 kWp monocrystalline photovoltaic panels; (4) induction boiler 21 kW + 28.7 kWp monocrystalline photovoltaic panels; (5) 11.6 kW air-to-water heat pump + 2.2 kW wind turbine with a horizontal axis of rotation; (6) 11.6 kW air-to-water heat pump + 2 kW wind turbine with a vertical axis of rotation; (7) induction boiler 21 kW + wind turbine with a horizontal axis of rotation of 7.5 kW; (8) induction boiler 21 kW + wind turbine with a vertical axis of rotation 6 kW. Source: own study.

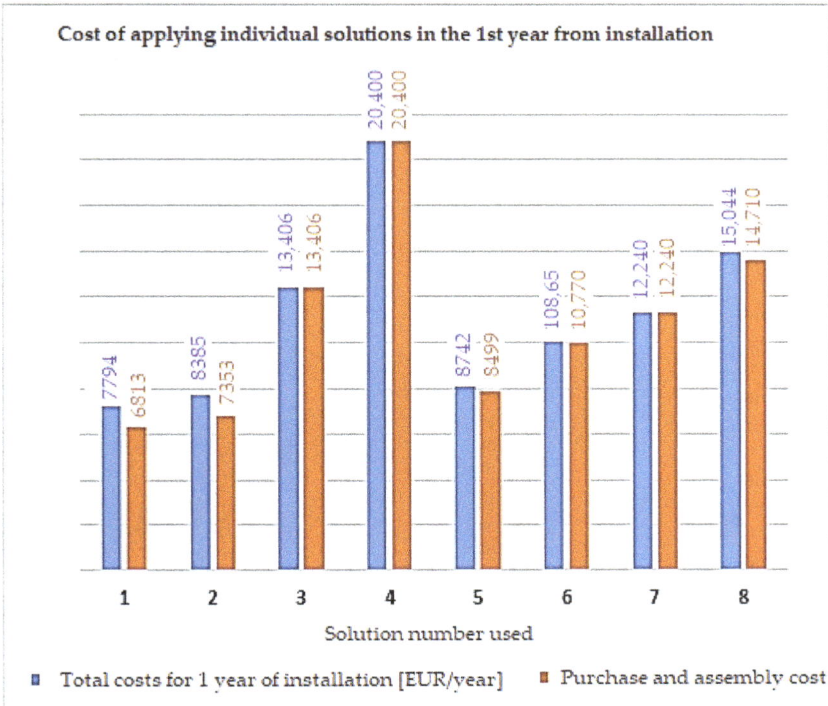

Figure 5. The sum of the costs of applying individual solutions in the first year of installation: (1) 11.6 kW air-to-water heat pump; (2) ground source heat pump 12.5 kW; (3) 11.6 kW air-to-water heat pump + 8.4 kWp monocrystalline photovoltaic panels; (4) induction boiler 21 kW + 28.7 kWp monocrystalline photovoltaic panels; (5) 11.6 kW air-to-water heat pump + 2.2 kW wind turbine with a horizontal axis of rotation; (6) 11.6 kW air-to-water heat pump + 2 kW wind turbine with a vertical axis of rotation; (7) induction boiler 21 kW + wind turbine with a horizontal axis of rotation of 7.5 kW; (8) induction boiler 21 kW + wind turbine with a vertical axis of rotation 6 kW. Source: own study.

Figure 3 shows the total installation costs of a given device in the building (the cost of purchasing the device, the cost of assembly, the cost of purchasing additional elements necessary for the installation, and the cost of purchasing energy, which may have been insufficient to cover the household's electricity demand). The highest costs (EUR 20,368.25) occurred with the use of a 21 kW induction boiler and 28.7 kWp monocrystalline photovoltaic panels. Such a high price is mainly due to the high power of the photovoltaic installation, and thus the need for more individual modules. In addition, the purchase of an induction boiler, which is not currently a popular solution, also involves a high cost. All three options with the use of an induction boiler are at the forefront of the presented options in terms of costs. The cheapest option is to choose one heat source: a heat pump. The costs for an air source heat pump or a ground source heat pump are similar level and amount to 7774–8206 EUR within the first year of installation. The solutions where only the purchase cost of the device itself and additional elements should be covered are:

- 11.6 kW air source heat pump + 8.4 kWp monocrystalline photovoltaic panels;
- 21 kW induction boiler + 28.7 kWp monocrystalline photovoltaic panels;
- 21 kW induction boiler + 7.5 kW wind turbine with a horizontal axis of rotation.

The power of these devices has been selected so that there is no need to buy the missing energy and thus expose the consumer to additional costs.

Lowering the energy needs of a building by increasing the requirements for thermal insulation of its housing and introducing high-performance modern heating devices and energy-saving electrical devices translates into measurable ecological and economic benefits. Characteristics of energy demand in the analyzed residential building, in which the domestic hot water demand plays a dominant role, amounting to almost 60% of the energy balance, makes it practically impossible to achieve the required EP level without modifying the heat source. As can be seen on the basis of the analysis, the use of alternative energy sources may be more advantageous than conventional solutions, both in terms of the amount of investment outlays and costs in the life cycle. It therefore confirms that the level of requirements specified in the technical requirements for buildings can promote the widespread use of alternative energy sources.

5. Conclusions

Promoting and recommending the use of installations based on renewable energy resources in construction, in addition to improving the thermal insulation of building partitions, has a significant impact on increasing energy security and achieving the required standards inside buildings at a lower operating cost. Such a procedure gives optimal and economically profitable effects in the long-term effects of respecting the resources of non-renewable primary energy. Targeted and well-thought-out measures to rationalize the use of final energy for the needs of buildings should no longer pose a challenge but should be a necessary task in a sustainable low-carbon economy.

At the initial heat source (old-type solid fuel boiler), CO_2 emissions to the atmosphere amounted to 28,123.2 kg during the 270-day heating period from 1 September to 31 May. This is a significant amount, considering that only one household releases so many harmful substances per year. It is also worth noting that the solid fuel boiler heated the rooms and hot water. Powering household appliances such as a washing machine or refrigerator as well as room lighting was associated with additional costs of about 410 EUR/year. The household's demand for electricity necessary to power household appliances, lighting rooms, central heating, and hot water is 22,907.72 kWh/year. The presented solutions have been adjusted so as to fully or partially cover the building's electricity needs. On the basis of the performed tests, the following conclusions were drawn:

1. Even using traditional technologies, that is, two-layer walls of constructing external partitions, it is possible to obtain a heat transfer coefficient U through the partition, appropriate for an energy-saving building. The availability of materials on the market is not a problem.
2. A three-layer wall is the most optimal solution for a single-family house built in energy-saving technology. Despite the higher investment costs, with such a solution the coefficient of demand for primary energy for heating is the lowest. Due to the cover layer, this solution is also the most durable and easy to maintain.
3. When using a heat pump instead of a solid fuel boiler, the amount of CO_2 released into the atmosphere is reduced by almost 90% per year and amounts to 2935.83 kg CO_2/year with an air source heat pump and 3196.79 kg CO_2/year with a ground source heat pump.
4. "Zero emission" occurs in solutions where the consumer himself produces the electricity needed to power the heat pump or induction boiler. Variants with the generation of electricity by means of a turbine with a vertical and a horizontal axis of rotation, as well as with the use of a photovoltaic installation, were presented.
5. The combination of two solutions using renewable energy sources is characterized by higher costs than that with one source of energy, but then the household becomes energy self-sufficient and is not affected by increases in electricity prices and power cuts.
6. Taking into account the co-financing programs, the most cost-effective solution for the analyzed single-family building is to install a heat pump in combination with photovoltaic panels. Then, it is possible to obtain funding of up to EUR 6480 under a government program called the "Clean Air Program" [57] for the replacement of

the current high-emission furnace with an environmentally friendly device and for the installation of photovoltaic panels. In the event that the consumer decides not to introduce photovoltaics, the surcharge may amount to a maximum of EUR 5400. Currently, there is a huge selection of photovoltaic panels on the market. The popularity of this solution is constantly growing, and the programs of subsidies or tax deductions for purchase costs encourage more people to install photovoltaics in their home. The aforementioned heat pump is also relatively popular.

7. By using two solutions using renewable energy sources from a highly energy-consuming building with energy class F, a house has become passive with energy class A^+. The household is then energy self-sufficient and is not affected by possible increases in energy prices or power outages.

In order to encourage building owners to use renewable energy sources, it is necessary for the state to create a broader support system and to modify the existing regulations enabling them to conduct prosumer activity on the electricity and heat market. Renewable energy installations, as an alternative to conventional systems, should become a standard in the design of nearly zero-energy buildings.

This type of research is necessary to generate more detailed scientific information that architectural engineers can use to design these types of buildings. Additional studies should be carried out to develop more complex models with external weather data as input data and to explore the benefits of passive technologies in different climatic regions. In addition, the algorithms for assessing the impact of thermomodernization of heating installations with the use of systems based on RES resources on the energy performance of the building and its certificate should be examined.

Author Contributions: Conceptualization, M.N.; methodology, M.N.; software, M.N.; validation, M.N.; formal analysis, M.N. and P.W.; investigation, M.N. and P.W.; resources, M.N.; data curation, M.N. and P.W.; writing—original draft preparation, M.N. and P.W.; writing—review and editing, P.W.; visualization, M.N.; supervision, P.W.; project administration, M.N. and P.W.; funding acquisition, M.N. All authors have read and agreed to the published version of the manuscript.

Funding: This research received no external funding.

Institutional Review Board Statement: Not applicable.

Informed Consent Statement: Not applicable.

Data Availability Statement: Not applicable.

Conflicts of Interest: The authors declare no conflict of interest.

References

1. Directive 2009/28/EC of the European Parliament and of the Council of 23 April 2009 on the Promotion of the Use of Energy from Renewable Sources and Amending and Subsequently Repealing Directives 2001/77/EC and 2003/30/EC (Text with EEA relevance). Available online: https://eur-lex.europa.eu/legal-content/PL/TXT/?uri=celex%3A32009L0028 (accessed on 15 December 2021).
2. Wróblewski, P.; Lewicki, W. A Method of Analyzing the Residual Values of Low-Emission Vehicles Based on a Selected Expert Method Taking into Account Stochastic Operational Parameters. *Energies* **2021**, *14*, 6859. [CrossRef]
3. Wróblewski, P. Analysis of Torque Waveforms in Two-Cylinder Engines for Ultralight Aircraft Propulsion Operating on 0W-8 and 0W-16 Oils at High Thermal Loads Using the Diamond- Like Carbon Composite Coating. *SAE Int. J. Engines* **2021**, *15*, 2022. [CrossRef]
4. Dziubak, T.; Bąkała, L.; Karczewski, M.; Tomaszewski, M. Numerical research on vortex tube separator for special vehicle engine inlet air filter. *Sep. Purif. Technol.* **2020**, *237*, 116463. [CrossRef]
5. Dziubak, T.; Dziubak, S.D. Experimental Study of Filtration Materials Used in the Car Air Intake. *Materials* **2020**, *13*, 3498. [CrossRef] [PubMed]
6. Dziubak, T.; Bąkała, L. Computational and Experimental Analysis of Axial Flow Cyclone Used for Intake Air Filtration in Internal Combustion Engines. *Energies* **2021**, *14*, 2285. [CrossRef]
7. Wróblewski, P.; Koszalka, G. An Experimental Study on Frictional Losses of Coated Piston Rings with Symmetric and Asymmetric Geometry. *SAE Int. J. Engines* **2021**, *14*, 853. [CrossRef]

8. Wróblewski, P. An Innovative Approach to Data Analysis in the Field of Energy Consumption and Energy Conversion Efficiency in Vehicle Drive Systems—The Impact of Operational and Utility Factors. In Proceedings of the 37th International Business Information Management Association (IBIMA), Cordoba, Spain, 1–2 April 2021; ISBN 978-0-9998551-6-4.
9. Kubińska-Jabcoń, E.; Kubiński, W.; Niekurzak, M. Analysis of the Economical, Ecological and Quality Advantages Consequent Resulting from Initiation of Technology Integrated in Metallurgy. *J. Mech. Eng. Autom.* **2015**, *5*, 549–553. [CrossRef]
10. Janoś, K. Polski Prąd Najdroższy w UE już 7 m-c z Rzędu. Tak Drogo Może być 15 Lat. Available online: https://www.money.pl/gospodarka/polski-prad-najdrozszy-w-ue-juz-7-miesiecy-z-rzedu-tak-drogo-moze-byc-15-lat-6576934583294752a.html (accessed on 12 December 2021).
11. Olczak, P.; Matuszewska, D.; Kryzia, D. "Mój Prąd" as an example of the photovoltaic one off grant program in Poland. *Energy Policy J.* **2020**, *23*, 123–138. Available online: https://epj.min-pan.krakow.pl/-Moj-Prad-as-an-example-of-the-photovoltaic-one-off-grant-program-in-Poland,122482,0,2.html (accessed on 15 October 2021). [CrossRef]
12. Niekurzak, M. The Potential of Using Renewable Energy Sources in Poland Taking into Account the Economic and Ecological Conditions. *Energies* **2021**, *14*, 7525. [CrossRef]
13. Eurostat Statistics Explained, Electricity Price Statistics. Available online: https://ec.europa.eu/eurostat/statistics-explained/index.php/Electricity_price_statistics (accessed on 10 December 2021).
14. Niekurzak, M.; Mikulik, J. Modeling of Energy Consumption and Reduction of Pollutant Emissions in a Walking Beam Furnace Using the Expert Method—Case Study. *Energies* **2021**, *14*, 8099. [CrossRef]
15. Niekurzak, M.; Kubińska-Jabcoń, E. Analysis of the Return on Investment in Solar Collectors on the Example of a Household: The Case of Poland. *Front. Energy Res.* **2021**, *9*, 224. [CrossRef]
16. Olczak, P.; Olek, M.; Kryzia, D. The Ecological Impact of Using Photothermal and Photovoltaic Installations for DHW Preparation. Polityka Energetyczna. *Energy Policy J.* **2020**, *23*, 65–74. [CrossRef]
17. Olczak, P.; Olek, M.; Matuszewska, D.; Dyczko, A.; Mania, T. Monofacial and Bifacial Micro PV Installation as Element of Energy Transition-The Case of Poland. *Energies* **2021**, *14*, 499. [CrossRef]
18. Borowski, P. Adaptation strategy on regulated markets of power companies in Poland. *Energy Environ.* **2019**, *30*, 3–26. [CrossRef]
19. Kochanek, E. *Wielowymiarowość Interesów Energetycznych w Dobie Transformacji Systemowej*; WAT: Warsaw, Poland, 2021.
20. *Polityka Energetyczna Polski 2040, Załącznik do Uchwały nr 22/2021 Rady Ministrów z Dnia 2 Lutego 2021 r*; Ministerstwo Klimatu i Środowiska: Warsaw, Poland, 2021.
21. Wiśniewski, G. *Europejskie Centrum Energii Odnawialnej, Ekonomiczne i Prawne Aspekty Wykorzystania Odnawialnych Zródeł Energii w Polsce*; Ministerstwa Środowiska: Warsaw, Poland, 2000. Available online: http://www.pga.org.pl/prawo/ekonomiczne_i_prawne_aspekty.pdf (accessed on 10 December 2021).
22. Szczerbowski, R. The forecast of Polish power production sector development by 2050—Coal scenario. *Energy Policy J.* **2018**, *19*, 5–18.
23. Młynarski, T.; Tarnawski, M. *Źródła Energii i ich Znaczenie dla Bezpieczeństwa Energetycznego w XXI Wieku*; Difin: Warsaw, Poland, 2016.
24. *Wnioski z Analiz Prognostycznych na Potrzeby Polityki Energetycznej Polski do 2050 Roku, Załącznik 2*; Ministerstwo Klimatu i Środowiska: Warsaw, Poland, 2021.
25. Gabryś, H. Elektroenergetyka w Polsce 2020. *Energetyka* **2020**, *8*, 365–373.
26. Robaina, M.; Neves, A. Complete decomposition analysis of CO_2 emissions intensity in the transport sector in Europe. *Res. Transp. Econ.* **2021**, *87*, 101074. [CrossRef]
27. Polski Sektor Energetyczny 2050. *4 Scenariusze*; Forum Energii: Warsaw, Poland, 2018.
28. Šadauskienė, J.; Seduikyte, L.; Paukštys, V.; Banionis, K.; Gailius, A. The role of air tightness in assessment of building energy performance: Case study of Lithuania. *Energy Sustain. Dev.* **2016**, *32*, 31–39. [CrossRef]
29. Cardoso, V.E.M.; Pereira, P.F.; Ramos, N.M.M.; Almeida, R. The Impacts of Air Leakage Paths and Airtightness Levels on Air Change Rates. *Buildings* **2020**, *10*, 55. [CrossRef]
30. Ziółkowski, J.; Żurek, J.; Małachowski, J.; Oszczypała, M.; Szkutnik-Rogoż, J. Method for Calculating the Required Number of Transport Vehicles Supplying Aviation Fuel to Aircraft during Combat Tasks. *Sustainability* **2022**, *14*, 1619. [CrossRef]
31. Wang, L.; Chen, Q. Validation of a Coupled Multizone-CFD Program for Building Airflow and Contaminant Transport Simulations. *HvacR Res.* **2007**, *13*, 267–281. [CrossRef]
32. Chan, W.R.; Joh, J.; Sherman, M.H. Analysis of air leakage measurements of US houses. *Energy Build.* **2013**, *66*, 616–625. [CrossRef]
33. Montoya, M.I.; Pastor, E.; Carrié, F.R.; Guyot, G.; Planas, E. Air leakage in Catalan dwellings: Developing an airtightness model and leakage airflow predictions. *Build. Environ.* **2010**, *45*, 1458–1469. [CrossRef]
34. Pan, W. Relationships between air-tightness and its influencing factors of post-2006 new-build dwellings in the UK. *Build. Environ.* **2010**, *45*, 2387–2399. [CrossRef]
35. Fernández-Agüera, J.; Domínguez-Amarillo, S.; Sendra, J.J.; Suarez, R. An approach to modelling envelope airtightness in multi-family social housing in Mediterranean Europe based on the situation in Spain. *Energy Build.* **2016**, *128*, 236–253. [CrossRef]
36. Bramiana, C.; Entrop, A.; Halman, J. Relationships between Building Characteristics and Airtightness of Dutch Dwellings. *Energy Procedia* **2016**, *96*, 580–591. [CrossRef]
37. Szkutnik-Rogoż, J.; Ziółkowski, J.; Małachowski, J.; Oszczypała, M. Mathematical Programming and Solution Approaches for Transportation Optimisation in Supply Network. *Energies* **2021**, *14*, 7010. [CrossRef]

38. Relander, T.-O.; Heiskel, B.; Tyssedal, J.S. The influence of the joint between the basement wall and the wood-frame wall on the airtightness of wood-frame houses. *Energy Build.* **2011**, *43*, 1304–1314. [CrossRef]
39. Hallik, J.; Gustavson, H.; Kalamees, T. Air Leakage of Joints Filled with Polyurethane Foam. *Buildings* **2019**, *9*, 172. [CrossRef]
40. Kalamees, T.; Alev, Ü.; Pärnalaas, M. Air leakage levels in timber frame building envelope joints. *Build. Environ.* **2017**, *116*, 121–129. [CrossRef]
41. Pinto, M.; Viegas, J.; De Freitas, V. Air permeability measurements of dwellings and building components in Portugal. *Build. Environ.* **2011**, *46*, 2480–2489. [CrossRef]
42. Langmans, J.; Desta, T.Z.; Alderweireldt, L.; Roels, S. Durability of self-adhesive tapes for exterior air barrier applications: A laboratory investigation. *Int. J. Vent.* **2016**, *16*, 30–41. [CrossRef]
43. Ziółkowski, J.; Oszczypała, M.; Małachowski, J.; Szkutnik-Rogoż, J. Use of Artificial Neural Networks to Predict Fuel Consumption on the Basis of Technical Parameters of Vehicles. *Energies* **2021**, *14*, 2639. [CrossRef]
44. Fokin, K.F. *Строительная Теплотехника Ограждающих Частей Зданий [Building Heat Engineering Enclosing Parts of Buildings]*; АВОК-ПРЕСС: Moscow, Russia, 2006.
45. Hens, S.L.C.H. *Building Physics-Heat, Air and Moisture 3e Fundamentals and Engineering Methods with Examples and Exercises*; John Wiley & Sons: Hoboken, NJ, USA, 2017; pp. 135–141.
46. Niekurzak, M. Determining the Unit Values of the Allocation of Greenhouse Gas Emissions for the Production of Biofuels in the Life Cycle. *Energies* **2021**, *14*, 8394. [CrossRef]
47. Shaw-Williams, D.; Susilawati, C.; Walker, G.; Varendorff, J. Towards net-zero energy neighbourhoods utilising high rates of residential photovoltaics with battery storage: A techno-economic analysis. *Int. J. Sustain. Energy* **2020**, *39*, 190–206. [CrossRef]
48. Brown, H.S.; Vergragt, P.J. Bounded socio-technical experiments as agents of systemic change: The case of a zero-energy residential building. *Technol. Forecast. Soc. Chang.* **2008**, *75*, 107–130. [CrossRef]
49. Yi, H.; Srinivasan, R.S.; Braham, W.W.; Tilley, D.R. An ecological understanding of net-zero energy building: Evaluation of sustainability based on emergy theory. *J. Clean. Prod.* **2017**, *143*, 654–671. [CrossRef]
50. Szalay, Z.; Zöld, A. Definition of nearly zero-energy building requirements based on a large building sample. *Energy Policy* **2014**, *74*, 510–521. [CrossRef]
51. Srinivasan, R.; Braham, W.W.; Campbell, D.E.; Curcija, C.D. Re (De) fining Net Zero Energy: Renewable Energy balance in environmental building design. *Build. Environ.* **2012**, *47*, 300–315. [CrossRef]
52. Robert, A.; Kummert, M. Designing net-zero energy buildings for the future climate, not for the past. *Build. Environ.* **2012**, *55*, 150–158. [CrossRef]
53. Zhou, Z.; Feng, L.; Zhang, S.; Wang, C.; Chen, G.; Du, T.; Li, Y.; Zuo, J. The operational performance of "net zero energy building": A study in China. *Appl. Energy* **2016**, *177*, 716–728. [CrossRef]
54. Pikas, E.; Thalfeldt, M.; Kurnitski, J. Cost optimal and nearly zero energy building solutions for office buildings. *Energy Build.* **2014**, *74*, 30–42. [CrossRef]
55. Nematchoua, M.K.; Tchinda, R.; Orosa, J.A. Adaptation and comparative study of thermal comfort in naturally ventilated classrooms and buildings in the wet tropical zones. *Energy Build* **2014**, *85*, 321–328. [CrossRef]
56. Strzeszewski, M.; Wereszczyński, P. Norma PN-EN 12831-3:2017-08E Nowa Metoda Obliczania Projektowego Obciążenia Ceiplnego. Poradnik. Warszawa. 2009. Available online: https://www.purmo.com/docs/Poradnik-Purmo-nowa-metoda-obliczania_12831_01_2012.pdf (accessed on 8 December 2021).
57. Ministerstwo Klimatu i Środowiska, Program Czyste Powietrze, Warsaw, Poland. 2022. Available online: https://czystepowietrze.gov.pl (accessed on 10 January 2022).

energies

Article

Research on Optimization of the Thermal Performance of Composite Rammed Earth Construction

Shenwei Yu [1], Shimeng Hao [1,2], Jun Mu [1,*], Dongwei Tian [1] and Mosha Zhao [3]

1 School of Architecture and Urban Planning, Beijing University of Civil Engineering and Architecture, Beijing 100044, China; 1108130420001@stu.bucea.edu.cn (S.Y.); haoshimeng@bucea.edu.cn (S.H.); 1108130420002@stu.bucea.edu.cn (D.T.)
2 State Key Laboratory of Subtropical Building Science, South China University of Technology, Guangzhou 510006, China
3 Institute for Acoustics and Building Physics, University of Stuttgart, Pfaffenwaldring 7, 70569 Stuttgart, Germany; zhaoms_53@hotmail.com
* Correspondence: mujun@bucea.edu.cn

Abstract: Rammed earth (RE) is a low-tech recyclable building material with good heat storage and moisture absorption performance that can better maintain the stability of the indoor thermal environment and improve indoor comfort. With innovations in and the development of new technology, the field of rammed earth construction technology is gradually expanding. However, deficiencies in the thermal insulation of traditional rammed earth structures make it impossible for them to meet China's building energy codes in cold regions. This study constructs a comprehensive evaluation index of the thermal performance of rammed earth walls that is based on the heat transfer mechanism, optimizing the thickness of the boundary conditions of the building interior's design temperature, as well as the energy demand and economic efficiency. This research also offers a new design for the thermal insulation of rammed earth construction by combining the building energy savings design code with WUFI Pro software. This study demonstrates that the optimum thickness of rammed earth construction in Beijing is about 360 mm, the thickness of extruded polystyrene board (XPS) is 50 mm (for public buildings) and 70 mm (for residential buildings), and the structural form of external insulation offers the highest performance benefit. In addition, this work also evaluates the risk of condensation inside composite rammed earth construction, finding that there is a risk of condensation on the exterior side of the wall and at the interface between the insulation panels and rammed earth wall, thus requiring an additional moisture-proof layer. In this study, thermal mass and insulation are fully considered and a design strategy for rammed earth construction given quantitatively, providing a theoretical basis for the application of rammed earth materials in cold regions.

Keywords: rammed earth construction; thermal insulation; thermal mass; comprehensive evaluation index; optimization

Citation: Yu, S.; Hao, S.; Mu, J.; Tian, D.; Zhao, M. Research on Optimization of the Thermal Performance of Composite Rammed Earth Construction. *Energies* **2022**, *15*, 1519. https://doi.org/10.3390/en15041519

Academic Editor: Fabrizio Ascione

Received: 10 January 2022
Accepted: 16 February 2022
Published: 18 February 2022

1. Introduction

With the development of modern rammed earth (RE) technology, earthen buildings have returned to being a topic of active interest. RE is perceived as sustainable due to its recyclability and low embodied energy, which is 15% to 25% of the cost of the same volume of burnt clay brickwork [1]. Moreover, the production of RE materials can effectively reduce the amount of energy consumed for manufacturing and transportation, which accounts for about 35% of the embedded energy for building materials [2,3]. RE buildings are warm in winter and cool in summer, maintaining the thermal stability of the building's internal environment and reducing energy consumption for cooling and heating [4]. This is because RE constructions are typically 300 mm to 600 mm thick [5] and can provide a large thermal mass compared to ordinary concrete blocks [4], leading to a time lag between the indoor temperature wave peak on hot summer days and reducing the impact of periodic outdoor temperature waves on the indoor environment [6,7]. Furthermore, with their high thermal capacity and density, RE constructions absorb and retain heat from the environment during the day and release that heat back into the internal space during cool nights [8].

Although many benefits could be gained through the use of RE construction, the process has its shortcomings. Surveys conducted by Paul and Taylor found that RE construction alone cannot guarantee thermal comfort or energy conservation in cold areas because RE buildings do not necessarily provide better thermal performance than do conventional buildings unless the external walls are sufficiently thick and properly insulated [9–12]. In previous research, Adams and Jones [13] tested the thermal conductivity of stabilized RE bricks and found that thermal conductivity increased exponentially with dry density. Goodhew and Griffiths [14] made measurements of different types of RE constructions using an innovative thermal probe technique. Soebarto [4] conducted temperature tests on uninsulated raw earth dwellings and found that the indoor ambient temperature was only 8–12.5 °C during the cold season, making indoor thermal comfort undesirable. Some researchers used slag, water glass, and plant fibers to optimize the thermal performance of RE materials [15–21], but the range of thermal conductivity of RE materials was 0.3634 to 0.7838 W/(m K) [18,22,23], which does not meet the regulatory requirements in China for cold areas. Though some other scholars have improved the thermal performance of RE buildings by combining them with passive designs such as glazing, shading, and ventilation, yielding positive effects on building energy consumption [24,25], the poor thermal insulation of RE materials is a problem yet to be solved. Researchers such as Taylor, Hall, and others have studied the heat transfer of insulated RE constructions [13], effectively solving the problem of poor thermal performance. However, their research conclusions were not given in terms of specific design strategies for composite RE walls in particular climates.

The advantages in terms of heat storage offered by RE construction are well studied, but the disadvantages of insulation have to date been relatively ignored. The current state of research on insulation can be broadly divided into two groups. On the one hand, some scholars have used additives to improve the thermal performance of RE, though it is still unable to meet the requirements of national codes. On the other hand, researchers have adopted passive technology and composite insulation earth walls, but there is as yet no quantitative design strategy. The review studies show that few scholars have carried out thermal performance optimization studies on RE constructions based on the relationship between thermal mass and thermal insulation. Therefore, in this research, a design strategy for composite RE construction was studied and optimized. To find a balance between thermal mass and insulation in RE construction, this study proposes a comprehensive evaluation method. The thickness of the construction layers is also described herein, in combination with the code and simulation tools. Finally, the newly proposed design strategy was assessed for risk and specific strategies applicable to Beijing; both are described below.

This research optimized the structural layers of composite RE walls and developed quantitative design strategies for the thicknesses of different layers most suitable to the Beijing area. The authors hope that the research findings will provide a basis and reference for the future design application of RE construction in cold climate zones.

2. Research Methodology

To seek a balance between thermal performance and energy efficiency in RE construction, in the present research, the particle size distribution, density, specific heat capacity, thermal conductivity, and thermal storage coefficients of the soil were evaluated. In conjunction with WUFI Pro and WUFI Plus, the connection between the thermal performance indexes of new raw soil materials was also analyzed, a comprehensive multi-index evaluation method was established, and appropriate thickness and construction forms for new RE walls in Beijing were derived (see Figure 1) [26].

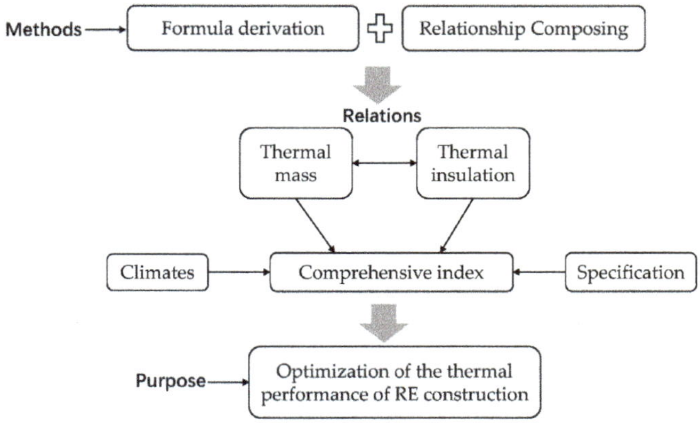

Figure 1. Technical route of the paper.

2.1. Research Region

In this work, Beijing was selected as a typical city in the cold region of China. It is the nation's capital and is located in the northern part of the North China Plain, backed by Yanshan Mountain and adjacent to Tianjin city and Hebei Province. According to an analysis of certain weather tools (see Figure 2), the summer in Beijing is hot and the winter cold and dry. The annual average temperature is about 12.9 °C, requiring indoor heating for up to four months. Thus, in this region, building insulation is an important component of building energy savings.

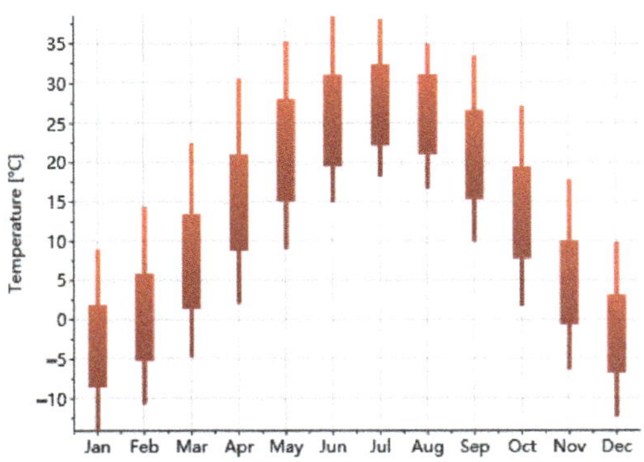

Figure 2. Annual temperatures in Beijing.

2.2. Research Materials

Earthen materials taken from Erlitou, Luoyang, and Henan were screened according to the engineering classifications outlined in the "Test Methods of Soils for Highway Engineering" JTG 3430—2020 [27]. Particle size was determined as shown in Table 1. The experimental results showed that 90% of the earthen particles were less than 8 mm in diameter.

Table 1. Granulometric Analysis: Sieving.

Diameter (mm)	1	0.4	0.2	0.08	<0.08	Waste	Total
Dry Weight with tare dwt (g)	1.44	1.44	5.44	33.93	390.46	1.35	436.02
Percentage (%)	0.33	0.33	1.25	7.78	90	0.31	100

2.3. Parameter Testing

The ratio of the test blocks of soil, sand, and gravel was 5:3:2, and the size was 300 mm × 300 mm × 30 mm. A model DRCD-3030 flat thermal conductivity tester was used to test the thermal conductivity of the blocks according to the method specified in the "Thermal insulation—Determination of steady-state thermal resistance and related properties—heat flow meter apparatus" GB/T10295-2008 [28]. The physical parameters of the text blocks were tested sequentially. The results are outlined below (see Table 2).

Table 2. Physical Parameters of the Test Blocks.

Name	Dry Density ρ (kg/m^3)	Specific Heat Capacity c KJ/(kg·K)	Thermal Conductivity λ (W/mK)	Thermal Storage Conductivity S (W/(m^2·K))
RE	2034	1.28	0.74	11.85
Aerated Concrete	700	1.05	0.18	3.10
Concrete	2500	0.92	1.74	17.20

2.4. Building Prototype

The scale of RE buildings in China is small, and the building shape coefficient generally does not exceed 0.3. This research used a demonstration project in Datong County, Qinghai Province, as a research prototype. An energy consumption model was established, and the impacts of different structural forms of composite RE walls on building energy demand were analyzed.

As seen in Figure 3, the length, width, and height of the building were 12.9 m, 7.7 m, and 6.3 m, respectively. The height of the ground floor was 3.6 m and the first floor was 3.3 m. The exterior walls of the building were made of RE, and the interior walls were brick. In addition, structural columns were used to increase the seismic performance of the building.

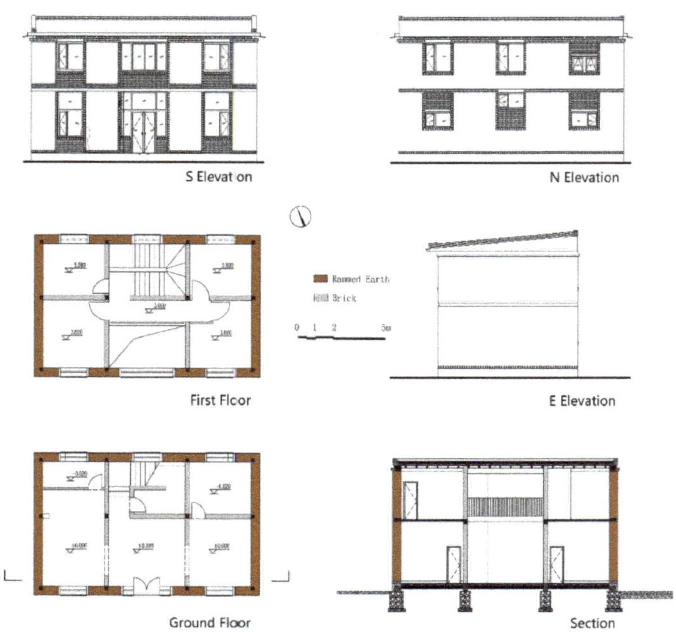

Figure 3. Metric of the building prototype.

2.5. Settings of WUFI Plus/Pro Models

The specifications of the base case model are seen in Table 3 [29].

Table 3. The specifications of the base case model.

Residential Houses				
Window-to-wall ratio (%)	South: 28.0	North: 10.8	East: 0.0	West: 0.0
People density (people/m^2)	0.025			
Wall U-value (W/m^2·K)	Value varies with thickness			
Window U-value (W/m^2·K)	2.0			
Floor U-value (W/m^2·K)	0.2			
Roof U-value (W/m^2·K)	0.25			
Heating setpoint (°C)	18			
Cooling setpoint (°C)	26			
HVAC schedule	Mon.–Fri.: 17:00–9:00; weekend: 0:00–24:00			
Ventilation (1/h)	0.8			

3. Numerical Analysis

As a porous material, the internal heat transfer of RE construction comes in three types: heat conduction, convection heat transfer, and radiation heat transfer. Among these, radiation heat transfer is only obvious when the temperature difference between solids is substantial and the pores are occupied by gas. In the present research, the earthen material was regarded as isotropic porous, and the effect of moisture migration on heat transfer was neglected. The formulation is only presented as a factor correlation analysis.

3.1. Heat Transfer

According to the Fourier formula, the heat flow q_{cond} for heat conduction and transfer within an RE construction can be written as:

$$q_{cond} = -\lambda(W,T)\Delta T \tag{1}$$

where λ is the thermal conductivity of raw earth materials, (W/mK); W is the relative humidity; and T is the temperature, °C. From Equation (1), it can be seen that the thermal conductivity of the material was related to the moisture content and temperature. Since the effect of temperature on the thermal conductivity of the material was very small, it was ignored here.

The convective heat flow q_{conv} within the RE construction includes both sensible and latent heat flows, expressed as follows:

$$q_{conv} = v\rho_{air}c_{air}T + J_V(L_V + c_V T) \tag{2}$$

where v is the airflow rate inside the wall, m/s; ρ_{air} is the density of dry air, kg/m^3; c_{air} is the specific heat capacity of dry air, KJ/(kg·K); JV is the water vapor flow rate, m^3/s; L_v is the evaporation/condensation enthalpy, kJ/kg; and c_v is the specific heat capacity of water vapor, KJ/(kg·K). As the airflow rate inside the earthen wall was very low, the water vapor sensible heat was neglected here, and Equation (2) can be simplified as:

$$q_{conv} = J_V L_V = -\delta_V \nabla P_V L_V \tag{3}$$

where δ_V is the water vapor penetration coefficient, kg/(m·s·Pa); and ∇P_V is the water vapor pressure, Pa. Combined with Equation (1), the heat transfer equation of the RE walls can then be described as in Equation (4):

$$\text{q} = q_{cond} + q_{conv} = -\delta_V L_V P_s \nabla\varphi - \left(\lambda + \delta_V L_V \varphi \frac{dP_s}{dT}\right)\nabla T \tag{4}$$

Processing Equation (4) led to the following results:

$$q = -\delta_V L_V (P_s \nabla \varphi + \varphi \frac{dP_s}{dT} \nabla T) - \lambda \nabla T \quad (5)$$

From Equation (5), it was determined that the heat transfer of RE walls could be divided into two main parts, one caused by the parameters related to thermal insulation (λ) and the other related to thermal mass (L_V). This led to the conclusion that it would be unscientific not to consider thermal mass when measuring the thermal performance of the walls.

3.2. *Subfactors*

As is commonly understood, thermal conductivity and the heat storage coefficient are used to evaluate the thermal insulation and mass of walls, where thermal conductivity affects the steady-state heat transfer of the wall as affected by the difference between indoor and outdoor temperatures, and the larger the thermal conductivity, the more unfavorable it is for the RE walls. The thermal storage coefficient positively affects the additional heat transfer caused by fluctuations of the outdoor environment. For the same material, the thermal conductivity thermal storage coefficients are constant. Thus, in the present research, it was necessary to convert the above two evaluation indicators into more detailed parameters to analyze how thermal insulation and mass affect building performance.

3.2.1. Thermal Insulation

This section reviews the related parameters of thermal insulation and mass in terms of their influence on the thermal performance of buildings. The relationship between the two was constructed in the form of the flow chart shown in Figure 4. The thermal insulation performance of walls mainly depends on the resistance of the material; the greater the thermal resistance, the better the insulation performance. Wall insulation performance is also affected by environmental humidity, wind speed, and solar radiation, so the measurement parameters of wall insulation performance require the heat transfer coefficient, which takes into account both the physical properties of the material and impact of the climate on the thermal performance of building walls.

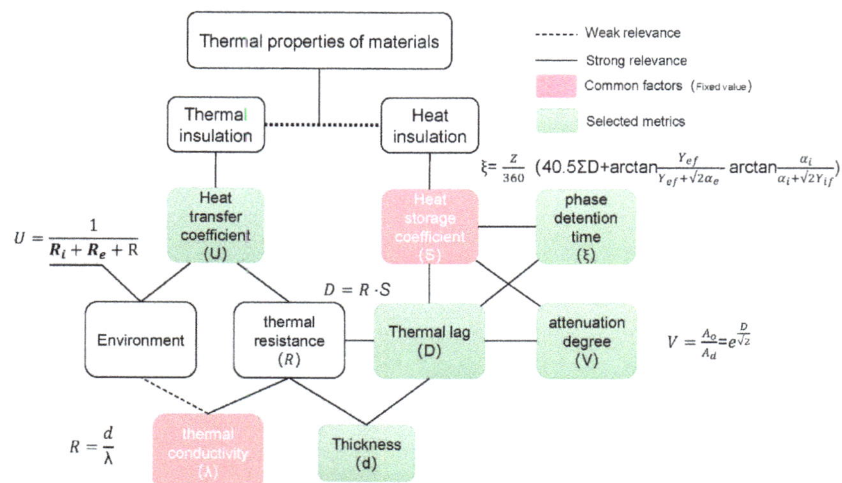

Figure 4. Relevance of the material's thermal parameters.

3.2.2. Thermal Mass

The factor most affecting thermal mass performance is the thermal storage coefficient. For the same wall materials, differences in thermal storage performance can be distinguished by the thermal inertia index, attenuation degree, and phase detention time. The greater the thermal inertia index and attenuation of the outdoor temperature wave, the longer the phase detention time and the better the thermal mass performance.

Therefore, using the wall thickness as the only variable, a comprehensive evaluation index was established to analyze the thermal performance of RE walls. Four parameters were employed: heat transfer coefficient, thermal inertia index, attenuation degree, and phase detention time.

3.3. Nondimensionalized Subfactors

To ensure the validity of the comprehensive evaluation index, it was necessary to reverse the value of a certain type of index through steered inversion, unifying the forms of all subfactors. In this research, we adopted the "quotient" transformation method to deal with the inverse indexes (U). After the transformation, all subfactors were isotropic, which was conducive to analyzing the composite index values. In addition, this section of the research adopted the mean value method for dimensionless evaluation indexes of RE walls to eliminate the adverse effects of the differences in physical meaning and scale units on the accuracy of evaluation results and to ensure comparability between subfactors.

4. Calculation and Analysis

As the thickness of the wall increases, the thermal insulation and mass increase. However, an increase in wall thickness seriously reduces the usable space and seismic performance of a building, especially since RE walls are heavy-weight structures. In real-world projects, it would be impractical to blindly increase the thickness of the exterior wall to improve the level of indoor thermal comfort. In the present research, WUFI Pro was used to dynamically simulate the heat and moisture transfer process of RE walls. Then, the variation tendency of the parameters (i.e., heat transfer coefficient, thermal inertia index, attenuation degree, and phase detention time) were fitted to the wall thickness to analyze the optimum effective thickness of the RE walls.

4.1. Thickness and Heat Transfer Coefficient

By analyzing the influential mechanism of wall heat transfer (as shown in Figure 5), it can be seen that heat transfer in RE walls can be described according to two aspects. From the macroscopic perspective, heat conduction and transfer occurring inside the wall are influenced by the temperature and humidity of the environment within which the internal and external surfaces of the wall are located. From the microscopic perspective, there is convective heat transfer occurring inside the RE wall under the action of air and vapor pressure. These two wall heat transfer processes are also influenced by solar radiation, as well as radiative heat transfer between objects. Therefore, measuring the thermal insulation performance of walls only from the perspective of thermal resistance would not be rigorous enough to accurately quantify the heat transfer mechanism of real-world RE walls. WUFI Pro software (Department Hygrothermics at the Fraunhofer IBP, Holzkirchen, Germany) was used to simulate and calculate the heat transfer coefficients of different thicknesses of RE walls, based on the climatic parameters of the Beijing area.

The heat transfer coefficient was calculated as outlined above. The results are shown in Table 4, which numerically illustrates the inverse function with thermal resistance.

The internal surface heat transfer resistance was taken to be Ri = 0.11 $(m^2{\cdot}K)/W$, and the external surface heat transfer resistance (Re) was about 0.04 $(m^2{\cdot}K)/W$ [30].

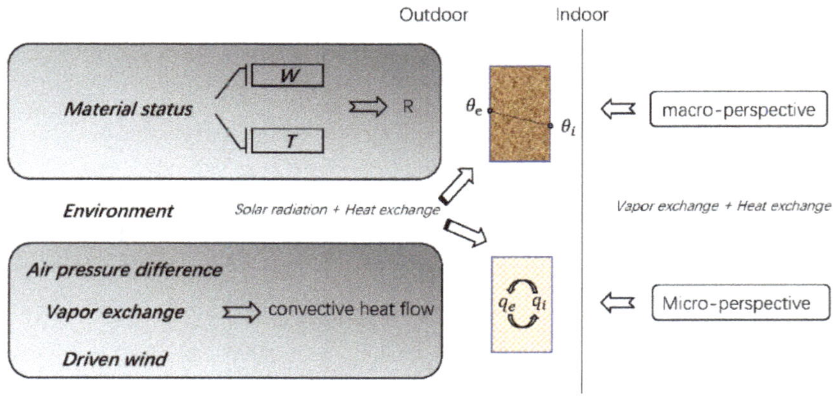

Figure 5. Analysis of heat transfer forms.

Table 4. Relationship between Thickness and the Heat Transfer Coefficient.

d mm	100	200	300	400	500	600	700	800	900	1000
U W/(m^2·K)	3.507	2.379	1.8	1.448	1.211	1.041	0.912	0.812	0.732	0.666
1/U	0.285	0.420	0.556	0.691	0.826	0.961	1.096	1.232	1.366	1.502

With an increase in thickness, the heat transfer coefficient decreases and the insulation performance of the wall improves. Based on these data, U = 112.9 × d^(−0.735) was obtained by fitting the heat transfer coefficient to the wall thickness. The influence factor of the heat transfer coefficient in the comprehensive evaluation index was about 0.7354. Moreover, to ensure the isotropy of the index, the heat transfer coefficient was converted using the quotient. The rotation parameters are shown in Table 4.

4.2. Thickness and Heat Storage Coefficient

As illustrated in Equation (6), the magnitude of the thermal storage coefficient depends on the specific heat capacity, density, and thermal conductivity of the material. For the same material, the effect of wall thickness on the thermal storage performance cannot be accurately interpreted through the study of S alone. Thus, in this research, indicators related to the thermal storage performance of walls were derived to assist with measurement.

$$S = \sqrt{c\rho\lambda} \tag{6}$$

As shown in Figure 6 ("-" here are minus signs), the indicator most directly related to the thermal storage coefficient is thermal inertia, which becomes larger with increases in wall thickness. Indirectly related to the thermal storage coefficient are the attenuation degree and delay time. Frequently, the larger the thermal inertia of the wall, the more remarkable the performance of the two indirect indicators. In addition, thermal inertia, temperature wave attenuation degree, and phase detention time can each be used to establish a direct relationship with wall thickness, and all showed a positive correlation.

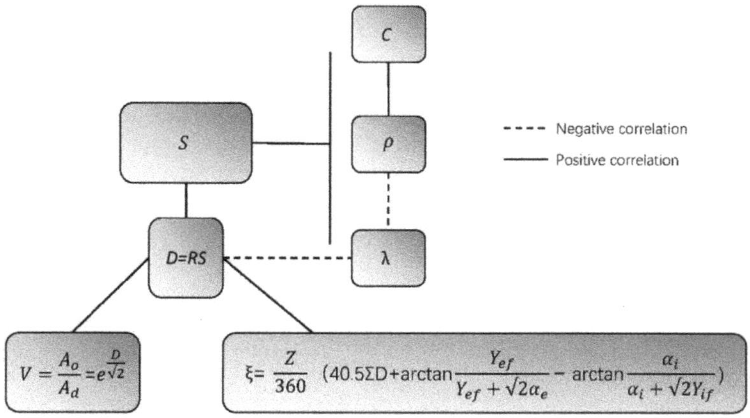

Figure 6. Relationships among the parameters of thermal storage performance.

4.2.1. Thickness and Thermal Inertia

The thermal inertia index (D) is a dimensionless index that reflects the decay rate of the fluctuating heat transfer process of the external wall. The larger the D value, the faster the surface temperature wave attenuates in the wall and the better the thermal stability of the wall. The thermal inertia index of an RE wall is numerically equal to the product of the material's heat storage coefficient and thermal resistance.

A thermal inertia index of earthen walls of different thicknesses could thus be calculated. The fitting function is shown in Figure 7. From the correlation of the functions, it can be seen that the thermal inertia index of earthen walls was directly proportional to wall thickness, and the influence factor of thermal inertia in the comprehensive evaluation index was 0.016.

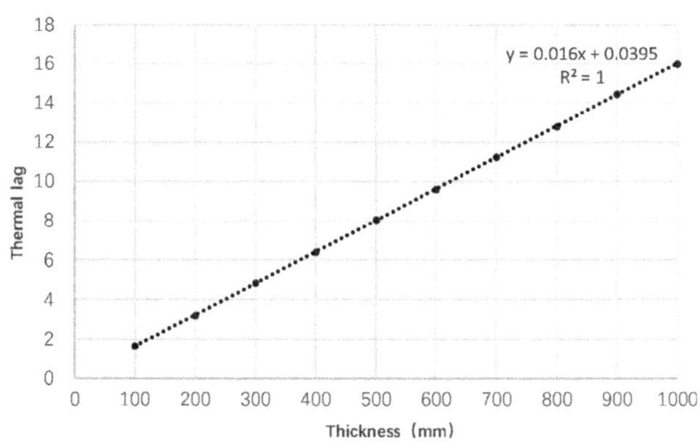

Figure 7. Thickness and thermal lag.

4.2.2. Thickness and Attenuation Degree

WUFI Pro was used to simulate the heat and moisture transfer process of the walls under different thicknesses. The external climate conditions set by the software were obtained from the weather tool (Beijing-hour.epw, Meteotest AG, Bern, Switzerland), Indoor ambient temperature was set at 18–26 °C, and relative humidity was 50%. The annual variation range of the internal and external surface temperatures of the walls were obtained by simulation (see Table 5, "-" in this table are minus signs). Here, the ratio of the harmonic amplitudes of the internal and external surface temperatures was used to measure the attenuation effect of the earthen walls of different thicknesses on the temperature wave.

The value was defined as the average annual attenuation of the temperature wave V_0 (see Equation (7)).

$$V_0 = A_0/A_d \tag{7}$$

where A_0 is the annual temperature amplitude of the external surface of the wall, A_0 = 41.8 °C, and A_d represents the annual temperature amplitude of the internal surface of the wall.

Table 5. Thickness and Attenuation Degree.

D (mm)	Section Temperature Fluctuation	A_0	A_d	V_0
100	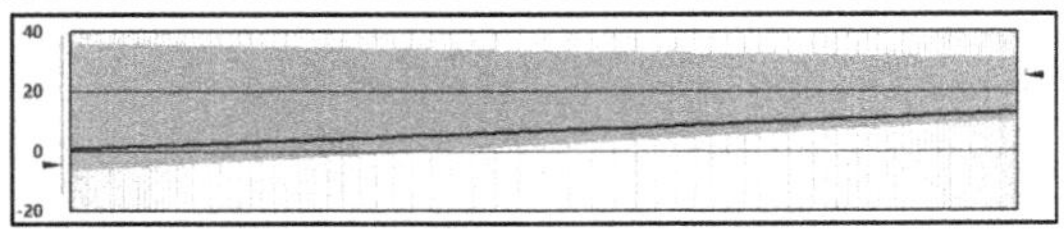		19.7	2.12
200	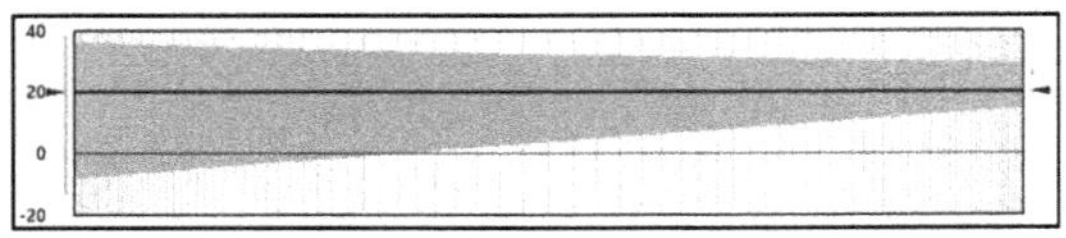		12.8	3.27
300	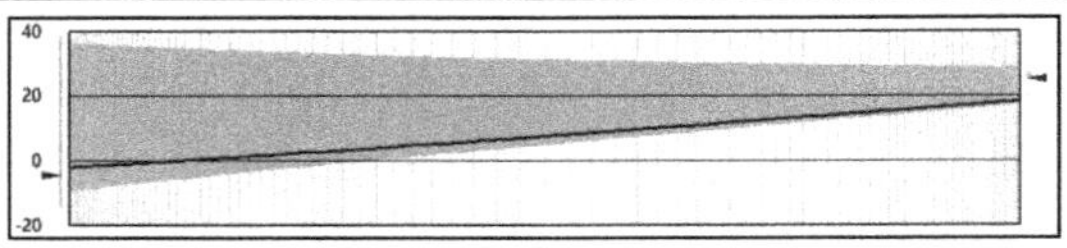		9.9	4.22
400	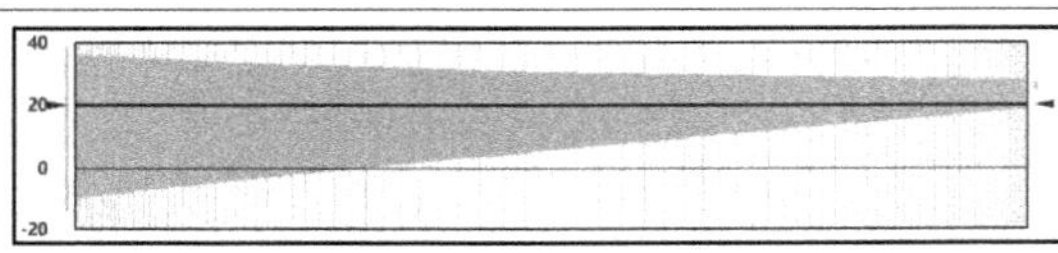		8.7	4.80
500	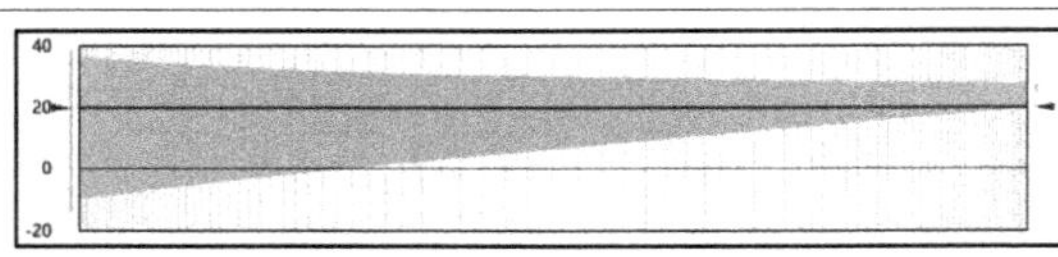	41.8	7.5	5.57
600	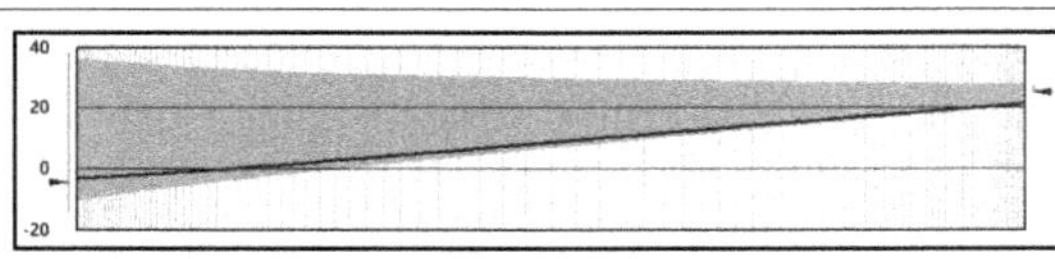		7.0	5.97
700	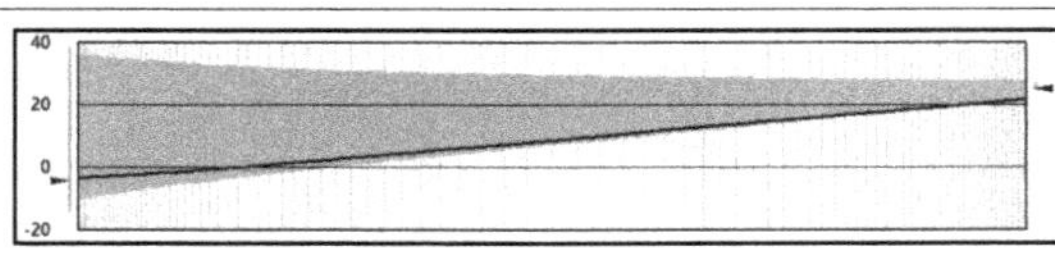		5.8	7.20
800	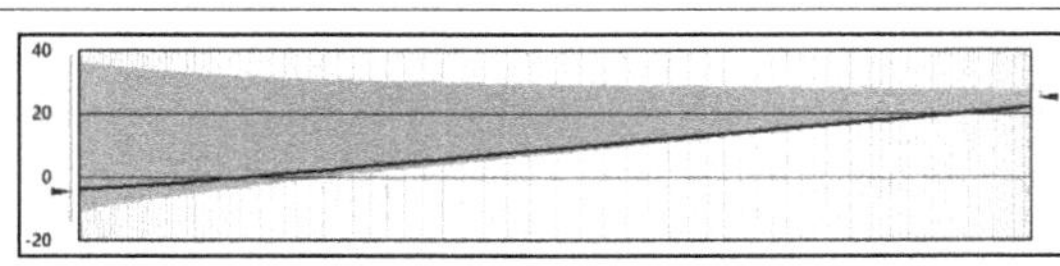		5.2	8.04
900	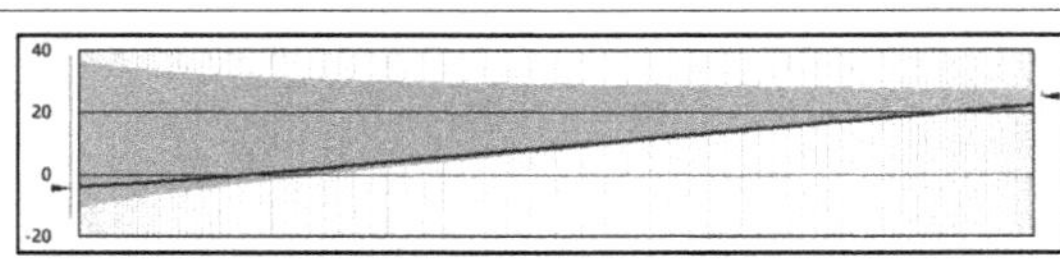		4.8	8.71

Table 5. *Cont.*

D (mm)	Section Temperature Fluctuation	A_0	A_d	V_0
1000	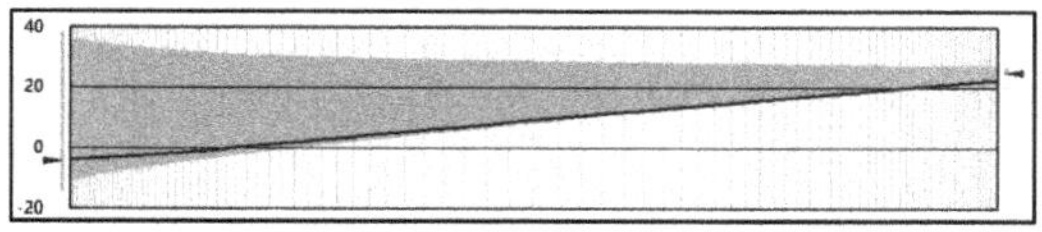		4.6	9.09

The values were fitted, and the results are shown in Figure 8. The correlation coefficient (R^2) between the attenuation and wall thickness was about 0.9932. It can be concluded that as the wall thickness increased, the attenuation degree increased, and the indoor environment fluctuation and temperature fluctuation range inside the wall became smaller. However, the growth trend of V_0 became slower, meaning that V_0 tended to stabilize regardless of the increase in thickness. The increase in wall thickness gradually reached saturation for the attenuation gain. According to the fitting function, the influence factor of V_0 was about 0.6332.

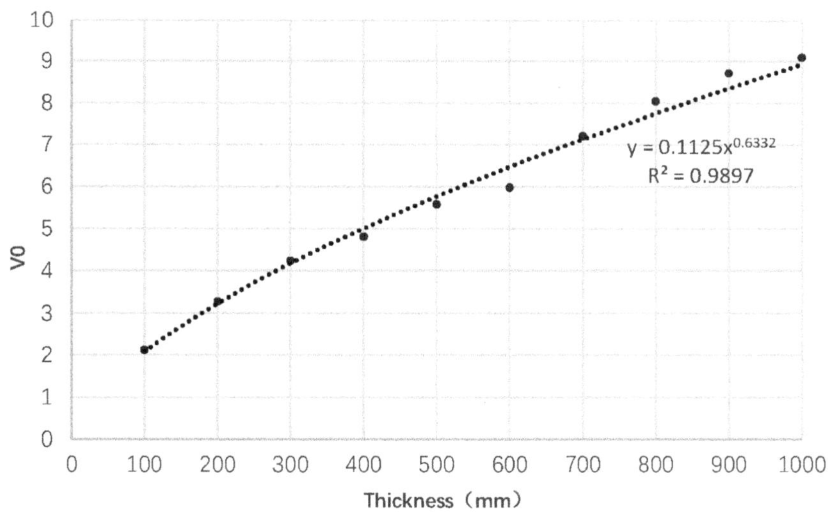

Figure 8. Fitting of the decay degree function.

4.2.3. Thickness and Time Lag

The time lag (ξ) refers to the interval between the appearance of the peak outdoor temperature wave and appearance of the peak temperature of the inner surface of the external wall. The larger the delay time, the smaller the impact of the outdoor temperature on the building's indoor environment. The delay times for the RE walls of different thicknesses were calculated, and a linear fit was performed, as shown in Figure 9 ("-" in here are minus signs).

The calculation period of the delay time was chosen to be 24 h. The image shows that the thicker the wall, the greater the delay time. However, when the delay time was greater than 24 h, it became meaningless to continue enhancing the delay time for building indoor environmental stability. The sensitivity factor of the delay time in the comprehensive evaluation index was calculated as 0.0434.

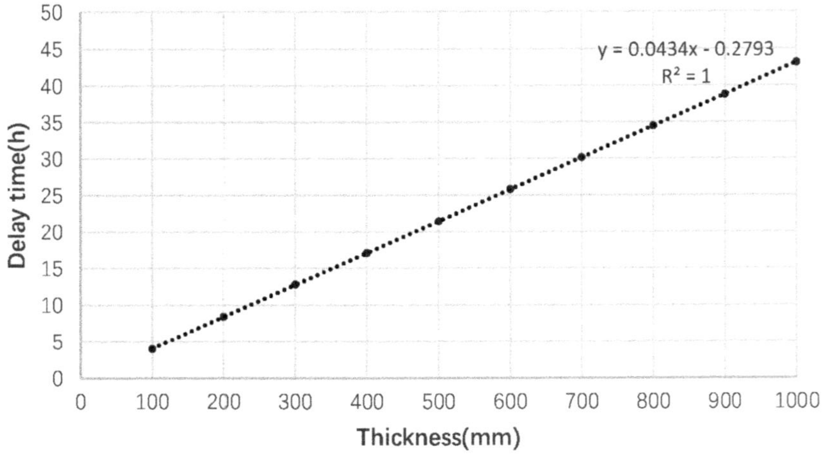

Figure 9. Fitting of the time lag.

4.2.4. Thickness and Comprehensive Evaluation Index

The comprehensive evaluation index was established according to the four-parameter influence factor fitted above, as shown in Equation (8). The comprehensive evaluation index under each wall thickness was then calculated and is shown in Figure 10.

$$M = 0.7354\beta_{1/U} + 0.016\beta_D + 0.6332V_0 + 0.0434\beta_\xi \quad (8)$$

From the data distribution pattern of the comprehensive evaluation index, it can be seen that the comprehensive thermal performance continued to increase with the increase in wall thickness. As the thickness increased, the thermal performance gain of the wall gradually decreased. Further, the calculation of the comprehensive index improvement rate revealed that it tended to be stable when the thickness of the rammed earth wall reached 600 mm. Increasing the wall thickness was not found to be significant for improving thermal performance.

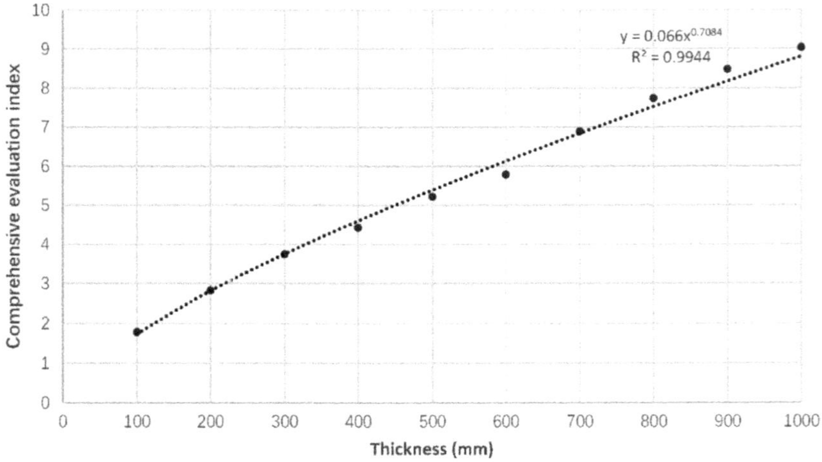

Figure 10. The relationship between thickness and M.

4.3. Defining Boundaries

This section describes the relationship between the wall thickness and thermal performance from the perspective of interior design temperature, building energy demand, and economic efficiency; it was used as a qualification for the wall thickness interval.

4.3.1. Design Temperature

Article 3.0.1 of the Code for Design of Heating, Ventilation, and Air Conditioning of Civil Buildings: GB 50736-2012 stipulates that the main room in cold and severely cold areas (e.g., in Beijing) should be 18–24 °C [31]. Here, the lower limit of the thickness interval of the RE wall was defined as the thickness that could meet the minimum heating temperature requirement in winter considering economic factors.

The effects of different thicknesses of RE walls on the lowest temperatures of the inner surface were simulated. The results are outlined in Table 6. As the wall thickness increased from 100 mm to 600 mm, the minimum temperatures on the inner surface of the RE walls were 11 °C, 15.5 °C, 17.6 °C, 19.7 °C, 20.3 °C, and 21.0 °C, respectively. From the data rules, it can be seen that as the thickness increased, the thermal performance improvement gain gradually decreased. In addition, when 18 °C was used as the thickness filter standard, the thickness of the RE wall needed to be between 300 mm and 400 mm.

Table 6. Minimum Temperatures of Inner Surfaces of Rammed Earth Walls of Different Thicknesses.

d (mm)	100	200	300	400	500	600
Minimum inner surface temperature (°C)	11.0	15.5	17.6	19.7	20.3	21.0

To further study the optimum thickness of RE walls for the Beijing area, the temperature variation range of the inner surfaces of walls of different thicknesses was simulated and the minimum temperature was numerically fitted (see Figure 11, "-" are minus signs). According to the restrictions put forth by local regulations and the fitting formula, the lower limit of the optimized wall thickness in Beijing was calculated to be about 325 mm, which is similar to Chen Mang's experimental tests of heavy clay walls in the cold region of China, with an error of 3.69% [32].

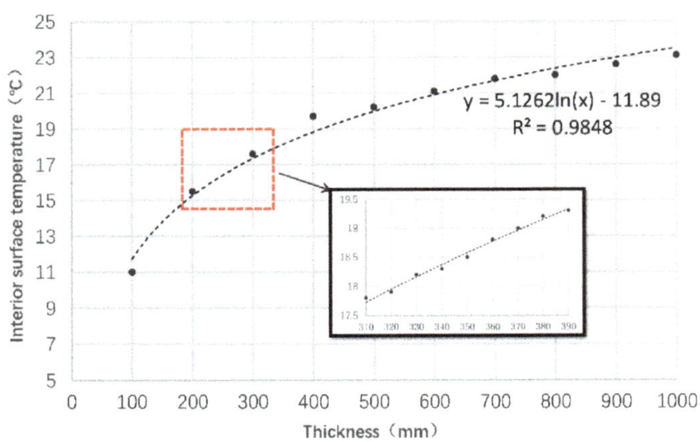

Figure 11. Relationship between the thickness and interior surface temperature.

In summary, the thickness interval of RE walls with a density of 2000 kg/m^3 in the Beijing area was determined to be approximately 325 mm to 600 mm, meeting the requirement of minimum heat transfer resistance for enclosures.

4.3.2. Energy Consumption

The thermal performance of RE walls can be visualized by the energy demand of the building. Therefore, the thickness of the RE wall was used as the only variable for studying the building energy demand under different conditions. Then, the effects of different thicknesses of RE walls (ranging from 100 mm to 1000 mm) on the energy demand of the building were simulated using WUFI Plus software (Department Hygrothermics at the Fraunhofer IBP, Holzkirchen, Germany). The results are shown in Figure 12. The energy consumption data were processed to derive energy savings rates for different thicknesses of walls and were compared with the index performance improvement rates. The following conclusions were drawn.

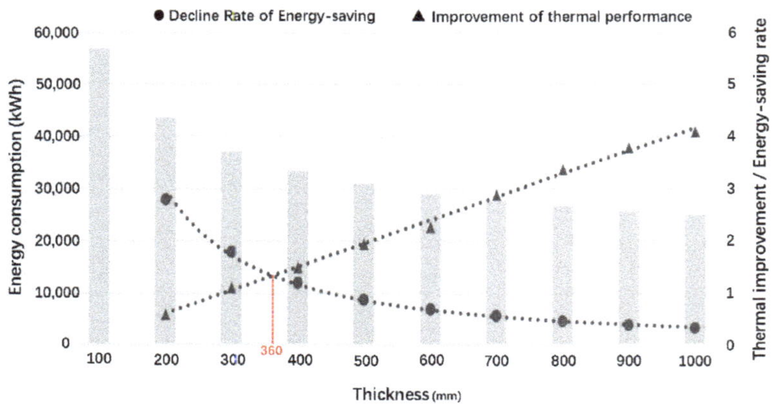

Figure 12. Energy demands of different wall thicknesses.

1. As the comprehensive performance of the RE wall improved, the rate of energy savings and energy consumption returns gradually decreased.
2. The performance improvement rate of the RE wall was defined as the investment and the energy savings rate as the benefit. It can be seen from the value change law that the investment benefit reached a balance when the wall thickness reached 360 mm; this was also when the wall performance benefit was the best.
3. Combined with the above research on the lower limits of wall thickness, the optimization interval of RE wall thickness in Beijing was found to be 325 mm to 360 mm.
4. The building energy consumption and construction economy were analyzed; the results are shown in Table 7. With an increase in wall thickness, the initial investment in construction increased, but the energy consumption demand of the building decreased under the gradient change of thickness. The payback period was about eight years, regardless of the increase in thickness (as compared to 325 mm), so the optimal wall thickness of RE walls in Beijing was determined to be about 360 mm.

Table 7. Economic Analysis of Energy Consumption and Construction.

Thickness (mm)	Energy Demand (kWh/m^2)	Electricity Consumption (¥0.72/ kWh)	Construction Cost (¥800/m^3)	Total Cost (¥/m^2)	Gain Payback (Years)
325	131.3	94.5	260	354.5	-
335	129.8	93.5	268	361.5	8
345	128.4	92.4	276	368.4	8
360	126.4	91.0	288	379.0	8

5. Optimization and Development

As described above, the optimum thickness of an RE wall was about 360 mm, but the heat transfer coefficient did not meet the requirements of local energy conservation regulations in Beijing. Therefore, it was necessary to improve the traditional RE wall by adding an insulation layer.

5.1. Basis for Insulation Improvement

This research used XPS to improve the insulation performance of RE walls. The structural layers of a standard composite RE wall is XPS insulation board and RE. The construction form of composite RE walls is mainly divided into three forms: external, internal, and sandwich insulation. The most economical thickness of insulation panels for energy savings was calculated according to the Beijing area specification standard (see Table 8).

Table 8. Thickness of XPS Insulation Board.

Normative Standard	U [W/m²·K]	Building Shape Coefficient	Thickness
Design Standard for Energy Efficiency of Public Buildings: DB11/687-2015	0.45	≤0.3	50 mm
Design Standard for Energy Efficiency of Residential Buildings: DB11/891-2012	0.35	≤0.3	70 mm

5.2. Energy Demand Simulations

WUFI Plus was used to simulate the building energy demand under different working conditions. The simulation was divided into three categories: external, sandwich, and internal insulation. Commonly used aerated concrete was also simulated for comparison under the same heat transfer coefficient.

Modeling and Simulations

The model was built according to the architectural prototype. Next, the gbXML format file was exported, and the parameters of each part of the building construction were set in the WUFI Plus software (For parameters see Table 3). Then, more than five cases were simulated, with the construction of the composite RE wall being the only variable for energy demand. The results are shown in Table 9.

Table 9. Energy Demands of Different Constructions.

Types	Structural Forms	Construction Details (U Is the Same)	Energy Demand (kWh)		
			Heating	Cooling	All
RE-1	External insulated rammed earth wall	50 mm + 360 mm	14,291.9	5512.0	19,803.9
		70 mm + 360 mm	13,306.4	5332.7	18,639.1
RE-2	Internal insulated rammed earth wall	360 mm + 50 mm	14,407.0	5466.1	19,873.0
		360 mm + 70 mm	13,402.1	5310.2	18,712.3
RE-3	Sandwich insulated rammed earth wall	180 mm + 50 mm + 180 mm	14,340.0	5492.0	19,832.0
		180 mm + 70 mm + 180 mm	13,339.2	5315.1	18,654.3
Aerated concrete	External insulated aerated concrete wall	15 mm + 200 mm	14,982.5	5698.6	20,681.1
		35 mm + 200 mm	13,709.9	5446.8	19,156.7
RE	Rammed earth wall	360 mm	24,247.7	7329.5	31,577.2
Concrete	Concrete wall	850 mm	31,045.7	3739.4	34,785.1
Aerated concrete	Aerated concrete wall	90 mm	29,676.0	4674.1	34,350.1

The following conclusions can be drawn from the simulation results:

1. When the heat transfer coefficient of the external building envelope was the same, RE buildings were found to be more energy efficient than concrete buildings.
2. By adding a certain thickness of XPS insulation board, the rate of energy savings of public buildings reached 37.3%, and that of residential buildings reached about 41.0%. This shows that the thermal performance of composite RE walls was greatly improved and can be adapted to cold climates.
3. When the thickness of the RE wall was 360 mm, the effects of different XPS thicknesses on building energy consumption were calculated, and the results are shown in Figure 13. From the data distribution pattern, it can be concluded that when the form of construction was external insulation, the energy demands of the building were the lowest and those of the internal insulation the highest. Thus, it would be appropriate to adopt the structural form of external insulation for composite RE walls in Beijing.
4. From the perspective of cooling and heating energy demands, it can be seen that external insulation had a greater impact on building heating energy demand, while internal insulation was more sensitive to building cooling energy consumption.

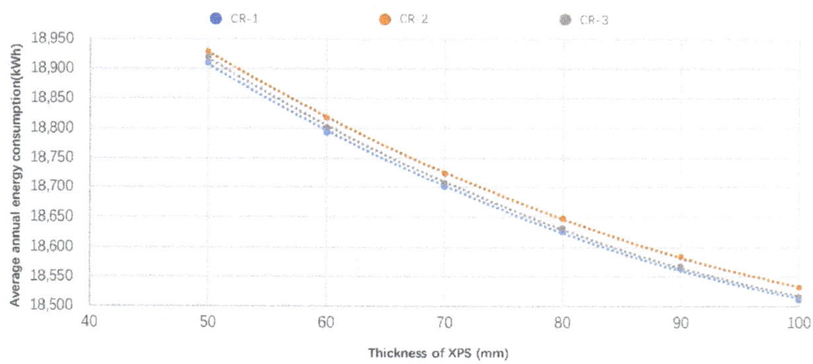

Figure 13. Energy demands of different XPS thicknesses.

5.3. Heat Flow Analysis

The results of the energy demand simulations showed that despite the same thermal design parameters of the wall, the influence of different constructions on the energy demand of the building was significantly different. This section focuses on an analysis of the causes of energy consumption variability. It is well known that the main form of energy exchange between a building and its surroundings is heat transfer through walls, expressed in terms of heat flow (kWh).

WUFI Pro was used to simulate the annual cumulative heat flow inside the three composite RE walls described above. The results are shown in Figure 14 ("-" are minus signs). Building heat loss was defined as a negative value and building heat gain as a positive value. During the period from May to October, the building was in a heat gain state. The rest of the year, the building was in a heat loss state. The annual average building heat loss in Beijing is about 1.85 times that of building heat gain. According to the energy demand simulation results, the following conclusions could be drawn. There was a positive correlation between heat flow size and energy demand. The external insulation wall had a low level of heat flow in winter and a substantial amount of heat flow in summer, while the internal insulation had a sizeable heat flow in winter and less heat flow in summer. This matched the simulation results for cooling and heating loads listed above. Thus, it was concluded that the structural form of internal insulation would be more suitable for areas where cooling is the dominant energy consumption and external insulation is more suitable for cold areas.

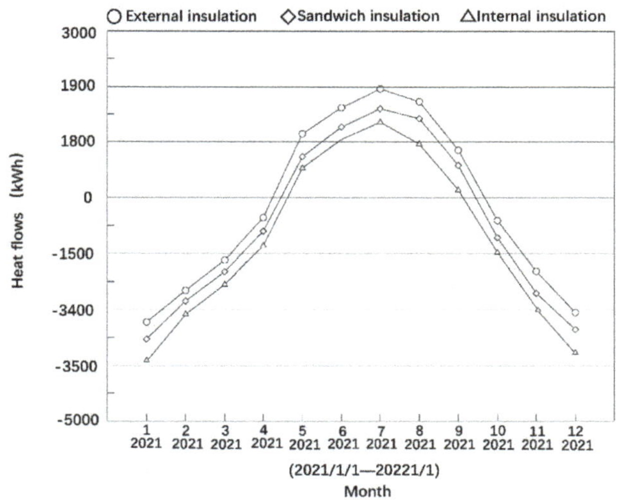

Figure 14. Average monthly heat flow for walls of different construction forms.

5.4. Condensation Risk Assessments

Through the above analysis, the best composite RE wall construction for the Beijing area was determined to be external insulation. For houses in areas with central heat, preventing condensation inside the walls is a priority. In this section, the condensation temperature of the longitudinal section of external insulated composite RE walls was analyzed in combination with WUFI Pro software. The model construction layers are described in Figure 15 ("-" here are hyphen). Monitoring points such as 1, 2, 3, and 4 were set to determine whether there was a risk of condensation inside the structure.

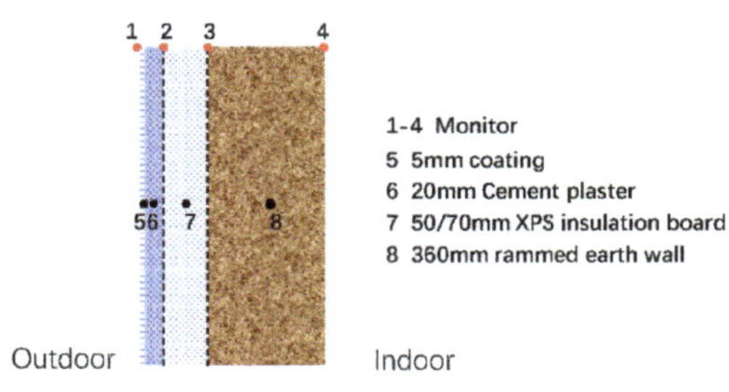

Figure 15. Schematic of the wall construction.

The dynamic moisture heat transfer process inside the wall was simulated, and the dewpoint and temperature were simulated for each monitoring point. When the temperature of the monitoring point is close to the condensation temperature, condensation is highly prone to occur and erode the building envelope, forming a safety hazard. In addition, wall condensation can also seriously affect the environmental quality of a building interior and thus adversely impact health.

5.4.1. Monitoring Point 1

Monitoring point 1 was placed on the outside of the wall; the simulation results are shown in Figure 16 ("-" are minus signs). It can be seen from the figure that the condensation temperature was close to the outdoor ambient temperature throughout the year, and the condensation phenomenon very easily occurred on the outside surface of the composite RE wall. This situation was more serious in summer. Therefore, the exterior wall surface needs some waterproofing treatment.

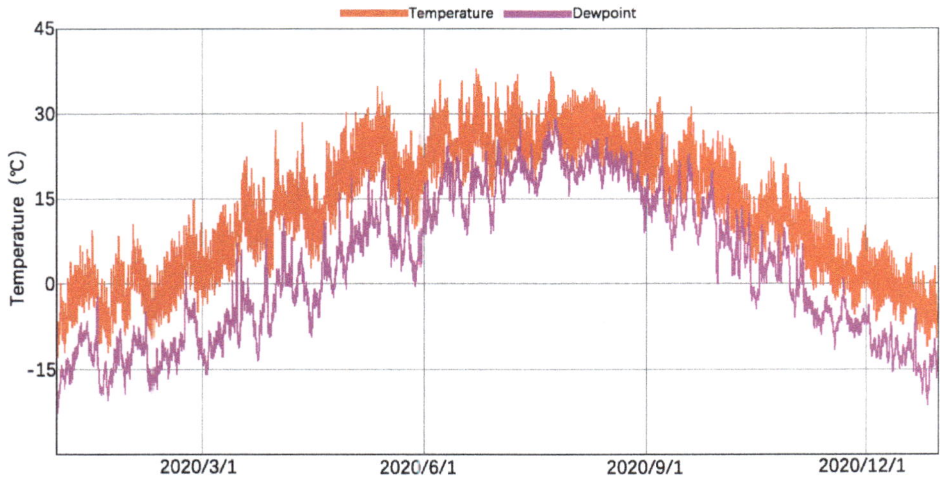

Figure 16. Dewpoint and temperature of monitoring point 1.

5.4.2. Monitoring Point 2

Monitoring point 2 was placed between the cement mortar and the XPS insulation board. As shown in Figure 17 ("-" are minus signs), there is a risk of condensation at this point, so a water-proof layer must be placed there.

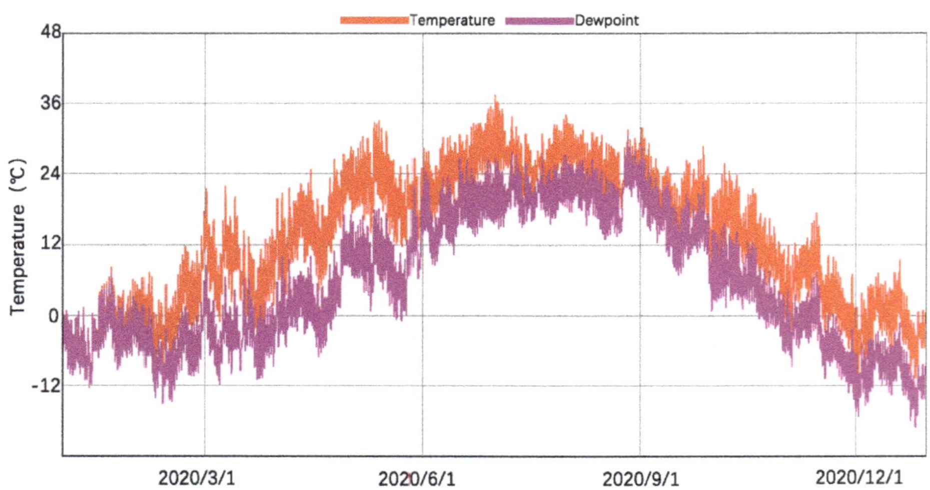

Figure 17. Dewpoint and temperature of monitoring point 2.

5.4.3. Monitoring Point 3

Monitoring point 3 was placed at the boundary between the insulation board and RE wall. The simulation results are shown in Figure 18 ("-" are minus signs). From the data distribution pattern of the intersection of winter (when there was no risk of condensation) and summer (when the ambient humidity was high), it can be seen that there was a certain risk of condensation (temperature and dewpoint differences of 3–5 °C), but given the high moisture adjustment capacity of the RE materials, no water-proofing measures are required at monitoring point 3.

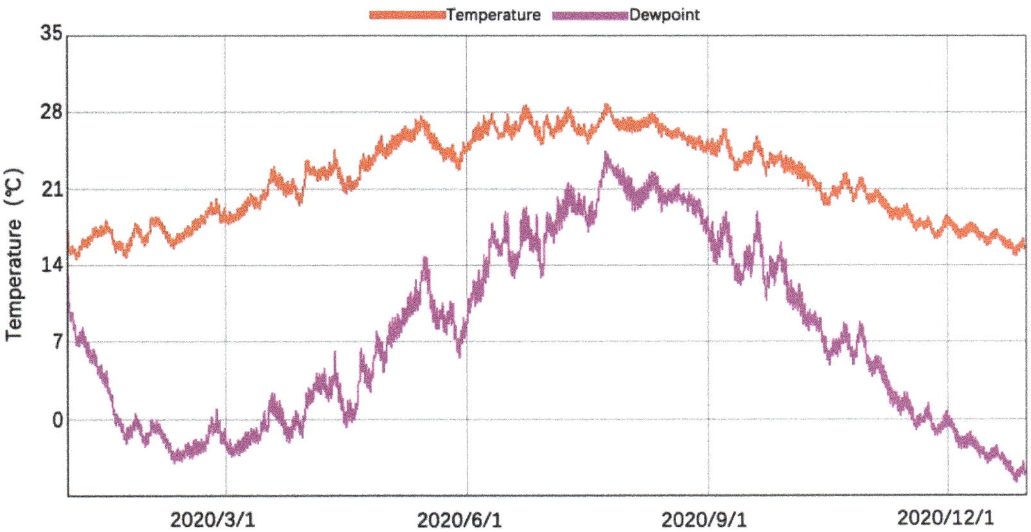

Figure 18. Dewpoint and temperature of monitoring point 3.

5.4.4. Monitoring Point 4

Monitoring point 4 was placed on the inside of the wall; the simulation results are shown in Figure 19. The statistics show that there was no risk of condensation on the inner surface of the composite RE wall.

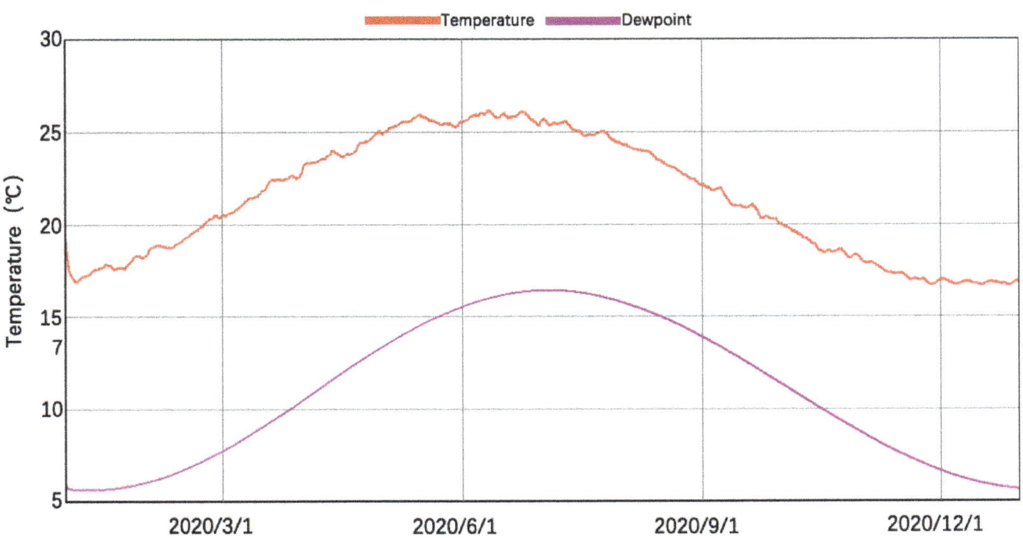

Figure 19. Dewpoint and temperature of monitoring point 4.

5.4.5. Condensation Optimization

According to the results of the above analysis, to prevent condensation in the wall, a waterproof layer would need to be added at monitoring point 2. The risk of condensation was analyzed. The results show that with the damp-proofing treatment, the risk of condensation has been removed at monitoring point 2 (see Figure 20, "-" are minus signs).

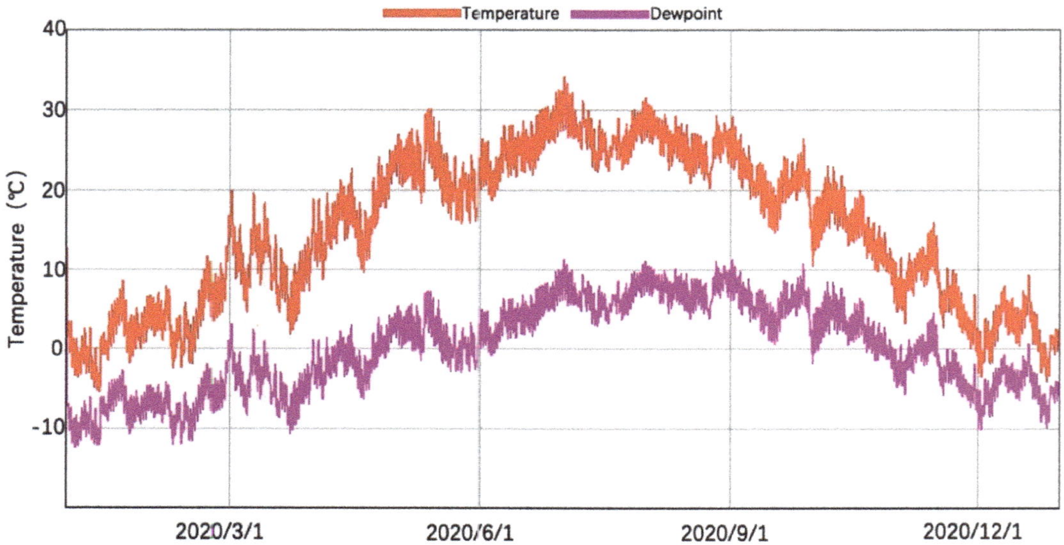

Figure 20. Optimized wall dewpoint and temperature of monitoring point 2.

5.5. *Strategy and Discussion*

Following the above study, the optimized wall construction layers are 5 mm coat-ing, water-proof layer, 20 mm cement plaster, bonding course, XPS insulation layer, bonding course, and RE construction in order (see Figure 21).

Figure 21. Schematic diagram of the composite RE wall construction.

The optimization solutions are also argued. The thickness for only RE walls is generally in the range of 400 mm–600 mm, and here 400 and 600 mm RE walls were selected for comparison to analyze their effects on building energy demands, and the simulation data are shown in Table 10.

Table 10. Comparative analysis of optimization strategies.

Types	Heating (kWh)	Cooling (kWh)	Energy Demand (kWh)	Energy Efficiency
RE-400	25,243.7	7470.3	32,714.0	-
RE-600	21,762.2	6812.1	28,574.3	12.66%
Optimization strategies (50 mm XPS)	14,291.9	5512.0	19,803.9	30.69%

From the data distribution rule, the energy saving rate of the optimized scheme reached 30.69%, and the optimized scheme can save a large amount of energy consumption for cooling and heating loads. Therefore, the optimization method adopted in this paper possesses certain scientificity and feasibility. However, the research in this paper is a preliminary exploration of the applicability of RE materials in the northern region of China, and the final evaluation of the optimized solution is still needed later in conjunction with actual construction and environmental measurements.

5.6. Applications

The research presented above suggests an optimized solution for the application of RE materials in the Beijing area. It also provides a reference for related applications in other cold regions. The overall application approach is designed with the regional climate in the context of the material's self thermal performance and the green building standards as the boundary conditions in order to find the optimal design strategy for RE materials under specific climate zones and working conditions. As shown in Figure 22 ("-" here are hyphen), the design of the application of RE materials is optimized on two main levels, one of which is the comprehensive thermal performance of the material itself and the other is the national and local design codes.

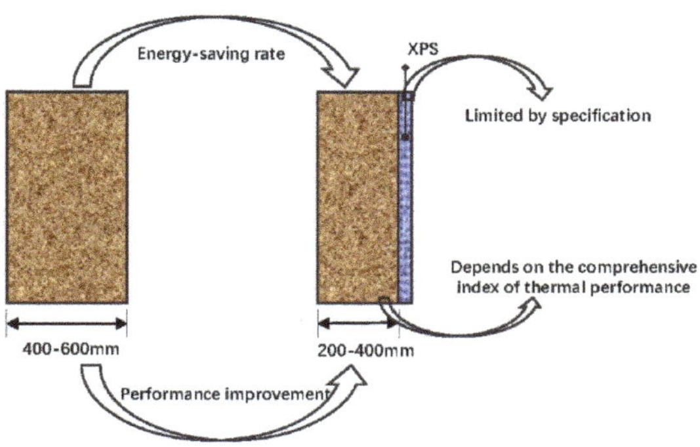

Figure 22. Applications ideas for RE materials.

6. Conclusions

This research used the mathematical model of heat transfer in porous material envelopes as a theoretical basis to study the influence of thermal insulation on the indoor thermal environment. This work also composed and derived the relationships among parameters affecting the thermal performance of buildings and constructed a comprehensive evaluation index for the thermal performance of RE walls. The findings suggest that after the thickness of an RE wall in Beijing exceeds 600 mm, the thermal performance improvement gradually decreases. The interior design temperature, energy consumption demand, and economic factors were used as boundary conditions to confine the thickness of the RE wall. It was found that the optimal thickness in Beijing would be about 360 mm. In addition, to meet China's building energy codes, an energy savings design for RE walls was presented in which XPS insulation board was added. Per the calculations, the final thickness of the insulation board for a public building with a building shape coefficient ≤ 0.3 was about 50 mm; for residential buildings, it was about 70 mm.

This study also found that external building insulation is conducive to reducing the winter heat flow in the envelope structure and plays a positive role in saving building heating energy. Internal building insulation has a greater impact on cooling energy consumption. Therefore, the external insulation composite RE wall was determined to be more suitable for the Beijing area. Moreover, this research evaluated the risk of condensation on composite RE walls, and a structural scheme applicable to the Beijing area was derived.

Author Contributions: Software, M.Z.; resources, J.M.; writing—original draft preparation, S.Y.; writing—review and editing, S.H.; data curation, D.T. All authors have read and agreed to the published version of the manuscript.

Funding: This research was funded by the State Key Project of National Natural Science of China (Grant number 51938002), the National Natural Science Foundation of China (Grant number 51878021), the General Project of Beijing Municipal Science & Technology Commission (Grant number KM201910 016016),and the State Key Laboratory of Subtropical Building Science (2019ZB09).

Institutional Review Board Statement: Not applicable.

Informed Consent Statement: Not applicable.

Data Availability Statement: Not applicable.

Conflicts of Interest: The authors declare that they have no known competing financial interests or personal relationships that could have appeared to influence the work reported in this paper.

References

1. Reddy, B.V.; Kumar, P.P. Embodied energy in cement stabilised rammed earth walls. *Energy Build.* **2010**, *42*, 380–385. [CrossRef]
2. Bribián, I.Z.; Capilla, A.V.; Usón, A.A. Life cycle assessment of building materials: Comparative analysis of energy and environmental impacts and evaluation of the eco-efficiency improvement potential. *Build. Environ.* **2011**, *46*, 1133–1140. [CrossRef]
3. Blengini, G.A. Life cycle of buildings, demolition and recycling potential: A case study in Turin, Italy. *Build. Environ.* **2009**, *44*, 319–330. [CrossRef]
4. Soebarto, V. Analysis of indoor performance of houses using rammed earth walls. In Proceedings of the Eleventh International IBPSA Conference, Glasgow, UK, 27–30 July 2009; p. 1530e7.
5. Hall, M.; Djerbib, Y. Rammed earth sample production: Context, recommendations and consistency. *Constr. Build. Mater.* **2004**, *18*, 281–286. [CrossRef]
6. Allinson, D.; Hall, M. Hygrothermal analysis of a stabilised rammed earth test building in the UK. *Energy Build.* **2010**, *42*, 845–852. [CrossRef]
7. Soudani, L.; Woloszyn, M.; Fabbri, A.; Morel, J.C.; Grillet, A.C. Energy evaluation of rammed earth walls using long term in-situ measurements. *Sol. Energy* **2017**, *141*, 70–80. [CrossRef]
8. Givoni, B. Characteristics, design implications, and applicability of passive solar heating systems for buildings. *Sol. Energy* **1991**, *47*, 425–435. [CrossRef]
9. Serrano, S.; de Gracia, A.; Cabeza, L.F. Adaptation of rammed earth to modern construction systems: Comparative study of thermal behavior under summer conditions. *Appl. Energy* **2016**, *175*, 180–188. [CrossRef]
10. El Fgaier, F.; Lafhaj, Z.; Brachelet, F.; Antczak, E.; Chapiseau, C. Thermal performance of unfired clay bricks used in construction in the north of France: Case study. *Case Stud. Constr. Mater.* **2015**, *3*, 102–111. [CrossRef]
11. Paul, W.; Taylor, P.A. A comparison of occupant comfort and satisfaction between a green building and a conventional building. *Build. Environ.* **2008**, *43*, 1858–1870. [CrossRef]
12. Taylor, P.; Fuller, R.; Luther, M. Energy use and thermal comfort in a rammed earth office building. *Energy Build.* **2007**, *40*, 793–800. [CrossRef]
13. Adam, E.; Jones, P. Thermophysical properties of stabilised soil building blocks. *Build. Environ.* **1995**, *30*, 245–253. [CrossRef]
14. Goodhew, S.; Griffiths, R. Sustainable earth walls to meet the building regulations. *Energy Build.* **2005**, *37*, 451–459. [CrossRef]
15. Goodary, R.; Lecomte-Nana, G.L.; Petit, C.; Smith, D.S. Investigation of the strength development in cement-stabilised soils of volcanic origin. *Constr. Build. Mater.* **2012**, *28*, 592–598. [CrossRef]
16. Zak, P.; Ashour, T.; Korjenic, A.; Korjenic, S.; Wu, W. The influence of natural reinforcement fibers, gypsum and cement on compressive strength of earth bricks materials. *Constr. Build. Mater.* **2016**, *106*, 179–188. [CrossRef]
17. Yusoff, M.Z.M.; Salit, M.S.; Ismail, N.; Wirawan, R. Mechanical properties of short random oil palm fibre reinforced epoxy composites. *Sains Malays.* **2010**, *39*, 87–92.
18. Achenza, M.; Fenu, L. On earth stabilization with natural polymers for earth masonry construction. *Mater. Struct.* **2006**, *39*, 21–27. [CrossRef]
19. Ciancio, D.; Beckett, C.T.S.; Carraro, J.A.H. Optimum lime content identification for lime-stabilised rammed earth. *Constr. Build. Mater.* **2014**, *53*, 59–65. [CrossRef]
20. Kariyawasam, K.K.G.K.D.; Jayasinghe, C. Cement stabilized rammed earth as a sustainable construction material. *Constr. Build. Mater.* **2016**, *105*, 519–527. [CrossRef]
21. Danso, H.; Martinson, D.B.; Ali, M.; Williams, J.B. Physical, mechanical and durability properties of soil building blocks reinforced with natural fibres. *Constr. Build. Mater.* **2015**, *101*, 797–809. [CrossRef]
22. Reddy, B.V.; Leuzinger, G.; Sreeram, V.S. Low embodied energy cement stabilised rammed earth building—A case study. *Energy Build.* **2014**, *68*, 541–546. [CrossRef]

23. Li, Q.; You, R.; Chen, C.; Yang, X. A field investigation and comparative study of indoor environmental quality in heritage Chinese rural buildings with thick rammed earth wall. *Energy Build.* **2013**, *62*, 286–293. [CrossRef]
24. Liu, J.; Hu, R.; Wang, R.; Yang, L. Regeneration of vernacular architecture: New rammed earth houses on the upper reaches of the Yangtze River. *Front. Energy Power Eng. China* **2009**, *4*, 93–99. [CrossRef]
25. Brambilla, A.; Jusselme, T. Preventing overheating in offices through thermal inertial properties of compressed earth bricks: A study on a real scale prototype. *Energy Build.* **2017**, *156*, 281–292. [CrossRef]
26. Lu, S.; Wang, Z.; Zhang, T. Quantitative analysis and multi-index evaluation of the green building envelope performance in the cold area of China. *Sustainability* **2020**, *12*, 437. [CrossRef]
27. Research Institute of Highway Ministry of Transport. *Test Methods of Soils for Highway Engineering (JTG 3430—2020)*; China Communications Press: Beijing, China, 2020. (In Chinese)
28. Nanjing Fiberglass Research and Design Institute Co., Ltd. *Thermal Insulation—Determination of Steady-State Thermal Resistance and Related Properties—Heat Flow Meter Apparatus (GB/T10295-2008)*; Standards Press of China: Beijing, China, 2008. (In Chinese)
29. Yu, S.; Hao, S.; Mu, J.; Tian, D. Optimization of Wall Thickness Based on a Comprehensive Evaluation Index of Thermal Mass and Insulation. *Sustainability* **2022**, *14*, 1143. [CrossRef]
30. Ministry of Housing and Urban-Rural Development of the People's Republic of China (MOHURD). *Code for Thermal Design of Civil Building (GB 50176-2016)*; China Architecture & Building Press: Beijing, China, 2016. (In Chinese)
31. Ministry of Housing and Urban-Rural Development of the People's Republic of China (MOHURD). *Design Code for Heating Ventilation and Air Conditioning of Civil Buildings (GB50736-2012)*; China Architecture & Building Press: Beijing, China, 2012. (In Chinese)
32. Chen, M. *Thermal Performance of Rammed Earth Wall*; Xi'an University of Architecture and Technology: Xi'an, China, 2017. Available online: https://kns.cnki.net/kcms/detail/detail.aspx?dbcode=CMFD&dbname=CMFD201901&filename=1018862061.nh&uniplatform=NZKPT&v=OPCxDjrz2B8pwdws4epbIGmOYwDe2xc1itEthiTkmONznMfGudEdqtuLJ0L4e5S0 (accessed on 26 October 2017).

energies

MDPI

Article

Impact of Hybrid Ventilation Strategies in Energy Savings of Buildings: In Regard to Mixed-Humid Climate Regions

Kyung-Yong Park [1], Deok-Oh Woo [2,*], Seung-Bok Leigh [1] and Lars Junghans [3]

1 Department of Architecture and Architectural Engineering, Yonsei University, 50, Yonsei-ro, Seodaemun-gu, Seoul 03722, Korea; kypark193@yonsei.ac.kr (K.-Y.P.); sbleigh@yonsei.ac.kr (S.-B.L.)
2 College of Engineering, Lawrence Technological University, 21000 W 10 Mile Rd., Southfield, MI 48075, USA
3 A.Alfred Taubman College of Architecture and Urban Planning, University of Michigan, 2000 Bonisteel Blvd, Ann Arbor, MI 48109, USA; junghans@umich.edu
* Correspondence: dwoo@ltu.edu

Abstract: It has been identified that improving building energy efficiency is an effective method to reduce greenhouse gas (GHG) emissions. Although standards have been established to satisfy a building's minimum energy demand while ensuring the comfort of its residents, they are difficult to implement in mixed-humid regions. This study proposes a hybrid ventilation strategy that can comprehensively reduce cooling, heating, and ventilation energy in mixed-humid climate regions to significantly decrease the primary energy demand and reduce the impact of buildings on the environment. This study evaluated the changes in energy saving potential and thermal comfort according to the extension of the natural ventilation period and passive strategies, such as decentralized ventilation. Changes in indoor air temperature, operative temperature, and PMV for each strategy were analyzed. As a result, extending the natural ventilation and the decentralized ventilation strategies can save 32% and 34% of the building's energy, respectively. Considering that electricity is the main energy source for cooling in Korea, the extension of the natural ventilation period was judged to be the best approach from the perspective of primary energy demand. The results can be used to predict changes in building energy demand and thermal comfort and select an appropriate ventilation strategy based on occupant information obtained using Internet of Things.

Keywords: mixed-humid climate; hybrid ventilation; heat recovery ventilator; adaptive thermal comfort

Citation: Park, K.-Y.; Woo, D.-O.; Leigh, S.-B.; Junghans, L. Impact of Hybrid Ventilation Strategies in Energy Savings of Buildings: In Regard to Mixed-Humid Climate Regions. *Energies* **2022**, *15*, 1960. https://doi.org/10.3390/en15061960

Academic Editors: Shi-Jie Cao and Wei Feng

Received: 25 January 2022
Accepted: 2 March 2022
Published: 8 March 2022

1. Introduction

Various efforts to reduce greenhouse gas (GHG) emissions in response to the global climate crisis have been made worldwide. South Korea has further declared its target of reducing GHG emissions by 37% compared to business as usual (BAU) by 2030 [1]. Building energy consumption can be reduced by improving building performance, which can lead to a significant reduction in GHG emissions from the building energy sector. Improving building performance is thus an effective method to reduce overall GHG emissions. In this regard, many countries have reinforced building design standards, such as building envelope performance. The European Union (EU) approved EPBD to promote improvements in building energy performance [2,3]. In Korea, a strong building envelope design standard was presented by benchmarking the passive house standards of Germany.

The passive house standard is the most widely used high-performance building standard in the world that satisfies the minimum building energy demand and indoor air quality (IAQ), without impacting the thermal comfort of the occupants [4]. The standard differs slightly depending on the climate zone, but the following requirements must be met universally:

1. The energy demand of the space must be under 15 kWh/m^2 of net living space (conditioned floor area) of 10 W/m^{2}year peak energy demand.

2. The renewable primary energy demand (PER, according to PHI method), all of domestic energy, such as heating, hot water, and domestic electricity, must be under 60 kWh/m^{2}year of conditioned floor area for Passive House Classic.
3. In terms of airtightness, only under 0.6 air change per hour (ACH) is allowed at 50 Pa pressure (ACH50), in both pressurized and depressurized states for pressure tests.
4. During winter as well as in summer, thermal comfort in all living areas must be met, with under 10% of the hours being over 25 °C [5,6].

In a mixed-humid climate zone, however, it is difficult to satisfy the passive house standard due to high humidity. High humidity causes discomfort due to various reasons. First, under high-humidity conditions, a lower indoor air temperature is required to satisfy thermal comfort because the latent heat loss from the surface of the human body is relatively small [7]. Second, dampness due to high humidity may give rise to mold, which can cause health problems, such as allergies and respiratory diseases [8]. Third, in terms of energy, a large amount of energy is required for the dehumidification process, as the conventional dehumidification process lowers the temperature of wet air to the dew point and increases it back to the setpoint temperature. As is generally known, the traditional dehumidification process requires a large amount of energy due to the cooling, dehumidification, and reheating processes.

Natural ventilation is one of the best passive strategies to improve IAQ and the thermal comfort level, while avoiding the use of mechanical devices. Natural ventilation, however, is difficult to adopt in a mixed-humid climate zone because it is applicable only when outdoor air conditions fall within comfort ranges. To overcome this limitation, Gail S. Brager et al. proposed an adaptive thermal comfort model that makes it possible to use natural ventilation even in humid climate through an improvement in air velocity [9]. The main concept of the model is to allow comfort under a higher temperature and humidity through an improvement in air velocity. Improving air velocity may slightly increase building energy consumption but can reduce total energy consumption by reducing the use of mechanical dehumidification, which requires a considerable portion of energy consumption.

Several studies used natural ventilation to allow occupants to live comfortably under a wider range of humidity conditions in a humid climate zone [10–13]. Lee et al. [10] analyzed the effect of natural ventilation using parallel double-windows in Korea. They proved that approximately 60% of cooling energy can be reduced by adopting an appropriate ventilation model. Pingel et al. [11] compared the thermal comfort in six buildings where natural ventilation was applied in a warm and humid climate zone. Several passive strategies were adopted in these buildings, and it was confirmed that exposed cavity wall, night ventilation, and optimized building forms are the most effective strategies. Hesaraki et al. [12] compared IAQ and energy saving according to the natural ventilation level. They found that sufficient air was provided even at low ventilation levels. Schulze et al. [13] calculated the energy saving effect of selecting an appropriate ventilation strategy through energy simulation. They showed that well-designed natural ventilation systems, such as those in Stuttgart, Turin, and Istanbul, could save cooling energy by 13–44 kWh/m^2 per year. These studies showed that natural ventilation is effective at reducing building cooling energy. In addition, well-designed ventilation strategies and an increase in air velocity increased the effect of natural ventilation in humid climate zones.

In addition to cooling energy, heating and ventilation loads in winter are one of the factors that make it difficult to satisfy the passive house standard in humid climate zones. Heating and ventilation losses mostly occur through openings and infiltration between conditioned indoor air and unconditioned outdoor air. In this case, however, it is difficult for occupants to control the airflow. Therefore, minimizing the supply air volume is an effective method to save energy.

For the traditional control method, however, the adjustment of the air volume is inefficient because the indoor environment is air conditioned even in the absence of occupants in the same way as when it is occupied [14]. In addition, since the building design standards of Korea already require a sufficiently high insulation envelope performance, residential buildings do not require a large-scale HVAC system. Therefore, IAQ can be satisfied just by adding a small cooling and heating unit to HRV, and energy can be saved through heat exchange between exhaust and returning air through the continuous supply of outdoor air. It was generally difficult, however, to apply HRV to residential buildings because of the high installation cost and operator expertise required. In addition, HRV has been overdesigned due to the conventional operation method of centralized control. Therefore, the installation of a small HRV in each room makes it possible to supply fresh air through DV, simultaneously saving heating and ventilation energy, even in winter.

Therefore, demand-based ventilation was used in several studies to find the optimal air supply volume. Nguyen et al. [15] divided demand-based control into distributed control and decentralized control. They mentioned that the primary difference between the two methods is whether the agents communicate with each other. Decentralized control adjusts local controllers without communication. More specifically, Kim et al. [16] described fan-assisted natural ventilation as one of the DV strategies. Kim compared the energy saving potentials of buildings with decentralized ventilation located in European climate zones. Among various distributed systems, radiant panel distributed ventilation (RPDV) was found to require the lowest initial and operating costs. Silva et al. [17] performed POE through surveys and measurements in residential buildings with centralized and decentralized ventilation systems. A total of 80% of the occupants responded that they were satisfied with the building environment after remodeling to a decentralized ventilation system. Based on these studies, heating and ventilation losses are expected to be reduced through DV, one of the demand control methods.

In recent years, many studies have been conducted on hybrid ventilation, a concept that combines passive ventilation and mechanical ventilation. These studies, however, proposed energy saving and comfort improvement in individual areas, such as cooling energy, heating energy, and ventilation energy, but suggested no specific strategy, protocol, or standard [18]. In addition, the effect of the existing hybrid ventilation strategy was proven in hot and dry climate zones, but a different ventilation strategy is required in climate zones where it is hot and humid in summer and cold and dry in winter as in Korea. Therefore, this study proposes a ventilation strategy that can comprehensively reduce cooling, heating, and ventilation energy in mixed-humid climate regions. Figure 1 shows a concept diagram of the proposed ventilation strategy in this study.

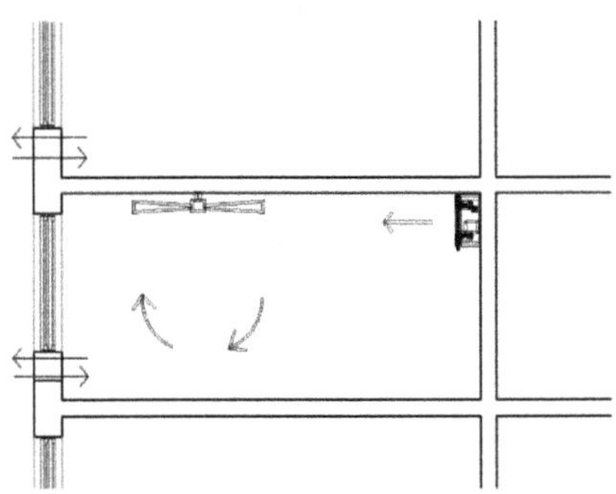

Figure 1. Concept diagram of proposed ventilation strategy.

In Korea, most HVAC systems are packaged air conditioners or split-unit systems, and electricity is used as the main energy source. Since electricity undergoes several conversion processes, the proportion of the primary energy demand is larger compared to other energy sources, such as thermal energy. Therefore, in this study, a hybrid ventilation strategy was proposed to significantly decrease the primary energy demand and reduce the impact of buildings on the environment.

2. Simulation Strategies

2.1. Natural Ventilation

Natural ventilation is a passive strategy traditionally used to reduce the cooling loads of buildings and secure IAQ [10]. Natural ventilation, however, is significantly affected by outdoor air conditions, and its effect is insignificant in a humid climate zone because it is applicable only when outdoor air conditions fall within the comfort ranges for temperature and humidity [12]. As a matter of fact, it is difficult to define the comfort ranges of outdoor air due to non-quantitative factors. Nevertheless, many studies have followed the criteria of ASHRAE Standard 55, which considers the PMV model [19]. The natural ventilation criteria of ASHRAE consider the temperature and absolute humidity of the outdoor air. Figure 2 shows the specific natural ventilation criteria suggested by ASHRAE Standard 55. When the period that allows natural ventilation was calculated using the climate data of Seoul, it was found that 2092 h makes up 22.4% of the year. Thus, the effect of natural ventilation is not significant. Therefore, additional energy saving can be achieved by extending the natural ventilation period and reducing the mechanical ventilation period. In this study, the concept of the adaptive thermal comfort model was used to extend the natural ventilation period.

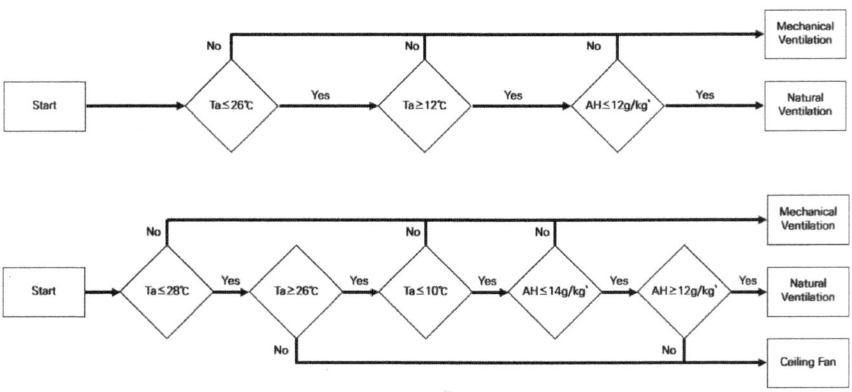

Figure 2. Conventional (**Top**) and proposed (**Bottom**) natural ventilation algorithms.

2.2. Increasing Air Velocity

The concept of the adaptive thermal comfort model is based on the supposition that it is possible to expand the comfort ranges of temperature and humidity that people generally feel by increasing indoor air velocity. In many previous studies, a cooling effect and comfort improvement were achieved using this concept [20–25]. In this study, ceiling fans were used as devices to increase the indoor air velocity. A ceiling fan can control a larger area than other airflow-generating devices, and can generate various air velocities. As the electric power varies depending on the generated air velocity, the energy demand was calculated in this study based on level 6 electric power (11 W), which generates a velocity of 1.5 m/s, by referring to a study by Zhai et al. [20]. Table 1 shows the fan power depending on air speed. Since the target building had four zones that require air-conditioning, the minimum requirement of four ceiling fans was met.

Table 1. Fan power and air speed at each setting.

Level	Power [W]	Mean Air Speed [m/s]
0	1 *	0.05
1	2	0.44
2	3	0.57
3	4	0.69
4	7	1.27
5	9	1.39
6	11	1.59

* Plug load.

2.3. Heat Recovery Ventilator

Since it is hot and humid in summer and cold and dry in winter in Korea, HRVs that can exchange both sensible heat and latent heat are required when heat exchangers are to be used. Existing HRVs, however, are large and require high installation costs. In addition, the condensation problem has continuously been raised. Currently, in Korea, the minimum standard for HRV performance is presented by KS B 6879, which is shown in Table 2 [26]. This standard stipulates an air-to-air heat recovery ventilator with heat exchange elements and filter units installed between indoor and outdoor spaces, with under 600 V and 3000 ACH. In this study, energy and thermal comfort were calculated while a small HRV that meets the minimum standard was used. It is possible to operate ventilation systems according to the number of occupants by installing small HRVs in the window of each room. In this process, the ventilation loss can be reduced by adjusting the air supply volume. Based on this, the energy efficiency of the entire building can be improved.

Table 2. Minimum standard of heat recovery ventilator based on KS B 6879.

	Ventilation Volume	Heat Recovery Rate		Energy Coefficient	
		Cooling	Heating	Cooling	Heating
Standard	Over 90%	Over 45%	Over 70%	Over 8.0	Over 15.0

3. Simulation

3.1. Simulation Description

In this study, an Energyplus engine was used to calculate the building energy demand. This can reflect the design elements and physical properties of a building through Rhino/Grasshopper. The energy simulation results were calculated on an hourly basis. Recently, Energyplus has been used in various studies because it is able to calculate building energy even in a complex energy model [27].

3.2. Climate Analysis

The climate of Korea is classified as a mixed-humid climate according to the Koppen Geiger classification criteria. In this study, the TMY weather data of Seoul, the capital city of Korea with the largest population, were used. Figure 3 shows the annual temperature and humidity distribution of outdoor air. Based on the criteria of Figure 2 (Bottom), 3055 h per year is available to extend the natural ventilation period in Korean climate.

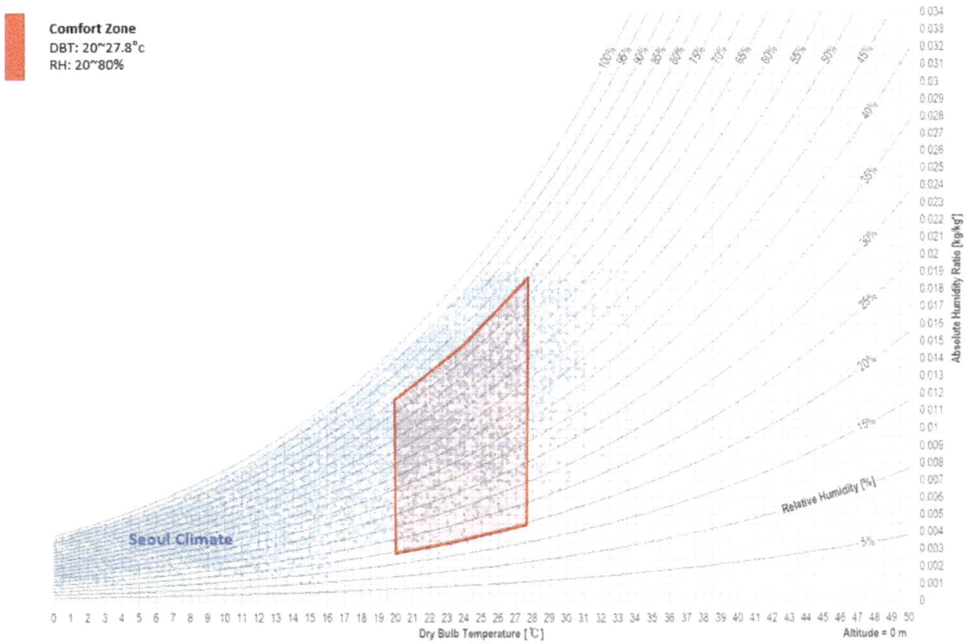

Figure 3. Annual temperature and relative humidity in Seoul, Korea.

3.3. Building Statement

Since Seoul has a high population density, most people live in high-rise apartments. The Korean government provides typical apartment drawings, such as floor plans, section plans, and system plans, to provide a high-quality residential environment. Figure 4 is the standard floor plan of apartments given by the Korean government. This suggests that the criteria for insulation performance and airtightness performance by region are used to secure a zero-energy building performance for new buildings. In this study, an energy simulation was performed based on the standard apartment drawings and building performance in Seoul. The input variables for simulation are shown in Table 3 [28].

Table 3. Input variable for energy simulation based on Korean building code of multi-stories apartment.

Number of Floors	Gross Area [m^2]	Total Height [m]	Volume [m^3]	WWR [%]				U-Value [W/m^{2}K]				ACH	Num of Occupants
				S	N	E	W	Roof	Wall	Floor	Win		
20	5954	56	16,671	60	10	0	0	0.15	0.17	0.17	1.00	0.3	240

Residential buildings in Korea mostly adopt hot water pipe floor heating for heating and packaged air conditioners or split-unit systems for cooling. The distribution energy of heating and cooling was calculated considering the characteristics of the media air and water. The pressure loss according to the medium is calculated using the following formula [29]:

$$P_{drop} = \frac{1000 \cdot f \cdot \rho \cdot v^2}{2D \cdot (1.6 + 1.1 L_{duct})} \tag{1}$$

where:

P_{drop} = Pressure drop (Pa);
f = Friction factor;

ρ = Density of mass (kg/m^3);
v = Velocity (m/s);
D = Hydraulic diameter (m), generally 1.204;
L_{duct} = Duct length (m).

The designed duct lengths (L) for each room in the unit household are shown in Table 4.

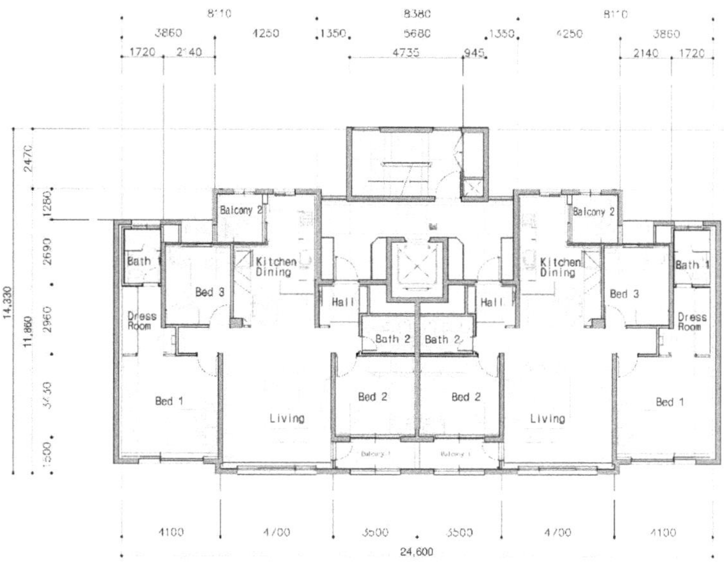

Figure 4. Typical floor plan for multi-story apartment in Korea.

Table 4. Designed duct length of each room from air handling unit.

	Bedroom 1	Bedroom 2	Bedroom 3	Kitchen	Living Room
Length (m)	11.0	4.5	9.4	7.0	6.5

3.4. Occupancy Schedule

In the conventional HVAC control method that involves the introduction of outdoor air, the required air supply volume is calculated based on the maximum number of occupants. Thus, energy can be wasted due to overdesign. The calculation of the number of occupants for energy simulation generally follows the basic occupancy schedules of EN and ISO. In this study, the basic occupancy schedule of Ahmed et al. [30], calculated based on the EN16798-1 and ISO/FDIS 17772-1 standards, was reflected in the energy simulation [31,32]. The occupancy schedule for each zone was calculated by applying the ratio obtained in the study by Ahmed et al. based on four people per household. Figure 5 shows the occupancy schedule for energy simulation with a conventional HVAC system and decentralized ventilation strategy in each room.

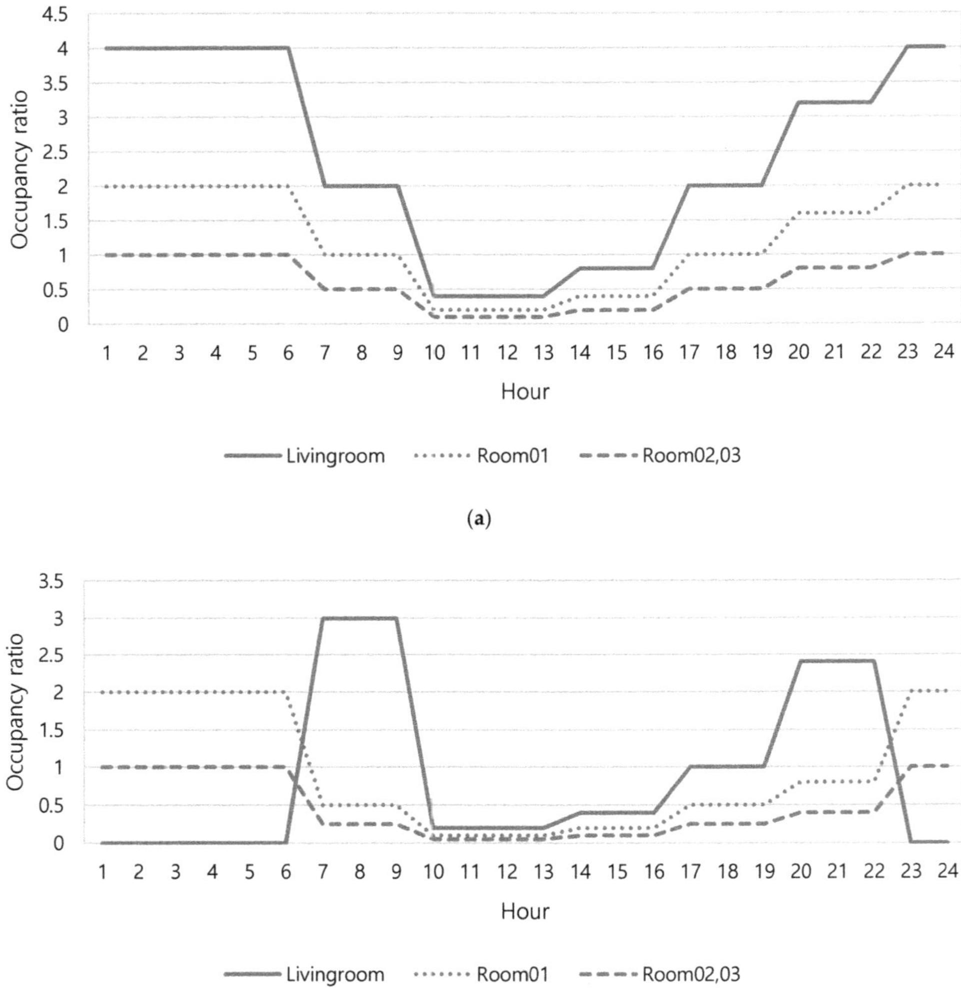

(a)

(b)

Figure 5. (**a**) Occupancy schedule for conventional HVAC system. (**b**) Occupancy schedule for decentralized ventilation.

4. Results

4.1. Energy Saving Potential

To evaluate the energy saving potential via the two passive strategies presented above, the energy demand, primary energy demand, and CO_2 emissions were compared for each strategy. Each strategy was divided into the following four cases and the effects of each strategy were compared: Case 1—baseline; Case 2—extended natural ventilation; Case 3—decentralized ventilation; and Case 4—combined passive strategies (proposed solution). Table 5 shows the description of each simulation case and Figure 6 shows the annual energy demand per unit area for each of the four strategies.

Table 5. Simulation cases description.

Operation Scenario	Case 1	Case 2	Case 3	Case 4
Name of case	Baseline case	Extended natural ventilation case	Decentralized ventilation case	Proposed solution case
Description	Basic ventilation strategy	Extending natural ventilation period with increasing air velocity using ceiling fan	Decentralized control for ventilation based on occupancy demand	Combined two strategies (Case 2 and Case 3)
Ventilation type	Centralized	Centralized	Decentralized	Decentralized
Control type	2-position Control	Floating Control	2-position Control	Floating Control

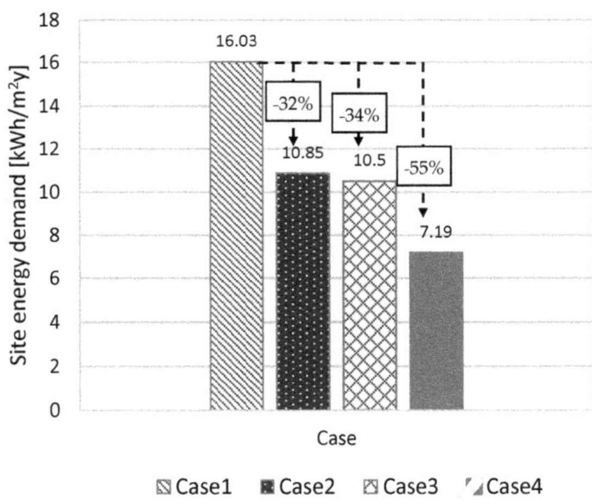

Figure 6. Site energy demand according to each strategy.

In Case 1, where the existing ventilation strategy is used, 16.03 kWh/m^2 of energy is required per year for HVAC. If the natural ventilation period can be extended in Case 1 by improving the air velocity, the energy demand is reduced by approximately 5.18 kWh/m^{2}year. The energy demand of Case 3 in which DV is used is similar to that of Case 2. The annual energy demands of Cases 2 and 3 were calculated to be 10.85 and 10.50 kWh/m^2, respectively. Conversely, in Case 4 where all of the proposed strategies are used, the energy demand can be further reduced to 7.19 kWh/m^2. Figure 7a–c show the energy consumption patterns of each strategy on a random date.

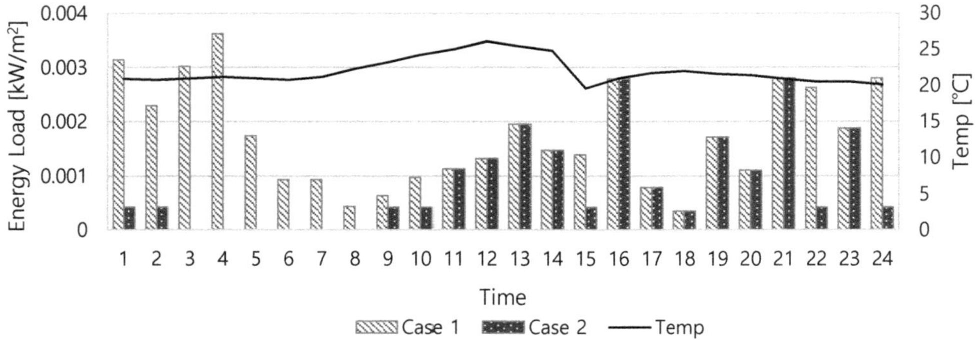

(a)

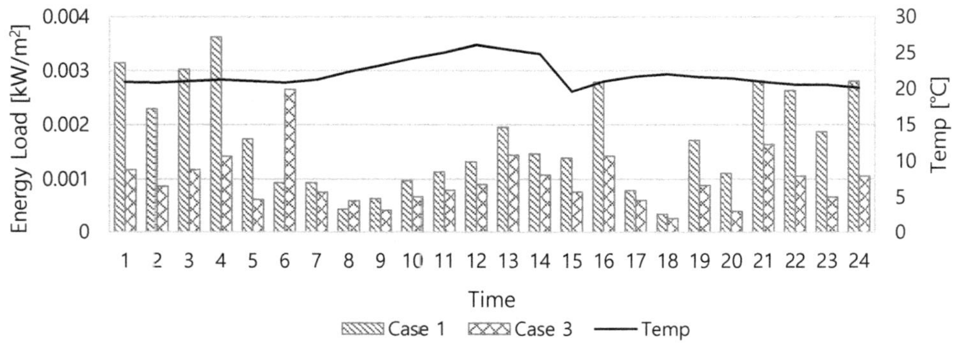

(b)

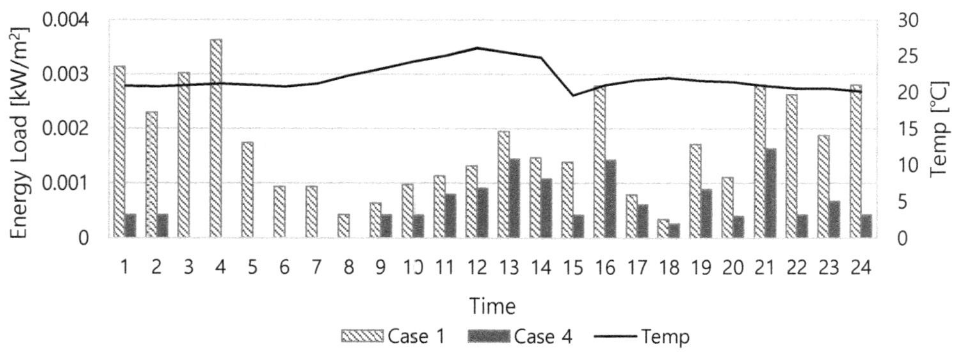

(c)

Figure 7. (**a**) Energy pattern comparison of Case 1 with Case 2. (**b**) Energy pattern comparison of Case 1 with Case 3. (**c**) Energy pattern comparison of Case 1 with Case 4.

The primary energy demand was calculated based on the actual energy demand. In the target building, a heat pump is used for heating and a compression chiller is used for cooling. Since the cooling and heating heat sources are different, the primary energy demand is also different from the energy demand calculated above. In Korea, coal gas has been traditionally used as a heating heat source and electricity as a cooling heat source. The

Korea Energy Agency suggests 1.1 and 2.75 as the conversion factors of gas and electricity, respectively, for primary energy calculation. Table 6 shows the primary energy factor depending on the energy source in Korea.

Table 6. Primary energy factor depending on energy source based on Building Energy Efficiency Certification.

Energy Source	Primary Energy Factor
Fuel	1.1
Electricity	2.75
District heating system	0.728
District cooling system	0.937

For this reason, the primary energy demand for each case was found to be 31.06 kWh/m^2 for Case 1, 19.15 kWh/m^2 for Case 2, 19.35 kWh/m^2 for Case 3, and 12.23 kWh/m^2 for Case 4. Figure 8 summarizes the primary energy demand for each strategy.

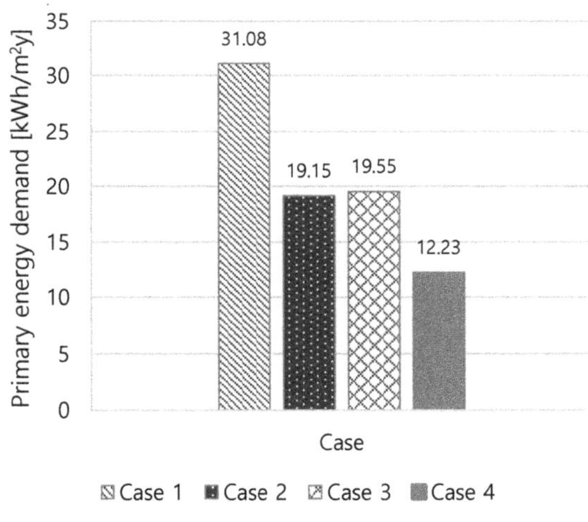

Figure 8. Primary energy demand of each case.

To examine the impact on the environment, the primary energy demand was converted into CO_2 emissions. Despite the same primary energy demand, the impact on the environment may vary depending on the energy production method. The energy production method differs by region. In Korea, which is the target region, energy is produced based on the proportions shown in Table 7. Figure 9 shows the results of calculating CO_2 emissions based on these proportions and the primary energy demand. CO_2 emissions were calculated to be 12.87 kgCO_2/m^{2}year for Case 1, 7.93 kgCO_2/m^{2}year for Case 2, 8.09 kgCO_2/m^{2}year for Case 3, and 5.06 kgCO_2/m^{2}year for Case 4.

Table 7. Energy source accountability to produce electricity in Korea.

Energy Source	Ration of Total Energy Production
Coal	42%
Liquefied Natural Gas (LNG)	27%
Nuclear	23%
Renewable	6%
Oil	1%
Etc.	1%

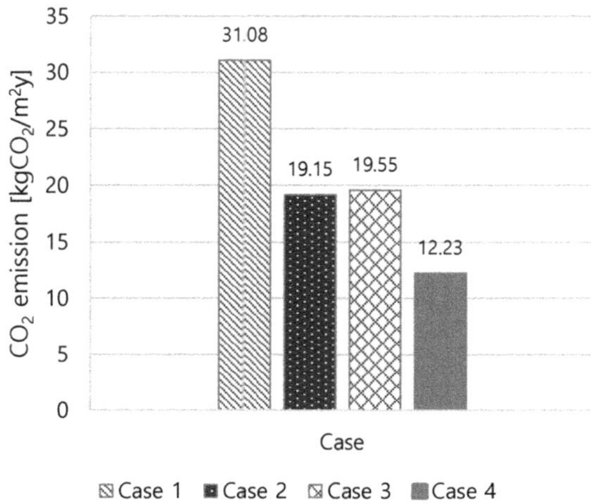

Figure 9. CO_2 emission of each case.

4.2. Thermal Comfort

To examine the comfort of occupants and the energy saving potential under each strategy, the indoor air temperature, operative temperature, and PMV were compared as thermal comfort indices. Thermal comfort was analyzed for Bedroom 01 with the highest density (two out of four occupants) in the household. Figure 10 shows the changes in the indoor air temperature of Bedroom 01 in a sample week.

In Case 1, the air temperature ranged from 20 to 26 °C due to the constant operation of HVAC. In Cases 2 and 3, similar indoor temperature distributions were observed. In Case 4, however, the temperature exceeded 26 °C or dropped below 20 °C in some instances, but the temperature difference was not severe and was found to be acceptable for most people.

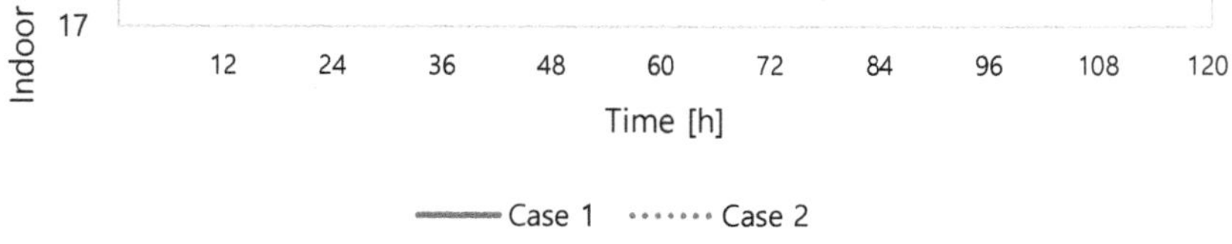

(**a**)

Figure 10. *Cont.*

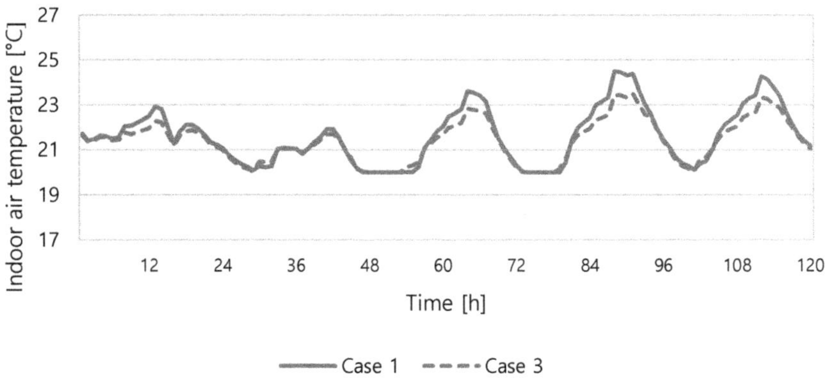

(**b**)

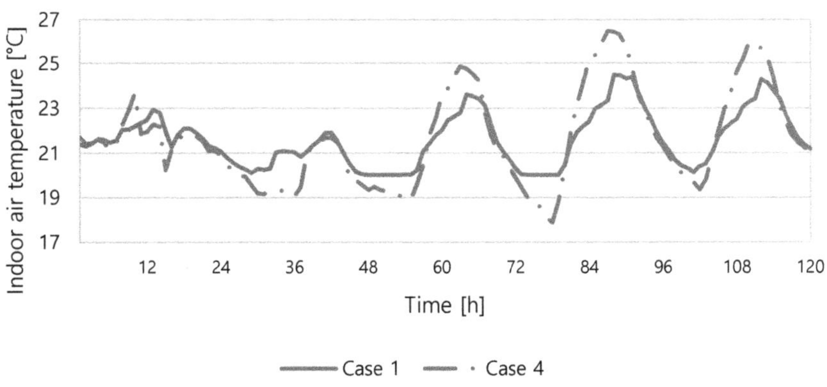

(**c**)

Figure 10. (**a**) The room air temperature comparison in a sample week of Case 1 with Case 2 (Bedroom 01). (**b**) The room air temperature comparison in a sample week of Case 1 with Case 3 (Bedroom 01). (**c**) The room air temperature comparison in a sample week of Case 1 with Case 4 (Bedroom 01).

Second, the operative temperatures, one of the indicators of the thermal comfort of occupants, were compared and analyzed. When the indoor air velocity is 0.2 m/s or less, the operative temperature is calculated as the average of the indoor air temperature (T_a) and the mean radiant temperature (MRT) [19,33].

$$T_o = \frac{T_a + T_{mrt}}{2} \quad (2)$$

In this study, however, the operative temperature was calculated using Equation (3) because the case of extending the natural ventilation period through an improvement in indoor air velocity was considered [19].

$$T_o = A\ T_a + (1 - A)\ T_{mrt} \quad (3)$$

For A, the following values in Table 8 were used according to the average indoor air velocity.

Table 8. A value for calculating operative temperature at more than 0.2 m/s air velocity.

V_a (Average Air Speed)	<0.2 m/s	0.2–0.6 m/s	0.6–1.0 m/s
A	0.5	0.6	0.7

ASHRAE standard 55 suggests acceptable operative temperature ranges as shown in Figure 11. This graph compares the monthly average outdoor air temperature with the indoor operative temperature, showing acceptability limits of 80% and 90%. In this study, the acceptable hours of a year in which the traditional control method falls within the 80% acceptability limits were examined, and their change for each strategy was evaluated to compare the thermal comfort. In Case 1, the acceptable hour limit was found to be 4933 h, at which point the building satisfied thermal comfort using HVAC systems. For other strategies, the acceptable hours slightly decreased due to the mixing of indoor air and outdoor air. Figure 12a,b show 80% acceptable hours for each strategy.

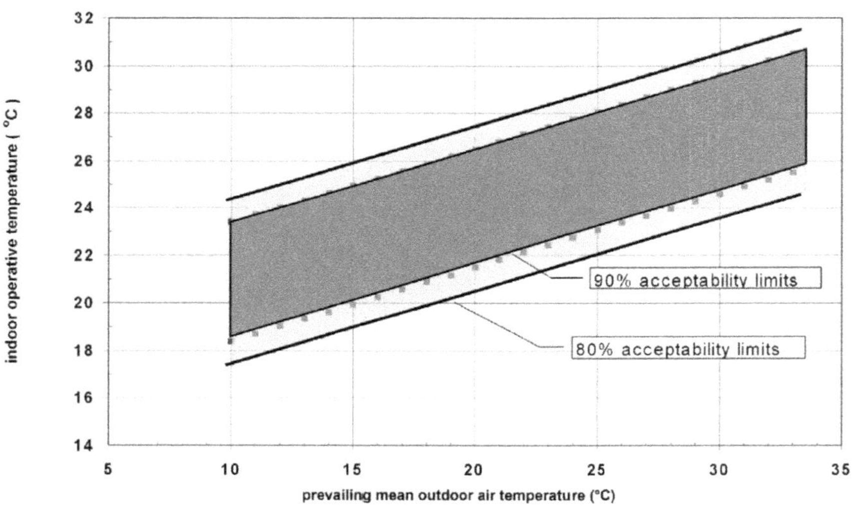

Figure 11. Acceptable operative temperature 6(T_o) ranges for naturally conditioned spaces.

Finally, PMV, which is most widely used as a thermal comfort index, was analyzed. PMV is calculated based on the following six variables: the air temperature, relative humidity, MRT, air velocity, amount of clothing, and metabolic rate. ASHRAE suggests thermal comfort ratings using this index, as shown in Figure 13. PMV is calculated using the following equation.

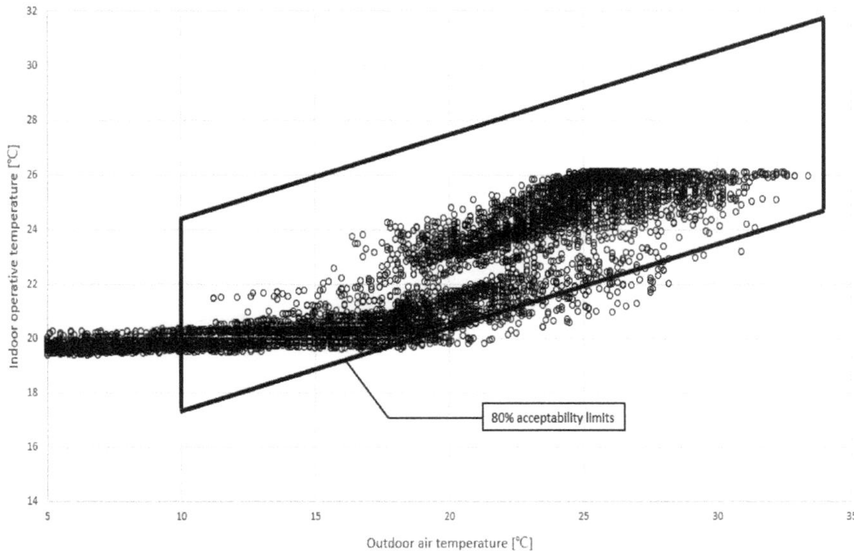

(a)

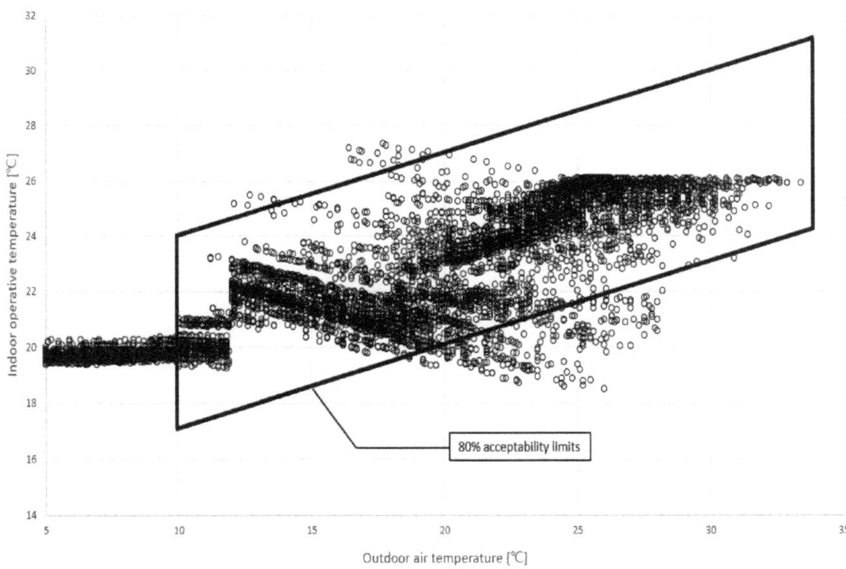

(b)

Figure 12. (**a**) Acceptability limits (80%) of operative temperature for naturally conditioned space in Case 1 (basic ventilation). (**b**) Acceptable hour (80%) of a year in case 4 (proposed solution).

$$PMV = \left(0.303e^{-0.036M} + 0.28\right)L_{\text{body}} \quad (4)$$

where:

PMV = Predicted Mean Vote;
M = Metabolic rate;
L_{body} = Thermal load on the body.

In this study, *PMV* was calculated every hour to compare the thermal comfort for each strategy. Non-quantitative factors, such as the amount of clothing and metabolic rate, were assumed to be different for summer and winter. The amount of clothing was set to 0.6 for summer and 1.0 for winter. Consequently, the average *PMV* values were found to be −0.32, −0.62, −0.66, and −0.70 for Cases 1–4, respectively. The average *PMV* was slightly different for each strategy, although it was generally similar. Figure 14 shows the *PMV* according to each case.

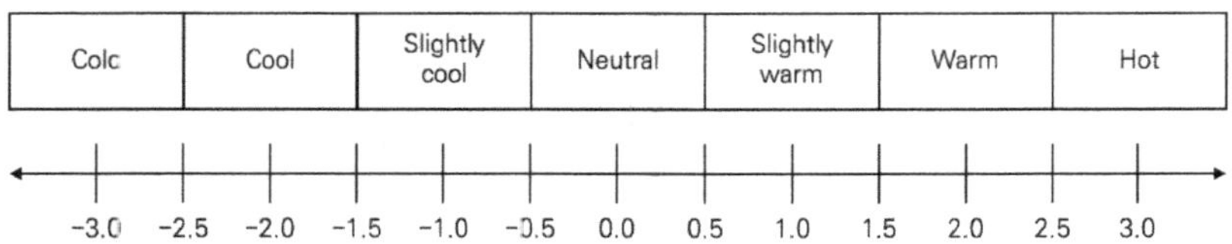

Figure 13. ASHRAE thermal sensation scale.

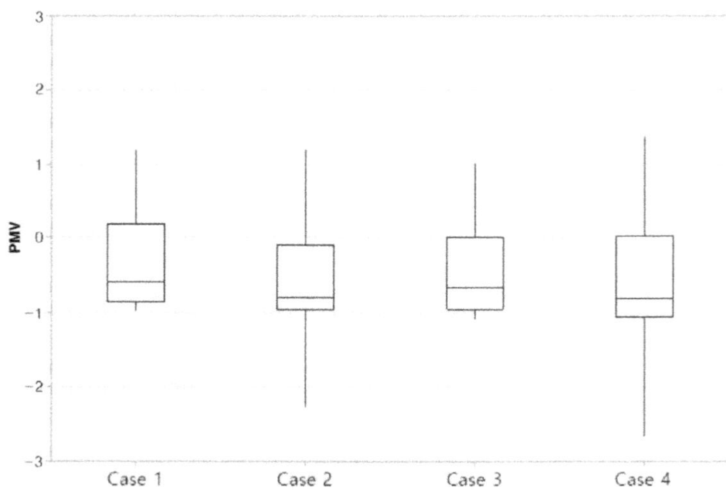

Figure 14. *PMV* variation by each case.

5. Discussion

5.1. Energy Saving Potential

This study evaluated the energy saving and changes in comfort through the existing passive strategy. Since the main strategy of this study is to extend the natural ventilation period by facilitating an increase in air velocity and to apply DV, cooling energy saving was important. The energy pattern analysis results showed that cooling energy could be saved by extending the natural ventilation period using ceiling fans. Conversely, when the outdoor air conditions were suitable for natural ventilation, the energy demand was the same as that of the existing control method. DV could also save energy by reducing heating and ventilation losses. The overall energy demand was reduced by DV, but the energy demand increased during time periods when occupants were concentrated, such as commuting hours. Total energy demand was significantly reduced by integrating the proposed strategies.

In terms of the primary energy demand, however, the results differed. Although the energy demand of the DV strategy was smaller than that of the ENV strategy, the primary energy demand of the DV strategy was found to be larger. This is because cooling and heating energy sources are different in Korea. Since electricity, which is the main energy source of cooling, goes through many conversion processes, its primary energy conversion factor is higher than that of gas. For this reason, CO_2 emissions showed the same pattern as the primary energy demand. Since the cooling load is the main load in a mixed-humid climate zone, the strategy to reduce the cooling load by extending the natural ventilation period is judged to be an appropriate passive strategy that can reduce the impact on the environment as well as the energy demand.

5.2. Thermal Comfort

From a thermal comfort perspective, energy saving sometimes adversely affects thermal comfort. Energy saving, however, can be sufficiently selected as an alternative strategy if changes in thermal comfort are insignificant or negligible. In this study, changes in thermal comfort were examined by analyzing three thermal comfort indices: the indoor air temperature, operative temperature, and *PMV*. The indoor air temperature of each strategy fluctuated in a wider range than that of the existing ventilation strategy. The temperature, however, always fell within the comfort range as cooling and heating systems were operated when the outdoor air was overly hot or cold. In addition, the acceptable hours in which the operative temperature fell within the 80% acceptability limits suggested by ASHRAE standard 55 were examined to compare the operative temperature for each strategy. The acceptable hours were found to be 4933 h a year for the existing strategy, and they were not significantly different for the strategies of Cases 2 to 4. Case 4 was the most uncomfortable strategy based on the operative temperature, but its acceptable hours were approximately 4300 h a year, which is not significantly different from that of the existing ventilation strategy. Finally, *PMV* was calculated every hour for each strategy. Since clo and met, which are non-quantitative factors, were assumed for the calculation of PMV, a comparison between the *PMV* values is not appropriate, but comparing the thermal comfort due to the change in passive strategy is significant. As shown in Figure 13, there was no significant difference in the annual distribution and average value of *PMV* for each strategy. In addition, the average *PMV* of Case 4 is somewhat smaller than that of the existing strategy. This appears to be because natural ventilation has a direct cooling effect, while stopping mechanical HVAC.

6. Conclusions

This study focused on reducing the latent heat cooling load for the spread of zero-energy buildings in mixed-humid climate zones without the use of additional machines. It was difficult to apply many existing passive strategies to mixed-humid climate zones, such as Korea, due to high humidity. Although it is impossible to reduce the latent heat load without mechanical dehumidification, an increase in air velocity allows higher humidity and temperature to be accepted. This makes it possible to reduce the use of mechanical cooling and heating and to perform natural ventilation and reduce the overall building energy demand even when outdoor air conditions exceed the comfort ranges. In addition, under existing HVAC systems, air-conditioning is performed before the air is dispersed to each room. To overcome these two limitations, this study evaluated the changes in energy saving potential and thermal comfort according to the extension of the natural ventilation period and passive strategies, such as decentralized ventilation.

Consequently, the extension of the natural ventilation period could save 5.18 kWh/m^2year of energy compared to the existing strategy, and cooling energy saving was important. The decentralized ventilation method could save 5.53 kWh/m^2year of energy compared to the existing strategy. The total amount of energy saved was larger for the decentralized ventilation method compared to the strategy involving extension of the natural ventilation period. Considering that electricity is the main energy source of cooling in Korea, however, the extension of the natural ventilation period was more efficient than the decentralized ventilation strategy from a perspective of primary energy demand. Therefore, the strategy of extending the natural ventilation period through an improvement in air velocity is judged to be the most effective in mixed-humid climate zones considering the impact on the environment.

In terms of thermal comfort, changes in indoor air temperature, operative temperature, and *PMV* for each strategy were analyzed in this study. The indoor air temperature fluctuated in a wider range depending on the strategy, but such fluctuations were within the comfort range as mechanical HVAC was operated when the outdoor air was extremely hot or cold. To compare the operative temperature, the acceptable hours in which the operative temperature falls within the 80% acceptability limits suggested by the ASHRAE standard were examined. For the existing control method, the acceptable hours were 4933 h a year. The acceptable hours for other strategies were not significantly different as they were also more than 4000 h. When *PMV* was compared, there was no significant difference for each strategy. For Case 4, where the proposed strategies were integrated, the PMV value was lower than those of other strategies due to the cooling effect of natural ventilation.

Although this study was limited by the lack of consideration of the combination of increasing the air velocity and mechanical ventilation, this study is significant in that the energy saving potential of typical residential buildings through the adaptive thermal comfort model and passive strategies, such as the decentralized ventilation strategy, in a mixed-humid climate was quantitatively evaluated. To the best of the authors' knowledge, no literature was found on the energy saving and the impact of thermal comfort by using hybrid ventilation strategy in the humid climate of South Korea. In addition, due to the recent development of measuring instruments, such as IoT sensors, it is now possible to identify the number of indoor occupants and their activities more easily and accurately than in the past. Therefore, the results of this study can be used to predict changes in building energy demand and thermal comfort according to the selection of an appropriate ventilation strategy based on the accurate information on occupants obtained using simple IoT equipment.

Author Contributions: Writing—original draft preparation, K.-Y.P.; writing—review and editing, D.-O.W.; project ad-ministration, S.-B.L.; Conceptualization, L.J. All authors have read and agreed to the published version of the manuscript.

Funding: This work was supported by the "Human Resources Program in Energy Technology" of the Korea Institute of Energy Technology Evaluation and Planning (KETEP), granted financial resource from the Ministry of Trade, Industry & Energy, Republic of Korea (No. 20194010000070).

Institutional Review Board Statement: Not applicable.

Informed Consent Statement: Not applicable.

Conflicts of Interest: The authors declare no conflict of interest.

Nomenclature

ACH	Air change per hour
D	Hydraulic diameter [m]
L_{body}	Thermal load on the body
L_{duct}	Duct Length [m]
M	Metabolic rate
MRT	Mean radiant temperature
P_{drop}	Pressure drop [Pa]
PMV	Predicted mean vote
T	Temperature [°C]
WWR	Window to wall ratio [%]
f	Friction factor
ρ	Density of mass [kg/m^3]
v	Velocity [m/s]

Subscripts	
a	Air
mrt	Mean radiant temperature
o	Operative temperature
Acronym	
ASHRAE	American society of heating, refrigerating, and air conditioning engineering
BAU	Business as usual
DV	Decentralized ventilation
ENV	Extended natural ventilation
EPBD	Energy performance of building directive
GHG	Greenhouse gas
HRV	Heat recovery ventilator
HVAC	Heating, ventilation, and air conditioning
IAQ	Indoor air quality
IoT	Internet of things
PER	Renewable primary energy demand
PHI	Passive house institute
POE	Post occupant evaluation
RPDV	Radiant panel distributed ventilation
TMY	Typical meteorological year

References

1. Kafle, S.; Parajuli, R.; Bhattarai, S.; Euh, S.H.; Kim, D.H. A review on energy systems and GHG emissions reduction plan and policy of the Republic of Korea: Past, present, and future. *Renew. Sustain. Energy Rev.* **2017**, *73*, 1123–1130. [CrossRef]
2. EPBD. Directive 2002/91/EC of the European Parliament and the Council, 16th December 2002, Concerning the Energy Efficiency of the Buildings. Available online: https://hoxe.vigo.org/pdf/valedorcidadan/AO_7ENG.pdf (accessed on 23 January 2022).
3. Figueiredo, A.; Figueira, J.; Vicente, R.; Maio, R. Thermal comfort and energy performance: Sensitivity analysis to apply the Passive House concept to the Portuguese climate. *Build. Environ.* **2016**, *103*, 276–288. [CrossRef]
4. Feist, W.; Schnieders, J.; Dorer, V.; Hass, A. Re-inventing air heating: Convenient and comfortable within the frame of the Passive House concept. *Energy Build.* **2005**, *37*, 1186–1203. [CrossRef]
5. Passive House Requirements, Passivhaus Institut (PHI), (Retrieved 11 December 2017). Available online: https://passiv.de/en/02_informations/02_passive-house-requirements/02_passive-house-requirements.htm (accessed on 23 January 2022).
6. Schnieders, J.; Hermelink, A. CEPHEUS result: Measurements and occupants satisfaction provide evidence for Passive Houses being an option for sustainable building. *Energy Policy* **2006**, *34*, 151–171. [CrossRef]
7. Tsutsumi, H.; Tanabe, S.; Harigaya, J.; Iguchi, Y.; Nakamura, G. Effect of humidity on human comfort and productivity after step changes from warm and humid environment. *Build. Environ.* **2007**, *42*, 4034–4042. [CrossRef]
8. Tang, S.; Zhi, C.; Fan, Y.; Ye, W.; Su, X.; Zhang, X. Unhealthy indoor humidity levels associated with ventilation rate regulations for high-performance buildings in China. *Build. Environ.* **2020**, *177*, 106839. [CrossRef]
9. de Dear, R.; Brager, G.S. Thermal comfort in naturally ventilated buildings: Revisions to ASHRAE Standard 55. *Energy Build.* **2002**, *34*, 549–561. [CrossRef]
10. Lee, S.-W.; Leigh, S.-B.; Kim, T.; Cheong, C.H.; Cho, S. Cooling energy reduction effect of parallel double-window system operation in residential buildings in South Korea. *Indoor Built Environ.* **2019**, *28*, 636–658. [CrossRef]
11. Doctor-Pingel, M.; Vardhan, V.; Manu, S.; Brager, G.; Rawal, R. A study of indoor thermal parameters for naturally ventilated occupied buildings in the warm-humid climate of southern India. *Build. Environ.* **2019**, *151*, 1–14. [CrossRef]
12. Hesaraki, A.; Myhren, J.A.; Holmberg, S. Influence of different ventilation levels on indoor air quality and energy savings: A case study of a single-family house. *Sustain. Cities Soc.* **2015**, *19*, 165–172. [CrossRef]
13. Schulze, T.; Eicker, U. Controlled natural ventilation for energy efficient buildings. *Energy Build.* **2013**, *56*, 221–232. [CrossRef]
14. Goyal, S.; Barooah, P.; Middelkoop, T. Experimental study of occupancy-based control of HVAC zones. *Appl. Energy* **2015**, *140*, 75–84. [CrossRef]
15. Nguyen, D.H.; Funabashi, T. Decentralized Control Design for User Comfort and Energy Saving in Multi-zone Buildings. *Energy Procedia* **2019**, *156*, 172–176. [CrossRef]
16. Kim, M.K.; Baldini, L. Energy analysis of a decentralized ventilation system compared with centralized ventilation systems in European climates: Based on review of analyses. *Energy Build.* **2015**, *111*, 424–433. [CrossRef]
17. Silva, M.F.; Maas, S.; de Souza, H.A.; Gomes, A.P. Post-occupancy evaluation of residential buildings in Luxembourg with centralized and decentralized ventilation systems, focusing on indoor air quality (IAQ). Assessment by questionnaires and physical measurements. *Energy Build.* **2017**, *148*, 119–127. [CrossRef]
18. Salcido, J.C.; Raheem, A.A.; Issa, R.R. From simulation to monitoring: Evaluating the potential of mixed-mode ventilation (MMV) systems for integrating natural ventilation in office buildings through a comprehensive literature review. *Energy Build.* **2016**, *127*, 1008–1018. [CrossRef]

19. *ANSI/ASHRAE Standard 55*; Thermal Environmental Conditions for Human Occupancy. American Society Heating, Refrigerating, and Air Conditioning Engineering: Peachtree Corners, GA, USA, 2017.
20. Zhai, Y.; Zhang, H.; Zhang, Y.; Pasut, W.; Arens, E.; Meng, Q. Comfort under personally controlled air movement in warm and humid environments. *Build. Environ.* **2013**, *65*, 109–117. [CrossRef]
21. Yang, B.; Schiavon, S.; Sekhar, C.; Cheong, D.; Tham, K.W.; Nazaroff, W.W. Cooling efficiency of a brushless direct current stand fan. *Build. Environ.* **2015**, *85*, 196–204. [CrossRef]
22. Ashoori, M.; Sheikhzadeh, G.; Sadripour, S. The effects of using ceiling fans on human thermal comfort in a three-dimensional room with centralized heating including an occupant. *Int. J. Numer. Methods Heat Fluid Flow* **2019**, *29*, 3976–3993. [CrossRef]
23. Arens, E.; Turner, S.; Zhang, H.; Paliaga, G. *Moving Air for Comfort*; UC Berkeley: Berkeley, CA, USA, 2009.
24. Schmidt, K.; Patterson, D.J. Performance results for a high efficiency tropical ceiling fan and comparisons with conventional fans: Demand side management via small appliance efficiency. *Renew. Energy* **2001**, *22*, 169–176. [CrossRef]
25. Ochiai, T.; Oda, S.; Sakai, M.; Ishiguro, S. *Thin Ceiling Circulator to Enhance Thermal Comfort and Cabin Space*; WCX SAE World Congress Experience: Detroit, MI, USA, 2019. [CrossRef]
26. KS B 6879. Heat Recovery Ventilators. 2020. Available online: https://www.standard.go.kr (accessed on 23 January 2022).
27. Li, S.; Liu, L.; Peng, C. A Review of Performance-Oriented Architectural Design and Optimization in the Context of Sustainability: Dividends and Challenges. *Sustainability* **2020**, *12*, 1427. [CrossRef]
28. Building Energy Code, Ministry of Land, Infrastructure and Transport. 2021. Available online: https://law.go.kr/LSW/eng/engLsSc.do?menuId=2&query=#liBgcolor11 (accessed on 23 January 2022).
29. ASHRAE Fundamental, C. *21 Duct Design*; American Society of Heating, Refrigerating, and Air conditioning Engineering: Peachtree Corners, GA, USA, 2017.
30. Ahmed, K.; Akhondzada, A.; Kurnitski, J.; Olesen, B. Occupancy schedules for energy simulation in new prEN16798-1 and ISO/FDIS 17772-1 standards. *Sustain. Cities Soc.* **2017**, *35*, 134–144. [CrossRef]
31. prEN16798-1; Energy Performance of Buildings—Part 1: Indoor Environmental Input Parameters for Design and Assessment of Energy Performance of Buildings Addressing Indoor Air Quality, Thermal Environment, Lighting and Acoustics—Module M1-6. 2016. Available online: https://global.ihs.com/doc_detail.cfm?document_name=PREN%2016798%2D1&item_s_key=00651620 (accessed on 23 January 2022).
32. ISO 17772-1; Energy Performance of Buildings—Indoor Environmental Quality—Part 1: Indoor Environmental Input Parameters for the Design and Assessment of Energy Performance of Buildings. 2017. Available online: https://www.iso.org/standard/60498.html (accessed on 23 January 2022).
33. Woo, D.-O.; Junghans, L. Framework for model predictive control (MPC)-based surface condensation prevention for thermo-active building systems (TABS). *Energy Build.* **2020**, *215*, 109898. [CrossRef]

buildings

MDPI

Article

Feasibility Study and Passive Design of Nearly Zero Energy Building on Rural Houses in Xi'an, China

Han Chang [1,*], Yaolong Hou [2,*], Inhee Lee [3,*], Tianye Liu [1] and Tri Dev Acharya [4]

1 Department of Architecture, School of Human Settlement and Civil Engineering, Xi'an Jiaotong University, Xi'an 710049, China; liu.tianye@xjtu.edu.cn
2 College of Railway Engineering, Zhengzhou Railway Vocational and Technical College, Zhengzhou 451460, China
3 Department of Architecture, Pusan National University, Busan 46241, Korea
4 Institute of Transportation Studies, University of California Davis, Davis, CA 95616, USA; tdacharya@ucdavis.edu
* Correspondence: changhan@xjtu.edu.cn (H.C.); houyaolong@zzrvtc.edu.cn (Y.H.); samlih@pusan.ac.kr (I.L.)

Abstract: Since the advent of reforms and opening-up of China, the focus has been on urban development. However, rural development has garnered attention in recent years. This research explores energy performance improvement methods for rural houses in Xi'an, China. It aims to discuss the feasibility of designing a nearly zero-energy building (nZEB), based on typical residential rural housing in Xi'an, through proposing new construction methods and examining the strategies for the refurbishment of an existing house. Initially, a typical rural house was modelled based on data collected from a field survey and historical documents. Subsequently, suitable passive design strategies were explored in the rural house design both in terms of proposing new construction methods and examining the refurbishment strategies of an existing house. After implementation of the passive design, the annual energy demand was reduced from 112 kWh/m^2 to 68 kWh/m^2 (new construction) and from 112 kWh/m^2 to 85 kWh/m^2 (refurbished). Even though the passive design significantly reduced the energy demand of the house, it could not achieve the Chinese nZEB standard. Therefore, a photovoltaic (PV) system and a storage battery were incorporated to meet the standard. Eighty per cent of the south roof area of the newly constructed and refurbished house was installed with a PV system and a storage battery with a capacity of 50 kWh and 52 kWh, respectively. After installation of the proposed renewable energy, the annual energy demand from the house was decreased to 35 kWh/m^2 (new construction) and 51 kWh/m^2 (refurbished), which both achieved the Chinese nZEB standard (equal to or below 55 kWh/m^2). The study shows the effectiveness of the methods used to design the nZEB and can be used to instruct the residents to build the nZEB in rural villages like Xi'an in China.

Keywords: rural house; passive design; nearly zero energy building; renewable energy

Citation: Chang, H.; Hou, Y.; Lee, I.; Liu, T.; Acharya, T.D. Feasibility Study and Passive Design of Nearly Zero Energy Building on Rural Houses in Xi'an, China. *Buildings* **2022**, *12*, 341. https://doi.org/10.3390/buildings12030341

Academic Editors: Shi-Jie Cao, Wei Feng and Alessandro Cannavale

Received: 27 December 2021
Accepted: 9 March 2022
Published: 11 March 2022

1. Introduction

Traditional buildings have high energy demands, requiring the consumption of fossil fuels. Therefore, architects worldwide have been focused on designing green buildings and advocating using renewable energy in buildings. There are various definitions of net Zero Energy Building (ZEB) according to source, site, cost and emission factors [1]; because the reduction of the delivered energy (coal, electricity from the public power plant and natural gas) is the main focus in this study, the ZEB definition proposed by U.S. Department of Energy was more suitable, which defines the ZEB as an energy-efficient building where, on a source energy basis, the actual annual delivered energy is less than or equal to the on-site renewable exported energy [2]. According to the Energy Performance of Buildings Directive (EPBD), a "Nearly Zero-Energy Building" (nZEB) means a building that has a very high energy performance and the nearly zero or very low amount of energy required should be covered to a very significant extent by energy from renewable sources, including energy from renewable sources produced on-site or nearby [3]. However, many countries have defined their own ZEB and nZEB standards according to the domestic situation. According to the technical standard in China [4], when the per-area energy consumption of a residential building is equal to or below 55 kWh/m^2, the building is classified as an nZEB.

Passive design started early in developed countries such as the US, which held its first passive design conference in May 1976. According to the description by the US Department of Energy [5], passive solar design takes advantage of the site, climate, and materials of a building to minimize energy use. Based on recent research on passive design, Song [6] reviewed evaluation methods for passive houses. Qi [7] classified passive design methods according to different weather conditions in different climate zones. Li [8] investigated the impediments of passive design applications in buildings in China; one crucial point is that there is no specific guidance. Yin [9] evaluated passive design methods for buildings from an economic perspective. The recent research also declares that passive design and renewable energy are two key methods to achieve nZEB and ZEB [2]. Therefore, according to the literature review of passive design, passive design should consider the energy-saving effect, evaluation method and economic factors. This study discussed the passive design strategies mainly considered the energy-saving effect of solar gain, insulation, appropriate structure and configuration of the village house in Xi'an. In terms of the specifications of the village house in Xi'an, appropriate local insulation materials and suitable strategies were explored; in the meantime, the basic economic factors were also discussed based on minimizing the construction and insulation materials. Although the passive design is an effective way to improve the energy performance of a house, it is difficult to meet the nZEB standard. Therefore, renewable energy systems are often incorporated in nZEB designs. According to the literature review, the long-term application of passive design in developed countries and some developing countries has led to the conversion of traditional buildings to nZEBs to some extent. Therefore, the next step in optimizing the energy performance of buildings is to pursue the renovation of buildings to design Positive Energy Buildings (PEB). In PEB, nZEBs support each other by connecting renewable energy with the grid [10–12]. Several researchers have evaluated the various effects of nZEBs: Dartevelle et al. [13] evaluated the long-term summer thermal comfort of nZEBs in Wallonia. Marta [14] evaluated the decarbonization effects of nZEBs in Southern Europe. Masi [15] performed a sensitivity analysis of nZEBs from a life cycle approach; this research also suggested that installing a storage system along with a PV system is more profitable. Presently, optimization methods for nZEB designs are also being investigated. Ferrara et al. [16] developed a model based on machine learning to optimize the design of renewable energy for nZEBs. Vujnovic et al. [17] proposed a cost-optimal energy performance calculation model for nZEBs. Herrero et al. [18] reviewed the modelling methods, simulation methods, and control tools of nZEBs. Kuivjõgi et al. [19] studied the phenomena in which commercial buildings are not converted to nZEBs in Estonia. From the literature review of the ZEB, nZEB and the PEB, the developed countries already comprehensively researched and practised energy-efficient buildings. However, as a developing country, the widespread application of nZEBs is not present in China, especially in rural areas. The methods to design an nZEB in developed countries cannot be simply implemented in developing countries due to the different environments, economies and policies.

North-western China has a population of 96.7 million, including a 55.5 million rural population [20]. In 2010, the delivered energy used (natural gas, coal and electricity from the grid, etc.) from rural houses in China was 1.38 trillion kWh, accounting for 26.1% of the total building energy consumption in China; the electricity consumption of rural houses in China was 136 billion kWh [21]. In winter, most rural residents in north-western China burn coal in their stoves for heating. Burning coal can adversely affect the health of rural residents and is harmful to the environment. Moreover, several cases of death from carbon monoxide poisoning because of coal-burning are reported every winter. Therefore, passive design and clean energy systems are a necessity for rural houses in north-western China.

This research explores energy performance improvement methods for typical rural houses in Xi'an in north-western China by exploring appropriate passive design strategies and applying renewable energy. This research aims to discuss the feasibility of designing a residential nZEB in a village in Xi'an.

This paper explored the appropriated passive design strategies, and with mature renewable energy, the study demonstrates that constructing an nZEB or refurbishing existing houses to nZEBs is feasible in the villages in Xi'an. The specific strategies and methods presented in this study were summarized in this paper, and it can be practical guidance for the residents to design nZEBs in the villages of Xi'an.

2. Materials and Methods

2.1. Study Area

Figure 1 shows the geographical information of Xi'an City in China.

Figure 1. Regions in north-western China under administrative classification.

Figure 2 presents the monthly average weather data of Xi'an. Figure 2a shows the monthly average dry bulb temperature in Xi'an for one year. Temperature data show that Xi'an has four distinct seasons. Figure 2b presents the monthly average relative humidity in Xi'an for one year; it ranges from 58% to 77%. Figure 2c shows the monthly average wind speed in Xi'an; the data reveal that the wind speed in winter is relatively higher than in summer. Figure 2d shows the monthly average direct normal solar radiation in Xi'an, revealing that direct normal solar radiation in summer is much higher than in winter; particularly in August, the average direct normal solar radiation almost reaches 160 Wh/m^2. Weather data are obtained from the American Society of Heating, Refrigerating and Air-Conditioning Engineers (ASHRAE) code [22].

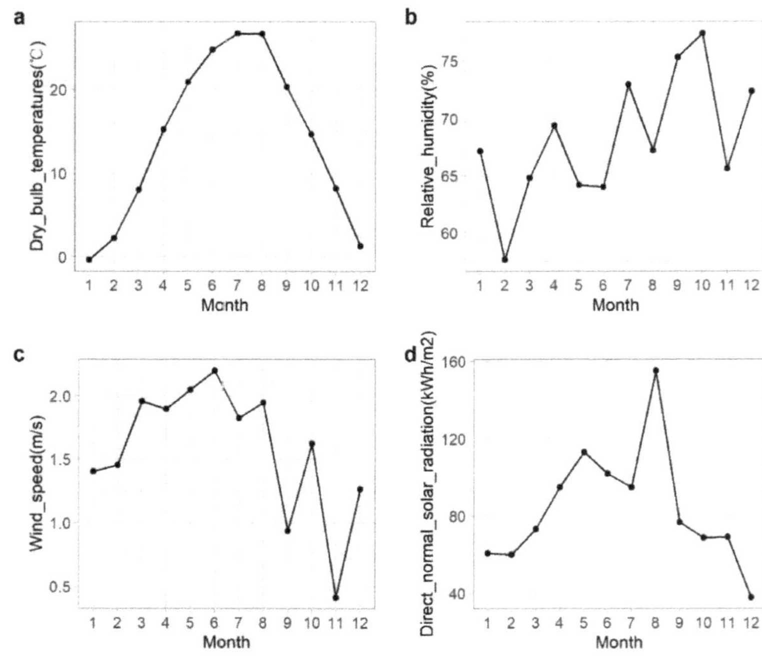

Figure 2. Monthly average weather data of Xi'an: (**a**) Dry bulb temperature; (**b**) Relative humidity; (**c**) Wind speed; (**d**) Direct normal solar radiation.

2.2. Typical Rural House and Sensors

To characterize typical rural housing in Xi'an, two field surveys were conducted and historical documents were reviewed. Figure 3 shows a picture of the village house in Xi'an. Once the typical rural house was characterized, it was selected and modelled. To incorporate nZEB designs in the typical rural house, the following two steps were taken: first, appropriate passive design strategies for the typical rural house were explored. DesignBuilder, a calibrated software from England based on EnergyPlus, was used to design the model of a typical rural house. This software was used to predict the energy consumption of different design varieties to find appropriate passive design strategies, such as the appropriate insulation material. The calculation engine of DesignBuilder is EnergyPlus and the EnergyPlus is developed in the U.S. used for analyzing the heating, cooling, lighting and ventilation of buildings. Even DesignBuilder is a calibrated commercial software, to verify the suitability of the software for this study, it was tested by comparing measured and simulated indoor temperature and relative humidity before applying this software. Secondly, as an application of passive design strategies is not sufficient to meet nZEB standards, renewable energy systems are used in passive design houses. The renewable energy system model, including photovoltaic cells and a storage battery, was created using the programming language Python. After the simulation, the results were organized and analyzed.

Figure 3. A village house in Xi'an.

In addition, Xi'an was classified as a humid subtropical climate under the Köppen climate classification, and the rural houses in the regions with the same climate condition can also refer to the achievements in this research to improve the energy performance of their houses.

In this research, Onset HOBO sensors were used, as seen in Figures 4 and 5. As shown in Figure 4, the outdoor sensor was installed in a solar radiation shield to ensure that outdoor temperature detection was not affected by the outdoor environment, such as solar radiation.

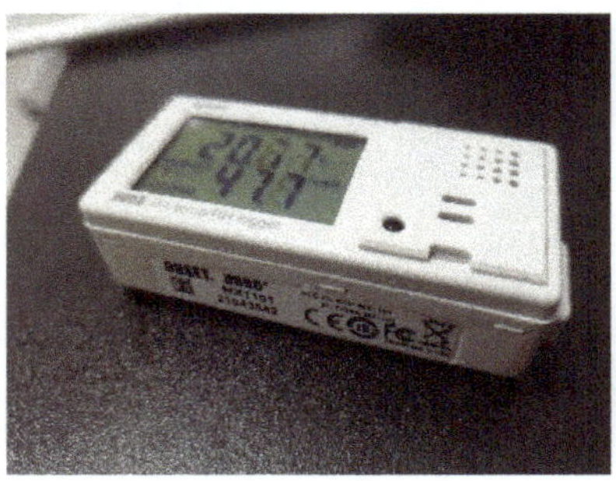

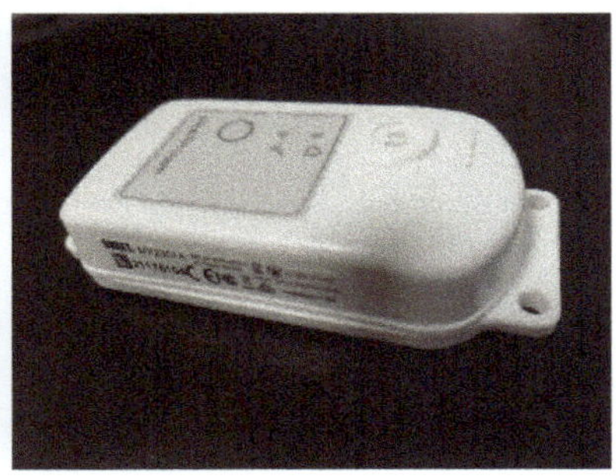

Figure 4. Temperature and relative humidity sensors for indoor (**left**) and outdoor (**right**) environments.

Figure 5. Outdoor temperature and relative humidity sensors (with solar radiation shield).

3. Energy Performance of the Typical Rural House

3.1. House Modelling and Data Measurement

Based on the field survey and topological study of rural houses in north-western China, a typical rural house in Xi'an was selected and modelled using DesignBuilder, as shown in Figure 6. Li et al. conducted four times filed surveys at the villages in North-western China including Shaanxi province to complete his doctoral thesis; for this study, three times field surveys were conducted at three villages (with 128 residential houses) in Xi'an. Table 1 concludes the distributions of the different village house structures in Shaanxi province and Xi'an based on the field surveys of this study and the research conducted by Li et al. [23]. As shown in Table 1, the proportion of the houses with brick-concrete structures is 73.6% and 95.8 in Shaanxi province and Xi'an, respectively. According to this study's survey results, all the occupied houses were brick-concrete structures (95.8%) and only a few unoccupied houses with dilapidated adobe-wood structures were found. Table 2 shows the distribution of the house construction time at the villages in Shaanxi province and Xi'an; 70% of the village houses in Shaanxi province and 81.6% of the village houses in Xi'an were constructed within 10 years. 36 houses were specifically studied during the surveys of this research. The detailed structure, construction materials and methods of the village house were surveyed through measurement and interviewing the senior workman (Due to the residents at the villages in Xi'an usually design and build their houses by referring to existing houses in the village, the houses in the villages are highly homogeneous. The material and structure of the occupied houses are the same; only the total area and window to wall ratio are slightly different; we selected the typical village house according to the most used window to wall ratio and area). The thermal coefficients of the materials were referred to from the related research [23]. The typical rural house had one and a half floors. The first floor was primarily for living. The top half-floor was used for storage and acted as a thermal buffer to improve the insulation effect of the roof. Before using the software for analysis, it was tested by comparing the measured data and simulated data. The devices mentioned in Section 2 were used to measure the indoor temperature, outdoor temperature and relative humidity under the condition of non-operational heating and cooling in the house. As shown in Figure 7, the outdoor sensor was hung from the tree, and the indoor sensor was installed 1.5 m above ground in the middle of the living room. The hourly indoor temperatures, outdoor temperatures and relative humidity from the afternoon on 14 October 2020 to the morning on 16 October 2020.

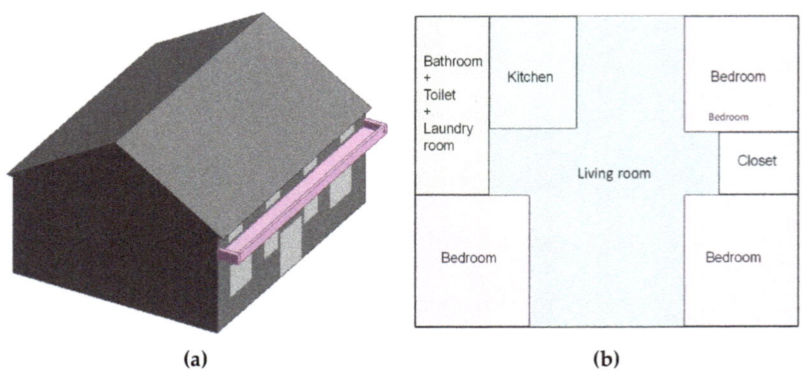

Figure 6. Typical rural house model: (**a**) 3D model; and (**b**) Configuration.

Table 1. Distribution of different house structures in the villages.

Structure	Shaanxi Province (Based on Li et al.'s Filed Surveys)	Xi'an (Based on Field Surveys of This Study)
Brick-concrete structure	73.7%	95.8%
Brick-wood structure	5.26%	0%
Adobe-wood structure	21.04%	4.2%

Table 2. Distribution of the house construction time at the villages.

Construction Time	Shaanxi Province (Based on Li et al.'s Filed Surveys)	Xi'an (Based on Field Surveys of This Study)
Within 10 years	70%	81.6%
Within 10~20 years	26.67%	8.9%
Within 20~30 years	3.33%	9.5%

(a) **(b)**

Figure 7. Temperature and relative humidity measurement: (**a**) Indoor; and (**b**) Outdoor.

3.2. *Software Test and Energy Demand Prediction*

The construction materials of a typical rural house are listed in Table 3. Considering the thermal design code for civil buildings in China [24], the specific characteristics of construction materials of the typical rural house in Xi'an are also listed in Table 3. As shown in Table 3, the original house did not have any insulation material. The specific characteristics of the materials were set in the model created using DesignBuilder. Table 4 summarizes the construction-specific characteristics of the typical rural house in Xi'an. The ventilation rate was obtained from the Chinese indoor air quality standard [25]. The infiltration rate of the rural house was the average measured data from [26]. Specific weather data, which were the input for the model, were obtained from the ASHRAE code [22].

Table 3. Materials and their thermal coefficients of a typical rural house in Xi'an, China.

Part Applied	Material	Thickness (mm)	Density (kg/m^3)	Specific Heat (J/kg·K)	Thermal Transmittance (W/m^2·K)
Brick wall	Cement sand mortar	15	1800	1050	62.00
	Brick	240	1800	1050	3.38
	Cement sand mortar	15	1800	1050	62.00
Floor (Semi-Roof)	Cement sand mortar	20	1800	1050	46.50
	Steel reinforced concrete	120	2500	920	14.50
	putty	5	1500	1050	152.00
Ground floor	Facing brick	10	2500	850	240.00
	Concrete	60	2300	920	25.17
	Spodosol	150	1600	1010	5.40
	Rammed earth	300	1796	884	2.40
Roof	Tile	20	1699	1050	37.00
	Cob	30	1400	1010	19.33
	Wood panel	10	500	2510	17.00

Table 4. Specific construction characteristics of the typical rural house in Xi'an.

Parameters	Specific Characteristics
No. of floors	1
Total area	120.5 m^2
Ceiling height	3.5 m
Structure	Brick mixed
Envelope	Brick
Roof	Pitched roof
Orientation	North to south
Glazing type	5.7 W/m^{2}K (Single no shading, SHGC: 0.82, VT: 0.88)
Window-to-wall ratio (South)	0.2
Window-to-wall ratio (North)	0.2
Window-to-wall ratio (East)	0
Window-to-wall ratio (West)	0
Wooden door	4.3 W/m^{2}K
Infiltration rate	0.55 AC/h
Ventilation rate	1 AC/h

Table 4. *Cont.*

Parameters	Specific Characteristics
Occupancy	0.02 person/m^2
Lighting	4.1 W/m^2
Equipment	13.3 W/m^2
HVAC/Heating	Natural gas (Gas boiler: Efficiency = 0.89)
HVAC/Cooling	Electricity from grid (Air conditioner: CoP = 3.9)
Area with heating and cooling system	Bedroom, lounge
Set point temperature	Summer 26 °C, Winter 18 °C

To test the software, the outdoor and living room were set sensors to measure the temperature and relative humidity with 5 min intervals from the afternoon on 14 October 2020 to the morning on 16 October 2020. The hourly average outdoor temperature and relative humidity data and other hourly climate data from the ASHRAE code were used for the software test. The hourly simulated results were compared with the measured results in Figure 8. The simulated results had a relatively small deviation from the measured data (0–3.4 °C for indoor temperature; 0–4.2% for relative humidity). The disagreement between the simulated and measured data is because the occupancy, ventilation rate cannot be precisely controlled in actual life; and unmeasured climate data (wind speed, solar radiation) can have a deviation from the historical climate data from ASHRAE. With enough devices and time, the limitation of the test will be carefully considered in future work. Although the measured data and simulated data have relatively low deviation, the trends of the simulated and measured data were the same. Additionally, a recent study conducted by Mazzeo [27] demonstrated that EnergyPlus is a sophisticated engine for the energy demand prediction of buildings. Thus, this software is appropriate for the analysis in this research.

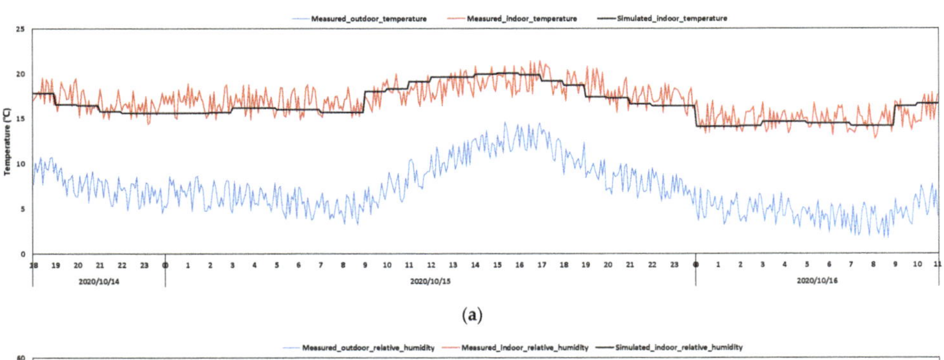

(**a**)

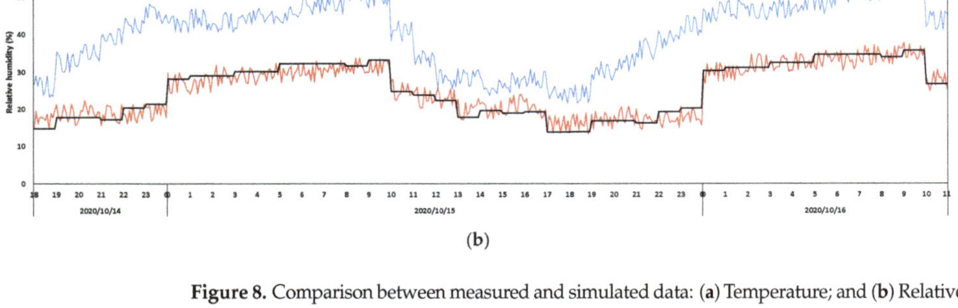

(**b**)

Figure 8. Comparison between measured and simulated data: (**a**) Temperature; and (**b**) Relative humidity.

The annual energy consumption of each category of the typical rural house was predicted by modelling the house and setting the parameters according to the data presented in Tables 3 and 4, as shown in Figure 9. From Figure 9, it can be inferred that the heating energy demand of the typical rural house is high, owing to the extreme cold weather in winter in the northwest region of China. In contrast, the cooling energy demand in the summer of the house was relatively low because of the relatively cool weather in the northwest region of China during summer. The energy demands from other aspects were calculated based on the occupancy density of the house, as summarized in Table 4.

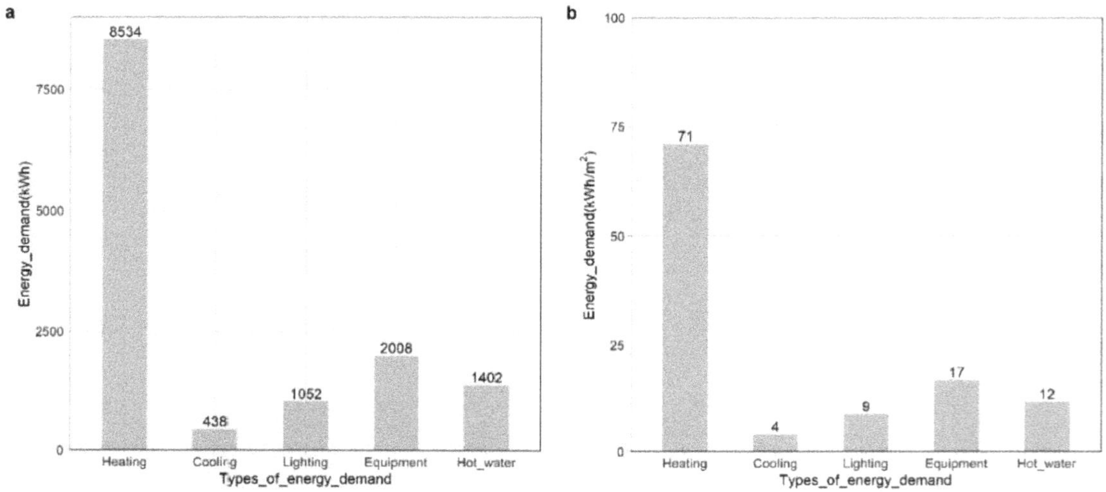

Figure 9. The annual energy demand of typical rural houses: (**a**) Annual total energy demand; (**b**) Annual per-square meter energy demand.

4. Application of the Passive Designs

4.1. Passive Designs for Newly Constructed Rural House

In this study, two methods were followed to create an nZEB based on the model of a typical rural house in Xi'an. The first method involved implementing the appropriate passive designs, and the second involved establishing a suitable renewable energy system in the village house. Two scenarios were considered in this study: the first scenario was the construction of a new passive house, and the second was the refurbishing of the existing house. In this section, suitable passive design strategies for the construction of a new passive house are explored. Passive designs consist of envelope improvement (orientation, wall and roof insulation, glazing type, window-to-wall ratio and sunroom application) and internal condition optimization (ceiling height and optimizing the configuration of the house). Table 5 summarizes the passive design steps. For each step, the possible alternatives to the original condition were evaluated, and an appropriate solution was proposed. Figure 10 presents the specific passive design exploration simulations of the possible alternatives, presented in alphabetical order. Once the results were obtained from the current step, the appropriate solution was selected and applied as the initial condition of the next step. Table 5 and Figure 11 also summarize the selected appropriate solutions for each step.

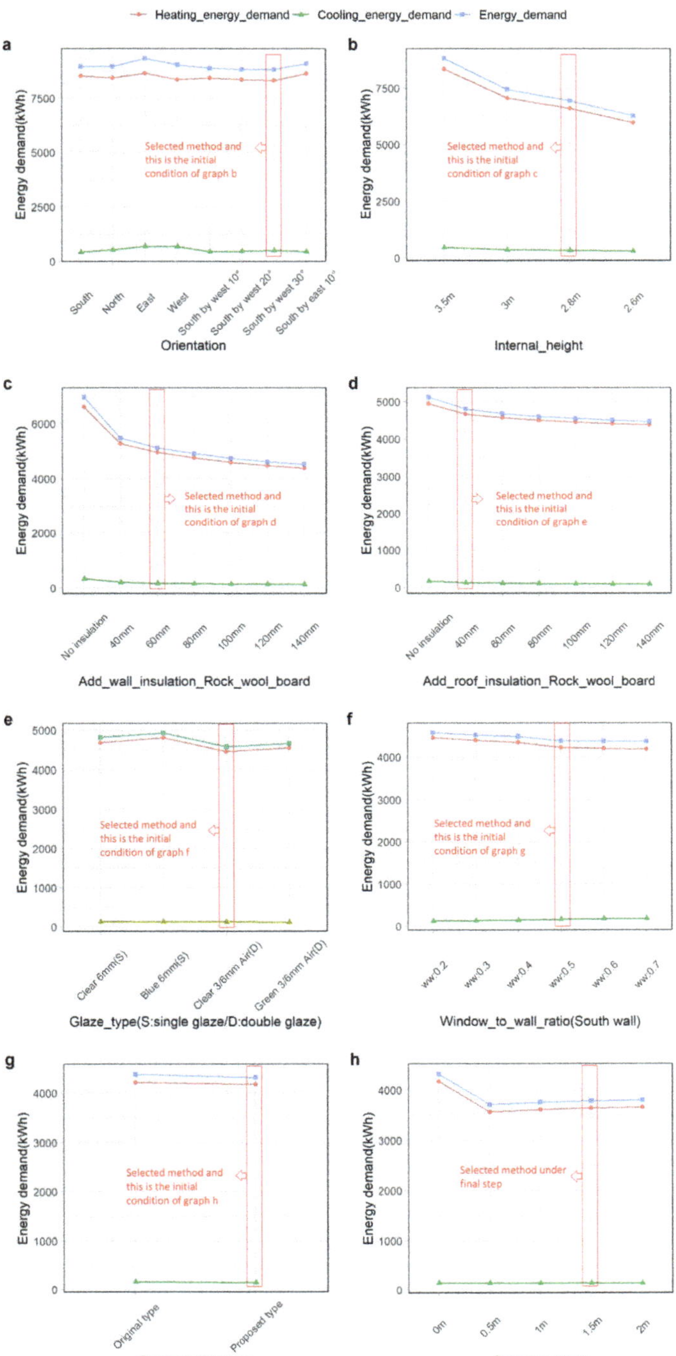

Figure 10. Simulation tests for exploring suitable passive design strategies: (**a**) Analysis of the orientation; (**b**) Analysis of the internal height; (**c**) Analysis of the wall insulation; (**d**) Analysis of the roof insulation; (**e**) Analysis of the glaze type; (**f**) Analysis of the window to wall ratio; (**g**) Analysis of the configuration type; (**h**) Analysis of sunroom effect.

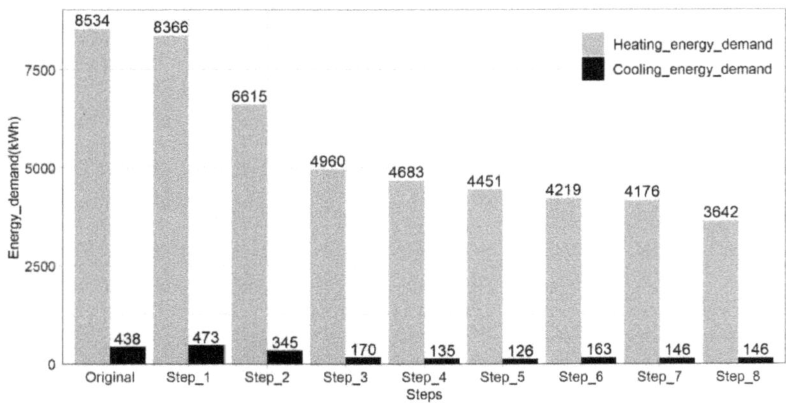

Figure 11. Annual heating and cooling energy demand according to the design of each step.

Figure 10a showed the energy-saving effect of the house with different orientations. Because the different orientations can help, the house benefits more from the solar gain; however, the larger solar gain can reduce the heating energy demand and increase the cooling energy demand; considering the total energy demand, the optimal orientation of the house was decided as south to west 20°. Figures 10 and 11 show that steps 2 and 3 reduced the energy demand of the house significantly. In step 2, the annual heating energy demand was reduced from 8366 kWh to 6615 kWh, and the annual cooling energy demand was reduced from 435 kWh to 345 kWh; in step 3, the annual heating energy demand was reduced from 6615 kWh to 4856 kWh, and the annual cooling energy demand was reduced from 345 kWh to 174 kWh. The economic condition of the villages in north-western China including Xi'an was very poor in the 20th century, but it has been significantly improved in the 21st century, and most of the villagers built their new houses. In the villagers' conventional view, the higher house represents the greater wealth of the family. Therefore, the ceiling height of the village house in Xi'an is mainly around 3.5 to 3.8 m. However, high ceiling height causes large heating and cooling energy demand due to the large volume of the space, step 2 demonstrated that the energy demand could be significantly reduced when the house ceiling height was decreased from 3.5 m to 2.8 m, which can also save a part of the construction material and reduce the labour cost. Considering the spatial comfort of the residents, the appropriate ceiling height was referred to the Chinese Residential Design Specification, which recommends a residential building with a ceiling height of 2.8 m [28]. According to the analysis results of step 3 as Figure 10c, the heating and cooling energy demand reduction were relatively rapid when the width of the insulation material was below or equal to 60 mm; the energy demand reduction was relatively smooth when the width of the insulation material above 60 mm; therefore, the insulation material of the external wall was selected with 60 mm width and the residents can choose wider width to pursue more significant energy performance of the house. The annual energy demand was decreased by 312 kWh when insulation material with 40 mm was implemented in the house; the reduction is relatively smooth when the width of the roof insulation is above 40 mm. Figure 12 shows the 3D structure of the typical village house. According to the roof system in Figure 12, there is a 120 mm reinforced concrete layer between the first floor and the half floor. Even the half-floor has the insulation effect, the insulation effect can be strengthened by adding an insulation material below the reinforced concrete. Figure 10e discussed the insulation effect of different glazing types, results showed that the clear glazing has better energy performance due to the clear glazing has the benefit to get solar gain; therefore, Figure 10f also discussed the energy performance of different window to wall ratios. As a result, the energy demand was decreased 241 kWh when the glazing was changed from single glazing to double glazing; the heating energy demand was decreased by 195 kWh when the window to wall ratio was changed from 0.2 to 0.5 and the cooling energy demand has a very small increase; the change of the heating energy demand was very small when the window to wall ratio was above 0.5. The effect of the double glazing and the window to wall ratio is relatively low because the solar gain only can significantly benefit the house in the day-time; however, the main energy demand from the residential house is in the morning and evening on the weekdays; the energy demand is high in the day-time only at weekends.

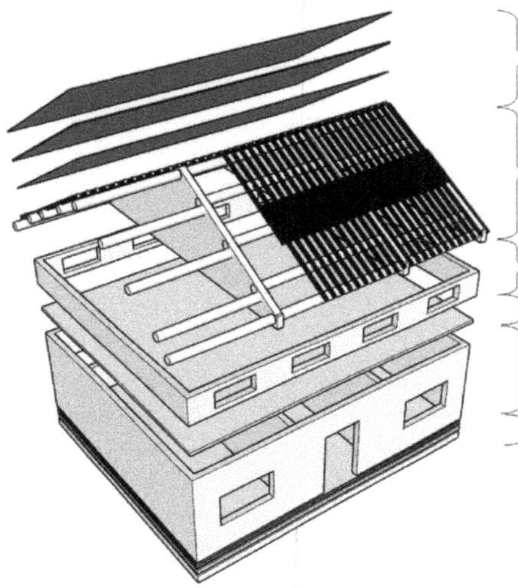

Figure 12. Three-dimensional structure of the typical village house in Xi'an, China.

Table 5. Passive design steps are applied to a typical rural house in Xi'an, China.

Steps	Designs
Original	Original condition
Step 1 (a)	Orientation change (North to south → South to west 20 degrees)
Step 2 (b)	Ceiling height change (3.5 m → 2.8 m)
Step 3 (c)	Add wall insulation (60 mm Rock wool board)
Step 4 (d)	Add roof insulation (60 mm Rock wool board)
Step 5 (e)	Glazing type change (Single glazing → Double glazing)
Step 6 (f)	Window to wall ratio change (South wall (0.2 → 0.5))
Step 7 (g)	Configuration change (Figure 13)
Step 8 (h)	Add sunroom (1.5 m)

Figure 13 shows the 3D model and the configuration of the house after applying passive design strategies. The configuration changes were made because the spaces with heating and cooling systems should be set next to each other to maintain proper heating or cooling. In this research model, spaces such as kitchens, bathrooms, toilets, laundry rooms and closets did not have heating and cooling systems. As shown in Figure 13, a 1.5 m sunroom was added in front of the house. The depth of the sunroom was selected based on the solar gain effect in winter and the functional consideration of the space; the sunroom can be completely opened or disassembled in summer to avoid solar gain. Figure 11 shows the annual energy-saving effect of the optimization design at each step. The village house in Xi'an does not have a basement, and there are four layers for ground insulation inheriting from the traditional method as shown in Figure 12. Therefore, the ground insulation was considered sufficient in this study.

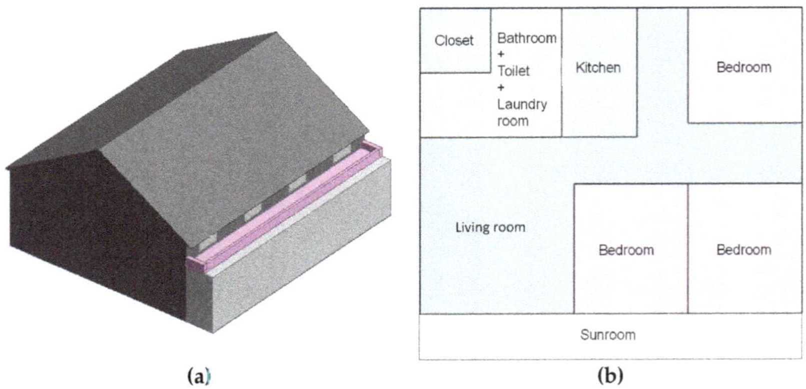

Figure 13. Model of the passive design rural house: (**a**) 3-D model; and (**b**) configuration.

Instead of using sophisticated insulation materials such as EPS boards, using local materials can meet the insulation demand of the house at a relatively low cost. Therefore, in step 3, insulation material called rock wool board made from local materials was applied, which reduced the heating and cooling energy demand significantly. In this research, the insulation materials with various sizes were discussed, the authors selected the insulation material with a specific size in terms of the extent of the energy-saving effect; however, the residents also can select the material with larger size according to the family's economic condition. Although changing the glazing type from single glazing to double glazing can be relatively expensive, the economic situation in Chinese villages has improved rapidly in recent years. Most new houses in the villages of Xi'an have adopted double glazing; therefore, incorporating double glazing in the design of the nZEB in step 5 is reasonable.

Table 6 summarizes the specific characteristics of the construction changes after the application of passive design strategies. Table 7 shows the materials and their thermal coefficients of the improved rural house. The annual energy demand of each category of the passive design rural house is shown in Figure 11. Comparing the energy demand of the original rural house with that of the passive design rural house, the annual heating energy demand decreased from 8534 kWh to 3642 kWh. Additionally, the cooling energy demand decreased from 438 kWh to 146 kWh, and the annual per-area energy consumption of the house was 68 kWh/m^2. However, according to the nZEB standard of China, the annual per-area energy consumption of the building should be equal to or below 55 kWh/m^2. Therefore, although the energy-saving effect of the passive design was significant, it still could not achieve the nZEB standard. Thus, a renewable energy system should be incorporated to achieve the nZEB standard.

Table 6. Specific characteristics of the construction changes in the passive design rural house.

Parameters	Specific Characteristics
Ceiling height	2.8 m
Orientation	South by west 20°
Glazing type	Double glazing; Clear 3 mm/6 mm air; 3.16 W/m^{2}K (Single no shading, SHGC: 0.76, VT: 0.81)
Window-to-wall ratio (South)	0.5

Table 7. Materials and their thermal coefficients of the passive design rural house.

Part Applied	Materials	Thickness (mm)	Density (kg/m^3)	Specific Heat (J/kg·K)	Thermal Transmittance (W/m^2·K)
Brick wall	15 mm cement sand mortar	15	1800	1050	62.00
	240 mm brick	240	1800	1050	3.38
	15 mm cement sand mortar	15	1800	1050	62.00
	Rock wool board	60	15	1400	1.00
	15 mm cement sand mortar	15	1800	1050	62.00
Floor (Semi-Roof)	20 mm cement sand mortar	20	1800	1050	46.50
	Rock wool board	40	15	1400	1.5
	120 mm steel reinforced concrete	120	2500	920	14.50
	5 mm putty	5	1500	1050	152.00
Ground	10 mm facing brick	10	2500	850	240.00
	60 mm concrete	60	2300	920	25.17
	150 mm spodosol	150	1600	1010	5.40
	300 mm rammed earth	300	1796	884	2.40
Roof	20 mm tile	20	1699	1050	37.00
	30 mm cob	30	1400	1010	19.33
	10 mm wood panel	10	500	2510	17.00

4.2. Passive Designs for Refurbished Rural House

In the previous section, possible passive designs for constructing a new house were discussed. In this section, possible passive design strategies for the existing house are proposed based on the passive design strategies for the first scenario. Due to changes in the house orientation, a decrease in the internal height and changes to the configuration of the existing house is not possible, and these passive design strategies were not considered. Thus, the passive design strategies of steps 3, 4, 5, 6 and 8 were chosen from Table 5 to refurbish the existing house. From Figure 14, the amount of energy reduction of each step can be obtained. The results also show that the addition of insulation material to the external wall in step 3 had a significant effect on reducing the heating and cooling energy demand. After applying all the proposed passive designs, the heating energy demand decreased from 8534 kWh to 5498 kWh, and the cooling energy demand decreased from 438 to 266 kWh. Considering the energy demand for hot water, lighting and equipment, the total annual energy demand of the passive refurbished house was 10226 kWh, and the annual per-area energy demand of the house was 85 kWh/m^2, which does not meet the nZEB standard of China. Therefore, the simulation of solar energy application was conducted to verify whether the passive refurbished house can meet the nZEB standard by incorporating a renewable energy system.

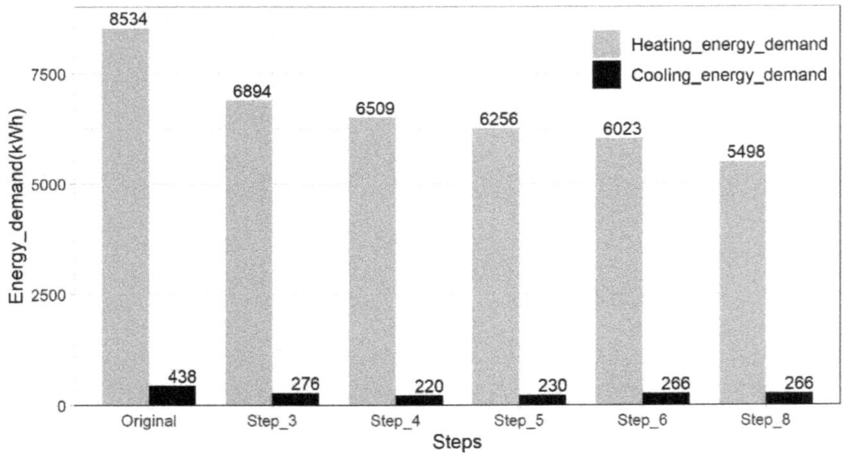

Figure 14. Annual heating and cooling energy demand according to each selected passive design.

5. Application of Renewable Energy

To achieve the nZEB standard, solar energy was solar energy. PV and storage batteries were considered to be installed in the house; even if the residents need to pay the initial cost of the PV and storage battery, the national government already has the policy to give financial support to the residents who install solar energy in their house; the government also plans to achieve the carbon peak in the year of 2030 and carbon neutralization in the year of 2060; to achieve this goal, more policies to encourage the residents to install renewable energy can be expected. Therefore, this research mainly discussed the feasibility of nZEB design in the village through the energy-saving prospect. To analyze the energy-saving effect of the PV and storage battery, models of a PV system and a storage battery were created and applied in the passive design for the rural house. Additionally, it was assumed that natural gas will be used for heating in the rural house as natural gas pipes are already installed in most villages in Xi'an. Therefore, electricity generation from the PV system and storage batteries will be used for other energy demands of the house. To conduct this analysis, a mathematical model of the PV system was developed. In this study, the mathematical model of the PV system was referred to a previous study by Yoza et al. [29]. Table 8 lists the necessary input parameters for the PV program. The amount of electricity generated by the PV system can be calculated using Equation (1):

$$P_{PV} = \eta n S_{PV} I_{PV}(1 - 0.005(t_{CR}\text{-}25)), \quad (1)$$

Table 8. Input parameters of the program (PV).

Input Data	Unit/Value
Conversion efficiency	14.4%
Coefficient of loss by changing direct current to alternating current	0.95
Coefficient of other losses	0.95
Number of panels	38 (80% area of the south roof)
Area of each panel	1.3 m^2
Ambient temperature	Data from ASHRAE (°C)
Solar radiation	Data from ASHRAE (W)

In Equation (1), η is the conversion efficiency of the PV panel, n is the number of PV panels, S_{PV} is the area of a single PV panel, I_{PV} is the effective solar radiation on an inclined panel surface, and $t_{ambient}$ is the ambient temperature. The recommended installation angle of a PV panel in Xi'an, China, is 26° [30]. Moreover, the effective solar radiation on an inclined panel surface was calculated using the method proposed by Erbs [31].

In this study, lead-acid storage batteries were selected because of their relatively low price and high energy performance. The specifications of the selected lead-acid storage batteries are listed in Table 9. The storage capacity of the storage batteries can be determined by dividing the depth of charge by the average daily load and multiplying by 3. As the heating energy was supplied by natural gas, the average electricity daily load of the new passive house and the passive refurbished house was 12.6 kWh and 13 kWh, respectively [32]. Additionally, the storage capacity of the storage batteries was determined to be 50 kWh for the newly constructed passive house and 52 kWh for the passive refurbished house. Figure 15 shows the program flowchart of the PV and storage battery systems.

Table 9. Input parameters of the program (Storage Battery).

Device Specification	Value
Storage capacity	50/52 kWh
Depth of discharge	75%
Maximum output capacity	6 kW
Maximum charge capacity	6 kW
Rate of charge loss	10%
Rate of output loss	10%
Rate of time loss	5%/month

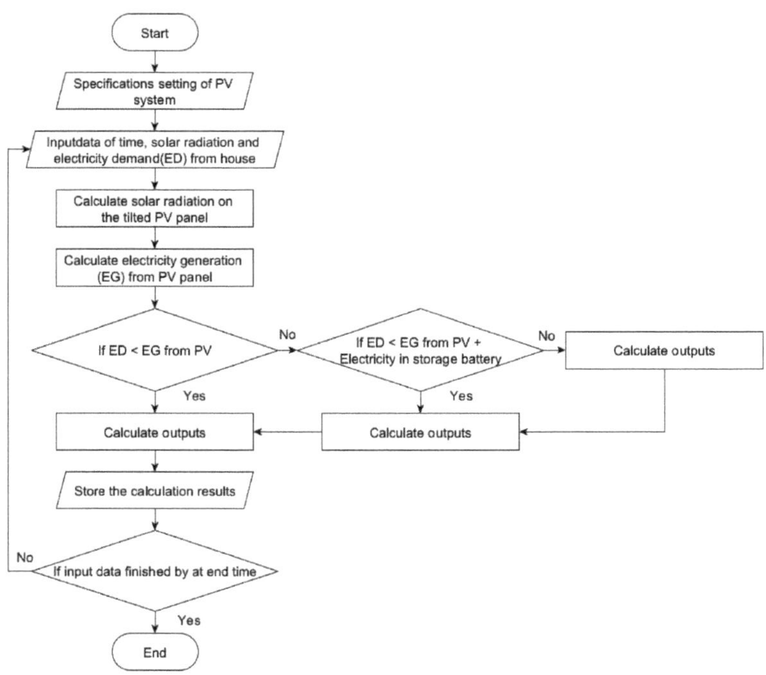

Figure 15. Program flowchart of PV and storage battery system.

The hourly simulation results of energy usage in the newly constructed passive house from the PV system and storage batteries on specific dates in the summer and winter seasons are shown in Figure 16. Although the electricity demand and usage patterns during the winter and summer seasons are the same, the amount of electricity generation and usage from renewable energy is different because the extent of solar radiation is different. According to Figure 16, electricity generation and usage from the PV system is mainly around noon. However, the energy demand is relatively low around noon because the residents are usually at work during the daytime. Therefore, the storage batteries store the surplus electricity generated by the PV system around noon and provide electricity in the morning and evening. Renewable energy can cover all the energy demand on most days in summer and it needs to purchase the electricity from the grid at several hours on most days in winter as the data shown on representative dates in winter and summer in Figure 16; the reason is that the solar radiation in summer is strong and in winter is medium. The energy performance of renewable energy in the intermediate seasons is the same as in summer. In the newly constructed house, the annual electricity demand from the public power plant was 561 kWh, the electricity usage from the storage batteries was 2961 kWh, and the electricity usage directly from the PV system was 1086 kWh. Owing to the similar electricity demand pattern, the size of the PV system and storage batteries were similar in the newly constructed passive house and refurbished passive house; furthermore, the energy performance of the PV system and the storage batteries were similar for the two cases. In the passive refurbished house, the storage batteries provided 2989 kWh, and the PV system provided 1091 kWh of the annual electricity demand of the house. Thus, in both cases, the storage batteries played a crucial role because the electricity usage from the storage batteries was around 2.7 times the electricity used directly from the PV panel. After the application of the renewable energy system, the annual per-area energy demand of the newly constructed passive house was 35 kWh/m^2, and the annual per-area energy demand of the refurbished passive house was 51 kWh/m^2. According to the nZEB standard of China, the annual per-area energy demand should be equal to or below 55 kWh/m^2. Thus, the nZEB standard was achieved through the proposed methods for both cases (new construction and refurbishing of the rural house).

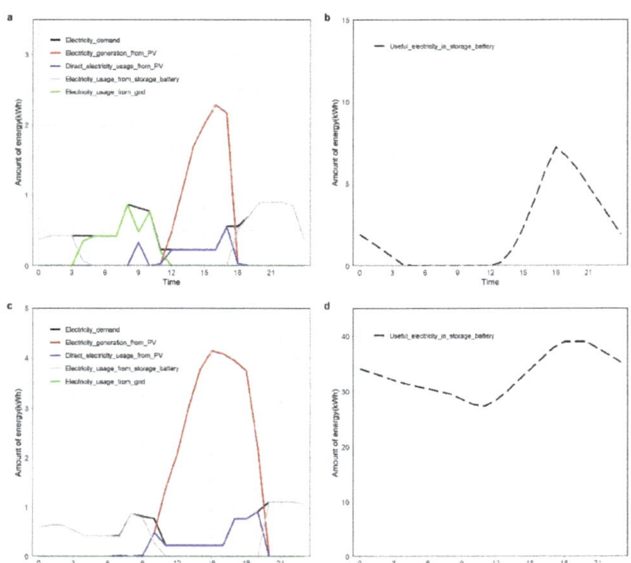

Figure 16. Energy performance of renewable energy system on specific dates in spring and summer for the newly constructed passive house (**a**) and (**b**): data in 1/12; (**c**) and (**d**): data in 8/16.

6. Discussion

Table 10 lists the energy demand reduction ratio of each step of the passive design strategies. The data show that the addition of the external wall insulation material and decreasing the internal height reduced the energy demand of the house significantly. The villagers prefer the higher houses due to the height of the house is related to the impression of wealth in the village social community; however, the higher village house could cause a large amount of heating and cooling energy demand. Therefore, only decreasing the ceiling height without sacrificing spatial comfort can not only reduce the energy demand from the house but also save on the material and labour costs. Due to the area of the external wall being large, the implementation of the insulation materials to the external wall has a significant effect to reduce the energy demand from the house. Therefore, the energy reduction ratio of the passive house is high at steps 2 and 3. On the other hand, as the traditional roof system already has an insulation effect and the window to wall ratio is low, the energy reduction ratio at steps 4, 5 and 6 is relatively small.

Table 10. Energy demand reduction ratio of each passive design strategy.

Steps	Designs	Energy Demand Reduction Ratio of Passive House	
		Newly Constructed	Refurbished
1 (a)	Orientation change (North to south → South to west 20 degrees)	0.99%	-
2 (b)	Ceiling height change (3.5 m → 2.8 m)	13.99%	-
3 (c)	Add wall insulation (60 mm Rock wool board)	13.62%	13.41%
4 (d)	Add roof insulation (60 mm Rock wool board)	2.32%	3.28%
5 (e)	Glazing type change (Single glazing → Double glazing)	1.79%	1.81%
6 (f)	Window to wall ratio change (South wall (0.2 → 0.5))	1.45%	1.47%
7 (g)	Configuration change (Figure 13)	0.45%	-
8 (h)	Add sunroom (1.5 m)	4.00%	0.39%

As the application of passive design strategies to the rural house could not achieve the Chinese nZEB standard, incorporating renewable energy systems was considered. According to the electricity demand of the houses, the appropriate size of the PV system and storage batteries were simulated for application in the newly constructed passive house and refurbished passive house. Table 11 presents the energy demand reduction ratio after applying the PV system and storage batteries to the newly constructed passive house and refurbished passive house. According to Table 11, the energy demand reduction ratio of the PV system for the newly constructed passive design house and refurbished passive design house was 8.1% and 8.2%, respectively. Similarly, the energy demand reduction ratio of the storage batteries for the newly constructed passive house and refurbished passive house was 22% and 22.3%, respectively. Therefore, the application of storage batteries had a significant effect on the energy demand of the houses. In addition to the energy demand reduction ratio of passive design strategies, the application of renewable energy also played a crucial role in reducing the energy demand of the houses. According to the analysis presented in Section 4, the newly constructed passive house and refurbished passive house achieved the nZEB standard of China after applying the passive design strategies and renewable energy system. Hence, it is feasible to refurbish an existing rural house to an nZEB or construct a new nZEB through the methods proposed in this study in the villages in Xi'an, China.

Table 11. Energy demand reduction ratio of applying renewable energy system.

Device Type	Size of Each Device	Energy Demand Reduction of Passive House New Constructed	Refurbished
PV	49.4 m^2	8.1%	8.2%
Storage battery (new constructed passive house)	50 kWh	22%	-
Storage battery (refurbished passive house)	52 kWh	-	22.3%

Due to the limitation of the funding, the measurement time and the devices are not enough to precisely do the software test. In the future, the software will be precisely tested including checking the infiltration rate of the house, surveying the occupancy of the village house and situation of hot water usage at different times.

7. Conclusions

In this study, the feasibility of designing an nZEB based on a typical rural house in Xi'an was explored. Appropriate passive design strategies and renewable energy systems were considered for the design of the nZEB. Two scenarios were considered: the first scenario was the construction of a new house, and the second was the refurbishing of an existing house. Based on the model of the typical rural house, two methods were followed to achieve the goal of this research. Initially, the appropriate passive design strategies were explored and applied in the newly constructed house; subsequently, the possible passive design strategies were selected for refurbishing the existing house. Furthermore, PV and storage battery systems were applied to the newly constructed passive house and refurbished passive house. The conclusions of this study are summarized below.

a. By applying the appropriate passive designs and renewable energy system in the typical rural house, the annual per-area energy demand of the newly constructed house was decreased to 35 kWh/m^2; for the refurbished house it decreased to 51 kWh/m^2. Considering the Chinese nZEB standard in which the annual per-area energy consumption should be equal to or below 55 kWh/m^2 confirmed that the proposed methods can transform a typical rural house in Xi'an, north-western China, into an nZEB through new construction or refurbishing.
b. Among all the passive strategies proposed in this study, decreasing the ceiling height to a reasonable size was a simple method, which reduced the energy demand of the house significantly. Moreover, it also saved construction materials and reduced labour costs. The insulation material made from local materials was simulated to be applied on the external wall of the house; the simulation results show that the energy demand of the house decreased significantly without incurring high costs.
c. Exclusively using passive design strategies in the village house could not meet the Chinese nZEB standard. Thus, the PV and storage battery system were simulated for application in the newly constructed passive house and the refurbished passive house. The simulation results indicate that the electricity usage from the storage battery was 2.7 times higher than the direct electricity usage from the PV panel. Therefore, storage batteries play a crucial role in solar energy systems in residential houses.

Furthermore, considering the energy demand reduction ratio of applying passive design strategies and renewable energy systems, the two methods reduced the energy demand of the rural house in Xi'an significantly. In addition, the achievements in this paper also can be used to instruct the residents to build the nZEB in rural villages like Xi'an in China.

Author Contributions: Conceptualization, H.C.; data curation, T.L.; formal analysis, H.C. and T.D.A.; investigation, H.C. and Y.H.; methodology, H.C. and Y.H.; project administration, H.C. and I.L.; resources, H.C., Y.H. and I.L.; software, I.L. and T.L.; supervision, Y.H., I.L. and T.D.A.; visualization, H.C.; writing—original draft, H.C.; writing—review and editing, T.D.A. All authors have read and agreed to the published version of the manuscript.

Funding: This work was supported by the Key Scientific Research Projects of Colleges and Universities in Henan Province (Grant No. 22B560020).

Institutional Review Board Statement: Not applicable.

Informed Consent Statement: Not applicable.

Data Availability Statement: Data available on request to the first author.

Conflicts of Interest: The authors declare no conflict of interest.

References

1. Torcellini, P.; Pless, S.; Deru, M. *Zero Energy Buildings: A Critical Look at the Definition*; CP-550-39833; Conference Paper National Renewable Energy Laboratory: Golden, CO, USA, 2006.
2. Feng, W.; Zhang, Q.; Ji, H.; Wang, R.; Zhou, N.; Ye, Q.; Hao, B.; Li, Y.; Luo, D.; Lau, S.S.Y. A Review of Net Zero Energy Buildings in Hot and Humid Climates: Experience Learned from 34 Case Study Buildings. *Renew. Sustain. Energy Rev.* **2019**, *114*, 109303. [CrossRef]
3. ZEBRA2020. Official nZEB Definition. Available online: https://zebra-monitoring.enerdata.net/overall-building-activities/nzeb-definitions-by-country.html (accessed on 16 August 2020).
4. *GB/T 51350-2019*; Technical Standard for Nearly Zero Energy Building. China Architecture & Building Press: Beijing, China, 2019.
5. U.S. Department of Energy. Passive Solar Home Design. Available online: http://energy.gov/energysaver/articles/passive-solar-home-design (accessed on 6 June 2021).
6. Song, Q. A Research on the Basic Theory and Method of Passive Building. Ph.D. Dissertation, Department of Architectural Technology and Science of Xi'an University of Architecture and Technology, Xi'an, China, 2015.
7. Qi, F. Passive Climate Partition of Architectural Design. Master's Dissertation, Department of Architecture of Xi'an University of Architecture and Technology, Xi'an, China, 2015.
8. Li, E. Passive Design Strategy on Residential Buildings for Sustainable Development of Developing Cities. Ph.D. Dissertation, Architecture Graduate School of Human-Environment Studies of Kyushu University, Fukuoka, Japan, 2012.
9. Yin, B.G. Benefit Evaluation and Policy Support of Passive Building. Master's Dissertation, Department of Administrative Management of Xi'an University of Architecture and Technology, Xi'an, China, 2014.
10. Magrini, A.; Lentini, G.; Cuman, S.; Bodrato, A.; Marenco, L. From Nearly Zero Energy Buildings (Nzeb) to Positive Energy Buildings (Peb): The Next Challenge—The Most Recent European Trends with Some Notes on the Energy Analysis of a Forerunner Peb Example. *Dev. Built Environ.* **2020**, *3*, 100019. [CrossRef]
11. Kotarela, F.; Kyritsis, A.; Papanikolaou, N.; Kalogirou, S.A. Enhanced Nzeb Concept Incorporating a Sustainable Grid Support Scheme. *Renew. Energy* **2021**, *169*, 714–725. [CrossRef]
12. Arif, S.; Taweekun, J.; Ali, H.M.; Yanjun, D.A.I.; Ahmed, A. Feasibility Study and Economic Analysis of Grid Connected Solar Powered Net Zero Energy Building (Nzeb) of Shopping Mall for Two Different Climates of Pakistan and Thailand. *Case Stud. Therm. Eng.* **2021**, *26*, 101049. [CrossRef]
13. Dartevelle, O.; van Moeseke, G.; Mlecnik, E.; Altomonte, S. Long-Term Evaluation of Residential Summer Thermal Comfort: Measured Vs. Perceived Thermal Conditions in Nzeb Houses in Wallonia. *Build. Environ.* **2021**, *190*, 107531. [CrossRef]
14. Monzón-Chavarrías, M.; López-Mesa, B.; Resende, J.; Corvacho, H. The Nzeb Concept and Its Requirements for Residential Buildings Renovation in Southern Europe: The Case of Multi-Family Buildings from 1961 to 1980 in Portugal and Spain. *J. Build. Eng.* **2021**, *34*, 101918. [CrossRef]
15. De Masi, R.F.; Gigante, A.; Vanoli, G.P. Are Nzeb Design Solutions Environmental Sustainable? Sensitive Analysis for Building Envelope Configurations and Photovoltaic Integration in Different Climates. *J. Build. Eng.* **2021**, *39*, 102292. [CrossRef]
16. Ferrara, M.; della Santa, F.; Bilardo, M.; de Gregorio, A.; Mastropietro, A.; Fugacci, U.; Vaccarino, F.; Fabrizio, E. Design Optimization of Renewable Energy Systems for Nzebs Based on Deep Residual Learning. *Renew. Energy* **2021**, *176*, 590–605. [CrossRef]
17. Vujnović, N.; Dović, D. Cost-Optimal Energy Performance Calculations of a New Nzeb Hotel Building Using Dynamic Simulations and Optimization Algorithms. *J. Build. Eng.* **2021**, *39*, 102272. [CrossRef]
18. Baiduwenku. Population of China. Available online: https://wenku.baidu.com/view/63f1c908941ea76e58fa04fd.html (accessed on 28 December 2020).
19. Energy Conservation Research Center of Tsinghua University. *China Energy Conservation Annual Report*; China Architecture & Building Press: Beijing, China, 2012.

20. Santos-Herrero, J.M.; Lopez-Guede, J.M.; Flores-Abascal, I. Modeling, Simulation and Control Tools for Nzeb: A State-of-the-Art Review. *Renew. Sustain. Energy Rev.* **2021**, *142*, 110851. [CrossRef]
21. Kuivjõgi, H.; Uutar, A.; Kuusk, K.; Thalfeldt, M.; Kurnitski, J. Market Based Renovation Solutions in Non-Residential Buildings—Why Commercial Buildings Are Not Renovated to Nzeb. *Energy Build.* **2021**, *248*, 111169. [CrossRef]
22. ASHRAE. *Energy Efficient Design for Low-Rise Residential Buildings, American Society of Heating, Refrigeration, and Air-Conditioning Engineers*; ASHRAE Standard 90.1; ASHRAE: Atlanta, GA, USA, 2016.
23. Li, Y. Study on Heating Pattern of Rural Residences. Ph.D. Dissertation, Department of Architectural Design and Theory of Xi'an University of Architecture and Technology, Xi'an, China, 2014.
24. Ministry of Construction of People's Republic of China. *Thermal Design Code for Civil Building*; China Planning Press: Beijing, China, 1993.
25. State Environment Protection Administration of China. *Indoor Air Quality Standard*; State Environment Protection Administration of China: Beijing, China, 2002.
26. Chen, D.L. Research on house infiltration in Architecture Energy Saving Area. *Archit. Energy Sav. Green Archit.* **2014**, *18*, 71–73. [CrossRef]
27. Mazzeo, D.; Matera, N.; Cornaro, C.; Oliveti, G.; Romagnoni, P.; De Santoli, L. EnergyPlus, IDA ICE and TRNSYS predictive simulation accuracy for building thermal behaviour evaluation by using an experimental campaign in solar test boxes with and without a PCM module. *Energy Build.* **2020**, *212*, 109812. [CrossRef]
28. Ministry of Housing and Urban-Rural Construction of the People's Republic of China. *Residential Design Specification*; China Architecture & Building Press: Beijing, China, 2008.
29. Yoza, A.; Yona, A.; Senjyu, T.; Funabashi, T. Optimal capacity and expansion planning methodology of PV and battery in smart house, Renew. *Energy* **2014**, *69*, 25–33. [CrossRef]
30. China Energy Storage. Best Angle of PV. Available online: http://www.escn.com.cn/news/show-442615.html (accessed on 15 March 2021).
31. Erbs, D.G.; Klein, S.A.; Duffie, J.A. Estimation of the diffuse radiation fraction for hourly, daily and monthly-average global radiation. *Sol. Energy* **1982**, *28*, 293–302. [CrossRef]
32. Chen, W.; Shen, H.; Deng, Y.J. Appling research of storage batteries in photovoltaic system. *Storage Battery* **2006**, *1*, 21–27.

buildings MDPI

Article

Experimental and Numerical Study on the Insulation Performance of a Photo-Thermal Roof in Hot Summer and Cold Winter Areas

Ying Zhang [1], Hongfa Sun [1,*], Jibo Long [1], Li Zeng [2] and Xiaohang Shen [3]

[1] College of Civil Engineering and Mechanics, Xiangtan University, Xiangtan 411105, China; 202021002620@smail.xtu.edu.cn (Y.Z.); hngzlong@xtu.edu.cn (J.L.)
[2] Wenzhou Industrial Design Institute Co., Ltd., Wenzhou 325000, China; kg3@163.com
[3] College of Civil Engineering, Hunan University, Changsha 410082, China; b1201s006@hnu.edu.cn
* Correspondence: sunhf1212@xtu.edu.cn

Abstract: The use of a solar architecture system is a feasible way to reduce the energy consumption of a building. The system also has important significance to the "Dual-carbon" plan. In this study, the heat transfer characteristics of a photo-thermal roof were analyzed in hot summer and cold winter zones; a model to calculate insulation performance was established. In the summer climate, the thermal performances of the photo-thermal roof and an ordinary roof were explored through experiments and simulations. The results showed that the heat transfer and temperature of the photo-thermal roof were lower than those of the ordinary roof. Heat transfer through a photo-thermal roof can be changed by adjusting the water flow of collectors. The water saturation of insulation materials is an important factor that affects the insulation performance of a roof. Compared with the ordinary roof, the change in water saturation was shown to have less impact on the insulation performance of the photo-thermal roof. The water saturation increased from 0 to 30%, while the heat transfer per unit area of the photo-thermal roof only increased by 0.9 W/m^2; 97.3% lower than that of the ordinary roof. The effect of reducing the insulation material thickness was less for the photo-thermal roof than for the ordinary roof. When the insulation material thickness was reduced from 100 mm to 0 mm, the average temperature in the indoor non-working area reached 38.5 °C and 27.1 °C in the ordinary roof and the photo-thermal roof, respectively. The insulation thickness of the photo-thermal roof had little effect on the indoor air temperature. The research results provide a reference for the roof energy-saving design of new buildings and the roof energy-saving transformation of existing buildings.

Keywords: building energy consumption; photo-thermal roof; water saturation; insulation performance; temperature distribution

Citation: Zhang, Y.; Sun, H.; Long, J.; Zeng, L.; Shen, X. Experimental and Numerical Study on the Insulation Performance of a Photo-Thermal Roof in Hot Summer and Cold Winter Areas. *Buildings* **2022**, *12*, 410. https://doi.org/10.3390/buildings12040410

Academic Editors: Shi-Jie Cao and Wei Feng

Received: 22 February 2022
Accepted: 26 March 2022
Published: 28 March 2022

1. Introduction

In order to ensure the sustainable development of human society, the "Dual-carbon" plan was proposed. The "Dual-carbon" plan is a "carbon emission peak and carbon neutrality" plan. The main way to achieve the goals of the "Dual-carbon" plan is to reduce the consumption of fossil energy to reduce greenhouse gas emissions [1]. Energy consumption by buildings accounts for more than one-third of global energy consumption, and the carbon dioxide emissions of buildings account for 40% of the total emissions worldwide [2,3]. Moreover, with the continuous improvement of people's living standards, building energy consumption will continue to grow [4,5]. Therefore, it is important to achieve the goals of the "Dual-carbon" plan in the construction field as soon as possible, reduce the energy consumption of buildings and develop renewable energy utilization technology through building energy conservation technology research [6].

In energy-saving building technology, the photovoltaic/thermal system has attracted extensive attention because of its excellent characteristics, such as its renewable and pollution-free nature [7]. Photovoltaic panels, solar collectors, and hybrid systems are the most suitable technologies for energy-saving building applications [8,9]. Most previous research in this field has focused on how to improve the energy efficiency of photovoltaic/thermal systems or has evaluated the economy of these systems [10–13]. Since the 1990s, many scholars have studied the design of solar heat collectors, photovoltaic panels, and hybrid systems integrated with buildings [14–17]. Solar collectors and photovoltaic panels combined with roofs have been shown to reach to about 120 °C in summer [18,19]. High temperature collectors and photovoltaic panels can enhance heat transfer to a building through convection, radiation, and other means, thereby affecting building energy consumption and indoor thermal comfort [20–22]. Unlike the overheating of photovoltaic panels, the temperature of the collector used to prepare domestic hot water is generally stable at 50–80 °C, and most of the solar radiation energy is used to heat the working medium and is then transferred with the working medium. If the fluid temperature of the solar collector changes, the temperature of the enclosure may change, affecting the indoor thermal environment [23]. Sourek [24] studied the interaction between the collector coupled with the building façade and the building and found that the indoor temperature increased by no more than 1 K in all investigated configurations. Buonomano [25] observed the summer superheating effects by detecting the passive thermal behavior effect of a light building with solar collectors coupled with the building façade in Milan. However, there are many diverse photo-thermal system and architecture combinations. Most previous studies only considered the passive effect of the collector in close contact with the building façade, and ignored the impact of the air gap between the photo-thermal system and the building's roof [26]. A study conducted in China pointed out that solar radiation absorption by the roof surface can account for more than 40% of the power consumption of the top-floor rooms during hot summer periods [27]. Inhibiting the heat absorption of the roof surface is very important to reduce the energy consumption of the top floor rooms of a building during the summer [28]. Solar collectors assembled on the roof also screen most of the solar rays, but the passive impact is often ignored [29].

China's dual carbon goals were proposed in 2020, which put forward higher requirements for the utilization of solar energy. Photo-thermal technology is an important technical mean to reduce building energy consumption and carbon emission. It is necessary to understand the influence of photo-thermal system on a building's thermal environment. Based on the existing literature, it can be found that the current utilization of solar energy in buildings is mainly photothermal utilization. However, the use of solar thermal insulation in buildings is not yet uncommon. Thus, this paper compared the thermal performance of a roof equipped with a field of thermal solar collectors with an air gap and a traditional bare roof in hot summer and cold winter zones. Combined with the influence of water saturation of roof insulation materials on roof heat transfer, the insulation performance of the photo-thermal roof was explored under different insulation layer thicknesses and different collector water temperatures. This research provides a theoretical basis for the application of photo-thermal roofs in the field of building energy conservation.

2. Methods

2.1. Experimental Model

The two experimental rooms were on the top floor of the civil engineering experimental building of Xiangtan University located at 112.86° longitude and 27.88° north latitude. The schematic structure of the experimental roof is shown in Figure 1. The structure of the roof from top to bottom was as follows: 40 mm reinforced concrete, 100 mm rock wool board, 20 mm cement mortar, 120 mm reinforced concrete, 20 mm cement mortar. The area used for the photo-thermal roof experiment was 5.6 m × 5 m, and 10 solar collectors 2000 mm × 1000 mm × 106 mm in size were used. The water temperature was adjusted by controlling the flow. There were no solar collectors on the ordinary roof, but its area, orientation, and roof structure were the same as for the photo-thermal roof. Both experimental test rooms were air-conditioned and cooled from 9:00 to 21:00 every day.

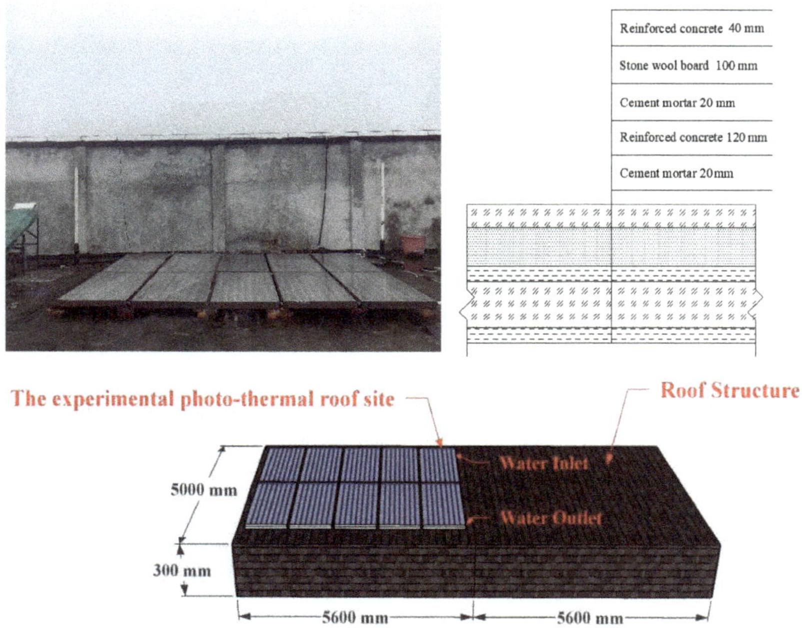

Figure 1. The experimental roof situation.

2.2. Experimental Measurements

In the experiment, temperature signals were collected by a k-type thermocouple with an accuracy level of 1 and a measurement error of less than 0.1%. The room studied under summer conditions was is an air-conditioned room, with the area 1 m from the inner surface of the roof being a non-working area. As shown in Figure 2, temperature detectors were placed on the roof surfaces, in the non-working area, and in the water inlet and outlet. The main temperature signals collected were: the internal and external roof surface temperature and the inlet and outlet water temperatures. Additionally, detectors were used to record the outdoor air temperature. The data acquisition unit (XSR-70A) is used to record the temperature data in real time. The radiation recorder (TBQ-2) was used to measure the solar irradiance of the test site. Test equipment and parameters are shown in Table 1.

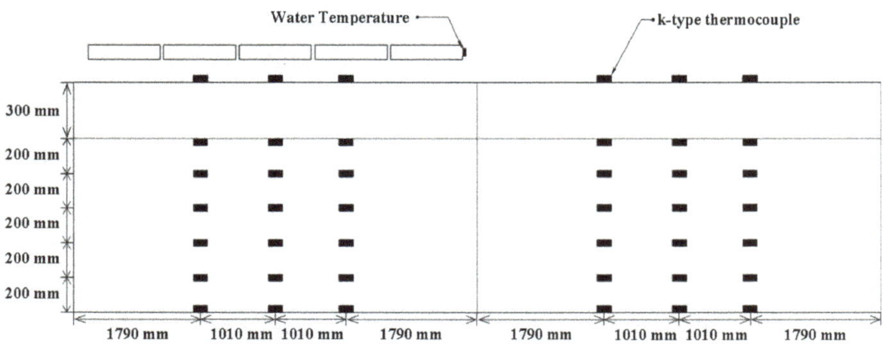

Figure 2. Locations of detectors.

Table 1. Test instruments and types and test parameters.

Test Parameters	Test Instrument	Type	Accuracy
Solar radiation intensity	Solar pyrometer	TBQ-2	≤0.2%
Temperature	Thermocouple	Ni-Cr and Ni-Si	±0.5 °C
	Data acquisition unit	XSR-70A	≤0.2%

2.3. Mathematical Model

2.3.1. Heat Transfer Analysis

As shown in Figure 3a, solar rays reached the photo-thermal roof, and a small portion were reflected by the glass cover of the collector, while the rest were absorbed by the collector. The collector converted solar radiation into heat. Most of the heat was used to heat the heat-carrying fluid and was carried away by the heat-carrying fluid. Some heat caused the temperature of the collector to increase. The collector then dissipated heat to the surrounding environment through convection and radiation. During heat dissipation by the collector, some of the heat was transferred to the room through the roof. The heat transfer conducted by the ordinary roof is shown in Figure 3b. The solar radiation directly reached the outer surface of the roof and increased the temperature of the roof, and the heat directly entered the room through the roof.

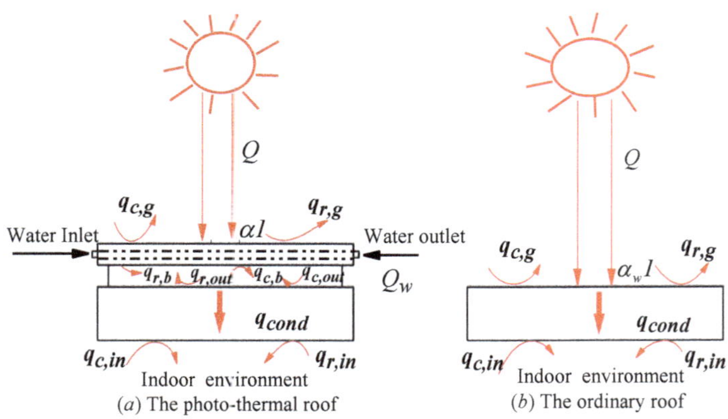

Figure 3. Roof heat transfer model.

The energy balance equation of the heat transfer model can be expressed using Equation (1) [30].

$$\frac{\partial}{\partial \tau}(\rho E) + \nabla[\vec{u}(\rho E + P)] = \nabla[\lambda_e \nabla T - \sum_j h_j \vec{J_j} + (\overline{\tau_{eff}} \vec{u})] + S_h \tag{1}$$

Since the thicknesses of each solid material layer of the solar collectors and roofs were much smaller than their length and widths, their heat conduction can be regarded as one-dimensional along the z-axis direction (height direction). Ignoring the diffusion component and viscosity dissipation, the above formula can be transformed into Equation (2).

$$\frac{\partial}{\partial \tau}(\rho E) + \nabla \vec{u} \rho E = \nabla(\lambda_e \nabla T) + S_h \tag{2}$$

The equation of energy balance was established with the collector as the research object as shown in Equation (3).

$$\rho C \frac{\partial T}{\partial \tau} = (1 - \rho_g) I - \lambda_p \frac{\partial^2 T}{\partial z^2} - S \tag{3}$$

In Equation (3), the left-hand side of the equation represents the internal energy of the collector, whereas the right-hand side of the equation represents the heat gain term, the heat conduction term, and the heat loss term. The heat loss of the collector includes two main parts. One part is the heat emitted by the collector and the external environment through radiation and convection, and the other part is the heat loss that results from heating the fluid in the collector. As the distance between the collector and the outer surface of the roof was very small, only 100 mm, and the coverage area of the collector was large, there was little natural wind between the collector and the roof. In the hot summer condition, the temperature of the rear cover plate of the normal working collector was higher than that of the outer surface of the roof, making it difficult for natural convection between them. Therefore, it was assumed that the air between the collector and the roof was almost stagnant. The convective heat transfer between the collector's rear cover and the roof was not considered. S_h can be expressed using Equation (4).

$$S_h = h_g(T_g - T_{o,a}) + \sigma\varepsilon_g(T_g^4 - T_{sky}^4) + \frac{\sigma\varepsilon_g(T_b^4 - T_{o,w}^4)}{\frac{1}{\varepsilon_b} + \frac{1}{\varepsilon_{o,w}} - 1} + q_w \tag{4}$$

Taking the outside surface of the photo-thermal roof as the research object, the heat balance equation can be expressed as shown in Equation (5).

$$\lambda_a \frac{\partial^2 T}{\partial z^2} + \frac{\sigma\varepsilon_g(T_b^4 - T_{o,w}^4)}{\frac{1}{\varepsilon_b} + \frac{1}{\varepsilon_{o,w}} - 1} = q_{cond} \tag{5}$$

Equation (6) is the energy balance equation for the photo-thermal roof.

$$\rho C \frac{\partial t}{\partial \tau} = S_h - \lambda_e \frac{\partial^2 T}{\partial z^2} - h_i(T_{i,w} - T_{i,a}) - \sigma\varepsilon_{i,w}(T_{i,w}^4 - T_n^4) \tag{6}$$

where S_h represents the heat gain of the photo-thermal roof, determined using (5). S_h can be expressed using Equation (7).

$$S_h = \lambda_a \frac{\partial^2 T}{\partial z^2} + \frac{\sigma\varepsilon_g(T_b^4 - T_{o,w}^4)}{\frac{1}{\varepsilon_b} + \frac{1}{\varepsilon_{o,w}} - 1} \tag{7}$$

Different from the photo-thermal roof, heat gain by the ordinary roof mainly came from direct solar irradiation. Taking the outer surface of the ordinary roof as the research object, the heat balance equation was established, as shown in Equation (8).

$$\alpha_w I - \sigma\varepsilon_{o,w}(T_{o,w}^4 - T_{sky}^4) - h_{o,w}(T_{o,w} - T_{o,a}) = q_{cond} \tag{8}$$

Equation (9) is the heat transfer energy balance equation for the ordinary roof.

$$\rho C \frac{\partial T}{\partial \tau} = \mathrm{S_h} - \lambda_\mathrm{e} \frac{\partial^2 T}{\partial z^2} - h_{i,w}(T_{i,w} - T_{i,a}) - \sigma\varepsilon_{i,w}(T_{i,w}^4 - T_n^4) \tag{9}$$

where S_h represents heat gain by the ordinary roof, determined using Equation (8). S_h can be expressed using Equation (10).

$$S_h = \alpha_w I - h_o(T_{o,w} - T_{o,a}) - \sigma\varepsilon_{o,w}(T_{o,w}^4 - T_n^4) \tag{10}$$

Although the collectors performed strongly in terms of heat accumulation, most of this heat energy was used to produce domestic hot water. The heat gained by the photo-thermal roof mainly came from radiant heat transfer between the collectors and the roof. However, the ordinary roof directly absorbed solar irradiation, converted it into heat energy, and transferred it to the rooms.

According to Equations (6) and (9), when the heat received on the roof outer surface and the roof structure is constant, the heat entering the room through the roof is only related to the effective thermal conductivity of the roof's insulation materials, and the more effective the thermal conductivity is, the more heat that is transferred into the room through the roof.

2.3.2. The Effective Thermal Conductivity of the Roof's Insulation Material

Building thermal insulation materials are mostly porous media filled with air and moisture. The volume ratio of air and water changes depending on the environmental temperature and humidity. Therefore, building insulation materials are actually a mixture of solid-phase insulation materials, liquid-phase water, and air, which can be expressed using Equation (11).

$$\varepsilon_s + \varepsilon_l + \varepsilon_a = 1 \tag{11}$$

Among these three substances, the volume of the solid-phase insulation material does not change, while the volume of air decreases as the liquid water content increases. The volume of liquid water can be expressed as water saturation using Equation (12).

$$\varphi = \frac{\varepsilon_l}{\varepsilon_l + \varepsilon_a} \tag{12}$$

The thermal conductivity of water is greater than that of air. When the liquid water content in the pores increases, the volume of air will reduce, and the effective thermal conductivity of the material will increase. When the pores are filled with air and water, due to the hydrophilicity of the solid surfaces of the pores, the volume of liquid water, and other factors, the distribution of liquid water varies widely. It is difficult to accurately calculate the effective thermal conductivity of the insulation material. If the heat conduction of the porous medium satisfies the requirements of the parallel heat conduction model, its effective thermal conductivity can be expressed using Equation (13) [31].

$$\lambda_{\mathrm{ep}} = \lambda_s \varepsilon_s + \lambda_l \varepsilon_l + \lambda_a \varepsilon_a \tag{13}$$

If the heat conduction of the porous medium satisfies the requirements of the serial heat conduction model, its effective thermal conductivity can be calculated using Equation (14) [31].

$$\lambda_{\mathrm{ec}} = \frac{1}{\frac{\varepsilon_s}{\lambda_s} + \frac{\varepsilon_l}{\lambda_l} + \frac{\varepsilon_a}{\lambda_a}} \tag{14}$$

The above two methods of calculating the effective thermal conductivity represent the extreme conditions. In reality, the situation is somewhere in between.

$$\lambda_{ec} \le \lambda_e \le \lambda_{ep} \tag{15}$$

In this study, stone wool board with a porosity of 95% was used as the roof insulation material. This material has a monofilament fiber thermal conductivity of about 1.25 W/(m·K) [32]. When the rock wool board was used as the roof insulation material, heat transfer was directed in the direction of its thickness. Some studies have pointed out that the interior part of the rock wool board was mainly composed of a large number of rock wool monofilament fibers and a small number of slag balls [32]. In the thickness direction, the inclined lapping of monofilament fibers of the rock wool board at different angles did not represent single horizontal or vertical placement, and a small number of slag balls were also randomly scattered inside, leading to the coexistence of serial heat conduction and parallel heat conduction in the rock wool board. In addition, the distribution of liquid water in the pores of the porous media was irregular. In order to simplify the analysis, the effective thermal conductivity was calculated by Equation (16).

$$\lambda_e = \frac{\lambda_{ep} + \lambda_{ec}}{2} \quad (16)$$

The above calculation method shows the variation law of effective thermal conductivity for the roof insulation material used in this study, and this is also shown in Figure 4.

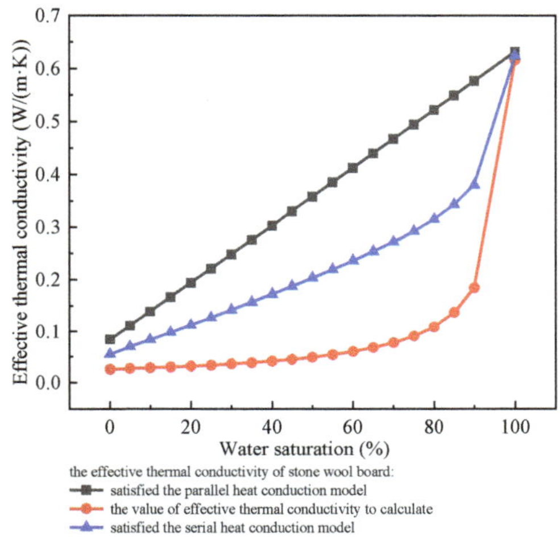

Figure 4. The effect of water saturation on the effective thermal conductivity of the insulation material.

As shown in Figure 4, the effective thermal conductivity calculated by the parallel heat conduction model changes linearly with the water saturation. In the serial heat conduction model, when the water saturation is less than 80%, the effective thermal conductivity increases slightly as the water saturation increases, from 0.025 W/(m·K) to 0.108 W/(m·K). When the water saturation is more than 80%, the change in the effective thermal conductivity relative to the water saturation increases sharply, from 0.108 W/(m·K) to 0.616 W/(m·K).

The varied relationship between the effective thermal conductivity and the water saturation calculated according to Equation (16) is shown in Figure 4. The water saturation can affect the effective thermal conductivity of rock wool board over a large range. With an increase in the water saturation, the effective thermal conductivity of rock wool board also increases. When the water saturation increases from 0% to 100%, the effective thermal conductivity increases by 933.3%, in theory. With an increase in water saturation from 0% to 30%, the effective thermal conductivity increases from 0.06 W/(m·K) to 0.14 W/(m·K), an increase of 60.7%.

2.4. Simulation Conditions

In order to determine the similarities and differences in thermal performance between the photo-thermal roof and the ordinary roof, a numerical simulation was used to create two physical models. One was a simplified three-dimensional model of an air-conditioned room with a photo-thermal roof, and the other was an ordinary air-conditioned room with the same bare roof structure and materials. The numerical model was calculated with FLUENT software. In order to simplify the model, a water pipe and its surrounding components in the center of the collector of the photo-thermal roof were considered to have the smallest width unit.

Considering the limited cooling capacity of the air conditioner, the air area of the top floor room was divided into two parts according to height. One part was the area of air 1000 mm below the roof, the non-working area, and the other part was the personnel working area. The air conditioner was able to maintain a constant temperature of 26 °C in the working area. Due to close contact with the roof, the temperature of the non-working area was affected by heat transfer from the roof and the temperature of the working area. In order to further simplify the model, only 0.5 m of the indoor air-conditioned area and its upper space components were modeled. In order to further simplify the model, only the non-working area and the area 0.5 m below it were modeled. The minimum size of the numerical model was 134 mm × 2000 mm × (δ + 1906) mm, where δ is the thickness of the insulation layer of the roof, which represents the amount of change.

The size of the water pipes in the solar collector was 8 × 2000 × 8 mm, and the distance between the two water pipes was 126 mm. The components used in the model were all general materials with general dimensions, and the dimensions and materials are shown in Table 2.

Table 2. Physical properties of each material layer used in the model.

Material Layers	Size x × y × z (mm^3)	Density (kg/m^3)	Specific Heat Capacity (J/(kg·K))	Thermal Conductivity (W/(m·K))
Glass cover	1000 × 2000 × 4.5	2500	840	0.76
Air layer in collector	1000 × 2000 × 45	1.205	1005	0.026
Absorber plate	1000 × 2000 × 1.5	2719	871	202.4
Copper pipe	8 × 2000 × 8	8978	381	387.6
Insulation in collector	1000 × 2000 × 55	50	1380	0.04
Air layer	4000 × 5000 × 100	1.205	1005	0.026
Reinforced concrete layer	1000 × 2000 × 160	2500	860	1.73
Insulation layer	1000 × 2000 × δ	120	750	0.04
Cement mortar layer	1000 × 2000 × 40	2000	840	0.87

This simulation introduced the solar ray tracing model. The geographical location was set to 27.88° N, 112.87° E with a time zone of +8. Ideal weather conditions were set as clear and cloudless, and the time was adjusted to obtain different solar radiation conditions. The DO radiation model was chosen, and the value of angular discretization was increased appropriately to obtain a more accurate simulation result. For the turbulence model, *k*-*ε* equation models with high accuracy were used. The boundary conditions of the model are shown in Table 3.

Table 3. Simulation conditions.

Boundary	Type	Value	Remarks
Glass cover	Convective heat transfer surface	$\alpha_g = 0.12$ $\tau_g = 0.8$	Introduction of solar radiation from the solar ray tracing model.
Absorber plate	Coupling surface	$\alpha_p = 0.95$	-
Copper pipe	Temperature surface/Coupling surface	50 °C/-	The temperature of the pipe was kept constant by adjusting the water flow.
Insulation in collector	Coupling surface	-	-
Outside surface of the photo-thermal roof	Coupling surface	-	-
Outside surface of the ordinary roof	Mixed	$\alpha_w = 0.6$	-
Inner surface of roofs	Coupling surface	-	-
Contact surface between air-conditioned area and non-air-conditioned area	Coupling surface	-	-

The simulated values of the meteorological parameters varied according to the purpose of the simulation. When using experimental data to verify the model, the meteorological parameters needed to be adjusted to be consistent with the experimental data. When using the model to explore practical problems, the meteorological parameters used were the typical meteorological parameters for Xiangtan City in summer [33], and the specific values are listed in Table 4.

Table 4. Typical meteorological parameters in summer in Xiangtan City, China.

Solar Radiation Intensity (W/m^2)	Ambient Temperature (°C)	Wind Speed (m/s)	Sky Temperature (K)
1000	35.8	2.6	296

The air density is affected by temperature. Therefore, the Boussinesq hypothesis was set to represent air density to simplify the solution. The coefficient of thermal expansion of air was set to 0.0037/k. The SIMPLEC algorithm was used to calculate the incompressible flow field. In the solving process, when the variable residuals of the continuity equation, momentum equation, k equation, and e equation were less than 10^{-3} and the variable residuals of energy equations and the radiation equation were less than 10^{-6}, convergence of the calculation was considered to be achieved.

2.5. Simulation Verification

Simulation verification was carried out using experimental data collected on July 12. Figure 5 shows that the solar radiation intensity was basically stable at about 930 W/m^2, and the air temperature was stable at about 37.8 °C during the period from 10:00 to 14:30. When the solar radiation intensity and air temperature remained constant, roof heat transfer reached a steady state. The outside surface temperature of the ordinary roof continued to rise from 10:00 to 14:00, and the temperature did not rise after reaching the maximum value of 53.0 °C at 14:30. At this time, the outside surface temperature of the ordinary roof reached the corresponding steady-state value. Compared with the ordinary roof, the rise in the outside surface temperature of the photo-thermal roof was slightly delayed. The total phase delay can be calculated using Equation (17), and the delay time of the outside surface of the photo-thermal roof was 2 h. Therefore, the outside surface temperature of the photo-thermal roof reached a steady-state value of 37.0 °C at 16:30. According to Equation (17) [34], the delay time of the inner surface of the roof was 5.5 h. Therefore, under a stable external thermal environment, the inner surface temperature of the ordinary roof reached its corresponding steady-state value of 43.2 °C at 20:00, and the inner surface temperature of the photo-thermal roof reached its corresponding steady-state value of 36.8 °C at 23:00.

$$\phi = \frac{\sum D}{\sqrt{2}} \times 57.3 - \text{arctg}\frac{1}{1+\frac{S_1\sqrt{2}}{\alpha_1}} + \text{arctg}\frac{1}{1+\frac{S_n\sqrt{2}}{\alpha_n}} \tag{17}$$

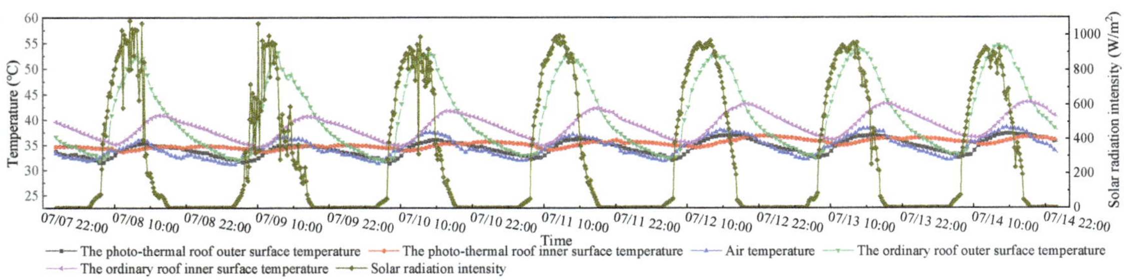

Figure 5. Variation of the temperature and solar radiation intensity on seven consecutive days.

Under the experimental conditions, the solar radiation intensity was 930 W/m^2, the air temperature was 37.8 °C, the thickness of the roof insulation layer was 100 mm, and the average water temperature in the collector was 50 °C. These values were substituted into the numerical calculation model. Considering that the experimental building was a new building, the roof had an excellent waterproofing performance. Therefore, the saturation rate was set to 0. It was calculated that the outer and inner surface temperatures of the ordinary roof were 54.3 °C and 38.7 °C, respectively, and the outer and inner surface temperatures of the photo-thermal roof were 30.7 °C and 28.1 °C, respectively. In general, the values simulated by the photo-thermal system were slightly smaller than the experimental values. This is because the radiation heat transfer between the wall and roof was not considered in the simulation calculation. In order to analyze the error between the experimental value and the simulated value, the relative error calculation results are shown in Table 5.

Table 5. Relative mean error for the simulation and experimental results.

Comparison of Simulation and Experimental Results	Ordinary Roof Outer Surface Temperature	Ordinary Roof Inner Surface Temperature	Photo-Thermal Roof Outer Surface Temperature	Photo-Thermal Roof Inner Surface Temperature
Experimental values	53.0	43.2	37.0	36.8
Simulation values	51.3	39.7	33.5	32.7
RE (%)	3.2%	8.1%	9.5%	11.1%
RMSE	3.3			

Table 5 shows that the simulated value for the ordinary roof was relatively close to the experimental value, and the maximum value of relative error (RE) was 8.1%. The simulated value for the photo-thermal roof was slightly different from the experimental value, and the maximum RE was 11.1%. The root-mean-squared error (RMSE) was 3.3. This suggests that the experimental values are in good agreement with the simulated values.

3. Results and Discussion

3.1. Experimental Results and Analysis

The thermal insulation performance of the roofs was tested during the hot season in 2021. Experimental data from seven consecutive sunny days from 8 to 14 July 2021 were used. The first three days were cloudy, and the influence of clouds on the solar radiation intensity value was greater on those three days than on the next four days. The relationships between the roofs' internal and external surface temperatures and the solar radiation intensity and air temperature are shown in Figure 5.

As can be seen from Figure 5, the temperature and solar radiation intensity values measured in the experiment showed periodic changes. The solar radiation intensity began to rise from about 6:00, reaching a maximum of about 900 W/m^2 from 10:00 to 14:30, and then gradually decreasing before dropping to 0 W/m^2 after 19:00. The outdoor air temperature fluctuated between 31.1 and 38.6 °C and reached its maximum value at 14:00–15:30 every day.

The roof temperature changed periodically under the influences of the solar radiation intensity and air temperature. However, the peaks of different temperature curves occurred at different times and the magnitude of the temperature fluctuation differed. The outer surface temperature of the ordinary roof reached a maximum value of about 53.3 °C from 13:00 to 14:30 and then gradually decreased until the next wave cycle began at 7:30 the next day. There was a delay in the inner surface temperature fluctuations of the ordinary roof, whereby a maximum temperature of about 42.3 °C occurred between 18:00 and 19:30 every day, before gradually decreasing and reaching a minimum value at about 10:30 the next day. The outer surface temperature of the photo-thermal roof reached a maximum value of about 36.5 °C from 15:30–16:30 daily and then gradually decreased until the minimum value was reached at about 8:30 the next day. The inner surface temperature of the photo-thermal roof reached a minimum value at 12:30–13:30 daily and a maximum value of about 36.0 °C at 22:00–24:00.

The temperature fluctuations were roughly similar, but the peaks and peak times differed slightly over different periods due to fluctuations in the solar radiation intensity and air temperature during the day. The solar radiation intensity fluctuated greatly in the first three days of the experiment, the air temperature was lower, the peak of each temperature was slightly lower, and the peak appeared slightly earlier than on the last four days.

As can be seen from Figure 5, on July 12, the weather was fine with few solar rays shielded by clouds, and the meteorological conditions were in line with the typical local summer climate. Therefore, July 12 was selected as a typical day and used to analyze the relationships among various temperatures. To cover the full cycle of each curve, the timeline for typical days to 0:30 on July 12 until 12:30 on July 13. Figure 6 shows the experimental data for the typical day.

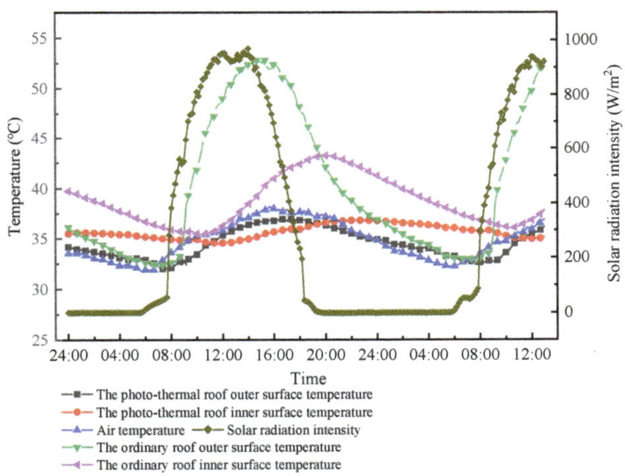

Figure 6. Variation in the temperature and solar radiation intensity on the typical day.

During the typical day, the outdoor air temperature was 31.7–38.2 °C, the solar radiation intensity increased from 6:00 to a maximum of 930 W/m^2 from 11:00 to 14:30 and then gradually decreased to 0 W/m^2 after 19:00.

As shown in Figure 6, the outer and inner surface temperatures of the photo-thermal roof were significantly lower than those of the ordinary roof for most of the study period. On the typical day, the outer and inner surface temperature peaks of the ordinary roof were 53.0 °C and 43.3 °C, while those of the photo-thermal were 37.0 °C and 36.8 °C, respectively. Under the same thermal disturbance conditions, the outer and inner surface temperature peaks of the photo-thermal roof reduced by 30.2% and 15.0%, respectively, compared with those of the ordinary roof. The heat gain of the roofs was composed of solar radiation and heat transfer from the surrounding air. The decrease in the outer surface temperature peak of the photo-thermal roof indicates a decrease in heat gain on the roof surface and a decrease in the amount of heat transferred to the inner surface of the roof through the roof structure.

This occurred because the solar irradiation on the photo-thermal roof was partly used to heat the water flow in the collector and partly used to increase the temperature of the collector, and then the heat from the collector was transferred to the outer surface of the roof. The air layer between the collector and the roof also weakened the amount of heat transfer. However, the ordinary roof was exposed directly to solar rays. When the outdoor air temperature was the same, the temperature peak of the outer surface of the ordinary roof was higher than that of the photo-thermal roof.

Figure 6 also shows that, for the ordinary roof, the outer surface temperature peak time was 14:30, the inner surface temperature peak time was 19:50, and the temperature wave delay time was 5 h and 20 min. The outer surface temperature of the photo-thermal roof reached its maximum value at 16:30 and the inner surface temperature reached its maximum value at 22:00. Compared to the ordinary roof, the temperature wave on the outer surface of the photo-thermal roof was delayed by 2 h and the inner surface temperature wave was delayed by 4 h.

In terms of the amplitude of the roof surface temperature, the order, from small to large, was as follows: the inner surface of the photo-thermal roof, the outer surface of the photo-thermal roof, the inner surface of the ordinary roof, the outer surface of the ordinary roof. On the typical day, the amplitudes were 2.0 °C, 4.5 °C, 7.3 °C, and 20.5 °C, respectively.

The outer surface temperature amplitude of the photo-thermal roof was 78.0% lower than that of the ordinary roof, and the amplitude of the surface temperature inside the photo-thermal roof was reduced by 72.6%. Thus, the photo-thermal roof was shown to have better thermal stability.

As shown in Figure 7, the change trend for the interior and exterior surface temperature difference between the ordinary roof and the photo-thermal roof was similar to that observed for solar radiation intensity. Although there was a delay in the increase in the inner surface temperature of the roofs, the time which the maximum temperature difference between the roof interior and exterior surface occurred was basically the same as that of the solar radiation intensity. When the solar radiation intensity was maximal, the roof outer surface received the most heat, while the inner surface released heat to the room at a low level, so the roof heat storage was maximal, as was the inner and outer surface temperature difference.

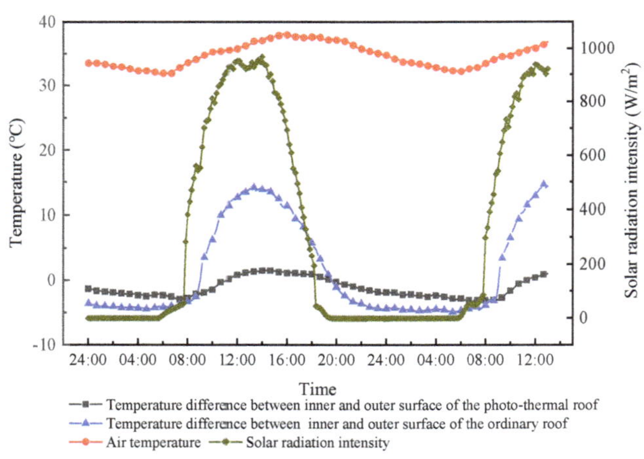

Figure 7. Variation in the temperature difference between internal and external surfaces and the solar radiation intensity on the typical day.

The maximum temperature difference between the inner and outer surfaces was 14.2 °C for the ordinary roof and 1.5 °C for the photo-thermal roof, as shown in Figure 7. Thus, in the photo-thermal roof, heat transfer was only one-ninth that in the ordinary roof under the experimental conditions. This shows that the photo-thermal roof has better insulating properties than the ordinary roof.

3.2. Effect of Water Saturation

In order to explore the influence of changes in the water saturation of insulation materials on the insulation performance of the photo-thermal roof, simulation studies were carried out when the water saturation level was 0, 10%, 20%, and 30%. The solar radiation intensity was set to 1000 W/m^2, the air temperature to 35.8 °C, the roof insulation thickness was set to 100 mm, and the water flow was adjusted to keep the pipe wall temperature constant at 50 °C to determine the relationship between the temperature, heat flux, and water saturation for the top room model with the photo-thermal roof and the ordinary room, as shown in Figures 8 and 9.

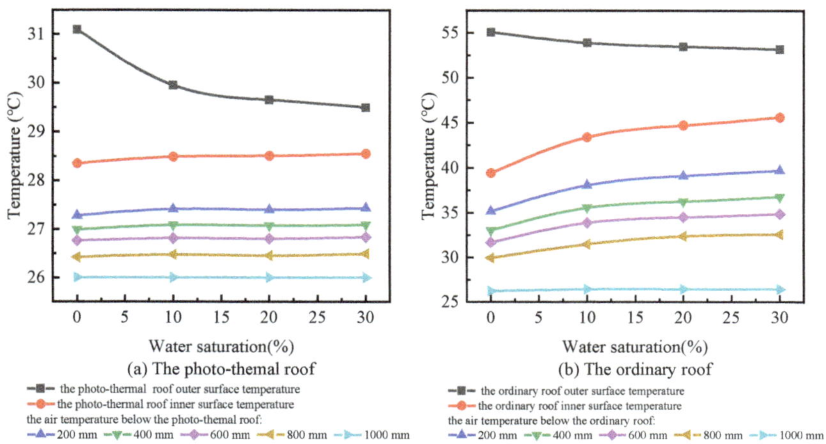

Figure 8. The effect of water saturation on temperature distribution in the rooms.

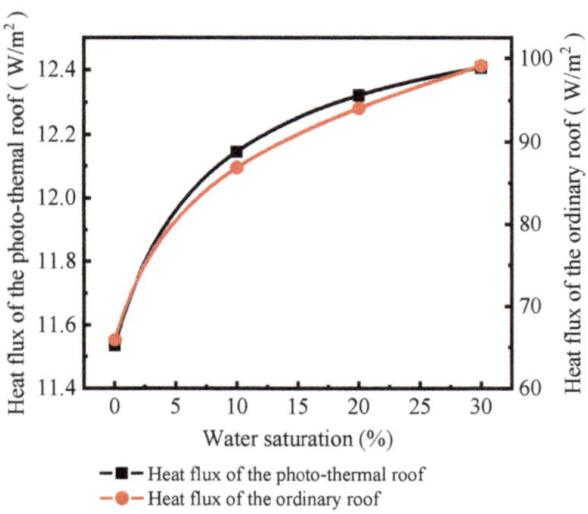

Figure 9. The influence of water saturation on the heat flux of the roofs.

An increase in water saturation led to an increase in the effective thermal conductivity of the roof, which enhances the heat transfer of the roof. The data presented in Figures 8 and 9 show that when the water saturation rose from 0 to 30%, the heat flux per unit area of the ordinary roof rose from 65.9 W/m^2 to 99.1 W/m^2, an increase of 50.4%; the inner surface temperature of the ordinary roof rose from 39.4 °C to 45.6 °C, an increase of 15.7%; and the air temperature 400 mm below the roof increased from 33.1 °C to 36.8 °C, an increase of 11.1%. Owing to the enhanced heat transfer, the outer surface temperature of the roofs decreased. When the water saturation rose from 0 to 30%, the outer surface temperature of the ordinary roof dropped from 55.1 °C to 53.2 °C, and the temperature gradient of the roof in the vertical direction dropped from 52.3 °C/m to 25.3 °C/m. Thus, as the water saturation of the roof insulation material increased, the heat transfer of the roof increased and the inner surface temperature and indoor air temperature increased, even though the temperature difference between the inner and outer surfaces of the roof may have reduced. The heat radiation from the inner surface of the roofs to the rooms was enhanced.

When the water saturation increased from 0 to 10%, the inner surface temperature and heat transfer of the roofs changed sharply. The inner surface temperature of the ordinary roof increased by 4 °C, and the heat flux per unit area increased by 21.0 W/m^2. However, the inner surface temperature of the photo-thermal roof only increased by 0.1 °C, and the heat flux of the roof only increased by 0.6 W/m^2.

With an increase in the water saturation, the change pattern of the photo-thermal roof temperature was similar to that of the ordinary roof. However, the photo-thermal roof showed a smaller temperature fluctuation. When the water saturation increased from 0 to 30%, the inner surface temperature of the photo-thermal roof rose from 28.4 °C to 28.6 °C, an increase of 0.2 °C. Compared with the ordinary roof, the temperature rise was 96.8% lower. The heat flux per unit area of the photo-thermal roof increased from 11.5 W/m^2 to 12.4 W/m^2, an increase of 0.9 W/m^2. Compared with the ordinary roof, the increase was 97.3% lower. As the water saturation increased, the temperature of the non-working area in the photo-thermal room did not show obvious fluctuations, and the indoor air temperature distribution was more uniform.

In summary, compared with the ordinary roof, the photo-thermal roof has smaller heat transfer, lower roof temperature and indoor non-air-conditioned area temperature. An increase in water saturation of insulation materials will enhance the heat transfer of roofs. However, an increase in water saturation has less negative impact on photo-thermal buildings than on ordinary buildings.

3.3. Effect of the Thickness of Roof Insulation Materials

In order to obtain an excellent thermal insulation performance, roof materials with low thermal conductivity are usually used. There are certain requirements regarding the thickness of thermal insulation materials. In hot summer and cold winter zones, 100 mm thick rock wool boards are usually used as roof insulation materials.

To determine the relationships among the temperature, heat flux, and roof insulation material thickness with the photo-thermal roof and the ordinary roof, the solar radiation intensity was set to 1000 W/m^2, the air temperature was set to 35.8 °C, the water saturation level was set to 0, and the water flow was adjusted to keep the pipe wall temperature constant at 50 °C, as shown in the Figures 10 and 11.

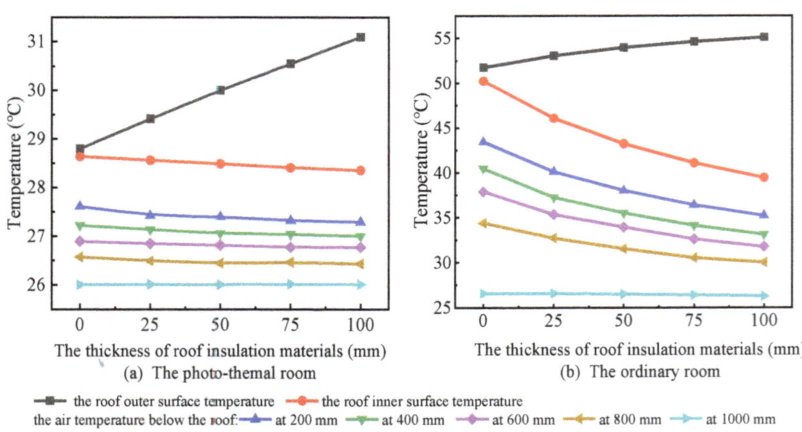

Figure 10. The influence of the insulation material thickness on the room temperature distribution.

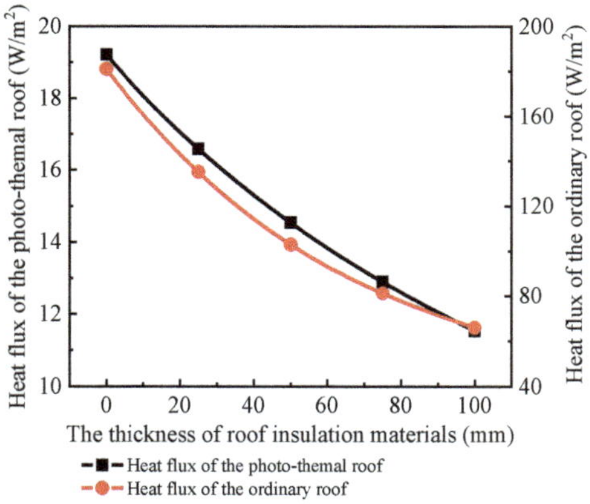

Figure 11. The influence of the insulation material thickness on the heat flux in the roofs.

As shown in Figures 10 and 11, the inner surface temperature and the indoor non-working area temperature of the roofs increased as the thickness of the insulation material decreased, while the temperature of the roof outer surface decreased. The heat flux of the roofs increased as the thickness of the insulation material decreased. When the thickness of insulation material was 100 mm, in contrast with the high temperature of 55.1 °C measured on the outside surface of the ordinary roof, the temperature of the photo-thermal roof was only 31.1 °C, a decrease of 43.2%. The inner surface temperature of the photo-thermal roof was 28.4 °C, which is 27.9% lower than that of the ordinary roof. The heat transfer per unit area of the photo-thermal roof also decreased from 65.9 W/m^2 to 11.5 W/m^2, a decrease of 82.5%.

When the insulation material thickness was reduced to 0, the inner surface temperature of the ordinary roof was 50.2 °C and the heat flux was 181.1 W/m^2. Under the same conditions, the inner surface temperature of the photo-thermal roof was 28.6 °C, a reduction of 38.0% compared with that of the ordinary roof, and the heat flux was 19.2 W/m^2, a reduction of 89.4% compared with the ordinary roof.

As shown in Figures 10 and 11, the indoor non-working area temperature of the ordinary roof reached to 40.1 °C when the thickness of the insulation material was reduced from 100 mm to 0 mm. In the same situation, the maximum temperature of the indoor non-working area of the photo-thermal room was only 28.1 °C, slightly higher than the temperature of the working area. Therefore, even when the thickness of the insulation layer was 0 mm, the photo-thermal roof still had an excellent thermal insulation performance.

Under actual conditions, the water saturation of the roof insulation materials may change periodically and cannot always be maintained in a completely dry state. If the thickness and water saturation of the roof insulation material changes, the inner surface temperature and heat flux will change under the above environmental conditions. Changes are shown in Figures 12 and 13.

A change in water saturation affects the effective thermal conductivity of the roof: the greater the water saturation, the greater the heat transfer of the roof, the lower the outer surface temperature of the roof, and the higher the inner surface temperature.

As shown in Figures 12 and 13, a decrease in the thickness of the roof insulation material or an increase in water saturation will promote heat transfer in the roof. When the thickness of the insulation material was constant, the heat transfer of the two roofs continuously increased, and the inner surface temperature of the roofs increased constantly as the water saturation level increased. When the water saturation level was 30% and the thickness of the roof insulation material was 100 mm, the inner surface temperature of the ordinary roof was 45.6 °C and the heat flux was 99.1 W/m^2. The inner surface temperature of the photo-thermal roof was 28.6 °C, 41.5% lower than that of the ordinary roof. The heat flux of the photo-thermal roof was 12.4 W/m^2, 87.5% lower than that of the ordinary roof. When the water saturation level was 30% and the thickness of the roof insulation material was 25 mm, the inner surface temperature of the ordinary roof was 48.9 °C and the heat flux was 155.6 W/m^2. The inner surface temperature of the photo-thermal roof was 28.5 °C, 41.7% lower than that of the ordinary roof. The heat flux of the photo-thermal roof was 16.9 W/m^2, 89.1% lower than that of the ordinary roof. It can also be seen from Figures 11 and 12 that the inner surface temperature of the photo-thermal roof was about 28.5 °C, which is only about 2.5 °C higher than the working area temperature as the thickness of the roof insulation layer changed from 100 mm to 25 mm and the water saturation of the insulation material increased from 0 to 30%. Thus, even when the insulation layer thickness was 25 mm and the water saturation level was high, the insulation performance of the photo-thermal roof was still good.

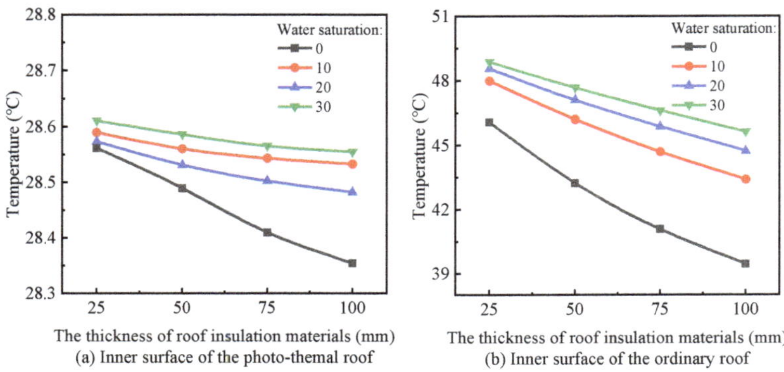

Figure 12. The influence of the thickness of insulation materials on the roof inner surface temperature when the water saturation changes.

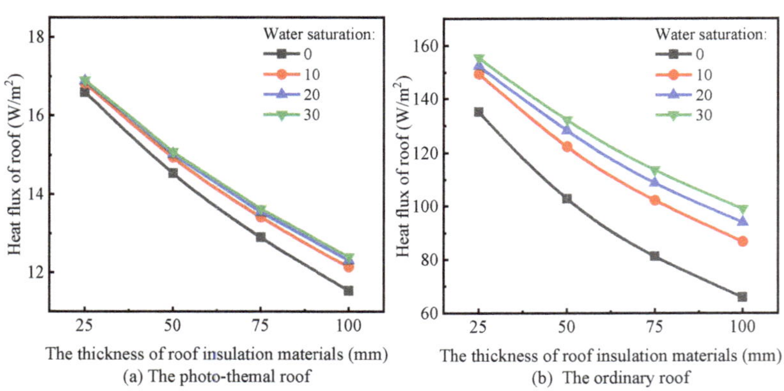

Figure 13. The influence of the thickness of insulation materials on the heat flux of the roof when the water saturation changes.

In summary, a decrease in the thickness of insulation material will lead to an increase in cooling load. In order to reduce the energy consumption of an air conditioner, the thickness of insulation material should be appropriately increased. The roof insulation layer is often made of porous materials. Water entering the pores of porous materials will weaken the thermal insulation performance of the roof. With high water saturation and thin thickness of the insulation layer, the photo-thermal roof can still maintain a good thermal insulation performance.

3.4. Variation in the Average Water Temperature of the Collectors

In practical projects, a certain level of moisture exists in a building's thermal insulation materials, and the thermal conductivity is corrected. Generally, the corrected effective thermal conductivity of rock wool board is 0.07 W/(m·K). Figure 14 shows the influence of the water temperature in the collector on the temperature distribution of the room when the solar radiation intensity was 1000 W/m^2, the air temperature was 35.8 °C, and the thickness of the rock wool board was 100 mm.

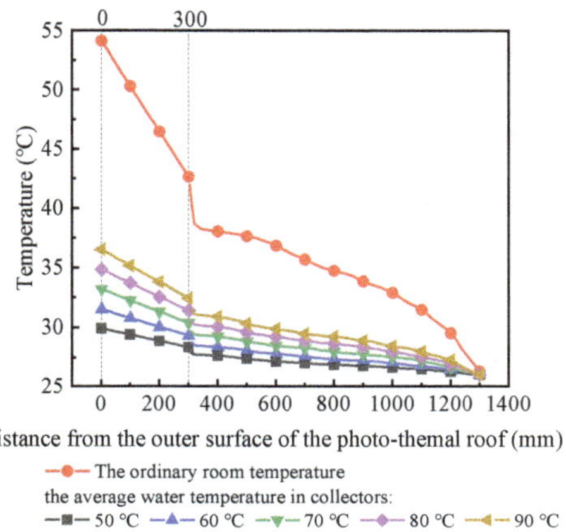

Figure 14. Temperature distribution of the roofs and non-air-conditioned area at different water temperatures.

As shown in Figure 14, the temperature of the photo-thermal roof and the non-working area increased as the average water temperature in the collectors increased. When the average water temperature rose from 50 °C to 90 °C, the outer surface temperature of the photo-thermal roof rose from 29.9 °C to 36.5 °C, an increase of 6.6%. The inner surface temperature of the photo-thermal roof increased from 28.4 °C to 32.4 °C, an increase of 14.1%. The average temperature of the indoor non-working area increased from 26.9 °C to 29.1 °C, an increase of 8.2%. However, compared with the temperature of the ordinary roof, the photo-thermal roof still showed an excellent insulation effect. Even when the average water temperature in the collectors was 90 °C, compared with the temperature of the ordinary roof, the outer surface temperature of the photothermal roof decreased by 32.7%, the inner surface temperature decreased by 23.9%, and the average temperature in the indoor non-working area decreased by 15.2%. The increase of the average water temperature also increases the roof temperature gradient. When the average water temperature rose from 50 °C to 90 °C, the vertical temperature gradient of the photo-thermal roof rose from 5.0 W/m to 13.6 W/m. However, when the average water temperature was 90 °C, the temperature gradient of the photo-thermal roof was still 64.8% lower than that of the ordinary roof.

When the solar radiation intensity was 1000 W/m^2, the air temperature was 35.8 °C, the thickness of the rock wool board was 100 mm, and the effective thermal conductivity was 0.07 W/(m·K), the heat transfer of the photo-thermal roof was affected by the change in the average water temperature in the collectors, as shown in Figure 15.

It can be seen from Figure 15 that an increase in the average water temperature in the collectors led to an increase in the photothermal roof heat transfer. When the water temperature increased from 50 °C to 90 °C, the roof heat transfer increased from 11.4 W/m^2 to 29.7 W/m^2, an increase of 160.5%. This occurred because the radiant heat transfer between the collector's rear cover and the roof is the main source of heat received by the photo-thermal roof. The increase in the average water temperature also increased the temperature of the collector's rear cover and enhanced the radiant heat transfer of the collector to the photo-thermal roof.

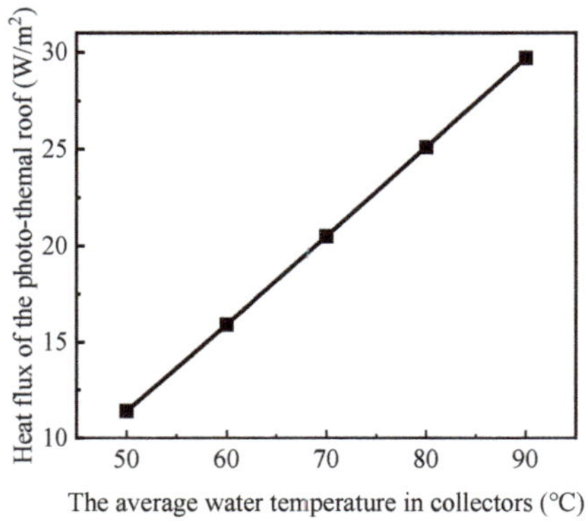

Figure 15. Effect of average water temperature in collectors on heat transfer of the photo-thermal roof.

Under the same environmental conditions, the heat transfer of the ordinary roof was 82.1 W/m^2. Even when the average water temperature was 90 °C, the heat transfer of the photo-thermal roof was 65.0% lower than that of the ordinary roof.

In order to explore the possibility of reducing the thickness of the thermal insulation layer of the photo-thermal roof under high water temperature, this paper also simulated a photo-thermal roof without insulation materials.

When the solar radiation intensity was 1000 W/m^2, the air temperature was 35.8 °C and the average water temperature in the collectors was 90 °C, and the thickness of the insulation layer was adjusted; the temperature distribution in the photo-thermal room is shown in Figure 16.

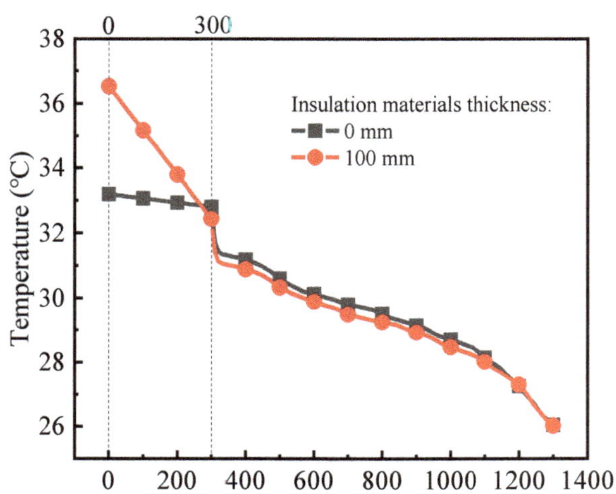

Figure 16. Effect of insulation material thickness on the temperature of the photo-thermal roof and the indoor non-air-conditioned area under high water temperatures.

As shown in Figure 16, when the thickness of the insulation layer decreased from 100 mm to 0 mm, the vertical temperature gradient of the roof decreased from 13.7 °C/m to 1.3 °C/m, the outer surface temperature of the roof decreased from 36.5 °C to 3.2 °C, the inner surface temperature increased from 32.4 °C to 32.8 °C, and the average temperature of the indoor non-working area increased from 29.1 °C to 29.3 °C. Under the above simulated conditions, when the thickness of the insulation layer changed from 100 mm to 0 mm, the roof heat transfer increased from 29.7 W/m^2 to 31.5 W/m^2. Therefore, the vertical temperature gradient of the photo-thermal roof is greatly reduced without insulation materials, but the increase of roof heat transfer is small. The roof insulation layer is not set, which increases the heat transfer coefficient and enhances the heat transfer of the photo-thermal roof. However, most of the heat by solar irradiation is absorbed by the water flow in the collector, and the heat transmitted to the roof is small. Increasing the heat transfer coefficient of the roof can increase the heat transfer less. From the above data, the heat transfer of the photo-thermal roof without the insulation layer is only increased by 0.8 W/m^2, the inner surface temperature is increased by 0.4 °C, and the average temperature of the indoor non-working area is increased by 0.2 °C. When the water temperature is 90 °C, the cancellation of the insulation layer of the photo-thermal roof has little impact on the thermal insulation performance of the roof.

In summary, since the water flow in the collector takes away the heat energy converted by solar energy, the thermal insulation performance of photo-thermal roof is better. An increase in the average water temperature of the collector will lead to an increase in the heat transfer of the photo-thermal roof, the roof temperature, and the indoor non-working area temperature. However, when the average water temperature is 90 °C, compared with ordinary roofs, the photo-thermal roof still has smaller heat transfer, lower roof temperature, and indoor non-working area temperature. At this time, even if the insulation material thickness was reduced to 0 mm, the photo-thermal roof still has good thermal insulation performance.

4. Conclusions

This paper studies the thermal insulation performance of a building roof combined with photothermal technology. In the summer climate of a hot summer and cold winter area, the thermal performance of the photo-thermal roof and the ordinary roof is simulated, and the following conclusions are obtained:

(1) Compared with the ordinary roof, the photo-thermal roof has smaller heat transfer, and lower roof temperature and indoor non-working area temperature. Adjusting the water flow of collectors can change the heat transfer through the roof. The photo-thermal roof has better insulation performance than the ordinary roof. Under the action of outdoor calculated temperature of air conditioning in summer, when the solar radiation intensity is 1000 W/m^2, the heat transfer per unit area of the photo-thermal roof is 82.5% lower than that of the ordinary roof.

(2) The water saturation of roof insulation materials is an important factor affecting insulation performance, and the influence of water saturation on the insulation performance of the ordinary roof is greater than that of the photo-thermal roof. Under the simulated conditions, the water saturation increases from 0 to 30%, the inner surface temperature of the ordinary roof rise is 6.2 °C, the photo-thermal roof temperature rise is only 0.2 °C, a decrease of 96.8%; the heat transfer per unit area of the ordinary roof increases by 33.2 W/m^2, and the photo-thermal roof increases by 0.9 W/m^2, a decrease of 97.3%.

(3) The influence of the insulation material thickness change on the insulation performance of the photo-thermal roof is less than the ordinary roof. When the thickness is very small, the photo-thermal roof still has excellent insulation performance. When the roof insulation layer is cancelled, the inner surface temperatures of the ordinary roof and the photo-thermal roof are 50.2 °C and 28.6 °C, respectively, and the average temperature of the indoor non-working area can reach 38.5 °C and 27.1 °C, respectively. The insulation thickness of the photo-thermal roof can be reduced or even cancelled.

Author Contributions: Conceptualization, Y.Z. and H.S.; methodology, H.S.; software, Y.Z.; validation, J.L.; data curation, L.Z. and X.S.; writing—original draft preparation, Y.Z., H.S., L.Z. and X.S.; writing—review and editing, Y.Z., H.S. and J.L; supervision, H.S.; project administration, H.S.; funding acquisition, H.S. All authors have read and agreed to the published version of the manuscript.

Funding: This work was supported by the National Natural Science Foundation of China (Grant No.52108099) and the Excellent Youth Project of Education Bureau of Hunan Province, China (Grant No.21B0134).

Institutional Review Board Statement: The study did not require ethical approval.

Informed Consent Statement: Informed consent was obtained from all subjects involved in the study.

Data Availability Statement: Not applicable.

Conflicts of Interest: The authors declare no conflict of interest.

Nomenclature

C	Specific heat capacity of materials (J/(kg·K))
D	Index of thermal inertia
E	The total energy composed of thermodynamic and kinetic energy (J/kg)
h_g	Convective heat transfer coefficient of the glass cover (W/m^2·K)
$h_{i,w}$	convective heat transfer coefficient of the inner surface of each material layer (W/m^2·K)
h_j	Enthalpy of matter (kJ/kg)
$h_{o,w}$	convective heat transfer coefficient of the outer surface of each material layer (W/m^2·K)
I	Solar radiation intensity (W/m^2)
J_J	Diffusion flux (kg/(m^2·s))
P	Pressure (Pa)
$q_{c,b}$	Heat convection between the collector's back surface and the air (W/m^2)
$q_{c,g}$	Heat convection between the glass cover and the external environment (W/m^2)
$q_{c,in}$	Heat convection between the roof inner surface and the indoor air (W/m^2)
$q_{c,out}$	Heat convection between the photo-thermal roof and the air (W/m^2)
q_{cond}	Conductive heat flux (W/m^2)
Q_w	Released heat by hot water (W)
$q_{r,b}$	Heat radiation of the collector's back surface to the roof (W/m^2)
$q_{r,g}$	Heat radiation of the glass cover to the external environment (W/m^2)
$q_{r,in}$	Heat radiation of the roof inner surface to the indoor environment (W/m^2)
$q_{r,out}$	Heat radiation of the roof to the collector's back surface (W/m^2)
S_i	Thermal effusivity of wall inner surface (W/m^2·K)
S_o	Thermal effusivity of wall outer surface (W/m^2·K)
S_h	Heat source (W/m^3)
T_b	Temperature of the back surface of the collector (°C)
T_g	The temperature of the glass cover (°C)
T_n	other surface temperatures in the room (°C)
T_{sky}	Sky temperature (°C)
$T_{i,a}$	The temperature of air in contact with the inner surface of the material layer (°C)
$T_{o,a}$	The temperature of air in contact with the outer surface of the material layer (°C)
$T_{i,w}$	The roof inner surface temperature (°C)
$T_{o,w}$	The roof outer surface temperature (°C)
α_g	Absorptivity of the glass cover
α_p	Absorptivity of the absorber plate
α_w	Absorptivity of roof outer surface
$\alpha_{i,w}$	Heat transfer coefficient of wall inner surface (W/m^2·K)
$\alpha_{o,w}$	Heat transfer coefficient of wall outer surface (W/m^2·K)
λ_a	Thermal conductivity of air (W/(m·k))
λ_e	Thermal conductivity of roof materials (W/(m·k))
λ_{ec}	Effective thermal conductivity of porous media satisfying serial heat conduction model (W/(m·k))
λ_{ep}	Effective thermal conductivity of porous media satisfying parallel heat conduction model (W/(m·k))
λ_l	Thermal conductivity of liquid phase in porous media (W/(m·k))
λ_p	Thermal conductivity of solar collectors (W/(m·k))
λ_s	Thermal conductivity of solid phase in porous media (W/(m·k))

ε_a	The volume percentage of the gas phase per unit volume (%)
ε_b	Emissivity of the collector rear cover
ε_g	Emissivity of the glass cover
$\varepsilon_{i,w}$	Emissivity of the roof inner surface
$\varepsilon_{o,w}$	Emissivity of the roof outer surface
τ	Unit time
τ_g	Transmissivity of the glass cover
φ	Water saturation (%)
σ	The Boltzmann constant, $\sigma = 5.67 \times 10^{-8}$ W/(m^2·K^4)
ϕ	Time lag

References

1. Zhou, N.; Price, L.; Yande, D.; Creyts, J.; Khanna, N.; Fridley, D.; Lu, H.; Feng, W.; Liu, X.; Hasanbeigi, A.; et al. A roadmap for China to peak carbon dioxide emissions and achieve a 20% share of non-fossil fuels in primary energy by 2030. *Appl. Energy* **2019**, *239*, 793–819. [CrossRef]
2. IEA. Online Data Services. 2019. Available online: https://www.iea.org/buildings (accessed on 20 February 2022).
3. Lu, X.; Memari, A.M. Comparison of the Experimental Measurement Methods for Building Envelope Thermal Transmittance. *Buildings* **2022**, *12*, 282. [CrossRef]
4. International Energy Agency. *Energy and Climate Change: World Energy Outlook Special Report 2015*; International Energy Agency: Paris, France, June 2015.
5. Nikkho, S.K.; Heidarinejad, M.; Liu, J.; Srebric, J. Quantifying the impact of urban wind sheltering on the building energy consumption. *Appl. Therm. Eng.* **2017**, *116*, 850–865. [CrossRef]
6. Xin, Z.; Xw, M.; By, C.; Yp, S.; Ml, S. Challenges toward carbon neutrality in China: Strategies and countermeasures. *Resour. Conserv. Recycl.* **2021**, *176*, 105959.
7. Li, H.; Zheng, R. Energy and Economic Performance of Solar Cooling Systems in the Hot-Summer and Cold-Winter Zone. *Buildings* **2018**, *8*, 37. [CrossRef]
8. Notton, G. *Tu1205 Projet Cost Overview of BISTS State of the Art, Models and Applications. COST Action TU1205 (BISTS): Building Integration of Solar Thermal Systems*; European Cooperation in Science and Technology (COST): Brussels, Belgium, 2015.
9. Celadyn, W.; Filipek, P. Investigation of the Effective Use of Photovoltaic Modules in Architectu re. *Buildings* **2020**, *10*, 145. [CrossRef]
10. Piratheepan, M.; Anderson, T. Performance of a Building Integrated Photovoltaic/Thermal Concentrator for Facade Applications. *Sol. Energy* **2017**, *153*, 562–573. [CrossRef]
11. Fekete, I.; Farkas, I. Numerical and experimental study of building integrated solar tile collectors. *Renew. Energy* **2018**, *137*, 45–55. [CrossRef]
12. Deng, C.; Chen, F. Preliminary investigation on photo-thermal performance of a novel embedded building integrated solar evacuated tube collector with compound parabolic concentrator. *Energy* **2020**, *202*, 117706. [CrossRef]
13. Lin, W.; Ma, Z.; Wang, S.; Sohel, M.; Cascio, E.L. Experimental investigation and two-level model-based optimisation of a solar photovoltaic thermal collector coupled with phase change material thermal energy storage. *Appl. Therm. Eng.* **2021**, *182*, 116098. [CrossRef]
14. Pugsley, A.; Zacharopoulos, A.; Mondol, J.D.; Smyth, M. BIPV/T facades—A new opportunity for integrated collector-storage solar water heaters? Part 1: State-of-the-art, theory and potential. *Sol. Energy* **2020**, *207*, 317–335. [CrossRef]
15. Harmim, A.; Boukar, M.; Amar, M.; Haida, A. Simulation and experimentation of an integrated collector storage solar water heater designed for integration into building facade. *Energy* **2018**, *166*, 59–71. [CrossRef]
16. Gautam, K.R.; Andresen, G.B. Performance comparison of building-integrated combined photovoltaic thermal solar collectors (BiPVT) with other building-integrated solar technologies. *Sol. Energy* **2017**, *155*, 93–102. [CrossRef]
17. Luo, K.; Ji, J.; Xu, L.; Li, Z. Seasonal experimental study of a hybrid photovoltaic-water/air solar wall system. *Appl. Therm. Eng.* **2019**, *169*, 114853. [CrossRef]
18. Rahmani, F.; Robinson, M.A.; Barzegaran, M.R. Cool roof coating impact on roof-mounted photovoltaic solar modules at texas green power microgrid. *Int. J. Electr. Power Energy Syst.* **2021**, *130*, 106932. [CrossRef]
19. Christoph, M.; Christoph, C.; Tilmann, E.K. Progress in building-integrated solar thermal systems. *Sol. Energy* **2017**, *154*, 158–186.
20. Ahmed, S.; Li, Z.; Javed, M.S.; Ma, T. A review on the integration of radiative cooling and solar energy harvesting. *Mater. Today Energy* **2021**, *21*, 100776. [CrossRef]
21. Yu, G.; Yang, H.; Yan, Z.; Ansah, M.K. A review of designs and performance of facade-based building integrated photovoltaic-thermal (BIPVT) systems. *Appl. Therm. Eng.* **2021**, *182*, 116081. [CrossRef]
22. Shao, N.; Ma, L.; Zhang, J. Experimental study on electrical and thermal performance and heat transfer characteristic of PV/T roof in summer. *Appl. Therm. Eng.* **2019**, *162*, 114276. [CrossRef]
23. Ponechal, R.; Barňák, P.; Ďurica, P. Comparison of Simulation and Measurement in a Short-Term Evaluation of the Thermal Comfort Parameters of an Office in a Low-Carbon Building. *Buildings* **2022**, *12*, 349. [CrossRef]
24. Sourek, M.B. Façade solar collectors. *Sol. Energy* **2006**, *80*, 1443–1452.

25. Buonomano, A.; Forzano, C.; Kalogirou, S.A.; Palombo, A. Building-façade integrated solar thermal collectors: Energy-economic performance and indoor comfort simulation model of a water-based prototype for heating, cooling, and DHW production. *Renew. Energy* **2018**, *137*, 20–36. [CrossRef]
26. Chang, H.; Hou, Y.; Lee, I.; Liu, T.; Acharya, T.D. Feasibility Study and Passive Design of Nearly Zero Energy Building on Rural Houses in Xi'an, China. *Buildings* **2022**, *12*, 341. [CrossRef]
27. Gao, Y.; Shi, D.; Levinson, R.; Guo, R.; Lin, C.; Ge, J. Thermal performance and energy savings of white and sedum-tray garden roof: A case study in a Chongqing office building. *Energy Build.* **2017**, *156*, 343–359. [CrossRef]
28. Tong, S.; Li, H.; Zingre, K.T.; Wan, M.P.; Chang, W.C.; Wong, S.K.; Toh, W.B.T.; Lee, I.Y.L. Thermal performance of concrete-based roofs in tropical climate. *Energy Build.* **2014**, *76*, 392–401. [CrossRef]
29. Barone, G.; Buonomano, A.; Forzano, C.; Giuzio, G.F.; Palombo, A. Passive and active performance assessment of building integrated hybrid solar photovoltaic/thermal collector prototypes: Energy, comfort, and economic analyses. *Energy* **2020**, *209*, 118435. [CrossRef]
30. Long, J.; Jiang, M.; Lu, J.; Du, A. Vertical temperature distribution characteristics and adjustment methods of a trombe wall. *Build. Environ.* **2019**, *165*, 106386. [CrossRef]
31. Bowen, B.D. "Heat transfer-A basic approach", by M. Necati Öuzisik, 1985, 780 pages, mcgraw-hill book company, (U.S.). *Can. J. Chem. Eng.* **1988**, *66*, 1036–1037. [CrossRef]
32. Zhou, J. Study on Microstructure and Energy Saving Characteristics of Rock Wool Board Used for Building External Wall Insulation System. Master's Thesis, Chang'an University, Xi'an, China, 2018.
33. *GB50736-2012*; Design Code for Heating Ventilation and Air Conditioning of Civil Buildings: GB50736-2012. China Architecture & Building Press: Beijing, China, 2012.
34. *Heat Transfer under Periodic Heat Action*; China Architecture & Building Press: Beijing, China, 1964.

buildings

MDPI

Article

Spatiotemporal Analysis of Influencing Factors of Carbon Emission in Public Buildings in China

Zhuoqun Du *, Yisheng Liu and Zhidong Zhang

School of Economics and Management, Beijing Jiaotong University, Beijing 100044, China; yshliu1@bjtu.edu.cn (Y.L.); davidzhang7@163.com (Z.Z.)
* Correspondence: duzhu0qun@163.com

Abstract: The rapid development of public buildings has greatly increased the country's energy consumption and carbon emissions. Excessive carbon emissions contribute to global warming. This paper aims to measure the carbon emissions in the operation of public buildings, and to identify the multiple influencing factors of carbon emissions in operational public buildings. First, the spatial and temporal variation characteristics of carbon emissions from public buildings in 30 provinces of China from 2008–2019 are analyzed. Second, a green building index is constructed, and the STIRPAT (Stochastic Impacts by Regression on Population, Affluence, and Technology) model is utilized to explore the relationship between each influencing factor and carbon emissions, using spatial and temporal geographically weighted regression analysis. The results show that the effects of population, urbanization rate, GDP per capita, green building index, and industrial structure on carbon emissions from public buildings all show spatial correlation and differences. There are east-west differences in the operational carbon emissions of public buildings in China's provinces. Cluster evolution shows a spatially increasing trend from west to east. To some extent, policymakers can develop appropriate policies for different provinces through the findings.

Keywords: GTWR; public buildings' carbon emission; spatiotemporal analysis

Citation: Du, Z.; Liu, Y.; Zhang, Z. Spatiotemporal Analysis of Influencing Factors of Carbon Emission in Public Buildings in China. *Buildings* **2022**, *12*, 424. https://doi.org/10.3390/buildings12040424

Academic Editors: Shi-Jie Cao and Wei Feng

Received: 22 February 2022
Accepted: 28 March 2022
Published: 31 March 2022

1. Introduction

Climate change has become one of the most important globally recognized issues. Reducing CO_2 emissions helps mitigate climate change, and carbon emission and environmental protection issues have aroused the attention of various countries. Many countries have begun to measure carbon emissions and take action to reduce them [1,2]. Recently, the Chinese government announced that it strives to achieve carbon peaking by 2030, and carbon neutrality by 2060. More detailed plans have also been specified to reach this goal. By 2030, carbon dioxide emissions per unit of GDP will drop 65% compared to 2005, the proportion of non-fossil energy consumption will reach about 25%, and the total installed capacity of solar and wind power generation will reach more than 1.2 billion kilowatts. The Ministry of Ecology and Environment has proposed that during the 14th and 15th Five-Year Plan periods, China will carry out CO_2 emission peaking actions and specify the peaking targets and action plans for localities and industries. China's total construction carbon emissions were 4.93 billion tons in 2018, accounting for 51% of the national carbon emissions. Carbon emissions from the production phase of building materials account for 28% of the total national carbon emissions, the construction phase accounts for 1%, and the building operation phase is 22% [3].

According to the data in 2018, the existing stock of public buildings in China's urban and rural areas is 12.8 billion square meters, accounting for 21.3% of the total civil construction area. From the annual data, the energy consumption for the construction of public buildings accounted for 44% of the total building construction energy consumption in 2018, and the total energy consumption for the operation of public buildings excluding northern heating accounted for 33% of the total building operation energy consumption [4].

In 2015, the energy consumption of public buildings in China was 34.1 billion tons of standard coal equivalent, accounting for about 40% of the total energy consumption of civil buildings. However, the public building area accounted for only 18% of the total area of civil buildings. Furthermore, a study showed that the energy intensity of public buildings in China was four times that of residential buildings from 2000 to 2015. Public buildings are important places for human activities, and their construction, operation, renewal, and demolition processes all generate significant energy consumption. Therefore, the CO_2 emissions of public buildings in China have attracted more attention [5].

Green buildings play a very important role in reducing carbon emissions. Wu et al. (2017) [6] found that with commercial buildings in China, green buildings are lower in carbon emissions than non-green buildings, in the operational phase. The Chinese government proposed the Green Building Creation Initiative in 2020, which aims to reach 70% of the green building area in new urban buildings in that year by 2022. The effect of green building on carbon reduction in the building sector and its spatial evolution is the focus of this study.

2. Literature Review

2.1. Spatiotemporal Analysis of Carbon Emission

Spatiotemporal analysis is a method that considers both temporal data and spatial position, and is mainly used to solve how coherent entities change over time. Scholars have performed many researches with different methods to study the spatiotemporal analysis of carbon emission and its influencing factors in different fields. For example, Chen et al. (2021) analyzed the temporal and spatial characteristics of industrial carbon emissions in four regions of Guangdong province from 2005 to 2015, and concluded that industrial carbon emissions have a trend of eastward expansion [7]. The spatial dynamic analysis model (SDDM) was used to study the impact of different technological progress factors on carbon emissions [8]. Cui et al. (2021) explored spatiotemporal dynamic evolution of carbon emission intensity and per capita carbon emissions from planting industry in 31 provinces in China across 20 years. The spatial inequality is measured by Theil index and its contribution rate [9]. Wang et al. (2020) employed the standard deviation ellipse method and tapio decoupling method to reveal the spatiotemporal characteristics of the relationship between carbon emissions from transportation industry and economic growth [10]. Hu et al. (2020) studied the spatial and temporal evolution relationship between economic growth and carbon emissions in Belt and Road countries [11]. Han et al. (2021) revealed the spatiotemporal characteristics of carbon intensity of 20 industries by extending the spatial weight matrix and spatial dubin model [12]. Falahatkar et al. (2020) quantified the relationship between carbon dioxide emission and urban form in 15 Iranian cities, and believed that carbon dioxide emission level was positively correlated with urban area growth and urban complexity increase [13]. Some scholars have studied the carbon emission spatiotemporal effect in the construction industry. Bai et al. (2021) estimated the building inventory and carbon emissions embodied by buildings in 31 provinces of China from 1997 to 2016, and proposed a spatiotemporal decomposition model to identify driving forces [14]. To sum up, in the construction industry, there are few spatial analyses on carbon emissions during the operation of public buildings.

2.2. Influencing Factors of Buildings Carbon Emission

The extraction of raw materials, on-site construction activities, and building operations produced the majority of carbon emissions of the construction sector [4]. Hard coal and its derivatives were the largest carbon dioxide emitters in China's construction industry [15]. In view of the significant impact of the construction industry, prior studies have been carried out to investigate influencing factors in order to develop mitigation strategies. For example, Lu et al. (2016) have analyzed the influencing factors of carbon emissions from construction activities in China, including energy intensity, energy structure, unit cost, level of construction automation, and machine efficiency [16]. Similarly, Zhang, Yan et al. (2019)

stated that building scale, building structure type, and production efficiency of material are the three main driving factors [4]. Wu, shen et al. (2019) used the STIRPAT model and found that the impact of population size, per capita GDP, energy intensity, and industrial structure on carbon emissions were heterogeneous among regions [17]. Mostafavi, Tahsildoost et al. concluded that strengthening the design parameters of envelope structure, optimizing the layout, and utilizing natural ventilation are conducive to reducing energy consumption of high-rises [18]. Tan, Lai, Gu, Zeng, & Li constructed a carbon emission prediction model including population, urbanization rate, and urban building area [19]. Wang et al. explored the driving forces of energy-related CO_2 emissions in the construction industry by implementing the comprehensive decomposition method, and finally found that technological progress of industrial output was the leading factor that suppressed CO_2 emissions [20]. Huang et al. propounded increased energy efficiency design for new buildings and energy-saving retrofit for existing buildings to carbon emission [21].

2.3. Research Gap

Based on a critical review of relevant studies, as well as substantive surveys and interviews with Chinese building industry professionals, we conclude that the spatial and temporal effects of carbon emissions from public buildings still require further research. Because using the STIRPAT model to decompose influence factor is more comprehensive, it is still worthwhile to use this model to study the factors that drive the carbon emission of operational public buildings.

3. Data Source and Methodologies

3.1. Study Area

China has 34 provincial districts. Four provincial districts (Tibet, Macao, Hong Kong, and Taiwan) are excluded due to data unavailability, so this study selected a total of 30 provincial districts. The study divided 30 provincial districts into four regions (East region, Central region, West region, and Northeast region) according to the National Bureau of Statistics (Table 1).

Table 1. Four regions and their provincial districts.

Regions	Provincial District
East region	Beijing, Tianjin, Hebei, Shanghai, Jiangsu, Zhejiang, Fujian, Shandong, Guangdong, Hainan
Central region	Shanxi, Anhui, Jiangxi, Henan, Hubei, Hunan
West region	Inner Mongolia, Guangxi, Chongqing, Sichuan, Guizhou, Yunnan, Shaanxi, Gansu, Qinghai, Ningxia, Xinjiang
Northeast region	Liaoning, Jilin, Heilongjiang

3.2. Data Source

The data of green buildings from 2008 to 2016 and 2018 to 2019 are obtained from the Chinese Green Building Evaluation Label Network. Green building data in 2017 are obtained through compilation of public information of green building projects on the website of provincial Construction Department. The remaining indicators are from China Energy Statistical Yearbook and China Statistical Yearbook for 2009–2020. The map data comes from the National Geomatics Center of China.

3.3. Methodology

3.3.1. Hot Spot Analysis

G_i^* statistic is used to analyze the spatial aggregation degree of carbon emissions during the operation of provincial public buildings, as shown in Equation (1).

$$G_i^*(d) = \sum_{i=1}^{n} w_{ij}(d)x_i / \sum_{i=1}^{n} x_i \quad (1)$$

x_i is the attribute value of the unit i, and x_i is the carbon emission in ith province public buildings during its operation; w_{ij} is the spatial weight matrix.

G_i^* is standardized by Equation (2) and the result is $Z(G_i^*)$. The larger the value $Z(G_i^*)$ is, the higher the spatial clustering in the region, indicating that it belongs to the hot spot area; the smaller it is, the lower the spatial clustering in the region, indicating it belongs to the cold spot area.

$$Z(G_i{}^*) = \frac{G_i^* - E(G_i^*)}{\sqrt{Var(G_i^*)}} \tag{2}$$

where $E(G_i^*)$, $Var(G_i^*)$ are the expectation and variance of G_i^*, respectively.

3.3.2. Geographically and Temporally Weighted Regression Model

The geographically and temporally weighted regression (GTWR) model is a deepening of the geographically weighted regression model, as shown in Equation (3). By using regional panel data for spatial regression, the temporal attributes are linked to the spatial attributes in the GTWR model, which better reflects the spatial and temporal change information of the study area, and makes the estimation results more effective.

$$y_i = \beta_0(u_i, v_i, t_i) + \sum_{k=1}^{p} \beta_k(u_i, v_i, t_i)x_{ik} + \varepsilon_i \tag{3}$$

where y_i is the dependent variable of sample i, x_{ik} is the kth independent variable at the sample point i, u_i, v_i are the latitude and longitude coordinates of the center of gravity, respectively, (u_i, v_i, t_i) are the spatial and temporal coordinates of the ith sample, $\beta_k(u_i, v_i, t_i)$ is the regression coefficient on the kth independent variable at the ith sample point, $\beta_0(u_i, v_i, t_i)$ is the space-time intercept of the ith sample point, and ε_i is the residual term.

3.3.3. Model Specification

Enrlich and Holdren first put forward the classic IPAT model in the early 1970s, which stipulates the influence of external factors on the environment. External factors include population size (P), affluency (A), and technology (T). The IPAT model was improved and transformed into the nonlinear random STIRPAT model, which is often used to analyze influencing factors of carbon emissions in different industries [22]. For example, Ma et al. surveyed the driving factors of carbon dioxide emission from public buildings in a country [23,24]; Yang and Jia explored the spatial effects of technology progress channels on CO_2 emissions for the agricultural, industrial, construction, transportation, and wholesale sectors [25]. The STIRPAT model is expressed as Equation (4):

$$I_i = aP_i^b A_i^c T_i^d \varepsilon_i \tag{4}$$

where i denotes the regional unit. I_i, P_i, A_i and T_i represent the impacts on the environment owing to population, affluence (per capita GDP), and technology factors in region i, respectively. Constant a represents the scale of the model. Meanwhile, b, c, and d are the estimated coefficients of population, affluence (per capita GDP), and technology, respectively. ε is the random error term. We take the logarithm of the STIRPAT model, obtaining the following Equation (5):

$$\ln I_i = \ln a + b \ln P_i + c \ln A_i + d \ln T_i + \ln \varepsilon \tag{5}$$

The STIRPAT equation allows the addition of plenty of relevant variables, and the transformation of the model into an extended version, as long as the dimensionality of these variables is reasonable [26,27].

In order to deeply explore the mechanism of carbon emission of public buildings, considering green buildings' specific characteristics, and looking for supporting references from a great deal of relevant previous studies, this study developed an extended version of the STIRPAT model using several meaningful variables retrieved from the population, affluence, and technology levels, respectively. The extended STIRPAT model is expressed in Equation (6):

$$\ln C_{it} = \ln a + \beta_0(u_i, v_i, t_i) + \beta_1(u_i, v_i, t_i) * \ln P_{it} + \beta_2(u_i, v_i, t_i) * \ln U_{it} + \beta_3(u_i, v_i, t_i) * \ln IS_{it} + \beta_4(u_i, v_i, t_i) * \ln G_{it} + \beta_5(u_i, v_i, t_i) * \ln IGB_{it} + \ln \varepsilon_i \quad (6)$$

where C_{it} refers to the carbon emission in the public building sector in province i over time t. (u_i, v_i, t_i) represents spatial coordinates of province i (i = 1, 2, 3, ..., 30). β_k (k = 1, 2, 3, 4, 5) denotes the kth regression coefficient in the ith province. The meaning and units of the variables are shown in Table 2.

Table 2. Declaration of the model variables.

Nomenclature	Variable	Unit	Type	Supporting References
C	Carbon emission of public buildings	tCO_2	dependent variable	
P	Population	Ten thousand people	explained variable	[28]
U	Urbanization level	%	explained variable	[24,29,30]
IS	Industrial structure	1	explained variable	[24,30,31]
G	GDP per capita	Yuan	explained variable	[32]
IGB	Index of green buildings		explained variable	

3.3.4. Index Calculation

1. Calculation methods of CO_2 emission

The operational energy consumption of public buildings includes heating, air conditioning, ventilation, lighting, elevators, cooking, domestic hot water, office electrical equipment, and comprehensive service equipment and facilities. Corresponding energy types include electricity, gas (natural gas, gas, and LPG), fuel oil (diesel), and coal combustion. This study uses a macro model for measuring carbon emissions from buildings based on energy balance sheets.

This paper mainly studies the operational stage of carbon emissions in public buildings. Because China's energy statistics yearbook does not provide building energy consumption directly [33], we need to select the energy consumption as public buildings' operational consumption. The specific accounting boundaries are shown in Table 3.

Table 3. Specific accounting boundary of public building.

Chinese Region Terminal Energy Category in Statistics Yearbook	Chinese Region Terminal Energy Category in Statistics Yearbook
Wholesale, retail trade and hotel, restaurants, and others	Coal, electricity, heat, liquefied petroleum gas, natural gas

This study mainly measures carbon emissions during the use of public buildings. Carbon emission in Chinese public buildings is measured by the end-use consumption of energy in each region in the China Energy Statistics Yearbook. The industries involved in public buildings are Transport, Storage and Post, Wholesale and Retail Trades, Hotels and Catering Services, and Other.

Energy type measurement includes coal, electricity, natural gas, LPG, and thermal power. Oil is not counted because it is mostly used in public buildings for transportation involving cars, and is not counted as energy consumption inside buildings for the time being. To obtain more meaningful and comprehensive results, we included three types of energy sources, such as coal, natural gas, and liquefied petroleum gas. According to the calculation method provided by IPCC (Equation (7)), coal, electricity, and heat consumed in the operation of public buildings are taken as the sources of carbon emissions.

$$C = \sum_{j}^{30} (\sum_{i}^{3} C_{ij} + C_{ej} + C_{hj}) = \sum_{j}^{30} (\sum_{i}^{3} E_{ij} \times O_{ij} \times LCV_{ij} \times CF_{ij} \times \frac{44}{12} + E_{ej} \times \delta e + E_{hj} \times \delta_h) \quad (7)$$

where C denotes the total carbon emissions from public building operation in each province, C_{ij} refers to carbon emission of the consumption of fossil energy i in the j province, and C_{ej} and C_{hj} represent the carbon emissions from the secondary energy consumption of electricity and heat in the j province. E_{ij} denotes the consumption of fossil energy i in the j province; O_{ij} refers to the oxidation rate of the fossil energy i in the province j; LCV_{ij} represents the average low-level calorific value of fossil energy i in province j; CF_{ij} denotes the carbon emission factor of fossil energy i in province j; factor $\frac{44}{12}$ refers to the ratio of CO_2 molecules to carbon atoms by weight; carbon emissions can be converted into CO_2 emissions by multiplying by this coefficient, E_{ej} and E_{hj} denote the electricity consumption and heat consumption in province j, respectively, δ_{ej} and δ_{hj} denote the carbon emission factor of electricity and heat consumption in province j.

The carbon emissions generated during the use of public buildings are estimated by referring to the low level calorific value, carbon emission factor, and carbon oxidation rate provided by IPCC.

Since the carbon emission factor of coal is not directly provided in IPCC, coal is considered as raw coal for calculation. The carbon emission factor of coal is 25.8 TC/TJ, the low calorific value is 20.908 GJ/T, and the carbon oxidation rate is 0.899. LNG is converted to natural gas volume for calculation, depending on its density as 0.42~0.46 g/cm^3.

The average CO_2 emission factors (kg-CO_2/kWh) of the national regional power grids in 2011 and 2012, as queried by the NDRC and the Guidelines for Provincial Greenhouse Gas Inventories, are shown in Table 4.

Table 4. Electricity carbon emission factors.

Regional Grid	Coverage of Provinces and Cities	2011	2012	Average Value
North China Regional Grid	Beijing, Tianjin, Heibei, Shanxi, Shandong, Western Inner Mongolia	0.8967	0.8843	0.8905
East China Regional Grid	Liaoning, Jilin, Heilongjiang, Eastern Inner Mongolia	0.8189	0.7769	0.7979
Northeast Regional Grid	Shanghai, Jiangsu, Zhejiang, Anhui, Fujian	0.7129	0.7035	0.7082
Central China Regional Grid	Henan, Hubei, Hunan, Jiangxi, Sichuan, Chongqing	0.5955	0.5257	0.5606
Northwest Regional Grid	Shaanxi, Gansu, Qinghai, Ningxia, Xinjiang	0.6860	0.6671	0.67655
Southern Regional Grid	Guangdong, Guangxi, Yunnan, Guizhou, Hainan	0.5748	0.5271	0.55095

Electricity carbon emission coefficients obtained based on public data query are generally measured by the government or relevant departments in a unified manner, which is easily accessible, and their data source is authoritative. However, the data are not published annually, which is not conducive to the measurement of time series of building carbon emission data. The average value of these two years was used in this study. Inner Mongolia power emission factor is taken as the average value of 0.8442 in the east and west. Table 5 shows Coefficient Thermal CO_2 emission [34].

2. Index of green building

Green buildings in China are classified as one-star, two-star, and three-star. Three stars are the highest level of green building. The index of green building is calculated through Equation (8).

$$IGB = D_1 \times 1 + D_2 \times 2 + D_3 \times 3 \quad (8)$$

where D_1, D_2, D_3 denote the number of one-star, two-star and three-star public green buildings in China, respectively.

Table 5. Coefficient Thermal CO_2 emission (tCO_2/MWh).

Provinces and Cities	Thermal CO_2 Emission Coefficient	Provinces and Cities	Thermal CO_2 Emission Coefficient	Provinces and Cities	Thermal CO_2 Emission Coefficient	Provinces and Cities	Thermal CO_2 Emission Coefficient
Anhui	116	Guizhou	292	Hunan	110	Sichuan	105
Beijing	88	Hainan	57	Jilin	132	Tianjin	108
Fujian	112	Hebei	122	Jiangsu	109	Xinjiang	109
Gansu	110	Henan	124	Jiangxi	134	Yunnan	149
Guangdong	93	Heilongjiang	155	Liaoning	130	Zhejiang	104
Guangxi	153	Hubei	122	Inner Mongolia	160	Chongqing	98
Ningxia	120	Qinghai	245	Shandong	114	Shanxi	116
Shaanxi	149						

3. Industrial Structure

We use the percentage of added value of the tertiary industry to GDP to describe industrial structure [35].

4. Empirical Results

4.1. Spatial Distribution of Carbon Emission in Different Areas

The regional energy balance of the China Energy Statistics Yearbook for 2009–2020 was used to estimate the carbon emissions from the operation of public buildings in each province of the country using end-use energy consumption. The specific measurement results, in accordance with the previously stated zoning, are shown in Figure 1.

In general, the eastern region has more carbon emissions than the central, western, and northeastern regions. The top three provinces generating carbon emissions from public buildings from 2008–2018 were Guangdong, Jiangsu, and Beijing. In 2019, Shandong surpassed Beijing among the top three.

Among the eastern regions, Guangdong Province has the most carbon emissions from public buildings and Hainan Province has the least. All provinces show an increasing trend year by year. Jiangsu Province and Hebei Province have a faster growth rate. In the central region, Henan Province is has the highest carbon emissions, except for 2017, and Jiangxi Province has the lowest carbon emissions. Other regions are steadily increasing, however, not as much as the vast majority of the eastern region's emissions.

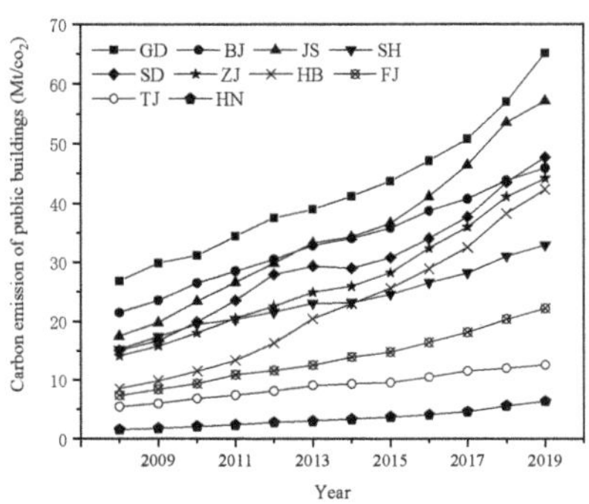

(a)

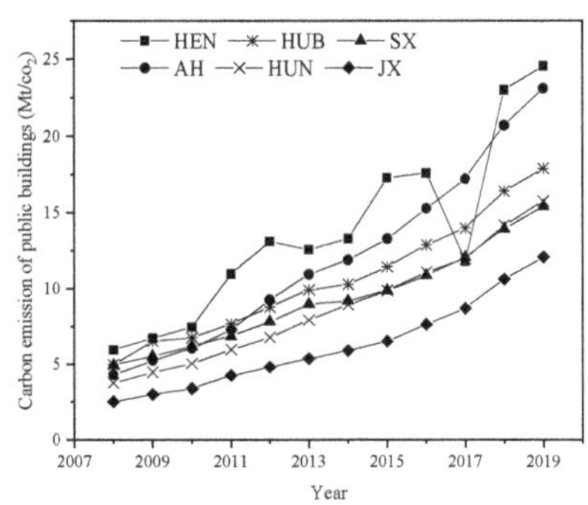

(b)

Figure 1. *Cont.*

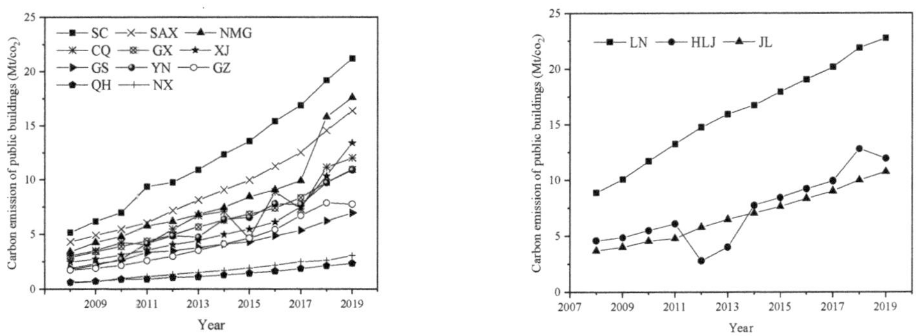

(c) (d)

Figure 1. Carbon emission of public buildings in China among provinces during 2008 to 2019: (**a**) Eastern Region; (**b**) Central Region; (**c**) Western Region; (**d**) Northeast Region. GD—Guangdong, BJ—Beijing, JS—Jiangsu, SH—Shanghai, SD—Shandong, ZJ—Zhejiang, HB—Hebei, FJ—Fujian, TJ—Tianjin, HN—Hainan, HEN—Henan, HUB—Hubei, SX—Shanxi, AH—Anhui, HUN—Hunan, JX—Jiangxi, SC—Sichuan, SAX—Shaanxi, NMG—Inner Mongolia, CQ—Chongqing, GX—Guangxi, XJ—Xinjiang, GS-Gansu, GZ—Guizhou, QH—Qinghai, NX—Ningxia, LN—Liaoning, HLJ—Heilongjiang, JL—Jilin.

Among the western regions, Sichuan Province consistently has the highest carbon emissions from public buildings, and Qinghai Province steadily has the lowest. Chongqing fluctuated more, with carbon emissions decreasing in 2015 and 2017. The growth rate is larger in Inner Mongolia and Xinjiang. Carbon emissions in Inner Mongolia province were always lower than Shaanxi province between 2008 and 2017, and exceeded Shaanxi province after 2018. Xinjiang Province surpassed Chongqing, Yunnan Province, and Guangxi Province in 2018.

In Northeast China, Liaoning Province has the highest carbon emissions from public buildings. Carbon emissions from public buildings in Heilongjiang Province were higher than those in Jilin Province, except in 2012 and 2013.

4.2. Hot Spot Analysis of Carbon Emissions from Public Building Operations

Hot spot analysis can reflect the spatial aggregation effect of carbon emissions from public buildings. The variation of the aggregation can be seen by plotting the hot spot and cold spot areas in different years. The natural interruption point grading method was used to classify the values of each year, calculated by Equations (1) and (2) into high, subhigh, sublow, and low value cluster areas in order of largest to smallest. The study area of this research is in the years 2008–2019, and the clustering results of 2010, 2013, 2016, and 2019 are drawn equally spaced for analysis, as shown in Figure 2.

From Figure 2, it can be seen that evolution of provincial carbon emission clustering in China shows a spatially increasing trend from west to east. Overall, the high-value clustering areas are mainly concentrated in Zhejiang, Jiangsu, Anhui, and Shanghai. The low-value clustering areas are mainly concentrated in Qinghai and Sichuan. In 2016, the number of provinces with high-value clustering areas increased, then decreased in 2019. Over time, the high value cluster areas first expanded and then contracted.

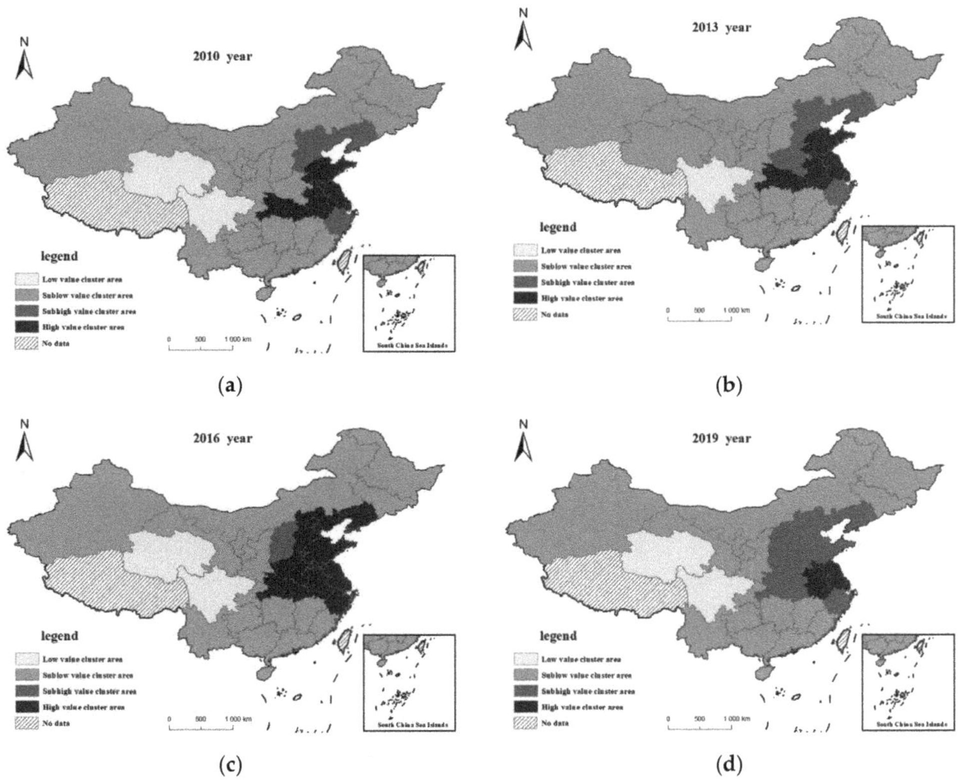

Figure 2. Spatial agglomeration pattern of public building carbon emissions in China from 2008 to 2019: (**a**) Spatial agglomeration pattern of public building carbon emissions in China in 2010; (**b**) Spatial agglomeration pattern of public building carbon emissions in China in 2013; (**c**) Spatial agglomeration pattern of public building carbon emissions in China in 2016; (**d**) Spatial agglomeration pattern of public building carbon emissions in China in 2019.

4.3. Spatial Effects of the Influencing Factors of Public Buildings Carbon Emissions

Carbon emissions from public building operations are the dependent variable; population, urbanization rate, industrial structure, GDP per capita, and green index are the explanatory variables in the STIRPAT model; the time range is 2008–2019; and the X and Y coordinates are the geographic coordinates of each province. With these factors, the runs can be entered into the spatio-temporal geographically weighted model to obtain the influence size of the five explanatory variables. In order to unify the comparison of influence size, the influence size values are arranged in descending order and divided by equal spacing, and the positive and negative influence are divided into six levels. Positive influence includes weak (WPI), medium (MPI), and strong positive influence (SPI), which indicates that the influence factor positively contributes to the carbon emission of public building operation; the negative influence includes weak (WNI), medium (MNI), and strong negative influence (SNI), which indicates that the influence factor negatively inhibits the carbon emission of public building operation [36]. Due to the number of years, this study selected 2010, 2013, 2016, and 2019 for spatial presentation and analysis. The GTWR model was run with an adjusted R^2 of 0.956 and an AICc of 199.253, indicating a better model effect.

Based on the regression results of the GTWR model, the spatial and temporal variability of the five influencing factors of carbon emissions of public buildings is analyzed one by one.

1. Spatial and temporal variation in the effect of population on carbon emissions

Population is a positive influence on the carbon emissions of public buildings in each province. The maximum value of the population regression coefficient is 8.135 and the minimum value is 0.082, which can be divided into WPI, MPI, and SPI according to the influence level. From the results, the provinces with the highest impact are Shanghai, Zhejiang, Jiangsu, Shandong, Anhui, Tianjin, Beijing, and Fujian. Mainly with the growth of population, the service industry activities in public buildings are frequent, thus increasing the carbon emissions from public buildings. As shown in Figure 3, the impact of population on carbon emissions from public buildings is increasing year by year. Spatially, the SPI of population size is gradually spreading from the northeast and eastern coastal regions to the central and western regions. By 2016, population size has reached a SPI within 30 provinces in the study area.

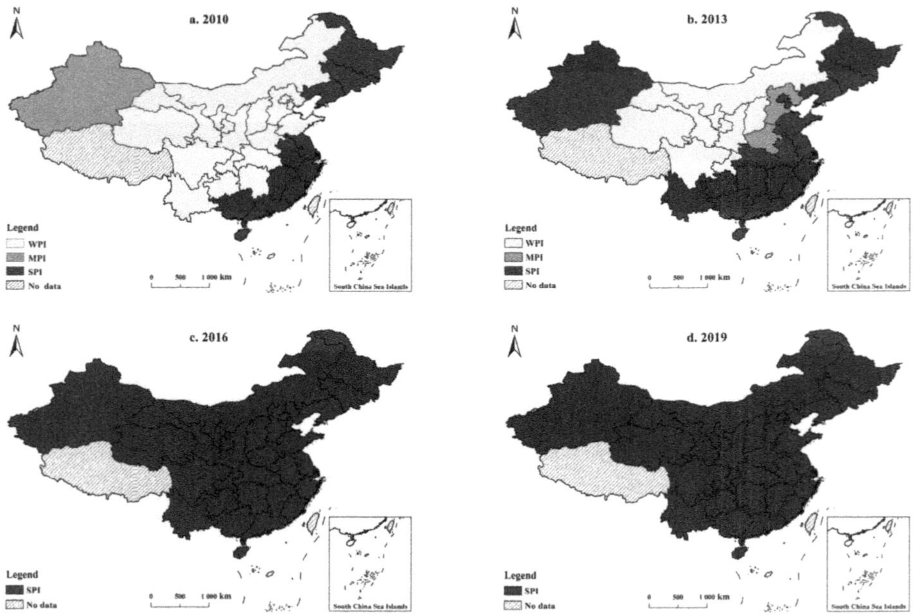

Figure 3. Spatial distribution of the regression coefficients of population: (**a**) Spatial distribution of the regression coefficients of the population in 2010; (**b**) Spatial distribution of the regression coefficients of the population in 2013; (**c**) Spatial distribution of the regression coefficients of the population in 2016; (**d**) Spatial distribution of the regression coefficients of the population in 2019.

2. Spatial and temporal variation in the effect of urbanization on carbon emissions

During the study period, the maximum value is 30.386 and the minimum value is −8.649. According to the influence level, it can be divided into SNI, MNI, WNI, WPI, and SPI. Overall, the urbanization rate has a predominantly negative impact on carbon emissions from public buildings. In 2010, WNI dominates, occupying 18 provinces, followed by WPI, occupying 11 provinces. In 2013, 2016, and 2019, WNI dominates, occupying 15 provinces, followed by WPI, occupying 14 provinces. It indicates that the urbanization rate has a small impact on the carbon emissions of public buildings. The scale of carbon emissions from public buildings does not increase with the increase in urbanization level; instead, it decreases with the optimization of energy consumption structure and the improvement of energy utilization efficiency. During the study period, the number of provinces with negative impact levels shows a decreasing trend, and spatially, the area of positive effect gradually expands from the northeast and some southeastern provinces to the whole eastern region, and then gradually shifts to the central region and Xinjiang province (Figure 4). Public buildings are mostly located in urban areas and less in rural areas. The increase in urbanization rate will promote the development of tertiary industry, which will further promote the generation and operation of public buildings. When the urbanization rate reaches a certain level, it will curb the carbon emissions of running public buildings because the stability of urbanization will make people start to raise the awareness of energy saving, not only limited to the use of more focus on energy efficiency.

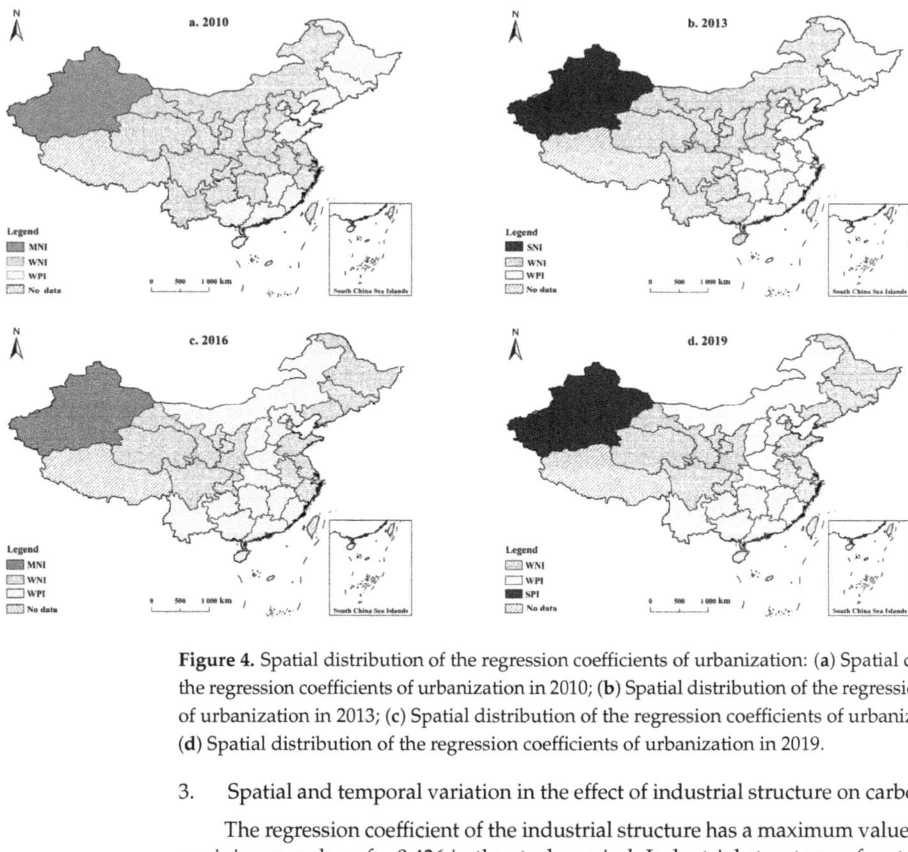

Figure 4. Spatial distribution of the regression coefficients of urbanization: (**a**) Spatial distribution of the regression coefficients of urbanization in 2010; (**b**) Spatial distribution of the regression coefficients of urbanization in 2013; (**c**) Spatial distribution of the regression coefficients of urbanization in 2016; (**d**) Spatial distribution of the regression coefficients of urbanization in 2019.

3. Spatial and temporal variation in the effect of industrial structure on carbon emissions

The regression coefficient of the industrial structure has a maximum value of 0.315 and a minimum value of −0.426 in the study period. Industrial structure refers to the ratio of the value added of the tertiary sector to the total value added of production. It is classified as SPI, WPI, WNI, and MNI according to the influence level. As shown in Figure 5, it is generally a positive impact. 2010, 2013, 2016, and 2019 are dominated by weak positive impact, occupying 16, 21, 18, and 15 provinces, respectively. The positive effect of industrial structure is shifted from the east to the center. In 2010, the positive effect of industrial structure is in the eastern coastal region, northeastern region, and Xinjiang province, and that positive effect is partially shifted to the central region in 2019.

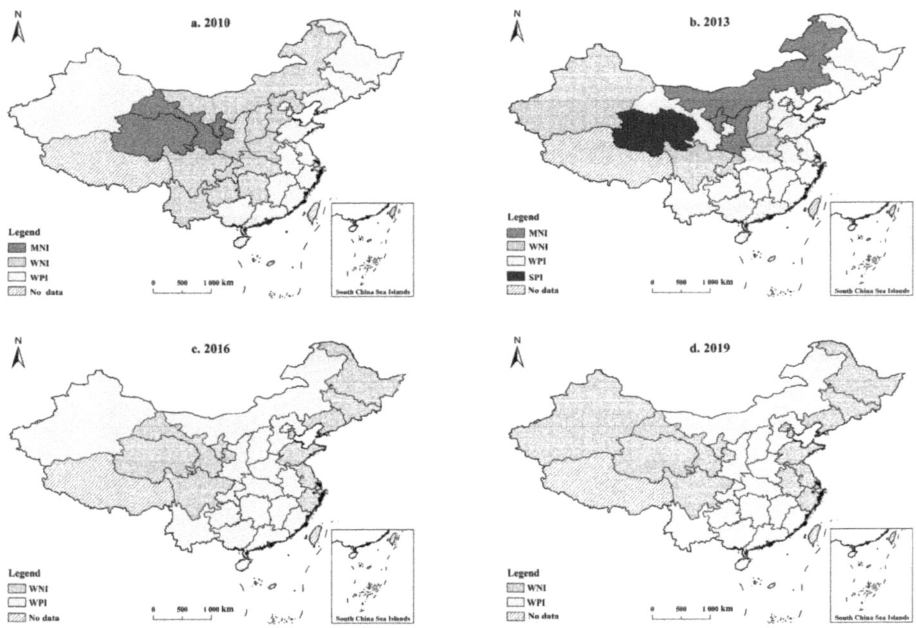

Figure 5. Spatial distribution of the regression coefficients of industrial structure: (**a**) Spatial distribution of the regression coefficients of industrial structure in 2010; (**b**) Spatial distribution of the regression coefficients of industrial structure in 2013; (**c**) Spatial distribution of the regression coefficients of industrial structure in 2016; (**d**) Spatial distribution of the regression coefficients of industrial structure in 2019.

4. Spatial and temporal variation in the effect of GDP per capita on carbon emissions

The maximum value of the regression coefficient for GDP per capita is 1.177 and the minimum value is −1.184. In 2010, the positive effect is dominant, with a total of 16 provinces in WPI and MPI. In 2013, the negative effect is dominant, with a total of 22 provinces in SNI, MNI, and WNI. In 2016, the positive effect is more pronounced, with 24 provinces in WPI. In 2019, the positive effect is pronounced, with a total of 22 provinces in WPI and MPI. As shown in Figure 6, the influence of GDP per capita shows a trend from positive to negative and then positive again. By 2019, the only regions with a negative effect are Shanghai, Jiangsu, Zhejiang, Anhui, Fujian, Jiangxi, Hubei, and Hunan. It indicates that population affluence is promoting the increase in carbon emissions from public buildings.

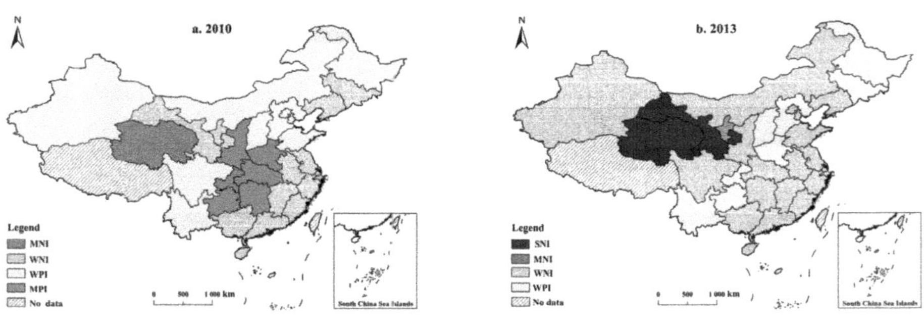

Figure 6. *Cont.*

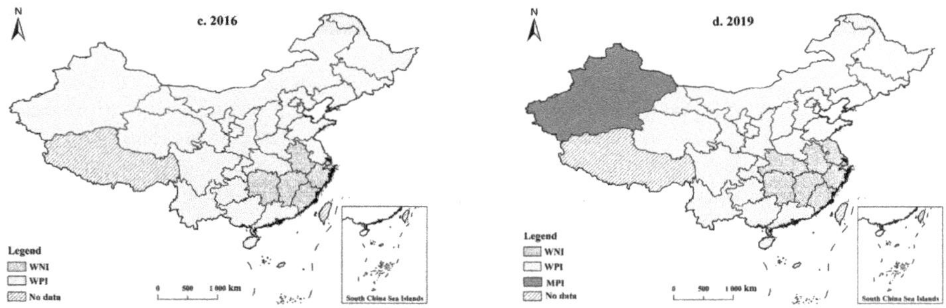

Figure 6. Spatial distribution of the regression coefficients of GDP per capita: (**a**) Spatial distribution of the regression coefficients of GDP per capita in 2010; (**b**) Spatial distribution of the regression coefficients of GDP per capita in 2013; (**c**) Spatial distribution of the regression coefficients of GDP per capita in 2016; (**d**) Spatial distribution of the regression coefficients of GDP per capita in 2019.

5. Spatial and temporal variation in the effect of the green building index on carbon emissions

The maximum value of the regression coefficient of the green building index is 0.832 and the minimum value is −0.098. As shown in Figure 7, in 2010, all 30 provinces have a positive effect. In 2013, the WNI dominated, occupying 16 provinces. In 2016, WPI occupied 17 provinces. In 2019, WPI occupies 21 provinces. In 2010, the green building index has a positive effect on carbon emissions from public buildings in all regions. However, by 2019, the inhibitory effect has 9 provinces. According to previous studies, green buildings contribute to carbon emission reduction than non-green buildings, but this study finds that the large-scale effect of carbon emission reduction from green buildings has not yet been developed.

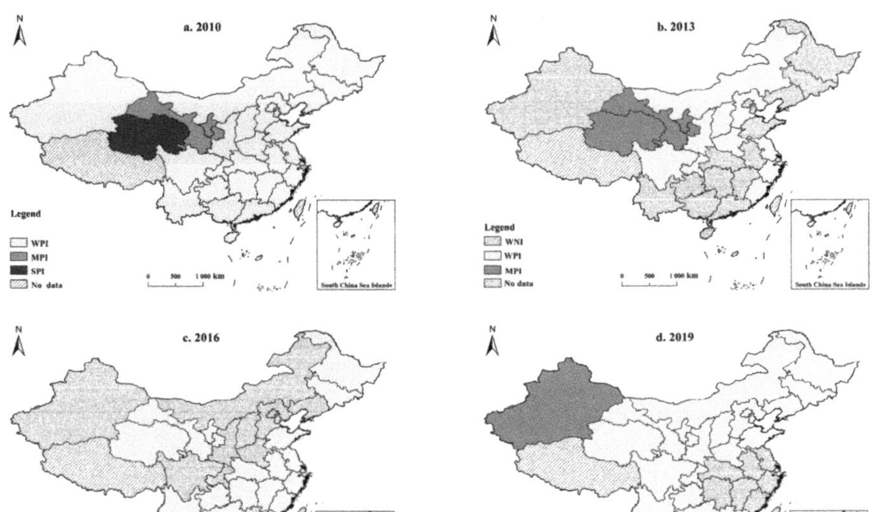

Figure 7. Spatial distribution of the regression coefficients of the green building index: (**a**) Spatial distribution of the regression coefficients of the green building index in 2010; (**b**) Spatial distribution of the regression coefficients of the green building index in 2013; (**c**) Spatial distribution of the regression coefficients of the green building index in 2016; (**d**) Spatial distribution of the regression coefficients of the green building index in 2019.

5. Conclusions and Policy Suggestions

5.1. Conclusions

Using the carbon emission of public buildings in operation in China from 2008 to 2019, the GTWR model was used to detect the spatial distribution of the influence coefficients of population, urbanization, industrial structure, GDP per capita, and index of green buildings. We found significant spatial heterogeneity changes between the five factors. Most importantly, the factor of the index of green buildings.

The empirical results of the green building index and carbon emissions of public building operation help to determine the direction and intensity of green building development, and help to enrich the study of the impact of green buildings on the carbon emissions of public buildings from a regional perspective.

In terms of carbon emissions from public building operations, the top three public building carbon emissions from 2008–2018 were Guangdong, Jiangsu, and Beijing, and in 2019, Shandong surpassed Beijing among the top three. Overall, the total carbon emissions from public buildings in the eastern region, except Hainan, Fujian, and Tianjin, are greater than those in the central, western, and northeastern regions, and the growth rate is obvious.

The results of the hotspot analysis show that there are east-west differences in the operational carbon emissions of public buildings in Chinese provinces. The evolution of clustering shows a spatially increasing trend from west to east.

The STIRPAT model shows that population has a positive influence on public building carbon emissions in each province, and the positive influence of population gradually spreads from the northeast and the eastern coastal regions to the central and western regions; urbanization rate has a predominantly negative influence on public building carbon emissions; industrial structure has a positive influence; the influence of GDP per capita and green building index shows a trend of positive to negative and then positive; the large-scale effect of green building carbon emission reduction has not yet been formed.

5.2. Suggestions

According to the above conclusions, we can make the following recommendations.

First, total carbon emission control should be carried out at the regional level under the constraints of other indicators such as socio-economic development rate and industry economic growth. Focus on controlling carbon emission hotspot areas and emission reduction measures should be formulated for cold spot areas, according to development needs to avoid generating large amounts of carbon emissions due to rapid economic development. Cooperation between provinces can be strengthened to develop inter-provincial carbon emission trading policies to balance provincial carbon emissions.

Second, for provinces with large populations, public buildings should be retrofitted with energy efficiency. It would be advisable while developing the economy to adjust the energy consumption structure, change the economic development mode, adhere to the path of low carbon development, and slow down the rapid growth of carbon emissions caused by the rapid growth of the regional economic development level.

Third, the development of green buildings still needs to continue to improve, and the current growth of green buildings has not yet formed a scale effect on the carbon emission aspect of public buildings. Because of the high cost of green building construction, there is a need to support the construction and development of green buildings in economically disadvantaged areas.

However, this study is subject to several limitations. Firstly, this paper uses China as an example, and the analysis for green buildings can be extended to other emerging economies and developing countries. Secondly, temperature has a different impact on the heating and cooling of public buildings under different climate backgrounds. Further study can take a closer look at the micro-level of the impact of temperature on the energy consumption of public buildings. Finally, the empirical results of this paper are helpful for policy makers to develop differentiated emission reduction strategies for high and low carbon emission provincial administrative regions.

Author Contributions: Conceptualization, Z.D. and Y.L.; methodology, Z.D. and Y.L.; validation, Z.D.; formal analysis, Z.D.; data curation, Z.D. and Z.Z.; writing—original draft preparation, Z.D.; writing—review and editing, Z.D.; visualization, Z.D. and Y.L.; project administration, Z.D. and Z.Z.; funding acquisition, Y.L. All authors have read and agreed to the published version of the manuscript.

Funding: This research was funded by National Natural Science Foundation of China, grant number 71871014.

Data Availability Statement: The data used to support the results in this article are included within the paper. In addition, some of the data in this paper are supported by the references mentioned in the manuscript. If you have any queries regarding the data, the data of this study would be available from the correspondence upon request.

Conflicts of Interest: All the authors declare no conflict of interest.

References

1. Nässén, J.; Holmberg, J.; Wadeskog, A.; Nyman, M. Direct and indirect energy use and carbon emissions in the production phase of buildings: An input–output analysis. *Energy* **2007**, *32*, 1593–1602. [CrossRef]
2. Zhu, W.; Feng, W.; Li, X.; Zhang, Z. Analysis of the embodied carbon dioxide in the building sector: A case of China. *J. Clean. Prod.* **2020**, *269*, 122438. [CrossRef]
3. *China Building Energy Consumption Annual Report 2020*; China Association of Building Energy Efficiency: Beijing, China, 2020.
4. Zhang, Y.; Yan, D.; Hu, S.; Guo, S. Modelling of energy consumption and carbon emission from the building construction sector in China, a process-based LCA approach. *Energy Policy* **2019**, *134*, 110949. [CrossRef]
5. Xiao, H.; Wei, Q.; Wang, H. Marginal abatement cost and carbon reduction potential outlook of key energy efficiency technologies in China's building sector to 2030. *Energy Policy* **2014**, *69*, 92–105. [CrossRef]
6. Wu, X.; Peng, B.; Lin, B. A dynamic life cycle carbon emission assessment on green and non-green buildings in China. *Energy Build.* **2017**, *149*, 272–281. [CrossRef]
7. Chen, L.; Xu, L.; Cai, Y.; Yang, Z. Spatiotemporal patterns of industrial carbon emissions at the city level. *Resour. Conserv. Recycl.* **2021**, *169*, 105499. [CrossRef]
8. Huang, J.; Chen, X.; Yu, K.; Cai, X. Effect of technological progress on carbon emissions: New evidence from a decomposition and spatiotemporal perspective in China. *J. Environ. Manag.* **2020**, *274*, 110953. [CrossRef]
9. Cui, Y.; Khan, S.U.; Deng, Y.; Zhao, M. Regional difference decomposition and its spatiotemporal dynamic evolution of Chinese agricultural carbon emission: Considering carbon sink effect. *Environ. Sci. Pollut. Res.* **2021**, *28*, 38909–38928. [CrossRef]
10. Wang, L.; Fan, J.; Wang, J.; Zhao, Y.; Li, Z.; Guo, R. Spatio-temporal characteristics of the relationship between carbon emissions and economic growth in China's transportation industry. *Environ. Sci. Pollut. Res.* **2020**, *27*, 32962–32979. [CrossRef]
11. Hu, M.; Li, R.; You, W.; Liu, Y.; Lee, C.-C. Spatiotemporal evolution of decoupling and driving forces of CO_2 emissions on economic growth along the Belt and Road. *J. Clean. Prod.* **2020**, *277*, 123272. [CrossRef]
12. Han, Y.; Jin, B.; Qi, X.; Zhou, H. Influential Factors and Spatiotemporal Characteristics of Carbon Intensity on Industrial Sectors in China. *Int. J. Environ. Res. Public Health* **2021**, *18*, 2914. [CrossRef] [PubMed]
13. Falahatkar, S.; Rezaei, F. Towards low carbon cities: Spatio-temporal dynamics of urban form and carbon dioxide emissions. *Remote Sens. Appl. Soc. Environ.* **2020**, *18*, 100317. [CrossRef]
14. Bai, J.; Qu, J. Investigating the spatiotemporal variability and driving factors of China's building embodied carbon emissions. *Environ. Sci. Pollut. Res.* **2021**, *18*, 19186–19201. [CrossRef]
15. Huang, L.; Krigsvoll, G.; Johansen, F.; Liu, Y.; Zhang, X. Carbon emission of global construction sector. *Renew. Sustain. Energy Rev.* **2018**, *81*, 1906–1916. [CrossRef]
16. Lu, Y.; Cui, P.; Li, D. Carbon emissions and policies in China's building and construction industry: Evidence from 1994 to 2012. *Build. Environ.* **2016**, *95*, 94–103. [CrossRef]
17. Wu, Y.; Shen, L.; Zhang, Y.; Shuai, C.; Yan, H.; Lou, Y.; Ye, G. A new panel for analyzing the impact factors on carbon emission: A regional perspective in China. *Ecol. Indic.* **2019**, *97*, 260–268. [CrossRef]
18. Mostafavi, F.; Tahsildoost, M.; Zomorodian, Z. Energy efficiency and carbon emission in high-rise buildings: A review (2005–2020). *Build. Environ.* **2021**, *206*, 108329. [CrossRef]
19. Tan, X.; Lai, H.; Gu, B.; Zeng, Y. Carbon emission and abatement potential outlook in China's building sector through 2050. *Energy Policy* **2018**, *118*, 429–439. [CrossRef]
20. Wang, M.; Feng, C. Exploring the driving forces of energy-related CO_2 emissions in China's construction industry by utilizing production-theoretical decomposition analysis. *J. Clean. Prod.* **2018**, *202*, 710–719. [CrossRef]
21. Huang, W.; Li, F.; Cui, S.-H.; Huang, L.; Lin, J.-Y. Carbon Footprint and Carbon Emission Reduction of Urban Buildings: A Case in Xiamen City, China. *Procedia Eng.* **2017**, *198*, 1007–1017. [CrossRef]
22. Rose, T.D.E.A. Effects of population and affluence on CO_2 emissions. *Ecology* **1997**, *94*, 175–179.
23. Ma, M.; Shen, L.; Ren, H.; Cai, W.; Ma, Z. How to Measure Carbon Emission Reduction in China's Public Building Sector: Retrospective Decomposition Analysis Based on STIRPAT Model in 2000–2015. *Sustainability* **2017**, *9*, 1744. [CrossRef]

24. Ma, M.; Yan, R.; Cai, W. An extended STIRPAT model-based methodology for evaluating the driving forces affecting carbon emissions in existing public building sector: Evidence from China in 2000–2015. *Nat. Hazards* **2017**, *89*, 741–756. [CrossRef]
25. Yang, X.; Jia, Z.; Yang, Z.; Yuan, X. The effects of technological factors on carbon emissions from various sectors in China—A spatial perspective. *J. Clean. Prod.* **2021**, *301*, 126949. [CrossRef]
26. Dietz, T.; Rosa, E.A. Rethinking the Environmental Impacts of Population, Affluence and Technology. *Hum. Ecol. Rev.* **1994**, *1*, 277–300.
27. York, R.; Rosa, E.A.; Dietz, T. STIRPAT, IPAT and ImPACT: Analytic tools for unpacking the driving forces of environmental impacts. *Ecol. Econ.* **2003**, *46*, 351–365. [CrossRef]
28. Yan, H.; Liu, H.; Qiu, R.; Zhang, Y. Influencing Factors Analysis of Construction Industry Carbon Emissions Based on Stepwise Regression. *J. Eng. Manag.* **2021**, *35*, 16–21.
29. Zhang, S.; Wang, K.; Yang, X.; Xu, W. Research on Emission Goal of Carbon Peak and Carbon Neutral in Building Sector. *Build. Sci.* **2021**, *37*, 189–198.
30. Xie, J. *Discussion on Peak Carbon Emissions and Energy Saving-Emission Reduction of Public Buildings in Chongqing Based on LEAP Model*; Chongqing University: Chongqing, China, 2019; p. 103.
31. Jiang, X. *Study on Forecast and Factor Decomposition of Carbon Emissions about Chinese Large-Scale Public Buildings*; Ocean University of China: Chongqing, China, 2012; p. 74.
32. Xiao, H.; Yi, D. Empirical Study of Carbon Emissions Drivers Based on Geographically Time Weighted Regression. *Model. Stat. Inf. Forum.* **2014**, *29*, 83–89.
33. Yang, D.; Liu, B.; Ma, W.; Guo, Q.; Li, F.; Yang, D. Sectoral energy-carbon nexus and low-carbon policy alternatives: A case study of Ningbo, China. *J. Clean. Prod.* **2017**, *156*, 480–490. [CrossRef]
34. Lin, F.J.Z. Spatiotemporal distribution and provincial contribution decomposition of carbon emissions for the construction industry in China. *Resour. Sci.* **2019**, *41*, 897–907.
35. Cheng, Z.; Li, L.; Liu, J. Industrial Structure, Technical Progress and Carbon Intensity in China's Provinces. *Renew. Sustain. Energy Rev.* **2018**, *81*, 2935–2946. [CrossRef]
36. Wang, H.; Zhang, B.; Liu, Y.; Liu, Y.; Xu, S.; Deng, Y.; Zhao, Y.; Chen, Y.; Hong, S. Muti-dimensional analysis of urban expansion patterns and their driving forces based on the center of gravity-GTWR model: A case study of the Beijing-Tianjin-Hebei urban agglomeration. *Acta Geogr. Sin.* **2018**, *73*, 1076–1092.

energies

Article

Thermal Parameters Calibration and Energy-Saving Evaluation of Spectral Selective Absorption Film Coated Glazing System Based on Heat Transfer Simulation

Xu Chen [1,*], Saihong Zhu [1] and Tianyi Chen [2]

[1] School of Civil and Transportation Engineering, Hebei University of Technology, Tianjin 300401, China; zhusaihong@hebut.edu.cn
[2] School of Architecture, Tianjin University, Tianjin 300072, China; tianyi_chen94@tju.edu.cn
* Correspondence: 201611601014@stu.hebut.edu.cn

Abstract: Spectral selective absorption film (SSAF), a solar control film, has a special energy-saving mechanism. In the previous studies of SSAF coated glazing systems, thermal parameters (global thermal transmittance (U) and solar heat gain coefficient (SHGC)) calculated by traditional algorithms were not verified. In order to evaluate the energy-saving effect of SSAF coated glazing systems accurately, U and the SHGC were calibrated and then used for energy consumption simulation. Firstly, the simulation models of the heat transfer process of SSAF coated glazing systems were established by COMSOL Multiphysics, considering simplified linear attenuation of radiative transfer. After being validated, the simulation models were used for the calibration of U and SHGC by the Multiple Linear Regression (MLR) model. As a result, the calibration coefficients of U and SHGC are 1.126 and 1.689, respectively. Secondly, the thermal parameters of SSAF coated glazing systems calibrated by the calibration coefficients were used for a building energy consumption simulation case. The result showed that the inner surface is the best coating position for single glazing systems (SG), while the outer surface is the best coating position for double glazing systems (DG) in hot summer and cold winter zone, hot summer and warm winter zone and the moderate zone of China.

Keywords: spectral selective absorption film; glazing system; heat transfer simulation; thermal parameters; building energy consumption

Citation: Chen, X.; Zhu, S.; Chen, T. Thermal Parameters Calibration and Energy-Saving Evaluation of Spectral Selective Absorption Film Coated Glazing System Based on Heat Transfer Simulation. *Energies* **2022**, *15*, 2780. https://doi.org/10.3390/en15082780

Academic Editors: Shi-Jie Cao and Wei Feng

Received: 3 March 2022
Accepted: 5 April 2022
Published: 11 April 2022

1. Introduction

As one of the most vital industries in all walks of life in many countries, the construction industry has been consuming the major energy in operation, such as for heating, air-conditioning and refrigeration. In 2019, the energy consumption in building operation was 1.03 billion kW·h in China, according to "Research Report on Building Energy Consumption and Carbon Emission" in China (2021), accounting for 21.2% of the total national energy consumption [1]. A considerable part of the energy consumption in building operation is caused by the heat exchange between the building envelope and the outdoor environment. The external window (including glass curtain wall) is weak for energy saving and thermal insulation in building envelopes. Although the external windows of new buildings have realized satisfactory energy-saving progress because of the emergence and application of new transparent materials/structures such as multiglazing systems with Low-E coating, insulating glazing systems with PCM, insulating glazing systems with aerogels, etc., the single clear glazing system and double clear glazing system installed on the large amounts of existing buildings are not energy efficient. It demonstrates that the energy consumption caused by external windows of existing buildings accounted for about 40~50% of total energy consumption of buildings [2] and even as high as 65~80% in where priority is given to cooling loads [3].

However, the single clear glazing system and the double clear glazing system after installation would rarely be retrofitted due to high retrofitting costs, long construction time and influences on the use of existing buildings during construction. It has been reported that these problems can be solved by solar control films, a type of energy-saving materials directly coated on glazing whether in a liquid-applied [4,5] or solid-applied [6] manner. They not only can play the role of saving energy by regulating solar radiation and surface emissivity but also have the advantages of simple construction, low cost and environmentally friendly performances. Solar control films mainly include heat mirror film, low emissivity film, spectral selective absorption film (SSAF), etc. [7]. The energy-saving mechanism of SSAF is different from other films. Its main functional components include transparent conductive oxides including antimony tin oxide (ATO), indium tin oxide (ITO), tungsten bronzes (M_xWO3, M including alkali metal elements Li, Na, K, Rb and CS) and its composites, which enable SSAFs to strongly absorb solar near-infrared radiation, blocking near-infrared radiation [8,9] off the room.

It is very necessary to clarify the energy-saving effect of SSAF, which is one of the most important issues for determining the prospects of its application. Currently, there is much research on it. Chamber tests with glazing systems were used to visually investigate the thermal insulation ability of SSAF. In a chamber test, Zhu et al. [10] confirmed that SSAF reduced the indoor temperature (T_{in} (K)) of the chamber by 14 °C. Full-scale case study could characterize the impact of SSAF on the indoor thermal environment. By conducting field research of typical buildings, Moretti et al. [11] confirmed that indoor temperatures decreased by 2–3 °C when SSAF reduced the incoming solar radiation by about 60% during sunny days in moderate climate. To better understand the thermal performance of SSAF from the mechanism, Alvarez et al. [12] and Xamán et al. [13,14] established the heat transfer models of an SSAF coated single glazing system (SG+F) and an SSAF coated double glazing system (DG+F). The temperature distribution, the boundary heating flux of SSAF coated glazing systems and T_{in} were numerically calculated by using FORTRAN computer language. In order to quantitatively evaluate the impact of SSAF on energy-saving, building performance softwares such as EnergyPlus, eQUEST and ESP-r were used to simulate energy consumption [15–21], which put forward the need to verify the model of energy consumption. Gijón-Rivera et al. [15], Li et al. [18], Moretti et al. [19], Teixeira et al. [16] and Pereira et al. [17,20] validated or calibrated entire models with experimental data.

As the decisive parameters determining the energy performance of glazing systems, global thermal transmittance (U) and solar heat gain coefficient (SHGC) usually together describe the applicability of a glazing system to different climatic conditions and its different energy-saving performance from other glazing systems [22]. However, the U and the SHGC of SSAF have not been verified. The U and the SHGC input into building performance simulation software were still calculated by traditional algorithms (as described in an ISO standard, ISO 9050: 2003 Glass in Building—Determination of Light Transmission, Solar Direct Transmission, Total Solar Energy Transmission, Ultraviolet Transmission and Related Glazing Factors and a Chinese industry standard, JGJ/T 151-2008 Calculation specification for thermal performance of windows, doors and glass-curtain walls). There was no proof to demonstrate exactly whether the values calculated by this method can also be applied to SSAF, both in the above standards and previous studies. It is clear that neither the results of traditional algorithms can apply to the U and the SHGC of an SSAF coated glazing system, nor the simulations of building energy consumption can work accurately. Although Cornaro [23] obtained the U and SHGC of an SSAF coated glazing system from a set of experiments with solar test boxes, the systematic error shall be eliminated later by a mass of experiments.

In this paper, the simulation model for the heat transfer process of SSAF-coated glazing systems will be established by numerical simulation software. After contrasting the results of heat transfer calculated by simulation results with that by U and SHGC, U and SHGC will be corrected by statistical analysis if necessary. The validated or calibrated U and

SHGC will be helpful for effective building energy simulation and the accurate evaluation of energy-saving effect of SSAF-coated glazing systems.

2. Heat Transfer Simulation of SSAF-Coated Glazing Systems

2.1. Description of Physical Model

As two typical forms of glazing systems in existing buildings, single glazing systems and double glazing systems can be coated with SSAF on the inner surface or the outer surface of them, as shown in Figure 1. Six millimeters of clear glass (SG) is a common form for single glazing systems; when coated with SSAF on the inner and the outer surface, it can be referred to as SG+F-in and SG+F-out, respectively, for the sake of simplicity; 6 mm clear glass + 12 mm air + 6 mm clear glass (DG) is a common form of double glazing systems, and when coated with SSAF on the inner surface and the outer surface, it can be referred to as DG+F-in and DG+F-out, respectively, for the sake of simplicity. These simplified expressions were used for the following studies.

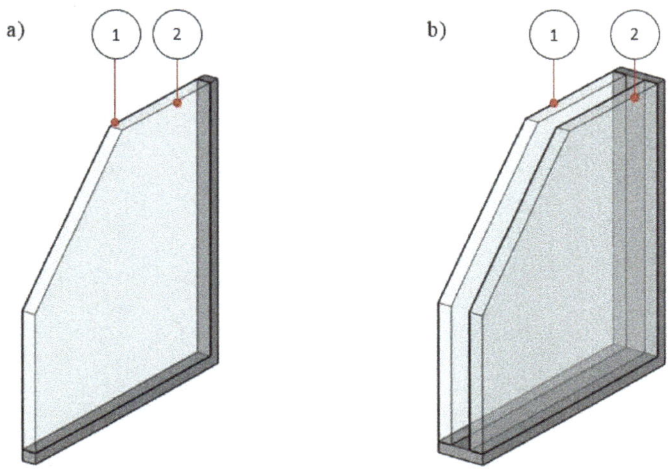

Figure 1. The positions of SSAF for existing buildings' glazing systems: (**a**) single glazing system (① the inner surface and ② the outer surface); (**b**) double glazing system (① the inner surface and ② the outer surface).

The heat transfer processes of SG+F-in and DG+F-in (as the examples of SG+F and DG+F, respectively) in the daytime (with solar radiation) are shown in Figure 2. Solar energy (G (W/m^3)) is assumed to strike perpendicularly to the glazing system from outdoor areas. For SG+F-in, some G directly enters the room through the coated glass (τ*G), some of G is absorbed by the SSAF and the glass and the rest is reflected, where τ is the transmissivity of the coated glazing system. The absorbed radiation energy is transformed into the internal energy of the coated glazing. Under the coupling effects of solar radiative absorption and the temperature difference between indoor and outdoor environments (outdoor temperature, T_{out} (K)), the heat exchange between the coated glazing system and the indoor environment is conducted by radiation (Q_r) and convection (Q_c). τ*G, Q_c and Q_r make up the total energy into a room through the glazing system. For DG+F, the difference between glazing temperatures on both sides of the cavity will cause a natural convection of internal fluid, and radiation will also be repeatedly reflected and transmitted on each surface.

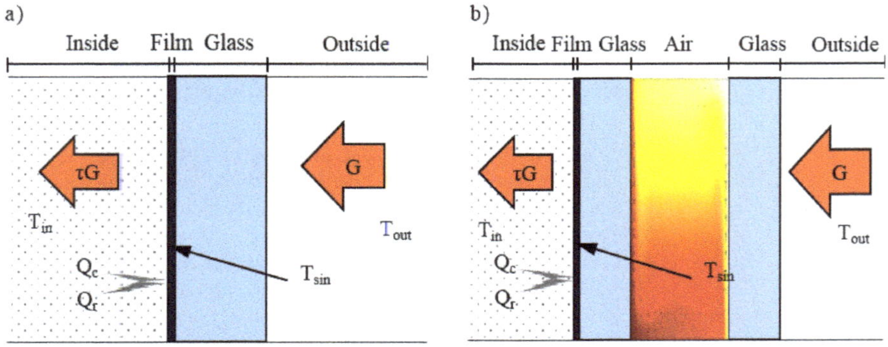

Figure 2. Heat transfer process of SSAF coated glazing systems: (**a**) SG+F [21]; (**b**) DG+F (T_{sin}: the inner surface temperature).

2.2. Heat Transfer Simulation Modeling

Based on the heat transfer mechanism of SG+F and DG+F, two-dimensional heat transfer simulation models were established by numerical simulation software. With the continuous improvement of computer performance and numerical methods, many forms of numerical simulation software have been more widely used [24,25]. Compared with traditional numerical methods, numerical simulation software will provide more convenient tools for the in-depth study on the heat transfer process of SSAF-coated glazing systems, characterized by simpler and faster simulation process and more intuitive graphical simulation results. COMSOL Multiphysics is a numerical simulation software characterized by an open architecture that supports adding and modifying control equations in the graphical interface. Similarly to other heat transfer simulation software, COMSOL Multiphysics also has various built-in common physical models and can complete process operations such as process modeling, parameter and condition setting, calculation and post-processing through the built-in model tree. The coupling of transfer processes of the SSAF coated glass, including heat transfer, natural convection and radiation heat transfer, was achieved by the physical fields of solid heat transfers and surface-to-surface radiation in COMSOL Multiphysics.

In simulation models, the part of glazing was set as a body heat source. Although the radiative transfer in semi-transparent materials considering radiation absorption is considered to decay exponentially [26], it was simplified as linear decay, considering the high thermal conductivity and the thin thickness of the glazing. Then, the heat flow intensity of the body heat source ($Q_g(W/m^2)$) is as follows:

$$Q_g = G_0 \times \alpha_g / L_g \tag{1}$$

where G_0 is the total solar radiation intensity on the glass surface (W/m^2); L_g is the glass thickness (m); and α_g is the radiation absorptivity of the glass.

SSAF was set as a boundary heat source, and its heat flow intensity ($Q_{SSAF}(W/m^2)$) is described as follows:

$$Q_{SSAF} = G_0 \times \tau_g \times \alpha_f \tag{2}$$

where τ_g is the solar radiation transmittance of glass; and α_f is the radiation absorptivity of SSAF.

When the coating position changes, the simulation model can be established and solved by adaptively adjusting the boundary conditions mentioned above.

2.3. Validation of Simulation Model

In order to verify the effectiveness of simulation models, the optical and thermophysical properties of SSAF and the clear glass (as shown in Table 1) in the research of G must be examined. Data from Alvarez et al. [12] were used as the parameter conditions of the simulation model in this part.

Table 1. Optical and thermophysical properties of the SSAF and the clear glass.

Parameters	L_g (mm)	α_g	τ_g	α_f
Values	6	0.14	0.78	0.64

Under the same boundary conditions (G = 750 W/m^2, T_{in} = 21 °C, T_{out} = 0, 10, 20, 30, 40 and 50 °C, respectively), the transient-state temperature distribution of SG+F-in was simulated and solved in COMSOL Multiphysics. The difference between these simulation results and the numerical calculation results of G. Alvarez et al. is 0.49–0.86%, which might be caused by the difference of parameter accuracy and meshing. Considering that the above results of G. Alvarez et al. have been verified by experiments [27], it is obvious that the effectiveness of the heat transfer simulation model in COMSOL Multiphysics is acceptable.

2.4. Heat Transfer Characteristics of SSAF Coated Glazing System in Typical Conditions

Depending on the cooling condition in summer or the heating condition in winter, the needed thermal performance of the SSAF coated glazing system is different. Therefore, the heat transfer simulations of SG+F and DG+F were conducted under typical conditions in summer (G = 500 W/m^2, T_{in} = 25 °C, T_{out} = 30 °C) and winter (G = 300 W/m^2, T_{in} = 20 °C, T_{out} = −20 °C), which were taken from JGJ/T 151-2008. The simulation results of SG+F and DG+F are shown in Figures 3 and 4, respectively.

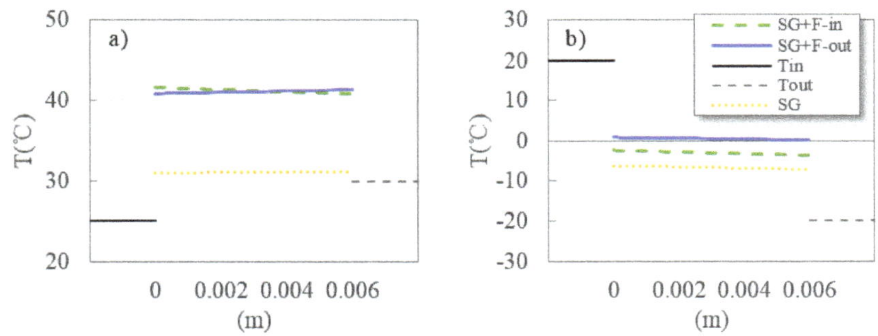

Figure 3. Temperature distribution of SG+F under typical conditions: (**a**) in summer; (**b**) in winter.

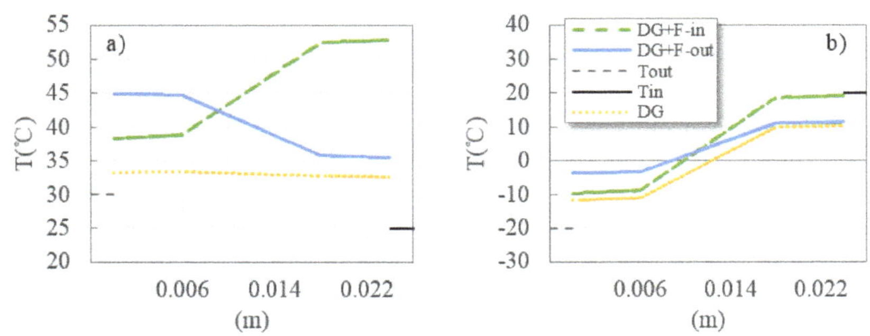

Figure 4. Temperature distribution of DG+F under typical conditions: (**a**) in summer; (**b**) in winter.

It can be seen from Figures 3 and 4 that the heat transfer simulation results can clearly and quantitatively reflect the special heat transfer mechanism in which SSAF strongly absorbs solar radiation. The absorbed solar radiation is converted into internal energy of the SSAF-coated glazing system, bringing it higher temperature. The temperature increase in SG+F is the most obvious under the typical conditions in summer. For SG+F-in and SG+F-out, the inner surface temperature (T_{sin}), which directly affects the heat exchange between the coated glazing system and the indoor environment, is 41.59 °C and 40.92 °C, respectively. T_{sin} of SG+F-in and T_{sin} of SG+F-out are separately 8.31 °C and 7.65 °C higher than that of SG. This means that there will be more Q_c, Q_r and indoor heat gain. It seems that τ*G of SG or DG should be paid more attention, because they intuitively characterize the solar radiation directly entering the room. However, the changes of Q_c and Q_r should also be fully considered in view of the particularity of the heat transfer mechanism of the SSAF coated glazing. More accurate Q_c and Q_r will help to obtain more accurate indoor heat gains when using the SSAF-coated glazing system. Under the typical conditions in winter, T_{sin} of SG+F-in and T_{sin} of SG+F-out are higher than that of SG by 3.86 °C and 7.15 °C, respectively. It not only indicates that the changes of Q_c and Q_r caused by SSAF are different under different weather conditions but also highlights that different coating positions will bring significant differences in indoor heat gain through SSAF-coated glazing systems.

The difference of thermal performance caused by the coating positions is more obvious in DG+F. Under the typical conditions in summer and winter, although the T_{in} of DG+F is still higher than DG, the difference of T_{in} between DG+F-in and DG+F-out is 17.25 °C and 7.71 °C respectively, which means a more significant impact on indoor heat.

3. Applicability Verification of Classical Algorithm for Thermal Parameters of SSAF-Coated Glazing System

The applicability of the classical algorithm for the U and SHGC of SSAF-coated glazing systems is evaluated by comparing the theoretical heat transfer with the calculated heat transfer. The theoretical heat transfer was calculated by the temperature distribution of SSAF coated glass obtained by simulation, while the calculated heat transfer was obtained by U and SHGC of SSAF-coated glazing systems, which was calculated by the classical thermal parameters algorithm.

3.1. Calculation Method of Theoretical Heat Transfer

Based on T_{sin} of the coated glazing systems obtained by simulation, Q_c can be obtained according to Newton's law of cooling:

$$Q_c = h_i \times (T_{sin} - T_{in}) \tag{3}$$

where h_i is the convective heat transfer coefficient of the inner surface of the coated glazing system (W/(m^2·K)).

Q_r can be obtained according to Stefan–Boltzmann's law:

$$Q_r = \sigma \times \varepsilon_f \times \left({T_{sin}}^4 - {T_{in}}^4\right) \tag{4}$$

where σ is Boltzmann constant, 5.67×10^{-8} W/(m^2·K^4), and ε_f is the surface emissivity of SSAF.

The theoretical total heat flux Q_t into a room through the glazing system is as follows.

$$Q_t = \tau \times G + Q_c + Q_r \tag{5}$$

3.2. Calculation Method of Calculated Heat Transfer

According to the physical meaning of U and SHGC, the calculated total heat flux Q_{ca} into a room through a glazing system is as follows.

$$Q_{ca} = G \times SHGC + U \times (T_{out} - T_{in}) \tag{6}$$

3.3. Error Analysis of Calculated Heat Transfer

In order to minimize the influence of weather conditions used for heat transfer comparison, the temperature difference between indoor and outdoor was set to −40 °C (T_{out} = −20 °C, T_{in} = 20 °C), −10 °C (T_{out}= −10 °C, T_{in} = 20 °C), 5 °C (T_{out} = 30 °C, T_{in} = 20 °C) and 15 °C (T_{out} = −35 °C, T_{in} = 20 °C), and the solar radiation intensities were set to 200 W/m^2, 500 W/m^2 and 800 W/m^2. The difference between Q_{ca} and Q_t was calculated, and the results are shown in Figure 5.

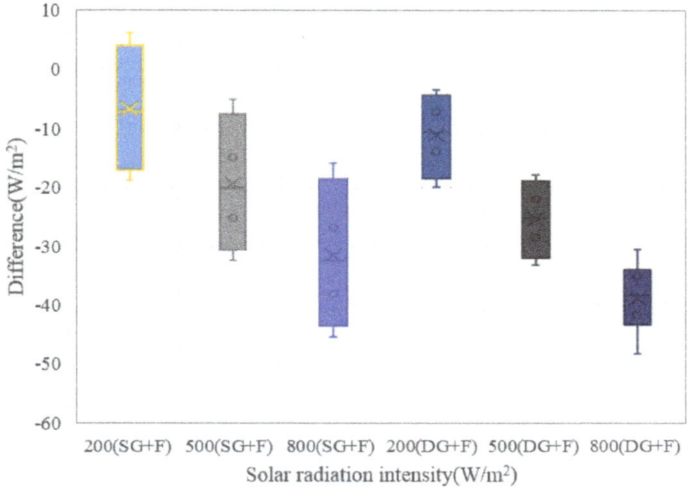

Figure 5. The difference between Q_{ca} and Q_t.

It can be seen that there are obvious differences between Q_{ca} and Q_t, which vary under different weather conditions. It is worth noting that this difference between SG+F and DG+F is not significant, but the difference gradually rises with the increase in solar radiation intensity. It shows that the U and the SHGC obtained directly by the classical algorithm do not fully consider the special heat transfer mechanism that SSAF strongly absorbs solar radiation. The U and the SHGC used by this method may further bring errors to building energy simulation and the energy-saving evaluation of SSAF. Therefore, it is necessary to calibrate the U and the SHGC of the coated glazing systems.

4. Calibration of Thermal Parameters of SSAF Coated Glazing Systems

4.1. Mechanism of Thermal Parameters Calibration

There are two kinds of calibration models for quantitative analysis parameters: multivariate linear calibration model and nonlinear calibration model. Multivariate linear correction models include Multiple Linear Regression (MLR), Principal Component Regression and Partial Least Squares Regression; nonlinear correction models include ANN, SVM, Kernel Partial Least Squares, etc. [28]. In this study, the MLR model was selected to calibrate the thermal parameters of SSAF-coated glazing systems, because it is simple to calculate and easy to understand.

Compared with U and SHGC calculated through traditional algorithms, the calibrated U and SHGC were set as U′ and SHGC′:

$$\text{s.t.}\begin{cases} U' = \alpha \times U \\ SHGC' = \beta \times SHGC \end{cases} \tag{7}$$

where α and β are calibration coefficients.

According to Equation (6), the total heat flux (Q_{ca}' (W/m^2)) into a room through the glazing system calculated by U′ and SHGC′ is as follows.

$$Q'_{ca} = (T_{in} - T_{out}) \times U' + G \times SHGC' = (T_{in} - T_{out}) \times U \times \alpha + G \times SHGC \times \beta \tag{8}$$

According to Equations (7) and (8), the α and β can be expressed as follows.

$$\begin{bmatrix} \alpha \\ \beta \end{bmatrix} = \left([U \times (T_{in} - T_{out}), G \times SHGC]^T [U \times (T_{in} - T_{out}), G \times SHGC]\right)^{-1} \times [U \times (T_{in} - T_{out}), G \times SHGC] \times Q'_{ca} \quad (9)$$

4.2. Solution and Verification of Calibration Coefficients

The calibration model of thermal parameters was established according to the mechanism of MLR and sampled by Latin Hypercube Sampling Method to generate 90 sample points. The parameters of these sample points included T_{in} and T_{out} (variable range $[-20, 40]$ °C), solar radiation intensity (variable range [0, 1000] W/m^2) and the types of the coated glazing systems (SG+F-in, SG+F-out, DG+F-in and DG+F-out). The Q_t, the U and SHGC of these sample points were calculated and fed into MATLAB with the code of MLR. Consequently, α and β, the calibration coefficients as the results of thermal parameters calibration, are 1.126 and 1.689, respectively.

The determination coefficient (R^2) of the calibration model was used to measure the fit degree between Q'_{ca} calculated by U′, SHGC′ and Q_t. R^2 is 0.97883, as shown in Figure 6, which indicates that the U and the SHGC calibrated by α and β have a good regressive relationship with theoretical heat transfer and embody the special heat transfer mechanism of SSAF completely.

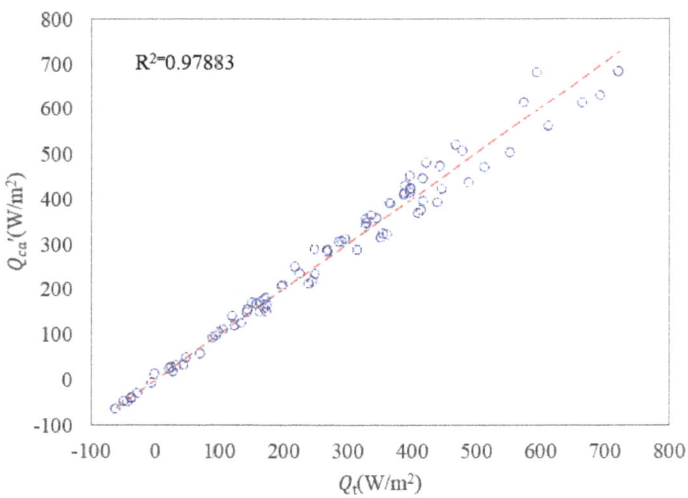

Figure 6. R^2 of the calibration model.

5. Energy Performance of the SSAF Coated Glazing System

As shown in Table 2, there are the U′ and the SHGC′ of SG+F-in, SG+F-out DG+F-in and DG+F-out, which were calibrated by α and β and the U and the SHGC of SG and DG. These thermal parameters were input into building performance software to evaluate the energy-saving effect of SSAF by conducting a building energy consumption simulation of the coated and non-coated glazing system.

China has a vast territory and complex terrains, causing great differences in climate among regions. According to the difference of T_{out}, the country is divided into five climatic zones: severe cold zone, cold zone, hot summer and cold winter zone, moderate zone and hot summer and warm winter zone. In order to study the applicability of SSAF in different climate zones, energy consumption simulation was conducted in the representative cities of these five climatic zones: Harbin, Tianjin, Chongqing, Kunming and Guangzhou in severe cold zone, cold zone, hot summer and cold winter zone, moderate zone and hot summer and warm winter zone, respectively. The monthly temperatures and the global horizonal solar radiation intensity of these representative cities based on CSWD (Chinese Standard Weather Data) are shown in Figure 7, and vertical solar radiation closely related to the glazing was automatically calculated by the building performance software.

Table 2. The thermal parameters of glazing systems.

The Type of Glazing Systems	U(W/(m^2·K)	SHGC	U′ (W/(m^2·K)	SHGC′
SG	5.70	0.81		
DG	2.80	0.75		
SG+F-in			4.40	0.58
SG+F-out			5.26	0.60
DG+F-in			2.44	0.72
DG+F-out			2.65	0.47

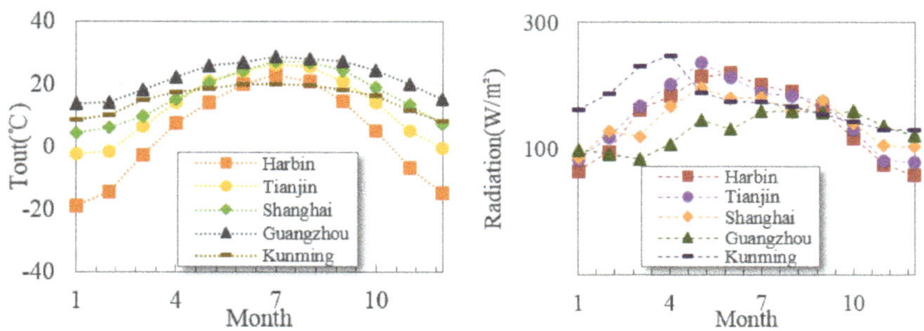

Figure 7. The monthly temperature and solar radiation intensity of the representative cities.

5.1. Building Model and Parameter Setting

The model of energy consumption simulation was established based on a typical existing office building. The model and its plan are shown in Figure 8. The total area of the office building is 1192 m^2, mainly including offices, toilets, corridors, staircases and other functional spaces. Three-hundred and seventy millimeters of brick–concrete structure is used in the exterior wall system; 100 mm reinforced waterproof concrete slab is used in the roof system and 100 mm concrete slab is used in the floor system. The other corresponding thermal parameters were taken within the conventional value range of the corresponding structure. The annual energy consumption per unit area of buildings evaluated by EnergyPlus was taken as the scheme performance evaluation index in this study. In order to facilitate data comparison, the room air temperature in all energy demand assessments was set to 20 °C and 26 °C for heating and cooling, respectively, and the operating time of HVAC systems was set from 6 am to 6 pm. Other parameters were set by the general setting of the office building.

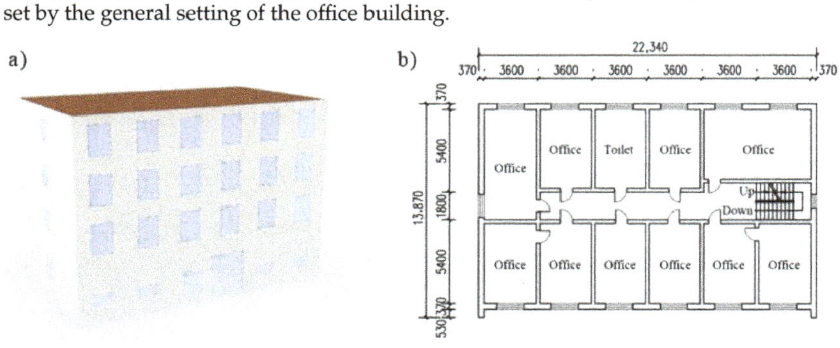

Figure 8. The typical building model and its plan: (**a**) the model; (**b**) the plan.

5.2. The Energy Performance of SSAF Coated Glazing System

It can be seen from Figure 9 that SSAF can reduce the building energy consumption of existing buildings with SG+F and DG+F in any climatic zones, but it is necessary to consider the optimum position of SSAF coating on the glazing system. In Harbin, compared with SG and DG, the building energy consumption of SG+F-out and DG+F-out increased by 1.80 W/m^2 and 6.23 W/m^2, which was the only case where SSAF did not save energy. For SG+F-in, the energy savings under the climatic conditions of Harbin, Tianjin, Shanghai, Guangzhou and Kunming were 3.96 W/m^2, 3.68 W/m^2, 3.51 W/m^2, 6.28 W/m^2 and 3.17 W/m^2, respectively. Particularly in Tianjin, Shanghai, Guangzhou and Kunming, the energy savings were higher than SG+F-out by 2.60 W/m^2, 1.18 W/m^2, 0.99 W/m^2 and 0.70 W/m^2, respectively. Therefore, the inner surface is the best coating position to SG, when SSAF is applied in existing buildings of severe cold zone, cold zone, hot summer and cold winter zone, moderate zone and hot summer and warm winter zone. The energy-saving reached 3.68 W/m^2~6.28 W/m^2 in the situations above. The energy-saving potential of SSAF is the greatest in hot summer and warm winter zone.

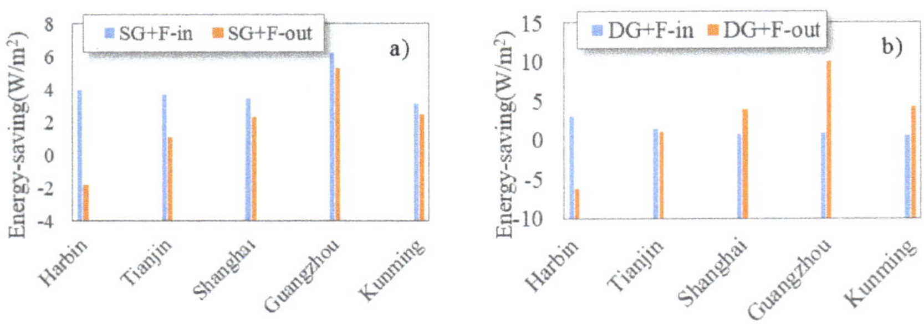

Figure 9. The energy-saving potential of SSAF: (**a**) SG+F; (**b**) DG+F.

The energy consumption of DG+F-in was lower than that of DG, but the limited energy saving was only 0.56 W/m^2~3.10 W/m^2. In Shanghai, Guangzhou and Kunming, when SSAF was coated on the outer surface of the glazing system, the energy saving of DG+F-out reached 3.89 W/m^2, 10.00 W/m^2 and 4.19 W/m^2, respectively. It shows that when SSAF is applied to the double-glazing systems of the existing buildings in hot summer and cold winter zone, moderate zone and hot summer and warm winter zone, the outer surface is the best position. Among them, SSAF has the greatest energy-saving potential in hot summer and warm winter zones. However, in Harbin, a typical city in the severe cold zone, the energy-saving performance of SSAF was obtained only when applied on the inner surface. It should be noted in particular that although in Tianjin, a typical city in cold zone, coating SSAF on whether inner or outer surfaces could save energy, energy saving was only 1.54 and 1.01 W/m^2. Therefore, whether SSAF is recommended or not, the economic benefits need to be considered, such as the investment payback period in this climate zone.

After the U and SHGC of the coated glazing systems are calibrated, the evaluation of energy performance of the coated glazing systems becomes easier and can better reflect the special mechanism of SSAF. Even without comparative experiments, it becomes convenient for SSAF-relevant energy performance comparison with other transparent materials/structures.

6. Conclusions

The heat transfer simulation models of SG+F and DG+F were established in COMSOL Multiphysics, where the radiative transfer in SSAF coated glazing systems was simplified as linear decay, the glass was set as a body heat source, and SSAF was set as a boundary heat source. Compared with previous numerical calculation results, the heat transfer simulation model proved acceptable. As the result of heat transfer simulation, temperature distributions of SG+F and DG+F clearly and quantitatively reflect the special heat transfer mechanism that SSAF strongly absorbs solar radiation and converts it into internal energy of the SSAF-coated glazing system. The heat transfer simulation models were used for the calibration of thermal parameters, including U and SHGC, by the MLR model. α and β, the calibration coefficients of U and SHGC, are 1.126 and 1.689, respectively, which have a good fit degree with theoretical heat transfer. Then, the thermal parameters of SSAF coated glazing system calibrated by α and β were used for a building energy consumption simulation case. The result showed that the inner surface is the best coating position of SSAF for the single glazing system (SG), while the outer surface is the best coating position for the double glazing system (DG) in hot summer and cold winter zone, hot summer and warm winter zone and moderate zone of China.

Author Contributions: X.C.: Methodology, software, validation and formal analysis; writing—original draft and visualization; S.Z.: conceptualization and resources; writing—review and editing, supervision and project administration; T.C.: formal analysis and data curation; writing—review and editing and visualization. All authors have read and agreed to the published version of the manuscript.

Funding: This research received no external funding.

Data Availability Statement: Not applicable.

Conflicts of Interest: The authors declare no conflict of interest.

References

1. Research Report on Building Energy Consumption and Carbon Emission in China. 2021. Available online: https://mp.weixin.qq.com/s/tnzXNdft6Tk2Ca3QYtJT1Q (accessed on 28 December 2021).
2. Gustavsen, A.; Grynning, S.; Arasteh, D.; Jelle, B.P.; Goudey, H. Key elements of and material performance targets for highly insulating window frames. *Energy Build.* **2011**, *43*, 2583–2594. [CrossRef]
3. Huang, Y.; Niu, J.; Chung, T. Study on performance of energy-efficient retrofitting measures on commercial building external walls in cooling-dominant cities. *Appl. Energy* **2013**, *103*, 97–108. [CrossRef]
4. Curcija, C.; Goudey, H.; Mitchell, R.; Manes, L.; Selkowitz, S. *Liquid-Applied Absorbing Solar Control Window Film Retrofit*; Lawrence Berkeley National Laboratory—Windows and Envelope Materials Group: San Francisco, CA, USA, 2014.
5. Berardi, U. Light transmittance characterization and energy-saving analysis of a new selective coating for in situ window retrofit. *Sci. Technol. Built Environ.* **2019**, *25*, 1152–1163. [CrossRef]
6. Calama-González, C.M.; León-Rodríguez, Á.L.; Suárez, R. Daylighting Performance of Solar Control Films for Hospital Buildings in a Mediterranean Climate. *Energies* **2019**, *12*, 489. [CrossRef]
7. Khaled, K.; Berardi, U. Current and future coating technologies for architectural glazing applications. *Energy Build.* **2021**, *244*, 111022. [CrossRef]
8. Yin, X.; Zhang, Z.; Wu, M.; Zhang, J.; Xu, G. Toward transparent composite films with selective solar spectral, flame retardant and thermal insulation functions. *Mater. Chem. Phys.* **2018**, *216*, 365–371. [CrossRef]
9. Dalapati, G.K.; Kushwaha, A.K.; Sharma, M.; Suresh, V.; Shannigrahi, S.; Zhuk, S.; Masudy-Panah, S. Transparent heat regulating (THR) materials and coatings for energy saving window applications: Impact of materials design, micro-structural, and interface quality on the THR performance. *Prog. Mater. Sci.* **2018**, *95*, 42–131. [CrossRef]
10. Zhu, S.; Chen, X.; Chen, T. Evaluation of thermal and energy performance of spectral selective absorbing film using outdoor measurements in cold climate. *Arab. J. Geosci.* **2021**, *14*, 1672. [CrossRef]
11. Moretti, E.; Belloni, E. Evaluation of energy, thermal, and daylighting performance of solar control films for a case study in moderate climate. *Build. Environ.* **2015**, *94*, 183–195. [CrossRef]
12. Alvarez, G.; Estrada, C.A. Transient heat conduction in a glass with chemically deposited SnS–Cu_xS solar control coating. *Renew. Energy* **1995**, *6*, 1023–1027. [CrossRef]
13. Xamán, J.; Jiménez-Xamán, C.; Álvarez, G.; Zavala-Guillén, I.; Hernández-Pérez, I.; Aguilar, J.O. Thermal performance of a double pane window with a solar control coating for warm climate of Mexico. *Appl. Therm. Eng.* **2016**, *106*, 257–265. [CrossRef]
14. Xamán, J.; Olazo-Gómez, Y.; Zavala-Guillén, I.; Hernández-Pérez, I.; Aguilar, J.O.; Hinojosa, J.F. Thermal evaluation of a Room coupled with a Double Glazing Window with/without a solar control film for Mexico. *Appl. Therm. Eng.* **2017**, *110*, 805–820. [CrossRef]
15. Gijón-Rivera, M.; Álvarez, G.; Beausoleil-Morrison, I.; Xamán, J. Appraisal of thermal performance of a glazed office with a solar control coating: Cases in Mexico and Canada. *Build. Environ.* **2011**, *46*, 1223–1233. [CrossRef]
16. Teixeira, H.; Gomes, M.G.; Moret Rodrigues, A.; Pereira, J. Thermal and visual comfort, energy use and environmental performance of glazing systems with solar control films. *Build. Environ.* **2020**, *168*, 106474. [CrossRef]

17. Pereira, J.; Glória Gomes, M.; Moret Rodrigues, A.; Almeida, M. Thermal, luminous and energy performance of solar control films in single-glazed windows: Use of energy performance criteria to support decision making. *Energy Build.* **2019**, *198*, 431–443. [CrossRef]
18. Li, C.; Tan, J.; Chow, T.; Qiu, Z. Experimental and theoretical study on the effect of window films on building energy consumption. *Energy Build.* **2015**, *102*, 129–138. [CrossRef]
19. Moretti, E.; Belloni, E.; Lascaro, E. The Influence of Solar Control Films on Energy and Daylighting Performance by Means of Experimental Data and Preliminary Unsteady Simulations. *Energy Procedia* **2015**, *78*, 340–345. [CrossRef]
20. Pereira, J.; Rivero, C.C.; Gomes, M.G.; Rodrigues, A.M.; Marrero, M. Energy, environmental and economic analysis of windows' retrofit with solar control films: A case study in Mediterranean climate. *Energy* **2021**, *233*, 121083. [CrossRef]
21. Chen, X.; Zhu, S.; Chen, T. Performance of a Solar Spectrum-Selective Absorption Film as a Building Energy-Saving Retrofit in China. *ACS Omega* **2021**, *6*, 31457–31468. [CrossRef]
22. Singh, M.C.; Garg, S.N. Energy rating of different glazings for Indian climates. *Energy* **2009**, *34*, 1986–1992. [CrossRef]
23. Cornaro, C.; Bucci, F.; Pierro, M.; Bonadonna, M.E.; Siniscalco, G. A new method for the thermal characterization of transparent and semi-transparent materials using outdoor measurements and dynamic simulation. *Energy Build.* **2015**, *104*, 57–64. [CrossRef]
24. Merlin, S.; Macmanus, C.; Yann, R. Modelling and numerical simulation of hygrothermal transfer through a building wall for locations subjected to outdoor conditions in Sub-Saharan Africa. *J. Build. Eng.* **2019**, *26*, 100901.
25. Merlin, S.; Romain, R.; Razika, K.; Lyes, B.; Macmanus, N.; Yann, R. Modeling, numerical simulation and validation of the hygrothermal transfer through a wooden building wall in Nancy, France. *Therm. Sci. Eng. Prog.* **2022**, *22*, 100808.
26. Modest, M.F. *Radiative Heat Transfer*; Elsevier: Amsterdam, The Netherlands, 2013.
27. Alvarez, G.; Jiménez, D.N.; Estrada, C.A. Thermal performance of solar control coatings: A mathematical model and its experimental verification. *J. Phys. D Appl. Phys.* **1998**, *31*, 2249. [CrossRef]
28. Tan, C.; Li, M.; Qin, X. Random Subspace Regression Ensemble for Near-Infrared Spectroscopic Calibration of Tobacco Samples. *Anal. Sci. Int. J. Jpn. Soc. Anal. Chem.* **2008**, *24*, 647. [CrossRef]

energies
MDPI

Review

Review of Residential Air Conditioning Systems Operating under High Ambient Temperatures

Mubarak Alawadhi and Patrick E. Phelan *

School for Engineering of Matter, Transport & Energy, Arizona State University, Tempe, AZ 85287-6106, USA; awadhi91@gmail.com
* Correspondence: phelan@asu.edu; Tel.: +1-480-965-1625

Abstract: This article provides an overview of residential vapor-compression air conditioners operating under high ambient temperatures (HAT). For the purpose of this article, a minimum temperature criterion, 40 °C and above, was developed to evaluate studies that were conducted at HAT. Several HAT organizations and projects were launched with the purpose of assessing the performance of low-GWP (GWP = global warming potential) refrigerants when operating under HAT and accelerating the transition to such refrigerants. Previous studies of air conditioner improvements (i.e., for condensers, evaporators, compressors, and refrigerants) were discussed under HAT conditions. This article also explores the challenges, the possible design modifications, and several limitations of air conditioners operating under HAT. Condenser improvements showed an 18 to 50% higher coefficient of performance (*COP*) and an 8 to 30% higher cooling capacity. Only one study was found for evaporator enhancement under HAT which improved the *COP* by ~7% and cooling capacity by ~10%. Experimental compressor improvements achieved 2 to 17 °C lower discharge temperature and up to 15% higher cooling capacity, whereas the *COP* ranged from −4% to +3% of the baseline values. Under HAT conditions, several A2L refrigerants exhibited an attractive performance compared to R-410A while none outperformed R-22 in terms of both cooling capacity and *COP*. Considering R-22 alternatives, all A1 refrigerants exhibited lower *COP*, A2L refrigerants achieved comparable *COP*, and A3 refrigerants reached higher *COP*.

Keywords: high ambient temperature; hot ambient; hot climate; air conditioning; global warming; climate change; cooling

Citation: Alawadhi, M.; Phelan, P.E. Review of Residential Air Conditioning Systems Operating under High Ambient Temperatures. *Energies* **2022**, *15*, 2880. https://doi.org/10.3390/en15082880

Academic Editors: Shi-Jie Cao and Wei Feng

Received: 20 February 2022
Accepted: 10 April 2022
Published: 14 April 2022

1. Introduction

The purpose of this article is to evaluate previous reports related to residential air conditioning systems under high ambient temperatures (HAT) and provide an overview of what has been carried out in this area to date. Most studies investigated residential air conditioners at a wide range of temperatures and hence it can be difficult and time-consuming to evaluate the results only at HAT, especially since there is no common temperature cut-off. Therefore, this article can be helpful for researchers who are interested in this topic for the following reasons: (i) it defines a criterion that segregates the moderate and high ambient temperatures, (ii) it shows the improvements of each technique compared to the baseline under HAT, (iii) it helps in avoiding repeated research work, (iv) it illustrates how A2L and A3 refrigerants operate under HAT compared with baseline refrigerants (e.g., R-410A and R-22), and (v) it recommends promising techniques for further development of AC units with A2L refrigerants under HAT. Lastly, to the best of the authors' knowledge, there is no review article that particularly discusses residential air conditioners operating under HAT. Section 1 reviews the main issues associated with HAT and their impacts on different regions of the world. Section 2 explains the approach followed to find interdisciplinary articles related to this study. Section 3 provides a brief description of different HAT organizations and projects. Section 4 discusses the literature studies of the main components of residential vapor compression systems operating under high ambient temperatures. In Section 5, various design modifications for residential air conditioners under high ambients are explored. Limiting factors of air conditioners operating under HAT conditions corresponding to Kuwait and Phoenix, Arizona, USA are investigated in Section 6. Section 7 summarizes the primary findings of the Section 4 literature studies.

Air conditioners operating under HAT must overcome high summer loads, high discharge temperature, and performance degradation. Consequently, to reduce both direct and indirect emissions, regulatory jurisdictions have added more requirements such as minimum energy efficiency and environmentally friendly refrigerants. The use of air conditioners in hot regions will be increasingly popular as a result of population and economic growth, which may affect the environment adversely, directly and indirectly if no mitigation measures are taken. Therefore, the problem becomes more complex and needs design modifications to alleviate the unit performance degradation, accelerate the use of environmentally friendly refrigerants, and comply with the minimum energy requirements under HAT conditions.

1.1. Climate Change

Since the discovery of modern electric air conditioners by Willis Carrier, there are now ~2 billion air conditioning units that have been installed globally [1]. The use of air conditioning has improved living standards by reducing heat exposure and providing comfortable indoor conditions. Hence, it became a necessity rather than a luxury in hot climates. In 2018, the percentage of households with air conditioners reached ~90% in some countries (i.e., USA and Japan) and less than 10% in other countries [1]. Moreover, ~69% of the installed units worldwide are attributed to China, the USA, Japan, and Korea, respectively, from highest to lowest. However, this is expected to change substantially in the next 30 years. The use of air conditioners will become increasingly popular as a result of population and economic growth, especially in hot regions [1]. By 2038, installed units are expected to double, reaching 4 billion units. India is expected to be the second-largest country in installed household air conditioners. The expansion of installed units is expected to improve human wellbeing in hot countries but can adversely affect the climate. In 2018, the International Energy Agency (IEA) stated that the energy consumption of air conditioners, including electric fans, accounted for 10% of global energy consumption and is expected to triple in the next 30 years [1]. The expected boom of air conditioners will cause a high electricity demand that utilities must be able to provide without harming the environment. "Growing demand for air conditioners is one of the most critical blind spots in today's energy debate", according to the IEA Executive Director [2].

Air conditioning systems interact with the environment mainly in three ways: electricity consumption, refrigerant leakage to the atmosphere, and heat rejected to the ambient. Electricity is produced primarily by burning fossil fuels, which represent ~63% of global electricity sources [3]. This process involves the releasing of greenhouse gases (GHG), mainly CO_2, into the atmosphere and eventually contributes to global warming. In fact, 25% of global GHG is due to electricity and heat production, which are the largest global sources of GHG [4]. Hence, the increase in the number of air conditioners indirectly increases the amount of CO_2 in the atmosphere. Moreover, refrigerators and air conditioners have been using several types of working fluids since the last century including: chlorofluorocarbons (CFCs), hydrochlorofluorocarbons (HCFCs), and hydrofluorocarbons (HFCs).

A leakage of those fluids can occur multiple times during the lifecycle of an air conditioner, for several reasons. Unfortunately, the environmental effects of those substances were not always well perceived when they were introduced. Later on, researchers discovered the environmental effect of these refrigerants, which are depleting the ozone layer and acting as potent GHG, up to 1000 times stronger than CO_2. Moreover, the way that air conditioners work is to pump heat from the indoors to the outdoors, which can worsen the heat island effect and increase the cooling demand [5,6]. For example, a study conducted in Phoenix, Arizona, USA found that while the effect of waste heat was negligible during the day it was substantial at night. Up to 1 °C of ambient temperature increase was observed at night, which may lead to a negative feedback loop between energy consumption and the heat island effect [7,8].

In 1984, a catastrophic impact of refrigerants was discovered—an ozone hole over Antarctica, and was attributed to the released chemicals (i.e., CFCs and HCFCs). As a result, the Vienna Convention was held in 1985 and was followed by the adoption of the Montreal Protocol in 1987, which mandated a total phase-out of HCFCs and CFCs [9]. HFCs were used as replacements for the banned refrigerants but as mentioned above they have a high effect on global warming. The world also started to observe that the earth's average temperature has been increasing since the pre-industrial era due to human activities that involved high emissions of GHGs [10], especially in the last 4 decades. Consequently, in 1997, the Kyoto Protocol was adopted to limit and reduce GHG emissions, especially in industrialized countries. There is a continuation of the United Nations Framework Convention on Climate Change (UNFCCC) meetings since the Kyoto Protocol with the goal of preventing harmful human activities from interfering with the climate system, including the Paris Agreement in 2015, which aims to limit global warming to well below 2 °C compared to pre-industrial levels. In fact, the earth's temperature was recorded in 2016 as the highest on record and was estimated to be ~1 °C warmer than the pre-industrial level [11]. In 2016, an important update to the Montreal Protocol was made in the Kigali Amendment and ratified by ~170 countries. This amendment aims to phase down the production and consumption of HFCs by more than 80% by 2047, and eventually limit the earth's temperature increase to below 1.5 °C.

Now that we have better scientific evidence on the environmental effects of human activities and the expected population growth, several measures can be taken to protect the environment. Further reductions in CO_2 emissions can be achieved by using alternative low global warming potential (GWP) refrigerants, energy-efficient systems, and more clean energy sources. Various alternative low-GWP refrigerants were found as suitable replacements but the search is still ongoing for high ambient temperature (HAT) countries. Furthermore, air conditioning manufacturers are improving the systems through several component changes (i.e., fans, compressors, etc.). Higher energy efficiency can also be promoted by governments in several ways, such as by setting minimum efficiency requirements and giving incentives to those who adopt them. For example, the Department of Energy (DOE) in the USA has imposed a minimum energy efficiency standard on appliances while the USA Environmental Protection Agency (EPA) developed the ENERGY STAR program to provide a label on efficient appliances in the market [12,13]. According to the IEA, a reduction of ~40% in energy demand can be achieved, in 2040, by following efficient cooling strategies [1].

1.2. High Ambient Temperature Regions

The high ambient temperature (HAT) exemption of the Kigali Amendment included 34 countries where suitable alternative refrigerants did not appear to exist [14]. Fifteen of those countries are recognized by the United Nations (UN) as least developed countries (LDC), contributing ~2.5% of the world population [15]. According to the UN, LDCs are at high risk of any disaster (i.e., economic, natural, epidemic) and have more than 75% of their population living in poverty [15]. Therefore, they need special consideration when implementing changes related to global warming. Gulf Cooperation Council (GCC) countries are also included in the exemption due to extreme temperatures in the summer.

The ANSI/ASHRAE standard 169-2021 [16] classified thermal climate zones based on Heating or Cooling Degree Days (HDD or CDD). The use of only average temperatures to classify thermal climate zones, however, can be misleading by averaging out an important factor—temperature variations. The CDD Equation (1), shown below, measures how hot the temperature is during a period of days relative to a reference temperature [17]:

$$\mathrm{CDD} = \sum\left[\frac{\mathrm{T_{High}} - \mathrm{T_{Low}}}{2} - \mathrm{T_{Reference}}\right] \quad (1)$$

Figure 1 shows the Cooling Degree Days (CDD) and average high temperature of the hottest month in a number of HAT countries [18,19]. The included HAT countries were selected based on location, average high temperature, and economic condition. Since all exempted HAT countries lie in the Northern Hemisphere, 12 countries from various regions of Africa (i.e., Western, Central, Eastern, and Northern) and 7 countries from Western Asia (i.e., the Arabian Peninsula) were selected. Omitted HAT countries are not expected to show significantly different results given that their average high temperatures are within the range of the included HAT countries. Furthermore, Phoenix, Arizona, USA was included in the analysis since it has considerably high summer temperatures, yet is not included in the exemption of the Kigali Amendment. It must be noted that there is more than one climate zone for large countries, which can give different CDD and temperature values. Moreover, CDD or average high temperature does not show the humidity levels in which some coastal countries may have mildly hot temperatures but extremely high humidity values. However, in this study, we are more concerned with the ambient temperature and its effect on the air conditioning system.

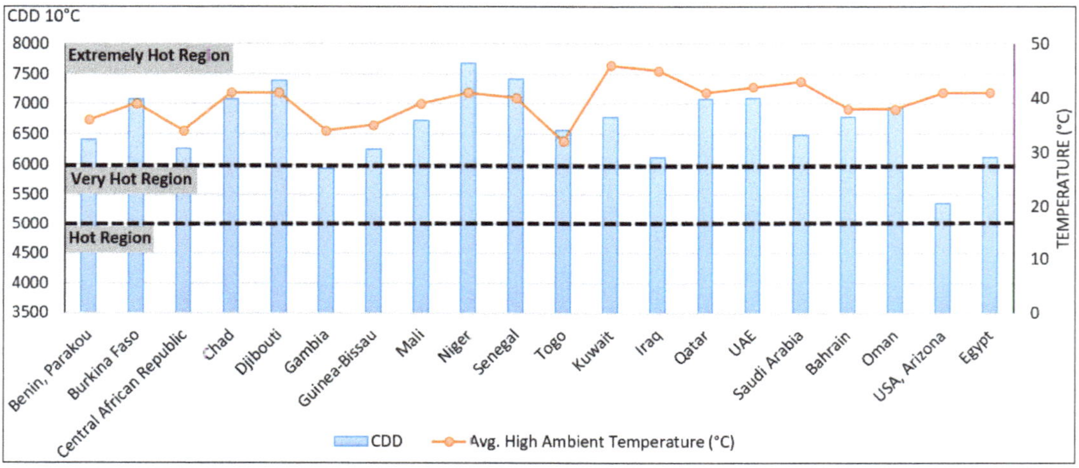

Figure 1. Average high ambient temperature for the hottest month of the year [18] and CDD°10 [19] for a number of high ambient temperature countries and regions.

ASHRAE classification of climate zones is shown in Table 1, where Climate Zone 0, or Extremely Hot, was introduced in 2013 [20]. Despite the fact that approximately 10% of USA counties have shifted to a warmer climate zone, no location falls under the Extremely Hot zone [20,21]. "For the HAT exemption to apply, a country must have an average of at least two months per year over ten consecutive years with a peak monthly average temperature above 35 °C", according to the exemption criteria of the Montreal Protocol [14]. Table 2 shows that almost all HAT countries defined by the Montreal Protocol fall under the Extremely Hot zone, while Phoenix, Arizona, USA lies in the Very Hot zone. The average high temperatures for the hottest month of the year range from 32 °C to 46 °C for Togo and Kuwait, respectively. In comparison, the average high temperature for Phoenix, Arizona, USA is relatively high, but the CDD is the lowest, which implies that the summer season can reach very high temperatures, yet the yearly values are tolerable. On the other hand, Togo has the lowest average high temperature but a relatively high CDD value due to moderate summer temperatures and almost constant ambient temperature throughout the year, ranging approximately from 23 to 32 °C [18].

Table 1. Thermal climate zone definitions from ASHRAE STANDARD 169-2021 [16].

Thermal Zone	Name	SI Units
0	Extremely hot	$6000 < CDD10\ ^\circ C$
1	Very hot	$5000 < CDD10\ ^\circ C \leq 6000$
2	Hot	$3500 < CDD10\ ^\circ C \leq 5000$
3	Warm	$CDD10\ ^\circ C \leq 3500$ and $HDD18.3\ ^\circ C \leq 2000$
4	Mixed	$CDD10\ ^\circ C < 3500$ and $2000 < HDD18.3\ ^\circ C \leq 3000$
5	Cool	$CDD10\ ^\circ C \leq 3500$ and $3000 < HDD18.3\ ^\circ C \leq 4000$
6	Cold	$4000 < HDD18.3\ ^\circ C \leq 5000$
7	Very cold	$5000 < HDD18.3\ ^\circ C \leq 7000$
8	Subarctic/arctic	$7000 < HDD18.3\ ^\circ C$

Table 2. Average high temperature for the hottest month of the year [18] and CDD°10 [19].

Country	Station [1]	CDD°10 [1]	Average High Ambient Temperature (°C) [2]	Climate Zone
Benin	Parakou, BJ (2.61 E, 9.36 N)	6412	36	Extremely Hot
Burkina Faso	Ouagadougou, BF (1.51 W, 12.35 N)	7081	39	Extremely Hot
Central African Republic	Bangui, CF (18.52 E, 4.40 N)	6270	34	Extremely Hot
Chad	Ndjamena, TD (15.03 E, 12.13 N)	7086	41	Extremely Hot
Djibouti	Camp Lemonier, DJ (43.15 E, 11.55 N)	7387	41	Extremely Hot
Gambia	Banjul/Yundum, GM (16.63 W, 13.20 N)	5933	34	Very Hot
Guinea-Bissau	Bissau Aeroport, GW (15.65 W, 11.89 N)	6258	35	Extremely Hot
Mali	Bamako/Senou, ML (7.95 W, 12.53 N)	6730	39	Extremely Hot
Niger	Niamey-Aero, NE (2.17 E, 13.48 N)	7675	41	Extremely Hot
Senegal	Tambacounda, SN (13.68 W, 13.77 N)	7415	40	Extremely Hot
Togo	Lome, TG (1.25 E, 6.17 N)	6574	32	Extremely Hot
Kuwait	Kuwait International Airport, KW (47.98 E, 29.22 N)	6784	46	Extremely Hot
Iraq	Al Najaf International Airport, IQ (44.40 E, 31.99 N)	6124	45	Extremely Hot
Qatar	Doha International Airport, QA (51.57 E, 25.26 N)	7087	41	Extremely Hot
UAE	Abu Dhabi International Airport, AE (54.65 E, 24.43 N)	7100	42	Extremely Hot
Saudi Arabia	King Khaled International Airport, SA (46.72 E, 24.93 N)	6492	43	Extremely Hot
Bahrain	Bahrain International Airport, BH (50.63 E, 26.27 N)	6784	38	Extremely Hot
Oman	Seeb International Airport, OM (58.28 E, 23.59 N)	6946	38	Extremely Hot
USA, Arizona	Phoenix Sky Harbor International Airport, AZ, US (112.01 W, 33.43 N)	5360	41	Very Hot
Egypt	Luxor, EG (32.71 E, 25.67 N)	6118	41	Extremely Hot

[1] Obtained from [19] over the period of 1 Decemebr 2020 to 31 November 2021. [2] Obtained from [18] over the period of 2014 to 2021.

From the previous examples of Togo and Phoenix, Arizona, USA it was clear that applying either the ASHRAE classification or the HAT definition of the Montreal Protocol may result in regions with mild temperatures, such as Togo, or omit regions with high ambients such as Phoenix, Arizona, USA. Furthermore, since there was no temperature cut-off for HAT countries in the previously mentioned methods, it is essential to define a temperature threshold for the purpose of this study. Based on the mean of average high temperatures of Extremely Hot countries exempted by the Montreal Protocol, in Table 2, high ambient temperature (HAT) conditions are defined here to be 40 °C and above, for the subsequent sections.

1.3. Challenges at High Ambient Temperatures

1.3.1. Energy Consumption and AC Performance

The energy efficiency and cooling capacity of air conditioners degrade as outdoor temperature increases. In most studies mentioned in this article, the degradation becomes substantial at high ambient temperature (HAT) conditions (i.e., 40 °C and above). The energy efficiency of an air conditioner can be described by its coefficient of performance (*COP*), which equals cooling capacity divided by energy consumption. Hence, the decrease in *COP* is observed with not only lower cooling capacity but even higher energy consumption. The ideal *COP*, $COP_{carnot} = T_L/(T_H - T_L) - in\ Kelvin$, can be used to explore the effect of outdoor temperatures on a system's performance. For example, when the indoor temperature is held constant at 18 °C, while the outdoor temperature is increased from 35 °C to 55 °C, the degradation of ideal *COP* is ~54%. In addition, the performance of an air conditioner can be greatly affected by refrigerant type, therefore, alternatives not only must have low GWP but show similar performance to the replaced refrigerants. As a result, an exemption of 34 developing countries was given to delay the phase-down of HFCs due to the apparent absence of suitable alternatives under HAT conditions [14].

An increase in energy consumption and demand during elevated ambient temperatures imposes high loads on electricity providers to satisfy the peak load demand. For example, household air conditioners in Kuwait account for 67% of residential electricity consumption and 72% of residential peak demand [22]. The ubiquitous usage of air conditioners poses heavy loads on electricity grids and increases the risk of power outages. This may even require investments in new systems. Consequently, fossil fuel power stations will release even higher amounts of GHGs which, in turn, increase global warming. Lundgren-Kownacki et al. [8] expect the increase in cooling demands to be more focused on fast-growing dense cities in tropical and subtropical regions. Moreover, as the global warming effect is increasing, hot regions are expected to have higher temperatures for longer periods and even more frequent heat waves [23]. The USA EPA stated that if the USA climate warms by 1 °C, an increase in electricity demand for cooling is expected to be 5 to 20% higher. This may strain the electricity grid and lead to power outages [24]. Since there appears to be no viable alternative refrigerant available for HAT countries, they are expected to move from HCFC to HFC refrigerants which have significant Global Warming Potential (GWP) values. As discussed in the previous section, all HAT countries are considered developing countries in Article 5 of the Montreal Protocol, and 15 countries are recognized as least developed countries (LDC) [15,25]. It is clear that the majority of those countries do not have the cutting-edge technology or advanced research and development (R & D) capability to handle this issue. Therefore, the support of developed countries is much needed to overcome this obstacle and transition to efficient air conditioning using low-GWP refrigerants.

1.3.2. Socio-Economic Impacts

The overall cost of an air conditioner includes the initial purchase price, maintenance costs, and energy cost. Hence, all of which must be considered in addition to the system efficiency, when comparing multiple brands. A higher efficiency system is usually more expensive upfront but yields lower energy costs and lower rates of degradation [26]. Furthermore, some low-GWP refrigerants are significantly expensive, such as R-1234yf, which reached prices up to 10 times higher than the baseline R-134a [27]. Owing to the Kigali Amendment, the phase down of HFCs may increase their prices due to lower global production, especially since Non-Article 5 countries (i.e., developed nations) are scheduled to reduce 70% of HFC production and consumption by 2024, whereas Article 5 countries (i.e., developing nations) are scheduled to reduce only 10% until 2035 [28].

The increase in energy consumption and demand during peak hours can lead to extending power plants or even building new ones, to keep up with the demand. As a result, energy prices can be affected and hence consumers may pay part of this cost, especially in the summer [29]. Looking at the total accrued expenses of an air conditioner, richer segments of society are more likely to afford the costs than poorer segments causing thermal inequities and inequalities of heat mortality [8]. This can be even more difficult for less developed countries (LDC) where they have less access to electricity and may end up living in uncomfortable conditions.

1.3.3. Human Health Impact

The human body has a core temperature of 37 °C, on average, and needs a comfortable surrounding environment to avoid health problems related to temperature. Exposure to high ambient temperatures for prolonged periods can cause the body to overheat leading to heat stroke, which can affect organs (i.e., brain, heart, kidney, etc.) and may lead to death [30]. Therefore, the International Organization for Standardization (ISO) introduced the Wet Bulb Globe Temperature (WBGT) index to assess the heat stress on individuals who are working in hot environments [31]. The WBGT index considers four environmental factors that can contribute to human heat stress: air temperature, radiant temperature, air speed, and humidity [32].

Sanderson et al. [33] performed an extensive study of mortality due to high ambient temperatures and found that cases are expected to increase under global warming. Studies have also shown that the human body can acclimatize to hot climates and thus reduce the risk of heat exposure. However, it is not known to what extent [34]. A study conducted in 16 countries has shown the temperature of increased mortality risk varies in different regions and was found to be higher in hot regions [35]. Moreover, Thompson et al. found that higher ambient temperatures increased suicide risk and recommended that mental health problems should be included in health response plans for high ambient temperatures [36].

1.3.4. Possible Mitigations

The challenges of high ambient temperatures are complex and there is no single solution to overcome those concerns. There are several solutions that can help in reducing the impact of high ambient temperatures which should be implemented by individuals and governments. Awareness campaigns to educate people about the current climate change dilemma and how to improve their energy consumption behavior should be a top priority for regulatory authorities. Additionally, average individuals must understand the benefit of transferring to a more efficient system (i.e., lower energy bills, longer life span, better performance, less environmental impacts, etc.). Governments must implement effective strategies for supporting consumers and encouraging them to use efficient systems. Those strategies include subsidizing efficient air conditioners, subsidizing new low-GWP refrigerants, setting minimum thermostat temperatures during peak hours, and imposing minimum energy standards. Successful implementation of such solutions will save energy for individuals and utilities as well as reduce CO_2 emissions.

2. Research Methodology

The purpose of this article is to evaluate previous studies related to residential air conditioning systems under high ambient temperatures and provide an overview of what has been carried out in this area so far. Initially, a minimum temperature criterion was developed, as mentioned in Section 1.2, to evaluate studies that focused only on hot ambient temperatures. A systematic search was developed in a multidisciplinary manner to include all related topics including engineering, industrial design, health, environment, and renewable energy. The database search was conducted mainly using the "ASU Library One Search" engine, which at first assessed any air conditioning system operating under high ambient temperatures. After careful review of each article, only those meeting the 40 °C and above criterion were included. Lastly, the goal was changed to focus only on high ambient temperatures for residential air conditioning systems. It is noted that throughout the review process, new topics were discovered to be related to the purpose of this study, and hence further research was conducted.

3. High Ambient Temperature Organizations and Projects

As mentioned in Section 1.3.1 the performance of an air conditioner degrades substantially at higher ambient temperatures. Hence, a major concern is how low global warming (GWP) alternative refrigerants perform under high ambient temperatures (HAT). To tackle this problem, the US Department of Energy, in cooperation with Oak Ridge National Laboratory, initiated an alternative refrigerant evaluation program specifically under HAT [37,38]. The program goal was to study the performance of low-GWP alternatives and compare them with R-410 and R-22, as baselines. International experts from various areas guided the program to ensure successful implementation. Moreover, the United Nations Environment Programme (UNEP) and United Nations Industrial Development Organization (UNIDO) sponsored two more programs with the same objective. The first was Promoting Low-GWP Alternative Refrigerants in the Air Conditioning Industry for High Ambient Conditions (PRAHA), while the second was the Egyptian Program for

Promoting Low-GWP Refrigerants (EGYPRA) [39,40]. Lastly, the Air Conditioning, Heating, and Refrigeration Institute (AHRI) conducted the Low-GWP Alternative Refrigerants Evaluation Program (Low-GWP AREP), through AHRI participants [41]. The main findings of the ORNL and AHRI Low-GWP AREP programs are summarized in Section 7.

3.1. *Oak Ridge National Laboratory (ORNL)*

The ORNL high ambient temperature program aimed to find low-GWP alternatives for CFCs and HCFCs in mini-split and rooftop air conditioners [37,38]. The program evaluated alternative low-GWP refrigerants in ~136 tests using soft-optimized mini-split air conditioners, provided by Carrier, and drop-in rooftop units, provided by SKM and Petra, under ambient temperatures up to 55 °C. Each test (i.e., drop-in and soft-optimized) included units designed for R-22 and R-410A as baselines. The alternative refrigerant selection was guided by an expert panel that included subject experts from different countries and UNEP and UNIDO personnel. Program results at high ambient temperatures (HAT) are shown in Section 7.

It must be observed that each testing program had different procedures, units, testing conditions, and levels of testing (i.e., drop-in, soft-optimized, purpose-built). Therefore, the results are not meant for direct comparison but, altogether, the findings can be used to understand the overall behavior of each alternative refrigerant. More comprehensive insights can be obtained by including additional factors such as metering device adjustments, charge quantity, and compressor speed to avoid any misinterpretation when comparing different systems [42].

3.2. *AHRI Low-GWP Alternative Refrigerants Evaluation Program (AREP)*

In response to the global warming effect of HFCs and other refrigerants, AHRI launched an industry-wide program (Low-GWP AREP) to assess new refrigerants and accelerate the transition to low-GWP refrigerants [41]. The evaluation of alternative refrigerants was performed in different applications including air conditioners, ice makers, and chillers. The program included a major category of tests under high ambient temperatures (HAT). Those reports include drop-in and soft-optimized testing with wide capacities of R-410A residential systems. AHRI required participating companies to conduct the tests at their facility using their own equipment, except for measuring the heat transfer coefficient. Nevertheless, AHRI played an important role in coordinating companies to avoid any duplicative work. All test reports were published and can be found on the AHRI website. Program results at high ambient temperatures (HAT) are shown in Section 7.

3.3. *Promoting Low-GWP Refrigerants for Air-Conditioning Sectors in High-Ambient Temperature Countries (PRAHA)*

PRAHA was a project that aimed to support the assessment of alternative refrigerants for air conditioners in high ambient temperature (HAT) countries [40,43,44]. The project was implemented by UNEP and UNIDO in two phases: PRAHA-I and PRAHA-II. PRAHA had seven local participating manufacturers from Saudi Arabia, Bahrain, Kuwait and the United Arab Emirates, and six international technology providers. PRAHA-I offered collaborative work in the regional air conditioning industry which involved building prototypes and testing alternative refrigerants and identifying minimum energy requirements for new systems. In addition, the project was able to coordinate with various component manufacturers to design compressors that are able to work efficiently with alternative refrigerants under HAT. The main findings of PRAHA-I are: (i) viable alternative refrigerants do exist at HAT but their optimization required some design modifications, and (ii) risk assessment is needed in order to safely use flammable refrigerants in HAT countries.

The second phase, PRAHA-II, focused on three elements: capacity building, design optimization, and risk assessment. The first element was able to create a platform that facilitated cooperation and exchange of knowledge among governments, research institutes, and industry associations. As a result of this awareness, the local industry started testing potential alternative refrigerants to boost the selection process. The second element focused on assessing and optimizing the prototype units. The design assessment involved some modeling work to evaluate the system performance with different components. It was highlighted that a major performance degradation occurred at ambients higher than 46 °C, and hence units were tested at this temperature. The aim of the last element was to provide a risk assessment of using flammable alternative refrigerants in HAT countries. An example of a risk assessment model was provided to test its applicability throughout the life cycle of the equipment. Furthermore, PRAHA recommended that HAT countries expand the assessment by incorporating several actual factors (e.g., local practices, cultural aspects). This can achieve a better model of tailored risk assessment, especially since the criteria for acceptable tolerances may differ among different countries.

3.4. Egyptian Program for Promoting Low-GWP Refrifèrants (EGYPRA)

EGYPRA is a project implemented by UNEP and UNIDO and initiated by the Arab Republic of Egypt [39]. The program aimed to test 19 purpose-built prototypes with dedicated compressors from several manufacturers, and 16 base units using eight alternative refrigerants under high ambient temperatures (HAT), while R-410A and R-22 are baselines. The systems were split and central units were tested at OEM labs at four different indoor/outdoor temperatures according to the Egyptian Organization for Standardization and Quality (EOS). It must be noted that testing conditions were different than ORNL and AHRI-AREP and hence the refrigerants had different behaviors. Higher cooling capacities were obtained at T_{high} than at T_3 outdoor temperatures (i.e., T_{high} = 50 °C and T_3 = 46 °C. This was attributed to the difference between indoor and outdoor temperatures, which affects the system efficiency, as explained in Section 1.3.1.

4. Residential Vapor Compression Systems at High Ambient Temperatures

Table 3 summarizes the studies on air conditioner improvements under HAT conditions. Further details on each study are provided below.

Table 3. Summary of studies related to air conditioner improvements at HAT conditions.

Authors	Objective	Results
Hajidavalloo and Eghtedari [45]	Analyzed the improvement of direct . evaporative cooling (DEC) on a 1.5-ton split-unit condenser	Achieved higher *COP*, higher cooling capacity, lower compression ratio, lower electric current consumption, and higher mass flow rate
T. Wang et al. [46]	Investigated the enhancements of direct evaporative cooling (DEC) applied to the condenser inlet air using R-410A	Lower compressor work, higher *COP*, and sub-cooling improvements
Shen and Bansal [47]	Studied the performance improvement for window air conditioners when using submerged sub-cooler and/or slinger	Submerged sub-cooler achieved up to ~5% higher *COP* while using both sub-cooler and slinger achieved up to ~7% higher *COP*
Eidan et al. [48]	Investigated the effect of direct evaporative cooling (DEC) on the condenser inlet air in dry–hot climates	Achieved higher cooling capacity and higher *COP*
Bahman and Groll [49]	Experimentally evaluated an interleaved evaporator circuitry using a 17.6 kW_{th} environmental control unit (ECU) with R-407C	Achieved higher *COP* and higher cooling capacity

Table 3. *Cont.*

Authors	Objective	Results
Al-Bakri and Ricco [50]	Investigated the heat transfer performance of a horizontal microchannel condenser (i.e., local heat transfer coefficient) using R-410A and a specifically designed test facility	The condensation HTC increases with (i) higher mass flow rate per unit area, (ii) higher vapor quality, (iii) smaller hydraulic diameter, and (iv) lower ambient temperature
López-Belchí [51]	Assessed the performance of R-134a, R-513A, and R-1234yf using a mini-channel condenser at condensing temperatures of 40 °C, 50 °C, and 60 °C	The pressure drop is greatest at higher quality, mass flow rate, and ambient temperature
		The HTCs are highest at high quality and mass flow rate but lower ambient temperatures
Ketwong et al. [52]	Studied three factors affecting air temperature when implementing DEC: FW temperature, mass ratio of water to air, and air wet-bulb temperature (WBT) for both humid and dry climates.	In hot–dry climates lower water-to-air mass ratio is recommended to achieve higher *COP* improvements while in hot–humid climates, a higher water-to-air mass ratio is recommended
Yang et al. [53]	Investigated an atomization cooling element (ACE) that uses condensate water to reduce the ambient air before entering the condenser	Achieved higher cooling capacity, lower power consumption decreased, and higher *COP*
Ding et al. [54]	Introduced a modified air source heat pump (ASHP) system that aims to provide additional sub-cooling and compressor injection, at medium pressure, by the means of an auxiliary circuit	Improved cooling capacity, increased power consumption, and lowered discharge temperatures.
Kang et al. [55]	Investigated the effect of liquid injection technique using an accumulator heat exchanger (AHX)	At constant valve opening, there was a substantial discharge pressure increase beyond the 15% injection ratio.
		At constant mass flow rate, the cooling capacity and *COP* decreased as the injection ratio increased
X. Wang et al. [56]	Investigated two vapor-injection methods in an R-410A 11 kW_{th} residential heat pump	The cooling capacity increased by 14 and 15%, and *COP* increased by 4 and 2%, for IHXC and FTC, respectively
Bahman et al. [57]	Investigated the effect of two technologies on compressors: liquid flooded with regeneration and vapor injection with economizing	The vapor injection method, at HAT, had higher *COP* improvements for all refrigerants except R-1234yf and provided lower discharge temperatures for all refrigerants except R-32, compared to oil flooding
		The vapor injection method showed the best performance when using R-410A while oil flooding had better performance using R-1234yf
Ribeiro [58]	Investigated the feasibility of using a novel refrigerating circuit to cool the outer shell of a compressor in a compact vapor compression unit used at telecommunication stations	The compressor shell temperature was always within 10 °C from the condensing temperature
		The cooling loop allowed the unit to work reliably under a higher temperature range

Table 3. *Cont.*

Authors	Objective	Results
Bahman et al. [59]	Assessed the improvement of retrofitting economized vapor injection (EVI) in a 17.6 kW_{th} environmental control unit (ECU)	The EVI system improved the cooling capacity and *COP* for both superheated injections by 12.7 and 3.1%, and saturated injections by 11.8 and 1.3%, respectively. The discharge temperature was lower by 5 and 1.7 °C for saturated and superheated injections, respectively.
J. Wu et al. [60]	Investigated the reliability of an R-290 rotary compressor under various conditions using mineral oil and synthetic oil (i.e., PAG)	Higher ambient temperatures decreased viscosities but did not have a major effect on solubility Higher suction temperatures increased viscosities and decreased solubility
C. Wang et al. [61]	Investigated the variations of dynamic pressure and oil viscosity of an R-290 rotary compressor using a 2.82 kW_{th} room air conditioner	Higher ambients decreased bearing film thickness At surface roughnesses of 0.2 and 0.6 μm, the peak contact forces were 3 and 174 N, and the minimum oil film thicknesses were 0.44 and 0.6 μm, respectively
Motta et al. [62]	Simulated the performance of a vapor compression system with R-22 and four alternatives: R-134a, R-290, R-410A and R-407C	Fluids with low critical temperature exhibited a large reduction in cooling capacity, while the compressor power increase was unaffected Adding a liquid-line/suction-line heat exchanger improved the *COP* for all refrigerants
Payne et al. [63]	Compared the performance of R-22 and R-410A in a split air conditioning system	The performance degradation was higher for R-410A as outdoor temperature increased At 54.4 °C outdoor temperature, R-410A exhibited lower cooling capacity and *COP* compared to R-22
Devotta et al. [64]	Assessed the performance of R-407C in a 5.28 kW_{th} window air conditioning unit designed for R-22	R-407C had lower cooling capacity, lower *COP*, and higher power consumption, compared to R-22 Discharge pressure of R-407C was higher than R-22
Devotta et al. [65]	Assessed the performance of R-290 as a drop-in using a 5.13 kW_{th} window air conditioner designed for R-22	R-290 had lower cooling capacity, higher *COP*, and lower power consumption, compared to R-22 The discharge pressure of R-290 was lower than R-22
Westphalen [66]	Investigated the possibility of using alternative drop-in refrigerants instead of R-407C in an environmental control unit (ECU)	R-1270 had higher cooling capacity, and higher *COP* compared to R-407C The ECU size could be reduced by 50 mm (i.e., condenser height) and still maintain the baseline refrigerant performance
Y. Wu et al. [67]	Investigated the feasibility of R-161 in a residential air conditioner	R-161 exhibited lower cooling capacity, higher COP, and lower discharge temperature compared to R-22

Table 3. *Cont.*

Authors	Objective	Results
Barve and Cremaschi [68]	Compared the drop-in performance of R-32 and R-1234yf using a 17.6 kW_{th} split heat pump with R-410A in residential applications	R-32 had higher cooling capacity, similar *COP*, and 20 to 30 °C higher discharge temperature compared to R-410A
		R-1234yf had 50% lower cooling capacity, higher COP, and lower discharge temperature compared to R-410A
Biswas and Cremaschi [69]	Assessed the characteristics of new low-GWP refrigerants DR-4 and DR-5 as a drop-in, using a 17.6 kW_{th} split heat pump with R-410A in residential applications	DR-5 achieved higher cooling capacity and higher *COP* compared to R-410A, while DR-4 exhibited lower cooling capacity and higher *COP* compared to R-410A
		Optimization increased the cooling capacity of DR-4 by 5 to 8% and *COP* by 2 to 6%, with respect to values from the drop-in test
J.H. Wu et al. [70]	Investigated the performance of R-290 and R-1270 in a 2.4 kW_{th} wall room air conditioner designed for R-22	R-290 had lower cooling capacity and higher *COP* compared to R-22, while R-1270 achieved higher cooling capacity and higher *COP*, compared to R-22
		After retrofitting the larger displacement compressor, the cooling capacity increased by approximately 9 and 15% while *COP* decreased by about 9 and 2%, for R-290 and R-1270, respectively
Joudi and Al-Amir [71]	Compared the performance of R-22 and three alternatives: R-410A, R-407C, and R-290, in 3.52 and 7.03 kW_{th} residential split air conditioners	R-290 system had the smallest optimum charge, power consumption, condensing temperature, pressure ratio, and highest *COP*
		Results showed that R-290 had the smallest value of TEWI
Sethi et al. [72]	Evaluated alternative low-GWP refrigerant of R-22 in a 6.2 kW_{th} mini-split air conditioner	The cooling capacity and COP for R-444B were within 2% of R-22, while R-407C had 2 to 3% lower cooling capacity and 4 to 7% lower *COP* compared to R-22
		R-444B had the lowest direct environmental impact due to its low GWP, low charge, and energy efficiency
Abdelaziz et al. [37]	Evaluated alternative low-GWP refrigerants of both R-410A and R-22, using soft optimized 5.25 kW_{th} mini-split air conditioners	R-22 alternatives: A2L had slightly higher discharge temperatures, cooling capacity within 5%, and *COP* within ~10%, while A3 had lower discharge temperatures, ~8% higher *COP* but within 10% lower cooling capacity
		All R-410A alternatives were A2L and showed promising results as alternatives at high ambients
Abdelaziz et al. [38]	Evaluated drop-in alternative low-GWP refrigerants for R-410A and R-22, using 27.2 and 38.7 kW_{th} roof top units	L41z(R-447B) and ARM-71a achieved higher *COP* and higher cooling capacity, compared to R-410A
		ARM-20a achieved about 1% higher *COP* and L-20A(R-444B) achieved about 2% higher cooling capacity, compared to R-22

Table 3. *Cont.*

Authors	Objective	Results
Taira et al. [73]	Investigated the performance of low-GWP HFO-mix refrigerant R-32/R-125/R-1234yf (67/7/26) and R-32 using a 7.1 kW mini-split air conditioner	At the same cooling capacity, the HFO mix had higher compressor speed, higher power input, and lower *COP* compared to R-32.
		The discharge temperature of R-32 was higher by 3.5 to 5.5 °C than the HFO mix
Oruç et al. [74]	Evaluated R-22 drop-in alternatives, R-422A, R-422D, R-417A, and R-424A using a 2.05 kW_{th} split air conditioner with a rotary compressor and capillary tube expansion device	R-22 had the highest cooling capacity, lowest compression ratio, and highest *COP*, while R-424A had comparable performance
		R-424A discharge temperature, cooling capacity, and *COP* were lower than the baseline by 19 °C, 20%, and 2.5%, respectively

4.1. Evaporators and Condensers

4.1.1. Introduction

Condensers are heat exchangers responsible for rejecting heat from the vapor compression system and are placed downstream from the compressor (i.e., operating at high pressures). In residential air conditioning systems, condensers are usually air-cooled and have aluminum fins for heat transfer enhancements (i.e., due to larger surface area). In a typical condenser, the refrigerant enters as a superheated vapor and leaves as a sub-cooled liquid, which can be factored into three stages: de-superheating, condensation, and subcooling. During the two-phase condensation stage, refrigerant velocity decreases due to the density increase when the refrigerant state changes from vapor into liquid [75]. The sub-cooled liquid at the exit sustains the metering device performance and helps increase the cooling capacity by allowing larger amounts of liquid to enter the evaporator (i.e., if there is no sub-cooling, there is a chance of flashing due to liquid-line pressure drop). For the condenser to reject heat, its temperature must be sufficiently higher than ambient air and any increase in ambient temperature requires higher condensing temperatures. With that being said, high ambient temperatures can bear higher loads on the compressors, lower the *COP* significantly, and increase the risk of compressor shutdown [46,48].

Microchannel heat exchangers are increasingly being used due to their lighter weight, smaller size, higher contact surface-area-to-volume ratio, reduced refrigerant charge, and providing similar heat transfer compared to conventional designs [50,51,76]. Their effectiveness is increased through internal and external fins and they can be used as condensers and evaporators [76]. The channels in micro and mini channel heat exchangers have hydraulic diameters less than 3 mm [77]. Fluid flow characteristics and heat transfer can be different when using microchannels instead of conventional size channels [78]. Al-Bakri and Ricco [50] found that condensation at near-critical pressure resulted in unique heat transfer behavior and could not be predicted by the literature studies. López-Belchí [51] attributed the increase in the heat transfer coefficient (HTC) of mini-channels to the flow pattern that reduces the liquid film leading to lower liquid resistance between the core gas and tube wall. In addition, he also mentioned the small hydraulic diameter effect on internal shear stress which can lead to high frictional pressure drop.

Many researchers evaluated the improvement resulting from pre-cooling air before entering the condenser using direct evaporative cooling (DEC). This cost-effective method showed promising enhancements in *COP* and cooling capacity, especially in hot–dry climates [45,46,52]. A common configuration is with a cooling pad placed at the inlet of the condenser, causing additional pressure drop, and injecting feedwater (FW) at the top of the media pad. Another study used a disk type atomization cooling element (ACE) to spray water droplets radially into air entering the condenser (i.e., this system required modifications that are explained in the literature section) with the use of condensate water from the evaporator, which can be questioned in desert climates (i.e., very low humidity).

On the other hand, evaporators are heat exchangers responsible for absorbing heat into the vapor compression system and placed upstream of the compressor (i.e., operating at low pressures). When the low-temperature two-phase refrigerant enters the evaporator, heat transfers from the return air and boils off the refrigerant into a vapor state. The refrigerant is usually superheated before entering the compressor to avoid liquid compression. Unlike condensers, evaporators deal with external sensible and latent heat transfer (i.e., from the air side) and they must be designed properly based on the climate and building demand. The evaporator coil must be sufficiently lower than the dew-point temperature (DPT) of the return air to make sure condensation is taking place, and evaporator temperature fluctuations can adversely affect the supply air condition. During refrigerant boiling, the kinetic energy is increased leading to higher pressure drops along the coil which reduce the evaporator saturation temperature. Uniform fluid distributions (i.e., refrigerant side and air side) are critical to achieving the desired cooling capacity without degrading system performance. Refrigerant maldistribution in parallel flow channels can be caused by (i) coils of different lengths or diameters, and (ii) uneven distribution of liquid–vapor at channel inlets, while air maldistribution can be affected by (i) evaporator geometry, (ii) fans, and (iii) dirty coils [79].

Several studies were conducted to assess the effect of air maldistribution and found a reduction in heat exchanger performance by up to 30% [80]. Another study by Choi et al. [81] found maximum cooling capacity degradations of 8.7% and 30% for air and refrigerant maldistribution, respectively. It must be noted that the effects of non-uniform distributions may lead to different conclusions based on the combinations of different factors (i.e., either exacerbate or outbalance the effects) [79,80,82]. Bahman and Groll [49] discussed passive and active controls (i.e., active control regulates the mass flow rate to control the exit superheat while passive control controls the exit superheat through design modification) to improve the evaporator performance under HAT. The passive control (i.e., interleaved circuitry) was preferred due to its reliability and lower implementation cost, and attractive improvements for both *COP* and cooling capacity at HAT. Studies related to condenser or evaporator improvements at HAT are discussed in Section 4.1.2 and summarized in Table 3.

4.1.2. The Literature

Hajidavalloo and Eghtedari [45] experimentally analyzed the improvement of direct evaporative cooling (DEC) on a 5.28 kW_{th} (1.5 RT) split-unit condenser under high ambient temperatures of 35 °C, 44 °C, and 49 °C. Experiments were conducted in two runs, with and without DEC, at each ambient temperature. Results showed significant improvements at ambient temperatures of 35 to 49 °C with 31.7 to 50.6% higher *COP*, 16.4 to 20.1% higher cooling capacity, 13 to 17% lower compression ratio, 11.6 to 20.3% lower electric current consumption, and 9.7 to 6% higher mass flow rate. It must be noted that greater improvements were observed at higher ambient temperatures.

T. Wang et al. [46] experimentally investigated the enhancements of direct evaporative cooling (DEC) applied to the condenser inlet air of a purpose-built system (i.e., similar to residential AC) using R-410A. The experiment was conducted in two runs with and without DEC under various outdoor temperatures including 44.5 °C. The evaporator and condenser air velocities were set to 1.6 m/s and 1.14 m/s, respectively, while the evaporative cooling pad was made from porous cellulosic paper. Experimental results showed improvements in sub-cooling, reduced compressor work, and increased *COP* by 18%, at 44.5 °C. Furthermore, the authors conducted a cost analysis based on USA water and electricity prices and showed that DEC is economically viable at high ambients.

Shen and Bansal [47] experimentally analyzed the performance of a window air conditioner (WAC) using a modified heat pump design model (HPDM) and evaluated the effectiveness of both a submerged sub-cooler and slinger (i.e., a ring placed around the condenser fan blades to collect and spray water in the ambient air entering the condenser). The unit had a cooling capacity of 2.93 kW_{th} (0.8 RT) with a single-speed rotary compressor and fin-tube evaporator and condenser. Moreover, the fin-tube sub-cooler was submerged in a water container, which used condensed water from the evaporator coil, placed downstream to provide additional sub-cooling. Results showed *COP* improvements when using a submerged sub-cooler of up to ~5% while using both sub-cooler and slinger achieved up to ~7%, over the range of ambient temperatures ~32.2 to 43.3 °C.

Eidan et al. [48] experimentally investigated the effect of direct evaporative cooling (DEC) on the condenser inlet air of a purpose-built system with a capacity of 7.0 kW_{th} (2.0 RT) under dry–hot climates (i.e., 10% RH = relative humidity, and 45 °C, 50 °C, 55 °C DBT = dry-bulb temperature). Three air velocities were imposed across the evaporative pads (1 m/s, 2 m/s, and 3 m/s) in which the highest velocity showed the highest pressure drop yielding the lowest evaporative cooling efficiency. Results showed significant improvements when applying evaporative cooling for all runs: lower compression ratio, higher refrigerant mass flow rate, higher cooling capacity by ~27 to 33%, and higher *COP* by ~17 to 33%.

Bahman and Groll [49] experimentally evaluated an interleaved evaporator circuitry using a 17.6 kW_{th} (5.0 RT) environmental control unit (ECU) with R-407C under outdoor temperatures up to 51.7 °C. Local air velocities were measured at the face of the evaporator to determine the maldistribution of the airflow. Based on the percentage of airflow at different locations, the arrangement of the interleaved circuitry was determined (i.e., refrigerant exiting a circuit with high airflow was redirected to a circuit with low airflow and so on) which yielded an additional refrigerant pressure drop of ~20 kPa compared to an unmodified evaporator. Results found that refrigerant superheat distribution was uniform across the interleaved evaporator circuitry (i.e., differences ranged from 1 °C to 5 °C). Moreover, the *COP* improved by 5.9 to 7.7% while cooling capacity improved by 8.4 to 10.6%, at outdoor temperatures of 40.6 to 51.7 °C.

Al-Bakri and Ricco [50] experimentally investigated the heat transfer performance of a horizontal microchannel condenser (i.e., local heat transfer coefficient) using R-410A and a specifically designed test facility. Ambient temperatures were 35 and 45 °C while condensing pressures were 70 and 80% of refrigerant critical pressure (i.e., near-critical). It was found that the condensation HTC increases with (i) higher mass flow rate per unit area, (ii) higher vapor quality, (iii) smaller hydraulic diameter, and (iv) lower ambient temperature. Results were validated using literature correlations and found unsatisfactory discrepancies which are attributed to the high operating conditions (i.e., ambient temperature and condensing pressure). The authors recommended further experimental analysis using R-410A at critical conditions to have a better understanding of heat transfer in microchannel condensers.

López-Belchí [51] experimentally and numerically assessed the performance of R-134a, R-513A, and R-1234yf using a mini-channel condenser at condensing temperatures of 40 °C, 50 °C, and 60 °C. The effect of variable flow rate, saturation pressure, vapor quality, and tube geometry was evaluated by studying both (i) local HTC, and (ii) frictional pressure drop, for the three refrigerants. Experimental results showed that the pressure drop is greatest at higher quality, mass flow rate, and ambient temperature. The HTCs are highest at high quality and mass flow rate but lower ambient temperatures. After validating the experimental data with model predictions from the literature, it was found that R-134a had the best thermal performance. Total equivalent warming impact (TEWI) analysis found R-513 applicable for very limited conditions while R-134a is better in most cases.

Ketwong et al. [52] theoretically studied the performance of direct evaporative cooling (DEC) for dry and humid conditions under ambient temperatures up to 40 °C. The feedwater (FW) temperature was varied in three scenarios: (i) less than air inlet wet-bulb temperature (WBT), (ii) equals air inlet WBT, and (iii) higher than air inlet WBT. Moreover, the mass ratio (MR: water mass flow rate to air mass flow rate) varied from 0.2 to 2.0. Numerical results showed that DEC was more effective in lowering ambient air temperatures, in hot–dry climates than in hot–humid climates. At 40 °C ambient temperature and 30 °C FW temperature, hot–dry conditions (30% RH) required lower MR to reduce ambient air temperature since the FW temperature was higher than air inlet WBT, while hot–humid conditions (70% RH) required higher MR to reduce the ambient air temperature since the FW temperature was lower than the air inlet WBT. Furthermore, simulation has shown that the lower the FW temperature the higher the improvement that could be reached for both climates.

Yang et al. [53] experimentally investigated a method that uses condensate water to reduce the ambient air before entering the condenser, under ambient temperatures, up to 43 °C. A 2.65 kW_{th} (0.75 RT) split unit, with R-22, was modified by changing the fan orientation (i.e., blowing towards the condenser instead of sucking air) and installing an atomization cooling element (ACE) disk on the fan. The effect of installing the ACE and changing the fan orientation was explored. Results at 43 °C showed that when using condensed water with an ACE disk, cooling capacity increased by 8.1%, power consumption decreased by 9.5%, and *COP* increased by 20%. The only concern with this method is when the condensate water is insufficient (i.e., in dry climates), the performance is expected to degrade and become lower than the baseline unit.

4.2. Compressors

4.2.1. Introduction

Compressors (i.e., positive displacement) are mechanical devices responsible for raising the pressure of the refrigerant by decreasing its volume and circulating the refrigerant in the cycle from high to low pressure. The low pressure-temperature refrigerant undergoes a compression process exiting as superheated with a high pressure-temperature state. It is imperative that the inlet refrigerant state is vapor otherwise the compressor (sometimes called a vapor pump) can be damaged since liquids are incompressible. There are several methods in real air conditioning systems that can provide proper superheating upstream of the compressor including an electronic expansion valve (EEV), thermostatic expansion valve (TXV), and an accumulator. The compression ratio (CR), absolute discharge pressure divided by absolute suction pressure, is key in identifying the compressor load. When the load on a compressor increases from higher ambient temperature or lower indoor temperature, the CR increases accordingly, indicating higher power consumption and lower performance [54,56,83]. Compressors in residential air conditioners are usually cooled by the refrigerant itself, hence proper mass flow rate and suction temperature must be maintained to provide sufficient cooling. Another important variable is the lubricant viscosity, which changes according to different operational parameters. The oil type must be compatible with the refrigerant and should maintain acceptable viscosities under various conditions. Yokozeki [84] discussed the behavior of viscosity and solubility of refrigerant-oil mixtures and the importance of choosing suitable oil viscosity and degree of solubility to avoid lowering system performance. When the ambient temperature increases, the discharge temperature and pressure will also increase so that the condenser is at a sufficiently higher temperature than the ambient, and hence heat can be rejected from the system. As a result, the CR will increase and the unit performance will decrease (i.e., lower cooling capacity, higher energy consumption, and lower *COP*). Therefore, compressors play a critical role in determining the cycle behavior and must be given special attention when developing improvements needed for HAT. Residential air conditioning compressors currently used are positive displacement types (i.e., reciprocating, rotary, and scroll).

Several methods of improving the compressor performance were initially developed for extreme low ambient temperatures and were found to be applicable for high ambient temperature applications in which cooling capacity can be greatly enhanced [57,85]. Two techniques were investigated for improving compressor performance: (i) liquid flooded compression (i.e., using oil), and (ii) refrigerant vapor injection. The latter was found to be effective in enhancing cooling capacity and increasing *COP* in certain conditions. It can be implemented by injection at intermediate pressure using an economizer: flash tank (i.e., injecting saturated vapor or liquid–vapor mixture), or an internal heat exchanger (i.e., injecting superheated vapor) [56,57,59]. On the other hand, (i) oil-flooded compression, and (ii) liquid refrigerant injection techniques, at the accumulator inlet, were evaluated to identify possible improvements in preventing high discharge temperature and compressor overheating [55,57]. In addition, exploiting a heat exchanger between the compressor inlet and condenser outlet (i.e., regenerator) prevents wet-compression, increases the sub-cooling, and hence improves the cooling capacity as well [55,57]. Many improvements were evaluated in the literature for compressors in residential air conditioners but for the purpose of this study, only feasible studies at HAT are discussed in Section 4.2.2 and summarized in Table 3.

4.2.2. The Literature

Ding et al. [54] introduced a modified air source heat pump (ASHP) system that aims to provide additional sub-cooling and compressor injection, at medium pressure, by the means of an auxiliary circuit (i.e., economizer heat exchanger, thermostatic expansion valve TXV, and solenoid valve). The auxiliary circuit starts from the condenser outlet to the compressor inlet, which can be used by activating the solenoid valve based on ambient temperatures. The condensing temperature reached 60 °C while the evaporating temperature was maintained at 2 °C. Results showed that using the auxiliary circuit improved cooling capacity, increased power consumption, and lowered the discharge temperatures. It must be noted that the compressor power increase was more significant at higher ambients (i.e., higher condensing temperatures), while the increase in cooling capacity was at the expense of a lower *COP*. Despite the lower *COP*, the supplementary circuit increased the operating range and the compressor reliability at HAT.

Kang et al. [55] investigated the effect of the liquid injection technique using an accumulator heat exchanger (AHX) in a 9 kW_{th} (2.6 RT) vapor compression cycle at 43 °C ambient temperature. The first test, without liquid injection, was conducted by varying the mass flow rate and measuring the superheat and sub-cooling effects. It was observed that the sub-cooling and superheat effects decreased with higher mass flow rate and higher evaporating pressure. The second test, with liquid injection, was conducted by varying the liquid injection ratio at two operating conditions: (i) constant valve opening, and (ii) constant flow rate. At condition (i) there was a substantial increase in the discharge pressure beyond the 15% injection ratio, while at condition (ii) the cooling capacity and *COP* decreased as the injection ratio increased. Therefore, the authors recommended optimizing the system performance by using higher flow rates at lower injections while limiting the flow rates at high injection ratios to ensure discharge pressure values are within an acceptable range.

X. Wang et al. [56] experimentally investigated two vapor injection methods in an R-410A 11 kW_{th} (3.13 RT) residential heat pump at 46.1 °C outdoor DBT where the baseline compressor was replaced with a vapor-injected scroll compressor. The two methods were the flash tank cycle (FTC) and the internal heat exchanger cycle (IHXC). The injection ratio for both methods varied to reach optimum performance, in which IHXC allowed for a wider range of ratios than FTC due to the use of a thermostatic expansion device. Moreover, the second-stage expansion valve used in the FTC was replaced with a larger one, rated for 18 kW_{th} (5.1 RT), and showed an improved performance. It was found that the IHXC and FTC had comparable performance improvement at 46.1 °C compared to the baseline, in which cooling capacity increased by up to ~14% and *COP* increased by up to ~4%.

Bahman et al. [57] numerically investigated the effect of two technologies on compressors: (i) liquid flooded with regeneration using a polyolester (POE) oil as a flooding agent, and (ii) saturated vapor injection with a flash tank economizer. A parametric study was conducted for the four refrigerants, R-410A (baseline), R-290 (propane), R-32, and R-1234yf, using the two technologies at different ambient temperatures of 25 to 55 °C. At HAT (i.e., 40 °C and above), the vapor injection method achieved higher *COP* improvements for all refrigerants except R-1234yf, compared to oil flooding. Moreover, vapor injection provided lower discharge temperatures for all refrigerants except R-32, compared to oil flooding. It must be noted that R-32 had extreme discharge temperatures of ~110 °C and hence, special attention must be paid when used at high ambients. Finally, the vapor injection method showed the best performance when using R-410A, while oil flooding had better performance using R-1234yf.

Ribeiro [58] experimentally investigated the feasibility of using a novel refrigerating circuit to cool the outer shell of a compressor in a compact vapor compression unit used at telecommunication stations, at ambient DBT up to 55 °C. The experiment was conducted using a linear compressor with a maximum operating temperature of 85 °C due to glued parts. It was found that the compressor shell temperature was always within 10 °C from the condensing temperature, in which at 55 °C ambient the condensing temperature was ~70 °C and the shell was ~79 °C. The cooling capacity was observed to increase at higher ambient temperatures due to the greater effect of increased evaporator temperature than increased condenser temperature.

Bahman et al. [59] experimentally assessed the improvement of retrofitting economized vapor injection (EVI) in a 17.6 kW_{th} (5.0 RT) environmental control unit (ECU) designed for military applications in extreme weather. The unit used R-407C as a working fluid and was tested under ambient temperatures up to 51.7 °C. A plate heat exchanger (PHX) economizer with EEV was used instead of a flash tank, due to the easier control of injected mass vapor flow rate. Experimental results showed that superheated and saturated vapor had better performance than the baseline case. Additionally, at HAT, the EVI system improved cooling capacity and *COP* for superheated injection by up to 12.7 and 3.1%, while for saturated injection, improvements were up to ~11.8 and 1.3%, respectively. On the other hand, the discharge temperature was lower by 5 and 1.7 °C for saturated and superheated injections, respectively.

J. Wu et al. [60] investigated the reliability of an R-290 rotary compressor under various conditions including ambients of 35 °C, 46 °C, and 55 °C, using mineral oil (MO) and synthetic oil (i.e., PAG = polyalkylene glycol). The performance of the R-290 rotary compressor, using PAG oil, was compared with different refrigerants—R-32, R-22, and R-410A—at similar cooling capacities (i.e., using different stroke volume compressors). Under various testing conditions, the following was observed: (i) higher ambient temperatures decreased viscosities but did not have a major effect on solubility, (ii) higher suction temperatures increased viscosities and decreased solubility, and (iii) higher ambients decreased bearing film thickness. Therefore, from tests (ii) and (iii), higher suction superheat can be used to compensate for the lower viscosity values (i.e., mineral oil but not synthetic oil) and hence avoid a too low oil film thickness that may lead to metallic contact.

C. Wang et al. [61] experimentally investigated the variations of dynamic pressure and oil viscosity of an R-290 rotary compressor using a 2.82 kW_{th} (0.8 RT) room air conditioner under ambient temperatures up to 50 °C. Results found that as outdoor temperature increased, the compression process took longer to reach higher discharge pressures. Moreover, the load on the crank part at 50 °C was nearly double the load at 30 °C. As the ambient temperature increased, there was a decrease in cooling capacity, *COP*, and mineral oil viscosity. Numerical simulations, at a 50 °C outdoor temperature, showed that at a surface roughnesses of 0.2 and 0.6 μm, the peak contact forces were 3 and 174 N, and the minimum oil film thicknesses were 0.44 and 0.6 μm, respectively. Hence, controlling the oil sump viscosity or using higher viscosity oil should be considered to maintain acceptable oil film thickness at high ambients.

4.3. *Refrigerants*

4.3.1. Introduction

Refrigerants play a substantial role in determining the performance of vapor compression cycles. They are used as working fluids to transfer heat between indoor and outdoor environments essentially by exploiting their latent heat characteristic. The most common refrigerants groups discussed in this section include chlorofluorocarbons (CFCs), hydrochlorofluorocarbons (HCFCs), hydrofluorocarbons (HFCs), hydrofluoroolefins (HFOs), and hydrocarbons (HCs). The history of refrigerants has evolved in four generations, which is well explained by Calm [86]. In 1985, the Vienna Convention was held with the objective of protecting the environment from harmful effects of ozone layer depletion [87], while in 1987 the Montreal Protocol was held with the aim of controlling and ultimately eliminating ozone-depleting substances (ODS) which created an ozone hole over Antarctica [86,88]. The Montreal Protocol is considered the most successful environment protection agreement, ratified by 198 countries, which provided a different ODS phase-out timetable for developed and developing countries [89]. In developed countries, CFCs and HCFCs were totally phased out by 1996 and 2020, respectively [90]. In developing countries, CFCs were phased out in 2010 while HCFCs are delayed until 2030 [90]. As a result, HFCs were introduced as a third-generation refrigerant due to their zero ozone depletion potential (ODP), relatively low flammability, and suitable thermophysical properties [91].

In 1997, the Kyoto Protocol was adopted with the goal to limit and reduce greenhouse gases (GHG), in which HFCs were designated as one of them [92]. In fact, HFCs are considered strong GHG and their impact on global warming can be 100 to 1000 times larger than CO_2 per unit mass [93]. This raised a concern about their impacts as they become increasingly used as alternatives to ODS in air conditioning and refrigeration applications [94]. Researchers suggested that HFC emissions are projected to reach 9 to 19% of global CO_2 emissions by 2050 if no mitigation measures are taken [95]. Therefore, the Kigali Amendment (2016) was introduced with the goal of protecting the ozone layer and the climate and was agreed upon by more than 170 countries [96]. The amendment, which came into force in 2019, aims to phase down the consumption of HFCs by more than 80% over the next 30 years [96]. Consequently, fourth-generation refrigerants are meant to replace the high-GWP HFCs with low-GWP alternatives (i.e., HFO/HFC blends, HFOs, HCs) [9,97]. Concerns were expressed about whether new alternatives are suitable under high ambient temperature (HAT) conditions, and an exemption was given to HAT countries which allowed for a delay in HFC reduction [14,98]. Since developed countries, such as the USA, have already phased out HCFCs, they are transitioning from high to low-GWP refrigerants [99]. On the other hand, HAT organizations are putting efforts into helping HAT countries bypass the high-GWP refrigerants and instead transition from HCFCs directly to low-GWP refrigerants [97].

Total equivalent warming impact (TEWI) is a measure of a refrigerant's impact on global warming, which takes into account both direct, due to leakages, and indirect emissions, from electricity consumption. It is worth mentioning that direct emissions represent about 2.9% of global GHG while indirect emissions represent about 4.9% of global GHG [100]. Hence, it is imperative for alternative refrigerants to also demonstrate adequate energy efficiency and cooling capacity, especially under HAT conditions. Furthermore, the performance of air conditioners (i.e., *COP* and cooling capacity) degrades substantially at high ambient temperatures. This adds complexity for HAT countries where low GWP might not be sufficient to reduce TEWI if the energy performance of the system is significantly penalized [68].

There are several organizations that have evaluated air conditioning systems (i.e., ORNL, PRAHA, AHRI, and EGYPRA) at various testing conditions under HAT, as described in Section 3. Depending on the system type and alternative refrigerant, different testing levels can be conducted, as defined in detail by AHRI low-GWP AREP. A drop-in test is the simplest type that only allows for minor changes: (i) charge optimization, (ii) expansion valve adjustment, and (iii) compressor speed adjustment [101]. A soft-optimized test allows additional modifications including (i) compressor displacement and/or motor size, (ii) use of variable speed compressor motor, (iii) flow control, (iv) lubricant, (v) size of tubing, and (vi) ratio of heat transfer area of condenser and evaporator, at constant total area [101]. Fully optimized types of equipment are those built specifically for certain refrigerants which is a more complex and time-consuming process. Despite the potential of higher efficiency levels for the fully optimized method, most tests conducted by organizations and companies are either drop-in or soft-optimized. Abdelaziz et al. [37] suggested that depending on measurement uncertainties, the performance is expected to improve slightly with further soft-optimization and substantially with additional engineering work. The engineering work can be any change to the unit beyond soft-optimization and drop-in modifications such as compressor cooling technologies and increasing the total area of heat exchangers (i.e., evaporator and condenser).

Refrigerants are classified in the ANSI/ASHRAE Standard 34-2019 according to two hazards involved, toxicity and flammability [102]. Lower toxicity is denoted by class A while higher toxicity is denoted by class B. Flammability is divided into four groups: high flammability denoted by class 3, low flammability denoted by class 2, mildly flammable denoted by subclass 2L, and non-flammable denoted by class 1. For example, an A2L refrigerant is non-toxic and mildly flammable (i.e., low burning velocity). An alternative refrigerant selection must consider several factors including environmental impact, thermo-physical properties, chemical properties, and safety. Despite there being no upper limit for GWP when nominating refrigerants, the candidate must show a significant reduction in GWP relative to the baseline [101]. Moreover, the performance of the alternative low-GWP refrigerant must be evaluated to limit indirect emissions. Desirable thermo-physical properties include high enthalpy of vaporization, high thermal conductivity, high critical temperature, and low vapor and liquid viscosities [71]. Low-GWP refrigerants include pure HCs, inorganics, HFOs, and HFO/HFC blends; blends can be either zeotropic or azeotropic. Many alternative refrigerants that have attractive (i.e., low) GWP and ODP values are categorized as flammable refrigerants, hence adding another challenge related to personal safety and risks associated with using flammable refrigerants [103]. Mildly flammable refrigerants (A2L) have been used in Japan for about a decade in residential air conditioners [104]. Furthermore, the EPA has approved the use of low-GWP hydrocarbon refrigerants, subject to use conditions, in various air conditioning and refrigeration applications (e.g., vending machines, room air conditioning units) [99]. Studies carried out for alternative refrigerants in residential air conditioning systems under high ambient temperature conditions are discussed in Section 4.3.2 and summarized in Table 3.

4.3.2. The Literature

Motta et al. [62] simulated the performance of a vapor compression system with R-22 (HCFC-type) and four alternatives: R-134a (HFC-type), R-290 (HC-type), R-410A (HFC-type), and R-407C (HFC-type), at an outdoor DBT of 25 to 55 °C. The results showed that R-410A had a higher degradation than other refrigerants, as the outdoor temperature increased. It was concluded that fluids with low critical temperature exhibited a large reduction in cooling capacity, while the compressor power increase was unaffected. The author performed a similar analysis after adding a liquid-line/suction-line heat exchanger to the cycle. This modification improved the *COP* for all refrigerants which varied depending on the different refrigerants' molar heat capacity.

Payne et al. [63] experimentally compared the performance of R-22 and R-410A in a split air conditioning system. The outdoor DBT ranged from 27.8 to 54.4 °C, while the indoor condition remained unchanged. Refrigerants had a comparable performance at low outdoor temperatures but the performance degradation was higher for R-410A as the outdoor temperature increased. The cooling capacity was similar at a 35 °C outdoor temperature but the R-410A *COP* was lower than R-22 by 4%. When the outdoor temperature increased to 54.4 °C, the cooling capacity and *COP* of R-410A were lower than R-22 by approximately 9 and 15%, respectively. The authors used identical evaporators and condensers, similar design compressors, and different lubricants. Furthermore, the authors conducted a higher outdoor temperature test at 68.3 °C using a customized compressor. R-410A reached a supercritical condition at the condenser inlet without noticeable changes in noise level or system operation.

Devotta et al. [64] experimentally assessed the performance of R-407C in a 5.28 kW_{th} (1.5 RT) window air conditioning unit designed for R-22. Outdoor DBT/WBT varied from 35 °C/30 °C to 46 °C/24 °C, while indoor DBT/WBT ranged from 27 °C/19 °C to 29 °C/19 °C. R-407C had approximately 2 to 8% lower cooling capacity, 8 to 13% lower

COP, and 6 to 7% higher power consumption, compared to R-22. The discharge pressure of R-407C was higher by 11 to 13% than R-22. In this study they also simulated the finned-tube heat exchangers, using the EVAP-COND model developed by the National Institute of Standards and Technology (NIST), USA [105]. The model accuracy was within ±3% of experimentally measured cooling capacities. It was used to further investigate the performance of heat exchangers for each fluid. Simulation results showed that both refrigerants had the lowest evaporator and condenser capacity at the highest outdoor temperatures. Simulated pressure drops of R-407C in the evaporator and condenser were lower than R-22 by up to ~16 and 41%, respectively. Despite the lower performance of R-407C, using it as a retrofit can extend the R-22 unit's life. It should be noted that retrofitting R-407C involved procedures for changing the unit's oil.

Another similar study was conducted by the same authors [65] to experimentally assess the performance of R-290 (the HC propane) as a drop-in using a 5.13 kW_{th} (1.5 RT) window air conditioner designed for R-22. Identical test conditions were applied. R-290 had an approximately 6 to 10% lower cooling capacity, 3 to 8% higher *COP*, and 12 to 13% lower power consumption, compared to R-22. The discharge pressure of R-290 was lower by ~13 to 18% than R-22. The authors used the same model, which was within ±4% of the experimentally measured cooling capacities, to simulate the performance of heat exchangers. The worst performance of the heat exchangers was observed at the highest outdoor temperature for both fluids. Simulated pressure drops of R-290 in the evaporator and condenser were lower than R-22 by up to ~48%. Moreover, R-290 had up to 22% lower condensing pressure and up to 3% lower evaporating pressure compared to R-22.

Westphalen [66] investigated the possibility of using alternative drop-in refrigerants instead of R-407C in a 17.6 kW_{th} (5.0 RT) environmental control unit (ECU). Different HC refrigerants were investigated based on their characteristics and R-1270 (propylene) was selected for the experimental test. Baseline and alternative refrigerants had similar characteristics except that R-1270 did not exhibit temperature glide. This was a concern for the microchannel heat exchangers design used in the ECU, which did not easily allow optimization for high-glide refrigerants. Experimental findings at a 51.7 °C outdoor DBT showed that R-1270 had a 12% higher cooling capacity and a 10% higher *COP* compared to R-407C. This improvement was partially attributed to the elimination of temperature glide when using R-1270. The author concluded that the condenser height could be reduced by 50 mm and still maintain the baseline refrigerant performance.

Y. Wu et al. [67] investigated the feasibility of R-161 (HFC-type) in a residential air conditioner. Theoretical simulations were conducted to investigate the performance of three refrigerants: R-161, R-22, and R-290 under various conditions. Theoretical results showed that R-161 had a better thermodynamic performance than R-290, a lower cooling capacity than R-22, and the highest *COP*. R-290 had the lowest discharge temperature. Furthermore, experimental analyses were conducted to verify the findings for R-161 and R-22 using a 3.5 kW_{th} (1.0 RT) residential air conditioning system where outdoor DBT ranged from 27 to 48 °C. At 48 °C, R-161 had a lower cooling capacity, higher *COP*, and lower discharge temperature than R-22 by 5.1%, 10.0%, and 3 °C, respectively.

Barve and Cremaschi [68] experimentally compared the drop-in performance of R-32 (HFC-type) and R-1234yf (HFO-type) using an R-410A 17.6 kW_{th} (5.0 RT) split heat pump in residential applications. The unit was tested at various outdoor DBTs including 43 °C and 46 °C. Experimental findings showed that R-32 had about a 10% higher cooling capacity, similar *COP*, and a 20 to 30 °C higher discharge temperature than R-410A. On the other hand, R-1234yf had a 50% lower cooling capacity, higher *COP*, and a lower discharge temperature. The unsatisfactory cooling capacity of R-1234yf, even after optimizing the thermostatic expansion valve (TXV), eliminated the refrigerant as a possible drop-in replacement.

Biswas and Cremaschi [69] experimentally assessed the characteristics of low-GWP refrigerants DR-4 and DR-5 as drop-in replacements, using an R-410A 17.6 kW_{th} (5.0 RT) split heat pump in residential applications. The unit was tested at various outdoor DBTs including 43 °C and 46 °C. The experimental assessment showed that DR-5 had up to 4% higher cooling capacity, up to 7% higher *COP*, and a higher discharge temperature than R-410A. At the same testing conditions, DR-4 showed up to 18% lower cooling capacity, up to 6% higher *COP*, and a lower discharge temperature than R-410A. Optimization of refrigerant charge and TXV increased the cooling capacity of DR-4 by 5 to 8% and *COP* by 2 to 6% with respect to values from the drop-in test, however, there was no significant improvement for DR-5.

J. H. Wu et al. [70] experimentally investigated the performance of R-290 and R-1270 in a 2.4 kW_{th} (0.7 RT) wall room air conditioner designed for R-22 under ambient temperatures up to 40 °C. At 35 °C, the experimental results showed that R-290 had a 5% lower cooling capacity and about a 10% higher *COP* compared to R-22. On the other hand, R-1270 had about a 2% higher cooling capacity and about a 1% higher *COP* compared to R-22. It was observed that, as the outdoor temperature increased, the cooling capacity of R-290 and R-1270 showed higher degradation rates than R-22. Moreover, the study also carried out refrigerant charge distribution tests since the practical charge was always higher than the allowable charge of flammable HC refrigerants. It was found that for both refrigerants, up to 63% of the charge was within the condenser and about 18% in the compressor. Hence, the authors suggested further investigations are needed on how to reduce the charges within the condenser, compressors, and liquid lines.

Joudi and Al-Amir [71] experimentally compared the performance of R-22 and three alternatives: R-410A, R-407C, and R-290, in 3.52 and 7.03 kW_{th} (1.0 and 2.0 RT) residential split air conditioners. Outdoor DBT varied from 35 to 55 °C to replicate the hot arid climate of Iraq. Experimental results showed that the R-290 system had the smallest optimum charge, power consumption, condensing temperature, pressure ratio, and highest *COP*. TEWI analysis was conducted for all refrigerants using the two air conditioning systems at variable outdoor temperatures. Results showed that R-290 had the smallest value of TEWI. Hence, R-290 is a very attractive candidate for R-22 replacement under high ambient temperatures.

Sethi et al. [72] evaluated alternative low-GWP replacement refrigerants for R-22 in a 6.2 kW_{th} (1.8 RT) mini-split air conditioner under high ambient temperatures. Experimental results under ambients up to 52 °C found that R-444B (HFC-type) was 5% more efficient than R-407C across most of the operating range. Life cycle climate performance (LCCP) was carried out to determine the environmental impacts of each refrigerant under summer ambient conditions in Kuwait. The LCCP analysis showed that the indirect impacts of each refrigerant showed significantly higher contributions to global warming than the direct impacts (i.e., ranging from 98.4 to 99.8% of the total contributions). Nevertheless, future electricity generation is likely to be less carbon-intensive, hence, the indirect impacts should be less important and the benefit of using low-GWP refrigerants would be greater.

Abdelaziz et al. [37] conducted an extensive evaluation of alternative low-GWP replacements for both R-410A and R-22, using soft optimized 5.25 kW_{th} (1.5 RT) mini-split air conditioners provided by Carrier. Outdoor DBT varied from 27.8 to 55 °C while indoor DBT/WBT ranged from 26.7 °C/19.4 °C to 29 °C/19 °C. The alternative refrigerant selection was guided by an expert panel, consisting of members of various nations, UNEP, and UNIDO personnel. R-22 alternative refrigerants were N-20B, DR-3, ARM-20B, L-20B(R-444B), R-290, and DR-93, while R-410A alternatives were R-32, DR-55, L41(R-447A), ARM-71A, and HPR-2A. R-22 alternatives showed promising results at high ambients where two of the A2L refrigerants had slightly higher discharge temperatures, cooling capacity within 5%, and efficiency approximately within 10%; A3 refrigerants had lower discharge temperatures, about 8% higher efficiency but within 10% lower cooling capacity. On the other hand, all R-410A alternatives were A2L and showed significant potential as alternatives at high ambients. R-32 had consistently better efficiency and capacity but 12 to 21 °C higher discharge temperatures.

Another similar study was conducted by ORNL [38] to evaluate drop-in alternative low-GWP refrigerants for R-410A and R-22, but this time using two roof top units where the first was 27.2 kW_{th} (7.7 RT) provided by SKM and the second was 38.7 kW_{th} (11.0 RT) provided by Petra. R-22 alternatives were L-20A (R-444B), ARM-20b, DR-7 (R-454A), and ARM-20a, while R-410A alternatives were DR-55, L41z (R-447B), ARM-71a, and R-32. Outdoor and indoor testing conditions were similar to the mini-split experiment mentioned previously. The experimental results found that at high ambient, L41z (R-447B) and ARM-71a had more than 7% higher *COP* and 3% higher cooling capacity compared to R-410A. However, all R-410A alternatives exhibited higher discharge temperatures. For the R-22 alternatives, at high ambients, ARM-20a had 0.8% higher *COP* and L-20A (R-444B) had 1.8% higher cooling capacity compared to R-22. It was noted that all refrigerants including the baseline showed a substantial efficiency degradation at higher ambient temperatures. Moreover, the testing units were designed for R-22 or R-410A, hence alternative refrigerants should not be expected to outperform the baseline refrigerants. Nevertheless, their performance could be expected to improve when manufacturers implement design modifications.

Taira et al. [73] explained the refrigerant market status for residential AC in Japan. In fact, R-410A had been phased out and R-32 was selected as a better alternative. In this study, the authors investigated the performance of low-GWP HFO-mix refrigerant R-32/R-125/R-1234yf (67/7/26) and R-32 using a 7.1 kW_{th} (2.0 RT) mini-split air conditioner under temperatures up to 52 °C. A variable frequency drive compressor was used to evaluate the system performance at different speeds, and electrical expansion was also used to regulate the mass flow rate as needed. It was found that at the same cooling capacity, the HFO mix had a higher compressor speed, lower discharge temperature, 72 W higher power input, and 6.3% lower *COP* compared to R-32. The authors explored the possible reasons for the HFO mix behavior and concluded that R-32 is superior to the HFO mix because of its latent heat characteristic, especially under high ambient conditions.

Oruç et al. [74] experimentally evaluated R-22 drop-in alternatives, R-422A (HFC-type), R-422D (HFC-type), R-417A (HFC-type), and R-424A (HFC-type) using a 2.05 kW_{th} (0.6 RT) split air conditioner with a rotary compressor and capillary tube expansion device. Ambient DBT ranged from 35 to 41 °C, while indoor DBT was set to 18 °C. Results showed that R-22 had the highest cooling capacity, lowest compression ratio, and highest *COP*, while R-424A had comparable performance. Discharge temperatures were highest for R-22 whereas R-424A had the lowest discharge temperatures. Finally, at an ambient of 41 °C, R-424A discharge temperature, cooling capacity, and *COP* were lower than the baseline by 19 °C, 20%, and 2.5%, respectively. Therefore, R-424A was the best alternative refrigerant, however, its GWP is higher than R-22 which makes it unfavorable due to its higher GWP and cost.

5. Practical Design Modifications for High Ambient Temperatures

Air conditioner manufacturers are usually obligated to certify their products before selling them in the markets and the certification requirements vary based on the minimum energy efficiency of local regulations. Policymakers are following this win-win strategy to curtail electrical demand, mitigate environmental impacts, and reduce utility bills for consumers. Davis et al. [106] conducted a study in Mexico City to evaluate the effect of replacing house appliances with efficient models and found higher energy consumption for replaced air conditioners. This was attributed to the consumer behavior that must be considered carefully in the evaluation of energy-efficiency programs. The performance of air conditioners is usually rated in two methods: (i) Energy Efficiency Ratio (*EER*), "the ratio of cooling capacity in Btu/h to the total power in Watts at 95 /75 outdoor and 80 /67 indoor DBT/WBT", and (ii) Seasonal Energy Efficiency Ratio (*SEER*), "the total heat removed from the conditioned solace during the annual cooling season in Btu divided by the total electrical energy in Watt*hours consumed by the air-conditioner during the same season" [107]. Both *COP* and *EER* represent the ratio of the cooling capacity to the input power, where EER has units of Btu h^{-1} W^{-1} and COP is unitless. Equation (2) can be used to convert between the two quantities:

$$EER = COP * 3.41 \quad (2)$$

In the USA, there are different minimum efficiency requirements for residential central AC depending on the region (i.e., North, Southeast, and Southwest) that must be met when selling an AC unit [108]. Nowadays, the minimum cooling efficiency of the Southern and Northern regions are 14 *SEER* and 13 *SEER*, and both are going to rise in 2023 to ~15 *SEER* and 14 *SEER*, respectively [109]. It must be noted that an additional minimum *EER* is required for the Southwest region only (i.e., Arizona, California, New Mexico, and Nevada) [110]. On the other hand, Kuwait has a minimum of 9.6 *EER* for direct-expansion (DX) units with and without inverters in 2022, and 10 *EER* starting in 2024 while phasing out DX units without inverters [111]. It must be noted that the previously mentioned *EER* value for Kuwait is measured at 46 °C, not 35 °C. The Air Conditioning, Heating, and Refrigeration Institute (AHRI) provide certification programs for air conditioners to ensure the products perform according to manufacturers' published claims, in which tests are conducted by an approved independent third-party laboratory and administered by AHRI. Those globally recognized certification programs help (i) manufacturers to distinguish their products in the market, sell more products and comply with local regulations, and (ii) customers to confidently compare different units' efficiencies (i.e., that are tested under similar conditions) [112]. In addition to the standard test condition T1 at 35 °C outdoor DBT, AHRI provides a wider range for high ambient countries, such as GCC countries with T3 being at 46 °C and T4-Kuwait being at 48 °C [113,114].

A motivation to know the practical modifications followed by AC manufacturers led to comparing units that are currently available in different markets. Four USA air conditioning companies were compared with three GCC companies to determine the main differences among nineteen air conditioners (i.e., 10 GCC units and 9 US units) working under high and mild ambient temperatures. The units include split, mini-split, packaged, and wall-mount types with a cooling capacity range from 10.55 to 26.38 kW_{th} (3.0 to 7.5 RT). The selection was random but the availability of detailed unit specifications was very limited and hence reduced the number of companies and units. The following observations were made:

1. The average energy efficiency ratio (*EER*) for the USA market is ~6% higher than in the GCC. This can be attributed to the more stringent regulations in the USA.
2. Almost all companies provided the *EER* and cooling capacity at the AHRI T1 condition.
3. A very limited number of AC companies provided face areas of evaporators and condensers.
4. The average condenser face area per ton and indoor air cubic feet per minute (CFM) per ton were inconclusive. This could be due to the use of mini-channel heat exchangers and the implementation of other design modifications.
5. The average condenser CFM per ton values were in the range from ~800 to 1100 CFM/ton.
6. Only GCC companies provided performance tests at 52 °C. This could be due to local regulations. For example, Kuwait requires the unit to work at 52 °C for at least two hours without tripping or overheating [111].
7. Both USA and GCC companies provided allowable operating temperatures of 52 °C while two ducted mini-split units of a GCC company had a 55 °C allowable temperature. In general, USA companies are providing a wide temperature range to meet the Southwest ambient temperatures, however, units are not allowed for installation unless they meet the minimum criteria for both *SEER* and *EER* ratings [115].
8. No GCC company provided the *SEER* value of any unit. This explains the higher importance of using *EER* when rating units operating under high ambient temperatures for long periods [108].

Due to the limited results found from the previous survey, further investigations were conducted to find common manufacturers' design modifications for high ambient temperatures (HAT). The USA regulations have set a minimum *EER* and *SEER* for the hot climate region (the Southwest) and alerted manufacturers to adhere to specific installation requirements. Some of the requirements can be critical in hot ambients such as (i) an outdoor and indoor unit combination must match the minimum certification ratings in the region, (ii) when the outdoor unit is not matching the indoor unit, the details of the indoor coil must be mentioned including face area, fin density, fin and tube materials, (iii) the outdoor temperature that locks out the low capacity operation when using a two-capacity compressor must be mentioned, and (iv) only a 5% variation is allowed in the face area and total fin surface area of the outdoor coil [116–118]. The two-capacity (or two-stage) compressor is a single or group of compressors operating with only two stages of capacity (i.e., a full compressor stage and a low compressor stage), according to AHRI 210/240 [107]. These specific requirements can be very helpful in assuring only efficient units are installed and can also help in future research and development purposes. A study conducted by the Proctor Engineering Group and AMAD explored the possible modifications of air conditioner components at a 46 °C ambient temperature, which are illustrated in Table 4 [119].

Table 4. Possible design modifications of residential air conditioners for high ambient conditions (46 °C). [119].

Component	Modification	Note
Evaporators/Condensers	Increase face area Increase number of tube rows	Space constraint Uniform airflow distribution is critical Higher pressure drop in air-side and refrigerant-side
	Improved refrigerant circuit configuration	Can be predicted via simulations and optimized based on airflow distribution Can raise manufacturing challenges
	Improved fin design	Geometry and higher fin density Higher air-side pressure drop and susceptible to trapping more dust
	Flat tube micro-channel	Higher ratio of surface area to volume Lower air-side pressure drop Issues with moisture drainage if used as evaporator
	Desuperheater	Heat exchanged with ambient air or suction-line refrigerant
Compressor	Customized models	Increased durability under high ambient temperatures Can increase the unit cost significantly
Fans and Motors	Indoor/outdoor fans	Indoor: centrifugal type blowers with a molded Styrofoam housing, cross flow fans Outdoor: propeller type fans In dry climates, running indoor fan after compressor shut-off can provide further air cooling by evaporating coil moisture for limited periods Higher airflow can improve the heat transfer through larger diameter or higher RPM Higher flow rates can increase fan power draw, increase noise, and raise concerns related to condensate drainage
	Electronically Commutated Motor (ECM)	Higher efficiency even at reduced speeds, compared to conventional permanent split capacity (PSC) motors
	Fan cycling	Ability to circulate air using the fan when the compressor is off
Controls	Variable speed controls	Inverter-driven variable speed compressor can lower consumption at lower demands The benefit can be eliminated during periods of high demand

Design modifications are expected to achieve better annual performance when optimized based on the dominant climate conditions. Thus, when a unit is designed for the highest ambient temperatures that rarely occur, the system can be oversized for long periods of the year leading to unnecessary penalties [120]. For example, oversized condensers can reduce the head pressure and compression ratio to some extent, yet the pressure should not be too low to ensure sufficient refrigerant mass flow rate is circulated and hence components work properly. Therefore, the optimized design depends on several factors and there is no specific change followed by manufacturers. Table 5 shows different design features found in units sold in the GCC market.

Table 5. Air conditioner design features found in HAT units sold in the GCC market.

Component	Feature
Compressor	Internal protection from high discharge temperature Overcurrent protection Low pressure and high pressure protection Voltage protection Cooled by refrigerant to limit discharge temperature Two-stage compressor and variable speed Short cycling protection Limiting liquid refrigerant in the compressor for improved durability
Evaporator/Condenser	Copper tubes with copper or aluminum fins Tubes: internally grooved, rifled, and ripple finned Fin: corrugated, lanced, and louvered Integral sub-cooler for condensers
Motor/Fan	Electronically Commutated Motor (ECM) Motor thermal protection Motor protection from water, particles, and solid objects
Housing	Thermal insulation for evaporator section

Under HAT, the condensing temperature can reach high values so that enough heat rejection is taking place (i.e., it must be sufficiently higher than the ambient temperature), which can lead to excessive discharge temperatures and compressor tripping. The condenser size must be suitable to lower the temperature difference (i.e., between the refrigerant and the ambient air) and hence reduce the discharge temperatures. Design modification toward a high efficiency can be achieved through the use of a larger heat exchanger surface area (i.e., condenser and evaporator), which helps in reducing the temperature difference, discharge temperature, and compression ratio, leading to improved compressor performance. In certain situations where both extreme and mild temperatures can occur, the condenser can have variable speed fans which can be regulated based on the ambient temperature. Moreover, at high humidity levels, the evaporator coil should be sufficiently low for dehumidification to take place, but it can be slightly higher in dry climates. The idea is to trade the latent cooling capacity for sensible cooling capacity and hence increase the sensible cooling efficiency [121]. Bhatia [122] explored the heat rejection of air-cooled condenser coils and the factors affecting their performance, including the condenser fan designs. The condenser fans were designed for either (i) draw-through airflow which has uniform air distribution across the coil but passes the hot discharge air over the fan and drive motor, or (ii) blow-through airflow which passes ambient air over the fan and drive motor but has less uniform air distribution across the coil. Noting that different refrigerants yield different behaviors depending on their characteristics, as mentioned in Section 4.3, the modification must also consider the type of refrigerant used. For example, in the AHRI-AREP test [123], evaluating R-32 as an alternative refrigerant for R-410A required a lower compressor speed to match the baseline cooling capacity and when the ambient temperature increased to ~49 °C, its discharge temperature increased significantly to its maximum allowable limit (i.e., 121 °C) and the unit stopped working.

6. Air Conditioner Limitations at High Ambient Temperatures

6.1. Climate

Scorching summer ambient temperatures can reach above 50 °C in the shade in the GCC region [49,72], leading to a limited unit performance in conjunction with higher cooling needs. A comparison between Kuwait and Phoenix, Arizona, USA climates was conducted to investigate the potential unit degradation resulting from high ambient temperatures. The hottest period of the day, for both locations, was found to be from approximately 9 a.m. to 5 p.m. (Figure 2) [124]. Subsequently, annual high ambient temperature (HAT) hours were evaluated for Kuwait and Phoenix (i.e., ASHRAE climate zones 0 and 1, respectively) to analyze the necessity for more efficient air conditioners in such climates. In 2017, Phoenix and Kuwait's annual hours, where ambient temperatures are 40 °C and above, were found to be ~423 and ~1430 h, respectively [124]. They are distributed over 2 months in Phoenix (~7.1 h/day) while over 5 months in Kuwait (~9.5 h/day). Figure 3 shows the average annual temperatures, from 9 a.m. to 5 p.m., for Kuwait and Phoenix. Therefore, it is essential to use efficient air conditioners (i.e., rated and certified for high temperatures) in Kuwait, but for Phoenix, a meticulous evaluation must be conducted to avoid any penalties for oversizing the unit when operating under mild temperatures. Moreover, an evaluation of average temperature increase in both locations, from 2010 to 2021, was conducted to assess the climate change effect on HAT seasons (i.e., June and July for Phoenix, Arizona, USA; May to September for Kuwait) [125]. Figure 4 shows the increased values to always be higher for the Kuwait HAT season, except in 2013 and 2016. The temperature increase in the Kuwait HAT season has always been higher than 1.2 °C over the past 6 years, raising another concern of global warming that is aggravating even the HAT levels of Kuwait.

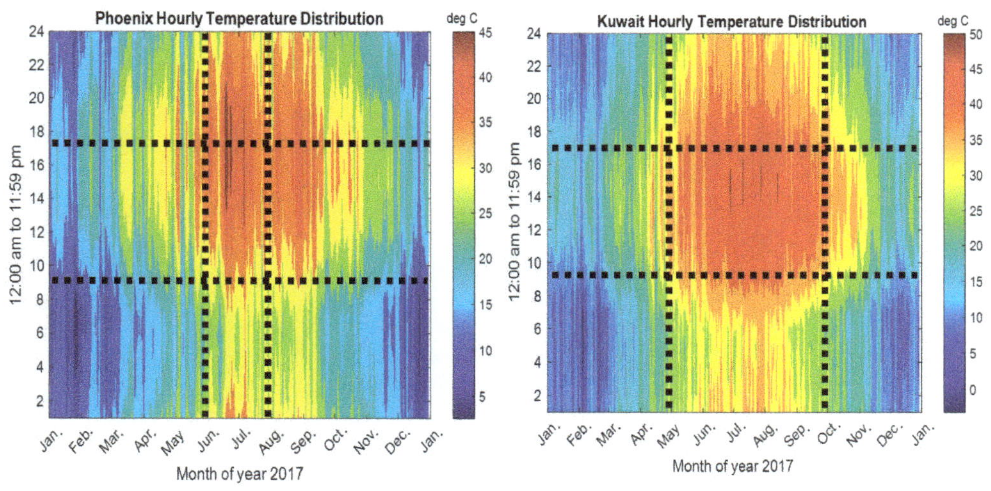

Figure 2. Hourly temperature distribution for Phoenix, Arizona, USA and Kuwait in 2017 [124].

The air conditioning process of residential air conditioners (i.e., split, mini-split, window and packaged) requires both cooling and dehumidification, which makes the indoor coil temperature a very critical factor in providing the proper thermal comfort. The dehumidification process takes place when the indoor return air passes over the evaporator cooling coil at a temperature lower than the air dew-point temperature (DPT), yet it must not be set very low (i.e., below 0 °C) to avoid coil freezing. In hot–humid climates, the dehumidification needs can be as critical as sensible cooling, especially in regions with very high relative humidities. The total air heat content can be attributed to sensible and latent heat, in which latent heat represents the moisture content. Hence, hot–humid climates bear additional cooling loads on the evaporator, which are attributed to increased moisture removal, leading to higher energy consumption. In addition, evaporators impose a very large pressure drop in the vapor compression cycle, and increased load fluctuations can adversely affect the cycle performance. As a result, the evaporating temperature is particularly limited in hot–humid climates than in hot–dry climates.

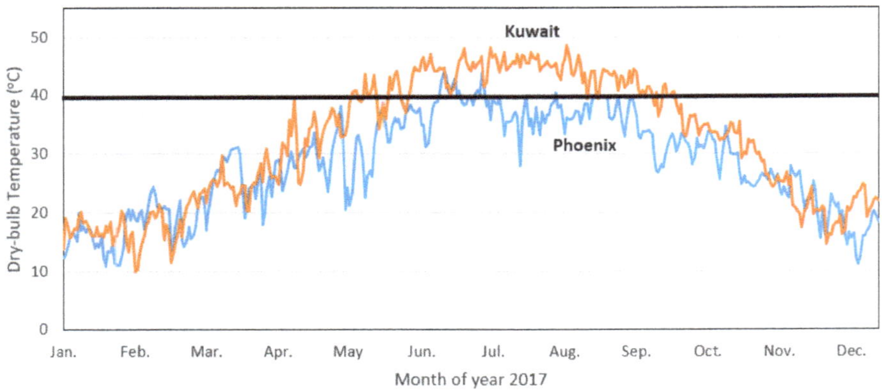

Figure 3. Annual high ambient temperature (HAT), averaged from 9 a.m. to 5 p.m., for Kuwait and Phoenix, Arizona, USA [124].

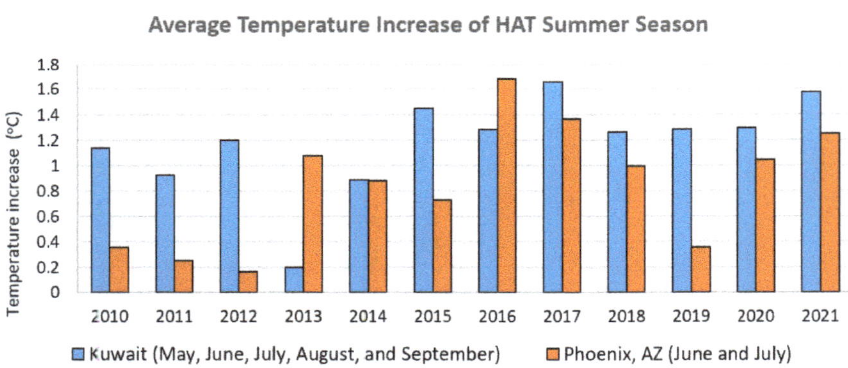

Figure 4. Average temperature increase for Kuwait and Phoenix, Arizona, USA with respect to the average temperatures for the period from 1981 to 2010 [125].

6.2. Components

Before discussing the performance limitations of an air conditioning unit under HAT, it is essential to understand how the unit efficiency is determined. As we mentioned in Section 5, there are some design modifications followed by manufacturers to improve the unit performance under HAT conditions, and there is no specific single strategy. Engineering judgment is the primary determinant of what modifications are needed and if the unit will reach the desired efficiency considering those changes [126]. The modifications are carefully optimized based on several factors including space, application type, ambient conditions, availability, economic feasibility, reliability, and mandatory regulations. A good design philosophy is to consider the annual peak temperatures but mainly design the system based on the ambient temperatures that occur for the longest periods throughout the year. If a scroll compressor is operating with a higher compression ratio than optimal, the gas is over-compressed leading to higher energy consumption than needed. While if it is operating with a lower compression ratio than optimal, the gas will not reach the desired discharge pressure until it floods back to the scroll pockets and some of it is then compressed twice, leading to higher energy consumption.

An air conditioning system consists of main and auxiliary components in series, where each component has its own limitations (e.g., compressor, condenser, and evaporator). The compressor is the heart of the unit and it is critical to know how it performs based on the selection of the remaining main parts. Proper condenser design at high ambient temperatures is crucial in determining how the compressor will function, where higher rates of heat rejection lead to lower compression ratio and hence lower work. Evaporators usually have high-pressure drops due to the increase in refrigerant velocity as evaporation is taking place leading to unwanted losses and contributing to a higher compression ratio. On the other hand, compressors have several limitations that must not be exceeded to avoid performance degradation or compressor damage. Suction superheat at the compressor suction or return gas temperature is usually specified by compressor manufacturers to be maintained at a minimum value to prevent liquid refrigerant flood back that can cause serious lubrication issues (e.g., oil dilution). Since many HAT compressors are refrigerant cooled, the superheat should not be too high to avoid the compressor overheating. Moreover, compressors also must have a limited discharge temperature to avoid premature compressor failures resulting from excessive wear and oil breakdown.

Favorable refrigerants are to be used especially under high temperatures, as discussed in the refrigerants Section 4.3.1. Another limitation is the use of A2L (mildly flammable) or A3 (highly flammable) refrigerants under high ambients, which can improve the system performance but are subject to additional design requirements to minimize the chance of system failure leading to a refrigerant leak and hence creating a fire hazard. Environmental impacts also limit the options of refrigerants since the new trend is towards zero ODP and extremely low (or zero) GWP.

6.3. System Performance

The compressor's operating envelope represents the allowable simultaneous condensing and evaporating temperatures that are determined based on refrigerant and oil mixtures. For high ambient temperatures it is vital to use compressors with an extended operating range to ensure smooth operation at normal evaporator temperatures, but high condensing temperatures. If the unit is not guaranteed to operate within those limits, compressor protective measures must be added, which are often seen in high ambient temperature AC units (Section 5). To assess the effect of changing evaporating or condensing temperatures on the pressure ratio, the evaporating temperature was varied by 7 °C whereas the condensing temperature was held constant, and vice versa. The saturation temperatures in the evaporator and condenser were assumed to be 10 °C different from the indoor and outdoor temperatures, respectively. Three refrigerants were compared, as shown in Figure 5, and they display the same behavior. The increase in compression ratio from changing the indoor temperature, 26 °C to 19 °C, was higher than from changing the outdoor temperature from 39 °C to 46 °C for all refrigerants (Table 6) [127–129]. Therefore, it is crucial to maintain the evaporator temperature at the design condition during high ambient temperatures to avoid an unwanted rise in the compression ratio leading to higher compressor work [71].

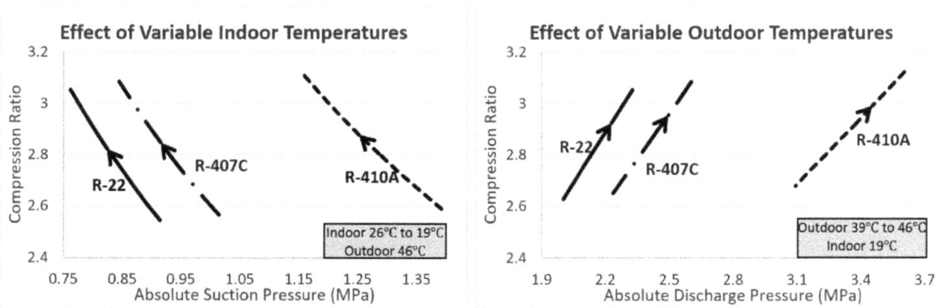

Figure 5. Effects of variable outdoor and indoor temperatures on compression ratio.

Table 6. Effects of variable outdoor and indoor temperatures on the compression ratio (CR).

		R-22	R-410A	R-407C
Variable Indoor	CR increase %	19.9%	20.5%	20.2%
	Magnitude	0.51	0.53	0.52
Variable Outdoor	CR increase %	16.2%	16.5%	16.3%
	Magnitude	0.43	0.44	0.43

6.4. Irreversibilities

At high ambient temperatures, the need for cooling increases since more heat enters the houses. However, unfortunately, the *COP* and cooling capacity of the unit drops significantly at high temperatures, presenting a limitation in the operating range of the unit, and a need for using more efficient units that can operate under HAT. The Carnot *COP*, mentioned in Section 1.3.1, represents the system's maximum theoretical efficiency according to the first law of thermodynamics and depends only on indoor and outdoor temperatures. Actual *COP*s are always less than ideal and most of the experimental studies showed drastic declines at higher ambients. The reason can be attributed to the irreversibilities of the air conditioner components, especially when operating at high ambient temperatures. Yumruta et al. [130] numerically analyzed how different condensing and evaporating temperatures would affect the exergy losses (i.e., irreversibilities) in a vapor compression system. It was found that at higher temperature differences between the evaporator and indoor air, exergy losses are increased in the evaporator. Similar behavior was observed for the condenser, where higher temperature differences led to higher exergy losses in the condenser. Kalaiselvam and Saravanan [131] experimentally tested scroll compressors and recommended some operating conditions to minimize the compressor irreversibilities: 4 °C evaporating temperature, 35 to 40 °C condensing temperature, within 65 °C discharge temperature, and 14 °C suction temperature. Ahamed et al. [132] reviewed different studies of exergy analysis for vapor compression systems and found that decreasing the condenser temperature or increasing the evaporating temperature improved the system *COP*, whereas the total system irreversibility was decreased. Bahman and Groll [133] experimentally investigated the components' irreversibilities of Environmental Control Units (ECUs), at capacities of 5.28, 10.55, and 17.58 kW_{th} (1.5, 3.0, and 5.0 RT), at a 51.7 °C outdoor DBT. The analysis showed that irreversibility contributions were significant for three system components: the compressor up to 42.5%, the evaporator up to 32.9%, and the condenser up to 22.4%. The 5-ton ECU's evaporator showed the highest irreversibility due to high refrigerant pressure drop (i.e., at the evaporator distributor) and air maldistribution.

6.5. Operation and Maintenance

People's behavior can also limit the performance of air conditioners from operation and maintenance aspects. A study conducted in Florida found that unmaintained air conditioners operating for long periods (i.e., more than ~1500 h/year) are susceptible to increased degradation rates, especially for larger capacity systems. The air conditioning unit degradation is another major AC-limiting factor leading to higher energy consumption with less ability for cooling, where the reasons for continuous degradation can be attributed, but not limited to, coil fouling, filter clogging, and refrigerant charge problems [26].

6.6. Possible Improvements

1. Proper control of indoor DBT during high ambient temperatures can compensate for the performance degradation due to high outdoor DBT. For example, setting an indoor thermostat at moderate temperatures can provide higher cooling capacity and higher *COP* of the unit, which leads to lower energy consumption and cost.
2. Applying optimized design modifications, Section 5, to maintain the overall system performance and achieve stable operation under harsh conditions.

3. Performing routine maintenance as recommended by local manufacturers to sustain the system performance and avoid higher operational costs.
4. Positioning the outdoor unit to face North or East can reduce the amount of direct sunlight as opposed to West or South facing.
5. Proper unit sizing, installation procedures, air distribution, and duct designs also have significant effects but are not included in this study.

7. Results and Discussion

Air conditioners operating under high ambient temperatures (HAT) must overcome high summer loads, high discharge temperatures, and performance degradation (i.e., lower cooling capacity and *COP*). Regulatory jurisdictions have added more requirements such as minimum energy efficiency and environmentally friendly refrigerants. Considering all previous factors, the problem becomes more complex and requires design modifications to comply with them. The climate change motivated the transition to low-GWP (A2L) refrigerants and ongoing efforts are underway to ensure the workability under high ambient conditions. In addition to the research completed by different entities, four main organizations conducted extensive work to investigate different A2L refrigerants under HAT (Section 3). The following results include two of those organizations (i.e., ORNL and AHRI Low-GWP AREP) and all related research articles found in the literature, where all of them comply with our definition of high ambient temperature, 40 °C and above, as mentioned in Section 3.1. It was found that the definition of the Montreal Protocol for hot countries, Section 3.1, was in compliance with the hottest ASHRAE climate zone. Nevertheless, a more stringent temperature was defined for the purpose of investigating the air conditioner performance and degradation, especially since most studies detected higher performance degradation at temperatures above 40 °C. Furthermore, due to the limited comprehensive resources in such areas, it was decided to focus on electric residential systems to investigate the drawbacks of each individual component.

7.1. Condensers and Evaporators

There were several studies conducted for the improvement of heat exchangers (i.e., evaporators and condensers) under HAT (Figure 6 and Table 7). Condenser improvements in the range from ~18 to 50% for *COP* and 8 to 30% for cooling capacity were achieved through direct evaporative cooling (DEC). Moreover, two studies tested mini-channel condensers at HAT where Al-Bakri and Ricco [50] found unique condensation heat transfer coefficients (HTC) at near-critical pressure and recommended further testing at critical pressures. Nevertheless, both studies [50,51] agreed that local HTC increases at higher quality and mass flow rate, but decreases at lower ambient temperature. On the other hand, only one study was found describing evaporator improvements at HAT [49], which showed improvements of ~7% in *COP* and ~10% in cooling capacity. This method implemented passive controls to improve the refrigerant circuitry based on the external air flow.

A mini-channel heat exchanger shows attractive characteristics in increasing the surface area and decreasing refrigerant charge, which are essential considerations for HAT air conditioners. The mini-channel heat exchangers could also be tested under HAT with A2L refrigerants (i.e., mildly flammable) since charge limitations are applied for flammable refrigerants. Direct evaporator cooling has shown significant improvements, especially under hot–dry conditions, but water availability is a major concern. Few studies explored the possibility of utilizing condensate water, recognizing that the amount can be insufficient in dry climates, and hence it is not recommended to depend solely upon it. The literature also revealed only a very limited number of studies conducted on evaporator improvements. Bahman and Groll [133] discussed the irreversibility of an environmental control unit (ECU) and showed that evaporator losses can be very significant under HAT. Hence, an active control method, where the refrigerant can be regulated to control the exit superheat, could be investigated under HAT and compared with passive control improvements [49].

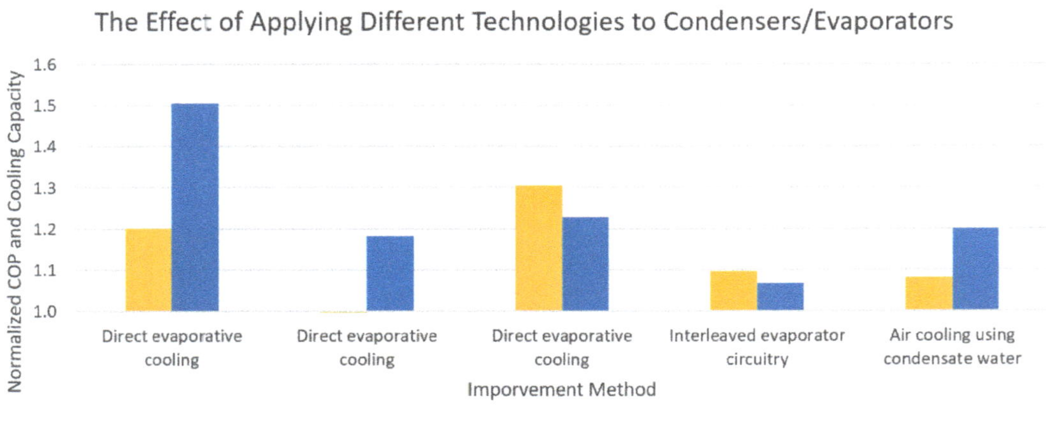

Figure 6. Improvements of different technologies applied to either evaporator or condenser [45,46,48,49,53].

Table 7. Details of condenser or evaporator improvements for each method.

Method	Refrigerant	Size	Test Type	Outdoor Condition	Cooling Capacity	COP	Note	Reference
Direct evaporative cooling	N/A	1.5 ton [5.3 kW_{th}]	Experimental	49.0 °C	+20.1%	+50.6%	-	[45]
Direct evaporative cooling	R-410A	N/A	Experimental	44.5 °C	N/A	+18.2%	-	[46]
Direct evaporative cooling	R-22	2.0 ton [7.0 kW_{th}]	Experimental	50.0 °C	+30.3%	+22.9%	-	[48]
Interleaved evaporator circuitry	R-407C	5.0 ton [17.6 kW_{th}]	Experimental	40.6 to 51.7 °C	+9.5%	+6.8%	At average values for tests 1, 2, and 3	[49]
Air cooling using condensade water (ACE)	R-22	0.8 ton [2.8 kW_{th}]	Experimental	43.0 °C	+8.1%	+20.0%	-	[53]

7.2. Compressors

Many compressor improvements found at high ambient temperatures were essentially induced from studies related to low ambients. Experimental and numerical improvement techniques found in the literature are (i) vapor injection, saturated or superheated, at the intermediate stage, (ii) liquid flooding, (iii) accumulator heat exchanger (AHX), and (iv) external shell cooling. Moreover, some studies analyzed the compressor reliability and the behavior of lubricants under HAT conditions.

Table 8 shows the experimental vapor injection methods [54–56] achieved up to 3% higher *COP*, 15% higher cooling capacity, and 4.5 °C lower discharge temperature. This method requires installing an additional economizer, which is either an internal heat exchanger or flash tank, where both types showed comparable performance, but the former provides wider mass flow rate control [56,59]. The cooling capacity improvement is attributed to the higher sub-cooling resulting from using an economizer. Additionally, the optimized numerical analysis showed potential for significant improvements using a new compressor (i.e., larger envelope and designed for higher condensing temperature) and properly sized economizer, as shown in Figure 7 [59]. Liquid flooding, using POE oil, was also numerically investigated and demonstrated that only R-1234yf had high improvements in *COP*, yet higher discharge temperatures, and hence needs further experimental testing [57]. Suction-line liquid injection through AHX, Figure 7, showed significant discharge temperature reductions of up to ~17 °C at the cost of lowering the *COP* by ~4.2% when using R-22 [55]. An external shell cooling method, using refrigerant, was conducted experimentally for a compact size compressor and led to attractive improvements [58]. This method is not included in Figure 7 and Table 8 since it was applied to a compact vapor-compression unit and was compared to thermoelectric coolers which yielded about 75% *COP* improvement under HAT. Moreover, it needs further investigation using residential size compressors, since the heat generated inside the compressor can be far from the surface and hence make it more difficult to apply such external cooling.

Table 8. Details of compressor improvements for each method.

Method	Injection	Refrigerant	Size	Test Type [1]	Outdoor Condition	Cooling Capacity	COP	Discharge Temperature	Note	Reference
Economizer heat exchanger	Vapor injection	R-22	NA	Experimental	N/A	+5.3%	0	−4.5 °C	60 °C condensing temperature	[54]
Accumulator heat exchanger	Suction-line refrigerant injection	R-22	2.6 ton [9.1 kW_{th}]	Experimental	43.0 °C	+0.2%	−4.2%	−16.9 °C	Injection ratio 15% using open EEV 36%	[55]
Flash tank	Saturated vapor injection	R-410A	3.1 ton [10.9 kW_{th}]	Experimental	46.1 °C	+15.0%	+2.2%	N/A	At 26% injection ratio	[56]
Internal heat exchanger	Superheated vapor injection	R-410A	3.1 ton [10.9 kW_{th}]	Experimental	46.1 °C	+13.9%	+3.0%	N/A	At 20% injection ratio	
Plate heat exchanger	Saturated vapor injection	R-407C	5.0 ton [17.6 kW_{th}]	Experimental	46.1 °C	+10.0%	+0.1%	−2.5 °C	[2]	[59]
	Superheated vapor injection	R-407C	5.0 ton [17.6 kW_{th}]	Experimental	46.1 °C	+10.8%	+1.9%	−1.8 °C	[2]	
	Superheated vapor injection	R-407C	5.0 ton [17.6 kW_{th}]	Numerical	46.1 °C	+33.3%	+8.7%	−7.1 °C	Optimized model [2]	

[1] All compressors used are scroll types. [2] Shown data are at average values for tests 1, 2, and 3.

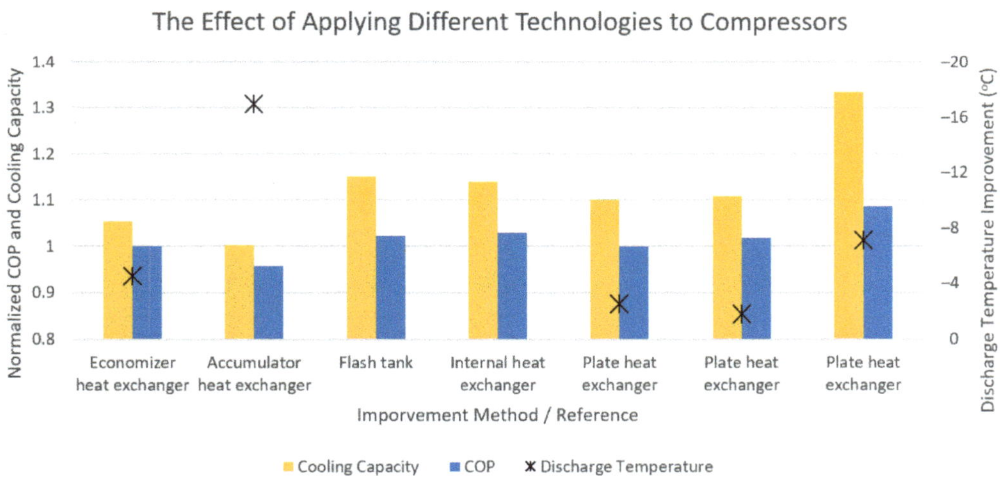

Figure 7. Improvements of different technologies applied to compressors under HAT conditions [54–56,59].

The compressor improvement techniques mentioned above can be extended to experimentally assessing mildly flammable refrigerants (A2L) under HAT, especially for the high discharge temperature challenge of R-32. Since numerical analysis showed R-32 could reach high discharge temperatures when using a vapor injection technique, two-phase injection within allowable compressor limits could be tested. Of note, this technique has higher costs, is difficult to control, and provides small improvements when compared to vapor injection [57]. In addition, increasing the injection ports at the compressor may lead to only slight improvements and hence it is not always feasible [57]. Utilizing an accumulator heat exchanger showed significant discharge temperature reductions for R-22 and could also be used for assessing R-32 under HAT.

Other studies explored the effects of high ambient temperature (HAT) on the compressor internals and oil viscosity. It was found that HAT decreased oil viscosity and the film thickness of bearings and increased the risk of metallic contact [61]. Moreover, at 50 °C ambient temperatures, the load on the crank of a rotary compressor was found to be double the load at 30 °C, while the compression process took longer periods to reach discharge pressure [61]. Therefore, the following techniques can be used to lower the risk of damaging the compressor internals: (i) maintain relatively higher suction temperatures,

(ii) use relatively higher viscosity oils, (iii) limit the compression ratio, (iv) use bearings with lower surface roughness, and (v) shut-off controls at excessive temperatures.

7.3. Refrigerants

Many studies were conducted to test different refrigerant performances at high ambient temperature (HAT) for the purpose of replacing R-22 and R-410A. The test levels are mainly drop-in and soft-optimized, as explained in Section 4.3.1. The data presented here include: (i) all related literature studies including ORNL results for both R-22 and R-410A and (ii) the AHRI low-GWP AREP program. A comparison among different refrigerants is shown in the following data based on: (i) cooling capacity, (ii) *COP*, (iii) discharge temperature, (iv) refrigerant classification, (v) indoor and outdoor temperatures, and (iv) unit size. The refrigerants are not meant to be compared directly since each test was subject to different test levels and conditions, but the data can help in determining the general behavior of each refrigerant under certain test conditions. The selection of new refrigerants is essentially dictated by the climate protocols that aim to lower the environmental impacts. Hence, using low GWP and zero ODP will most likely lead to higher flammability, which adds complexity to the system and more safety requirements. With that being said, the mildly flammable refrigerants (A2L) show attractive performance and are being tested by several researchers.

Table 9 shows all related studies found in the literature with the aim of replacing R-22 with different refrigerants (i.e., A1, A2L, and A3 types). It is obvious that there is no A1 refrigerant that outperformed R-22, and they all had lower *COP* with a wide range of cooling capacities (Figure 8). R-424A showed a similar *COP* but at the cost of a significantly lower cooling capacity. Moreover, A2L R-444B refrigerant reached a similar *COP* and slightly higher cooling capacity than R-22, while ARM-20a had a slightly higher *COP* but lower cooling capacity. ARM-20a showed a substantially lower discharge temperature by 22 °C. In general, most A2Ls are within 4% of the R-22 cooling capacity and within *COP* reductions of up to 15% (Table 9). Most A3 refrigerants showed higher *COP* than R-22 but almost all of them had a lower cooling capacity. The interesting behavior of R-1270 was observed when a larger displacement compressor was used (i.e., higher mass flow rate), which significantly improved the cooling capacity at comparable *COP* and had a slightly lower optimum refrigerant charge [70]. For the same study, R-290 showed cooling capacity improvements but much lower *COP* values, which explains the importance of using an optimum-sized compressor for each refrigerant. It is noted that discharge temperatures of both R-290 and R-1270 were not given in their respective studies (Table 9).

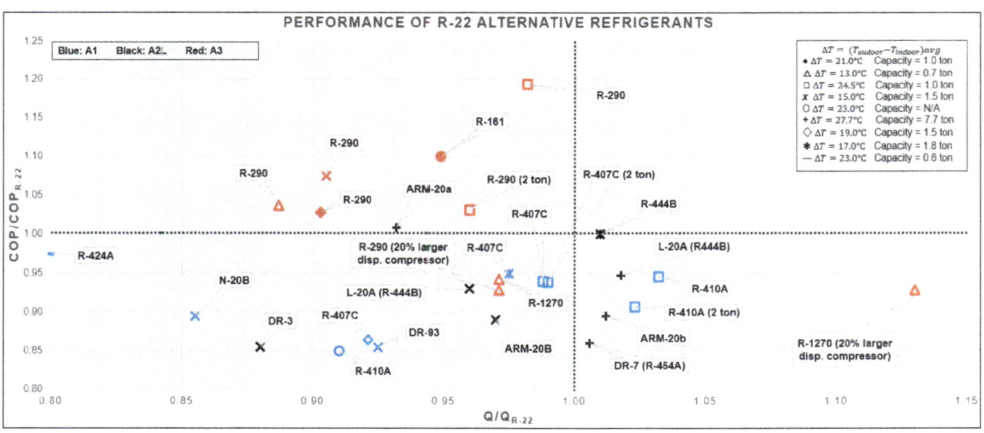

Figure 8. Experimental results for R-22 alternative refrigerants at HAT conditions, where Q is cooling capacity.

Table 9. Experimental results for R-22 alternative refrigerants at HAT conditions.

System Type	Size	Baseline Refrigerant	Alternative	Test Level	Baseline [1] Cooling Capacity/COP	Indoor Condition	Cooling Capacity [1]	COP [1]	Discharge Temperature [1]	Outdoor Condition [2]	Legend [3]	Reference
Residential heat pump AC	1.0 ton [3.5 kW_{th}]	R-22	R-161	N/A	2.9 kW_{th}/0.68	DBT/WBT 27 °C/19 °C	−5.1%	+10.0%	−3 °C	DBT/WBT 48 °C/30 °C	•	[67]
Residential split AC	1.0 ton [3.5 kW_{th}] 2.0 ton [7.0 kW_{th}]	R-22	R-290 (2.0 ton) R-407C (2.0 ton) R-410A (2.0 ton)	N/A	3.0 kW_{th}/2.32 (6.7 kW_{th}/2.60)	DBT/WBT 25 °C/19 °C	−1.8% (−4.0%) −1.2% (−1.0%) +3.2% (+2.3%)	+19.4% (+3.1%) −6.0% (−6.1%) −5.4% (−9.3%)	N/A	DBT 40 °C to 55 °C	□	[71]
Split AC system	N/A	R-22	R-410A	N/A	9.3 kW_{th}/2.28	DBT/WBT 26.7 °C/19.4 °C	−9.0%	−15.0%	N/A	DBT 54.4 °C	O	[63]
Window AC	1.5 ton [5.3 kW_{th}]	R-22	R-407C R-290	N/A Drop-in	4.2 kW_{th}/1.84 4.1 kW_{th}/1.76	DBT/WBT 29 °C/19 °C	−7.9% −9.7%	−13.5% +2.8%	N/A	DBT/WBT 46 °C/24 °C	◇	[64,65]
Wall room AC	0.7 ton [2.5 kW_{th}]	R-22	R-290 (20% larger disp. compressor) R-1270 (20% larger disp. compressor)	N/A	2.4 kW_{th}/0.87	DBT/WBT 27 °C/19 °C	−11.3% (−2.9%) −2.9% (+13.0%)	+3.7% (−7.1%) −5.7% (−7.1%)	N/A	DBT/WBT 40 °C/24 °C	△	[70]
Mini split AC	1.5 ton [5.3 kW_{th}]	R-22	N-20B DR-3 ARM-20B L-20A (R-444B) R-290 DR-93	Soft optimized	4.9 kW_{th}/1.9	DBT/WBT 29 °C/19 °C	−14.5% −12.0% −3.0% −4.0% −9.5% −7.5%	−10.5% −14.5% −11.0% −7.0% +7.5% −14.5%	−7 °C −4 °C +3 °C +6 °C −13 °C −2 °C	DBT 52 °C, 55 °C	X	[37]
Rooftop AC	7.7 ton [27.1 kW_{th}]	R-22	L-20A (R-444B) ARM-20b DR-7 (R-454A) ARM-20a	Drop-in	19.8 kW_{th}/1.84	DBT/WBT 29 °C/19 °C	+1.8% +1.2% +0.6% −6.8%	−5.3% −10.5% −14.0% +0.8%	−2 °C −8 °C −11 °C −22 °C	DBT 52 °C	+	[38]
Mini split AC	1.8 ton [6.3 kW_{th}]	R-22	R-444B R-407C	N/A	5.2 kW_{th}/1.80	DBT/WBT 32 °C/13 °C 29 °C/19 °C	+1.0% −2.5%	0 −5.0%	N/A	DBT/WBT 53 °C/32 °C 46 °C/29 °C	✱	[72]
Split AC	0.6 ton [2.1 kW_{th}]	R-22	R-424A	Drop-in	1.7 kW_{th}/2.41	DBT 18 °C	−20%	−2.5%	−19 °C	DBT 41 °C	—	[74]

[1] Values are approximated from graphs at the mentioned average outdoor conditions unless they are numerically presented by author(s). [2] Only high ambient temperatures were selected (i.e., 40 °C and above). [3] Symbols are illustrated in Figure 8.

All the literature studies evaluated only A2L refrigerants as R-410A alternatives. In Table 10, R-1234yf showed extremely low cooling capacity which was attributed to the need for additional system modifications, beyond charge and TXV optimization, which led to lower flow rates and pressure ratios [68]. Most of the remaining alternatives showed improvements in both cooling capacity and *COP*, however, all of them had higher discharge temperatures than R-410A, except for refrigerant DR-4 (i.e., had lower discharge temperatures at the cost of considerably lower cooling capacity). Therefore, the behavior of any nominated refrigerant must be carefully observed under high ambient temperatures. The discharge temperatures of R-32 are higher than R-410A, ranging from ~16 °C to 30 °C, which makes it imperative to utilize compressor cooling technologies and large condensers. Moreover, after soft optimizing the split system using drop-in DR-5 refrigerant, there was a negligible change in performance but higher discharge temperatures which makes it a better drop-in alternative. All R-410A replacement refrigerants in Table 10 are plotted in Figure 9 excluding R-1234yf [37,38,68,69].

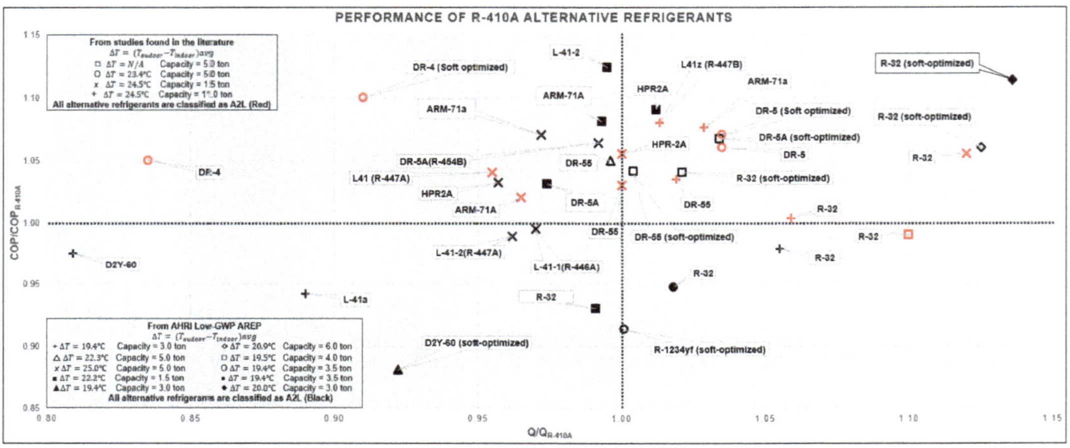

Figure 9. Experimental results for R-410A alternative refrigerants at HAT conditions, where Q is cooling capacity.

The AHRI low-GWP program [123,134–142], discussed in Section 3.2, conducted several tests under high ambient temperatures, Figure 9 and Table 11. All R-410A alternatives are A2L types, and many of these refrigerants showed higher *COP* and cooling capacity than R-410A. R-32 with soft-optimization achieved the highest performance among all refrigerants, however, the discharge temperature was higher by ~16 °C. R-32 discharge temperatures for different experiments ranged from 16 to 26 °C. Various refrigerants had higher *COP* and similar cooling capacity to R-410A with only a small increase in discharge temperature of less than 10 °C. Hence, A2L refrigerants are capable of replacing R-410A but their flammability requires further risk assessment under HAT to decide which level of replacement is recommended (i.e., soft optimized or new equipment).

Table 10. Experimental results (except for AHRI Low-GWP AREP) for R-410A alternative refrigerants at HAT conditions.

System Type	Size	Baseline Refrigerant	Alternative	Test Level	Baseline [1] Cooling Capacity COP	Indoor Condition	Cooling Capacity [1]	COP [1]	Discharge Temperature [1]	Outdoor Condition [2]	Legend [3]	Reference
Residential split heat pump	5.0 ton [17.6 kW_{th}]	R-410A	R-32 R-1234yf	Drop-in	N/A	N/A	+10.0% −49.0%	−1.0% +7.0%	+30 °C −29°C	DBT 43 °C, 46 °C	□	[68]
			DR-4 (Soft optimization) DR-5 (Soft optimization)	Drop-in	N/A	DBT 21.1 °C	−16.5% (−9.0%) +3.5% (+3.0%)	+5.0% (+10.0%) +6.0% (+7.0%)	−9 °C (−2 °C) +3 °C (+8 °C)	DBT 43 °C, 46 °C	O	[69]
Mini split AC	1.5 ton [5.3 kW_{th}]	R-410A	R-32 DR-55 L41 (R-447A) ARM-71A HPR-2A	Soft optimized	3.4 kW_{th}/1.97	DBT/WBT 29 °C/19 °C	+12.0% 0 −4.5% −3.5% 0	+5.5% +3.0% +4.0% +2.0% +5.5%	+21 °C +8 °C +14 °C +8 °C +14 °C	DBT 52 °C, 55 °C	X	[37]
Rooftop AC	11 ton [38.7 kW_{th}]	R-410A	DR-55 L41z (R-447B) ARM-71a R-32	Drop-in	30.7 kW_{th}/1.80	DBT/WBT 29 °C/19 °C	+1.9% +1.3% +2.9% +5.9%	+3.5% +8.0% +7.6% 0	+7 °C +10 °C +7 °C +19 °C	DBT 52 °C, 55 °C	+	[38]

[1] Values are approximated from graphs at the mentioned average outdoor conditions unless they are numerically presented by the author(s). [2] Only high ambient temperatures were selected (i.e., 40 °C and above). [3] Symbols are illustrated in Figure 9.

Table 11. AHRI Low-GWP AREP results for R-410A alternative refrigerants at HAT conditions.

System Type	Size	Baseline Refrigerant	Alternative	Test Level	Baseline [1] Cooling Capacity/COP	Indoor Condition	Cooling Capacity [1]	COP [1]	Discharge Temperature [1]	Outdoor Temperature [2]	Legend [3]	Reference
Rooftop AC	4.0 ton [14.1 kW_{th}]	R-410A	DR-55 R-32 DR-5A	Soft-optimized	12.1 kW_{th}/2.25	DBT/WBT 26.6 °C/19.4 °C	+0.4% +2.1% +3.4%	+4.1% +4.0% +6.7%	+4.6 °C +20.2 °C +6.2 °C	DBT 40.5 °C to 51.7 °C	□	[123]
Rooftop AC	6.0 ton [21.1 kW_{th}]	R-410A	R-32	Soft-optimized	18.1 kW_{th}/2.48	DBT/WBT 26.6 °C/19.4 °C	+12.5%	+6.0%	+20.7 °C	DBT 40.5 °C to 54.4 °C	◇	[134]
Rooftop AC	5.0 ton [17.6 kW_{th}]	R-410A	DR-55	Drop-in	14.5 kW_{th}/2.29	DBT/WBT 26.6 °C/19.4 °C	−0.4%	+5.0%	+4.6 °C	DBT 46.1 °C, 51.7 °C	△	[135]
Split system heat pump	3.0 ton [10.6 kW_{th}]	R-410A	ARM-71a DR-5A (R-454B) HPR2A L-41-1 (R-446A) L-41-2 (R-447A)	Drop-in	7.4 kW_{th}/1.87	DBT/WBT 26.7 °C/15.6 °C	−2.8% −0.8% −4.3% −3.0% −3.8%	+7.0% +6.4% +3.2% −0.5% −1.1%	+7.9 °C +9.0 °C +1.6 °C +14.7 °C +8.2 °C	DBT 51.7 °C	x	[136]
Rooftop AC	5.0 ton [17.6 kW_{th}]	R-410A	L-41-2 ARM-71A HPR2A DR-5A R-32	Drop-in	14.5 kW_{th}/2.29	DBT/WBT 26.7 °C/19.4 °C	−0.5% −0.7% +1.2% −2.6% −0.9%	+12.4% +8.1% +9.0% +3.1% −7.0%	+6.4 °C +6.9 °C +13.5 °C +9.2 °C +26 °C	DBT 46.1 °C, 51.7 °C	■	[137]
Split AC system	3.0 ton [10.6 kW_{th}]	R-410A	R-32	Soft-optimized	8.4 kW_{th}/2.02	DBT/WBT 28.9 °C/18.9 °C	+13.6%	+11.4%	+15.9 °C	DBT/WBT 46.1 °C/25.7 °C 51.7 °C/27.7 °C	✱	[138]
Air source heat pump	3.0 ton [10.6 kW_{th}]	R-410A	D2Y-60	Soft-optimized	8.5 kW_{th}/2.4	DBT/WBT 26.7 °C/19.4 °C	−7.8%	−11.9%	−1.6 °C	DBT 46.1 °C	▲	[139]
Air source heat pump	3.0 ton [10.6 kW_{th}]	R-410A	R-32 D2Y-60 L-41a	Drop-in	8.5 kW_{th}/2.4	DBT/WBT 26.7 °C/19.4 °C	+5.5% −19.1% −11.0%	−2.1% −2.5% −5.8%	+18.1 °C −11.5 °C +6.7 °C	DBT 46.1 °C	+	[140]
Split system heat pump	3.5 ton [12.3 kW_{th}]	R-410A	R-1234yf	Soft-optimized	10.9 kW_{th}/2.64	DBT/WBT 26.7 °C/19.4 °C	+0.1%	−8.7%	−17.8 °C	DBT 46.1 °C	O	[141]
Split system heat pump	3.5 ton [12.3 kW_{th}]	R-410A	R-32	Drop-in	10.9 kW_{th}/2.64	DBT/WBT 26.7 °C/19.4 °C	+1.8%	−5.3%	+23.2 °C	DBT 46.1 °C	•	[142]

[1] Values are approximated from graphs at the mentioned average outdoor conditions unless they are numerically presented by author(s). [2] Only high ambient temperatures were selected (i.e., 40 °C and above). [3] Symbols are illustrated in Figure 9.

8. Conclusions

This article provides an overview of residential vapor-compression air conditioners operating under high ambient temperatures (HAT). For the purpose of this article, a minimum temperature criterion, 40 °C and above, was developed to evaluate studies that were conducted at HAT. Several HAT organizations and projects (i.e., ORNL, AHRI AREP, EGPYRA, and PRAHA) were launched with the purpose of assessing the performance of low-GWP refrigerants when operating under HAT and accelerating the transition to such refrigerants. Previous studies of air conditioner improvements (i.e., for condensers, evaporators, compressors, and refrigerants) were discussed under HAT conditions. This article also discussed the challenges, the possible design modifications, and several limitations of air conditioners operating under HAT. The main challenges at high ambient temperature are (i) the reduced efficiency, (ii) the higher energy consumption, (iii) the high discharge temperature associated with promising A2L refrigerants, (iv) the higher minimum energy efficiency requirements, and (v) the ambient temperature increase due to global warming.

Climate limitations were explored for Kuwait and Phoenix, Arizona, USA, and the following was observed:

- The yearly HAT season of Kuwait was found to be ~5 months and therefore designing efficient air conditioners for such a climate is necessary.
- The yearly HAT season in Phoenix, Arizona, USA was found to be ~2 months, and hence a careful evaluation must be conducted to avoid any penalties for oversizing the unit when operating under mild temperatures.
- Over the past 6 years in both Kuwait and Phoenix, Arizona, USA, the average temperatures during the HAT seasons increased by up to ~1.7 °C over the period 1981 to 2010, indicating already high ambient temperatures are increasing even further.
- After surveying the units sold in the USA and GCC regions, it was found that both provided allowable operating temperatures of up to 52 °C.
- In Kuwait, the units are certified by meeting the minimum value of *EER*, while in the Southwest region of the USA, the unit must meet the minimum values of both *EER* and *SEER*.

Main findings and suggested improvements:

- Refrigerant R-32 showed the highest improvements in terms of cooling capacity and *COP*, however, the discharge temperature was too high for a conventional unit to run continuously without tripping.
- Refrigerants DR-5, DR-5a, and ARM-71a showed better performance than R-410A but a slightly higher discharge temperature of +2 to +9 °C, which is expected to be mitigated using compressor and condenser enhancements techniques.
- The compressor cooling techniques that yielded the highest improvements (i.e., lowest discharge temperatures) are saturated vapor injection and accumulator heat exchanger at the suction line.
- Condenser improvements were mainly related to evaporative cooling techniques which showed significant improvements in cooling capacity and *COP*.
- The irreversibilities of evaporators were found to be significant at HAT and only one study of evaporator improvement was found under such conditions. The evaporator circuitry was interleaved based on the air maldistribution, which showed considerable improvements in cooling capacity and *COP* but needs to be implemented at the design stage.
- The TEWI and LCCP analyses showed that the indirect impact poses a significantly higher effect on the environment than the direct impact for an air conditioner operating under HAT conditions and hence, the energy efficiency remains the dominant obstacle that needs further improvement. Therefore, to meet the purpose of lowering global warming and CO_2 emissions, HAT countries should only allow the transition to similar or better energy efficiency A2L units.

Recommended research paths:

- Several studies assessed microchannel condensers under HAT using a bespoke test facility where one of them, at near-critical pressure, found a unique heat transfer behavior and recommended further testing at critical pressure to have a better understanding of the heat transfer behavior.
- Evaporators can be assessed using active control methods which control the refrigerant flow to obtain a uniform exit superheat.
- The literature studies with the purpose of evaluating the performance of different A2L and A3 refrigerants compared to the baseline R-410A provided sufficient data and hence further assessments are not recommended.
- Risk assessments represent a major concern for the next transition to mildly flammable refrigerants and hence, any further development in this area can significantly accelerate the transition to A2L or even A3 refrigerants.

9. Future Work

This article was primarily focused on residential air conditioners only, while the following ideas could be further investigated at HAT:

1. Applying active controls of evaporator refrigerant flow to regulate the exiting superheat.
2. Investigating the condensation behavior of a microchannel condenser at critical pressure since it showed unique findings at near-critical pressure.
3. Applying external cooling techniques to the compressor shell using an additional heat exchanger (e.g., coolant can be refrigerant or phase change material).
4. Investigating the energy savings when using a two-stage compressor compared to a single-stage under real HAT conditions.

Author Contributions: Conceptualization, M.A. and P.E.P.; methodology, M.A.; software, M.A.; validation, M.A. and P.E.P; formal analysis, M.A. and P.E.P.; investigation, M.A. and P.E.P.; resources, M.A. and P.E.P.; data curation, M.A.; writing—original draft preparation, M.A.; writing—review and editing, P.E.P.; visualization, M.A.; supervision, P.E.P. All authors have read and agreed to the published version of the manuscript.

Funding: This research received no external funding.

Institutional Review Board Statement: Not applicable.

Informed Consent Statement: Not applicable.

Data Availability Statement: Not applicable.

Acknowledgments: M.A. gratefully acknowledges the support of The Public Authority for Applied Education and Training—College of Technological Studies, Kuwait.

Conflicts of Interest: The authors declare no conflict of interest.

References

1. The Future of Cooling—Analysis—IEA. Available online: https://www.iea.org/reports/the-future-of-cooling (accessed on 6 January 2022).
2. Air Conditioning Use Emerges as One of the Key Drivers of Global Electricity-Demand Growth—News—IEA. Available online: https://www.iea.org/news/air-conditioning-use-emerges-as-one-of-the-key-drivers-of-global-electricity-demand-growth (accessed on 20 January 2022).
3. Electricity Mix—Our World in Data. Available online: https://ourworldindata.org/electricity-mix (accessed on 6 January 2022).
4. Global Greenhouse Gas Emissions Data | US EPA. Available online: https://www.epa.gov/ghgemissions/global-greenhouse-gas-emissions-data (accessed on 6 January 2022).
5. Learn about Heat Islands | US EPA. Available online: https://www.epa.gov/heatislands/learn-about-heat-islands (accessed on 6 January 2022).
6. Excess Heat from Air Conditioners Causes Higher Nighttime Temperatures | ASU News. Available online: https://news.asu.edu/content/excess-heat-air-conditioners-causes-higher-nighttime-temperatures (accessed on 6 January 2022).
7. Salamanca, F.; Georgescu, M.; Mahalov, A.; Moustaoui, M.; Wang, M. Anthropogenic heating of the urban environment due to air conditioning. *J. Geophys. Res.* **2014**, *119*, 5949–5965. [CrossRef]

8. Lundgren-Kownacki, K.; Hornyanszky, E.D.; Chu, T.A.; Olsson, J.A.; Becker, P. Challenges of using air conditioning in an increasingly hot climate. *Int. J. Biometeorol.* **2018**, *62*, 401–412. [CrossRef] [PubMed]
9. Groll, E.A. Update on Refrigerants: Past, Present and Future. ASHRAE Chapter Meetings. 2021. Available online: https://www.ashrae.org/professional-development/learning-portal/instructor-led-training/ashrae-instructors/eckhard-a-groll (accessed on 6 January 2022).
10. Causes of Climate Change | US EPA. Available online: https://www.epa.gov/climatechange-science/causes-climate-change (accessed on 6 January 2022).
11. Climate Change: Global Temperature | NOAA Climate.gov. Available online: https://www.climate.gov/news-features/understanding-climate/climate-change-global-temperature (accessed on 6 January 2022).
12. U.S. Environmental Protection Agency; U.S. Department of Energy. ENERGY STAR®. Available online: https://www.energystar.gov/ (accessed on 7 January 2022).
13. Energy Efficiency—Site. Available online: https://www.regulations.doe.gov/eecompass/energy-efficiency (accessed on 6 January 2022).
14. OzonAction Kigali Fact Sheet No. 9—Technical Issues: High Ambient Temperature | Ozonaction. Available online: https://www.unep.org/ozonaction/resources/factsheet/ozonaction-kigali-fact-sheet-no-9-technical-issues-high-ambient-temperature (accessed on 7 January 2022).
15. Least Developed Countries | UNCTAD. Available online: https://unctad.org/topic/least-developed-countries (accessed on 7 January 2022).
16. *ANSI/ASHRAE Standard 169-2021*; Climatic Data for Building Design Standards. American Society of Heating, Refrigerating and Air-Conditioning Engineers: Atlanta, GA, USA, 2021.
17. Degree-Days—U.S. Energy Information Administration (EIA). Available online: https://www.eia.gov/energyexplained/units-and-calculators/degree-days.php (accessed on 7 January 2022).
18. The Weather Year Round Anywhere on Earth—Weather Spark. Available online: https://weatherspark.com/ (accessed on 7 January 2022).
19. Heating & Cooling Degree Days—Free Worldwide Data Calculation. Available online: https://www.degreedays.net/ (accessed on 4 December 2021).
20. Pacific Northwest National Laboratory for the U.S. Department of Energy. Building Energy Codes Program (ANSI/ASHRAE/IES Standard 90.1-2016: HVAC). 2017. Available online: www.energycodes.gov (accessed on 8 January 2022).
21. Athalye, R.; Taylor, T.; Liu, B. Impact of Ashrae Standard 169-2013 on Building Energy Codes and Energy Efficiency. Available online: https://www.semanticscholar.org/paper/IMPACT-OF-ASHRAE-STANDARD-169-2013-ON-BUILDING-AND-Athalye-Taylor/320bcf954de579fab3c2e2e680f4b45a2bd40ebf (accessed on 8 January 2022).
22. Alajmi, T.; Phelan, P. Modeling and forecasting end-use energy consumption for residential buildings in Kuwait using a bottom-up approach. *Energies* **2020**, *13*, 1981. [CrossRef]
23. US EPA and Climate Change Division. Climate Change Indicators in the United States. 2014. Available online: www.epa.gov/climatechange/indicators (accessed on 8 January 2022).
24. Climate Change and Heat Islands | US EPA. Available online: https://www.epa.gov/heatislands/climate-change-and-heat-islands (accessed on 7 January 2022).
25. Article 5: Special Situation of Developing Countries | Ozone Secretariat. Available online: https://ozone.unep.org/treaties/montreal-protocol/articles/article-5-special-situation-developing-countries (accessed on 7 January 2022).
26. Fenaughty, K.; Parker, D. Evaluation of Air Conditioning Performance Degradation: Opportunities from Diagnostic Methods. 2018. Available online: www.fsec.ucf.edu (accessed on 8 January 2022).
27. Sherry, D.; Nolan, M.; Sherry, N.; Associates-NSA; Seidel, S. HFO-1234yf: An Examination of Projected Long-Term Costs of Production I. Why is the Production Cost of Hydrofluoroolefin (HFO)-1234yf Important? Available online: https://www.1234facts.com (accessed on 8 January 2022).
28. OzonAction Kigali Fact Sheet No. 5—HFC Baselines and Phase-Down Timetable | Ozonaction. Available online: https://www.unep.org/ozonaction/resources/factsheet/ozonaction-kigali-fact-sheet-no-5-hfc-baselines-and-phase-down-timetable (accessed on 7 January 2022).
29. Prices and Factors Affecting Prices—U.S. Energy Information Administration (EIA). Available online: https://www.eia.gov/energyexplained/electricity/prices-and-factors-affecting-prices.php (accessed on 8 January 2022).
30. Heatstroke—Symptoms and Causes—Mayo Clinic. Available online: https://www.mayoclinic.org/diseases-conditions/heat-stroke/symptoms-causes/syc-20353581 (accessed on 8 January 2022).
31. ISO—ISO 7243:2017—Ergonomics of the Thermal Environment—Assessment of Heat Stress Using the WBGT (Wet Bulb Globe Temperature) index. Available online: https://www.iso.org/standard/67188.html (accessed on 8 January 2022).
32. Parsons, K. Heat Stress Standard ISO 7243 and Its Global Application. 2006. Available online: https://pubmed.ncbi.nlm.nih.gov/16922180/ (accessed on 9 January 2022).
33. Sanderson, M.; Arbuthnott, K.; Kovats, S.; Hajat, S.; Falloon, P. The use of climate information to estimate future mortality from high ambient temperature: A systematic literature review. *PLoS ONE* **2017**, *12*, e180369. [CrossRef] [PubMed]
34. Hanna, E.G.; Tait, P.W. Limitations to thermoregulation and acclimatization challenge human adaptation to global warming. *Int. J. Environ. Res. Public Health* **2015**, *12*, 8034–8074. [CrossRef] [PubMed]

35. Vicedo-Cabrera, A.M.; Scovronick, N.; Sera, F.; Royé, D.; Schneider, R.; Tobias, A.; Astrom, C.; Guo, Y.; Honda, Y.; Hondula, D.M.; et al. The burden of heat-related mortality attributable to recent human-induced climate change. *Nat. Clim. Change* **2021**, *11*, 492–500. [CrossRef]
36. Thompson, R.; Hornigold, R.; Page, L.; Waite, T. Associations between high ambient temperatures and heat waves with mental health outcomes: A systematic review. *Public Health* **2018**, *161*, 171–191. [CrossRef]
37. Abdelaziz, O.; Shrestha, S.; Munk, J.; Linkous, R.; Goetzler, W.; Guernsey, M.; Kassuga, T. Alternative Refrigerant Evaluation for High-Ambient-Temperature Environments: R-22 and R-410A Alternatives for Mini-Split Air Conditioners. 2015. Available online: https://www.energy.gov/eere/buildings/downloads/alternative-refrigerant-evaluation-high-ambient-temperature-environments-r (accessed on 25 October 2021).
38. Abdelaziz, O.; Shrestha, S.; Shen, B.; Elatar, A.; Linkous, R.; Goetzler, W.; Guernsey, M.; Bargach, Y. Alternative Refrigerant Evaluation for High-Ambient-Temperature Environments: R-22 and R-410A Alternatives for Rooftop Air Conditioners. 2016. Available online: https://www.energy.gov/eere/buildings/downloads/alternative-refrigerant-evaluation-high-ambient-temperature-environments--0 (accessed on 25 October 2021).
39. United Nations Environment Programme; United Nations Industrial Development Organization. EGYPRA—Promotion of Low-GWP Refrigerants for the Air Conditioning Industry in Egypt—Report. 2019. Available online: https://wedocs.unep.org/xmlui/handle/20.500.11822/32719 (accessed on 8 January 2022).
40. Promoting Low-GWP Refrigerants for Air-Conditioning Sectors in High-Ambient Temperature Countries (PRAHA) | UNEP—UN Environment Programme. Available online: https://www.unep.org/resources/report/promoting-low-gwp-refrigerants-air-conditioning-sectors-high-ambient-temperature (accessed on 8 January 2022).
41. AHRI Low-GWP Alternative Refrigerants Evaluation Program. Available online: https://www.ahrinet.org/arep (accessed on 8 January 2022).
42. Wang, X.; Amrane, K. AHRI Low Global Warming Potential Alternative Refrigerants Evaluation Program (Low-GWP AREP) Summary of Phase II Testing Results. 2016. Available online: http://docs.lib.purdue.edu/iracc/1586 (accessed on 9 January 2022).
43. United Nations Environment Programme; United Nations Industrial Development Organization. Promoting Low-GWP Refrigerants for Air-Conditioning Sectors in High-Ambient Temperature Countries Phase II: Project Report. 2019. Available online: https://wedocs.unep.org/handle/20.500.11822/32721;jsessionid=F650AA2B1F5332A47B0F4515C62A073B (accessed on 9 January 2022).
44. United Nations Environment Programme; United Nations Industrial Development Organization. Promoting Low-GWP Refrigerants for Air-Conditioning Sectors in High-Ambient Temperature Countries (PRAHA): Project Report. 2016. Available online: https://www.unido.org/our-focus/safeguarding-environment/implementation-multilateral-environmental-agreements/montreal-protocol/finding-climate-friendly-ways-cool-down (accessed on 9 January 2022).
45. Hajidavalloo, E.; Eghtedari, H. Performance improvement of air-cooled refrigeration system by using evaporatively cooled air condenser. *Int. J. Refrig.* **2010**, *33*, 982–988. [CrossRef]
46. Wang, T.; Sheng, C.; Nnanna, A.G.A. Experimental investigation of air conditioning system using evaporative cooling condenser. *Energy Build.* **2014**, *81*, 435–443. [CrossRef]
47. Shen, B.; Bansal, P. Assessment of Environmentally Friendly Refrigerants for Window Air Conditioners Assessment of Environmentally Friendly Refrigerants for Window Air Conditioners. 2014. Available online: http://docs.lib.purdue.edu/iracc/1425 (accessed on 23 December 2021).
48. Eidan, A.A.; Alwan, K.J.; Alsahlani, A.; Alfahham, M. Enhancement of the Performance Characteristics for Air-Conditioning System by Using Direct Evaporative Cooling in Hot Climates. *Energy Procedia* **2017**, *142*, 3998–4003. [CrossRef]
49. Bahman, A.M.; Groll, E.A. Application of interleaved circuitry to improve evaporator effectiveness and COP of a packaged AC system. *Int. J. Refrig.* **2017**, *79*, 114–129. [CrossRef]
50. Al-Bakri, B.A.R.; Ricco, P. Condensation heat transfer coefficient for rectangular multiport microchannels at high ambient temperature. *Int. J. Heat Mass Transf.* **2019**, *138*, 866–878. [CrossRef]
51. López-Belchí, A. Assessment of a mini-channel condenser at high ambient temperatures based on experimental measurements working with R134a, R513A and R1234yf. *Appl. Therm. Eng.* **2019**, *155*, 341–353. [CrossRef]
52. Ketwong, W.; Deethayat, T.; Kiatsiriroat, T. Performance enhancement of air conditioner in hot climate by condenser cooling with cool air generated by direct evaporative cooling. *Case Stud. Therm. Eng.* **2021**, *26*, 101127. [CrossRef]
53. Yang, H.; Pei, N.; Liu, L.; Fan, M.; Qin, Y. Experimental study on the effect of condensate water on the performance of split air conditioning system. *Energy Rep.* **2021**, *7*, 840–851. [CrossRef]
54. Ding, Y.; Chai, Q.; Ma, G.; Jiang, Y. Experimental study of an improved air source heat pump. *Energy Convers. Manag.* **2004**, *45*, 2393–2403. [CrossRef]
55. Kang, H.; Lee, S.; Kim, Y. Effects of liquid refrigerant injection on the performance of a refrigeration system with an accumulator heat exchanger. *Int. J. Refrig.* **2008**, *31*, 883–891. [CrossRef]
56. Wang, X.; Hwang, Y.; Radermacher, R. Two-stage heat pump system with vapor-injected scroll compressor using R410A as a refrigerant. *Int. J. Refrig.* **2009**, *32*, 1442–1451. [CrossRef]
57. Bahman, A.M.; Groll, E.A.; Horton, W.T.; Braun, J.E. Technologies to Improve the Performance of A/C Systems in Hot Climate Regions Technologies to Improve the Performance of A/C Systems in Hot Climate Regions. 2014. Available online: http://docs.lib.purdue.edu/iracc/1431 (accessed on 25 October 2021).

58. Ribeiro, G.B. A novel vapor-compression refrigeration loop for high ambient temperatures. *Int. J. Refrig.* **2016**, *70*, 103–107. [CrossRef]
59. Bahman, A.M.; Ziviani, D.; Groll, E.A. Vapor injected compression with economizing in packaged air conditioning systems for high temperature climate. *Int. J. Refrig.* **2018**, *94*, 136–150. [CrossRef]
60. Wu, J.; Shi, H.; Li, J. Analysis on the reliability of R290 rotary compressor in the high ambient temperature. *Int. J. Refrig.* **2020**, *110*, 132–141. [CrossRef]
61. Wang, C.; Wu, J.; Du, Y.; Lei, B. Lubricating condition evaluation of the rotary compressor under high ambient temperature. *Int. J. Refrig.* **2020**, *118*, 220–229. [CrossRef]
62. Motta, S.F.Y.; Domanski, P.A. Performance of R-22 and Its Alternatives Working at High Outdoor Temperatures. 2000. Available online: https://docs.lib.purdue.edu/iracc/464/ (accessed on 10 January 2022).
63. Payne, W.V.; Domanski, P.A. A Comparison of An R22 and An R410A Air Conditioner Operating at High Ambient Temperatures. 2002. Available online: http://docs.lib.purdue.edu/iracc/532 (accessed on 25 October 2021).
64. Devotta, S.; Padalkar, A.; Sane, N. Performance assessment of HCFC-22 window air conditioner retrofitted with R-407C. *Appl. Therm. Eng.* **2005**, *25*, 2937–2949. [CrossRef]
65. Devotta, S.; Padalkar, A.; Sane, N. Performance assessment of HC-290 as a drop-in substitute to HCFC-22 in a window air conditioner. *Int. J. Refrig.* **2005**, *28*, 594–604. [CrossRef]
66. Westphalen, D. Converting a Military Environmental Control Unit to Propylene Refrigerant. 2006. Available online: http://docs.lib.purdue.edu/iracc/840 (accessed on 11 January 2022).
67. Wu, Y.; Liang, X.; Tu, X.; Zhuang, R. Study of R161 Refrigerant for Residential Air-conditioning Applications. 2012. Available online: http://docs.lib.purdue.edu/iracc/1189 (accessed on 11 January 2022).
68. Barve, A.; Cremaschi, L. Drop-in Performance of Low GWP Refrigerants in a Heat Pump System for Residential Applications. 2012. Available online: http://docs.lib.purdue.edu/iracc/1211 (accessed on 10 January 2022).
69. Biswas, A.; Cremaschi, L. Performance and Capacity Comparison of Two New LGWP Refrigerants Alternative to R410A in Residential Air Conditioning Applications. 2012. Available online: http://docs.lib.purdue.edu/iracc/1210 (accessed on 11 January 2022).
70. Wu, J.; Yang, L.; Hou, J. Experimental performance study of a small wall room air conditioner retrofitted with R290 and R1270. *Int. J. Refrig.* **2012**, *35*, 1860–1868. [CrossRef]
71. Joudi, K.A.; Al-Amir, Q.R. Experimental Assessment of residential split type air-conditioning systems using alternative refrigerants to R-22 at high ambient temperatures. *Energy Convers. Manag.* **2014**, *86*, 496–506. [CrossRef]
72. Sethi, A.; Becerra, E.V.; Motta, S.F.Y.; Spatz, M.W. Low GWP R22 replacement for air conditioning in high ambient conditions. *Int. J. Refrig.* **2015**, *57*, 26–34. [CrossRef]
73. Taira, S.; Haikawa, D.T.; Minamida, T.; Haikawa, T.; Ota, F. Performance Evaluation of Heat pump System using R32 and HFO-Mixed Refrigerant in High Ambient Temperature. 2016. Available online: https://docs.lib.purdue.edu/iracc/1736/ (accessed on 11 January 2022).
74. Oruç, V.; Devecioğlu, A.G.; Berk, U.; Vural, I. Experimental comparison of the energy parameters of HFCs used as alternatives to HCFC-22 in split type air conditioners. *Int. J. Refrig.* **2016**, *63*, 125–132. [CrossRef]
75. Al-Bakri, B.A.R. Micro-Channel Air Cooled Condenser Performance with Two-Phase Flow of Zeotropic Refrigerant at High Ambient Temperatures. Ph.D. Thesis, The University of Sheffield, Sheffield, UK, 2018.
76. Parmar, S.K.; Patel, K.N.; Jejurkar, A.N. A Review of Air Conditioning with the Use of Mini-Channel and Micro-Channel Heat Exchanger. *Int. J. Sci. Eng. Res.* **2017**, *8*, 86–90. Available online: http://www.ijser.org (accessed on 9 January 2022).
77. Kandlikar, S.G.; Li, D.; King, M.R.; Garimella, S.; Colin, S. *Heat Transfer and Fluid Flow in Minichannels and Microchannels*; Elsevier: Amsterdam, The Netherlands, 2006.
78. Kandlikar, S.G. A roadmap for implementing minichannels in refrigeration and air-conditioning systems—Current status and future directions. *Heat Transf. Eng.* **2007**, *28*, 973–985. [CrossRef]
79. Mader, G.; Palm, B.; Elmegaard, B. Maldistribution in air–water heat pump evaporators. Part 1: Effects on evaporator, heat pump and system level. *Int. J. Refrig.* **2015**, *50*, 207–216. [CrossRef]
80. Blecich, P. Experimental investigation of the effects of airflow nonuniformity on performance of a fin-and-tube heat exchanger. *Int. J. Refrig.* **2015**, *59*, 65–74. [CrossRef]
81. Choi, J.M.; Payne, W.V.; Domanski, P.A. Effects of Non-Uniform Refrigerant and Air Flow Distributions on Finned-Tube Evaporator Performance. 2003. Available online: https://tsapps.nist.gov/publication/get_pdf.cfm?pub_id=860899 (accessed on 9 January 2022).
82. Timoney, D.J.; Foley, P.J. Some Effects of Air Flow Maldistribution on Performance of a Compact Evaporator with R134a. *Heat Recovery Syst. CHP* **1994**, *14*, 517–523. [CrossRef]
83. Guo, W.; Wang, G.J.H.Z.D.; Ji, G.; Zhan, H.; Wang, D. R32 Compressor for Air Conditioning and Refrigeration Applications in China. 2012. Available online: http://docs.lib.purdue.edu/icec/2098 (accessed on 9 January 2022).
84. Yokozeki, A.M. Solubility and Viscosity of Refrigerant-Oil Mixtures. 1994. Available online: https://docs.lib.purdue.edu/icec/1002/ (accessed on 9 January 2022).
85. Xu, X.; Hwang, Y.; Radermacher, R. Refrigerant injection for heat pumping/air conditioning systems: Literature review and challenges discussions. *Int. J. Refrig.* **2011**, *34*, 402–415. [CrossRef]

86. Calm, J.M. The next generation of refrigerants—Historical review, considerations, and outlook. *Int. J. Refrig.* **2008**, *31*, 1123–1133. [CrossRef]
87. The Vienna Convention for the Protection of the Ozone Layer | Ozone Secretariat. Available online: https://ozone.unep.org/treaties/vienna-convention (accessed on 9 January 2022).
88. The Montreal Protocol on Substances that Deplete the Ozone Layer | Ozone Secretariat. Available online: https://ozone.unep.org/treaties/montreal-protocol (accessed on 9 January 2022).
89. Montreal Protocol on Substances that Deplete the Ozone Layer—DAWE. Available online: https://www.awe.gov.au/environment/protection/ozone/montreal-protocol (accessed on 9 January 2022).
90. United Nations Environment Programme. *Ozone Secretariat. Handbook for the Montreal Protocol on Substances that Deplete the Ozone Layer*, 14th ed.; United Nations Environment Programme, Ozone Secretariat: Nairobi, Kenya, 2020.
91. Miyara, A. Developments of next generation refrigerants and heat transfer. In *MATEC Web of Conferences*; EDP Sciences: Ulys, France, 2018; Volume 204, p. 00004. [CrossRef]
92. What is the Kyoto Protocol? | UNFCCC. Available online: https://unfccc.int/kyoto_protocol (accessed on 9 January 2022).
93. Hydrofluorocarbons (HFCs) | Climate & Clean Air Coalition. Available online: https://www.ccacoalition.org/fr/slcps/hydrofluorocarbons-hfcs (accessed on 9 January 2022).
94. Reducing Hydrofluorocarbon (HFC) Use and Emissions in the Federal Sector through SNAP | US EPA. Available online: https://www.epa.gov/snap/reducing-hydrofluorocarbon-hfc-use-and-emissions-federal-sector-through-snap (accessed on 9 January 2022).
95. Velders, G.J.M.; Fahey, D.W.; Daniel, J.S.; Mcfarland, M.; Andersen, S.O. The large contribution of projected HFC emissions to future climate forcing (Factsheet). 2009. Available online: https://www.epa.gov/sites/default/files/documents/factsheet_velders_hfc.pdf (accessed on 10 January 2022).
96. The Montreal Protocol Evolves to Fight Climate Change | UNIDO. Available online: https://www.unido.org/our-focus-safeguarding-environment-implementation-multilateral-environmental-agreements-montreal-protocol/montreal-protocol-evolves-fight-climate-change (accessed on 9 January 2022).
97. Abdelaziz, O.; Shrestha, S.; Munk, J.; Member, J.M.; Member, M.; Linkous, R.L. Evaluation of Alternative Refrigerants for High Ambient Applications in a Mini-Split AC Unit. 2016. Available online: https://www.osti.gov/biblio/1237149-evaluation-alternative-refrigerants-high-ambient-applications-mini-split-ac-unit (accessed on 9 January 2022).
98. Makhnatch, P.; Mota-Babiloni, A.; López-Belchí, A.; Khodabandeh, R. R450A and R513A as lower GWP mixtures for high ambient temperature countries: Experimental comparison with R134a. *Energy* **2019**, *166*, 223–235. [CrossRef]
99. Substitutes in Residential and Light Commercial Air Conditioning and Heat Pumps | US EPA. Available online: https://www.epa.gov/snap/substitutes-residential-and-light-commercial-air-conditioning-and-heat-pumps (accessed on 9 January 2022).
100. IIF-IIR; Morlet, V.; Coulomb, D.; Dupont, J.L. The Impact of the Refrigeration Sector on Climate Change, 35th Informatory Note on Refrigeration Technologies. 2017. Available online: https://iifiir.org/en/fridoc/the-impact-of-the-refrigeration-sector-on-climate-change-141135 (accessed on 9 January 2022).
101. Participants' Handbook: Ahri Low-Gwp Alternative Refrigerants Evaluation Program (Low-Gwp Arep). 2015. Available online: https://www.ahrinet.org/app_content/ahri/files/research/participants_handbook2015-04-17.pdf (accessed on 9 January 2022).
102. *ANSI/ASHRAE Standard 34-2019*; Designation and Safety Classification of Refrigerants. American Society of Heating, Refrigerating and Air-Conditioning Engineers: Atlanta, GA, USA, 2019.
103. Abu-Heiba, A.; Patel, V.K.; Baxter, V.D.; Abdelaziz, O.; Elatar, A.F. Flammable Refrigerant Charge Limits: Can or Should These Limits Be Higher? (Journal Article) | OSTI.GOV. *ASHRAE J.*. 2018. Available online: https://www.osti.gov/biblio/1459281-flammable-refrigerant-charge-limits-can-should-limits-higher (accessed on 9 January 2022).
104. The Japan Refrigeration and Air Conditioning Industry Association. 2013. Available online: https://www.jraia.or.jp/english/about/letter/Vol.1304%20NL.pdf (accessed on 9 January 2022).
105. EVAP-COND, Version 5.0 | NIST. Available online: https://www.nist.gov/services-resources/software/evap-cond-version-50 (accessed on 5 January 2022).
106. Davis, L.W.; Fuchs, A.; Gertler, P. Cash for coolers: Evaluating a large-scale appliance replacement program in Mexico. *Am. Econ. J. Econ. Policy* **2014**, *6*, 207–238. [CrossRef]
107. AHRI. AHRI Standard 210/240, Performance Rating of Unitary Air-conditioning & Air-Source Heat Pump Equipment. 2017. Available online: https://www.ahrinet.org/App_Content/ahri/files/STANDARDS/AHRI/AHRI_Standard_210-240_2017.pdf (accessed on 10 January 2022).
108. Air Conditioner Regional Standards Brochure | Department of Energy. Available online: https://www.energy.gov/gc/downloads/air-conditioner-regional-standards-brochure (accessed on 10 January 2022).
109. Efficiency Requirements for Residential Central AC and Heat Pumps to Rise in 2023—Today in Energy—U.S. Energy Information Administration (EIA). Available online: https://www.eia.gov/todayinenergy/detail.php?id=40232 (accessed on 10 January 2022).
110. Federal Register: Energy Conservation Program: Energy Conservation Standards for Residential Central Air Conditioners and Heat Pumps. Available online: https://www.federalregister.gov/documents/2017/01/06/2016-29992/energy-conservation-program-energy-conservation-standards-for-residential-central-air-conditioners (accessed on 10 January 2022).

111. Ministry of Electricity and Water. *Energy Conservation Code for Buildings, MEW/R-6/2018*; Ministry of Electricity and Water: Kuwait City, Kuwait, 2018.
112. AHRI Certification. Available online: https://www.ahrinet.org/certification (accessed on 10 January 2022).
113. AHRI Certification Program Expands Test Conditions to Enable Greater Global Efficiency. Available online: https://www.ahrinet.org/news-events/news-and-shipping-releases?A=1608 (accessed on 10 January 2022).
114. AHRI MENA. Available online: https://www.ahrinet.org/mena (accessed on 10 January 2022).
115. Federal Register: Energy Conservation Program: Enforcement of Regional Standards for Central Air Conditioners. Available online: https://www.federalregister.gov/documents/2016/07/14/2016-16441/energy-conservation-program-enforcement-of-regional-standards-for-central-air-conditioners (accessed on 10 January 2022).
116. Code of Federal Regulations: 10 CFR 430.2—Definitions. Available online: https://www.ecfr.gov/current/title-10/chapter-II/subchapter-D/part-430/subpart-A/section-430.2 (accessed on 10 January 2022).
117. Code of Federal Regulations: 10 CFR Part 429—Certification, Compliance, and Enforcement for Consumer Products and Commercial and Industrial Equipment. Available online: https://www.ecfr.gov/current/title-10/chapter-II/subchapter-D/part-429 (accessed on 10 January 2022).
118. Code of Federal Regulations: 10 CFR 429.16—Central Air Conditioners and Central Air Conditioning Heat Pumps. Available online: https://www.ecfr.gov/current/title-10/chapter-II/subchapter-D/part-429/subpart-B/section-429.16 (accessed on 10 January 2022).
119. Engineering Proctor; AMAD. High EER at 46 °C Kingdom of Saudi Arabia Air Conditioner Project Draft Final Report. Available online: https://www.proctoreng.com/dnld/I104EERFinalReportPEG012113.pdf (accessed on 10 January 2022).
120. Djunaedy, E.; van den Wymelenberg, K.; Acker, B.; Thimmana, H. Oversizing of HVAC system: Signatures and penalties. *Energy Build.* **2011**, *43*, 468–475. [CrossRef]
121. Shapiro, C.; Aldrich, R.; Arena, L. Retrofitting Air Conditioning and Duct Systems in Hot, Dry Climates Consortium for Advanced Residential Buildings (CARB). 2012. Available online: http://www.osti.gov/bridge (accessed on 10 January 2022).
122. Bhatia, A. Heat Rejection Options in HVAC Systems. Available online: https://www.cedengineering.com/userfiles/Heat%20Rejection%20Options%20R1.pdf (accessed on 10 January 2022).
123. Schultz, K.; Perez-Blanco, M.; Kujak, S. AHRI Low-GWP AREP System Soft-Optimization Tests of Refrigerant R-32, DR-5A, and DR-55 in a R-410A 4-ton Unitary Rooftop Heat Pump-Cooling Mode Performance—TEST REPORT #56. 2015. Available online: https://www.ahrinet.org/App_Content/ahri/files/RESEARCH/AREP_Final_Reports/AHRI_Low_GWP_AREP_Rpt_056.pdf (accessed on 11 January 2022).
124. White Box Technologies Weather Data. Available online: http://weather.whiteboxtechnologies.com/ (accessed on 11 January 2022).
125. NOAA National Centers for Environmental information. Climate at a Glance: Global Time Series. Available online: https://www.ncdc.noaa.gov/cag/ (accessed on 13 January 2022).
126. Application Engineering Bulletin. Copeland 17-1251 Design Considerations for High Ambient Conditions. 1978. Available online: http://www.hvacrinfo.com/ae_index.htm (accessed on 11 January 2022).
127. Freon™ 22 (R-22) Refrigerant | Hydrochlorofluorocarbon HCFC. Available online: https://www.freon.com/en/products/refrigerants/r22 (accessed on 11 January 2022).
128. Freon™ 407C (R-407C) Refrigerant | R-22 Retrofit Refrigerant. Available online: https://www.freon.com/en/products/refrigerants/407c (accessed on 11 January 2022).
129. Freon™ 410A (R-410A) Refrigerant | HFC Replacement for R-22. Available online: https://www.freon.com/en/products/refrigerants/410a (accessed on 11 January 2022).
130. Yumrutas, R.; Kunduz, M.; Kanoglu, M. Exergy Analysis of Vapor Compression Refrigeration Systems. *Exergy Int. J.* **2002**, *2*, 266–272. [CrossRef]
131. Kalaiselvam, S.; Saravanan, R. Exergy analysis of scroll compressors working with R22, R407c, and R417a as refrigerant for HVAC system. *Therm. Sci.* **2009**, *13*, 175–184. [CrossRef]
132. Ahamed, J.U.; Saidur, R.; Masjuki, H.H. A review on exergy analysis of vapor compression refrigeration system. *Renew. Sustain. Energy Rev.* **2011**, *15*, 1593–1600. [CrossRef]
133. Bahman, A.M.; Groll, E.A. Application of second-law analysis for the environmental control unit at high ambient temperature. *Energies* **2020**, *13*, 3274. [CrossRef]
134. Abbadi, E.M.; Khawaldeh, E.M. AHRI Low-GWP AREP System Soft-Optimization Tests of Refrigerant R-32 in a 6-Ton Rooftop Packaged Air-Conditioner—TEST REPORT #55. 2015. Available online: https://www.ahrinet.org/App_Content/ahri/files/RESEARCH/AREP_Final_Reports/AHRI_Low_GWP_AREP_Rpt_055.pdf (accessed on 12 January 2022).
135. Uselton, D.; Crawford, T. AHRI Low-GWP AREP Drop-in Test of Refrigerant Blend DR-55 in a Five-Ton R-410A Rooftop Packaged Unit—TEST REPORT #53 System. 2015. Available online: https://www.ahrinet.org/App_Content/ahri/files/RESEARCH/AREP_Final_Reports/AHRI_Low_GWP_AREP_Rpt_053.pdf (accessed on 12 January 2022).
136. Burns, L.; Chen, C. AHRI Low-GWP AREP System Drop-in Tests of Refrigerant Blends ARM-71a, DR-5A (R-454B), HPR2A, L-41-1 (R-446A), L-41-2 (R-447A) in a R-410A Split System Heat Pump—TEST REPORT #52. 2015. Available online: https://www.ahrinet.org/App_Content/ahri/files/RESEARCH/AREP_Final_Reports/AHRI_Low_GWP_AREP_Rpt_052.pdf (accessed on 12 January 2022).

137. Uselton, D.; Crawford, T. AHRI Low-GWP AREP System Drop-in Test of R-32 and Refrigerant Blends ARM-71a, HPR2A, L-41-2 and DR-5A in a Five-Ton R-410A Rooftop Packaged Unit—TEST REPORT #47," 2015. Available online: https://www.ahrinet.org/App_Content/ahri/files/RESEARCH/AREP_Final_Reports/AHRI_Low_GWP_AREP_Rpt_047.pdf (accessed on 12 January 2022).
138. Li, H. Robert by Goodman Manufacturing. AHRI Low-GWP AREP System Soft-Optimization Tests of Refrigerant R-32 in a 3-ton Split System Air-Conditioner—Test Report #42. 2015. Available online: https://www.ahrinet.org/App_Content/ahri/files/RESEARCH/AREP_Final_Reports/AHRI_Low-GWP_AREP-Rpt-042.pdf (accessed on 13 January 2022).
139. Alabdulkarem, A.; Hwang, Y.; Radermacher, R. AHRI Low-GWP AREP System Soft-Optimized Test of Refrigerant D2Y60 in Air Source Heat Pump—TEST REPORT #32. 2013. Available online: https://www.ahrinet.org/App_Content/ahri/files/RESEARCH/AREP_Final_Reports/AHRI-Low-GWP-AREP-Rpt-032.pdf (accessed on 12 January 2022).
140. Alabdulkarem, A.; Hwang, Y.; Radermacher, R. AHRI Low-GWP AREP System Drop-In Tests of Refrigerants R-32, D2Y-60, and L-41a in Air Source Heat Pump—TEST REPORT #20. 2013. Available online: https://www.ahrinet.org/App_Content/ahri/files/RESEARCH/AREP_Final_Reports/AHRI%20Low-GWP%20AREP-Rpt-020.pdf (accessed on 12 January 2022).
141. Uselton, D. AHRI Low-GWP AREP System Soft-Optimized Test of Refrigerant HFO-1234yf (R-1234yf) in a Split System Heat Pump (with Addendum)—TEST REPORT #10. 2013. Available online: https://www.ahrinet.org/App_Content/ahri/files/RESEARCH/AREP_Final_Reports/AHRI%20Low-GWP%20AREP-Rpt-010_with%20addendum.pdf (accessed on 12 January 2022).
142. Crawford, T.; Uselton, D. AHRI Low-GWP AREP System Drop-in Test of Refrigerant R-32 in Split System Heat Pump (with Addendum)—TEST REPORT #4. 2012. Available online: https://www.ahrinet.org/App_Content/ahri/files/RESEARCH/AREP_Final_Reports/AHRI%20Low-GWP%20AREP-Rpt-004_with%20addendum.pdf (accessed on 12 January 2022).

energies

MDPI

Article

Research on Ecological Compensation Mechanism for Energy Economy Sustainable Based on Evolutionary Game Model

Yingbo Qin * and Wenping Wang

School of Economics and Management, Southeast University, Nanjing 211189, China; 230149149@seu.edu.cn
* Correspondence: qinyingbo@seu.edu.cn

Abstract: Energy and the environment are important foundations for sustainable economic development. In order to realize the sustainable development of energy economy, clarify the respective responsibilities of the government and enterprises, and explore the internal mechanism of ecological compensation, this paper constructs an evolutionary game model by means of social benefits, ecological compensation, supervision costs, government subsidies, enterprise punishment, enterprise additional income, enterprise emission reduction costs and other means, through the government's supervision and non-supervision behavior, as well as the cooperation and non-cooperation behavior of enterprises, and analyzes the model parameters and game results. The research shows that: (1) due to the delay effect of technical progress of pollutant discharge, in the long run, the cost of enterprise emission reduction must be less than its benefits. (2) Social benefits brought by government regulation must exceed half of ecological compensation. (3) Government subsidy should not be higher than ecological compensation. Then, taking Inner Mongolia coal mine as an example, the field investigation is carried out from the three aspects of atmosphere, water, and soil, and the ecological environment loss of the mining area is preliminarily calculated based on the national technical specification for ecological environment assessment. Based on the above results, it shows that: (1) From the perspective of the government, social benefit must exceed USD 10.69 million annually; (2) from the perspective of enterprises, government subsidy should be lower than USD 21.38 million annually. In short, ecological compensation mechanism for coal resource development should include two parts: resource consumption reduction and environmental restoration; take the ecological compensation threshold as the standard, strictly eliminating declining enterprises, stimulating technological innovation, weakening the government subsidy, relying on non-governmental organizations to enlarge the social benefits brought by government supervision.

Keywords: evolutionary game model; ecological compensation; energy economy sustainable

Citation: Qin, Y.; Wang, W. Research on Ecological Compensation Mechanism for Energy Economy Sustainable Based on Evolutionary Game Model. *Energies* **2022**, *15*, 2895. https://doi.org/10.3390/en15082895

Academic Editor: Adam Smoliński

Received: 12 February 2022
Accepted: 11 April 2022
Published: 14 April 2022

Publisher's Note: MDPI stays neutral with regard to jurisdictional claims in published maps and institutional affiliations.

1. Introduction

Energy production and consumption can be regarded as "synchronous indicators" to measure the economic and social development of a country or region, which can accurately and directly reflect the economic operation [1–3]. There is a high correlation between energy consumption and economic growth [4–6]. According to the statistics of the International Energy Agency, China has always been in the forefront of the world in the field of energy production and consumption, which not only effectively promotes the development process of industrialization, urbanization, and modernization, but also plays an important role in improving people's living standards. China's latest energy statistical yearbook shows that coal resources are still dominant in the energy structure. From 2000 to 2020, although its proportion decreased from 68.5% to 56.8%, the total consumption is growing at a high speed. With the continuous increase of total energy consumption, China's environmental quality is also facing severe challenges, and various environmental pollution problems occur frequently, which has attracted great attention [7,8]. Comprehensively promoting the construction of a safe and efficient energy system is not only a powerful starting point for

China's high-quality economic and social development in the new era, but also an important embodiment of coping with global climate change and fulfilling the responsibility of a big country [9,10]. Therefore, China should start with coal resources and urgently solve the contradiction between economic growth and ecological environment deterioration.

Coal resource mining is a complex system with diversified operation modes and information collection tools [11,12], which is compatible with economy, society, and environment, and provides power for social and economic development [13]. At the same time, it also brings some negative effects, which are closely intertwined. In addition, the current method is to control after it is polluted [14]. The instability of its management has greatly increased the loss process of ecological environment and resource exploitation, and the loss of ecological services and ecological compensation cost have also increased [15]. An effective ecological compensation mechanism must be established to fundamentally change the stage of environmental deterioration in the mining area and put the ecological environment management on the track of the virtuous circle. The negative impact of modern human activities on the natural environment is becoming more and more serious, resulting in the gradual decline of the ecological environment [16]. Therefore, restoring the ecological environment through various means and methods is the requirement of modern society [17]. In view of the deterioration of the ecological environment caused by the development of coal resources, the current common practice is to make necessary ecological compensation. The specific accounting methods are mainly reflected in the damaged value of the ecosystem, the damage of the residents in the mining area, and the cost of environmental governance in the mining area [18]. The development of mineral resources will cause serious damage to the ecosystem, so the damaged value of the ecosystem can be calculated as the compensation standard [19]. The loss of eco-environmental value caused by the development of mineral resources covers many aspects: for example, the loss of pollution load value is mainly calculated according to the pollution control cost of air, water, and solid waste in the mining area. The loss of land stress value mainly refers to the loss of opportunity cost caused by soil collapse in the mining area and the loss of nutrient elements such as nitrogen, phosphorus, and potassium caused by erosion. This paper will calculate ecological compensation based on the above methods.

Ecological compensation, mainly includes compensation scope, compensation subject, compensated subject, etc., [20]. Since the research core of this paper is to solve how to reduce the negative impact of coal energy production on the ecological environment, the following will focus on the compensation subject. As far as the government is concerned, on the one hand, it is the direct beneficiary of mineral resources development. On the other hand, a specific institution that can represent the regional public interest is needed to perform the duties of supervision and compensation. Therefore, the government is the responsible compensation subject [21,22]. As far as mining enterprises are concerned, they are not only the direct destroyer of the ecological environment, but also obtain huge profits from it. There is no doubt that they should bear the responsibility of compensation to ensure that the damaged ecological environment can be restored to the state before development [23,24]. So, in the process of ecological compensation, the problem is how to clarify the respective responsibilities of the government and enterprises, and how to grasp the strength of ecological compensation? How to formulate the elimination standards of enterprises that do not meet the requirements? Is there a restrictive relationship between different policies, such as government subsidies and ecological compensation? The solution of the above problems is very important to build a reasonable ecological compensation mechanism and promote a virtuous cycle of energy economy.

As ecological compensation involves many stakeholders, there is both competition and cooperation among various stakeholders, and the relationship is complex. How to clarify the responsibilities of various stakeholders is closely related to the final implementation effect of ecological compensation. Based on this, game theory has become the preferred tool for scholars. For example, Wei and Luo [25] study how local governments balance the sustainable development of local economy and the effective protection of water resources from an ecological perspective, and how to maximize the profits of local enterprises in the ecological compensation system. Xu and Yi [26] studied the corresponding proportional compensation between the upstream and downstream of the same basin and the feedback Nash equilibrium strategy of investment compensation from the different compensation modes between the government and enterprises. Shen and Gao [27] studies the decision-making behavior and influencing factors of ecological compensation in its basin, and believes that the increase of environmental tax rate has a significant impact on the decision-making behavior of local governments with low initial probability, and the improvement of supervision ability can promote local governments and polluting enterprises to reach a stable state faster.

It can be found from the literature summary that the research objects of ecological compensation mostly focus on water resources and the upstream and downstream of rivers, and less on the ecological compensation in the process of coal resource development. In addition, environmental pollution has a delay effect, and the existing literature mostly studies the problem from a static perspective, thus the dynamic balance of ecological compensation is often ignored. More importantly, there is little case on field research, and the combination of theory and practice is not really achieved. Most of the proposed schemes are general suggestions, which are not practical and targeted, so there are still some doubts about whether those problems can be really solved. Based on this, this paper explores the dynamic equilibrium point of ecological compensation by studying the dynamic game process between the government and coal resource enterprises, and puts forward targeted ecological compensation suggestions with the help of field research data, so as to provide important reference value for building a reasonable ecological compensation mechanism.

2. Materials and Methods

2.1. Establishment of Evolutionary Game Model

2.1.1. Participants and Strategy Selection

Evolutionary game model can explain why the group reaches the current state by analyzing the dynamic evolution process, and can ensure the equilibrium of evolutionary stability strategy. In order to deeply analyze strategic choice behavior, the participants of this model are the government and enterprises. The government refers to the complex of the central government and local governments, including the Environmental Protection Bureau, Environmental Protection Stations, and other institutions, which formulate and implement environmental protection policies and pollution discharge indicators, collect fines and grant subsidies through supervision. Enterprises refer to various industrial units that explore coal resources and produce various pollutants.

In economics, market failure determines the indispensable and important position of the government in environmental protection. There is a certain cost in supervision, but sometimes the government will relax supervision for pursuing economic growth. Therefore, the government has two strategies: regulation and non-regulation. The goal of enterprises is to maximize economic benefits, achieve emission indicators through technological progress, and obtain certain subsidies. If not, a fine will be imposed for ecological compensation. Therefore, enterprises also have two strategies: cooperation and non-cooperation.

2.1.2. Model Assumptions

Hypothesis 1 (H1). *In this model, the government and enterprises are the two subjects of the game, and there is no information asymmetry between them.*

Hypothesis 2 (H2). *In this model, the government implements supervision to achieve environmental protection or it can only focus on economic growth and ignore supervision. In order to maximize benefits, enterprises may take cooperative or uncooperative action feedback.*

Hypothesis 3 (H3). *If the probability of government regulation is x, the probability of non-regulation is 1-x. If the probability of enterprises cooperating with government supervision is y, the probability of non-cooperating is 1-y. The values of x and y are between 0 and 1.*

Hypothesis 4 (H4). *The government's supervision cost is C, and the social benefit is B, but it is -B when the supervision does not work. The government collects fines P from enterprises for ecological compensation.*

Hypothesis 5 (H5). *Under the condition of government supervision, if the enterprise cooperates with the government to implement environmental protection, it will receive subsidy R. The cost of reducing pollutant discharge by introducing advanced equipment is D, and the sum of production cost saved by adopting pollutant discharge technology and additional benefits brought by good reputation is S. If the enterprise does not take environmental protection measures, it will be imposed a fine P for ecological compensation.*

According to the above assumptions, the following game matrix is obtained, as shown in Table 1.

Table 1. Game matrix between the government and enterprises.

Government	Enterprises	
	Cooperation	Non-Cooperation
Regulation	(B-C, R+S-D)	(P-B-C, -P)
Non-regulation	(0, S-D)	(0, 0)

2.1.3. Model Derivation

According to the model hypothesis and game matrix, the replication dynamic equations of the government and enterprises can be obtained respectively.

(1) Expectation and replication dynamic equation under different strategies adopted by the government

Expectation of the government under the regulation:

$$U_{G1} = y(B - C) + (1 - y)(P - B - C) = 2By - Py - B - C \quad (1)$$

Expectation of the government under the non-regulation:

$$U_{G2} = y \times 0 + (1 - y) \times 0 = 0 \quad (2)$$

Average expectation of the government:

$$\overline{U_G} = xU_{G1} + (1 - x)U_{G2} \quad (3)$$

The replication dynamic equation of the government:

$$\begin{aligned} F(x) = \tfrac{dx}{dt} = x(U_{G1} - \overline{U_G}) = x(1 - x)(U_{G1} - U_{G2}) \\ = x(1 - x)(2By - Py - B - C) \end{aligned} \quad (4)$$

(2) Expectation and replication dynamic equation under different strategies adopted by enterprises

Expectation of enterprises under the cooperation:

$$U_{E1} = x(R + S - D) + (1 - x)(S - D) = Rx + S - D \quad (5)$$

Expectation of enterprises under the non-cooperation:

$$U_{E2} = x \times (-P) + (1 - x) \times 0 = -Px \quad (6)$$

Average expectation of enterprises:

$$\overline{U_E} = yU_{E1} + (1-y)U_{E2} \tag{7}$$

The replication dynamic equation of enterprises:

$$\begin{aligned} F(y) &= \tfrac{dy}{dt} = y(U_{E1} - \overline{U_E}) = y(1-y)(U_{E1} - U_{E2}) \\ &= y(1-y)(Rx - Px + S - D) \end{aligned} \tag{8}$$

By combining Equations (4) and (8), the replicated dynamic equations of the government and enterprises can be obtained:

$$\begin{cases} F(x) = x(1-x)(2By - Py - B - C) \\ F(y) = y(1-y)(Rx - Px + S - D) \end{cases} \tag{9}$$

2.2. *Accounting Method of Ecological Compensation (P)*

Based on the field investigation of manlailiang coal mine in Inner Mongolia, its type mainly is coking coal, with low ash content, medium sulfur content and medium selectivity. The total area of the mining area is about 19 square kilometers, and the vertical and horizontal changes of coal seams and coal quality are obvious, and the metamorphic degree is gradually deepened from top to bottom in the vertical zoning. From the perspective of feasibility, the ecological compensation of the mining area mainly involves several aspects such as atmosphere, water, and soil.

2.2.1. Layout of Monitoring/Sampling Points

Monitoring and sampling points mainly includes atmosphere, surface water, groundwater, soil and so on. In this paper, geographic information system (GIS) is used to mark each sampling point, as shown in Figure 1 below.

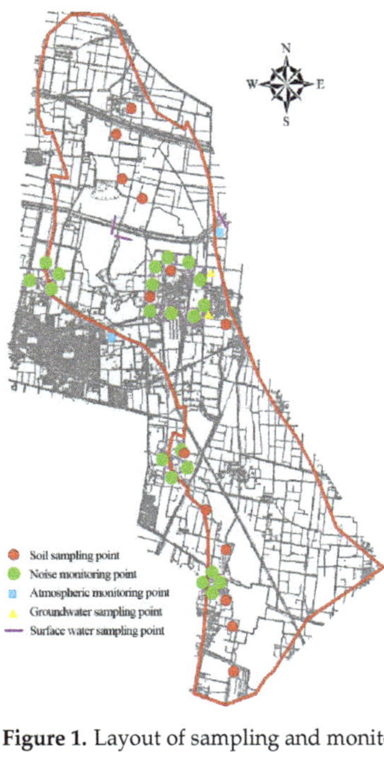

Figure 1. Layout of sampling and monitoring sample points.

Atmosphere: Two atmospheric environment monitoring points are set up around the industrial square in the mining area.

Surface water: According to the distribution of the water system in the study area and the discharge of coal mine wastewater, the water quality of the collapsed water area and the impact of coal mine wastewater are studied [28]. A monitoring section is set up before the outlet of the collapsed area that receives wastewater enters the outer river, in the collapsed area after the water outlet entered the outer river. One cross-section was set up at 500 m in the upstream and downstream of the river channel to monitor the water quality of the subsidence area, and the upstream monitoring point data were used for comparison [29].

Groundwater: Samples of shallow groundwater in the coal mine field were taken for monitoring. In the industrial square, waste rock hill, and nearby villages, one civil water well was selected for sampling, making a total of three sampling points.

Soil: One soil sampling point was set up at the bottom of the gangue dump and nearby villages to analyze the heavy metal content of the soil, making a total of two sampling points; along the time sequence of the surface subsidence of coal mining, soil samples were taken at the edge of the subsidence area to analyze the soil. In terms of physical properties, a total of 11 sampling points were set; in addition, another 38 monitoring sampling points were set up in other farmlands within the boundary of the minefield to analyze the content of heavy metals in the soil. There were a total of 51 sampling points in the soil.

2.2.2. Test Items and Methods

Atmosphere: According to the "Ambient Air Quality Standards" and the characteristics of mine waste discharge, the monitoring items of the ambient air current situation in the industrial square and its vicinity are determined as SO_2 daily average concentration and SO_2 hourly average concentration. Air samples were collected in the industrial square and its vicinity for seven consecutive days, from 23–29 July 2020. The time is in accordance with the time specified in the "Ambient Air Quality Standard".

Soil: Bulk density can reflect the degree of soil tightness and is the most important indicator of soil subsidence and deformation. Therefore, physical properties mainly analyze soil bulk density and water content, and heavy metals mainly analyze six types of Cd, Cr, Pb, As, Cu, and Zn. Sampling was carried out in three times, one on 1 August 2020, a small amount of comprehensive monitoring, one sample for each type, a total of 3 points; the second time was conducted on 24 April 2020, focusing on sampling and analyzing the physical properties of the soil. A total of 11 points were set up along the direction of mining subsidence; the third time was conducted on 7 August 2020, focusing on the analysis of heavy metal content, and 51 points were relatively evenly distributed within the mining area.

Surface and Groundwater: The monitoring element refers to the "Surface Water Quality Standard" (GB3838-2002), which combines the characteristics of coal mine wastewater with known pollution factors in surface water bodies, as well as confirmed test factors, including the output of domestic sewage and the concentration of BOD_5, COD, NH_3-N, sampling for two consecutive days, mixed sampling once a day.

2.2.3. Test Instrument Selection

Main instruments: multi-parameter water quality analyzer (Pro2030, the United States), X-diffraction fluorescence spectrometer (BRUKER S1 TURBOSD, Karlsruhe, Germany), multi-function noise tester (AR.27-628 series, China), soil profile moisture rapid tester (IMKO-PICO-BT, Ettlingen, Germany), and so on.

2.2.4. Design of Ecological Compensation of Coal Resources Mining

Generally speaking, compensation must be paid first, and the degree of damage is the calculation assumption of the environmental compensation standard (Sergeev et al., 2017). Ecological destroyers must make up for the losses caused by the destruction of the ecological environment. The ecological compensation standard in the proposal is to compensate the losses caused by ecological damage and environmental pollution. According to the technical specifications for ecological environment assessment issued by China in 2015, the value of ecological environment loss of atmosphere, water, and soil is estimated.

Ecological compensation P:

$$P = EC_{air} + EC_{water} + EC_{soil} \quad (10)$$

Air pollution loss:

$$EC_{air} = \sum (A_i \times d_i) \quad (11)$$

In the Equation (11), A_i represents the total amount of air pollution in type i and d_i represents the unit treatment cost of the corresponding pollution type.

Water pollution loss:

$$EC_{water} = \sum (W_i \times c_i) + \sum W_i \times m \quad (12)$$

In the Equation (12), W_i represents the total amount of water pollution in type i and c_i represents the unit treatment cost of the corresponding pollution type; m represents the market price of water resources transaction.

Soil value loss:

$$EC_{soil} = \theta \times O/(10^4 \times r \times h) \quad (13)$$

In the Equation (13), θ represents the total amount of soil conservation and O represents opportunity cost per unit soil area; r represents soil bulk density and h represents soil depth.

3. Results

3.1. Equilibrium Points and Strategy Selection of Evolutionary Game Model

If Equation (9) is 0, five equilibrium points can be obtained, that are (0, 0), (0, 1), (1, 0), (1, 1), (x*, y*). By calculating the partial derivatives of x and y respectively, the corresponding Jacobean matrix can be obtained, and further different stability strategy points can be chosen.

$$\det A = \begin{vmatrix} (1-2x)(2By - Py - B - C) & x(1-x)(2B-P) \\ (R-P)y(1-y) & (1-2y)(Rx - Px + S - D) \end{vmatrix} \quad (14)$$

By substituting the five equilibrium points into Equation (10) and combining their positive and negative characteristics, the following results can be obtained, as shown in the Table 2.

Table 2. Property of equilibrium points.

Equilibrium Points	Value of Determinant	Sign Characteristics	Trace of Matrix	Sign Characteristics	Stability
(0, 0)	(B+C)(D-S)	+	−(B+C+D-S)	−	ESS
(0, 1)	(B-C-P)(D-S)	+	B-C-P+D-S	+	Instable
(1, 0)	(B+C)(R-P+S-D)	+	B+C+R-P+S-D	+	Instable
(1, 1)	(B-C-P)(R-P+S-D)	+	−(B-C-P+R-P+S-D)	−	ESS
(x*, y*)	(B+C)(D-S) (B-C-P)(R-P+S-D)/ (2B-P)(R-P)	+	0	0	Saddle-point

3.2. Calculation of Ecological Compensation for Coal Resource Development

3.2.1. The Impact of Coal Resource Development on the Atmospheric Environment

SO_2 is monitored four times a day, and the statistical results are shown in Table 3. Due to the influence of temperature and airflow, the overall change of SO_2 in a day shows an increasing trend during the day and decreasing trend at night, as shown in Figure 2. The content of gangue hillside is significantly higher than that outside the mine boundary. Although it does not reach the level of pollution, it already had an impact on the environment. To simplify the calculation, the average value of concentration change is 0.01 mg/L, and the air sampling depth is 1.5 m, and the treatment cost of sulfur dioxide per kilogram is USD 0.19. Therefore, air pollution loss is USD 0.21 million annually.

Table 3. Evaluation results of the hourly average concentration range of SO_2.

Monitoring Points	Concentration Range	Single Factor Index	Number of Monitoring
Mining East	0.018–0.042	0.036–0.084	28
Gangue hillside	0.037–0.052	0.064–0.104	28
Mining West	0.014–0.037	0.028–0.074	28

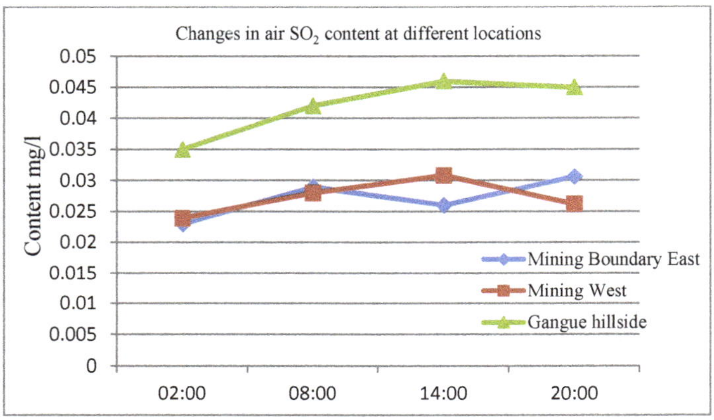

Figure 2. Changes in air SO_2 content at different locations.

3.2.2. The Impact of Coal Resource Development on the Soil and Surface Subsidence

Soil bulk density is an index that reflects the degree of soil tightness and indirectly expresses soil porosity, which is the main physical property affected by coal mining. The soil moisture content comprehensively reflects the physical properties of the soil. Therefore, 11 sampling points were set up along the edge of the collapsed basin to determine the soil moisture content and bulk density. The average water content of the monitoring points is 10.12%, and the average soil bulk density is 1.16 g/cm^3.

Figure 3 shows the change in bulk density of each point the sampling points were in the order of coal mining collapse, with the first collapse in the middle near the industrial square, and the shorter the collapse time toward the north and south. It can be seen from the figure that the longer the time collapses, the lower the soil water content at the edge of the collapsed basin, and the soil bulk density tends to increase. It shows that the impact of coal mining subsidence on the physical properties of soil is mainly reflected in the fact that the soil at the edge of the basin tends to be compact, and the soil moisture content decreases, and it tends to be significant over time.

MSPS mining subsidence prediction software is used to predict the evolution process of the surface subsidence of coal mine A, and the surface subsidence area statistics are shown in Table 4. The results show that the mining process of coal resources produces a small amount of disturbance to various environmental factors, but nonetheless, both are within the national control standards. The main influence is the slight influence of pH value and f tunnel water on the surface water. The spontaneous combustion of coal gangue SO_2 and silicified ore has little influence on the surrounding atmosphere; the noise near the wind farm has little influence; the minor influence on the surrounding soil and farmland is flying Pb field ash; surface subsidence leads to an increase in soil volume and a decrease in water content. The biggest problem with the impact of coal mining on the ecological environment is that the mountain crust causes a change in the shape of the ground, which greatly increases the pit surface and reduces the arable land.

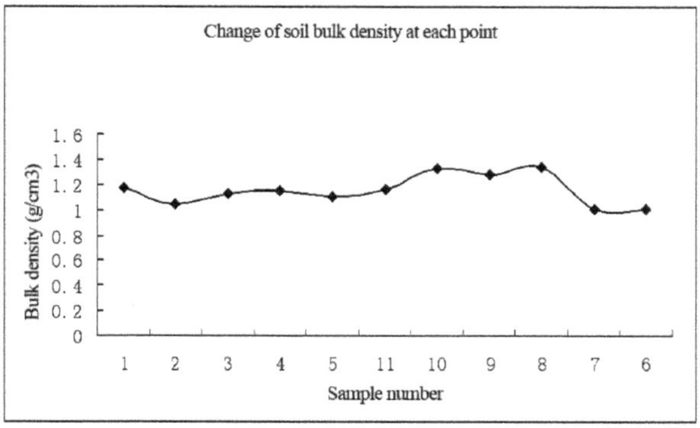

Figure 3. Changes in soil bulk density.

Table 4. Statistics of collapsed area of mine A in each period (m^2).

Years	Collapse Depth (mm)			
	10–1500	1500–3000	3000–5000	>5000
2013	5,918,660.59	873,725.24	457,703.63	1,012,279.10
2015	7,481,752.09	1,543,751.71	1,041,739.20	1,280,927.35
2017	7,427,833.44	1,943,836.27	1,497,840.37	1,830,113.19
2019	9,093,214.39	1,964,655.57	1,575,652.78	2,400,149.78
2020	8,708,807.66	2,174,142.65	2,265,369.82	2,804,318.26
Finally	8,762,543.81	2,251,783.35	2,193,314.57	3,001,731.36

According to the statistical bulletin of national economic and social development, the output value of grass per square kilometers is USD 0.15 million. The annual soil conservation is 65,000 tons, and the average soil bulk density is 1.2 t/m^3. Therefore, soil value loss is USD 0.53 million annually.

3.2.3. The Impact of Coal Resource Development on the Water

The output of domestic sewage in coal mine is 717 m^3/d, and the main pollutants are BOD_5, COD, NH_3-N, etc., whose discharge concentrations are 80 mg/L, 20 mg/L, and 150 mg/L respectively. The price of industrial water is 0.65 USD/m^3, and the charging standard for sewage discharge is USD 0.11 per pollution equivalent. Therefore, water pollution loss is USD 0.36 million annually. Therefore, the total ecological environment loss is about USD 1.09 million annually.

In order to eliminate the impact of COVID-19 on the price of coking coal, this paper takes the average market price of USD 188.6 per ton in 2015–2020 years. The annual coal output of Inner Mongolia is about 1 million 200 thousand tons, and the coal resource tax rate is 9%, so the mining area needs to pay USD 20.28 million per year. Coupled with the loss of ecological value, it can be concluded that the total cost of ecological compensation is about USD 21.38 million per year.

4. Discussion

4.1. Analysis of Parameters and Results in Evolutionary Game Model

4.1.1. Analysis of Parameters

(1) Social benefit B must exceed half of ecological compensation P.

For the government, if y = (B + C)/(2B − P), F(x) is 0, which is means the model is always in a stable state. Otherwise, there are two stable points x = 0 and 1. Since y is greater than 0 and less than 1, B is greater than C + P, whose economic significance is that the social benefits of government supervision must be greater than the sum of supervision cost and ecological compensation. At the same time, since B + C is greater than zero, B must be greater than half of P, that is, the social benefits of government supervision must exceed half of ecological compensation. If the cost of government regulation is higher than social benefits, the government tends not to regulate in the long-term game.

(2) Considering the delayed effect of technological progress of pollutant discharge, in the long run, the cost D of reducing pollutant discharge must be less than the additional benefit S.

For enterprises, if x = (D − S)/(R − P), F(y) is 0, which means the model is always in a stable state. Otherwise, there are two stable points y = 0 and 1. Since x is greater than 0 and less than 1, R + S are greater than D + P, whose economic significance is that the sum of government subsidies and additional benefits brought by the cooperation of enterprises must be greater than the sewage cost and ecological compensation.

(3) Government subsidy R should not be higher than ecological compensation P.

Generally, S is greater than D, so P must be greater than R, that is, ecological compensation should be greater than government subsidies to enterprises. If the emission cost and punishment of enterprises are greater than government subsidies and cooperation benefits, enterprises tend to implement non-cooperation strategies in the long-term game.

4.1.2. Analysis of Evolutionary Game Results

Based on the results in Table 2, the game evolution trajectory of government and enterprises can be obtained, as shown in Figure 4.

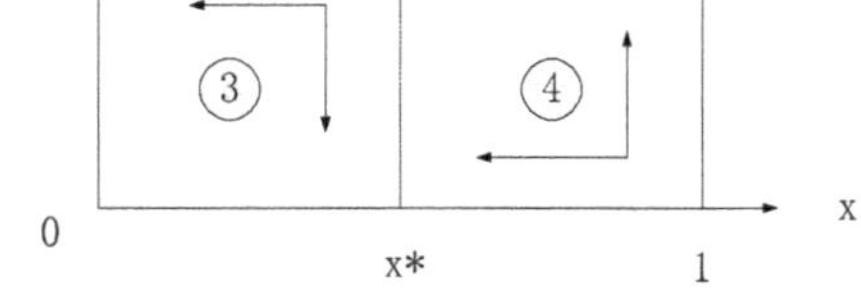

Figure 4. Evolutionary game trajectory.

Figure 4 shows the evolution direction of the game when the initial state falls in different regions. When it is in region③, the game converges to (0, 0), which is non-regulation and non-cooperation. When it is in region①, the game converges to (1, 1), which is regulation and cooperation. When it is in region②, the game converges to (1, 0), which is regulation and non-cooperation. When it is in region④, the game converges to (0, 1), which is non-regulation and cooperation.

The initial state of the game determines the final convergence result. Therefore, in order to achieve the optimal regulation and cooperation state, it is bound to expand the probability that the initial state falls in region 1, that is, the smaller the value of x* or y*, the better. That is: (1) regulatory cost C↓, ecological compensation P↓, social benefit B↑; (2) government subsidy R↑, enterprise additional income S↑, enterprise sewage improvement cost D↓, ecological compensation or fine P↓. In short, from the perspective of the government, its regulatory cost C and ecological compensation P should be reduced as much as possible, and the social benefit B should be expanded at the same time. From the perspective of enterprises, government subsidy R and additional income s can stimulate enterprises to promote emission reduction technology, so as to reduce emission reduction cost D and avoid punishment P.

4.2. Equilibrium Results of Game Model in Specific Cases

Based on the above calculation results, the total cost of ecological compensation is about USD 21.38 million per year in manlailiang coal mine of Inner Mongolia. Combined with the parameter analysis of evolutionary game model, social benefit B must exceed USD 10.69 million annually and government subsidy R should be lower than USD 21.38 million annually.

5. Conclusions

The ecological compensation mechanism in the process of coal resources development should include both compensation for resource consumption and compensation for ecological environment damage.

5.1. Ecological Compensation Types

5.1.1. Resource Depletion Compensation Mechanism

At present, the mining compensation of coal resources is mainly realized through the resource tax system, which has the nature of development compensation to adjust the differential income of developers, and promote the price of resource market to be reasonable. To improve the compensation mechanism for resource consumption, it needs to start from the following aspects. First, a flexible and appropriate tax rate market mechanism should be established. To improve the flexibility of oil and gas resource tax rate, it should not only have a long-term and stable resource tax rate system, but also make dynamic adjustment to determine the short-term optimal tax rate and optimal development volume. Second, it needs to optimize the collection method of special coal tax, carry out classified collection, guide sustainable development, and improve utilization efficiency. Third, it must locate the resource tax to compensate the consumption cost of contemporary resources, establish the coal resource depletion compensation fund system according to local conditions, reasonably allocate the extraction proportion of resource tax, and achieve the purpose of earmarking.

5.1.2. Environmental Restoration Compensation Mechanism

First, an ecological environment deposit system should be established. For the possible environmental pollution risks caused by coal resource development and utilization enterprises, a certain compensation deposit should be charged, assessed according to the production and environmental maintenance of the enterprise, and then deducted or returned according to the actual situation. Second, the source of ecological compensation can be provided through transfer payment. The government has the identity of both coal resource owner and social manager. It needs to promote production through supervision and publicity, and maintain the environment and ecology through compensation and restoration. Third, in order to avoid occupying a large amount of exploration land without investment, the marketization of mining rights should be implemented, that is, responsible and powerful mining enterprises should be introduced through bidding, auction, and other forms.

5.2. Ecological Compensation Threshold

5.2.1. Strictly Implement the Elimination Mechanism of Declining Industries

The backward technology of emission reduction in China's coal industry is still serious: the overall technical level is low and the production equipment is backward; the waste of coal resources is serious and the recovery rate is low; the insufficient washing rate of raw coal is also a prominent problem. While eliminating backward production capacity, enterprise should pay more attention to the development and application of high and new technology. Through fiscal and tax support, capital and project support and other policy guidance methods, the government will guide the declining industries to gradually withdraw from the existing leading industrial structure layout, so as to promote the optimal allocation of production factors. This mechanism is conducive to solving the problems of shrinking main business scale, insufficient operation, and decline in equipment utilization and so on.

5.2.2. Promote Technological Innovation Compensation Mechanism

The government should encourage enterprises to increase scientific research funds, ensure technological innovation in emission reduction, accelerate the transformation of enterprise achievements, promote new technologies and products, and improve the emission reduction efficiency of enterprises. Establishing a pollutant emission trading market as soon as possible so that "emissions" and "emission rights" can flow in the market and have the characteristics of commodities to some extent. Enterprises can reduce emissions and then sell the remaining emissions to the market to obtain additional economic benefits; this kind of transaction belongs to the flow and exchange of special commodities, so the emissions must flow to enterprises with high production capacity and economic efficiency, so as to achieve the purpose of better allocation of environmental resources.

5.2.3. Weaken the Government Subsidy Mechanism and Strengthen the Initiative of Enterprise for Ecological Compensation

Ecological compensation should be based on the technological innovation of enterprise emission reduction. On this basis, the government should give enterprises certain technical subsidies through administrative means, but the proportion of subsidies should not be too high. It is necessary to introduce more market mechanisms and guide public opinion to stimulate the enthusiasm of enterprises for ecological compensation by improving the brand value of enterprises and expanding product sales.

5.2.4. Rely on Non-Governmental Organizations to Enlarge the Social Benefits Brought by Government Regulation

Healthy Nongovernmental Organizations are an international trend to improve environmental compensation and environmental protection. According to Arrow's Impossibility Theorem, it is difficult to reach many people who have suffered environmental damage, and it is difficult to take coordinated actions. Therefore, from the perspective of environmental victims, there are problems and difficulties in limiting environmental degradation and compensating for the environment. These environmental groups have broad, diverse, and complete social components, representing the collective interests of many different environments and pollution, and focusing on environmental degradation and environmental control measures. They can combine small-scale and insignificant forces to form a powerful and comprehensive ecological balance and environmental protection force in order to effectively deal with ecological and environmental pollution at the social level. This is the way our country should try and use coal production.

Author Contributions: Conceptualization, Y.Q. and W.W.; methodology, Y.Q.; software, Y.Q.; validation, Y.Q.; formal analysis, Y.Q. and W.W.; investigation, Y.Q.; resources, Y.Q.; data curation, Y.Q.; writing—original draft preparation, Y.Q.; writing—review and editing, Y.Q. and W.W.; visualization, Y.Q.; supervision, W.W.; project administration, W.W.; funding acquisition, W.W. All authors have read and agreed to the published version of manuscript.

Funding: This research was funded by Chinese National Social Sciences Fund (Major Research Plan 12&ZD207); Chinese National Natural Sciences Fund (71973023); Later support of Chinese National Social Science Fund (19FGLB010); Major Program of Jiangsu Social Science Foundation in China (19GLA003).

Institutional Review Board Statement: Not applicable.

Informed Consent Statement: Not applicable.

Data Availability Statement: All data, models generated or used during the study appear in the submitted article.

Conflicts of Interest: The authors declare that they have no competing interest.

References

1. Acheampong, A.O.; Boateng, E.; Amponsah, M. Revisiting the economic growth–energy consumption nexus: Does globalization matter? *Energy Econ.* **2021**, *102*, 105472. [CrossRef]
2. Kan, S.; Chen, B.; Chen, G. Worldwide energy use across global supply chains: Decoupled from economic growth? *Appl. Energy* **2019**, *250*, 1235–1245. [CrossRef]
3. Sekerin, V.; Dudin, M.; Gorokhova, A.; Bank, S.; Bank, O. Mineral resources and national economic security: Current features. *Min. Miner. Depos.* **2019**, *13*, 72–79. [CrossRef]
4. Xie, F.; Liu, C.; Chen, H.; Wang, N. Threshold Effects of New Energy Consumption Transformation on Economic Growth. *Sustainability* **2018**, *10*, 4124. [CrossRef]
5. Chontanawat, J. Dynamic Modelling of Causal Relationship between Energy Consumption, CO_2 Emission, and Economic Growth in SE Asian Countries. *Energies* **2020**, *13*, 6664. [CrossRef]
6. Niu, S.; Zhang, X.; Zhao, C.; Niu, Y. Variations in energy consumption and survival status between rural and urban households: A case study of the Western Loess Plateau, China. *Energy Policy* **2012**, *49*, 515–527. [CrossRef]
7. Li, X.Q.; Hu, Z.Y. Interaction between economic growth energy consumption and environmental pollution. *Int. J. Environ. Pollut.* **2020**, *68*, 162–177. [CrossRef]
8. Han, X.; Sun, T.; Feng, Q. Study on environmental pollution loss measurement model of energy consumption emits and its application in industrial parks. *Sci. Total Environ.* **2019**, *668*, 1259–1266. [CrossRef]
9. Kang, H.; Wang, G.; Wang, S. High-Quality Development of China's Coal Industry. *Eng. Sci.* **2021**, *23*, 130–138. [CrossRef]
10. He, J.; Fu, C.; Long, Y. Promoting the High-Quality Development of Central China through Energy Revolution: Theoretic Analysis and Implementation Paths. *Chin. J. Eng. Sci.* **2021**, *23*, 60–67. [CrossRef]
11. Falshtynskyi, V.S.; Dychkovskyi, R.O.; Saik, P.B.; Lozynskyi, V.H.; Cabana, E.C. Formation of thermal fields by the energy-chemical complex of coal gasification. *Nauk. Visnyk Natsionalnoho Hirnychoho Universytetu* **2017**, *5*, 36–42.
12. Skrypnyk, O.; Shapar, A.; Taranenko, O. Determining local wetness conditions within the mined lands using GIS. *Min. Miner. Depos.* **2020**, *14*, 53–58. [CrossRef]
13. Cooper, G.; Cowan, D. *Terracing Potential Field Data*; ASEG Extended Abstracts; ASEG: Crows Nest, Australia, 2009.
14. Guo, X.; Niu, D.; Xiao, B. Assessment of Air-Pollution Control Policy's Impact on China's PV Power: A System Dynamics Analysis. *Energies* **2016**, *9*, 336. [CrossRef]
15. Song, W.; Deng, X. Effects of Urbanization-Induced Cultivated Land Loss on Ecosystem Services in the North China Plain. *Energies* **2015**, *8*, 5678–5693. [CrossRef]
16. Lee, M.; Morris, W.; Leblanc, G.; Harris, J. Curvature analysis to differentiate magnetic sources for geologic mapping. *Geophys. Prospect.* **2012**, *61*, 572–585. [CrossRef]
17. Li, X. Curvature of a geometric surface and curvature of gravity and magnetic anomalies. *Geophysics* **2015**, *80*, G15–G26. [CrossRef]
18. Chang, Z.; Qiao, J.; Zhao, J.; Dong, Z.; Wang, J. Review on the key issues of basis and criteria of ecological compensation in China. *Ecol. Sci.* **2020**, *39*, 248–255.
19. Li, S.; Wang, J.; Zhang, Z. Ecological compensation for mineral resources exploitation: A review. *Chin. J. Ecol.* **2019**, *38*, 1551–1559.
20. Dong, Q.; Liu, X.Z. The Legal Norms of Ecological Compensation in Coastal Cities under Regional Cooperation. *J. Coast. Res.* **2020**, *8*, 552–555. [CrossRef]
21. Gao, X.; Shen, J.; He, W.; Sun, F.; Zhang, Z.; Zhang, X.; Yuan, L.; An, M. Multilevel Governments' Decision-Making Process and Its Influencing Factors in Watershed Ecological Compensation. *Sustainability* **2019**, *11*, 1990. [CrossRef]
22. Gao, X.; Shen, J.; He, W.; Sun, F.; Zhang, Z.; Guo, W.; Zhang, X.; Kong, Y. An evolutionary game analysis of governments' decision-making behaviors and factors influencing watershed ecological compensation in China. *J. Environ. Manag.* **2019**, *251*, 109592. [CrossRef] [PubMed]
23. Xu, Z.; Liu, B.; Li, S.; Chen, Y.; Meng, Z. A standard compensation system on the environmental influences of Chinese coal resource development. *Arab. J. Geosci.* **2021**, *14*, 1–17. [CrossRef]
24. Abuya, W.O. Mining conflicts and Corporate Social Responsibility: Titanium mining in Kwale, Kenya. *Extr. Ind. Soc.* **2016**, *3*, 485–493. [CrossRef]

25. Wei, C.; Luo, C. A differential game design of watershed pollution management under ecological compensation criterion. *J. Clean. Prod.* **2020**, *274*, 122320. [CrossRef]
26. Chen, Z.; Xu, R.; Yi, Y. A Differential Game of Ecological Compensation Criterion for Transboundary Pollution Abatement under Learning by Doing. *Discret. Dyn. Nat. Soc.* **2020**, *2020*, 1–13. [CrossRef]
27. Shen, J.; Gao, X.; He, W.; Sun, F.; Zhang, Z.; Kong, Y.; Wan, Z.; Zhang, X.; Li, Z.; Wang, J.; et al. Prospect theory in an evolutionary game: Construction of watershed ecological compensation system in Taihu Lake Basin. *J. Clean. Prod.* **2021**, *291*, 125929. [CrossRef]
28. Scofield, N.; Roggenthen, W.M. Petrologic evolution of plagioclase-rich cumulates from the Wichita Mountains, Oklahoma: Effects upon magnetic remanence properties. *Geology* **1986**, *14*, 908–911. [CrossRef]
29. Seisuma, Z.; Kulikova, I. Distribution of metal concentrations in sediments of the coastal zone of the Gulf of Riga and open part of the Baltic Sea. *Oceanology* **2012**, *52*, 780–784. [CrossRef]

buildings

MDPI

Article

The Impacts of Greenery Systems on Indoor Thermal Environments in Transition Seasons: An Experimental Investigation

Xiaoli Hao [1,2,*], Liping Liu [1], Hang Tan [1], Yaolin Lin [3,*], Jinhua Hu [1,2] and Wei Yin [1,2]

1 College of Civil Engineering, Hunan University of Science and Technology, Xiangtan 411201, China; lpliu@hnust.edu.cn (L.L.); tanhang@capol.cn (H.T.); hujinhua@hnust.edu.cn (J.H.); yinwei@hnust.edu.cn (W.Y.)
2 Hunan Engineering Research Center for Intelligently Prefabricated Passive House, Hunan University of Science and Technology, Xiangtan 411201, China
3 School of Environment and Architecture, University of Shanghai for Science and Technology, Shanghai 200093, China
* Correspondence: haoxiaoli2002@aliyun.com (X.H.); yaolinlin@gmail.com (Y.L.)

Abstract: The impacts of greenery systems (GSs) on microclimate conditions and building energy performance have been frequently investigated using experiments and simulations during the past decades, especially in summer and winter. However, few studies have focused on the performance of GSs in transition seasons. The ambient weather conditions vary with great fluctuations during transition seasons, which may result in severe oscillations in indoor environments. To investigate the impacts of GSs on indoor environments, an experiment was conducted using a contrastive test platform, which consisted of two experimental rooms, one equipped with a GS and the other without, from 1 April 2019 to 31 May 2019 in Hunan, China. Both rooms were free-running. The experimental results showed that the GS had the ability to reduce the oscillations in the indoor environment. The oscillations in indoor dry-bulb temperature (DBT) and relative humidity (RH) were reduced by 39.3% and 28.8%, respectively. The maximum daily DBT and RH ranges were, respectively, cut down by 3.5 °C and 12.4%. The maximum reductions in external and internal surface temperatures were 29.5 °C and 9.4 °C, respectively, for the GS, while the average reductions were 1.6~4.1 °C and 0.2~1.3 °C, respectively, depending on the orientation of the surfaces. The operative temperature (OT) during the daytime on sunny days was also lowered by the GS. The differences in OT between the two rooms ranged from −1.8 °C to 8.2 °C, with an average of 1.0 °C. The GS can improve the indoor thermal comfort during transition seasons. The thermal dissatisfaction was decreased by 7.9%. This lengthened the thermal comfort time by 15% across the whole day and by 28% during the daytime. This indicates reductions in air-conditioning system operating times, leading to energy savings.

Keywords: green wall; green roof; indoor thermal environment; thermal comfort; energy saving

Citation: Hao, X.; Liu, L.; Tan, H.; Lin, Y.; Hu, J.; Yin, W. The Impacts of Greenery Systems on Indoor Thermal Environments in Transition Seasons: An Experimental Investigation. *Buildings* **2022**, *12*, 506. https://doi.org/10.3390/buildings12050506

Academic Editors: Shi-Jie Cao and Wei Feng

Received: 5 March 2022
Accepted: 16 April 2022
Published: 19 April 2022

1. Introduction

The latest report from the Intergovernmental Panel on Climate Change (IPCC) [1] shows that the global surface temperature was 1.09 °C higher in 2011–2020 than in 1850–1900. It is an unequivocal fact that the global climate is warming due to pollution emissions from human activities. It is quite urgent for humans to take measures to mitigate global climate change. The rapid process of urbanization and urban densification is responsible for global climate warming [2]. According to statistics from the United Nations [3], 60–80% of energy consumption and 75% of carbon emissions in the world are from cities at present, although the area of land occupied by cities is only 3% of that of the earth. Furthermore, it is predicted that the proportion of the urban population around the world population will increase from 54% in 2014 to 66% by 2050 [4]. More fossil energy will be consumed and more pollution will be released by cities if sustainable strategies are not adopted in the future.

To accommodate for this population growth, a large amount of buildings and structures have been built in cities, resulting in large amounts of permeable land, such as bare soil and vegetation, being replaced by impervious concrete surfaces [5,6]. Increases in impervious land cause higher urban temperatures than in the surrounding rural areas. This phenomenon is called the urban heat island (UHI) effect [7]. The UHI can lead to many socio-economic problems, such as increasing the incidence of heat-related mortality and the cooling energy consumption in summer [8,9]. Therefore, strategies for mitigating the UHI effect are needed to achieve the goal of sustainable development.

Tian et al. [10] reviewed mitigation strategies for dealing with the UHI effect. Among these strategies, the construction of green infrastructure was considered a feasible approach to mitigate the UHI effect and global climate warming, as well as other problems. Green infrastructure can be classified into urban green spaces (UGS), green roofs (GR), and green walls (GW) [10]. UGS include the green spaces provided by urban roads, urban parks, residences, and workplaces. However, in crowded urban areas, the land is very precious and the UGS is very limited. Green roofs, also called as eco-roofs, vegetated roofs, or living roofs, contain plants vegetation on in areas that are usually idle. However, green wall plants and green vegetation on the vertical surfaces of buildings, such as the walls, façades, and blind walls, are also called vertical greening systems (VGS) [11]. Due to the occupation of urban land, GR and GW have wider application potential as strategies to mitigate the UHI and global warming compared with UGS. Both GW and GR can be classified as greenery systems (GSs) in buildings [12].

The environmental benefits [13] provided by GW and GR include improvements in indoor and outdoor thermal comfort and air quality [14–18]; reductions in energy consumption via enhancing building thermal performance [19–27]; the mitigation of the UHI effect and global climate change by cooling the urban area and lessening of GHG emissions [28–35]; decreased urban noise pollution [36,37]; relief of urban drainage pressures via storm water management, which is mainly provided by GR (the role of GW on storm water management is limited) [38,39]; and the promotion of biodiversity in urban environment [40,41]. In addition, GW and GR can also bring about many social benefits, such as improving a city's image [42], enhancing the well-being of citizens [43], and increasing property values [44].

Due to the significant environmental and social benefits of GW and GF, much research has been focused on this topic, especially regarding the impacts on microclimate conditions and building energy consumption [45]. Experimental and simulation methods have been adopted in previous studies [46,47]. A systematic review on the influence of GW and GF on building environments and energy was presented by Seyam [48]. He found that six parameters, including solar radiation, ambient/indoor dry-bulb temperatures (DBT), ambient relative humidity (RH), and internal/external surface temperatures of walls and roofs, were frequently measured, while the indoor RH received little attention.

The impacts of GS on temperature reductions (TR) of external and internal surfaces [48], which are defined as temperature differences between external and internal surfaces of bare and vegetated walls and roofs, were most frequently investigated. The magnitude of the TRs of external and internal surfaces varied depending on the building envelope, the installation location of the GS, the outdoor climate conditions, and the design of the GS. Human occupancy also has a significant impact on the TRs of external and internal surfaces [49,50]. For a given GS, the TR value may be positive or negative, depending on the time of the day, season, and solar radiation intensity [48].

The reductions in temperature differences of external and internal surfaces between envelopes with and without GS have often been the focus, because these are closely associated with heat flux reductions through walls or roofs, thereby affecting the energy savings achieved by GSs. In most studies, the energy consumption was reduced via the use of a GS, while the energy saving rates varied across a wide range, depending on the vegetation coverage, outdoor climate conditions, and GS design [48].

Compared with TR values of external and internal surfaces and the energy savings achieved by GSs, the impacts on indoor building environments have been less investigated. For an air-conditioned space, the indoor DBT and RH are controlled by the air-conditioning system and are almost unaffected by the GS. However, the experimental results from Hao et al. [51] showed that the GS can reduce the indoor operative temperature (OT) of air-conditioned spaces by 0.4 °C on average and 2.1 °C in summer. In addition, it can also reduce the oscillations in indoor OT by 1.1 °C. According to the four seasons (spring, summer, autumn, and winter), experiments by Mangone et al. [52] with indoor plants improved the thermal comfort of an air-conditioned office by 12% compared to an office with an identical indoor temperature set-point but without plants. Lately, research results have revealed that small indoor green walls can help relieve the mental stress of the occupants [53].

The impacts of GSs on indoor thermal environments in non-air-conditioned spaces are more significant compared with air-conditioned spaces. The experimental results from Olivieri et al. [49] showed that a vegetation layer reduced the indoor DBT by 4 °C on average in a continental Mediterranean climate zone under summer conditions, while a lower OT in a chamber with vegetation was achieved compared with a chamber without vegetation due to a 5 °C lower internal surface temperature. To investigate the thermal performance of a living wall system (LWS) in a hot and humid climate, Chen et al. [54] constructed two experimental chambers of identical size, materials, and structure, except that one contained the LWS and the other did not. The experimental results for the two chambers revealed that the indoor DBT of the chamber with the LWS was reduced by 1.1 °C in summer. An LWS was applied on a school building in a hot and arid climate zone and the indoor DBT in the unconditioned school was measured in the peak summer month of July by Haggag et al. [55]. It was found that the indoor DBT with the LWS was always lower than that without the LWS, with reductions ranging from 2 °C at night to 6 °C during the daytime. Yang et al. [56] investigated the impact of a vertical GS on an indoor thermal environment in summer without the operation of an air-conditioning system. A double-skin green façade (DSGF) was added to a campus building and the indoor thermal environmental parameters were recorded. The experimental data indicated that the indoor DBT values were reduced on average by 0.6–1.2 °C, while the indoor OT values were lowered on average by 0.6–1.1 °C and 1.9–2.7 °C at maximum owing to the installation of the DSGF.

The transition seasons, which are between winter and summer, show different climatic condition compared with winter and summer, during which the impact of the outdoor environment on the indoor environment has its own features. To create a more comfortable indoor environment, Long et al. [57] performed an experimental investigation on the natural ventilation performance of a solar chimney system with an earth-to-air heat exchanger during the transition seasons. Cao et al. [58] investigated the window opening behavior of infant families during transition seasons. Yu et al. [59] experimentally and numerically analyzed the soil temperatures using a ground heat exchanger in transition seasons. However, there are few studies on the impacts of GSs on indoor environments during the transition seasons.

A literature review showed that less attention has been paid to the impacts of GSs on indoor environments compared with those on energy performance. However, this subject deserves particular attention because humans are spending more time indoors, especially following the coronavirus disease outbreak in 2019. Indoor environments have effects not only on indoor thermal comfort and resident health, but also on energy consumption. In addition, most of the existing studies have been conducted in summer or winter [48], while few have been conducted in transition seasons, during either spring or autumn. However, the weather fluctuates a lot in transition seasons, which could lead to severe oscillations in indoor environments, causing human discomfort, short-cycling of air-conditioning systems, and switching between heating and cooling demands. These factors not only increase a building's energy consumption, but also shorten the service life of the air-conditioning system. Thus, technologies that can reduce the oscillations in indoor environments and decrease the energy consumption in transition seasons are preferred. The aim of this paper is to investigate the effects of a GS on an indoor environment and energy consumption in transition seasons through experimentation.

2. Materials and Methods

In order to investigate the influence of GS on indoor thermal environments and building energy performance, an experimental test platform, which consisted of two experiment rooms, both with sensors for environmental parameter measurements and a data collection and recording system, was set up on the top of an office building on the campus of Hunan University of Science and Technology, as shown in Figure 1. The size, structure, and materials of the walls and roof, as well as the orientations of the two experiment rooms, were identical, except that one room was equipped with GS while the other one was not. The room equipped with GS was called the VGRoom. The other room was used as a reference and was referred to as the RefRoom. The experimental setup had been used to investigate the effects of GS on indoor environments and energy savings in summer and winter [19,51]. In this paper, it was used to investigate the performance of GS during transition seasons.

Figure 1. Location on the map (**a**) and photograph (**b**) of the experimental setup.

The experimental room measured 3 m × 2.5 m with a height of 3 m. It had lightweight walls and a roof, with an overall heat transfer coefficient of 1.09 W/(m^2·K). Two windows measuring 0.9 m × 1.2 m and an insulated door measuring 0.9 m × 2 m were installed in each room. The U-values of the windows and doors were, respectively, 4.1 W/(m^2·K) and 2.1 W/(m^2·K). A VGS with a living wall structure and GR with modular trays were installed on the VGRoom. Schefflera octophylla (Lour.) Harms and Sedum linear plants were selected for the green walls and roofs, respectively. These are evergreen plants with the characteristics of light favorability, drought tolerance, and easy maintenance.

To record the indoor and outdoor environmental parameters during the experiment, many sensors were installed. Two self-recording sensors (accuracy levels of ±0.5 °C for temperature and ±3% for RH) were used to record the indoor DBT and RH values. The temperatures of interior and exterior wall and roof surfaces of each room were measured using twenty T-type thermocouples, with an accuracy level of ±0.5 °C. The indoor black-globe temperature (BGT) values were measured using two black-globe thermometers, with a globe probe diameter of 150 mm. The accuracy of the black-globe thermometer was ±0.4 °C. The solar radiation was measured using a pyranometer, with an accuracy level of ±5%. The instrument used for measuring outdoor DBT and RH values was the same as that used for indoor measurements. The accuracy levels, measurement ranges, and sampling intervals for all abovementioned instruments are listed in Table 1. Figure 2 shows the locations where the sensors were installed in detail. An Agilent data collector was used to automatically collect the experimental data expect for the indoor and outdoor DBT and RH and the BGT, which were automatically recorded by the measuring instruments themselves. More details about the experimental setup can be found in the references [19,51].

Table 1. Monitoring data and instrumentation.

Parameter	Sampling Interval	Measuring Device	Measuring Range	Accuracy
Air temperature (°C)	1 min	Temp self-recording meter	−40~100 °C	±0.5 °C
Relative humidity (%)	1 min	RH self-recording meter	0~100%	±3%
Black globe temperature (°C)	1 min	BGT self-recording meter	−20 °C~+80 °C	±0.4 °C
Surface temperature of wall (°C)	1 min	T-type thermocouple	−200 °C~300 °C	±0.5 °C
Solar radiation intensity (W/m^2)	1 min	Pyranometer	0~2000 W/m^2	±5%

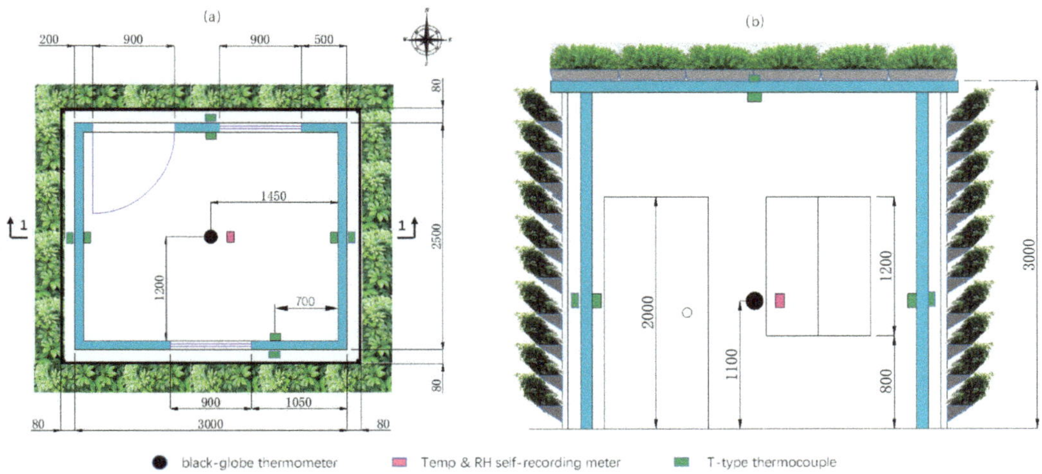

Figure 2. Installation locations of sensors: (**a**) plan and (**b**) cross-section 1-1 (the unit of dimensions is mm).

The experiment was carried out in Xiangtan, a city in Hunan Province in China at 28° N latitude and 112° E longitude. It features typical hot summer and cold winter weather conditions with four distinct seasons. The rainfall in Xiangtan is abundant, especially in spring and summer. It belongs to the Cfa category according to the Köppen–Geiger climate classification [60]. The transition seasons in Xiangtan are quite long and include two segments, spring (from March to May) and autumn (from October to November). The daily averaged outdoor temperatures in the transition seasons usually range from 12 °C to 26 °C. The experiment was conducted from 1 April 2019 to 31 May 2019. All data were automatically recorded at intervals of 10 min. The recorded data were hourly-averaged and were used for analysis.

During the experiment, both rooms were kept under free-running mode and no air-conditioner or heater was used. The rooms remained closed and unoccupied, except occasionally people entered the rooms to collect experimental data in order to avoid the influence of occupants.

The indoor environment is significantly affected by the outdoor climatic conditions on free-running mode. In the transition seasons, the indoor environment fluctuates a lot due to large fluctuations in the outdoor weather conditions. Measures to mitigate the oscillations of the indoor environment without running an air-conditioning system are preferred, as this leads to energy savings. To evaluate the effects of GS on reducing indoor environment oscillations, two indices, the DBT oscillation weakening rate (*TOWR*) and RH oscillation weakening rate (*HOWR*), were defined as:

$$TOWR = \frac{ADSDT_{Ref} - ADSDT_{VG}}{ADSDT_{Ref}} \times 100\% \quad (1)$$

$$HOWR = \frac{ADSDH_{Ref} - ADSDH_{VG}}{ADSDH_{Ref}} \times 100\% \quad (2)$$

where $ADSDT$ and $ADSDH$ are, respectively, the average daily standard deviations (SD) of DBT (in °C) and RH (in %). The subscripts Ref and VG denote the RefRoom and VGRoom, respectively. The $ADSDT$ and $ADSDH$ for RefRoom and VGRoom during the experiment will be analyzed and compared in the Discussion.

For the indoor thermal environmental evaluation, OT, which combines the effects of both convective and radiative heat transfers, is a more reasonable indicator than the DBT. It can be calculated using Equation (3):

$$t_{op} = \frac{h_{con}t_{air} + h_{rad}\bar{t}_{rad}}{h_{con} + h_{rad}} \quad (3)$$

where t_{op} is the operative temperature in °C; h_{con} and h_{rad} are, respectively, the convective and linear radiative heat transfer coefficients in $W/(m^2 \cdot °C)$; t_{air} is the indoor DBT in °C. $\bar{t}_{rad}$ is the mean radiant temperature (MRT) in °C; $\bar{t}_{rad}$ can be determined using Equation (4) [61].:

$$\bar{t}_{rad} = \left\{ \left(t_{bg} + 273\right)^4 + 0.4 \times 10^8 \left|t_{bg} - t_{air}\right|^{1/4} \times \left(t_{bg} - t_{air}\right) \right\}^{1/4} - 273 \quad (4)$$

where t_{bg} is the indoor BGT in °C. With the experiment data for indoor DBT and BGT, the MRT and OT can be determined using Equations (3) and (4). According to the suggestions from ASHRAE [62], values of 3.1 and 4.5 $W/(m^2 \cdot K)$ were adopted for h_{con} and h_{rad}, respectively. The impacts of GS on indoor OT will be discussed later.

3. Results

3.1. Outdoor Weather Conditions

The experiment data were collected from 4:00 p.m. on April 1 to 7:00 a.m. on 31 May and lasted for 1432 h. The data were recorded every ten minutes and then hourly-averaged. Figure 3 shows the variations in hourly-averaged outdoor DBT, RH, and solar radiation values during the experiment. Table 2 presents the statistical values for outdoor environmental parameters. Figure 3 and Table 2 show significant fluctuations in DBT and RH in outdoor air during the experiment. The outdoor DBT values ranged from 12.5 °C to 36.3 °C, covering the heating and cooling periods. The average outdoor DBT was 21.9 °C, which is a thermally comfortable temperature. The outdoor air RH values fluctuated between 36.6% and 96.8%, with an average value of 78%. This illustrated the humid climate in Xiangtan in spring. During the experiment, the maximum solar radiation intensity was 901.9 W/m^2, with an average value of merely 112.7 W/m^2. It was mostly rainy or cloudy during the experiment, with low solar radiation intensity, low outdoor DBT, and high outdoor RH values. It can also be seen from Figure 3 that a high outdoor DBT was usually accompanied by high solar radiation intensity.

Table 2. Environmental parameters during the experiment.

	Maximum	Minimum	Average
Outdoor DBT (°C)	36.3	12.5	21.9
Outdoor RH (%)	96.8	36.6	78.0
Solar radiation (W/m^2)	901.9	0	112.7

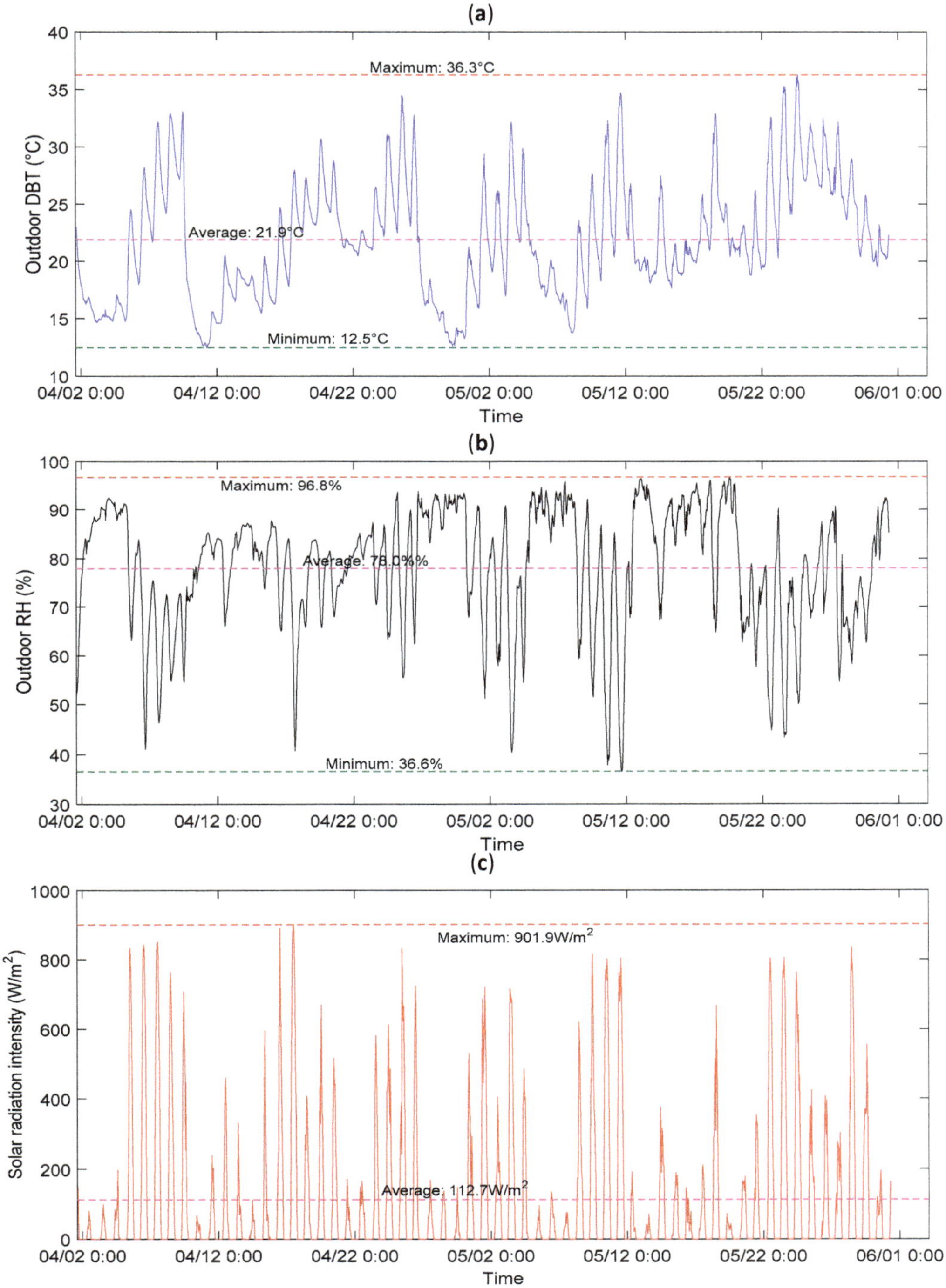

Figure 3. Outdoor DBT (**a**), RH (**b**), and solar radiation (**c**) values during the experiment.

Figure 4 displays the daily averaged outdoor DBT values during the experiment. The daily averaged DBT is the arithmetic mean of the 24-haveraged DBT. Most of the daily averaged outdoor temperatures were between 12 °C and 26 °C, except for a few days. According to the Chinese standard for climatic division [63], the transition seasons (spring and autumn) are when the daily average outdoor temperature is between 10 °C and 22 °C. Summer arrives if the daily average outdoor temperatures for five consecutive days are all higher than 22 °C. From Figure 4, it can be seen that the date for switching from spring to summer in meteorological terms was 22 May in 2019, which was close to the normal switching date of May 19 in Xiangtan. Therefore, the period from 1 April 2019 to 22 May 2019 was the spring transition season. However, there were 21 days in which the daily average temperature was higher than 22 °C, temperatures at which cooling may be required to maintain thermal comfort.

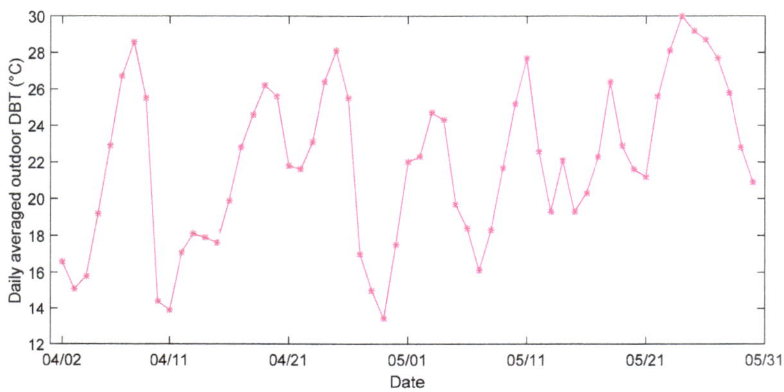

Figure 4. Daily averaged outdoor DBT values during the experiment.

3.2. Indoor DBT

The indoor DBT values for the two experimental rooms are shown in Figure 5, along with the outdoor DBT values. The statistical results for the indoor DBT and RH are presented in Table 3. Due to the drastic changes in outdoor climatic conditions during the experiment, large indoor DBT oscillations were observed, especially in the RefRoom. The maximum, minimum, and averaged indoor DBTs of the RefRoom during the experiment were 37.8 °C, 11.9 °C and 23.0 °C, respectively. The maximum daily temperature range (MDTR), which is defined as the maximum difference between daily maximum and minimum DBTs, was 18 °C in the RefRoom. In contrast, the indoor DBT of the VGRoom showed less oscillation than that of the RefRoom. The maximum, minimum, and averaged indoor DBTs for the VGRoom were 32.5 °C, 13.0 °C, and 22.1 °C, respectively, and the MDTR of the VGRoom was 14.5 °C during the experiment. The indoor DBT oscillation was reduced by 3.5 °C with the GS, although the averaged DBT was reduced by only 0.9 °C. This indicates that the VGRoom achieved a more stable indoor thermal environment than the RefRoom in the transition season. During the experiment, the hourly indoor DBT difference between the RefRoom and VGRoom ranged from −1.8 °C to 7.9 °C, with an average of 0.9 °C.

To investigate the daily variations in indoor DBT in detail, two typical days, 7 April and 14 April, were selected as the representatives of two kinds of weather conditions in spring in Xiangtan. April 7 is a typical sunny day with averaged outdoor DBT and RH values of 26.7 °C and 64.2%, respectively. The averaged and maximum solar radiation intensities were 256.3 W/m^2 and 851.6 W/m^2, respectively. The indoor DBT values of the experimental rooms on April 7 are shown in Figure 6a. It can be observed that the indoor and outdoor DBT variations are notable and share similar variation trends. The DBTs rose in the morning and dropped off in the afternoon. However, the magnitudes of temperature oscillations were different. The indoor DBT oscillation for the RefRoom was much higher than that of the VGRoom. The indoor DBT of the RefRoom was obviously higher than that outdoors, while the indoor DBT of the VGRoom was lower than that outdoors during daytime. However, at night the indoor DBT of the VGRoom was higher than that of the RefRoom. Figure 6b shows the indoor and outdoor DBT variations on 14 April, which was a typical cloudy day. During the day, the averaged outdoor DBT, RH, and solar radiation intensity values were, respectively, 17.9 °C, 84.7%, and 15.8 W/m^2, while the maximum solar radiation intensity was only 111.8 W/m^2. Compared with sunny days, the temperature oscillation on cloudy day was small. The indoor DBT of the VGRoom was always slightly higher than those of the RefRoom and the outdoor DBT on cloudy days. In general, the temperature difference between indoor and outdoor environments was small. The indoor DBT of the VGRoom was almost the same as that of the outdoor. It can be found from Figure 6 that the solar radiation had a great impact on indoor DBT values during the transition season, while the GS mitigated this impact significantly.

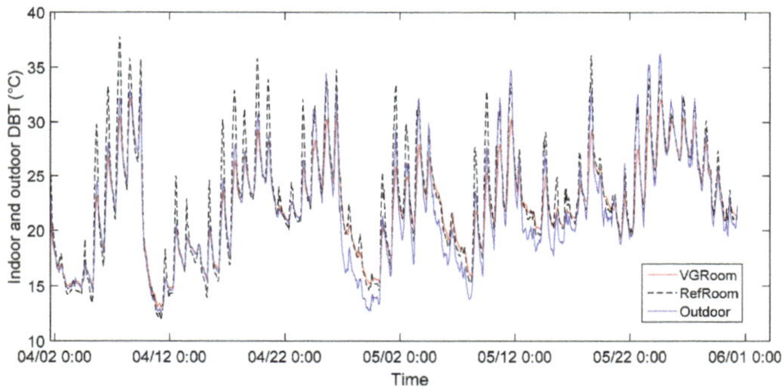

Figure 5. Indoor DBT values for the experimental rooms and the outdoor DBT.

Table 3. Statistical results for the indoor DBT and RH.

	DBT (°C)				RH (%)			
	Maximum	**Minimum**	**Average**	**MDTR**	**Maximum**	**Minimum**	**Average**	**MDRHR**
VGRoom	32.5	13.0	22.1	14.5	93.2	42.8	77.7	38.1
RefRoom	37.8	11.9	23.0	18.0	93.3	34.7	75.7	50.5

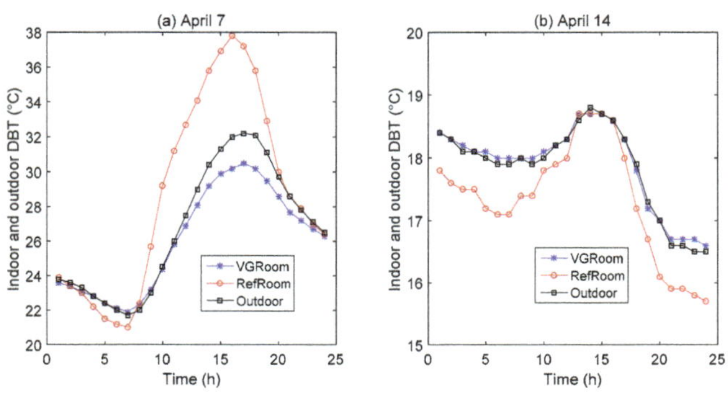

Figure 6. Indoor DBT values for the experimental rooms and the outdoor DBT. (**a**) The indoor DBT values on April 7. (**b**) The indoor and outdoor DBT variations on 14 April.

3.3. Indoor RH

Figure 7 shows the indoor RH of the two experimental rooms during the experiment. The oscillations in RH were also drastic. Table 3 presents the statistical results for the indoor

RH. The RH of the VGRoom during the experiment ranged from 42.8% to 93.2%, with an average of 77.7%, while for the RefRoom the values from 34.7% to 93.3%, with an average of 75.7%. The maximum daily RH ranges (MDRHRs), defined as the maximum difference between the daily maximum and minimum relative humidity, were 50.5% and 38.1% in the RefRoom and VGRoom, respectively. Similar to the indoor DBT, the oscillations in indoor RH for the VGRoom were also less noticeable than for the RefRoom. The smaller oscillations in RH in the VGRoom may be due to the smaller indoor DBT oscillations. During the experiment, the hourly indoor RH differences between the RefRoom and VGRoom ranged from −16.5% to 8.3%, with an average of −2.0%.

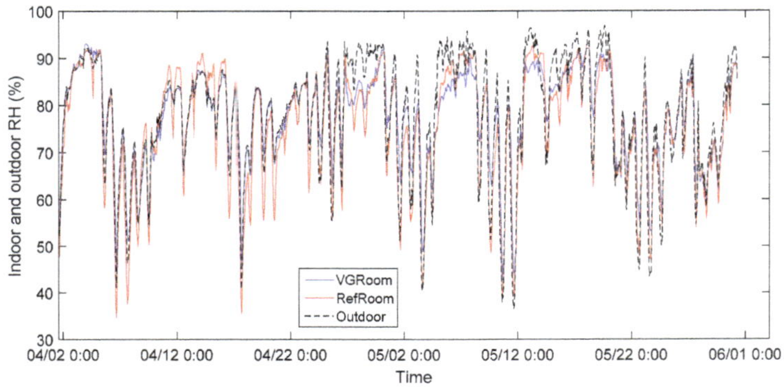

Figure 7. Indoor RH values for the experimental rooms and the outdoor RH.

3.4. Indoor BGT

The BGTs for both experimental rooms are shown in Figure 8. It can be seen that the indoor BGT values for RefRoom were significantly higher for the VGRoom during the daytime, especially on sunny days. During the experiment, the BGT values in the RefRoom ranged from 11.9 °C to 37.3 °C, while in the VGRoom the values from 12.7 °C to 32.6 °C. The BGT variation range in the VGRoom was obviously smaller than that in the RefRoom. The average BGT values in VGRoom and RefRoom were, respectively, 22.0 °C and 23.1 °C, while a 1.1 °C reduction was achieved due to the GS. Figure 9 shows the BGT values over two typical days. The variation trend for BGT values on typical days is similar to that for DBT values shown in Figure 6. On sunny days, the BGT values in the RefRoom were much higher than in the VGRoom during the daytime, while the values were slightly lower late at night. On cloudy days, the VGRoom showed slightly higher BGT values most of the time, except for a short period in the afternoon. In general, the two rooms show similar indoor BGT values on cloudy days, without notable differences.

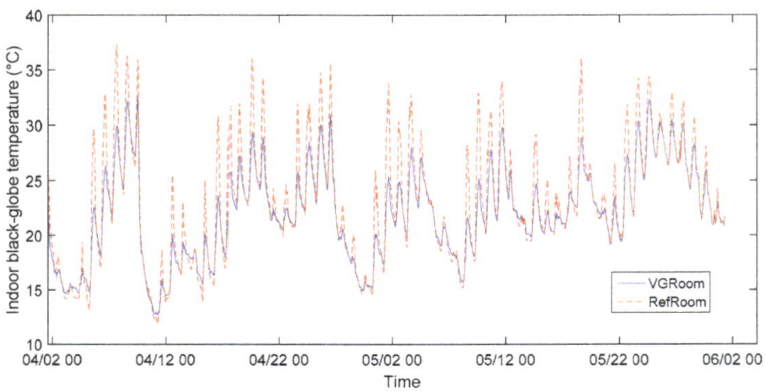

Figure 8. Indoor BGT values in the experimental rooms.

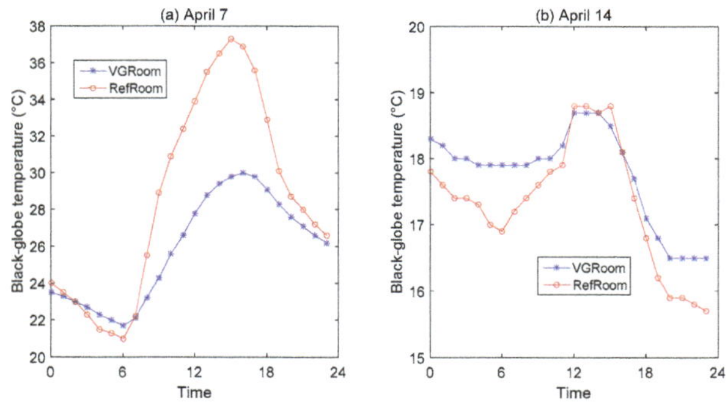

Figure 9. Indoor BGT variations over two typical days.

3.5. Internal and External Surface Temperatures of Building Walls and Roofs

The surface temperatures of the west wall are shown in Figure 10. The variation trends for the other walls and roofs are similar as for the west wall. Table 4 summarizes the variation ranges of surface temperatures for all walls and roofs during the experiment. From Figure 10, it can be seen that the external surface temperatures of the walls of the RefRoom were very high during daytime and much higher than that of the VGRoom. This verified that the shading effect of the GS can help to reduce the external wall and roof surface temperatures significantly during the daytime. However, the external surface temperatures of the RefRoom were slightly lower than for the VGRoom at night. This was due to the blockage of the GS in terms of radiative heat transfer between the external wall surfaces and the cold night sky, leading to lower heat loss at night. Benefiting from lower external surface temperatures during the daytime and higher external surface temperatures at night, the internal surface temperatures of the VGRoom were more stable than those of the RefRoom. This trend can also be seen in Table 4. During the experiment, the maximum and averaged surface temperatures for all walls and the roof of the RefRoom were higher than those of the VGRoom, while the minimum surface temperatures for all the walls and the roof were even lower. The external surface temperatures of the RefRoom and VGRoom were, respectively, in the ranges of 7.6~68.1 °C and 9.1~38.6 °C, while the internal surface temperatures varied across ranges of 10.7~43.2 °C and 11.8~34.1 °C, respectively. Both the temperature ranges of internal and external surfaces of the VGRoom were narrower than those of the RefRoom. Due to the smaller oscillations for external surfaces, the variations in indoor BGT values for the VGRoom were smaller, as shown in Figure 8. Compared with the RefRoom, the external and internal surface temperatures of the VGRoom were reduced by 1.6~4.1 °C and 0.2~1.3 °C on average, respectively. The maximum temperature reductions were 29.5 °C and 9.4 °C for external and internal surfaces, respectively. Among the walls and the roof, the maximum surface temperature of the roof and west wall were higher than the others. The maximum surface temperatures of the roof and west wall were, respectively, 68.1 °C and 63.8 °C for the RefRoom and 38.6 °C and 34.7 °C for the VGRoom.

Table 4. Summary of internal and external surface temperatures of walls and roofs (°C).

		East Wall		West Wall		South Wall		North Wall		Roof	
		Int. surf.	**Ext. surf.**	**Int. surf.**	**Ext. surf.**	**Int. surf.**	**Ext. surf.**	**Int. surf.**	**Ext. surf.**	**Int. surf.**	**Ext. surf.**
VGRoom	Maxi.	33.6	32.8	33.5	34.7	34.1	32.8	33.5	34.0	33.8	38.6
	Mini.	11.8	9.6	11.9	9.4	12.8	9.8	11.8	9.4	12.0	9.1
	Aver.	21.7	20.6	21.7	20.7	22.3	20.7	21.6	20.7	21.6	19.9
RefRoom	Maxi.	38.6	54.1	41.4	63.8	39.0	46.0	39.3	40.9	43.2	68.1
	Mini.	11.1	9.1	11.2	8.7	10.9	9.1	10.9	8.9	10.7	7.6
	Aver.	22.7	22.9	22.7	23.1	22.5	22.3	22.6	22.3	22.9	24.0

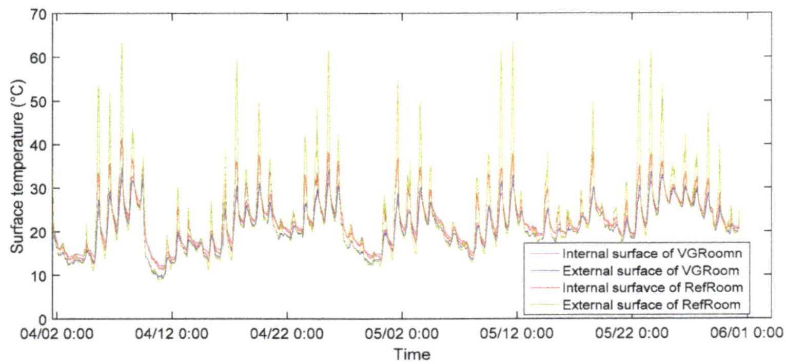

Figure 10. Internal and external surface temperatures of the west walls of the two experimental rooms.

4. Discussion

4.1. Impacts of GS on Indoor DBT and RH

Figure 11 shows the effects of the outdoor environment on the indoor environment in the two experiment rooms. It can be seen from Figure 11 that both the indoor DBT and RH vary linearly with the outdoor DBT and RH. Table 5 presents the results of the linear fitting process. The coefficients p1 and p2 are the slope and intercept of the linear fitting, respectively. The statistic R^2 for all fits has a value close 1, which indicates the high fitness. It can be seen from Table 5 that the slopes of fitting curves for VGRoom are lower than for the RefRoom. A smaller slope indicates less influence from the outdoor environment. It was verified that GS can reduce the impacts of outdoor DBT and RH on the indoor climate.

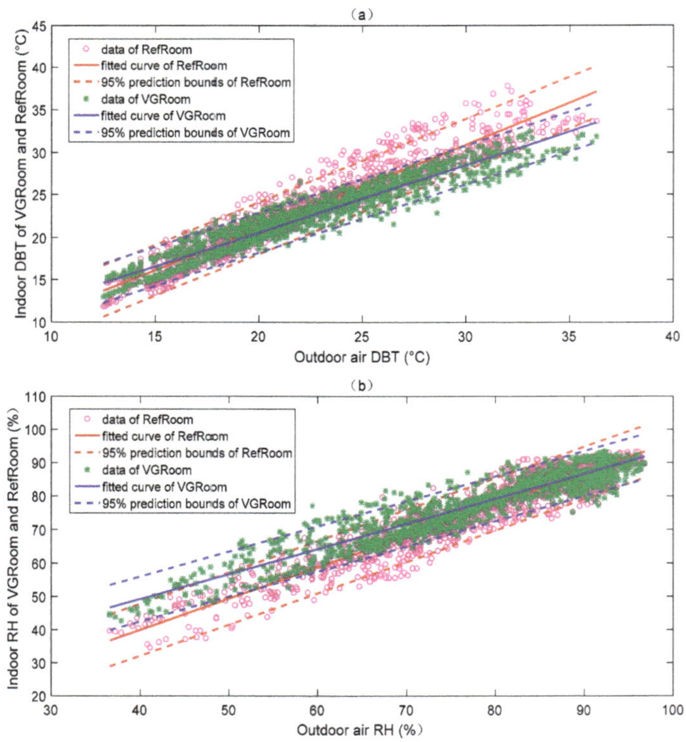

Figure 11. Relationships between (**a**) indoor and outdoor DBT and (**b**) indoor and outdoor RH values.

Table 5. Results of the linear fitting.

	DBT			RH		
	p1	p2	R^2	p1	p2	R^2
VGrRoom	0.7942	4.666	0.9215	0.7492	19.28	0.8822
RefRoom	0.9859	1.338	0.9141	0.9392	2.414	0.8970

Figure 12 shows the daily averaged indoor DBT values and the SD for the two rooms. It can be seen that the daily averaged indoor DBT for the RefRoom is higher than that of the VGRoom most of the time, except for some cloudy or rainy days. The SD of indoor DBT for the VGRoom is always lower than that of the RefRoom. Differing from the indoor DBT values, the daily averaged indoor air RH of the VGRoom is higher than that of the RefRoom, as shown in Figure 13a. However, Figure 13b shows a lower SD for the indoor air RH in the VGRoom than in the RefRoom. SD is a statistic characterizing the degree of variation in data, whereby a larger SD indicates a greater variation in data. The lower SD of the DBT and RH values in the VGRoom indicates less oscillation in indoor DBT and RH values during experiment. This also verified that GS had a good effect on reducing the oscillation of the indoor environment in the transition season. To investigate the role of GS on reducing the oscillation in indoor environments in the transition season, the experiment results of the two rooms were statistically analyzed. From the statistical results, the $ADSDT$ and $ADSDH$ for the RefRoom and VGRoom were, respectively, 2.8 °C and 7.3%, 1.7 °C and 5.2%. Therefore, $TOWR$ and $HOWR$ were, respectively, 39.3% and 28.8%. This means that the indoor DBT and RH oscillations were, respectively, decreased by 39.3% and 28.8% due to the use of GS, and a more stable indoor environment was achieved in the VGRoom in the transition season.

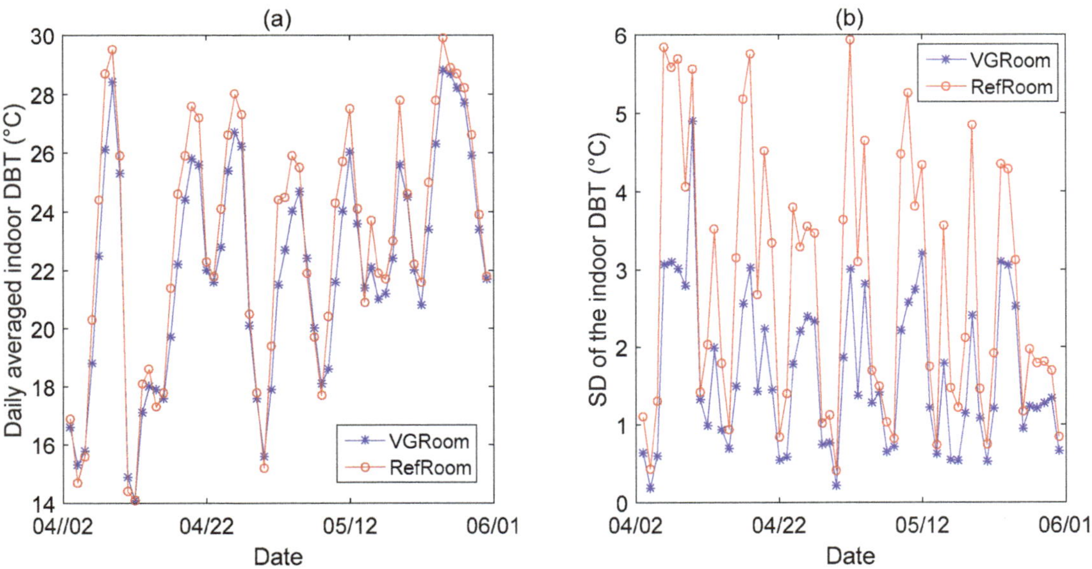

Figure 12. Daily averaged indoor DBT values (**a**) and their daily SD (**b**) during the experiment.

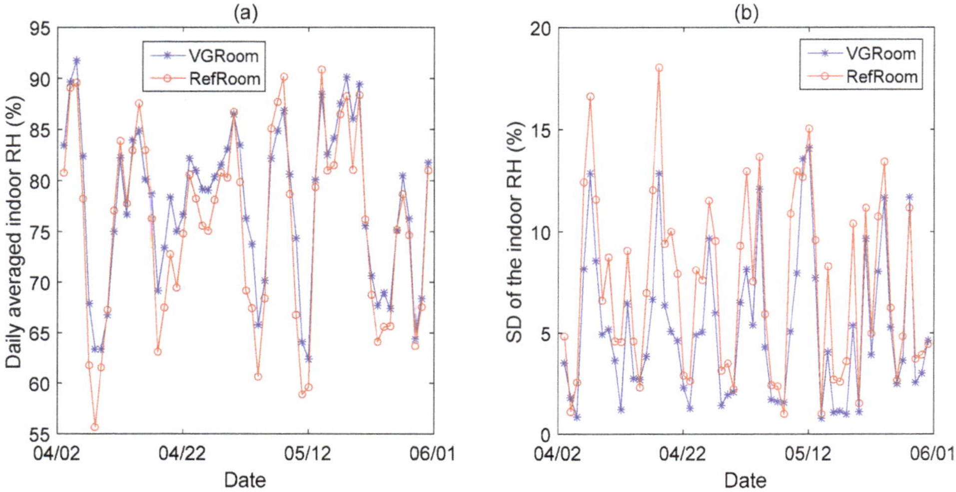

Figure 13. Daily averaged indoor air RH values (**a**) and their daily SD (**b**) during the experiment.

4.2. *Impacts of GS on Indoor OT*

The OT was calculated using Equation (3), and Figure 14 shows the indoor OT values for the two rooms. During the experiment, the OT values in RefRoom varied from 11.9 °C to 37.4 °C with an average of 23 °Cm while the values in VGRoom ranged from 12.8 °C to 32.6 °C with an average of 22 °C. It can be seen from Figure 14 that the OT values in the RefRoom were much higher than in the VGRoom during daytime on sunny days. However, the OTs during nighttime and on cloudy or rainy days were about the same. The differences in OT values between the RefRoom and VGRoom ranged from −1.8 °C to 8.2 °C with an average of 1.0 °C. This verified that the GS can significantly reduce the indoor OT fluctuations in the transition season. The difference was larger than that obtained by Yang et al. [56] and Hao et al. [51]. Yang et al. [56] found maximum and average reductions in the indoor OT of 1.9–2.7 °C and 0.6–1.1 °C, respectively, for a free-running room in summer (August). The experimental results from Hao et al. [51] for air-conditioned rooms showed an averaged reduction of 0.4 °C and a maximum reduction of 2.1 °C in indoor OT in summer. The reason could be due to the greater fluctuations in outdoor climatic conditions in the transition season. This also illustrates that the regulating effect of GS on indoor thermal environment is affected by the outdoor climate conditions and the air-conditioning system.

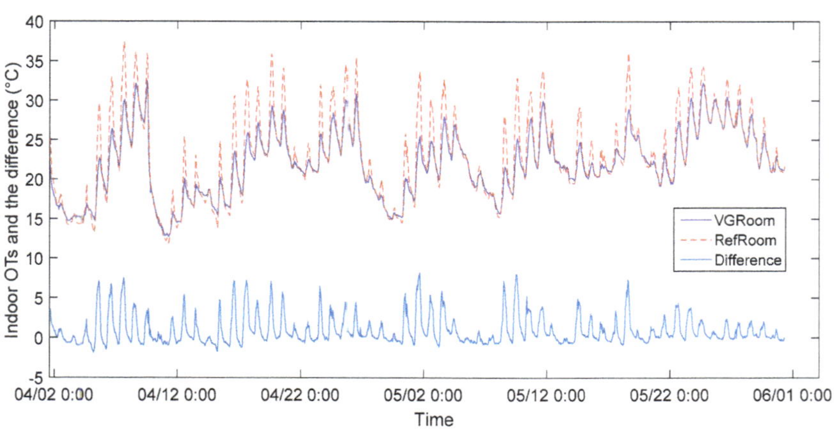

Figure 14. Indoor OTs for two experiment rooms and the differences.

4.3. Impacts of GS on Indoor Thermal Comfort

The hourly indoor climate data for the experimental rooms during the experiment are presented in Figure 15, with the comfort zone marked in the figure. The comfort zone was determined with the method provided by ASHRAE Standard 55-2017 [64]. In the calculation, the metabolic rate of a person with sedentary activity in an office, which is 1.2 met (70 W/m^2), was adopted. The thermal insulation of clothing was set at 0.75 clo. for the transition season, representing the mean of the thermal insulation levels of clothing for summer (cooling season, 0.5 clo.) and winter (heating season, 1.0 clo.), as suggested by ISO 7730 [65]. The air velocities measured were lower than 0.1 m/s in both rooms because the doors and windows were closed during the experiment and there was no rapid indoor air movement. The predicted mean vote (PMV) was set in the range of −0.7~+0.7. From Figure 15, it can be seen that there is a longer time period during which the indoor climate conditions were within the comfort zone in the VGRoom than in the RefRoom. The indoor DBT and RH oscillations of the VGRoom were also smaller than in the RefRoom.

The PMV and the predicted percentage dissatisfied (PPD) recommended by the international standard ISO 7730 [65] are often used for predicting the thermal sensation and the degree of thermal dissatisfaction of people in a thermal environment. From the measured data, the PMV and PPD were calculated for the two experimental rooms. A computer program provided by ISO 7730 was used to calculate the PMV and PPD of the two rooms during the experiment, and the results are shown in Figures 16 and 17.

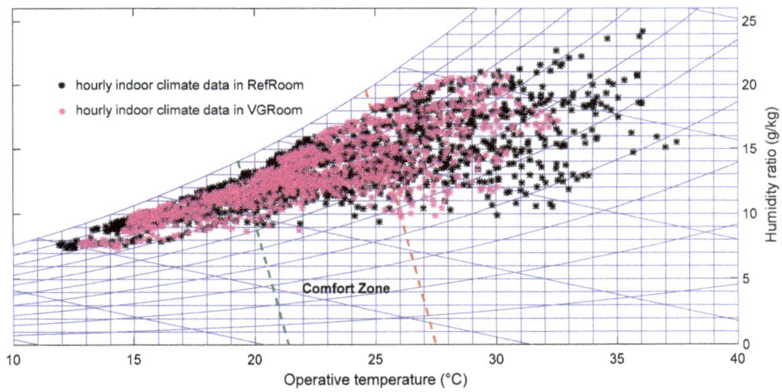

Figure 15. Hourly indoor climate data shown as a psychrometric chart of moist air.

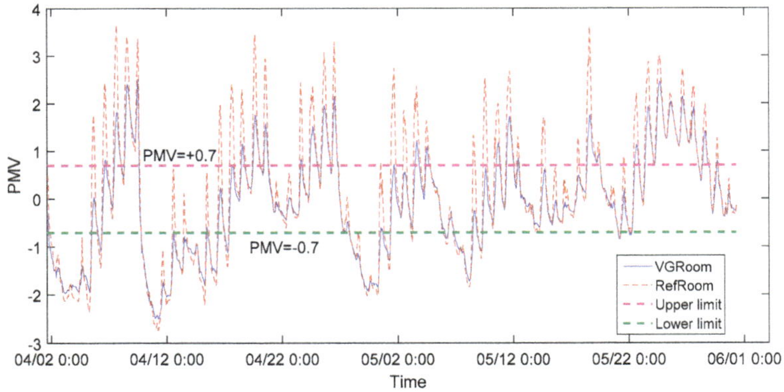

Figure 16. PMV values during the experiment.

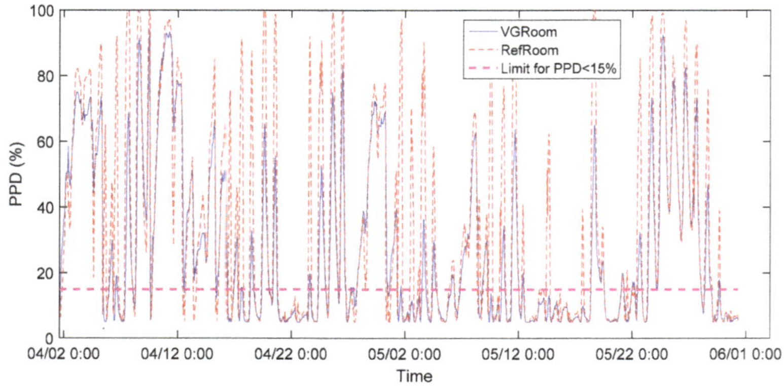

Figure 17. PPD values during the experiment.

From Figures 16 and 17, it can be seen that both the PMV and PPD values of the two rooms varied over a wide range, although smaller oscillations in PMV and PPD were observed in the VGRoom than in the RefRoom. This illustrates that the thermal comfortable condition was not met for some of the time. According to ISO 7730 [65], three categories of indoor environment criteria, A, B, and C, were defined for the indoor thermal environmental design. Each category prescribes the range of PMV and the maximum PPD that should be achieved in the indoor thermal environment. For category C, the indoor PMV should be in the range of −0.7 to +0.7 and the PPD should be no higher than 15%. The experimental result showed that there were 676 h in which the category C criterion was met during the 1432 h of experiment in the VGRoom, and the average PPD was 27.5%. However, for the RefRoom, there were only 587 h that the requirements for category C criterion were met, and the average PPD was 35.4%. This indicates that the GS can lengthen the period of indoor thermal comfort without the need for air-conditioning by 15% and can decrease the thermal dissatisfaction by 7.9% in the transition season. During the daytime (from 7:00 a.m. to 7:00 p.m.), there were 342 h for the VGRoom and 267 h for the RefRoom in which the indoor thermal comfort criterion was met. The GS improved the thermal comfort during the daytime by 28%. This shows the notable effect of the GS on improving indoor thermal comfort in the transition season, especially during the daytime. If air-conditioning systems are used to maintain the indoor thermal comfort, the GS can decrease the operation time of air-conditioning system by 15% for a whole-day-occupied building or 28% for a daytime-occupied building, meaning energy savings can be achieved.

4.4. Impacts of GS on Heat Transfer through the Walls or Roof

To investigate the direction of heat transfer through the walls, the surface temperatures of all walls were examined. Figure 18 shows the average temperature differences between the external and internal surfaces (Det_Ts) of all walls and the roof. It reveals a negative temperature gradient from external surfaces to internal surfaces for all walls and the roof of the VGRoom. This phenomenon was also found by Yang et al. [56]. Therefore, heat transfer through the walls and roof occurred from inside to outside during the whole day for the VGRoom. The cooling load through opaque walls and roofs can be estimated using the temperature difference between the external and internal surfaces. Thus, it indicates a cooling effect of the GS and significant energy saving potential on warm days. However, for the RefRoom, the average envelope temperature differences between the surfaces of east and west walls and the roof were positive, while for the south and north walls they were slightly less than zero. In addition, the temperature differences during the daytime were much higher than for the whole day. This indicates a higher cooling load during the daytime than at night. Compared to the RefRoom, discrepancies and temperature differences in the VGRoom were lower during the whole day and daytime. Among the walls and the roof, the largest temperature difference occurred for the roof, followed by the west wall. The smallest temperature difference was observed for the north wall.

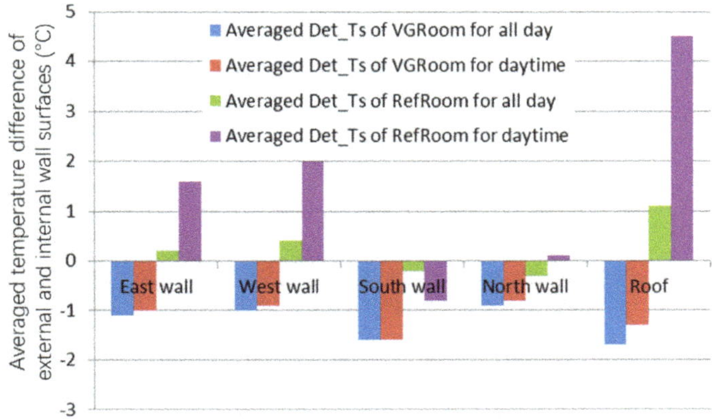

Figure 18. Average temperature differences of external and internal surfaces of the walls and roof.

5. Conclusions

In this study, the effects of GS on the indoor thermal environment and building energy performance during transitional seasons were explored via experimentation. The experiment was conducted in Xiangtan in China, which has a long transition season between the cold winter and hot summer. The experiment lasted for two months in the spring of 2019. The experimental results for two rooms, one with the GS and the other without the GS, were compared and analyzed. During the experiment, no air-conditioning system was run in either room. The findings are summarized below.

The indoor DBT and RH values of both rooms oscillated significantly in the transition season due to great fluctuations in outdoor climatic conditions when the air-conditioning system was not operated. The experimental results indicate an obvious effect of the GS in reducing the oscillation of the indoor environment in the transition season. The maximum daily DBT and RH variations were, respectively, depressed by 3.5 °C and 12.4% due to the application of the GS, although the average values during the experiment were only reduced by 0.9 °C and –2.0%. Two indices, TOWR and HOWR, were defined for evaluating the effects of GS on reducing indoor environment oscillation. The TOWR and HOWR results showed 39.3% and 28.8% oscillation reductions in indoor DBT and RH with the GS. This verified the ability of the GS to maintain a more stable indoor environment.

For various orientations, the external and internal surfaces temperatures were reduced by 1.6~4.1 °C and 0.2~1.3 °C on average by the GS, respectively. The maximum reductions in external and internal surface temperatures were 29.5 °C and 9.4 °C, respectively. For the room with the GS, the average internal surface temperatures of all walls and the roof were higher than for the corresponding external surfaces, indicating an outgoing heat flux throughout the whole day. However, the average temperature differences of external and internal surfaces for the RefRoom were either positive or negative, depending on the orientation of the wall. The temperature difference during the daytime was much higher than that throughout the whole day.

The OT for the room with the GS was lower than that without the GS during the daytime on sunny days. However, almost identical OT values were observed at night and on cloudy or rainy days. The differences in OT between the two rooms ranged from −1.8 °C to 8.2 °C, with an average of 1.0 °C. This differences were higher than those found by Yang et al. [56] for a non-air-conditioned room and by Hao et al. [51] for an air-conditioned room in summer. This reveals that the regulating effect of the GS is affected by the outdoor climate conditions and the air-conditioning system.

The experimental result showed that the GS can improve the indoor thermal comfort and decrease the thermal dissatisfaction by 7.9% in the transition season. A 15% longer time that the indoor conditions can meet the thermal comfort criterion was achieved by the VGRoom. The role of the GS in improving indoor thermal comfort was more significant in the daytime. The time that the indoor conditions satisfied the thermal comfort criterion was lengthened by 28% during the daytime. This indicates a reduction in air-conditioning operating time and that energy savings can be achieved.

In this paper, we presented the results of research on the impacts of the GS on an indoor microclimate. It will be helpful for designers and building owners to learn about the role of the GS in improving indoor environments and reducing energy consumption. However, the influence of a building's dynamic properties, such as the thermal capacity of the outer envelope and the air tightness, was not considered in this paper. For future research, an investigation on the dynamic properties of the test object and a comparison and comparative analysis of CO_2 concentrations in the tested rooms and the external environment shall be included.

Abbreviation

BGT	Black-globe temperature (°C)
DBT	Dry-bulb temperature (°C)
DSGF	Double-skin green façade
GHG	Greenhouse gas
GR	Green roof
GS	Greenery system
GW	Green wall
HOWR	Relative humidity oscillation weakening rate (%)
IPCC	Intergovernmental Panel on Climate Change
LWS	Living wall system
MDRHR	Maximum daily relative humidity ranges (%)
MDTR	Maximum daily dry-bulb temperature range (°C)
MRT	Mean radiant temperature (°C)
OT	Operative temperature (°C)
PMV	Predicted mean vote
PPD	Predicted percentage dissatisfied (%)
RH	Relative humidity (%)
RefRoom	Experimental room without greenery system
SD	Standard deviations
TOWR	Dry-bulb temperature oscillation weakening rate (%)
TR	Temperature reductions (°C)
UGS	Urban green spaces
UHI	Urban heat island
VGRoom	Experimental room equipped greenery system
VGS	Vertical greening systems

Author Contributions: X.H and Y.L. contributed to the conception of the study and the development of the methodology. Investigation, L.L., H.T. and J.H.; Writing—review and editing, X.H., W.Y. and Y.L. All authors have read and agreed to the published version of the manuscript.

Funding: This work is financially supported by scientific research fund of Hunan Provincial Education Department (Funding Numbers: 19A180 and 21B0460).

Institutional Review Board Statement: Not applicable.

Informed Consent Statement: Not applicable.

Data Availability Statement: Not applicable.

Conflicts of Interest: The authors declare no conflict of interest.

References

1. IPCC. Summary for Policymakers. In *Climate Change 2021: The Physical Science Basis. Contribution of Working Group I to the Sixth Assessment Report of the Intergovernmental Panel on Climate Change*; Cambridge University Press: Cambridge, UK, 2021.
2. Meulen, S.H.V.D. Costs and benefits of green roof types for cities and building owners. *J. Sustain. Dev. Energy Water Environ. Syst.* **2019**, *7*, 57–71. [CrossRef]
3. U.N. *Sustainable Development Goal 11: Sustainable Cities and Communities, Make Cities and Human Settlements Inclusive, Safe, Resilient and Sustainable*. Available online: https://armenia.un.org/index.php/en/sdgs/11 (accessed on 1 March 2022).
4. Morakinyo, T.E.; Lai, A.; Lau, K.K.-L.; Ng, E.Y.Y. Thermal benefits of vertical greening in a high-density city: Case study of Hong Kong. *Urban For. Urban Green.* **2019**, *37*, 42–55. [CrossRef]
5. Mathew, A.; Khandelwal, S.; Kaul, N. Spatial and temporal variations of urban heat island effect and the effect of percentage impervious surface area and elevation on land surface temperature: Study of Chandigarh city, India. *Sustain. Cities Soc.* **2016**, *26*, 264–277. [CrossRef]
6. Dutta, D.; Rahman, A.; Paul, S.; Kundu, A. Impervious surface growth and its inter-relationship with vegetation cover and land surface temperature in peri-urban areas of Delhi. *Urban Clim.* **2021**, *37*, 100799. [CrossRef]
7. Oke, T.R. City size and the urban heat island. *Atmos. Environ.* **1973**, *7*, 769–779. [CrossRef]
8. Tan, C.L.; Wong, N.H.; Jusuf, S.K. Outdoor mean radiant temperature estimation in the tropical urban environment. *Build. Environ.* **2013**, *64*, 118–129. [CrossRef]
9. Kim, S.W.; Brown, R.D. Urban heat island (UHI) variations within a city boundary: A systematic literature review. *Renew. Sustain. Energy Rev.* **2021**, *148*, 111256. [CrossRef]
10. Tian, L.; Li, Y.; Lu, J.; Wang, J. Review on Urban Heat Island in China: Methods, Its Impact on Buildings Energy Demand and Mitigation Strategies. *Sustainability* **2021**, *13*, 762. [CrossRef]
11. Manso, M.; Teot'onio, I.; Silva, C.M.; Cruz, C.O. Green roof and green wall benefits and costs: A review of the quantitative evidence. *Renew. Sustain. Energy Rev.* **2021**, *135*, 110111. [CrossRef]
12. Besir, A.B.; Cuce, E. Green roofs and facades: A comprehensive review. *Renew. Sustain. Energy Rev.* **2018**, *82*, 915–939. [CrossRef]
13. Daemei, A.B.; Shafiee, E.; Chitgar, A.A.; Asadi, S. Investigating the thermal performance of green wall: Experimental analysis, deep learning model, and simulation studies in a humid climate. *Build. Environ.* **2021**, *205*, 108201. [CrossRef]
14. Charoenkit, S.; Yiemwattana, S. Living walls and their contribution to improved thermal comfort and carbon emission reduction: A review. *Build. Environ.* **2016**, *105*, 82–94. [CrossRef]
15. Perini, K.; Ottel'e, M.; Haas, E.M.; Raiteri, R. Vertical greening systems, a process tree for green façades and living walls. *Urban Ecosyst.* **2013**, *16*, 265–277. [CrossRef]
16. Weerakkody, U.; Dover, J.W.; Mitchell, P.; Reiling, K. Particulate matter pollution capture by leaves of seventeen living wall species with special reference to rail-traffic at a metropolitan station. *Urban For. Urban Green.* **2017**, *27*, 173–186. [CrossRef]
17. Pugh, T.A.M.; MacKenzie, A.R.; Whyatt, J.D.; Hewitt, C.N. Effectiveness of Green Infrastructure for Improvement of Air Quality in Urban Street Canyons. *Environ. Sci. Technol.* **2012**, *46*, 7692–7699. [CrossRef]
18. Ottelé, M.; van Bohemen, H.D.; Fraaij, A.L. Quantifying the deposition of particulate matter on climber vegetation on living walls. *Ecol. Eng.* **2010**, *36*, 154–162. [CrossRef]
19. Xing, Q.; Hao, X.; Lin, Y.; Tan, H.; Yang, K. Experimental investigation on the thermal performance of a vertical greening system with green roof in wet and cold climates during winter. *Energy Build.* **2018**, *183*, 105–117. [CrossRef]
20. Yuan, S.; Rim, D. Cooling energy saving associated with exterior GSs for three US Department of Energy (DOE) standard reference buildings. *Build. Simul.* **2018**, *11*, 625–631. [CrossRef]
21. Kalani, K.W.D.; Dahanayake, C.; Chow, L. Studying the potential of energy saving through vertical greenery systems: Using EnergyPlus simulation program. *Energy Build.* **2017**, *138*, 47–59.
22. Foustalieraki, M.; Assimakopoulos, M.; Santamouris, M.; Pangalou, H. Energy performance of a medium scale green roof system installed on a commercial building using numerical and experimental data recorded during the cold period of the year. *Energy Build.* **2017**, *135*, 33–38. [CrossRef]
23. Coma, J.; Pérez, G.; de Gracia, A.; Burés, S.; Urrestarazu, M.; Cabeza, L.F. Vertical greenery systems for energy savings in buildings: A comparative study between green walls and green facades. *Build. Environ.* **2017**, *111*, 228–237. [CrossRef]
24. Perini, K.; Bazzocchi, F.; Croci, L.; Magliocco, A.; Cattaneo, E. The use of vertical greening systems to reduce the energy demand for air conditioning. Field monitoring in Mediterranean climate. *Energy Build.* **2017**, *143*, 35–42. [CrossRef]
25. Pérez, G.; Coma, J.; Sol, S.; Cabeza, L.F. Green facade for energy savings in buildings: The influence of leaf area index and facade orientation on the shadow effect. *Appl. Energy* **2017**, *187*, 424–437. [CrossRef]
26. Ran, J.; Tang, M. Passive cooling of the green roofs combined with night-time ventilation and walls insulation in hot and humid regions. *Sustain. Cities Soc.* **2018**, *38*, 466–475. [CrossRef]
27. Tan, H.; Hao, X.; Long, P.; Xing, Q.; Lin, Y.; Hu, J. Building envelope integrated green plants for energy saving. *Energy Explor. Exploit.* **2019**, *38*, 222–234. [CrossRef]
28. Shafiee, E.; Faizi, M.; Yazdanfar, S.-A.; Khanmohammadi, M.-A. Assessment of the effect of living wall systems on the improvement of the urban heat island phenomenon. *Build. Environ.* **2020**, *181*, 106923. [CrossRef]
29. Andric, I.; Kamal, A.; Al-Ghamdi, S.G. Efficiency of green roofs and green walls as climate change mitigation measures in extremely hot and dry climate: Case study of Qatar. *Energy Rep.* **2020**, *6*, 2476–2489. [CrossRef]

30. Sierra-P´erez, J.; Rodríguez-Soria, B.; Boschmonart-Rives, J.; Gabarrell, X. Integrated life cycle assessment and thermodynamic simulation of a public building's envelope renovation: Conventional vs. Passivhaus proposal. *Appl. Energy* **2018**, *212*, 1510–1521. [CrossRef]
31. Daemei, A.B.; Azmoodeh, M.; Zamani, Z.; Khotbehsara, E.M. Experimental and simulation studies on the thermal behavior of vertical greenery system for temperature mitigation in urban spaces. *J. Build. Eng.* **2018**, *20*, 277–284. [CrossRef]
32. Battista, G.; Evangelisti, L.; Guattari, C.; Vollaro, E.D.L.; Vollaro, R.D.L.; Asdrubali, F. Urban heat Island mitigation strategies: Experimental and numerical analysis of a university campus in Rome (Italy). *Sustainability* **2020**, *12*, 7971. [CrossRef]
33. Mutani, G.; Todeschi, V. The Effects of Green Roofs on Outdoor Thermal Comfort, Urban Heat Island Mitigation and Energy Savings. *Atmosphere* **2020**, *11*, 123. [CrossRef]
34. Guattari, C.; Evangelisti, L.; Asdrubali, F.; Vollaro, R.D.L. Experimental Evaluation and Numerical Simulation of the Thermal Performance of a Green Roof. *Appl. Sci.* **2020**, *10*, 1767. [CrossRef]
35. Todeschi, V.; Mutani, G.; Baima, L.; Nigra, M.; Robiglio, M. Smart Solutions for Sustainable Cities—The Re-Coding Experience for Harnessing the Potential of Urban Rooftops. *Appl. Sci.* **2020**, *10*, 7112. [CrossRef]
36. Wong, N.H.; Kwang Tan, A.Y.; Tan, P.Y.; Chiang, K.; Wong, N.C. Acoustics evaluation of vertical GSs for building walls. *Build. Environ.* **2010**, *45*, 411–420. [CrossRef]
37. Pérez, G.; Coma, J.; Barreneche, C., de Gracia, A.; Urrestarazu, M.; Burés, S.; Cabeza, L.F. Acoustic insulation capacity of Vertical Greenery Systems for buildings. *Appl. Acoust.* **2016**, *110*, 218–226. [CrossRef]
38. Sims, A.W.; Robinson, C.E.; Smart, C.C.; O'Carroll, D.M. Mechanisms controlling green roof peak flow rate attenuation. *J. Hydrol.* **2019**, *577*, 123972. [CrossRef]
39. Liu, W.; Wei, W.; Chen, W.; Deo, R.C.; Si, J.; Xi, H.; Li, B.; Feng, Q. The impacts of substrate and vegetation on storm water runoff quality from extensive green roofs. *J. Hydrol.* **2019**, *576*, 575–582. [CrossRef]
40. Madre, F.; Clergeau, P.; Machon, N.; Vergnes, A. Building biodiversity: Vegetated façades as habitats for spider and beetle assemblages. *Glob. Ecol. Conserv.* **2015**, *3*, 222–233. [CrossRef]
41. Chiquet, C.; Dover, J.W.; Mitchell, P. Birds and the urban environment: The value of green walls. *Urban Ecosyst.* **2012**, *16*, 453–462. [CrossRef]
42. Theodoridou, I.; Karteris, M.; Mallinis, G., Papadopoulos, A.; Hegger, M. Assessment of retrofitting measures and solar systems' potential in urban areas using Geographical Information Systems: Application to a Mediterranean city. *Renew. Sustain. Energy Rev.* **2012**, *16*, 6239–6261. [CrossRef]
43. Magliocco, A.; Perini, K. The perception of green integrated into architecture: Installation of a green facade in Genoa, Italy. *AIMS Environ. Sci.* **2015**, *2*, 899–909. [CrossRef]
44. Ichihara, K.; Cohen, J.P. New York City property values: What is the impact of green roofs on rental pricing? *Lett. Spat. Resour. Sci.* **2010**, *4*, 21–30. [CrossRef]
45. Bustami, R.A.; Belusko, M.; Ward, J.; Beecham, S. Vertical greenery systems: A systematic review of research trends. *Build. Environ.* **2018**, *146*, 226–237. [CrossRef]
46. Abdo, P.; Huynh, B.P. An experimental investigation of green wall bio-filter towards air temperature and humidity variation. *J. Build. Eng.* **2021**, *39*, 102244. [CrossRef]
47. Kenai, M.-A.; Libessart, L.; Lassue, S.; Defer, D. Impact of green walls occultation on energy balance: Development of a TRNSYS model on a brick masonry house. *J. Build. Eng.* **2021**, *44*, 102634. [CrossRef]
48. Seyam, S. The impact of greenery systems on building energy: Systematic review. *J. Build. Eng.* **2019**, *26*, 100887. [CrossRef]
49. Olivieri, F.; Neila, J. Experimental study of the thermal-energy performance of an insulated vegetal façade under summer conditions in a continental mediterranean climate. *Build. Environ.* **2014**, *77*, 61–76. [CrossRef]
50. Tseng, Y.-C.; Lee, D.-S.; Lin, C.-F.; Chang, C.-Y. A Novel Sensor Platform Matching the Improved Version of IPMVP Option C for Measuring Energy Savings. *Sensors* **2013**, *13*, 6811–6831. [CrossRef]
51. Hao, X.; Xing, Q.; Long, P.; Lin, Y.; Hu, J.; Tan, H. Influence of vertical GSs and green roofs on the indoor operative temperature of air-conditioned rooms. *J. Build. Eng.* **2020**, *31*, 101373. [CrossRef]
52. Mangone, G.; Kurvers, S.; Luscuere, P. Constructing thermal comfort: Investigating the effect of vegetation on indoor thermal comfort through a four season thermal comfort quasi-experiment. *Build. Environ.* **2014**, *81*, 410–426. [CrossRef]
53. Yeom, S.; Kim, H.; Hong, T. Psychological and physiological effects of a green wall on occupants: A cross-over study in virtual reality. *Build. Environ.* **2021**, *204*, 108134. [CrossRef]
54. Chen, Q.; Li, B.; Liu, X. An experimental evaluation of the living wall system in hot and humid climate. *Energy Build.* **2013**, *61*, 298–307. [CrossRef]
55. Haggag, M.; Hassan, A.; Elmasry, S. Experimental study on reduced heat gain through green façades in a high heat load climate. *Energy Build.* **2014**, *82*, 668–674. [CrossRef]
56. Yang, F.; Yuan, F.; Qian, F.; Zhuang, Z.; Yao, J. Summertime thermal and energy performance of a double-skin green facade: A case study in Shanghai. *Sustain. Cities Soc.* **2018**, *39*, 43–51. [CrossRef]
57. Long, T.; Zhao, N.; Li, W.; Wei, S.; Li, Y.; Lu, J.; Huang, S.; Qiao, Z. Natural ventilation performance of solar chimney with and without earth-air heat exchanger during transition seasons. *Energy* **2022**, *250*, 123818. [CrossRef]
58. Cao, Y.; Pan, S.; Liu, Y.; Yu, H.; Wang, X.; Chang, L.; Ni, M.; Liu, H. The window opening behavior of infant families: A case study during transition season in the cold region of China. *Energy Build.* **2021**, *254*, 111588. [CrossRef]

59. Yu, T.; Wang, D.; Zhao, X.; Liu, J.; Kim, M.K. Experimental and Numerical Study of an Active Solar Heating System with Soil Heat Storage for Greenhouses in Cold Climate Zones. *Buildings* **2022**, *12*, 405. [CrossRef]
60. Kottek, M.; Grieser, J.; Beck, C.; Rudolf, B.; Rubel, F. World map of the Köppen-Geiger climate classification updated. *Meteorol. Z.* **2006**, *15*, 259–263. [CrossRef]
61. *ISO 7726:1998*; Ergonomics of the Thermal Environment—Instruments for Measuring Physical Quantities. ISO: Geneva, Switzerland, 1998.
62. American Society of Heating. Refrigerating and Air-Conditioning Engineers (ASHRAE). In *ASHRAE Handbook-Fundamentals*; ASHRAE: Atlanta, GA, USA, 2009.
63. *QX/T 152:2012*; China Meteorological Administration. Standard for Division of Climatic. Meteorological Press: Beijing, China, 2012.
64. *ASHRAE Standard 55:2017*; Thermal Environmental Conditions for Human Occupancy. ASHRAE: Atlanta, GA, USA, 2017.
65. *ISO 7730:2005*; Ergonomics of the Thermal Environment—Analytical Determination and Interpretation of Thermal Comfort Using Calculation of the PMV and PPD Indices and Local Thermal Comfort Criteria. ISO: Geneva, Switzerland, 2005.

energies

MDPI

Review

An Overview of Extrusion as a Pretreatment Method of Lignocellulosic Biomass

Delon Konan [1], Ekoun Koffi [2], Adama Ndao [1], Eric Charles Peterson [1,3], Denis Rodrigue [4] and Kokou Adjallé [1,*]

[1] Centre Eau, Terre, Environnement (ETE), Institut National de la Recherche Scientifique (INRS), 490 Rue de la Couronne, Quebec City, QC G1K 9A9, Canada; behibro_ange-delon.konan@inrs.ca (D.K.); adama.ndao@inrs.ca (A.N.); eric_peterson@sifbi.a-star.edu.sg (E.C.P.)

[2] Département de Génie Mécanique et Énergétique, Institut National Polytechnique Felix Houphouët Boigny (INPHB), Yamoussoukro P.O. Box 1093, Côte d'Ivoire; ekoun.koffi@inphb.ci

[3] Singapore Institute of Food and Biotechnology Innovation (SIFBI), Agency for Science Technology and Research (A*STAR), 31 Biopolis Way, Singapore 138669, Singapore

[4] Département de Génie Chimique, Université Laval, Pavillon Adrien-Pouliot, 1065 Avenue de la Médecine, Quebec City, QC G1V 0A6, Canada; denis.rodrigue@gch.ulaval.ca

* Correspondence: kokou.adjalle@inrs.ca

Abstract: Lignocellulosic biomass is both low cost and abundant, and unlike energy crops, can escape associated ethical dilemmas such as arable land use and food security issues. However, their usage as raw material in a biorefinery implies an inherent upstream pretreatment step to access compounds of interest derived from lignocellulosic biomass. Importantly, the efficiency of this step is determinant for the downstream processes, and while many pretreatment methods have been explored, extrusion is both a very flexible and promising technology. Extrusion is well-known in both the polymer and pharmaceutical industries and has been used since the 18th century. However, as a pretreatment method for lignocellulosic biomass, extrusion is relatively new. The first use for this purpose dates back to the 1990s. Extrusion enjoys a high degree of flexibility due to the many available parameters, but an understanding of extrusion requires a knowledge of these parameters and the different relationships between them. In this paper, we present a concise overview of lignocellulosic biomass extrusion by reviewing key extrusion parameters and their associated extruder design components and operating conditions.

Keywords: biomass pretreatment; lignocellulosic biomass; extrusion; reactive extrusion

Citation: Konan, D.; Koffi, E.; Ndao, A.; Peterson, E.C.; Rodrigue, D.; Adjallé, K. An Overview of Extrusion as a Pretreatment Method of Lignocellulosic Biomass. *Energies* **2022**, *15*, 3002. https://doi.org/10.3390/en15093002

Academic Editors: Shi-Jie Cao and Wei Feng

Received: 26 March 2022
Accepted: 18 April 2022
Published: 20 April 2022

1. Introduction

Petroleum, its derivatives, and more generally fossil materials, have found deep-rooted applications in all sectors of modern life. Gasoline, kerosene, sanitizers, fertilizers, asphalt, textiles, cosmetics, pharmaceuticals, solvents, diluents, plastics, printing inks, vaseline, and rust removers are some of the products that have become an integral part of today's lifestyles [1,2]. Among all these products, those with energy applications (fuels) are of particular importance because they enter into the production process of almost everything produced on an industrial scale.

The problems associated with the use of fossil fuels are well known and their consequences on the environment are increasingly obvious. However, getting out of this dependence on fossil fuels means finding competitive alternatives. Among the renewable energies available today, lignocellulosic biomass is one of those capable of replacing fossil materials in many applications, including energy production [3–6]. Long considered useless or of little interest, lignocellulosic biomass (LCB) is one of the most abundant resources on earth. Global lignocellulosic biomass production is estimated at several billion dry tons per year. In Canada, lignocellulosic residues (forest and agricultural) are estimated between 64 and 561 million dry tons per year, and less than 30 million tons are used in the industry [7].

LCB is an important source of renewable energy. However, many difficulties hinder its use as a raw material in the industry. LCBs must be pretreated before their utilization in a biorefinery process. The goal of this step is to deconstruct the lignocellulosic structure to get access to the desired compound (i.e., glucose, xylose, etc.). It is well documented that pretreatment is the limiting step in the biorefinery context [8–10], at least for two reasons. First, LCBs are recalcitrant to pretreatment. Lignocellulose is a complex matrix, and as the main constituent of plants cells walls, lignocellulose acts in nature as a defense system against microbial, chemical, and physical attacks. This matrix is mainly comprised of cellulose, hemicellulose, and lignin linked to each other by a diversity of strong and weak bonds (ester, ether, hydrogen, Van der Waals, etc.). The second reason is a corollary of the first since several pretreatments methods have been developed to address the complex nature of LCBs, but still need improvements regarding efficiency, cost, and environmental aspects.

LCB pretreatments are classified into four classes: (a) chemical, (b) physical, (c) biological, and (d) physicochemical [11,12]. Class (a) pretreatments include acid hydrolysis, alkaline hydrolysis, ozonolysis, organosolvation, oxidative delignification, ionic liquids, deep eutectic solvents, and natural deep eutectic solvents. Class (b) includes physical treatments such as extrusion, milling, irradiation, microwave, ultrasound, pyrolysis, and pulsed electric fields. Biological pretreatments (i.e., class (c)) are named according to the type of organism involved: fungi, bacteria, and archaea. Class (d) includes physicochemical treatment methods such as steam explosion, liquid hot water, SPORL (sulfite pretreatment to overcome recalcitrance of lignocellulose), AFEX (ammonia fiber explosion), CO_2 explosion, and wet oxidation.

Extrusion, from class (b), is a promising pretreatment method. It presents many key advantages for biomass pretreatment in a biorefinery context and is a complex technology with a simple core principle. It consists of destructuring LCB under high shearing forces through contact with one or two rotative screws into a barrel, or more specifically, an extruder. This technology is particularly adaptable, can be used for diverse purposes outside biomass pretreatment, and possesses several parameters that can be modified according to the desired goal [13]. Short residence time is another advantage of extrusion, usually requiring only a few minutes [14]. Concerning operating conditions, extrusion can be run in batch, fed-batch, and continuous processing, and can be run at a mild temperature with low energy consumption and high solid loadings. This technology is also known for rapid heat transfer and effective mixing. Moreover, extrusion offers the possibility to be coupled with other pretreatments methods, and is also a scalable technology possible to achieve comparable results when transferring from a laboratory scale to pilot and industrial scales [15]. Extrusion has been used in food, polymer, and many other industries for a long time (since 1797), but as a pretreatment method for LCB, extrusion is quite recent (the 1990s), and is receiving increasing attention [16].

The great flexibility of Lignocellulosic Biomass Extrusion (LBE) is an advantage. However, at the same time, this flexibility adds a layer of complexity because of the great number of parameters available. Those parameters are important to better understand how LBE works and for scaling up purposes. Thus, the main purposes of this paper are to present an overview of relevant LBE parameters, to show the influence of extrusion setups on the efficiency of the pretreatment, to give core information about typical operational practices, and to highlight R&D needs.

2. Lignocellulosic Biomass

2.1. Biomass Composition

Since the beginning of biomass extrusion, it has been used for various purposes such as furfural recovery [17], lipid extraction (with microalgae) [18], pigment extraction [19], torrefaction/pelletization [20], biomass briquettes making [21], and composite materials formation [22]. For lignocellulosic biomass pretreatment, sugars recovery (monosaccharides, oligosaccharides, and polysaccharides) remain the preponderant goal so far [9,23,24]. The reasons are that sugars (cellulose and hemicellulose) represent 50 to 80% of LCB and also because downstream processes utilizing sugar are well mastered today as they have been studied since the beginnings of first-generation biorefinery.

Figure 1 illustrates lignocellulose. Cellulose is the principal constituent of plant cells wall and the most abundant polymer from living organisms [25]. It is a linear D-glucose polymer with β(1–4) glycosidic bonds. In LCB, cellulose occupies between 20% and 50% of all components [26]. Hemicellulose, like cellulose, is a biopolymer. It consists of about 15–35% of LCB on a dry basis [26]. While cellulose is a hexose polymer composed of only one type of monomer, hemicellulose is a heteropolymer (mixtures of pentoses and hexoses). The most abundant monomers in hemicellulose by order are xylose and arabinose for pentoses and mannose, glucose, and galactose for hexoses. Hemicellulose is also a nonlinear polymer with significant short branching sidechains that contribute to the overall cohesion of lignocellulosic structures. This biopolymer is embedded between cellulose fibers and lignin and plays the role of a binder via covalent bonds. Compared to cellulose, the molecular weight of hemicellulose is low, and its structure is easily hydrolyzed.

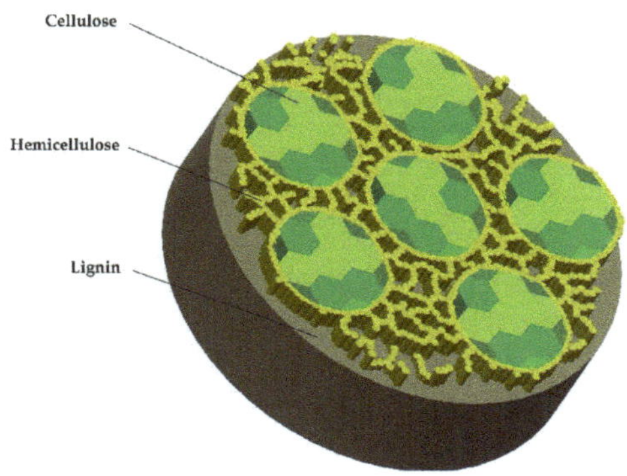

Figure 1. Lignocellulose.

After cellulose, lignin is the most abundant biopolymer on earth and is counted for about 5% to 30% in the composition of LCB [27]. Lignin is a three-dimensional aromatic biopolymer [28]. It is also a plant cell wall component like hemicellulose and cellulose in woody plant tissues. The main role of lignin is to prevent the cell from exterior threats such as microorganism attacks. It is made of three monomers (monolignols): *p*-coumaryl alcohol, coniferyl alcohol, and sinapyl alcohol, which respectively appear in the lignin polymer as Hydroxyphenyl (H lignin), Guaiacyl (G lignin), and Syringyl (S lignin). Softwood is mainly composed of G lignin units while hardwood has essentially both S and G lignin units [29]. Monolignols are linked one to another by alkyl-aryl, alkyl-alkyl, and aryl-aryl bonds. The relative abundance of one of these linkages over the others determines the physicochemical and biological properties of the lignin [30]. Lignin and cellulose are linked both by hydrogen (weak) and covalent (strong) bonds. Figure 2 shows an overview of some LCB compositions (details about data sources are presented in Appendix A). Cellulose, hemicellulose, and lignin form a complex and resistant material (i.e., lignocellulose) whose structure can vary depending on many factors (type of biomass, sources, stage of maturity, plant part, etc.). Generally, agricultural residues require less harsh pretreatment conditions than forest residues. During lignocellulosic biomass extrusion (LBE), the mechanical action of the screws on the extrudate disrupts the lignocellulose material. Covalent and hydrogen bonds are altered and weakened, while the degree of polymerization of cellulose is technically reduced and a part of the lignin layer is removed. As Table 1 shows, so far, studies do not permit to state clearly whether or not there is a significant difference between biomass composition before and after extrusion. The differences observed can be for diverse reasons: A structural change in the biomass during extrusion, the denaturation of certain compound according to the severity of the pretreatment conditions, the fact of bias related to precision and accuracy of the protocol used for biomass composition estimation, etc.

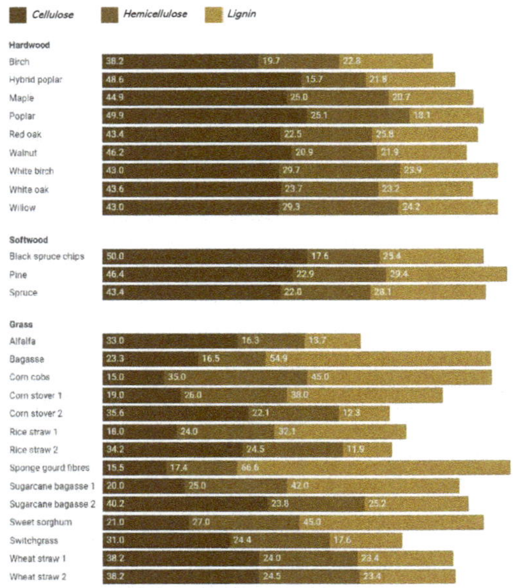

Figure 2. Composition of some lignocellulosic biomass.

Table 1. Lignocellulosic biomass composition before and after extrusion.

Biomass	Compounds	Composition (%)		References
		Before Extrusion	After Extrusion	
Bulgur bran	Glucose Hemicellulose Total lignin	36.38 ± 0.32 29.42 ± 0.13 12.54 ± 0.14	30.86 ± 0.64 33.18 ± 0.53 16.24 ± 0.31	[31]
Eucalyptus	Cellulose Hemicellulose Lignin Ash	46.90 ± 1.21 12.87 ± 0.35 31.15 ± 0.40 0.86 ± 0.00	44.90 ± 1.86 13.71 ± 0.32 32.97 ± 0.86 0.57 ± 0.05	[32]
Olive stone	Cellulose Hemicellulose Lignin	20.8 ± 0.2 25.9 ± 0.1 35.5 ± 0.6	18.3 ± 2.8 22.4 ± 0.4 39.0 ± 0.2	[33]
Barley straw	Glucose Hemicellulose Lignin Ash	32.9 26.1 18.8 3.9	32.80 15.53 15.71 2.17	[34]
Corn stover	Cellulose Hemicellulose Lignin	32.75 ± 0.32 31.08 ± 0.57 10.07 ± 0.91	33.98 ± 0.14 30.20 ± 0.28 9.89 ± 0.43	[35]
Oat hull	Cellulose Hemicellulose Lignin	31.16 ± 1.15 28.72 ± 0.25 18.12 ± 0.63	34.32 ± 2.06 26.40 ± 0.53 15.00 ± 1.30	[36]
Wheat straw	Cellulose Hemicellulose Lignin Ash	37.8 ± 1.9 28.2 ± 0.5 19.8 ± 0.3 3.7 ± 0.0	46.9 ± 0.1 28.7 ± 0.1 15.4 ± 0.1 3.3 ± 0.0	[37]
Corn cob	Cellulose Hemicellulose Neutral detergent soluble	42.0 ± 0.15 45.9 ± 0.90 9.3 ± 0.95	34.8 ± 0.23 38.9 ± 0.52 19.0 ± 0.60	[13]

2.2. Crystallinity

Usually, in LBE, crystallinity refers to cellulose. Natural cellulose polymers contain both crystalline (D-glucose monomers ordered) and amorphous (D-glucose monomers disordered) sequences [38,39], and the crystallinity index is the overall percentage of the crystalline fraction. That is, it is the relative quantity of crystalline sequence in cellulose. Crystallinity is determined by X-ray diffraction and the following formula:

$$CI = \frac{I_{200} - I_{am}}{I_{200}}, \quad (1)$$

where I_{200} represent the height of the (200) peak and I_{am} is the minimum intensity between the (200) and the (110) peaks [40].

Hemicellulose and lignin are considered non-crystalline polymers (amorphous polymers) and are both heteropolymers. However, hemicellulose can also be highly crystalline because of multiple ramifications of homopolymers with a crystalline structure (xylans, mannans, arabinans, and galactans) attached to the principal heteropolymer chain [41]. The crystallinity of cellulose is particularly important in LBE when enzymes are involved before, during, or after the processing for enzymatic hydrolysis. This is because the amorphous part of cellulose is more susceptible to saccharification compared to the crystalline part, and can be degraded between five and thirty times more quickly [42,43].

Kuster et al. [44] recorded a decrease from 57% to 54% of the crystallinity index after extrusion of sugarcane bagasse. It could be obvious that the crystallinity index might decrease after extrusion, but many studies showed that it is not always so. For example, Vandenbossche et al. [45] extruded four types of biomass (i.e., barley straw, sweet corn, blue agave, and oil palm empty fruit bunch) and found that the crystallinity of all extrudates was higher than for the raw materials (Table 2). Marone et al. [46] came to the same conclusion with corn stover. Fu et al. [47] observed the crystallinity of Douglas fir residuals and found that after extrusion, the crystallinity index slightly increased, which was attributed to effects from both heat and moisture content. Recrystallization can occur in cellulose because, under high temperatures, hydrogen atoms in the amorphous region undergo a realignment [48].

As an indicator of enzymatic hydrolysis yield, cellulose crystallinity is also controversial. Some authors reported a strong correlation between crystallinity and glucose and xylose/mannose yield, while others showed that crystallinity index is not accurate to predict sugar yield [48,49].

Table 2. Crystallinity index.

Biomass	Substrate Crystallinity (%)	Extrudate Crystallinity (%)	References
Banana fibers	39	-	[50]
Sugarcane bagasse	48	-	[50]
Sponge gourd fibers	50	-	[50]
Sweet corn	41 ± 3	47 ± 6	[45]
Barley straw	44 ± 8	46 ± 2	[45]
Blue agave bagasse	27 ± 7	52 ± 1	[45]
OPEFB	50 ± 8	51 ± 7	[45]
Corn stover	48 ± 4	51.2 ± 3.4	[46]
Sugarcane bagasse	57.3 ± 1.3	54.0 ± 0.23	[44]

2.3. Particle Size

Usually, biomass will undergo a size reduction before its application to the extrusion process. A grinder is used in that case, and this step involves energy consumption and must be included in the energy balance of LBE process. At pilot and industrial scales, biomass size reduction can seriously affect the economic profitability of the LBE. However, particle size plays an important role in lignin removal, reaction kinetics, hydrolysis rate, rheological properties of the substrate inside the barrel, and sugar yield. A strong correlation between the particle size and the extrusion Specific Mechanical Energy (SME) (−0.786), the torque (−0.788), the glucose recovery yield (−0.813), and the xylose/mannose recovery yield (−0.787) has been observed during extrusion of Douglas-fir forest residuals, all with p-value inferior to 0.01 [48]. This means that when the particle size decreased, the SME, torque, glucose recovery yield, and xylose-mannose recovery yield increased. Additionally, many authors reported a size reduction of the extrudate relatively to the substrate after the extrusion process [51–53]. The reduction of the extrudate particle size increases their specific surface, which has clear advantages with respect to improvements in enzymatic saccharification.

2.4. Morphology

After extrusion, a visual inspection of the extrudate allows a first appreciation of the impact of the extrusion pretreatment on the biomass. Particle size is reduced, the extrudate looks rough, crumbly, and has a broken surface to the touch [46,54]. LCBs' microstructure can be observed by Scanning Electron Microscopy (SEM). Usually, extrudates show a disruptive surface with a lot of clear exfoliations compared to substrates which are compact (bundled) and have smooth surfaces. Important fibrillation in the extrudate has also been reported in the literature. Extrudate microfibrils are twisted, untied, and untangled [55–58]. The disruptive and fibrillation effect of extrusion can be remarkably enhanced when chemicals are used during the extrusion process (reactive extrusion). For example, Han et al. [59] observed a significant disruption and fibrillation in the microstructure of the extrudate (wood powder of pussy willow) when [EMIM]Ac (1-ethyl-3-methylimidazolium acetate) and DMSO (dimethyl sulfoxide) were used as additive during the extrusion process. The chemicals reacted with the water molecules inside the substrate and then created voids (porosity) in the biomass. Byun et al. [58] experienced a similar microstructure with Amur silvergrass. Porosity is also created by water evaporation under mild and high-temperatures extrusion (above 100 °C), but this effect is significantly enhanced with hydrophilic chemical additives. The increase in porosity results in an increase in the specific surface of the extrudate, which is highly beneficial for enzymatic hydrolysis [53]. Karunanithy and Muthukumarappan [60] demonstrated that the efficiency of enzymatic hydrolysis strongly depends on the accessibility of sugars to the enzymes. The greater the accessible surface, the higher the rate of enzymatic hydrolysis. Size reduction also participates in the increase of the specific surface, as highlighted in the preview section. Cellulose microfibrils in Han's extrudate were less than 500 nm in diameter [59].

2.5. Moisture

Substrate moisture is a key parameter for LBE. Most of the time, LCBs after harvest are not immediately pretreated, but rather undergo preparation before extrusion. The storage conditions (i.e., temperature and time) determine the biomass moisture content. Ambient temperature is preferred for storage to reduce energy consumption, and storage can last from a few hours to many months. The biomass is stored until the desired moisture for extrusion is reached. This moisture ranges from 6% to 50% according to the type of biomass (see Table 3). It is important to note that materials above 50% are not sufficiently consistent to be extruded and behave more like a liquid than a solid. Moisture content around 25% seems to be an optimum for high (above 70%) sugar recovery from barley and wheat straw [37,61], but more investigations are required.

2.6. Biomass Preparation before Pretreatment

Biomass preparation is a necessary step for successful extrusion. We investigated twenty-seven LBE studies in order to identify common practices during biomass preparation before extrusion (Table 3). Biomass preparation steps consist of [a] sorting/washing, [b] drying, [c] grinding/milling, [d] sieving, [e] mixing with additives, [f] storing, and

[g] extrusion. The first step (sorting) is an inspection of the sample collected, to remove contaminants (plastic, sand, etc.). Sometimes washing is necessary to remove the contaminants [53]. The drying step has at least two goals. The first one is to restrict microbial activity in the biomass, especially if the biomass is very wet, while the second is to lower the moisture content of the raw material [44]. Grinding/milling steps are for the size reduction of the substrate, and sieving ensures a desired particles size for the substrate [13]. There are two ways to use additives in LBE: after and during the extrusion process. Some prefer the former and run step [e] [59]. Then, the biomass is directly extruded or stored until extrusion [32].

Authors freely adapt these steps to their material and their goal. Table 3 shows that some omit certain steps or change the order. For example, Liu et al. [62] used only [d] (milling) for corn stover preparation; Kuster et al. [44] used steps [b] (drying), [c] (milling), [d] (sieving), and [e] (mixing with additive) to prepare sugarcane bagasse and sugarcane straw for extrusion; while for eucalyptus tree, Duque et al. [32] opted for a [c]-[b]/[f]-[d] sequence. However, generally speaking, all the above steps mentioned remain important for best practices for biomass preparation before extrusion.

Table 3. Practices of biomass preparation before extrusion.

Substrate	Source	Steps	Size (mm)	Storage Time	Tempera	Moisture (%)	Additives before Extrusion	Reference
Barleystraw	Research centre	[b]-[c]-[d]-[g]	5	-	-	-	No	[45]
Barley straw	Research centre	[c]-[d]-[f]-[g]	5	Stored until use	-	6	No	[61]
Big bluestem	Farm	[c]-[d]-[f]-[g]	0.4–0.8	Stored until use	RT	-		[63]
Big bluestem	Farm	[c]-[d]-[e]/[f]-[g]	2, 4, 6, 8, 10	~8 h	RT	10, 20, 30, 40, 50	Water	[64]
Blue agave	Manufacture	[b]-[c]-[d]-[g]	2	-	-	-	No	[45]
Corn cob	Farm	[b]-[c]-[d]-[g]	2	-	RT	-	No	[13]
Corn stover	Farm	[c]-[b]-[e]-[g]	2	-	RT	22.5, 25, 27.5	No	[49]
Corn stover	Farm	[c]-[b]-[e]-[f]-[g]	2	8 h	RT	50	NaOH	[15]
Corn stover	Farm	[c]-[g]	2–5	0 h	-	-	No	[62]
Corn stover	Farm	[c]-[d]-[e]/[f]-[g]	2, 4, 6, 8, 10	~8 h	RT	10, 20, 30, 40, 50	Water	[64]
Eucalyptus trees	Research centre	[c]-[b]/[f]-[d]-[g]	60–190	2 months	-	20	No	[32]
Hardwood biomass (oak, fir, and pine sawdust)	-	[e]-[g]	1	-	RT	21–28	NaOH	[55]
Miscanthus	Farm	[c]-[d]-[g]	3	-	7	-	No	[65]
Olive tree pruning	Farm	[b]-[c]-[e]-[g]	1–4	-	RT	10	No	[66]
OPEFB	Manufacture	[b]-[c]-[d]-[f]	2	-	-	-	No	[45]
Prairie cordgrass	Farm	[c]-[d]-[e]/[f]-[g]	2, 4, 6, 8, 10	~8 h	RT	10, 20, 30, 40, 50	Water	[64]
Rape straw	Research centre	[c]-[d]-[g]	1.4–2.36	24 h	45 ± 5	6.44	No	[67]
Rice hull	Manufacture	[a]-[b]-[c]-[d]-[g]	25.4	24 h	60	-	No	[53]
Soybean hulls	Manufacture	[b]-[d]-[g]	1.041	24 h	RT	40, 45, 50	No	[68]
Sugarcane bagasse	Mill	[b]-[c]-[d]-[e]-[f]-[g]	0.2–2	24 h	Cold room	10.4 ± 0.36 8.9 ± 0.30	Water, Glycol, Ethylene glycol, Tween 80	[44]
Sugarcane bagasse	Manufacture	[c]-[d]-[e]-[g]	0.425–1.000	-	40	10	[EMIM]Ac	[69]
Sugarcane straw	Mill	[b]-[c]-[d]-[e]-[f]-[g]	0.2–2	24 h	Cold room	12.05 ± 0.36 10.34 ± 0.26	Water, Glycol, Ethylene glycol, Tween 80	[44]

Table 3. *Cont.*

Substrate	Source	Steps	Size (mm)	Storage Time	Tempera	Moisture (%)	Additives before Extrusion	Reference
Sweet corn	Manufacture	[b]-[c]-[d]-[g]	6	-	-	-	No	[45]
Switchgrass	Farm	[c]-[d]-[e]/[f]-[g]	2, 4, 6, 8, 10	~8 h	RT	10, 20, 30, 40, 50	water	[64]
Switchgrass (matured)	Farm	[c]-[d]-[f]-[g]	0.3–1.2	Stored until use	RT	-		[63]
Wheat straw	-	[c]-[f]-[g]	5	Stored until use	40	6	No	[37]
Wood powder of pussy willow	-	[d]-[b]-[f]-[e]-[g]	25.4	24 h	40	-	[EMIM]Ac, DMSO, [EMIM]Ac/DMSO	[59]

3. Extruder

An extruder is a thermomechanical device composed of different parts, with the most important being the barrel (inside which are one or more screws) and the die. These two parts are generally temperature controlled by a system of heating and cooling. Most often, extruders are equipped with one or more liquid injection points (Figure 3). The first patent of an extruder was filled by Joseph Bramah in 1797. Today, several types of extruders are available according to the number of screws. However, single-screw extruders and twin-screw extruders are both widely used for LBE, although twin-screw designs are more common. These screws rotate around their axis thanks to a drive motor and exert a significant mechanical force on the biomass, which is caught between the screws and between the screws and the wall.

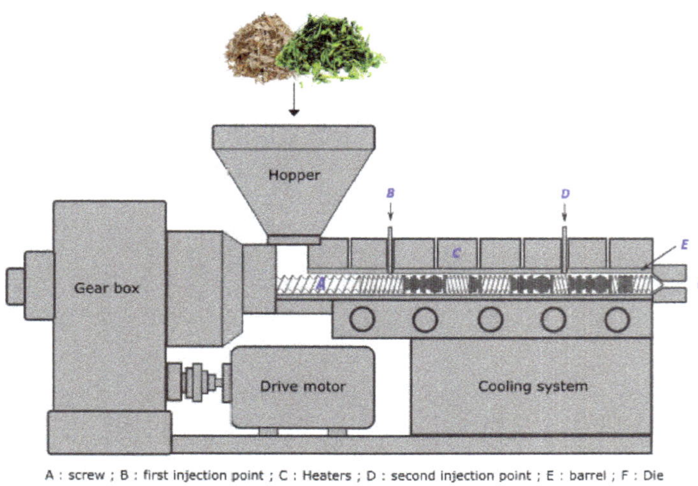

Figure 3. Extruder.

3.1. Screw Type

Screws have two principal functions: convey and disrupt. The lignocellulosic substrate is conveyed from the feeding zone (zone under the hopper) to the die. During transport, the substrate undergoes high shearing forces as a consequence of protrusions of the screws, which results in the disorganization of the lignocellulose complex, with a part of the lignin layer removed while the cellulose crystallinity is technically assumed to decrease.

An extruder screw is made of a non-corrosive and resistant (high shearing forces) metal [67], and consists of a shaft surmounted by different shapes of protrusion, with two typical screw types: the one-piece screw and the modulated screw. The one-piece screw is a full bar on which protrusions are made directly on the shaft (Figure 4). In the case of a modulated screw, this consists of a bar ridged lengthwise on which modules (screw elements) are mounted (Figure 5). Contrary to one-piece screws, modulated screws offer more flexibility because the configuration of the screw can be changed by using different modules [55]. In the case of LBE, modulated screws are better suited, as most of the time the screw configuration must be changed according to the type of biomass.

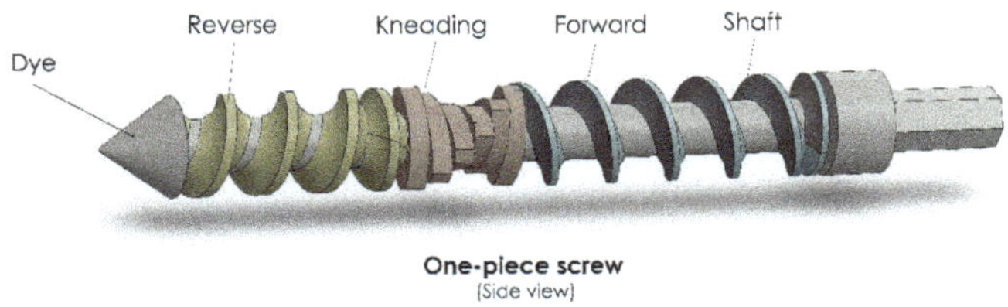

Figure 4. One-piece screw.

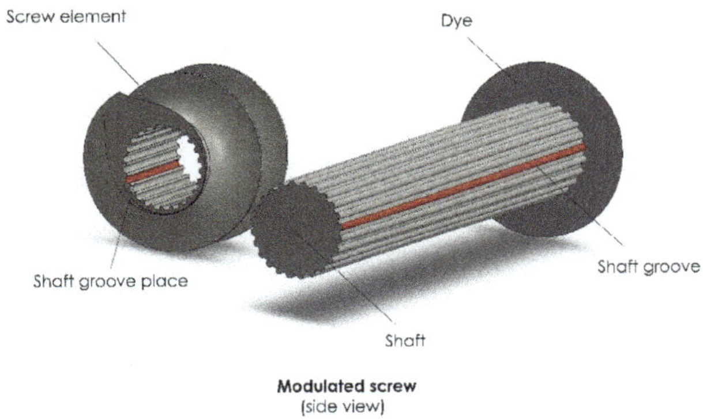

Figure 5. Modulated screw.

Similarly, twin-screws have more than one configuration and can be co-rotative (turn in the same direction) or counter-rotative (turn in opposite directions). Furthermore, counter-rotative screws can be intermeshing or non-intermeshing (Figure 6). Conversely, co-rotative screws are always intermeshing and provide better mixing than counter-rotative configurations [70,71].

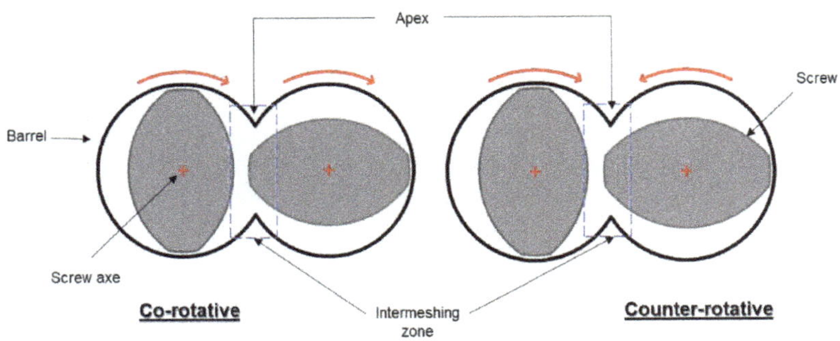

Figure 6. Co- and counter-rotative extruder screws.

Screws can also take different longitudinal geometries. Cylindrical, conical, and mixed-shape screw configurations are the best known. For a typical cylindrical screw design, the diameter is the same along the screw from the beginning to the end of the screw, and this is the most common design for LBE. For conical screw designs, the diameter constantly decreases (or increases) from one end to the other, and this kind of design is used for biomass briquetting [72]. Mixed-shape screw designs feature two different screw diameters linked by a conical compression zone, which facilitates a transition from one diameter to the next. Such compression zone designs can be both considered soft (long) or strong (short). Conical and mixed-shape screws have been explored in studies for materials such as thermoplastic polyurethane, polyvinyl chloride (PVC), and plastic composites [73–75].

The choice of the screw type is made when the extruder is purchased, and generally, the operators do not have much freedom for modification afterwards.

3.2. Screw Configuration

Lignocellulosic Biomass Extrusion (LBE) efficiency is strongly dependent on the screw configuration [76]. Screw configuration is the final layout obtained from the arrangement of screw elements on the shaft. Contrary to screw type, authors have full control of overall screw configuration.

LBE screw configuration is formed by transport elements (forward and reverse) and mixing elements (kneading). Gatt and Vandenbossche [14] proposed the following screw configuration for LBE: F-T-M-R-M-R-M-T; where F = forward transport element (with more spaced spirals than T), T = transport element (forward), M = mixing element (kneading), and R = reverse transport element. Although screw configuration differs from one user to another, the T-M pattern is almost always present at the beginning of a screw configuration [14,77–79].

Wahid et al. [80] investigated the effect of screw configuration on the pretreatment of wheat straw and deep litter in order to produce biogas. They tested many screw configurations by using a starting screw configuration only composed of forward screw elements, and they changed some of these forward screw elements by kneading or reverse screw elements to get a new configuration. Five screw configurations were then obtained. These are (a) mild kneading (medium length kneading block replacing some forward screw elements); (b) long kneading (a long block of kneading screw elements replacing some forward screw elements); (c) reverse (a block of reverse screw elements replacing some forward screw elements); (d) kneading and reverse (a block of kneading screw element and a block of reverse screw elements replacing some forward screw element on the same shaft. However, these two blocks are separated from each other by some forward screw elements; and (e) kneading with reverse (the same configuration as the previous but here the two blocks are contiguous). Configuration (a) was found suitable for deep litter (soft texture) and configuration (d) for wheat straw because they gave the best compromise between energy consumption, sugar availability, and methane yield. As for configuration (b), it was found unproductive because of important energy consumption for both LCBs. The authors have also demonstrated that the energy consumption increases as reverse and/or kneading elements are added to the screw configuration and at the same time, these elements enhance the disruptive effect of the screw on the biomass (like with the (d) configuration). In the same perspective, Kuster et al. [44] pre-treated sugarcane biomass and observed that the glucose recovery yield was improved when reverse elements are placed just after the last kneading zone. With a similar screw configuration, Negro et al. [66] reported an increase in the overall sugar yield with olive-tree prunings.

Thus, a screw configuration starting with T-M followed by a reverse element after one or two kneading elements, including the last kneading element, should be optimal to improve the sugar recovery yield. However, more investigations are required to confirm this assertion.

3.3. Screw Elements

Each screw element type has a geometry that defines its function, and this geometric variation will systematically affect extrusion performance. Furthermore, lignocellulose composition differs from one type of biomass to another (wood, agricultural residues, etc.) as well as variability within a specific biomass type according to different factors (age, maturity stage, etc.), and this also affects screw element selection. Thus, an ideal geometry exists according to each specific biomass to be pre-treated. However, from a practical perspective, this can prove to be difficult because of downtime associated with reconfiguration, which can limit productivity.

3.3.1. Forward Screw Element

The Forward Screw Element (FSE) is an elliptical screw element designed to convey the substrate forward while turning around its axis on a rotor force. It appears at the beginning of the screw, under the feeding zone of the extruder. FSE are selected for extrusion processing according to their depth, length of the pitch, and flight angle. Figure 7 shows a side view of FSE. The geometry and orientation of the design is important for performance. For instance, as the tip angle increases, the speed at which the substrate is conveyed also increases. Similarly, increasing pitch, in turn, translates to a larger available volume in the FSE. Finally, increasing the screw tip width increases the clearance surface (between tip and barrel) and reduces the available volume in the FSE.

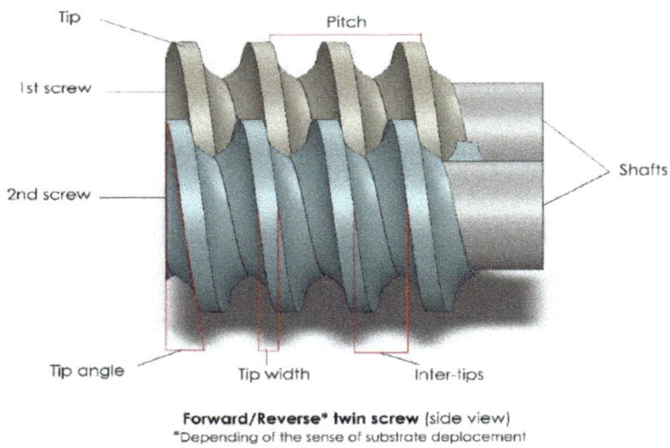

Figure 7. Intermeshing forward/reverse twin-screw elements.

FSE has an impact on the resulting extrudate properties, and this was demonstrated by Djuric and Kleinebudde [81] by wet granulation of lactose monohydrate with a twin-screw extruder. Wet granulation is one of the ways to make solid oral forms (tablets, capsules) in the pharmaceutical industry [82], where fine powder particles are agglomerated together to form larger compounds. Djuric and Kleinebudde [81] tested the porosity and the friability of the extrudate obtained after using different FSE pitches and found that the friability of the extrudate increased with the pitch length. With respect to LBE, depending on the rheological behavior of the substrate, excessively small pitches can lead to extra flow resistance in the barrel, while increasing the FSE pitch may lead to substrate friability; the substrate has insufficient viscosity to ensure a suitable fluidity inside the barrel. Usually, FSE with greater pitch are set directly under the feeding zone, while those with lower pitch are placed downstream from the feeding zone. Kohlgrüber et al. [83] considered a pitch range 1.5–2 times that of the screw diameter as the most suitable for FSE under the feeding zone.

3.3.2. Reverse Screw Element

A Reverse Screw Element (RSE) has the same design as a FSE, but with opposite flights (Figure 7). Set together on the same shaft as an FSE, an RSE is an obstacle to the forward displacement of the substrate, and thus an RSE represents a high zone of resistance. The goal of RSE in LBE is to increase pressure on the substrate and also to reach a steady state, especially with small pitches [84]. Similar to FSE, a side view section of an RSE shows the same behavior with respect to tip angle and tip width, with an additional particularity: as the pitch decreases, the resistance generated by RSE highly increases, which controls the back pressure and increases the specific mechanical energy.

RSE has an impact on the LBE efficiency, as Gu et al. [48] have shown through their investigation of glucose and xylose/mannose yield obtained during an LBE of Douglas-fir residues. Using a twin-screw divided into six zones, they found a significant increase in the glucose and xylose/mannose yield next to the RSE due to high shearing forces. Kuster et al. [44] reported similar results on sugarcane biomass, with the insertion of a RSE increasing the yield of lignocellulose hydrolysis. Zheng et al. [78] investigated height screw configuration to find the best one for xylose separation from steam-exploded corncobs and found that xylose recovery was higher using configurations containing one or more RSE. They also found that while xylose yield varied with configurations containing RSE, these outcomes were always superior to a configuration without RSE. However, regardless of how a RSE improves LBE pretreatment, attention must be paid to the specific mechanical energy.

3.3.3. Kneading Element

Kneading elements (KE) play a disruptive and distributive effect on the substrate, and can also act as mild flow-restricting elements [14,85]. A screw configuration for LCB pretreatment will typically contain at least one kneading block comprised of two or more juxtaposed KE. During LBE a kneading block is ideally set immediately downstream of the first FSE [14,77,78]. Furthermore, Kuster et al. [44] demonstrated that the best place for a RSE is just after a kneading block because of the additional back pressure and resistance provided by the RSE.

Kneading blocks geometry depends on the angles between KE, KE staggering, tip thickness, and clearance. Figure 8 presents both facing and lateral views of a kneading block. As KE tip thickness increases, the kneading surface also increases, while reducing the available volume in the kneading block. Furthermore, creating an offset angle between the KE will improve the distributive function of the kneading block. The optimum offset angle (α) as a function of the number of KE (nKE) is given by:

$$\alpha = \frac{180}{n\text{KE}} \tag{2}$$

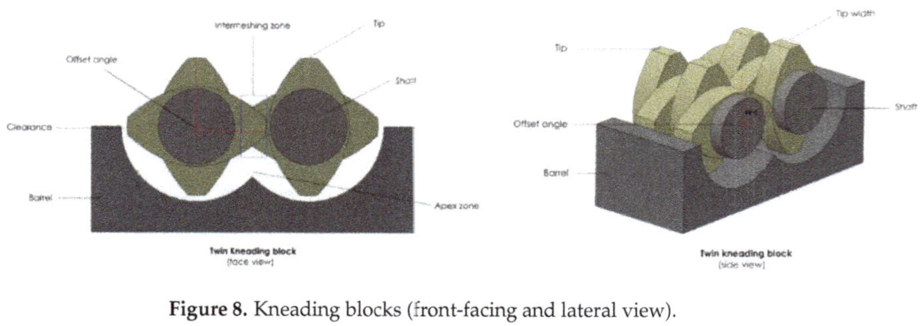

Figure 8. Kneading blocks (front-facing and lateral view).

All the KE presented in Figure 8 are vertical and then have no conveying function, only disruptive and distributive function. However, sometimes a KE is staggered either because a conveying function or an increase or decrease of shearing forces on the substrate in the kneading zone is needed. Shearing forces increase when the staggering of the kneading block is opposite to the displacement direction of the substrate and decrease when the staggering is in the same direction [55,86]. The choice of the staggering angle of the kneading block should be related to the extrusion purpose [44].

Usually, kneading blocks are composed of 4–8 KE. The length of a kneading block influences the LBE. During their experiment on wheat straw and deep litter, Wahid et al. [80] found that a longer kneading block improved the biodegradability of the extruded biomass better than a shorter one. As for Fu et al. [47], they recorded a 7% increase in the glucose yield when the number of KE was increased. Therefore, it may seem interesting to lengthen the kneading block to maximize the biodegradability of the extrudate. However, the authors pointed out that long kneading blocks increase the temperature, the residence time, and the specific mechanical energy, and this must be taken into consideration before lengthening the kneading block, especially for bioextrusion, as enzyme degradation can result in excessive kneading effects or a rise in temperature. A good alternative is to use both KE and RSE instead of a long kneading block, which is more favorable. The first solution is better than the second in terms of temperature control (due to reduced shearing forces), specific mechanical energy saving, and screw length shortening [80].

3.4. Die Shape

The die is the end of the screw through which the substrate exits the extruder, and its diameter is consistently lower than the inner barrel diameter. Similar to the screws and the barrel, the die is generally heated. It is an important part of the extruder because it influences the back pressure inside the barrel and in turn the overall efficiency of the LBE process [48]. The die entry is a high-pressure zone, as the substrate inside the barrel is conveyed by the screws and forced to pass through the die which has a smaller diameter [75,87].

Different shapes of dies are available, but a typical common design for a LBE is a cone entry followed by a cylinder at the end (Figure 3). Patil et al. [88] studied the influence of this shape over the pressure in the barrel, and both the entry angle (2α) and the length-to-die diameter ratio (L/D_{die}) were found to be correlated with the internal pressure. For entry angles (2α) up to 30°, the pressure linearly increased with a slope of 0.5. With respect to L/D_{die} ratio, the relation has a slope of 0.6. Understanding this relationship, a given die design can be used to regulate the extrusion pressure [55]. Moreover, a larger die requires a lower specific mechanical energy than a smaller one because the pressure at the die entry for a larger die entry is lower and requires less mechanical energy.

Sometimes extrusion is run without a die for many different reasons. The principal reason for LBE operation without a die is reports of serious packing at the die entrance due to insufficient fluidity of the substrate (lack of solvent or catalyst) [59,89].

3.5. Torque

The torque (i.e., moment or moment of a force) is the capacity of a force to turn an object around its axis. For an extruder, the torque is the aptitude of the screws to turn around their axis, and it is an indicator of the efficiency of the extruder [90]. Torque also plays a role in the determination of the specific mechanical energy and is correlated to other extrusion parameters. For example, substrate moisture is inversely correlated with torque [68]. The torque increases when the barrel temperature and the screw speed are lowered [86,91,92]. Adding Reverse Screw Elements (RSE) to the screw configuration tends to increases the torque [93,94]. Concerning the particle size of the substrate, there is no evidence about its impact on the torque [64].

Importantly, torque influences the sugar recovery yield. Higher torque leads to sugar recovery improvement. Gu et al. [48] recorded an increase from 27% up to 43% of glucose yield and from 13% to 21% for xylose/mannose yield when the torque was increased from 15 Nm to 70 Nm. However, there is no specific torque range for LBE extrusion because it can differ from one extruder to another, according to the type of biomass and the extrusion conditions [86]. However, one approach to lower the torque and still reach good sugar recovery yield is to use additives (solvent or catalyst), especially those with a great affinity towards cellulose such as ethylene glycol and glycerol [40].

3.6. Specific Mechanical Energy

The specific mechanical energy (SME) is an input parameter that is expressed in Watt-hour per kilogram (Wh/kg) or Joule per kilogram (J/kg). The SME is the energy supplied for one kilogram of extrudate obtained. It is an indicator of the stability and capacity of the extrusion process, as a rapid change of the SME usually relates to instability in the flow [14,32].

SME is a function of the torque, the mass flow, the power of the extruder motor, and the screw speed, as outlined in the following formula:

$$\text{SME} = \frac{(Total\ torque - friction\ Torque) \times N \times (P_m)}{(max_t) \times (max_{ss}) \times m_f}, \tag{3}$$

where N is the screw speed (rev/min), m_f is the mass flow rate (kg/s), max_t is the maximum allowable torque, max_{ss} is the maximum allowable screw speed, and P_m is the power of the drive motor at a rated speed of max_{ss}. Gu et al. [48] found that the SME is correlated with the median particles size and the crystallinity of the substrate respectively with $r = -0.79$ and $r = -0.87$. Furthermore, it has been reported that the viscosity of the substrate influences the SME as less viscous substrates require higher SME [95].

Zheng et al. [78] studied the role of the SME in xylose recovery yields and found that mass flow higher than 1.45 kg/h negatively affected the xylose recovery yield. However, when the additive flow (water) was increased, they recorded an improvement in the xylose recovery yield while the SME decreased concurrently, which was attributed to lower friction in the barrel due to increased moisture content. These results show that additives can be used to lower the SME in LBE and improve the sugar recovery results.

Energy consumption is one of the main concerns of biomass pretreatment. The goal is to recover the highest among of the desired compound under the least energy consumption possible. Thus, the SME should be set in the optimum range for a given LBE. For example, Lamsal et al. [79] tested SME values from 222 to 639 Wh/kg and found that 416,6 Wh/kg was optimum for wheat bran. Figure 9 gives an overview of some SME for LBE (details about data sources are presented in Appendix B). In cases where the SME is an output, the value can be predicted with a highly accurate model ($R^2 = 0.978$) developed by Lei et al. [93] for a twin-screw extruder.

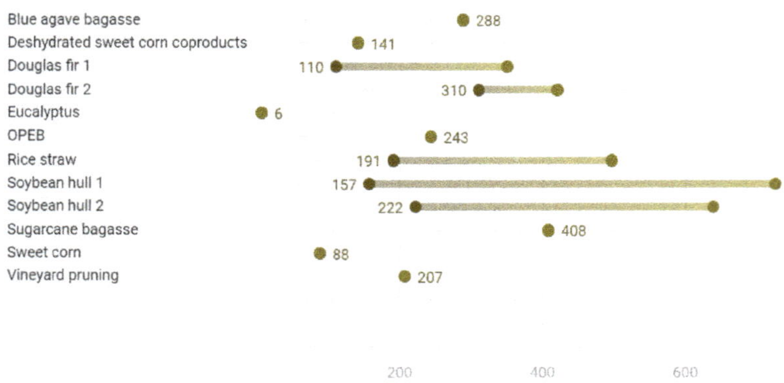

Figure 9. Specific mechanical energy for some lignocellulosic biomass extrusion (Wh/kg).

Comparing power consumption between pretreatment methods is complicated and somewhat unnecessary. Indeed, the pre-treatment method must be evaluated with regard to the profitability of the product which allows it to manufacture at the end of the chain (ethanol, biogas, biodiesel, enzymes, resin, etc.). Kazi et al. [96] used an ASPEN Plus model to simulate the profitability (on short-term economic viability) of four LCB pretreatment methods for ethanol production. The pretreatment methods were dilute acid, 2-stage dilute acid, ammonia fiber explosion (AFEX), and hot water. Corn stover was the raw biomass for all of them. Results showed the dilute acid pretreatment as the best pretreatment method as it gave the lowest product value (1.36 dollars of gasoline-equivalent). On the other hand, Yoo et al. [97] compared the profitability (for a year) of dilute acid and extrusion pretreatment in the production of ethanol. The Monte Carlo model was used for this purpose and soybean hulls were the substrate. Extrusion pretreatment was the best pretreatment as it produced 23.4% more ethanol than the dilute acid pretreatment. The main reason was the high conversion of cellulose to glucose, achieved with extrusion pretreatment [98].

In practice, in order to lower the energy balance of the whole extrusion pretreatment, good practices are:

- Avoid the use of a thermal source during biomass preparation, instead privilege room temperature or solar heat.
- Use kneading screw elements and reverse screw elements sparingly in the screw configuration. As highlighted in Section 3.2, these two elements enhance the disruptive effect of screws on biomass, but at the same time, they increase the energy consumption [66,80]. The operator must find a compromise according to the objectives of their extrusion pretreatment.
- Opt for continuous extrusion to avoid unnecessary energy consumption and also because starting up the extruder is time-consuming and energy-intensive. Therefore, plan each extrusion well and prepare everything before starting.
- Make sure the moisture of the substrate is sufficient to ensure smooth transport of the substrate in the barrel, as dry matter content and extruder electricity consumption are strongly linked ($R^2 = 0.73$) [99]. This practice also helps to avoid the overloading of the barrel and the jamming of the screws.
- Limit to the strict minimum the number of passes of the biomass in the extruder. This number may vary from one type of biomass to another. For this, preliminary tests are necessary. As highlighted in Section 5.2, several studies have shown that beyond a certain number of passes, there is no longer any significant improvement in the sugar recovery rate [44,69].

4. Additives

Reactive extrusion is performed in an extruder where one or more additives chemically react with the biomass to achieve a change in composition and structure of the lignocellulose. Reactive extrusion is very common in LBE. Usually, the additive is a catalyst, a solvent, an enzyme (bioextrusion), or a combination of them. Additives can also be acid, alkali, organosolv, mineral, etc. Water, sodium hydroxide (NaOH), ethylene glycol, lime (CaOH), sulphuric acid (H_2SO_4), Tween 80 (polysorbate 80) as surfactant, and [EMIM]Ac (1-Ethyl-3-methylimidazolium acetate) as ionic liquid are the most used. Reactive extrusion presents a key advantage because LCBs have poor flow capabilities. The aim of using additives is to enhance the flowability of the substrate in the barrel and facilitate saccharification via hydrolysis of the chemicals over the substrate. Reactive extrusion can be performed via two methods: mixing the additive with the substrate during the preparation step or adding the additive during the extrusion process.

4.1. Addition before Extrusion

Adding additives to the biomass before extrusion (i.e., during biomass preparation) modifies the biomass moisture and this has certain advantages for reactive extrusion. In particular, this approach allows for a more precise control of the moisture content and mixing is optimal for maximum contact with the additive. Furthermore, when the mixture is stored in the presence of reactive additives, delignification can start during this period, weakening the lignocellulose structure and subsequently facilitating extrusion flow. Many studies have demonstrated that using additives before the extrusion process can be a very good practice, and the application of different kinds of additives has been investigated. Kuster et al. [44] studied the effect of water, glycerol, Tween 80, and ethylene glycol on sugar recovery with sugarcane bagasse and sugarcane straw, and water was found to be the best additive for both biomasses as evidenced by sugar recovery yield. However, this also led to problems with flow during extrusion. Under the pretreatment conditions tested (i.e., long residence time of the substrate inside the barrel, high shearing forces, and temperature), the substrate began to dry and then blocked the screws. Glycerol, as an additive, achieved a slightly lower glucose recovery yield, but substantially improved the flow conditions. On the other hand, with willow and [EMIM]Ac as additives, Han et al. [59] achieved recovery yield for glucose and xylose of 99% and 99.5%, respectively. Da Silva et al. [69] applied the same solvent on sugarcane bagasse, and achieved 90% glucose recovery yield at 25 wt.% for 8 min extrusion. Zhang et al. [15] obtained 86.8% of glucose recovery yield and 50.5% of xylose recovery yield with dry corn stover combined with NaOH.

4.2. Addition during Extrusion

Adding reagents during extrusion involves the use of a pump with a controllable flow rate. Some extruders are equipped with one or two additive pumps, wherein a first additive is injected in the screw zone located after the biomass hopper (Figure 3). Subsequently, another reagent is added downstream in the extruder to neutralize the first additive (in the case of alkali). The two flows and their respective concentration must be correctly adjusted. If not, the first additive action will not be optimal, or a poor neutralization can occur, negatively affecting further treatment of the extrudate. Extrusion performed this way saves time and energy compared to the case when the biomass and additives are mixed before extrusion and is well adapted to continuous extrusion. These advantages are crucial, especially at pilot and industrial scales. However, the mixing time for biomass and additive is reduced, and the sugar recovery yield can be affected [59,77,100]. Choi and Oh [67] pretreated ripe straw with sulphuric acid without a neutralization reagent. Only 43% of glucan (glucose) at 3.5% w/v H_2SO_4 was recovered. Thus, the application of additives before or during the extrusion process must consider the objectives of the experiments, as well as energy consumption, and scale up implications. Sometimes, LBE is coupled with other pretreatments methods [101]. In that case, the second pretreatment method must be taken into account during the decision-making process.

5. Working Parameters

5.1. Temperature

Extrusion is defined as a high-temperature technology [102]. In fact, there are three ranges of temperature for LBE: under 100 °C (low temperatures), between 100 °C and 150 °C (mild temperatures), and above 150 °C (high temperatures) [47,103].

The temperature inside the extruder barrel results from the heat generated by both external and internal sources. The external source is coming from the heating system of the extruder, while the internal source is the heat generated by the effect of shear forces inside the extruder (viscous dissipation) [104]. Some extruders offer the possibility to impose a temperature profile along the screw. For example, Montiel et al. [100] pretreated blue agave bagasse using an extruder with four screw sections with different temperatures: 22 °C in the feeding zone, 50 °C in deconstruction zone, 25 °C in the neutralization zone, and 25 °C in the filtration zone. In this case, a higher temperature in the neutralization zone, such as 50 °C in the deconstruction zone, can denature the neutralization agent. This is a good example of how a temperature profile across the extruder design is advantageous for setting the optimum temperature for each screw zone.

It is unclear which temperature range (low, mild, or high) is suitable for a better sugar recovery. For Karunanithy et al. [105], single-extruded pine wood ran at different temperatures (100 °C, 150 °C, and 180 °C) achieved bests recovery results at 180 °C with 66.1% of total sugar recovery. Zheng et al. [78] experienced similar results after a twin-screw extrusion of sweet corn, with xylose recovery yield increasing with temperature (65–100 °C). At higher temperatures, biomass moisture loss is important, which can cause a powerful disturbance in the biomass structure due to shearing forces and elevated thermal action. Higher temperatures can have additional negative impacts on the extrusion process, as the substrate releases volatile organic compounds which can hinder downstream processes (enzymatic saccharification, fermentation, etc.) [104,106,107]. Gu et al. [48] used a twin-screw extruder to pre-treat Douglas fir residuals. The screws had five sections (T1 to T5) along with the screws, with the following temperature profile: T1 and T2 (25 °C), T3 (50 °C), T4 and T5 (50, 100, or 150 °C). The results showed that glucose and xylose/mannose yield decreased when the temperature increased in sections T4 and T5, where the conditions ranged from 50 °C to 150 °C. They attributed this result to the fact that moisture evaporation is pronounced in T4 and T5 at high temperatures, leading to particle agglomeration, an increase in particle size, and cellulose recrystallization.

5.2. Residence Time

The biomass residence time in extrusion is considered a particular advantage for this process, as it is very short compared to other pretreatment methods. For LBE, the timespan is on the scale of minutes, with residence times around 1 min 30 s being achieved by Karunanithy et al. [105] and Vaidya et al. [56]. However, there are no standard residence times for laboratory studies, as residence times between 1 and 10 min are common [86]. On the other hand, a short residence time could be a problem, especially in the case of bioextrusion (extrusion with enzymes) or in reactive extrusion, when additives are added during the processing, both of which would require longer times for the necessary reactions to take place.

The residence time is the consequence of many factors. For example, screw design can play a role, as cylindrical screws generate longer residence time than conical screws for the same screw length [108]. Depending on screw speed and the screw configuration, the residence time can be lengthened or shortened. Screw configurations containing more KE, RSE, and short pitches elements lengthen the residence time, while more FSE and larger pitches elements tend to shorten the residence time [80,109,110]. On the other hand, many authors found that the screw speed is inversely proportional to the residence time [32,48,111]. Generally speaking, flow resistance inside the barrel translates into longer residence times. Furthermore, an extruder without a die at the end of the barrel results in shortened residence times. The length to diameter ratio (L/D) of the screws also influences the residence time, with higher ratios increasing the residence time [86].

So far, there is no evidence about the role of other parameters such as liquid/solid ratio on the residence time. Based on current knowledge, it can be assumed that a higher ratio will shorten the residence time because adding additives enhances the substrate flowability.

It has been reported that long residence times enhance sugar recovery yield as the effects of the shearing forces and all the other pretreatment conditions over the substrate are exerted over a longer period. However, long residence times also raise the SME [78,112]. Usually, operators increase the residence time by recirculating the extrudate into the extruder as many times as needed (i.e., number of passes). The number of passes can be up to ten or more. Da Silva et al. [69] investigated the effect of the number of passes on saccharification yield for sugarcane bagasse with an ionic liquid as the additive. The results showed that the glucose and xylose recovery increased after the first pass, but additional passes did not significantly increase the yields of glucose and xylose recovery compared to the first pass. Kuster et al. [44] experimented with 10 extrusion passes with both sugarcane bagasse and straw. As previously mentioned, a slight improvement of the glucose recovery yield was observed for each pass (after the first). However, for bagasse after 3 passes and 7 for straw, no improvement was recorded. Additionally, multi passes did not affect the crystallinity index, as no significant variation of the index was observed after the first pass.

5.3. Screw Speed

From the initial development of extrusion as a processing step, even in the case of LCB pretreatment, screw speed has been considered as an important parameter [80,113,114]. Screw speed is measured in rotations per minute (rpm) and usually ranges from 30 to 200 rpm in laboratory LBE settings. Screw speeds less than 100 rpm are considered low and those above 120 rpm are considered high. In particular cases, it can be set very low (down to 5 rpm), or very high (up to 420 rpm) [59,68,115,116]. As noted in the preview sections, screw speed influences the torque, the SME, the barrel temperature, the residence time, and the substrate flow rate.

Screw speed is one of the most documented parameters in LBE. Like temperature, screw speed alone cannot guarantee the efficiency of the extrusion pretreatment [49,79,86]. It must be related to other extrusion parameters. For example, Karunanithy and Muthukumarappan [117] extruded switchgrass, while Heredia et al. [118] extruded Brewers' spent grain. Both studies recorded opposite behavior about screw speed over the glucose recovery yield. While Karunanithy and Muthukumarappan [117] found the best result by decreasing the screw speed from 150 to 50 rpm, Heredia et al. [118] obtained their bests results by increasing the screw speed from 100 up to 200 rpm. However, it is important to notice that Heredia et al. [118] ran the extrusion under 20% moisture at 50 °C, while Karunanithy and Muthukumarappan [117] worked at 15% moisture content and 150 °C. It is also possible to vary the screw speed between high and low settings during the extrusion process. In that case, a way to overcome the rise of the SME is to lower the torque at high screw speeds [64,78]. Nevertheless, this processing requires more complex design and controls to operate.

6. Challenges, Limitations, and Future Prospects

Reactive extrusion is an interesting technology due to the advantages it has over other pretreatment methods, and the traditional disadvantages from which it is freed. These are: the high risk of corrosion of the equipment, the use of large quantities of water during or after pretreatment, the appearance of inhibitors, the pollution and toxicity linked to the use of certain chemical products, and the length of the pre-treatment time which can last several days for example in the case of biological pre-treatments (fungi, bacteria, termites, etc.) [24,119]. On the other hand, the challenges related to extrusion as a method of LCB pretreatment can essentially be summarized in four points: the initial investment cost, the energy consumption, the post-purchase flexibility of the design parameters, and the process scale-up.

Although on a medium and long-term basis extrusion is a commercially attractive solution and far better than many other pretreatment methods, the investment costs in this technology are high [97,98]. The high prices of extruders hinder the democratization of their use. In this case, it might be interesting to diversify the use of the extruder. For example, the same extruder could be used to manufacture composite materials whose commercialization could allow a faster return on investment. Concerning energy consumption, extruders need a heat source and a cooling system in addition to a power supply. These are made possible through the use of electricity. The consequence is a non-negligible energy consumption. It is possible from several practical techniques, such as those presented in Section 3.6, to save energy or improve the energy efficiency of the extruder, but less energy-consuming extruders are of essential needs to accelerate the return on investment in the case where the extruder is exclusively used for LCB pretreatment. Another important aspect is the relatively small leeway of the extruder holders for the modification of the screw elements. Extruder owners in most cases have to refer to the equipment supplier for modifications, which add delays and affect the productivity of the extruder. Although technically very difficult, the design of adaptable screw elements according to the desired shapes, inclinations, and diameters or the development of an extruder capable of variably housing one, two, or three screws could revolutionize the use of extruders for maximum destruction of the lignocellulosic complex but also for many other applications. The limits of extrusion are those of mechanical pretreatments in general. They necessarily require an external energy source. Moreover, with mechanical pretreatments, it is impossible to be selective and to target, for example, the types of chemical bonds in the biomass that we would like to break, as this is the case with most chemical pretreatments. Thus, biomass fractionation (separation into its three major components) with extrusion requires coupling with another pretreatment method such as Organosolv [120].

Regarding future prospects, in addition to solving the challenges mentioned above, it is about finding an interesting coupling of extrusion with another method of pretreatment. The idea would be to benefit from the many advantages offered by extrusion while mitigating its disadvantages. So far, several coupling ideas have been studied in the literature. For example, extrusion has been coupled with liquid hot water for eucalyptus and aspen pretreatment [121], with Steam explosion for corncob pretreatment [122], Ultrasonication for rice hull pretreatment [53], Ionic Liquid (IL) for pussy willow and sugar bagasse pretreatment [59], and with Organosolv method for prairie cordgrass pretreatment [122]. As extrusion is one of the most used methods in pretreatment couplings, it is of great interest to investigate, in a review, each of the couplings extrusion has been implicated in, in order to highlight their efficiency, their advantages and disadvantages, their need for improvement, and if possible to advise possible interesting coupling ideas based on lessons learned from existing couplings.

Extrusion optimization and scaling up is also an aspect that is attracting more and more interest. Indeed, experimental designs with response surfaces have been and continue to be used for optimization. The problem is that they take time and are sometimes very expensive. Today, new computational techniques (therefore faster and less expensive) are in development. A genetic algorithm method is one that currently focuses attention. Nastaj and Wilczynski's work [123] entitled "optimization and scale-up for polymer extrusion" is a rich source of information on this subject.

7. Conclusions

Extrusion is a very flexible method of lignocellulosic biomass pretreatment due to the many parameters available, with optimal conditions for a given process, and can include extruder design, biomass type, additives, and operating conditions. While some parameters related to the extruder design are limited to initial design plans, several other parameters can be adopted as needed by operators to customize for a process' given needs. Most of these parameters are correlated, and clear identification of the purpose of the extrusion and the downstream treatments, as well as the possibility to scale up the process, are important when selecting the extrusion settings. Particular attention should be paid to the energy consumption during the biomass preparation and the extrusion process, with settings leading to satisfactory sugar recovery with the lowest energy consumption as a focus.

Funding: This work was supported by Institut National de la Recherche Scientifique (INRS) (Grant No. 121486) and the Natural Sciences and Engineering Research Council of Canada (NSERC) (Grant No. RGPIN-2020-05720).

Data Availability Statement: Not applicable.

Conflicts of Interest: The authors declare no conflict of interest.

Appendix A

Table A1. Specific mechanical energy for some lignocellulosic biomass extrusion (Wh/kg).

Substrate	Value or Range (Wh/kg)	References
Blue agave bagasse	288	[84]
Deshydrated sweet corn coproducts	141	[84]
Douglas fir 1	110–350	[78]
Douglas fir 2	310–420	[47]
Eucalyptus	6	[84]
OPEB	243	[84]
Rice straw	191–496	[92]
Soybean hull 1	157–726	[68]
Soybean hull 2	222–639	[79]
Sugarcane bagasse	408	[84]
Sweet corn	88	[84]
Vineyard pruning	207	[84]

Appendix B

Table A2. Composition of some lignocellulosic biomass.

N°	Substrate	Composition (%)			References
		Cellulose	Hemicellulose	Lignin	
		Hardwood			
1.	Birch	38.2	19.7	22.8	[124]
2.	Hybrid poplar	48.6	15.7	21.8	[42]
3.	Maple	44.9	25	20.7	[42]
4.	Poplar	49.9	25.1	18.1	[42]
5.	Red oak	43.4	22.5	25.8	[42]
6.	Walnut	46.2	20.9	21.9	[42]
7.	White birch	43	29.7	23.9	[39]
8.	White oak	43.6	23.7	23.2	[42]
9.	Willow	43	29.3	24.2	[124]
		Softwood			
1.	Black spruce chips	50	17.6	25.4	[39]
2.	Pine	46.4	22.9	29.4	[124]
3.	Spruce	43.4	22	28.1	[124]
		Grasses			
1.	Alfalfa	33	16.3	13.7	[39]
2.	Bagasse	23.33	16.52	54.87	[50]
3.	Corn cobs	15	35	45	[125]
4.	Corn stover 1	19	26	38	[126]
5.	Corn stover 2	35.6	22.1	12.3	[124]
6.	Rice straw 1	18	24	32.1	[125]
7.	Rice straw 2	34.2	24.5	11.9	[124]
8.	Sponge gourd fibres	15.46	17.44	66.59	[50]
9.	Sugarcane bagasse 1	20	25	42	[127]
10.	Sugarcane bagasse 2	40.2	23.8	25.2	[42]
11.	Sweet sorghum	21	27	45	[127]
12.	Switchgrass	31.0	24.4	17.6	[42]
13.	Wheat straw 1	38.2	24	23.4	[124]
14.	Wheat straw 2	38.2	24.5	23.4	[42]

References

1. Gupta, R.; Sharma, M.; Jangir, V.; Gautam, A.K.; Chandra, A.; Arya, R.K. A theoretical and mathematical study on the importance of petroleum derivatives in the production of sanitizer based products. *Mater. Today Proc.* **2022**, *in press*. [CrossRef]
2. Al-Samhan, M.; Al-Fadhli, J.; Al-Otaibi, A.M.; Al-Attar, F.; Bouresli, R.; Rana, M.S. Prospects of refinery switching from conventional to integrated: An opportunity for sustainable investment in the petrochemical industry. *Fuel* **2022**, *310*, 122161. [CrossRef]
3. Braga, A.; Faria, N. Biotechnological production of specialty aromatic and aromatic-derivative compounds. *World J. Microbiol. Biotechnol.* **2022**, *38*, 80. [CrossRef] [PubMed]
4. Reshmy, R.; Paulose, T.A.P.; Philip, E.; Thomas, D.; Madhavan, A.; Sirohi, R.; Binod, P.; Kumar Awasthi, M.; Pandey, A.; Sindhu, R. Updates on high value products from cellulosic biorefinery. *Fuel* **2022**, *308*, 122056. [CrossRef]
5. Liu, K.; Du, H.; Zheng, T.; Liu, H.; Zhang, M.; Zhang, R.; Li, H.; Xie, H.; Zhang, X.; Ma, M.; et al. Recent advances in cellulose and its derivatives for oilfield applications. *Carbohydr. Polym.* **2021**, *259*, 117740. [CrossRef]
6. Peter, S.; Lyczko, N.; Gopakumar, D.; Maria, H.J.; Nzihou, A.; Thomas, S. Nanocellulose and its derivative materials for energy and environmental applications. *J. Mater. Sci.* **2022**, *57*, 6835–6880. [CrossRef]
7. Antar, M.; Lyu, D.; Nazari, M.; Shah, A.; Zhou, X.; Smith, D.L. Biomass for a sustainable bioeconomy: An overview of world biomass production and utilization. *Renew. Sustain. Energy Rev.* **2021**, *139*, 110691. [CrossRef]
8. Hoang, A.T.; Nizetic, S.; Ong, H.C.; Mofijur, M.; Ahmed, S.F.; Ashok, B.; Bui, V.T.V.; Chau, M.Q. Insight into the recent advances of microwave pretreatment technologies for the conversion of lignocellulosic biomass into sustainable biofuel. *Chemosphere* **2021**, *281*, 130878. [CrossRef]
9. Zhang, J.; Zou, D.; Singh, S.; Cheng, G. Recent developments in ionic liquid pretreatment of lignocellulosic biomass for enhanced bioconversion. *Sustain. Energy Fuels* **2021**, *5*, 1655–1667. [CrossRef]

10. Xu, H.; Kong, Y.; Peng, J.; Song, X.; Liu, Y.; Su, Z.; Li, B.; Gao, C.; Tian, W. Comprehensive analysis of important parameters of choline chloride-based deep eutectic solvent pretreatment of lignocellulosic biomass. *Bioresour. Technol.* **2021**, *319*, 124209. [CrossRef]
11. Kumar, P.; Barrett, D.M.; Delwiche, M.J.; Stroeve, P. Methods for pretreatment of lignocellulosic biomass for efficient hydrolysis and biofuel production. *Ind. Eng. Chem. Res.* **2009**, *48*, 3713–3729. [CrossRef]
12. Galbe, M.; Zacchi, G. Pretreatment of lignocellulosic materials for efficient bioethanol production. *Adv. Biochem. Eng. Biotechnol.* **2007**, *108*, 41–65. [CrossRef] [PubMed]
13. Pérez-Rodríguez, N.; García-Bernet, D.; Domínguez, J.M. Extrusion and enzymatic hydrolysis as pretreatments on corn cob for biogas production. *Renew. Energy* **2017**, *107*, 597–603. [CrossRef]
14. Gatt, E.; Rigal, L.; Vandenbossche, V. Biomass pretreatment with reactive extrusion using enzymes: A review. *Ind. Crops Prod.* **2018**, *122*, 329–339. [CrossRef]
15. Zhang, S.; Keshwani, D.R.; Xu, Y.; Hanna, M.A. Alkali combined extrusion pretreatment of corn stover to enhance enzyme saccharification. *Ind. Crops Prod.* **2012**, *37*, 352–357. [CrossRef]
16. Morales-Huerta, J.C.; Hernández-Meléndez, O.; Hernández-Luna, M.G.; Manero, O.; Bárzana, E.; Vivaldo-Lima, E. An experimental and modeling study on the pretreatment and alkaline hydrolysis of blue agave bagasse in twin-screw extruders. *Ind. Eng. Chem. Res.* **2021**, *60*, 12449–12460. [CrossRef]
17. Zhang, H.; Yang, W.; Roslan, I.I.; Jaenicke, S.; Chuah, G.-K. A combo Zr-HY and Al-HY zeolite catalysts for the one-pot cascade transformation of biomass-derived furfural to γ-valerolactone. *J. Catal.* **2019**, *375*, 56–67. [CrossRef]
18. Li, Q.; Zhou, Z.; Zhang, D.; Wang, Z.; Cong, W. Lipid extraction from nannochloropsis oceanica biomass after extrusion pretreatment with twin-screw extruder: Optimization of processing parameters and comparison of lipid quality. *Bioprocess Biosyst. Eng.* **2020**, *43*, 655–662. [CrossRef]
19. Pan-Utai, W.; Iamtham, S. Physical extraction and extrusion entrapment of C-phycocyanin from Arthrospira platensis. *J. King Saud Univ.—Sci.* **2019**, *31*, 1535–1542. [CrossRef]
20. Sarker, T.R.; Azargohar, R.; Stobbs, J.; Karunakaran, C.; Meda, V.; Dalai, A.K. Complementary effects of torrefaction and pelletization for the production of fuel pellets from agricultural residues: A comparative study. *Ind. Crops Prod.* **2022**, *181*, 114740. [CrossRef]
21. Ahiduzzaman, M.; Islam, A.S. Development of biomass stove for heating up die barrel of rice husk briquette machine. *Procedia Eng.* **2013**, *56*, 777–781. [CrossRef]
22. Rowell, R.M. Challenges in biomass–Thermoplastic composites. *J. Polym. Environ.* **2007**, *15*, 229–235. [CrossRef]
23. Guiao, K.S.; Tzoganakis, C.; Mekonnen, T.H. Green mechano-chemical processing of lignocellulosic biomass for lignin recovery. *Chemosphere* **2022**, *293*, 133647. [CrossRef] [PubMed]
24. Zhang, Y.; Ni, S.; Wu, R.; Fu, Y.; Qin, M.; Willför, S.; Xu, C. Green fractionation approaches for isolation of biopolymers and the critical technical challenges. *Ind. Crops Prod.* **2022**, *177*, 114451. [CrossRef]
25. Li, T.; Chen, C.; Brozena, A.H.; Zhu, J.Y.; Xu, L.; Driemeier, C.; Dai, J.; Rojas, O.J.; Isogai, A.; Wagberg, L.; et al. Developing fibrillated cellulose as a sustainable technological material. *Nature* **2021**, *590*, 47–56. [CrossRef] [PubMed]
26. Kumar, D.; Singh, V. Bioethanol Production From Corn. In *Corn*; AACC International Press: Washington, DC, USA, 2019; pp. 615–631.
27. Niu, N.; Feng, L.; Lin, Y.; Li, X.; Zhang, D.; Yao, S. The sources of dechlorane plus (DP) in surface sediment from Bohai sea and the northern part of the yellow sea, China: Evidence from the fractional abundance of anti-DP (fanti) combined with lignin biomarker. *Reg. Stud. Mar. Sci.* **2020**, *39*, 101437. [CrossRef]
28. Fernández-Rodríguez, J.; Erdocia, X.; Sánchez, C.; González Alriols, M.; Labidi, J. Lignin depolymerization for phenolic monomers production by sustainable processes. *J. Energy Chem.* **2017**, *26*, 622–631. [CrossRef]
29. Ahvazi, B.; Wojciechowicz, O.; Ton-That, T.M.; Hawari, J. Preparation of lignopolyols from wheat straw soda lignin. *J. Agric. Food Chem.* **2011**, *59*, 10505–10516. [CrossRef]
30. Garlapati, V.K.; Chandel, A.K.; Kumar, S.P.J.; Sharma, S.; Sevda, S.; Ingle, A.P.; Pant, D. Circular economy aspects of lignin: Towards a lignocellulose biorefinery. *Renew. Sustain. Energy Rev.* **2020**, *130*, 109977. [CrossRef]
31. Yağcı, S.; Sutay Kocabaş, D.; Çalışkan, R.; Özbek, H.N. Statistical investigation of the bioprocess conditions of alkali combined twin-screw extrusion pretreatment to enhance fractionation and enzymatic hydrolysis of bulgur bran. *J. Sci. Food Agric.* **2022**. *early view*. [CrossRef]
32. Duque, A.; Manzanares, P.; González, A.; Ballesteros, M. Study of the application of alkaline extrusion to the pretreatment of eucalyptus biomass as first step in a bioethanol production process. *Energies* **2018**, *11*, 2961. [CrossRef]
33. Doménech, P.; Duque, A.; Higueras, I.; Iglesias, R.; Manzanares, P. Biorefinery of the olive tree—Production of sugars from enzymatic hydrolysis of olive stone pretreated by alkaline extrusion. *Energies* **2020**, *13*, 4517. [CrossRef]
34. Duque, A.; Doménech, P.; Álvarez, C.; Ballesteros, M.; Manzanares, P. Study of the bioprocess conditions to produce bioethanol from barley straw pretreated by combined soda and enzyme-catalyzed extrusion. *Renew. Energy* **2020**, *158*, 263–270. [CrossRef]
35. Wang, Z.; He, X.; Yan, L.; Wang, J.; Hu, X.; Sun, Q.; Zhang, H. Enhancing enzymatic hydrolysis of corn stover by twin-screw extrusion pretreatment. *Ind. Crops Prod.* **2020**, *143*, 111960. [CrossRef]
36. Debiagi, F.; Faria-Tischer, P.C.S.; Mali, S. A green approach based on reactive extrusion to produce nanofibrillated cellulose from oat hull. *Waste Biomass Valoriz.* **2021**, *12*, 1051–1060. [CrossRef]

37. Coimbra, M.C.; Duque, A.; Saéz, F.; Manzanares, P.; Garcia-Cruz, C.H.; Ballesteros, M. Sugar production from wheat straw biomass by alkaline extrusion and enzymatic hydrolysis. *Renew. Energy* **2016**, *86*, 1060–1068. [CrossRef]
38. Park, S.; Baker, J.O.; Himmel, M.E.; Parilla, P.A.; Johnson, D.K. Cellulose crystallinity index: Measurement techniques and their impact on interpreting cellulase performance. *Biotechnol. Biofuels* **2010**, *3*, 10. [CrossRef]
39. Chen, J.; Adjallé, K.; Lai, T.T.; Barnabé, S.; Perrier, M.; Paris, J. Effect of mechanical pretreatment for enzymatic hydrolysis of woody residues, corn stover and alfalfa. *Waste Biomass Valorization* **2019**, *11*, 5847–5856. [CrossRef]
40. Lee, S.H.; Teramoto, Y.; Endo, T. Enzymatic saccharification of woody biomass micro/nanofibrillated by continuous extrusion process I–effect of additives with cellulose affinity. *Bioresour. Technol.* **2009**, *100*, 275–279. [CrossRef]
41. Ansell, M.P.; Mwaikambo, L.Y. The structure of cotton and other plant fibres. In *Handbook of Textile Fibre Structure*; Woodhead Publishing: Sawston, UK, 2009; pp. 62–94.
42. Zhao, X.; Zhang, L.; Liu, D. Biomass recalcitrance. Part I: The chemical compositions and physical structures affecting the enzymatic hydrolysis of lignocellulose. *Biofuels Bioprod. Bioeng.* **2012**, *6*, 465–482. [CrossRef]
43. Ghaemi, F.; Abdullah, L.C.; Ariffin, H. Lignocellulose structure and the effect on nanocellulose production. In *Lignocellulose for Future Bioeconomy*; Elsevier: Amsterdam, The Netherlands, 2019; pp. 17–30.
44. Moro, M.K.; Teixeira, R.S.S.; da Silva, A.S.; Fujimoto, M.D.; Melo, P.A.; Secchi, A.R.; da Silva Bon, E.P. Continuous pretreatment of sugarcane biomass using a twin-screw extruder. *Ind. Crops Prod.* **2017**, *97*, 509–517. [CrossRef]
45. Vandenbossche, V.; Brault, J.; Vilarem, G.; Hernández-Meléndez, O.; Vivaldo-Lima, E.; Hernández-Luna, M.; Barzana, E.; Duque, A.; Manzanares, P.; Ballesteros, M.; et al. A new lignocellulosic biomass deconstruction process combining thermo-mechano chemical action and bio-catalytic enzymatic hydrolysis in a twin-screw extruder. *Ind. Crops Prod.* **2014**, *55*, 258–266. [CrossRef]
46. Marone, A.; Trably, E.; Carrère, H.; Prompsy, P.; Guillon, F.; Joseph-Aimé, M.; Barakat, A.; Fayoud, N.; Bernet, N.; Escudié, R. Enhancement of corn stover conversion to carboxylates by extrusion and biotic triggers in solid-state fermentation. *Appl. Microbiol. Biotechnol.* **2018**, *103*, 489–503. [CrossRef] [PubMed]
47. Fu, Y.; Gu, B.-J.; Wang, J.; Gao, J.; Ganjyal, G.M.; Wolcott, M.P. Novel micronized woody biomass process for production of cost-effective clean fermentable sugars. *Bioresour. Technol.* **2018**, *260*, 311–320. [CrossRef] [PubMed]
48. Gu, B.J.; Dhumal, G.S.; Wolcott, M.P.; Ganjyal, G.M. Disruption of lignocellulosic biomass along the length of the screws with different screw elements in a twin-screw extruder. *Bioresour. Technol.* **2019**, *275*, 266–271. [CrossRef] [PubMed]
49. Zhang, S.; Xu, Y.; Hanna, M.A. Pretreatment of corn stover with twin-screw extrusion followed by enzymatic saccharification. *Appl. Biochem. Biotechnol.* **2012**, *166*, 458–469. [CrossRef] [PubMed]
50. Guimarães, J.L.; Frollini, E.; da Silva, C.G.; Wypych, F.; Satyanarayana, K.G. Characterization of banana, sugarcane bagasse and sponge gourd fibers of Brazil. *Ind. Crops Prod.* **2009**, *30*, 407–415. [CrossRef]
51. Zheng, J.; Choo, K.; Rehmann, L. The effects of screw elements on enzymatic digestibility of corncobs after pretreatment in a twin-screw extruder. *Biomass Bioenergy* **2015**, *74*, 224–232. [CrossRef]
52. Kang, K.E.; Han, M.; Moon, S.-K.; Kang, H.-W.; Kim, Y.; Cha, Y.-L.; Choi, G.-W. Optimization of alkali-extrusion pretreatment with twin-screw for bioethanol production from miscanthus. *Fuel* **2013**, *109*, 520–526. [CrossRef]
53. Zhang, Y.; Li, T.; Shen, Y.; Wang, L.; Zhang, H.; Qian, H.; Qi, X. Extrusion followed by ultrasound as a chemical-free pretreatment method to enhance enzymatic hydrolysis of rice hull for fermentable sugars production. *Ind. Crops Prod.* **2020**, *149*, 112356. [CrossRef]
54. Rashed, G.I.; Chong, L.; Zhiyuan, S.; Hongzhen, F.; Chen, C.; Junjie, L.; Qiao, L.; Jianhui, D.; Huazheng, C.; Hongmei, W.; et al. Effect of a promising CSESE pretreatment on the morphological structure and properties of jute fibers. *E3S Web Conf.* **2021**, *252*, 02049. [CrossRef]
55. Senturk-Ozer, S.; Gevgilili, H.; Kalyon, D.M. Biomass pretreatment strategies via control of rheological behavior of biomass suspensions and reactive twin screw extrusion processing. *Bioresour. Technol.* **2011**, *102*, 9068–9075. [CrossRef] [PubMed]
56. Vaidya, A.A.; Gaugler, M.; Smith, D.A. Green route to modification of wood waste, cellulose and hemicellulose using reactive extrusion. *Carbohydr. Polym.* **2016**, *136*, 1238–1250. [CrossRef] [PubMed]
57. Debiagi, F.; Faria-Tischer, P.C.S.; Mali, S. Nanofibrillated cellulose obtained from soybean hull using simple and eco-friendly processes based on reactive extrusion. *Cellulose* **2019**, *27*, 1975–1988. [CrossRef]
58. Byun, J.; Cha, Y.-L.; Park, S.-M.; Kim, K.-S.; Lee, J.-E.; Kang, Y.-G. Lignocellulose pretreatment combining continuous alkaline single-screw extrusion and ultrasonication to enhance biosugar production. *Energies* **2020**, *13*, 5636. [CrossRef]
59. Han, S.-Y.; Park, C.-W.; Endo, T.; Febrianto, F.; Kim, N.-H.; Lee, S.-H. Extrusion process to enhance the pretreatment effect of ionic liquid for improving enzymatic hydrolysis of lignocellulosic biomass. *Wood Sci. Technol.* **2020**, *54*, 599–613. [CrossRef]
60. Karunanithy, C.; Muthukumarappan, K. Thermo-mechanical pretreatment of feedstocks. In *Green Biomass Pretreatment for Biofuels Production*; Gu, T., Ed.; Springer: Dordrecht, The Netherlands, 2013; pp. 31–65.
61. Duque, A.; Manzanares, P.; Ballesteros, I.; Negro, M.J.; Oliva, J.M.; Saez, F.; Ballesteros, M. Optimization of integrated alkaline–extrusion pretreatment of barley straw for sugar production by enzymatic hydrolysis. *Process Biochem.* **2013**, *48*, 775–781. [CrossRef]
62. Liu, C.; Van Der Heide, E.; Wang, H.; Li, B.; Yu, G.; Mu, X. Alkaline twin-screw extrusion pretreatment for fermentable sugar production. *Biotechnol. Biofuels* **2013**, *6*, 97. [CrossRef]

63. Karunanithy, C.; Muthukumarappan, K.; Gibbons, W.R. Sequential extrusion-microwave pretreatment of switchgrass and big bluestem. *Bioresour. Technol.* **2014**, *153*, 393–398. [CrossRef]
64. Karunanithy, C.; Muthukumarappan, K. Influence of extruder and feedstock variables on torque requirement during pretreatment of different types of biomass—A response surface analysis. *Biosyst. Eng.* **2011**, *109*, 37–51. [CrossRef]
65. Cha, Y.-L.; Yang, J.; Seo, S.-i.; An, G.H.; Moon, Y.-H.; You, G.-D.; Lee, J.-E.; Ahn, J.-W.; Lee, K.-B. Alkaline twin-screw extrusion pretreatment of Miscanthus with recycled black liquor at the pilot scale. *Fuel* **2016**, *164*, 322–328. [CrossRef]
66. Negro, M.J.; Duque, A.; Manzanares, P.; Sáez, F.; Oliva, J.M.; Ballesteros, I.; Ballesteros, M. Alkaline twin-screw extrusion fractionation of olive-tree pruning biomass. *Ind. Crops Prod.* **2015**, *74*, 336–341. [CrossRef]
67. Choi, C.H.; Oh, K.K. Application of a continuous twin screw-driven process for dilute acid pretreatment of rape straw. *Bioresour. Technol.* **2012**, *110*, 349–354. [CrossRef] [PubMed]
68. Yoo, J.; Alavi, S.; Vadlani, P.; Amanor-Boadu, V. Thermo-mechanical extrusion pretreatment for conversion of soybean hulls to fermentable sugars. *Bioresour. Technol.* **2011**, *102*, 7583–7590. [CrossRef]
69. Da Silva, A.S.A.; Teixeira, R.S.S.; Endo, T.; Bon, E.P.S.; Lee, S.-H. Continuous pretreatment of sugarcane bagasse at high loading in an ionic liquid using a twin-screw extruder. *Green Chem.* **2013**, *15*, 1991. [CrossRef]
70. Shah, A.; Gupta, M. Comparision of the flow in co-rotating and counter-rotating twinscrew extruders. *ANTEC* **2004**, *5*, 443–447.
71. Cantine, O. PEO Hot Melt Extrudates for Controlled Drug Delivery. Ph.D. Thesis, Universite Lille, Lille Cedex, France, 2017.
72. Solano, D.; Vinyes, P.; Arranz, P. *The Biomass Briquetting Process: A Guideline Report*; UNDP-CEDRO Publication: Beirut, Lebanon, 2016.
73. Zhang, X.; Zhang, W.; Su, J.; Wang, M.; Lu, C. Preparation, characterization, and properties of polyethylene composites highly filled with calcium carbonate through co-rotating conical twin-screw extrusion. *J. Vinyl Addit. Technol.* **2014**, *20*, 108–115. [CrossRef]
74. Leng, J.; Wu, J.; Chen, N.; Xu, X.; Zhang, J. The development of a conical screw-based extrusion deposition system and its application in fused deposition modeling with thermoplastic polyurethane. *Rapid Prototyp. J.* **2019**, *26*, 409–417. [CrossRef]
75. Vera-Sorroche, J.; Kelly, A.L.; Brown, E.C.; Gough, T.; Abeykoon, C.; Coates, P.D.; Deng, J.; Li, K.; Harkin-Jones, E.; Price, M. The effect of melt viscosity on thermal efficiency for single screw extrusion of HDPE. *Chem. Eng. Res. Des.* **2014**, *92*, 2404–2412. [CrossRef]
76. Kelly, A.L.; Brown, E.C.; Coates, P.D. The effect of screw geometry on melt temperature profile in single screw extrusion. *Polym. Eng. Sci.* **2006**, *46*, 1706–1714. [CrossRef]
77. Duque, A.; Manzanares, P.; Ballesteros, I.; Negro, M.J.; Oliva, J.M.; Gonzalez, A.; Ballesteros, M. Sugar production from barley straw biomass pretreated by combined alkali and enzymatic extrusion. *Bioresour. Technol.* **2014**, *158*, 262–268. [CrossRef] [PubMed]
78. Zheng, J.; Choo, K.; Rehmann, L. Xylose removal from lignocellulosic biomass via a twin-screw extruder: The effects of screw configurations and operating conditions. *Biomass Bioenergy* **2016**, *88*, 10–16. [CrossRef]
79. Lamsal, B.; Yoo, J.; Brijwani, K.; Alavi, S. Extrusion as a thermo-mechanical pre-treatment for lignocellulosic ethanol. *Biomass Bioenergy* **2010**, *34*, 1703–1710. [CrossRef]
80. Wahid, R.; Hjorth, M.; Kristensen, S.; Møller, H.B. Extrusion as pretreatment for boosting methane production: Effect of screw configurations. *Energy Fuels* **2015**, *29*, 4030–4037. [CrossRef]
81. Djuric, D.; Kleinebudde, P. Impact of screw elements on continuous granulation with a twin-screw extruder. *J. Pharm. Sci.* **2008**, *97*, 4934–4942. [CrossRef]
82. Fülöp, G.; Domokos, A.; Galata, D.; Szabó, E.; Gyürkés, M.; Szabó, B.; Farkas, A.; Madarász, L.; Démuth, B.; Lendér, T.; et al. Integrated twin-screw wet granulation, continuous vibrational fluid drying and milling: A fully continuous powder to granule line. *Int. J. Pharm.* **2021**, *594*, 120126. [CrossRef]
83. Kohlgrüber, K.; Ullrich, M.; Hepperle, J.; Rudolf, R.; König, T.; Liesenfelder, U.; Bierdel, M.; Kirchhoff, J.; Lechner, F.; Sämann, H.-J.; et al. *Co-Rotating Twin-Screw Extruders Fundamentals, Technology, and Applications*; Hanser Publications: Munich, Germany, 2008. [CrossRef]
84. Vandenbossche, V.; Brault, J.; Vilarem, G.; Rigal, L. Bio-catalytic action of twin-screw extruder enzymatic hydrolysis on the deconstruction of annual plant material: Case of sweet corn co-products. *Ind. Crops Prod.* **2015**, *67*, 239–248. [CrossRef]
85. Choudhury, G.S.; Gautam, A. Screw configuration effects on macroscopic characteristics of extradates produced by twin-screw extrusion of rice flour. *J. Food Sci.* **1999**, *64*, 479–487. [CrossRef]
86. Duque, A.; Manzanares, P.; Ballesteros, M. Extrusion as a pretreatment for lignocellulosic biomass: Fundamentals and applications. *Renew. Energy* **2017**, *114*, 1427–1441. [CrossRef]
87. Abeykoon, C.; Kelly, A.L.; Brown, E.C.; Coates, P.D. The effect of materials, process settings and screw geometry on energy consumption and melt temperature in single screw extrusion. *Appl. Energy* **2016**, *180*, 880–894. [CrossRef]
88. Patil, P.D.; Feng, J.J.; Hatzikiriakos, S.G. Constitutive modeling and flow simulation of polytetrafluoroethylene (PTFE) paste extrusion. *J. Non-Newton. Fluid Mech.* **2006**, *139*, 44–53. [CrossRef]
89. Ai, B.; Li, W.; Woomer, J.; Li, M.; Pu, Y.; Sheng, Z.; Zheng, L.; Adedeji, A.; Ragauskas, A.J.; Shi, J. Natural deep eutectic solvent mediated extrusion for continuous high-solid pretreatment of lignocellulosic biomass. *Green Chem.* **2020**, *22*, 6372–6383. [CrossRef]
90. Guha, M.; Ali, S.Z.; Bhattacharya, S. Twin-screw extrusion of rice flour without a die: Effect of barrel temperature and screw speed on extrusion and extrudate characteristics. *J. Food Eng.* **1997**, *32*, 251–267. [CrossRef]

91. Akdogan, H. Pressure, torque, and energy responses of a twin screw extruder at high moisture contents. *Food Res. Int.* **1996**, *29*, 423–429. [CrossRef]
92. Chen, W.H.; Xu, Y.Y.; Hwang, W.S.; Wang, J.B. Pretreatment of rice straw using an extrusion/extraction process at bench-scale for producing cellulosic ethanol. *Bioresour. Technol.* **2011**, *102*, 10451–10458. [CrossRef]
93. Lei, H.; Fulcher, R.G.; Ruan, R.; van Lengerich, B. Empirical modeling of die pressure, shaft torque, sme, and product temperature of rice flour in a corotating twin-screw extruder. *Cereal Chem. J.* **2005**, *82*, 582–587. [CrossRef]
94. Harmann, D.V.; Harper, J.M. Effect of extruder geometry on torque and flow. *Trans. ASAE* **1973**, *16*, 1175–1178. [CrossRef]
95. Godavarti, S.; Karwe, M.V. Determination of specific mechanical energy distribution on a twin-screw extruder. *J. Agric. Eng. Res.* **1997**, *67*, 277–287. [CrossRef]
96. Kazi, F.K.; Fortman, J.A.; Anex, R.P.; Hsu, D.D.; Aden, A.; Dutta, A.; Kothandaraman, G. Techno-economic comparison of process technologies for biochemical ethanol production from corn stover. *Fuel* **2010**, *89*, S20–S28. [CrossRef]
97. Yoo, J. *Technical And Economical Assessment of Thermo-Mechanical Extrusion Pretreatment for Cellulosic Ethanol Production*; Kansas State University: Manhattan, KS, USA, 2011.
98. Zheng, J.; Rehmann, L. Extrusion pretreatment of lignocellulosic biomass: A review. *Int. J. Mol. Sci.* **2014**, *15*, 18967–18984. [CrossRef]
99. Hjorth, M.; Gränitz, K.; Adamsen, A.P.S.; Møller, H.B. Extrusion as a pretreatment to increase biogas production. *Bioresour. Technol.* **2011**, *102*, 4989–4994. [CrossRef]
100. Montiel, C.; Hernández-Meléndez, O.; Vivaldo-Lima, E.; Hernández-Luna, M.; Bárzana, E. Enhanced bioethanol production from blue agave bagasse in a combined extrusion–saccharification process. *BioEnergy Res.* **2016**, *9*, 1005–1014. [CrossRef]
101. Beisl, S.; Biermair, F.; Friedl, A.; Mundigler, N.; Miltner, A. Sequential extrusion and organosolv pretreatment for wheat straw valorization. *Chem. Eng. Trans.* **2017**, *61*, 853–858.
102. Merci, A.; Urbano, A.; Grossmann, M.V.E.; Tischer, C.A.; Mali, S. Properties of microcrystalline cellulose extracted from soybean hulls by reactive extrusion. *Food Res. Int.* **2015**, *73*, 38–43. [CrossRef]
103. Appels, L.; Degreve, J.; Van der Bruggen, B.; Van Impe, J.; Dewil, R. Influence of low temperature thermal pre-treatment on sludge solubilisation, heavy metal release and anaerobic digestion. *Bioresour. Technol.* **2010**, *101*, 5743–5748. [CrossRef]
104. Formela, K.; Zedler, L.; Hejna, A.; Tercjak, A. Reactive extrusion of bio-based polymer blends and composites—Current trends and future developments. *Express Polym. Lett.* **2018**, *12*, 24–57. [CrossRef]
105. Karunanithy, C.; Muthukumarappan, K.; Gibbons, W.R. Extrusion pretreatment of pine wood chips. *Appl. Biochem. Biotechnol.* **2012**, *167*, 81–99. [CrossRef]
106. Formela, K.; Wołosiak, M.; Klein, M.; Wang, S. Characterization of volatile compounds, structural, thermal and physico-mechanical properties of cross-linked polyethylene foams degraded thermo-mechanically at variable times. *Polym. Degrad. Stab.* **2016**, *134*, 383–393. [CrossRef]
107. Espert, A.; de las Heras, L.A.; Karlsson, S. Emission of possible odourous low molecular weight compounds in recycled biofibre/polypropylene composites monitored by head-space SPME-GC–MS. *Polym. Degrad. Stab.* **2005**, *90*, 555–562. [CrossRef]
108. Seifert, S. Comparison of parallel and conical twin screw extruders from the processing point of view. *Plast. Rubber Compos.* **2013**, *34*, 134–142. [CrossRef]
109. Raquez, J.-M.; Narayan, R.; Dubois, P. Recent advances in reactive extrusion processing of biodegradable polymer-based compositions. *Macromol. Mater. Eng.* **2008**, *293*, 447–470. [CrossRef]
110. Gogoi, B.K.; Oswalt, A.J.; Choudhury, G.S. Reverse screw element(s) and feed composition effects during twin-screw extrusion of rice flour and fish muscle blends. *J. Food Sci.* **1996**, *61*, 590–595. [CrossRef]
111. da Silva, A.S.; Sobral Teixeira, R.S.; de Oliveira Moutta, R.; Santana, V.; de Barros, R.D.R.O.; Antonieta, M.; da Silva Bo, E.P. Sugarcane and woody biomass pretreatments for ethanol production. In *Sustainable Degradation of Lignocellulosic Biomass—Techniques, Applications and Commercialization*; Books on Demand: Norderstedt, Germany, 2013.
112. Duque, A.; Álvarez, C.; Doménech, P.; Manzanares, P.; Moreno, A.D. Advanced bioethanol production: From novel raw materials to integrated biorefineries. *Processes* **2021**, *9*, 206. [CrossRef]
113. Guo, Y.; Chung, C.I. Dependence of melt temperature on screw speed and size in extrusion. *Polym. Eng. Sci.* **1989**, *29*, 415–419. [CrossRef]
114. Chung, C.I. Maximum pressure developed by solid conveying force in screw extruders. *Polym. Eng. Sci.* **1975**, *15*, 29–34. [CrossRef]
115. Choi, W.-I.; Oh, K.-K.; Park, J.-Y.; Lee, J.-S. Continuous sodium hydroxide-catalyzed pretreatment of empty fruit bunches (EFB) by continuous twin-screw-driven reactor (CTSR). *J. Chem. Technol. Biotechnol.* **2014**, *89*, 290–296. [CrossRef]
116. Endersen, P.G.; Lechner, F. Co-rotating fully intermeshing twin-screw compounding: Advancements for improved performance and productivity. *Plast. Eng.* **2012**, *64*, 32–38. [CrossRef]
117. Karunanithy, C.; Muthukumarappan, K. Effect of extruder parameters and moisture content of switchgrass, prairie cord grass on sugar recovery from enzymatic hydrolysis. *Appl. Biochem. Biotechnol.* **2010**, *162*, 1785–1803. [CrossRef]
118. Heredia-Olea, E.; Pérez-Carrillo, E.; Serna-Saldívar, S.O. Effect of extrusion conditions and hydrolysis with fiber-degrading enzymes on the production of C5 and C6 sugars from brewers' spent grain for bioethanol production. *Biofuel Res. J.* **2015**, *2*, 203–208. [CrossRef]

119. Zhao, L.; Sun, Z.-F.; Zhang, C.-C.; Nan, J.; Ren, N.-Q.; Lee, D.-J.; Chen, C. Advances in pretreatment of lignocellulosic biomass for bioenergy production: Challenges and perspectives. *Bioresour. Technol.* **2022**, *343*, 126123. [CrossRef]
120. Brudecki, G.; Cybulska, I.; Rosentrater, K. Integration of extrusion and clean fractionation processes as a pre-treatment technology for prairie cordgrass. *Bioresour. Technol.* **2013**, *135*, 672–682. [CrossRef] [PubMed]
121. Tian, D.; Shen, F.; Yang, G.; Deng, S.; Long, L.; He, J.; Zhang, J.; Huang, C.; Luo, L. Liquid hot water extraction followed by mechanical extrusion as a chemical-free pretreatment approach for cellulosic ethanol production from rigid hardwood. *Fuel* **2019**, *252*, 589–597. [CrossRef]
122. Zhang, H.-J.; Fan, X.-G.; Qiu, X.-L.; Zhang, Q.-X.; Wang, W.-Y.; Li, S.-X.; Deng, L.-H.; Koffas, M.A.G.; Wei, D.-S.; Yuan, Q.-P. A novel cleaning process for industrial production of xylose in pilot scale from corncob by using screw-steam-explosive extruder. *Bioprocess Biosyst. Eng.* **2014**, *37*, 2425–2436. [CrossRef] [PubMed]
123. Nastaj, A.; Wilczyński, K. Optimization and Scale-Up for Polymer Extrusion. *Polymers* **2021**, *13*, 1547. [CrossRef]
124. Jørgensen, H.; Kristensen, J.B.; Felby, C. Enzymatic conversion of lignocellulose into fermentable sugars: Challenges and opportunities. *Biofuels Bioprod. Bioeng.* **2007**, *1*, 119–134. [CrossRef]
125. Prasad, S.; Singh, A.; Joshi, H.C. Ethanol as an alternative fuel from agricultural, industrial and urban residues. *Resour. Conserv. Recycl.* **2007**, *50*, 1–39. [CrossRef]
126. Zhu, Y.; Lee, Y.Y.; Elander, R.T. Optimization of dilute-acid pretreatment of corn stover using a high-solids percolation reactor. *Appl. Biochem. Biotechnol.* **2005**, *124*, 1045–1054. [CrossRef]
127. Kim, M.; Day, D.F. Composition of sugar cane, energy cane, and sweet sorghum suitable for ethanol production at Louisiana sugar mills. *J. Ind. Microbiol. Biotechnol.* **2011**, *38*, 803–807. [CrossRef]

energies

Article

Handling Load Uncertainty during On-Peak Time via Dual ESS and LSTM with Load Data Augmentation

Jin Sol Hwang, Jung-Su Kim * and Hwachang Song

Department of Electrical and Information Engineering, Seoul National University of Science and Technology, Seoul 01811, Korea; 16100665@seoultech.ac.kr (J.S.H.); hcsong@seoultech.ac.kr (H.S.)
* Correspondence: jungsu@seoultech.ac.kr; Tel.: +82-2-970-6475

Abstract: This paper proposes a scheduling method of dual ESSs (Energy Storage Systems) for the purpose of reducing the peak load when there are sudden loads or generation changes during the on-peak time. The first ESS is scheduled once a day based on a day-ahead load prediction, and the second ESS is scheduled every 15 min during on-peak time based on a short-term load prediction by LSTM (Long Short-Term Memory). Special attention is paid to training the LSTM for the short-term load prediction by using the augmented past load data which is generated by adding possible uncertainties to the past load and temperature data. Based on the load forecast, optimization problems for the scheduling are formulated. The proposed scheduling method is validated using load and temperature data from a real building. In other words, when the proposed method is applied to the real building energy data in the case study, it not only shaves the peak load during on-peak time interval effectively but also results in lower electricity price although there are sudden load or temperature changes during the time interval.

Keywords: building energy management; deep learning; energy storage system; load forecast; real-time control

Citation: Hwang, J.S.; Kim, J.-S.; Song, H. Handling Load Uncertainty during On-Peak Time via Dual ESS and LSTM with Load Data Augmentation. *Energies* **2022**, *15*, 3001. https://doi.org/10.3390/en15093001

Academic Editors: Shi-Jie Cao and Wei Feng

Received: 23 February 2022
Accepted: 16 April 2022
Published: 20 April 2022

1. Introduction

As the number of big buildings has increased, there has been an increase not only in their energy consumption and CO_2 emission but also in their proportion of global total energy consumption and CO_2 emissions. For instance, electricity consumption in buildings was about 55% of global total electricity consumption in the year 2019 [1]. Hence, it is of the utmost importance to devise building energy management systems (BEMS) in order to manage the total energy consumption efficiently.

In recent BEMS's research, a large amount of effort is directed to integrating renewable energy sources (RES) systematically [2]. Since RES is intermittent by nature and the load is unknown, inevitably BEMS has to be devised in such a way that it can deal with uncertainties in both the load and RES. ESS (Energy Storage Systems) are known to be the most efficient method to handle this problem. Consequently, one of the most important BEMS's tasks is to have an algorithm to charge and discharge ESS in such a way that the management of the building's energy is carried out successfully in terms of supply–demand balance, low electricity prices, and lower peak loads, etc. BEMS charges ESS when the demand is low and discharges it when the demand is high for the purpose of reducing electricity prices or the peak load during on-peak times [3,4].

The optimal ESS charging and discharging during a day has to be performed while taking the load into account. However, the load is unknown in advance, which means that BEMS has to be able to predict the load [5]. Various artificial intelligence-based methods for load prediction have been developed using ANN (Artificial Neural Network) [6,7] and RNN (Recurrent Neural Network) for sequential data prediction [8]. To overcome the weak long-term dependency of RNN, the LSTM (Long Short-Term Memory) network has been made and popularly applied to forecast loads [9–11]. Recently, GRU (Gated Recurrent

Units) have been developed in order to reduce the number of parameters in LSTM and improve the convergence speed of training, and they are also employed to predict the load [12,13].

Usually, there are two approaches for ESS operation in BEMS. In the first approach, a day-ahead load prediction is made at the beginning of a day, and the schedule of charging and discharging is determined for the day considering the prediction. When a day starts, BEMS charges or discharges ESS following the determined schedule. In the offline scheduling approach, ESS works such that supply–demand balance is maintained and the electricity price according to ToU (Time-of-Use) is minimized [14,15]. This approach cannot deal with any uncertainties in the load or the temperature during the on-peak time since charging and discharging are scheduled at the beginning of a day. The other possible approach is to use another ESS, which is used online using real-time information on load and temperature to achieve the objectives of the BEMS, such as peak reduction. This second ESS is used under the assumption that the first ESS operates in accordance with the schedule made at the beginning of the day. As a compromise of these two approaches, in [16,17], only one ESS is employed, but it is re-scheduled at a certain time during a day based on real-time information.

Recently, renewable generation and EV (Electrical Vehicles) charging stations have been installed in buildings. Although they provide more electrical power and convenience, it is difficult to predict how much power must be generated by renewable generation sources and how much will be consumed by EVs during a day. If during on-peak times renewable generation produces less electricity than forecast or many EVs try to charge simultaneously, then the peak load can be very large. Hence, online real-time monitoring and ESS scheduling are important in order to deal with such situations. For instance, a real-time ESS operation method is developed to handle uncertainties [18–20]. A multi-time scale coordination is devised to reduce the effect of uncertainties [21,22] in BEMS operation. An optimal online ESS operation method to take uncertainties in solar generation and load variation into account using stochastic programming is designed in [23]. Furthermore, real-time energy management for apartment buildings using MPC (Model Predictive Control) [24], and energy management considering random events such as EV charging have been an emerging research area [25,26].

Along this line of research, this paper focuses on real-time ESS scheduling for peak load reduction when there is a large discrepancy in supply and demand during on-peak times. To this end, it is assumed in this paper that there are two ESSs: one (called ESS_{off} hereafter) is scheduled offline using a day-ahead load prediction and the other (called ESS_{on} hereafter) is scheduled online and is charged or discharged during on-peak time based on real-time short-term load predictions. ESS_{on} is necessary because the scheduled plan for ESS_{off} might not be effective to reduce the peak load due to the uncertainties during on-peak time period. Hence, the performance of ESS_{on} is heavily dependent on real-time short-term load prediction. Roughly speaking, ESS_{off} deals with a slowly varying deficient load and ESS_{on} works for a rapidly varying deficient load.

The main contribution of the paper is twofold. First, for the purpose of scheduling ESS_{on}, a short-term prediction based on LSTM and past temperature and load is developed. Especially, the training data for the LSTM is augmented such that the possible uncertainties during on-peak time are added to the past load and temperature data. The trained LSTM is used to predict the load during the on-peak time period. Second, using the short-term prediction by the LSTM, an optimization is formulated to make a plan for charging and discharging of ESS_{on}. In the optimization, various requirements on, for example, SoC (State of Charge) limit of the amount of charging and discharging at a time are modeled as constraints in the optimization. A case study using real temperature and load data of a building shows that the proposed scheduling method for ESS_{on} successfully reducing the peak load and thereby reducing the electricity cost. This means that the proposed method deals with uncertainties during the on-peak time efficiently.

In this paper, the variables with subscript 'off' mean that they are determined offline while the variables with subscript 'on' are determined online.

The paper is organized as follows. In Section 2, the configuration and objective of the paper are introduced. In Section 3, the proposed online ESS scheduling method is presented, which is followed by a case study based on real building data in Section 4. Section 5 concludes the paper.

2. Objective and Description of the Proposed Method

In this section, the problem under consideration and the structure of the proposed BEMS are described. The proposed BEMS schedules a dual ESS (ESS_{off} and ESS_{on}) by solving optimization problems. The optimization problems rely on load predictions made by LSTM (Long Short-Term Memory). Figures 1 and 2 describes the proposed BEMS. The mathematical symbols in Figure 2 are explained in the next section.

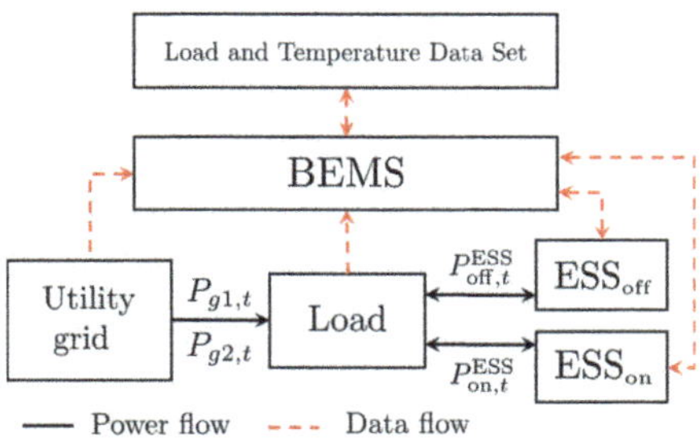

Figure 1. Structure of the proposed BEMS.

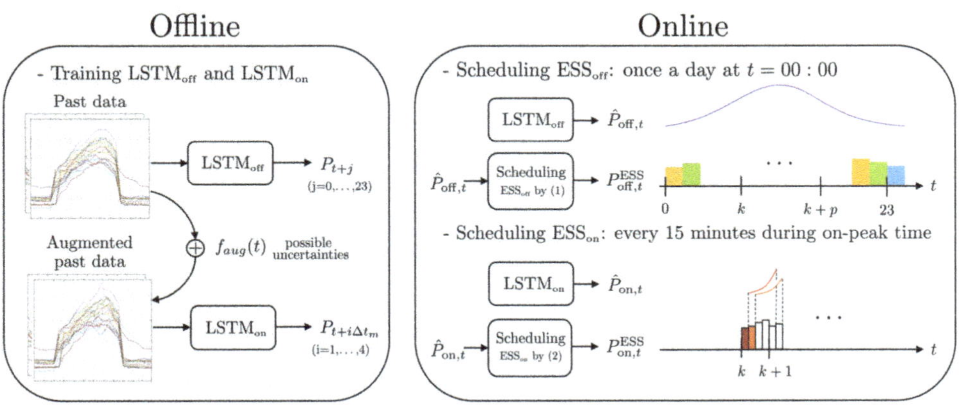

Figure 2. The structure of the proposed BEMS method.

To be specific, in offline mode, the two-deep learning networks $LSTM_{off}$ and $LSTM_{on}$ are trained using the past temperature and load data set. Then, at the beginning of a day, the trained $LSTM_{off}$ computes a day-ahead load prediction and the prediction is used to define an optimization problem with decision variables $P^{ESS}_{off,0}, \cdots, P^{ESS}_{off,23}$ for scheduling ESS_{off} where $P^{ESS}_{off,t}$ denotes the amount of charging or discharging for ESS_{off} every hour (i.e., 24 times of charging or discharging a day). In addition, during an on-peak time period, $LSTM_{on}$ generates the load prediction every 15 min for the next 1 h to consider uncertain situations which are not taken into account in $LSTM_{off}$ (or scheduling ESS_{off}). Based on short-term predictions by $LSTM_{on}$ including current load (i.e., $P_0, \hat{P}_{on,1}, \cdots, \hat{P}_{on,4}$), an optimization problem with decision variables $P^{ESS}_{on,0}, \cdots, P^{ESS}_{on,4}$ (i.e., 5×15 min) is defined for charging and discharging plan for ESS_{on}. Then, only the first element $P^{ESS}_{on,0}$ of the optimal solution is applied to ESS_{on}, and this procedure is repeated every 15 min similar to the receding horizon implementation in MPC (Model Predictive Control) [27].

The main objective of the BEMS design in this paper is to reduce the peak load during on-peak times by scheduling ESS_{on} based on short-term load prediction by $LSTM_{on}$, which is trained using the augmented load data to deal with abrupt large load changes. In achieving the objective, physical constraints such as SoC bounds or the limit of ESS output are taken into account.

3. Proposed Method

In this section, details of the proposed method are described. The results consist of offline ESS scheduling and online scheduling. The offline ESS scheduling is a modified result of that in [17] and the focus of this paper is mainly placed on the online ESS scheduling. Hence, offline ESS scheduling is briefly reviewed for the self-sufficient presentation of the main result and then the proposed online ESS scheduling is presented.

3.1. ESS_{off} Scheduling

3.1.1. Training $LSTM_{off}$ for a Day-Ahead Load Forecast Using Past Load and Temperature Data

In building energy management, a day-ahead load prediction is indispensable for optimization-based ESS scheduling. Since LSTM is efficient at predicting time series data [28], it is employed to forecast the load demand of the building. Since LSTM is already a well-known deep learning technique, it is not explained here. For details, see [17,28].

For a day-ahead prediction, $LSTM_{off}$ is trained using past building loads. In this work, the load data are assumed to be measured and saved every hour. Moreover, to enhance the prediction accuracy, hourly measured outdoor temperature data are also used. Hence, the input data to $LSTM_{off}$ for training is of the form $\{P_{t-j}, T_{t-j}\}$ $(j = 1, \cdots, 24)$ and $\{P_t, P_{t+1}, \cdots, P_{t+23}\}$ is used as the output of the network. When $LSTM_{off}$ is trained, if $\{P_{t'-j}, T_{t'-j}\}$ $(j = 1, \cdots, 24)$ is injected into the trained $LSTM_{off}$, $LSTM_{off}$ generates $\{\hat{P}_{\mathrm{off},t'}, \hat{P}_{\mathrm{off},t'+1}, \cdots, \hat{P}_{\mathrm{off},t'+23}\}$ as the load prediction for the next 24 h.

3.1.2. Scheduling ESS_{off} via Optimization

For the purpose of scheduling ESS_{off}, the trained $LSTM_{off}$ generates the load prediction $\hat{P}_{\mathrm{off},t}$ at midnight for the next 24 h using the load and temperature data from the previous day. Then, to decide the amount of charging and discharging $P^{\mathrm{ESS}}_{\mathrm{off},t}$ for the next 24 h, an optimization problem is formulated on the basis of the prediction as follows:

$$\min_{\substack{P^{\mathrm{ESS}}_{\mathrm{off},t} \\ t=0,\cdots,23}} \sum_{t=0}^{23} \left\{ C_{g,t}\hat{P}_{g1,t} + w_1 P_{gL1,t} + w_2 \left(P^{\mathrm{ESS}}_{\mathrm{off},t} - P^{\mathrm{ESS}}_{\mathrm{off},t-1} \right)^2 + w_3 {P^{\mathrm{ESS}}_{\mathrm{off},t}}^2 \right\} \tag{1a}$$

subject to $\forall t \in \{0, 1, \cdots, 23\}$

$$\hat{P}_{g1,t} = \hat{P}_{\mathrm{off},t} + P^{\mathrm{ESS}}_{\mathrm{off},t} \tag{1b}$$

$$P_{gL1,t} = \begin{cases} P_{g,min} - \hat{P}_{g1,t}, & \hat{P}_{g1,t} < P_{g,min} \text{ and } t \in \{\text{on-peak time}\} \\ 0, & \text{otherwise} \end{cases} \tag{1c}$$

$$SoC_{\mathrm{off},t} = SoC_{\mathrm{off},t-1} + \frac{\eta}{E_{c1}} P^{\mathrm{ESS}}_{\mathrm{off},t} \tag{1d}$$

$$SoC_{min} \leq SoC_{\mathrm{off},t} \leq SoC_{max} \tag{1e}$$

$$\mid P^{\mathrm{ESS}}_{\mathrm{off},t} \mid \leq P^{\mathrm{ESS}}_{\mathrm{off},max} \tag{1f}$$

$$\mid \sum_{t=0}^{23} P^{\mathrm{ESS}}_{\mathrm{off},t} \mid \leq \alpha \tag{1g}$$

where $C_{g,t}$ is the constant denoting the electricity price at time t and $\hat{P}_{g1,t}$ is the estimated amount of electricity to be used. Hence, the first term $C_{g,t}\hat{P}_{g1,t}$ in the objective function denotes the estimated electricity cost that will be paid. w_i $(i = 1,2,3)$ are weights that make each term contribute similarly to the entire objective function. $P_{g,min}$ in (1c) represents the contracted minimum amount of electricity from the main grid during on-peak time. It is assumed that the operator of the building's energy provides the main grid operator with this information. It is useful to define $P_{g,min}$ for not only the energy manager of the building's energy but also the main grid operator since the existence of such a value can help make a long-term energy plan for both the building and the main grid. $P_{gL1,t}$ is the difference between $P_{g,min}$ and $\hat{P}_{g1,t}$ at time t during the on-peak time. Since the contract is made between the building energy operator and the grid operator such that the building spends at least $P_{g,min}$ electricity during the on-peak time, a penalty has to be imposed on $P_{gL1,t}$, which is expressed using the second term $w_1 P_{gL1,t}$. $P^{\text{ESS}}_{\text{off},t}$ is the amount of charging or discharging from ESS_{off} at time t. Hence, the third and fourth terms in the objective function imply that the variation and amount of charging or discharging have to be small. Such consideration is helpful for both health and lifetime of the ESS. Note that $P^{\text{ESS}}_{\text{off},t} \geq 0$ implies charging and $P^{\text{ESS}}_{\text{off},t} < 0$ means discharging. (1d) denotes how $SoC_{\text{off},t}$ changes according to $P^{\text{ESS}}_{\text{off},t}$ where η and E_{c1} denote the efficiency of $P^{\text{ESS}}_{\text{off},t}$ and the capacity of ESS_{off}. (1e,f) are the constraints on $SoC_{\text{off},t}$ and $P^{\text{ESS}}_{\text{off},t}$, respectively. The last constraint (1g) is used to maintain the initial value of $SoC_{\text{off},t}$ to a constant value at the beginning of a day by setting the α small enough. Note that $\alpha \equiv 0$ leads to an equality constraint that can restrict the feasibility of the problem. The optimization problem is a modified version of that in [17].

After computing the prediction $\hat{P}_{\text{off},t}$ by LSTM_{off}, if the operator solves the optimization problem (1), the result can be depicted by Figure 3. In Figure 3, k is the start point of the on-peak time period and p is the length of the on-peak time period, and $\hat{P}_{\text{off},t}$ and $\hat{P}_{g1,t}$ denote the load power prediction before and after ESS_{off} is applied, respectively. Note that the peak of $\hat{P}_{\text{off},t}$ is reduced thanks to ESS_{off}. Mostly, ESS_{off} discharges during on-peak time $\in [k, k+p]$ to reduce the electricity cost with $P_{g,min}$ being the minimum, and charges during off-peak time to satisfy (1g). Therefore, comparing $\hat{P}_{\text{off},t}$ with $\hat{P}_{g1,t}$ employing ESS_{off}, peak reduction is achieved, thereby resulting in lower cost. Note that reducing the peak of $\hat{P}_{g1,t}$ can bring about a reduction of the real consumed power.

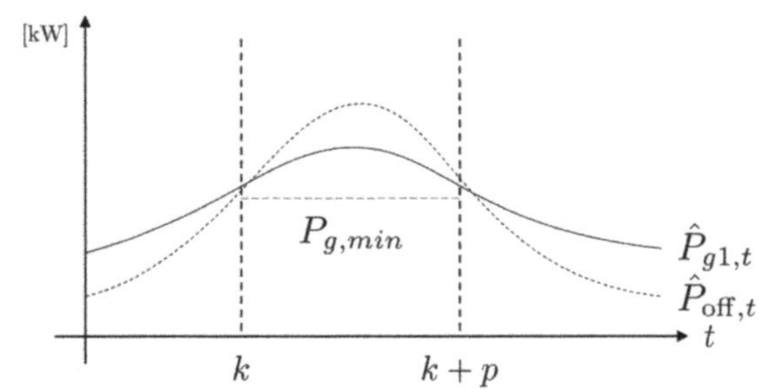

Figure 3. Offline ESS optimization.

On the other hand, if the load uncertainty in real-time is denoted by φ_t, then the real load can be expressed by $P_{g1,t} = \hat{P}_{\text{off},t} + P^{\text{ESS}}_{\text{off},t} + \varphi_t$. This means that the cost for buying electricity from the main grid can vary significantly depending on the load uncertainty from the offline forecast. If the uncertainty is small, the cost variation becomes acceptable, but the cost will not be trivial if the uncertainty is significant due to, for example, abrupt large load changes such as charging electric vehicles. To deal with these situations, we present strategies on how to reduce the effect of the uncertain load during the on-peak times by improving the performance of the short-term load prediction, and how to operate ESS_{on} based on the enhanced short-term load prediction.

3.2. ESS_{on} Scheduling

Under the assumption that ESS_{off} is scheduled, this section presents a scheduling method for ESS_{on} which works mainly during the on-peak time period. LSTM_{on} is trained using augmented past load and temperature data for online short-term load forecasting first, and then a charging and discharging strategy is proposed using the online load forecast by using LSTM_{on} and convex optimization.

3.2.1. Online Short-Term Load Forecast via LSTM with Data Augmentation

ESS_{on} is employed in this work for the purpose of handling sudden large variations in load or temperature during the on-peak time period and is scheduled based on short-term load forecasts by LSTM_{on}. Hence, LSTM_{on} has to be trained such that it can generate short-term load forecasts even when there are unexpected load or temperature variations that do not exist in the past data. To this end, the past load and temperature data can be augmented [29]. The augmented data can be generated by adding possible values of sudden uncertainties (synthetic uncertainties) considering the possible situations in the building to past load and temperature data. To generate the synthetic uncertainties systematically, a probability density function (PDF) such as a Gaussian distribution can be used [30].

In other words, the synthetic uncertainties are generated by sampling data using a PDF, and it is added to the past data. For details, see the next section.

With this augmented data, LSTM_{on} is trained such that its input $\{P_{t-i\Delta t_m}, T_{t-i\Delta t_m}\}(i = 0, \cdots, 7)$ and its output $\{P_{t+\Delta t_m}, \cdots, P_{t+4\Delta t_m}\}$, where Δt_m is 15 min. When the training is complete, the trained LSTM_{on} generates $\{\hat{P}_{\text{on},t'+\Delta t_m}, \cdots, \hat{P}_{\text{on},t'+4\Delta t_m}\}$ when $\{P_{t'-i\Delta t_m}, T_{t'-i\Delta t_m}\}$ is given as the input to LSTM_{on}. Hence, the trained LSTM_{on} forecasts the next one-hour load and its resolution is 15 min. Such LSTM_{on} is expected to generate more accurate short-term load forecasts compared with the day-ahead load forecast when there are nontrivial variations in load or temperature.

3.2.2. Online ESS Operation

This subsection presents a scheduling ESS_{on} via convex optimization based on the short-term load forecast $\hat{P}_{\text{on},t}$ by LSTM_{on}.

During on-peak times, a short-term load prediction is made every 15 min for the next hour. In other words, at time k during the on-peak time period, the short-term load forecast $\{\hat{P}_{\text{on},k+\Delta t_m}, \cdots, \hat{P}_{\text{on},k+4\Delta t_m}\}$ is generated every Δt_m. Then, the following optimization problem is solved with the forecast.

$$\min_{\substack{P^{\text{ESS}}_{\text{on},t} \\ t=k,\cdots,k+4\Delta t_m}} \sum_{t=k}^{k+4\Delta t_m} \left\{ w_4 P_{gU2,t} + w_5 P_{gL2,t} + w_6 {P^{\text{ESS}}_{\text{on},t}}^2 \right\} \tag{2a}$$

$$\text{subject to } \forall t \in \{k, k+\Delta t_m, \cdots, k+4\Delta t_m\}$$

$$\hat{P}_{g2,t} = \hat{P}_{\text{on},t} + P^{\text{ESS}}_{\text{off},t} + P^{\text{ESS}}_{\text{on},t} \tag{2b}$$

$$P_{gU2,t} = \begin{cases} \hat{P}_{g2,t} - (\hat{P}_{g1,t} + \delta), & \hat{P}_{g2,t} > \hat{P}_{g1,t} + \delta \\ 0, & else \end{cases} \tag{2c}$$

$$P_{gL2,t} = \begin{cases} P_{g,min} - \hat{P}_{g2,t}, & \hat{P}_{g2,t} < P_{g,min} \\ 0, & else \end{cases} \tag{2d}$$

$$SoC_{\text{on},t} = SoC_{\text{on},t-\Delta t_m} + \frac{\eta}{E_{c2}} P^{\text{ESS}}_{\text{on},t} \Delta t_m \tag{2e}$$

$$SoC_{min} \leq SoC_{\text{on},t} \leq SoC_{max} \tag{2f}$$

$$|\ P^{\text{ESS}}_{\text{on},t}\ | \leq P_{\text{on},max} \tag{2g}$$

where $P^{\text{ESS}}_{\text{on},t}$ denotes the amount of charging or discharging from ESS_{on} at time t. In the cost function (2a), $w_4 P_{gU2,t}$ penalizes the power when it is higher than $\hat{P}_{g1,t} + \delta$. The first term makes ESS_{on} work only when the difference between the required load $\hat{P}_{g2,t}$ computed online and $\hat{P}_{g1,t}$ computed offline is higher than δ. Since ESS_{on} is usually expensive equipment, it is used only when there are large uncertainties. As depicted in Figure 4, ESS_{on} makes $\hat{P}_{g2,t}$ be between the two red dashed lines, $\hat{P}_{g1,t} + \delta$ and $P_{g,min}$, during on-peak time. In other words, if the uncertainties predicted by the online load forecast are small, ESS_{on} does not do anything, which is helpful for the lifetime of ESS_{on}. On the contrary, if the estimated uncertainty is nontrivial, ESS_{on} tries to reduce the effect of the uncertainty.

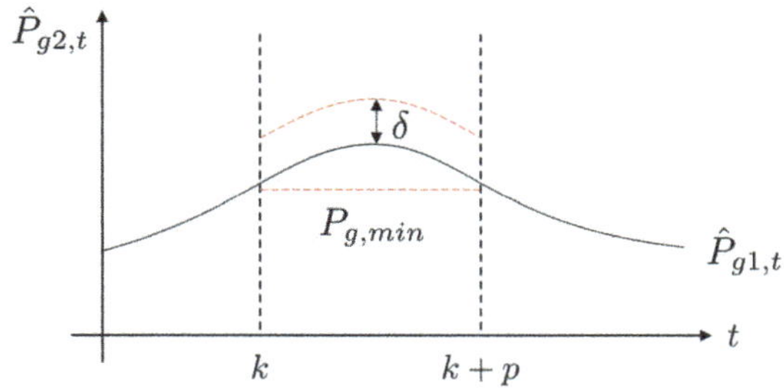

Figure 4. Online ESS optimization.

In the first term of the cost function, $\hat{P}_{g2,t}$ means estimates of the required load power since it is the sum of the estimated power $\hat{P}_{on,t}$ and the outputs of the two ESSs. The other terms in the cost are similar to the cost function (1).

Figure 5 summarizes how the online load forecast and ESS_{on} work during on-peak time. At time $t = k$, $LSTM_{on}$ generates short-term load forecast $\{\hat{P}_{on,k+\Delta t_m}, \cdots, \hat{P}_{on,k+4\Delta t_m}\}$. Based on this short-term forecast, the optimization problem (2) is solved to determine $P^{ESS}_{on,k}, P^{ESS}_{on,k+\Delta t_m}, \cdots, P^{ESS}_{on,k+4\Delta t_m}$. $P^{ESS}_{on,k}$ is applied for 15 min, and this procedure is repeated during on-peak time.

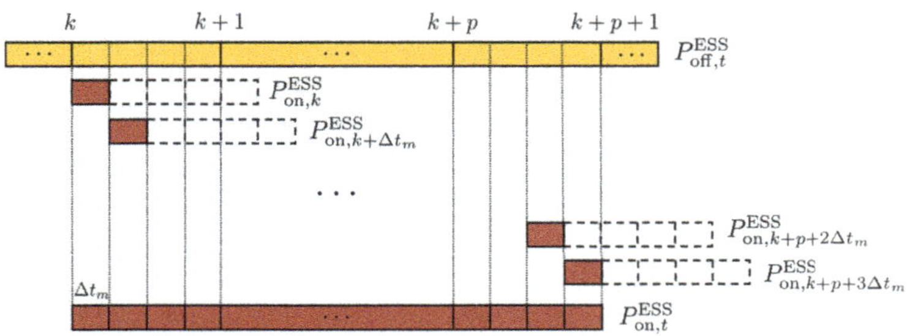

Figure 5. Online ESS operation process.

After the on-peak time, to maintain the initial value of $SoC_{on,t}$, the following optimization problem is solved. This procedure can be seen as (1g) in ESS_{off} scheduling. The cost function is similar to the third and fourth terms in the objective function (1), considering the health and lifetime of the ESS_{on}. At time $t = k + p + 1$, right after the on-peak time, ESS_{on} charges or discharges for two hours considering $P^{ESS}_{on,t}$ during the on-peak time. Note that γ is a small constant.

$$\min_{P^{ESS}_{on,t}} \sum_{t=k+p+1}^{k+p+2+3\Delta t_m} \left\{ w_6 {P^{ESS}_{on,t}}^2 + w_7 \left(P^{ESS}_{on,t} - P^{ESS}_{on,t-\Delta t_m} \right)^2 \right\} \tag{3a}$$

$$\text{subject to } \forall t \in \{k+p+1, k+p+1+\Delta t_m, \cdots, k+p+2+3\Delta t_m\}$$

$$\left| \sum_{t=p}^{k+p+2+3\Delta t_m} P^{ESS}_{on,t} \right| \leq \gamma \tag{3b}$$

Algorithm 1 summarizes the proposed energy management method using LSTM with the augmented dataset, dual ESSs, and convex optimization. Moreover, Figure 2 illustrates Algorithm 1.

Algorithm 1: Proposed energy management method.

```
Offline
  Train LSTM_off using past load and temperature data set
  Train LSTM_on using augmented past load and temperature data set
Online
  Repeat at t = 00:00
    A day-ahead load forecast using LSTM_off
    Make charging and discharging scheduling of ESS_off by solving the
     optimization (1)
    /* Repeat the following every 15 min during on-peak time   */
    for t ∈ on-peak time do
      Short-term load forecast using LSTM_on
      Make charging and discharging scheduling of ESS_on by
       solving the optimization (2)
    end for
    Initialize SoC_on,t by solving the optimization (3)
```

4. Case Study

This section shows the application of the proposed energy management scheme using data from a real building. It is shown that the online short-term forecast-based ESS scheduling can reduce the peak load effectively even when there are nontrivial load uncertainties during the on-peak time. This is because the trained $LSTM_{on}$ using augmented datasets generate better load forecasts, which is not easy to accomplish for a day-ahead load forecast.

4.1. Training Data and Data Augmentation

For training $LSTM_{off}$, load and outdoor temperature datasets are taken from [31] and they are measured for a commercial building located in Richland, WA, during the summer season (June–September) from 2009 to 2011.

For training the $LSTM_{on}$, the dataset is augmented for the purpose of obtaining a better load prediction when there are severe load uncertainties during on-peak time. To be specific, the dataset is augmented by adding the synthetic uncertainties to the original dataset. To augment the dataset systematically, the value of the synthetic uncertainty is generated using the normal distribution in (4) with a mean μ and a standard deviation σ as follows:

$$f_{aug}(t) = \frac{\beta}{\sigma\sqrt{2\pi}}\exp\left(-\frac{(t-\mu)^2}{2\sigma^2}\right) \quad (4)$$

where $\beta \in [-40, 40]$ is a scaling factor. $\mu \in [10, 18]$ and $\sigma \in [1, 3]$ are chosen to create a variety of uncertainties. Several examples of synthetic uncertainties generated by $f_{aug}(t)$ are shown in Figure 6.

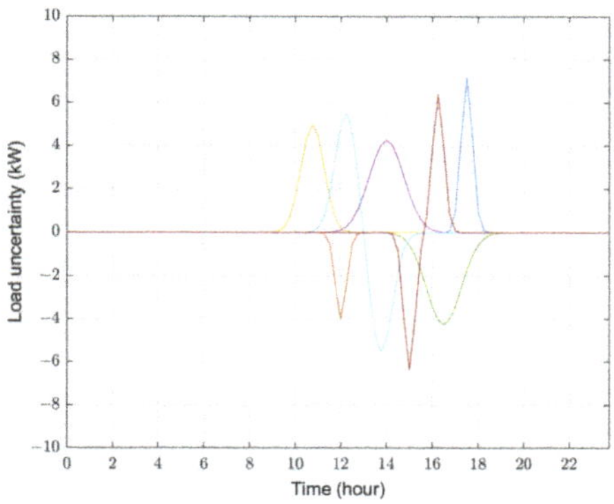

Figure 6. Load uncertainty data.

Figure 7 shows the original load and temperature dataset and the augmented dataset using the synthetic uncertainties calculated by (4).

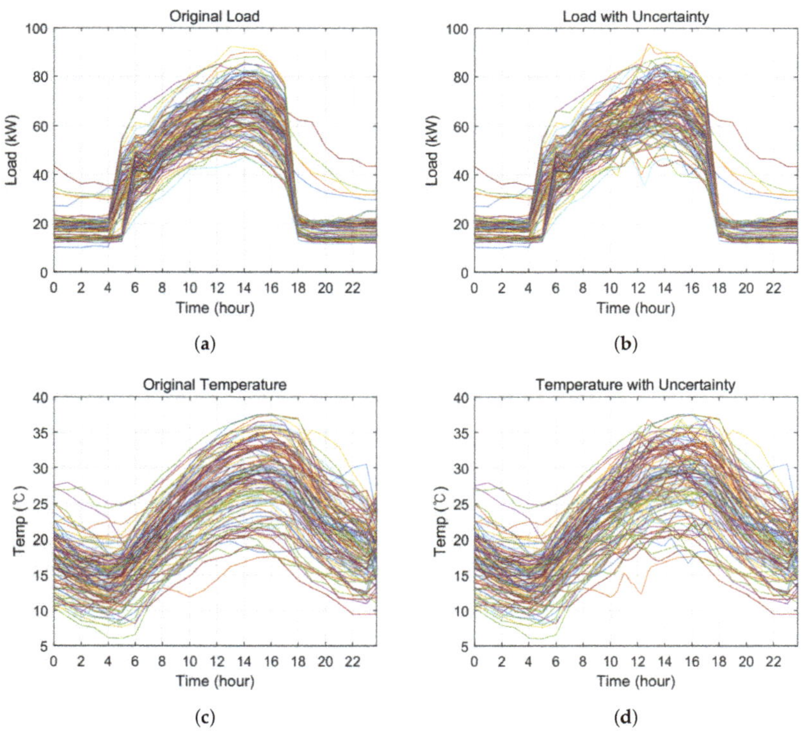

Figure 7. Original data and data with synthetic uncertainties. (**a**) Original load data, (**b**) Load with uncertainty data, (**c**) Original temperature data, (**d**) Temperature data with uncertainty.

4.2. Offline and Online Load Forecast by the Trained LSTMs

This section shows how accurately the trained $LSTM_{off}$ using the past dataset and $LSTM_{on}$ using the augmented past dataset generate load forecasts when there are uncertainties. Table 1 summarizes the structures and hyperparameters of $LSTM_{off}$ and $LSTM_{on}$. For training the LSTMs, Tensorflow 2.0 in Intel(R) Core(TM) i7-4790 with 8GB memory was used [32].

Table 1. LSTM Network parameters for training.

Parameter	$LSTM_{off}$	$LSTM_{on}$
Number of layers	3	3
Number of neurons	128 × 128 × 24	128 × 128 × 4
Batch size	128	64
Number of epochs	100	100
Learning rate	0.001	0.001
Loss function	MAE	MAE
Optimizer	ADAM	ADAM

Figures 8–10 show load forecast results by $LSTM_{off}$ and $LSTM_{on}$ for positive, negative, and sign indefinite synthetic uncertainties, respectively. In the figures, the red dotted lines denote the original load data and the black solid lines show the augmented load data. On the left of Figures 8–10, the blue solid lines are the load forecast $\hat{P}_{off,t}$ by $LSTM_{off}$. As seen in the figures, the load forecast by $LSTM_{off}$ is not accurate. This is natural since $LSTM_{off}$ is trained using the red lines but actually used the black lines (i.e., input to the trained $LSTM_{off}$) for the forecast. On the other hand, on the right of Figures 8–10, the short lines with various colors denote the one-hour load forecast by $LSTM_{on}$. Note that the online load forecast by $LSTM_{on}$ for the next one hour period is carried out every 15 min repeatedly during the on-peak time.

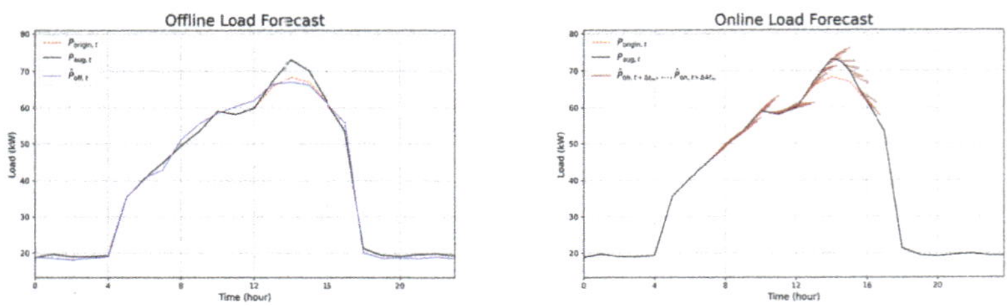

(a) (b)

Figure 8. Offline and online load forecasts with positive synthetic uncertainties. (**a**) Offline load forecast, (**b**) Online load forecast.

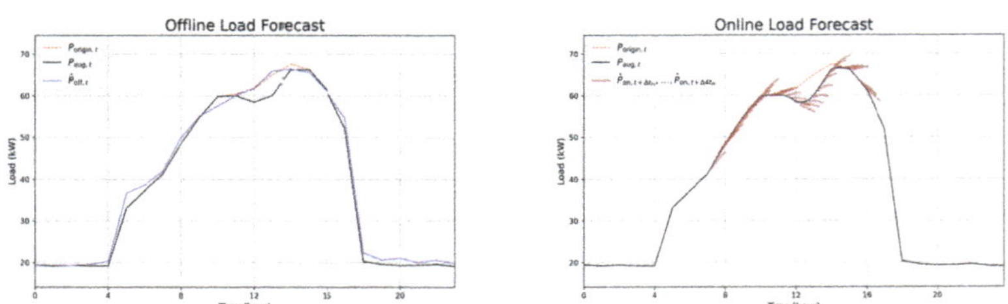

(a) (b)

Figure 9. Offline and online load forecasts with negative synthetic uncertainties. (**a**) Offline load forecast, (**b**) Online load forecast.

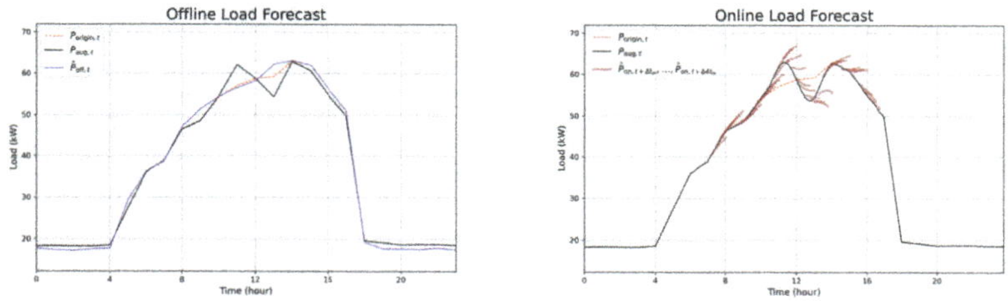

(a) (b)

Figure 10. Offline and online load forecasts with sign indefinite synthetic uncertainties.(**a**) Offline load forecast, (**b**) Online load forecast.

In view of the forecast results in Figures 8–10, LSTM_{on} generates better load forecasts for a load with uncertainties than LSTM_{off}. For quantitative comparison, the forecast errors are computed using Root Mean Square Error (RMSE) for the test data. Each RMSE for offline and online is given in (5a,b) considering the different resolutions. N is the number of test data and only the forecast during the on-peak time is calculated. Note that the online load forecast is repeated every 15 min. LSTM_{off} results in 5.611 kW and LSTM_{on} does 2.022 kW. Hence, the load forecast by LSTM_{on} can be used for ESS scheduling in scenarios where there are sudden load changes during on-peak times.

$$\text{RMSE}_{\text{off}} = \sqrt{\frac{1}{N} \cdot \frac{1}{p+1} \sum_{i=1}^{N} \sum_{t=k}^{k+p} (P^i_{\text{aug},t} - \hat{P}^i_{\text{off},t})^2} \tag{5a}$$

$$\text{RMSE}_{\text{on}} = \sqrt{\frac{1}{N} \cdot \frac{1}{4p+1} \cdot \frac{1}{4} \sum_{i=1}^{N} \sum_{t=k}^{k+p} \sum_{j=0}^{3} (P^i_{\text{aug},t+j\Delta t_m} - \hat{P}^i_{\text{on},t+j\Delta t_m})^2} \tag{5b}$$

4.3. ESS Scheduling Based on Online Load Forecast

Using the offline load estimate $\hat{P}_{\text{off},t}$ and online estimate $\hat{P}_{\text{on},t}$, $P^{\text{ESS}}_{\text{off},t}$ and $P^{\text{ESS}}_{\text{on},t}$ are determined by solving the convex optimization (1) and (2), respectively. For optimization, the CVX MATLAB toolbox is employed [33] and the tuning parameters for the optimization problems are given in Table 2. Each weight w_i is chosen such that each term in the objective function has a similar influence on the entire cost function.

Table 2. Parameters for ESS optimization.

Offline $P^{\text{ESS}}_{\text{off},t}$		Online $P^{\text{ESS}}_{\text{on},t}$		Common	
Parameter	**Value**	**Parameter**	**Value**	**Parameter**	**Value**
w_1, w_2, w_3	5, 0.05, 0.1	w_4, w_5, w_6, w_7	3, 50, 50, 0.1	η	0.95
E_{c1}	120 kWh	E_{c2}	40 kWh	SoC_{min}	0.1
α	3 kW	δ, γ	1 kW, 5 kW	SoC_{max}	0.9
$P_{\text{off},max}$	30 kW	$P_{\text{on},max}$	20 kW	$P_{g,min}$	54 kW

For both ESS_{off} and ESS_{on}, the initial SoC are set to 0.5, and $P_{g,min}$ is 0.8 times the peak of the average real load data. During the on-peak time, when the real power $P_{g1,t}$ and $P_{g2,t}$ bought from the main grid are smaller than $P_{g,min}$, then the penalty is calculated as

$$C_{\text{off}}^{pen} = \sum_{t \in j} C_{p,t}(P_{g,min} - P_{g1,t}); \{j | P_{g1,j} < P_{g,min}\}, \tag{6a}$$

$$C_{\text{on}}^{pen} = \sum_{t \in i} C_{p,t}(P_{g,min} - P_{g2,t}); \{i | P_{g2,i} < P_{g,min}\} \tag{6b}$$

where the penalty price $C_{p,t}$ is adjusted as double the value of $C_{g,t}$. Hence, the resulting costs for offline ESS scheduling and online ESS operations are given by

$$C_{\text{off}} = \sum_{t=0}^{23} C_{g,t} P_{g1,t} + C_{\text{off}}^{pen}, \tag{7a}$$

$$C_{\text{on}} = \sum_{t=0}^{k-1} C_{g,t} P_{g1,t} + \sum_{t=k}^{k+p+2+3\Delta t_m} C_{g,t} P_{g2,t} + \sum_{t=k+p+3}^{23} C_{g,t} P_{g1,t} + C_{\text{on}}^{pen} \tag{7b}$$

where period $t \in [k, k+p+2+3\Delta t_m]$ is the time when ESS_{on} is operated.

The simulation results are given in Figures 11–13 when $P_{\text{off},t}^{\text{ESS}}$ and $P_{\text{on},t}^{\text{ESS}}$ are applied to the cases with positive, negative, and sign indefinite synthetic uncertainties. Scheduling results are given on the left column in Figures 11–13 when only ESS_{off} (i.e., $P_{\text{off},t}^{\text{ESS}}$) is used. On the other hand, scheduling results are given on the right column in Figures 11–13 when both ESS_{off} and ESS_{on} (i.e., $P_{\text{off},t}^{\text{ESS}}, P_{\text{on},t}^{\text{ESS}}$) are used.

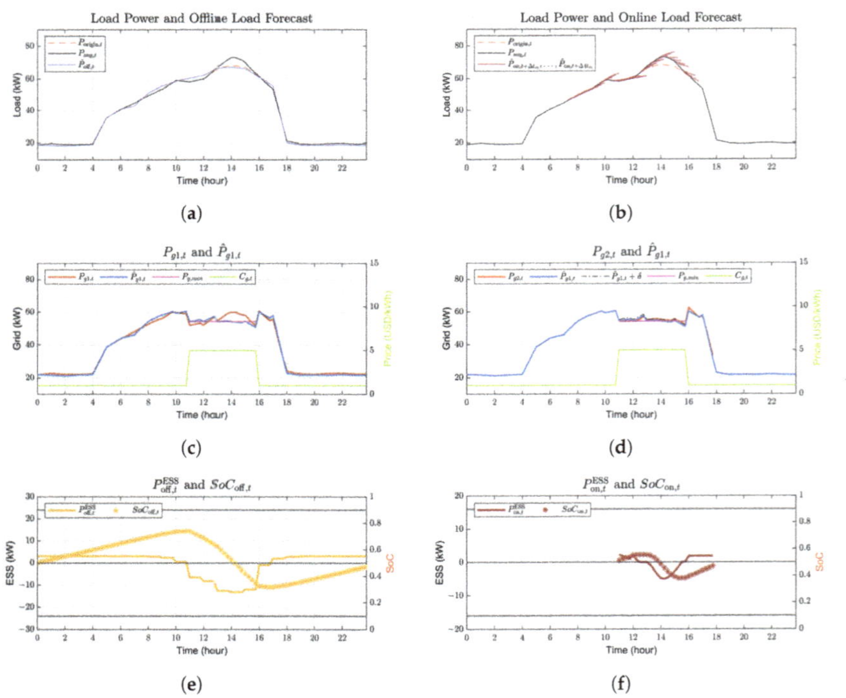

Figure 11. Offline ESS scheduling and online ESS operations with positive synthetic uncertainties. (**a**) Offline load forecast, (**b**) Online load forecast, (**c**) The estimated power $\hat{P}_{g1,t}$ and real power $P_{g1,t}$ from the main grid by using only ESS_{off}, (**d**) The estimated power $\hat{P}_{g1,t}$ and real power $P_{g2,t}$ from the main grid by using both ESS_{off} and ESS_{on}, (**e**) The amount of charging or discharging of ESS_{off} and $SoC_{\text{off},t}$, (**f**) The amount of charging or discharging of ESS_{on} and $SoC_{\text{on},t}$.

Figure 11 is the operation result when the uncertainty is positive. In Figure 11a,b, $\hat{P}_{\text{off},t}$ and $\hat{P}_{\text{on},t}$ are given, which are also presented in Figure 8. In Figure 11c,d, the power $P_{g1,t}$ and $P_{g2,t}$ bought from the main grid are depicted together with corresponding estimate

$\hat{P}_{g1,t}$ and $P_{g,min}$. The green solid line denotes the electricity price $C_{g,t}$; ToU pricing, where the start point of on-peak time k is 11:00; and the peak period p is 4 h. The price for off-peak time is 1 USD/kWh and on-peak time is 5 USD/kWh. In Figure 11e,f, $P^{ESS}_{off,t}$, $P^{ESS}_{on,t}$ (i.e., amount of charging and discharging of ESS_{off} and ESS_{on}), and corresponding SoC (i.e., $SoC_{off,t}$, $SoC_{on,t}$) are presented. The grey solid line denotes the constraints of SoC, SoC_{min} and SoC_{max}, respectively.

When Figure 11a,c are compared, the estimated power $\hat{P}_{g1,t}$ denoted by the blue line is reduced during on-peak time based on the optimization (1) but the real power $P_{g1,t}$ denoted by the brown line becomes higher than $\hat{P}_{g1,t}$ due to uncertain load that is not predicted by $\hat{P}_{off,t}$. In the case of offline scheduling, since using only ESS_{off} based on a day-ahead load forecast cannot consider the uncertainties during on-peak times, this results a higher cost.

On the other hand, when the brown lines denoting $P_{g1,t}$ in Figure 11c and the red lines denoting $P_{g2,t}$ in Figure 11d are compared, it is verified that the proposed scheduling method for ESS_{on} effectively reduces the peak load during the on-peak time. To be more specific, the effect of the uncertain load is eliminated by optimization (2c) based on the online short-term load forecast $\hat{P}_{on,t}$, which has better performance than $\hat{P}_{off,t}$. By adding ESS_{on}, peak load reduction is achieved when considering the uncertain load and reducing the required payment. This can be confirmed by the final cost paid. For Figure 11c, it costs 2072 but only 2042 for Figure 11d.

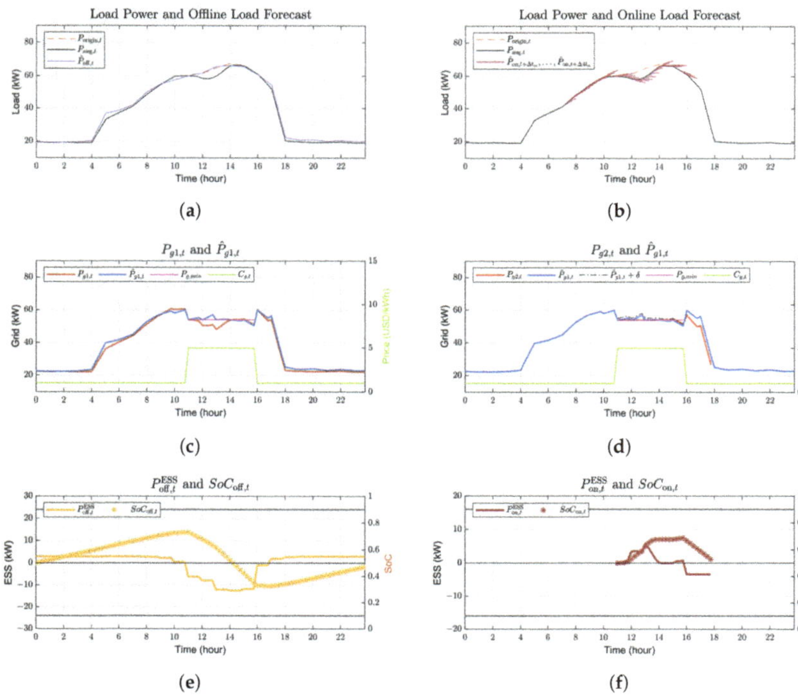

Figure 12. Offline ESS scheduling and online ESS operations with negative synthetic uncertainties. (**a**) Offline load forecast, (**b**) Online load forecast, (**c**) The estimated power $\hat{P}_{g1,t}$ and real power $P_{g1,t}$ from the main grid by using only ESS_{off}, (**d**) The estimated power $\hat{P}_{g1,t}$ and real power $P_{g2,t}$ from the main grid by using both ESS_{off} and ESS_{on}, (**e**) The amount of charging or discharging of ESS_{off} and $SoC_{off,t}$, (**f**) The amount of charging or discharging of ESS_{on} and $SoC_{on,t}$.

Figure 12 shows the operation results obtained by the proposed scheduling method when the uncertainty is negative. Similar observations to those in Figure 11 are possible. The brown line denoting $P_{g1,t}$ in Figure 12c becomes lower than $P_{g,min}$ during the on-peak time due to the negative uncertain load, resulting in a penalty, but in the case of $P_{g2,t}$ as denoted by the red line in Figure 12d, it does not deviate from $P_{g,min}$ constraints through ESS_{on}. It is verified that the proposed scheduling for ESS_{on} can effectively reduce the effect of the negative variations. Quantitatively, for Figure 12c, it costs 2079 but only 2042 for Figure 12d.

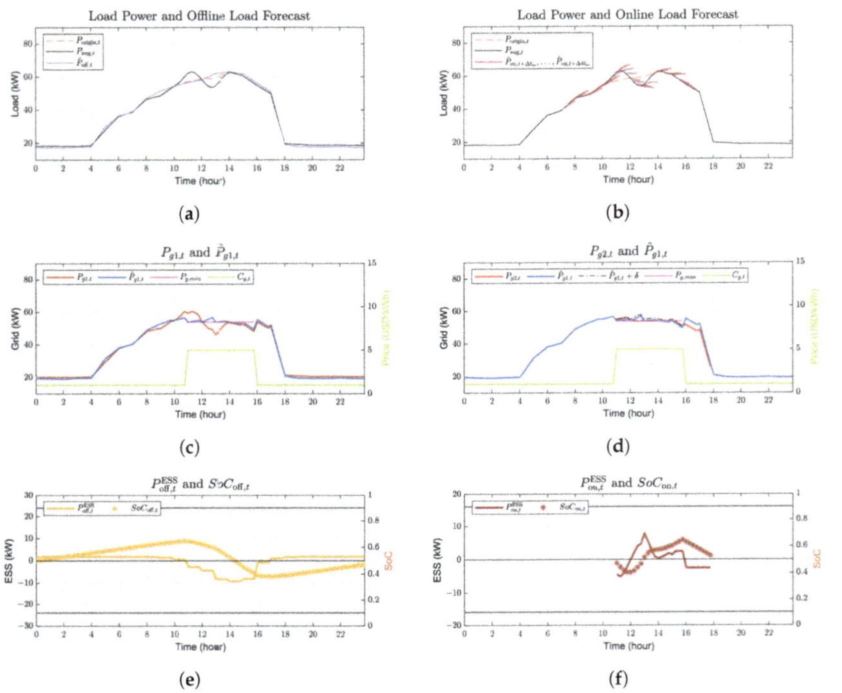

Figure 13. Offline ESS scheduling and online ESS operations with indefinite synthetic uncertainties. (**a**) Offline load forecast, (**b**) Online load forecast, (**c**) The estimated power $\hat{P}_{g1,t}$ and real power $P_{g1,t}$ from the main grid by using only ESS_{off}, (**d**) The estimated power $\hat{P}_{g1,t}$ and real power $P_{g2,t}$ from the main grid by using both ESS_{off} and ESS_{on}, (**e**) The amount of charging or discharging of ESS_{off} and $SoC_{\text{off},t}$, (**f**) The amount of charging or discharging of ESS_{on} and $SoC_{\text{on},t}$.

As the last case study, Figure 13 presents the operation results of the proposed ESS scheduling when the variation can be indefinite (i.e., either positive or negative). In Figure 13, both positive and negative variations are used. Again, similar observations to Figures 11 and 12 are also possible. Quantitatively, for Figure 13c, it costs 2031 but only 1962 for Figure 13d.

5. Discussion

This section analyzes the effect of parameters δ and the initial value of $SoC_{\text{on},t}$ on $P^{\text{ESS}}_{\text{on},t}$ (i.e., amount of charging or discharging of ESS_{on}).

5.1. Effect of δ

Figures 14 and 15 represent the operation results with both ESS for the same condition as in Figure 11 with different values of δ. $\delta = 1$ is used in Figure 11. Figure 14 corresponds to $\delta = 0$ and results in a cost value of 2035, and Figure 15 shows $\delta = 4$ and cost of 2062. As the value of δ becomes smaller, ESS_{on} has to deal with more uncertainties according to the optimization (2c). This is verified by comparing Figure 11 ($\delta = 1$) with Figure 14 ($\delta = 0$) since $P^{\text{ESS}}_{\text{on},t}$ discharges more in Figure 14. On the other hand, $P^{\text{ESS}}_{\text{on},t}$ discharges less in Figure 15 since $\delta = 4$ makes ESS_{on} cover less uncertainties compared with the previous cases. Therefore, the proper choice of δ must be used since ESS_{on} has to consider the constraints in (2f).

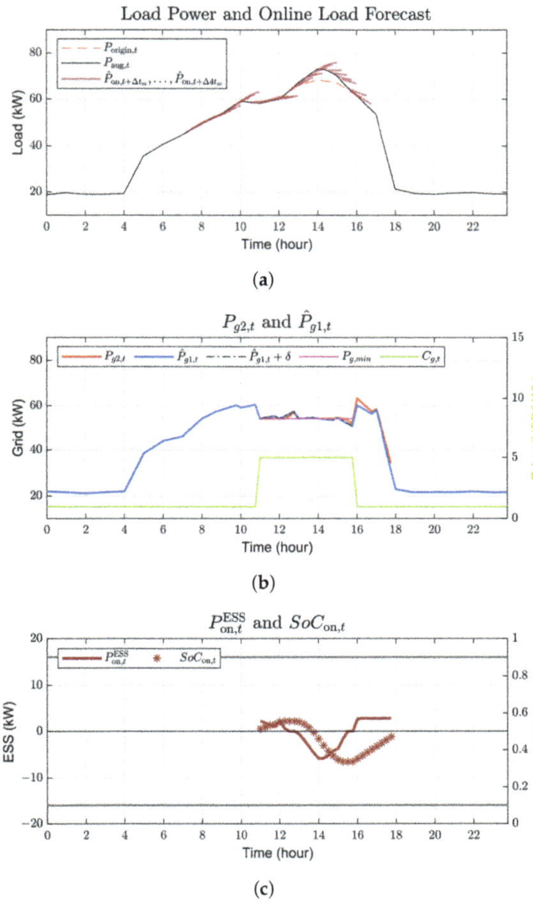

Figure 14. Online ESS operation at $\delta = 0$. (**a**) Online load forecast, (**b**) The estimated power $\hat{P}_{g1,t}$ and real power $P_{g2,t}$ from the main grid by using both ESS_{off} and ESS_{on}, (**c**) The amount of charging or discharging of ESS_{on} and $SoC_{on,t}$.

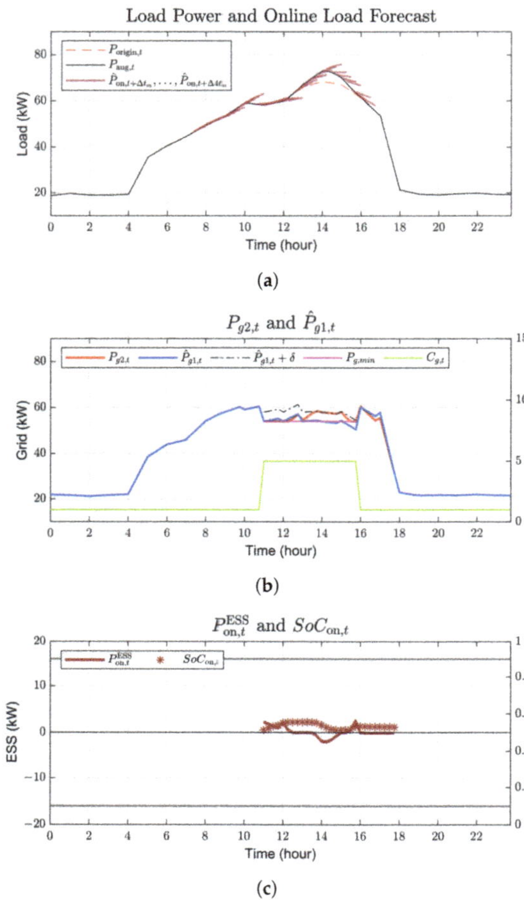

Figure 15. Online ESS operation at $\delta = 4$. (**a**) Online load forecast, (**b**) The estimated power $\hat{P}_{g1,t}$ and real power $P_{g2,t}$ from the main grid by using both ESS_{off} and ESS_{on}, (**c**) The amount of charging or discharging of ESS_{on} and $SoC_{on,t}$.

5.2. Effect of $SoC_{on,t}$ on the Initial Value

Figures 16 and 17 validate why it makes sense to set the initial value of $SoC_{on,t}$ to 0.5.

The configuration of Figure 16 is the same as that of Figure 11 except for the initial value of $SoC_{on,t} = 0.2$. As seen in Figure 16, ESS_{on} starts to discharge in order to handle the uncertainties but stops discharging after a short time due to the SoC constraints. This makes $P_{g2,t}$ violate the constraint on $\hat{P}_{g1,t} + \delta$, thereby leading to the value of the cost being 2050, which is larger than the case in Figure 11 with the initial value of $SoC_{on,t}$ 0.5.

Conversely, in Figure 17, the initial value of $SoC_{on,t}$ is set to 0.8 and all the other settings are the same as those in Figure 12. At this time, ESS_{on} has to charge to deal with the negative uncertainties, but it reaches the upper limit soon. Hence, it can not charge any more. As a result, $P_{g2,t}$ becomes smaller than $P_{g,min}$ sometimes. This results in the value of the cost being 2028, which is larger than that in Figure 12 with the initial value of $SoC_{on,t}$ set to 0.5.

Hence, since it is unknown which uncertainties occur during on-peak times, it is reasonable to set the initial value of $SoC_{on,t}$ to 0.5.

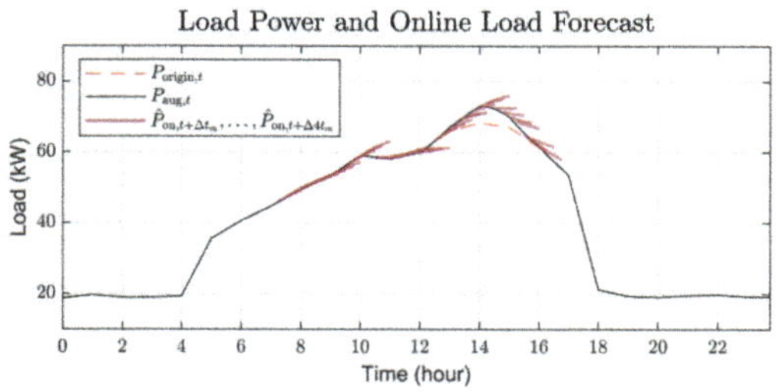

(a)

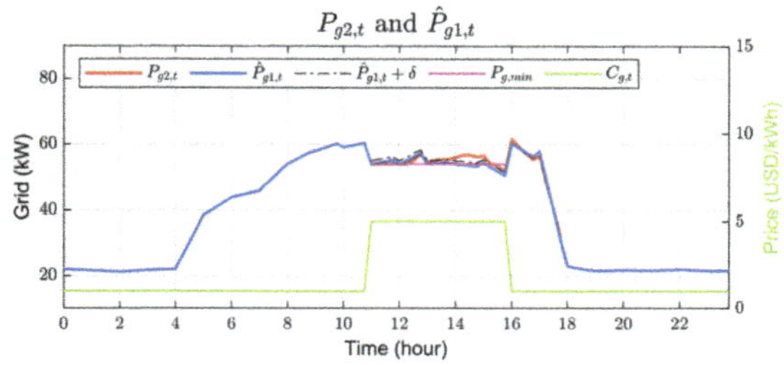

(b)

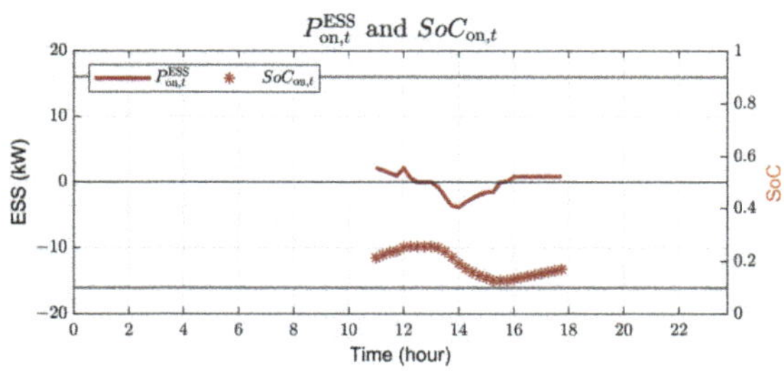

(c)

Figure 16. Online ESS operations with initial $SoC_{on,t}$ value being 0.2. (**a**) Online load forecast, (**b**) The estimated power $\hat{P}_{g1,t}$ and real power $P_{g2,t}$ from the main grid by using both ESS_{off} and ESS_{on}, (**c**) The amount of charging or discharging of ESS_{on} and $SoC_{on,t}$.

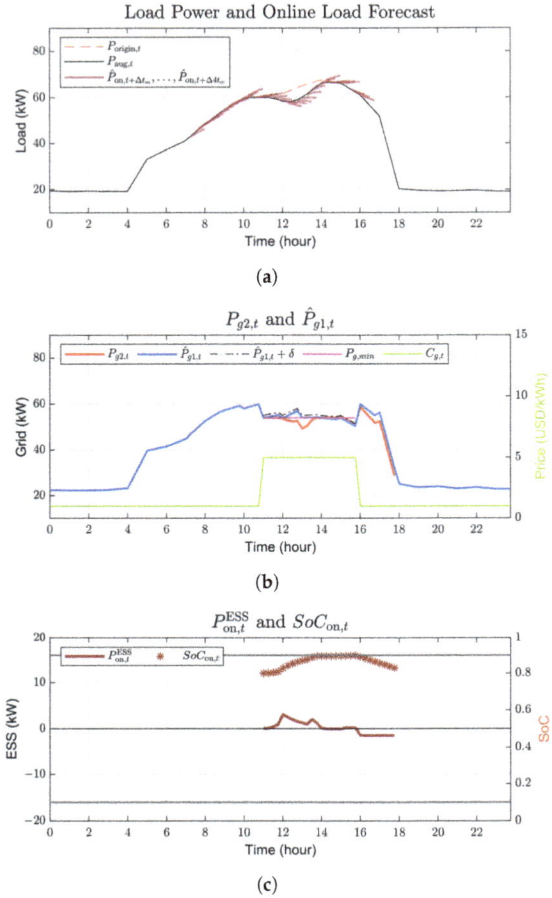

Figure 17. Online ESS operations with initial $SoC_{\text{on},t}$ value being 0.8. (**a**) Online load forecast, (**b**) The estimated power $\hat{P}_{g1,t}$ and real power $P_{g2,t}$ from the main grid by using both ESS_{off} and ESS_{on}, (**c**) The amount of charging or discharging of ESS_{on} and $SoC_{\text{on},t}$.

6. Conclusions

In this paper, a method for building energy management through real-time ESS operations is presented for a case where a sudden load variation occurs. For this, the LSTM network-based load forecast method was used for both the offline and online forecasts. In addition, to increase the prediction accuracy, the load of the building and the outdoor temperature were selected as input variables for the network. For the online load forecast, the LSTM network was trained using augmented past load and temperature data. Based on the day-ahead load forecast, the offline ESS was scheduled. On top of this, the online ESS is scheduled to reduce the effect of load uncertainties during on-peak times by using short-term load forecasts. For the scheduling of the two ESSs, optimization problems were formulated considering various physical constraints. In a case study, it was confirmed that the proposed method can reduce the effect of the uncertainties during on-peak times, which then leads to lower costs.

Author Contributions: J.S.H. surveyed the backgrounds of this research, designed the preprocessing data, designed the deep learning network, and performed the simulations and experiments showing the benefits of the proposed method. J.-S.K. and H.S. supervised and supported this study. All authors have read and agreed to the published version of the manuscript.

Funding: This study was supported by the Research Program funded by SeoulTech (Seoul National University of Science and Technology).

Institutional Review Board Statement: Not applicable.

Informed Consent Statement: Not applicable.

Conflicts of Interest: The authors declare no conflict of interest.

Abbreviations

The following abbreviations are used in this manuscript:

BEMS	Building Energy Management System
ESS	Energy Storage System
LSTM	Long Short-Term Memory
SoC	State of Charge
ToU	Time-of-Use
MAE	Mean Absolute Error
RMSE	Root Mean Square Error

References

1. United Nations Environment Programme. *Global Status Report for Buildings and Construction-Towards a Zero-Emissions, Efficient and Resilient Buildings and Construction Sector*; Technical Report; United Nations Environment Programme: Nairobi, Kenya, 2020.
2. Vasilj, J.; Gros, S.; Jakus, D.; Sarajcev, P. Multi-market Scheduling of Battery Storages Within Renewable Portfolios. In Proceedings of the 2018 IEEE PES Innovative Smart Grid Technologies Conference Europe (ISGT-Europe), Sarajevo, Bosnia and Herzegovina, 21–25 October 2018; pp. 1–6. [CrossRef]
3. Rahimi, A.; Zarghami, M.; Vaziri, M.; Vadhva, S. A simple and effective approach for peak load shaving using Battery Storage Systems. In Proceedings of the 2013 North American Power Symposium (NAPS), Manhattan, KS, USA, 22–24 September 2013; pp. 1–5. [CrossRef]
4. Joshi, K.A.; Pindoriya, N.M. Day-ahead dispatch of Battery Energy Storage System for peak load shaving and load leveling in low voltage unbalance distribution networks. In Proceedings of the 2015 IEEE Power Energy Society General Meeting, Denver, CO, USA, 26–30 July 2015; pp. 1–5. [CrossRef]
5. Kodaira, D.; Jung, W.; Han, S. Optimal Energy Storage System Operation for Peak Reduction in a Distribution Network Using a Prediction Interval. *IEEE Trans. Smart Grid* **2020**, *11*, 2208–2217. [CrossRef]
6. Deb, C.; Eang, L.S.; Yang, J.; Santamouris, M. Forecasting diurnal cooling energy load for institutional buildings using Artificial Neural Networks. *Energy Build.* **2016**, *121*, 284–297. [CrossRef]
7. Oprea, S.V.; Bâra, A. Machine Learning Algorithms for Short-Term Load Forecast in Residential Buildings Using Smart Meters, Sensors and Big Data Solutions. *IEEE Access* **2019**, *7*, 177874–177889. [CrossRef]
8. Shi, H.; Xu, M.; Li, R. Deep Learning for Household Load Forecasting—A Novel Pooling Deep RNN. *IEEE Trans. Smart Grid* **2018**, *9*, 5271–5280. [CrossRef]
9. Zheng, J.; Xu, C.; Zhang, Z.; Li, X. Electric load forecasting in smart grids using Long-Short-Term-Memory based Recurrent Neural Network. In Proceedings of the 2017 51st Annual Conference on Information Sciences and Systems (CISS), Piscataway, NJ, USA, 22–24 March 2017; pp. 1–6. [CrossRef]
10. Bouktif, S.; Fiaz, A.; Ouni, A.; Serhani, M.A. Optimal Deep Learning LSTM Model for Electric Load Forecasting using Feature Selection and Genetic Algorithm: Comparison with Machine Learning Approaches. *Energies* **2018**, *11*, 1636. [CrossRef]
11. Kong, W.; Dong, Z.Y.; Jia, Y.; Hill, D.J.; Xu, Y.; Zhang, Y. Short-Term Residential Load Forecasting Based on LSTM Recurrent Neural Network. *IEEE Trans. Smart Grid* **2019**, *10*, 841–851. [CrossRef]
12. Zheng, J.; Chen, X.; Yu, K.; Gan, L.; Wang, Y.; Wang, K. Short-term Power Load Forecasting of Residential Community Based on GRU Neural Network. In Proceedings of the 2018 International Conference on Power System Technology (POWERCON), Guangzhou, China, 6–8 November 2018; pp. 4862–4868. [CrossRef]
13. Sajjad, M.; Khan, Z.A.; Ullah, A.; Hussain, T.; Ullah, W.; Lee, M.Y.; Baik, S.W. A Novel CNN-GRU-Based Hybrid Approach for Short-Term Residential Load Forecasting. *IEEE Access* **2020**, *8*, 143759–143768. [CrossRef]
14. Gibilisco, P.; Ieva, G.; Marcone, F.; Porro, G.; Tuglie, E.D. Day-ahead operation planning for microgrids embedding Battery Energy Storage Systems. A case study on the PrInCE Lab microgrid. In Proceedings of the 2018 AEIT International Annual Conference, Bari, Italy, 3–5 October 2018; pp. 1–6. [CrossRef]
15. Ross, M.; Abbey, C.; Bouffard, F.; Joós, G. Microgrid Economic Dispatch With Energy Storage Systems. *IEEE Trans. Smart Grid* **2018**, *9*, 3039–3047. [CrossRef]
16. Choi, S.; Min, S.W. Optimal Scheduling and Operation of the ESS for Prosumer Market Environment in Grid-Connected Industrial Complex. *IEEE Trans. Ind. Appl.* **2018**, *54*, 1949–1957. [CrossRef]

17. Hwang, J.S.; Rosyiana Fitri, I.; Kim, J.S.; Song, H. Optimal ESS Scheduling for Peak Shaving of Building Energy Using Accuracy-Enhanced Load Forecast. *Energies* **2020**, *13*, 5633. [CrossRef]
18. Shi, W.; Li, N.; Chu, C.C.; Gadh, R. Real-Time Energy Management in Microgrids. *IEEE Trans. Smart Grid* **2017**, *8*, 228–238. [CrossRef]
19. Zhang, Z.; Wang, J.; Ding, T.; Wang, X. A Two-Layer Model for Microgrid Real-Time Dispatch Based on Energy Storage System Charging/Discharging Hidden Costs. *IEEE Trans. Sustain. Energy* **2017**, *8*, 33–42. [CrossRef]
20. Li, S.; Yang, J.; Song, W.; Chen, A. A Real-Time Electricity Scheduling for Residential Home Energy Management. *IEEE Internet Things J.* **2019**, *6*, 2602–2611. [CrossRef]
21. Sharma, S.; Verma, A.; Xu, Y.; Panigrahi, B.K. Robustly Coordinated Bi-Level Energy Management of a Multi-Energy Building Under Multiple Uncertainties. *IEEE Trans. Sustain. Energy* **2021**, *12*, 3–13. [CrossRef]
22. Sharma, S.; Xu, Y.; Verma, A.; Panigrahi, B.K. Time-Coordinated Multienergy Management of Smart Buildings Under Uncertainties. *IEEE Trans. Ind. Inform.* **2019**, *15*, 4788–4798. [CrossRef]
23. Hafiz, F.; Awal, M.A.; Queiroz, A.R.d.; Husain, I. Real-Time Stochastic Optimization of Energy Storage Management Using Deep Learning-Based Forecasts for Residential PV Applications. *IEEE Trans. Ind. Appl.* **2020**, *56*, 2216–2226. [CrossRef]
24. Rezaei, E.; Dagdougui, H. Optimal Real-Time Energy Management in Apartment Building Integrating Microgrid With Multizone HVAC Control. *IEEE Trans. Ind. Inform.* **2020**, *16*, 6848–6856. [CrossRef]
25. Tushar, M.H.K.; Zeineddine, A.W.; Assi, C. Demand-Side Management by Regulating Charging and Discharging of the EV, ESS, and Utilizing Renewable Energy. *IEEE Trans. Ind. Inform.* **2018**, *14*, 117–126. [CrossRef]
26. liu, z.; Wu, Q.; Shahidehpour, M.; Li, C.; Huang, S.; Wei, W. Transactive Real-Time Electric Vehicle Charging Management for Commercial Buildings With PV On-Site Generation. *IEEE Trans. Smart Grid* **2019**, *10*, 4939–4950. [CrossRef]
27. Rawlings, J.B.; Mayne, D.Q.; Diehl, M.M. *Model Predictive Control: Theory, Computation, and Design*, 2nd ed.; Nob Hill Publishing: Madison, WI, USA , 2020.
28. Hochreiter, S.; Schmidhuber, J. Long Short-Term Memory. *Neural Comput.* **2006**, *9*, 1735–1780. [CrossRef]
29. Shorten, C.; Khoshgoftaar, T.M. A survey on Image Data Augmentation for Deep Learning. *J. Big Data* **2019**, *6*, 60. [CrossRef]
30. Aprillia, H.; Yang, H.T.; Huang, C.M. Statistical Load Forecasting Using Optimal Quantile Regression Random Forest and Risk Assessment Index. *IEEE Trans. Smart Grid* **2021**, *12*, 1467–1480. [CrossRef]
31. OpenEI. Available online: https://openei.org/datasets/dataset/consumption-outdoor-air-temperature-11-commercial-buildings (accessed on 30 September 2019).
32. Tensorflow.org: Deep Learning Library Developed by Google. Available online: https://www.tensorflow.org/ (accessed on 20 November 2020).
33. CVX: Matlab Software for Disciplined Convex Programming. Available online: http://cvxr.com/ (accessed on 20 January 2021).

designs

Article

Impacts of Location on Designs and Economics of DIY Low-Cost Fixed-Tilt Open Source Wood Solar Photovoltaic Racking

Nicholas Vandewetering [1], Koami Soulemane Hayibo [2] and Joshua M. Pearce [2,3,*]

1 Department of Civil & Environmental Engineering, Western University, London, ON N6A 3K7, Canada; nvandew@uwo.ca
2 Department of Electrical & Computer Engineering, Western University, London, ON N6A 3K7, Canada; khayibo@uwo.ca
3 Ivey School of Business, Western University, London, ON N6A 3K7, Canada
* Correspondence: joshua.pearce@uwo.ca

Abstract: Although small solar photovoltaic (PV) systems avoid most soft costs, they still have a relatively high \$/W value due to racking costs. In order to fulfill the promise of small-scale plug-and-play solar, a do-it-yourself PV rack design is provided and analyzed here for six criteria: (1) made from locally-accessible renewable materials, (2) 25-year lifetime to match PV warranties, (3) able to be fabricated by average consumers, (4) able to meet Canadian structural building codes, (5) low cost and (6) that it is shared using an open-source license. The open-source wood-based fixed-tilt ground-mounted bifacial photovoltaic rack design evaluated here was found to be appropriate throughout North America. Economic analysis of the bill of materials showed the racking system ranges from 49% to 77% less expensive compared to commercial proprietary racking in Canada. The racking design, however, is highly dependent on the cost of lumber that varies widely throughout the world. Even for an absolute lower-cost design in Togo due to a lower fixed tilt angle and lower loads from lack of snow, it was not found to be economic because of the relatively high cost of wood. The recent volatile lumber market warrants local evaluation from those considering the use of the open-source design. This design, however, provides for a PV rack that can be manufactured with distributed means throughout most of the world enabling more equitable access to solar energy to support a circular bioeconomy.

Keywords: open-source; photovoltaic; racking; solar energy; biomaterials; wood; photovoltaic; mechanical design; balance of systems; renewable energy

Citation: Vandewetering, N.; Hayibo, K.S.; Pearce, J.M. Impacts of Location on Designs and Economics of DIY Low-Cost Fixed-Tilt Open Source Wood Solar Photovoltaic Racking. *Designs* **2022**, *6*, 41. https://doi.org/10.3390/designs6030041

Academic Editors: Shi-Jie Cao and Wei Feng

Received: 14 March 2022
Accepted: 18 April 2022
Published: 21 April 2022

1. Introduction

Solar photovoltaic (PV) technology is a naturally distributed renewable energy technology that is well established as a leading sustainable energy source [1] because of an excellent ecological balance sheet [2–4]. The last remaining barrier to widespread PV deployment has been economic costs [5], but PV prices have dropped 60% in the last decade [6–10]. This has brought the levelized cost of solar electricity [11] is often the lowest cost option on a large scale [12,13]. Not surprisingly, PV is the most rapidly expanding electricity generation source [13,14]. Even when economies of scale are not in play, Levin & Thomas [15] showed that small solar home systems can play an important role in achieving U.N. 'Sustainable Energy for All' goals. In the developed world, most PV systems are grid-tied and there has been a surge of interest among consumers because they can effectively lower their electric utility bills with lower-cost solar electricity [16,17].

Even with clear lifetime economic benefits, however, the capital cost of a PV systems can be challenging for many consumers, particularly the poor, both in the developing [18] and the developed countries [19–22]. One approach to overcoming this challenge is to start with a small do-it-yourself (DIY) [23] or use plug-and-play solar, where PV modules are connected through microinverters directly to the household circuits by consumers.

This latter approach is legal throughout some of Europe and is technically compliant with regulations in the U.S., although there is a widespread disparity in interpretations between the states and utilities [24,25]. Several studies have proposed methods to streamline the technology and regulations for such systems [25–28] because it would open up a large new market for PV to a wide range of consumers and potentially save Americans alone $13 billion/year [29].

Although these small DIY or plug-and-play systems avoid most of the soft costs associated with PV systems, they still have a relatively high $/W cost. This is because the majority of the cost declines in PV have come in the form of reduced PV module costs, while the relative cost of the balance of systems (BOS) has become more important [4,6,10]. The BOS consists of racking, electronics, and wiring. Racking, in particular, has been largely ignored in the peer-reviewed literature while the PV industry focused on various proprietary and costly aluminum extrusion profiles. Until recently, this was not important due to the fact that the relative costs of PV racking were marginal for the complete system and thus only modest progress was made in reducing PV racking materials and costs [30]. Because of this, current PV racking components can often dominate the costs of a PV system—particularly for smaller systems. For example, the low spot price for a PV module is currently US$0.19/W [31] while racking for a 3-module system costs US$535 (list price US$635) [32] which is US$1.78–0.45/W for 100 W and 400 W modules, respectively. Similarly, a 3-module pole mount is selling for US$1194, even with 400 W modules this is equivalent to about US$1/W for racking excluding the foundation [33].

Recently, several types of plastic-based PV racking have been proposed to reduce costs for low-tilt angle arrays, including: small-scale mobile PV arrays [34], flat roofs [35], ground-mounted systems at the equator in the developing world [36]. The vast majority of PV systems, however, have a much greater tilt angle, (e.g., approximately equivalent to the latitude). In addition, conventional racking materials for ground-mount systems (metal and concrete) have high embodied energies and play a major role in the environmental impact of a PV system [37]. Thus, in order to fulfill the promise of distributed DIY and plug-and-play solar, what is needed is a PV rack design that is; (1) made from locally-accessible sustainable renewable materials, (2) can be fabricated using simple hand tools by the average consumer, (3) has a 25-year life time to be equivalent to common PV warranties, (4) is structurally sound in order to weather high wind speeds and major snow loads (depending on the region), (5) has a low cost and (6) that it is shared using an open-source license so that many people can fabricate it themselves, or companies can make versions to offer in their local markets.

In order to overcome these challenges, this study reports on the technical and economic viability of an open-source DIY wood-based racking system. Specifically, a full structural analysis is completed for fixed-tilt wood-based PV racks developed following Canadian building codes for two locations: Ontario, Canada (to represent a northern latitude with heavy snow loads) and Togo (to represent a low-tilt angle system with no snow loads for locations close to the equator). The complete designs and bill of materials (BOM) of the racks are provided along with basic instructions and are released with an open-source license that will enable anyone to fabricate the rack system. The BOM costs are compared to the cost of proprietary commercial PV racks for both locations. The results of this study are discussed in the context of using an open-source DIY design for increasing PV deployments both in the global north and the global south.

2. Materials and Methods

All abbreviations are detailed in Abbreviations section.

2.1. Renewable Materials Selection for Racking: Wood

Wood was selected as a building material because it is locally available throughout most of the world. Choosing wood for the construction of PV racking can have both economic and environmental advantages. Unlike other construction materials, responsibly-sourced wood has the advantage of being sustainable [38], renewable, and comprised of approximately half carbon, which was recently taken from the atmosphere. When combined with lower energy needs for processing, wood actually has a negative combined embodied energy and carbon over alternative racking construction materials. For instance, aluminum (even with 31% recycled content), which is the most common PV racking material, has over 5 times the embodied CO_2e/kg of wood [39], giving wood a distinctive advantage.

There are many choices, however, when it comes to wood species and how they might be treated for decay resistance. These are somewhat governed by local availability. Since the most common and easily available choice is treated softwood species it is reviewed here. Pressure-treating greatly extends the service life of wood and is commonly used on the fast-growing and more economical softwood species. Micronized copper azole is among the latest generation of wood preservatives and is considered safer than other preservative systems for humans, animals, and the environment [40]. Now, a common treatment used in residential settings, micronized copper azole, goes by many brand names, (e.g., MicroPro/LifeWood and Wolmanized Outdoor wood, Yellawood, and SmartSense). Micronized copper azole is also less corrosive on fasteners and can come into direct contact with aluminum, which is typically used in solar panel frames. Pressure-treated SPF (Spruce, Pine, Fir) lumber was selected because of its low cost, high availability, and overall durability in outdoor conditions. It is the most common wood used to construct decks, fences, gazebos, and other outdoor structures in Canada. Depending on the weather conditions, pressure-treated lumber can stay up for up to 40 years without signs of decaying [41].

2.2. Material Properties

There is a limited variety of dimensional lumber that can be used in building wooden structures. The dimensional properties of common structural lumber are summarized in Table 1. It should be noted that in all cases, the base should be less than the height so that the member is loaded in its strong axis, thus producing the optimal moment of inertia and the first moment of area values.

Table 1. Dimensional properties of common sizes of construction lumber.

Lumber	Base b [mm]	Height h [mm]	Area A [mm^2]	Moment of Inertia I [mm^4]	First Moment of Area Q [mm^3]
2 × 4	38	89	3382	2,232,402	37,625
2 × 6	38	140	5320	8,689,333	93,100
2 × 8	38	184	6992	19,726,763	160,816
2 × 10	38	235	8930	41,096,604	262,319
2 × 12	38	286	10,868	74,079,911	388,531
4 × 4	89	89	7921	5,228,520	88,121
6 × 6	140	140	19,600	32,013,333	343,000

Where the cross-sectional area, A, in mm^2 is calculated by,

$$A = bh \tag{1}$$

The moment of inertia, I, in mm^4, for rectangular cross-sections is calculated by,

$$I = \frac{1}{12} bh^3 \tag{2}$$

The first moment of area, Q, in mm^3, for rectangular cross-sections is given by,

$$Q = \frac{hA}{8} \tag{3}$$

2.3. Case Studies

Two case studies were selected to demonstrate the wide range of potential PV system design situations. London, Ontario, Canada (42.9849° N, 81.2453° W) was selected as an example of a high-latitude location in the global north that would be expected to experience snow loading and need a substantial tilt angle. Lomé, Togo (6.1256° N, 1.2254° E) located near the equator was selected as a global south-based installation that would not have any snow-related losses nor necessitate designing for snow loads. The open-source System Advisory Model (SAM) [42–44] was used to determine the optimal fixed-tilt angle for both locations on an annual basis of energy production.

The optimal angle for London, Ontario for a fixed-tilt solar PV system according to SAM is 34° as shown in Figure 1a and for Lomé Togo is 10° as shown in Figure 1b.

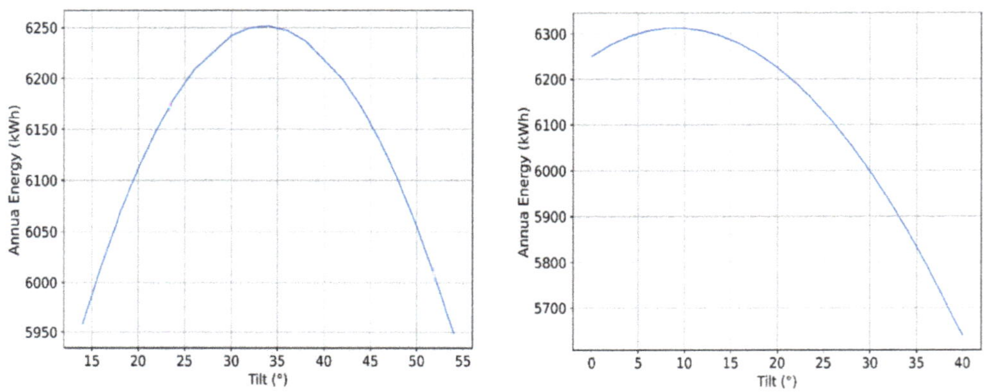

(**a**) (**b**)

Figure 1. Annual energy production for a solar module in (**a**) London Ontario and (**b**) Lomé Togo as a function of tilt angle.

These locations represent the extreme cases of a system in a location of high latitude and a system located by the equator. SAM can be used to determine the optimal angle for a location of any latitude. The optimal angle for locations with most of the human population is expected to be between these two extremes.

2.4. PV Racking Basic Design Parameters

The PV rack was designed to do a small system of ~1 kW so that it could be used for plug-and-play PV [25] and represent an initial cost that would be accessible to a greater population. A 400 W LG 400 W NeON2 BiFacial Solar Panel [45], was selected to take advantage of the bifacial PV rear surface solar absorption, which not only enhances electricity production [45–47] but also increases snow clearing on the front [48,49]. This racking system was specially designed for bifacial PV, but the design is still useful for any type of solar module. In regions where snow clearing or rear reflection are not of interest, bifacial modules can be substituted for any rigid framed module, which may further reduce the cost of the system. When selecting modules, it is important to ensure the module's rear and front load capacities are greater than the design load calculated in Appendix A. The lumber structural capacities are shown in Appendix B. The approximate dimensions of the LG 400 W NeON2 modules are 1 m × 2 m. If modules with different dimensions are to be used, then the specifications in the assembly instructions in Section 3.1.2 can be scaled up or down to meet the given module's requirements. Resizing the system will modify the design load calculated in Appendix A, which means smaller systems reduce costs by selecting smaller members such that no limits are exceeded in the structural analysis outlined in Appendix C.

In order for the racking to be widely applicable, they are designed to be ground-mounted and have a 500 mm ground clearance to avoid obstructing snow slide-off even in extreme northern environments [50]. The tilt angle is 34° for London, Ontario, and 10° for Lomé, Togo. Both systems are designed to withstand 80 mph winds and the London racking system must be able to handle typical snow load as well. These were chosen as extreme conditions so that if the system could withstand the mechanical loads in these conditions, it would be able to handle any loads in less severe environments. Following open hardware design guidelines [51–53] the design is released under CERN Open Hardware License V2 strong reciprocal variant (CERN-OHL-S) [54].

2.5. Design Analysis Assumptions

The snow, wind, and dead loads are calculated and combined for both case study designs following Appendix A. The structural analysis calculations are shown in Appendix B. Many common assumptions must be made to simplify and idealize the structural analysis. It should be noted that these assumptions are conservative, meaning that they further ensure that the structure will not fail unexpectedly.

- All loads shall act perpendicular to the face of the modules, so joists are experiencing the worst-case flexural load.
- All members are idealized as pins connected with no fixed end moments since joist hangers and brackets still allow for rotation [55].
- The wind load and snow load will only be applied to the surface of the modules because the accumulation of snow on wooden members is practically negligible.
- The wind load and snow load are assumed to be distributed evenly throughout the surface of the modules because snow and wind accumulation is only considered for large structures as per NBCC 4.1.6 [56].
- The modules can be idealized as a one-way slab since the length to span ratio is 2 [57].
- The modules used for these systems are the LG NeON 2 from Volts Energies in Quebec. According to the supplier, the modules can endure a front load of up to 5400 Pa, and a rear load of up to 4000 Pa [45]. Since the design loads will be much less than these values, the modules will have sufficient structural capacity.

2.6. Economic Analysis

The economic cost of the solar PV wood racking equipment for Togo is obtained locally. The dimensions of the wood pieces are converted into metric units and used for pricing. In Togo, and the Western Africa region in general, one type of wood that is widely available and suitable for use in racking is teak wood [58,59]. Togolese teak is not pressure-treated, but it is naturally resistant to water. Furthermore, the lumber is usually treated using a water-resistant varnish that allows it to be used in outdoor applications. Because of the lack of strict enforcement of the regulations on local wood costs, the price varies from one vendor to another, and an average price has been used in this study. Additionally, the cost of the connection equipment is obtained from local vendors in Togo.

The comparison between the cost of the different racking systems is done on a 3-module basis, a per W basis, and extended to the levelized cost of electricity (LCOE). The percent differences for each are given. The costs of these DIY wooden racking systems are compared to the cost of both commercial and residential DIY metal racking in Canada and Togo. This comparison is important because in Togo, for non-utility grade solar PV systems, the racking is usually manufactured locally using steel that is treated with water-resistant paint. It should be noted that commercial systems are typically not sold in three module systems and will usually include the modules in their pricing. Thus, the cost of commercial metal racking systems, minus the commercial cost of modules, is converted to a cost per W basis, and then multiplied by the 1200 W in 3 bifacial modules [volts.ca] to calculate a cost per 3-module basis. The LCOE of the racking is obtained by dividing the cost of the racking by the lifetime energy production of the system. The lifetime energy production of the system is obtained by performing a SAM simulation using the optimal tilt angles for

Canada and Togo, respectively, as shown in Figure 1. The SAM simulation is performed by considering bifacial solar modules, a system lifetime of 25 years, and a module degradation rate of 0.5%/year [60].

After the calculation of the cost of wood racking for London, Ontario, the result is compared to that of a design with no snow load. The no snow load case is provided as the projections for snow losses are substantially decreased with climate change future projections even over the relatively medium-term (e.g., 2040) [61].

3. Results

Following the design procedure outlined in the previous section, two wood-based fixed-tilt wood-based racks were designed for both case study locations.

3.1. London, Ontario: 34 Degree Fixed Racking System

3.1.1. Bill of Materials

The bill of materials (BOM) of the London, Ontario system is shown in Table 2 in Canadian dollars sourced from Copp's Build-All, London, or Home Depot, London.

Table 2. 34-Degree Fixed Rack List of Materials.

Member Name	Piece [1]	Cost per Piece [2]	Quantity	Cost
Outside Joists	2 × 6 × 8	$16.12	2	$32.24
Inside Joists	2 × 8 × 8	$22.75	2	$45.50
Beams	2 × 8 × 10	$28.50	2	$57.00
Back Posts	6 × 6 × 8	$47.05	2	$94.10
Front Posts	4 × 4 × 10	$21.95	1 [3]	$21.95
Lateral Bracing	2 × 4 × 8	$9.99	2	$19.98
Lateral Bracing	2 × 4 × 10	$12.48	1	$12.48
Joist to Beam Connection	2 × 4 Fence Bracket	$0.36	8	$2.88
Bracing to Post Connection	2 × 4 Fence Bracket	$0.36	6	$2.16
Beam to Post Connection	$\frac{1}{2}$″ Carriage Bolt (6″ & 8″), Nut, & Washer	$4.44 [4]	8	$35.52
Tension Based Connections	2-1/2″ Brown Deck Screws	$9.99	100 Pack	$9.99
Shear Based Connections	1-1/2″ Joist Hanger Nails	$3.62	1 lb	$3.62
Module to Block Connections	1/4″ Carriage Bolt (2-1/2″), Nut, & Washer	$0.48 [4]	24	$11.52
			Total Cost with No Concrete	$348.94
Concrete for Posts	30 MPa Quikrete concrete	$4.98	8 bags	$39.84
			Total Cost:	$388.78

[1] All lumber is to be pressure treated, and all hardware is to be hot-dipped galvanized. [2] All costs are in Canadian Dollars as of 13 December 2021, before tax. [3] 1 piece to be cut to serve as 2 front posts. [4] Cost per connection (1 bolt, 1 nut, 1 washer).

3.1.2. London Assembly Instructions

The system requires at least two builders to install. Refer to Table 3 for the typical time spent completing each component per two builders.

To begin, four holes at least 250 mm in diameter are dug at least 1.2 m into the ground to prevent frost heaving of any soil type according to Table 9.12.2.2 in the National Building Code of Canada (NBCC) [56]. The holes are spaced according to Figure 2 from center to center. The front posts are made by cutting one 4 × 4 × 10 into two pieces at 1.6 m. The back posts are made by cutting each 6 × 6 × 8 to a length of 2.7 m. If concrete is being used for the footings, mix two bags of 30 kg Quikrete ready-mix concrete with water in a wheelbarrow. The mix should be evenly distributed under and around the posts. Once the hole is filled, dug-up topsoil should be used to cap the dug hole to ensure water slopes away from the footing as shown in Figure 3. Refer to instructions provided on the bag to ensure the mix cures to a serviceable hardness before continuing construction.

Table 3. Forces and deflections of structural members in the fixed angle rack.

Task	Typical Time Spent [1]
Digging holes and post installation	2.5 h [2]
Beam Installation	1.0 h
Brace Installation	0.5 h
Joist Installation	2.0 h
Block Installation	0.5 h
Module Installation	1.0 h
Total Time Spent	7.5 h

[1] Assuming 2 builders with some construction experience. Not including time to gather materials, acquire equipment, etc. [2] Not including curing time for concrete/footing mixture. Refer to supplier's instructions for suitable curing time before continuing to construct.

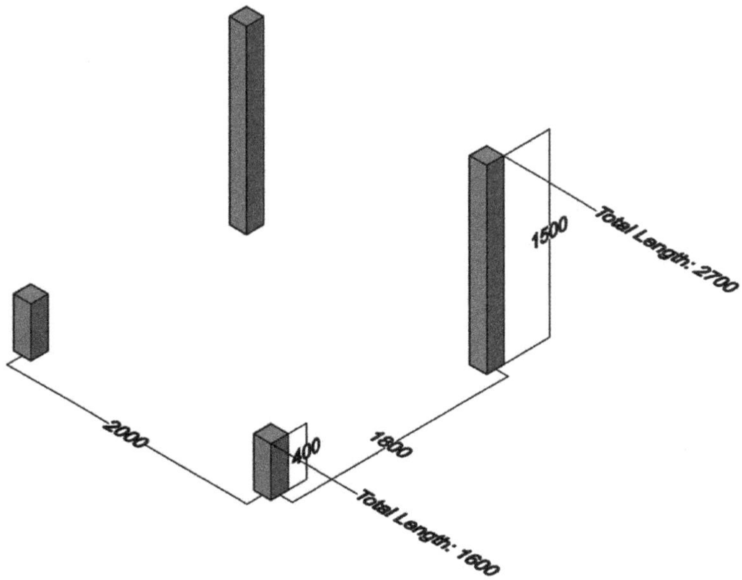

Figure 2. Center to center spacing of vertical posts for the 34-degree rack.

Two 2 × 8 × 10 beams cut into 3 m pieces are to be installed as shown in Figure 4a. Use 2-1/2″ brown deck screws to tighten the connection between the beams and the posts. Two 1/2″ holes are drilled through the posts and beams as seen in Figure 4b, and 6″ and 8″ carriage bolts are inserted for the 4 × 4 and 6 × 6 posts, respectively. All carriage bolts are to be tightly secured with a washer and nut as shown in Figure 4c.

A 2 × 4 bracing must be installed to enhance the system's stability. Two 2 × 4 × 8 s are cut to 1.6 m to install from front post to back post, 300 mm above the ground as seen in Figure 5a. Another 2 × 4 × 8 is cut to 2 m to be installed between the first 2 × 4 s in the middle of the system, about 750 mm from the front posts. All 2 × 4 s are installed as shown in Figure 5b with 2 × 4 galvanized fence brackets and 1-1/2″ joist hanger nails for stronger and more ductile connections. 2-1/2″ brown deck screws should be used to further sink the 2 × 4 into the post, and to improve the connection's resistance to pullout.

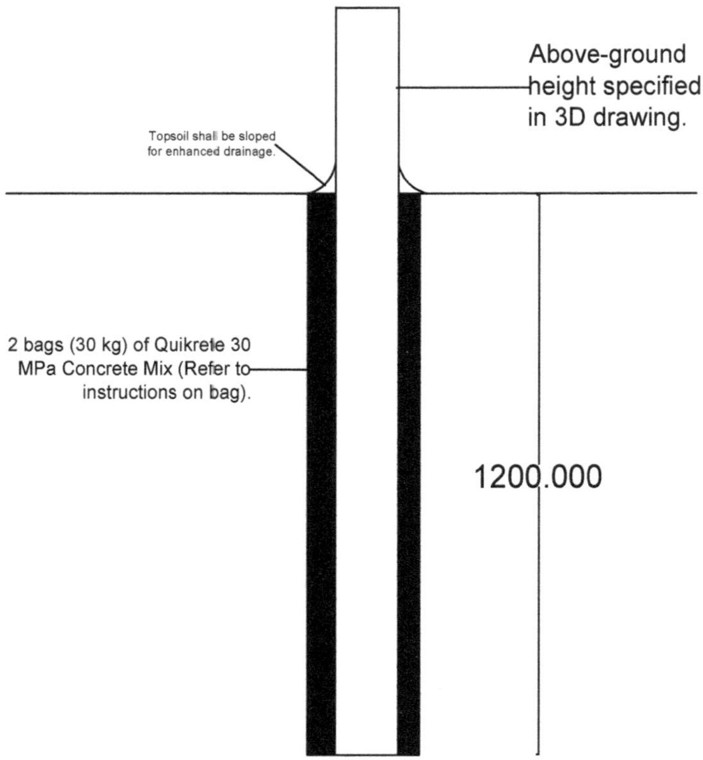

Figure 3. Foundational installment of vertical posts.

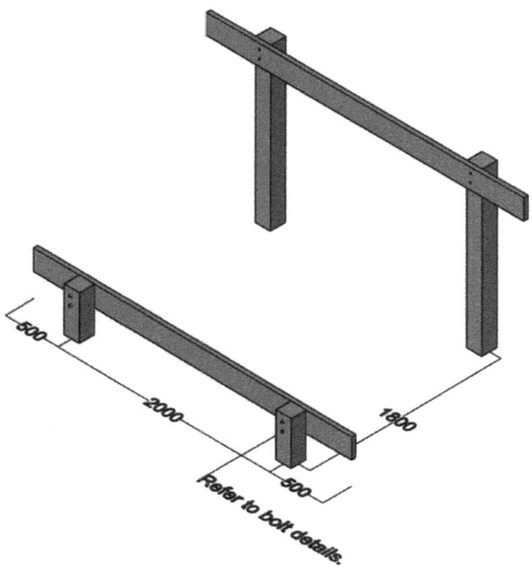

(a)

Figure 4. *Cont.*

(**b**) (**c**)

Figure 4. (**a**) 2 × 8 horizontal beam installment onto 4 × 4 and 6 × 6 posts, (**b**) $\frac{1}{2}''$ holes are drilled, and carriage bolts are inserted, and (**c**) connections are tightly secured using a nut and washer.

(**a**) (**b**)

Figure 5. (**a**) Spacing for 2 × 4 bracing, and (**b**) 2 × 4 fence brackets, joist hanger nails, and screws used for connections.

The outside joists are made from 2 × 6 × 8 s, and the inside joists are made from 2 × 8 × 8 s as shown in Figure 6a. A miter saw is adjusted to cut the joists to an angle of 34 degrees, and pieces are cut to 1.95 m. The joists are spaced 1 m from each other as shown in Figure 6b. Additional fence brackets are used for the bracing to install the joist to beam connections. The bottom lip of the bracket can be bent to align with the 34-degree angle.

Four brown deck screws per beam should be installed as shown in Figure 6c to further sink the joist into the beam, and to improve load transfer between the members.

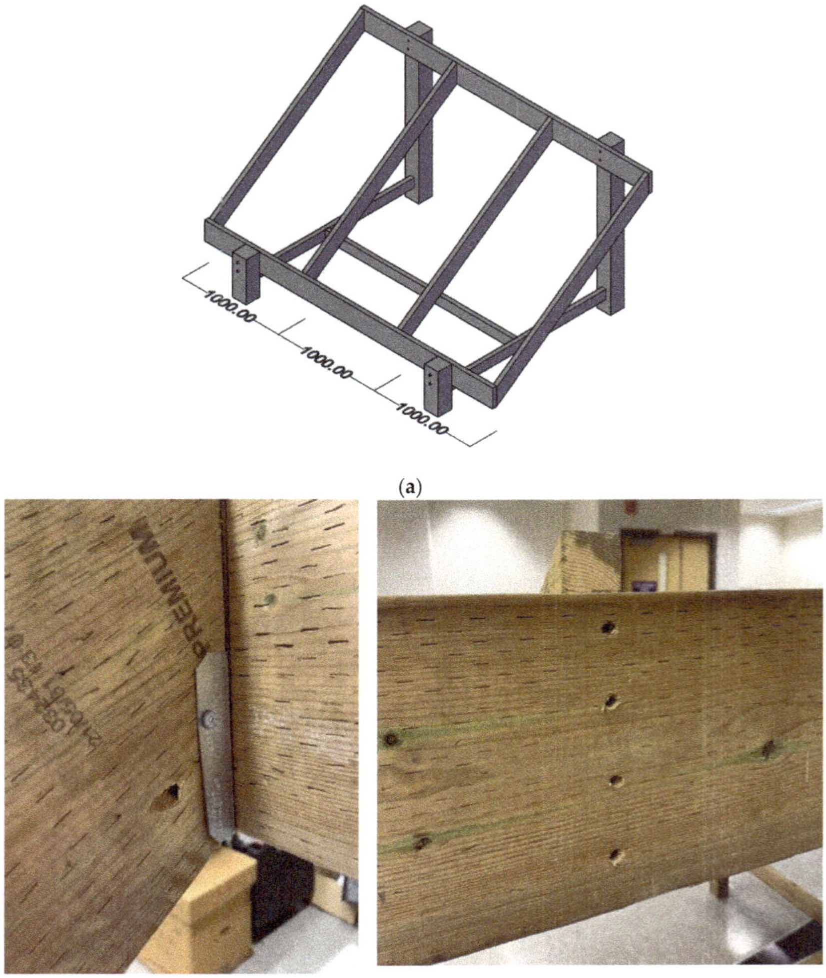

(**a**)

(**b**) (**c**)

Figure 6. (**a**) Joist spanning specifications between the beams, (**b**) fence brackets with joist hanger nails to be installed at the bottom of the joists, and (**c**) 4 brown deck screws installed from behind the beam.

Once the joists are installed, scrap pieces of lumber can be cut into blocks and installed onto the joists with two screws as shown in Figure 7a. These blocks serve as the connection between the module and the lumber and can be adjusted to match the holes of the module frame. The overhang of these blocks shall not exceed 100 mm. Once these blocks are installed, the modules can be placed onto the blocks. Drill a $\frac{1}{4}''$ hole through the bottom of the block, and insert a $\frac{1}{4}'' \times 2 - 1/2''$ galvanized bolt from under the system. Then, place the module onto the bolt, and secure the connection with a nut and washer as shown in Figure 7b. To enhance the load transfer to the joist, place another block under the overhanging block, and screw the second block into the joist as shown in Figure 7c.

(**a**) (**b**) (**c**)

Figure 7. (**a**) Extra lumber made into blocks that line up with the module's holes, (**b**) a bolt inserted from under and secured with a nut and washer, and (**c**) extra blocks placed under the overhanging block to enhance load transfer to the joist.

Once all connections are secured, the build is complete (Figure 8). The system can then be disassembled in the reverse order it was initially constructed.

Figure 8. Finished system.

Following the calculations shown in Appendix B for the structural analysis, the forces and deflections of the fixed angle system specifically for the London, Ontario system have been summarized in Table 4. When constructing a system, it is important to follow the structural design process in Appendix B to ensure the system can withstand the design load outlined in Appendix A. Depending on the design load, smaller members can be selected, and thus the net cost of the system can be reduced such that the maximum shear, moment, deflection, and axial forces are less than the capacities shown in Table A7 of Appendix B.

Table 4. Forces and deflections of structural members in the fixed angle rack.

Member Name	Shear [kN]	Moment [kNm]	Deflection [mm]	Tension/Compression [kN]
Outside Joists	0.95	0.55	3.30	N/A
Inside Joists	1.90	1.10	1.65	N/A
Beams	1.90	0.50	2.73	N/A
Back Posts	3.86	0.90	0.6	−2.70 [1]
Front Posts	3.86	0.31	1.6	−2.70 [1]
Bracing Support	N/A	N/A	N/A	−1.59

[1] For a 250 mm diameter hole, this load induces a bearing pressure of 55 kPa.

3.2. Lomé Togo 10 Degree Fixed Rack

3.2.1. BOM 10 Degree Fixed Rack

The bill of materials (BOM) of the Lomé Togo system is shown in Table 5 in Canadian dollars (CAD) to compare to Table 2.

Table 5. 10 Degree Fixed Rack List of Materials for *Lomé Togo if purchased in Canada.*

Member Name	Piece [1]	Cost per Piece [2]	Quantity	Total
Joists	2 × 6 × 8	$16.12	4	$64.48
Beams	2 × 6 × 10	$20.15	2	$40.30
Back Posts	4 × 4 × 8	$17.49	2	$34.98
Front Posts	4 × 4 × 10	$21.95	1 [3]	$21.95
Lateral Bracing	2 × 4 × 8	$9.99	3	$29.97
	2 × 4 × 10	$12.48	1	$12.48
Joist to Beam Connection	2 × 4 Fence Bracket	$0.36	8	$2.88
Bracing to Post Connection	2 × 4 Fence Bracket	$0.36	6	$2.16
Beam to Post Connection	$\frac{1}{2}$″ Carriage Bolt (6″), Nut, & Washer	$4.44 [4]	8	$35.52
Tension Based Connections	2″ Brown Deck Screws	$9.99	100 Pack	$9.99
Shear Based Connections	1-1/2″ Joist Hanger Nails	$3.62	1 lb	$3.62
Module to Block Connections	1/4″ Carriage Bolt (2-1/2″), Nut, & Washer	$0.48 [4]	24	$11.52
			Total with no concrete	$269.85
Concrete for Posts	30 MPa Quikrete concrete	$4.98	8 bags	$39.84
			Total Cost:	309.69

[1] All lumber is to be pressure treated, and all hardware is to be hot-dipped galvanized. [2] All costs are in Canadian Dollars as of 13 December 2021, before tax. [3] 1 piece to be cut to serve as 2 front posts. [4] Cost per connection (1 bolt, 1 nut, 1 washer).

The values in Togo found in Table 6 are converted to $CAD where 1 $CAD = 457.800 XOF [62] so the total with no concrete is $507.63 and with concrete, it is $540.10, which is 39% percent more than building in Canada.

3.2.2. Lomé, Togo Fixed Rack Installation Instructions

The post installation process for the Togo system is the same as the fixed-angle process in Canada. The spacing of posts changes with a fixed tilt angle so that it shall correspond to the center to center spacing shown in Figure 9.

The beam, bracing, joist, and block installation are the same process as the 34-degree fixed rack. The miter saw should be set at 10-degree cuts when cutting the joists.

The 10-degree (Figure 10) structural analysis process is identical to the 34-degree structural analysis. Lomé, Togo's design load is half of London, Ontario's design load, thus, analysis results were half of London Ontario's analysis results.

Table 6. 10 Degree Fixed Rack List of Materials for *Lomé Togo if purchased in Togo* [1].

Member Name	Piece [2]	Cost per Piece [3]	Quantity	Total
Joists	2 × 6 × 8	$28.14	4	$112.56
Beams	2 × 6 × 10	$34.63	2	$69.26
Back Posts	4 × 4 × 8	$32.47	2	$64.94
Front Posts	4 × 4 × 10	$37.88	1 [4]	$37.88
Lateral Bracing	2 × 4 × 8	$27.06	3	$81.18
	2 × 4 × 10	$32.47	1	$32.47
Joist to Beam Connection	2 × 4 Fence Bracket	$5.41	8	$43.28
Bracing to Post Connection	2 × 4 Fence Bracket	$5.41	6	$32.46
Beam to Post Connection	$\frac{1}{2}$″ Carriage Bolt (6″), Nut, & Washer	$1.30	8	$10.40
Tension Based Connections	2″ Brown Deck Screws	$16.23	250 Pack	$16.23
Shear Based Connections	1-1/2″ Joist Hanger Nails	$4.33	1 lb	$4.33
Module to Block Connections	1/4″ Carriage Bolt (2-1/2″), Nut, & Washer	$0.11	24	2.64
			Total with no concrete	$507.63
Concrete for Posts	Concrete (Cement + Sand + Gravel)	$32.47	528 lbs	$32.47
		Total Cost:		$540.10

[1] It should be noted that lumber price has doubled in Togo during 2021 [63]. [2] All the lumbers are varnish-treated, and all hardware is to be hot-dipped galvanized. [3] All costs are in Canadian Dollars as of 28 January 2022, before tax. [4] 1 piece to be cut to serve as 2 front posts.

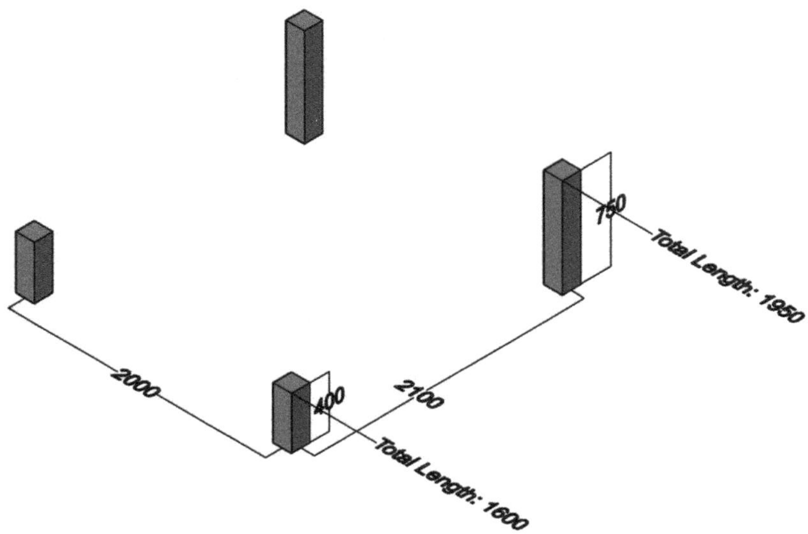

Figure 9. Foundational installment and spacing of 4 × 4 posts.

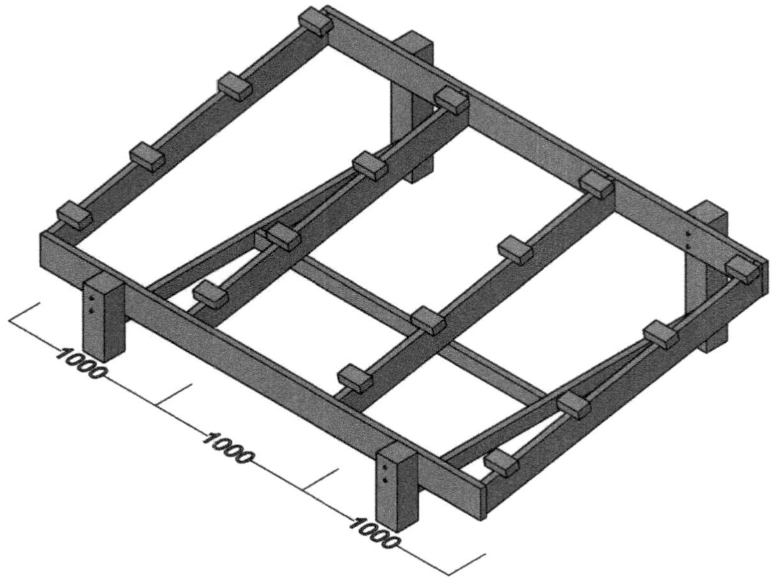

Figure 10. Completed 10-degree fixed rack.

3.3. Economic Analysis

The cost per 3-module basis, cost per W, and LCOE for each system have been summarized in Table 7.

Table 7. Cost per 3 modules, Cost per W, and LCOE for each system in $CAD.

System	Cost per 3-Modules [CAD$]	Cost per W [CAD$/W]	LCOE [CAD$/kWh]
London Canada Design DIY Wood purchased in Canada	388.78	0.32	0.010
Lomé, Togo Design/No snow DIY Wood Purchased in Canada	309.69	0.26	0.008
Lomé, Togo Design/No snow DIY Wood Purchased in Togo	540.10	0.45	0.014

The material costs for the open-source wood rack for London, ON Latitude purchased in Canada, is CAD$388.78. The cost per installed W for this system is CAD$0.32/W, which corresponds to an LCOE of CAD$0.010/kWh. Subtracting the cost of the footings (which do not come with any commercial systems) the cost is CAD$348.94, which is CAD$0.29/W. Using a US$ to $CAD of 1.28 [64] for comparison, 3-module system costs CAD$682.83 [32] using the same PV is CAD$0.57/W and a 3-module pole mount rack is CAD$1523.93 [33] is CAD$1.27/W. Other Canadian racking systems available on the market fall between this range. Even when scaled up to a 12-module system (4 of these OS wood designs) the cost is CAD$0.32/W, while commercial systems cost CAD$3430.75 [65] or CAD$0.71/W. Thus, the percent savings for small systems range from 49% to 77% and this also leaves the open-source wood rack less expensive (17% savings) than residential roof racking costs of CAD$0.35/W [66].

The 10-degree design that does not need to handle snow loads is CAD$309.69; however, if the wood is purchased in Togo, it increases to CAD$540.10. The impact that wood prices have on this design is substantial. In Togo, the availability of an extremely inexpensive DIY metal rack (CAD$165) coupled with the much higher cost of wood, makes it uneconomic.

It should be noted there is no structural stability study behind the design of the locally manufactured steel racking in Togo.

For a no snow load design, when following the design load procedure outlined in Appendix A, a London, Ontario system with a fixed tilt angle of 34 degrees will have a design load of approximately 0.83 kPa, which is practically the same design load as Lomé, Togo's design load of 0.81 kPa. Thus, the no snow system can be built with the same BOM as the wooden Togo system at a cost of CAD$309.69. A no snow load design results in a price reduction of about 20%.

The Canada and Togo systems represent the two extreme cases of a system being used in a location of high latitude and a location near the equator. Therefore, a system anywhere in the world will have a cost roughly between these two extreme cases. The cost of any given system is dependent on the region's wind and snow load, which can be calculated using Appendix A, and the cost of lumber in the region, which is further discussed in Section 4.2.

4. Discussion

4.1. Limitations

For the wood members, regardless of the preservative system chosen or available, the buried wood posts must be rated for a below-ground application, while the upper structure can be rated for ground contact or lower. The higher the rating, the more preservative chemicals are in the wood, and normally the higher the cost. More information and resources for homeowners can be found American Wood Protection Association (awpa.org). A treated wood solar rack can last a lifetime common to the PV warranties of 25 years if installed in accordance with manufacturers' instructions. This typically includes hand treating exposed board cuts and drilled holes with a 2% copper naphthenate solution (commonly available at home improvement stores, lumber yards, or online). In addition, particular care must be taken if the installation is in a higher wood deterioration zone. The Forestry Chronicle provides a decay hazard map showing high hazards of decay off the coast of British Colombia [67]. The American Wood Protection Association (AWPA) provides a decay hazard zone map showing high hazard decay in California, and the south [68]. As all materials come to an end of their service lives, treated wood must be disposed of properly, which usually means taking it to a landfill or transfer station. The material should not be burned or reused as mulch so there is a need for future work to find an environmentally benign method of recycling treated wood such as low-temperature pyrolysis [69].

To avoid the use of treated wood, there are species that offer natural decay resistance. White oak (*Quercus alba*) offers impressive strength and decay resistance [70] and would be suitable for long life in most solar racking applications. White oak also has slightly better mechanical properties [71] than red oak. Additionally, white cedar and western red cedar are commonly available at hardware stores for approximately 2.5 times the cost of pressure-treated wood. This again pushes the wood rack to become uneconomical with the current cost of metal.

The footings of the system can be made with concrete, or any other type of ground-rated mixture to hold the posts in place. However, the mixture should have a compressive strength of at least 30 MPa to ensure it can withstand ground stresses due to freeze–thaw cycles if put in cold environments [72]. For warm regions in which freeze–thaw cycles will not be problematic, a minimum compressive strength of 20 MPa is required [72]. To make a 20 MPa mix, use a water to cement ratio of 0.70 instead of the typical ratio of 0.55. Additionally, the air content of the mixture should not exceed 8% to ensure water does not induce too much internal stress in the mix. These compressive strength requirements lend the opportunity to work with alternative building materials, which are left for future work. All hardware used in the system, such as nails, screws, brackets, bolts, etc., shall be either hot-dipped galvanized, electrogalvanized, or stainless steel to ensure a lifetime of 25 years.

Although common zinc plated hardware is inexpensive and easily accessible, it cannot be used for outdoor applications.

4.2. Wood Price Sensitivity

Although the open-source wood-based PV rack provided substantial savings in Canada and was able to be constructed with no special tools, the same was not found in Togo. The system is highly sensitive to the price of lumber, which recently has been volatile and increasing as shown in Figure 11. Once a sustainable equilibrium is reached, the costs of this design could decrease by a factor of 2–3 as shown by the earlier years in Figure 11. This would make the substantial savings observed in the Canadian market even more striking and allow for the system to compete with the low-cost metal racks in Togo.

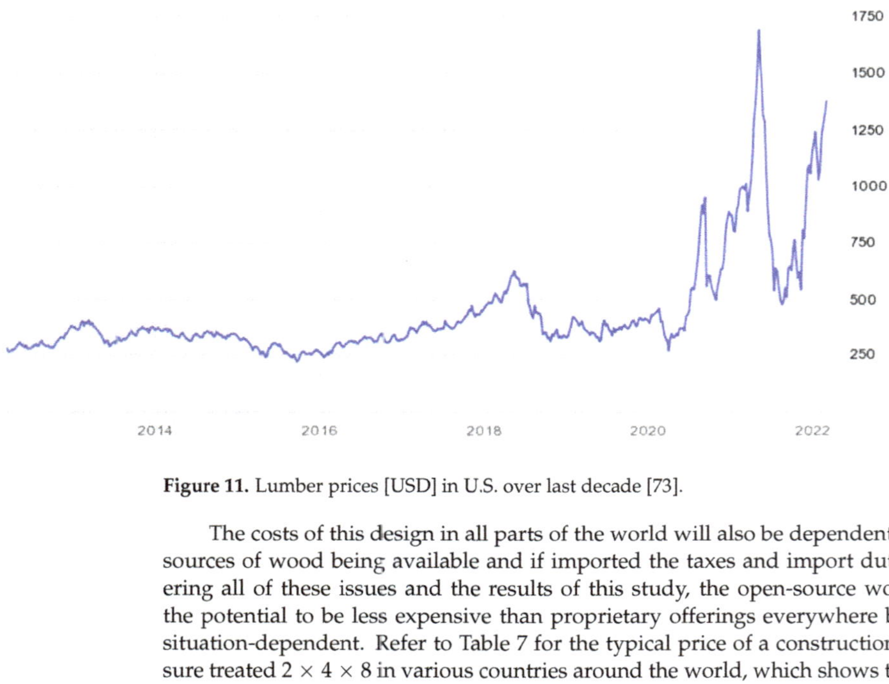

Figure 11. Lumber prices [USD] in U.S. over last decade [73].

The costs of this design in all parts of the world will also be dependent on the local sources of wood being available and if imported the taxes and import duties. Considering all of these issues and the results of this study, the open-source wood rack has the potential to be less expensive than proprietary offerings everywhere but is highly situation-dependent. Refer to Table 7 for the typical price of a construction grade pressure treated 2 × 4 × 8 in various countries around the world, which shows the impact of international import duties and material availability. As can be seen in Table 8—the two case studies selected represented both the high (Togo) and relatively low (Canada) costs of wood globally.

Table 8. The typical price of a pressure treated 2 × 4 × 8 in various countries converted to USD as.

Country	Price [USD] [1]	Source [2]
Canada	$8.46	The Home Depot
USA	$9.68	The Home Depot
Togo	$21.67	Tao Bentho and Komi Jacques Gakli-Gaka
United Kingdom	$15.70	B & Q
Netherlands	$10.32	Woodvision
Australia	$13.92	Bunnings
Brazil	$12.03	Fremade Madeiras
India	$4.96 [3]	IndiaMart

[1] Priced as of 2 April 2022. [2] Prices at each source's competition are approximately the same. [3] Priced before pressure treating. The cost is expected to at least double to treat.

The cost of pressure-treated lumber is noticeably higher in other continents compared to the cost in North America. This makes it more difficult to observe the economic benefits of using wood for PV racking globally, but once sustainable equilibrium is reached, the cost of a wood system can prove to be competitive with metal systems on a global level.

As per Section 1.3.1.1 Div. A of the Ontario Building Code [56], a building inspection and permit for a ground-mounted system is not required. A grid-connected PV system may require an inspection from the Electrical Safety Authority [74]. These rules depend on locality and may change as PV obtains high grid penetration rates [75].

Future work can further refine the design to make a rack specific for: (1) monofacial PV (which would use less wood as the cross members could go behind the module back surface and decrease lumber lengths), (2) multi-tilt angle capability, (3) vertical racks and (4) investigate at the impact of the designs on a greater number of woods and recycled plastic lumber. In addition, the use of strategic bracing or wire tension can also be explored for reducing the materials needed. Wood racking should also be evaluated for viability for use in spray cooling PV systems [76].

In addition, a full life cycle analysis is needed to determine if this design, although able to be made with sustainably harvested wood, is indeed ecologically superior to more traditional metal-based racking designs. This analysis can include all the externality costs and evaluate the system for on-grid, off-grid, and microgrid [77]. All of this work supports the potential for further reducing the costs and applications of open-source PV racks.

5. Conclusions

Wood can be a sustainable source of photovoltaic system racking materials to support a circular bioeconomy. The open-source wood-based fixed-tilt ground-mounted bifacial photovoltaic rack design evaluated here is appropriate throughout North America. The results found the open-source wood rack contributes only 1.1 US cents to the LCOE and can be fabricated for roughly 49% to 77% less than proprietary small-scale metal racks. Although the design of the system needs fewer materials in regions that do not experience snowfall, this may or may not equate to lower capital costs. This is because the system is highly dependent on the costs of lumber. For designers attempting to optimize both the LCOE as well as the environmental impact of PV systems, wood may be a viable lower-cost option than conventional racking materials. As wood costs come back towards historic norms, wood PV racking provides a promising method to further improve the economical footprint for PV systems anywhere in the world. Lastly, this design provides for a rack that can be manufactured with a distributed means throughout most of the world enabling more equitable access to solar energy.

Author Contributions: Conceptualization, J.M.P.; methodology, N.V., K.S.H.; software, K.S.H.; validation, N.V., K.S.H.; formal analysis, J.M.P., N.V., K.S.H.; investigation, N.V., K.S.H.; resources, J.M.P.; data curation, N.V., K.S.H.; writing—original draft preparation J.M.P., N.V., K.S.H.; writing—review and editing, J.M.P., N.V., K.S.H.; visualization, N.V., K.S.H.; supervision, J.M.P.; funding acquisition, J.M.P. All authors have read and agreed to the published version of the manuscript.

Funding: This research was funded by Thompson Endowment.

Institutional Review Board Statement: Not applicable.

Informed Consent Statement: Not applicable.

Data Availability Statement: Data will be made available upon request.

Acknowledgments: The authors would like to thank Paul Vandewetering of Paul's Build-All for assistance in the construction of the 34-degree rack. In addition, the authors would like to thank Tao Bentho and Komi Jacques Gakli-Gaka for providing equipment costs for Togo.

Conflicts of Interest: The authors declare no conflict of interest. The funders had no role in the design of the study; in the collection, analyses, or interpretation of data; in the writing of the manuscript, or in the decision to publish the results.

Abbreviations

Structural Analysis	
w	Uniform distributed load
L	Member length
$Vmax$	Maximum Shear Force
$Mmax$	Maximum Bending Moment
Δmax	Maximum Deflection
Design Loads	
S	Design Snow Load
I_S	Snow Importance Factor
S_S	Ground Snow Load Factor
C_b	Basic Roof Snow Load Factor
C_W	Wind Exposure Factor
C_S	Slope Factor
C_a	Accumulation Factor
S_r	Rain Factor
W	Design Wind Load
I_w	Wind Importance Factor
q	1/50 Year Wind Pressure
C_e	Exposure Factor
C_pC_g	Pressure and Gust factor
C_{ei}	Internal Exposure Factor
C_{gi}	Internal Gust Factor
C_{pi}	Internal Pressure Factor
D	Design Dead Load
Material Properties	
b	Cross-Sectional Base
h	Cross-Sectional Height
A	Cross-Sectional Area
I	Moment of Inertia
Q	First Moment of Area
fb	Flexural Resistance
fv	Shear Stress Resistance
ft	Tensile Stress Resistance
fc	Compression Stress Resistance Parallel to the grain
E	Elastic Modulus
$Emin$	Minimum Elastic Modulus
C_d	Duration Factor
C_t	Temperature Factor
C_L	Beam Stability Factor
C_M	Wet Service Factor
C_{fu}	Flat Use Factor
C_i	Incising Factor
C_f	Member Size Factor
$fb^*, fv^*, ft^*, fc^*, E^*$	Factored Resistances
Mr	Factored Bending Moment Resistance
Vr	Factored Shear Force Resistance
Tr	Factored Tensile Force Resistance
Cr	Factored Compression Force Resistance
$D_{footing}$	Diameter of ground footing

Appendix A. Specified Loads

Appendix A.1. Snow Load NBCC 4.1.6

The governing snow load [56] is composed of multiple factors shown in Equation (A1),

$$S = I_s[S_s(C_bC_wC_sC_a) + S_r] \quad \text{(A1)}$$

The Importance Factor, I_S, is taken as 0.8 because the failure of a PV system has little to no risk of the loss of life.

The Ground Snow Load Factor, S_S, is dependent on the location of the structure. It is the 1-in-50-year snow load found in Table C-2 in NBCC 4.1.6 For London, Ontario, a value of 1.90 must be taken [56].

The Basic Roof Snow Load Factor, C_b, is 0.80 for small structures.

The Wind Exposure Factor, C_w, is taken as 0.75 if the structure is exposed to wind in all directions.

The Slope Factor, C_s, is dependent on the tilt angle of the system. It can be calculated using the following equation from NBCC 4.1.6.

$$C_S = \frac{60 - angle}{45} \quad \text{(A2)}$$

The Accumulation Factor, C_a, is taken as 1.00 for small monoslope structures.

The Rain Factor, S_r, depends on the location of the structure. For London, Ontario, a value of 0.40 must be taken according to Table C-2 in NBCC 4.1.6 [56].

The factors and calculated net snow load are summarized in Table A1.

Table A1. Design Snow Load. Note that the minimum specified snow load is to be taken as 1.00 kPa.

Coefficient	Value
I_s	0.80
S_s	1.90 [1]
C_b	0.80
C_W	0.75
C_S	0.58 [2]
C_a	1.00
S_r	0.40 [1]
S	0.85 [3]

[1] This value is specific to London, Ontario. Refer to Table C-2 in NBCC 4.1.6 for other locations. [2] This value is specific to 34-degree systems. Refer to Equation X for other angles. [3] As per NBCC, the minimum specified design snow load shall not be less than 1.00 kPa.

Appendix A.2. Wind Load

The specified wind load is adapted from the National Building Code of Canada 2015, Division B, 4.1.7, and is composed of both an external wind pressure, and internal wind pressure,

$$W = p + p_i \quad \text{(A3)}$$

where the external wind pressure is composed of the following factors,

$$p = I_w q C_e C_t C_p C_g \quad \text{(A4)}$$

The internal wind pressure is composed of the following factors,

$$p_i = I_w q C_{ei} C_t C_{gi} C_{pi} \quad \text{(A5)}$$

The Wind Importance Factor, I_W, is taken as 0.80 because the failure of a PV system has little to no risk of the loss of life.

The Hourly Wind Pressure Factor, q, is dependent on the location of the structure. It is the 1-in-50-year wind load found in Table C-2 in NBCC 4.1.6 For London, Ontario, a value of 0.47 must be taken.

The Exposure Factor, C_e, shall be taken as 0.90 for structures less than 10 m in height.

The topographic factor, C_t, is to be taken as 1.00.

The Pressure and Gust Factors, C_p and C_g, are combined and found using the table from Figure 4.1.7.6-A from NBCC [56]. From the figure, the governing building for 34-degree systems is 2, which corresponds to a C_pC_g value of -1.30.

The Internal Exposure Factor, C_{ei}, is the same as the External Exposure Factor, C_e since the structure has a "dominant opening", meaning that wind can attack the inside of the system just as easily as it can attack the outside.

The Internal Gust Factor, C_{gi}, is taken as 2.00 since is not a large unpartitioned volume, such as an arena.

The Internal Pressure Factor, C_{pi}, will be taken as −0.70 because the wind has easy access to push against the back of the system.

The factors and calculated net snow load are summarized in Table A2.

Table A2. Design Wind Load as per NBCC 4.1.7.3.

Coefficient	Value
I_W	0.80
q	0.47
C_e	0.90
C_t	1.00
C_pC_g	−1.30
C_{ei}	0.90
C_{gi}	2.00
C_{pi}	−0.70
p	−0.44
p_i	−0.47
W	−0.91 kPa

Appendix A.3. Specified Dead Load

The dead load, or the self-weight of the structure, will consist of the weight of the PV modules, and the weights of the wooden members. The weight of brackets and fasteners is much smaller than the design load and can be considered negligible. According to Natural Resources Canada's CanmetENERGY research centre [78], a dead load of PV systems, also known as the superimposed dead load, shall be taken as 0.24 kPa. The self-weight of the lumber depends on what dimensions of lumber has sufficient capacity to carry the load. The self-weight of lumber is variable due to changing moisture content and amount of knots. For analysis purposes, it is best practice to use the lumber weights provided by the supplier, and translate the given weight into a uniform distributed load in kN/m.

Appendix A.4. Load Combinations

Since many simple assumptions are made in the analysis process, safety factors must be applied to the specified loads to reduce the probability of failure. These factored loads are then combined as principle loads and companion loads via the load combinations found in NBCC 4.1.3.2 [56]. The combination of principal loads and companion loads that creates the largest net load shall be used in the analysis. Principle loads are the mandatory loads that must be checked, and companion loads are only added if they act in the same direction as the principal loads. Since the design wind load acts in the negative direction and the governing snow load is in the positive direction, they shall not be combined as doing so will reduce the net load. The combinations that produced the largest positive and negative are found in Table A3.

Table A3. Load Combinations as per NBCC 4.1.3.2. Based on Design Dead, Snow, and Wind Loads.

Combination	Net Load [kPa]
0.9D [1] + 1.4W	−1.06
1.25D + 1.5S	1.80

[1] A factor of either 0.9 or 1.25 can be used for the dead load D. Since the wind load W acts in the opposite direction as D, 0.9 is selected to create the maximum negative net load.

It should be noted that the load path for the positive case is the same as the negative case, the only difference being the direction of the load. Since all connections can resist loads in both directions, and all members have the same material properties in both directions, the analysis for the negative case is the same as the positive case. Thus, the negative case can be ignored, and the positive case will be used for the analysis.

The same process will be applied to a system located in Lomé, Togo. For this system, it is justified to assume that the snow load is 0 kPa. All wind coefficients can be the same as the London, Ontario system with the exception of q and C_pC_g because they depend on the wind speed and tilt angle, respectively. The hourly wind pressure q for Lomé, Togo, is not readily tabulated, but can be calculated using the wind pressure Equation (A6) shown below,

$$q = \frac{1}{2}\rho V^2 \tag{A6}$$

where ρ is the density of air, and V is the 1-in-50 year wind speed. NBCC suggests taking a ρ value of 1.29 kg/m^3 [56]. According to the European Conference on Severe Storms 2019, V for Lomé, Togo will be taken as 20 m/s [79]. Plugging these values into the equation, q becomes 0.26 kPa.

C_pC_g for 10 degrees turns out to be −1.30 as well, thus producing a net wind load of −0.42 KPa. The governing load combination for this system is 0.9D + 1.4W, which produces a specified design load for this system to be −0.81 KPa.

Appendix B. Lumber Structural Capacity

The material properties of lumber are extremely variable due to the multiple different species of wood, the existence of knots that alter the load path in the member, and the fluctuating moisture content carried inside the wood. The National Design Speciation for Wood Construction 2018 [80] provides engineers with trustworthy design values for the vast majority of wood species. Almost all pressure-treated lumber in Canada is made of Spruce Pine Fir grades number 1 and 2 [81], which possess the structural capacities summarized in Table A4.

Table A4. Unfactored capacities for No. 1/2 Spruce Pine Fir Lumber.

Capacity	Value [MPa]
fb	6.03
fv	0.93
ft	3.10
fc	7.93
E	9652.60
Emin	3516.30

Although these capacities have proven to be reliable, resistance factors must be applied to the capacities to account for unexpected weaknesses and to ensure a safe and serviceable design.

The load duration factor, C_d, is selected as 1.15 since the snow load is the governing load on the structure.

The temperature factor, C_t, is 1.00 since the structure will not be exposed to temperatures over 100 degrees Fahrenheit.

The beam stability factor, C_L, is calculated as 0.98 as per the guidelines outlined in Appendix C of the NDS.

The wet service factor, C_M, is calculated as 1.00 because the product of fb and C_f is less than 7.9.

The flat use factor, C_{fu}, is 1.00 because all members are to be loaded upon their strong axis in which h > b.

The incising factor, C_i, is selected as 0.8 as per Table 4.3.8 in the NDS [80].

The repetitive member factor, C_r, is 1.00 since the structure cannot be considered a floor or roof system with multiple of the same member.

Finally, the size factor, C_f, is selected as 1.00 because lumber greater than 2 × 12 will not be used for this design.

These factors have been summarized in Table A5,

Table A5. Resistance factors as per the National Design Specification for Wood Construction.

Coefficient	Value
C_d	1.15
C_t	1.00
C_L	0.98
C_M	1.00
C_{fu}	1.00
C_i	0.80
C_r	1.00
C_f	1.00

As per the National Design Specification for Wood Construction, the factored bending capacity is calculated as,

$$fb^* = fb\left[C_D C_t C_L C_M C_{fu} C_i C_r C_F\right] \tag{A7}$$

The factored shear stress is calculated as,

$$fv^* = fv[C_D C_M C_i] \tag{A8}$$

The factored tensile stress is calculated as,

$$ft^* = ft\left[C_D C_M C_f C_i\right] \tag{A9}$$

The factored compressive stress is calculated as,

$$fc^* = fc\left[C_D C_M C_t C_f C_i\right] \tag{A10}$$

Finally, the factored elastic modulus is calculated as,

$$E^* = E[C_M C_t C_i] \tag{A11}$$

These factored capacities have been summarized in Table A6.

Table A6. Factored capacities for No. 1/2 Spruce Pine Fir Lumber.

Factored Capacity	Value [MPa]
fb^*	5.44
fv^*	0.86
ft^*	2.85
fc^*	7.29
E^*	9169.97

Using the factored capacities and dimensional properties, resistance values can be computed using the following equations,

The resisting bending moment, Mr, can be calculated as,

$$Mr = \frac{2fb^* I}{h} \tag{A12}$$

The resisting shear force, *Vr*, can be calculated as,

$$Vr = \frac{fv^* Ib}{Q} \tag{A13}$$

The resisting tensile force, *Tr*, can be calculated as,

$$Tr = ft^* A \tag{A14}$$

The resisting compressive force, *Cr*, can be as,

$$Cr = fc^* A \tag{A15}$$

Resisting values for each dimensional lumber have been summarized in Table A7.

Table A7. Resisting values for No. 1/2 Spruce Pine Fir Lumber.

Lumber	*Mr* [kN·m]	*Vr* [kN]	*Tr* [kN]	*Cr* [kN] [1]
2 × 4	0.27	1.93	9.65	24.66
2 × 6	0.67	3.03	15.18	38.79
2 × 8	1.17	3.99	19.95	50.98
2 × 10	1.90	5.09	25.48	65.12
2 × 12	2.82	6.20	31.01	79.25
4 × 4	0.64	4.52	22.60	57.76
6 × 6	2.49	11.18	55.92	142.92

[1] Compression resistances are for loads parallel to the wood grain.

To ensure that the structure does not fail, the following conditions must be met for each member:

The resisting bending moment must be greater than or equal to the maximum applied bending moment,

$$Mr \geq Mmax \tag{A16}$$

The resisting shear force must be greater than or equal to the maximum applied shear force,

$$Vr \geq Vmax \tag{A17}$$

The resisting tensile force must be greater than or equal to the maximum applied tensile force,

$$Tr \geq Tmax \tag{A18}$$

The resisting compressive force must be greater than or equal to the maximum applied compressive force,

$$Cr \geq Cmax \tag{A19}$$

Finally, the maximum deflection cannot exceed the member length divided by 360 as per NBCC 9.4.3 [NRC],

$$L/360 \geq \Delta max \tag{A20}$$

Now that a design load and material properties are given, structural analysis can be conducted to determine the optimal dimensions of lumber needed to build a serviceable system.

Appendix C. 34 Degree Fixed System Structural Analysis

The net load is distributed evenly throughout the surface of the modules. As per the supplier of the modules [volts.ca], it is assumed that the panels have sufficient capacity to carry these loads. The load is then transferred from the panels to the joists. Each joist carries its own weight as a uniform distributed load, w, and 4-point loads that represent the block connections. w is calculated using Equation (A21),

$$w = 1.25(OW) \tag{A21}$$

OW represents the own weight of the member, which needs to be multiplied by a factor of 1.25 because it is a dead load [82]. Since the required dimensions of lumber to carry the load is unknown, an assumption needs to be made (for example, assume 2×8) to carry out the analysis. If the assumption results in the maximum applied value being greater than the resistance values, then a larger member needs to be used.

The point loads can be calculated by dividing each joist's tributary loading into four points because it is assumed that the load is distributed evenly throughout the modules. The tributary area represents how much width of the panels each joist is responsible for carrying. For example, in this three-module system, which is 3 m wide, the middle joists have a tributary width of 1 m (0.5 m on each side), and the end joists have a tributary width of 0.5 m (only one side). The value for each point load on the joists is calculated as,

$$Point\ Load = \frac{Design\ Load \times Tributary\ Area}{4} \tag{A22}$$

Once w and the point loads are calculated, the free body diagram for each joist can be made (Figure A1),

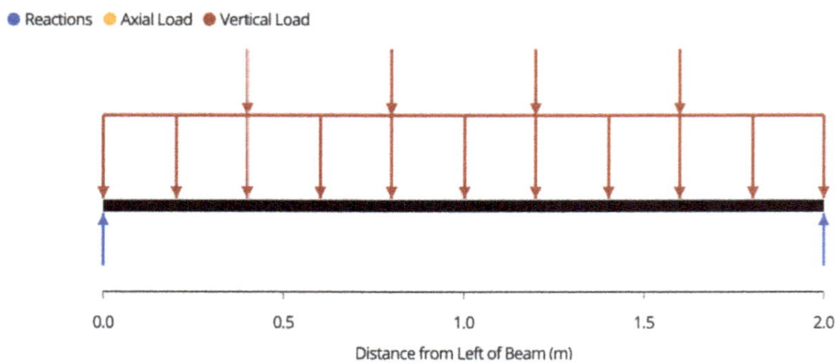

Figure A1. Free body diagram of joists. Note that the two outside joists will have half the tributary area of the inside joists, and thus will carry approximately half of the load.

Each joist is supported by a beam on each end. The reaction, and thus the load that each joist transfers to each beam is the following,

$$Reaction = \frac{4 * Point\ Load + wL}{2} \tag{A23}$$

The shear force diagram throughout each joist is the seen as Figure A2,

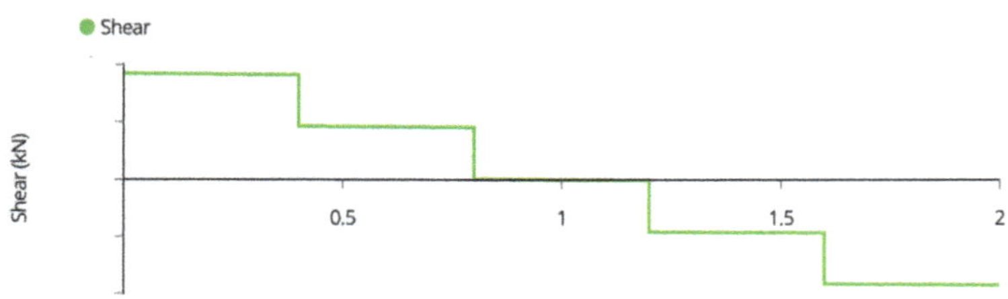

Figure A2. Joist Shear Force Diagram.

The maximum shear force occurs at the supports, and thus the maximum shear is calculated as the reaction shown above.

The bending moment diagram in each joist is seen as Figure A3,

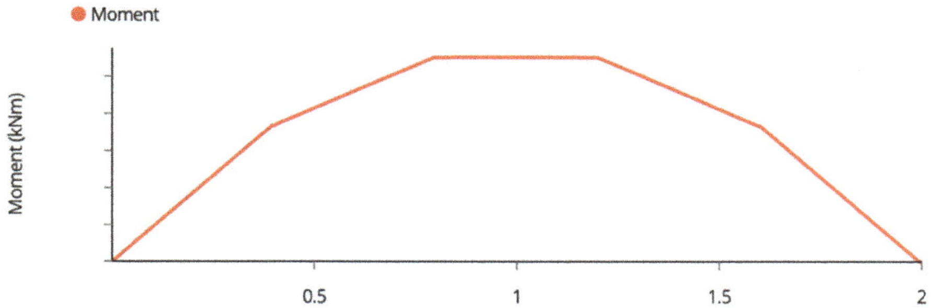

Figure A3. Joist Bending Moment Diagram.

The maximum bending moment of the joists occurs at the midspan. The maximum bending moment can easily be calculated by

$$Mmax = \frac{wL^2}{8} + \frac{3(Point\ Load)L}{5} \quad \text{(A24)}$$

The deflection diagram throughout each joist is seen as Figure A4,

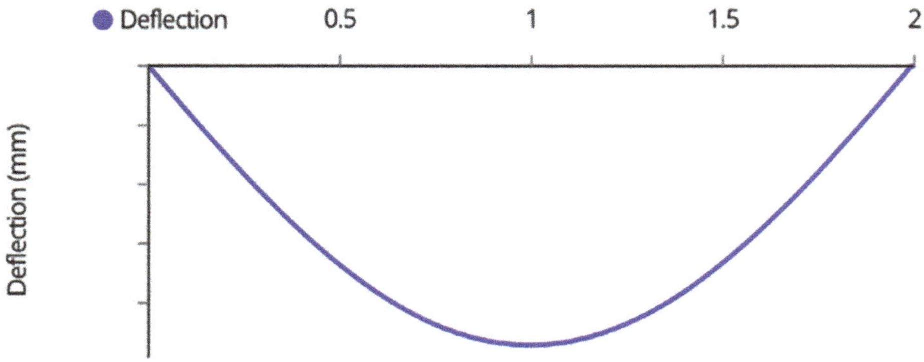

Figure A4. Joist Deflection Diagram.

The maximum deflection of the joists occurs at the midspan. For simplicity of analysis, assume the 4-point loads serve as a uniform distributed load and calculate the maximum deflection using the following equation,

$$\Delta max = \frac{5(w + 2(Point\ Load)L^4}{384EI} \quad \text{(A25)}$$

The beams then carry the point loads of each joist and the factored own weight of the member. The free-body diagram is described in Figure A5,

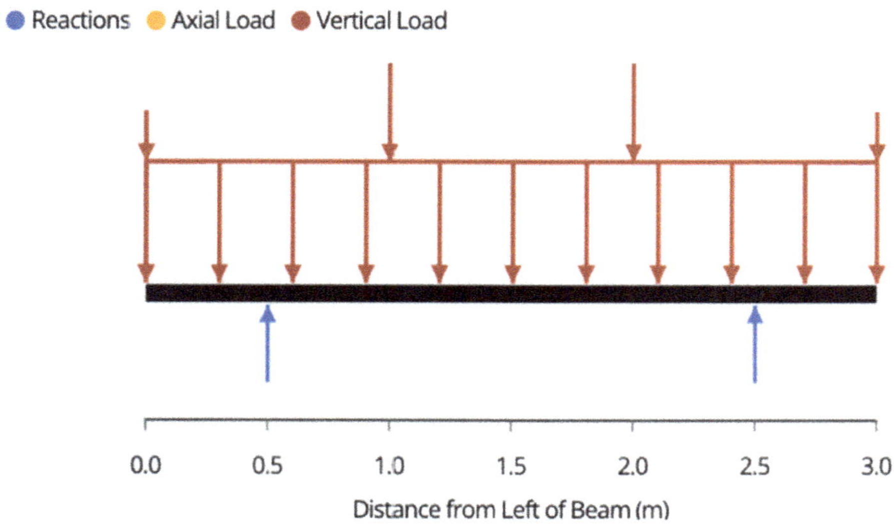

Figure A5. Beam Free-Body Diagram.

Due to the symmetric loading of the beams, the post loads, or the support reactions are described as,

$$Reaction = \frac{(\sum_{k=1}^{4} Joist\ Reactions) + wL}{2} \tag{A26}$$

The shear force diagram for the beams is described in Figure A6,

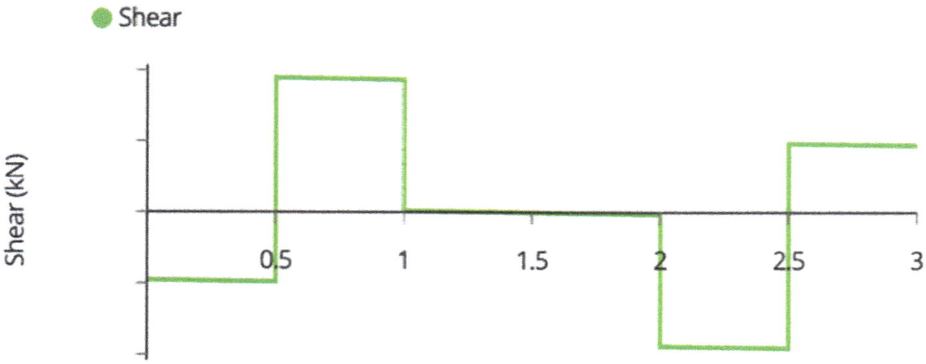

Figure A6. Beam Shear Force Diagram.

The maximum shear forces occur on the inside of the reactions with a value of,

$$Vmax = Reaction - Joist\ Load_1 - \frac{wL}{6} \tag{A27}$$

The bending moment diagram for the beam is shown in Figure A7,

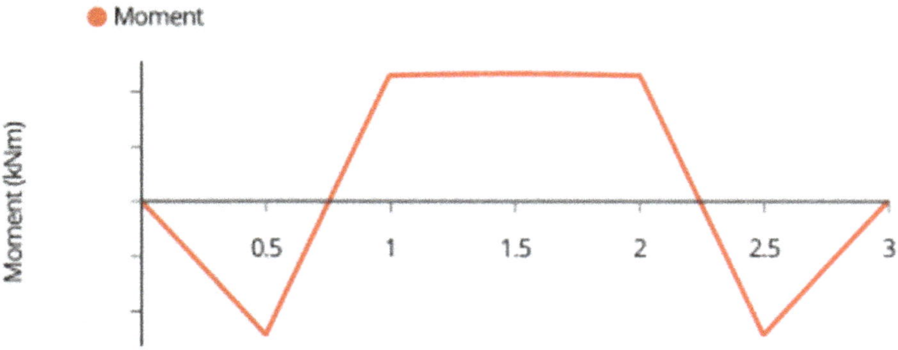

Figure A7. Beam Bending Moment Diagram.

The maximum moment occurs at the supports and can be found by integrating the shear force throughout the first sixth of the beam as described in Equation (A28), or by simply finding the area under the shear force diagram.

$$Mmax = \int_0^{L/6} V(x)dx = (Joist\ Load_1)\frac{L}{6} + \frac{wL^2}{72} \quad \text{(A28)}$$

The deflection diagram is shown in Figure A8,

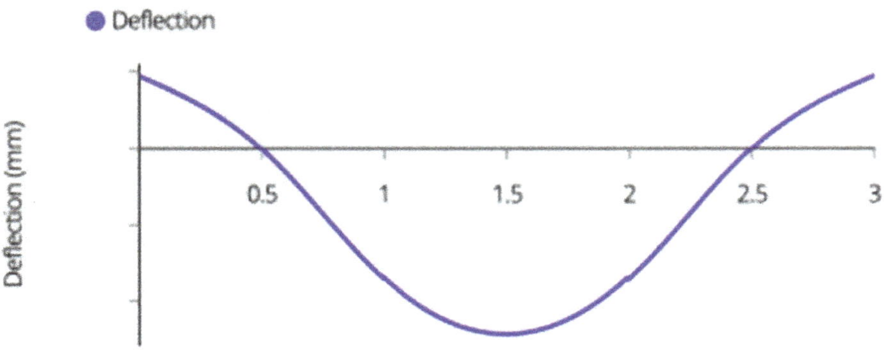

Figure A8. Beam Deflection Diagram.

The maximum deflection occurs at midspan and can be solved using the differential equation and initial conditions below, or by using the moment area theorem or virtual work method described in many structural engineering textbooks.

$$\begin{aligned} \frac{d^2\Delta}{dx^2} &= \frac{M(x)}{EI} \\ \Delta\left(\frac{L}{3}\right) &= 0 \\ \Delta'\left(\frac{L}{2}\right) &= 0 \end{aligned} \quad \text{(A29)}$$

The load is then transferred to the posts. It should be noted that the posts are not loaded purely in compression; an eccentricity described in the free body diagram in Figure A9 induces a bending moment. The post can be idealized as a cantilever with a support reaction from a 2 × 4 brace.

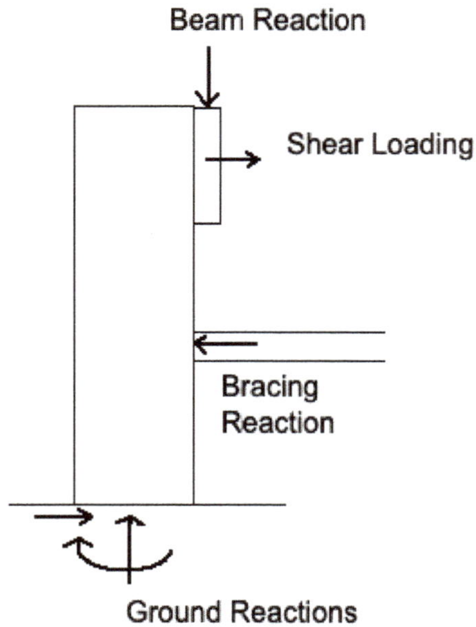

Figure A9. Post Free Body Diagram.

The compressive load of the column is equal to the beam reaction solved above. Along with this compressive load comes a shear loading that is induced by wind and snow loads. This loading can act in either the left or right direction, but the load should be analyzed in the direction that induces bending in the same direction as the beam reaction for assessment of the critical case. The magnitude of this shear load in each post is described in Equation (A30),

$$Shear\ Load = \frac{(Design\ Load)\cos(\theta)}{4} \tag{A30}$$

where θ is the tilt angle of the system. It should be noted that this cantilever with bracing support is an indeterminate structure, meaning that it has too many supports to be solved with static analysis, and thus can not be expressed with generalized equations. The structure can be solved by using finite element analysis, or by an analytical method such as the moment distribution or slope-deflection method.

The shear force diagram of the post is seen in Figure A10,

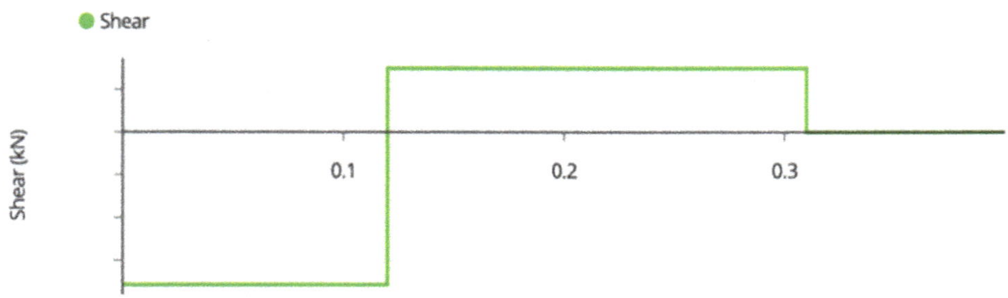

Figure A10. Post Shear Force Diagram.

The bending moment diagram of the post is seen in Figure A11,

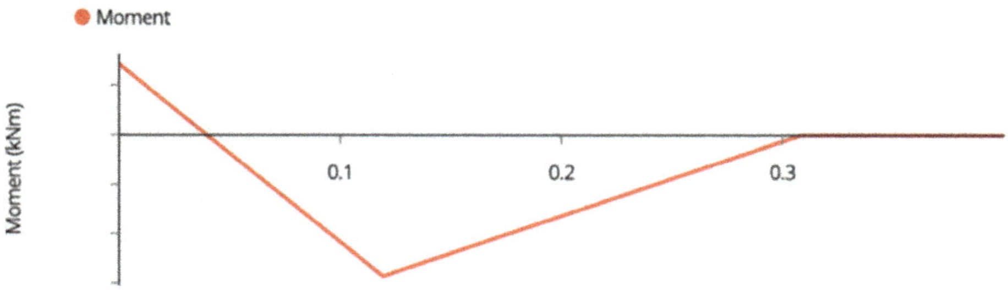

Figure A11. Post-Bending Moment Diagram.

The deflection diagram of the post is seen in Figure A12,

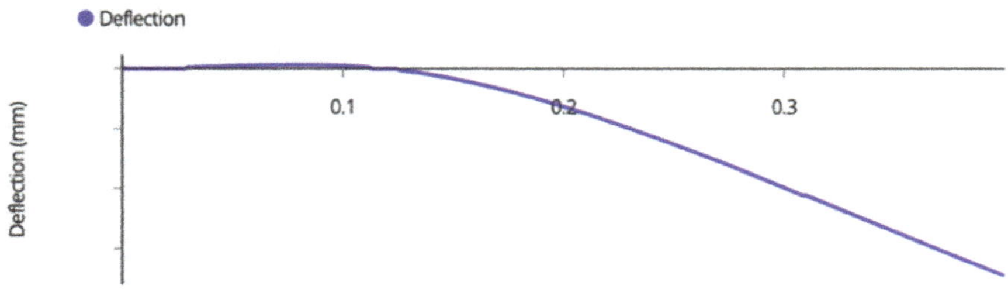

Figure A12. Post Deflection Diagram.

The 2 × 4 brace support is a slender compression member. Not only should the applied force be less than the compressive resistance, but should be less than the buckling resistance governed by the following Euler Buckling equation,

$$Cbuckling = \frac{\pi^2 E I_{weak}}{L^2} \tag{A31}$$

$$where\ I_{weak} = \frac{1}{12} h b^3 \tag{A32}$$

Once all components of the post have been analyzed, the load will finally transfer itself to the ground. Table 9.4.4.1 of the NBCC provides maximum allowable bearing pressures for different types of soil and rock [56]. In the worst case, soft clays support a maximum bearing pressure of 75 kPa. To ensure, that the ground is not overloaded and settles, the bearing pressure can be calculated with the following equation,

$$Bearing\ Pressure = \frac{Post\ Compression}{\frac{\pi}{4}(D_{Footing})^2} \tag{A33}$$

If the applied pressure is more than the allowable, 150 mm of compacted clear stone gravel can be added to the bottom of the footing, or the footing diameter can be increased.

Throughout the system, each connection transfers the load from one member to another via a shear force within the fasteners that compose that connection. For bolts complying with ASTM A307A, the shear resistance of a $\frac{1}{2}''$ carriage bolt holding the beams is about 23.8 kN, and the shear resistance of a $\frac{1}{4}''$ carriage bolt holding the modules is 5.21 kN [83], both of which are beyond the demand of these systems, and thus will not be critical to the design.

References

1. Pearce, J.M. Photovoltaics—A path to sustainable futures. *Futures* **2002**, *34*, 663–674. [CrossRef]
2. Fthenakis, V.M.; Moskowitz, P.D. Photovoltaics: Environmental, health and safety issues and perspectives. *Prog. Photovolt. Res. Appl.* **2000**, *8*, 27–38. [CrossRef]
3. Pearce, J.; Lau, A. Net energy analysis for sustainable energy production from silicon based solar cells. In *ASME Solar 2002: International Solar Energy Conference*; American Society of Mechanical Engineers Digital Collection: Little Falls, NJ, USA, 2002; pp. 181–186.
4. Fthenakis, V.; Alsema, E. Photovoltaics energy payback times, greenhouse gas emissions and external costs: 2004—Early 2005 status. *Prog. Photovolt. Res. Appl.* **2006**, *14*, 275–280. [CrossRef]
5. Barbose, G.L.; Darghouth, N.R.; LaCommare, K.H.; Millstein, D.; Rand, J. *Tracking the Sun: Installed Price Trends for Distributed Photovoltaic Systems in the United States-2018 Edition*; Electricity Markets & Policy: Berkeley, CA, USA, 2018.
6. Feldman, D.; Barbose, G.; Margolis, R.; Bolinger, M.; Chung, D.; Fu, R.; Seel, J.; Davidson, C.; Darghouth, N.; Wiser, R. *Photovoltaic System Pricing Trends: Historical, Recent, and Near-Term Projections 2015 Edition*; National Renewable Energy Lab.: Golden, CO, USA, 2015.
7. Fu, R.; Feldman, D.J.; Margolis, R.M. *US Solar Photovoltaic System Cost Benchmark: Q1 2018*; No. NREL/TP-6A20-72399; National Renewable Energy Lab (NREL): Golden, CO, USA, 2018.
8. Barron, A.R. Cost reduction in the solar industry. *Mater. Today* **2015**, *18*, 2–3. [CrossRef]
9. Reuters. Solar Costs to Fall Further, Powering Global Demand—Irena. Reuters 2017. Available online: https://www.reuters.com/article/singapore-energy-solar-idUSL4N1MY2F8 (accessed on 13 April 2020).
10. Matasci, S. Solar Panel Cost: Avg. Solar Panel Prices by State in 2019: EnergySage. Solar News, EnergySage, 5 June 2019. Available online: news.energysage.com/how-much-does-the-average-solar-panel-installation-cost-in-the-u-s/ (accessed on 28 January 2022).
11. Branker, K.; Pathak, M.J.M.; Pearce, J.M. A review of solar photovoltaic levelized cost of electricity. *Renew. Sustain. Energy Rev.* **2011**, *15*, 4470–4482. [CrossRef]
12. Dudley, D. *Renewable Energy Will Be Consistently Cheaper Than Fossil Fuels By 2020, Report Claims [WWW Document]*; Forbes: Jersey City, NJ, USA, 2019. Available online: https://www.forbes.com/sites/dominicdudley/2018/01/13/renewable-energy-cost-effective-fossil-fuels-2020/ (accessed on 13 April 2020).
13. Solar Industry Research Data. Available online: https://www.seia.org/solar-industry-research-data (accessed on 13 April 2020).
14. Vaughan, A. Time to Shine: Solar Power Is Fastest-Growing Source of New Energy. Available online: https://www.theguardian.com/environment/2017/oct/04/solar-power-renewables-international-energy-agency/ (accessed on 28 January 2022).
15. Levin, T.; Thomas, V.M. Can developing countries leapfrog the centralized electrification paradigm? *Energy Sustain. Dev.* **2016**, *31*, 97–107. [CrossRef]
16. Lang, T.; Ammann, D.; Girod, B. Profitability in absence of subsidies: A techno-economic analysis of rooftop photovoltaic self-consumption in residential and commercial buildings. *Renew. Energy* **2016**, *87*, 77–87. [CrossRef]
17. Prehoda, E.; Pearce, J.; Schelly, C. Policies to Overcome Barriers for Renewable Energy Distributed Generation: A Case Study of Utility Structure and Regulatory Regimes in Michigan. *Energies* **2019**, *12*, 674. [CrossRef]
18. Agenbroad, J.; Carlin, K.; Ernst, K.; Doig, S. Minigrids in the Money: Six Ways to Reduce Minigrid Costs by 60% for Rural Electrification. Rocky Mountain Institute. 2018. Available online: https://rmi.org/insight/minigrids-money/ (accessed on 28 February 2022).
19. Alafita, T.; Pearce, J.M. Securitization of residential solar photovoltaic assets: Costs, risks and uncertainty. *Energy Policy* **2014**, *67*, 488–498. [CrossRef]
20. Rai, V.; Reeves, D.C.; Margolis, R. Overcoming Barriers and Uncertainties in the Adoption of Residential Solar PV. *Renew. Energy* **2016**, *89*, 498–505. [CrossRef]
21. Horváth, D.; Szabó, R.Z. Evolution of Photovoltaic Business Models: Overcoming the Main Barriers of Distributed Energy Deployment. *Renew. Sustain. Energy Rev.* **2018**, *90*, 623–635. [CrossRef]
22. Yousaf, H.; Shakeel, S.R.; Rajala, A.; Raza, Z. Addressing Financial Barriers Influencing the Adoption of Solar PV: The Role of Business Models. In *Proceedings of the Advances in Human Factors, Business Management and Leadership*; Kantola, J.I., Nazir, S., Salminen, V., Eds.; Springer International Publishing: Cham, Switzerland, 2021; pp. 42–49.
23. Grafman, L.; Pearce, J.M. *To Catch the Sun*; Humboldt State University Press: Arcata, CA, USA, 2021; ISBN 978-1-947112-62-9.
24. Renewables International, 2013. Photovoltaics after Grid Parity Plug-and-Play PV: The Controversy 2013. Renewables. Available online: http://www.renewablesinternational.net/plug-and-play-pv-the-controversy/150/452/72715/ (accessed on 18 December 2015).
25. Mundada, A.S.; Nilsiam, Y.; Pearce, J.M. A review of technical requirements for plug-and-play solar photovoltaic microinverter systems in the United States. *Sol. Energy* **2016**, *135*, 455–470. [CrossRef]
26. Khan, M.T.A.; Norris, G.; Chattopadhyay, R.; Husain, I.; Bhattacharya, S. Autoinspection and Permitting with a PV Utility Interface (PUI) for Residential Plug-and-Play Solar Photovoltaic Unit. *IEEE Trans. Ind. Appl.* **2017**, *53*, 1337–1346. [CrossRef]
27. Khan, M.T.A.; Husain, I.; Lubkeman, D. Power electronic components and system installation for plug-and-play residential solar PV. In Proceedings of the 2014 IEEE Energy Conversion Congress and Exposition (ECCE), Pittsburgh, PA, USA, 14–18 September 2014; pp. 3272–3278.

28. Lundstrom, B.R. *Plug and Play Solar Power: Simplifying the Integration of Solar Energy in Hybrid Applications*; Cooperative Research and Development Final Report, CRADA Number CRD-13-523; National Renewable Energy Lab. (NREL): Golden, CO, USA, 2017.
29. Mundada, A.S.; Prehoda, E.W.; Pearce, J.M. U.S. market for solar photovoltaic plug-and-play systems. *Renew. Energy* **2017**, *103*, 255–264. [CrossRef]
30. Feldman, D.; Barbose, G.; Margolis, R.; Wiser, R.; Darghout, N.; Goodrich, A. *Photovoltaic (PV) Pricing Trends: Historical, Recent, and Near-Term Projections, Sunshot*; National Renewable Energy Laboratory: Washington, DC, USA, 2012.
31. PVinsights. PVinsights 2022. Available online: http://pvinsights.com/ (accessed on 16 January 2022).
32. Alt E Store. Tamarack Solar Top of Pole Mounts for Large Solar Panels. Available online: https://www.altestore.com/store/solar-panel-mounts/top-of-pole-solar-panel-mounts/tamarack-solar-top-of-pole-mounts-6072-cell-solar-panels-p40745/ (accessed on 16 January 2022).
33. TPM3 Pole Mount for Three 60/72 Cell Solar Modules. Available online: https://www.off-the-grid-solar.com/products/tpm3-pole-mount-for-three-60-72-cell-solar-modules (accessed on 11 March 2022).
34. Wittbrodt, B.; Laureto, J.; Tymrak, B.; Pearce, J.M. Distributed Manufacturing with 3-D Printing: A Case Study of Recreational Vehicle Solar Photovoltaic Mounting Systems. *J. Frugal Innov.* **2015**, *1*, 1. [CrossRef]
35. Wittbrodt, B.; Pearce, J.M. 3-D Printing Solar Photovoltaic Racking in Developing World. *Energy Sustain. Dev.* **2017**, *36*, 1–5. [CrossRef]
36. Wittbrodt, B.T.; Pearce, J.M. Total U.S. Cost Evaluation of Low-Weight Tension-Based Photovoltaic Flat-Roof Mounted Racking. *Sol. Energy* **2015**, *117*, 89–98. [CrossRef]
37. Beylot, A.; Payet, J.; Puech, C.; Adra, N.; Jacquin, P.; Blanc, I.; Beloin-Saint-Pierre, D. Environmental impacts of large-scale grid-connected ground-mounted PV installations. *Renew. Energy* **2014**, *61*, 2–6. [CrossRef]
38. Lehmann, S. Sustainable Construction for Urban Infill Development Using Engineered Massive Wood Panel Systems. *Sustainability* **2012**, *4*, 2707–2742. [CrossRef]
39. Hammond, G.P.; Jones, C.I. Inventory of (Embodied) Carbon & Energy (ICE) Ver 3.0 Beta. University of Bath, UK. 2019. Available online: https://circularecology.com/embodied-energy-and-carbon-footprint-database.html (accessed on 8 May 2020).
40. Freeman, M.H.; McIntyre, C.R. Comprehensive Review of Copper-Based Wood Preservatives. *For. Prod. J.* **2008**, *58*, 6–27.
41. Adpearance, I. What You Need to Know about Pressure Treated Wood | AIFP | PDX, OR. Available online: https://www.lumber.com/blog/what-you-need-to-know-about-pressure-treated-wood (accessed on 17 February 2022).
42. Gilman, P. *SAM Photovoltaic Model Technical Reference*; NREL/TP-6A20-64102; National Renewable Energy Lab. (NREL): Golden, CO, USA, 2015; p. 63. [CrossRef]
43. SAM Open Source—System Advisor Model (SAM). Available online: https://sam.nrel.gov/about-sam/sam-open-source.html (accessed on 11 March 2022).
44. *System Advisor Model (SAM)*; National Renewable Energy Laboratory: Golden, CO, USA, 2022. Available online: https://github.com/NREL/SAM (accessed on 28 January 2022).
45. LG 400W NeON2 BiFacial Solar Panel | LG400N2T-J5—Volts Energies. Available online: https://volts.ca/collections/solar-panels/products/lg400n2t-j5-solar-panel (accessed on 10 January 2022).
46. Molin, E.; Stridh, B.; Molin, A.; Wäckelgård, E. Experimental Yield Study of Bifacial PV Modules in Nordic Conditions. *IEEE J. Photovolt.* **2018**, *8*, 1457–1463. [CrossRef]
47. Riedel-Lyngskær, N.; Ribaconka, M.; Pó, M.; Thorseth, A.; Thorsteinsson, S.; Dam-Hansen, C.; Jakobsen, M.L. The Effect of Spectral Albedo in Bifacial Photovoltaic Performance. *Sol. Energy* **2022**, *231*, 921–935. [CrossRef]
48. Burnham, L.; Riley, D.; Walker, B.; Pearce, J.M. Performance of Bifacial Photovoltaic Modules on a Dual-Axis Tracker in a High-Latitude, High-Albedo Environment. In Proceedings of the 2019 IEEE 46th Photovoltaic Specialists Conference (PVSC), Chicago, IL, USA, 16–21 June 2019; pp. 1320–1327.
49. Hayibo, K.; Petsiuk, A.; Mayville, P.; Brown, L.; Pearce, J.M. Monofacial vs Bifacial Solar Photovoltaic Systems in Snowy Environments. Available online: https://ssrn.com/abstract=4056922 (accessed on 10 January 2022).
50. Heidari, N.; Gwamuri, J.; Townsend, T.; Pearce, J.M. Impact of Snow and Ground Interference on Photovoltaic Electric System Performance. *IEEE J. Photovolt.* **2015**, *5*, 1680–1685. [CrossRef]
51. Gibb, A.; Abadie, S. *Building Open Source Hardware: DIY Manufacturing for Hackers and Makers*; Pearson Education: Upper Saddle River, NJ, USA, 2014; ISBN 978-0-321-90604-5.
52. Definition (English)—Open Source Hardware Association. Available online: https://www.oshwa.org/definition/ (accessed on 13 April 2020).
53. Oberloier, S.; Pearce, J.M. General Design Procedure for Free and Open-Source Hardware for Scientific Equipment. *Designs* **2018**, *2*, 2. [CrossRef]
54. CERN OHL version 2·Wiki·Projects/CERN Open Hardware License. Available online: https://www.ohwr.org/project/cernohl/wikis/Documents/CERN-OHL-version-2 (accessed on 13 April 2020).
55. Joist Hangers and End Moments—Structural Engineering General Discussion—Eng-Tips. Available online: https://www.eng-tips.com/viewthread.cfm?qid=339938 (accessed on 10 January 2022).
56. Canada, N.R.C. National Building Code of Canada 2015. Available online: https://nrc.canada.ca/en/certifications-evaluations-standards/codes-canada/codes-canada-publications/national-building-code-canada-2015 (accessed on 17 February 2022).

57. One Way Slab Design. DAILY CIVIL 2018. Available online: https://dailycivil.com/one-way-slab-design-how-to-design-one-way-slab-1/ (accessed on 28 January 2022).
58. Kokutse, A.D.; Akpenè, A.D.; Monteuuis, O.; Akossou, A.; Langbour, P.; Guibal, D.; Tomazello, M.F.; Gbadoe, E.; Chaix, G.; Kokou, K. Selection of plus trees for genetically improved teak varieties produced in benin and togo. *Bois For. Des Trop.* **2017**, *328*, 55. [CrossRef]
59. Hounlonon, M.C.; Kouchade, C.A.; Kounouhewa, B.B. Propriétés physiques et méca-niques du bois de teck de provenances tanzanienne et locale au bénin. *Bois For. Des Trop.* **2017**, *331*, 45–53. [CrossRef]
60. Phinikarides, A.; Kindyni, N.; Makrides, G.; Georghiou, G.E. Review of Photovoltaic Degradation Rate Methodologies. *Renew. Sustain. Energy Rev.* **2014**, *40*, 143–152. [CrossRef]
61. Ryan, A.; Williams Daniel, J.; Lizzadro-McPherson, J.; Pearce, M. The Impact of Snow Losses on Solar Photovoltaic Systems in North America in the Future. Unpublished work. 2022.
62. 1 CAD to XOF—Canadian Dollars to CFA Francs Exchange Rate. Available online: https://www.xe.com/currencyconverter/convert/?Amount=1&From=CAD&To=XOF (accessed on 28 January 2022).
63. Sociétéle, P.L.R. Togo: Quand le Prix du Bois Touche le Plafond. L-FRII 2021. Available online: https://l-frii.com/togo-quand-le-prix-du-bois-touche-le-plafond/ (accessed on 28 January 2022).
64. 1 USD to CAD—US Dollars to Canadian Dollars Exchange Rate. Available online: https://www.xe.com/currencyconverter/convert/?Amount=1&From=USD&To=CAD (accessed on 13 March 2022).
65. 12 Solar Panel Ground Mounting Kit IronRidge. Available online: https://sunwatts.com/12-solar-panel-ground-mounting-kit-ironridge/ (accessed on 13 March 2022).
66. Solcan. Personal Communication. 3 November 2021.
67. Setliff, E.C. Wood Decay Hazard in Canada Based on Scheffer's Climate Index Formula. *For. Chron.* **1986**, *62*, 456–459. [CrossRef]
68. Forest Products Laboratory. Fungal Decay Hazard Map. Available online: https://www.apawood.org/data/sites/1/documents/technicalresearch/rip/fplrip-4723-022.pdf (accessed on 7 March 2022).
69. Helsen, L.; Van den Bulck, E.; Hery, J.S. Total recycling of CCA treated wood waste by low-temperature pyrolysis. *Waste Manag.* **1998**, *18*, 571–578. [CrossRef]
70. Wiemann, M.C. Characteristics and Availability of Commercially Important Woods. 46. Available online: https://www.fpl.fs.fed.us/documnts/fplgtr/fplgtr190/chapter_02.pdf (accessed on 28 January 2022).
71. Green David, W.; Winandy Jerrold, E.; Kretschmann David, E. *Mechanical Properties of Wood. Wood Handbook: Wood as an Engineering Material*; General Technical Report FPL; GTR-113; USDA Forest Service, Forest Products Laboratory: Madison, WI, USA, 1999; pp. 4.1–4.45.
72. Kessy, J.G.; Alexander, M.G.; Beushausen, H. Concrete Durability Standards: International Trends and the South African Context. *J. S. Afr. Inst. Civ. Eng.* **2015**, *57*, 47–58. [CrossRef]
73. Lumber—2022 Data—1978–2021 Historical—2023 Forecast—Price—Quote—Chart. Available online: https://tradingeconomics.com/commodity/lumber (accessed on 2 March 2022).
74. Norfolk County, Solar Panels Mounted on Buildings 2013. Available online: https://www.norfolkcounty.ca/download/living/building/faq/SOLAR%20PANELS%20MOUNTED%20ON%20BUILDINGS%202013.pdf (accessed on 28 January 2022).
75. Cuello-Polo, G.; O'Neill-Carrillo, E. Power System Modeling for the Study of High Penetration of Distributed Photovoltaic Energy. *Designs* **2021**, *5*, 62. [CrossRef]
76. Benato, A.; Stoppato, A.; De Vanna, F.; Schiro, F. Spraying Cooling System for PV Modules: Experimental Measurements for Temperature Trends Assessment and System Design Feasibility. *Designs* **2021**, *5*, 25. [CrossRef]
77. Mothilal Bhagavathy, S.; Pillai, G. PV Microgrid Design for Rural Electrification. *Designs* **2018**, *2*, 33. [CrossRef]
78. CanmetENERGY (Canada). *Solar Ready Guidelines for Solar Domestic Hot Water and Photovoltaic Systems*; CanmetENERGY: Ottawa, ON, Canada, 2012; ISBN 978-1-100-20633-2. Available online: https://www.nrcan.gc.ca/sites/www.nrcan.gc.ca/files/canmetenergy/files/pubs/SolarReadyGuidelines_en.pdf (accessed on 28 January 2022).
79. Kettle, A.J. Extreme 50 Year Return Wind Speeds from the USAF Data Set. 2019, p. 3. Available online: https://meetingorganizer.copernicus.org/ECSS2019/ECSS2019-218-3.pdf (accessed on 28 January 2022).
80. NDS. 2018. Available online: https://awc.org/publications/2018-nds/ (accessed on 17 February 2022).
81. MicroProSPEC-SpecifierGuide. Available online: https://microprosienna.com/wp-content/uploads/2020/07/MicroProSPEC-SpecifierGuide.pdf (accessed on 28 January 2022).
82. Dead Loads. Available online: https://www.designingbuildings.co.uk/wiki/Dead_loads (accessed on 17 February 2022).
83. Load Calculator | Fastenal. Available online: https://www.fastenal.com/en/84/load-calculator (accessed on 17 February 2022).

energies

Article

Influence of a Better Prediction of Thermal Satisfaction for the Implementation of an HVAC-Based Demand Response Strategy

Jongyeon Lim [1,2] and Wonjun Choi [3,*]

[1] Department of Architectural Engineering, Kangwon National University, Kangwon-do 24341, Korea; jongyeon.lim@kangwon.ac.kr
[2] Department of Integrated Energy and Infra System, Kangwon National University, Kangwon-do 24341, Korea
[3] School of Architecture, Chonnam National University, Gwangju 61186, Korea
* Correspondence: wonjun.choi@jnu.ac.kr

Abstract: Building system operation faces the challenge of reducing energy use and implementing a demand response, which can be defined as a temporary modification in energy loads affecting dynamic energy price and reliability information. The heating, ventilation, and air-conditioning (HVAC) system in buildings provides an opportunity for implementing demand response strategies due to the thermal inertia in building zones. However, an HVAC-based demand response is not a prevalent strategy in actual facility management due to the lack of understanding among building operators of their facilities and occupants. Herein, we focus on developing a better understanding of the occupant side by obtaining a reliable prediction of occupants' thermal satisfaction. We evaluate the prediction performance of a probabilistic model provided in our previous paper using a case study with a subset of the ASHRAE Global Thermal Comfort Database II. The influence of a better prediction of thermal satisfaction on the implementation of the HVAC-based demand response strategy is further discussed. The conventional method overestimates productivity deterioration due to changes in the thermal environment, making it challenging to implement an HVAC-based demand response strategy aggressively. A robust prediction model using a probabilistic approach can solve this problem, allowing building operators to adopt an aggressive stance for implementing a demand response. The results of this study offer fresh insight into the impact of a probabilistic model in the prediction of thermal satisfaction for establishing an HVAC-based demand response strategy.

Keywords: thermal comfort; thermal satisfaction; demand response strategy; thermal sensation; occupant performance; predicted mean vote

Citation: Lim, J.; Choi, W. Influence of a Better Prediction of Thermal Satisfaction for the Implementation of an HVAC-Based Demand Response Strategy. *Energies* **2022**, *15*, 3094. https://doi.org/10.3390/en15093094

Academic Editors: Shi-Jie Cao and Wei Feng

Received: 14 March 2022
Accepted: 21 April 2022
Published: 23 April 2022

1. Introduction

The purpose of heating, ventilation, and air-conditioning (HVAC) systems is to maintain desired environmental conditions in a specific physical space. These conditions are collectively referred to as the conditions for human thermal comfort, which is defined as a mental state of satisfaction with the thermal environment [1]. The term satisfaction is often synonymously used with acceptability [2]. A thermally comfortable environment ensures that the conditions are satisfactory/acceptable to most occupants (i.e., more than 80%, according to ASHRAE Standard 55 [3]) within a specific physical space. HVAC systems are responsible for efficiently maintaining the desired service level in accordance with ASHRAE Standard 55.

Recently, energy use in buildings has emerged as an important issue. Building operators and facility managers face the challenge of reducing energy costs to temporarily reduce the load in response to the spike in electricity prices [4,5], which is called a demand response. HVAC systems, along with lighting, are most commonly adjusted to achieve energy savings in demand response strategies in buildings. This is because HVAC systems are well-suited to a demand response strategy that evenly distributes the burden across the facility, which is least likely to have negative effects on building occupants. Furthermore, large thermal inertia in an occupied space allows HVAC systems to be temporarily unloaded without an immediate negative impact on the occupants. However, despite this potential for energy saving, the HVAC-based demand response is not a prevalent strategy in actual facility management. One reason is the lack of automated HVAC control systems required during demand response events [6]. Because of a lack of hardware, simple strategies are typically prioritized, such as the global temperature adjustments of spaces (e.g., turning the cooling thermostat up by 3 °C) [7]. However, applying the demand response strategy also requires software that determines proper operating points. In this study, we focused on a software approach to develop a successful HVAC-based demand response strategy (e.g., a setpoint temperature adjustment), for example a predictive model capable of accurately predicting occupants' thermal satisfaction under varying thermal conditions.

In implementing HVAC-based demand response strategies, building operators and facility managers should consider meeting energy saving targets while minimizing the negative impacts on occupants' thermal comfort, because thermal discomfort due to reduced service levels can harm occupants' productivity, leading to economic losses [8–12]. In other words, changes in thermal satisfaction due to setpoint temperature adjustments not within acceptable boundaries can result in the further deterioration of the occupants' well-being and lead to an expected loss of productivity, which is the most significant concern for building operators. To this end, a multi-objective optimizer for HVAC control is sometimes installed and tested. Homod et al. [13] proposed a fuzzy forward control strategy to simultaneously balance energy saving and achieve occupant satisfaction. Schito et al. [14] demonstrated the multi-objective optimization of the HVAC control in museums to achieve visitors' comfort and energy savings without compromising the integrity of the artwork. Reena et al. [15] and Turley et al. [16] presented a framework for energy and comfort management in buildings. Although these previous works are meaningful advances in this field, there are limitations associated with simulation-based approaches and with using the existing thermal comfort index, such as the predicted mean vote (PMV) and predicted percentage of dissatisfied (PPD) proposed by Fanger [17]. In reality, even a simple strategy, such as setpoint adjustment, is difficult to aggressively implement for achieving energy saving goals because building operators and facility managers are generally afraid of not meeting the desired service level and facing complaints from occupants as a result [18]. To successfully implement an HVAC control that meets both the thermal comfort and energy saving requirements, a technique that can offer reliable data on occupants' thermal satisfaction should be developed.

For over 50 years, PMV and PPD have been widely employed to assess the indoor climate and thermal satisfaction of occupants. As addressed by Benton et al. [19], several studies have validated the relationship between indoor climate and occupants' thermal satisfaction provided by the PMV–PPD model. However, several studies, such as those by Schiller [20] and Xavier and Lamberts [21], reported discrepancies between the PPD and occupant dissatisfaction in practical scenarios. Recent advancements in data science have facilitated overcoming this prediction failure. Katić et al. [22] and Ghahramani et al. [23] focused on individuals' thermal comfort responses and developed a personal comfort prediction model by adopting machine learning algorithms such as support vector machines and ensemble algorithms. Li et al. [24] proposed a high accuracy comfort prediction method using an artificial neural network with three physiological input parameters. Although the intrinsic objective of these studies and our study was to improve prediction performance, we mainly focused on strategies to reflect the stochastic characteristics of thermal satisfaction in the prediction model.

We believe that the reported failure in the prediction by Fanger's model may be a result of inherent limits in the deterministic approach used to provide a link between environmental conditions and human sensations. Additionally, nonthermal factors (such as race, age, gender, ethnicity, and region), which are not considered in typical deterministic PMV models but make the prediction highly uncertain, may also cause a prediction failure.

In the authors' previous study [25], a stochastic model which is distinct from the existing simple linear regression methods and can probabilistically reproduce dispersed occupants' response to thermal sensation, was developed. Compared with the deterministic method based on Fanger's PMV–PPD model, our model adequately provides the stochastic characteristics of dispersed thermal sensation votes across occupants and a robust prediction of thermal satisfaction.

In this study, we present an argument that the reliable prediction of thermal satisfaction can assist building operators and facility managers in aggressively implementing demand response strategies, ever since determining that proper operating points can reduce energy use while minimizing deterioration in productivity and thermal comfort. To support this argument, we compared a conventional prediction model (i.e., Fanger's PMV–PPD model [17]) for thermal satisfaction with a data-driven probabilistic prediction model proposed in our previous research [25]. In addition, we also discussed the influence of the differences in prediction accuracy for the implementation of an HVAC-based demand response strategy.

The remainder of this paper is structured as follows. In Section 2, the field survey data used in the case study are briefly presented. Thereafter, in Sections 3 and 4, the prediction performance is tested with publicly available data on occupants' response to thermal sensation (ASHRAE Global Thermal Comfort Database II [26]), emphasizing thermal satisfaction, and compared with that of the conventional method. In addition, the impact of the better prediction of occupants' thermal satisfaction on the prediction of occupants' productivity is quantified. The significance of implementing HVAC-based demand response strategies is also discussed. Finally, the limitations of this study and the conclusions, along with directions for future studies, are presented in Sections 5 and 6, respectively.

2. Data Description

A subset of ASHRAE Global Thermal Comfort Database II [26] was used to base the discussion on thermal satisfaction and productivity on real-world data. Although the database is a collection of field surveys performed under various conditions (climate, building type, experimental range, etc.), we extracted and used only specific field survey data collected from an air-conditioned office in the Midlands, UK (number of data: $n = 4316$; monitoring data measured continuously for one week in summer), which was also used by Oseland [27]. Table 1 lists the contents of the dataset used. Three additional indices related to thermal comfort, i.e., the PMV, PPD, and operative temperature (OT), were then calculated for each observed value by guidance in engineering references [1,3,28].

Table 1. Dataset contents.

Variable	Description
Thermal sensation vote (TSV)	The seven-point scaled thermal sensation −3: cold, −2: cool, −1: slightly cool, 0: neutral, 1: slightly warm, 2: warm, 3; hot
Thermal acceptability	0: unacceptable, 1: acceptable
Clothing insulation [clo]	Intrinsic clothing ensemble insulation of the subject
Metabolic rate [met]	Average activity level of the subject
Air temperature [°C]	Air temperature 1.1 m above the floor
Globe temperature [°C]	Globe temperature 1.1 m above the floor
Relative humidity [%]	Relative humidity
Air velocity [m/s]	Airspeed 1.1 m above the floor

3. Limitations of the Conventional TSV Model

A field survey is considered in which participants are requested to vote their thermal sensation on a seven-point scale, and the measured OT value defines the indoor thermal condition.

In surveys, the measured PMV value often disagrees with the thermal sensation perceived by the occupants because the psychological and behavioral factors influencing the occupants' perceptions are not fully considered in the PMV model. Consequently, the actual TSVs are often biased towards the warm or cool sides compared with the measured PMV. Figure 1 shows the OT–TSV and OT–PMV relationships obtained from the survey. These relationships show a bias between TSV and PMV that often exceeds 1 scale unit. Fortunately, a least-squares line can provide a good approximation of the mean TSV for each OT level ($r^2 \approx 0.94$).

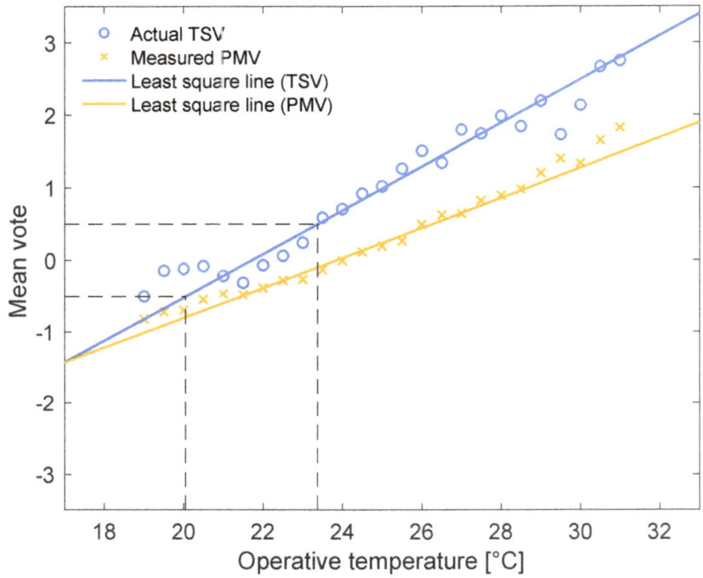

Figure 1. OT−TSV and OT−PMV relationships. The mean TSV and mean PMV are plotted against OTs binned in 0.5 °C increments. Least-squares lines are generated using weighted regression to account for the number of votes in each OT bin.

A comfort zone [3] can be deduced as the OT range from 20.05 to 23.37 °C using a least-squares line (the blue line in Figure 1). Table 2 lists the statistics of the TSVs made in this range. Although this range should be equivalent to 90% satisfaction, the actual percentage of thermally satisfied occupants in this range is approximately 76%. This is attributed to the discrepancy in the deterministic estimation approach that solely produces the mean vote and ignores its variation. This limitation is closely associated with the conventional method of predicting thermal satisfaction.

Table 2. TSV statistics obtained at the comfort zone identified in the survey data.

Thermal Sensation	Cold	Cool	Slightly Cool	Neutral	Slightly Warm	Warm	Hot
Counts	46	118	251	561	280	137	39
Share	3.2%	8.2%	17.5%	39.2%	19.6%	9.6%	2.7%

Generally, a simple linear regression model based on constant variance is used to derive the dose–response relationship between a given thermal condition and the consequent thermal sensation in occupants, presumably owing to its mathematical simplicity. However, this is an optional method, and its results may not necessarily be linked to the prediction of thermal satisfaction.

Figure 2 shows the percentage of dissatisfied occupants as determined using the field survey dataset, wherein the data points are organized in terms of OT. The following components are listed in the legend:

(a) OPD: Observational thermal dissatisfaction (percentage of occupants voting for a thermal sensation of cold, cool, warm, or hot for each data bin divided into 0.5 °C OT intervals).

(b) PPD: Percentage of dissatisfied occupants calculated using Fanger's PMV–PPD equation.

(c) mPPD: Percentage of dissatisfied occupants calculated using Fanger's PMV–PPD equation modified with the mean TSV (the blue line in Figure 1). Many studies on the modification of the thermal comfort model based on field surveys have used this approach [29–32].

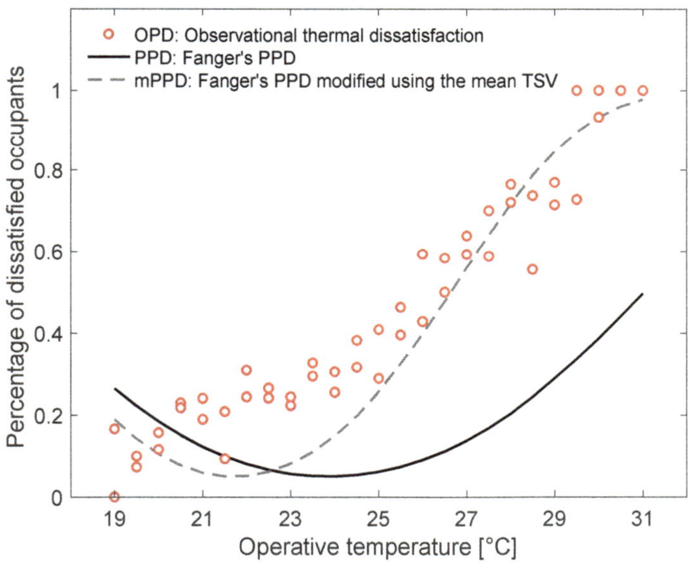

Figure 2. Relationship between the operative temperature and the percentage of dissatisfied occupants.

The following discussion is only concerned with the results between the OTs of 20 and 30 °C because the data from the other ranges were insufficient.

In Figure 2, the observed thermal satisfaction was the highest (i.e., the lowest OPD) at an OT of ~22 °C. The mPPD captured the OT range wherein the lowest thermal dissatisfaction was observed. However, the PPD did not reflect this observation, as it was biased to the higher OT side. Furthermore, both the PPD and mPPD predicted an extremely low dissatisfaction rate (approximately 5%) in the thoroughly conditioned environment. However, in a field survey, more than 20% of the occupants remained thermally dissatisfied even within the OT range, where the lowest thermal dissatisfaction appeared.

4. Results

This section aims to emphasize the performance of our model in predicting thermal sensation and thermal satisfaction and discuss its importance in implementing HVAC-based demand response strategies. First, the predicted results of applying the probabilistic model proposed in our previous study [25] are presented; this is based on a case study with the data presented in Section 3. The used model was validated by k-fold cross-validation ($k = 5$) to prevent overfitting to the specific training data. Then, the influence of a better prediction of thermal satisfaction by using the probabilistic prediction approach for the implementation of an HVAC-based demand response strategy is discussed. Although the essence of the prediction method is briefly described below, the details of the prediction method are omitted here to avoid obscuring the focus of this paper (for details, see Lim et al. [25]).

The prediction model comprises a statistical framework for estimating model parameters and a regression method for considering the impact of the measured (thermal) and non-measured (nonthermal) factors on thermal sensation. Our model, including the variance parameter defined as a thermal-condition-dependent variable, provides a reliable prediction of the distribution of thermal sensations in a given indoor climate. This consideration can generate reliable information on the thermal satisfaction of occupants.

4.1. Model-Predicted Thermal Sensation

Figure 3 shows the observed probability of occupants responding to each category of the seven-point scale TSV in given OT conditions. Figure 4 presents the prediction results obtained from the conventional and proposed models in given thermal conditions. The predictions shown in Figure 4a were deterministically established as 0 or 1 according to the pseudocode listed in Algorithm 1. Figure 4b shows the probability distribution of the occupant response in each TSV category. A comparison of Figures 3 and 4 indicates that the proposed model reflects the observations better compared to the prediction by a simple linear regression. Therefore, the proposed model can provide a thermal sensation profile reflecting the inherent stochastic characteristics of the actual TSV.

Algorithm 1. Pseudocode for deterministically establishing the response probability for each TSV category.

```
mean TSV = −6.5368 + 0.3011 × OT (based on the blue line in Figure 1).
If mean TSV < −2.5 then "thermal sensation of cold"
    elseif mean TSV < −1.5 then "thermal sensation of cool"
    elseif mean TSV < −0.5 then "thermal sensation of slightly cool"
    elseif mean TSV < 0.5 then "thermal sensation of neutral"
    elseif mean TSV < 1.5 then "thermal sensation of slightly warm"
    elseif mean TSV < 2.5 then "thermal sensation of warm"
    elseif "thermal sensation of hot"
end
```

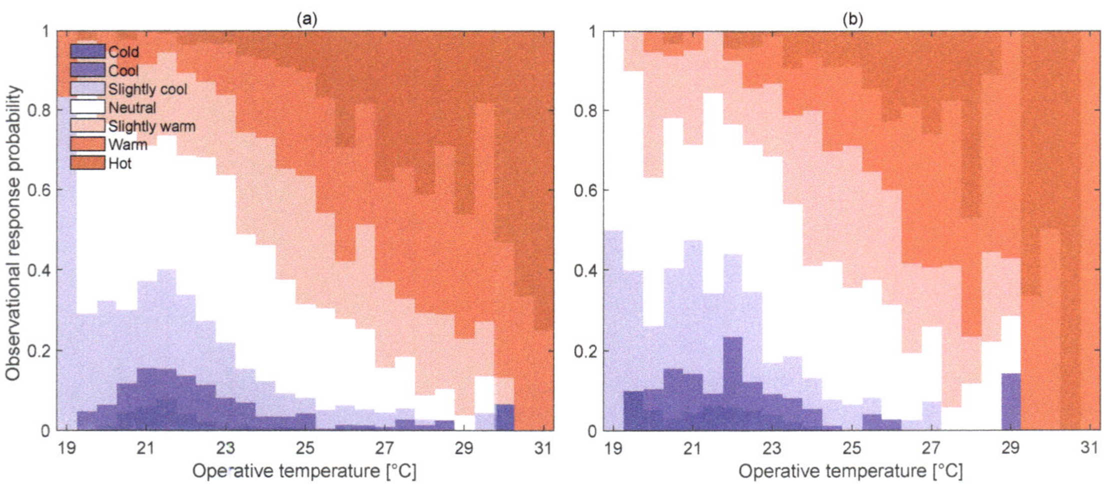

Figure 3. Observational percentage of occupants responding to each TSV category from the (**a**) training and (**b**) validation datasets. Each stacked bar was drawn using bin width of 0.5 °C.

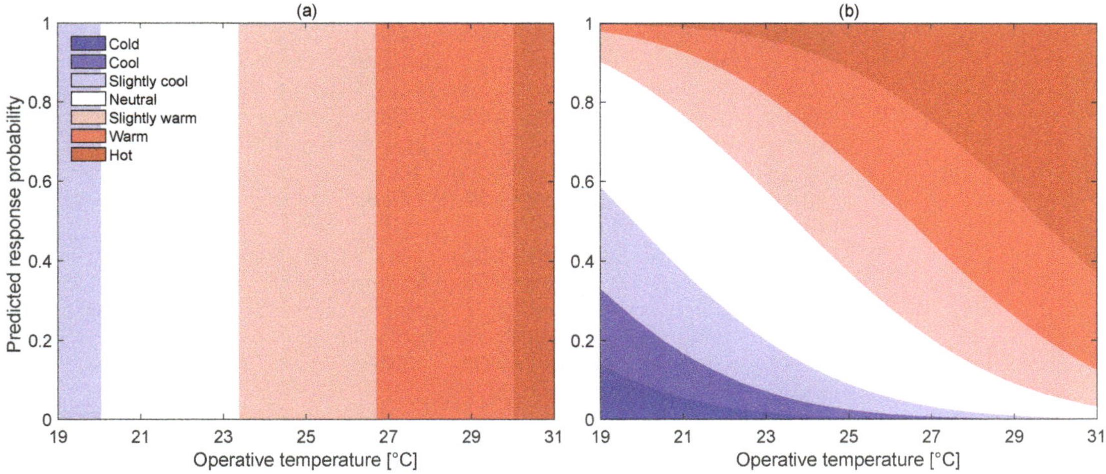

Figure 4. Thermal sensation profiles based on the (**a**) conventional and (**b**) probabilistic models.

4.2. Model-Predicted Thermal Dissatisfaction

In Figure 5, the model-predicted thermal dissatisfaction (MPD) data are superimposed on the observed data, and the data predicted using the conventional methods plotted in Figure 3. The MPD, the percentage of dissatisfied occupants, was predicted using our model proposed in [33], which was generated with points estimated from the MAP and a 99% credible interval (CI). In our model, the MPD predicted thermal satisfaction (i.e., the percentage of thermally satisfied occupants) more accurately (without overestimation) and captured the thermal conditions that yielded the highest thermal satisfaction. In addition, the probabilistic prediction results using the 99% CI of the estimated model parameters showed good agreement with the observations compared to other prediction results.

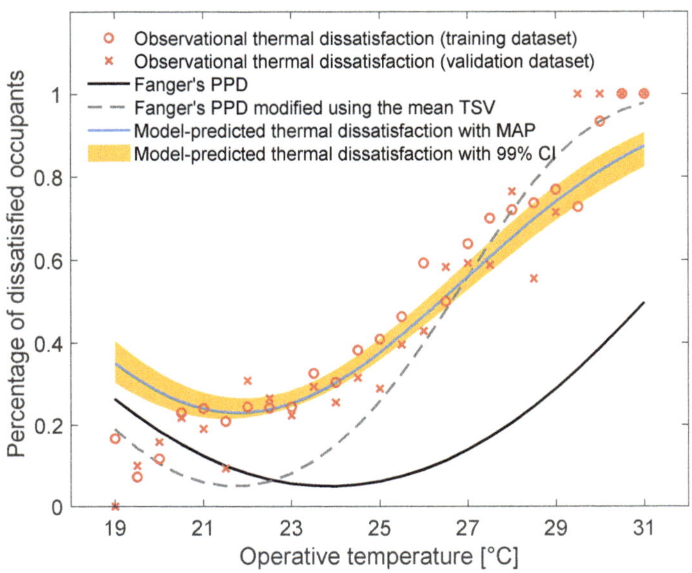

Figure 5. Relationship between the operative temperature and the percentage of dissatisfied occupants.

4.3. Impact on the Implementation of the HVAC-Based Demand Response Strategy

As indicated by the scenario mentioned in Section 1, HVAC engineers must determine a demand response strategy that temporarily decreases energy use. For example, HVAC engineers must tune the indoor climate such that a suboptimal environment is obtained. The advantages (the financial incentives earned by limiting and/or shifting power demands) and disadvantages (the economic losses due to the deterioration of the occupants' performance resulting from a sub-optimal indoor environment) of this choice must be balanced. In this section, the implications of a better prediction of thermal satisfaction while achieving this balance are discussed.

Such discussions can be ineffective because of the variation in the occupants' performance with the quality of the indoor environment, which often depends on indirect evidence [1]. In such cases, measurement results such as [9,10,34,35] serve as references. Herein, the discussion presented is based on a meaningful relationship between the relative occupant performance (RP) and TSV, which was reported by Jensen et al. [9] and is expressed as follows:

$$\mathrm{RP} = -0.0069 \times \mathrm{TSV}^2 - 0.0123 \times \mathrm{TSV} + 0.9945 \tag{1}$$

It is concluded that the initial approximation of the prediction of practical office work performance and the generated OT–RP relationship shown in Figure 6 is based on Equation (1). When deterministically approaching the TSV, the RP is calculated by applying the TSV–OT relationship based on the blue line shown in Figure 1. When probabilistically approaching the TSV using the proposed model, the RP at a given OT is calculated as follows:

$$\mathrm{RP} = \begin{bmatrix} \mathrm{P}(\mathrm{TSV} = -3) \\ \mathrm{P}(\mathrm{TSV} = -2) \\ \vdots \\ \mathrm{P}(\mathrm{TSV} = 2) \\ \mathrm{P}(\mathrm{TSV} = 3) \end{bmatrix}^{\mathrm{T}} \times \begin{bmatrix} -0.0069 \times (-3)^2 - 0.0123 \times (-3) + 0.9945 \\ -0.0069 \times (-2)^2 - 0.0123 \times (-2) + 0.9945 \\ \vdots \\ -0.0069 \times (2)^2 - 0.0123 \times (2) + 0.9945 \\ -0.0069 \times (3)^2 - 0.0123 \times (3) + 0.9945 \end{bmatrix} \tag{2}$$

where each element in the left vector is the probability of the TSV being $-3, \ldots, 3$, which can be obtained from Figure 4b.

As shown in Figure 6, the deterministic approach overestimates the occupant performance around the comfort zone. The overestimation level decreases with increasing OT and eventually proceeds to an underestimation after a certain OT point. Further investigation is required to quantitatively prove the gap between the two curves shown in Figure 6 because Equation (1) is one of the numerous indicators of occupants' performance over a wide range of tasks in indoor environments, for which there are different scientific arguments. It is noteworthy that, despite the lack of quantitative agreement, it is agreed that lowering indoor environmental quality decreases productivity [1], which indicates the wider implications of the proposed model. The proposed method predicts that occupant performance would more smoothly decrease with increasing OT than the occupant performance predicted using the conventional method. In addition, the proposed method provides more accurate predictions. This allows HVAC engineers to aggressively tune the indoor climate to meet energy-saving goals. Using such temporary demand response strategies, HVAC engineers and building operators can permanently improve the energy efficiency of HVAC systems while maintaining acceptable levels of occupant productivity.

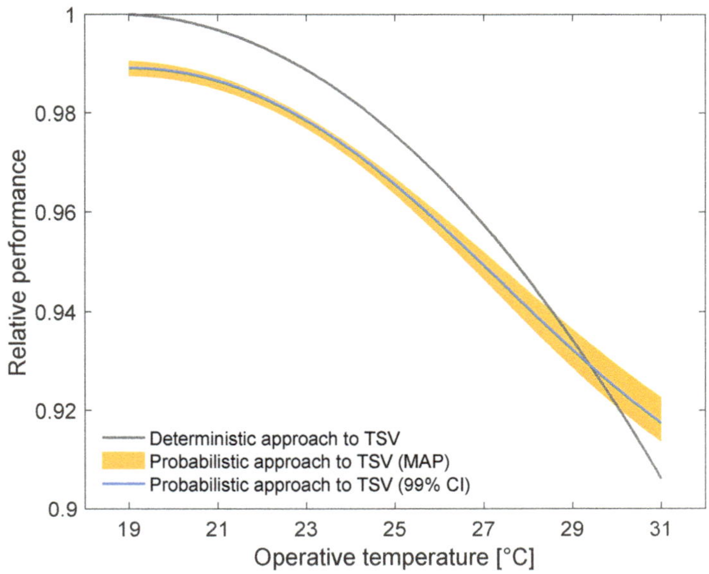

Figure 6. Relationship between the operative temperature and the relative performance of occupants carrying out regular office work.

5. Discussion

The results suggest that building operators and facility managers can more actively utilize demand response strategies if the prediction accuracy of thermal sensation and thermal satisfaction is improved by using a data-driven probabilistic approach. Therefore, the probabilistic prediction method proposed in this study can not only improve the understanding of thermal comfort exhibited by occupants in a specific space but also be used as an HVAC control technology considering the trade-off relationship between the energy use of building facilities and the service level.

Despite our attempt to provide a comprehensive description of the impact of thermal satisfaction prediction on implementing an HVAC-based demand response strategy, this study should be regarded as a case study using in situ experimental data. The results of this study (Figure 6) reveal an apparent change in the productivity level due to environmental variations. However, further research is required to determine how well the relative difference value represents the actual differences. Herein, we adopted the claims of the existing research (i.e., Equation (1)) to quantify relative occupant performance; however, quantitative evidence on productivity deterioration due to the changes in the thermal environment by demand response control is scarce.

The probabilistic prediction method for thermal satisfaction proposed in this study was verified using subset data of the ASHRAE Global Thermal Comfort Database II. To generalize the prediction performance of this model, validations using various field survey data on changes in thermal comfort and occupant performance should be accompanied in the future.

6. Conclusions

Thermal satisfaction contributes to productivity in daily life. Therefore, it is an important criterion that designers and engineers in charge of building projects must consider. Recently, as energy use in buildings has emerged as a significant issue, the real challenge is balancing thermal satisfaction with energy use for HVAC systems. HVAC-based demand response control is an important technology for reducing energy use while minimizing the negative impacts on occupants' thermal satisfaction, and this can be achieved using both hardware and software development. As it is costly to replace the control hardware of existing buildings, improvement using software will attract attention in the future.

This study provides new insights into the important issue faced by building operators and facility managers, which is achieving energy savings targets while maintaining the desired level of service. An improved thermal satisfaction prediction, a type of software improvement, allows building operators and facility managers to operate their systems more flexibly, which helps to aggressively implement HVAC-based demand response strategies, such as a setpoint temperature adjustment. In addition, the results of this study will be helpful for studies that require a deep understanding of thermal comfort, such as PMV-based HVAC controls, and future research on effective strategies for implementing HVAC-based demand responses.

Author Contributions: Conceptualization, J.L. and W.C.; Methodology, J.L. and W.C.; Software, J.L.; Writing—Original Draft Preparation, J.L.; Writing—Review and Editing, J.L. and W.C.; Visualization, J.L. and W.C. All authors have read and agreed to the published version of the manuscript.

Funding: This work was partly supported by the 2020 Research Grant from Kangwon National University and the National Research Foundation of Korea (NRF) grant funded by the Korean Government (MSIT) (No. NRF-2020R1F1A1068197).

Data Availability Statement: Data sharing is not applicable.

Conflicts of Interest: The authors declare no conflict of interest.

References

1. ASHRAE. *ASHRAE Handbook: Fundamentals*; ASHRAE: Atlanta, GA, USA, 2013.
2. de Dear, R.J.; Brager, G.S. Thermal Comfort in Naturally Ventilated Buildings: Revisions to ASHRAE Standard 55. *Energy Build.* **2002**, *34*, 549–561. [CrossRef]
3. ASHRAE. *ASHRAE Standard 55—Thermal Environmental Conditions for Human Occupancy*; ASHRAE: Atlanta, GA, USA, 2017.
4. Guo, H.; Qiao, W.; Liu, J. Dynamic Feedback Analysis of Influencing Factors of Existing Building Energy-Saving Renovation Market Based on System Dynamics in China. *Sustainability* **2019**, *11*, 273. [CrossRef]
5. Hoyo-Montaño, J.; Valencia-Palomo, G.; Galaz-Bustamante, R.; García-Barrientos, A.; Espejel-Blanco, D. Environmental Impacts of Energy Saving Actions in an Academic Building. *Sustainability* **2019**, *11*, 989. [CrossRef]
6. Goldman, C.; Hopper, N.; Sezgen, O.; Moezzi, M.; Bharvirkar, R. Does Real-Time Pricing Deliver Demand Response? A Case Study of Niagara Mohawk's Large Customer RTP Tariff. In Proceedings of the ACEEE 2004 Summer Study on Energy Efficiency in Buildings; 2004. Available online: https://www.aceee.org/files/proceedings/2004/data/papers/SS04_Panel5_Paper10.pdf (accessed on 20 April 2022).
7. Watson, D.S.; Kiliccote, S.; Motegi, N.; Piette, M.A. Strategies for Demand Response in Commercial Buildings. In Proceedings of the ACEEE Summer Study on Energy Efficiency in Buildings, Pacific Grove, CA, USA, 13–18 August 2006.
8. Lan, L.; Wargocki, P.; Lian, Z. Quantitative Measurement of Productivity Loss Due to Thermal Discomfort. *Energy Build.* **2011**, *43*, 1057–1062. [CrossRef]
9. Jensen, K.L.; Toftum, J.; Friis-Hansen, P. A Bayesian Network Approach to the Evaluation of Building Design and Its Consequences for Employee Performance and Operational Costs. *Build. Environ.* **2009**, *44*, 456–462. [CrossRef]
10. Seppänen, O.; Fisk, W.J.; Lei, Q.H. *Room Temperature and Productivity in Office Work*; Lawrence Berkeley National Laboratory: Berkeley, CA, USA, 2006; Available online: https://escholarship.org/uc/item/9bw3n707 (accessed on 20 April 2022).
11. Adams, J.N.; Bélafi, Z.D.; Horváth, M.; Kocsis, J.B.; Csoknyai, T. How Smart Meter Data Analysis Can Support Understanding the Impact of Occupant Behavior on Building Energy Performance: A Comprehensive Review. *Energies* **2021**, *14*, 2502. [CrossRef]
12. Ali, Q.; Thaheem, M.J.; Ullah, F.; Sepasgozar, S.M.E. The Performance Gap in Energy-Efficient Office Buildings: How the Occupants Can Help? *Energies* **2020**, *13*, 1480. [CrossRef]
13. Homod, R.Z.; Sahari, K.S.M.; Almurib, H.A.F.; Nagi, F.H. Gradient Auto-Tuned Takagi–Sugeno Fuzzy Forward Control of a HVAC System Using Predicted Mean Vote Index. *Energy Build.* **2012**, *49*, 254–267. [CrossRef]
14. Schito, E.; Conti, P.; Urbanucci, L.; Testi, D. Multi-Objective Optimization of HVAC Control in Museum Environment for Artwork Preservation, Visitors' Thermal Comfort and Energy Efficiency. *Build. Environ.* **2020**, *180*, 107018. [CrossRef]
15. Mary Reena, K.E.; Mathew, A.T.; Jacob, L. A Flexible Control Strategy for Energy and Comfort Aware HVAC in Large Buildings. *Build. Environ.* **2018**, *145*, 330–342. [CrossRef]
16. Turley, C.; Jacoby, M.; Pavlak, G.; Henze, G. Development and Evaluation of Occupancy-Aware HVAC Control for Residential Building Energy Efficiency and Occupant Comfort. *Energies* **2020**, *13*, 5396. [CrossRef]
17. Fanger, P.O. *Thermal Comfort*, 1st ed.; McGraw-Hill: New York, NY, USA, 1972.

18. Chen, J.; Augenbroe, G.; Wang, Q.; Song, X. Uncertainty Analysis of Thermal Comfort in a Prototypical Naturally Ventilated Office Building and Its Implications Compared to Deterministic Simulation. *Energy Build.* **2017**, *146*, 283–294. [CrossRef]
19. Benton, C.C.; Bauman, F.S.; Fountain, M.E. A Field Measurement System for the Study of Thermal Comfort. *ASHRAE Trans.* **1990**, *96*, 623–633.
20. Schiller, G.E. A Comparison of Measured and Predicted Comfort in Office Buildings. *ASHRAE Trans.* **1990**, *96*, 609–622.
21. de Paula Xavier, A.A.; Lamberts, R. Indices of Thermal Comfort Developed from Field Survey in Brazil. *ASHRAE Trans.* **2000**, *106*, 45–58.
22. Katić, K.; Li, R.; Zeiler, W. Machine Learning Algorithms Applied to a Prediction of Personal Overall Thermal Comfort Using Skin Temperatures and Occupants' Heating Behavior. *Appl. Ergon.* **2020**, *85*, 103078. [CrossRef]
23. Ghahramani, A.; Tang, C.; Becerik-Gerber, B. An Online Learning Approach for Quantifying Personalized Thermal Comfort via Adaptive Stochastic Modeling. *Build. Environ.* **2015**, *92*, 86–96. [CrossRef]
24. Li, D.; Menassa, C.C.; Kamat, V.R. Personalized Human Comfort in Indoor Building Environments under Diverse Conditioning Modes. *Build. Environ.* **2017**, *126*, 304–317. [CrossRef]
25. Lim, J.; Choi, W.; Akashi, Y.; Yoshimoto, N.; Ooka, R. Bayesian Prediction Model of Thermally Satisfied Occupants Considering Stochasticity Due to Inter- and Intra-Individual Thermal Sensation Variations. *J. Build. Eng.* **2022**, *52*, 104414. [CrossRef]
26. Földváry Ličina, V.; Cheung, T.; Zhang, H.; de Dear, R.; Parkinson, T.; Arens, E.; Chun, C.; Schiavon, S.; Luo, M.; Brager, G.; et al. Development of the ASHRAE Global Thermal Comfort Database II. *Build. Environ.* **2018**, *142*, 502–512. [CrossRef]
27. Oseland, N.A. Acceptable Temperature Ranges in Naturally Ventilated and Air-Conditioned Offices. *ASHRAE Trans.* **1998**, *104*, 1018–1030.
28. *ISO 7730:2005*; Ergonomics of the Thermal Environment: Analytical Determination and Interpretation of Thermal Comfort Using Calculation of the PMV and PPD Indices and Local Thermal Comfort Criteria. International Organization for Standardization: Geneva, Switzerland, 2005.
29. Majewski, G.; Orman, Ł.J.; Telejko, M.; Radek, N.; Pietraszek, J.; Dudek, A. Assessment of Thermal Comfort in the Intelligent Buildings in View of Providing High Quality Indoor Environment. *Energies* **2020**, *13*, 1973. [CrossRef]
30. Lin, T.-P. Thermal Perception, Adaptation and Attendance in a Public Square in Hot and Humid Regions. *Build. Environ.* **2009**, *44*, 2017–2026. [CrossRef]
31. Hwang, R.-L.; Lin, T.-P. Thermal Comfort Requirements for Occupants of Semi-Outdoor and Outdoor Environments in Hot-Humid Regions. *Archit. Sci. Rev.* **2007**, *50*, 357–364. [CrossRef]
32. Spagnolo, J.; de Dear, R. A Field Study of Thermal Comfort in Outdoor and Semi-Outdoor Environments in Subtropical Sydney Australia. *Build. Environ.* **2003**, *38*, 721–738. [CrossRef]
33. Lim, J.; Akashi, Y.; Yoshimoto, N. Bayesian Inference for Predicting an Inter-Individual Variation of Thermal Sensation Votes. In Proceedings of the Building Simulation 2019: 16th Conference of IBPSA, Rome, Italy, 2–4 September 2019.
34. Roelofsen, P. The Design of the Work Place as a Strategy for Productivity Enhancement. In Proceedings of the 7th REHVA World Congress (CLIMA 2000), Napoli, Italy, 15–18 September 2001.
35. Berglund, L.G.; Gonzales, R.R.; Gagge, A.P. Predicted Human Performance Decrement from Thermal Discomfort and ET*. In Proceedings of the Indoor Air '90, Toronto, ON, Canada, 29 July–3 August 1990.

energies

Article

Quantifying of Vision through Polymer Dispersed Liquid Crystal Double-Glazed Window

Mohammed Lami [1,*], Faris Al-naemi [1], Hameed Alrashidi [2] and Walid Issa [1]

[1] Industry & Innovation Research Institute, Sheffield Hallam University, Sheffield S1 1WB, UK; acesfa@exchange.shu.ac.uk (F.A.-n.); aceswi@exchange.shu.ac.uk (W.I.)
[2] Kuwait Fund for Arab Economic Development, Mubarak Al-Kabeer St., Safat, P.O. Box 2921, Kuwait City 13030, Kuwait; basicolor@hotmail.com
* Correspondence: b8018683@my.shu.ac.uk

Abstract: The visual linking of a building's occupants with the outside views is a basic property of windows. However, vision through windows is not yet a metricized factor. The previous research employs a human survey methods to assess the vision through conventional windows. The recently fabricated smart films add a changeable visual transparency feature to the windows. The varied operating transparency challenges the evaluation of vision. Therefore, surveying human preferences is no longer a feasible approach for smart windows. This paper proposes an image-processing-based approach to quantify the vision quality through smart windows. The proposed method was experimentally applied to a polymer dispersed liquid crystal (PDLC) double-glazed window. The system instantaneously determines the available contrast band of the scenes seen through the window. The system adjusts the excitation of the PDLC film to maintain a desired vision level within the determined vision band. A preferred vision ratio (*PVR*) is proposed to meet the requirements of occupant comfort. The impact of the *PVR* on vision quality, solar heat gain, and daylight performance was investigated experimentally. The results show that the system can determine the available vision comfort band during daytime considering different occupant requirements.

Keywords: vision through windows; visual comfort; PDLC smart film; smart windows; image processing

Citation: Lami, M.; Al-naemi, F.; Alrashidi, H.; Issa, W. Quantifying of Vision through Polymer Dispersed Liquid Crystal Double-Glazed Window. *Energies* **2022**, *15*, 3196. https://doi.org/10.3390/en15093196

Academic Editors: Alessandro Cannavale and Dimitrios Katsaprakakis

Received: 26 February 2022
Accepted: 25 April 2022
Published: 27 April 2022

1. Introduction

Windows are considered to be an essential part in the envelope of buildings as they play an important role in providing the daylight for the internal environments and maintaining vision comfort for the occupants.

The major tendency to conduct large studies on windows has resulted in associated challenges such as vision comfort. Vision comfort not only concerns the quantity of daylight in indoor environments, but also refers to the vision quality through the window. Understanding the quality and quantity of daylight in buildings is thus crucial to satisfying the indoor occupant comfort. Daylight glare (DG), correlated color temperature (CCT), and color rendering index (CRI) are the three crucial components to understand visual comfort [1]. However, these components describe the comfort in a static scenario where glazing has a constant transparency. Using switchable glazing, such as the newly developed thin films, in which the transparency changes, results in more challenges to assess the vision comfort.

Studies have been carried out to investigate the impact of daylight, and evaluate or improve the daylight performance and visual comfort in current and future buildings [2–4]. For example, Wang et al. [5] studied the influence of outdoor views provided by the window on patients who had undergone surgery in Taiwan. The authors concluded that vision comfort through the window dramatically reduced the usage of medicines to treat aches.

Pilechiha et al. [6] introduced a new method to optimize energy consumption through windows of office buildings. The author emphasized that the proposed method should not impact the quality of vision through the window or the quantity of the daylight inside the building.

Dušan et al. [7] proposed a special approach to design a lighting system for industrial buildings. The author stated that, in addition to the availability of adequate daylight in the working area, the workers must have contact with the external environment to improve their performance.

In addition to the studies mentioned above, other papers analyzed the function of windows and their influence on different kinds of occupants, such as employees, workers, and patients. These studies affirmed that the essential functions of windows are providing daylight and maintaining good vision of the outdoor views [8–10].

Ma et al. [11] surveyed the preferences of a set of students regarding the vision comfort and discomfort glare of a study area for six months. The authors used the output data to create a building information model (BIM), which was translated into an algorithm for a machine learning system, and the resultant algorithm was used to control a proposed window shading system.

Day et al. [12] surveyed 5031 people in the United States regarding their satisfaction in their visual comfort. They were distributed in three different buildings. The buildings employed automated blinds, electrochromic glazing, and roller shades techniques, respectively, to control the accessed daylight in the buildings. The people in buildings 1 and 2, who occupied the most luminous environments, were satisfied, with higher levels of output and higher levels of contentment. Moreover, the satisfaction level of occupants was directly proportional to the seating position inside the building. The most satisfied people where those sitting closer to the windows. Negative feedback was provided by the occupants of building 3, which had the lowest level of luminance due to the use of electrochromic glazing.

Lim et al. [13] examined the influence of an artificial light-controlled system on energy consumption and occupants' behavior in two different workplaces. Access to daylight was provided in the first workplace, but not in the second. The results showed that utilization of the authors' proposed lighting system in the first area contributed to a reduction of 78% in the lighting energy consumption. Additionally, the occupants in both workplaces were satisfied with the brightness of the environment.

The cases cited above, in addition to other studies [14–16], depended on individuals' observations to evaluate the vision comfort in different environments, including the quantity of the received daylight and the quality of vision through the window towards external views. However, questionnaires are a time-consuming method to gather human preferences during different seasons. In addition, this approach cannot ensure that the gathered data are sufficient to cover all expected user needs in practical applications.

New technologies have been developed recently in the windows industry, which have introduced innovative materials to replace or add to the glazing, such as semi-transparent photovoltaic panels (STPVs), suspended particle device (SPD) films, electrochromic films, and polymer dispersed liquid crystal (PDLC) films. All have different optical and thermal characteristics compared to normal glass. Integrating these panels and films into glazing introduces other dimensions to be investigated, i.e., some have controllable optical and thermal properties [17–22].

Another study by Ghosh et al. [1] analyzed the daylight performance of building-integrated photovoltaic (BIPV) windows in an office building in Saudi Arabia. The result shows that a transparency of 50 to 70 percent can ensure a comfort glare level during winter, and 70% transparency in summer. By comparison, increasing the transparency to 90% resulted in a high daylight level inside the building; however, the discomfort due to glare was indicated to be high.

Detsi et al. [23] evaluated the energy consumption of lighting, cooling, and heating loads in office buildings in Athens and Stockholm. The study proposed a triple-glazed window of thermochromic, electrochromic, and low-e coatings in Insulated Glass Units (IGUs). Overall annual energy savings of 18.5% and 8.1% were achieved in Athens and Stockholm, respectively. The study concluded that the proposed window provides good visual comfort for employees in terms of the daylight glare index. The visible transparency of the proposed window ranged from 0.46 to 0.75; however, the study did not evaluate the quality of the external views within the given range of transparency.

Ko et al. [24] investigated the performance of an SPD window in an office building in South Korea. The research method relied on varying the visible transparency of the SPD film by changing the applied voltage; thus, the solar heat gain coefficient (SHGC) was modified accordingly (the SHGC controllable range was 0.11 to 0.45). An annual energy saving of 4.1% was realized; however, reducing the visible transparency compromised the clarity of the vision toward the outside environment for the employees.

Oh et al. [25] analyzed the feasibility of attaching different types of PDLC thin film to the windows and the façade walls in terms of energy consumption and daylight performance. Utilization of different kinds of PDLC thin films with existing windows showed energy savings of between 3.1% and 17%; in addition, the discomfort glare ratio of the internal environment was enhanced. Again, the study ignored the effect of reducing the visible transmission, after adding the PDLC film, on the quality of vision through the window.

Hemaida et al. [26] investigated the influence of employing PDLC windows in office buildings on cooling consumption in Riyadh, Saudi Arabia, and heating consumption in London, UK. The study concluded that the proposed window was more efficient, reducing cooling capacity by 12.8% in Riyadh, while the same window led to a reduction in heating capacity of 4.9% in London. The study also deduced that the PDLC window provides an acceptable level of daylight for interior environments. The results showed that the window accomplished 75.8% of the annual required daylight glare in Riyadh and approximately 63% in London. The study mentioned that the visible transmittance of the window ranged from 44% (OFF state) to 79% (ON state); however, the study neglected to assess the impact of decreasing the glazing visible transmission on the vision through the window.

Previous research, especially that utilizing dynamic windows, has focused on evaluating vision comfort by quantifying the daylight in internal environments. The assessment usually relies on measuring the available luminance. However, visual access to the external environment through the window has been studied only by surveying and analyzing human preferences. With introducing the switchable films and dynamic optical and thermal properties of glazing, there are many settings that can bring challenges to be surveyed or investigated to assess the vision comfort properly. If artificial intelligence is used to dominate how windows are controlled, it is important to quantify visual comfort and develop algorithms to parameterize it.

Human visual perception is based on the contrast of images [27–29]. The contrast property enables human eyes to distinguish between two adjacent objects in an image [30,31]. Therefore, quantifying vision through glazing can be achieved using an artificial vision system to derive the real-time contrast value, which reflects the quality of human vision.

Previous studies have employed image processing to determine the daylight glare index, discomfort glare index, and other factors that reflect the quantity of daylight in a specific area. For example, Fan et al. [32] used image-based simulation to determine the daylight level in a gymnasium. In addition, Hegazy et al. [33] quantified the daylight in a virtual environment using luminosity histogram analysis. This method uses digital images of the environment, in which each pixel is represented in the luminosity histogram according to its brightness level, which reflects the luminance of a particular point in a given area. Futagami et al. [34] utilized an image-based motion sensor to determine the luminance level of a workplace. The images were taken by a visible light camera. Day et al. [12] evaluated the daylight illuminance in three office buildings. The research team captured HDR images and then processed them using the Evaglare algorithm [35] to determine the daylight glare probability. None of these studies evaluated the vision to exterior views through PDLC glazing.

To summarize, the previous studies emphasized that visual linking of occupants to the external views is an essential function for windows. Unfortunately, assessment of vision through windows was not considered by most of the previous research, although some papers relied on surveying of individuals' preferences to assess the vision through windows. However, surveys usually require a large amount of time to cover all seasonal conditions; in addition, the outcome data may not be adequate to cover the wide variety of preferences of users in real life. Furthermore, following the integration of thin-film technologies with the window industry, and the addition of variable-transparency features to windows, surveying is no longer a feasible method to evaluate the vision through windows.

Therefore, this paper proposes an online image-processing-based method to quantify the vision through a PDLC double-glazed window. This approach can be used to establish levels of occupants' visual comfort, which are fed into a controller to constrain the transparency setting of the PDLC to prevent discomfort. The proposed system adds observability and controllability features for buildings' controllers, i.e., artificial intelligence systems, to mathematically define the visual comfort of smart windows and to ensure a minimum level of viewing quality towards external views for the occupants. Overall, the proposed method will provide an alternative way to assess vision through windows, and save time in window design, considering all the expected configurations of smart windows.

2. Materials and Methods

Figure 1 demonstrates the proposed online vision quantification methodology. It has two components: (a) an image-processing model and (b) a mathematical model.

a. To develop the image-processing model, a vision sensor, such as an HD camera, captures a set of photos from inside an environment towards the external views through the PDLC double-glazed window. A group of 16 photos is captured every 30 min. Each photo is taken with different levels of PDLC transparency, i.e., from an opaque state to a transparent state. Thereafter, the contrast value of each image is extracted individually based on a developed algorithm that sends a combination of transparency and contrast.
b. To develop the mathematical model, an instantaneous contrast curve is obtained between contrast values and the operating visible transmission for each time slot during the daytime. The vision comfort band is derived with the lowest and highest allowed values of image contrast at a given time. This is fed back to the PDLC transparency setting to further limit the transparency range.

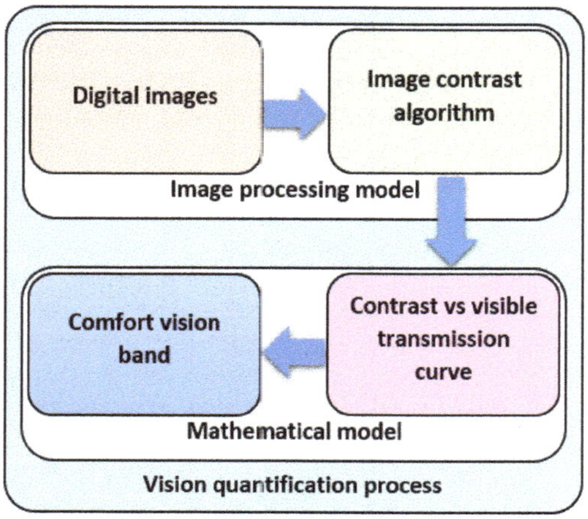

Figure 1. Visual comfort quantification block diagram.

2.1. Methodology Setup

The experimental setup includes a test room, sensors, and data loggers required for indoor and outdoor measurements, as shown in Figure 2. The optical properties of the glazing were measured by a spectrometer device in the laboratory. In addition, a weather data file was necessary to apply the experimental outdoor conditions to the simulated model; however, three kinds of data were required to prepare the weather data. Firstly, outdoor weather data include the temperature, humidity, atmosphere pressure, wind speed, and wind direction. Secondly, solar radiation data encompass global horizontal irradiance, diffused horizontal irradiance, and global vertical irradiance. The direct normal irradiance (*DNI*) is also needed for the weather data file; however, Equation (1) was used to estimate it [36]:

$$DNI = \frac{GHI - DHI}{\cos\theta} \quad (1)$$

where DNI is the direct normal irradiance in W/m^2, GHI is the global horizontal irradiance in W/m^2, DHI is the diffused horizontal irradiance in W/m^2, and θ is the zenith angle.

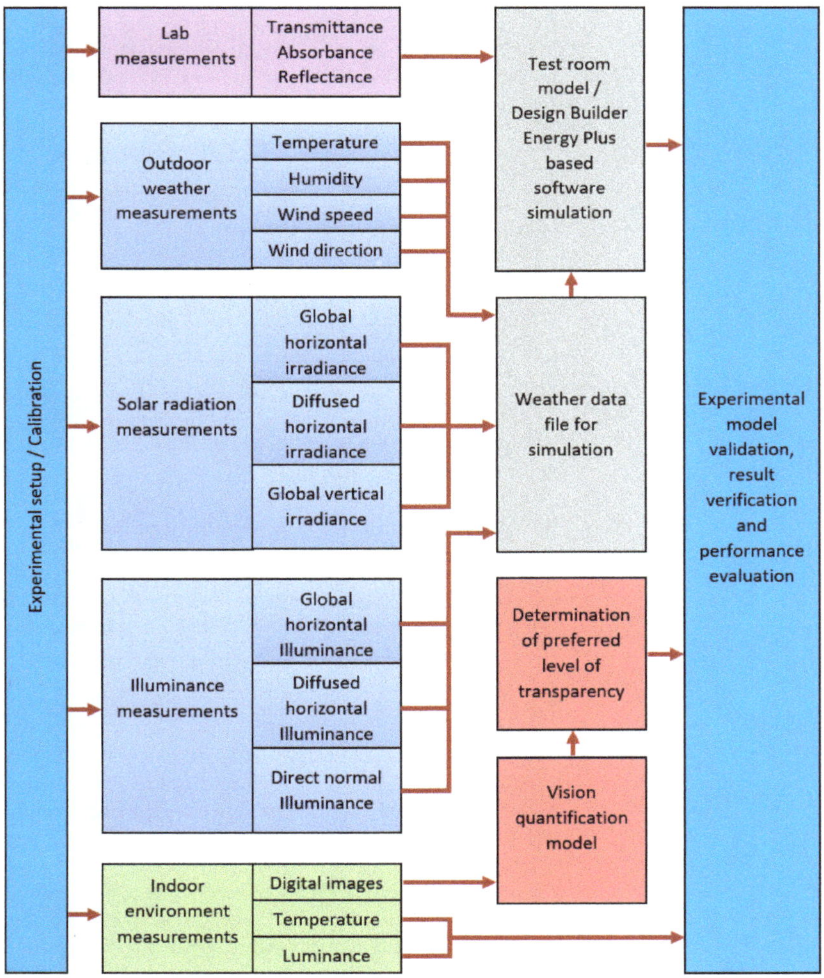

Figure 2. Methodology setup.

Thirdly, the illuminance data comprise global horizontal illuminance, diffused horizontal illuminance, and direct normal illuminance. The Illuminance Engineering Society of North America (IESNA) approach provides an appropriate approximation for the available illuminance in solar radiation. The following equations elucidate the illuminance calculation procedures of the IESNA method, as proposed by Kandilli and Ulgen [37]:

$$E_{XT} = G_{SC}\ \{1 + 0.034 \cos[\frac{2\pi}{365}(n-2)]\} \tag{2}$$

where E_{XT} is the extraterrestrial illuminance, G_{SC} is the solar illuminance constant, and n is the day of the year. Then, the direct normal illuminance E_{DN} can be determined using Equation (3):

$$E_{DN} = E_{XT}\exp(-cm) \tag{3}$$

where m is the air mass, which can be expressed as follows:

$$m = \frac{1}{\sin(\propto_S)} \tag{4}$$

where $\propto_S$ is the angle of altitude in degrees, and the small "c" is a constant, which can be quoted from Table 1 according to the sky condition. Thereafter, direct horizontal illuminance E_{DH} is calculated using Equation (5):

$$E_{DH} = E_{DN}\sin(\propto_S) \tag{5}$$

Table 1. The constants of sky parameters.

Sky Condition	c	A, Klux	B, Klux	C, Klux
Clear	21	0.8	15.5	0.5
Partly cloudy	0.8	0.3	45	1
Cloudy	No direct sun	0.3	21	1

Based on the condition of the sky, the diffuse (or sky) horizontal illuminance E_{KH} can be computed using Equation (6):

$$E_{KH} = A + B\sin^C(\propto_S) \tag{6}$$

The constants A, B, and C categorize the condition of the sky into cloudy, partly cloudy, and clear. Table 1 depicts the value of sky parameters for each category; it is noted that, on a cloudy day, no direct sun is available; therefore, $E_{DN} = 0$.

Finally, the total illuminance E falling on the horizontal plane is given in Equation (7):

$$E = E_{KH} + E_{DH} \tag{7}$$

The experimental indoor measurements involve the temperature and illuminance level inside the test room, which are used to evaluate the performance and validate the system. Additionally, digital images are captured from inside the test room towards the external environment through the window. These images are applied to the proposed vision quantification model to determine the comfort vision band in real time. The outdoor measurements were employed to prepare the weather data file of the commercial DESIGN BUILDER software. DESIGN BUILDER software was used to derive the energy model. The weather and optical data were applied to the model to achieve the simulation. The simulated and experimental data, in addition to the outcome of the quantification process, were used to validate the experimental model, verify the experimental result, and evaluate the system performance.

2.2. Image Processing

For contrast measurement, the *RMS* contrast method in Equation (8) was adopted [38]:

$$RMS\ Contrast = \sqrt{\frac{1}{M*N}\sum_{i=0}^{M-1}\sum_{j=0}^{N-1}\left(I_{ij}-\overline{I}\right)^2} \quad (8)$$

where M and N are the numbers of columns and rows of the image I, respectively; I_{ij} is the value of the pixel in the ith column and the jth row; and $\overline{I}$ is the average brightness of the image I, which is given in Equation (9) in the range [0, 1]:

$$\overline{I} = \frac{1}{M*N}\sum_{i=0}^{M*N-1} I_i \quad (9)$$

The images in Figure 3 were captured in an external environment during daytime on a sunny day through a double-glazed PDLC window with different visible transparencies. Each image was taken after setting the applied AC voltage of the PDLC within the rated voltage (0 to 200 V), as shown in the figure.

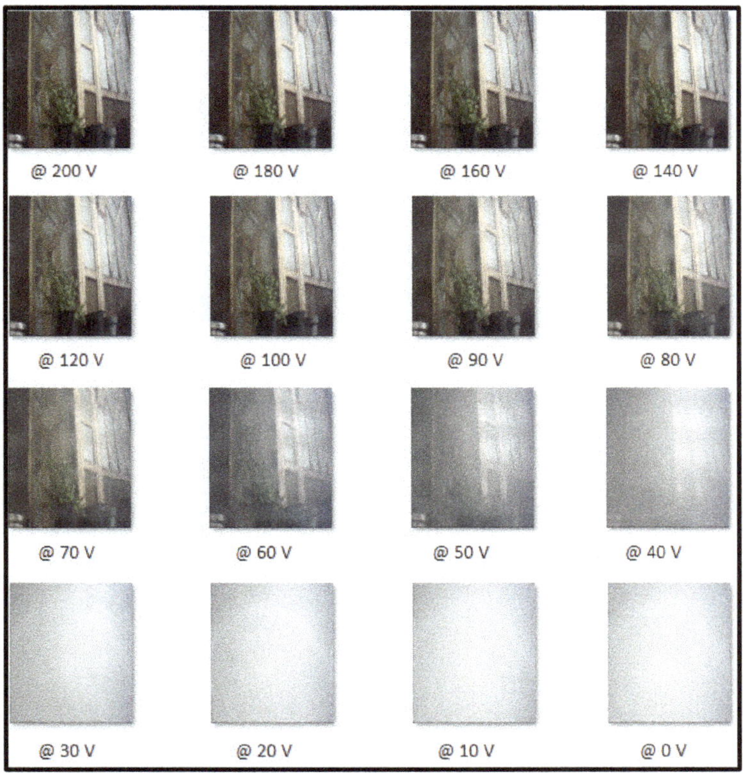

Figure 3. Digital images with different applied voltages.

The relationship of the contrast with the applied voltage was obtained, as shown in Figure 4. Three regions can be recognized in the curve. The first region is between the minimum contrast point (C_{min}) and point (C_y); obviously, the contrast has low values as the PDLC film is in its opaque state. The second region is between (C_y) and the threshold contrast point (C_{th}), in which the contrast value is dramatically increased with the increase in the applied voltage. The third region is between (C_{th}) and the maximum contrast point (C_{max}), in which the contrast begins to show a saturation response to the applied voltage.

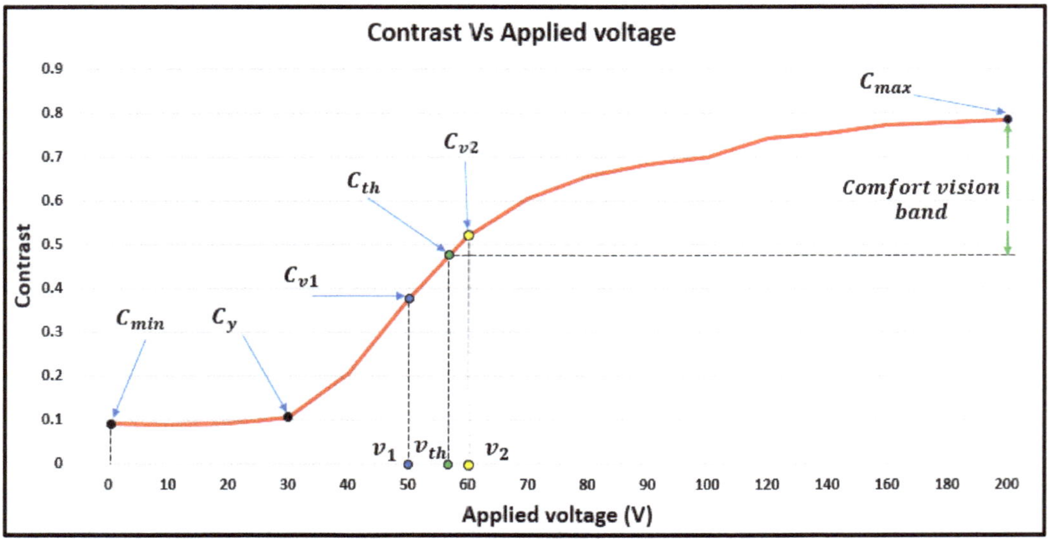

Figure 4. Contrast curve.

The vision is considered to be comfortable if the contrast is above approximately 0.48; this value is the threshold value of the contrast, and was calculated graphically from the contrast–voltage curve. However, this point does not correspond to a fixed transparency and voltage in all cases. The threshold contrast point (v_{th}, C_{th}) represents a proposed minimum comfort band that limits the transparency. In Figure 4, the threshold contrast value lies between two given points, (C_{v1}, v_1) and (C_{v2}, v_2); hence, the threshold contrast value can be expressed mathematically as below:

$$\text{Threshold Contrast } (C_{th}) = 0.707\ (C_{max} - C_{min}) \tag{10}$$

To estimate the threshold voltage (v_{th}), the segment of the curve between (C_{v1} and C_{v2}) is assumed to be a linear relationship; this assumption leads to:

$$\frac{C_{v2} - C_{th}}{C_{v2} - C_{v1}} = \frac{v_2 - v_{th}}{v_2 - v_1},$$

or:

$$v_{th} = v_2 - \left[(v_2 - v_1) \left(\frac{C_{v2} - C_{th}}{C_{v2} - C_{v1}} \right) \right] \tag{11}$$

Consequently, the system chooses one operating point within the comfort vision band based on the user's preference. However, the user's preference is also referenced to v_{th}. The range that the user can set is from 0 to 100%, which maps to the range from v_{th} to v_{max}. In our paper, a preferred vision ratio (*PVR*) is proposed to quantify the user setting. Thus, if the *PVR* = 0%, then the operating point is at the threshold value (C_{th}), whereas the system works at (C_{max}) if the *PVR* = 100%. Consequently, the preferred contrast value (C_{PVR}) can be determined as below:

$$\text{The preferred contrast value } (C_{PVR}) = PVR\ (C_{max} - C_{th}) + C_{th} \tag{12}$$

The corresponding applied voltage at the preferred contrast v_{PVR} can be defined by substituting Equation (12) into Equation (11), as shown below:

$$v_{PVR} = v_b - \left[(v_b - v_a) \left(\frac{C_b - (PVR(C_{max} - C_{th}) + C_{th})}{C_b - C_a} \right) \right] \ldots\ldots 0 \le PVR < 1 \tag{13}$$

3. Experiment

3.1. Experimental Setup

Two small-scaled test rooms of the same size were made from a triple-layer insulated board. The board consists of a thick layer of polystyrene with thermal conductivity of 0.046 W/m.K. The polystyrene is sandwiched between two thin layers of steel. The first room had a conventional double-glazed window having the specifications shown in Table 2. This room was considered as the reference model in the experiments. The second room had the same window, except that a PDLC thin film was attached to the inner pane of the double-glazed window. The second room was used to evaluate the proposed vision quantification process. Both rooms were equipped with a temperature sensor to measure the temperature of the enclosure; in addition, a flux sensor was centered inside each test room at a height of 15 cm. Table 2 provides more details about the test room structure, double-glazed window, and the PDLC film.

Table 2. Specifications of the experimental setup.

Item	Part	Details
Board	Polystyrene thickness	4.9 cm
	Steel thickness	0.05 cm
Wall dimension of the room		55×55 cm^2
Room dimensions		$45 \times 45 \times 45$ cm^3
Double glazed window	Room area	30×30 cm^2
	WWR	44%
	Glazing type	Clear
	Glazing thickness	0.4 cm
	Thermal conductivity for the glazing	0.9 W/m.K
	Glazing transmittance	0.83
	Glazing reflectance	0.12
	Glazing absorbance	0.039
	Depth of the air layer	2 cm
	Outside reveal depth	0 cm
PDLC film	Rated voltage	220 AC volt
	Thickness	0.36 mm
	Visible transparency (OFF/ON)	0.05/0.83

For outdoor weather conditions, a weather station was utilized to record the temperature, humidity, pressure, wind speed, and wind direction. Additionally, three pyranometers were employed to measure global horizontal irradiance (GHI), diffused horizontal irradiance (DHI), and global vertical irradiance (GVI). All pyranometers were interfaced with a datalogger device to collect the data. More details about the sensors used in the experiment are provided in Table 3.

Table 3. Technical specifications of the sensors.

Sensor	Technical Specifications
Pyranometer (for global horizontal irradiance measurement)	Veinasa-ZFS Solar total radiation sensor, sensitivity 7~14 μV per w/m^2, and measurement range 0~2000 W/m^2
Pyranometer (for diffused horizontal irradiance measurement)	Veinasa-SFS Scattered radiation sensor, sensitivity 7~14 μV per w/m^2, and measurement range 0~2000 W/m^2
Pyranometer (for global vertical irradiance measurement)	Davis Instruments 6450 Solar Radiation Sensor, sensitivity 1.67 mV per W/m^2, measurement range 0~1800 W/m^2
Lux meter (for indoor illuminance measurement)	OPT3001 Ambient Light Sensor, measurement range 0.01 lux to 83 k lux
Temperature sensor (for indoor temperature measurement)	AM2302 humidity and temperature sensor, sensitivity 0.1 Celsius, and measurement range −40~80 Celsius

In the second test room, because the experiments required a variable AC voltage for the PDLC film, a 300 V/500 VA VARIAC transformer was used to adjust the desired voltage during the experiment. Furthermore, a high-definition camera was fixed in the room 5 cm from the inner pane of the window. The camera was horizontally oriented toward the window to capture the external views through the window. The digital camera was connected to a PC via an IP network; therefore, the images were directly transferred to the PC, and Windows-based software was used to manage the image-capturing operation. Finally, MATLAB code was developed to compute the contrast value of the captured images during the execution of the experiments. The frequency of this process was 30 min. Figure 5 shows the experimental setup, including the reference test room, PDLC test room, three pyranometers, and weather station.

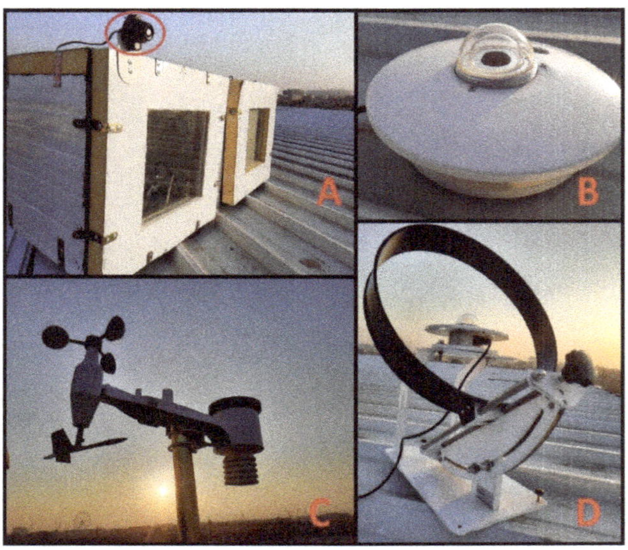

Figure 5. Experimental setup. (**A**): Reference room with double-glazed window (closer), PDLC room (further), and the pyranometer for vertical global irradiance measurement (circled). (**B**): Pyranometer for global horizontal irradiance measurement. (**C**): The weather station. (**D**): Pyranometer for diffuse horizontal irradiance measurement.

3.2. Experiment Scenarios

The two test rooms were installed in Basra, Iraq (latitude: 30.5 and longitude: 47.8) on the roof of the Basra oil training institute. The window for each room was oriented to the south. The experiment was continuously carried out for four days during daytime, from 11 to 14 August 2021. The experiment was executed every day with a different *PVR*. Table 4 shows the four scenarios, 1 to 4, that were implemented during the first to the fourth day of the experiment, respectively. It is worth mentioning here that the *PVR* was determined only for the window of the second test room to show the influence of the *PVR* on vision quality, in addition to the other parameters of the system, such as internal temperature, internal daylight, and solar heat gain.

Table 4. The scenarios of the experiment.

Scenario	Day	Test Room 1	Test Room 2	
		Window Type	*PVR*	Window Type
1	1st	Conventional double-glazed window	0	Double-glazed PDLC window
2	2nd		0.3	
3	3rd		0.65	
4	4th		1	

The experiment was executed from 5:30 AM to 6:30 PM. The process shown in Figure 6 was run 27 times daily (every 30 min) on the second room. The process required 2 min to be executed manually each time. The algorithm started by setting the PDLC film voltage to a specific value from among 16 values in the range 0 to 200 V. One image was taken each time for each voltage value. MATLAB then obtained the voltage versus contrast relation and the threshold value C_{th}.

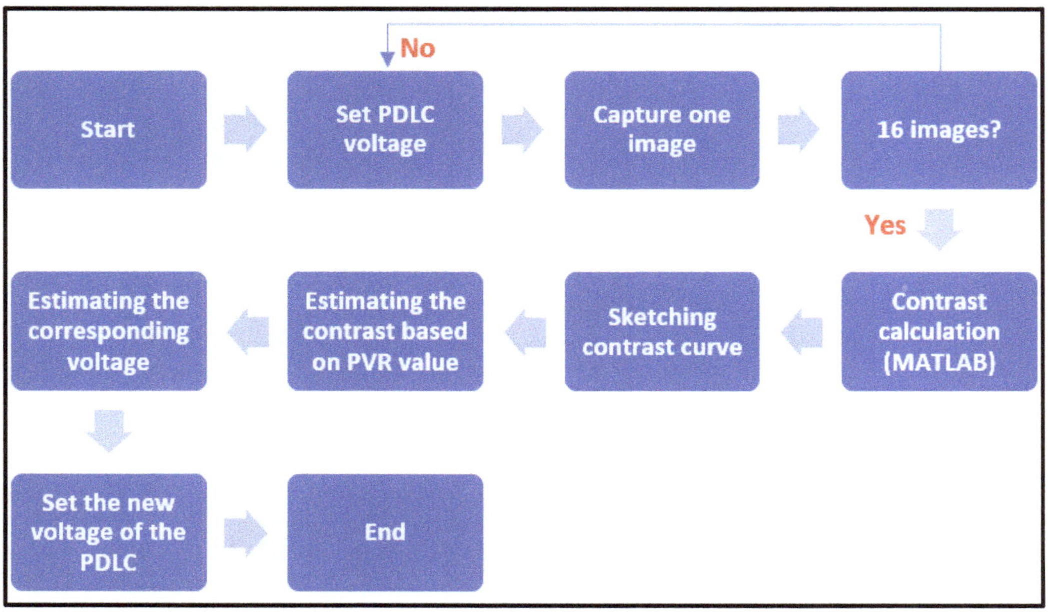

Figure 6. Algorithm for adjusting the excitation of the PDLC film used in the experiment.

Thereafter, the instantaneous contrast value was calculated using Equation (12) based on the preferred value of the *PVR*. The corresponding voltage was graphically determined from the instantaneous contrast–voltage curve, or from Equation (13). Finally, VARIAC output was set to the new voltage value from Equation (13) to realize the instantaneous preferred transparency. This value should ensure visual comfort.

3.3. Simulation

The two experimental test rooms, namely, the first room with the conventional double-glazed window and the second room with the PDLC double-glazed window, were modeled in DESIGN BUILDER software based on the properties listed in Table 2. Simultaneously, the weather data file required for the simulation was prepared using the computed illuminance data, experimental solar irradiance data, and weather conditions recorded by the weather station. Since the visible transparency of the PDLC film was changed during the experiments, during the simulation, it was necessary to adjust the visible transmission of the PDLC film accordingly at different times. This task was achieved by referring to the transparency–voltage curve of the film. A spectrometer was used to measure the visible transmission of the utilized PDLC film against different input AC voltages; the resultant curve is shown in Figure 7.

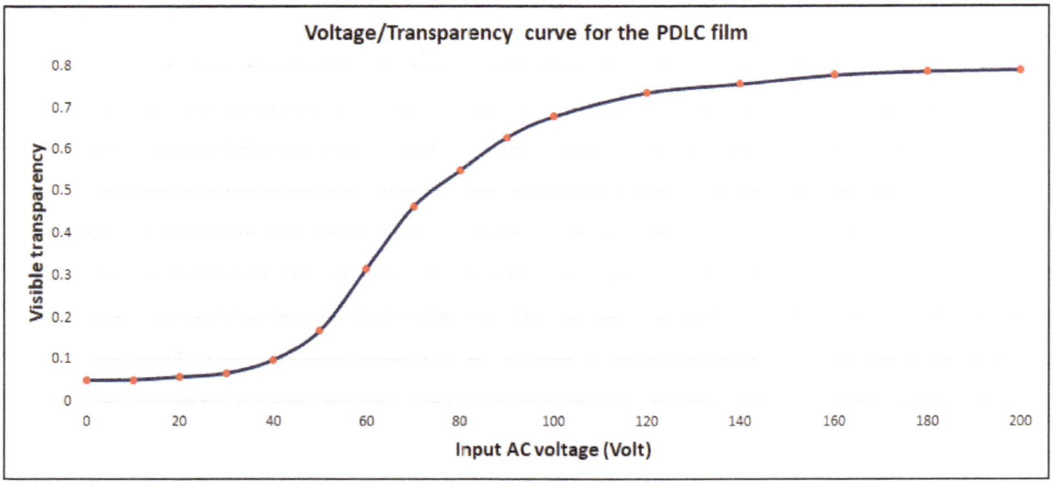

Figure 7. Transparency–voltage curve of the PDLC thin film.

4. Results, Validation, and Discussion

The objectives of this section are:

a. To demonstrate the experimental and simulation results of the implemented scenarios mentioned in Table 4.
b. To compare the experimental and simulation outcomes to validate the system.
c. To investigate the impact of the *PVR* on different factors of the system.

According to the IESNA approach mentioned in Equations (2)–(7), the illuminance data were calculated for 11 to 14 August 2021, as shown in Figure 8, including direct normal illuminance E_{DN}, diffuse horizontal illuminance E_{KH}, and the total illuminance E. The effective daylight period was from 5:00 am and 7:00 pm; however, the calculation showed that the available estimated illuminance during the four days was approximately equal. There are two reasons for this: first, since the days of the experiment were consecutive, the sun elevation angle was almost the same; second, because the four days had the same sky clearness parameters (sunny days).

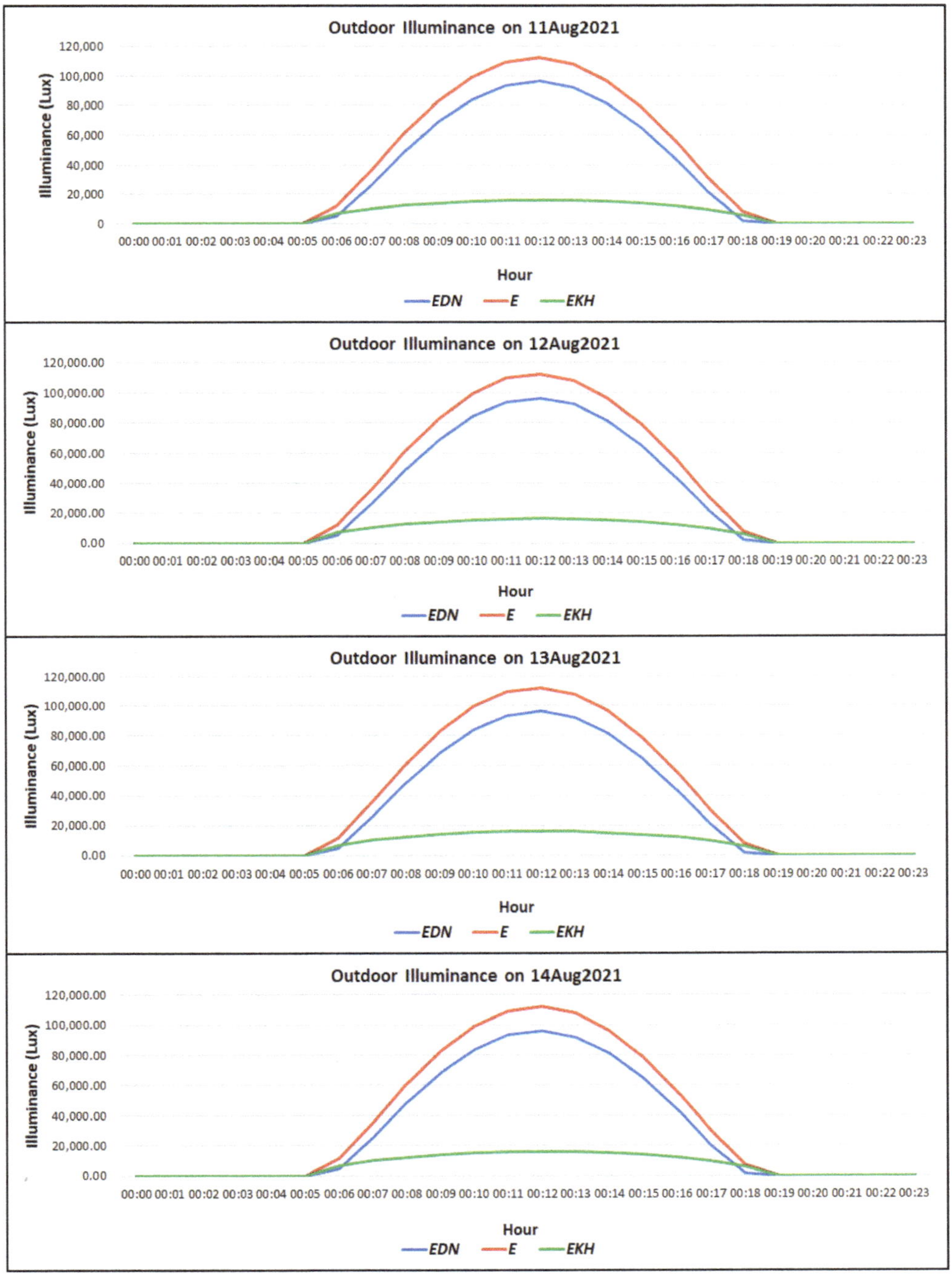

Figure 8. Calculated outdoor illuminance data during experiment days based on IESNA approach.

The solar irradiance data during the four days are shown in Figure 9, including global horizontal irradiance GHI and diffuse horizontal irradiance DHI; solar radiation slightly varied from 11 to 14 August 2021.

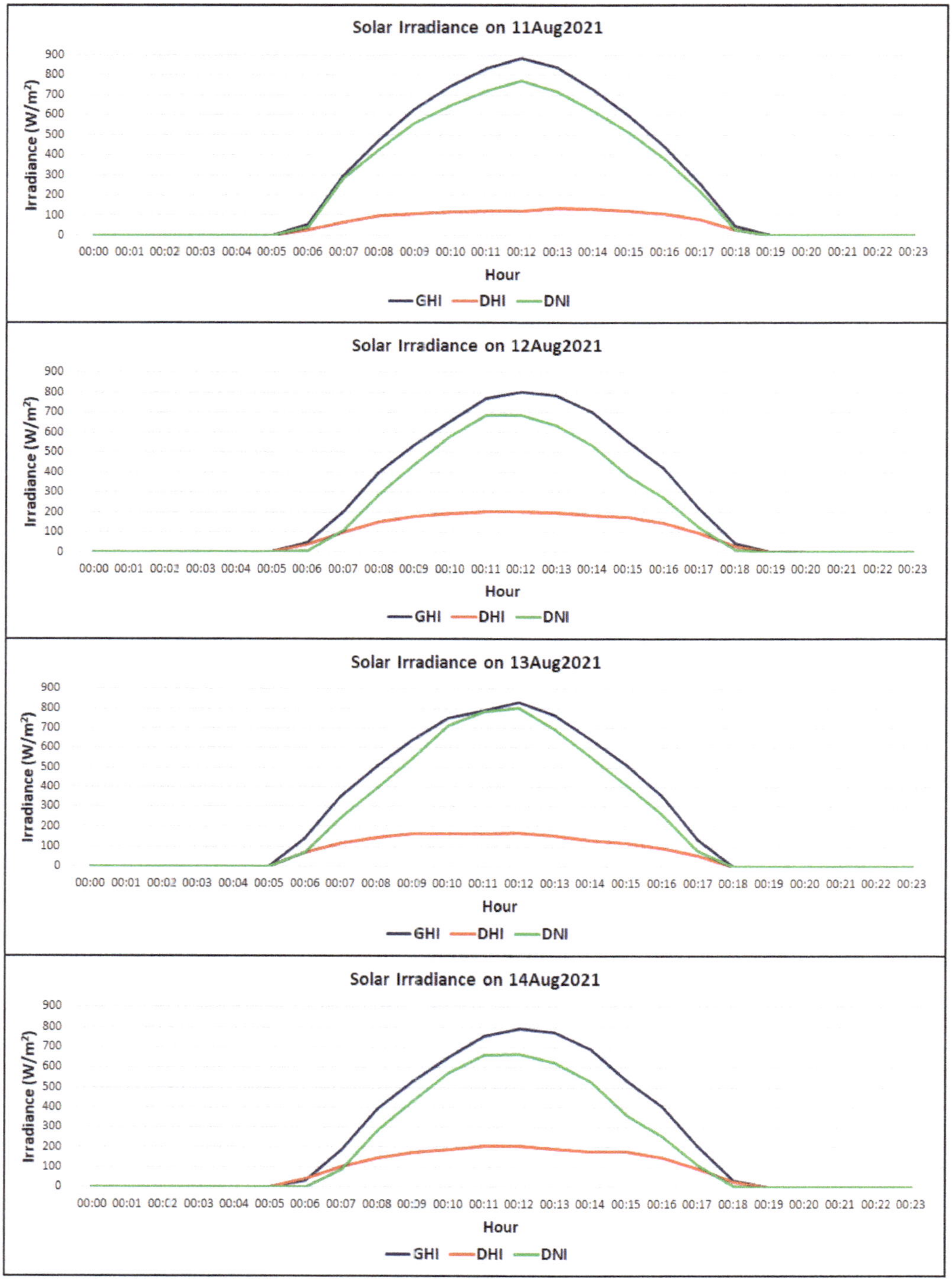

Figure 9. Solar irradiance data.

The indoor and outdoor temperatures of the test rooms are plotted against the days of the experiment in Figure 10. During the first day, when the *PVR* was adjusted to 0, which corresponds to the minimum available contrast in the comfort vision band, the experimental and simulated indoor temperatures responded to the variation in the outdoor temperature with good correlation. The maximum recorded temperature inside the reference model at 2:00 p.m. was 58.8 °C, and the average temperature was 41 °C, whereas the highest and average temperatures for the PDLC model were 56 and 38 °C, respectively. The average deviations between the experimental and simulated temperatures inside the double-glazed window room and the PDLC room were 0.12 and 0.37 °C, respectively.

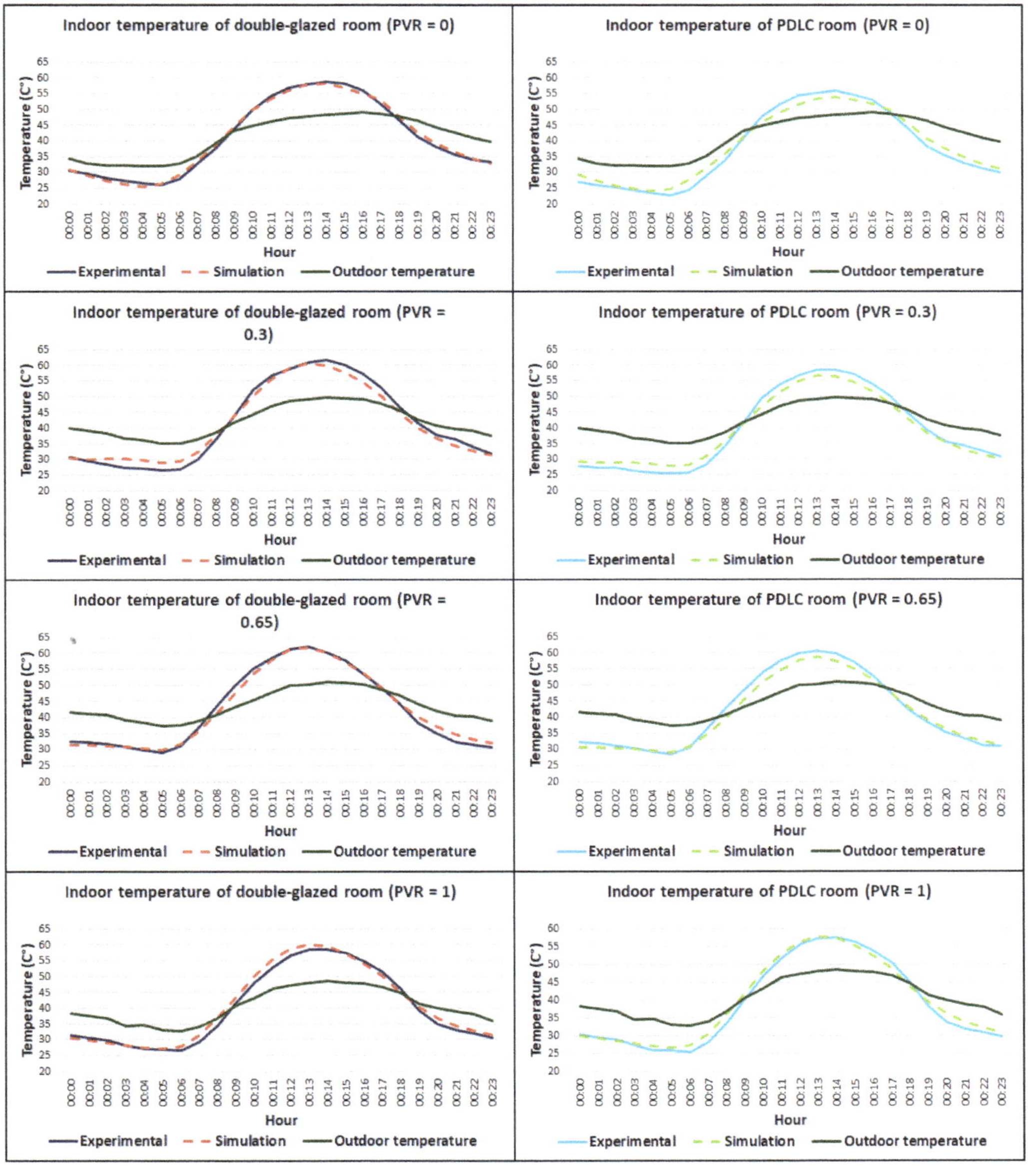

Figure 10. Indoor temperature for the test rooms versus *PVR*.

During the second day, the *PVR* was equal to 0.3, and the experimental and simulated indoor temperatures varied in the same pattern as that of the outdoor temperature. The experimental and simulated indoor temperatures in both rooms were approximately equal. Moreover, the maximum observed temperature inside the reference model at 2:00 pm was 61.7 °C, and the average temperature was 41.5 °C, whereas the highest and average temperatures for the PDLC model were 58.5 and 39.4 °C, respectively. The average differences between the experimental and simulated temperatures inside the reference room and the PDLC room were 0.19 and 0.22 °C, respectively.

Similar observations were made on the third day, when the *PVR* was set to 0.65. The greater temperature noted inside the reference model at 2:00 pm was 62 °C, and the average temperature was 42.3 °C, whereas the highest and average temperatures for the PDLC model were 60.7 and 40.8 °C, respectively. The average variances between experimental and simulated temperatures inside the reference window room and the PDLC room were 0.1 and 0.3 °C, respectively.

For the fourth day, when the *PVR* was set to 1, the highest recorded temperature inside the reference model was at 2:00 pm, and was 58.6 °C, and the average temperature was 40 °C, whereas the highest and average temperatures for the PDLC model were 57.5 and 38.8 °C respectively. The average disparities between the experimental and simulated temperatures inside the reference room and the PDLC room were 0.2 and 0.24 °C, respectively.

Evidently, increasing the value of the preferred vision ratio (*PVR*) led to an increase in the PDLC room temperature; for instance, for *PVR* of 0, 0.3, 0.65, and 1, the average temperature inside the reference room exceeded that of the PDLC room by 3, 2.1, 1.5, and 1.2 °C respectively. Alternatively, with the rise in the *PVR*, the temperature of the PDLC model approached the temperature inside the reference room; this is because the increase in the *PVR* allows more solar irradiance to enter the PDLC test room as the transparency becomes higher.

Figure 11 shows the experimental and simulated daylight values, in lux; these represent the illuminance intensity at the point inside the test rooms where the flux sensor was positioned. For the simulated daylight curves, DESIGN BUILDER software provides a range of illuminance values for a given point during the simulation; therefore, the values of simulated indoor illuminance are sketched as ranges in different hours during the daytime. Figure 11 also elucidates vertical global irradiance (*VGI*).

Starting with day 1 and a *PVR* of 0, the experimental and simulated indoor illuminance values respond to the variation in the outdoor vertical global irradiance with good correlation. The maximum recorded illuminance inside the reference model at 12:00 pm (at the highest value of *VGI*) was 4.1 Klux, and the average illuminance was 2.6 Klux, whereas the highest and average illuminance values for the PDLC model were 3 and 1.9 Klux, respectively. Similar trends can be observed for other *PVR* settings on other days.

As a result of increasing the *PVR*, more daylight reached the internal environment. For example, when the *PVR* was set to 0 during the first day of the experiment, the average daylight in the PDLC test room was 73% of that in the reference test room. Thereafter, the change in daylight increased to 76, 80, and 85% for *PVR* values of 0.3, 0.65, and 1, respectively. As mentioned previously, raising the *PVR* offers higher visible transparency and permits more daylight to enter the PDLC test room. The average deviations between experimental and simulated daylight values inside the reference room and the PDLC room were 20 lux and 39 lux on day 1, 57 lux and 69 lux on day 2, 81 lux and 29 lux on day 3, and 39 lux and 108 lux on day 4, respectively.

As the PDLC window has different transparency compared to the conventional double-glazed window, the window has a different solar heat gain coefficient (*SHGC*). Consequently, the solar heat gain values inside the two test rooms were not identical. Figure 12 depicts the experimental solar irradiance on the window and the simulated solar heat gain for both test rooms during the days of the experiment.

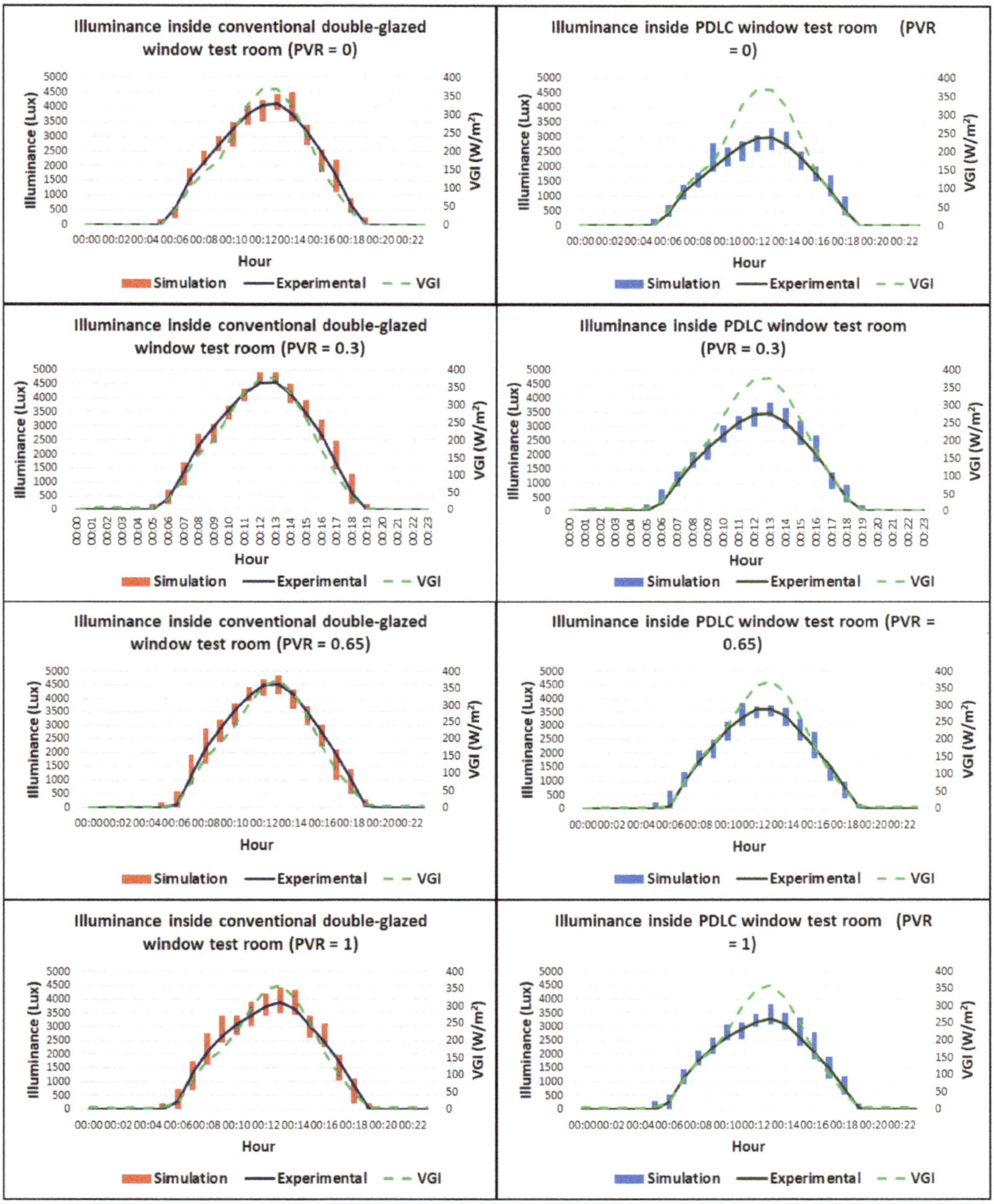

Figure 11. Indoor illuminance for the test rooms versus *PVR*.

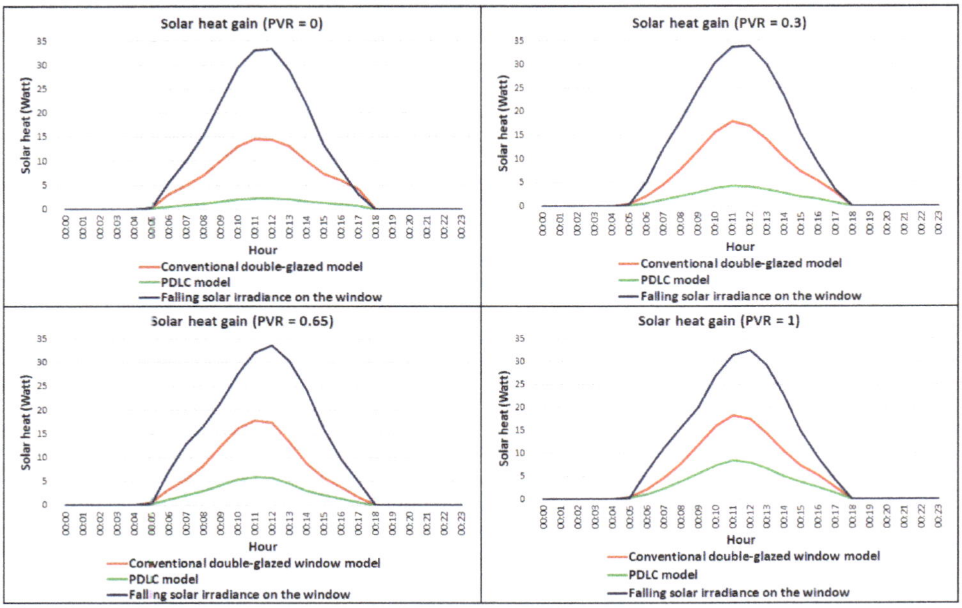

Figure 12. Solar heat gain for the test rooms versus *PVR*.

Overall, the effective solar heat gain occurred between 5:00 am and 6:00 pm, with the highest gain at 12:00 pm. The heat gain mimics the envelope of the solar irradiance on the window. During the first day of the experiment, with a *PVR* of 0, the maximum heat gain in the reference test room was 14.6 W and the average value was 8.32 W. By comparison, the highest recorded heat gain in the PDLC test room was 2.23 W and the average value was 1.38 W. Furthermore, with a *PVR* of 0.3, the greater heat gain noted inside the reference test room was 17.85 W and the average value was 8.95 W, whereas for the PDLC test room, the highest recorded heat gain was 4.25 W and the mean value was 2.24 W. Additionally, when the value of the *PVR* was set to 0.65, the superior observed heat gain in the reference test room was 17.8 W, and the average value was 8.81 W; similarly, the highest registered value in the PDLC test room was 5.88 W and the median value was 3.05 W. Finally, during day 4 of the experiment, with a *PVR* of 1, the heat gain in the reference room had a maximum value of 18.28 W and an average value of 9.09 W, whereas the highest observed value in the PDLC test room was 8.33 W and the mean value was 4.28 W.

Clearly, raising the preferred vision ratio resulted in an increase in a heat gain inside the PDLC test room; for instance, with a *PVR* of 0, the average heat gain inside the PDLC test room was only 16.6% of that in the reference room; this percentage was 25.1% with a *PVR* of 0.3, 34.7% with a *PVR* of 0.65, and 47.1% with a *PVR* of 1. Moreover, the solar heat gain coefficients computed by the DESIGN BUILDER software were 0.42, 0.49, 0.56, and 0.78 based on *PVR*s of 0, 0.3, 0.65, and 1, respectively.

Modifying the *PVR* has a direct impact on the quality of vision through the PDLC window, as elucidated earlier, since it determines the operating point on the comfort vision band of the contrast–voltage curve, which represents the contrast value of the instantaneous external scene seen through the window. Consequently, the visible transparency was estimated accordingly. Figure 13 depicts the contrast values during daytime, which were determined by the proposed algorithm during the days of the experiment. Obviously, choosing a higher *PVR* prompts the system to increase the contrast; thus, the average contrast values were 0.18, 0.25, 0.31, and 0.37 for *PVR*s of 0, 0.3, 0.65, and 1, respectively.

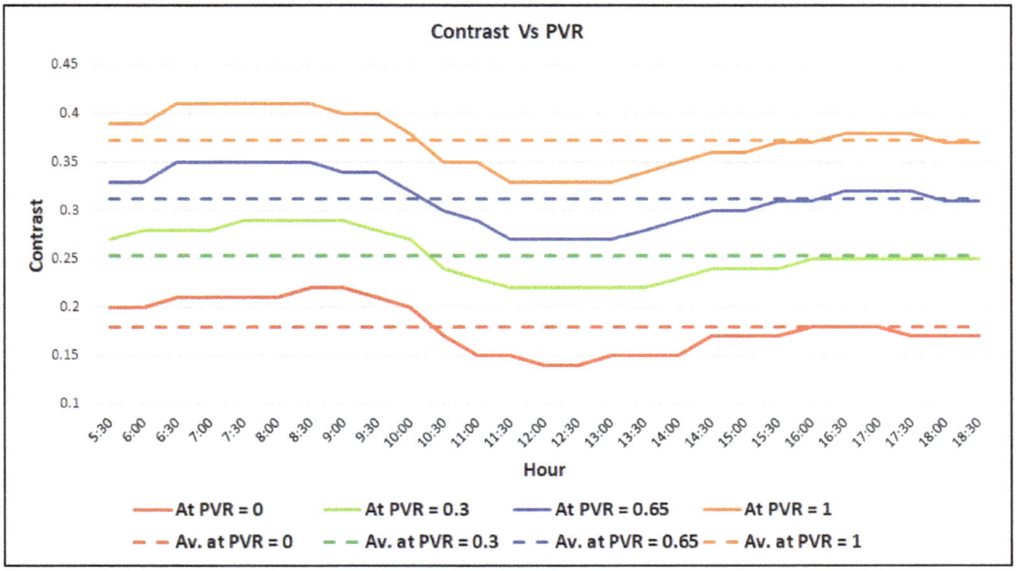

Figure 13. Contrast versus the *PVR*.

However, Figure 13 shows a fluctuation in the contrast value during the daytime. The proposed system increases the contrast value during the early and late hours of daytime (low illuminance hours). In contrast, the system lessens the contrast during midday hours when the available illuminance in the external environment is too luminous. Hence, the lowest contrast occurred at the highest outdoor illuminance (at 12:00 pm).

During the experiment, the contrast–voltage curves were inflated during low daylight hours or reduced during high daylight hours; thus, the comfort vision band was expanded or limited accordingly. This modification in the contrast–voltage curve led to a change in the *PVR*; therefore, the system tried to keep the *PVR* constant by simultaneously updating the contrast value.

The interaction of the system with the outdoor illuminance mimics the behavior of the human eye. The pupil in the human eye is responsible for controlling the quantity of light that enters the eye to maintain a certain level of vision, and has a dilatable diameter to achieve a light-controlling action [39,40].

The proposed algorithm was executed every 30 min during daytime, as mentioned earlier; however, the experiment shows that the achieved contrast values were in the vicinity of the setpoint. To achieve a more precise contrast value, the processing frequency can be increased to 10 min.

In practical applications, the present study assumed that the scene was fixed during image capturing. However, if the view instantaneously changes (e.g., the sudden appearance of dirt, dust, or rain on the window), two consequences can be expected. First, if the sudden change is small then it will not affect the contrast calculation. Second, if the sudden change is large, then it may lead to a nonstandard form of the contrast curve. In this case, the quantification process should be modified to recognize this change and to avoid this limitation in real time using image-processing techniques.

Figure 14 shows the overall assessment of the system in terms of the daylight entering the indoor environment, indoor temperature, solar heat gain, and vision contrast through the window, which are directly proportional to the *PVR*.

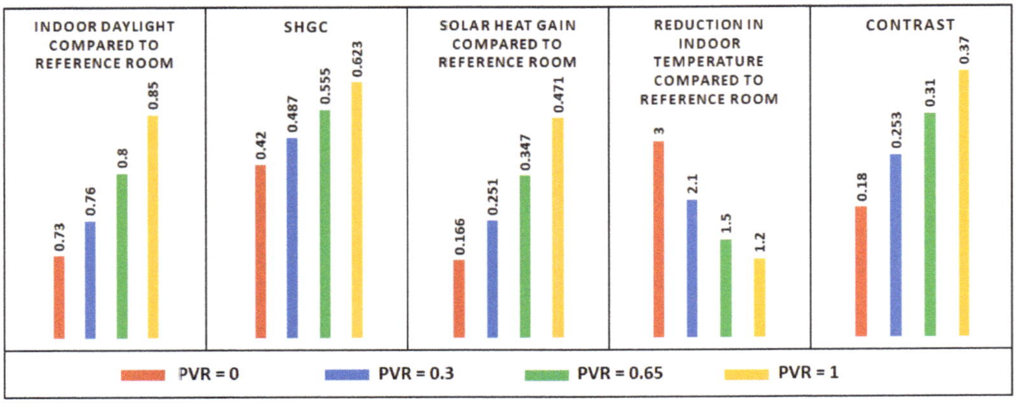

Figure 14. Results summary.

It can be concluded that setting a higher *PVR* leads to better vision quality through the window and enhances the daylight performance in indoor environments. Unfortunately, this choice results in an increase in the heat gain inside buildings, and raises the average indoor temperature; as a result, more energy consumption is required for cooling loads in hot regions. However, providing good vision quality is undoubtedly an essential challenge and a trade-off can be considered. By comparison, in a temperate climate, such as that of the UK, this challenging trade-off may be more easily addressed because the energy consumption is expected to increase due to the lower temperature. Utilizing the newly developed thin films must not impact the vision through windows, but it is necessary to achieve a balance between energy consumption, vision quality, and daylight performance to obtain optimum overall performance.

The conducted experiment showed that utilizing normal LDR images is a suitable approach for the contrast–voltage curve. Using HDR images can generate more precise contrast curves. However, HDR images have greater color representation (16 bit instead of 8 bit in LDR images); this increases the complexity of the image processor, which effectively inflates the cost. At the stage of implementing the processor, it will be possible to compare the efficiency of the processing using LDR and HDR images, and thereby evaluate the precision of the contrast values in both cases.

If a commercial building is equipped with the proposed system, based on the procedures of the quantification algorithm, the visible transparency of the windows would change from an opaque to a translucent state every 30 min. This would not be acceptable to the occupants. Therefore, it is proposed to conduct the quantification procedures on a reference window located in a building that has fewer people, and distribute the data and transparency settings to other glazing in the building.

It is worth mentioning that the proposed quantification method is not restricted to PDLC windows, but can also be employed for other kinds of conventional and smart windows, whenever it is necessary to evaluate the quality of vision. However, conventional windows provide a high quality of external scenes, mostly due to the high visible transparency of their glass; therefore, the proposed method may be more useful in the case of smart windows.

5. Conclusions

This study aimed to quantify the vision through smart windows using an image-processing approach to achieve visual comfort for the occupants. A small-scale room was equipped with a PDLC double-glazed window and used as an example to realize the proposed system. Online image processing was proposed to derive the available comfort vision from the contrast–voltage curve of the PDLC film in real time, and to constrain the

PDLC transparency range to achieve occupant satisfaction. The main contributions were as follows: first, mathematically defining the comfort vision band for PDLC double-glazed windows; and second, introducing the preferred vision ratio (*PVR*), which ensures a desired level of vision for the occupants. The main outcomes of this work can be summarized as follows:

1. Vision through smart windows can be quantified in terms of the contrast of the scenes measured through the window toward the external views.
2. The proposed *PVR* allows users to choose the preferable level of vision through the window.
3. The experiment showed that varying the *PVR* from 0 to 1 can modify the average contrast value from 0.18 to 0.37 in a clear sky environment.
4. The proposed algorithm ensures that the transparency of the window will not be less than a threshold value, in turn ensuring a minimum level of vision quality.
5. The experiment showed that varying the *PVR* of the PDLC window from 0 to 1 increased the solar heat gain coefficient (SHGC) from 0.42 to 0.623.
6. The experiment showed that modifying the *PVR* of the PDLC window from 0 to 1 allowed more daylight to enter the internal environment (73% to 85% compared to conventional double-glazed windows).

The test room was simulated in the DESIGN BUILDER software under the experimental weather conditions, and the simulation results were compared with the experiment outcome to validate the experimental results. In commercial buildings, employing the proposed system constrains the vision through smart windows within the comfort vision band. The system also helps to replace the conventional human preference surveying methods, thus saving time and effort. Moreover, future studies can utilize the proposed *PVR* to evaluate the vision level through smart windows. Overall, the proposed system adds observability and controllability capabilities for building controllers, thus helping achieve comfortable living environments for the occupants. In future, the proposed system may use HDR images for more precise contrast curves. Moreover, employing image-processing techniques enables the system to deal with instantaneous changes that may occur in the scenes during image capturing. In addition, an embedded system can be designed to implement the proposed quantification algorithm. This would help to reduce the time taken in the process and obtain more precise real-time visible transparency values for the glazing.

Author Contributions: Conceptualization, M.L., F.A.-n. and W.I.; Data curation, M.L.; Formal analysis, M.L., F.A.-n. and W.I.; Investigation, M.L.; Methodology, M.L.; Resources, M.L. and H.A.; Software, M.L.; Supervision, F.A.-n. and W.I.; Validation, M.L.; Visualization, M.L.; Writing—original draft, M.L.; Writing—review & editing, F.A.-n., W.I. and H.A. All authors have read and agreed to the published version of the manuscript.

Funding: This research received no external funding.

Institutional Review Board Statement: Not applicable.

Informed Consent Statement: Not applicable.

Data Availability Statement: Not applicable.

Conflicts of Interest: The authors declare no conflict of interest.

References

1. Ghosh, A.; Mesloub, A.; Touahmia, M.; Ajmi, M. Visual comfort analysis of semi-transparent perovskite based building integrated photovoltaicwindow for hot desert climate (riyadh, saudi arabia). *Energies* **2021**, *14*, 1043. [CrossRef]
2. Rabani, M.; Madessa, H.B.; Nord, N. Achieving zero-energy building performance with thermal and visual comfort enhancement through optimization of fenestration, envelope, shading device, and energy supply system. *Sustain. Energy Technol. Assess.* **2021**, *44*, 101020. [CrossRef]
3. Basinska, M.; Kaczorek, D.; Koczyk, H. Building thermo-modernisation solution based on the multi-objective optimisation method. *Energies* **2020**, *13*, 1433. [CrossRef]

4. Cesari, S.; Valdiserri, P.; Coccagna, M.; Mazzacane, S. The energy saving potential of wide windows in hospital patient rooms, optimizing the type of glazing and lighting control strategy under different climatic conditions. *Energies* **2020**, *13*, 2116. [CrossRef]
5. Wang, C.H.; Kuo, N.W.; Anthony, K. Impact of window views on recovery—an example of post-cesarean section women. *Int. J. Qual. Health Care* **2019**, *31*, 798–803. [CrossRef] [PubMed]
6. Pilechiha, P.; Mahdavinejad, M.; Rahimian, F.P.; Carnemolla, P.; Seyedzadeh, S. Multi-objective optimisation framework for designing office windows: Quality of view, daylight and energy efficiency. *Appl. Energy* **2020**, *261*, 114356. [CrossRef]
7. Katunský, D.; Dolníková, E.; Doroudiani, S. Integrated lighting efficiency analysis in large industrial buildings to enhance indoor environmental quality. *Buildings* **2017**, *7*, 47. [CrossRef]
8. Batool, A.; Rutherford, P.; McGraw, P.; Ledgeway, T.; Altomonte, S. Window Views: Difference of Perception during the COVID-19 Lockdown. *LEUKOS-J Illum. Eng. Soc. North Am.* **2021**, *17*, 380–390. [CrossRef]
9. Dodo, Y.A.; Kandar, M.Z.; Ossen, D.R.; Jibril, J.D.a.; Bornoma, A.H.; Abubakar, A.I. Importance of a view window in rating green office buildings. *Adv. Mater. Res.* **2013**, *689*, 180–183. [CrossRef]
10. Roetzel, A.; Tsangrassoulis, A.; Dietrich, U.; Busching, S. On the influence of building design, occupants and heat waves on comfort and greenhouse gas emissions in naturally ventilated offices. A study based on the EN 15251 adaptive thermal comfort model in Athens, Greece. *Build. Simul.* **2010**, *3*, 87–103. [CrossRef]
11. Ma, G.; Pan, X. Research on a visual comfort model based on individual preference in china through machine learning algorithm. *Sustainability* **2021**, *13*, 7602. [CrossRef]
12. Day, J.K.; Futrell, B.; Cox, R.; Ruiz, S.N. Blinded by the light: Occupant perceptions and visual comfort assessments of three dynamic daylight control systems and shading strategies. *Build. Environ.* **2019**, *154*, 107–121. [CrossRef]
13. GLim, H.; Keumala, N.; Ghafar, N.A. Energy saving potential and visual comfort of task light usage for offices in Malaysia. *Energy Build.* **2017**, *147*, 166–175. [CrossRef]
14. Meresi, A. Evaluating daylight performance of light shelves combined with external blinds in south-facing classrooms in Athens, Greece. *Energy Build.* **2016**, *116*, 190–205. [CrossRef]
15. Aries, M.B.C.; Veitch, J.A.; Newsham, G.R. Windows, view, and office characteristics predict physical and psychological discomfort. *J. Environ. Psychol.* **2010**, *30*, 533–541. [CrossRef]
16. Jakubiec, J.A. The Use of Visual Comfort Metrics in the Design of Daylit Spaces. Massachusetts Institute of Technology. 2014. Available online: http://hdl.handle.net/1721.1/91295 (accessed on 1 September 2021).
17. Setyantho, G.R.; Park, H.; Chang, S. Multi-criteria performance assessment for semi-transparent photovoltaic windows in different climate contexts. *Sustainability* **2021**, *13*, 2198. [CrossRef]
18. Basher, M.K.; Alam, M.N.E.; Alameh, K. Design, development, and characterization of low distortion advanced semitransparent photovoltaic glass for buildings applications. *Energies* **2021**, *14*, 3929. [CrossRef]
19. Gardymova, A.P.; Krakhalev, M.N.; Zyryanov, V.Y.; Gruzdenko, A.A.; Alekseev, A.A.; Rudyak, V.Y. Polymer dispersed cholesteric liquid crystals with a toroidal director configuration under an electric field. *Polymers* **2021**, *13*, 732. [CrossRef]
20. Chen, G.; Hu, W.; Zhou, L.; Zhang, H.; Wang, L.; Zhang, C.; Sun, J.; Zhang, L.; Yuan, X.; Zhu, S. The electro-optical properties and adhesion strength of epoxy-polymercaptan-based polymer dispersed liquid crystal films. *Crystals* **2021**, *11*, 567. [CrossRef]
21. Shchegolkov, A.V.; Jang, S.H.; Shchegolkov, A.V.; Rodionov, Y.V.; Sukhova, A.O.; Lipkin, M.S. A brief overview of electrochromic materials and related devices: A nanostructured materials perspective. *Nanomaterials* **2021**, *11*, 2376. [CrossRef]
22. Brzezicki, M. A systematic review of the most recent concepts in smart windows technologies with a focus on electrochromics. *Sustainability* **2021**, *13*, 9604. [CrossRef]
23. Detsi, M.; Manolitsis, A.; Atsonios, I.; Mandilaras, I.; Founti, M. Energy savings in an office building with high WWR using glazing systems combining thermochromic and electrochromic layers. *Energies* **2020**, *13*, 3020. [CrossRef]
24. Ko, Y.; Oh, H.; Hong, H.; Min, J. Energy consumption verification of spd smart window, controllable according to solar radiation in south korea. *Energies* **2020**, *13*, 5643. [CrossRef]
25. Oh, M.; Lee, C.; Park, J.; Lee, K.; Tae, S. Evaluation of energy and daylight performance of old office buildings in South Korea with curtain walls remodeled using polymer dispersed liquid crystal (PDLC) films. *Energies* **2019**, *12*, 3679. [CrossRef]
26. Hemaida, A.; Ghosh, A.; Sundaram, S.; Mallick, T.K. Simulation study for a switchable adaptive polymer dispersed liquid crystal smart window for two climate zones (Riyadh and London). *Energy Build.* **2021**, *251*, 111381. [CrossRef]
27. Bourbakis, N.G. *Emulating Human Visual Perception for Measuring Difference in Images Using an SPN Graph Approach*; IEEE: New York, NY, USA, 2002.
28. Sasaki, H.; Todorokihara, M.; Ishida, T.; Miyachi, J.; Kitamura, T.; Aoki, R. Effect of noise on the contrast detection threshold in visual perception. *Neurosci. Lett.* **2006**, *408*, 94–97. [CrossRef]
29. Butler, P.D.; Silverstein, S.M.; Dakin, S.C. Visual Perception and Its Impairment in Schizophrenia. *Biol. Psychiatry* **2008**, *64*, 40–47. [CrossRef]
30. Veth, M.J.; Raquet, J.F. Image-Aided Navigation—Concepts and Applications. In *Position, Navigation, and Timing Technologies in the 21st Century: Integrated Satellite Navigation, Sensor Systems, and Civil Applications*; IEEE: New York, NY, USA, 2021; pp. 1571–1595. [CrossRef]
31. Driggers, R.; Jacobs, E.; Young, S. Signal Processing and Performance Analysis for Imaging Systems. Artech. 2008. Available online: http://ieeexplore.ieee.org/document/9100126 (accessed on 1 September 2021).

32. Fan, Z.; Liu, M.; Tang, S. A multi-objective optimization design method for gymnasium facade shading ratio integrating energy load and daylight comfort. *Build. Environ.* **2021**, *207*, 108527. [CrossRef]
33. Hegazy, M.; Yasufuku, K.; Abe, H. Evaluating and visualizing perceptual impressions of daylighting in immersive virtual environments. *J. Asian Archit. Build. Eng.* **2021**, *20*, 768–784. [CrossRef]
34. Futagami, T.; Yano, T.; Huang, C.; Enohara, T. Experimental evaluation for daylight-linked gradation lighting control using image-based motion sensors. *IEEJ Trans. Electr. Electron. Eng.* **2020**, *15*, 723–732. [CrossRef]
35. Suk, J.; Schiler, M. Investigation of Evalglare software, daylight glare probability and high dynamic range imaging for daylight glare analysis. *Lighting Res. Technol.* **2013**, *45*, 450–463. [CrossRef]
36. Kandilli, C.; Ulgen, K. Numerical Assessment of Downward Incoming Solar Irradiance in Smoke Influenced Regions—A Case Study in Brazilian Amazon and Cerrado. *Remote Sens.* **2021**, *13*, 4527. [CrossRef]
37. Kandilli, C.; Ulgen, K. Solar illumination and estimating daylight availability of global solar irradiance. *Energy Sources Part A Recovery Util. Environ. Eff.* **2008**, *30*, 1127–1140. [CrossRef]
38. Ionescu, C.; Fosalau, C.; Petrisor, D. A study of changes in image contrast with various algorithms. In Proceedings of the 2014 International Conference and Exposition on Electrical and Power Engineering, Iasi, Romania, 16–18 October 2014; pp. 100–104. [CrossRef]
39. Wyatt, H.J. The Form of the Human Pupil. *Vis. Res.* **1995**, *35*, 2021–2036. [CrossRef]
40. Hanke, F.D.; Hanke, W.; Scholtyssek, C.; Dehnhardt, G. Basic mechanisms in pinniped vision. *Exp. Brain Res.* **2009**, *199*, 299–311. [CrossRef] [PubMed]

buildings

MDPI

Article

Heating and Cooling Primary Energy Demand and CO_2 Emissions: Lithuanian A+ Buildings and/in Different European Locations

Kęstutis Valančius [1], Monika Grinevičiūtė [2] and Giedrė Streckienė [1,*]

[1] Department of Building Energetics, Vilnius Gediminas Technical University, Sauletekio ave. 11, 10223 Vilnius, Lithuania; kestutis.valancius@vilniustech.lt
[2] Creativity and Innovation Centre "Linkmenų fabrikas", Vilnius Gediminas Technical University, Linkmenų g. 28, 08217 Vilnius, Lithuania; monika.grineviciute@vilniustech.lt
* Correspondence: giedre.streckiene@vilniustech.lt

Citation: Valančius, K.; Grinevičiūtė, M.; Streckienė, G. Heating and Cooling Primary Energy Demand and CO_2 Emissions: Lithuanian A+ Buildings and/in Different European Locations. *Buildings* **2022**, *12*, 570. https://doi.org/10.3390/buildings12050570

Academic Editors: Shi-Jie Cao and Wei Feng

Received: 1 April 2022
Accepted: 26 April 2022
Published: 28 April 2022

Abstract: National legal and political regulation in the field of energy efficiency is closely connected to minimizing energy consumption in buildings. Within the framework of implementing Directive 2018/844/EU on the energy performance of buildings in Europe, the practice of its application differs from country to country. This study aims to reveal the differences in the energy indicators of an energy-efficient building in European states. To that end, an analysis was made to compare the results of a single-family home model in 11 city locations with different climatic conditions (from the Mediterranean to Nordic) and appropriate national regulations in place for the past three years. The simulation was done using IDA Indoor Climate and Energy software, EQUA Simulation AB, Stockholm, Sweden. The demand for primary energy is based on primary energy factors. A comparison of overall heat transfer coefficients for walls and windows in an energy-efficient building in different locations was made to reveal the differences in applicable national regulations. The results showcase the primary energy demand depending on the different climatic conditions for building heating and cooling purposes, as appropriate, and on CO_2 emissions. The study has shown the energy demand for cooling to increase significantly—by 65% in the case of Vilnius, whereas only a slight decrease in the demand for heating. Furthermore, a Lithuanian energy class A+ building is singled out as an individual case, its energy indicators determined for a different location under analysis.

Keywords: energy class; energy efficiency; primary energy; heating and cooling; CO_2 emissions

1. Introduction

Modern buildings have come to be the third-largest group of consumers of fossil fuel after industry and agriculture. What is more, 30–40% of all key global resources are consumed by none other than the construction sector [1]. The building sector is considered the main and largest energy consumer on the global scale, and 40% of all primary energy (PE) generated in the US and the European Union (EU) is consumed in buildings [2,3]. With the economy growing and urban development picking up pace, one can look forward to a further increase in the building sector and a parallel surge in energy consumption, to be boosted by the accelerating climate change as well. The primary objective of this study is to assess the energy performance of energy-efficient buildings in European countries, to analyze the relationship between different climatic conditions and the building's primary energy needs, and to determine how climate change contributes to a building's energy consumption for heating and cooling.

It is a widely known fact that greenhouse gas (GHG) emissions are the number one reason behind climate change and the global warming that follows, as well as extreme weather [4]. Obviously, buildings also have a role in driving climate change: 19% of all GHG emissions occur through energy processes decarbonization in buildings [5]. It is the matter of building decarbonization that has been the focal point of attention over the past few decades, as achieving energy efficiency in buildings and switching to renewable energy has a large potential in reducing GHG emissions in the future [6].

In a bid to improve the energy performance of buildings, in 2018 the EU adopted Directive 2018/844/EU, partially amending Directive 2010/31/EU on the energy performance of buildings [7]. The latest EU directive is geared towards improving the energy efficiency of buildings, ensuring the right indoor climate, and reducing the use of fossil fuel whilst increasing the availability of renewable energy. This directive aims to contribute to the EU's goal of decarbonizing the building sector by 2050 [8]. The results of an analysis conducted by the European Commission (EC) in 2011 have shown that GHG emissions in the building sector can be cut by a staggering 90% by 2050 [9]. The EU goal of becoming a climate-neutral zone is also seeing a contribution from the building sector with nearly zero-energy buildings (NZEB), which were supposed to become the benchmark in the EU residential building market as of 2021 [7].

EU member state documents provide varying definitions of energy-efficient buildings, yet their underlying feature is defined as follows: these buildings consume little energy, and any energy consumption is done in an efficient way [10]. Some of Europe's first high-energy class buildings were introduced in German: these are the so-called passive houses and buildings bearing the Minergie seal of quality in Switzerland [1].

Following the adoption of the updated Building Energy Performance Directive [7], European countries and Member States had to amend their national legislation to include legal solutions in connection with NZEB. The main requirements for the primary energy consumption, total annual heating, and cooling demands, envelope heat transfer indicators, airtightness, and infiltration of buildings differ from one European country to another. However, these are just some of the differences that affect an effective entrenchment of NZEB in Europe.

A review by the Institute for Energy Efficiency has backed the information presented in the amendment to the European Commission regulation (Directive (EU) 2018/844) that differences in climatic conditions preclude the application of a single NZEB efficiency value suitable for all European countries [11]. Therefore, in the EU (and Norway) these buildings are covered by different national regulations and requirements, which makes a consistent increase in the availability of such buildings in different economies more difficult.

The latest review of political strategies by the Building Performance Institute Europe (BPIE) has highlighted the key differences when it comes to implementing the requirements of Directive 2018/844/EU among the EU states. According to the BPIE, (1) the timing of hands-on application of the NZEB concept varies among the states (some of the Member States have complied with the requirements for implementation ahead of time while others are lagging behind); (2) Member States use different definitions and approaches to determine national NZEB definitions; (3) approaches to calculation and performance levels to be achieved by NZEBs under construction are variegated; (4) a portion of the energy consumed to be replaced by renewable energy varies from country to country. Another important aspect is that some of the Member States had developed (and have never updated) their approaches to NZEB years before such buildings became mandatory. As a result, the national standards of these buildings are not aligned with the EU's goal of becoming a climate-neutral zone by [12].

One stipulation of Directive 2018/844/EU is the mandatory inclusion of the numerical indicator of primary energy (kWh/m^2/year) in the national plans of Member State NZEB strategies. Considering that the energy efficiency of a building is affected by different climatic conditions and the building's typology, geometry, location, engineering mechanical systems, and so on, many Member States (with the exception of Austria, Flanders, Germany, Italy, Luxembourg, and Portugal where PE values are calculated on the basis of benchmark buildings) have set a certain range of primary energy consumption [12]. In its 2016 recommendations and guidelines for ensuring the good NZEB practice in the Member States, the EC indicated the comparable limit values of primary energy differentiated by four key climatic zones: Mediterranean, Oceanic, Continental, and Nordic [13,14]. According to a report by the International Energy Agency (2018), there is a global slowdown in the progress of energy policy, indicating that the evolution of building energy codes is failing to keep up with the growth of the economies of rapidly developing countries [14]. In 2018, two-thirds of countries worldwide were short on building energy efficiency codes and legal regulations. It means that in 2018 more than three billion square meters of useful building area were built without any mandatory energy performance requirements.

1.1. Climate Change on Residential Buildings in Europe

To achieve the sustainable development scenario, all countries of the world must switch to mandatory building energy efficiency laws by the year 2030 [14]. Improvements in the energy efficiency of buildings and sustainable development of renewable energy are a must in terms of overcoming the ever-growing energy consumption and the consequences of climate change [8]. Climate change and building energy processes share a paradoxical bond: these days, the processes that take place in buildings contribute to climate change; according to a number of studies (Table 1), the consequences of climate change will drive the energy consumption for building cooling purposes up. Scientific studies conducted decades ago noted that climate change would have a direct impact on the energy and thermal properties of buildings [15,16].

Table 1. A summary of previous studies pertaining to the effect of climate change on residential buildings in Europe.

Country	Period	Climate Scenario	Conclusion	Reference
Sweden	2050–2100	RCP [1] scenario 4.5 (the radiative forcing of GHG is reduced to 4.5 W/m^2) and RCP scenario 8.5 (GHG increases, its radiative forcing going up to 8.5 W/m^2) [17]	A 13–22% drop in the demand for heating, a 33–49% increase in the demand for cooling	[18]
Finland	2030–2050–2100	Drafted on the basis of the CMIP3 global climate model [19]	A 20–40% drop in the demand for heating, a 40–80% increase in the demand for cooling	[20]
Switzerland	2100	It is assumed that the average annual air temperature will increase by 4.4 °C compared to the climatologic standards of 1961–1990	A 33–44% drop in the demand for heating (cooling is not considered)	[21]
Germany	2060	It is assumed that the average annual air temperature will increase by 1–3 °C	A 44–75% drop in the demand for heating and a 28–59% increase in the demand for cooling	[16]
Greece	2100	Three scenarios by the Intergovernmental Panel on Climate Change are used [15]	A 44–75% drop in the demand for heating and a 28–59% increase in the demand for cooling	[22]

[1] RPC—Representative Concentration Pathway.

All kinds of research have been undertaken around the globe over the past few decades in order to analyze the effect climate change has on buildings. Table 1 shows the results of simulation studies of residential buildings in Europe highlighting the impact of climate change on the energy needs of buildings.

Depending on the climatic data of different countries, climate change scenarios, and other assumptions, the summary of studies in Table 1 shows that when the building's demand for heating drops by roughly one-half, the demand for cooling may go up by a massive 80%. Isaac and van Vuuren have estimated that climate change will drive the need for heating energy by more than 30% worldwide by 2010, while the demand for cooling energy will go up by nearly 80% [23].

1.2. The Current Situation in the European Zones Covered by the Analysis

Based on the values established by the EC, countries with prevalently milder (Mediterranean) climates must ensure the lowest demand for net primary energy and the largest share of energy from renewable sources [12]. Still, considering the primary energy of a building and notwithstanding whether it is supplied from renewable sources, the range of primary energy across all four European climate zones is much narrower: the PE demands of a single-family home must fall within the recommended 50–90 kWh/m^2/year (Table 2) [13].

Table 2. European Commission standards of building performance and renewable energy resources for different climate zones [13].

Climate Zone	Demand for Net PE, kWh/m^2/year	Energy from Renewable Energy Sources, kWh/m^2/year	PE Ceiling, Including Energy from Renewable Sources, kWh/m^2/year	Renewable Energy Sources as a Percentage of Total PE
Mediterranean	0–15	50	50–65	87%
Oceanic	15–30	35	50–65	61%
Continental	20–40	30	50–70	50%
Nordic	40–65	25	65–90	32%

Based on a review by the BPIE (2021), regulations of 13 Member States point to primary energy values that fall within the limit of 50–90 kWh/m^2/year as recommended by the Commission. Denmark, Croatia, and Ireland are more stringent in their requirements, and their recommended values are below those laid down in the EC guidelines. Whereas countries such as Bulgaria, Latvia, Cyprus, Hungary, the Czech Republic, Finland, and Romania are disregarding the guideline recommendations and have set primary energy values above those recommended by the EC [12]. The differences in PE demand in European states and the gap between the national values and the EC requirement for the demand in countries covered by the analysis are shown in Figure 1.

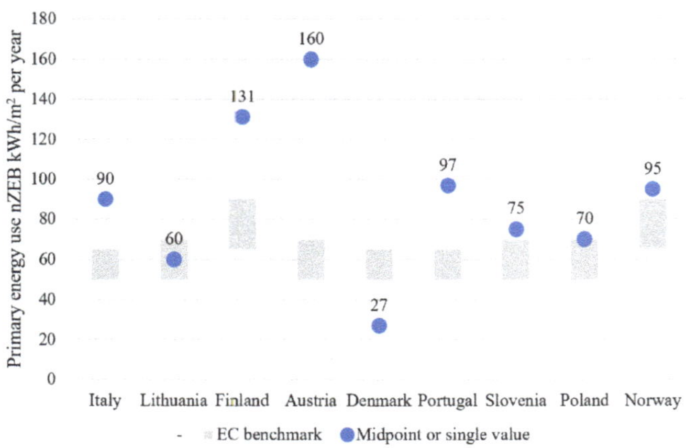

Figure 1. NZEB values (kWh/m^2/year) for single-family homes in European countries analyzed.

A more precise breakdown of the values established by the countries analyzed and those recommended by the EC are presented in Table 3.

Evidently, only Lithuania and Poland make it to the EC's brackets of primary energy demand. Denmark's national regulations stipulate a PE value that is nearly three times below the EC recommendations. For Norway (it is assumed that Norway is appraised on a par to the EU countries) and Slovenia, the PE limits are close to what the EC recommends. Therefore, the aim of this paper is to analyze and compare differences in the energy indicators of an energy-efficient building in different European states (with climate from Mediterranean to Nordic). This reveals differences in both climatic conditions and country legislation. At the same time, this contributes to the application of the NZEB concept. A single-family home is selected as a case study.

Table 3. PE values in the countries covered by the analysis and those recommended by the European Commission.

Country Analyzed	PE Value as Recommended by the EC, kWh/m²/year [1]	Source
Italy	65	PE value determined based on an assumption (considering the results of projects completed) [24]
Lithuania	70	[25]
Finland	90	[12]
Austria	70	[26]
Denmark	65	[12]
Portugal	65	[27]
Slovenia	70	[12]
Poland	70	[12]
Norway	90	[28]

[1] Ref [13] is the source of PE value as recommended by the EC. If the subject countries provide several limit values, the average PE values are specified for the purposes of this comparison.

2. Methodology and Case Study

This case study analyses the following European countries and cities: Lithuania (Vilnius), Finland (Jyväskylä), Italy (Palermo and Bologna), Denmark (Copenhagen), Portugal (Lisbon and (Bragança), Slovenia (Ljubljana), Poland (Warsaw), Norway (Oslo), Austria (Bregenz). This choice has been driven by the desire to analyze and compare the data and results for northern and southern, as well as eastern, western, and central states alike. A comparison of the heating transfer ratios applicable to energy-efficient buildings is followed by an analysis of building energy indicators in different climatic conditions conducted on the basis of the methodology developed by Sartori et al. and D'Agostino and Parker [29,30].

The methods used in the study are presented in Figure 2. They are based on the data gathered from Guidelines on Energy System Analysis and Cost Optimality in Early Design of ZEB, a report by Sartori et al. and the methodology developed by D'Agostino and Parker designed to analyze NZEB buildings in terms of price optimization [29,30].

2.1. *Study-Case Building Characteristics*

The modeled building is a single-apartment, one-floor residential home with a gross useful area of 100 m^2. The home is assumed to be inhabited by a family of four. The façade of the simulated building is facing south, nearly 68% of the façade (17.60 m^2) is a panoramic window. The key data of the building's envelopes are presented in Table 4.

Table 4. Envelope areas of the modeled building.

Building Envelope	Envelope Area, m²
Walls	72.20
Doors	1.60
Windows	30.2
Floors	100.0
Roof	100.0

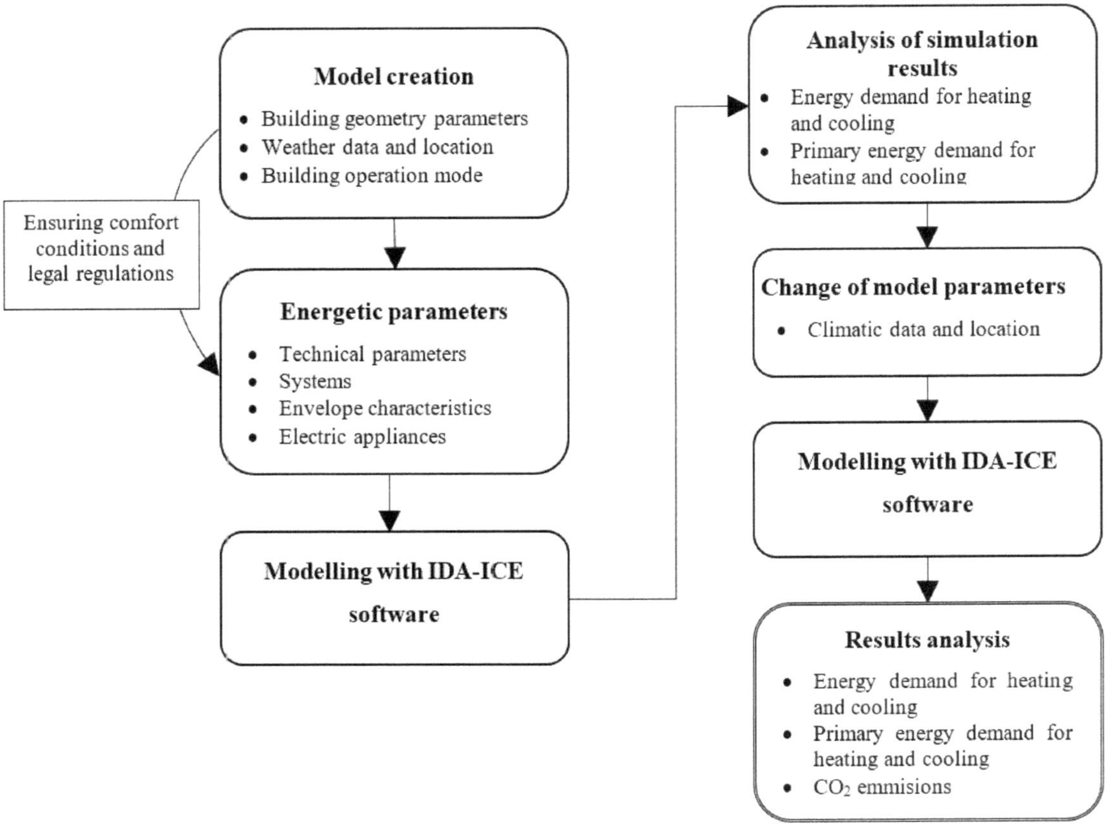

Figure 2. Study method (adapted by the authors in reliance on [29,30]).

The key data of the building's windows are presented in Table 5. The ratios were obtained from the IDA-ICE database. The type of window chosen is a quad-glazed window with in-between areas 90% filled with argon gas.

Table 5. Data of see-through building envelopes.

See-Through Envelopes	Solar Heat Inflow Ratio, g	Solar Permeability Ratio, T	Visible Permeability Ratio, T_{vis}	Interior Emission Ratio	Exterior Emission Ratio
Windows	0.60	0.55	0.74	0.837	0.837

The number and behavior of the residents have a large impact on the outcomes of energy calculations, which makes selecting the right data highly relevant. Table 6 contains the key usage mode parameters for the building model.

Notwithstanding the model of the state under analysis, these building usage parameters do not change in the course of modeling and remain the same for each building model. A visualization of the building in the environment of IDA-ICE 4.8 software used for dynamic simulation is shown in Figure 3.

Table 6. Building usage parameters.

Parameter	Comment
Energy consumption by electrical appliances	The annual consumption of electrical energy by household appliances in a residential district is 30 kWh/m^2 [31]
Lighting	630 kWh per residential home per year [31]
Energy for water heating	Hot water consumption per 24 h is assumed to be 75 L/person/24 h
Building occupancy	From 8 a.m. until 3 p.m. (or until 5 PM for one-half of the residents) on weekdays, weekends at home.
Resident activity level	0.77 MET. 1 MET = 58 W/m^2 of the body surface area (assumed to be 1.80 m^2) [31]

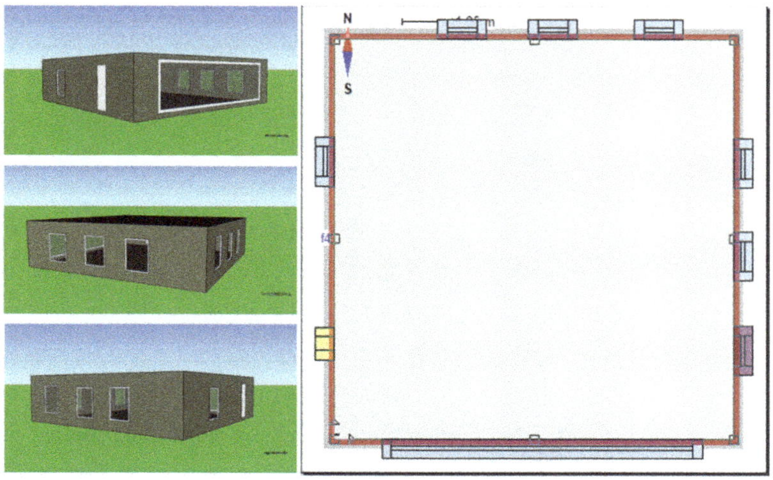

Figure 3. Visualization of the modeled residential home in the environment of IDA-ICE software.

2.2. Analysis of Energy Indicators of Buildings under Different Climatic Conditions

Each building model in IDA-ICE software is designed to conform to the principal requirements laid down in the building directives and standards of the subject country (Table 7) that apply to energy-efficient buildings, the main difference among the models being the envelope parameters and climatic data. Data such as building geometry, occupancy, number of residents, hot water consumption, and electrical energy consumption for electrical appliances remain the same across all building models covered by the analysis. A summary of the climatic data used in the models of the building analyzed is presented in Table 7, the key assumptions used for the purposes of the study are given in Table 8 and primary energy factors and CO_2 emission factors are presented in Table 9.

Table 7. Summary of study models.

Model No.	Climatic Data Used	Methodology Defining the Energy Characteristics of the Building, Which Is Used for Modelling Purposes
1	Vilnius, Lithuania	[25]
2	Jyväskylä, Finland	[32]
3	Palermo, Southern Italy	[33]
4	Bologna, Northern Italy	[33]
5	Copenhagen, Denmark	[34]
6	Lisbon, Southern Portugal	[35]
7	Bragança, Northern Portugal	[35]
8	Ljubljana, Slovenia	[36]
9	Warsaw, Poland	[37]
10	Oslo, Norway	[38]
11	Bregenz, Austria	[26]

Table 8. Study assumptions.

Indicator	Assumption
Scope of application	New buildings
Building type	A single-household residential home (100 m^2)
Climatic data	Vilnius, Jyväskylä, Palermo, Bologna, Copenhagen, Bragança, Lisbon, Ljubljana, Warsaw, Oslo, Bregenz
Established parameters	Envelopes (walls, floors, windows, roof, partitions), air flow, mechanical ventilation, heating, cooling, hot water, building usage mode, electrical energy consumption
Survey of energy indicators	Computer modelling using IDA-ICE software
Indicators analysed	Primary energy for heating and cooling, CO_2 emissions

Table 9. Primary energy factors (PEF) and CO_2 emission factors (EF) in subject countries.

Type of Energy	Country	PEF	kg CO_2/kWh	Comment
Central heating	Lithuania	0.62	0.100	Heat from heating systems, Lithuanian average
	Finland	0.50	0.195	EF average for Finland's central systems
	Italy	1.50	0.500	
	Denmark	0.60	0.260	
	Portugal	1.00	0.200	
	Slovenia	1.20	0.371	
	Poland	0.80	0.270	Central heating in Warsaw comes from a combined heat and power plant (fueled with coal)
	Norway	0.11	0.019	The CO_2 EF value was chosen with reference to the central heating technology in place in Oslo (based on wood pellet boilers and biofuel)
	Austria	0.19	0.200	
Central cooling	Finland	0.28	0.027	
	Italy	0.50	0.270	
Electricity	Lithuania	2.50	0.420	An average for different methods to generate electricity
	Finland	1.20	0.141	
	Italy	2.42	0.410	
	Denmark	1.80	0.420	
	Portugal	2.50	0.144	
	Slovenia	2.50	0.353	
	Poland	3.00	1.190	
	Norway	1.79	0.0022	
	Austria	1.91	0.176	
Natural gas	Lithuania	1.10	0.220	
	Finland	1.34	0.199	
	Italy	1.05	0.3696	
	Denmark	1.00	0.220	
	Portugal	1.00	0.202	
	Slovenia	1.20	0.200	
	Poland	1.20	0.201	
	Norway	1.244	0.203	
	Austria	0.19	0.200	

2.3. *Climatic Data and Potential Climate Change Effect*

Climate in Europe can be described as a mid-latitude climate with the following climate zones: Mediterranean, oceanic, continental, and Nordic [30]. Climate zones are defined on the basis of heating degree days (HDD) or cooling degree days (CDD). A heating/cooling degree day is a unit of measure for quantitative evaluation of the demand for heating/cooling in a building. A map of heating and cooling degree days in Europe is shown in Figure 4.

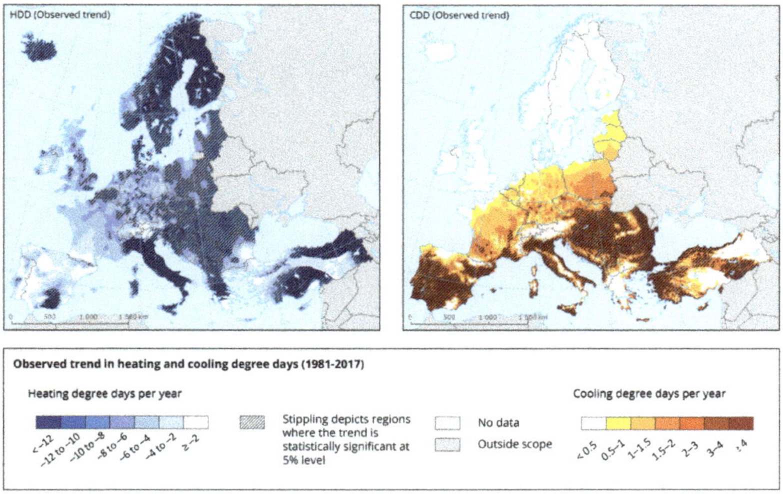

Figure 4. Heating and cooling degree days in Europe [39].

The climatic data used for the purposes of the modeling were obtained from the ASHRAE IWEC2 database using IDA-ICE software. ASHRAE stands for the American Society of Heating, Refrigerating, and Air-Conditioning Engineers. This society has initiated a collection of typical climatic data across over 3000 locations worldwide, to be used in energy calculation and modeling programs. Every climatic data package contains hourly climatic data for the selected location: the outside air temperature, relative humidity, wind speed, solar irradiance.

3. Results of the Analysis of Primary Energy for Building Heating and Cooling

Following the analysis of the differences in heat transfer ratios applicable to efficient buildings covered in this chapter, the study proceeds with four separate cases and produces an appropriate set of results for each of them. A summary of these phases is shown in Table 10.

Table 10. Summary of phases of the analysis of the study.

Case No.	Analysis Details
Case 1	The energy indicators of an energy-efficient building in the European states covered by the analysis (thermal energy consumption for heating and cooling; primary energy for heating and cooling; primary energy consumption in the building when all of the subject countries use a typical source of energy)
Case 2	The energy indicators of an energy-efficient building in the European states covered by the analysis (thermal energy consumption for heating and cooling; primary energy for heating and cooling; primary energy consumption in the building when all of the subject countries use the same source of energy (natural gas for heating, electricity for cooling (the average of different ways to generate electricity); identification of CO_2 emissions)
Case 3	The energy indicators of a Lithuanian building, energy class A+, in the European states covered by the analysis (thermal energy consumption for heating and cooling; primary energy for heating and cooling; primary energy consumption in the building when all of the subject countries use the same source of energy (natural gas for heating, electricity for cooling (the average of different ways to generate electricity); identification of CO_2 emissions)
Case 4	The energy indicators of a residential building conforming to the regulations for new buildings in the European states covered by the analysis (Italy, Finland, Austria, Denmark, Portugal, Slovenia, Poland, Norway) under Lithuanian climatic conditions (thermal energy consumption for heating and cooling; primary energy for heating and cooling; identification of CO_2 emissions)

3.1. Comparison of Heat Transfer Ratios Applicable to Energy-Efficient Buildings

To be able to highlight the key differences in the regulations governing energy-efficient buildings in the states covered by the analysis better, a comparison of heat transfer coefficients (of walls and windows by choice) is carried out, its results are presented in Figure 5.

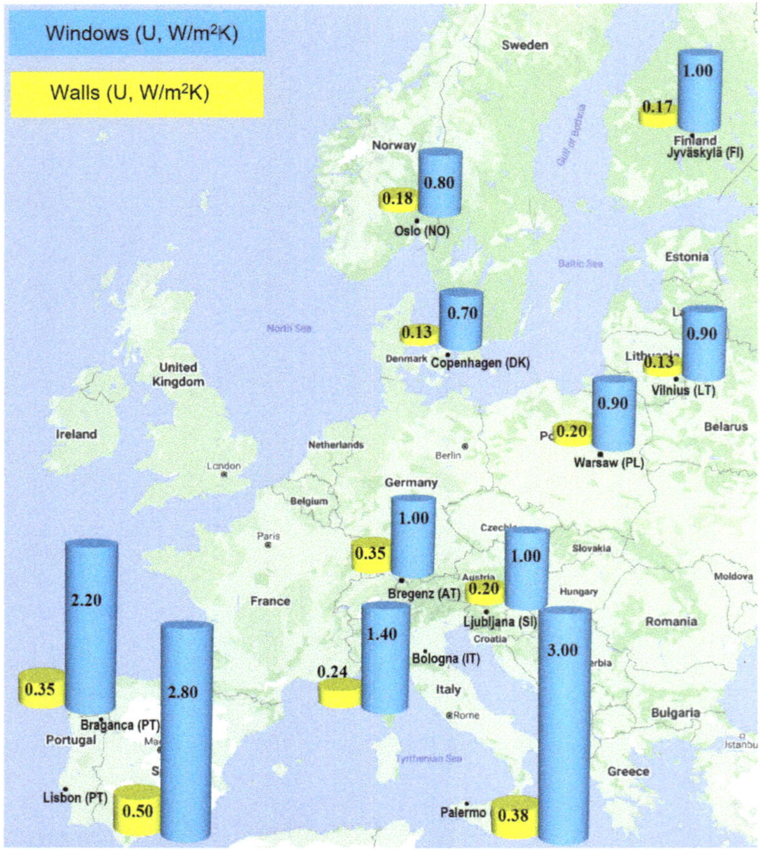

Figure 5. Heat transfer coefficient values of the exterior walls and windows of a building in the subject countries.

To achieve a set of standards on a par with an energy-efficient building with economically optimal measures, the focus must be placed first and foremost on different climatic conditions: insulation and airtightness of buildings are a must in colder climates, whereas high energy class equipment and efficient lighting are the number one measures to reduce the building's energy consumption in warmer European climate zones [30].

Figure 5 shows that the tightest requirements in relation to the values of building envelope heat transfer coefficients apply in countries with colder climates, such as Lithuania, Denmark, Finland, and Norway. In Denmark, the value of heat transfer coefficients in a class B2020 building is comparable with that which applies to a class A+ energy-efficient building in Lithuania. Higher heat transfer coefficient values for exterior walls and ceilings apply in southern and central Europe—Italy, Portugal, and Austria.

3.2. Case 1. Typical Source of Energy

The building model primary energy demands for heating and cooling in different European states using the typical sources of energy in those states and considering the different factors of primary energy conversion are presented first (Figure 6). The results of primary energy for heating and cooling of the calibrated model depending on the heat source LT A+ (similar to nZEB) building are 40–67 kWh/m^2/year (Figures 6 and 7), which shows that the model is suitable considering the requirements and recommendations (Figure 1).

The data presented in Figure 6 show that the highest level of consumption of primary energy for building heating and cooling purposes exists in Ljubljana, Slovenia, Warsaw, Poland, and Bragança, Portugal. Whereas the lowest aggregate primary energy demand for heating and cooling was obtained in buildings that conform to the national requirements for energy-efficient buildings that apply in Austria and Norway. It is important to mention here that the amount of primary energy used for building heating and cooling purposes in Denmark is similar to that in Lithuania (4.2 MWh and 4.0 MWh, appropriately). A building designed for the Lithuanian climate conforms to energy class A+, and for the Danish climate, to the B2020 label [34].

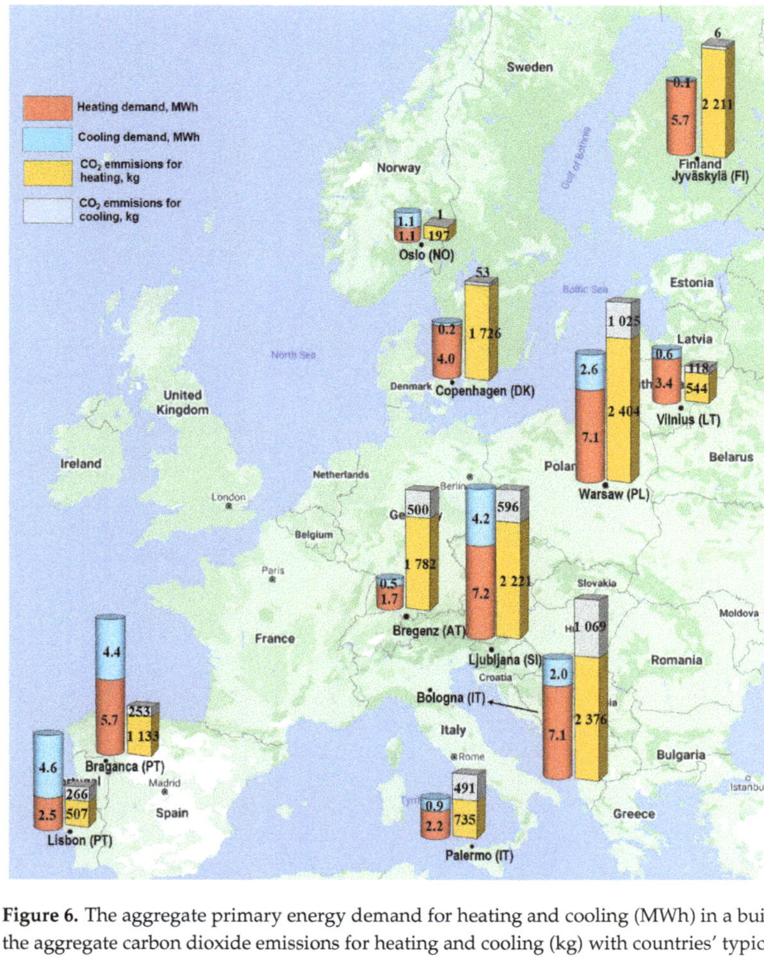

Figure 6. The aggregate primary energy demand for heating and cooling (MWh) in a building and the aggregate carbon dioxide emissions for heating and cooling (kg) with countries' typical sources of energy.

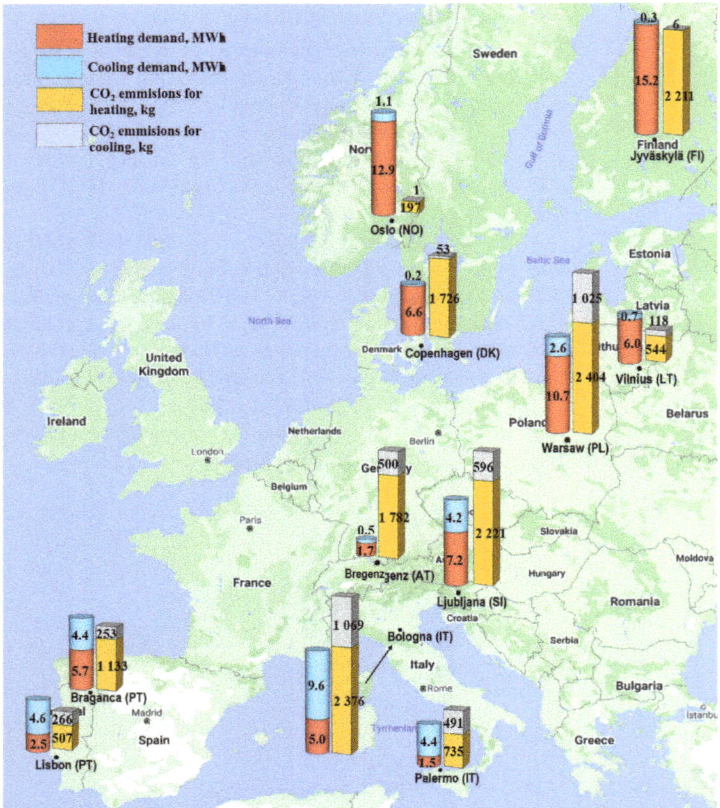

Figure 7. The aggregate primary energy demand for heating and cooling (MWh) in a building and the aggregate carbon dioxide emissions for heating and cooling (kg) using the same source of energy (natural gas for heating and electricity from RES and NER for cooling).

A building designed for the cold climate of Jyväskylä, Finland with its chilly winters and temperatures that drop to −27.9 °C (Tables 3 and 5) will consume 50% less primary energy compared to a building that exists in Ljubljana, Slovenia, and, for instance, 37% less primary energy compared to a building in Bologna, Italy. In Bologna, central heating is provided by a CHP plant (fueled with methane gas) [40], while the main fuel used in the central heating systems of Finland consists of peat and wood [41].

3.3. Case 2. The Same Source of Energy

The results shown in Figure 7 indicate the amount of primary energy (MWh) that would be required to satisfy the energy demands of a building if all models of the building use the same type of fuel (natural gas for heating and electricity from renewable energy sources (RES) and non-renewable energy resources (NER) for cooling).

The data shown in Figure 7 indicate that the highest demand for primary energy for heating and cooling with natural gas and electricity (from RES and NER) exists in a Finnish residential building. Compared to the original option (where the Finnish central systems use biofuel for heating and cooling purposes), the building's demand for primary energy goes up by 63%. In Lithuania, when the primary energy demands of a building are satisfied with natural gas, this gap stands at 40%. The largest difference can be observed in Norway: if Oslo used natural gas and electricity from NER instead of biofuel for heating and supply of electrical energy, the demand for primary energy would increase by 84%.

3.4. Case 3. Lithuania in Analysed European Countries

This analysis aims to determine the thermal consumption for heating and cooling in an energy class A+ building and the demands for primary energy in the European states covered by the analysis. To that end, the building model "Vilnius, Lithuania", its envelope heat transfer ratio values conforming to the values of energy class A+, is built on the bases of the climatic data of the subject states (for instance, the model "class A+ building in Palermo" shows results obtained by modeling an energy class A+ building with the climatic data for the city of Palermo, stating in the model that the building is located in the city of Palermo).

Figure 8 shows the aggregate demand for primary energy to secure the heating and cooling needs of an energy class A+ building in the subject countries. The same figure also shows the aggregate carbon dioxide emissions. The primary energy and the CO_2 emissions are calculated with PEF and EF values applied in the countries covered by the analysis (for instance, with the model "class A+ building in Palermo", primary energy and CO_2 emissions are calculated using PEF and EF as applicable in Italy).

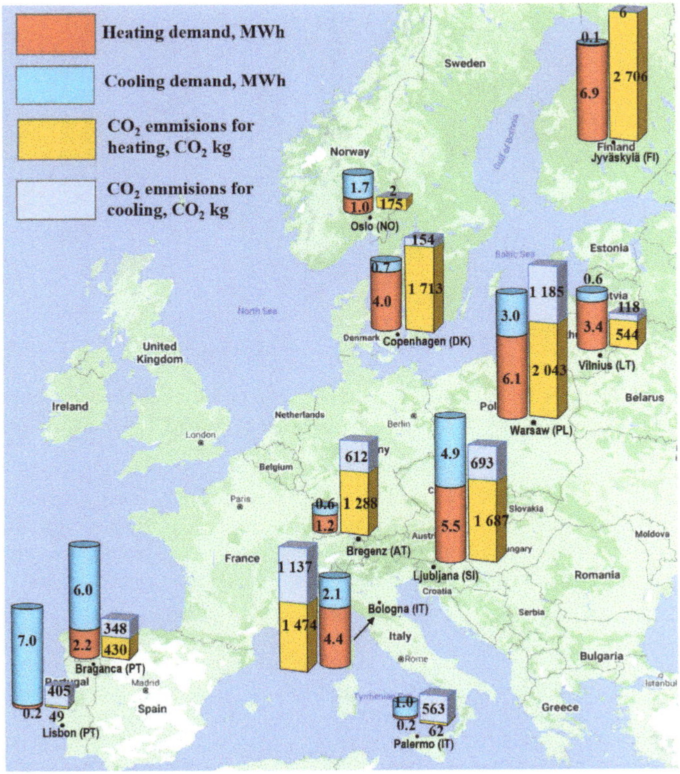

Figure 8. The aggregate primary energy demand for heating and cooling (MWh) in an energy class A+ building and the aggregate carbon dioxide emissions for heating and cooling (kg) in different European climate zones.

The data in Figure 8 show that the highest demand for primary energy in an energy class A+ building would exist if the building were located in typical central European climate in Ljubljana, Slovenia, where said demand would be about 61% more than if the building were located in Lithuania. The largest CO_2 emissions would occur if the building model in question were located in Warsaw, Poland, surpassing the CO_2 emissions in Lithuania by 79%, the smallest, in Lisbon, Portugal, falling 39% below the Lithuanian level.

3.5. Case 4. Analyzed European Countries' Buildings in Lithuania Climate

In this case, the modeling results covered by the analysis were obtained by modeling buildings in the subject countries with the climatic data for the city of Vilnius. The aim here is to identify the demands for primary energy if the building models of the subject countries were located in Vilnius and exposed to the Lithuanian climate.

Figure 9 shows the total demand for primary energy for heating and cooling in buildings of the European states covered by the analysis, as well as the total amount of carbon dioxide emissions under prevalent Lithuanian climatic conditions. The primary energy and the CO_2 emissions were calculated with the PEF and EF values that apply in Lithuania (heating from heating systems, the Lithuanian average, and the average of different methods to generate electricity; see Table 9 for details).

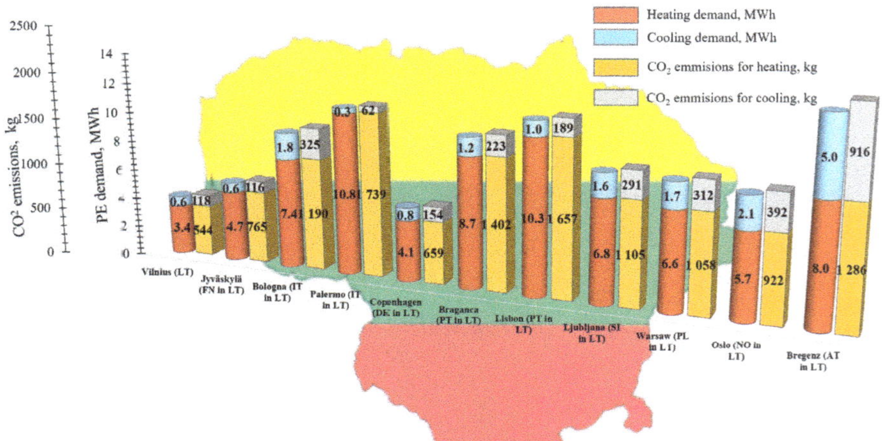

Figure 9. The aggregate primary energy demand for heating and cooling (MWh) in a building and the aggregate carbon dioxide emissions for heating and cooling (kg) in Lithuanian climatic conditions.

So, the data in Figure 9 indicate that under prevalent Lithuanian climatic conditions, the highest level of consumption of primary energy would exist in a building that matches the Austrian residential house model. With the Lithuanian climatic conditions prevalent, this building would consume 70% primary energy more compared to the Austrian consumption level (yet its CO_2 emissions would drop by 7%).

3.6. Discussion

From 2016 to 2022, Lithuania underwent significant changes in the regulation of energy efficiency in buildings. This has led to a reduction in the U-values of buildings and an increase in the use of renewable energy to meet energy needs. The study aimed to analyze what these changes look like at the European level. It is obvious that the tightest requirements for the heat transfer of building envelopes (exterior walls and windows) exist in northern Europe (such as Denmark and Lithuania), and the highest permissible heat transfer coefficients are in southern countries (such as Portugal and Italy).

Furthermore, it should be emphasized that, in determining the primary energy of electricity (which is very important when using heat pumps), the fuels used for production are often difficult to define due to common electricity networks. Therefore, it is most appropriate to compare the heating and cooling demands "up to" the heat source. This study did not address energy efficiency alternatives and the inclusion of additional RES to increase the sustainability of the building, so in the future it is planned to examine the energy performance of energy buildings in a broader perspective, taking into account not only climatic differences but also applied technological alternatives and architectural solutions.

Finally, going back to the beginning, not all buildings in the countries surveyed meet the required, recommended primary energy consumption values for heating and cooling. The purpose of this study was to compare and illustrate these differences and to highlight that good practice can be shared.

4. Conclusions

Four analyses with different viewpoints were performed in the study. This included such cases as the single-family building in subject countries used a typical source of energy; the same source of energy; Lithuanian building of A+ energy class is placed in other countries, and the simulated buildings are in Lithuanian climate. The outcomes of the simulations were as follows:

- The results of the different analyses performed within the scope of this study have shown the gap in the demand for primary energy in a building located in Denmark and one located in Lithuania to be the smallest, standing at about 9%. The biggest difference can be seen between the energy indicators of a high energy class building in Palermo and those of an energy class A+ building: if a Palermo residential building was operated in Lithuania, its primary energy demand for heating and cooling would surge by 88% compared to a building located in Palermo. The annual volume of thermal energy consumed for heating is the highest in the building model that mirrors the Finnish energy-efficient building. An energy class A+ building based in Lithuania requires 52% less thermal energy to satisfy its heating demands per year compared to Finland. The lowest value of thermal energy consumption for cooling is observed in a class B2020 building located in Copenhagen, Denmark. Compared to Lithuania, this building will use 55% less thermal energy for its cooling needs than an energy class A+ building in Lithuania.
- The largest amount of primary energy and CO_2 emissions to satisfy the heating and cooling demands of a building is consumed in Ljubljana, Slovenia (11 MWh), Warsaw, Poland (10 MWh), and Bragança, Portugal (9 MWh), whereas the lowest demand for primary energy and CO_2 emissions for heating and cooling was observed in buildings that conform to the Austrian and Norwegian national requirements for energy-efficient buildings, standing at about 2.0 MWh of primary energy.
- The highest demand for primary energy and CO_2 emissions for heating and cooling with natural gas and electricity (from RES and NER) exists in a residential building in Finland. Compared to the original option (where the Finnish central systems use biofuel for heating and cooling), the building's demand for primary energy rises by 63%. In Lithuania, if the demand for primary energy in a building is met with natural gas, this difference stands at 40%. The lowest degree of primary energy consumption with natural gas and electricity (from RES and NER) would exist in an Austria-based building: 2.2 MWh.
- Under prevalent Lithuanian climatic conditions, the highest level of primary energy consumption and CO_2 emissions would exist in a building that corresponds to the Austrian residential house model. Under such conditions, the building would consume nearly 70% more primary energy than under ordinary conditions, as they exist in Austria. In Lithuania, the smallest amount of primary energy under prevalent Lithuanian climatic conditions would be consumed by a Lithuanian energy class A+ building: 4 MWh of primary energy.

Author Contributions: Conceptualization, K.V. and M.G.; methodology, M.G.; software, M.G.; validation, K.V.; formal analysis, G.S.; investigation, M.G.; data curation, G.S.; writing—original draft preparation, M.G., K.V. and G.S.; writing—review and editing, K.V. and G.S.; visualization, G.S.; supervision, K.V. All authors have read and agreed to the published version of the manuscript.

Funding: This research received no external funding.

Institutional Review Board Statement: Not applicable.

Informed Consent Statement: Not applicable.

Data Availability Statement: Not applicable.

Conflicts of Interest: The authors declare no conflict of interest.

Nomenclature

U	overall heat transfer coefficient, W/(m^{2}K)
Abbreviations	
BPIE	Building Performance Institute Europe
CDD	Cooling Degree Days
EC	European Commission
EF	Emission Factor
EU	European Union
GHG	Greenhouse Gas
HDD	Heating Degree Days
NER	Non-renewable Energy Resources
NZEB	Nearly Zero-Energy Building
PE	Primary Energy
PEF	Primary Energy Factor
RES	Renewable Energy Sources
RPC	Representative Concentration Pathway

References

1. Chel, A.; Kaushik, G. Renewable energy technologies for sustainable development of energy efficient building. *Alex. Eng. J.* **2018**, *57*, 655–669. [CrossRef]
2. Allouhi, A.; El Fouih, Y.; Kousksou, T.; Jamil, A.; Zeraouli, Y.; Mourad, Y. Energy consumption and efficiency in buildings: Current status and future trends. *J. Clean. Prod.* **2015**, *109*, 118–130. [CrossRef]
3. Cao, X.; Dai, X.; Liu, J. Building energy-consumption status worldwide and the state-of-the-art technologies for zero-energy buildings during the past decade. *Energy Build.* **2016**, *128*, 198–213. [CrossRef]
4. Berardi, U.; Jafarpur, P. Assessing the impact of climate change on building heating and cooling energy demand in Canada. *Renew. Sustain. Energy Rev.* **2020**, *121*, 109681. [CrossRef]
5. Pachauri, R.K.; Allen, M.R.; Barros, V.R.; Broome, J.; Cramer, W.; Christ, R. *Climate Change 2014: Synthesis Report. Contribution of Working Groups I, II and III to the Fifth Assessment Report of the Intergovernmental Panel on Climate Change*; IPCC: Geneva, Switzerland, 2014.
6. Larsen, M.A.D.; Petrović, S.; Radoszynski, A.M.; McKenna, R.; Balyk, O. Climate change impacts on trends and extremes in future heating and cooling demands over Europe. *Energy Build.* **2020**, *226*, 110397. [CrossRef]
7. *Directive 2018/844 of the European Parliament and of the Council of 30 May 2018 amending Directive 2010/31/EU on the Energy Performance of Buildings and Directive 2012/27/EU on Energy Efficiency, 2018*; European Parliament: Strasbourg, France, 2018.
8. Tanasa, C.; Dan, D.; Becchio, C.; Corgnati, S.P.; Stoian, V. Cost-optimal and indoor environmental quality assessment for residential buildings towards EU long-term climate targets. *Energy Sustain. Dev.* **2020**, *59*, 49–61. [CrossRef]
9. A Roadmap for Moving to a Competitive Low Carbon Economy in 2050. In *Communication from the Commission to the European Parliament, the Council, the European Economic and Social Committee and the Committee of the Regions*; European Commission: Brussels, Belgium, 2011.
10. Andaloro, A.P.F.; Salomone, R.; Ioppolo, G.; Andaloro, L. Energy certification of buildings: A comparative analysis of progress towards implementation in European countries. *Energy Policy* **2010**, *38*, 5840–5866. [CrossRef]
11. D'Agostino, D.; Mazzarella, L. What is a Nearly zero energy building? Overview, implementation and comparison of definitions. *J. Build. Eng.* **2019**, *21*, 200–212. [CrossRef]
12. BPIE. Nearly Zero: A Review of EU Member State Implementation of New Build Requirements. 2021. Available online: https://www.bpie.eu/wp-content/uploads/2021/06/Nearly-zero_EU-Member-State-Review-062021_Final.pdf.pdf (accessed on 1 April 2022).
13. *Guidelines for the Promotion of Nearly Zero-Energy Buildings and Best Practices to Ensure that, by 2020, all New Buildings are Nearly Zero-Energy Buildings, C/2016/4392*; Publications Office of the European Union: Brussels, Belgium, 2016.
14. IEA. Available online: https://www.iea.org/reports/tracking-buildings-2020 (accessed on 1 April 2022).
15. Nakicenovic, N.; Alcamo, J.; Davis, G.; Vries, B.D.; Fenhann, J.; Gaffin, S. Special Report on Emissions Scenarios. 2000. Available online: https://escholarship.org/content/qt9sz5p22f/qt9sz5p22f.pdf (accessed on 1 April 2022).
16. Olonscheck, M.; Holsten, A.; Kropp, J.P. Heating and cooling energy demand and related emissions of the German residential building stock under climate change. *Energy Policy* **2011**, *39*, 4795–4806. [CrossRef]

17. Swedish Meteorological and Hydrological Institute Klimatscenarier (Climate Scenarios). Available online: https://www.smhi.se/en/climate/future-climate/advanced-climate-change-scenario-service/met/sverige/medeltemperatur/rcp45/2071-2100/year/anom (accessed on 1 April 2022).
18. Dodoo, A.; Gustavsson, L.; Bonakdar, F. Effects of future climate change scenarios on overheating risk and primary energy use for Swedish residential buildings. *Energy Procedia* **2014**, *61*, 1179–1182. [CrossRef]
19. Meehl, G.A.; Covey, C.; Delworth, T.; Latif, M.; McAvaney, B.; Mitchell, J.F. The WCRP CMIP3 multimodel dataset: A new era in climate change research. *Bull. Am. Meteorol. Soc.* **2007**, *88*, 1383–1394. [CrossRef]
20. Jylhä, K.; Jokisalo, J.; Ruosteenoja, K.; Pilli-Sihvola, K.; Kalamees, T.; Seitola, T.; Mäkelä, H.M.; Hyvönen, R.; Laapas, M.; Drebs, A. Energy demand for the heating and cooling of residential houses in Finland in a changing climate. *Energy Build.* **2015**, *99*, 104–116. [CrossRef]
21. Frank, T. Climate change impacts on building heating and cooling energy demand in Switzerland. *Energy Build.* **2005**, *37*, 1175–1185. [CrossRef]
22. Asimakopoulos, D.A.; Santamouris, M.; Farrou, I.; Laskari, M.; Saliari, M.; Zanis, G.; Giannakidis, G.; Tigas, K.; Kapsomenakis, J.; Douvis, C.; et al. Modelling the energy demand projection of the building sector in Greece in the 21st century. *Energy Build.* **2012**, *49*, 488–498. [CrossRef]
23. Isaac, M.; van Vuuren, D.P. Modeling global residential sector energy demand for heating and air conditioning in the context of climate change. *Energy Policy* **2009**, *37*, 507–521. [CrossRef]
24. Green Building Council Italia. Energy Efficiency of Buildings in Italy. 2019. Available online: https://c2e2.unepdtu.org/kms_object/green-building-council-italia-gbc-italia/ (accessed on 20 March 2022).
25. STR 2.01.02:2016 Pastatų Energinio Naudingumo Projektavimas ir Sertifikavimas 2019. Available online: https://e-seimas.lrs.lt/portal/legalAct/lt/TAD/15767120a80711e68987e8320e9a5185/sEHkqghNMX (accessed on 20 March 2022).
26. OIB Guidline Energy Conservation And Thermal Protection. *Cost Optimality OIB-330. 6-005/18-001*; Austrian Institute of Construction Engineering: Vienna, Austria, 2018.
27. Laustsen, J.; Warming, N.B. New Building Codes for NZEB. Key Elements and Overall Picture in EU Countries. Joint Workshop Presentations. Session 7, NZEB Building Code. 2019. Available online: https://www.ca-res.eu/fileadmin/cares/PublicArea/Joint_workshop_presentations/Session_7_NZEB_Building_Codes.pdf (accessed on 1 April 2022).
28. Kurnitski, J. NZEB requirements in Nordic countries. *REHVA Eur. HVAC J.* **2019**, *56*, 8–12.
29. Sartori, I.; Løtveit, S.V.; Skeie, K. Guidelines on Energy System Analysis and Cost Optimality in Early Design of ZEB. 2018. Available online: https://sintef.brage.unit.no/sintef-xmlui/bitstream/handle/11250/2498764/ZEB%2bprreport%2bno%2b41_Nett_ZEB%2bpr%2breport%2bno%2b41.pdf?sequence=2&isAllowed=y (accessed on 1 April 2022).
30. D'Agostino, D.; Parker, D. A framework for the cost-optimal design of nearly zero energy buildings (NZEBs) in representative climates across Europe. *Energy* **2018**, *149*, 814–829. [CrossRef]
31. Soleimani-Mohseni, M.; Nair, G.; Hasselrot, R. Energy simulation for a high-rise building using IDA ICE: Investigations in different climates. *Build. Simul.* **2016**, *9*, 629–640. [CrossRef]
32. National Building Code of Finland Energy Efficiency 1010/2017 Decree of the Ministry of the Environment on the Energy Performance of New Buildings. 2018. Available online: https://www.ym.fi/download/noname/%7BE12CDE2C-9C2B-4B84-8C81-851349E2880B%7D/140297. (accessed on 20 March 2022).
33. Italian Republic Decree of the Italian Ministry of Economic Development 26 June 2015 "Applicazione delle metodologie di calcolo delle prestazioni energetiche e definizione delle prescrizioni e dei requisiti minimi degli edifici" (Iin Italian) 2015. Available online: https://www.ambientesicurezzaweb.it/wp-content/uploads/sites/5/2015/07/Calcolo-completo.pdf (accessed on 20 March 2022).
34. Danish Building Research Institute Cost Optimal Levels of Minimum Energy Performance Requirements in the Danish Building Regulations. 2018. Available online: https://ens.dk/sites/ens.dk/files/Energibesparelser/bilag_3a_cost-optimal_levels_of_minimum_energy_performance_requirements_in_the_danish_building_regulations_-_sbi_2018.pdf (accessed on 20 March 2022).
35. Portaria n.° 379-A/2015. Ministry of Environment, Spatial Planning and Energy, Official Gazette No. 207/2015, 2nd Supplement, Series I of 2015-10-22. Available online: https://data.dre.pt/eli/port/379-a/2015/10/22/p/dre/pt/html (accessed on 20 March 2022).
36. Energetski zakon Uradni list RS, št. 60/19–Uradno Prečiščeno Besedilo). Energy Act (Uradni list RS, št. 17/14 in 81/15). Available online: http://www.pisrs.si/Pis.web/pregledPredpisa?id=ZAKO6665 (accessed on 20 March 2022).
37. The Energy Performance of Buildings Law The Act of 29 August 2014. Available online: http://prawo.sejm.gov.pl/isap.nsf/DocDetails.xsp?id=WDU20140001200 (accessed on 20 March 2022).
38. Norwegian Building Regulations Regulations on Technical Requirements for Construction Works, "Forskrift om tekniske krav til byggverk (Byggteknisk forskrift-TEK17)". 2017. Available online: https://dibk.no/globalassets/byggeregler/regulation-on-technical-requirements-for-construction-works--technical-regulations.pdf (accessed on 20 March 2022).
39. EEA European Environment Agency. Heating and Cooling Degree Days in Europe. Available online: https://www.eea.europa.eu/data-and-maps/figures/trend-in-heating-and-cooling-1 (accessed on 30 March 2022).

40. Cefla Engineering Hera District Heating Project. Available online: https://www.ceflaengineering.com/en/project/energy-hera-district-heating-renewable-sources (accessed on 20 March 2022).
41. Power Engineering Internationa District Heat Solution for Finland CHP Plants. Available online: www.powerengineeringint.com/decentralized-energy/district-energy/district-heat-solution-for-finland-chp-plants (accessed on 20 March 2022).

energies

MDPI

Article

Impact of Window Frames on Annual Energy Consumption of Residential Buildings and Its Contribution to CO_2 Emission Reductions at the City Scale

Younhee Choi [1], Akihito Ozaki [1] and Haksung Lee [2,*]

[1] Faculty of Human-Environment Studies, Kyushu University, 744 Motooka Nishi-ku, Fukuoka 819-0395, Japan; choi.youn.hee.692@m.kyushu-u.ac.jp (Y.C.); ozaki@arch.kyushu-u.ac.jp (A.O.)
[2] Department of Architecture, Chungbuk National University, 1 Chungdae-ro, Cheongju 28644, Korea
* Correspondence: leehaksung@chungbuk.ac.kr; Tel.: +82-43-261-2435; Fax: +82-43-261-2634

Abstract: Windows are among building components that have the strongest effect on thermal load. They play a significant role in heat loss in buildings because they usually have a largely higher thermal conductance than other components of the building envelope. Although many studies have highlighted the relevance of heat transfer through frames and aimed to improve their thermal performance, poorly insulated aluminum frames (thermal conductivity of aluminum is 160 W/m·K, while that of polyvinyl chloride [PVC] is 0.17 W/m·K) are still in use in Japan. Therefore, the U-values of different window frames were calculated, and annual thermal loads were calculated according to the window configurations, including the frame, glazing, and cavity. We focused on standard residential buildings in Japan with a total floor area of 120.6 m^2 (two-story building), and the number of newly built houses and the application rate of window configurations in 2019 were surveyed to estimate the CO_2 emissions by regions. CO_2 emissions were reduced by approximately 3.98–6.58% with the application of PVC frames. Furthermore, CO_2 emissions were converted into the amount of CO_2 gas absorbed by cedar trees, which cover nearly 18% of the total land area of Japan. In conclusion, analogous to the amount of CO_2 gas absorbed by cedar trees, the absorption effect was equivalent to 327,743–564,416 cedar trees. Changing the window frame material can facilitate a significant energy-saving effect as a considerable amount of energy is saved, especially at a city scale.

Keywords: window frame; thermal transmittance; CO_2 emissions; energy consumption; city scale

Citation: Choi, Y.; Ozaki, A.; Lee, H. Impact of Window Frames on Annual Energy Consumption of Residential Buildings and Its Contribution to CO_2 Emission Reductions at the City Scale. *Energies* **2022**, *15*, 3692. https://doi.org/10.3390/en15103692

Academic Editor: Luca Gonsalvi

Received: 15 April 2022
Accepted: 17 May 2022
Published: 18 May 2022

1. Introduction

A recent publication by the Intergovernmental Panel on Climate Change [1] indicated that the largest growth in carbon emissions originates from electricity generation, transport, industry, and building operations. The building sector accounts for approximately 40% of the total energy consumption in many countries; this indicates the importance of minimizing energy consumption in this industry [2,3].

Windows are typically responsible for a large fraction of heat loss in buildings because of the large differences between their thermal transmittance values (U-value) and other building components [3–6]. However, windows can contribute to heating via solar energy transmitted through glazing. Window frames cover a relatively small fraction of the entire building's surface; however, they can be responsible for a major part of heat loss depending on their insulation properties (compared to those of other elements of the building envelope) and solar transmittance [5,6]. However, although significant improvements have been made in designing highly insulating window frames [3,6–10], poorly insulated aluminum frames are still in use in many regions worldwide [11,12]. The thermal conductivity of aluminum is 160 W/m·K, while that of polyvinyl chloride (PVC) is 0.17 W/m·K.

In Japan, as of 2020, the application frequencies of relatively poorly insulated aluminum frames and composite frames (polyvinyl chloride with aluminum) were 10.1%

and 67.5%, respectively [13]. The installation rate of PVC frames was 10% in 2013 and increased to 16.5% and 22.5% in 2016 and 2020, respectively [13,14]. This implies that PVC frames have been applied rapidly in the Japanese construction industry in recent years. However, given that PVC frames are applied to more than 50% of buildings in America and Europe [15], the insulation level of window frames in Japan remains low, which attracted our attention. In many countries, including Japan, the thermal performance of window frames can be significantly improved, which warrants additional research. Therefore, it is necessary to quantify the effect of significant efforts to increase the thermal performance of window configurations—an ongoing trend in Japan and around the world—on the annual energy consumption and CO_2 emissions in the building sector. In addition, to evaluate the effect of improving the thermal performance of window frames nationwide, the energy-saving effect of window configurations should be quantified at the city scale and not at the scale of single residential buildings.

Although many studies have highlighted the relevance of heat transfer through frames with the aim of improving the thermal performance of frames, research on the impact of a typical frame material (e.g., aluminum frame, PVC frame, and aluminum–PVC composite frame) on the overall thermal performance of residential buildings or its impact on the energy consumption reduction rate at the city scale is lacking. Few studies have evaluated the impact of typical frames on the building's energy performance [16,17]. Certain investigations have focused on using low-conductivity materials to reduce the thermal transmittance of the frame and minimize the additional heat loss caused by glazing and spacer systems [15]. Appelfeld et al. [6] used three novel designs of thin glass fiber-reinforced polyester frames to calculate the impact of replacing aluminum frames without thermal breaks on the energy consumption of an office building. They found that the optimal design could reduce total energy consumption by 6.5 kWh/m^2 per year (almost 20%). However, the window design proposed in their study included the lately developed uncommon frame. Moreover, existing investigations have focused on single buildings; the embodied impacts of typical window frames on the annual energy consumption at the city scale have been overlooked.

The objectives of this study were (1) to identify the impact of differences in the frame material on the heating and cooling energy consumption of residential buildings, (2) quantify the energy-saving impact of the application of highly insulated frames for new residential buildings in various regions in Japan at the city scale, and (3) discuss the feasibility of national countermeasures to minimize the CO_2 emission rate by improving the thermal insulation level of window frames.

This paper is structured as follows. In Section 2, the methodology and research process for analyzing the impact of the frame material on the annual energy consumption, heating and cooling, and CO_2 emission rates are presented. In Section 3, the simulation configurations used to determine the U-values of the frames and the simulated results are presented. In Section 4, the impact of the window frame on the building's annual energy consumption and the energy conversion to CO_2 emissions at the city scale are analyzed. Finally, Section 5 summarizes the key findings of this study.

2. Methodology

To achieve the goals of sustainability, a multi-disciplinary approach covering a number of features in the whole life cycle assessment phase, which includes the manufacturing phase, use phase, and demolition phase, is required [18]. However, energy consumption on operating accounts for approximately 80–90% of total life cycle energy [19]. Therefore, in this study, energy consumption during the operation stage due to different window frames have been discussed. The thermal properties of window frames based on their material and effect on annual energy consumption and CO_2 emissions at the city scale were calculated in three steps: (1) calculation of the U-value of the frame by material, (2) quantification of the annual energy consumption with the application of different window frames, and (3) conversion of the effect of different window frames on energy consumption and CO_2

emission rate into the city scale. The three steps of the research methodology are explained in the following subsections.

2.1. Step 1: U-Value of the Frame

In the first step, we calculated the U-values of the frames based on the material. U-values were obtained using a simulation program called THERB for HAM [20], which was originally an unsteady computational simulation tool for evaluating the thermal environment of buildings. This software has been validated through standardized tests in Japan, such as the building energy simulation test procedure [21]. For the user-selectable option, THERB for HAM can simulate highly precise heat transfer through the building envelope using two-dimensional analysis based on the finite element method under unsteady-state conditions. The U-values of the frames were calculated using the method recommended in ISO 10077-2 [22]. The method is based on using a highly insulated panel that substitutes glazing and eliminates the effect of the thermal bridge using the glazing spacer [6,22].

2.2. Step 2: Impact of Window Frame on Building Thermal Load, Energy Consumption, and CO_2 Emissions

Based on the U-value of the frame calculated in Step 1, the effect of the material change in the frame on the annual building thermal load was evaluated. THERB for HAM was used to evaluate the thermal load of the target buildings.

The annual thermal load of the target residential building was calculated and converted into energy consumption in terms of electricity and CO_2 emission rate to determine the energy-saving effect based on the window configurations. The annual performance factor (APF) was used to convert the thermal load into energy consumption. The power consumption was calculated by dividing the thermal load by APF. APF was set at 5.8, which is the median value of the basic statistics on the sales performance of home appliances [23].

The CO_2 emission rate can be calculated by multiplying the power consumption by the CO_2 emission factor. For the CO_2 emission factor, the values of the local power company by region were sourced from the Ministry of the Environment [24].

2.3. Step 3: Impact of Window Frame on CO_2 Emissions at City Scale

To quantify the CO_2 emission rates according to the application of window combinations (including glazing, cavity, and frame) by region, data analysis was performed on the application rate of window combinations according to region and the number of newly built houses.

According to energy conservation standards [23], Japan is divided into eight regions. In this study, to analyze the effect of window frame materials on energy saving at the city scale, seven regions with thermal insulation regulations were selected [25]. One city was selected from each of the seven regions; Asahikawa, Sapporo, Morioka, Nagano, Toyama, Tokyo, and Fukuoka were selected as sample cities in regions 1 to 7, respectively.

Table 1 shows the frequency of window frame applications according to region [13] and the number of newly built households (last column) [26] in 2019. Frames can be fabricated using a variety of materials [27], but most installed window frames in Japan are made of aluminum, PVC, and composites of aluminum and PVC [13]. The application frequency of the PVC frame in the cold regions (regions 1 and 2) was approximately 99.00%, as shown in Table 1. However, the nationwide adoption rate of PVC frames in Japan was only 23.47%. The frequency of adoption of aluminum frames in window configuration is still quite high in Japan. In this study, wooden frames were excluded from the analysis because wooden frames are rarely used in houses in Japan [13].

A total of 151,569 newly built residential buildings were investigated in this study. Region 6, which includes cities with high population densities, such as Tokyo, had the highest number of newly built households, accounting for approximately 72.6% of the total.

The number of new households surveyed in this study (151,569) was approximately 17.2% of the total number of newly constructed households in Japan [28]. Therefore, the results of this study represent approximately 17% of the energy consumption of the entire residential building sector in Japan, which has considerable research significance.

Table 1. Frame application frequency and number of newly built households by region.

Region	Frame Application Frequency by Households (Number of Households, %)				Sum (Number of Newly Built Households, %)
	PVC	Composite	Aluminum	Wood	
Entire area	35,571 (23.47%)	97,709 (64.47%)	18,155 (11.98%)	24 (0.02%)	151,569 (100%)
Region 1 (Asahikawa)	1495 (99.00%)	0 (0.00%)	14 (0.90%)	2 (0.10%)	1510 (100%)
Region 2 (Sapporo)	15,839 (99.00%)	0 (0.00%)	144 (0.90%)	6 (0.10%)	15,999 (100%)
Region 3 (Morioka)	1063 (53.60%)	893 (45.00%)	26 (1.30%)	2 (0.10%)	1984 (100%)
Region 4 (Nagano)	433 (18.50%)	1729 (73.80%)	176 (7.50%)	5 (0.20%)	2343 (100%)
Region 5 (Toyama)	1102 (35.50%)	1983 (63.90%)	19 (0.60%)	0 (0.00%)	3103 (100%)
Region 6 (Tokyo)	12,982 (11.80%)	81,195 (73.80%)	15,733 (14.30%)	10 (0.10%)	110,020 (100%)
Region 7 (Fukuoka)	2658 (16.00%)	11,909 (71.70%)	2043 (12.30%)	0 (0.00%)	16,610 (100%)

3. Simulation for U-Value of the Frame (Step 1)

3.1. Description of the Examined Window Configurations

Figure 1 shows a representative cross-section for each window frame. In this study, PVC, composite (PVC and aluminum), and aluminum frames were selected for analysis. The geometry of each frame is that of a typical window frame in Japan. The width of the frame varied depending on the frame material. The width of the PVC frame was 103 mm, and the widths of the composite and aluminum frames were 61 mm.

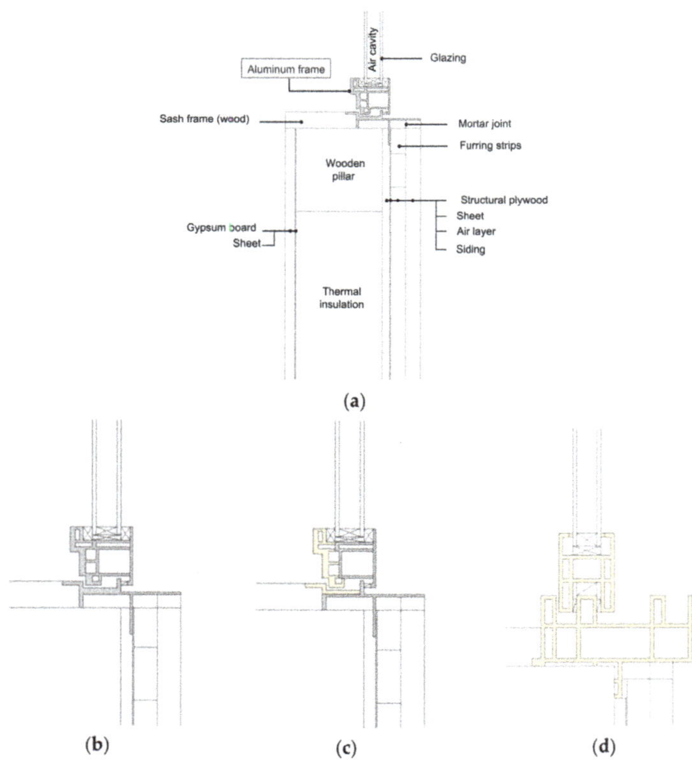

Figure 1. Cross-section of (**a**) aluminum frame including wall, (**b**) aluminum frame, (**c**) composite frame (aluminum and polyvinyl chloride [PVC]), and (**d**) PVC frame.

Glazing was configured in the same manner; the windows were double-glazed with a low-e coating and filled with air to a thickness of 16 mm. The PVC frame was set as white, and the composite and aluminum frames were gray to ensure that solar absorption rates were uniform among the frame materials. The effect of the frame material only on thermal performance was analyzed.

3.2. Simulation Conditions

This section presents the simulation conditions used to calculate the U-values of the frames. The temperature distribution was determined by calculating the two-dimensional heat transfer for the cross-sectional models. The thermal transmittance rates for all frame materials were calculated and compared according to the temperature distribution. The input conditions and material properties were based on ISO 1007-2 [22]. The indoor and outdoor temperatures were fixed at 20 °C and 0 °C, respectively, and the calculation was performed until a steady-state was achieved in the absence of external solar radiation and wind speed. The thermal conductivity of the window cavities was calculated in terms of the equivalent thermal conductivity according to ISO 1007-2 [22]. The thermal conductivity values of the frames were set as 0.17 W/(m·K) for PVC-Hard and 160 W/(m·K) for aluminum. The densities of the frames were set as 1390 kg/m^3 for PVC-Hard and 2800 kg/m^3 for aluminum.

3.3. Simulation Results: U-Value of Frames

Figure 2 shows the calculated temperature distributions for each reference frame. The indoor surface temperature of the frame was maintained at its highest value in the PVC frame (Figure 2c), followed by that of the composite (Figure 2b) and aluminum frames (Figure 2a). Additionally, the temperature distribution of the PVC frame was clearly divided between the indoor and outdoor sectors. However, low temperature values were distributed almost uniformly throughout the aluminum frame. Therefore, the PVC frame exhibits the lowest heat loss due to transmission, and the aluminum frame exhibits the highest heat loss. In conclusion, the thermal conductivity values of the PVC, composite, and aluminum frames were 1.88 W/m^2·K, 5.32 W/m^2·K, and 8.78 W/m^2·K, respectively.

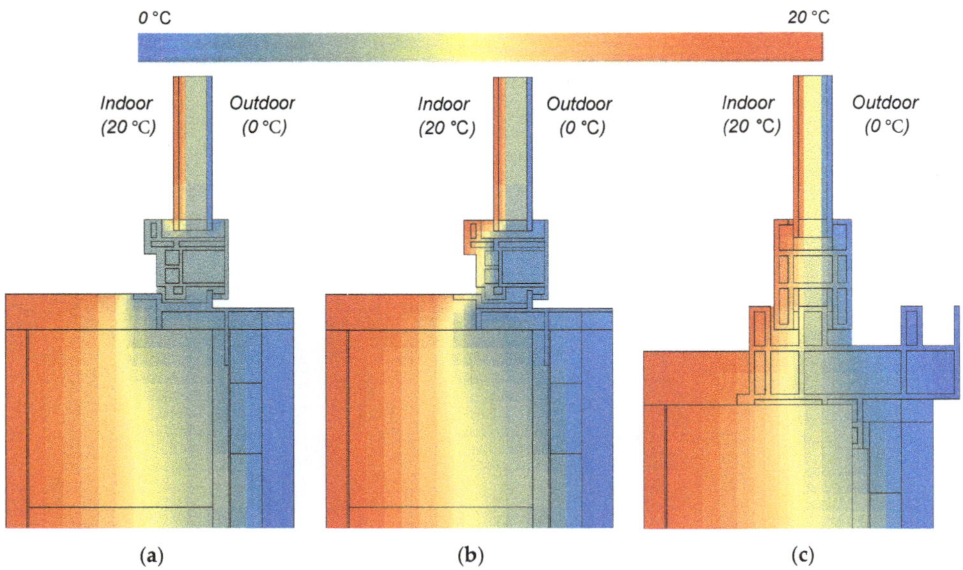

Figure 2. Temperature distribution for each reference frame. PVC, polyvinyl chloride. (**a**) Aluminum; (**b**) Composite; (**c**) PVC.

4. Impact of the Window Frame on Annual Building Thermal Load, Energy Consumption, and CO_2 Emissions (Steps 2 and 3)

4.1. Description of Reference Building and Window Configurations

To analyze the effect of the frame materials on the annual thermal load, energy consumption, and CO_2 emissions, a standard residential building in Japan was selected for analysis. Figure 3 illustrates a prototype of a residential building from the Institute for Building Environment and Energy Conservation (IBEC) [29]. This is a standard two-story building in Japan, with a total floor area of 120.6 m^2; that is, 67.9 m^2 for the first floor and 52.17 m^2 for the second floor [29].

Figure 3. Floor plan of a reference building.

The window-to-wall ratio—the total area of the glazing (window) divided by the total wall area—of the reference building varies according to climatic conditions. The window-to-wall ratio of the reference building for regions 1 and 2 was 14.7%, whereas that for regions 3–7 was 18.9% [29]. The cold area had a relatively small window area. Table 2 lists the thermal performances of the building envelope. The thickness of the thermal insulating material varied according to the energy conservation standard for each region [23].

The main purpose of this study was to analyze the effect of frame materials on energy performance. Therefore, examining the effect of the changes in the frame material under uniform conditions of glazing or air cavity is important. However, in the actual construction market, generally applied glazing specifications differ for each frame material. Therefore, seven different types of window combinations were determined, as shown in Table 3. Table 3 shows the window configurations based on the window frame material, filled-gas-type air cavity, and cavity thickness, with regard to the sales performance of Japanese manufacturers [14]. The application status of each window configuration, according to the region and based on the defined window configuration, is shown in Table 4. In regions 1 and 2, which are cold regions, the application of high-performance windows with PVC frames was approximately 99%. Conversely, in regions 6 and 7, the installation ratio of the aluminum frame, which has a low thermal performance, was approximately 12–15% owing to the relatively warm outdoor conditions. Among the window configurations adopting PVC frames, window type 3, in which cavities are filled by dry air, is not popular in the Japanese market. Nevertheless, this window type was selected to identify the U-value of the window frame based on its material and to analyze the effect of the frame material on the annual building thermal load.

Table 2. Description of material parameters.

Category	Layer	Thickness [m]	Thermal Conductivity [W/(m·K)]	Specific Heat [J/(kg·K)]	Specific Weight [kg/m^3]
Exterior wall	Gypsum board	0.009	0.220	870.0	706.0
	Thermal insulation (glass wool) *	0.130/0.080/0.040	0.045	840.0	56.0
	Structural plywood	0.012	0.160	1880.0	556.0
	Ventilated cavity	0.022	0.024	1005.0	1.2
	Siding	0.015	0.140	760.0	1110.0
Ceiling	Interior finishing materials	0.012	0.160	1880.0	556.0
	Ventilated cavity	0.030	0.024	1005.0	1.2
	Thermal insulation (glass wool) *	0.280/0.210/0.090	0.050	840.0	56.0
	Structural Plywood	0.012	0.160	1880.0	556.0
	Sheet	0.001	0.160	840.0	1270.0
	Roof finish	0.030	0.349	800.0	2000.0
Floor	Plywood	0.012	0.111	1880.0	550.0
	Thermal insulation (glass wool) *	0.150/0.110/0.050	0.045	840.0	56.0
	Plywood	0.012	0.160	1880.0	556.0

* Thickness of thermal insulation varied according to the region (Regions 1–2/Regions 3–4/Regions 5–7).

Table 3. Description of window configurations. AL, aluminum; PVC, polyvinyl chloride.

Classification	Type of Window						
	1	2	3 *	4	5 *	6 *	7
Frame	PVC	PVC	PVC	AL+PVC	AL+PVC	AL	AL
Glazing	Low-E triple	Low-E double	Low-E double	Low-E double	Low-E double	Low-E double	Single
Cavity	Argon 15 mm × 2	Argon 16 mm	Air 16 mm	Argon 16 mm	Air 16 mm	Air 16 mm	Air 16 mm

* Window specifications used to calculate U-value for each frame material in Section 3.

Table 4. Application frequency or window assembly according to region.

Region	Type of Window							Sum
	1	2	3 *	4	5 *	6 *	7	
Region 1 (Asahikawa)	881 (58.42%)	613 (40.65%)	0 (0.00%)	0 (0.00%)	0 (0.00%)	14 (0.93%)	0 (0.00%)	1508 (100%)
Region 2 (Sapporo)	9345 (58.47%)	6494 (40.63%)	0 (0.00%)	0 (0.00%)	0 (0.00%)	144 (0.90%)	0 (0.00%)	15,983 (100%)
Region 3 (Morioka)	239 (12.06%)	824 (41.57%)	0 (0.00%)	264 (13.32%)	629 (31.74%)	26 (1.31%)	0 (0.00%)	1982 (100%)
Region 4 (Nagano)	42 (1.80%)	391 (16.72%)	0 (0.00%)	294 (12.57%)	1435 (61.38%)	176 (7.53%)	0 (0.00%)	2338 (100%)
Region 5 (Toyama)	177 (5.71%)	924 (29.79%)	0 (0.00%)	508 (16.38%)	1475 (47.55%)	8 (0.26%)	10 (0.32%)	3102 (100%)
Region 6 (Tokyo)	2298 (2.09%)	10,684 (9.72%)	0 (0.00%)	21,029 (19.13%)	60,165 (54.74%)	5727 (5.21%)	10,006 (9.10%)	109,909 (100%)
Region 7 (Fukuoka)	691 (4.16%)	1967 (11.84%)	0 (0.00%)	4514 (27.17%)	7396 (44.52%)	276 (1.66%)	1767 (10.64%0	16,611 (100%)

* Window specifications used to calculate U-values for each frame material in Section 3.

4.2. Simulation Conditions

The calculated U-value of the frame for each material was applied as an input value for the unsteady-state energy analysis simulation to calculate the annual building load, power consumption, and CO_2 emission rate for each condition.

Weather data for the standard year of the expanded Automated Meteorological Data Acquisition System [30] were used for weather conditions. In all cases, the internal gain was generated constantly, and sensible heat of 260 W and latent heat of 220 W were set as the internal heat generations from four people in the house. The ventilation rate was set as 0.5 air change rate per hour and heating and cooling systems were operated continuously (27 °C in the cooling season and 20 °C in the heating season). The simulations were conducted for a year, and the simulation time step was 10 min.

4.3. Simulation Results: Impact of the Window Frame Material on Annual Energy Consumption and CO_2 Emissions (Step 2)

The heating, cooling, and total heat loads by region for the seven window configurations are presented in Table 5a–c. In regions 1–2, which are relatively cold regions, the heating load is lower than that of other regions owing to the strengthening of the thermal insulation performance of the building envelope (Table 5a). However, regions 4–7 show higher heating loads than the cold regions despite warm external conditions, owing to the low insulation performance of the buildings. The regions also exhibited a significant reduction in the heating load following an improved thermal performance of the window configurations (window types 1 to 7).

In the case of the cooling load in Table 5b, owing to the high level of thermal insulation of the building envelope, a low level of cooling load was observed in the warm climate regions. The cooling load decreases as the insulation performance of windows is improved in all regions. In cold regions, the cooling load is increased owing to a decrease in the heat transfer rate through the windows, attributable to the improved insulation performance of the window glazing (window types 1 and 2). Owing to the conflicting relationship between the insulation performance and cooling load, the identification of an appropriate insulation level appears to be necessary.

Regarding the annual thermal load in Table 5c, the total load decreased as the insulation performance of the window configurations was strengthened. This tendency was evident in all regions. The window configurations adopting the aluminum frame (window types 6–7) clearly require a higher annual building thermal load than windows adopting PVC frames (window types 1–3).

Figure 4 shows the results of window combinations (window types 3, 5, and 6) where the filled-gas-type air cavity and the thickness of the cavity are uniform, and only the frame material is changed. In this analysis, the net effect of the frame material alone on energy use is observed. The results of the annual energy consumption and CO_2 emission reductions for the three window combinations are shown according to region. When changing from an aluminum frame to a composite frame of aluminum and PVC, the annual energy consumption decreased by approximately 0.11–1.26%. On a regional average, the annual energy consumption decreased by approximately 0.75%. In addition, when the frame material was changed from aluminum to PVC, the annual energy consumption according to region decreased by approximately 1.49–3.55%, and on average, by approximately 2.62%.

Table 5. (**a**) Annual heating load by window configurations. (**b**) Annual cooling load by window configurations. (**c**) Annual total heat load by window configurations.

(a)

Region	Annual Heating Load [kWh/year]						
	Window Type 1	**Window Type 2**	**Window Type 3**	**Window Type 4**	**Window Type 5**	**Window Type 6**	**Window Type 7**
Region 1 (Asahikawa)	915.0	1655.7	1611.1	1896.5	1849.2	1751.0	1795.7
Region 2 (Sapporo)	695.1	1221.8	1183.5	1301.6	1259.3	1169.0	1296.4
Region 3 (Morioka)	869.1	883.3	872.7	880.2	867.8	828.2	916.4
Region 4 (Nagano)	2344.8	2585.0	2561.8	2666.0	2644.0	2577.0	3518.4
Region 5 (Toyama)	3097.6	3373.2	3352.7	3479.4	3457.2	3406.3	4469.3
Region 6 (Tokyo)	4165.3	4511.1	4495.2	4652.9	4635.2	4597.9	5963.1
Region 7 (Fukuoka)	4720.4	5087.1	5070.2	5223.4	5206.3	5165.3	6529.1

690 kWh/year — 6,530 kWh/year

(b)

Region	Annual Cooling Load [kWh/year]						
	Window Type 1	**Window Type 2**	**Window Type 3**	**Window Type 4**	**Window Type 5**	**Window Type 6**	**Window Type 7**
Region 1 (Asahikawa)	8762.1	8255.6	8478.4	8303.0	8546.6	8709.9	10,466.6
Region 2 (Sapporo)	7025.2	6539.3	6727.0	6583.2	6787.7	6952.7	8397.5
Region 3 (Morioka)	8784.1	9201.8	9340.3	9304.1	9487.7	9538.8	10,905.3
Region 4 Nagano)	6582.2	6689.0	6806.7	6729.8	6891.1	7030.8	7484.7
Region 5 (Toyama)	7477.7	7675.5	7829.4	7731.4	7902.0	8023.6	8756.8
Region 6 (Tokyo)	4705.6	4742.5	4859.2	4779.3	4907.4	5066.3	5325.2
Region 7 (Fukuoka)	4420.8	4492.0	4599.7	4533.7	4652.7	4789.3	5118.6

4,420 kWh/year — 11,000 kWh/year

(c)

Region	Annual Total Heat Load [kWh/year]						
	Window Type 1	**Window Type 2**	**Window Type 3**	**Window Type 4**	**Window Type 5**	**Window Type 6**	**Window Type 7**
Region 1 (Asahikawa)	9677.1	9911.2	10,089.5	10,199.5	10,395.8	10,460.9	12,262.3
Region 2 (Sapporo)	7720.3	7761.0	7910.5	7884.8	8047.0	8121.7	9693.9
Region 3 (Morioka)	9653.2	10,085.0	10,213.0	10,184.3	10,355.5	10,367.0	11,821.7
Region 4 (Nagano)	8927.0	9274.0	9368.6	9395.8	9535.1	9607.8	11,003.1
Region 5 (Toyama)	10,575.3	11,048.6	11,182.1	11,210.8	11,359.2	11,429.9	13,226.1
Region 6 (Tokyo)	8870.9	9253.6	9354.4	9432.2	9542.6	9664.2	11,288.4
Region 7 (Fukuoka)	9141.2	9579.1	9669.9	9757.1	9859.0	9954.6	11,647.7

7,720 kWh/year — 13,300 kWh/year

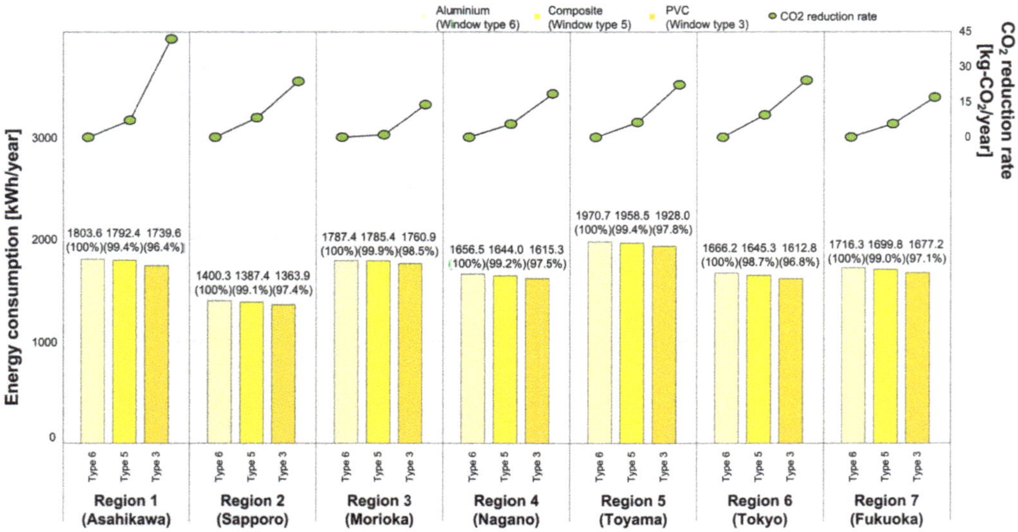

Figure 4. Annual energy consumption and CO_2 reduction rate of window frame according to region. PVC, polyvinyl chloride.

The energy-saving rates were the highest where the frame material was changed from aluminum to composite (0.6%) and PVC (3.6%) in the coldest region (region 1). Nevertheless, a similar energy-saving effect is expected in regions 2–7 regardless of the external temperature conditions.

The CO_2 emission rate was calculated by multiplying power consumption by the CO_2 emission factor. Based on the aluminum frame in the coldest region (region 1), the application of the PVC frame and the composite frame reduced the CO_2 emission rate by 42.01 kg and 7.36 kg, respectively, yielding the largest reduction among all regions. Furthermore, in a relatively warm region (region 6), when the aluminum frame was replaced with composite and PVC frames, CO_2 emission savings of 9.54 kg and 24.3 kg were achieved, respectively.

In summary, depending on the application of the composite frame, the reduction amounts were 1.04–9.54 kg of CO_2 per household per year, and 14.02–42.01 kg of CO_2 depending on the application of the PVC frame. On average, compared to the installation of aluminum frames, composite and PVC frames showed a reduction of 6.31 kg of CO_2 and 23.19 kg of CO_2, respectively. Furthermore, CO_2 emissions were converted into the amount of CO_2 gas absorbed by cedar trees, which are prevalent in Japanese forests and cover nearly 18% of the total land area of Japan [31,32]. The amount of CO_2 gas absorbed per year by a single cedar tree is approximately 14 kg of CO_2 [33]; therefore, it is possible to reduce the CO_2 gas emission equivalent to that absorbed by a single cedar tree by strengthening the thermal insulation performance of the frame. In particular, when the window frame was changed to PVC, a greenhouse gas reduction effect equivalent to the amount of CO_2 gas absorbed by 1–3 cedar trees per household for one year was observed; therefore, the effect of changing the window frame material on the annual energy consumption and CO_2 emissions throughout the study region was observed to be quite significant.

4.4. Simulation Results: Impact of the Window Frame on CO_2 Emissions at the City Scale (Step 3)

Figure 5 shows the annual CO_2 emissions according to the regions in 2019. Figure 5a–g compare CO_2 emissions under three conditions: (1) CO_2 emission status of newly built houses in each region as of 2019, (2) an alternative where the window compositions were changed to window types 1 and 2 (Alternative 1), and (3) an alternative where window compositions were all changed to window type 1 (Alternative 2). In Alternative

1, the frequency of window types 1 and 2 was based on the sales ratio of window frame manufacturers in 2019.

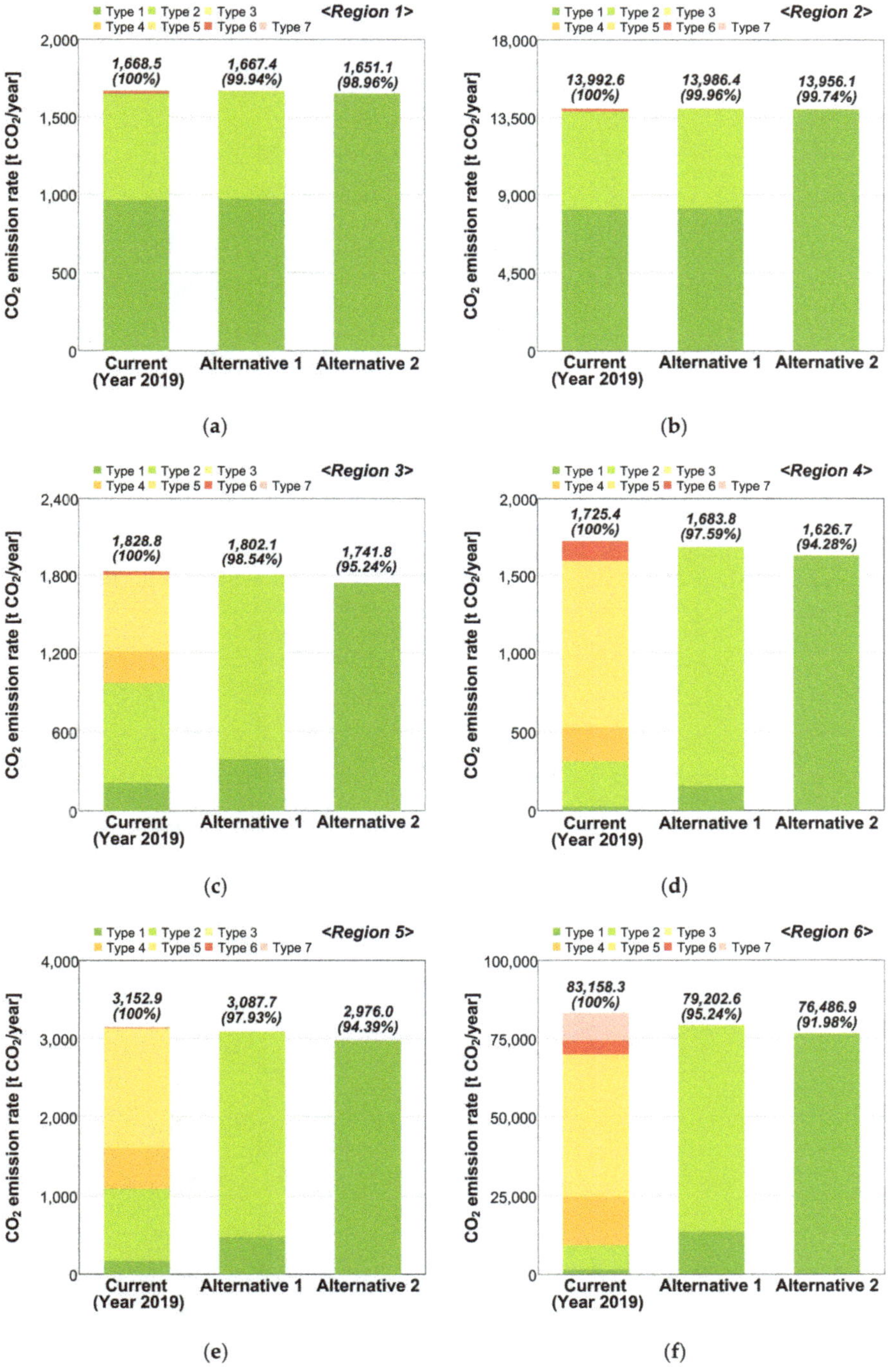

Figure 5. *Cont.*

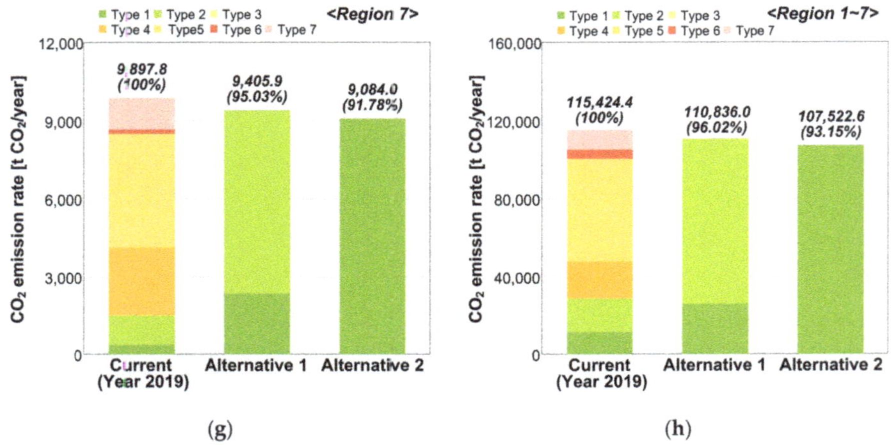

Figure 5. CO_2 emission rate by window configurations in (**a**) region 1, (**b**) region 2, (**c**) region 3, (**d**) region 4, (**e**) region 5, (**f**) region 6, (**g**) region 7, and (**h**) nationwide.

CO_2 emissions in all regions clearly decreased upon applying Alternatives 1 and 2. In the cold regions, the original PVC frame installation rate was high. Therefore, in regions 1 and 2, the reduction rates were approximately 1.06–17.43 t of CO_2 and 6.26–36.50 t of CO_2, respectively. In conclusion, in regions 1 and 2, a CO_2 reduction effect equivalent to the amount of CO_2 absorbed by 76–1245 cedar trees and 448–2607 trees per year can be achieved. In region 2, the reduction exceeded that in region 1 owing to the large number of newly constructed residential buildings. Consequently, the amount of the CO_2 emission reduction in cold regions throughout the region accounted for 0.16% and 0.68% following the implementation of Alternatives 1 and 2, respectively (Table 6).

Table 6. CO_2 reduction rate according to alternative window configurations.

Region	Alternative 1		Alternative 2	
	CO_2 Reduction Rate [t CO_2]	Frequency of CO_2 Reduction Rate [%]	CO_2 Reduction Rate [t CO_2]	Frequency of CO_2 Reduction Rate [%]
Region 1 (Asahikawa)	1.06	0.02%	17.43	0.22%
Region 2 (Sapporo)	6.26	0.14%	36.50	0.46%
Region 3 (Morioka)	26.65	0.58%	87.03	1.10%
Region 4 (Nagano)	41.56	0.91%	98.66	1.25%
Region 5 (Toyama)	65.17	1.42%	176.92	2.24%
Region 6 (Tokyo)	3955.81	86.21%	6671.42	84.43%
Region 7 (Fukuoka)	491.89	10.72%	813.87	10.30%
Sum	4588.41	100%	7901.82	100%

In regions 3–5, the CO_2 reduction rates of Alternatives 1 and 2 corresponded to 26.65–65.17 t of CO_2 and 87.03–176.92 t of CO_2, respectively, corresponding to the amount of CO_2 absorbed by 1904 to 12,637 cedar trees per year. Compared to cold regions (regions 1 and 2), the application rate of the aluminum frame was high; therefore, the amount of CO_2 reduction due to frame replacement was relatively large. The amount of CO_2 reduction in regions 3–5 compared to the reductions throughout the region were approximately 2.91% (Alternative 1) and 4.59% (Alternative 2) (Table 6).

Region 6, which has a relatively high aluminum frame adoption rate, has a high population density and a large number of new buildings; therefore, Alternatives 1 and 2 in region 6 accounted for 86.21% and 84.43% of the reduction rate of the entire region, respectively. In conclusion, as the thermal performance of the frame is improved, an effect equivalent to the CO_2 gas absorption effect of approximately 282,558–476,530 cedar trees in Tokyo alone can be achieved. Region 6 includes many cities with a high population density, such as Tokyo and Osaka; therefore, it is expected that PVC frames can significantly contribute to the reduction of greenhouse gas emissions, especially in areas such as region 6, which has a high population density and a large amount of new housing.

In region 7, the hottest region, CO_2 emissions of approximately 491.89 CO_2 and 813.87 CO_2 can be reduced under the conditions of Alternatives 1 and 2, respectively. In conclusion, in region 7, the CO_2 absorption is equivalent to 35.135 and 58.133 cedar trees, accounting for a relatively high rate of the total CO_2 emission reduction in the area (approximately 10.30–10.72%).

In all of the regions, it was confirmed that CO_2 gas emissions could be reduced by approximately 3.98% and 6.85% following the implementation of Alternatives 1 and 2, respectively (Figure 5h), corresponding to CO_2 absorption by 327,743 and 564,416 cedar trees, respectively. This confirms that changing the material of the window frame has a significant energy-saving impact, and a considerable amount of energy can be saved at the city scale.

5. Conclusions

This study examined the potential consequences of global warming on the energy performance of window frames in seven cities in Japan. The U-value of window frames was simulated according to the frame material: aluminum, composite (aluminum and PVC), and PVC. This study suggests the energy-saving effect of increasing the thermal performance of window frames at the city scale. The main results of this study are as follows:

(1) The U-values of different window frames were calculated using a two-dimensional simulation, and the heat-loss characteristics of the envelope were analyzed. The U-values of the PVC, composite, and aluminum frames were 1.88 $W/m^2{\cdot}K$, 5.32 $W/m^2{\cdot}K$, and 8.78 $W/m^2{\cdot}K$, respectively.

(2) The impact of the window frame material on only the annual energy consumption varied from approximately 0.11% to 3.55%, depending on the region. On a regional average, the annual energy consumption decreased by approximately 0.75%. Consequently, if the composite frame was used instead of the aluminum frame, the reduction amount was 1.04–9.54 kg of CO_2 per household per year. Reductions of 14.02–42.01 kg of CO_2 could be achieved following the application of a PVC frame instead of an aluminum frame. When the window frame was changed to PVC, a greenhouse gas reduction effect was observed to be equivalent to the amount of CO_2 gas absorbed by 1–3 cedar trees per household for one year.

(3) CO_2 gas emissions can be reduced by approximately 3.98–6.85% through the application of PVC frames in seven regions of Japan. If the reduction rate is converted to the amount of CO_2 gas absorbed by cedar, the absorption effect corresponds to 327,743 and 564,416 cedar trees. This confirms that changing the material of the window frame has a significant energy-saving impact, and a considerable amount of energy can be saved at the city scale.

(4) Even in areas where PVC frames are already used commonly, the city-level energy-saving effect is proportional to the number of new houses. Tokyo is one such area, and the energy-saving rate in Tokyo accounts for approximately 85% of all examined regions. This necessitates the continuous improvement of the insulation of window configurations, even in regions or countries that have already adopted numerous PVC frames.

The results of this study may provide a basis for predicting the future energy-saving potential of countries targeting the dealumination of windows. In addition, related studies and national policymakers can use our research methods as references. Research studies related to energy-efficient retrofits can refer to the effects of the window configurations reported here.

It is worth noting that this study analyzed only approximately 17.2% of all new housing units in Japan. Therefore, for accurate analysis and practical application, it is necessary to conduct analyses of all new housing units in Japan. This limitation will be addressed in future studies.

Author Contributions: Y.C.: Data curation, Visualization, Investigation, Writing—Original draft preparation. H.L.: Software, Writing—Reviewing and Editing. A.O.: Conceptualization, Methodology, Supervision. All authors have read and agreed to the published version of the manuscript.

Funding: This research received no external funding.

Institutional Review Board Statement: Not applicable.

Informed Consent Statement: Not applicable.

Data Availability Statement: Not applicable.

Acknowledgments: This work was supported by the research grant of the Chungbuk National University in 2021.

Conflicts of Interest: The authors declare no conflict of interest.

References and Note

1. IPCC. *Climate Change 2007: Synthesis Report, Summ. For Policymakers*; IPCC: Geneva, Switzerland, 2007.
2. Heydari, A.; Sadati, S.E.; Gharib, M.R. Effects of different window configurations on energy consumption in building: Optimization and economic analysis. *J. Build. Eng.* **2021**, *35*, 102099. [CrossRef]
3. De Gastines, M.; Correa, É.; Pattini, A. Heat transfer through window frames in EnergyPlus: Model evaluation and improvement. *Adv. Build. Energy Res.* **2019**, *13*, 138–155. [CrossRef]
4. Sinha, A.; Kutnar, A. Carbon Footprint versus Performance of Aluminum, Plastic, and Wood Window Frames from Cradle to Gate. *Buildings* **2012**, *2*, 542. [CrossRef]
5. Gustavsen, A.; Jelle, B.P.; Arasteh, D.; Kohler, C. *State-of-the-Art Highly Insulating Window Frames–Research and Market Review*; Norges Byggforskningsinstitutt: Oslo, Norway, 2007.
6. Appelfeld, D.; Hansen, C.S.; Svendsen, S. Development of a slim window frame made of glass fibre reinforced polyester. *Energy Build.* **2010**, *42*, 1918–1925. [CrossRef]
7. Misiopecki, C.; Bouquin, M.; Gustavsen, A.; Jelle, B.P. Thermal modeling and investigation of the most energy-efficient window position. *Energy Build.* **2018**, *158*, 1079–1086. [CrossRef]
8. Takada, K.; Hayama, H.; Mori, T.; Kikuta, K. Thermal insulated PVC windows for residential buildings: Feasibility of insulation performance improvement by various elemental technologies. *J. Asian Archit. Build. Eng.* **2020**, *20*, 340–355. [CrossRef]
9. Rahman, W.A.W.A.; Sin, L.T.; Rahmat, A.R. Injection moulding simulation analysis of natural fiber composite window frame. *J. Mater. Process. Technol* **2008**, *197*, 22–30. [CrossRef]
10. Khetib, Y.; Alotaibi, A.A.; Alshahri, A.H.; Rawa, M.; Cheraghian, G.; Sharifpur, M. Impact of phase change material on the amount of emission in the double-glazed window frame for different window angles. *J. Energy Storage* **2021**, *44*, 103320. [CrossRef]
11. International Energy Agency. *Technology Roadmap Energy Efficient Building Envelopes*; IEA: Paris, France, 2013.
12. El-Darwish, I.; Gomaa, M. Retrofitting strategy for building envelopes to achieve energy efficiency. *Alex. Eng. J.* **2017**, *56*, 579–589. [CrossRef]
13. Survey on the Use of Building Materials for Residential Buildings. 2021. Available online: https://www.jsma.or.jp/ (accessed on 16 May 2022).
14. Survey on the Use of Building Materials for Residential Buildings. 2013. Available online: https://www.jsma.or.jp/ (accessed on 16 May 2022).
15. Baldinelli, G.; Lechowska, A.; Bianchi, F.; Schnotale, J. Sensitivity Analysis of Window Frame Components Effect on Thermal Transmittance. *Energies* **2020**, *13*, 2957. [CrossRef]
16. Laustsen, J.B.; Svendsen, S. Improved windows for cold climates. In Proceedings of the 7th Nordic Symposium on Building Physics, Reykjavik, Iceland, 13–15 June 2005; pp. 987–994.
17. Zajas, J.; Heiselberg, P. Parametric study and multi objective optimization of window frame geometry. *Build. Simul.* **2014**, *7*, 579–593. [CrossRef]
18. Asdrubali, F.; Roncone, M.; Grazieschi, G. Embodied energy and embodied GWP of windows: A critical review. *Energies* **2021**, *14*, 3788. [CrossRef]
19. Ramesh, T.; Prakash, R.; Shukla, K.K. Life cycle energy analysis of buildings: An overview. *Energy Build.* **2010**, *42*, 1592–1600. [CrossRef]
20. Ozaki, A.; Tsujimaru, T. Prediction of hygrothermal environment of buildings based upon combined simulation of heat and moisture transfer and airflow. In Proceedings of the IBPSA 2005—International Building Performance Simulation Association, Montréal, QC, Canada, 15–18 August 2005.
21. Choi, Y.; Cho, W.; Ozaki, A.; Lee, H. Influence of the moisture driving force of moisture adsorption and desorption on indoor hygrothermal environment and building thermal load. *Energy Build.* **2021**, *253*, 111501. [CrossRef]

22. *BS EN ISO 10077-2:2017*; BSI Standards Publication Thermal Performance of Windows, Doors and Shutters—Calculation of Thermal Transmittance. B.S.I. Standards: London, UK, 2017.
23. Konisi, T.; Saito, H.; Isikawq, K. Energy Conservation Standards, Inst. Build. Environ. Energy Conserv. Available online: https://www.ibec.or.jp/ee_standard/build_standard.html (accessed on 16 May 2022).
24. Ministry of the Environment, Japan. CO_2 Emission Factor by Electric Power Company. 2020. Available online: https://www.env.go.jp/press/files/jp/115373.pdf (accessed on 16 May 2022).
25. Area Classification for Energy Saving, Japan. 2020. Available online: https://www.mlit.go.jp/jutakukentiku/shoenehou_assets/img/library/chiikikubun-sinkyuu.pdf (accessed on 16 May 2022).
26. Number of newly built households by region in Japan, Tokyo, n.d.
27. *ISO 15099:2003*; Thermal Performance of Windows, Doors and Shading Devices—Detailed Calculations. International Organization for Standardization (I.S.O.): Geneva, Switzerland, 2003.
28. *Number of Houses Built in Japan*; Nomura Research Institute (NRI): Tokyo, Japan, 2020. Available online: https://www.nri.com/jp/news/newsrelease/lst/2020/cc/0609_1 (accessed on 16 May 2022).
29. Nishizawa, S. *Explanation of Energy Consumption Calculation Method in Criteria of Housing Business*; Institute for Building Environment and Energy Conservation (IBEC): Tokyo, Japan, 2009; pp. 35–47.
30. Expanded AMeDAS Weather Data, Japan Meteorol. Agency. 2021. Available online: https://www.jma.go.jp/bosai/map.html#5/34.5/137/&elem=temp&contents=amedas&interval=60 (accessed on 16 May 2022).
31. Uetake, J.; Tobo, Y.; Kobayashi, S.; Tanaka, K.; Watanabe, S.; DeMott, P.J.; Kreidenweis, S.M. Visualization of the seasonal shift of a variety of airborne pollens in western Tokyo. *Sci. Total Environ.* **2021**, *788*, 147623. [CrossRef] [PubMed]
32. Yamada, T.; Saito, H.; Fujieda, S. Present state of Japanese cedar pollinosis: The national affliction. *J. Allergy Clin. Immunol.* **2014**, *133*, 632–639. [CrossRef] [PubMed]
33. *How Much Carbon Dioxide Does the Forest Absorb?* Ministry of Agriculture, Forestry and Fisheries: Tokyo, Japan, 2014. Available online: https://www.rinya.maff.go.jp/j/sin_riyou/ondanka/20141113_topics2_2.html (accessed on 16 May 2022).

energies

MDPI

Article

Study on the Variation in Heating Energy Based on Energy Consumption from the District Heating System, Simulations and Pattern Analysis

Soo-Jeong Kim [1] and Doo-Yong Park [2,*]

1 Department of Architecture, Korea University, Seoul 02841, Korea; crystal0213@hanmail.net
2 Energy Division, Korea Conformity Laboratories (KCL), Seoul 06711, Korea
* Correspondence: pdy0528@hanmail.net

Abstract: This study aims to analyze the actual heating energy consumption according to the location and size of apartment houses. The study shows the variation in heating energy consumption in accordance with the living pattern of residents in such apartments. By calculating the average annual heating energy consumption and distribution of the measured heating energy of two years, it was found that the outdoor temperature was inversely proportional to the average heating energy consumption. Moreover, the lowest/highest floors and corner houses were the most vulnerable since they had a lot of area exposed to the outside air and, thus, consume a huge amount of heating energy. According to this study, the heating load had relevance to the factors such as wall loss, window loss, ventilation loss, and solar radiation gain that were analyzed in accordance with the growth in house size. Based on the survey outcome on the living pattern and number of residents, a simulation was conducted to analyze the variation in heating energy consumption. Households consumed the average heating energy for 15.8 h/day and occupied for 16.4 h/day. Households consumed more than the average heating energy for 22.2 h/day and occupied for 21.2 h/day, meaning 6.4 extra hours than those consuming the average heating energy. Households consumed less than the average heating energy for 5.2 h/day and occupied for 10.9 h/day, meaning 10.6 less hours/day than those consuming the average heating energy and 17 less hours/day than those consuming more than the average heating energy.

Keywords: actual heating energy consumption; district heating system; EnergyPlus; variation; living pattern

Citation: Kim, S.-J.; Park, D.-Y. Study on the Variation in Heating Energy Based on Energy Consumption from the District Heating System, Simulations and Pattern Analysis. *Energies* **2022**, *15*, 3909. https://doi.org/10.3390/en15113909

Academic Editors: Shi-Jie Cao and Wei Feng

Received: 19 April 2022
Accepted: 23 May 2022
Published: 25 May 2022

1. Introduction

Building energy takes up the largest portion of greenhouse gas emission and energy consumption by 36%. As environmentally friendly low-energy building technology has come to the fore, the goal is to ensure every new apartment building that will be built from 2025 to be a zero-energy house, which is a 100% energy-independent building generating zero carbon emission. The government also intends to save building energy and cut down greenhouse emissions by enforcing the Green Building Creation Support Act. Although existing research examines energy reduction studies and application plans supported by changes in the building planning factors such as direction and floor of houses as well as the enhancement in building materials, they have not been realized due to economic feasibility [1,2]. In a recent study, an annual simulation and economic analysis on orientation of office buildings were conducted for four cities with different Iranian climates [3]. As a result of the analysis, in the east–west orientation, energy saving was up to 13.6%, and the value of the simple payback period was also evaluated to be lower than in the north–south orientation. However, analysis of residents' behavior was not included. According to the existing study, the behavior of residents is the main influence on the total energy consumption and the behavior of building dwellers is attributable to a twofold energy consumption deviation [4,5]. The study also revealed that the variation in energy consumption within the same building was caused by the difference in living pattern of residents, density of occupation, and building environment system [6,7]. In particular, Shin [8,9] developed an algorithm for occupant-based heating control and conducted a study applying it. The accuracy of the algorithm developed based on indoor CO_2 concentration and passive infrared (PIR) signals was evaluated as 83.5–98.9%. In addition, it was analyzed that the optimal heating start and stop time was found through simulation and the heating energy consumption was reduced by up to 3.1%. In addition, it was found that the thermal discomfort time was reduced from 62.5 to 8.3 h. The limitations of existing studies include that results were derived only by simulation and comparative verification with measured data, and the behavior of residents, a major influencing factor of heating energy, was not considered, and there is no comprehensive research.

The novelty of this study is divided into three. First, while most of the existing studies were simulated, this study aims to analyze the measured heating energy of the selected flat-type apartment, the most common type of apartment in Korea, and verify the results through simulation reflecting the resident survey. Second, the actual heating energy consumption of apartment houses is analyzed by location and size of the house, and the heating load factors are classified into ventilation loss, window loss, wall loss, solar heat increase, human heat generation, and facility heat generation. Finally, this study has novelty that was not present in previous studies by investigating patterns of residents and analyzing variation in heating energy. This study looked into a total of two years of measured heating energy in the winter season between November 2010 and March 2011 (hereinafter referred to as '2010 measured data') as well as November 2011 to March 2012 (hereinafter referred to as '2011 measured data'), and categorized the consumption by different house sizes (35, 84, and 164 m^2). The study also calculated the annual heating energy by accumulating the heating energy from November to March and compared between the measured average monthly heating energy and monthly outside air temperature measured by KMA (Korea Meteorological Administration). The measured heating energy from each household was identified by house size and location. Lastly, the study analyzed the variation in heating energy consumption according to the residents' living patterns to determine how the behavior of residents affects their energy consumption. Figure 1 shows the study flow chart.

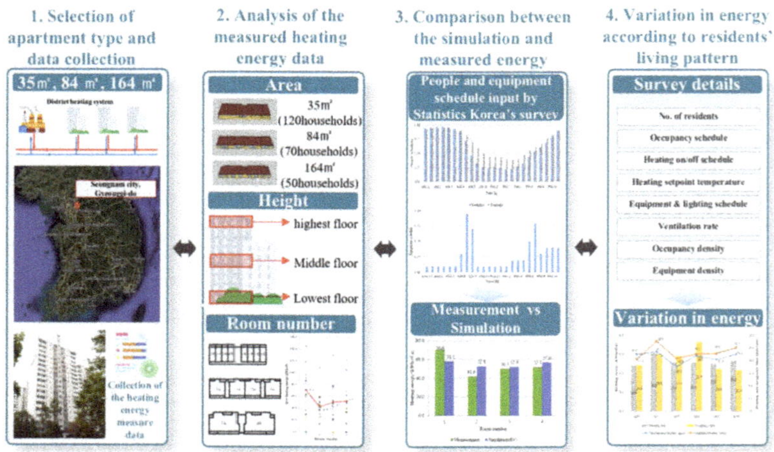

Figure 1. Study flow chart.

2. Materials and Methods

2.1. Subject and Method of Evaluation

Apartment Houses Subjected to Evaluation

This study analyzes the apartment houses located in Seongnam city, Gyeonggi Province, where the actual heating energy consumption in the winter season through district heating was obtained. For the study, three buildings with different exclusive private areas within the housing complex were selected. They were all flat-type apartment houses. The subjects are shown in Table 1 and were modeled and simulated by placing nearby buildings according to the apartment-housing layout in order to consider the impact of solar shading from those nearby buildings. A total 240 households composed of 120 households sized 35 m^2, 70 households sized 84 m^2, and 50 households sized 164 m^2, were evaluated

Table 1. Selection by type of apartment for evaluation.

Exclusive Private Area	Modeled Households	Modeled Nearby Buildings
35 m^2 (120 households)	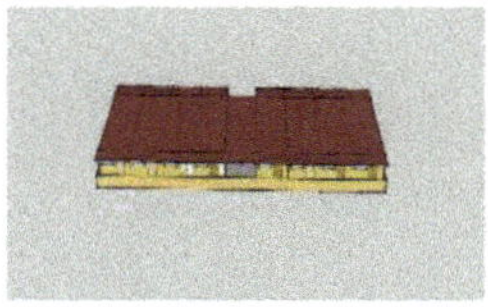	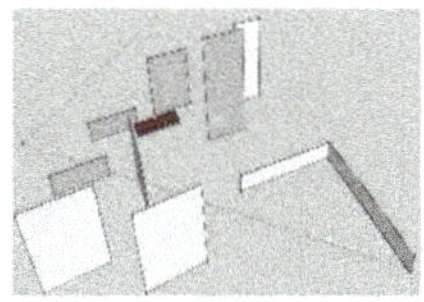
84 m^2 (70 households)	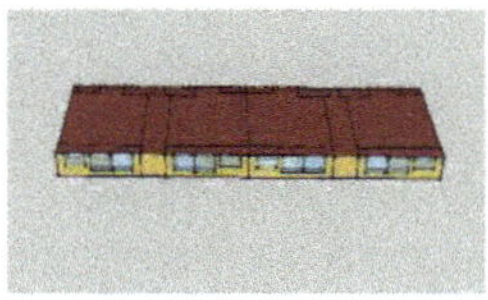	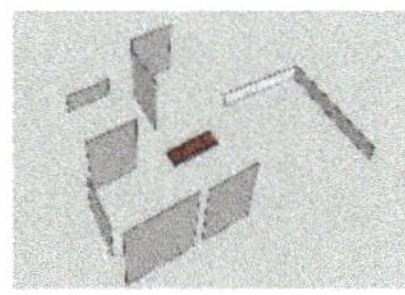
164 m^2 (50 households)	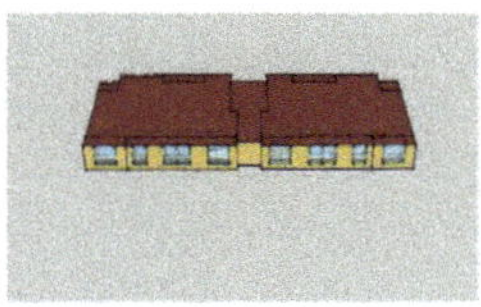	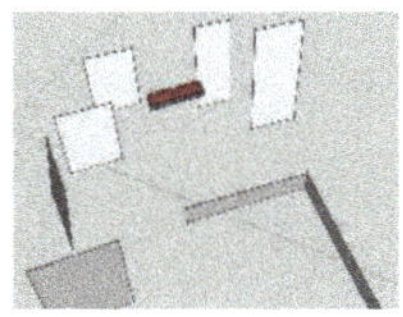

2.2. Simulation

This study compared between the simulation inputting the envelope composition materials of apartment houses and the measured heating energy. It examined the influence of loss or gain of heating load elements on the increasing heating energy consumption as the size of exclusive private area of households grew (Refer to Section 3.5). In addition, to verify that the residents' living patterns (reflecting the density of occupancy and hours of heating use) and the number of residents were the reasons behind the differences in heating energy consumption from the same building (Refer to Section 4.5), the study conducted a resident survey on living patterns and reflected such outcome into the simulation.

2.2.1. Evaluation Tools for Simulation

The heating energy analysis program used in this study is EnergyPlus, which enables text-based input and selection, and uses the same engine as the DOE-2 and BLAST programs. In addition, it can be linked with Google SketchUp and is a program that calculates the building energy using numeric analysis based on heat balance algorithms. Additionally, LBNL Window 7 program was used for calculating the g-value and visible transmittance value of windows that met the criteria of heat transmission coefficient being inputted.

2.2.2. Input Criteria for Simulation

In this study, regarding the outer wall, floor, roof, and outside window components, the heat transmission coefficient by each region constructed on the apartment house was applied as a material property. The below Table 2 shows the set value of material property for envelope composition.

Table 2. Set up value of material property for envelope composition of apartment houses subjected to evaluation.

Division	Input Data	Set Value of Material Property
Setpoint temperature	Heating	20 °C
Internal heat	Equipment	3.24 W/m^2
	People	1.44 W/m^2
	Number of residents	0.03 person/m^2
Air change per Hour (ACH)	Heating space	0.7 ACH
	Non-heating space	2.0 ACH
Wall U-Value	Exterior wall	0.58 W/m^2·K
	Interior wall	0.64 W/m^2·K
	Roof/floor	Adiabatic (insulation)
Windows U-Value	Exterior window	3.84 W/m^2·K
	Interior window	5.47 W/m^2·K
Windows properties	SHGC (Solar Heat Gain Coefficient)	0.613
	Visible transmittance	0.56

Regarding the material property from Table 2, the set temperature of heating, air change per hour, electric power heating value, and heat from people referred to the Operational Regulation on the Certification of Building Energy Efficiency [10]. The weather data for simulation utilized the KMA's data regarding Seoul region, which is the nearest to the studied apartment house. Additionally, the mid-floor houses were simulated as a benchmark. Since the mid-floors had both upper and lower neighbors to their border of roofs and floors, adiabatic process was set up.

3. Analysis of the Measured Heating Energy Data

The measured values of actual heating energy consumption targeting 240 households sized 35, 84, and 164 m^2 were analyzed. During the data analysis, households that consumed zero energy from November to March were excluded from the calculation of average consumption of heating energy.

3.1. Analysis of the Measured Heating Energy by Household Size

3.1.1. Measured Annual Average Heating Energy Consumption from Households Sized 35 m^2

The below Figure 2 shows the distribution of measured annual average heating energy consumption from 120 households sized 35 m^2.

The average heating energy consumption per unit area of 120 households sized 35 m^2 was measured at 69.7 kWh/m^2·a in 2010 and 58.6 kWh/m^2·a in 2011. The average heating energy consumption that was revised by excluding the households with zero consumption from November to March was measured at 83.0 kWh/m^2·a in 2010 and 70.8 kWh/m^2·a in 2011.

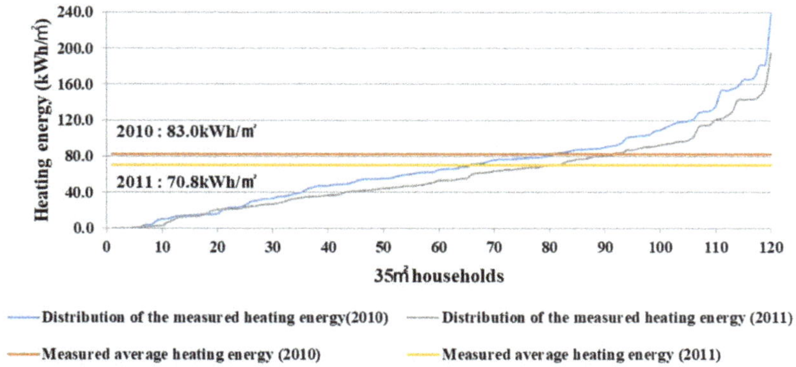

Figure 2. Analysis of the measured heating energy consumption from households sized 35 m^2.

3.1.2. Measured Annual Average Heating Energy Consumption from Households Sized 84 m^2

The below Figure 3 shows the distribution of annual average heating energy consumption from 70 households sized 84 m^2 in 2010 and 2011.

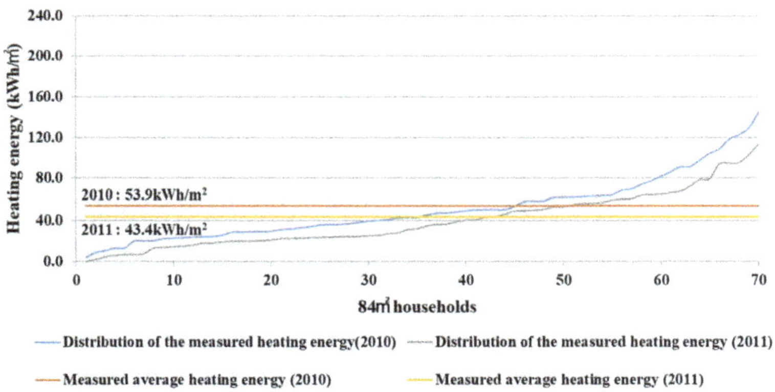

Figure 3. Analysis of the measured heating energy consumption of households sized 84 m^2 in 2010 and 2011.

The average heating energy consumption per unit area of 70 households sized 84 m^2 was measured at 51.0 kWh/m^2·a in 2010 and 39.6 kWh/m^2·a in 2011. The average heating energy consumption that excluded the households with zero consumption during November and March was measured at 53.9 kWh/m^2·a in 2010 and 43.4 kWh/m^2·a in 2011.

3.1.3. Annual Average Heating Energy Consumption by Households Sized 164 m^2

The below Figure 4 shows the distribution of annual average heating energy consumption from 50 households sized 164 m^2 in 2010 and 2011.

The average heating energy consumption per unit area of 50 households sized 164 m^2 was measured at 38.1 kWh/m^2·a in 2010 and 33.4 kWh/m^2·a in 2011. The average heating energy consumption that excluded the households with zero consumption during November and March was measured at 41.3 kWh/m^2·a in 2010 and 35.3 kWh/m^2·a in 2011.

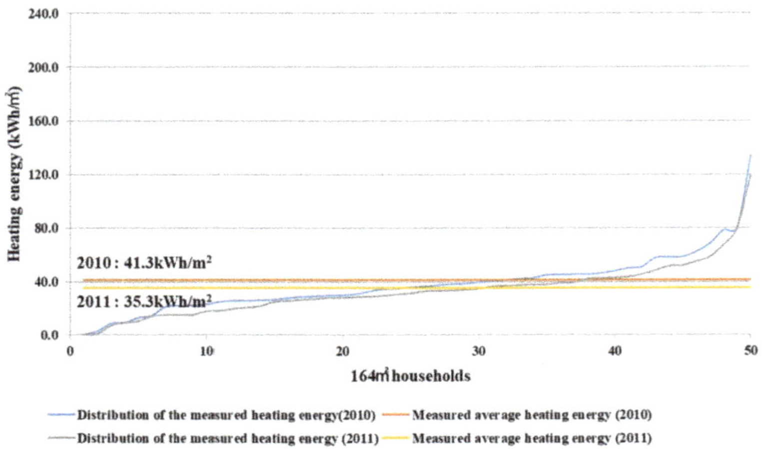

Figure 4. Analysis of the measured heating energy consumption of households sized 164 m^2 in 2010 and 2011.

3.2. Comparison between the Measured Monthly Heating Energy Consumption by Household Size and the Outside Temperature

Figure 5 classified the measured monthly average heating energy consumption from 2010 to 2011 by house size. Figure 5 shows a graph comparing the monthly outside temperature data of 2010 with that of 2011 provided by KMA.

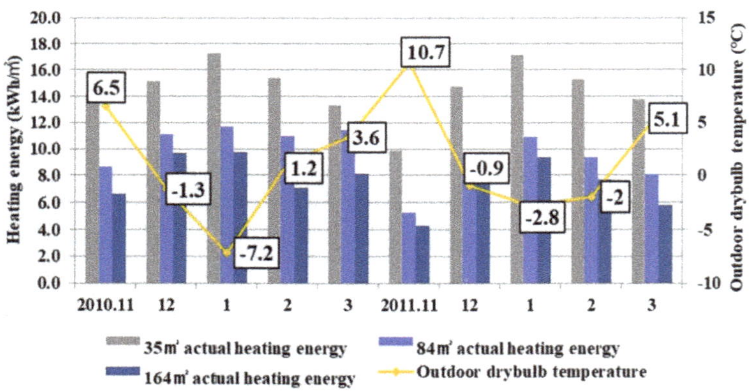

Figure 5. Comparison between the measured monthly average heating energy consumption by house size and the outside temperature.

The lowest outside temperature during the winter season of 2010 and 2011 was in January. From Figure 5, when the outside temperature of January 2010 was −7.2 °C, the measured average heating energy consumption was the highest of the year, recording 18.3 kWh/m^2·a from 35 m^2 houses, 11.7 kWh/m^2·a from 84 m^2 houses, and 9.8 kWh/m^2·a from 164 m^2 houses. From Figure 5, given that the lowest temperature of January 2011 was −2.8 °C, the measured average heating energy consumption was the highest of the year, recording 17.2 kWh/m^2·a from 35 m^2 houses, 10.9 kWh/m^2·a from 84 m^2 houses, and 9.4 kWh/m^2·a from 164 m^2 houses. Additionally, the highest outside temperature during the winter season was in November. From Figure 5, when the outside temperature of November 2010 was 6.5 °C, the measured average heating energy consumption was the lowest of the year, recording 14.4 kWh/m^2·a from 35 m^2 houses, 8.7 kWh/m^2·a from

84 m² houses, and 6.6 kWh/m²·a from 164 m² houses. From Figure 5, given that the highest temperature of November 2011 was 10.7 °C, the measured average heating energy consumption was the lowest of the year, recording 9.9 kWh/m²·a from 35m² houses, 5.3 kWh/m²·a from 84 m² houses, and 4.3 kWh/m²·a from 164 m² houses.

Since the monthly outside temperature of 2010 was lower than that of 2011, the monthly average heating energy consumption of 2010 was larger accordingly. The average heating energy consumption from 35 m² houses recorded 83.0 kWh/m²·a in 2010, larger than 70.8 kWh/m²·a recorded in 2011. A similar trend was witnessed from 84 and 164 m² houses, respectively. The graph in Figure 5 indicates that the average heating energy consumption is inversely proportional to the outside temperature.

3.3. *Maximum and Minimum Values of Measured Monthly Heating Energy Consumption by Household Size*

3.3.1. Maximum and Minimum Values of Measured Monthly Heating Energy Consumption from Households Sized 35 m²

The below Figure 6 shows the maximum and minimum values of measured monthly heating energy consumption from 35 m² houses in 2010 and 2011.

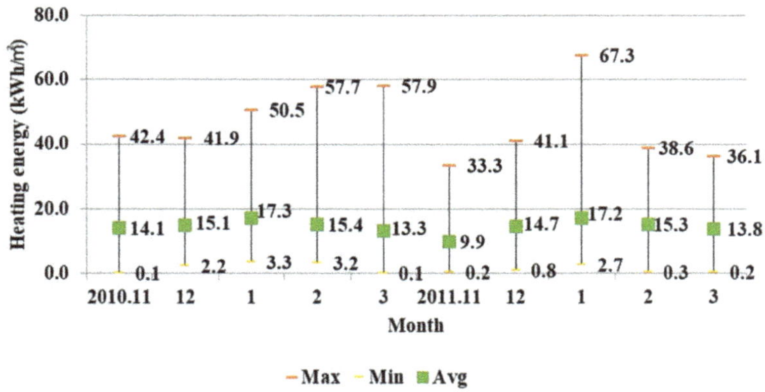

Figure 6. Maximum and minimum values of monthly heating energy from households sized 35 m².

The measured monthly average heating energy consumption in 2010 and 2011 reached the maximum in January when the temperature was the lowest and the minimum in November when the temperature was the highest during the winter season. Such findings imply that the outside temperature and the measured monthly average heating energy consumption are inversely proportional. Still, as shown in Figure 6, the maximum heating energy consumption in January 2010 from 35 m² houses was 50.5 kWh/m²·a, failing to be the highest of the year. Meanwhile, the maximum heating energy consumption in January 2011 was 67.3 kWh/m²·a, being the highest of the year.

3.3.2. Maximum and Minimum Values of Measured Monthly Heating Energy Consumption from Households Sized 84 m²

The below Figure 7 shows the maximum and minimum values of measured monthly heating energy consumption from 84 m² houses in 2010 and 2011.

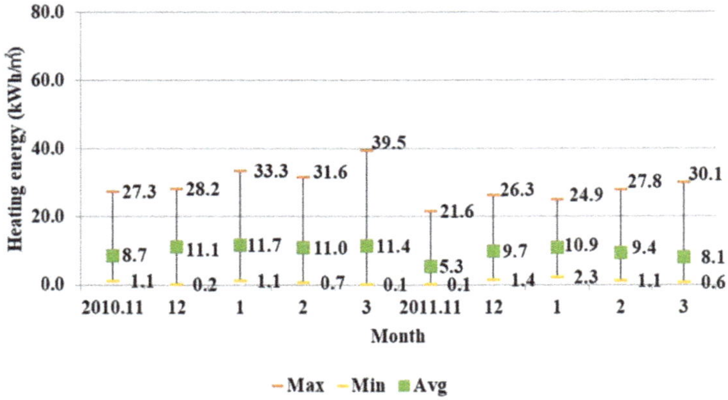

Figure 7. Maximum and minimum values of monthly heating energy from households sized 84 m^2.

From Figure 7, the measured average heating energy consumption in November 2010 and 2011 from 84 m^2 houses reached the lowest by 8.7 and 5.3 kWh/m^2·a, respectively, in reverse proportion to the highest outside temperature during the winter season. However, the measured minimum heating energy consumption was 1.1 kWh/m^2·a in November 2010, failing to be the lowest of the year, and 0.1 kWh/m^2·a in 2011, posting the lowest.

3.3.3. Maximum and Minimum Values of Measured Monthly Heating Energy Consumption from Households Sized 164 m^2

The below Figure 8 shows the maximum and minimum values of measured monthly heating energy consumption from 164 m^2 houses in 2010 and 2011.

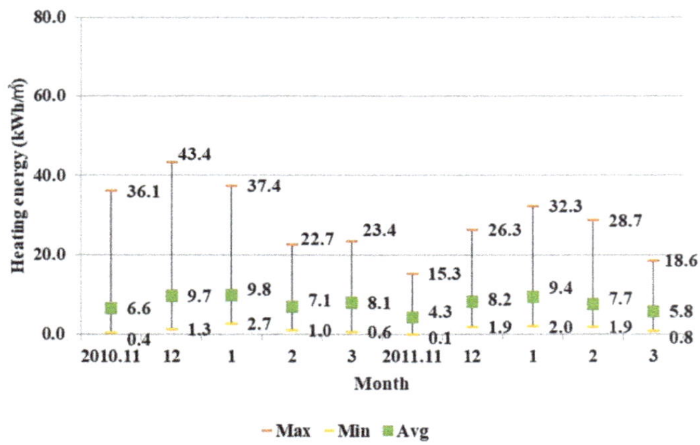

Figure 8. Maximum and minimum values of monthly heating energy from households sized 164 m^2.

From Figure 8, the measured maximum heating energy consumption from 164 m^2 houses recorded 43.4 kWh/m^2·a in December 2010, meaning that January was not the highest. However, in January 2011, the measured maximum heating energy consumption recorded 32.3 kWh/m^2·a, which was the highest of the year.

As shown in the graphs in Figures 6–8, it was found that the maximum and minimum values of measured heating energy consumption by household size were uniformly in reverse proportion to the outside temperature. Such analysis indicates that despite the same household size and outside condition, the energy consumption of households can differ by going above or below the average usage level depending on the residents' living patterns and number of residents.

3.4. Analysis of the Measured Heating Energy Consumption According to Household Location

Below the measured heating energy consumption from 84 m^2 houses as well as the variation in heating energy by floor location and room number is analyzed.

3.4.1. Comparison of the Measured Heating Energy Consumption from the Households Sized 84 m^2 by Floor Location

Figure 9 shows the distribution graph of measured heating energy consumption from all floors (by the height of household) in 2010 and 2011.

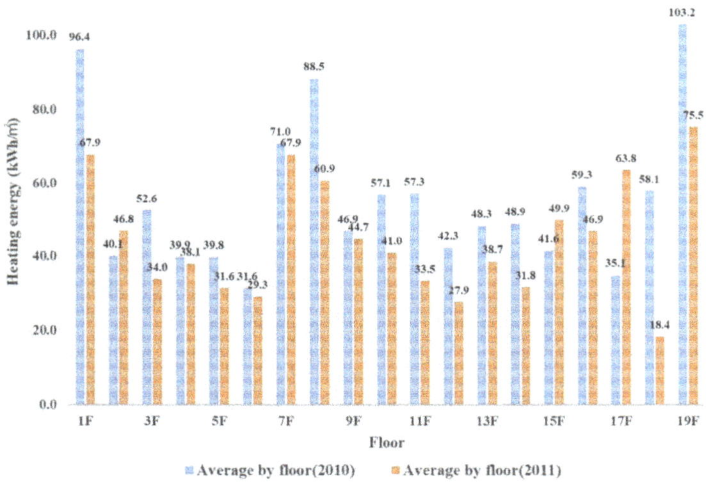

Figure 9. Distribution of measured heating energy consumption by the height of households.

Table 3 comparatively analyzes the measured heating energy consumption by each floor including the lowest, middle, and highest floors. The 1st floor was evaluated as the lowest floor since its surface bordered to the outside air. As for the mid-floors, the measured heating energy consumption from the 2nd to 18th floors was averaged. As for the highest floor, the measured heating energy consumption of the 19th floor whose roof touched the outside air was measured.

Table 3. Analysis of measured heating energy consumption by the height of households.

(Unit: $kWh/m^2{\cdot}a$)	2010	2011
Lowest floor (1st floor)	96	68
Mid-floors (average between 2nd and 18th)	50	41
Highest floor (19th floor)	103	75

As a result of comparing the measured heating energy consumption of 2010 and 2011 by floor location, the mid-floors were measured at 50 and 41 $kWh/m^2{\cdot}a$ whereas the lowest floor recorded 96 and 68 $kWh/m^2{\cdot}a$ and the highest floors posted 103 and 75 $kWh/m^2{\cdot}a$. The reason why the lowest and highest floors showed larger heating energy consumption than the mid-floors was assumed to be due to the heat loss occurring through the roof and floor.

3.4.2. Comparison of the Measured Heating Energy from the Households Sized 84 m^2 by Room Number

Figure 10 shows the distribution graph of measured heating energy consumption of all floors by room number in 2010 whereas Figure 11 shows that in 2011.

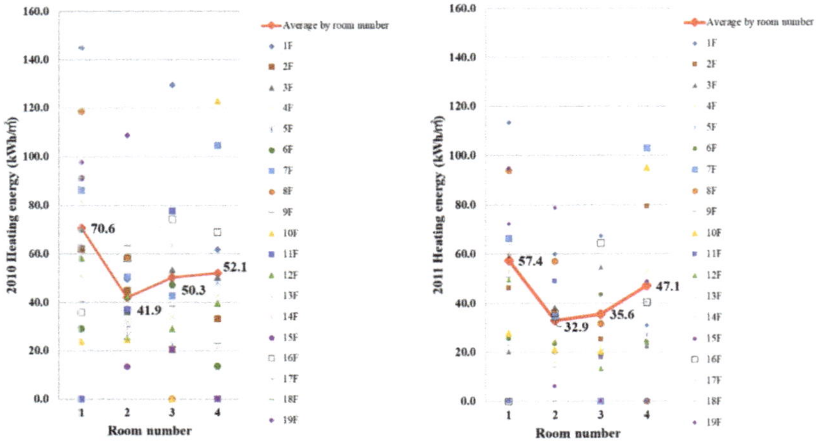

Figure 10. Distribution and measured average heating energy in 2010 and 2011 by room number.

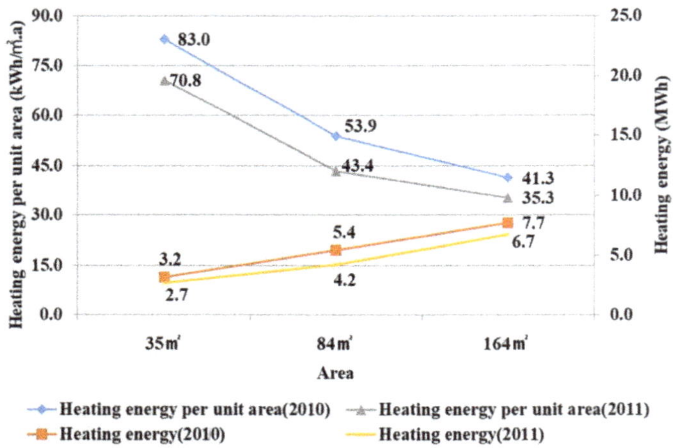

Figure 11. Measured heating consumption and heating energy per unit area.

When averaging out the distribution of measured heating energy in 2010 by room number, house no. 1 and no. 4 on the corner showed higher average heating energy consumption (70.6 and 52.1 kWh/m^2) than those from mid-floors (41.9 and 50.3 kWh/m^2·a). The reason why households on the corner-side showed higher heating energy consumption than those from mid-floors was because they had more space bordering the outside air and, thereby, leading to the rise in wall loss. The average heating energy consumption of house no. 1 and no. 4 on the corner in 2011 were 57.4 and 47.1 kWh/m^2·a, being larger than 32.9 and 35.6 kWh/m^2·a from mid-floors (household no. 2 and no. 3). Since the outside temperature was lower in 2010 than in 2011 (see Figure 5), the measured average heating energy consumption by room number was higher in 2010 than in 2011.

3.5. Analysis of the Measured Annual Heating Energy Consumption According to the Growth in Household Size

Table 4 classified the monthly heating energy consumption in 2010 from the households sized 35, 84, and 164 m^2 whereas Table 5 shows that in 2011.

Table 4. Measured annual heating energy consumption according to the growth in household size in 2010.

Size	2010~2011 (Unit: kWh)					Measured Heating Energy (Unit: MWh)
	November	December	January	February	March	
35 m^2	569	647	703	706	587	3.2
84 m^2	845	1095	1182	1099	1132	5.4
164 m^2	1152	1791	1875	1344	1491	7.7

Table 5. Measured annual heating energy consumption according to the growth in household size in 2011.

Size	2011~2012 (Unit: kWh)					Measured Heating Energy (Unit: MWh)
	November	December	January	February	March	
35 m^2	365	560	660	591	518	2.7
84 m^2	487	939	1068	915	744	4.2
164 m^2	792	1589	1782	1454	1099	6.7

As shown in Tables 4 and 5, when the household size becomes larger, the measured annual heating energy consumption increases to 3.2, 5.4, and 7.7 MWh in 2010 as well as 2.7, 4.2, and 6.7 MWh in 2011.

3.6. Analysis of the Measured Annual Heating Energy Consumption According to the Growth in Household Size

Table 6 shows the annual heating energy consumption per unit area of 2010 and 2011 as well as the flat surface of each household and building.

Table 6. Heating energy per unit area as well as flat surface of each household and building according to household size.

Household Size	Year of Measurement	Heating Energy ($kWh/m^2 \cdot a$)	Flat Surface of Each Building	Flat Surface of Each Household
35 m^2	2010	83.0	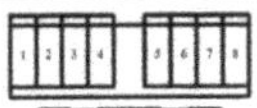	
	2011	70.8		
84 m^2	2010	53.9	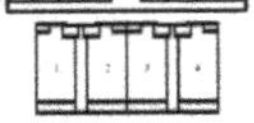	
	2011	43.4		
164 m^2	2010	41.3	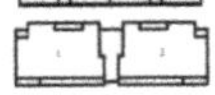	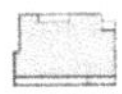
	2011	35.3		

Figure 11 shows the measured heating consumption and the heating energy per unit area.

As shown in Figure 11, when the household size becomes larger, the measured annual heating energy consumption increases whereas the heating energy per unit area decreases. To analyze such, a simulation was conducted.

4. Comparison between the Measured Average Energy Consumption and the EnergyPlus Simulation

4.1. Simulation for Analyzing the Heating Load and Heating Load Elements According to Household Size

EnergyPlus was operated for the simulation to analyze the cause of decreasing heating energy consumption per unit area according to the growth in house size. The material property of envelop composition in Table 2 was entered as an input for simulation while the KMA's 2010 Seoul region data were used to create weather data. Figure 12 shows the simulation results of heating energy consumption and heating load per unit area.

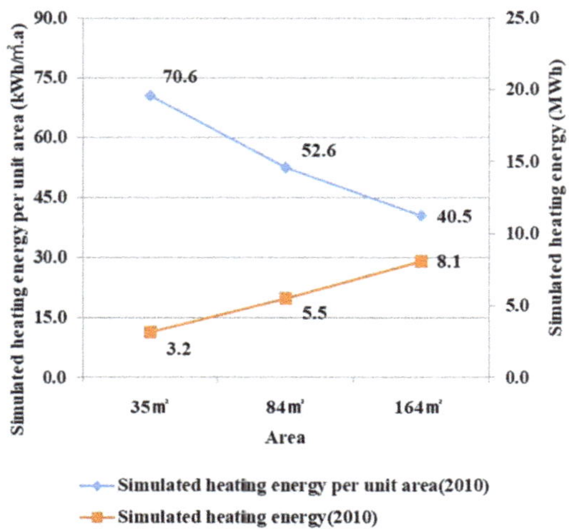

Figure 12. Simulated heating energy consumption and heating load per unit area (EnergyPlus).

According to the simulation results of Figure 12, as the house size grew, the annual heating energy consumption increased while the heating load per unit area decreased. Tables 7 and 8 examined the impact of heating load elements on the heating load as the house size grew. In the process of such analysis, the annual heating energy consumption and heating load were sorted by ventilation loss, wall loss, window loss, solar heat gain, heat from people, and heat from equipment.

Table 7. Annual heating load usage according to the growth in house size (EnergyPlus).

Simulation of Annual Heating Load Usage (Unit: MWh)							
Household Size	Ventilation Loss	Wall Loss	Window Loss	Solar Heat Gain	Heat from People	Heat from Equipment	Heating Energy Consumption
35 m^2	2.3	1.9	0.4	0.8	0.3	0.3	3.2
84 m^2	7.5	1.9	1.6	4.2	0.6	0.7	5.5
164 m^2	11.8	4.1	2.8	8.0	1.2	1.3	8.1

Table 8. Annual heating load per unit area according to the growth in house size (EnergyPlus).

Simulation of Heating Load and Elements per Unit Area (Unit: $kWh/m^2 \cdot a$)							
Household Size	Ventilation Loss	Wall Loss	Window Loss	Solar Heat Gain	Heat from People	Heat from Equipment	Heating Load
35 m^2	50.5	41.2	9.0	18.2	6.0	6.0	70.6
84 m^2	71.8	18.4	14.8	40.2	5.9	6.3	52.6
164 m^2	58.6	20.2	13.8	39.7	5.9	6.4	40.5

When splitting up the annual heating energy consumption per unit area, the heating load usage per unit area also saw equal values or increase in window loss, solar heat gain, heat from people, and heat from equipment as the house size grew. According to Table 8, as the house size grow by about 2.4 times from 35 to 84 m^2, the heating load decreases due to the declining wall loss and the rising solar heat gain. According to Table 9, the households sized 35 and 84 m^2 both show equal heating load usage (1.9 MWh) for wall loss regardless of the growth in house size. Therefore, when splitting up such households per unit area, Table 8 shows that the wall loss per unit area greatly decreases from 41.2 kWh/m^2·a for the 35 m^2 house to 18.4 kWh/m^2·a for the 84 m^2 house. If referring to the floor plan of different house sizes in Table 8, the 35 m^2 house is vertically long with a relatively large wall area whereas the 84 m^2 house is horizontally long with a large window area. Although the wall area is similar between two different house sizes, the window area greatly increases in case of the 84 m^2 house. Therefore, although the households sized 35 and 84 m^2 both show 1.9 MWh for wall loss, the solar heat gain jumps about fivefold from 0.8 MWh for the 35 m^2 house to 4.2 MWh for the 84 m^2 house, reducing the heating load from that of the 35 m^2 house. According to Table 7, as the house size doubles from 84 to 164 m^2, the increased rate of ventilation loss falls short of twofold from 7.5 to 11.8 MWh. Therefore, when splitting up such households per unit area, Table 8 shows that the ventilation loss per unit area reduces from 71.8 kWh/m^2·a for the 35 m^2 house to 58.6 kWh/m^2·a for the 84 m^2 house. In summary, as the house size grows, the ventilation loss and wall loss among the elements of annual heating load usage will remain the same or increase but to a lesser extent than the growth rate of house size. Such factors lead to a decline in heating load per unit area.

4.2. Ratio between Volume and Elevation Area (AV Value) According to the Growth in Household Size

Although the shape of the building is the same, the amount of load varies according to the W/D ratio while the length, width, and height of the building have an impact on the solar heat gain and the amount of external heat loss [11]. For this reason, the index for building volume ratio is used to design in a way to reduce the heat loss from the external building environment. Figure 13 shows the calculation of the ratio between volume and elevation area based on the following Equation (1).

$$\text{Ratio between volume and floor area (AV value)} = \text{Ratio between floor area(A)/volume(v)} \quad (1)$$

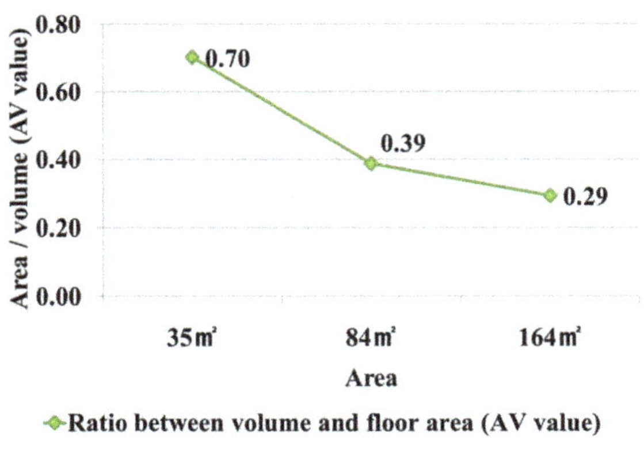

Figure 13. Ratio between volume and elevation area (AV value).

According to Figure 13, as the house size grows, the AV values decreases to 0.7, 0.39, and 0.29. The lower the ratio between volume and elevation area (AV value) is, the lesser the impact from the heat gain or heat loss from radiation and convection [11]. In summary, as the house size grows in Figure 12, the annual heating load per unit area decreases because the impact of heat gain or heat loss from convection becomes smaller.

4.3. EnergyPlus Simulation

For the overview of simulation, refer to Section 2 while the schedule for people and equipment is set up as follows.

4.3.1. People Schedule Setup

As shown in Figure 14, the ratios of occupancy by residents during weekdays and weekends were entered in EnergyPlus as an hourly schedule by using the 2009 time use survey data from Statistics Korea for simulation [12].

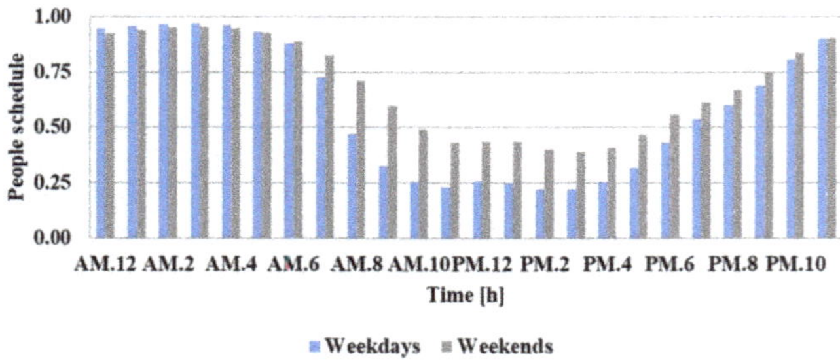

Figure 14. People schedule from Statistics Korea's survey.

4.3.2. Indoor Equipment Schedule Setup

The schedule of using indoor equipment from Figure 15 referred to the estimation method of housing sector electricity consumption using the time for living activities [13,14]. By referring to the estimation method of electricity consumption based on statistical data such as the population and housing survey and people's time use survey, this schedule of using indoor equipment was entered into EnergyPlus for simulation.

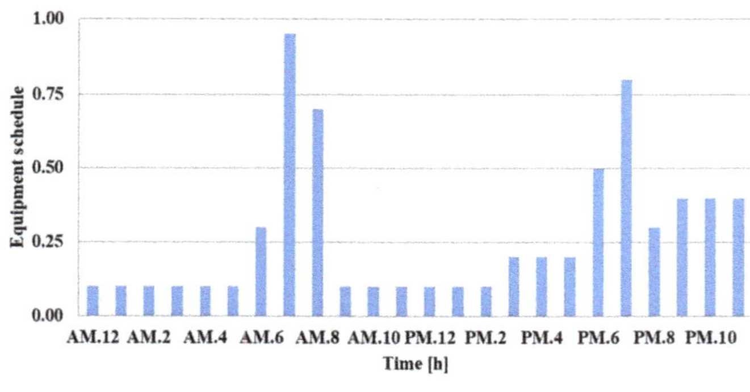

Figure 15. Equipment schedule from Statistics Korea's survey.

4.4. Comparison between the Measured Average Energy Consumption and the Simulation

Figure 16 shows the comparison of heating energy simulation results from mid-floor houses and 84 m^2 houses using the measured average heating energy consumption of 84 m^2 houses in 2010 and Seoul region's weather data in 2010.

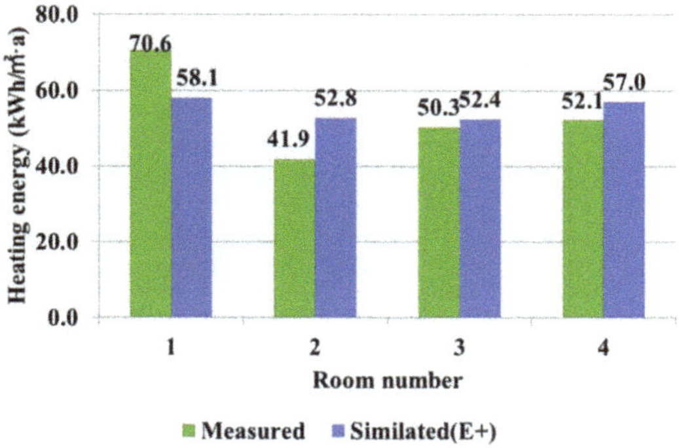

Figure 16. Comparison between the measured average heating energy consumption and the EnergyPlus simulation.

When comparing between the measured average heating energy consumption by room and the simulation results, 50.3 kWh/m^2·a from room No. 3 was similar to the simulation result of 52.4 kWh/m^2·a. Additionally, room No. 1 and 4 of corner houses that border to outside air showed larger heating energy consumption than mid-floor houses in terms of both simulation and actual measured results.

4.5. Analysis of the Variation in Heating Energy Consumption According to Residents' Living Patterns

Through the survey on measured heating energy, it was found that the heating energy consumption by each household differed despite the same size and envelope performance [15–18]. This was due to the residents' living patterns, occupancy patterns, usage of home appliances, and number of residents. The following Table 9 indicates the survey items and details used to analyze the impact on the heating energy consumption.

Table 9. Survey details and simulation overview.

Items	Survey Details		
No. of residents	Survey		
Occupancy schedule	Survey on weekdays and weekends		
Heating on/off schedule	Survey of heating on/off per hour during weekdays/weekends on a monthly basis (November, December, January, February, March)		
Heating setpoint temperature	23 °C (References [19–22])		
Equipment and lighting schedule	Survey from Statistics Korea (References [13,14])—Figure 16		
Air change per hour (ACH)	Heating Non-heating	0.7 ACH 2.0 ACH	Refer to the Operational Regulation on the Certification of Building Energy Efficiency [10]
Occupancy density	1.44 W/m^2		
Equipment density	3.24 W/m^2		

Among the 84-m^2-sized mid-floor houses of room No. 2 and 3, a total of 12 households (6 households showing average energy consumption, 3 households showing more than the average, and 3 households showing less than the average) were selected for the survey. The simulation reflecting such survey results was compared with the measured heating energy consumption.

4.6. Comparison between the Simulation and the Actually Measured Value from the Households Consuming the Average

Table 10 shows the survey results of 6 households consuming the average heating energy of 41.9~50.3 kWh/m^2·a among the 84-m^2-sized mid-floor houses of room No. 2 and 3 from Figure 16.

Table 10. Survey results of residents consuming the average heating energy.

Case	Selected Household	No. of Residents	Survey Results
1	Room No. 2C2	4	Salary worker, housewife, high school senior, high school junior
2	Room No. 3C3	3	Salary worker, salary worker, high school graduate
3	Room No. 5C3	4	Salary worker, housewife, elementary 1st grade, kindergartener
4	Room No. 6C3	3	Elderly couple, elementary 3rd grade
5	Room No. 1102	4	Salary worker, housewife, high school graduate, university student
6	Room No. 1303	4	Salary worker, salary worker, elementary 3rd grade and 6th grade

Table 11 and Figure 17 indicate the survey results of heating hour and occupancy hour of 6 households consuming a heating energy of 43.6~51.5 kWh/m^2·a as well as the simulation and actual measured values.

Table 11. Survey results of heating energy, heating hour and occupancy hour (Case 1–6).

Case	Heating Energy [kWh/m^2·a]		Heating Hour [Hours/Day]			Occupancy Hour [Hours/Day]		
	Actually Measured Value	EnergyPlus	Weekday	Weekend	Average	Weekday	Weekend	Average
1	49.9	47.4	14.4	14.4	14.4	14.0	16.0	14.6
2	51.5	56.9	18.4	18.4	18.4	17.8	18.5	18.0
3	43.6	47.9	12.0	24.0	15.4	16.7	19.7	17.5
4	49.3	50.3	15.4	15.4	15.4	21.0	24.0	21.9
5	47.3	50.4	15.0	15.0	15.0	12.8	14.8	13.3
6	50.2	53.5	15.2	18.0	16.0	12.5	14.0	12.9
	Survey average		15.1	17.5	15.8	15.8	17.8	16.4

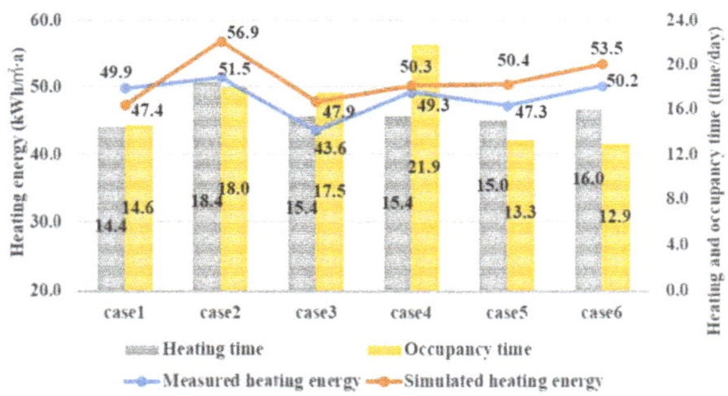

Figure 17. Comparison between the measured heating energy consumption and simulation as well as living pattern (case 1–6).

According to Figure 17, the simulation reflecting the survey results of households consuming the average heating energy amount showed similar results to the measured heating energy consumption. Except for case 1, cases 2~5 showed slightly higher simulation results than the measured heating energy consumption. Among the 6 cases in Table 11, the households consumed the average heating energy for 15.8 h/day and for 16.4 h/day. The households from cases 1, 5, and 6 occupied an average of 14.6, 13.3, and 12.9 h/day and had similar numbers of residents (4 people) as shown in Table 12. According to Table 13, the households from case 1 consumed the average heating for 14.4 h/day at 47.4 kWh/m^2·a of heating energy whereas the households from cases 5 and 6 consumed the average heating for 15 and 16 h/day at 50.4 and 53.5 kWh/m^2·a, respectively.

Table 12. Survey results of residents consuming more than the average heating energy.

Case	Selected Household	No. of Residents	Survey Results
7	Room No. 703	1	Freelancer
8	Room No. 902	3	Elderly couple, salary worker
9	Room No. 1603	4	Salary worker, housewife, baby 1 and 2

Table 13. Survey results of heating energy, heating hour, and occupancy hour (Case 7-9).

Case	Heating Energy [kWh/m²·a]		Heating Hour [Hours/Day]			Occupancy Hour [Hours/Day]		
	Actually Measured Value	EnergyPlus	Weekday	Weekend	Average	Weekday	Weekend	Average
7	63.5	65.9	21.6	21.6	21.6	22.1	21.0	21.8
8	63.3	59.6	21.0	21.0	21.0	20.0	21.7	20.5
9	70.2	61.4	24.0	24.0	24.0	21.0	22.3	21.4
Survey average			22.2	22.2	22.2	21.0	21.6	21.2

4.7. Comparison between the Simulation and the Actually Measured Value from the Households Consuming More Than the Average

Table 12 shows the survey results of 3 households consuming more than the average heating energy of 41.9~50.3 kWh/m²·a among the 84-m²-sized mid-floor houses from Figure 16.

Table 13 and Figure 18 indicate the survey results of heating hour and occupancy hour of 3 households consuming a heating energy of 63.3~70.2 kWh/m²·a as well as the simulation and actual measured values.

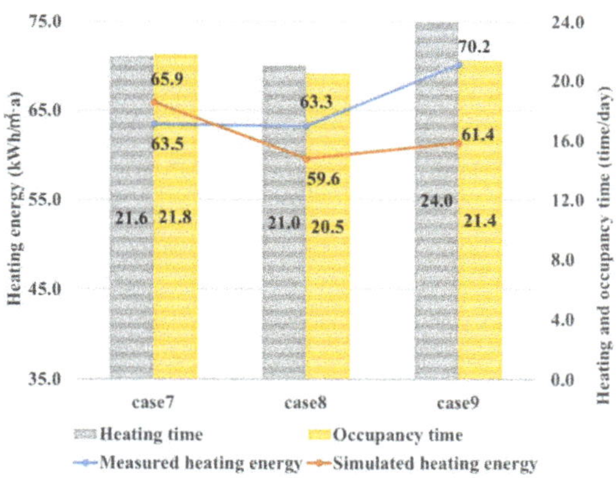

Figure 18. Comparison between the measured heating energy consumption and simulation as well as living pattern (case 7–9).

According to Figure 18, except for case 7, cases 8 and 9 showed slightly lower simulation results than the measured heating energy consumption. From Table 13, the households consumed more than the average heating energy for 22.2 h/day and occupied for 21.2 h/day. Accordingly, the heating hour and occupancy hour increased compared to the households from cases 1~6 from Table 11 that consumed the average heating energy. From Table 13, the households from cases 7 and 8 consumed the average heating energy for 21.6 and 21.0 h/day and occupied for 21.8 and 20.5 h/day. Although the heating hour and occupancy hour of the two households were similar, the heating energy recorded differently by 65.9 and 59.6 kWh/m²·a, respectively. Such difference was because the household from case 7 had 1 resident whereas the household from case 8 had 3 residents showing higher heat from people and thereby consumed less heating energy.

4.8. Comparison between the Simulation and the Actually Measured Value from the Households Consuming Less Than the Average

Table 14 shows the survey results of 3 households consuming less than the average heating energy of 41.9~50.3 kWh/m^2·a among the 84-m^2-sized mid-floor houses from Figure 16.

Table 14. Survey results of residents consuming less than the average heating energy.

Case	Selected Household	No. of Residents	Survey Results
10	Room No. 402	2	Salary worker, salary worker
11	Room No. 1002	4	Salary worker, salary worker, university student 1 and 2
12	Room No. 1403	5	Salary worker, salary worker, university student 1, 2, and 3

Table 15 and Figure 19 indicate the survey results of heating hour and occupancy hour of 3 households consuming a heating energy of 16.2~24.6 kWh/m^2·a as well as the simulation and actual measured values.

Table 15. Survey results of heating energy, heating hour, and occupancy hour (Case 10-12).

Case	Heating Energy [kWh/m^2·a]		Heating Hour [Hours/Day]			Occupancy Hour [Hours/Day]		
	Actually Measured Value	EnergyPlus	Weekday	Weekend	Average	Weekday	Weekend	Average
10	18	20.7	4.4	5.0	4.6	10.5	18.0	12.6
11	24.6	28.4	5.8	7.2	6.9	9.3	14.0	10.6
12	16.2	19.0	4.2	4.2	4.2	8.4	11.8	9.4
	Survey average		5.1	5.5	5.2	9.4	14.6	10.9

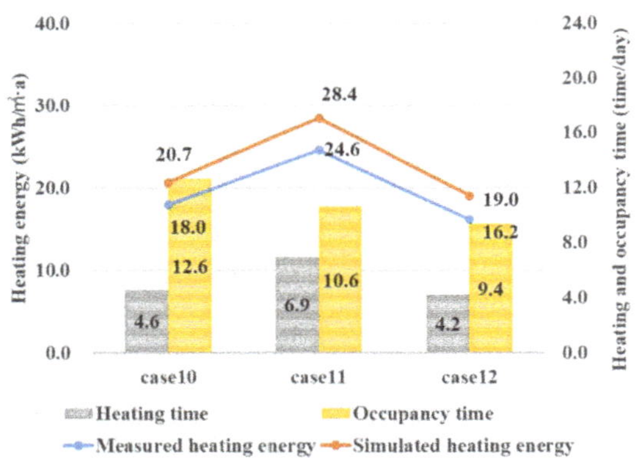

Figure 19. Comparison between the measured heating energy consumption and simulation as well as living pattern (case 10–12).

From Figure 19, the 3 households consuming less than the average heating energy showed a slightly higher simulation than the measured heating energy consumption. The households consumed less than the average heating energy for 5.2 h/day and occupied for 10.9 h/day. Accordingly, their heating hours and occupancy hours declined compared to the households from cases 1~6 that consumed the average heating energy for 15.8 h/day and occupied for 16.4 h/day and greatly differed from the households from cases 7~9 that consumed more than the average heating energy for 22.2 h/day and occupied for 21.2 h/day.

According to Table 14, the households from cases 11 and 12 had 4 and 5 residents, respectively, and their occupancy hours were similar by showing 10.6 and 9.4 h/day from Table 15. Meanwhile, the average heating hour declined from 6.9 h/day for case 11 to 4.2 h/day for case 12. Accordingly, as a result of simulation in Table 15, the heating energy declined from 28.4 kWh/m^2·a for case 11 to 19.0 kWh/m^2·a for case 12. The measured heating energy consumption also dropped from case 11 to case 12. According to the households from cases 10 and 12 from Table 15, the average heating hour was similar by showing 4.6 and 4.2 h/day. However, the occupancy hour of 12.6 h/day for case 10 was higher than 9.4 h/day for case 12. Still, the simulation results showed similar level of heating energy by posting 20.7 and 19.0 kWh/m^2·a, respectively. This implied that the impact from occupancy hour was small. If the heating hour was similar, the heating energy showed a similar level as well.

5. Conclusions

The results of this study are outlined as follows:

(1) Since the outside temperature in the winter season of 2010 was lower than that of 2011, the annual heating energy consumption was higher in 2010 than in 2011. As the average heating energy of 84 m^2 houses was the lowest in November (showing the highest outside temperature amid the winter season) by recording 8.7 kWh/m^2·a in 2010 and 5.3 kWh/m^2·a in 2011 whereas it was the highest in January (showing the lowest outside temperature amid the winter season) by recording 11.7 kWh/m^2·a in 2010 and 10.9 kWh/m^2·a in 2011, it was found that the outside temperature was inversely proportional to the heating energy consumption. Meanwhile, the maximum and minimum heating energy values were not uniformly in reverse proportion to the outside temperature. That was because the heating energy consumption of the households of the same size and condition could be more than or less than the average due to the residents' living pattern and number of residents.

(2) The average heating energy consumption of mid-floors was 50 kWh/m^2·a in 2010 and 41 kWh/m^2·a in 2011. However, the average heating energy consumption of the lowest floor was 96 kWh/m^2·a in 2010 and 68 kWh/m^2·a in 2011 while the highest floor was 103 and 75 kWh/m^2·a, respectively. The reason behind higher energy consumption of highest/lowest floors was that those floors suffered more heat loss through the roof and floor than mid-floors. Additionally, when analyzing the measured heating energy consumption of 2010 by room number, household room no. 1 and no. 4 from the corner-side having larger area exposed to the outside air showed higher average heating energy consumption (57.4 and 47.1 kWh/m^2·a) than mid-floors. (32.9 and 35.6 kWh/m^2·a) Research will be needed to reduce the difference by applying phase change materials, etc., in the future to the part where the heating load of the lowest and highest floor is larger than that of the mid-floors.

(3) As the house size grows, the measured annual heating energy of 2010 and 2011 increases whereas the heating energy per unit area decreases. As a result of analyzing the heating load elements per unit area through simulation, when the house size expands from 35 to 84 m^2 by about 2.4 times, the wall loss per unit area decreases and the solar heat gain increases and, thereby, reducing the heating load. When the unit area doubles from 84 to 164 m^2, the ventilation loss jumps from 7.5 to 11.8 MWh by not more than twofold. Accordingly, the ventilation loss per unit area of 84 m^2 houses declines. Among each element of annual heating load usage, the ventilation loss and wall loss remained the same or increased as the house size grew but at a lesser rate than the house size growth rate. Therefore, the heating load per unit area declined.

(4) As the house size increased, the ratio between volume and elevation area (AV value) dropped to 0.7, 0.39, and 0.29. The lower the ratio between volume and elevation area (AV value) is, the lesser the impact from the heat gain or heat loss from radiation and convection. In conclusion, as the house size grows, the annual heating load per unit area decreases.

(5) The following is the outcome of the simulation reflecting the survey on the living pattern of households that consumed the average, more than the average, and less than the measured average heating energy. Households consumed the average heating energy for 15.8 h/day and occupied for 16.4 h/day. Households consumed more than the average heating energy for 22.2 h/day and occupied for 21.2 h/day, meaning 6.4 extra hours than those consuming the average heating energy. Households consumed less than the average heating energy for 5.2 h/day and occupied for 10.9 h/day, meaning 10.6 less hours/day than those consuming the average heating energy and 17 less hours/day than those consuming more than the average heating energy.

This study conducted a comparative analysis on the variation in heating energy consumption according to the residents' living pattern and behavior of usage as a method of energy savings. The outcome will serve as basis data in efficiently managing actual energy users in consideration of their behavior of energy usage as well as establishing policies and educational directions. For in-depth analysis in future studies, verification through long-term measured data for more than two years, performing various climate-specific analyses for universal application, and an increase in the number of samples when investigating resident patterns are necessary.

Author Contributions: Conceptualization, S.-J.K.; methodology, D.-Y.P.; formal analysis, S.-J.K.; data curation, D.-Y.P.; writing—original draft preparation, S.-J.K.; writing—review and editing, D.-Y.P.; supervision, D.-Y.P. All authors have read and agreed to the published version of the manuscript.

Funding: This work was supported by Korea Institute of Planning and Evaluation for Technology in Food, Agriculture and Forestry (IPET) through Agri-Food Export Business Model Development Program, funded by Ministry of Agriculture, Food and Rural Affairs (MAFRA) (grant number: 321073-02).

Institutional Review Board Statement: Not applicable.

Informed Consent Statement: Not applicable.

Data Availability Statement: The data presented in this study are available on request from the corresponding author. The data are not publicly available due to the supporting project involving a confidentiality agreement.

Acknowledgments: This paper is based on the author's master's thesis.

Conflicts of Interest: The authors declare no conflict of interest.

References

1. Carlier, R.; Dabbagh, M.; Krarti, M. Energy Performance of Integrated Wall and Window Switchable Insulated Systems for Residential Buildings. *Energies* **2022**, *15*, 1056. [CrossRef]
2. Yarramsetty, S.; Rohullah, M.S.; Sivakumar, M.V.N. An investigation on energy consumption in residential building with different orientation: A BIM approach. *Asian J. Civ. Eng.* **2020**, *21*, 253266. [CrossRef]
3. Spanodimitriou, Y.; Ciampi, G.; Scorpio, M.; Mokhtari, N.; Teimoorzadeh, A.; Laffi, R.; Sibilio, S. Passive Strategies for Building Retrofitting: Performances Analysis and Incentive Policies for the Iranian Scenario. *Energies* **2022**, *15*, 1628. [CrossRef]
4. Gillott, M.; Hall, M. Domestic energy and occupancy: A novel post-occupancy evaluation study. *Int. J. Low-Carbon Technol.* **2010**, *5*, 148157.
5. Banker, N.; Steemers, K. *Energy and Environment in Architecture*; Taylor & Francis: London, UK, 2005. [CrossRef]
6. Yun, G.Y.; Kong, H.J.; Kim, J.T. A Field Survey of Occupancy and Air-Conditioner Use Patterns in Open Plan Offices. *Indoor Built Environ.* **2011**, *20*, 137147.
7. Buttitta, G.; Finn, D. A high-temporal resolution residential building occupancy model to generate high-temporal resolution heating load profiles of occupancy-integrated archetypes. *Energy Build.* **2020**, *206*, 109577. [CrossRef]
8. Shin, M.S.; Rhee, K.N.; Jung, G.J. Optimal heating start and stop control based on the inferred occupancy schedule in a household with radiant floor heating system. *Energy Build.* **2020**, *209*, 109737. [CrossRef]
9. Shin, M.S.; Rhee, K.N.; Lee, E.T.; Jung, G.J. Performance evaluation of CO2-based ventilation control to reduce CO_2 concentration and condensation risk in residential buildings. *Build. Environ.* **2018**, *142*, 451–463. [CrossRef]
10. Korea Energy Agency. *Regulations for the Operation of Building Energy Efficiency Certification System*; Korea Energy Agengy: Seoul, Korea, 2011.

11. Korea Presidential Commission on Architecture Policy. *Guidelines for Passive Design of Buildings*; Korea Presidential Commission on Architecture Policy: Seoul, Korea, 2012.
12. Statistics Korea (KOSTAT). *Life Time Statics*; Statistics Korea (KOSTAT): Daejeon, Korea, 2009.
13. Yoo, S.Y.; Kim, J.Y. Analysis of Internal Heat Gain and System Operation Patterns in Residential buildings. In Proceedings of the Society of Air-Conditioning and Refrigerating Engineers of Korea, Kangwon, Korea, 8 July 2011; pp. 683–686.
14. Yoo, J.H. Method for Estimating Electricity Consumption of Residential Sectors by National Time Use Survey. In Proceedings of the Journal of the Korean Architectural Association, Chungbuk, Korea, 23 October 2010; Volume 30, p. 501502.
15. ASHRAE. *ASHRAE Handbook: Fundamentals*; American Society of Heating, Refrigerating and Air-Conditioning Engineers, Inc.: Peachtree Corners, GA, USA, 2009.
16. United States Department of Energy. Energy Plus Manual Engineering Reference, Version 3.0. USA. Available online: https://bigladdersoftware.com/epx/docs/8-9/engineering-reference/ (accessed on 18 April 2022).
17. Wolfgang Feist, Passive House Planning Package, Passiv Haus Institut. 2007. Available online: https://www.google.com.hk/url?sa=t&rct=j&q=&esrc=s&source=web&cd=&cad=rja&uact=8&ved=2ahUKEwjGvvPDvfr3AhXOxosBHWsTCswQFnoECAMQAQ&url=http%3A%2F%2Fwookware.org%2Ffiles%2FPHPP.pdf&usg=AOvVaw0EASMx-Qz8TPqH3H0Jk_DR (accessed on 18 April 2022).
18. Cheung, C.K.; Fuller, R.J.; Luther, M.B. Energy-efficient envelope design for high-rise apartments. *Energy Build.* **2005**, *37*, 37–48. [CrossRef]
19. Bae, N.R.; Chun, C.Y. Changes of Residents' Indoor Environment Control Behavior as a Result of Provided Education and Environmental Information. *J. Korean Archit. Assoc.* **2008**, *24*, 285293.
20. Chun, C.Y.; Kim, H.J.; Bae, N.R. Research on Thermal Environment in Apartments' Living rooms and Residents' Control Behaviors of a Thermostat. *J. Archit. Inst. Korea Struct. Constr.* **2005**, *21*, 209216.
21. Choi, W.K.; Kim, H.J.; Lee, H.G.; Suh, S.J. Theoretical Study on the Patterns of Temperature Variation and Thermal Load According to Each Household Heating and Cooling in Apartment. *J. Korean Archit. Assoc.* **2007**, *23*, 279288.
22. Yu, J.; Yang, C.; Tian, L. Low-energy envelope design of residential building in hot summer and cold winter zone in China. *Energy Build.* **2008**, *40*, 15361546. [CrossRef]

designs

MDPI

Article

Open-Source Design and Economics of Manual Variable-Tilt Angle DIY Wood-Based Solar Photovoltaic Racking System

Nicholas Vandewetering [1], Koami Soulemane Hayibo [2] and Joshua M. Pearce [2,3,*]

1 Department of Civil & Environmental Engineering, Western University, London, ON N6A 3K7, Canada; nvandew@uwo.ca
2 Department of Electrical & Computer Engineering, Western University, London, ON N6A 3K7, Canada; khayibo@uwo.ca
3 Ivey School of Business, Western University, London, ON N6A 3K7, Canada
* Correspondence: joshua.pearce@uwo.ca

Abstract: Fixed-tilt mechanical racking, consisting of proprietary aluminum extrusions, can dominate the capital costs of small-scale solar photovoltaic (PV) systems. Recent design research has shown that wood-racking can decrease the capital costs of small systems by more than 75% in North America. To determine if wood racking provides enough savings to enable labor to be exchanged profitably for higher solar electric output, this article develops a novel variable tilt angle open-source wood-based do-it-yourself (DIY) PV rack that can be built and adjusted at exceptionally low costs. A detailed levelized cost of electricity (LCOE) production analysis is performed after the optimal monthly tilt angles are determined for a range of latitudes. The results show the racking systems with an optimal variable seasonal tilt angle have the best lifetime energy production, with 5.2% more energy generated compared to the fixed-tilt system (or 4.8% more energy, if limited to a maximum tilt angle of 60°). Both fixed and variable wooden racking systems show similar LCOE, which is only 29% of the LCOE of commercial metal racking. The results of this study indicate that the novel variable tilt rack, whether used as a small-scale DIY project or scaled up to fulfill larger energy demands, provides both the lowest cost option even when modest labor costs are included and also may provide specific advantages for applications such as agrivoltaics.

Keywords: open source; photovoltaic; racking; solar energy; biomaterials; wood; mechanical design; balance of systems; renewable energy; do-it-yourself

Citation: Vandewetering, N.; Hayibo, K.S.; Pearce, J.M. Open-Source Design and Economics of Manual Variable-Tilt Angle DIY Wood-Based Solar Photovoltaic Racking System. *Designs* **2022**, *6*, 54. https://doi.org/10.3390/designs6030054

Academic Editors: Shi-Jie Cao and Wei Feng

Received: 11 May 2022
Accepted: 9 June 2022
Published: 14 June 2022

1. Introduction

Solar photovoltaic (PV) technology is a well-established distributed sustainable energy technology [1]. There have been massive PV price declines in the last decade [2,3], which have reduced the levelized cost of electricity (LCOE) [4] to generally be the least-cost option on a large scale [5,6]. In turn, this has made PV the most rapidly expanding electricity generation source (notably the former dominant source, coal is in decline) [6,7]. Worldwide dominance of solar PV as a source of electricity continues to be an economic cost [8]. There is some evidence that at least part of this economic barrier continues to be the focus on large-scale PV deployments [9]. This is despite the fact that small solar home systems can play an important role in achieving the U.N.'s 'Sustainable Energy for All' goals [10]. For grid-tied PV systems that have a lower LCOE than the retail rate of grid electricity (e.g., surpassing grid parity), there is massive interest among consumers because they can save money with lower-cost solar electricity [11]. This is particularly true if real net metering is maintained (where prosumers are credited an equivalent economic amount for the electricity they use and send to the grid even if it is not the full value of solar electricity) [12]. Despite clear lifetime economic benefits, the capital expenditures (CAPEX) of PV systems can be challenging for many non-wealthy consumers, both in developing [13] and developed countries [14].

Historically, large-scale centralized PV costs were lower than small-scale distributed costs, and this type of growth continues to dominate the market [9]. The use of small open-source do-it-yourself (DIY) or plug-and-play solar [15,16] has lower costs per unit power than possible for large-scale systems [17]. Among residential systems in the U.S., 8–10 kW systems were roughly 16% less expensive than 2–4 kW systems, and non-residential systems over 1000 kW were 43% less expensive than systems under 10 kW [18]. This small-scale approach deserves appropriate regulation [16,19–21] as it could have a substantial economic impact [22]. (Part of the reason small-scale PV systems can be more economic than large-scale industrial PV deployments is that DIY or plug-and-play systems avoid most of the soft costs associated with PV systems. They still have a large CAPEX for many individuals (e.g., even with a ~9-year simple payback period for PV systems under warranty for 25 years that provide an internal rate of return in the double digits, some consumers cannot afford to buy 9 years of electricity upfront even if they receive 'free' electricity for another 16 years or more)).

The majority of PV system cost declines have focused on the modules themselves (historically the largest component cost), but now the relative cost of the balance of systems (BOS) made up of racking, electronics, and wiring has become more important [3,23,24]. For the smallest systems, the simple mechanical racking, made up of various proprietary and costly aluminum extrusion profiles that have barely reduced costs [25], can dominate the cost of a whole system. For example, PV module spot prices are currently USD 0.19/W [26], so three 400 W modules cost USD 228 while a three-module rack costs USD 535 (list price USD 635) [27] (e.g., ~2× the cost of the PV) and a three-module pole mount is selling for USD 1194 [28] (e.g., ~5× the cost of the PV).

To combat the distorted market for small-scale PV systems, there has recently been several open-source hardware-based PV racks described, including low-tilt angle arrays, small-scale mobile PV arrays [29], cable-based systems for flat roofs [30], ground-mounted systems at the equator in the developing world [31], tensegrity structures [32], and post-market building-integrated PV (BIPV) [33]. These are all fixed-tilt systems, which are common for industrial PV. This is because simply cleaning large-scale systems can cost anywhere from USD 15 to USD 35 per module [34], so additional labor-intensive work, such as adjusting the tilt of multiple arrays, may be cost-prohibitive. A system in Ontario that is manually adjusted to the optimal tilt angle twice a year generates revenue of USD 1.90 per module each year [35]. All of these systems are fixed-tilt, based on the assumption in the large-scale PV industry that it does not make economic sense to pay people to seasonally adjust the tilt angle of modules. Manual variable-tilt small-scale PV systems exist but have high racking costs and are limited in availability. Therefore, historically, it has generally not been worth hiring personnel to adjust tilt angles. Is this still true, however, if the CAPEX of racking can be radically reduced? More recently, a fixed-tilt wood-based PV DIY rack showed decreases in costs between 49% and 77% compared to proprietary small-scale metal racks [36].

To determine if there has been a shift in the economics of manual tilt adjustment for PV with lower-cost racking, this article develops a variable tilt angle open-source wood-based PV rack following the evolution of the design in [36]. The complete designs and bill of materials (BOM) of the variable tilt angle rack are provided along with basic instructions and are released with an open-source license that will enable anyone to fabricate this rack in their community. A detailed energy production analysis is performed after the optimal monthly tilt angles are determined for a case study location, London, Ontario. The economic analysis accounts for the cost of wood, purchased locally. Then a careful study of the energy production of DIY tilt angle adjustment is run over sensitivity of tilts per year. The energy analysis and economics analysis are combined to evaluate the levelized cost of electricity (LCOE) of the multi-tilt wood racking. The LCOE of the multi-tilt wood racking is then compared to the LCOE of fixed-tilt wood racking, fixed-tilt commercial metal racking, and single-axis tracking commercial metal racking. A sensitivity analysis is then run on the maximum achieved tilt angle as well as the labor cost involved in building the system.

The results of this study are discussed in the context of upending the assumption that large scale leads to better economic performance when using open-source DIY design for increasing PV deployments.

2. Materials and Methods

All variables in this report adopt the same abbreviations from Appendix A of the [36] design.

2.1. Material Properties

Construction-grade pressure-treated lumber and hardware purchased from typical hardware stores were used for this design. Material properties from the design in [36] were used for the build of this system. The resisting capacities of each of the wooden members can be calculated by following Appendix A of the design in [36]. Throughout the structural analysis of this design, the limitations outlined in [36] shall not be exceeded.

2.2. Material Stability

Most softwoods are natural organic material subject to deterioration under high moisture conditions, making them an inadequate building material for outdoor use. Some species of wood, such as mahogany and redwood, can naturally fight against decay but are too expensive to be used for outdoor structural use. The best alternative is pressure-treated SPF (Spruce, Pine, Fir) lumber, which offers effective moisture resistance, high availability, and low costs, thus making it an ideal material for structural use in wet climates. Pressure-treated lumber can last upwards of 40 years depending on the moisture conditions [37], making it suitable to use for PV modules with a warranty and expected life of 25 years.

Metal racks are also subject to material degradation that shorten their lifespan due to cross-sectional rusting and corrosion under moist conditions. To account for degradation, engineers apply specified resistance factors that reduce the capacities of both metal and wood structures based on material imperfections under wet conditions. The resistance factors for steel and aluminum are outlined in their respective engineering design handbooks, and the resistance factors for wood are detailed in Appendix A of the design in [36]. Compared to steel and aluminum, wood is a much weaker material, meaning that thicker structural members are required to carry the same loads. Therefore, by designing around the factored capacities of pressure-treated wood, it is possible to design a wooden rack to stand as long as a metal rack under the same weather conditions by building with larger structural wooden members. It should be noted, however, that it makes the most economic sense to only design a rack to match most modules' warranty and an approximate lifetime of 25 years.

2.3. Energy and Economic Analysis

The energy production analysis of the system is performed using the System Model Advisor (SAM) software [38,39]. A base model is created in SAM with the parameters in Table 1 for a multi-tilt system located in London, Ontario, Canada. A comparative analysis is run between the multi-tilt wood rack in this study and the fixed-tilt wood rack proposed by [36]. The purpose of the energy simulation is to determine the optimal energy production of the system if the tilt angle is adjusted with varying time steps throughout the year. Therefore, a first simulation is run to determine the best tilt angle for each month of the year. To find the best simulation parameters, all the parameters in Table 1 are maintained, except the value of the tilt angle, which is varied from 0° to 90°. The resulting hourly energy production for the first year is aggregated by month and for each tilt angle. The dependence of the maximum monthly energy production on the tilt angle is then analyzed to determine the best tilt angle for each month. The best monthly tilt angles are then used as seasonal tilt angle parameters to determine the lifetime energy production of the system.

Table 1. Parameters used in the PV system energy production simulation in SAM.

System Parameter	Value
Modules DC Power (W_{DC})	1200
Inverter AC Power (W_{AC})	1225
DC to AC Ratio	0.98
Azimuth (°)	180
Lifetime (years)	25
Annual PV Degradation Rate (%)	0.5

Three different energy simulation scenarios are analyzed. The first scenario considers that the racking system is adjusted monthly using the optimal angles. The second scenario explores the case where the maximum angle that can be reached is 60°. In the second scenario, any month with a tilt angle greater than 60° is capped at 60°. As a result, in the case of London, Ontario, the tilt angles of January, February, November, and December are set to 60° instead of the values. Lastly, the third case scenario is the optimal yearly fixed tilt angle (34°), as analyzed by [36].

A detailed economic analysis based on the bill of materials is conducted. The cost of the multi-tilt wood racking system is based on local purchases, and since the system is a DIY system, the labor cost is not factored into the base case study. The lifetime energy production of the PV system and the cost of the racking is used to determine the levelized cost of electricity (LCOE) of the multi-tilt wood racking system, which is then compared to the LCOE structural BCS in fixed-tilt wood racking, commercial fixed-tilt metal racking, and commercial single-axis tracking metal racking. As the system is designed to be DIY and only the racking cost is being analyzed, the discount factor is not included in the calculation of the *LCOE* (see Equation (1)).

$$LCOE\ (CAD\$/kWh) = \frac{Net\ Racking\ Cost\ (CAD\$)}{Lifetime\ Energy\ Production\ (kWh)} \quad (1)$$

Finally, sensitivity analyses are conducted to assess the power output and system cost in different regions of the world. Simulations are run for multiple locations with various latitudes to determine if a user in different regions can yield the same technical benefit. Furthermore, the cost of pressure-treated wood in different countries is compared to determine if this system can truly provide an economic benefit outside of Canada.

2.4. Design Analysis Assumptions

The same design analysis assumptions from the design in [36] were used for this system, including the specified snow, wind, and total design load.

3. Results

3.1. Bill of Materials

The bill of materials (BOM) of the multi-tilt system is shown in Table 2 in Canadian dollars sourced from Copp's Build-All and The Home Depot, London.

The modules used for this racking design are LG 400 W NeON2 modules. The approximate dimensions of the modules are 1 × 2 m. If modules with different dimensions are to be used, then the specifications in the assembly instructions in Section 3.2 can be scaled up or down to fulfill the given module size requirements. Resizing the system will change the loads applied to the wood, which means smaller systems reduce costs by selecting smaller structural members such that no limits are exceeded when conducting the structural analysis outlined in Appendix A.

Table 2. Variable angle rack list of materials.

Member Name	Piece [1]	Cost per Piece [2]	Quantity	Cost
Outside Joists	2 × 6 × 8	$16.12	2	$32.24
Inside Joists	2 × 8 × 8	$22.75	2	$45.50
Beams	2 × 8 × 10	$28.50	3	$85.50
Posts	4 × 4 × 10	$21.95	2 [3]	$43.90
Joist to Beam Connection	2 × 4 Fence Bracket	$0.36	8	$2.88
Back Supports	2 × 4 × 8	$9.99	3	$29.97
2 × 4 Hinges	8″ Gate Hinges	$11.96	6	$71.76
4 × 4 to Beam Hinges	4″ Gate Hinges	$9.51	2	$19.02
Beam to Post Connection	1/2″ Carriage Bolt (6″) Nut, & Washer	$4.44 [4]	2	$8.88
Tension Based Connections	3″ Brown Deck Screws	$9.99	100 Pack	$9.99
Shear Based Connections	1-1/2″ Joist Hanger Nails	$3.62	1 lb	$3.62
Module to Block Connections	1/4″ Carriage Bolt (2-1/2″), Nut, and Washer	$0.48 [4]	27	$12.96
			Total Cost with No Concrete	$366.22
Concrete for Posts	30 MPa Quikrete concrete	$4.98	8 bags	$39.84
			Total Cost:	$406.06

[1] All lumber is to be pressure treated, and all hardware is to be hot-dipped galvanized. [2] All costs are in Canadian Dollars as of 13 December 2021, before tax. [3] One piece to be cut to serve as two posts. [4] Cost per connection (1 bolt, 1 nut, 1 washer).

3.2. Variable Angle Installation Instructions

To begin, four holes at least 250 mm in diameter are dug at least 1.2 m into the ground, as shown in Figure 1a, to prevent frost heaving of any soil type, according to Table 9.12.2.2 in the National Building Code of Canada (NBCC) [40]. The 4 × 4 posts should have a center-to-center spacing, as described in Figure 1b, all cut to 1.825 m.

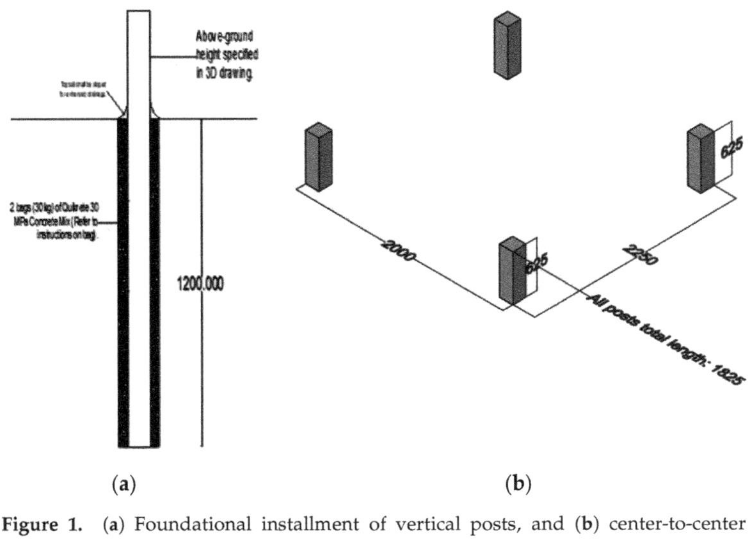

(a) (b)

Figure 1. (**a**) Foundational installment of vertical posts, and (**b**) center-to-center spacing of vertical posts.

A 2 × 8 × 10 beam can then be installed onto the back posts, as shown in Figure 2. The overhang beyond each post is 0.5 m. The beam is to be installed to the posts with 3″ brown deck screws and one galvanized nut, bolt, and washer per post. If the posts are subject to ground movement, temporary 2 × 4 bracing can be installed to hold the system up during construction, as shown below. These braces can then be removed once the footings are stiff and secure.

Figure 2. The 2 × 8 beam installed between the back posts, with temporary 2 × 4 lateral bracing installed between the left and right posts.

Once the base is installed, the frame can then be assembled, as shown in Figure 3a, using 2 × 6 outside joists and 2 × 8 inside joists connected to 2 × 8 beams. Connections, as shown in Figure 3b, should be assembled using fence brackets and 1-1/2″ joist hanger nails. Additionally, 4 3″ brown deck screws per joist should be installed to further sink the joist into the beam and to improve load transfer between the members.

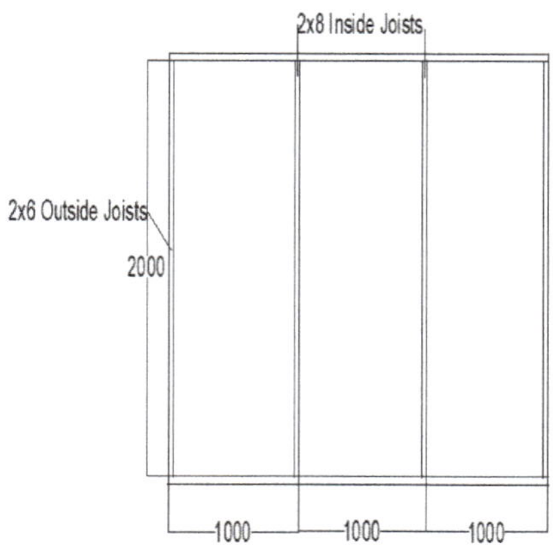

Figure 3. (**a**) Frame composition consisting of two 2 × 8 outside joists, two 2 × 8 inside joists, and two 2 × 8 beams, and (**b**) each joist connected with fence brackets, 1-1/2″ joist hanger nails, and 3″ brown deck screws.

The frame can then be connected to the two front posts using 3-1/2″ traditional gate hinges. The 2 × 8 of the frame should be connected to the top of the post, as shown in Figure 4.

Figure 4. Installation of the frame onto the front posts using galvanized gate hinges.

Three 2 × 4 s should then be cut to a length of 2 m. This will allow for a maximum tilt angle of 60 degrees. These 2 × 4 s will be used as the adjustable back supports for the system. Three galvanized gate T hinges will be connected between the joists on the frame, and another three will be connected on the beam, as shown in Figure 5.

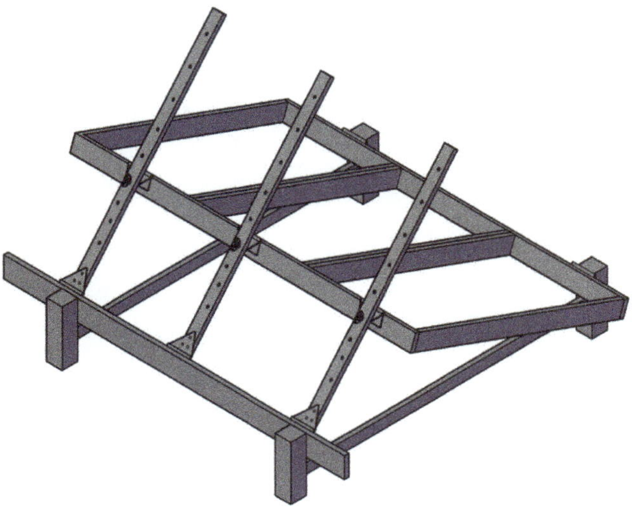

Figure 5. The spacing of 2 × 4 s on the system.

Holes of 1/4″ should be drilled in the 2 × 4 s based on the desired tilt angle, as shown in Figure 6a. Align the frame hinges with the hole for the given tilt angle, and secure the connection with a 1/4″ nut, bolt, and washer, as shown in Figure 6b. While drilling the 2 × 4 s, temporary supports or extra helpers should be used to hold the frame up to prevent damage.

(a) (b)

Figure 6. (**a**) 1/4″ holes drilled based on preferred tilt angle, and (**b**) the 2 × 4 connected to the frame via a nut, bolt, and washer.

Once the joists are installed, scrap pieces of lumber can be cut into blocks and installed onto the joists with two screws, as shown in Figure 7a. These blocks serve as the connection between the module and the lumber and can be adjusted to match the holes of the module frame. The overhang of these blocks shall not exceed 100 mm. Once these blocks are installed, the modules can be placed onto the blocks. Drill a $\frac{1}{4}$″ hole through the block and insert a $\frac{1}{4}$″ × 2-1/2″ galvanized bolt from under the system. Then, place the module onto the bolt, and secure the connection with a nut and washer, as shown in Figure 7b. To enhance the load transfer to the joist, place another block under the overhanging block, and screw the second block into the joist, as shown in Figure 7c.

(a) (b) (c)

Figure 7. (**a**) Extra lumber made into blocks that line up with the module's holes, (**b**) A bolt inserted from under and secured with a nut and washer, and (**c**) extra blocks placed under the overhanging block to enhance load transfer to the joist.

Once all connections are secured, the build is complete (Figure 8a). The system can then be disassembled in the reverse order it was initially constructed. Figure 8b provides a detailed back view to ensure all components of the system are present.

(a)

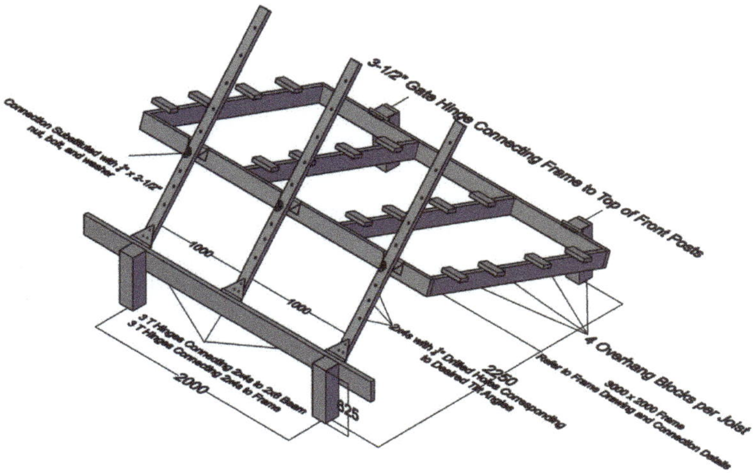

(b)

Figure 8. (**a**) Finished system, and (**b**) detailed back view of the final system.

Annually, the bolt connections should be checked to ensure loose nuts are retightened. When possible, it is best practice to brush off snow from the modules to minimize creep in the lumber.

3.3. *Build Time*

The system requires at least two builders for installation. Refer to Table 3 for the typical time to complete each component per two builders.

Table 3. Time invested in building the multi-tilt racking system.

Task	Typical Time to Complete [1]
Hole digging and post installation with temporary bracing	3.0 h [2]
Frame Installation	1.0 h
Back Support Drilling and Hinge Installation	3.0 h
Block Installation	0.5 h
Module Installation	1.0 h
Total Time	8.5 h

[1] Assuming 2 builders with some construction experience. Not including time to gather materials, acquire equipment, etc. [2] Not including curing time for concrete/footing mixture. Refer to the supplier's instructions for suitable curing time before continuing to construct.

To adjust the angle of the system, the nuts on the 2 × 4 s should be loosened, and the bolts should be taken off the hinges. It should be noted that once all the bolts are taken off, the system is free to fall and damage the racking frame. Approximately 150 lbs of force is required to keep the frame and modules from falling. Thus, temporary supports or extra help should be used to hold the frame up while repositioning the system to a new angle. Lift or drop the frame to the desired angle, then align the hinges with the drilled holes designated to that angle. The tilt angle can be found by using an angle level or by calculating the arctan of the frame's rise divided by the run. Secure the connection with the nut, bolt, and washer. For higher angles, it can be difficult to lift the system, and thus, more help may be required. Refer to Table 4 for approximate adjustment times and recommended number of workers based on the desired tilt angle.

Table 4. Number of workers and time required for changing the tilt angle in different cases.

Tilt Angle (Degrees)	Typical Time Spent (Minutes)	Number of Workers
0 to 35	3	2
35+	4	3

Following the calculations shown in Appendix A for the structural analysis, the forces and deflections of the system, specifically for the London, Ontario, system, have been summarized in Table 5. These values represent the worst-case scenario that governs the design of this system. When constructing a system, it is important to follow the structural design process in Appendix A to ensure the system can withstand the design load outlined in Appendix A. Depending on the design load, smaller members can be selected, and thus the net cost of the system can be reduced such that the maximum shear, moment, deflection, and axial forces are less than the capacities shown in Appendix A.

Table 5. Forces and deflections of structural members for the variable angle rack in London, Ontario, under the worst-case design load.

Member Name	Shear (kN)	Moment (kNm)	Deflection (mm)	Tension/Compression (kN)
Outside Joists	0.95	0.45	3.30	N/A
Inside Joists	1.90	0.90	1.65	N/A
Beams	1.90	0.50	2.73	N/A
2 × 4 Back Supports	N/A	N/A	N/A	−2.22
Middle Back Support	N/A	N/A	N/A	−1.31
Back Beam	0.70	0.70	0.65	N/A
Back Posts	0.45	0.43	0.80	−2.70 [1]
Front Posts	0.45	0.51	0.90	−2.70 [1]

[1] For a 250 mm diameter hole, bearing pressure is equal to 55 kPa.

3.4. Energy Simulation Results

The results of the base model simulation are displayed in Figure 9, which shows the monthly energy production of the system during the first year for different tilt angles ranging from 0° (horizontal modules) to 90° (vertical modules) in steps of 1° for the city of London, ON, Canada.

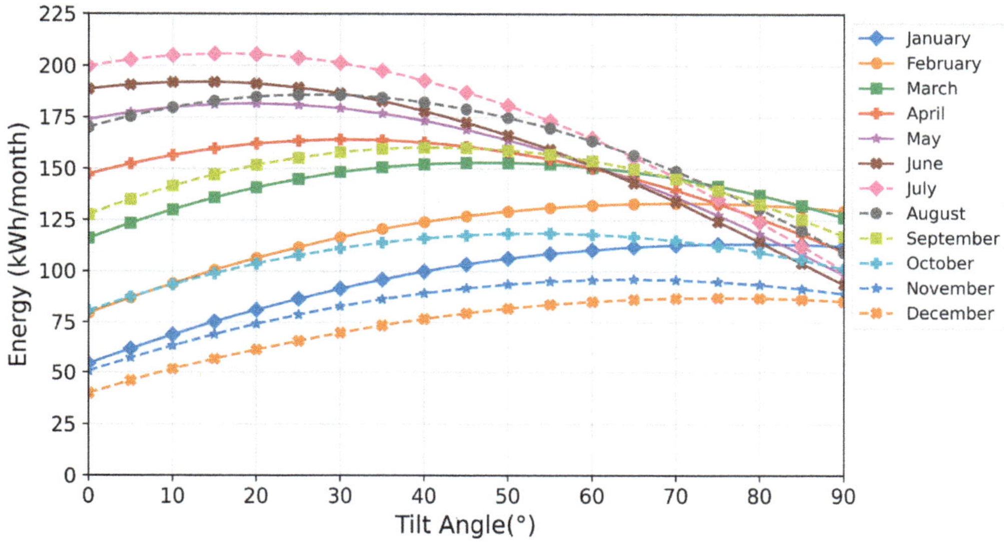

Figure 9. Monthly energy production during the first operational year of the 1200 W bifacial PV system for different tilt angles in London, Ontario.

The bell shapes of the monthly energy production curves show the value of the optimal angle for maximum energy generation each month. The monthly optima tilt angles are reported in Table 6. According to the values in Table 6, the system is at its highest tilt angle (79°) at the beginning of the year. The system has to be lowered every month until it reaches its lowest optimal tilt angle in July (16°) and has to be lifted again between August and January.

Table 6. Monthly optimal tilt angles in London, Ontario.

Month	Optimal Tilt (°)	Maximum Energy Generation during the First Operational Year (kWh/Month)
January	79	114
February	71	133
March	46	153
April	30	164
May	19	182
June	13	192
July	16	206
August	28	186
September	41	160
October	54	119
November	64	96
December	76	87

Three different energy simulation scenarios are analyzed. The first scenario results are shown in Table 6, the second scenario results cap the maximum adjustable tilt angle at 60°, and the third scenario fixes the optimal yearly tilt angle at 34°. The results of the simulations of the three scenarios are shown in Figure 10.

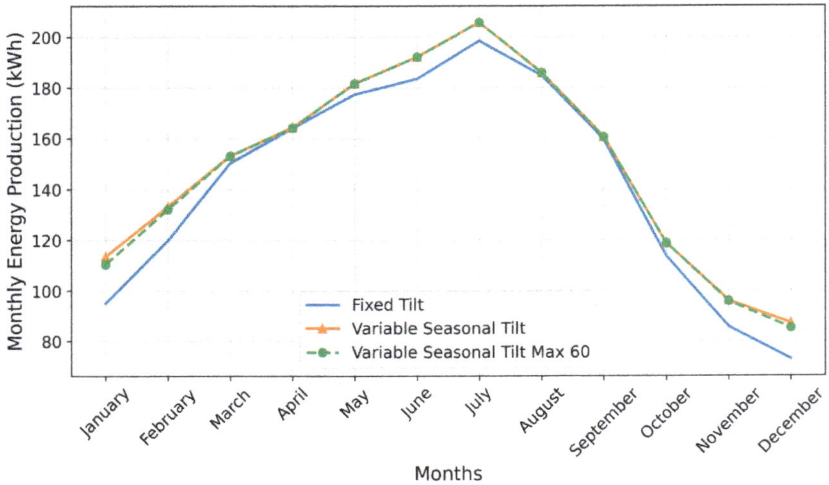

Figure 10. Monthly energy production during year 1 of the simulations' three scenarios.

The results clearly show that the multi-tilt system produces more energy every month than the fixed-tilt as it uses seasonal tracking to optimize the solar energy harvested throughout the year. The maximum energy production gain of the optimum tilt angle system compared to the fixed-tilt racking system was 17% in December. Even in April, where the graph shows a similar energy production trend for the two systems, there is a small production gain from the multi-tilt of 0.13%. Furthermore, capping the maximum angle of the multi-tilt system to 60° only has a noticeable impact on the energy production of the system during January and December in the case of London, Ontario. During these months, the production is expected to be lower because of snow-related losses in Canada [41] and Ontario in particular [42,43], even if somewhat mitigated with bifacial PV modules [44]. Although, it should be pointed out that with the impacts of climate change, these losses are expected to decrease in the future [45].

3.5. Economic Analysis Results

The economic analysis of the system is shown in Table 7. As can be seen in Table 7, the open-source design provided in this study costs less than a third of the commercial equivalent variable tilt angle racking system.

Table 7. Cost analysis of each racking system with a set of 3 PV modules of 400 W each, the total system installed DC power is 1200 W, assuming no construction labor costs.

Racking System (3 Modules)	Lifetime Energy (kWh)	Racking Cost (CAD)	LCOE (CAD/kWh)	Cost per W (CAD/W)
Wood Fixed Tilt [36]	40,065	388.78	0.0097	0.32
Wood Optimal Variable Tilt	42,152	406.06	0.0096	0.34
Wood Optimal Tilt with Max Angle Capped at 60°	41,997	406.06	0.0097	0.34
Metal Optimal Commercial Variable Tilt [46]	42,152	1399.99	0.0332	1.17

The results in Table 7 show the lifetime energy production of the PV system, the racking cost, the LCOE of the racking, and the cost of each racking system per Watt. The racking systems (metal or wood) with an optimal variable seasonal tilt angle show the best lifetime energy production (42.15 MWh), which is 5.2% more energy generated by the fixed-tilt system (40.07 MWh). Even when the maximum angle of the wooden racking with variable tilt is limited to 60°, it generates 4.8% more energy (42.0 MWh) than the fixed-tilt system. In terms of LCOE, the wooden racking systems show similar LCOE (~CAD 0.01/kWh), and their LCOE represents only 29% of the LCOE of a commercial metal racking (CAD 0.033/kWh) serving the same purpose.

4. Discussion

4.1. Wooden Racking for Agrivoltaic and Impact of Labor Cost

The impact of the increase in energy production from the fixed-tilt wooden racking system to the variable angle wooden racking is seen in the LCOE of the two systems. Despite the wooden fixed-tilt system having a lower capital cost than the variable tilt wooden racking, they both have similar LCOE (~CAD 0.01/kWh). This is a crucial argument in determining which racking system to adopt. Specifically, the variable angle wooden racking design proposed in this study and the fixed-tilt wooden racking proposed by [36] are suitable for providing standalone clean solar PV energy to remote locations.

When the variable tilt wooden racking is compared to a commercial metal racking serving the same purpose (variable seasonal tilt angle) and purchased in Canada [46], the two systems have the same lifetime energy production. Nevertheless, the wooden racking is less costly than the commercial metal racking. In this study, data have been collected regarding the manpower and time needed for changing the angle of the proposed system seasonally (see Table 4). When the numbers in Table 4 are applied to the case of London, Ontario, the total annual labor time for changing the tilt of the wooden racking system is 1.9 person-hours/year, representing 47.5 person-hours during the lifetime of the wood racking system (25 years). Therefore, even when the labor cost for changing the tilt angle is factored into the wooden racking system's cost, it remains economically more feasible than the commercial metal racking as long as the labor cost does not exceed CAD 21/hour. CAD 21/hour is 50% higher than CAD 14/hour, which is the average minimum wage across Canada [47]. Furthermore, this assumption is conservative as there is no information regarding the manpower required to change the tilt angle of the commercial metal racking. Hence, if the commercial metal racking requires more than one person to operate, the economic advantages of the variable tilt wooden racking system will be increased. On the other hand, when the labor cost is considered in the calculation of the variable angle wooden racking (as high as CAD 0.20), then the fixed-tilt wooden system is economically more viable. This does not, however, account for the additional benefits provided by the variable angle racking for specific applications such as agrivoltaics (see Section 4.2).

It should also be pointed out that there are several circumstances where the labor cost is effectively zero. For small-scale DIY systems [17], where prosumers are not charging themselves to vary the tilt angle of their own system or there is no opportunity cost to invest the few minutes once a month to change the tilt angle, the effective cost is zero, similar to businesses with salaried employees or even hourly employees where there is no cost to reassigning a few minutes once a month (e.g., secretaries manning a phone).

4.2. Agrivoltaics

Agrivoltaics in farmlands is a promising strategy for the co-development of land for both PV electrical generation and agriculture [48,49]. Services provided by agrivoltaics include: (i) renewable electricity generation, (ii) decreased greenhouse gas emissions, (iii) increased crop yield, (iv) plant protection from excess solar energy, (v) plant protection from inclement weather such as hail, (vi) water conservation, (vii) agricultural employment, (viii) local food, and (ix) increased revenue [50]. Benefits (i) and (ii) are well-established for all PV systems as they produce renewable electricity, which can offset greenhouse gas (GHG) emissions from fossil fuel-based electricity production [51]. From a farmer's perspective, agrivoltaics also ensure that land remains productive during the winter by generating electricity year-round. Strikingly, agrivoltaics actually increase crop yield for a variety of crops [52–56], which along with solar electricity generation, substantially increases land-use efficiency [57]. Crops grow better with some PV-related shading because the PV array creates a microclimate beneath the modules that alters air temperature, relative humidity, wind speed, wind direction, and soil moisture [58]. Agrivoltaics also protects crops not only from excess solar energy preventing heat stress but also from inclement weather such as hail, while the PV performance can increase because of lower operating temperatures caused by the plants [54,59,60]. Combining PV and agriculture has the potential to increase global land productivity by 35–73% [61] while minimizing agricultural displacement for energy production [49,61,62]. Additionally, the microclimate created by agrivoltaics provides more efficient use of water and water conservation [63–66]. Furthermore, the novel variable tilt racking system analyzed here is particularly useful from an agrivoltaic perspective. The variable angle system will allow having the same LCOE as the fixed-tilt but will offer additional advantages specific to agrivoltaic farms, such as raising the modules to the selected highest angle to facilitate harvesting, planting, or weeding. In addition, the tilt angle can be lowered to protect the plants, for example, if a hail warning is issued. The labor involved with this tilt angle changing is trivial compared to that of farming, and additional farm labor is not always seen as a negative. By maintaining the land for use in agriculture, employment of farmers remains intact, and these farmers provide local sources of food along with all of the concomitant benefits [67,68]. Thus, agrivoltaics is looked at favorably not only by the PV community [69] but also by farmers [70] and farming communities [71]. The combination of farming and PV electricity generation increases revenue per acre for the farmer, and the local community also benefits from protecting access to fresh food and renewable energy [72]. Advanced inverter management can also provide stability [73] to rural electric grids and improve their power quality [74–76] if used in agrivoltaic systems. Finally, if the agrivoltaic system also includes storage, it can be used to create emergency islanded power grids that can reduce outage impacts [77,78], which may be particularly useful for isolated communities. For all these reasons, the system described here is perhaps best suited for small-scale agrivoltaics.

Lastly, the impact of the microenvironment for this type of racking system should be considered in the performance of the PV. Considerable work has been conducted to show that cooling PV modules with water by humidification by placing wet sacks at the back end of the PV module, water cooling by flowing the water on the front end of the module, and the combination of these two strategies improve performance in warm and humid climates [79]. Similarly, it is well-established that agrivoltaics create a microclimate under the PV modules that have similar but lesser PV performance benefits from direct intervention measures [80–82]. Although the wet sack method is not easily integrated into agrivoltaic wood-based racking systems because of compromising the agricultural potential, there is a potential for future work to consider the use of front-side cooling of agrivoltaic PV by integrating it into the irrigation of the crops (e.g., water is first used to cool the PV and then collected and dispersed to water the crops. Future work is needed to explore this potential both technically and economically.

4.3. System Location Sensitivity

The energy production of the system is simulated for different locations in the northern hemisphere to assess and compare the energy production performance of the fixed-tilt wooden racking proposed by Vandewetering et al. [36] and the variable tilt wooden racking proposed in this study. The location considered for the simulation and the latitude of those locations are shown in Table 8.

Table 8. Cities and corresponding latitudes that were considered for the location sensitivity analysis.

Cities	Mexico	Cairo	Boulder, Colorado	L'Anse	Winnipeg	Helsinki
Latitude (°)	19.45	30.05	40	46.77	49.89	59.97

Figure 11 shows the comparison of the annual energy production during the first operational year for the locations shown in Table 8. The results in Figure 11 show clearly that wherever the two systems are located, the variable tilt system will produce more energy than the fixed-tilt racking for the same operational conditions. Nevertheless, it is necessary to mention that the actual LCOE of the two systems is highly dependent on the local wood prices. This sensitivity is briefly discussed in Section 4.4, and a more detailed wood price sensitivity analysis is available in previous work [36].

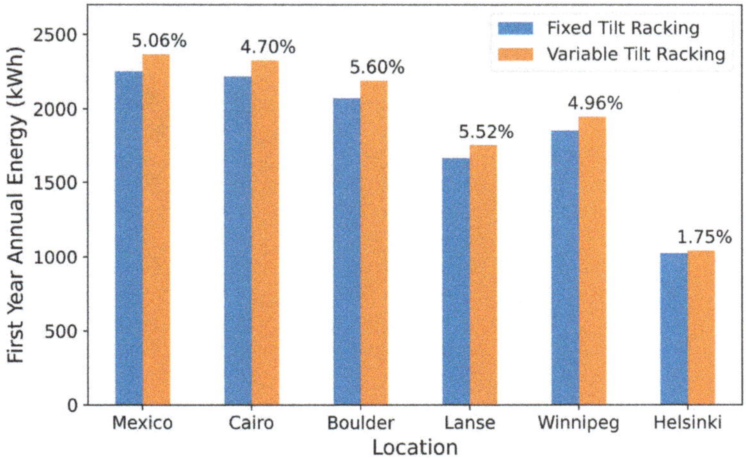

Figure 11. Comparison of the energy generated during the first year of operation by the fixed-tilt racking to the energy generated by the variable tilt racking for different locations in the northern hemisphere. The percentages on top of the bar plot represent the energy gain by using the variable tilt racking instead of the fixed-tilt racking.

4.4. *Wood Price Sensitivity*

The system is highly sensitive to the price of lumber, which has shown to be volatile in the sensitivity analysis in the design in [36].

The costs of this design will also be dependent on the local sources of wood being available and, if imported, the taxes and import duties. Refer to Table 9 for the typical price of a construction grade pressure-treated 2 × 4 × 8 in various countries.

Table 9. The typical price of a pressure-treated 2 × 4 × 8 in various countries converted to USD.

Country	Price (USD) [1]	Source [2]
Canada	$8.46	The Home Depot
USA	$9.68	The Home Depot
Togo	$21.67	Collected Locally
United Kingdom	$15.70	B and Q
Netherlands	$10.32	Woodvision
Australia	$13.92	Bunnings
Brazil	$12.03	Fremade Madeiras
India	$4.96 [3]	IndiaMart

[1] Priced as of 2 April 2022. [2] Prices at each source's competition are approximately the same. [3] Price before pressure-treating, which is expected to cost at least double to treat.

This open-source wood PV rack is (1) made from locally accessible, sustainable, renewable materials, (2) can be fabricated using simple hand tools by the average consumer, (3) has a 25-year lifetime to be equivalent to common PV warranties, (4) is structurally sound in order to weather high wind speeds and major snow loads (depending on the region), (5) has a low cost and (6) that is shared using an open-source license so that many people can fabricate it themselves, or companies can make versions to offer in their local markets.

4.5. Limitations

This system is entirely built with materials available at local hardware stores and requires a few hours of labor to complete the build. Since the multi-tilt system is not pre-assembled, this system is mainly targeted toward do-it-yourselfers (DIYers). The limitation to DIY multi-tilt racking is that many people, such as the elderly or physically disabled, may have to hire a general contractor to build the system, which adds a labor cost to the system build. Depending on the location and labor rates, this would diminish the economic advantages shown here. Nevertheless, metal racking systems also require initial labor costs if not built by the owner, thus making the cost comparison between wood and metal racking still valid.

The primary benefit of this design is to convert physical labor into electrical power, but on a three-panel array, the difference in power remains small. By making many replications of this three-panel array, the benefit can be scaled up to produce significantly more power for the owner. Table 10 summarizes the amount of extra energy generated in a lifetime from monthly tilt adjustments in many scenarios, where this system is scaled up to satisfy the energy demand of practical applications. By scaling up the number of three-panel arrays, both the generated power and economic benefit of monthly adjustments proportionally scale up. It should be noted that costs can further be reduced in scaled-up systems since many hardware stores honor a contractor discount to those who purchase orders in bulk. This makes all economic estimates used in this study extremely conservative for any form of scale up case.

Table 10. The additional energy generated over the lifetime of a system for scaled up applications, assuming monthly tilt adjustments, capped at 60°.

Application	System Size	Number of Systems Required	Additional Energy Generated over Lifetime
Base System	1.2 kW	1	1.93 MWh
Typical Residential Home in Ontario [1]	7.2 kW	6	11.58 MWh
Small Business [2]	21.6 kW	18	34.74 MWh
Agrivoltaics	100 kW	84	160.19 MWh

[1] Assuming an energy consumption of 9500 kWh/year [83]. [2] Assuming an energy consumption of 30,000 kWh/year [84].

Despite the technical and economic benefits the multi-tilt system provides, it is more difficult to install compared to fixed-tilt systems. Not only does it require more time to install, but it is also more labor-intensive, especially when measuring and drilling holes into the 2 × 4 s while holding the large frame in place. To mitigate this, extra lumber to serve as temporary bracing and support can be used to hold the frame and posts up while measurements and cuts are being made.

For agrivoltaic applications, the use of pressure-treated lumber can pose the threat of introducing toxic chemicals into the vegetation. Since the 1930s, pressure-treated lumber contained chromated copper arsenate to protect the wood from rotting, which caused major health and environmental concerns but has been discontinued and replaced with copper azole preservative since 2004 [85]. Copper azole is a much safer alternative since it removes exposure from chromium and arsenic, and copper is already abundant in soil and groundwater, but high doses of copper can result in severe liver and kidney damage [86].

Therefore, if this system is in close contact with vegetation, it is highly advised to prevent pressure-treated lumber from contacting the soil. Natural alternatives such as cedar lumber can be used, but the cost is at least 2.5 times that of pressure-treated. The best solution to eliminate soil contact is to extend the concrete footing up and above the ground. This may require using an extra bag of concrete in each footing to ensure enough mix reaches a significant level above the ground. Future work in sand and acrylonitrile styrene acrylate waste composites [87] or other plastic composites may serve as a less permeable alternative to concrete footings to further prevent soil contamination in some applications.

5. Conclusions

This study detailed the designs of a novel variable tilt angle open-source wood-based PV racking system. The system costs less than one-third of the CAPEX of variable tilt angle commercial racking solutions. These lower costs make it economical in some contexts for labor to be exchanged profitably for higher solar electric output. The results of this study show that the racking systems with an optimal variable seasonal tilt angle present the best lifetime energy production, with 5.2% more energy generated compared to the fixed-tilt system (or 4.8% more energy, if limited to a maximum tilt angle of 60°). Thus, a few minutes of labor per kW can be traded for about 5% more annual solar electricity production. Both fixed and variable wooden racking systems show similar LCOE of less than 1 cent, which is only 29% of the LCOE of commercial metal racking. The economic analysis found that in several contexts, the novel variable tilt rack provides the lowest cost option even when modest labor costs are included. Finally, the novel variable tilt racking design shown here has several specific advantages over fixed-tilt designs for applications such as agrivoltaics.

Author Contributions: Conceptualization, J.M.P.; methodology, N.V., K.S.H.; software, K.S.H.; validation, N.V., K.S.H.; formal analysis, J.M.P., N.V., K.S.H.; investigation, N.V., K.S.H.; resources, J.M.P.; data curation, N.V., K.S.H.; writing—original draft preparation J.M.P., N.V., K.S.H.; writing—review and editing, J.M.P., N.V., K.S.H.; visualization, N.V., K.S.H.; supervision, J.M.P.; funding acquisition, J.M.P. All authors have read and agreed to the published version of the manuscript.

Funding: This research was funded by the Thompson Endowment and the Natural Sciences and Engineering Research Council of Canada (NSERC).

Institutional Review Board Statement: Not applicable.

Informed Consent Statement: Not applicable.

Data Availability Statement: Data will be made available upon request.

Acknowledgments: The authors would like to thank Paul Vandewetering of Paul's Build-All for assistance in the construction of the variable angle rack.

Conflicts of Interest: The authors declare no conflict of interest. The funders had no role in the design of the study; in the collection, analyses, or interpretation of data; in the writing of the manuscript, or in the decision to publish the results.

Appendix A Variable Tilt System Structural Analysis

It can be difficult to determine the specified design load for this system because the design load is dependent on the tilt angle, which will vary throughout the system's life cycle. For the purpose of analysis, the angle that produces the heaviest combined wind and snow loads is any angle between 30 and 45 degrees, as suggested by the NBCC wind and snow design load procedure [40]. For London, Ontario, design loads of 1.80 and −0.81 kPa are yielded when following the procedure outlined in the design in [36]. To ensure that the system will not unexpectedly fail, the applied shear forces, bending moments, and deflections shall not exceed the limits of each structural member outlined in the National Design Specification for Wood Construction [88].

The net load is distributed evenly throughout the surface of the modules. As per the supplier of the modules [89], it is assumed that the panels have sufficient capacity to carry these loads. The load is then transferred from the panels to the joists. Each joist carries its own weight as a uniform distributed load, w, and four-point loads that represent the block connections. w is calculated using Equation (A1),

$$w = 1.25(OW) \tag{A1}$$

OW represents the own weight of the member, which needs to be multiplied by a factor of 1.25 because it is a dead load [90]. Since the required dimensions of lumber to carry the load is unknown, an assumption needs to be made (for example, assume 2×8) to carry out the analysis. If the assumption results in the maximum applied value being greater than the resistance values, then a larger member needs to be used.

The point loads can be calculated by dividing each joist's tributary loading into four points because it is assumed that the load is distributed evenly throughout the modules. The tributary area represents how much width of the panels each joist is responsible for carrying. For example, in this three-module system, which is 3 m wide, the middle joists have a tributary width of 1 m (0.5 m on each side), and the end joists have a tributary width of 0.5 m (only one side). The value for each point load on the joists is calculated using Equation (A2),

$$Point\ Load = \frac{Design\ Load \times Tributary\ Area}{4} \tag{A2}$$

Once w and the point loads are calculated, the free body diagram shown in Figure A1 for each joist can be made.

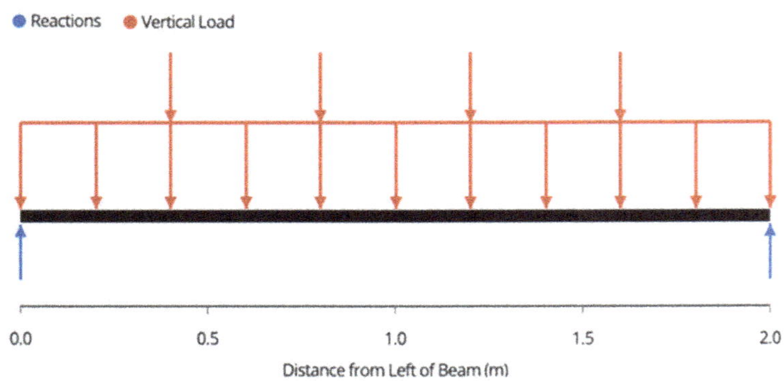

Figure A1. Free body diagram of joists. Note that the two outside joists will have half the tributary area of the inside joists and thus will carry approximately half of the load.

Each joist is supported a beam on each end. The reaction and thus the load that each joist transfers to each beam is calculated using Equation (A3),

$$Reaction = \frac{4 * Point\ Load + wL}{2} \tag{A3}$$

The shear force diagram throughout each joist is seen in Figure A2.

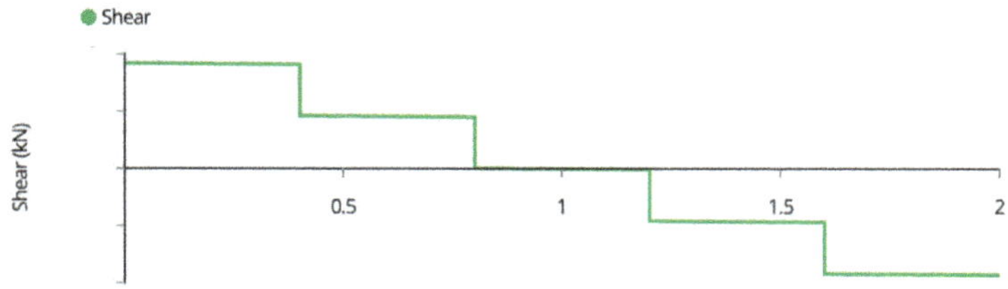

Figure A2. Joist shear force diagram.

The maximum shear force occurs at the supports, and thus, the maximum shear is calculated as the reaction shown above.

The bending moment diagram in each joist is seen in Figure A3.

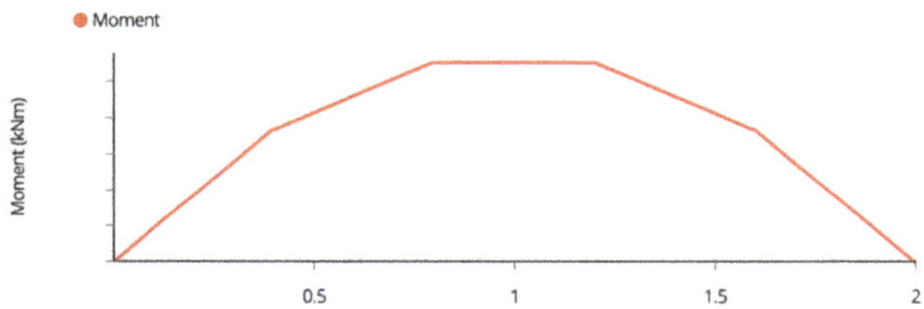

Figure A3. Joist bending moment diagram.

The maximum bending moment of the joists occurs at the midspan. The maximum bending moment can easily be calculated using Equation (A4),

$$M_{max} = \frac{wL^2}{8} + \frac{3(Point\ Load)L}{5} \tag{A4}$$

The deflection diagram throughout each joist is plotted in Figure A4.

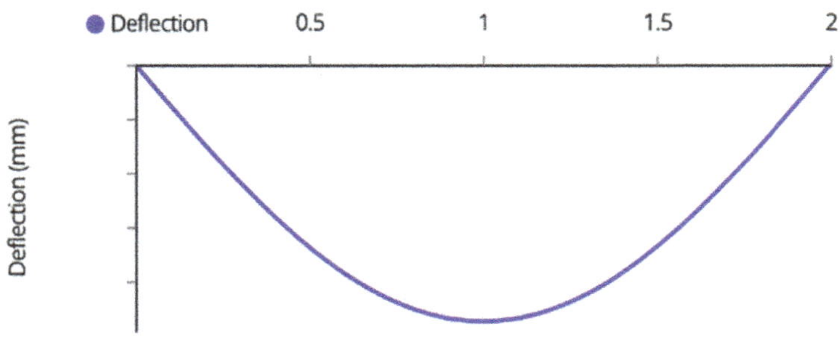

Figure A4. Joist deflection diagram.

The maximum deflection of the joists occurs at the midspan. For simplicity of analysis, assume the four-point loads serve as a uniform distributed load, and calculate the maximum deflection using Equation (A5),

$$\Delta_{max} = \frac{5(w + 2(Point\ Load)L^4}{384EI} \tag{A5}$$

In the frame, half of the load will be transferred to the front 2 × 8, and the other half to the back 2 × 8. The free body diagram for the front 2 × 8 is shown in Figure A5.

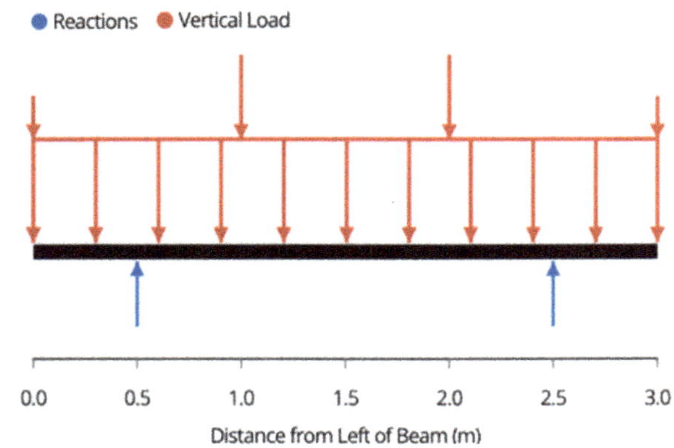

Figure A5. Beam free body diagram.

Due to the symmetric loading of the beams, the post loads or the support reactions are described in Equation (A6),

$$Reaction = \frac{(\sum_{k=1}^{4} Joist\ Reactions) + wL}{2} \quad \text{(A6)}$$

The shear force diagram for the beams is described in Figure A6.

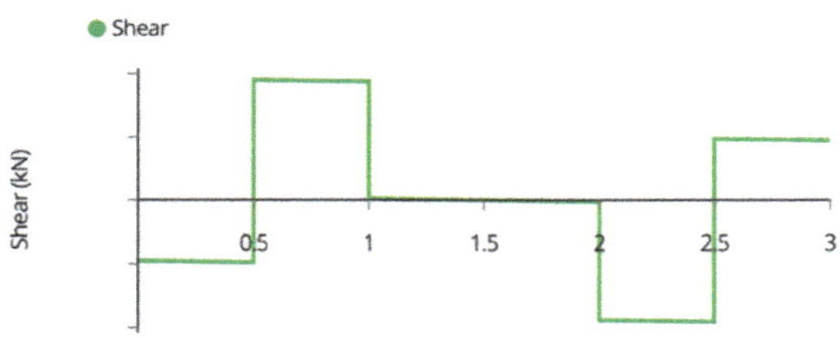

Figure A6. Beam shear force diagram.

The maximum shear forces occur on the inside of the reactions and are calculated using Equation (A7)

$$Vmax = Reaction - Joist\ Load_1 - \frac{wL}{6} \quad \text{(A7)}$$

The bending moment diagram for the beam is shown in Figure A7.

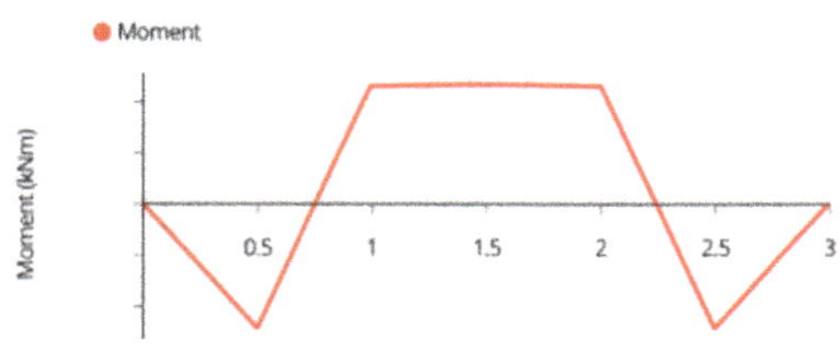

Figure A7. Beam bending moment diagram.

The maximum moment occurs at the supports and can be found by integrating the shear force throughout the first sixth of the beam, as described in Equation (A8) or by simply finding the area under the shear force diagram.

$$Mmax = \int_0^{L/6} V(x)dx = (Joist\ Load_1)\frac{L}{6} + \frac{wL^2}{72} \quad \text{(A8)}$$

The deflection diagram is shown in Figure A8.

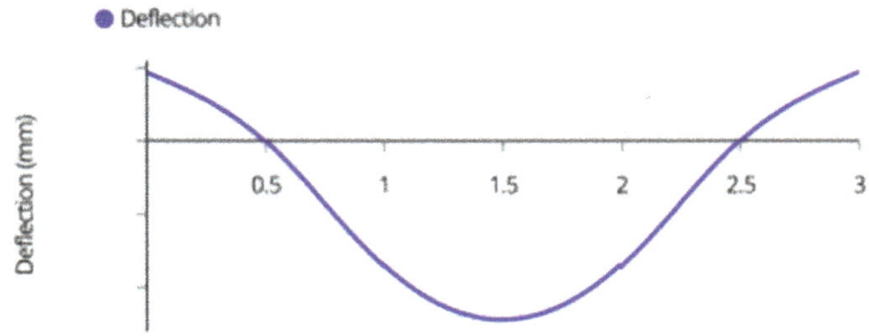

Figure A8. Beam deflection diagram.

The maximum deflection occurs at the midspan and can be solved using the differential Equation (A9) and initial conditions below or by using the moment area theorem or virtual work method described in many structural engineering textbooks.

$$\begin{aligned} \frac{d^2\Delta}{dx^2} &= \frac{M(x)}{EI} \\ \Delta\left(\frac{L}{3}\right) &= 0 \\ \Delta'\left(\frac{L}{2}\right) &= 0 \end{aligned} \quad \text{(A9)}$$

The back 2 × 8 in the frame is then supported by three hinges. The free body diagram for the beam is shown in Figure A9. This is an indeterminate structure, meaning that it has too many supports to be solved with static analysis and thus cannot be expressed with generalized equations. The structure can be solved by using finite element analysis or by an analytical method such as the moment distribution or slope-deflection method.

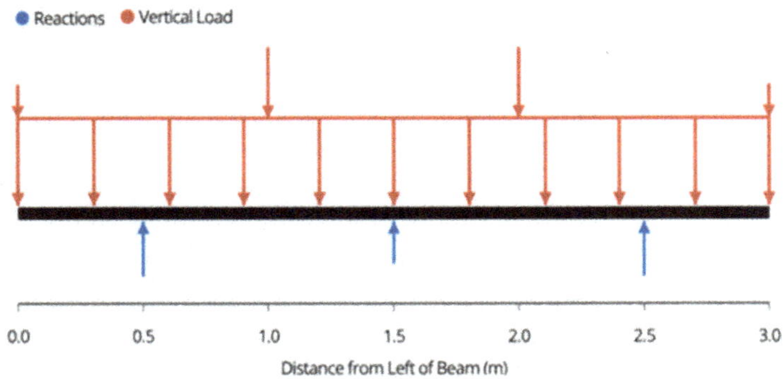

Figure A9. Beam free body diagram for variable angle rack.

The shear force diagram for each beam is shown in Figure A10.

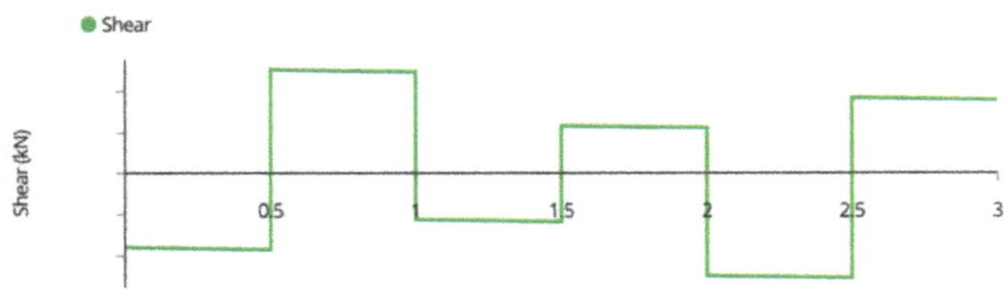

Figure A10. Beam shear force diagram.

The bending moment diagram for each beam is shown in Figure A11.

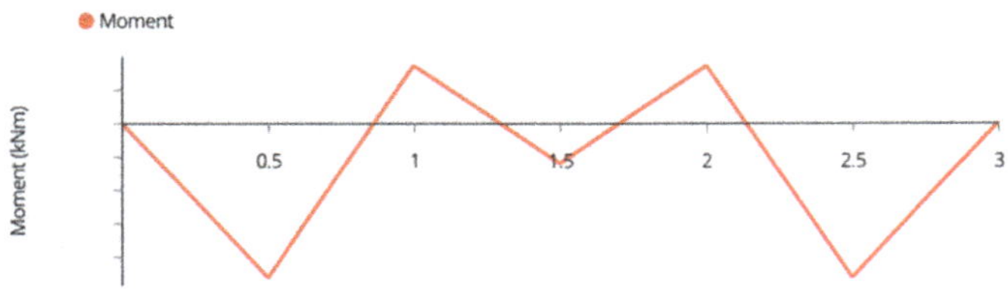

Figure A11. Beam bending moment diagram.

The deflection diagram for each beam is shown in Figure A12.

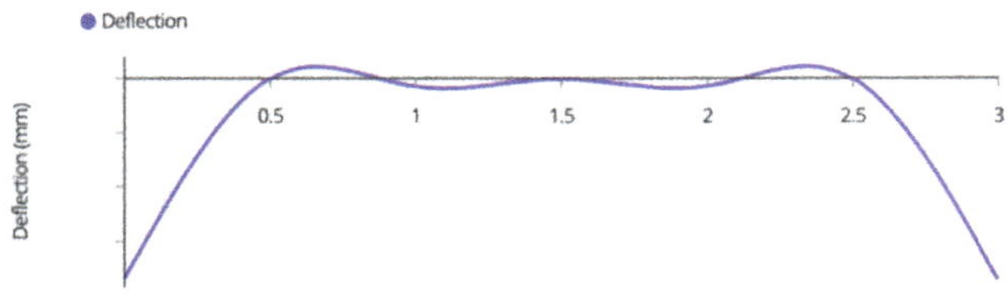

Figure A12. Beam deflection diagram.

The load is then transferred to the supporting 2 × 4 via the T hinges. Since these hinges are free to rotate, the 2 × 4 s can be idealized as truss members in pure compression (the moments are released). The compressive force is equal to the reaction supporting the beam. These back supports are slender and thus should be checked for buckling using Equations (A10) and (A11).

$$Cbuckling = \frac{\pi^2 E I_{weak}}{L^2} \tag{A10}$$

$$where\ I_{weak} = \frac{1}{12} h b^3 \tag{A11}$$

After confirming that both Cbuckling and Cmax are less than Cr, the load can then be transferred to the bottom beam with the free body diagram shown in Figure A13.

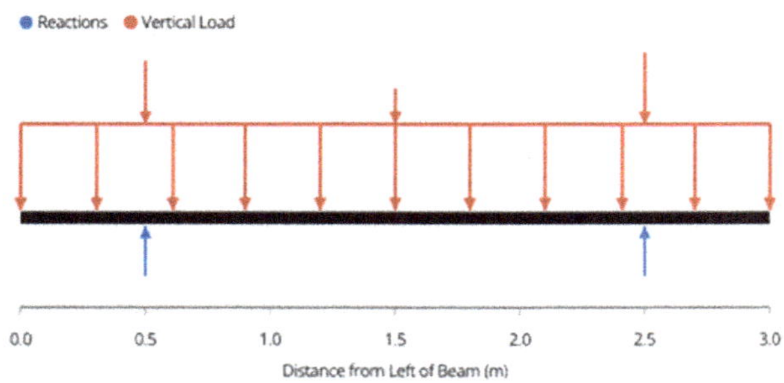

Figure A13. Bottom beam free force diagram.

The shear force diagram is shown in Figure A14.

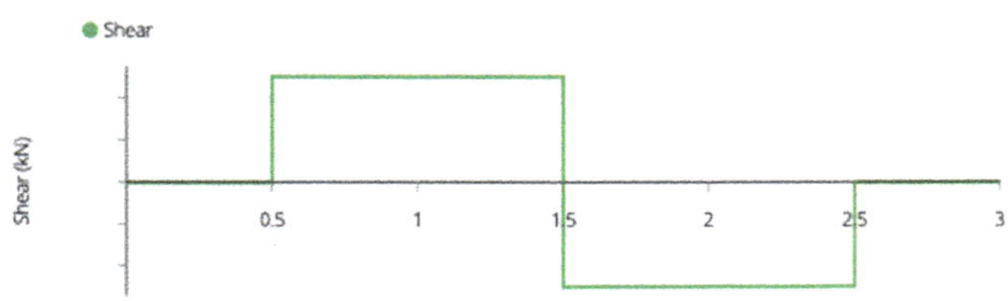

Figure A14. Bottom beam shear force diagram.

The bending moment diagram is shown in Figure A15.

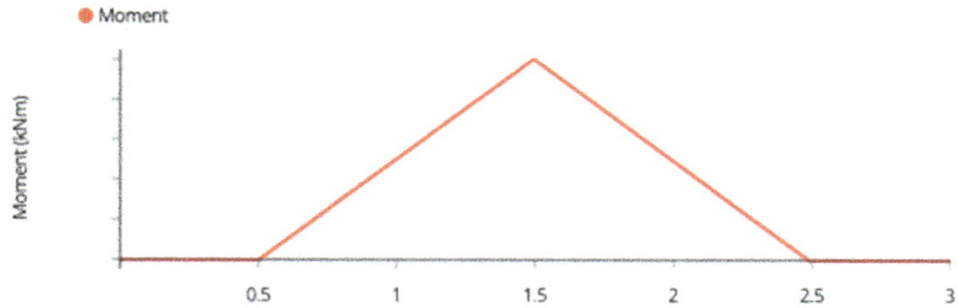

Figure A15. Bottom beam bending force diagram.

The deflection diagram is shown in Figure A16.

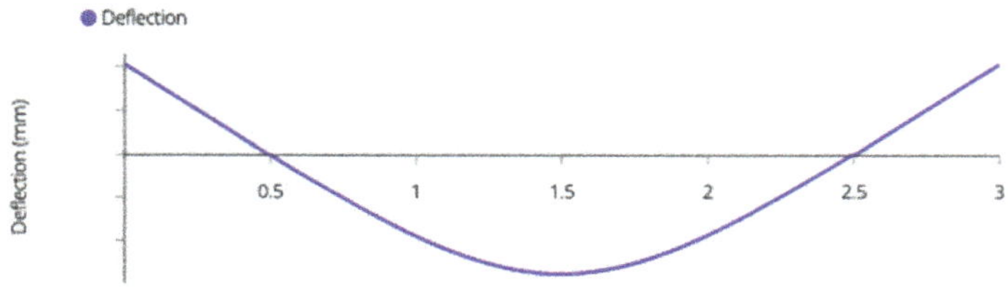

Figure A16. Bottom beam deflection diagram.

The load is then transferred to the posts. It should be noted that the posts are not loaded purely in compression; an eccentricity, e, described in the free body diagram in Figure A17, induces a bending moment.

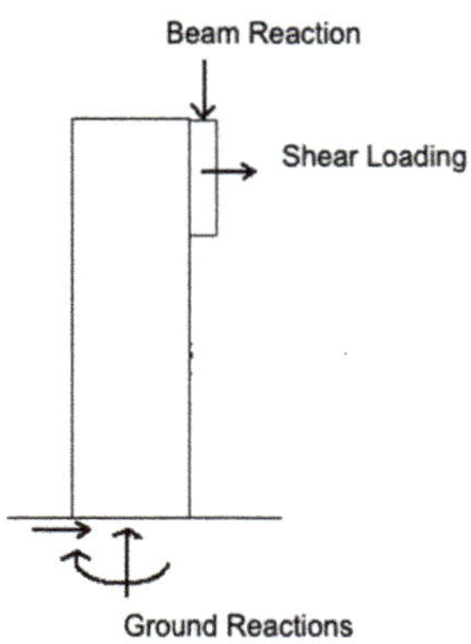

Figure A17. Post free body diagram.

The compressive load of the column is equal to the beam reaction solved above. Along with this compressive load comes a shear loading that is induced by wind and snow loads. This loading can act in either the left or right direction, but the load should be analyzed in the direction that induces bending in the same direction as the beam reaction. The magnitude of this shear load in each post is described in Equation (A12),

$$Shear\ Load = \frac{(Design\ Load)\cos(\theta)}{4} \tag{A12}$$

where θ is the tilt angle of the system. For conservative analysis, assume the lowest tilt angle of 0 degrees to maximize this load calculation. The post will serve as a determinant beam column. The shear force diagram is shown in Figure A18.

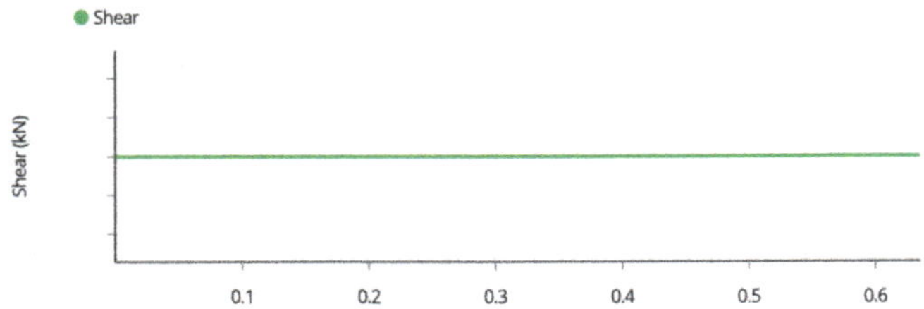

Figure A18. Post shear force diagram.

The maximum shear can be calculated using Equation (A13),

$$Vmax = Shear\ Loading \tag{A13}$$

The bending moment diagram can be found in Figure A19.

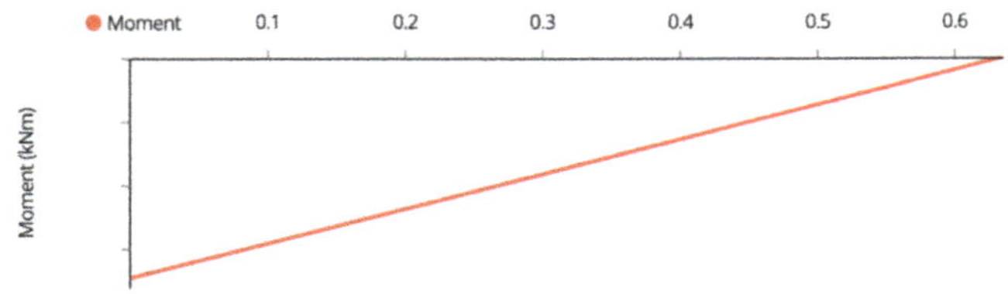

Figure A19. Post bending moment diagram.

The maximum bending moment can be calculated using Equation (A14).

$$M_{max} = Shear\ Loading * L + Beam\ Reaction * e \tag{A14}$$

The deflection diagram is shown in Figure A20.

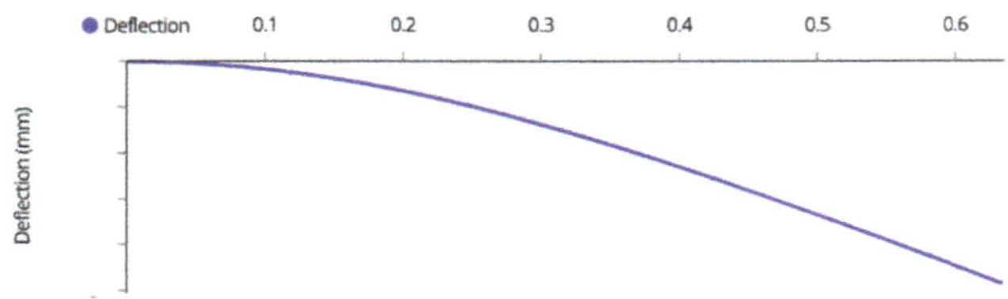

Figure A20. Post bending moment diagram.

The maximum deflection can be calculated using Equation (A15).

$$\Delta_{max} = \frac{Shear\ Loading * L^3}{3EI} + \frac{Beam\ Reaction * e * L^2}{2EI} \tag{A15}$$

Once all components of the post have been analyzed, the load will finally transfer itself to the ground. Table 9.4.4.1 of the NBCC provides maximum allowable bearing pressures for different types of soil and rock. In the worst case, soft clays support a maximum bearing pressure of 75 kPa [40]. To ensure, that the ground is not overloaded and settles, the bearing pressure can be calculated with Equation (A16),

$$Bearing\ Pressure = \frac{Post\ Compression}{\frac{\pi}{4}(D_{Footing})^2} \tag{A16}$$

If the applied pressure is more than the allowable, 150 mm of compacted clear stone gravel can be added to the bottom of the footing, or the footing diameter can be increased.

Throughout the system, each connection transfers the load from one member to another via a shear force within the fasteners that compose that connection. For bolts complying with ASTM A307A, the shear resistance of a $\frac{1}{2}''$ carriage bolt holding the beams is about 23.8 kN, and the shear resistance of a $\frac{1}{4}''$ carriage bolt holding the modules is 5.21 kN [91], both of which are beyond the demand of these systems, and thus will not be critical to the design.

References

1. Pearce, J.M. Photovoltaics—A path to sustainable futures. *Futures* **2002**, *34*, 663–674. [CrossRef]
2. Fu, R.; Feldman, D.J.; Margolis, R.M. *US Solar Photovoltaic System Cost Benchmark: Q1 2018*; No. NREL/TP-6A20-72399; National Renewable Energy Lab.: Golden, CO, USA, 2018.
3. How Much Do Solar Panels Cost? *2022 Guide*. Available online: https://news.energysage.com/how-much-does-the-average-solar-panel-installation-cost-in-the-u-s/ (accessed on 10 May 2022).
4. Branker, K.; Pathak, M.J.M.; Pearce, J.M. A review of solar photovoltaic levelized cost of electricity. *Renew. Sustain. Energy Rev.* **2011**, *15*, 4470–4482. [CrossRef]
5. Dudley, D. Renewable Energy Will Be Consistently Cheaper Than Fossil Fuels by 2020, Report Claims [WWW Document]. Forbes. 2019. Available online: https://www.forbes.com/sites/dominicdudley/2018/01/13/renewable-energy-cost-effective-fossil-fuels-2020/ (accessed on 13 April 2020).
6. Solar Industry Research Data. Available online: https://www.seia.org/solar-industry-research-data (accessed on 13 April 2020).
7. Vaughan, A. Time to shine: Solar power is fastest-growing source of new energy. *The Guardian*, 6 October 2017.
8. Barbose, G.L.; Darghouth, N.R.; LaCommare, K.H.; Millstein, D.; Rand, J. *Tracking the Sun: Installed Price Trends for Distributed Photovoltaic Systems in the United States-2018*; Berkeley Lab: Berkely, CA, USA, 2018.
9. IEA. Solar PV—Renewables 2020—Analysis. Available online: https://www.iea.org/reports/renewables-2020/solar-pv (accessed on 7 April 2022).
10. Levin, T.; Thomas, V.M. Can developing countries leapfrog the centralized electrification paradigm? *Energy Sustain. Dev.* **2016**, *31*, 97–107. [CrossRef]
11. Lang, T.; Ammann, D.; Girod, B. Profitability in absence of subsidies: A techno-economic analysis of rooftop photovoltaic self-consumption in residential and commercial buildings. *Renew. Energy* **2016**, *87*, 77–87. [CrossRef]
12. Hayibo, K.S.; Pearce, J.M. A Review of the Value of Solar Methodology with a Case Study of the U.S. VOS. *Renew. Sustain. Energy Rev.* **2021**, *137*, 110599. [CrossRef]
13. Agenbroad, J.; Carlin, K.; Ernst, K.; Doig, S. Minigrids in the Money: Six Ways to Reduce Minigrid Costs by 60% for Rural Electrification. Rocky Mountain Institute. 2018. Available online: https://rmi.org/insight/minigrids-money/ (accessed on 28 February 2022).
14. Alafita, T.; Pearce, J.M. Securitization of residential solar photovoltaic assets: Costs, risks and uncertainty. *Energy Policy* **2014**, *67*, 488–498. [CrossRef]
15. Renewables International. Photovoltaics after Grid Parity Plug-and-Play PV: The Controversy 2013. Renewables. 2013. Available online: http://www.renewablesinternational.net/plug-and-play-pv-the-controversy/150/452/72715/ (accessed on 18 December 2015).
16. Mundada, A.S.; Nilsiam, Y.; Pearce, J.M. A review of technical requirements for plug-and-play solar photovoltaic microinverter systems in the United States. *Solar Energy* **2016**, *135*, 455–470. [CrossRef]
17. Grafman, L.; Pearce, J.M. *To Catch the Sun*; Humboldt State University Press: Arcata, CA, USA, 2021; ISBN 978-1-947112-62-9.
18. Barbose, G.; Darghouth, N.; Millstein, D.; Cates, S.; DiSanti, N.; Widiss, R. *Tracking the Sun IX: The Installed Price of Residential and Non-Residential Photovoltaic Systems in the United States*; Lawrence Berkeley National Lab. (LBNL): Berkeley, CA, USA, 2016. Available online: https://www.osti.gov/servlets/purl/1345194 (accessed on 10 May 2022).
19. Khan, M.T.A.; Norris, G.; Chattopadhyay, R.; Husain, I.; Bhattacharya, S. Autoinspection and Permitting with a PV Utility Interface (PUI) for Residential Plug-and-Play Solar Photovoltaic Unit. *IEEE Trans. Ind. Appl.* **2017**, *53*, 1337–1346. [CrossRef]
20. Khan, M.T.A.; Husain, I.; Lubkeman, D. Power electronic components and system installation for plug-and-play residential solar PV. In Proceedings of the 2014 IEEE Energy Conversion Congress and Exposition (ECCE), Pittsburgh, PA, USA, 14–18 September 2014; pp. 3272–3278.
21. Lundstrom, B.R. *Plug and Play Solar Power: Simplifying the Integration of Solar Energy in Hybrid Applications*; Cooperative Research and Development Final Report; CRADA Number CRD-13-523; National Renewable Energy Lab. (NREL): Golden, CO, USA, 2017.
22. Mundada, A.S.; Prehoda, E.W.; Pearce, J.M. U.S. market for solar photovoltaic plug-and-play systems. *Renew. Energy* **2017**, *103*, 255–264. [CrossRef]
23. Fthenakis, V.; Alsema, E. Photovoltaics energy payback times, greenhouse gas emissions and external costs: 2004–early 2005 status. *Prog. Photovolt. Res. Appl.* **2006**, *14*, 275–280. [CrossRef]
24. Feldman, D.; Barbose, G.; Margolis, R.; Bolinger, M.; Chung, D.; Fu, R.; Seel, J.; Davidson, C.; Darghouth, N.; Wiser, R. *Photovoltaic System Pricing Trends: Historical, Recent, and Near-Term Projections 2015*; NREL: Golden, CO, USA, 2015.

25. Feldman, D.; Barbose, G.; Margolis, R.; Wiser, R.; Darghout, N.; Goodrich, A. *Photovoltaic (PV) Pricing Trends: Historical, Recent, and Near-Term Projections, Sunshot*; NREL: Golden, CO, USA, 2012.
26. PVinsights. PVinsights 2022. Available online: http://pvinsights.com/ (accessed on 16 January 2022).
27. Alt E Store. Tamarack Solar Top of Pole Mounts for Large Solar Panels. Available online: https://www.altestore.com/store/solar-panel-mounts/top-of-pole-solar-panel-mounts/tamarack-solar-top-of-pole-mounts-6072-cell-solar-panels-p40745/ (accessed on 16 January 2022).
28. TPM3 Pole Mount for Three 60/72 Cell Solar Modules. Available online: https://www.off-the-grid-solar.com/products/tpm3-pole-mount-for-three-60-72-cell-solar-modules (accessed on 10 March 2022).
29. Wittbrodt, B.; Laureto, J.; Tymrak, B.; Pearce, J.M. Distributed Manufacturing with 3-D Printing: A Case Study of Recreational Vehicle Solar Photovoltaic Mounting Systems. *J. Frugal Innov.* **2015**, *1*, 1. [CrossRef]
30. Wittbrodt, B.T.; Pearce, J.M. Total U.S. Cost Evaluation of Low-Weight Tension-Based Photovoltaic Flat-Roof Mounted Racking. *Solar Energy* **2015**, *117*, 89–98. [CrossRef]
31. Wittbrodt, B.; Pearce, J.M. 3-D Printing Solar Photovoltaic Racking in Developing World. *Energy Sustain. Dev.* **2017**, *36*, 1–5. [CrossRef]
32. Arefeen, S.; Dallas, T. Low-Cost Racking for Solar Photovoltaic Systems with Renewable Tensegrity Structures. *Solar Energy* **2021**, *224*, 798–807. [CrossRef]
33. Pearce, J.M.; Meldrum, J.; Osborne, N. Design of Post-Consumer Modification of Standard Solar Modules to Form Large-Area Building-Integrated Photovoltaic Roof Slates. *Designs* **2017**, *1*, 9. [CrossRef]
34. Fixr. 2022 Solar Panel Maintenance Costs. *Solar PV Maintenance Cost.* Available online: https://www.fixr.com/costs/solar-panel-maintenance (accessed on 26 April 2022).
35. Lubitz, W.D. Effect of Manual Tilt Adjustments on Incident Irradiance on Fixed and Tracking Solar Panels. *Appl. Energy* **2011**, *88*, 1710–1719. [CrossRef]
36. Vandewetering, N.; Hayibo, K.S.; Pearce, J.M. Impacts of Location on Designs of DIY Low-Cost Fixed-Tilt Open Source Wood Solar Photovoltaic Racking. *Designs* **2022**, *6*, 41. [CrossRef]
37. What You Need to Know about Pressure Treated Wood. Available online: https://www.lumber.com/blog/what-you-need-to-know-about-pressure-treated-wood (accessed on 17 February 2022).
38. Gilman, P. *SAM Photovoltaic Model Technical Reference*; NREL/TP-6A20-64102; National Renewable Energy Lab. (NREL): Golden, CO, USA, 2015; p. 63. [CrossRef]
39. *System Advisor Model (SAM)*; National Renewable Energy Laboratory: Golden, CO, USA, 2022; Available online: https://github.com/NREL/SAM (accessed on 10 May 2022).
40. Canada, N.R.C. National Building Code of Canada 2015. Available online: https://nrc.canada.ca/en/certifications-evaluations-standards/codes-canada/codes-canada-publications/national-building-code-canada-2015 (accessed on 17 February 2022).
41. Pawluk, R.E.; Chen, Y.; She, Y. Photovoltaic Electricity Generation Loss Due to Snow—A Literature Review on Influence Factors, Estimation, and Mitigation. *Renew. Sustain. Energy Rev.* **2019**, *107*, 171–182. [CrossRef]
42. Andrews, R.W.; Pollard, A.; Pearce, J.M. The Effects of Snowfall on Solar Photovoltaic Performance. *Solar Energy* **2013**, *92*, 84–97. [CrossRef]
43. Andrews, R.W.; Pearce, J.M. Prediction of Energy Effects on Photovoltaic Systems Due to Snowfall Events. In Proceedings of the 2012 38th IEEE Photovoltaic Specialists Conference, Austin, TX, USA, 3–8 June 2012; pp. 3386–3391.
44. Hayibo, K.S.; Petsiuk, A.; Mayville, P.; Brown, L.; Pearce, J.M. Monofacial vs Bifacial Solar Photovoltaic Systems in Snowy Environments. *Renew. Energy* **2022**, *193*, 657–668. [CrossRef]
45. Ryan, A.; Williams Daniel, J.; Lizzadro-McPherson, J.; Pearce, M. The Impact of Snow Losses on Solar Photovoltaic Systems in North America in the Future. to be published.
46. Trifecta 3-Panel Ground Mount KitDefault Title. Available online: https://www.thecabindepot.ca/products/trifecta-3-panel-ground-mount-kit (accessed on 22 April 2022).
47. Retail Council of Canada Minimum Wage by Province. Retail Council of Canada 2022. Available online: https://www.retailcouncil.org/resources/quick-facts/minimum-wage-by-province (accessed on 10 May 2022).
48. Mamun, M.A.A.; Dargusch, P.; Wadley, D.; Zulkarnain, N.A.; Aziz, A.A. A Review of Research on Agrivoltaic Systems. *Renew. Sustain. Energy Rev.* **2022**, *161*, 112351. [CrossRef]
49. Mavani, D.D.; Chauhan, P.M.; Joshi, V. Beauty of Agrivoltaic System regarding double utilization of same piece of land for Generation of Electricity & Food Production. *Int. J. Sci. Eng. Res.* **2019**, *10*, 118–148.
50. Pearce, J.M. Agrivoltaics in Ontario Canada: Promise and Policy. *Sustainability* **2022**, *14*, 3037. [CrossRef]
51. Fthenakis, V.M.; Kim, H.C.; Alsema, E. Emissions from Photovoltaic Life Cycles. *Environ. Sci. Technol.* **2008**, *42*, 2168–2174. [CrossRef]
52. Marrou, H.; Wery, J.; Dufour, L.; Dupraz, C. Productivity and radiation use efficiency of lettuces grown in the partial shade of photovoltaic panels. *Eur. J. Agron.* **2013**, *44*, 54–66. [CrossRef]
53. Valle, B.; Simonneau, T.; Sourd, F.; Pechier, P.; Hamard, P.; Frisson, T.; Ryckewaert, M.; Christophe, A. Increasing the Total Productivity of a Land by Combining Mobile Photovoltaic Panels and Food Crops. *Appl. Energy* **2017**, *206*, 1495–1507. [CrossRef]

54. Barron-Gafford, G.A.; Pavao-Zuckerman, M.A.; Minor, R.L.; Sutter, L.F.; Barnett-Moreno, I.; Blackett, D.T.; Thompson, M.; Dimond, K.; Gerlak, A.K.; Nabhan, G.P.; et al. Agrivoltaics Provide Mutual Benefits across the Food–Energy–Water Nexus in Drylands. *Nat. Sustain.* **2019**, *2*, 848–855. [CrossRef]
55. Hudelson, T.; Lieth, J.H. Crop Production in Partial Shade of Solar Photovoltaic Panels on Trackers. *AIP Conf. Proc.* **2021**, *2361*, 080001. [CrossRef]
56. Sekiyama, T. Performance of Agrivoltaic Systems for Shade-Intolerant Crops: Land for Both Food and Clean Energy Production. Master's Thesis, Harvard Extension School, Cambridge, MA, USA, 2019.
57. Trommsdorff, M.; Kang, J.; Reise, C.; Schindele, S.; Bopp, G.; Ehmann, A.; Weselek, A.; Högy, P.; Obergfell, T. Combining Food and Energy Production: Design of an Agrivoltaic System Applied in Arable and Vegetable Farming in Germany. *Renew. Sustain. Energy Rev.* **2021**, *140*, 110694. [CrossRef]
58. Adeh, E.H.; Selker, J.S.; Higgins, C.W. Remarkable Agrivoltaic Influence on Soil Moisture, Micrometeorology and Water-Use Efficiency. *PLoS ONE* **2018**, *13*, e0203256. [CrossRef]
59. Dupraz, C.; Marrou, H.; Talbot, G.; Dufour, L.; Nogier, A.; Ferard, Y. Combining Solar Photovoltaic Panels and Food Crops for Optimising Land Use: Towards New Agrivoltaic Schemes. *Renew. Energy* **2011**, *36*, 2725–2732. [CrossRef]
60. Schindele, S.; Trommsdorff, M.; Schlaak, A.; Obergfell, T.; Bopp, G.; Reise, C.; Braun, C.; Weselek, A.; Bauerle, A.; Högy, P.; et al. Implementation of Agrophotovoltaics: Techno-Economic Analysis of the Price-Performance Ratio and Its Policy Implications. *Appl. Energy* **2020**, *265*, 114737. [CrossRef]
61. Mow, B. Solar Sheep and Voltaic Veggies: Uniting Solar Power and Agriculture | State, Local, and Tribal Governments | NREL [WWW Document]. 2018. Available online: https://www.nrel.gov/state-local-tribal/blog/posts/solar-sheep-and-voltaic-veggies-uniting-solar-power-and-agriculture.html (accessed on 2 July 2020).
62. Adeh, E.H.; Good, S.P.; Calaf, M. Solar PV Power Potential is Greatest Over Croplands. *Sci. Rep.* **2019**, *9*, 11442. [CrossRef] [PubMed]
63. Elamri, Y.; Cheviron, B.; Lopez, J.-M.; Dejean, C.; Belaud, G. Water Budget and Crop Modelling for Agrivoltaic Systems: Application to Irrigated Lettuces. *Agric. Water Manag.* **2018**, *208*, 440–453. [CrossRef]
64. Al-Saidi, M.; Lahham, N. Solar energy farming as a development innovation for vulnerable water basins. *Dev. Pract.* **2019**, *29*, 619–634. [CrossRef]
65. Giudice, B.D.; Stillinger, C.; Chapman, E.; Martin, M.; Riihimaki, B. Residential Agrivoltaics: Energy Efficiency and Water Conservation in the Urban Landscape. In Proceedings of the 2021 IEEE Green Technologies Conference (GreenTech), Denver, CO, USA, 7–9 April 2021; pp. 237–244.
66. Miao, R.; Khanna, M. Harnessing Advances in Agricultural Technologies to Optimize Resource Utilization in the Food-Energy-Water Nexus. *Annu. Rev. Resour. Econ* **2019**, *12*, 65–85. [CrossRef]
67. Brain, R. *The Local Food Movement: Definitions, Benefits, and Resources*; Utah State University: Logan, UT, USA, 2012.
68. Feenstra, G.W. Local Food Systems and Sustainable Communities. *Am. J. Altern. Agric.* **1997**, *12*, 28–36. [CrossRef]
69. Pascaris, A.S.; Schelly, C.; Burnham, L.; Pearce, J.M. Integrating Solar Energy with Agriculture: Industry Perspectives on the Market, Community, and Socio-Political Dimensions of Agrivoltaics. *Energy Res. Soc. Sci.* **2021**, *75*, 102023. [CrossRef]
70. Pascaris, A.S.; Schelly, C.; Pearce, J.M. A First Investigation of Agriculture Sector Perspectives on the Opportunities and Barriers for Agrivoltaics. *Agronomy* **2020**, *10*, 1885. [CrossRef]
71. Pascaris, A.S.; Schelly, C.; Rouleau, M.; Pearce, J.M. Do Agrivoltaics Improve Public Support for Solar Photovoltaic Development? *Survey Says: Yes!* 2021. Available online: https://osf.io/preprints/socarxiv/efasx/ (accessed on 10 May 2022).
72. Dinesh, H.; Pearce, J.M. The potential of agrivoltaic systems. *Renew. Sustain. Energy Rev.* **2016**, *54*, 299–308. [CrossRef]
73. Zhang, P.; Li, W.; Li, S.; Wang, Y.; Xiao, W. Reliability Assessment of Photovoltaic Power Systems: Review of Current Status and Future Perspectives. *Appl. Energy* **2013**, *104*, 822–833. [CrossRef]
74. Jamil, E.; Hameed, S.; Jamil, B. Qurratulain Power Quality Improvement of Distribution System with Photovoltaic and Permanent Magnet Synchronous Generator Based Renewable Energy Farm Using Static Synchronous Compensator. *Sustain. Energy Technol. Assess.* **2019**, *35*, 98–116. [CrossRef]
75. Craciun, B.-I.; Sera, D.; Man, E.A.; Kerekes, T.; Muresan, V.A.; Teodorescu, R. Improved Voltage Regulation Strategies by PV Inverters in LV Rural Networks. In Proceedings of the 2012 3rd IEEE International Symposium on Power Electronics for Distributed Generation Systems (PEDG), Aalborg, Denmark, 25–28 June 2012; pp. 775–781.
76. Abdul Kadir, A.F.; Khatib, T.; Elmenreich, W. Integrating Photovoltaic Systems in Power System: Power Quality Impacts and Optimal Planning Challenges. *Int. J. Photoenergy* **2014**, *2014*, e321826. [CrossRef]
77. Singh, K.; Mishra, S.; Kumar, M.N. A Review on Power Management and Power Quality for Islanded PV Microgrid in Smart Village. *Indian J. Sci. Technol.* **2017**, *10*, 1–4. [CrossRef]
78. Saleh, M.S.; Althaibani, A.; Esa, Y.; Mhandi, Y.; Mohamed, A.A. Impact of Clustering Microgrids on Their Stability and Resilience during Blackouts. In Proceedings of the 2015 International Conference on Smart Grid and Clean Energy Technologies (ICSGCE), Offenburg, Germany, 20–23 October 2015; pp. 195–200.
79. Junaidh, P.S.; Vijay, A.; Mathew, M. Power Enhancement of Solar Photovoltaic Module Using Micro-Climatic Strategies in Warm-Humid Tropical Climate. In Proceedings of the 2017 Innovations in Power and Advanced Computing Technologies (i-PACT), Vellore, India, 21–22 April 2017; pp. 1–6.

80. Marrou, H.; Guilioni, L.; Dufour, L.; Dupraz, C.; Wery, J. Microclimate under Agrivoltaic Systems: Is Crop Growth Rate Affected in the Partial Shade of Solar Panels? *Agric. For. Meteorol.* **2013**, *177*, 117–132. [CrossRef]
81. Katsikogiannis, O.A.; Ziar, H.; Isabella, O. Integration of Bifacial Photovoltaics in Agrivoltaic Systems: A Synergistic Design Approach. *Applied Energy* **2022**, *309*, 118475. [CrossRef]
82. Gorjian, S.; Bousi, E.; Özdemir, Ö.E.; Trommsdorff, M.; Kumar, N.M.; Anand, A.; Kant, K.; Chopra, S.S. Progress and Challenges of Crop Production and Electricity Generation in Agrivoltaic Systems Using Semi-Transparent Photovoltaic Technology. *Renew. Sustain. Energy Rev.* **2022**, *158*, 112126. [CrossRef]
83. Residential Electricity and Natural Gas Plans. Available online: https://energyrates.ca/residential-electricity-natural-gas/ (accessed on 10 May 2022).
84. Average Business Energy Consumption. Energy Bills. Bionic. Available online: https://bionic.co.uk/business-energy/guides/average-energy-usage-for-businesses/ (accessed on 31 May 2022).
85. US EPA, O. Overview of Wood Preservative Chemicals. Available online: https://www.epa.gov/ingredients-used-pesticide-products/overview-wood-preservative-chemicals (accessed on 28 May 2022).
86. Haywood, S. The Effect of Excess Dietary Copper on the Liver and Kidney of the Male Rat. *J. Comp. Pathol.* **1980**, *90*, 217–232. [CrossRef]
87. Jin, D.; Meyer, T.K.; Chen, S.; Ampadu Boateng, K.; Pearce, J.M.; You, Z. Evaluation of Lab Performance of Stamp Sand and Acrylonitrile Styrene Acrylate Waste Composites without Asphalt as Road Surface Materials. *Constr. Build. Mater.* **2022**, *338*, 127569. [CrossRef]
88. NDS. 2018. Available online: https://awc.org/publications/2018-nds/ (accessed on 17 February 2022).
89. LG 400W NeON2 BiFacial Solar Panel. LG400N2T-J5—Volts Energies. Available online: https://volts.ca/collections/solar-panels/products/lg400n2t-j5-solar-panel (accessed on 10 January 2022).
90. Dead Loads. Available online: https://www.designingbuildings.co.uk/wiki/Dead_loads (accessed on 17 February 2022).
91. Load Calculator. Fastenal. Available online: https://www.fastenal.com/en/84/load-calculator (accessed on 17 February 2022).

designs

Article

Getting Fit in a Sustainable Way: Design and Optimization of a Low-Cost Regenerative Exercise Bicycle

Huy Pham [1], Aseesh Paul Bandaru [1] Pranav Bellannagari [2], Sohail Zaidi [2] and Vimal Viswanathan [1,*]

1 Mechanical Engineering Department, San Jose State University, San Jose, CA 95192, USA; huy.pham@sjsu.edu (H.P.); aseeshpaul.bandaru@sjsu.edu (A.P.B.)
2 Intelliscience Institute, San Jose, CA 95192, USA; pranavbella@gmail.com (P.B.); syed.zaidi@sjsu.edu (S.Z.)
* Correspondence: vimal.viswanathan@sjsu.edu

Abstract: With the increase in demand for more sustainable energy sources, recent researchers have been looking into harvesting energy spent by humans for various purposes. One of the available sources of such energy is exercise equipment. While a few products are available in the market to harvest the power expended during an exercise session, these products are costly, and the cost may prohibit a day-to-day user from purchasing those. Motivated by this challenge, this paper describes a long-running research project that uses a static exercise bicycle to sustainably harvest human energy. A regenerative spin bike that uses the friction between a flywheel and a BaneBots wheel was designed and deployed. For the motor mount, two methods are investigated: linear preloading and rotary preloading. A commercially available indoor static bicycle is modified to incorporate the flywheel and the motor attachment. The generated electricity is converted to DC using a three-phase rectifier. A car charger is used for charging any devices attached to the setup. The resulting configuration is very effective in operating small electronic devices. This setup, which uses only off-the-shelf components, can be considered a replacement for its expensive custom-made counterparts.

Keywords: energy harvesting; green energy; green gym; sustainable energy

Citation: Pham, H.; Bandaru, A.P.; Bellannagari, P.; Zaidi, S.; Viswanathan, V. Getting Fit in a Sustainable Way: Design and Optimization of a Low-Cost Regenerative Exercise Bicycle. *Designs* **2022**, *6*, 59. https://doi.org/10.3390/designs6030059

Academic Editors: Shi-Jie Cao and Wei Feng

Received: 6 May 2022
Accepted: 16 June 2022
Published: 18 June 2022

1. Introduction

Increasing demand for energy and awareness about CO_2 emissions [1] has made many companies invest more in researching and developing an environmentally sustainable method for energy harvesting. According to REN21(Renewable Energy Policy Network for the 21st Century) 's 2021 Report [2], renewable sources contributed 18.1% to our energy consumption and 26% to our electricity generation in recent years (7.5% from biomass, 4.2% from heat energy, 1% from biofuels, 3.6% from water, 2% combined from wind, solar, biomass, geothermal, and ocean power). However, several other sustainable energy sources are overlooked that can provide a solution for the inevitable energy crisis. Human energy is one such energy form.

Different approaches are proposed in the literature to harvest energy from human motion. Piezoelectric footwear and other piezoelectric devices have been gaining popularity recently, e.g., [3–5]. In the piezoelectric footwear design, the dynamic force generated by the human heel is harvested and amplified using a piezoelectric circuit embedded in footwear. While several researchers continue to work on this area, these devices face several challenges, including low power output [6]. A backpack for generating power using human kinetic energy has also been proposed by researchers [7]. This work uses a flexible mechanical motion rectifier (MMR) to harvest energy. A few other recent works have focused on generating power from human joints using wearable, lightweight systems that involve minimal user effort [8–10]. There are also works focusing on harvesting energy based on the position of the center of mass of the human body [11–13]. Recent efforts have also tried to generate electricity from mechanical motions and vibrations [14].

A typical human burns an average of 2000 kCal (8.5 MJ) of energy daily, equivalent to the energy stored in a car's battery. Many studies investigate the capture and reuse of this human energy e.g., [15–18]. However, only a few have promised sustainable ways to harvest the energy [19–22]. Gymnasiums (gyms) are the perfect example of the wastefulness of human energy. In a typical gym, powered exercise equipment consumes massive amounts of energy. Most gyms operate lights, TVs, fans, and speakers regardless of the presence of users. The possibility of capturing the energy burnt by the gym-goers sustainably to operate all these equipment has been mostly overlooked in the literature. If this human energy can be harvested efficiently, the energy consumption by gyms can be reduced to a great extent.

In recent years, several research studies have explored the generation of green energy from gym equipment. A survey of college students found that they were willing to use a campus gym more often if the energy they burnt in the exercise equipment could be used for charging their electronic devices [23]. The mechanical engineering researchers at the University of California at Berkeley have developed a facility within their campus gym to harvest energy and feed it back to the facility [24]. Another similar study from California [25] shows that using a four-switch buck-boost topology, human power can be harvested from gym equipment with more than 90% efficiency. The student researchers at Cal Poly Pomona have created an array of energy-harvesting elliptical machines in their recreation center to power the facility [26]. A similar study is also reported from the University of British Columbia, where energy-harvesting elliptical machines reduced the CO_2 production by 0.2 kg/day [27]. Similar efforts have been reported from other parts of the world [28–31].

Beyond research studies, there have been a few attempts to commercialize green gym technology. In the United States, Adam Boesel created a "green micro-gym" in Oregon [32] that uses a combination of human energy and solar energy to power the gym. A similar attempt has been reported in England by the "Great Outdoor Gym Company" [33]. In addition to feeding the gym equipment, the facility uses human power to light a display board to make the users aware of their contributions to powering the gym.

While many previous attempts to harvest human energy from gym equipment have been successful, most of them suffer from a considerable initial cost to retrofit equipment to produce electricity. The technology needs to address several challenges before it can go mainstream. First, a person cannot generate a good amount of electricity with a bicycle generator in a single workout [34]. Due to the initial cost of the equipment and the relatively minor return, it will take several years for the gyms to make this technology profitable. Many gyms advertise themselves as green gyms to attract environmentally cautious customers without taking any action to harvest human energy [35]. This technology requires significantly more research and refinement before being widely adaptable worldwide.

Motivated by the abovementioned factors, this study targets to develop cost-effective and efficient technology to harvest energy from an exercise bike-powered generator. The target is to use off-the-shelf components and existing technology to gather energy from an exercise bike (static bike) in a laboratory setting and optimize the design using design for manufacturability and assembly (DFMA) considerations. We have explored this concept in a previous study [36] done at the San Jose State University gym facility. While the prior study showed promising results, we found that the technology was not good enough to achieve sustainable results in the long term. Ideally, the device will be compatible with and will not change the footprint of the exercise bike and, at the same time, be easy to maintain. This type of electrical energy generation will help minimize environmental impacts and preserve the environment by reducing the emission of carbon dioxide into the atmosphere. Further, a detailed cost analysis is performed on the new design to calculate the years it will take for an investor to reach the return of their investment point.

2. Methodology

2.1. Design Specifications

To develop the most cost-effective system, all the available technologies in the literature are investigated and compared. Several previous works attempted to create a system to harvest power from exercise equipment. Unfortunately, most of these works only described their design without providing additional data for quantification. In our study, we compared the technology used in these designs and chose the best elements that work for such a system. e.g., the table shows the Pugh chart [37] comparing the energy conversion subsystems of various designs proposed in the literature and the ranking of those systems based on the requirements for our project. The target of this approach was to use the best elements in the existing literature to develop an energy harvesting system that could provide optimum performance.

Based on this preliminary analysis, a friction-driven mechanism has been identified as the ideal candidate. The traditional friction wheel mechanism has been updated with elements from other energy harvesting mechanisms to achieve the project's end goals. The target is to develop a device that can be retrofitted to any existing exercise bike. The device is also expected to be of high quality and low cost. Figure 1 shows the CAD model of an exercise spin bike, as envisioned, with an RC motor attached. The design uses the friction between a flywheel and a BaneBot wheel to achieve the drive and generate electricity. The BaneBot wheel is attached to the motor using a BaneBot Hub. Achieving the right amount of friction between the flywheel and the BaneBot wheel is critical. It is also crucial to size the BaneBot Hub correctly to fit the motor shaft.

Figure 1. CAD model of a spin bike with an RC motor attached to the flywheel.

2.1.1. Friction Drive Transmission

Compared to gear or belt-driven systems, the friction drive transmission possesses certain unique advantages. They can transfer torque with high mechanical efficiency and minimal backlash at high speeds. They are also known for quieter operations. Since their operation is smoother, high-speed motors can be utilized, which offers a higher power density [38]. These drives are also mechanically safer and more straightforward to assemble. This system is also simple to manufacture, making it an ideal candidate for our application.

When the user pedals to rotate the flywheel, the motor rotates. There are two methods to mount the motor, as shown in Figure 2.

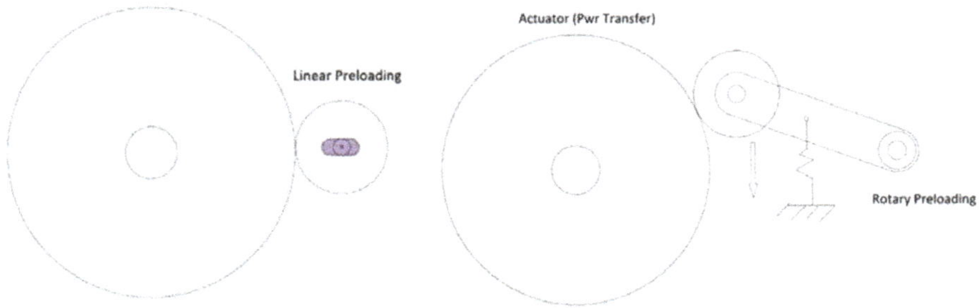

Figure 2. Two methods of preloading a friction drive transmission.

1. Linear Preloading: The linear preloading method utilizes an oblong hole to mount the motor using a screw and a nut. The motor wheel is pushed against the flywheel and then tightens the nut.
2. Rotary Preloading: The motor is mounted on a pivot arm, and a spring is attached to apply the pressure against the flywheel.

For the linear preloading, the tension must be set precisely for the drive to transmit the power. On the other hand, in the rotary preloading, we need enough force so that it does not slip. In addition, if the flywheel surface is imperfect, the spring system is more forgiving.

A friction drive is very similar to a geared transmission with no teeth. The computations for this drive type are the same as the standard geared transmission. The calculations performed on the friction drive system are shown in Figure 3 and the equations below.

$$Gear\ ratio = \frac{T_1}{T_2} = \frac{R_1}{R_2} = \frac{C_1}{C_2}$$

where T = torque, R = radii, d = diameter, C = circumference

$$Power,\ P = \frac{2\pi NT}{60}$$

where N = rotational speed (RPM), T = Torque (Nm)

$$Instantaneous\ velicuty\ of\ the\ point\ of\ contact = \frac{rev}{min} \cdot \frac{\pi d}{rev}$$

2.1.2. Choosing the Motor

The most critical task is determining a highly efficient and durable permanent magnet motor that generates power in a usable voltage range. The goal is to have a motor spin at a rate that will generate 12–15 V when a person pedals the exercise bike.

A Sunny Health SF-BD1423 Belt Drive Indoor Cycling Bike is used to build the prototype in this project. The Sunny Health SF-BD1423's flywheel diameter is measured at 18 inches. The following relationships are used to calculate the motor RPM (Figure 4).

$$Speed_1 \cdot 2\pi r_1 = \mathrm{V}$$

$$-Speed_2 \cdot 2\pi r_2 = \mathrm{V}$$

$$\frac{Speed_1}{Speed_2} = -\frac{r_2}{r_1} = -\frac{C_2}{C_1}$$

$$d_{flywheel} = 18$$

$$d_{Banebotwheel} = 2.875$$

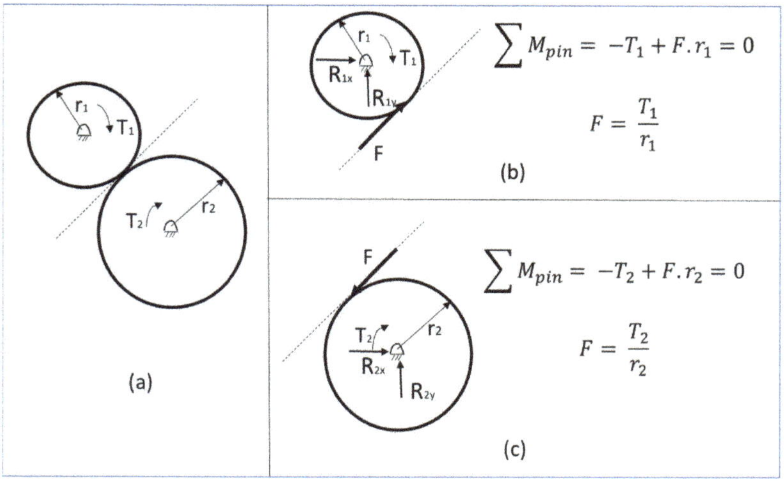

Figure 3. (**a**) Motor and flywheel free body diagram; (**b**) the free body diagram of the motor; and (**c**) the free body diagram of the flywheel.

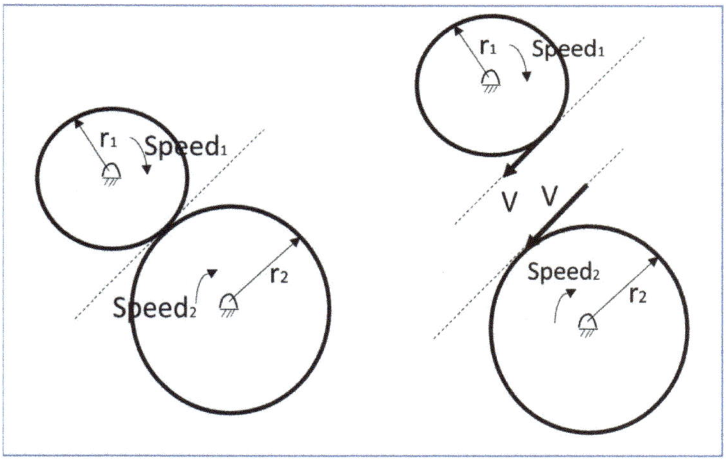

Figure 4. Flywheel and motor speed relationship (free body diagram).

Pedaling the spin bike at an average pace rotates the flywheel around 250 to 350 RPM. Taking the lower end of that, we have:

$$Speed_{Banebotwheel} = Speed_{flywheel} \cdot \frac{18}{2.875} = 1583 \text{ RPM}$$

The motor needs to produce 11–12 V at this RPM.

$$\frac{1583 \text{ RPM}}{11 \text{ V}} = 143.9$$

$$\frac{1583 \text{ RPM}}{12 \text{ V}} = 131.9$$

Therefore, the target motor should be around 140 KV.
P = 0.5 KW, Speed = 1583 RPM
$Torque_{motor}$ = P × 9.549/Speed = 0.5 kW × 9.549/1583 RPM = 3.02 Nm

We also have

$$\frac{T_1}{T_2} = \frac{R_1}{R_2} = \frac{C_1}{C_2}$$

Using this relationship, we can calculate the torque input by the flywheel.

$$Torque_{flywheel} = \frac{d_1}{d_2}\,Torque_{motor}$$

$$Torque_{flywheel} = \frac{18}{2.875}\cdot 3.02 = 18.91\text{ Nm}$$

2.1.3. Design of Shaft

One disadvantage of the friction drive is the internal forces transmitted to the shafts. The shafts need to be designed to handle these radial loads, which makes it a combined bending and torsion loading problem (Figure 5).

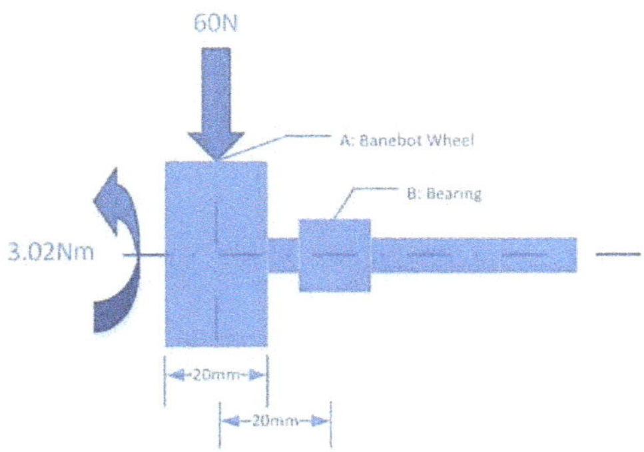

Figure 5. Motor shaft combined bending and twisting moment.

P = 0.5 kW = 5.0 × 10^2 W
Speed (N) = 1583 RPM
Radial loading (bending) is 60 N.
Shaft material is mild steel with safe stress τ = 60 MPA
Motor torque is calculated above T = 3.02 Nm or 3.02 × 10^3 N·mm
The bending moment at point B, M_B = 60 N × 20 mm = 1.2 × 10^3 N·mm
Therefore, the Equivalent twisting moment is given by

$$T_{eq} = \sqrt{T^2 + M^2} = \sqrt{3.02^2 + 1.20^2}$$

T_{eq} = 3.25 × 10^3 N·mm
Furthermore, $T_{eq} = \frac{\pi}{16} d^3 \cdot \tau$ (by Strength criteria for shaft design)
T_{eq} = 3.25 × 103 N·mm = $\frac{\pi}{16} d^3 \cdot 60$
$d^3 = \frac{3250\text{ N·mm} \cdot 16}{\pi \cdot 60}$
d = 6.5 mm

2.1.4. Components and Cost

Table 1 shows the components required and the estimated cost. If the cost of the spin bike and some of the optional parts are removed, the electricity harvesting system can be built for less than $200, not counting the cost of the tools. However, for prototyping, these items are purchased in small quantities. When purchased in bulk, the cost can be reduced significantly. In summary, this design costs way less than similar products in the market.

Table 1. A comparison of the energy harvesting subsystems of various designs in the literature. This method was used to select the best subsystems for our design.

			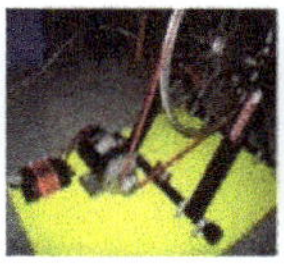			
Customer Needs	**DESIGN A**	**DESIGN B**	**DESIGN C**	**DESIGN D**	**DESIGN E**	**DESIGN F**
Power generation efficiency	S	0	0	0	0	0
Universality	S	0	0	−1	0	0
Portability and compactness	S	0	0	−1	1	1
Works with multiple power sources	S	0	0	0	0	0
Integrated USB for mobile devices	S	0	0	0	0	0
Low noise	S	0	0	0	0	0
Environment: home or commercial	S	0	0	0	0	0
Ergonomic	S	−1	−1	1	1	0
Aesthetics	S	−1	−1	1	1	1
Durability	S	−1	0	0	−1	0
Energy storage	S	0	0	0	0	0
Safety	S	0	0	1	0	1
Maintenance	S	−1	0	0	−1	0
Cost	S	0	0	0	0	1
TOTAL +	0	0	0	3	3	4
TOTAL −	0	−4	−2	−2	−2	0
SUM	0	−4	−2	1	1	4

2.2. Prototype Building

2.2.1. Motor Assembly

Figure 6 shows the motor assembly. Two flat sheet metal brackets are used to mount the motor to the spin bike. The bracket has four screw holes lining up with the hole patterns on the motor at 38 mm spacing. Four hex cap M4 screws are used to secure the motor to the bracket. To secure the Babebot wheel to the motor shaft, a T81 Hub was used. The hub comes with 2 set screws and a snap ring. First, the hub was secured with two set screws, and then the Babebot wheel was slid in and secured with the snap ring.

2.2.2. Mounting the Motor Assembly to the Spin Bike Frame

Modification to the spin bike frame is required to mount the motor assembly. The mount is shown in Figure 7. First, the metal frame is drilled through to mount the two parallel brackets with an all-threaded bolt. Access to the inside frame is obstructed, so a more extended drill bit is needed to drill the holes at an angle. It should be noted that this is the only modification needed on the original bike frame for the retrofit.

Figure 6. (**a**) The motor assembly as designed; (**b**) the exploded assembly view of the design.

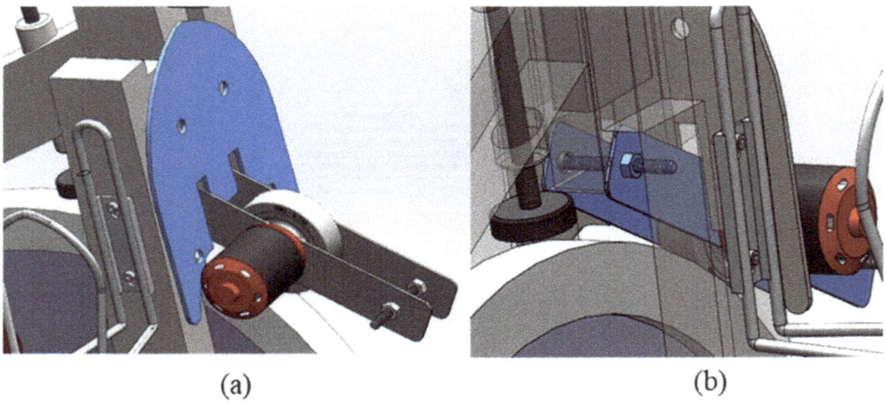

Figure 7. (**a**) The motor mounting and assembly as envisioned; (**b**) the bike frame modification needed to mount the motor.

The front steel plate must be cut out to avoid interference with the motor bracket, as shown in Figure 8. An all-threaded bolt and three hexagonal nuts are used to keep the two flat brackets parallel. A bungee cord applies a radial force to the flywheel so the Banebot wheel does not slip.

The assembly in the finished prototype is shown in Figure 8.

2.3. *Electrical Components and Connections*

The power generated can be converted to DC using a 3-phase rectifier. Figure 9 shows the simple schematic and components. The multimeter reads out the voltage generated when the motor is spun. The goal is to have 11–15 V recommended for car socket chargers and inverters. Devices can be charged or connected by using 12 V car chargers. The more devices are plugged in, the more power is required and hence more resistance.

A "RadioShack project box" is used to house the 3-phase bridge rectifier, multimeter, two 12 V sockets, and electrical wiring. A Dremel is used to cut out a rectangular slot of 45 mm × 86 mm for the multimeter. A 1.25" spade drill bit is used to make the two slots for the 12 V sockets. The wiring as completed on the prototype is shown in Figure 10.

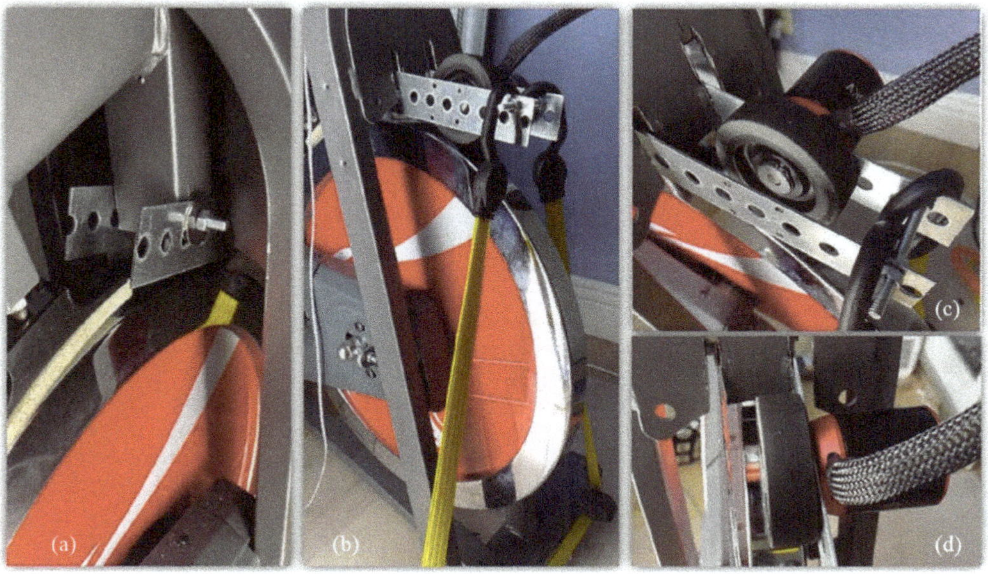

Figure 8. The motor assembly in the prototype (**a**) the modification of the bike frame to mount the motor; (**b**) the friction drive assembly; (**c**) the friction drive and the motor assembly; and (**d**) a closer view of the friction drive with the Motor.

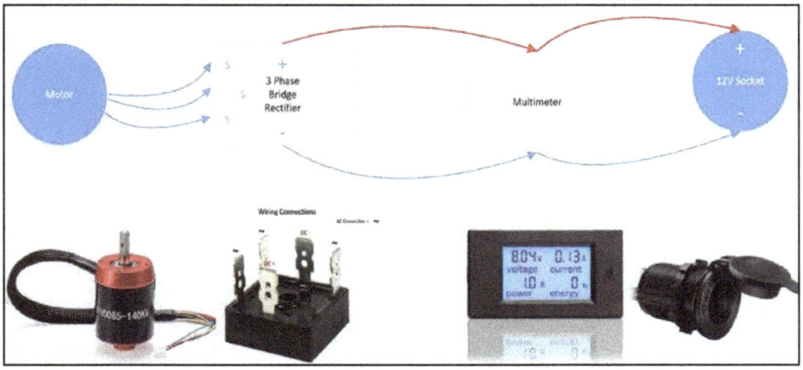

Figure 9. Convert AC to DC with a 3-phase bridge rectifier.

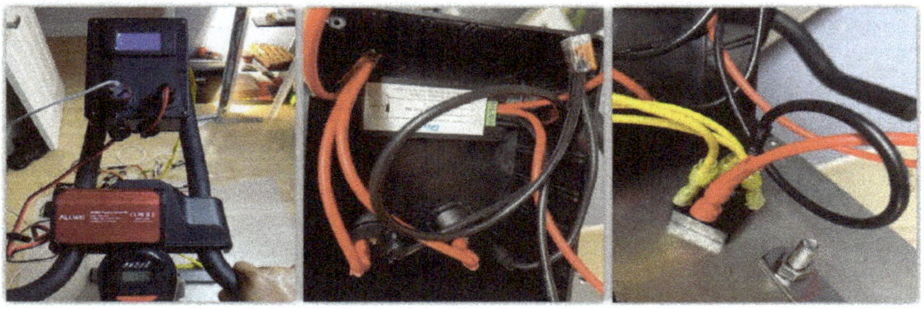

Figure 10. The electrical components on the prototype.

3. Data Characterization and Comparison with Similar Work

3.1. The Setup

To power up or charge devices, and Anker USB charger is used. This 24 W dual USB charger can output 12 W per port and charge any mobile device or tablet. A Jackery battery pack is added to the system to harvest the excess power for future use. This battery pack has a carport charger that can be plugged directly into the 12 V car sockets. Connecting directly into the 12 V car socket is important to avoid power loss through another connection, such as an inverter. The inverter is used to connect any appliance that requires 110 V AC.

3.2. Results and Discussion

An experiment was conducted with ten participants to evaluate the system's performance. The sample size was restricted primarily due to the global pandemic (COVID-19). All critical data parameters were recorded, such as exercise duration, input calories, output voltage, current, power produced, and devices percentage charge. All the data points are tabulated in Table 2. Of the eleven participants, ten rarely do or never do any cycling exercise. Fifty percent of the participants said they never go to the gym and rarely do any exercise. The other half of the participants do some kind of exercise but rarely do any cycling exercise. Only one participant (trial #11 and trial #12) is a weekend biker who just picked up the sport for six months.

Table 2. Components and cost (including the bike).

Items	Cost (USD)	
Sunny Health Spin Bike	299	
RC Motor	81.89	
35 Amp 3 phase bridge rectifier	12	
DC meter	20	
12 GA Wire	11	
Female Disconnects	8	
Female Bullet Connector	11	
Heat Shrink Tubing	7	
MT60 e wire bullet connectors	17	
All-Thread rods	21	
Nuts	5	
Lock Washer	6	
Loctite Threadlocker	6	
Washers	12	
BaneBots Hub	4	
BaneBots Wheel	3	
Project box	24	
12 V car sockets	36	
12 V Socket Splitter	12	
Wago Connectors	9	
12 V Jackery Battery	250	Optional
Power Inverter DC to AC	30	Optional
USB Car Charger Adaptor	10	Optional
	$1000.00	

A basic setup with one iPhone 11 Pro Max and one Jackery, 240 Wh Battery pack, is connected to the system for the first seven trials. The more devices are plugged in, the more power draw required and hence a higher resistance the peddler will feel. With one iPhone and one Jackery battery connected seems to be the sweet spot resistance that most participants think they can pedal for 15 to 30 min without taking a break.

The raw data obtained from the experiment were compared against the theoretical calculations using the Coulomb-count method. According to the Coulomb count method, the state of charge at time t ($SOC_{@t}$) is equal to the charge at the current time (t−1) divided by the total capacity. The rated charge capacity ($Q_{nominal}$) for an iPhone 11 Pro Max is 16.8 Ampere-hour (60,480 Coulombs).

$$SOC_{@t} = SOC_{@(t-1)} + I_{@t} \frac{\Delta t}{Q_{nominal}}$$

For example, assuming that the battery charge is at 0% at time $t = 0$,

$$SOC_{@15\ min} = SOC_{@\ 0\ min} + 4.34\ A * \frac{0.25\ \text{h}}{16.8\ A\ \text{h}} = 6.46\%$$

From Table 2, the battery is only charged to 3% at 12 min with the current shown on the multimeter. This translates to an efficiency of approximately 46% for the system. Please note that the charge percentage of the iPhone is offered only in whole numbers, and fractions are not available at any given time. Following the same method, theoretical calculations are performed for all the trials, and the results are shown in Table 3. The last column shows the experimental values. It is observed that the retrofit design can perform energy harvesting at 40–50% efficiency consistently.

Table 3. Raw data collected from the experiment.

Trial	Duration (min)	Average Voltage	Average Energy (WH)	Calories	Power (W)	Battery %	Phone %	Battery Accumulation	Phone Two %
1	15	12.00	13.25	105	53.00	3	15	3	-
2	14	11.50	11.43	110	49.00	3	16	6	-
3	13	13.00	11.05	108	51.00	4	25	10	-
4	24	11.50	23.20	176	58.00	3	24	13	-
5	16	14.00	15.73	150	59.00	4	18	17	-
6	18	11.00	13.96	70	46.53	1	9	18	-
7	15	11.20	11.90	120	47.60	2	3	20	-
8	18	11.50	11.39	60	37.95	4	N/A	24	-
9	18	11.40	11.29	140	37.62	3	N/A	27	-
10	18	13.00	12.71	120	42.38	4	N/A	31	-
11	30	14.00	31.85	260	63.70	10	35	41	30
12	30	13.45	57.75	315	115.50	9	30	52	25

It is observed that the USB charger always produces a constant output of 13 W. Measurements are taken for every 1 V output increment from 9 V to 15 V (Figure 11). If the voltage increases, the current draw decreases so that the total output is always around 13 W. The charging rate is the same even though the peddling speed is faster. So, the percentage charged is limited by the charger capacity. According to the specification, each USB port is rated at 12 W. However, the recorded data showed 13 W, 8.3 percent higher than the specification. With more advanced technology and newer mobile devices released with much higher battery capacity, it is necessary to have a portable charger with higher output.

It is recommended to add a higher-rated charger to take full advantage of our pedaling power for future design iterations.

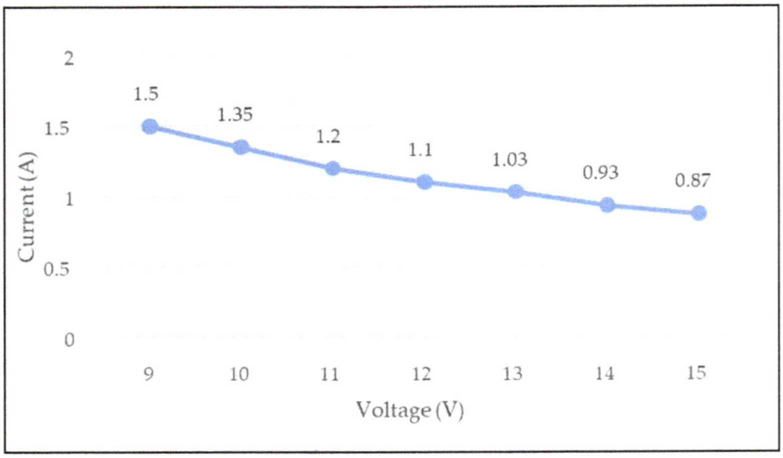

Figure 11. Experimental measurement with a 12 W charger.

A digital tachometer is used to record the flywheel's rotational speed, which is used to calculate the motor's rotational speed. Figure 12 shows the result of voltage generated vs. the motor RPM.

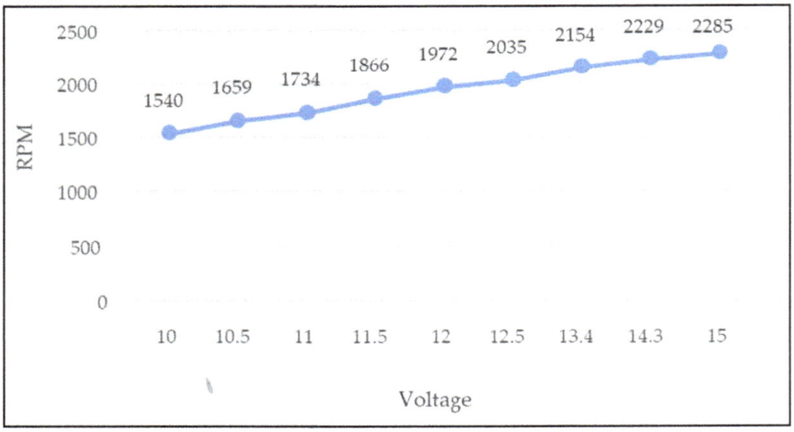

Figure 12. The voltage vs. motor RPM characteristics of the design.

Theoretically, to generate 12 V, the motor should spin at the speed of 1680 RPM. The recorded value at 12 V is 1972. This is expected due to the energy loss in the motor and 3-phase bridge rectifier. So, from the motor to the AC-DC conversion phase, there is a loss of 17.4%. Energy loss is present in every stage, from production to storage. We can expect another 5–10% loss at the 12 V socket connection, 15% in the battery storage, and another 15% if the inverter converts DC to AC. Losses can be minimized by reducing the number of electrical connections and using mechanical links wherever possible. One of the main reasons a Jackery battery storage is used is its 12 V car socking charging option. This eliminates the inverter from the equation, which saves 15% efficiency.

For the participants who do not exercise regularly, charging an iPhone and the Jackery battery pack is enough resistance to allow them to pedal for 15 to 30 min continuously. For one participant who bikes every weekend, charging two iPhone 11 Pro Max and the battery pack is an easy task. Two iPhone and the battery pack is drawing about 65 W. According to the participant who does cycling weekly, charging two iPhones and the battery pack is about the same as riding his bike on a leveled street. The more devices are plugged in, the heavier the resistance. To make it an actual workout for the subject, the two iPhones are replaced by a desktop electric fan at a "high" setting. This is drawing a total of 115 W. According to the subject, this setting is similar to going upwards on a steady hill. The subject was able to pedal for 30 minutes straight without any exhaustion. The participant reported the ability to continue for an hour in this setting.

Of all the 11 participants, 4 participants generated 13 to 14 V, while four others generated an average of 11.5 to 12.5 V. Three participants generated an average of 10–11.5 V. The voltage distribution is shown in Figure 13. If a person pedals at a speed that only generates at around 11 V, the energy is only enough to charge the iPhone. When the pedaling rate exceeds the one corresponding to 12 V, the excess energy will go to the battery pack. The iPhone charging speed is constant since the output wattage of the USB charger is maxed out at 13 W.

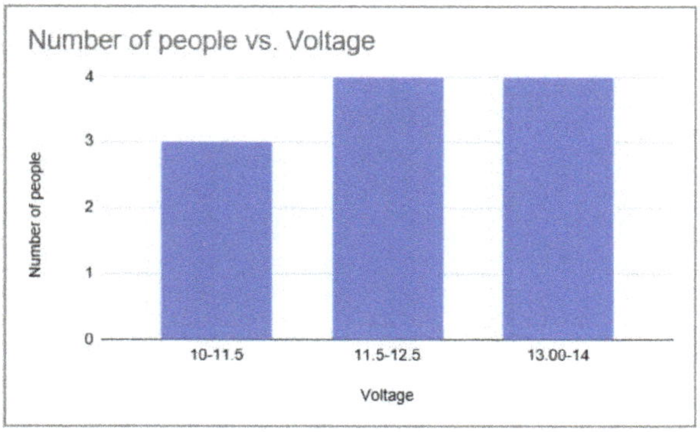

Figure 13. The average voltage generated as a function of the number of people using the device.

On the other hand, the faster-pedaling speed, the faster the battery pack is charged. Since the battery pack can accept a wide range of voltage from 12 V to 30 V, there is no risk of damaging the battery pack. However, if the inverter is used, the optimal voltage range when using the inverter is 12 to 15 V. An integrated circuit in the inverter protects the inverter from overcharge or undercharge devices. This theory is validated by pedaling at a breakneck speed that generated 20 V, and the inverter shut off the circuit temporarily. After about 30 seconds, the rate is reset to under 15 V and the inverter is turned back on.

Figure 14 and Table 4 show that in the first ten trials with an average of around 12 V, the Jackery battery pack can be fully charged in about 9 h, which is slower than the charging time of 7–8 h stated by the manufacturers if it was charged via a regular AC outlet. However, if we can pedal fast enough and consistently generate 13 to 14 V, the battery can be charged more quickly than plugging into the wall outlet at less than 6 h to completion (Table 5).

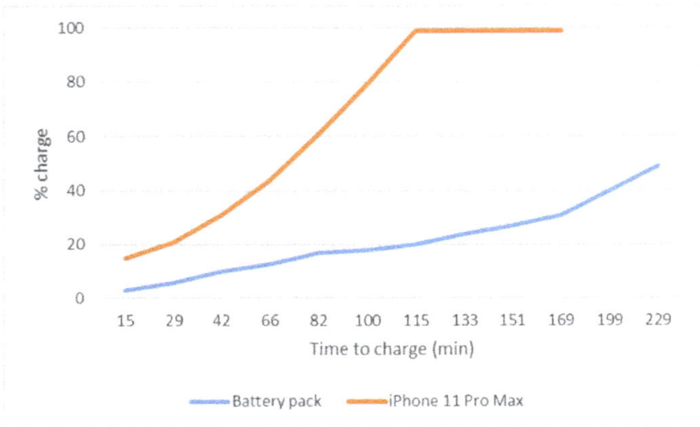

Figure 14. Percentage of charge as a function of time for the battery pack and the iPhone 11 Pro Max.

Table 4. Estimating the battery charge using the Coulomb-count method and comparison with the experimental data.

Trial	Trial Duration	Total Duration	Average Voltage	Average Current	Battery% Accumulation (Theoretical)	Battery% Accumulation (Experimental)
1	15	15	12	4.34	6.46	3
2	14	29	11.5	4.3	12.43	6
3	13	42	13	4.23	17.88	10
4	24	66	11.5	4.13	27.71	13
5	16	82	14	4.21	34.41	17
6	18	100	11	4.23	41.95	18
7	15	115	11.2	4.25	48.27	20
8	18	133	11.5	3.3	54.17	24
9	18	151	11.4	3.3	60.06	27
10	18	169	13	3.26	65.88	31
11	30	199	14	4.55	69.27	41
12	30	229	13.45	8.59	94.83	52

Table 5. Charging speed in comparison to a wall outlet.

Level of Experience	% Charge on the Battery Pack	Time to Charge (h)	Comparison with a Wall Outlet
Little or no exercise	31	2.82	Slower than the wall outlet
	100	9.10 (projected)	
Weekend biker	18	1	Faster than the wall outlet
	100	5.56 (projected)	

3.3. Comparison with Existing Studies

While multiple papers in the literature attempted to develop energy harvesting exercise systems, studies that presented numerical data to demonstrate their effectiveness are scarce. In this subsection, the performance of our Banebot-based exercise bike energy harvester is compared against two other systems developed previously: the design proposed by

Anyanwu and Anthony [39] (referred to as "design A") and the bicycle power generator design by Patowary, et al. [40] (referred to as "design B"). Both the designs use belt drive for transmission of rotation from the flywheel to the generator. In contrast, in the design of the Banebot, friction drive transmission has been employed with rotary preloading, ensuring adequate traction between the flywheel and the motor. Moreover, unlike in the case of Designs A and B, there is minimal energy loss in the BaneBot's transmission as there is no belt or chain drive.

Both designs A and B have the hind wheel of the setup modified into a flywheel by eliminating the rubber and other parts on the upper surface of the wheel's rim. The Rim's upper surface comes in contact with the belt for the pulley arrangement between the flywheel and generator. The flywheel operates like any bike's hind wheel, employing a chain and sprocket mechanism to transfer the movement from pedals into rotation of the flywheel, causing transmission losses both in the chain drive and belt drive. Moreover, these designs made use of traditional road bikes for their setups, unlike in the case of Banebot, for which a stationary exercise bike has been used, which consists of a heavier flywheel, making the design even more energy-efficient and stable. Figure 15 shows a comparison of the RPM-voltage characteristics of all three designs.

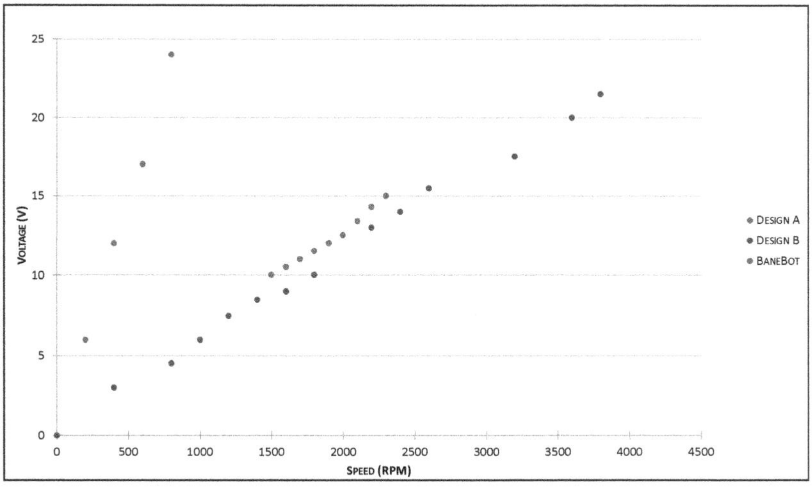

Figure 15. A comparison of the flywheel RPM vs. generated voltage characteristics of the three designs being compared.

For design A, the scatter plot is almost linear, whereas the variation in voltage is higher comparatively. The voltage in the battery changes from 0 to 24 Volts within 800 RPM of the flywheel. Moreover, the battery selected for Design A is a 12 V DC battery, while the voltage being produced to charge that battery is 24 Volts, which can cause the battery life to depreciate, and depending on the usage frequency, the battery can even die out in a few hours or days. Considering an upgrade to a bigger battery for Design A, the experimental data collected is still impractical for long-term usage, as the Voltage readings are extremely RPM sensitive, which indicates that an experienced biker who rides at higher speeds can wreck the battery in a matter of hours, causing flames or explosions.

For design B, the voltage readings are not as sensitive as those of design A. There is a steady increase in voltage in a linear fashion, which is optimal for battery charge. The Voltage recorded up to the maximum possible speed (4000 RPM) on that particular equipment is under 22 volts. The alignment of the dots in the series indicates a steadier battery charge comparatively. This closely resembles the plot of the experimental data from the BaneBot. It can be observed from the plotted data that the BaneBot produced higher voltages with lower fluctuation in comparison to the Design B. Considering the transmission losses, the battery charging fashion, and battery life assumptions from the results, in comparison, it can be concluded that the BaneBot is the optimal design of the three options.

4. Design Optimization

4.1. Modularity Design

A modular design approach was taken when redesigning the power generator system. The goal is to have the power generation and conversion features added to the existing spin bike without any modification to the spin bike. Modularity will also speed up the factory assembling process and reduce the overall cost of manufacturing the product. The final product will have two main modules: the motor assembly and the control box for AC-DC conversion. The modules will be connected via a single wire harness with an MT30 connector. Minimum assembly is required to mount the motor assembly to the flywheel and secure the control box to the spin bike frame. All the subassemblies and electrical wiring will be readily assembled and tested before packaging and shipping to customers.

4.2. Components to Be Manufactured

To lower the material and labor costs, the following list gives components that need to be designed and manufactured:

1. The control box housing will be an injection-molded ABS. ABS is used for its nonconductive property to prevent users from getting an electrical shock during operation.
2. The mounting plate allows the motor assembly to be easily attached to the spin bike frame and flywheel.
3. Motor mount bracket with holes' footprint matches that of the motor footprint.
4. The control box base will be made from sheet metal. This will help dissipate the heat from the 3-phase bridge rectifier.
5. MT30 Wire harness allows for modular connection between the motor assembly and the control box. This will save substantial assembly time for not having to solder the wires into the connector.

The rest of the components can be found off-the-shelf either from online retail stores or directly from the manufacturers.

4.3. Plastic Injection Molding Design Considerations

The control box houses the voltage meter, the 3-phase bridge rectifier, the electrical connections between them, and the 12 V sockets. This housing will be made of ABS using a plastic injection molding manufacturing process. The reasons for choosing plastic molding are:

- Injection molding is the cheapest process at scale for plastic
- Consistent, repeatable reproduction of a part
- Mid to high-volume manufacturing
- The greatest variety of materials, colors, and configurations
- Can handle a variety of sizes and shapes
- Custom cosmetics from polish to texturing

There are many design considerations to be considered when designing parts for plastic injection molding, including the tolerances, type of materials, surface finishes, draft angles on part surfaces, wall thickness, bosses, ribs, gussets, and fillets [41].

4.4. Design for Manufacturability and Assembly (DFMA)

Design for Manufacturability and Assembly (DFMA) is a tool that product designers use to eliminate waste and inefficiencies at various points of the product development process [42]. Boothroyd Dewhurst DFM Concurrent Costing tool [43] was used to understand each part's manufacturing costs. The software interface of this tool has three areas, as shown in Figure 16:

1. The process chart is in the upper left. This area shows the list of steps involved in manufacturing the part.
2. The question panel is on the right-hand side, where the part details are entered.
3. The result box on the lower left shows the cost results regarding materials, setup, process, and rejects. Depending on the manufacturing process, we can also calculate a tooling investment. We add up the four buckets of costs shown at the top to give a part cost and amortize the tooling investment over the specified manufacturing life volume. By adding those together, we can get the total price in DFM concurrent costing.

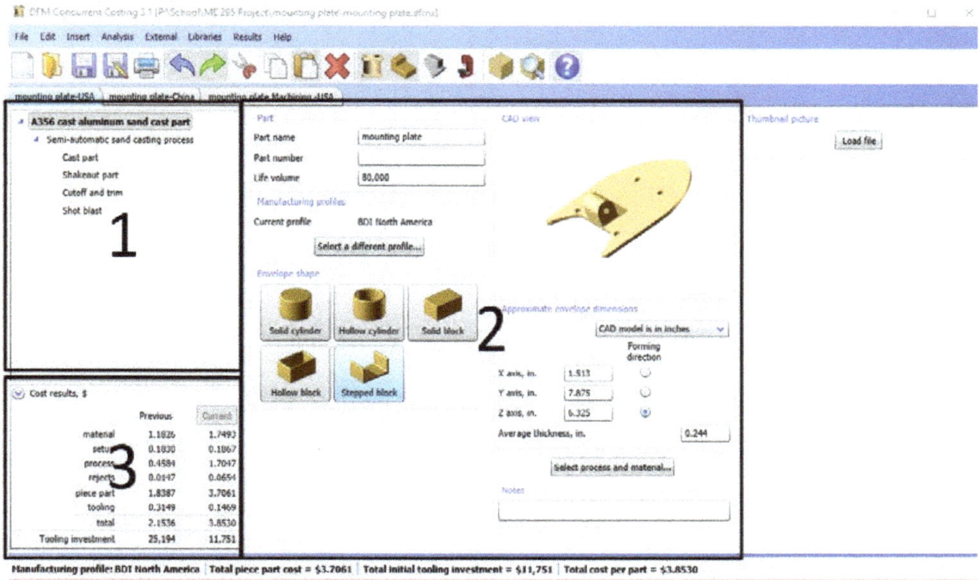

Figure 16. DFM of mounting plate.

5. Cost Analysis

5.1. Costed Bill of Materials (BOM)

After conducting DFM and DFA, a costed BOM (Table 6) is put together to estimate the part and the assembly cost. The total preliminary labor and materials cost is $104. Based on this information, a target price for the energy harvester bike can be aggressively set at $400. In the current market, the cheapest commercially available counterpart costs $3000. Customers can assemble this kit and operate the power generator bike in under 15 minutes.

Table 6. Costed BOM for the exercise bike energy harvester.

Part #	Description	SYS $	Cost	Cateo	QTY	Source	Total
30,001	PACKAGING	$3.00	$3.30	BOXES	1	DFM ESTIMATE	$3.30
30,002	FLAT MOUNTING BAR BRACKET REVB	$0.24	$0.27	BRACKET	2	DFM ESTIMATE	$0.53
30,003	14 GA HARNESS WITH T30 CONNECTOR	$1.08	$1.19	ELECTRONIC	1	DFM ESTIMATE	$1.19
30,004	35A 3 PHASE BRIDGE RECTIFIER	$2.66	$2.92	HARNESS	1	ALIEXPRESS	$2.92
30,005	DC VOLTAGE METER	$10.06	$11.07	ELECTRONIC	1	ALIEXPRESS	$11.07

Table 6. *Cont.*

Part #	Description	SYS $	Cost	Cateo	QTY	Source	Total
30,006	12 VOLT SOCKET	$1.73	$1.90	ELECTRONIC	2	ALIEXPRESS	$3.81
30,007	3.1A DUAL USB CAR CHARGER	$1.86	$2.05	ELECTRONIC	1	ALIEXPRESS	$2.05
30,008	5 PORT WAGO CONNECTOR	$0.18	$0.20	ELECTRONIC	2	ALIEXPRESS	$0.39
30,009	14 GA WIRES	$0.50	$0.55	ELECTRONIC	1	ALIEXPRESS	$0.55
30,010	WIRE HARNESS SPADE T30	$1.25	$1.38	ELECTRONIC	1	DFM ESTIMATE	$1.38
30,011	1/4″-20 HEX NUT	$0.05	$0.05	HARDWARE	8	MCMASTER	$0.43
30,012	Zinc Yellow-Chromate Plated Hex Head Screw, Grade 8 Steel, 1/4″-20 Thread Size, 2-1/2″ Long, Fully Threaded	$0.73	$0.80	HARDWARE	2	MCMASTER	$1.60
30,013	Zinc Yellow-Chromate Plated Hex Head Screw, Grade 8 Steel, 1/4″-20 Thread Size, 3/4″ Long	$0.12	$0.14	HARDWARE	4	MCMASTER	$0.55
30,014	M5 BOLT FOR CONTROL BOX AND RECTIFIER	$0.12	$0.13	HARDWARE	5	MCMASTER	$0.66
30,015	M5 NUTS	$0.02	$0.02	HARDWARE	1	MCMASTER	$0.02
30,016	M4X.7MM SOCKET HEAD SCREW	$0.09	$0.10	HARDWARE	4	MCMASTER	$0.39
30,017	UBOLT WITH MOUNTING PLATE	$1.42	$1.56	HARDWARE	1	MCMASTER	$1.56
30,018	CONTROL BOX	$1.07	$1.18	HOUSING	1	DFM ESTIMATE	$1.18
30,019	CONTROL BOX BASE	$0.40	$0.44	HOUSING	1	DFM ESTIMATE	$0.44
30,020	RACERSTAR BRH5065-140kV MOTOR	$46.95	$51.65	MOTOR	1	ALIEXPRESS	$51.65
30,021	BANEBOT T81 HUB, 8MM SHAFT 60A	$3.15	$3.47	MOTOR HUB	1	BANEBOTS.COM	$3.47
30,022	BANEBOTS WHEEL, 2-7/8″X0.8″, HUB MOUNT, 60A,	$2.45	$2.70	MOTOR WHEEL	1	BANEBOTS.COM	$2.70
30,023	RECEPTICAL MOUNTING PLATE	$1.93	$2.12	MOUNTING	1	DFM ESTIMATE	$2.12
30,024	BUNGEE CORD WITH HOOKS 1M	3.933	$4.33	MOUNTING	1	ALIEXPRESS	$4.33
30,025	6″ 14G Red Wire with spade connector-pre-stripped end	$0.20	$0.22	Harness	3	DFM ESTIMATE	$0.66
30,026	6″ 14G Blk Wire with spade connector-pre-stripped end	$0.20	$0.22	Harness	3	DFM ESTIMATE	$0.66
						MATERIALS COST	99.59
						LABOR COST	3.87
						TOTAL COST	$103.46

5.2. Return on Investment

The collected data concluded that people who do cycling exercises could pedal the spin bike at a higher speed and easily maintain 13 V to 15 V for 30 to 60 minutes. A person who does cycling exercise regularly can produce 200 Wh in every hour-long exercise session. If this spin bike power generator is placed in a gym, with an estimated using time of 8 h per day, this will generate 1600 Wh per day. Using a utility company cost rate of $0.159 per KWh would save a gym $93.68 per year. It would take 4.28 years to get the return on cost investment. On the other hand, it would take over 17 years to recover the cost for a household of four with two hours of usage per day. Table 7 summarizes the return of investment data for a regenerative exercise bicycle.

Table 7. Return of Investment calculation.

	Wh/Day	Wh/Year	kWh	Savings on Electricity	ROI (Years)
Regular cycling exercise can produce	200	73,000	73	11.68	34.25
In a gym with average usage time of 8 h/day	1600	584,000	584	93.44	4.28
In a regular household of 4 with 2 h of usage per day	400	146,000	146	23.36	17.12

6. Contributions of This Work

The idea of generating usable power from exercise equipment is not novel, but not many researchers have attempted to create a complete system that can achieve this functionality. Most of the existing research focuses on either understanding the need for such a device or developing individual components that can contribute to the design of such a system in the future (e.g., [11–17]). We attempted to create a device that can generate electricity from human energy using off-the-shelf components only. We also performed a design for manufacturability and assembly analysis to optimize the design for future iterations and manufacturing. Further, experimental data are collected to show the efficiency of the device. The novelty of this work is in developing such a working system and collecting data to show the system's working.

7. Conclusions and Future Work

Today, the energy crisis has become a problem worldwide, and renewable energy research has been trying to address this problem. One such alternate renewable resource, human power generation, is presented in this paper. The rotational energy of the flywheel in a stationary exercise bicycle, generated by pedaling, can be used to charge mobile devices and power small home appliances. This paper presents the design and methodology for developing a static-exercise bike-based energy harvesting system. The proposed system consists of the best elements from the previous designs in the literature. Experiments are performed to quantify the efficiency of the proposed approach. It was observed that for a person doing a little to no exercise, the energy harvester could be used to charge their phones at speed lower than the AC wall outlet. At the same time, a person with a regular exercise habit can charge a phone faster than the wall outlet. Further, this device's power outcome is compared against two other designs in the literature that provided quantitative measurements. The results show that the design proposed in this paper is more efficient compared to the previous designs.

We plan to continue working on the optimized design for future work. We plan to optimize the design by rebuilding a new prototype with the mounting plate method. More research is needed to tie multiple power generators together and feed the power back to the grid with a grid-tie inverter.

Author Contributions: Conceptualization, V.V. and H.P.; methodology, H.P., S.Z. and V.V.; validation, H.P., V.V. and P.B.; formal analysis, H.P. and P.B.; investigation, H.P., A.P.B. and P.B.; resources, H.P. and S.Z.; data curation, H.P. and A.P.B.; writing—original draft preparation, H.P., V.V., P.B., S.Z. and A.P.B.; writing—review and editing, V.V., S.Z. and A.P.B.; visualization, H.P., V.V. and A.P.B.; supervision, V.V.; project administration, S.Z.; funding acquisition, H.P. and S.Z. All authors have read and agreed to the published version of the manuscript.

Funding: This research received no external funding.

Institutional Review Board Statement: Ethical review and approval were waived for this study as the study was exempt from such a review. The study involved very minimum interventions with humans.

Informed Consent Statement: Informed consent was obtained from all subjects involved in the study.

Data Availability Statement: All the data presented in this study are available in this article.

Conflicts of Interest: The authors declare no conflict of interest.

References

1. Nazarko, Ł.; Žemaitis, E.; Wróblewski, Ł.K.; Šuhajda, K.; Zajączkowska, M. The Impact of Energy Development of the European Union Euro Area Countries on CO_2 Emissions Level. *Energies* **2022**, *15*, 1425. [CrossRef]
2. Murdock, H.E.; Gibb, D.; André, T.; Sawin, J.L.; Brown, A.; Ranalder, L.; Collier, U.; Dent, C.; Epp, B.; Hareesh Kumar, C.; et al. Renewables 2021 Global Status Report. In Proceedings of the Technical Report No. Renewable Energy Policy Network for the 21st Century, Online, 28 June 2021; Ren21: Paris, France, 2021.
3. Qian, F.; Xu, T.; Zuo, L. Design, Optimization and Testing of a Piezoelectric Footwear Energy Harvester. *Energy Convesrion Manag.* **2018**, *171*, 1352–1364. [CrossRef]
4. Wu, H.; Tang, L.; Yang, Y.; Sih, C. A Novel Two-degrees-of-freedom Piezoelectric Energy Harvester. *J. Intell. Mater. Syst. Struct.* **2014**, *24*, 357–368. [CrossRef]
5. Zhao, J.; You, Z. A Shoe-Embedded Piezoelectric Energy Harvester for Wearable Sensors. *Sensors* **2014**, *14*, 12497–12510. [CrossRef]
6. Toprak, A.; Tigli, O. Piezoelectric Energy Harvesting: State of the Art and Challenges. *Appl. Phys. Rev.* **2014**, *1*, 031104. [CrossRef]
7. Fan, K.; Xia, P.; Li, R.; Guo, J.; Tan, Q.; Wei, D. An Innovative Energy Harvesting Backpack Strategy through a Flexible Mechanical Motion Rectifier. *Energy Convers. Manag.* **2022**, *264*, 115731. [CrossRef]
8. Donelan, J.; Li, Q.; Naing, V.; Hoffer, J.A.; Weber, D.J.; Kuo, A.D. Biomechanical energy harvesting, generating electricity during walking with minimal user effort. *Science* **2008**, *319*, 807–810. [CrossRef]
9. Fan, J.; Xiong, C.H.; Huang, Z.K.; Wang, C.B.; Chen, W.B. A lightweight biomechanical energy harvester with high power density and low metabolic cost. *Energy Convers. Manag.* **2019**, *195*, 641–649. [CrossRef]
10. Zou, H.; Li, M.; Zhao, L.; Liao, X.; Gao, Q.; Yan, G.; Du, R.; Wei, K.; Zhang, W. Cooperative compliant traction mechanism for human-friendly biomechanical energy harvesting. *Energy Convers. Manag.* **2022**, *258*, 115523. [CrossRef]
11. Shi, H.U.; Luo, S.; Xu, J.; Mei, X. Hydraulic system based energy harvesting method from human walking induced backpack load motion. *Energy Convers. Manag.* **2021**, *229*, 113790. [CrossRef]
12. Shi, H.U.; Yue, Y.; Wang, H.; Xu, J.; Mei, X. Design and performance analysis of human walking induced energy recovery system by means of hydraulic energy conversion and storage. *Energy Convers. Manag.* **2020**, *217*, 113008. [CrossRef]
13. Wang, W.; Cao, J.Y.; Zhang, N.; Lin, J.; Liao, W.H. Magnetic-spring based energy harvesting from human motions: Design, modeling and experiments. *Energy Convers. Manag.* **2017**, *132*, 189–197. [CrossRef]
14. Zou, H.; Zhao, L.; Gao, Q.; Zuo, L.; Liu, F.; Tan, T.; Wei, T.; Zhang, W. Mechanical modulations for enhancing energy harvesting: Principles, methods and applications. *Appl. Energy* **2019**, *255*, 113871. [CrossRef]
15. Bowers, B.J.; Arnold, D.P. Spherical, Rolling Magnet Generators for Passive Energy Harvesting from Human Motion. *J. Micromech. Microeng.* **2009**, *19*, 094008. [CrossRef]
16. Rao, Y.; Cheng, S.; Arnold, D.P. An Energy Harvesting System for Passively Generating Power from Human Activities. *J. Micromech. Microeng.* **2013**, *23*, 114012. [CrossRef]
17. Sirigireddy, P.; Eladi, P.B. Numerical Design of Novel Piezoelectric Generating Structure that Effectively Utilizes the Force Generated from Human Motion. In Proceedings of the 2020 IEEE International Conference on Power Electronics, Smart Grid and Renewable Energy (PESGRE2020), Cochin, India, 2–4 January 2020.
18. Saha, C.; O'donnell, T.; Wang, N.; Mccloskey, P. Electromagnetic Generator for Harvesting Energy from Human Motion. *Sens. Actuators A Phys.* **2008**, *147*, 248–253. [CrossRef]
19. Oh, Y.; Kwon, D.-S.; Eun, Y.; Kim, W.; Kim, M.-O.; Ko, H.-J.; Kang, S.G.; Kim, J. Flexible Energy Harvester with Piezoelectric and Thermoelectric Hybrid Mechanisms for Sustainable Harvesting. *Int. J. Precis. Eng. Manuf.-Green Technol.* **2019**, *6*, 691–698. [CrossRef]
20. Su, M.; Brugger, J.; Kim, B. Simply Structured Wearable Triboelectric Nanogenerator Based on a Hybrid Composition of Carbon Nanotubes and Polymer Layer. *Int. J. Precis. Eng. Manuf.-Green Technol.* **2020**, *7*, 683–698. [CrossRef]
21. Seo, B.; Cha, Y.; Kim, S.; Choi, W. Rational Design for Optimizing Hybrid Thermo-Triboelectric Generators Targeting Human Activities. *ACS Energy Lett.* **2019**, *4*, 2069–2074. [CrossRef]
22. Visconti, P.; Bagordo, L.; Velázquez, R.; Cafagna, D.; De Fazio, R. Available Technologies and Commercial Devices to Harvest Energy by Human Trampling in Smart Flooring Systems: A Review. *Energies* **2022**, *15*, 432. [CrossRef]
23. Blechman, A.; Braker, G.; Chodnicki, B.; Dubow, E.; Pernia, K.; Sy, T.; Thompson, M.; Tucker, J. A Study of the Benefits of Retrofitting Cardiovascular Exercise Equipment of a Gym with Human Energy Harvesting Technology. Ph.D. Thesis, University of Maryland, College Park, MD, USA, 2009.
24. Haji, M.N.; Lau, K.; Agogino, A.M. Human Power Generation in Fitness Facilities. In Proceedings of the ASME 2010 4th International Conference on Energy Sustainability, Phoenix, AZ, USA, 17–22 May 2010.
25. Hilario, A.J. Energy Harvesting from Elliptical Machines Using Four-Switch Buck-Boost Topology. Master's Thesis, California Polytechnic State University, San Luis Obispo, CA, USA, 2011.

26. Lum, M.; Yuen, J. Energy Harvesting from Elliptical Machines: Dc Converter Troubleshooting. Master's Thesis, Department of Electrical Engineering, California Polytechnic University at San Louis Obispo, San Luis Obispo, CA, USA, 2009.
27. Barois, R.; Caverly, M.; Marshall, K. An Investigation into Using Electricity Harvesting Elliptical Machines as a Renewable Energy Source. *UBC Undergrad. Res.* **2010**. [CrossRef]
28. Chalermthai, B.; Sada, N.; Sarfraz, O.; Radi, B. Recovery of Useful Energy from Lost Human Power in Gymnasium. In Proceedings of the 2015 IEEE 15th International Conference on Environment and Electrical Engineering (EEEIC), Rome, Italy, 10–13 June 2015; IEEE: Piscataway, NJ, USA, 2015.
29. Thiruchelvam, V.; Hammad, C.; Medni, S. Design of Controls, Monitoring and a Energy Storage System for a Energy Harvesting Gymnasium Equipment. In Proceedings of the MATEC Web of Conferences, Cape Town, South Africa, 1–3 February 2016.
30. Suhalka, R.; Khandelwal, M.C.; Sharma, K.K.; Sanghi, A. Generation of Electrical Power Using Bicycle Pedal. *Int. J. Recent Res. Rev.* **2014**, *7*, 63–67.
31. Ullah, M.T.; Karim, M.A.B.; Uddin, M.H.; Tauseef, G.M. Harvesting Green Energy from Wastage Energy of Human Activities Using Gymnasium Bicycle at Chittagong City. In Proceedings of the 2015 3rd International Conference on Green Energy and Technology (ICGET), Dhaka, Bangladesh, 11 September 2015; IEEE: Piscataway, NJ, USA, 2015.
32. Boesel, A. Interview with Adam Boesel, Green Micro Gym, 2011 (Audio). 2011. Available online: https://pdxscholar.library.pdx.edu/sustainhist/32/ (accessed on 27 February 2022).
33. Tgo. The Great Outdoor Gym Company. 2021. Available online: https://www.tgogc.com/ (accessed on 27 February 2022).
34. Ahsan-Uz-Zaman, K.; Ullah, K.M.; Mishir, M.; Alam, M. Generation of Electrical Power Using Gymnasium Bicycle. In Proceedings of the 2017 IEEE Region 10 Humanitarian Technology Conference (R10-HTC), Dhaka, Bangladesh, 21–23 December 2017; IEEE: New York, NY, USA, 2017.
35. Gibson, T. These Exercise Machines Turn Your Sweat into Electricity. Available online: https://spectrum.ieee.org/these-exercise-machines-turn-your-sweat-into-electricity (accessed on 27 February 2022).
36. Ihsan, M.; Viswanathan, V. Self-Powering Gyms: A Case Study on Energy Harvesting from a Static Bicycle. In Proceedings of the ASME International Mechanical Engineering Congress and Exposition, Salt Lake City, UT, USA, 10–14 November 2019; IEEE: Piscataway, NJ, USA, 2019.
37. Otto, K.N.; Wood, K. *Product Design: Techniques in Reverse Engineering and New Product Development*; Tsinghua University Press: Beijing, China, 2003.
38. Ai, X.; Wilmer, M.; Lawrentz, D. Development of Friction Drive Transmission. *J. Tribol.* **2005**, *127*, 857–864. [CrossRef]
39. Anyanwu, S.I.; Anthony, A.E. Design and Fabrication of a Pedal Operated Power Generator. *Innov. Syst. Des. Eng.* **2016**, *7*, 11–13.
40. Patowary, T.; Hasan, M.; Karim, K. Generation of Electricity by Pedal Power. Master's Thesis, Department of Electrical and Electronic Engineering, Islamic University of Technology (IUT), Gazipur, Bangladesh, 2016.
41. Rosato, D.V.; Rosato, M.G. *Injection Molding Handbook*; Springer Science & Business Media: Berlin/Heidelberg, Germany, 2012.
42. Boothroyd, G. Design for Manufacture and Assembly: The Boothroyd-Dewhurst Experience. In *Design for X*; Springer: Dordrecht, The Netherlands, 1996.
43. Boothroyd, G. Design for Assembly—The Key to Design for Manufacture. *Int. J. Adv. Manuf. Technol.* **1987**, *2*, 3–11. [CrossRef]

buildings

MDPI

Article

Date Palm Surface Fibers for Green Thermal Insulation

Mohsin Raza [1], Hyder Al Abdallah [1], Ayah Abdullah [1] and Basim Abu-Jdayil [1,2,*]

1 Chemical and Petroleum Engineering Department, College of Engineering, United Arab Emirates University, Al Ain P.O. Box 15551, United Arab Emirates; 201990122@uaeu.ac.ae (M.R.); 201870175@uaeu.ac.ae (H.A.A.); 201540053@uaeu.ac.ae (A.A.)
2 National Water and Energy Center, United Arab Emirates University, Al Ain P.O. Box 15551, United Arab Emirates
* Correspondence: babujdayil@uaeu.ac.ae; Tel.: +971-50-7537300

Abstract: Some of the major challenges of the twenty-first century include the continued increase in energy consumption and environmental pollution. One approach to overcoming these challenges is to increase the use of waste materials and environmentally friendly manufacturing methods. The high energy consumption in the building sector contributes significantly to global climatic changes. Here, by using date palm surface fibers, a high-performance green insulation material was developed via a simple technique that did not rely on any toxic ingredients. Polyvinyl alcohol (PVA) was used as a binding agent. Four insulation samples were made, each with a different density within the range of 203 to 254 kg/m^3. Thermal conductivity and thermal diffusivity values for these four green insulators were 0.038–0.051 $W/(m\cdot K)$ and 0.137–0.147 mm^2/s, respectively. Thermal transmittance (U-value) of the four insulation composites was between 3.8–5.1 $W/m^2\cdot K$, which was in good comparison to other insulators of similar thickness. Thermogravimetric analysis (TGA) showed that insulating sample have excellent thermal stability, with an initial degradation temperature of 282 °C, at which just 6% of its original weight is lost. Activation energy (E_a) analysis revealed the fire-retardancy and weakened combustion characteristics for the prepared insulation composite. According to differential scanning calorimetry (DSC) measurements, the insulating sample has a melting point of 225 °C, which is extremely close to the melting point of the binder. The fiber-based insulating material's composition was confirmed by using Fourier transform infrared spectroscopy (FTIR). The ultimate tensile range of the insulation material is 6.9–10 MPa, being a reasonable range. Our study's findings suggest that developing insulation materials from date palm waste is a promising technique for developing green and low-cost alternatives to petroleum-based high-cost and toxic insulating materials. These insulation composites can be installed in building envelopes during construction.

Keywords: building insulation; date palm surface fibers; green insulation; insulation material; thermal conductivity; thermogravimetric analysis; waste management

Citation: Raza, M.; Abdallah, H.A.; Abdullah, A.; Abu-Jdayil, B. Date Palm Surface Fibers for Green Thermal Insulation. *Buildings* **2022**, *12*, 866. https://doi.org/10.3390/buildings12060866

Academic Editors: Shi-Jie Cao and Wei Feng

Received: 12 May 2022
Accepted: 17 June 2022
Published: 20 June 2022

1. Introduction

For decades, mostly all industrial applications have initiated energy efficiency and sustainability. Almost every field of engineering is seeing plans to reduce energy consumption and to improve its conservation [1]. Low energy consumption has also become a major impediment to achieving urban sustainability. Many countries have nonetheless pledged to reduce their energy usage and carbon emissions to combat global climate change [2]. This rising energy demand has prompted a refocusing of energy research towards the usage and development of renewable sources to replace non-renewable fossil fuels that are rapidly diminishing in supply and that emit harmful pollutants into the atmosphere [3]. According to the United Nations Environment Program, buildings consume around 40% of the world's energy, 25% of its water, and are responsible for 50% of total greenhouse gas emissions [4]. Therefore, several attempts have been made to improve energy efficiency in the building sector. Thermal insulation in building materials can help to reduce energy consumption significantly. Using efficient insulation materials can help to save energy by reducing heat losses and gains during building heating and cooling [5]. A pivotal characteristic of all insulation materials used in modern construction is their low thermal conductivity, this typically is <0.1 W/(m·K) [6]. If a material's thermal conductivity is <0.07 W/(m·K), it is generally called a thermal insulator [7].

In most building construction practices, petrochemical and inorganic insulation materials are being used, such as polyurethane, mineral wool, expanded polystyrene, and fiberglass. The manufacturing of these insulation materials consumes much energy, which greatly impacts the environment in adverse ways over their whole life cycle [8]. For example, making 1 kg of expanded polystyrene emits 3 kg of carbon dioxide [9]. Natural fibers are very appealing and have a huge potential as eco-friendly raw materials for thermal insulation. Moreover, natural fibers are biodegradable and have minimal ecological impact [10]. The market for insulation materials is highly competitive in terms of both performance and cost. Natural insulation materials are currently a niche market [11]. Green composites are a type of biocomposite in which natural fibers are used to strengthen a bio-based polymer [12,13] Matrix. The United Arab Emirates (UAE) is home to ca. 40 million date palm trees. Their leaves, stem, trunk, leaf stem, surface fibers, and date fruit seeds (Figure 1) are among the plant parts generated and considered waste materials [14]. The possibility of discovering uses for date palm wood in developing insulation materials could open new markets for what is typically regarded as waste or for its usage in low-value products. In the UAE, the construction industry consumes 35–40% of the country's total energy generated [15], the majority of which comes from fossil fuels [16]. The UAE is constantly extending its development projects as a global hub for the international community, with real estate being the fastest growing sector in recent decades [17]. Therefore, developing an insulation material from local waste material will bring a cost saving from waste disposal and availability of cheap thermal insulation material.

Figure 1. A date palm tree and its key material parts at the Al-Foah Experimental Farm of the UAE University, Al-Ain, United Arab Emirates.

Today, various studies have been carried out to develop building insulation materials using unconventional approaches. Therefore, researchers are focusing on the development of sustainable building insulators especially made from renewable/or waste materials. Piotr Kosinski et al. [18] reported the findings of research on raw hemp shives obtained from the Polish crop of industrial hemp as a loose-fill thermal insulating material. The measurements of the pore size distribution, thermal conductivity, and air permeability of the material were the main emphasis of the research. In the density range of 109–124 kg/m^3, thermal conductivity 0.049–0.052 W/(m·K) showed a slight tendency to rise with density. In this study, the least density that could be achieved was 109 kg/m^3. Yan Qiuhui et al. [19] performed an acid/base two-step catalytic procedure followed by ambient pressure drying to produce a hydrophobic and low-cost SiO_2 aerogel. The developed material fulfills well with insulation standards, with a thermal conductivity of 0.0212 W/(m·K), a specific surface area of 920 m^2/g, a density of 0.109 g/cm^3, a porosity of 95.05 percent, and a super hydrophobic and mesoporous structure. Srihanum et al. [20] developed low-density rigid polyurethane foam incorporated with renewable polyol. Palm olein-based polyol (PP) was employed as a partial substitute for conventional sucrose/glycerin-initiated polyether polyol (GP). When palm-based rigid polyurethane foams with up to 30 percentage PP were compared with GP foams, the compressive strength and strain were shown to be improved. The addition of 10 percentage PP resulted in the lowest thermal conductivity (0.0232 W/(m·K)) of low-density rigid polyurethane foam with a density below 30 kg/m^3. Abu-Jdayil et al. [21] conducted research in which they combined polyurethane dust waste with unsaturated polyester resin to develop a heat insulation composite material for building. Insulation composite materials were produced with polyurethane dust ratios ranging from 10% to 50%. With low thermal conductivity in the range of 0.076 to 0.10 W/(m·K), the findings demonstrated good thermal insulation performance. Furthermore, the composites had compression strengths of 56 to 100 MPa and tensile strengths of 10.3 to 28 MPa. Dissanayake DG et al. [22] developed an insulation material using textile waste and a natural rubber. Textile wastes are harmful when landfilled, therefore cotton/polyester mixed wastes were utilized to produce sound insulation material. The newly designed material's noise reduction coefficient values varied from 0.5 to 0.7. The sound insulation capabilities of the samples were found to be comparable to those of commercially available sound insulation panels. Waseem Hittini et al. [23] developed thermal insulation composite from devulcanized waste rubber tires. Using a melt extruder, a grounded devulcanized rubber tire was combined with polystyrene in various amounts (0–50 wt%). Insulation composites with less than 40 wt% filler content showed very good properties, with thermal conductivity ranging from 0.0502 to 0.07084 W/(m·K), density from 462.8 to 482.32 kg/m^3, and compressive strength from 11.66 to 7.47 MPa.

Hence, utilizing date palm wastes in the UAE and the Gulf region is of paramount importance. Limited research, however, has been performed on the utilization of date palm waste to develop thermal insulation materials. An overview of the studies that included date palm waste to develop building insulators is presented in Table 1. In most studies, date palm waste was utilized as a reinforcing agent with different polymers to develop insulation composites through a melt extrusion process. However, there are very few studies in which date palm waste has been utilized without melt extrusion or pretreatment process.

Table 1. Date palm waste-based insulators.

Date Waste Type	Treatments	Polymer	Thermal Conductivity (W/m·K)	Ref.
Pits (fruit seeds)	-	Unsaturated polyester	0.126–0.138	[24]
Fibers	-	Poly (β-hydroxybutyrate)	0.086–0.112	[25]
Fibers	Alkaline treatment (NaOH & KOH)	Polylactic acid	0.076–0.084	[26]
Wood	-	Polylactic acid	0.0692–0.0757	[27]
Pits	-	Polylactic acid	0.0794 to 0.0682	[28]
Fibers	silane-treated (APTES)	Polylactic acid	0.085–0.105	[29]
Leaflets		Expanded polystyrene	0.11–0.16	[30]
Leaves	Corn starch and wood adhesive as binder	-	0.045–0.065	[31]
Surface fibers	Corn starch as binder	-	0.0475–0.0697	[32]
Trunk wood	Isocyanate based polyurethane and polyvinyl acetate as binder	-	0.1357–0.14	[33]
Fibers	Corn starch, glue, and white cement as binder	-	04234–0.05291	[34]

Therefore, there exists a demand to devise thermal insulation materials that utilize date palm waste without any reliance on sophisticated processing. This study aims to develop such a novel thermal insulation material by utilizing the date palm waste coupled with a natural binder. In bypassing the extrusion process/or any pretreatment of the date fibers, this should further promote the insulation composite's renewability, lower its toxicity, and, most importantly, reduce its cost of production. Thus, our goal was to build a low-cost low-impact material with a good thermal performance that may be employed on a large scale in building applications. To the authors' best knowledge, this is the first empirical study to develop building insulation material using date palm surface fibers and polyvinyl alcohol. A US patent for this process has also been recently granted to the same authors (US 11255052) [35]. Therefore, utilizing date palm waste material for developing building insulation materials not only entails nil carbon emissions but its production also helps to reduce the huge amounts of biomass waste materials that arise yearly in MENA (Middle East/North Africa) regions.

2. Materials and Methods

2.1. Materials

The natural fibers used in this study were collected from date palm trees at the Al-Foah Experimental Farm, United Arab University (Figure 1), Al-Ain, United Arab Emirates. The natural fibers were cut from the tree by knife. The polyvinyl alcohol (PVA) used was bought from Sigma Aldrich (99% purity) and applied here without any further modification.

2.2. Methods

2.2.1. Sample Preparation

The date palm surface fibers (DPSFs) were first cleaned with water to remove any lingering dust or contaminants. The raw DPSFs were rough and rather inflexible; hence, these natural fibers were soaked in water for 24 h at room temperature to render them flexible for processing. The raw fibers were dried in an oven until constant weight was obtained. At 230 °C, an 8 wt.% PVA solution was made by stirring for 30–40 min until all the PVA granules had dissolved in deionized water and the solution turned transparent. Date palm surface fibers were immersed in the PVA solution (200 mL) for 10 min. This ensured that each thread of date palm fiber came into contact with the PVA binder. No specific weight percentage of DPFs/PVA was used. After that, the fibers were removed from the PVA solution and carefully inserted into the mold to completely fill the required space. Figure 2 shows the filled molds for the DPSF/PVA samples produced for the later thermal conductivity and mechanical testing measurements. The molds were then placed in a hot press machine (AUTOFOUR/3015 by Carver, Wabash, IN, USA) for 3 h at 70 °C under a load of 3 tons. Next, the compressed DPSF/PVA samples were oven dried for 48 h at 100 °C. After the water content had been completely vaporized, a coherent sample was obtained. Four samples (A, B, C, and D) of varying density (Figure 3) were created, then comprehensively analyzed [34] and systematically tested. The apparent (bulk) density of each sample was determined using the method of Centiner and Shea [36]. The total mass of each respective sample was divided by the volume of the mold. Similarly, a real density can also be measured using the area of the mold. The square mold used to make each sample for thermal conductivity measurement had dimensions of 11 cm × 11 cm × 3 mm (length × width × thickness). The dog-bone-shaped mold for making samples to test mechanical properties had dimensions of 100 mm × 14 mm × 3 mm (length × width × thickness). The density of each sample is based on three replicates. The average values have been used to represent the samples (Figure 4).

2.2.2. Thermal Conductivity Measurement

The thermal conductivity of the composite samples was measured using a LaserComp FOX-200, Tokyo, Japan. This instrument measures the thermal conductivity of a given sample according to the steady-state method. Its principal is that by measuring the temperature gradient and the power input, and by following the ASTM C1045-07 standard, one can calculate the thermal conductivity value. The thermal conductivity measurement was performed in triplicate. The standard deviation was very low. For example, the calculations of thermal conductivity at 25 °C for the four samples can be seen in Figure 5.

Figure 2. The two types of molds used for the DPSF/PVA composite fabrication.

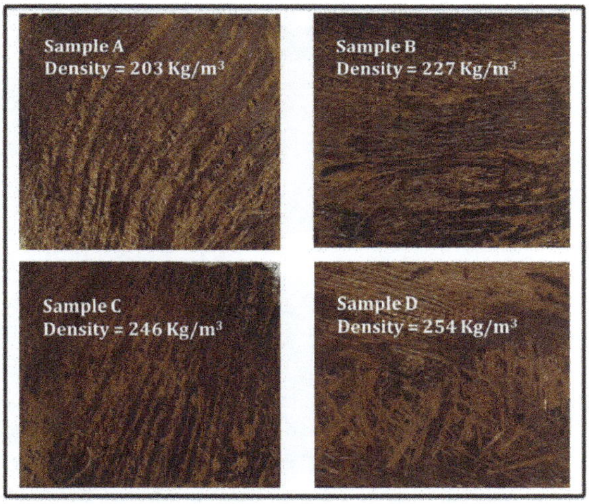

Figure 3. Images of the prepared DPSF/PVA insulation composites varying in density (four samples).

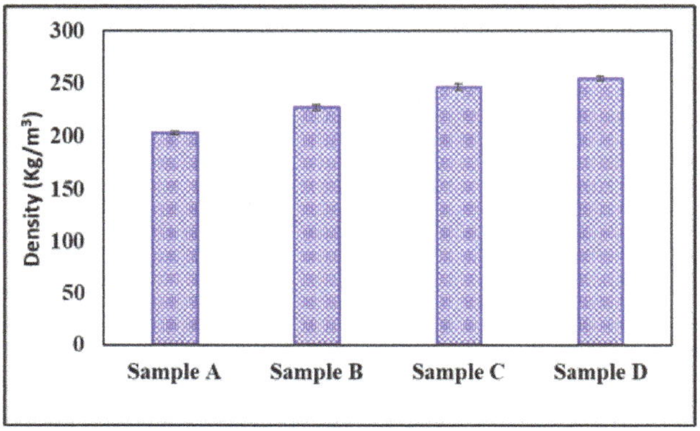

Figure 4. Density of four insulation composites based on three triplicates.

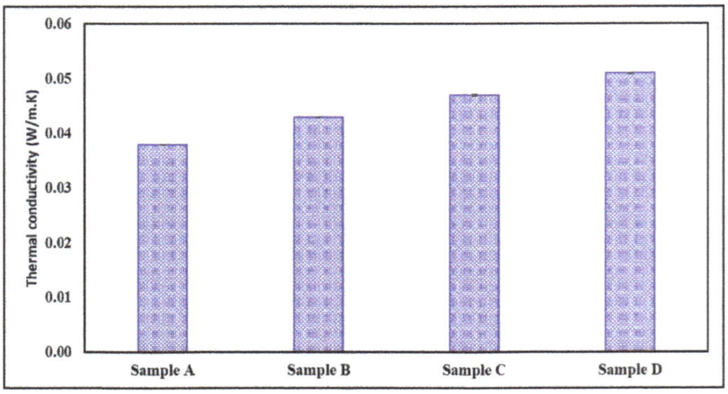

Figure 5. Thermal conductivity of four insulation composites based on three triplicates.

2.2.3. Thermogravimetric Analysis (TGA)

This investigation was carried out to determine the prepared composite's thermal stability. A thermogravimetric analyzer was used to test an insulation composite sample consisting of 5–10 mg (TGA Q500 by TA Instruments, New Castle, DE, USA). The insulation composite sample decomposition was studied under a heating rate of 10 °C/min spanning temperatures of 25–800 °C. The carrier gas was nitrogen, which flowed at a rate of 20 mL/min. To obtain least error and to have good accuracy the experiment was performed two times.

2.2.4. Kinetic Study

The activation energy (E_a) was estimated using the fundamental Coats–Redfern equation, as presented in Equation (1) [37]:

$$\ln\left[\frac{g(\alpha)}{T^2(K)}\right] = \ln\left[\frac{AR}{\beta E_a}\right]\left(1-\frac{2RT}{E_a}\right) - \frac{E_a}{RT} \quad (1)$$

where $g(\alpha)$ denotes the model for the reaction mechanism; β (°C/min) is the rate of heating; R is a universal constant, taking the value of 0.008321 kJ/mol; T (Kelvin) is the reaction temperature; and A (m^{-1}) is the pre-exponential or frequency factor.

The α term is the conversion amount, calculated using Equation (2) [38]:

$$\alpha = \frac{m_o - m_i}{m_o - m_f} \quad (2)$$

where m_o is the original mass of the respective sample at t = 0, m_i is the instantaneous mass of the respective sample at any time t, and m_f is the final mass of the respective sample.

Plotting the left side of Equation (1) ($\ln\left[\frac{g(\alpha)}{T^2(K)}\right]$) vs. $1/T$ will yield both E_a and A. The fitted straight line's slope will be equivalent to $-E_a/R$ and its intercept is $\ln\left[\frac{AR}{\beta E_a}\right]$.

2.2.5. Differential Scanning Calorimetry (DSC)

The differential scanning calorimetry (DSC) analysis was conducted using a TA instrument (model DSC25, TA instruments, New Castle, USA). A DSC analysis was performed on a 5-mg sub-sample across a temperature range of 25–250 °C. Nitrogen at a flow rate of 5 mL/min was used to maintain the system under inert conditions. The heating rate was set to 10 °C/min in the ramp mode. To obtain least error and to have good accuracy the experiment was performed two times.

2.2.6. Fourier Transformation Infrared (FTIR)

The FTIR analysis of the four prepared samples was performed on an IRTracer-100 FTIR spectrometer (Shimadzu, Kyoto, Japan). The FTIR results were used to examine the changes in the functional groups at different wavelengths. An advanced Fourier transform infrared (ATR-FTIR) spectrograph was obtained per sample, having a range of 500 to 4000 cm^{-1}, 34 scans on average, and a spectral resolution of 4 cm^{-1}.

2.2.7. Scanning Electron Microscopy (SEM)

A JEOL/EO scanning electron microscope (SEM), operating at 5 kV, was used to observe the surface morphology of the prepared composite samples. The samples were gold-coated before this analysis to avoid any electrostatic charges during their examination.

2.2.8. X-Ray Diffraction (XRD)

The index of crystallization of the samples was determined by the X-ray diffraction analysis method. Each sample was investigated using Cu K radiation, for which the working lamp parameters were set as follows: receiving slit = 0.15 mm, v = 40 kV, I = 30 mA. The SEGAL method was used to calculate the crystallinity index, as follows:

$$X_{cr} = \frac{I_{200} - I_{am}}{I_{200}} \tag{3}$$

where X_{cr} is the index of crystallinity; I_{200} corresponds to the diffracted intensity at the highest crystalline peak; and I_{am} denotes the diffraction intensity of the amorphous region.

2.2.9. Mechanical Testing

The tensile properties of the four fabricated composite samples were measured with an autograph testing machine (Shimadzu Co. AGS-20NJ Series, Kyoto, Japan). To do this, each sample was extended at a constant elongation velocity of 10 mm/min until it broke. Dog-bone-shaped samples, as shown in Figure 2, were prepared for this mechanical analysis.

3. Results and Discussion

3.1. Thermal Conductivity (k) and Diffusivity (α)

Figure 6 shows the thermal conductivity profiles of four composite samples, whose respective densities were 203 kg/m^3, 227 kg/m^3, 247 kg/m^3, and 254 kg/m^3. Evidently, thermal conductivity of the samples increases with their increasing density. At 25 °C, the thermal conductivity of Sample A with the lowest density (203 kg/m^3) was 0.038 W/(m·K), 0.043 W/(m·K) for Sample B with a density of (227 kg/m^3), 0.047 W/(m·K) for Sample C with a density of (246 kg/m^3), and 0.051 W/(m·K) for Sample D with the highest density (254 kg/m^3).

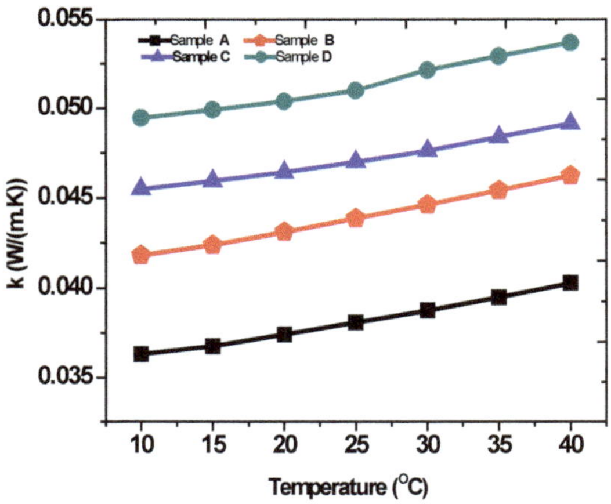

Figure 6. Thermal conductivity of the four prepared samples of insulation composite differing in density.

It should also be noted that for a given density of the composite, its thermal conductivity also increases linearly with temperature. For example, if the sample has a density of 203 kg/m^3 (Sample A), its thermal conductivity increases from 0.036 W/(m·K) to 0.040 W/(m·K) across a temperature range of 10–40 °C (5 °C increments).

Figure 7 shows that at every fixed temperature (from 10 to 40 °C), the thermal conductivity increases linearly as the sample density increases. At 30 °C, this being close to the normal application temperature for building insulating materials, the thermal conductivity increased from 0.03875–0.05215 W/(m·K) over a sample density spanning 203 kg/m^3 to 254 kg/m^3. Therefore, the thermal conductivity of dried samples is a function of both temperature and density.

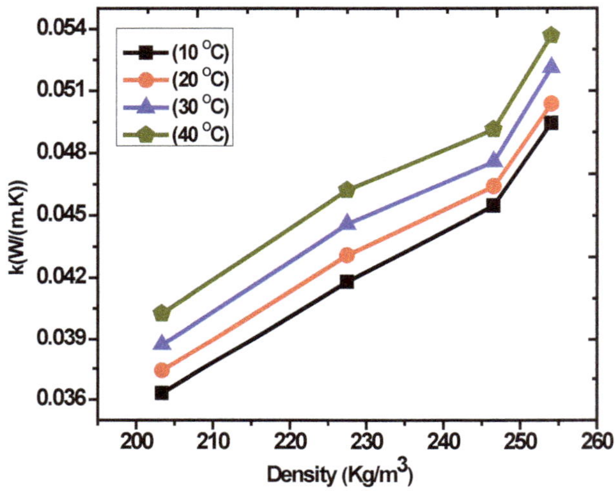

Figure 7. Thermal conductivity profiles of the insulation composite at fixed temperatures.

Figure 8 shows the linear fits of thermal conductivity against temperature. Very high regression coefficient values characterized all four samples (>0.98), whose slope values were very small (≈0.0001). This indicated that the thermal conductivity of these insulation samples changed little (low sensitivity) in response to a rising temperature. This is a fundamental characteristic of robust insulation materials.

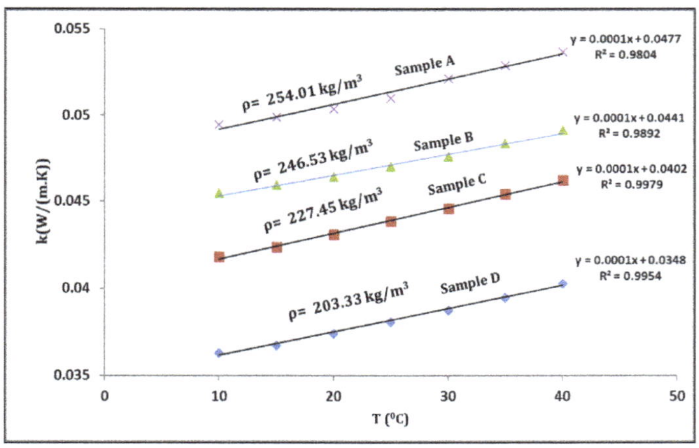

Figure 8. Linear regressions of thermal conductivity as a function of temperature.

In Table 2, thermal conductivity values of our prepared heat insulating samples are compared with those of other renewable insulation materials. Ali and Alabdulkarem [32] prepared a thermal insulation material using date palm surface fibers and corn starch as a binding agent; their lowest thermal conductivity was 0.0475 W/(m·K) at a density of 176 kg/m^3. In our study, the thermal conductivity of the date palm fibers bound with polyvinyl alcohol was 0.038 W/(m·K) at a density of 203 kg/m^3. Hence, although their density is lower, their thermal conductivity value is higher than that of date palm fibers bonded with PVA. In addition, the thermal conductivity of this novel composite material is comparable to that of conventional insulation materials. Cellular glass (0.041 W/(m·K) at 115 kg/m^3), glass mineral wool (0.035 W/(m·K) at ca. 20 kg/m^3), polyurethane foam (0.023–0.026 W/(m·K) at 30–40 kg/m^3), expanded polystyrene (0.034–0.038 W/(m·K) at 15–30 kg/m^3), and extruded polystyrene (0.033–0.035 W/(m·K) at 20–40 kg/m^3) are prominent examples of this similarity in performance [39].

Table 2. Thermal conductivity coefficients of the fiber-based insulation composites.

No.	Materials	Thermal Conductivity Coefficient, k (W/(m·K))	Ref.
1.	Date palm surface fibers bonded with PVA (polyvinyl alcohol)	0.038 W/(m·K) (25 °C) at ρ = 203.33 kg/m^3 0.043 W/(m·K) (25 °C) at ρ = 227.45 kg/m^3 0.047 W/(m·K) (25 °C) at ρ = 246.53 kg/m^3 0.051 W/(m·K) (25 °C) at ρ = 254.01 kg/m^3	This study
2.	Date palm surface fibers bonded with corn starch	0.047 W/(m·K) (25 °C) at ρ = 176 kg/m^3	[32]
3.	Date palm surface fibers, apple of Sodom fibers, and agave fibers bonded with corn starch, wood adhesive glue, and white cement	Range of 0.0418–0.056 W/(m·K) for all samples at ρ = 170–300 kg/m^3	[34]
4.	Waste wool and recycled polyester fibers	0.035 W/(m·K) (35 °C) at ρ = 62.50 kg/m^3	[40]
5.	Bamboo fibers bonded with protein-based bone glue	0.055 W/(m·K) (25 °C) at ρ = 246 kg/m^3	[41]

The heat capacity (C_p) of the date palm thermal insulation composite over various temperatures is presented in Table 2. Heat capacity clearly has a linear relation with temperature. Another important physical property of an insulating material is its thermal diffusivity. It is the rate of temperature distribution through a material that predicts the behavior in terms of thermal conduction relative to the heat storage capacity (C_p). High diffusion means a high heat transfer rate and vice versa. The thermal diffusivity (α) was calculated using Equation (4):

$$\textit{Thermal diffusivity}\ (\alpha) = \frac{\textit{Thermal conductivity}\ (k)}{\rho \times C_p} \quad (4)$$

where thermal diffusivity (α) is measured in units of m^2/s or mm^2/s, the ρ is the density of the sample, and C_p is the heat capacity of the sample.

Table 3 shows the thermal diffusivity values of the four prepared samples. At 25 °C, Sample A had the lowest thermal diffusivity, being 0.137 mm^2/s. In the earlier thermal conductivity analysis, we found the conductivity coefficient (k) rising predictably with temperature, albeit slightly, and considered better at lower temperatures. However, the thermal diffusivity attains higher values at lower temperatures and lower values as the temperature rises. This trend is linked to a greater heat capacity of the samples at higher temperatures that lowers the transport rate of heat through the material. Hence, thermal conductivity cannot be used alone as a measure of insulating material performance. Moreover, thermal diffusivity stayed nearly constant for all four composite samples, in that it changes little with respect to density. A similar relationship between thermal conductivity and thermal diffusivity was reported by Cetiner et al. [36]. Thermal diffusivity of our DPSF-based insulation material is comparable to commercially available insulation materials like cellular glass boards (0.42 mm^2/s) and hempcrete blocks (0.15 mm^2/s) [39].

Table 3. Thermal diffusivity values of the different insulation composite samples.

T (°C)	Heat Capacity C_p (J/g °C)	Thermal Diffusivity (α) (mm^2/s)			
		Sample A	Sample B	Sample C	Sample D
10	0.179	0.248	0.255	0.256	0.270
15	1.122	0.161	0.166	0.166	0.175
20	1.291	0.142	0.145	0.145	0.153
25	1.365	0.137	0.138	0.139	0.147
30	1.442	0.132	0.136	0.133	0.142
35	1.501	0.129	0.133	0.130	0.138
40	1.545	0.128	0.131	0.129	0.136

Although thermal conductivity and diffusivity are significant, we can additionally account for the thickness of the insulation in our analyses. The thicker the insulating layer the slower the rate of heat loss and the better the building's ability to retain heat. The U-value comes into play here. The U-value represents the amount of heat lost through a certain thickness of material. This allows you to compare insulation materials and thickness immediately. The calculation of thermal transmittance (U-value) is as follows [42]:

$$\textbf{\textit{Thermal transmittance}}\ (\textbf{U-value}) = \frac{\textbf{\textit{thermal conductivity}}}{\textbf{\textit{thickness}}} \tag{5}$$

where U-value is in $W/(m^2{\cdot}K)$, thermal conductivity in $W/(m{\cdot}K)$, and thickness of samples is taken in meters. When considering U-values, a lower number is better; a lower U-value implies that a material transfers less heat and thus is a good insulator [43]. Table 4 shows the U-values of four insulation samples. It can be seen that the U-value of date palm surface fibers based insulation samples is in close comparison to other insulation materials. However, the U-value comparison is more accurate when the thickness is the same between insulators. The general consensus is that the thicker the insulation the lesser the heat transmission. Thermal conductivity, on the other hand, is unaffected by insulation thickness, which instead influences thermal resistance [44]. Lakatos et al. [45] examined the relationships between thermal conductivity and expanded material thickness. They demonstrated that, contrary to U-values, thermal conductivity is independent of insulator thickness. Zack J. et al. [46] showed that U-values decreased with increasing the samples thickness from 40–80 mm. Therefore, the U-values of date palm surface fibers based insulation material can further be decreased at higher sample thickness. However, Mahlia T.M.I. et al. [47] suggested that there is a relation between thermal conductivity and thickness of insulation materials. Therefore, an optimum thickness can be evaluated.

3.2. *Thermal Analysis*

3.2.1. Thermogravimetric Analysis (TGA)

Thermal degradation behavior is also another crucial characteristic of insulation materials. A higher thermal degradation range with a lower weight loss is considered an excellent property of insulation materials. Thermogravimetric analysis (TGA) was performed here to analyze the degradation behavior of raw fibers of date palm and the insulation composite material prepared with them. Initial degradation temperature (T_{IN}), mass loss (%) at T_{IN}, major weight loss region, and major degradation temperature range (T) are the primary physical properties of interest for insulation materials. A multi-step thermal decomposition pattern was obtained from thermogravimetric analysis (Figure 9). These results are summarized in Table 5. Raw fibers started to degrade at a very high temperature of 255 °C, corresponding to T_{IN}. At this T_{IN}, the sample lost 11% of its original weight. The initial degradation temperature of the insulation composite further increased to 282 °C where the sample loss was only 6% of its original weight. The major degradation range of the insulation composite expands going from 282 to 383 °C. Major weight loss, which is ca. 84% of the original weight, thus falls within that major degradation range. Major degradation rates were also identified by DTG (Figure 10). The T_{IN} for the insulation composite (282 °C) is a very high temperature, one not yet realized for any building insulation material during the normal application period, and its corresponding weight loss is very small. This makes the DPSF/PVA composite highly suitable for use as a building insulating material.

Table 4. U-value of insulation materials.

No.	U-Value W/m^2·K	Ref.
Sample A (10 mm)	3.8	This study
Sample B (10 mm)	4.3	
Sample C (10 mm)	4.7	
Sample D (10 mm)	5.1	
Polystyrene (50 mm)	1.25	[48]
Celotex (50 mm)	2.25	
Glass wool (100 mm)	0.4	[49]
Concrete blocks (100 mm)	11.1	
Clay bricks (100 mm)	7.6	
Fiber glass (50 mm)	1	[50]

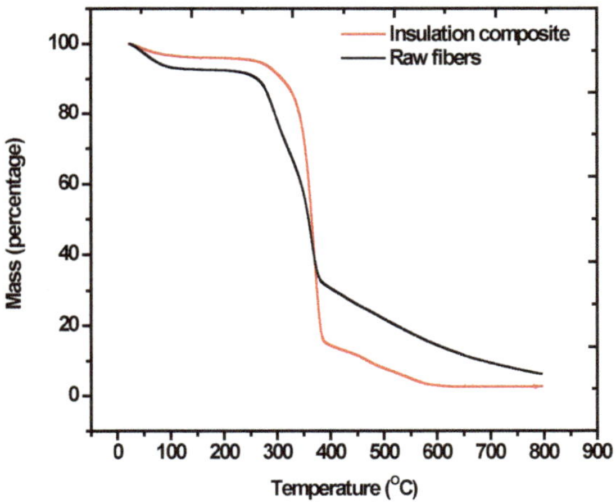

Figure 9. TGA analysis of raw fibers and the insulation composite.

Table 5. Thermal characteristics of raw fibers and insulation composite.

Thermal Property	Raw Fibers	Insulation Composite
Initial degradation temperature at (T_{IN}) (°C)	255	282
Mass loss (%) at T_{IN}	11	6
Major degradation temperature range (°C)	255–381	282–383
Maximum degradation temperature (T_{max})	365	368

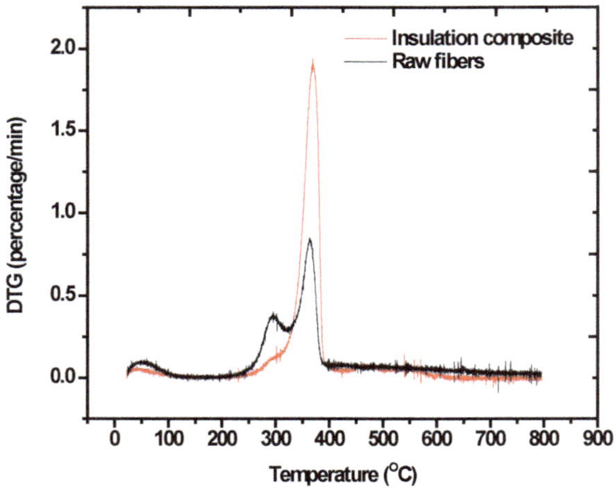

Figure 10. DTG of raw fibers and the insulation composite.

3.2.2. Kinetic Analysis

Many researchers have performed kinetic analyses of insulation materials. This type of analysis is helpful for understanding the thermal stability of insulation materials and establishing their aging and fire-retardant characteristics. Zheng et al. [51] produced a thermal insulation material using cellulose pulp, whose fire retardancy was developed using an intumescent fire retardant, and its activation energy E_a increased due to the formation of thermally stable char. Later, Lie et al. [52] conducted a kinetic analysis of thermal insulator waste in the form of extruded polystyrene to learn more about its fire retardancy and recycling potential. Earlier work by Jiao et al. [53] ran kinetic analyses of three insulation materials: rigid polyurethane foam, extruded polystyrene, and expanded polystyrene. The goal of that study was to look at the thermal degradation properties and volatile products of those insulation materials.

Activation energy (E_a) was calculated here for date palm raw fibers and insulation composite material by using the Coat–Redfern integral method. Kinetic analyses were applied to the major degradation region, using the thermogravimetric data. The major degradation region (i.e., active pyrolysis zone) is indicated in Figure 11.

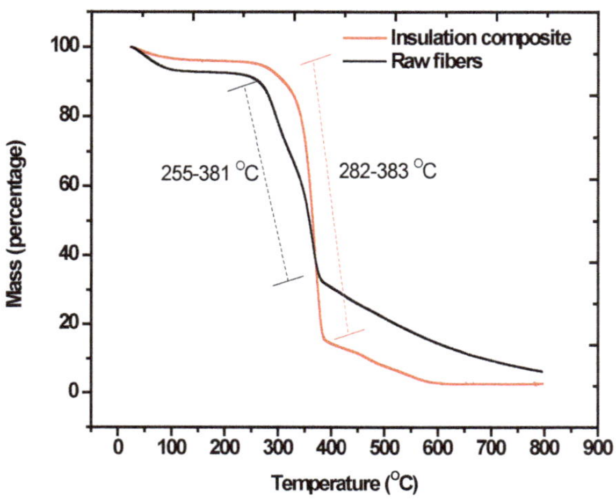

Figure 11. TGA curves of raw fibers and the insulation composite.

Two diffusion models, the anti-Jander equation and the Ginstling equation, yielded excellent linear fits to the thermogravimetric data of raw date palm fibers ($R^2 \approx 0.99$) and the insulation composite ($\approx$0.92). These results are summarized in Table 6. The E_a increased from 101 kJ/mol to a much higher value of 170 kJ/mol for the insulation composite. The 68% increase in the E_a value shows that the binding of date palm fibers with PVA augmented its fire retardancy and weakened its combustibility. The best-fitted diffusion models are also depicted in Figure 12.

Table 6. Activation energy (E_a) of date palm fibers and the insulation composite.

Equation Model	E_a (kJ/mol)	
	Raw fibers	*Insulation composite*
Anti-Jander	101.26	170.57
Ginstling	101.67	171.63

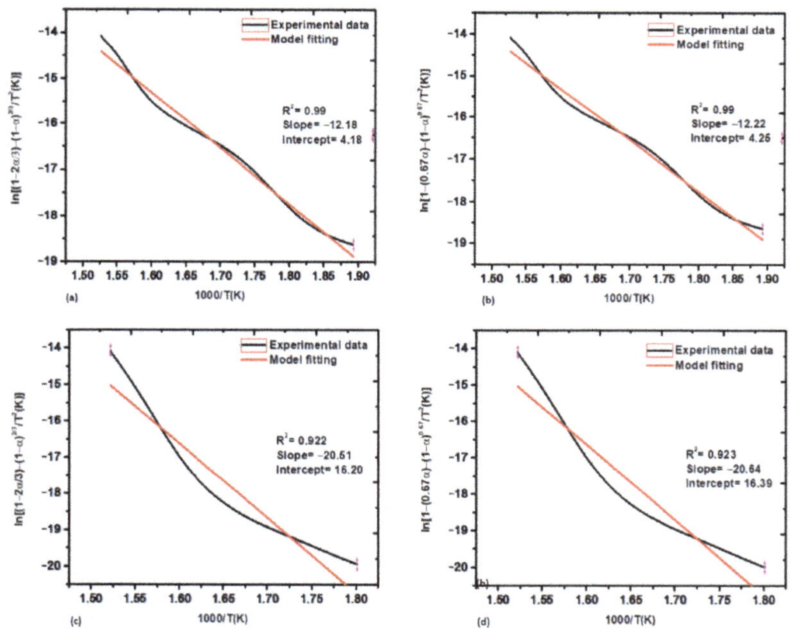

Figure 12. Fitted diffusion models. (**a**) The anti-Jander model for raw fibers, (**b**) the Ginstling model for raw fibers, (**c**) the anti-Jander model for insulation composite, and (**d**) the Ginstling model for the insulation composite.

3.2.3. Differential Scanning Calorimetry (DSC) Analysis

DPSFs were successfully bound with polyvinyl alcohol (PVA). Differential scanning calorimetry (DSC) is a technique employed to observe the thermal transitions of polymers upon heating. An insulation composite sample (one with least thermal diffusivity) was analyzed to identify its glass transition (T_g), cold crystallization (T_c), and melting point T_m. A second cycle isotherm for an insulation composite from 25–250 °C (10 °C/min) is depicted in Figure 13. Upon heating a sample at a certain temperature, its state changes from a hard glassy material into a soft rubbery material. This corresponds to the first step change in the DSC curve, better known as the T_g. The T_g value for the composite sample was 86.2 °C, slightly higher than the T_g value for pure PVA (80 °C). Upon further heating a sample above T_g, a polymer often acquires much greater mobility and, once reaching the proper threshold temperature, it gives off enough heat to attain an ordered structure. This heat release or dumping of heat can be recognized by an upward peak in the DSC plot. The area of this peak conveys the latent heat of crystallization, and the temperature at this stage is denoted T_c. Yet, most importantly, it tells us that the polymer in question can crystallize. In cases where we have amorphous or semi-crystalline polymers, this peak fails to appear, as we would expect in the DPSF/PVA prepared sample because its PVA component is a semi-crystalline polymer. When the material is further heated beyond T_c, the molecules will no longer retain their arranged pattern and a temperature point is reached when they fall apart and the polymer melts. This is known as the T_m and is discernible on a DSC curve as a steep downward curve. The melting point temperature (T_m) for the insulation composite sample was 225 °C. Therefore, the developed material has a very high melting point or dissociation from its binding material.

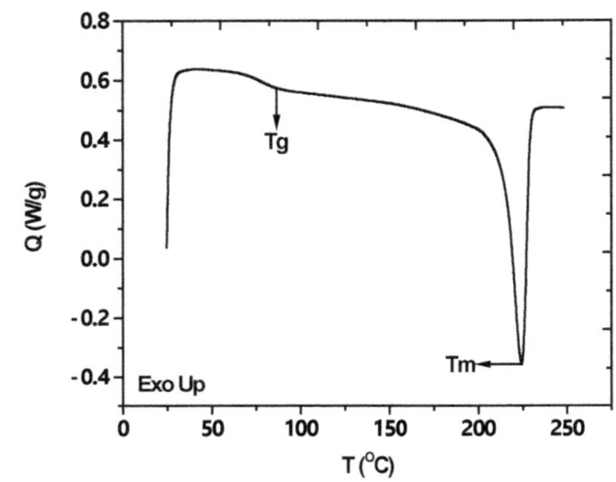

Figure 13. DSC thermograph of the DPSF/PVA sample.

We also used the DSC data to measure the crystallinity of the sample, using Equation (6) [28]:

$$\textbf{\textit{Degree of crystallinity}}\ (X_{cr}) = \frac{\Delta Hm - \Delta Hc}{\Delta H_m^{\infty}} \tag{6}$$

where ΔHm is the measured melting enthalpy from the sample (55.68 J/g); ΔHc is the measured enthalpy from the sample; ΔH_m^{∞} is the enthalpy of 100% crystalline PVA (138.6 J/g). Hence, the X_{cr} for the insulation sample is 40.17%.

3.3. Microstructure Properties

Figure 14 shows the FTIR analysis of the insulation composite sample (one with least thermal diffusivity). This analysis is useful for identifying the green nature of a composite's fabrication process. The FTIR spectra of insulation composite resembles closely with pure lignocellulose material exhibiting typical vibration spectrum of natural fibers, where the cellulose is the dominant component. The examination of the stretching peaks indicated that carbon, hydrogen, and oxygen were the core components present in the insulation composite. Hence, it is expected to be fully organic in nature. These functional groups are allocated different wavelength numbers. Carbon–carbon (C=C), hydrocarbon (C-H), carbonyl (C=O), and hydroxyl (OH) are the main fundamental groups. There is some medium stretching at 3321 cm^{-1}, due to an aliphatic primary amine (N-H) group that appears in the sample, and two strong alkane C-H stretching peaks appear at 2918 cm^{-1} and 2846 cm^{-1}. Then, at 1602 cm^{-1}, we can see a medium C=C stretching peak associated with the conjugated alkene group. The medium-bending peak appears at 1418 cm^{-1} because of an OH-bending alcohol group. At 1237 cm^{-1} there is a medium C-N stretching of an amine group. The peak at 1015 cm^{-1} arises from the medium C-N stretching of an amine group. Similar FTIR analysis has been reported in the literature for DPSFs [32] and agave fibers bonded with corn starch and wood adhesive. [34]. Mohammad Ali et al. [32] also reported the presence of similar spectrums when date palm fibers were bonded with corn starch. FTIR results showed that the fabrication process is expected to be a green process, with no toxic materials present.

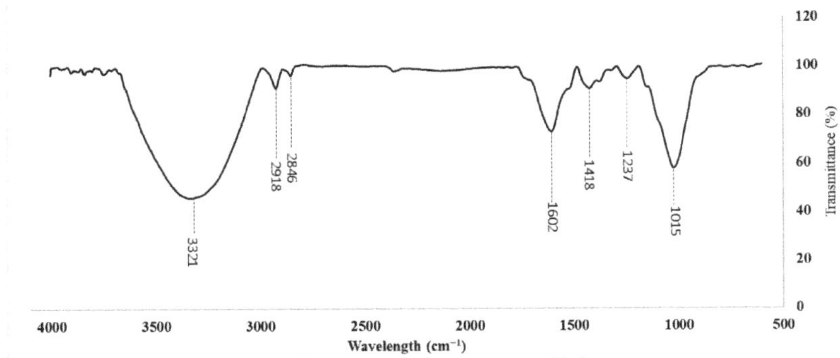

Figure 14. FTIR spectra of the insulation composite material.

Scanning electron microscopy is used to provide information on the microstructure of the composites and permits the observation of the interaction of the DPF fibers with the PVA solution. SEM images of the DPSFs and insulation composite can be seen in Figure 15. Evidently, the fibers are bonded together with PVA and display a complex binding structure. This analysis is helpful in showing, at the microstructure level, the bonding capacity of polyvinyl alcohol to hold together the surface fibers of date palm. The raw date palm fibers are long and needle shaped. In our recent study [14], it demonstrated that these DPSFs are a lignocellulosic biomass composed of 44 wt.% cellulose, 22 wt.% hemicellulose, 30 wt.% lignin, with the remainder of it comprising extractives, such as waxes or pectins, among others. Good compatibility between date palm fibers and PVA binder has been observed without any pretreatment of raw fibers. Moreover, a homogenous structure is shown in the insulation composite without any huge gaps and voids.

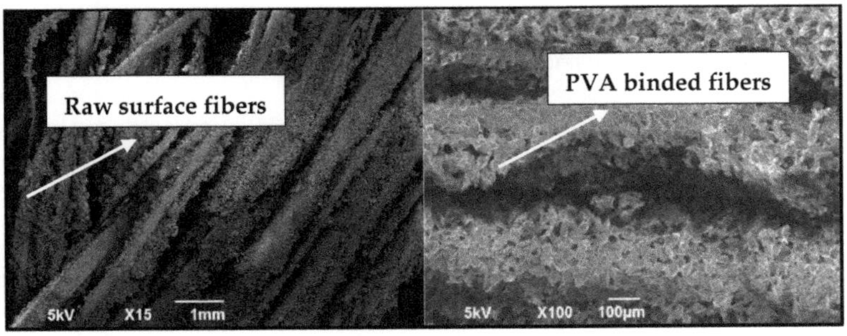

Figure 15. SEM image of raw date palm fibers (**left side**) and the prepared insulation composite (**right side**).

The X-ray diffraction analysis was performed on raw date palm fibers and insulation composite. Date palm surface fibers are basically composed of lignin, hemicellulose, and cellulose, of which cellulose is known as crystalline in nature, while lignin is the amorphous phase; therefore, the semi-crystalline behavior of date surface fibers can be seen in Figure 16. The crystallinity of insulation composites is an important parameter for evaluating their strength. A sharp peak appeared in the insulation composite XRD scan, indicating the presence of the crystalline phase and suggesting that the crystal structure of PVA was incorporated. The crystallinity index of the DPSFs is only 34.6%, but it increased to 41.7%

for the insulation composite. This increase in crystallinity is due to the binding of fibers with a crystalline PVA polymer binder. Due to the ordered PVA structure, the intermolecular bonding is significant, which results in a higher mechanical strength [54]. The degree of crystallinity found here matches the crystallinity index of kapok tree fiber (Xcr = 45%) [55]. The value of the crystallinity index also corroborates the value obtained from our DSC analysis of the insulation composite. A greater crystalline structure provides more hardness and strength to materials.

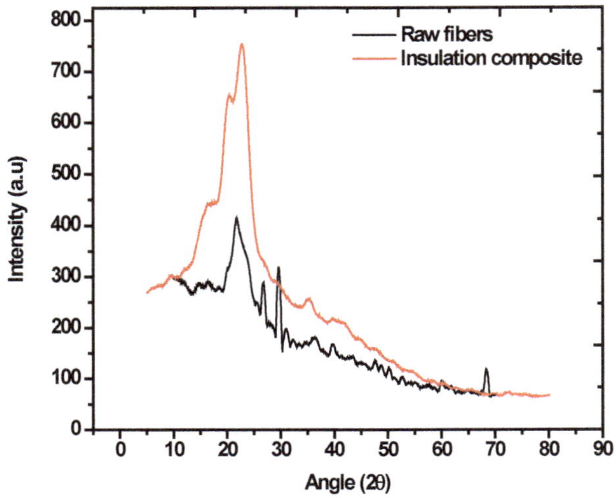

Figure 16. XRD analysis of raw fibers and insulation composite.

3.4. Mechanical Properties

Mechanical strength is an indispensable characteristic of any insulation material. Tensile strength and Young's modulus were measured for the four prepared insulation composite samples. Their stress–strain graphs are shown in Figure 17. It is noted that the tensile strength of the insulation composite increased for Sample B and Sample C as compared with Sample A. However, it again decreases for Sample D. The tensile strength of Sample B and Sample C is closely matched. It was expected to not have a clear relation between density and tensile, since the nature of fibers is uneven. During testing, if any single fiber breaks the fracture point is obtained, eliminating the effect of samples density. However, it should be noted that during tensile testing even when any single fiber breaks to give break point, the remaining sample is still in contact, unlike extruded polymeric samples that break in parts. The value of tensile strength ranged from 7 to 10 mPa. Additionally, presented in Table 7 are values for Young's modulus, which is a measure of tensile stiffness. We should note that the mechanical properties of fiber/binder composites depend on the properties of the binder as well as the fiber used. The mechanical properties of natural fiber-based or biodegradable thermal insulation materials are generally lower than those of conventional insulation materials. These findings for mechanical performance are comparable to reported values in recent studies. For example, Feng et al. [56] obtained a maximum tensile strength of 20.36 mPa for polypropylene composites when these were reinforced with Kenaf and pineapple leaf fibers and treated with varying concentrations of sodium hydroxide and 3-aminopropyltriethoxysilane. Pawlak et al. [57] used maleinized linseed oil to plasticize polylactic acid, which was then reinforced with sheep wool fibers recovered from the dairy industry; they recorded a tensile strength of ca. 25 mPa for their PLA composites whose content of sheep wool fibers was 10%. Moreover, in stark contrast, the tensile strength of polyurethane foam is just 0.110 mPa [58] and 0.120 mPa for the EPS [59].

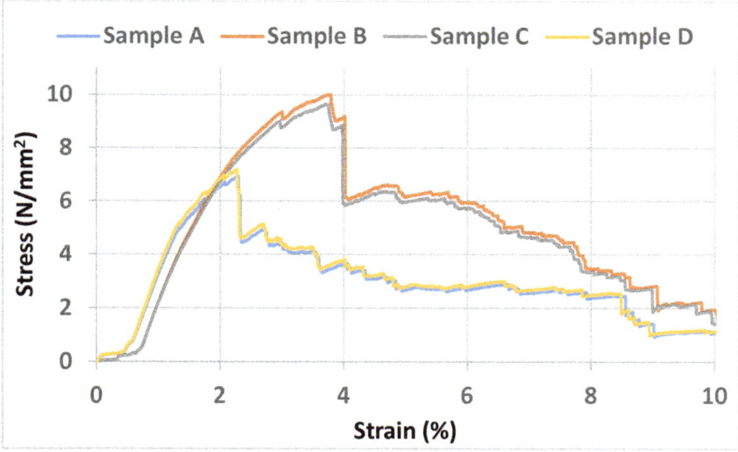

Figure 17. Stress–strain graphs of insulation composites.

Table 7. Mechanical properties of insulation composites.

Sample No.	Tensile Strength (MPa)	Young's Modulus (MPa)
Sample A	6.9	3.80
Sample B	10	3.56
Sample C	9.8	4.11
Sample D	7.2	5.98

4. Conclusions

Date palm surface fibers, an abundant form of lignocellulosic biomass waste, were utilized here to develop a thermal insulation material. The goal was to develop a more environmentally friendly biodegradable material for use as thermal insulation that could be used in buildings instead of conventional fossil fuel based insulators.

- Polyvinyl alcohol was successfully used as a binder in the synthesis of fiber-based heat insulation composites.
- Their thermal conductivity was found to be density- and temperature-dependent. Four thermal insulation composites with densities of 203–254 kg/m^3 had thermal conductivity and diffusivity values of 0.038–0.051 $W/(m{\cdot}K)$ and 0.137–0.147 mm^2/s, respectively. Crucially, both the thermal conductivity and diffusivity were on par with those of existing commercial insulators.
- Thermal transmittance (U-value) of the four insulation composites were between 3.8–5.1 $W/m^2{\cdot}K$, which was in good comparison to other insulators of similar thickness.
- TGA and DSC analyses confirm the higher thermal stability of the novel insulation composite. Initial degradation temperature and melting point were 282 °C and 225 °C, respectively.
- An E_a analysis was applied to predict the more resistive nature of the insulation material. An activation energy increased to 171 kJ/mol for the insulation composite compared with raw fibers (E_a = 101 kJ/mol).
- FTIR analysis showed that only carbon, hydrogen, and oxygen are the main constituents of the thermal composite and that it is organic in nature.
- Further, our SEM results confirmed the binding of PVA into the fiber structures.
- The prepared insulation composites had a tensile strength of 6.9–10 MPa.

- Therefore, date palm surface fiber based building insulation material exhibits excellent properties for use as a substitute for conventional insulation.

Author Contributions: M.R., formal analysis, data curation, writing—original draft; H.A.A., investigation, writing—review and editing; A.A., investigation, writing—review and editing; B.A.-J., conceptualization, methodology, validation, resource management, writing—review and editing. All authors have read and agreed to the published version of the manuscript.

Funding: The research work was funded by research grants (no. 12R014 and 31R272) from the United Arab Emirates University.

Institutional Review Board Statement: Not applicable.

Informed Consent Statement: Not applicable.

Data Availability Statement: Not applicable.

Conflicts of Interest: The authors state that they have no conflicting economic interests or personal connections that could be perceived to have impacted the research presented in this study.

References

1. Economidou, M.; Todeschi, V.; Bertoldi, P.; D'Agostino, D.; Zangheri, P.; Castellazzi, L. Review of 50 years of EU energy efficiency policies for buildings. *Energy Build.* **2020**, *225*, 110322. [CrossRef]
2. Miner, K.R.; Turetsky, M.R.; Malina, E.; Bartsch, A.; Tamminen, J.; McGuire, A.D.; Sweeney, C.; Elder, C.D.; Miller, C.E. Permafrost carbon emissions in a changing Arctic. *Nat. Rev. Earth Environ.* **2022**, *3*, 55–67. [CrossRef]
3. Pigliautile, I.; Chàfer, M.; Pisello, A.L.; Pérez, G.; Cabeza, L.F. Inter-building assessment of urban heat island mitigation strategies: Field tests and numerical modelling in a simplified-geometry experimental set-up. *Renew. Energy* **2020**, *147*, 1663–1675. [CrossRef]
4. Aly, N.M.; Seddeq, H.; Elnagar, K.; Hamouda, T. Acoustic and thermal performance of sustainable fiber reinforced thermoplastic composite panels for insulation in buildings. *J. Build. Eng.* **2021**, *40*, 102747. [CrossRef]
5. Gounni, A.; Alami, M.E. Experimental study of heat transfer in a reduced scale cavity incorporating phase change material into its vertical walls. *J. Therm. Sci. Eng. Appl.* **2018**, *10*, 011010. [CrossRef]
6. Berardi, U. The impact of aging and environmental conditions on the effective thermal conductivity of several foam materials. *Energy* **2019**, *182*, 777–794. [CrossRef]
7. Asdrubali, F.; D'Alessandro, F.; Schiavoni, S. A review of unconventional sustainable building insulation materials. *Sustain. Mater. Technol.* **2015**, *4*, 1–17. [CrossRef]
8. Yildiz, G.; Yahia, M.E. Comparative performance evaluation of conventional and renewable thermal insulation materials used in building envelops. *Teh. Vjesn.* **2020**, *27*, 283–289.
9. Hill, C.; Norton, A.; Dibdiakova, J. A comparison of the environmental impacts of different categories of insulation materials. *Energy Build.* **2018**, *162*, 12–20. [CrossRef]
10. Abedom, F.; Sakthivel, S.; Asfaw, D.; Melese, B.; Solomon, E.; Kumar, S.S. Development of natural fiber hybrid composites using sugarcane bagasse and bamboo charcoal for automotive thermal insulation materials. *Adv. Mater. Sci. Eng.* **2021**, *2021*, 2508840. [CrossRef]
11. Dickson, T.; Pavía, S. Energy performance, environmental impact and cost of a range of insulation materials. *Renew. Sustain. Energy Rev.* **2021**, *140*, 110752. [CrossRef]
12. Ilyas, R.A.; Aisyah, H.A.; Nordin, A.H.; Ngadi, N.; Zuhri, M.Y.M.; Asyraf, M.R.M.; Sapuan, S.M.; Zainudin, E.S.; Sharma, S.; Abral, H.; et al. Natural-Fiber-Reinforced Chitosan, Chitosan Blends and Their Nanocomposites for Various Advanced Applications. *Polymers* **2022**, *14*, 874. [CrossRef]
13. Asyraf, M.R.M.; Ishak, M.R.; Sapuan, S.M.; Yidris, N.; Ilyas, R.A.; Rafidah, M.; Razman, M.R. Potential application of green composites for cross arm component in transmission tower: A brief review. *Int. J. Polym. Sci.* **2020**, *2020*, 8878300. [CrossRef]
14. Raza, M.; Abu-Jdayil, B.; Al-Marzouqi, A.H.; Inayat, A. Kinetic and thermodynamic analyses of date palm surface fibers pyrolysis using Coats-Redfern method. *Renew. Energy* **2022**, *183*, 67–77. [CrossRef]
15. Inayat, A.; Raza, M. District cooling system via renewable energy sources: A review. *Renew. Sustain. Energy Rev.* **2019**, *107*, 360–373. [CrossRef]
16. Alzaabi, M.S.A.; Mezher, T. Analyzing existing UAE national water, energy and food nexus related strategies. *Renew. Sustain. Energy Rev.* **2021**, *144*, 111031. [CrossRef]
17. Rehman, M.S.U.; Shafiq, M.T.; Afzal, M. Impact of COVID-19 on project performance in the UAE construction industry. *J. Eng. Des. Technol.* **2021**, *20*, 245–266.
18. Kosiński, P.; Brzyski, P.; Tunkiewicz, M.; Suchorab, Z.; Wiśniewski, D.; Palczyński, P. Thermal Properties of Hemp Shives Used as Insulation Material in Construction Industry. *Energies* **2022**, *15*, 2461. [CrossRef]
19. Yan, Q.; Feng, Z.; Luo, J.; Xia, W. Preparation and characterization of building insulation material based on SiO_2 aerogel and its composite with expanded perlite. *Energy Build.* **2022**, *255*, 111661. [CrossRef]

20. Srihanum, A.; Tuan Noor, M.T.; Devi, K.P.; Hoong, S.S.; Ain, N.H.; Mohd, N.S.; Din Mat, N.S.M.N.; Kian, Y.S. Low density rigid polyurethane foam incorporated with renewable polyol as sustainable thermal insulation material. *J. Cell. Plast.* **2022**, *58*, 1–19. [CrossRef]
21. Abu-Jdayil, B.; Abdallah, H.A.; Mlhem, A.; Alkhatib, S.; Sayah, E.A.; Abdulsalam, H.; Asayel, A.; Alaydaroos, A. Utilization of Polyurethane Foam Dust in Development of Thermal Insulation Composite. *Buildings* **2022**, *12*, 126. [CrossRef]
22. Dissanayake, D.; Weerasinghe, D.; Thebuwanage, L.; Bandara, U. An environmentally friendly sound insulation material from post-industrial textile waste and natural rubber. *J. Build. Eng.* **2021**, *33*, 101606. [CrossRef]
23. Hittini, W.; Mourad, A.-H.I.; Abu-Jdayil, B. Utilization of devulcanized waste rubber tire in development of heat insulation composite. *J. Clean. Prod.* **2021**, *280*, 124492. [CrossRef]
24. Abu-Jdayil, B.; Mourad, A.-H.I.; Hussain, A.; Al Abdallah, H. Thermal insulation and mechanical characteristics of polyester filled with date seed wastes. *Constr. Build. Mater.* **2022**, *315*, 125805. [CrossRef]
25. Mlhem, A.; Abu-Jdayil, B.; Tong-Earn, T.; Iqbal, M. Sustainable heat insulation composites from date palm fibre reinforced poly (β-hydroxybutyrate). *J. Build. Eng.* **2022**, *54*, 104617. [CrossRef]
26. Al Abdallah, H.; Abu-Jdayil, B.; Iqbal, M.Z. The Effect of Alkaline Treatment on Poly(lactic acid)/Date Palm Wood Green Composites for Thermal Insulation. *Polymers* **2022**, *14*, 1143. [CrossRef]
27. Abu-Jdayil, B.; Barkhad, M.S.; Mourad, A.-H.I.; Iqbal, M.Z. Date palm wood waste-based composites for green thermal insulation boards. *J. Build. Eng.* **2021**, *43*, 103224. [CrossRef]
28. Barkhad, M.S.; Abu-Jdayil, B.; Iqbal, M.Z.; Mourad, A.-H.I. Thermal insulation using biodegradable poly (lactic acid)/date pit composites. *Constr. Build. Mater.* **2020**, *261*, 120533. [CrossRef]
29. Al Abdallah, H.; Abu-Jdayil, B.; Iqbal, M.Z. Improvement of Mechanical Properties and Water Resistance of Bio-based Thermal Insulation Material via Silane Treatment. *J. Clean. Prod.* **2022**, *346*, 131242. [CrossRef]
30. Masri, T.; Ounis, H.; Sedira, L.; Kaci, A.; Benchabane, A. Characterization of new composite material based on date palm leaflets and expanded polystyrene wastes. *Constr. Build. Mater.* **2018**, *164*, 410–418. [CrossRef]
31. Ali, M.; Alabdulkarem, A.; Nuhait, A.; Al-Salem, K.; Iannace, G.; Almuzaiqer, R. Characteristics of agro waste fibers as new thermal insulation and sound absorbing materials: Hybrid of date palm tree leaves and wheat straw fibers. *J. Nat. Fibers* **2021**, 1–9. [CrossRef]
32. Ali, M.E.; Alabdulkarem, A. On thermal characteristics and microstructure of a new insulation material extracted from date palm trees surface fibers. *Constr. Build. Mater.* **2017**, *138*, 276–284. [CrossRef]
33. Haseli, M.; Layeghi, M.; Hosseinabadi, H.Z. Characterization of blockboard and battenboard sandwich panels from date palm waste trunks. *Measurement* **2018**, *124*, 329–337. [CrossRef]
34. Alabdulkarem, A.; Ali, M.; Iannace, G.; Sadek, S.; Almuzaiqer, R. Thermal analysis, microstructure and acoustic characteristics of some hybrid natural insulating materials. *Constr. Build. Mater.* **2018**, *187*, 185–196. [CrossRef]
35. Abu-Jdayil, B.; Raza, M.; Al Abdallah, H. Thermal Insulating Material Made from Date Palm Surface Fibers. U.S. Patent 11255052b1, 22 July 2022.
36. Cetiner, I.; Shea, A.D. Wood waste as an alternative thermal insulation for buildings. *Energy Build.* **2018**, *168*, 374–384. [CrossRef]
37. Gayathri, K.; Rajesh, K.M.; Krishnan, P.V.; Anandan, K.; Rexalin, D.A.; Anbalagan, G. A study on kinetic properties of brucinium hydroge (s) malate pentahydrate single crystal by Coats Redfern method. In *AIP Conference Proceedings*; AIP Publishing LLC: Jodhpur, India, 2020; Volume 2265, p. 030425.
38. Naqvi, S.R.; Uemura, Y.; Osman, N.; Yusup, S. Kinetic study of the catalytic pyrolysis of paddy husk by use of thermogravimetric data and the Coats–Redfern model. *Res. Chem. Intermed.* **2015**, *41*, 9743–9755. [CrossRef]
39. GreenSpec. Insulation Materials and Their Thermal Properties. 2021. Available online: https://www.greenspec.co.uk/building-design/insulation-materials-thermal-properties/ (accessed on 11 May 2022).
40. Patnaik, A.; Mvubu, M.; Muniyasamy, S.; Botha, A.; Anandjiwala, R.D. Thermal and sound insulation materials from waste wool and recycled polyester fibers and their biodegradation studies. *Energy Build.* **2015**, *92*, 161–169. [CrossRef]
41. Nguyen, D.M.; Grillet, A.-C.; Diep, T.M.H.; Bui, Q.B.; Woloszyn, M. Influence of thermo-pressing conditions on insulation materials from bamboo fibers and proteins based bone glue. *Ind. Crops Prod.* **2018**, *111*, 834–845. [CrossRef]
42. Alana. Thermal Conductivity & U-Values. September 2020. Available online: https://ewistore.co.uk/thermal-conductivity-u-values/#login-modal (accessed on 11 May 2022).
43. Mahmoodzadeh, M.; Gretka, V.; Hay, K.; Mukhopadhyaya, C.S.P. Determining overall heat transfer coefficient (U-Value) of wood-framed wall assemblies in Canada using external infrared thermography. *Build. Environ.* **2021**, *199*, 107897. [CrossRef]
44. Sahu, D.K.; Sen, P.K.; Sahu, G.; Sharma, R.; Bohidar, S. A review on thermal insulation and its optimum thickness to reduce heat loss. *Int. J. Innov. Res. Sci. Technol* **2015**, *2*, 1–6.
45. Lakatos, Á.; Kalmár, F. Investigation of thickness and density dependence of thermal conductivity of expanded polystyrene insulation materials. *Mater. Struct.* **2013**, *46*, 1101–1105. [CrossRef]
46. Zach, J.; Korjenic, A.; Petránek, V.; Hroudová, J.; Bednar, T. Performance evaluation and research of alternative thermal insulations based on sheep wool. *Energy Build.* **2012**, *49*, 246–253. [CrossRef]
47. Mahlia, T.; Taufiq, B.; Masjuki, H. Correlation between thermal conductivity and the thickness of selected insulation materials for building wall. *Energy Build.* **2007**, *39*, 182–187. [CrossRef]

48. Cassidy, K. My Builder. 2014. Available online: https://www.mybuilder.com/questions/v/14947/using-50mm-polystyrene-instead-of-50mm-celotex-for-extension (accessed on 11 May 2022).
49. NBS. What Is A U-Value? 2022. Available online: https://www.thenbs.com/knowledge/what-is-a-u-value-heat-loss-thermal-mass-and-online-calculators-explained (accessed on 11 May 2022).
50. Kumar, D.; Alam, M.; Zou, P.X.; Sanjayan, J.G.; Memon, R.A. Comparative analysis of building insulation material properties and performance. *Renew. Sustain. Energy Rev.* **2020**, *131*, 110038. [CrossRef]
51. Zheng, C.; Li, D.; Ek, M. Mechanism and kinetics of thermal degradation of insulating materials developed from cellulose fiber and fire retardants. *J. Therm. Anal. Calorim.* **2019**, *135*, 3015–3027. [CrossRef]
52. Li, A.; Zhang, W.; Zhang, J.; Ding, Y.; Zhou, R. Pyrolysis kinetic properties of thermal insulation waste extruded polystyrene by multiple thermal analysis methods. *Materials* **2020**, *13*, 5595. [CrossRef]
53. Jiao, L.; Xu, G.; Wang, Q.; Xu, Q.; Sun, J. Kinetics and volatile products of thermal degradation of building insulation materials. *Thermochim. Acta* **2012**, *547*, 120–125. [CrossRef]
54. Song, F.; Wang, Q.; Wang, T. The effects of crystallinity on the mechanical properties and the limiting PV (pressure× velocity) value of PTFE. *Tribol. Int.* **2016**, *93*, 1–10. [CrossRef]
55. Mwaikambo, L.Y.; Ansell, M.P. Chemical modification of hemp, sisal, jute, and kapok fibers by alkalization. *J. Appl. Polym. Sci.* **2002**, *84*, 2222–2234. [CrossRef]
56. Feng, N.L.; Malingam, S.D.; Razali, N.; Subramonian, S. Alkali and silane treatments towards exemplary mechanical properties of kenaf and pineapple leaf fibre-reinforced composites. *J. Bionic* **2020**, *17*, 380–392. [CrossRef]
57. Pawlak, F.; Aldas, M.; Parres, F.; López-Martínez, J.; Arrieta, M.P. Silane-functionalized sheep wool fibers from dairy industry waste for the development of plasticized pla composites with maleinized linseed oil for injection-molded parts. *Polymers* **2020**, *12*, 2523. [CrossRef]
58. Ma, S.; Xiao, Y.; Zhou, F.; Schartel, B.; Chan, Y.Y.; Korobeinichev, O.P.; Trubachev, S.A.; Hu, W.; Ma, C.; Hu, Y. Effects of novel phosphorus-nitrogen-containing DOPO derivative salts on mechanical properties, thermal stability and flame retardancy of flexible polyurethane foam. *Polym. Degrad. Stab.* **2020**, *177*, 109160. [CrossRef]
59. Rydzkowski, T.; Reszka, K.; Szczypiński, M.; Szczypiński, M.M.; Kopczyńska, E.; Thakur, V.K. Manufacturing and evaluation of mechanical, morphological, and thermal properties of reduced graphene oxide-reinforced expanded polystyrene (EPS) nanocomposites. *Adv. Polym. Technol.* **2020**, *2020*, 3053471. [CrossRef]

energies

MDPI

Review

A Review of Thermally Activated Building Systems (TABS) as an Alternative for Improving the Indoor Environment of Buildings

María M. Villar-Ramos [1], Iván Hernández-Pérez [2,*], Karla M. Aguilar-Castro [2], Ivett Zavala-Guillén [3], Edgar V. Macias-Melo [2], Irving Hernández-López [4] and Juan Serrano-Arellano [5]

1 Doctorado en Ciencias en Ingeniería, División Académica de Ingeniería y Arquitectura, Universidad Juárez Autónoma de Tabasco (DAIA-UJAT), Carretera Cunduacán-Jalpa de Méndez km. 1, Cunduacán 86690, Tabasco, Mexico
2 División Académica de Ingeniería y Arquitectura, Universidad Juárez Autónoma de Tabasco (DAIA-UJAT), Carretera Cunduacán-Jalpa de Méndez km. 1, Cunduacán 86690, Tabasco, Mexico
3 Centro de Investigación Científica y de Educación Superior de Ensenada CICESE, Carretera Ensenada-Tijuana No. 3918, Zona Playitas, Ensenada 22860, Baja California, Mexico
4 Departamento de Ingeniería Química y Metalurgia, Universidad de Sonora (UNISON), Blvd. Luis Encinas y Rosales S/N, Col. Centro, Hermosillo 83000, Sonora, Mexico
5 División de Estudios de Posgrado e Investigación, Tecnológico Nacional de México/IT de Pachuca, Carretera Mexico-Pachuca km. 87.5, Colonia Venta Prieta, Pachuca de Soto 42080, Hidalgo, Mexico
* Correspondence: ivan.hernandezp@ujat.mx

Citation: Villar-Ramos, M.M.; Hernández-Pérez, I.; Aguilar-Castro, K.M.; Zavala-Guillén, I.; Macias-Melo, E.V.; Hernández-López, I.; Serrano-Arellano, J. A Review of Thermally Activated Building Systems (TABS) as an Alternative for Improving the Indoor Environment of Buildings. *Energies* **2022**, *15*, 6179. https://doi.org/10.3390/en15176179

Academic Editors: Shi-Jie Cao and Wei Feng

Received: 31 July 2022
Accepted: 21 August 2022
Published: 25 August 2022

Abstract: Among the alternatives for improving the thermal comfort conditions inside buildings are the thermally activated building systems (TABS). They are embedded in different building components to improve the indoor air temperature. In this work, a review and analysis of the state of the art of TABS was carried out to identify their potential to improve thermal comfort conditions and provide energy savings. Furthermore, this study presents the gaps identified in the literature so that researchers can develop future studies on TABS. The articles found were classified and analyzed in four sections, considering their implementation in roofs, walls, floors, and the whole envelope. In addition, aspects related to the configuration of the TABS and the fluid (speed, temperature, and mass flow rate) were analyzed. It was found that when TABS are implemented in roofs, walls, and floors, a reduction in the indoor temperature of a building of up to 14.4 °C can be obtained. Within the limitations of the TABS, the complexity and costs of their implementation compared to the use of air conditioning systems are reported. However, the TABS can provide energy savings of up to 50%.

Keywords: thermally activated building systems; thermal comfort; thermal mass; energy savings; radiant envelope; heat exchanger pipes

1. Introduction

According to experts from the Intergovernmental Panel on Climate Change (IPCC), climate change has impacted all countries [1]. The level of electricity consumption for thermal comfort has increased dramatically due to population growth. Moreover, in the last decade, significant changes in many meteorological phenomena and weather conditions have occurred in the world. It is estimated that the amount of energy consumed by the residential and construction sector in 2018 was 36%, of which 39% of the energy was related to CO_2 emissions worldwide [2].

A building is a construction or an enclosure made of different materials destined to be inhabited or destined to be used for conducting other activities. It is well known that most of the heat gains of the building envelope occur due to the received solar irradiance, the heat exchange with the outdoor environment, and its geometry and orientation. These heat gains or losses of the building envelope usually cause the inhabitants to use an air conditioning system to achieve thermal comfort. The scientific community has begun to search for and analyze construction alternatives that can reduce or increase thermal loads in buildings and reduce global electricity consumption through renewable energy [3,4]. Several construction alternatives to improve the indoor environment of a building are available, such as earth-to-air heat exchangers [5–7], ventilated roofs [8–10], reflective materials [11–13], passive design strategies [14–16], and thermally activated building systems (TABS) [17–19], among others. These design strategies also contribute to the proposal of intervention alternatives for the rehabilitation of spaces, considering the occupants, thermal adaptation, and energy use [14].

TABS can reduce heat gains or losses because of the heat exchange from embedded pipes installed in different building components. These pipes exchange heat directly with the thermal mass of the building and improve the temperature of the indoor environment [20]. Inside the pipes, water or air is generally circulated; these pipes are embedded in roofs, floors, or walls, depending on factors such as the climatic zone, orientation, and construction materials, among others. TABS are used to decrease or increase the temperature in the indoor space of an enclosure. The integration of the system with the construction structure allows the use of solar energy to be included, since the working fluid can be reused for other applications, helping to reduce pollution from greenhouse gases. According to the literature, the cost of implementing TABS is higher than traditional HVAC systems. However, the energy saving provided by TABS is 25% higher compared to the lifetime of a traditional air conditioning system. Using water as the working fluid in TABS increases energy savings because water can transport energy 3500 times more effectively than air. Among the advantages of using TABS over HVAC systems are: high indoor air quality, greater energy efficiency and smaller size, and low maintenance costs [21]. Although thermally activated systems have been studied, analysis of the envelope is required to obtain a better performance [22].

Various authors have analyzed the application of TABS, which have been referred to with different names depending on the application and the location in the building envelope. The literature is not consistent in labeling TABS. Table 1 summarizes the different names used for TABS.

Table 1. Other names used for TABS and their applications.

Name	Location	Mode
• Hollow core slab • Embedded tubes with hot/cold fluids • Slab cooling/heating system • Floor heating system • In slab heating floor • Radiant floor • Concrete core • Pipe-embedded envelope • Radiant slab cooling • Concrete core cooling slab • Thermally activated building constructions • Active building storage systems • Embedded hydronic pipe systems	• Floor • Roof • Wall • Whole envelope	• Heating • Cooling

TABS have been used for both heating and cooling and are located in different sections of a building envelope depending on their application. Previous review articles on TABS are available and discuss several aspects related to the thermal behavior of this technology. For instance, Rhee and Kim [23] analyzed the basic and applied literature on radiant heating and cooling systems embedded in the building envelope. The authors analyzed the main uses of radiant systems and thermal comfort, their cooling/heating capacity, obtained from different approaches such as computational fluid dynamics (CFD) analysis, energy simulation, system configuration, and control strategies for use at other times of the day. In the literature, TABS have also been analyzed according to their application, design, topology, and control strategies. Romaní et al. [24] analyzed TABS based on their application's simulation and control strategies. The authors studied the system's generalities and design and classified the TABS by mode of operation, position, and working fluid. Romaní et al. classified the TABS as radiant floor, radiant ceiling, hollow core slab, concrete core, and pipe-embedded envelope. Ma et al. [25] conducted a review of the state of the art of energy storage and dissipation in TABS. The authors focused on the extraction of energy from the indoor environment of buildings and how it can be reused in other systems. The authors concluded that by applying TABS correctly, an improvement in energy efficiency can be obtained. The possibilities and limitations of using TABS on walls were analyzed, such as in a work published by Krajčík and Šikula [26]. The authors examined the use of TABS and compared four types of wall cooling system. Krajčík et al. [27] carried out a review on TABS embedded in walls and their use as thermal barriers, with a focus on the reduction in thermal loads. The authors selected only systems with pipes embedded in the wall for heating and cooling. Krajčík et al. classified the wall system and the thermal barriers by their function, the configuration of material layers, and the location of the pipes. On the other hand, the analysis of TABS has also been considered in works incorporating insulation materials, such as phase change materials, as studied by Cai et al. [28], which contribute to improving the indoor environment and storing energy.

A bibliometric analysis was performed in order to explore the existing status of the literature on TABS. Figure 1 shows a visual map where some aspects of the selected articles were analyzed. To search for the term "Thermally Activated Building System", the Scopus database was used by prioritizing the publication period from 2015 to 2022. To input the collected results, the open-source software VOSviewer, was used. It was found that the most cited topics in the literature related to the term were: (1) thermally activated, (2) cooling systems, (3) energy efficiency, (4) heat storage, and (5) thermal comfort. Figure 2 shows the countries of origin where the most articles about TABS have been published for the last seven years. TABS are being researched extensively in China, the United States, Germany, Spain, Belgium, and India, among others.

The present study aims to explore the state of the art of TABS in buildings and to present information that would help researchers to develop new practices, technologies, and research directions. The reviewed and analyzed articles were obtained mainly from databases such as Scopus, Web of Science, and Google Scholar by prioritizing the publication period from 2015 to 2022. Within the criteria for selecting the articles, only articles in which the heat exchanger tubes were embedded in a building envelope component (either roof, wall, or floor), or even in the whole building envelope, were considered. This work is divided into four sections, with the main findings highlighted when TABS are installed on roofs, walls, floors, and the whole building envelope. In each section, a summary table describes the applications of TABS in the building envelope, the type of climate studied, the mode of operation, the TABS features, and the simulation methods.

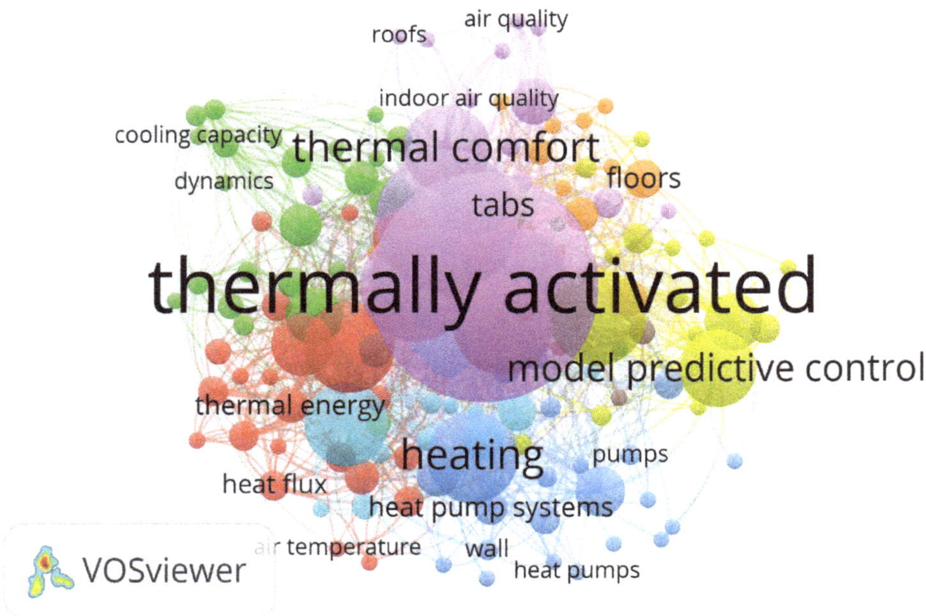

Figure 1. Network of keywords relating to thermally activated building systems.

Figure 2. Network of the origins of papers about thermally activated building systems.

2. TABS Embedded in Building Roofs

Building roofs are usually the building components with the most significant temperature fluctuations, and they receive solar energy for more hours than any other component. Thus, in zones with a warm climate, building roofs are sources of unwanted heat that affect indoor thermal comfort conditions. This section focuses on the research works related to TABS integrated into building roofs. The improvements in thermal comfort, combinations with other technologies, and the energy savings provided by this technology are given in this section.

2.1. Potential of TABS to Improve Thermal Comfort When Installed in Roofs

The thermal comfort conditions inside buildings depend on variables such as air relative humidity, air temperature, and air speed, among others [29,30]. Several studies were conducted to determine the influence of roofs with TABS on the indoor air temperature of buildings, which can be considered as a method of assessing the thermal comfort improvements provided by TABS.

One of the first studies was that presented by Gwerder et al. [31], who proposed a control algorithm for TABS to comply with comfort requirements. The proposed method incorporates the change between the heating and cooling modes of the TABS to satisfy thermal comfort. The algorithm was tested in a simulation example. The hourly temperature analysis demonstrated that the TABS maintained the indoor air temperature between 21 and 27 °C for the whole year. Another control strategy for TABS in which the operating mode (heating or cooling) is determined by the average indoor air temperature was reported by Wit and Wisse [32]. They analyzed the thermal behavior of TABS with different hydronic typologies integrated into the roof of two office buildings. After several experimental tests, the results demonstrated that TABS could maintain the comfort conditions of the two buildings because most of the measured indoor air temperatures fell within the 80–90% satisfaction zones during the testing period. In another study of a roof integrated with TABS, Rey Martínez et al. [33] analyzed the indoor air quality and thermal comfort of a building. The building had four floors, TABS powered by a water chiller, and a cooling tower. The authors found that the operating temperature remained between 23 and 25 °C and the CO_2 levels at 850 ppm during occupancy. A simulation study of a building incorporated with TABS in the roof is described in Chung et al. [34]. EnergyPlus simulation software was used to apply different control strategies in each area of the case study building. The authors varied the water supply temperature from 19 to 25 °C in the interior zone and the perimeter for heating and cooling, grouping the tests into three case studies. Chung et al. concluded that by separating the proposed building into zones with different control strategies according to each floor's needs, the thermal comfort improved by 5%. The experimental study presented by Dharmasastha et al. [35] analyzed the thermal behavior of a hybrid system integrated with a TABS coupled to a gypsum roof reinforced with fiberglass (TAGFRG). They built a test chamber with copper tubes of 0.01 m internal diameter embedded in the roof under hot and humid conditions in Chennai, India. The authors found that the TABS decreased up to 5.1 °C in the roof interior surface temperature and 6.7 °C in the indoor air of the test chamber. Saw et al. [36] studied the thermal behavior of a roof cooling system with a closed-loop pulsating heat pipe (CLPHP) and compared it with a bare metal roof system design. The authors proposed a rooftop CLPHP as an active cooling system for a tropical climate. This system consisted of a closed circuit of copper pipe, placed between two aluminum plates under a sheet roof and insulated on the lower surface. Methanol was used as a working fluid in the copper pipe circuit. They simulated solar radiation using two halogen lamps. Saw et al. found that a cool roof system with CLPHP reduced the indoor air temperature of the test cabin from 34 °C to 29.6 °C compared to the bare metal roof system.

Table 2 summarizes the studies presented in this section. The authors determined the influence of TABS installed in building roofs on thermal comfort by analyzing the indoor air temperature, satisfaction zone compliance, and comfort improvement percentage. The influence of TABS installed in roofs appears to be beneficial for increasing the thermal comfort in buildings.

Table 2. Improvements in thermal comfort of buildings with TABS embedded in roofs.

Reference	Weather	Mode *	TABS Features *	Model	Findings
[31]	-	H, C	F = Water, DBP = 0.2 m, ϕ = 0.015 m	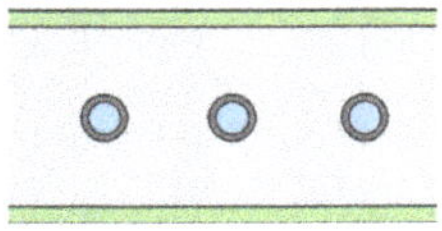	The thermal comfort can be maintained between 21 and 27 °C if TABS are used with intermittent operation.
[32]	Temperate	H, C	F = Water	-	TABS maintained the indoor air temperature within 80–90% of comfort satisfaction zone.
[33]	-	C	F = Water	-	TABS maintained the indoor air temperature in a range between 23 and 25 °C.
[34]	-	H, C	F = Water		A control strategy by zones improves the thermal comfort by 5% with the TABS installed in the roof.
[35]	Warm and humid	C	F = Water, DBP = 0.054 m, ϕ = 0.01 m		TABS decreased the indoor air temperature of the test chamber by up to 6.7 °C.
[36]	Tropical	C	F = Methanol, ϕ = 0.00635 m		A CLPHP coupled to a metal roof reduced the indoor air temperature by up to 13% compared with the bare metal roof system.

* H = heating, C = cooling, F = fluid, DBP = distance between the pipes, and ϕ = diameter of the pipes.

2.2. Combination of TABS with Other Technologies for Roofs

Several research works analyzed the combination of TABS with other technologies to improve the indoor temperature of buildings. In the reported studies, TABS was combined with solar collectors, ground heat exchangers, and materials that favor energy storage. Wu et al. [37] developed a numerical model to analyze the behavior of a combined heating system formed by solar air collectors connected to a TABS with intermittent operation. The authors observed that the solar air collector inlet temperature ranged from 17 to 24 °C, while the air collector outlet temperature ranged from 35 to 62 °C, with an average efficiency of 47.1%. They concluded that with the proposed system an acceptable thermal comfort temperature could be maintained inside the building ranging from 17 to 24 °C. In a recent study, Chung et al. [38] simulated the behavior of a coupled system (TABS + ground heat exchanger) considering 28 climatic conditions and varying the burial depth of the ground heat exchanger. The authors found that the coupled system removed the peak thermal loads by up to 75%, while a chiller cooling system removed it by 32%. They also found that in this coupled system, the climatic conditions caused variations in the load-handled ratio, obtaining better results in warm humid climates when the depth of the ground heat exchanger was buried at 2 m.

Other authors studied the combination of TABS with phase change material (PCM). A study of roofs with PCM and TABS is also available. Yu et al. [39] studied a roof with embedded tubes through which air circulated. They validated and compared through a CFD numerical simulation the thermal properties of the system with a PCM as insulation. The authors proposed a concrete roof with a thickness of 0.19 m and a layer of 0.03 m of paraffin as the PCM. The results show that the optimum phase transition temperature increases linearly by approximately 2 °C when the average temperature of the outdoor air rises. Compared to a roof without PCM, they found that the interior surface temperature decreases by between 3.7 and 4.0 °C in different regions of China. In a more recent study, the same authors [40] proposed a ventilated roof model with embedded pipes and a stabilized layer of PCM (VRSP). The authors developed a steady-state three-dimensional heat transfer model of the VRSP system in ANSYS FLUENT. The convective heat transfer coefficient on the interior surface of the roof was 8.72 W m^{-2} K^{-1} and 23.26 W m^{-2} K^{-1} on the exterior surface, and the indoor air temperature was set at 26 °C. The effect of the phase transition temperature, the thickness of the PCM layer, and the airflow rate in the tubes was studied. The researchers found that the optimal design of the roof had a phase transition temperature of 29–31 °C, a thickness of the PCM layer of 0.02–0.35 m, and an airflow rate of 1.4–2.5 m s^{-1}. It was shown that the optimum design reduced the average temperatures of the interior surface by a factor ranging between 0.4 and 3.2 °C compared to the non-ventilated roof. Moreover, the daily heat gain of the roof was reduced by a factor ranging between 9 and 82%. In a recent study, Heidenthaler et al. [41] performed a comparative analysis of TABS embedded in concrete and wooden roof slabs. The authors used the finite element analysis (FEA) simulation software HTflux. They analyzed four basic variants of fir and beech wood, of which they obtained five additional combinations. They also varied the depth at which the tubes were embedded (0.03 and 0.06 m). The authors concluded that by using wood in TABS, adequate heat flux densities can be achieved for heating in low-energy buildings, supplying the fluid at higher temperatures compared to concrete structures. The authors found that the basic combination of beech (radial/tangential) with 6 cm embedded pipes has a potential energy storage capacity 53% greater than a concrete structure. Other authors that analyzed the behavior of a roof TABS coupled with a ground source heat pump (GSHP) were Hu et al. [42]. The authors carried out an energetic and exergetic analysis of a building for summer and winter. The authors found that adding a cooling tower improves system performance with an efficiency of up to 16%, maintaining the indoor ambient temperature within the range of 18–26 °C.

Table 3 summarizes the studies presented in this section. The coupling of TABS with other technologies such as solar collectors or ground heat exchangers is expected to increase the cooling or heating effect that TABS provides. Such a combination of technologies demonstrates that TABS can be integrated into renewable energy sources and will help to reduce the emission of greenhouse gases. Moreover, as mentioned above, the combination of TABS with other technologies such as PCM increases the thermal mass of the building roofs, which enhances the peak indoor air temperature reduction.

Table 3. Studies that analyzed the combination of TABS with other roof technologies.

Reference	Weather	Mode *	TABS Features *	Model	Combination
[37]	Cold region	H	F = Air, DBP = 0.12 m, ϕ = 0.04 m, v = 2 m s^{-1}	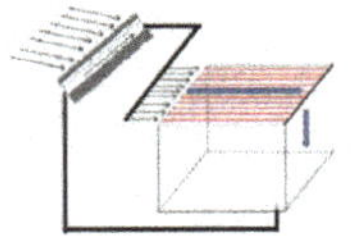	Solar air collector and TABS
[38]	Equatorial, arid, warm temperature, snow, polar	H, C	-	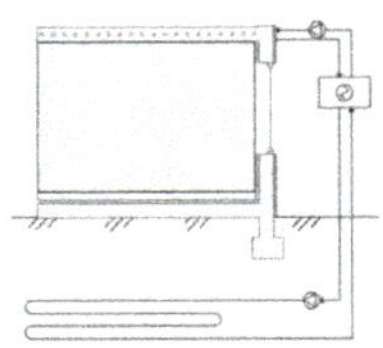	Horizontal ground heat exchanger and TABS
[39]	Cool, winter, hot summer, mild regions	-	F = Air, DBP = 0.024 m, ϕ = 0.08 m	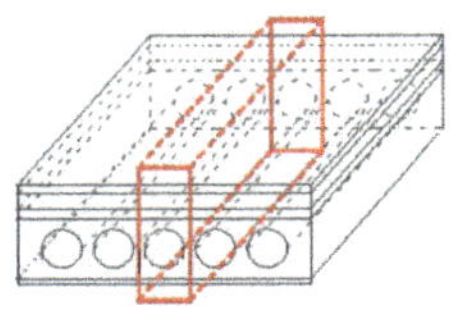	PCM and TABS

Table 3. *Cont.*

Reference	Weather	Mode *	TABS Features *	Model	Combination
[40]	-	-	F = Air, DBP = 0.024 m, ϕ = 0.08 m	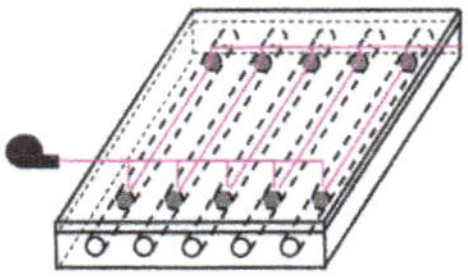	PCM and TABS
[41]	-	H	F = Water, DBP= 0.15 m, ϕ = 0.016 m	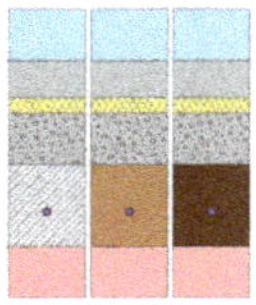	Wood and TABS
[42]	Cold winter	H, C	F = Water		GSHP with TABS.

* H = heating, C = cooling, F = fluid, DBP = distance between the pipes, ϕ = diameter of the pipes, and v = fluid velocity.

2.3. *Potential of TABS to Reduce the Energy Consumption When Installed in Roofs*

The reduction in the energy consumption of air-conditioned buildings due to the incorporation of TABS in building roofs was analyzed in two research works. In the first work, Lehmann et al. [43] investigated the functionality and application range of a TABS by simulating a typical office in TRNSYS. The authors analyzed thermal comfort aspects, maximum allowable heat gains in the room, and the re-cooling of the building mass. They studied a building that was 6 m long, 5 m wide, and 3 m high facing west. This room had pipes of 0.020 m internal diameter embedded in a 0.3 m-thick concrete roof slab and a 0.25 m separation between pipes. It was found that the maximum allowable total heat gains were 39 W m^{-2} with carpet in the room and 32 W m^{-2} with a false floor, with a room temperature between 21 and 24 °C. Furthermore, the authors found that the TABS reduced the energy consumption for cooling by 50% compared to the base case. The second study is a simulation study mentioned in Section 2.1. Chung et al. [34] also estimated the influence of the TABS installed on the roof on the thermal loads of the building prototype. They found that compared to the reference case, the heating load was reduced by 10%, the cooling load was reduced by 36%, and the total energy consumption decreased by 13% due to the TABS. Table 4 summarizes the studies presented in this section, where the authors demonstrated the potential of TABS for reducing energy consumption and reducing heating and cooling loads.

Table 4. Reductions in energy consumption provided by TABS embedded in building roofs.

Reference	Mode *	TABS Features *	Model	Findings
[34]	H, C	F = Water		The heating load was reduced by 10%, the cooling load was reduced 36%, and total energy consumption decreased by 13% with the TABS.
[43]	C	F = Water, DBP = 0.25 m, ϕ = 0.020 m, $\dot{m}$ = 13 kg h^{-1}	-	The TABS reduced the energy consumption by 50% in cooling mode.

* H = heating, C = cooling, F = fluid, DBP = distance between the pipes, ϕ = diameter of the pipes, and $\dot{m}$ = mass flow rate.

3. TABS Embedded in Building Walls

The walls of a building are another type of building envelope component that exchange energy with the outdoor environment because of their significant surface area. Several studies about TABS on building walls have been carried out. TABS embedded in walls is a potential solution to improve their thermal behavior by increasing or decreasing heat losses and saving energy. The aim of this section is to introduce the simulation methods to predict the behavior of TABS, the most suitable configuration, and other techniques to improve the design and construction of TABS.

3.1. Influence of the Flow Characteristics on the Thermal Behavior of TABS Embedded in Walls

Some authors have carried out experimental studies with TABS in walls, where they have varied the fluid parameters such as inlet temperature and fill ratio to analyze the thermal behavior of building walls with TABS. Zhu et al. [44] proposed a two-phase thermosyphon loop (TPTL) incorporated into a thermally activated wall and tested it under winter conditions. The authors carried out experimental tests using a test wall 1 m wide, 0.9 m, high, and 0.2 m thick with embedded tubes of 0.009 m internal diameter, using ethanol as a working fluid. The authors varied the fluid temperature from 25 to 65 °C and the fluid fill ratio from 60 to 144%. The authors found that the fill ratio between the volume of the working fluid and the evaporator volume has a critical impact on the thermal resistance and the starting behavior of the TPTL. They found that the optimal fill ratio is around 116%. A theoretical–experimental study of pipes embedded in a wall for analyzing the influence of pipe depth and spacing on the indoor temperature gradient was carried out by Romaní et al. [45]. They installed an experimental prototype in Spain, which measured 2.85 × 1.85 × 1.95 m. The prototype walls were made of alveolar brick, with 0.016 m-diameter polyethylene tubes embedded 0.036 m from the interior surface and a 0.15 m separation between each pipe. The experimental results obtained by the authors show that the indoor temperature near the east, west, and south walls remains between 25 and 31 °C.

Table 5 summarizes the articles about wall-embedded TABS and the influence that the working fluid has on thermal behavior. As can be seen, TABS does not only use water as the working fluid.

Table 5. Influence of the flow on TABS embedded in walls.

Reference	Mode *	TABS Features *	Model	Findings
[44]	H	F = Ethanol, DPB = 0.20 m, ϕ = 0.00822 m	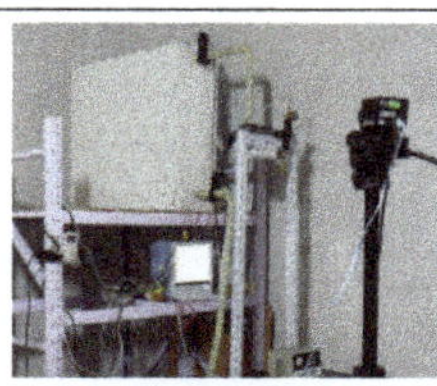	The system exchanges 25.5 W m^{-2} with the internal surface of the wall
[45]	-	F = Water, DPB = 0.05–0.30, ϕ = 0.016 m		The temperature difference between the inner and outer surface of the wall decreases by up to 20 °C

* H = heating, F = fluid, DBP = distance between the pipes, and ϕ = diameter of the pipes.

3.2. Prediction of the Behavior of TABS Embedded in Walls

Some works have analyzed the thermal behavior of building walls with TABS by modeling the system. The authors have analyzed the system using different methods such as resistance–capacitance (RC), the number of transfer units (NTU), and finite difference (FD). Fluid parameters such as inlet temperature, inlet velocity, and mass flow rate were analyzed. Some of these studies have been validated with experimental data, such as that of Todorović et al. [46]. The authors used the analytical expression of Faxen-Rydberg-Huber to determine the thermal characteristics of walls heated by embedded tubes. This expression was experimentally and theoretically verified using three heated wall panels with different structures and geometric characteristics. The panels operated in heating mode, the temperature of the water from feeding the pump was set at 40 °C, and the volumetric flow circulated at 2 L min^{-1}. The authors compared measurements of the average surface temperature of the panels, using a test contact, thermistors, and a thermal imaging camera. The differences between the average temperatures of the panel surfaces were 1.8 to 4.5% when measured using a non-contact and contact method. The authors concluded that the difference between the analytically calculated average temperature and the experimental measurements is 13.7 and 8.6% by contact and non-contact methods, respectively. A model of the frequency-domain finite difference (FDFD) of the thermal behavior of a building wall construction was developed by Xie et al. [47]. The researchers built a test room to validate the model, being 5.6 m long, 3.3 m wide, and 2.8 m high, divided into two test chambers by a 0.31 m-wide wall. The experimental test had embedded polypropylene tubes of 0.02 m diameter placed with a separation of 0.2 m. The authors varied the water inlet temperature to 17.5, 19, and 20 °C, while the water inlet velocity was set at 0.5 m s^{-1}. They found that by supplying the water in the tubes at 17.5 °C, a heat exchange with the wall internal surface of 25.5 W m^{-2} could be obtained. The results show that the finite difference model could predict the behavior of construction with embedded pipes. The relative errors were 6.5% and 4.4% between the measurement and the prediction by the FDFD model for the external surface heat flux and the pipe-embedded building envelope internal pipe surface heat flux, respectively. In other research by the same group, Zhu et al. [48] developed a semi-dynamic thermal model of an active pipe-embedded building. The model consists of construction with embedded pipes in a 3 m-high and 2 m-wide wall . This model was coupled with a resistance and capacitance model (RC) that predicts heat transfer along the width of the structure and a number of transfer units model (NTU) to evaluate heat transfer in the pipes. To assess the behavior of the semi-dynamic model, they developed a CFD model in FORTRAN that functioned as an experimental virtual test for comparison. They tested and verified three case studies where the water inlet temperature was set at 20 °C, varying the water inlet velocity from 0.5 to 0.7 m s^{-1} and the thermophysical properties of the wall; the pipe spacing was 0.02 m. The authors observed that the changes in the heat fluxes taken away by the water are not obvious with different velocities in the water. Meanwhile, an average difference of about 0.5 °C in the outlet temperature of the fluid was found throughout the day. The results demonstrate that the semi-dynamic model predicts the thermal behavior of a TABS with a relative error of 5%. Later, Zhu et al. [49] validated a simplified semi-dynamic model of a chamber with tubes embedded in the envelope. They built two chambers with a controlled environment to perform the validation, one with pipes embedded in the envelope and the other without embedded pipes as a reference. The walls of the chambers were made of alveolar brick, with a layer of cement mortar covering both surfaces of the walls, with polybutylene tubes of 0.020 m in diameter embedded in the layer of cement mortar. The water velocity was varied from 0.8 to 0.5 m s^{-1}, and the water temperature from 18 to 19 °C. The authors concluded that the difference between the model and the experimental validation was minimal. The average relative error to predict the outlet water temperature was less than 0.10 °C, while the heat transferred to the water had a difference of 11%. Other authors that varied the flow rate with a numerical model were Ibrahim et al. [50], who analyzed the behavior of the surface temperatures and the fluid of a chamber with TABS in the walls through which water circulated. To compare the experimental results with the numerical model, they used two chambers: a reference sample, and the other as a test. The experimental chambers measured 2.25 × 1.6 × 1.2 m (length, width, and height), composed of concrete walls with a thickness of 0.12 m with a layer of 0.04 m aerogel plaster. The copper pipes were embedded in the aerogel plaster and placed in a serpentine shape, with a separation between pipes of 0.10 m. The authors used a mixture of 60% water and 40% antifreeze as the working fluid, with a variable volumetric flow rate of 5.53–11.6 L h^{-1} controlled by a pump in a closed circuit. The authors found that the performance of the wall with TABS is affected by the weather, the indoor temperature, the solar absorptivity of the envelope, and the mass flow rate. Qu et al. [51] investigated the relationship between the design and the operating parameters of a thermally activated wall system (TAW) using a mathematical model developed in COMSOL and validated with experimental data from a test chamber. The variables analyzed were the separation between each tube, the area of the thermally activated wall, the flow rate, and the inlet temperature of the water. The authors proposed optimal design graphics for a thermally activated wall system for China's climatic zones. The test chamber measured 2 m × 2 m × 2 m, and was thermally activated on the south wall with embedded tubes, where three separations between tubes (0.01, 0.02, and 0.03 m) were tested. The water flow circulating through the TAW had a velocity of 0.2 m s^{-1}, and a heat pump supplied three different temperatures (15, 17, and 19 °C). The results indicate that the water inlet temperature and the indoor air temperature affected the heat transfer of the TAW. They found that the maximum inner wall surface temperature occurred for a separation between tubes of 0.02 m and a water velocity of 0.2 m s^{-1}, and the maximum and minimum values reached 1.78 °C and 1.80 °C during the cooling and heating mode.

Other studies have analyzed the effect that the arrangement, the separation and the distance between the pipes have on the indoor temperature. Jiang et al. [52] investigated the influence of the velocity and the type of arrangement on the performance considering the changes in the water temperature. They compared two TABS arrangements in a numerical study: a serial pipe-embedded wall (SPW) and a wall with embedded tubes connected in parallel (PPW). The authors found that the inlet water temperature had a more significant effect on the interior temperature than the sol-air temperature. The authors observed that reducing the water temperature below 26 °C in summer and increasing the temperature above 18 °C in winter reduced the cooling and heating thermal loads. Romaní et al. [45] made a numerical model of a radiant wall in 2D, validated with an experimental prototype. The radiant wall was simulated using the finite volume method (FVM). The parametric study showed that the separation and the depth at which the pipes are placed significantly influenced the walls' thermal behavior. The authors obtained better performance when placing the pipes at a depth of 0.045 and 0.065 m and with a separation of 0.0125 and 0.0150 m because the heat fluxes and the temperature inside are minimized.

Table 6 summarizes the articles about models developed for TABS embedded in walls. Among the variables analyzed are the fluid supply velocity, the fluid inlet temperature, and the configuration of the pipes that supply the fluid. As can be seen, TABS are mostly used for heating and use water as the working fluid. Furthermore, TABS have different arrangements and configurations depending on the building construction methods.

Table 6. Models developed to predict the behavior of TABS.

Reference	Mode *	TABS Features *	Model	Simulation Method/Model	Simulation Tool
[45]	-	F = Water, DPB = 0.05–0.30, ϕ = 0.016 m		FVM	-
[46]	H	F = Water, DBP = 0.07 and 0.10 m, ϕ = 0.016 m and 0.0116 m, $\dot{V}$ = 2 L h^{-1},		Faxen-Rydberg-Huber	-

Table 6. *Cont.*

Reference	Mode *	TABS Features *	Model	Simulation Method/Model	Simulation Tool
[47]	H, C	F = Water, DBP = 0.20 m, ϕ = 0.020 m, v = 0.5 m s^{-1}	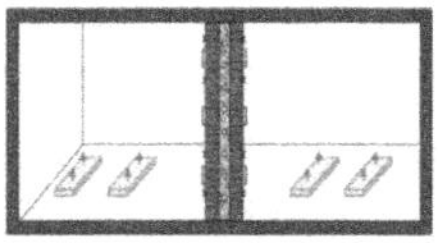	FDFD	FLUENT
[48]	-	F= Water, DBP = 0.20 m, ϕ = 0.02 m, v = 0.5–0.7 m s^{-1}	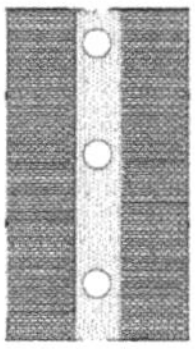	RC-NTU	Program written in FORTRAN Code
[49]	H, C	F = Water, DBP = 0.20 m, ϕ = 0.02 m, v = 0.8–0.5 m s^{-1}		RC-NTU	-
[50]	H	F= Water, antifreeze, DPB = 0.10 m, ϕ = 0.012 m, $\dot{V}$ = 5.3–116 L h^{-1}		-	TRNSYS
[51]	H, C	F = Water, DPB= 0.10, 0.20, and 0.30 m, ϕ = 0.02 m, v = 0.2 m s^{-1}		FEM	COMSOL
[52]	H, C	F = Water, DPB = 0.08 m, ϕ = 0.008 m, v = 2.7 m s^{-1}	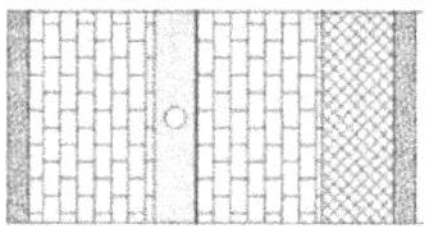	FVM	FLUENT

* H = heating, C = cooling, F = fluid, DBP = distance between the pipes, ϕ = diameter of the pipes, $\dot{V}$ = volumetric flow rate, and v = fluid velocity.

3.3. Heat Losses and Heat Dissipation of Walls Integrated with TABS

A building wall integrated with TABS can reduce the heat losses of buildings in winter. Ibrahim et al. [50] found that heat losses were reduced by between 9% and 35% in the Mediterranean climates when a wall with embedded pipes was used. On the other hand, a building wall integrated with TABS can dissipate heat more effectively than a conventional wall. Li and Zhang [53], analyzed the behavior of a wall implanted with heat pipes (WIHP). The authors compared the WIHP with a conventional wall in the summer months in Tianjin, China. The WIPH wall dimensions were 1.72 m long, 1.72 m wide, and 0.34 m thick, with 24 capillary tubes of 0.002 m in diameter and a length of 0.60 m. The authors concluded that the WIPH system had a greater heat dissipation effect in the summer. Its heat transfer capacity was 50.7 kW m^{-2}, and the average temperature of the WIPH was 2 °C lower than the conventional wall. Other authors, such as Iffa et al. [54], included the use of an insulating material for energy storage and saving. The authors coupled an active insulation system with a TABS embedded in a wall. They analyzed the behavior of the system through simulation and experimental tests. It was found that if both systems are coupled, a flux of up to 81.92 W m^{-2} can be transferred from the wall to the air. Thermal barriers were proposed by Krzaczek et al. [55] to maintain the changes in the internal energy of the walls at a level close to zero. They proposed a thermal barrier model in residential construction, which consisted of a system of tubes embedded in the walls to heat or cool a building, controlled by a fuzzy logic program. The pipes were supplied by water without antifreeze at 25.3 °C for summer and 20.5 °C for winter. The experimental test period was 17 months. They found that the control method through the thermal barrier system was efficient for maintaining a comfortable temperature inside, finding that the temperature variations in the exterior and interior wall of construction were less than 1 °C.

Table 7 summarizes the studies presented in this section. These research works indicate that walls integrated with TABS allow heat gains to the indoor of buildings to be reduced. However, it is important to analyze the most suitable configurations to increase heat dissipation, because a reduction of 2 °C in the case of the study reported by Li and Zhang [53] could be considered low if the complexity of the installation of the TABS on the walls is considered.

Table 7. Heat losses and heat dissipation of TABS embedded in walls.

Reference	Mode *	TABS Features *	Model	Findings
[50]	H	F = Water, antifreeze, DPB = 0.10 m, ϕ = 0.012 m, $\dot{V}$ = 5.3–116 L h^{-1}		With the system proposed the heat losses were reduced from 35% to 9%.
[53]	C	ϕ = 0.0042 m		The system has a heat transfer capacity of 50.7 kW m^{-2}. The temperature of the wall with TABS was 2 °C lower than a conventional wall.
[54]	H	F = Water, DPB = 0.076–0.152 m, ϕ = 0.019 m	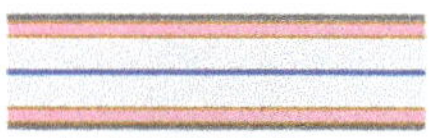	The TABS reduced the temperature of the wall by 10 °C.
[55]	H, C	F = Water, DPB = 0.20 m, ϕ = 0.025 m,		The control method through the thermal barrier system was able to maintain a comfortable temperature inside, with a temperature variation smaller than 1 °C.

* H = heating, C = cooling, F = fluid, DBP = distance between the pipes, ϕ = diameter of the pipes, and $\dot{V}$ = volumetric flow rate.

3.4. TABS Walls and Other Techniques for Energy Saving

The TABS is studied for its capacity to improve buildings' indoor thermal comfort; some authors have proposed integrating new insulating materials and techniques to control the system. Comparing two TABS arrangements in a numerical study, Jiang et al. [52] found that the energy load reduction rate of a serial pipe-embedded wall (SPW) system is higher (25.2%) than that of a wall with embedded tubes connected in parallel (PPW) (8.7%). The influence of the TABS on heating energy consumption in a typical Serbian home was determined by Stojanović et al. [56]. The authors simulated a TABS in EnergyPlus. The TABS was fed with groundwater, where three supply temperatures were used: 10, 14, and 18 °C. The authors concluded that when the TABS is used for heating, energy savings of up to 75% can be obtained with a supply temperature of 18 °C. Furthermore, they emphasized that all renewable sources can be used as an energy source for the TABS when it is used for heating. Guerrero et al. [57] proposed a new prefabricated panel for residential building facades. The authors proposed the integration of phase change materials (PCM) and concrete as structural elements. In this structure, water circulates through heat exchange pipes embedded in mortar cement, made of plastic material with an outer diameter of 0.01 m and a separation of 0.10 m between the pipes. The inlet water temperature and the distance between pipes were varied, from 30 to 45 °C and from 0.08 to 0.12 m, respectively. The authors concluded that the system design depends on the meteorological conditions; if it is designed for winter, a phase change temperature around 24 °C is required. If it is used for summer, the required phase change temperature is around 20 °C. The efficiency was reduced to 6% when the distance increased from 0.08 m to 0.12 m. On the other hand, the efficiency reached approximately 7% with the increasing inlet water temperature of 45 °C. Chen et al. [58] also proposed a thermo-activated PCM composite wall (TAPCW). The TAPCW consisted of placing an intermediate layer with tubes embedded in a macro-encapsulated PCM on the outer side. The authors used a validated numerical model to study the thermal and energy-saving performance of TAPCW under winter weather conditions in northern China. The authors analyzed different values for the spacing between each tube, the thickness of the PCM, and the orientation. The parametric study showed that the separation between pipes has a more significant influence on the system than the thickness of the PCM. They found that a separation between pipes of 0.01 m could be used for the thermal barrier function and a separation between tubes of 0.075 m for the heating function. The researchers also found that the TAPCW oriented to the north was more effective because it had an interior temperature increase of up to 1.8 °C and reduced energy consumption by 65%. Guerrero Delgado et al. [59] characterized and evaluated a panel designed for facades with an integrated TABS. As a first stage, the authors studied the behavior of the TABS through modeling in ANSYS FLUENT operated under different climatic zones. In the second stage, the authors integrated the TABS into a building using a simplified model to evaluate the energy demand and the system energy-saving potential. Guerrero Delgado et al. concluded that the proposed TABS is fully compatible with renewable energies, showing that energy savings of up to 40% for heating can be obtained. Kisilewicz et al. [60] present preliminary results of the thermal behavior of a thermally activated wall coupled with a ground heat exchanger. The authors compared an actively insulated wall against a reference wall under Hungarian climatic conditions. The thermally activated insulated wall construction consisted of a concrete layer on the outside, a layer of extruded polystyrene, tubes embedded in reinforced concrete, and an interior layer of extruded polystyrene. As working fluid in the embedded tubes, in summer, they used refrigerant at a lower temperature than that of the indoor air and a temperature higher than that of outdoor air in winter. The authors concluded that thermally activated insulation significantly improves the exterior wall's insulation parameters because, in the periods analyzed, they obtained a reduction in total energy loss through the external walls from 53 to 81%. To control the water supply temperature in the system, Qu et al. [61] proposed a model for the heat transfer of a TABS in walls under the climatic conditions of Beijing, China. The authors built a test chamber to validate the energy consumption and simulated indoor temperature in EnergyPlus. The test cabin had the following dimensions: 0.8 m long, 0.8 m wide, and 0.8 m high. It had embedded tubes of 0.02 m in diameter and separation between pipes of 0.05 m. The results indicate that pre-cooling a room overnight and reducing the water supply temperature can improve thermal comfort and reduce the unit capacity by over 35%. Kalús et al. [62] proposed the design of a thermally activated precast panel. The authors presented the development of a facade system, through calculations and a parametric study of the system for heating and cooling mode. They discovered that a number of variables, including pipe diameter, distance between the pipes, pipe dimensions, mean heat transfer medium temperature, and the heat storage capacity of building structures, affect the thermal and cooling performance of thermal insulation panels.

Table 8 shows the main works on thermally activated walls. These studies analyzed the behavior of TABS by changing parameters such as fluid velocity, temperature, and the location of the pipes, among others. These changes resulted in energy savings from 40 to 75%. Furthermore, by adding a layer of PCM to the TABS wall system, it is possible to obtain up to 65% energy savings.

Table 8. Energy savings of TABS embedded in walls and other techniques.

Reference	Mode *	TABS Features *	Model	Findings
[52]	H, C	F = Water, DPB = 0.08 m, ϕ = 0.008 m, v = 2.7 m s^{-1}	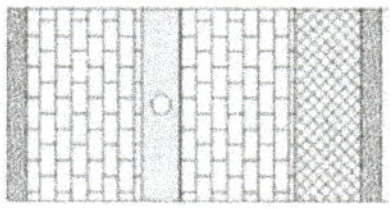	A serial pipe-embedded wall reduced energy load rate by 25.2% while a wall with embedded tubes connected in parallel reduced it by 8.7%.
[56]	H	F = Water	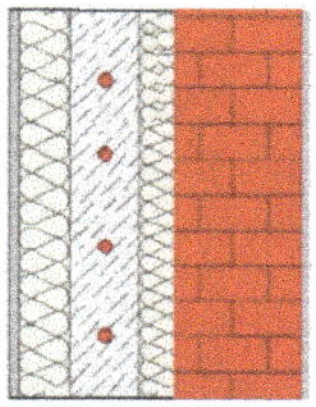	When the TABS is adapted for heating a home, it can provide energy savings of up to 75%.
[57]	H	F = Water, DPB = 0.08, 0.10, 0.12 m, ϕ = 0.01 m, v = 2.7 m s^{-1}		The efficiency of the TABS increases by up to 7% when the inlet temperature increases, and when increasing the distance between the pipes it decreases by up to 6%.
[58]	H	F = Water, DPB = 0.15 m, ϕ = 0.02 m	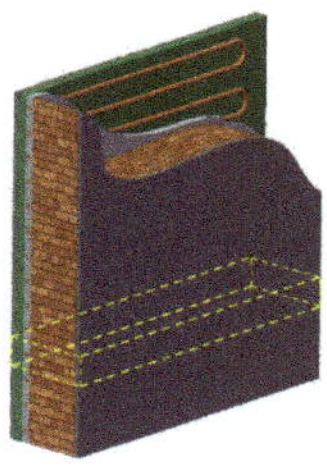	The thermo-activated PCM composite wall increased the indoor temperature and reduced energy consumption by 65%.
[59]	H	F = Water, DPB = 0.10 m, ϕ = 0.01 m, v = 1–2 m s^{-1}	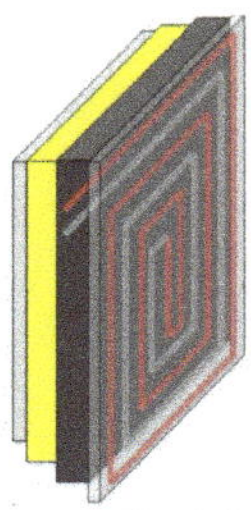	The TABS provided energy savings of up to 40% in heating mode.
[60]	H, C	F = Refrigerant, Water, DPB = 0.2 m, ϕ = 0.02 m		TABS improves the exterior wall's insulation parameters because it causes a reduction in total energy loss from 53 to 81%.

Table 8. *Cont.*

Reference	Mode *	TABS Features *	Model	Findings
[61]	H, C	F = Water, DPB = 0.15 m, ϕ = 0.014 m		The proposed system reduced the discomfort rate by over 35%.
[62]	H, C	-	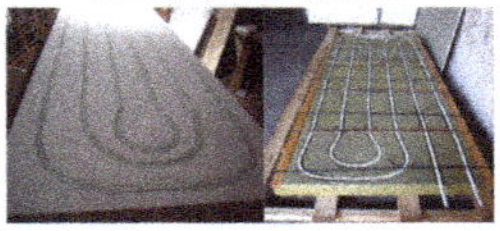	The proposed panel application is most suitable for buildings made with materials with good thermal conductivity and heat accumulation.

* H = heating, C = cooling, F = fluid, DBP = distance between the pipes, ϕ = diameter of the pipes, and v = fluid velocity.

4. TABS Embedded in Floor

This section focuses on TABS installed in the floor, their configurations, and materials used to improve the thermal comfort of the buildings.

4.1. Strategies for Improving the Behavior of TABS Embedded in Floors

Some authors have chosen to analyze floor TABS by simulating their behavior to improve thermal comfort [63]. Joe and Karava [64] developed a model predictive control (MPC) to optimize its behavior, reduce energy consumption and costs, and increase thermal comfort. The authors compared simulated and experimental data from three buildings in heating and cooling mode: (1) with a hydronic radiant floor system, (2) with a wall diffuser, and (3) with a roof diffuser. The authors found that significant energy and cost reductions were achieved compared to a traditional HVAC system. The cost savings were close to 34%, and the energy savings were 16%. Meanwhile, the building with the radiant floor system obtained energy savings of 50% and 29% compared to buildings 2 and 3. In the case of Feng et al. [65], they analyzed the impact of solar radiation on floor cooling in order to find the cooling load of the proposed system. The authors modeled the system in Energy Plus with a total of 864 simulations. The authors found the floor cooling capacity to be 35.6–44.0 W m^{-2} at an operating temperature of 24 °C. Yang et al. [66] analyzed the behavior of a radiant floor heating system embedded in concrete. The authors analyzed the behavior of the system by simulating different scenarios in Modelica. The separation between tubes (100–500 mm), the thickness of the concrete (50–190 mm) and the temperature of the water supply (35–60 °C) were varied. The authors found that by supplying a water temperature of 35 °C, the indoor temperature was kept below the comfort temperature. However, increasing the separation between tubes and supplying a higher temperature increased the thermal comfort. With respect to the thickness of the concrete, the authors found that by increasing the thickness, fewer disturbances were obtained in the indoor air temperatures, but the energy consumption of electricity increased.

Some authors have analyzed the behavior of TABS using different construction materials on the floor. To analyze its thermal behavior and the ability to store heat, Ma et al. [67] proposed a radiant floor with embedded pipes. The authors analyzed the thermal behavior of the radiant concrete panel experimentally and with a simplified model. They compared two concrete blocks with aluminum-plastic (XPAP) embedded pipes, where one block had aluminum fins attached to the bottom surface of the pipe and another block had embedded tubes without fins. Inside the pipes, water was circulated at three different temperatures, 25.0 °C, 29.8 °C and 34.6 °C. The authors found that the radiant floor with aluminum fins reduced the temperature through the concrete block and improved energy storage, increasing exponentially with increasing fin height. The authors concluded that the height and material of the fins integrated into the tubes have a significant effect on the energy storage rate. Other authors that analyzed the effect of varying construction materials on the behavior of TABS were Pardo et al. [68]. The authors developed an RC model of a TABS embedded in the floor and compared two types of materials as a cover, granite and wood. The authors analyzed 216 dwellings, where they varied the location, glazing, insulation, heat capacity and orientation. The authors found that wooden floors can offer a good performance when compared to materials with high thermal conductivity. With wood as floor covering, a 6.4% reduction in energy demand and a 1.4% reduction in comfort hours were obtained.

Table 9 summarizes the studies that analyzed the behavior of TABS in flats. The various authors have not only analyzed the behavior of TABS, but have also varied the construction materials. It can be seen that most of the works were carried out for heating, varying the configuration.

Table 9. Studies that analyzed strategies for improving the behavior of floor TABS.

Reference	Mode *	TABS Features *	Model	Findings
[64]	H, C	F = Water	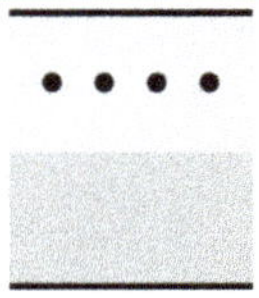	When applying the MPC, the indoor temperature was 22–26 °C for the cooling mode and 17–25 °C for the heating mode. Soil temperature was maintained in a range of 15 to 29 °C.
[65]	C	F = Water		The radiant floor increased its cooling capacity up to 140 W m^{-2}.
[66]	H	F = Water, DBP = 0.1–0.5 m, ϕ = 0.026 m		A tube spacing of 0.3 m maintained a comfort temperature of 21 to 25 °C
[67]	H	F = Water, DBP = 0.15 m, ϕ = 0.02 m	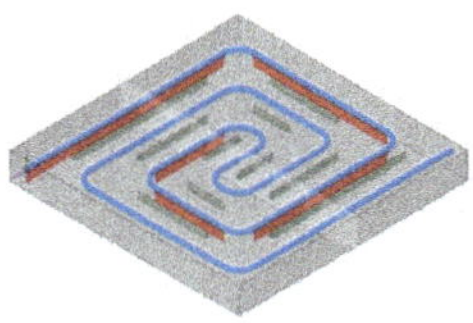	Implementing aluminum fins on the heat exchanger tubes improves the thermal behavior of a floor. Storage capacity increases with fin material embedded in exchanger tubes.
[68]	H	F = Water, DPB = 0.15 m, ϕ = 0.016 m, $\dot{V}$ = 200 L h^{-1}		By using the TABS on the floor, the energy demand decreased by 18% and thermal comfort increased by 14%.

* H = heating, C = cooling, F = fluid, DBP = distance between the pipes, ϕ = diameter of the pipes, v = fluid velocity, and $\dot{V}$ = volumetric flow rate.

4.2. *Combination of TABS with Other Technologies for Floors*

Floor TABS have also been coupled with other systems in order to improve system efficiency and thermal performance. Park et al. [69] conducted a study to estimate the thermal comfort and energy consumption of a TABS combined with a radiant floor heating system (RFHS) and an air conditioning system package (PAC). The authors performed the analysis using simulations from EnergyPlus of a conventional residence construction and a low-thermal-load residential construction, in which they proposed 17 different combinations. The authors found that combining TABS with other systems showed better thermal comfort. However, the configurations TABS-PAC and TABS in cooling mode maintained the indoor comfort conditions. The authors suggested that the TABS should be operated under pre-cooling conditions considering the occupancy and cooling load of the building. Cen et al. [70] experimentally compared the behavior of a radiant floor system with a fain coil. The authors analyzed the influence of the size of the space on thermal comfort, using a variable space height of 3, 5, 7 and 9 m. The authors found that the indoor air temperature is similar using either of the two systems when the height of the room is 5, 7 and 9 m, with a temperature of 20.8 to 25.4 °C. However, they found that the air velocity is lower at all heights when using a radiant floor system (0.03 to 0.04 m s^{-1}) than when using a fan coil (0.09 a 0.12 m s^{-1}). The authors concluded that the thermal comfort is better with a radiant floor system when the size of the space is lower. Zhang et al. [71] combined radiant floor cooling with an underfloor ventilation system (RFCUV). The manuscript focused on developing a dynamic model for radiant floor energy savings compared to the proportional integral derivative model. The water supplied to the radiant floor was 22 °C to 24 °C. The authors found that the system had an energy saving of 17.5%, 8.2% for the air-cooled chiller and 20.5% for the air handling unit. The behavior of a composite hybrid radiant floor was evaluated by Gu et al. [72] in a hot and humid climate. The authors proposed four case studies where they combined the floor coil, fan coil and outdoor air unit in an office divided by zones. The water was supplied to the tubes by means of an air source heat pump in a temperature range of 7 °C to 45 °C. The best case was the one with the combination of the four elements, with a surface temperature of the floor that dropped to 22.8 °C, a relative humidity from 52.1 to 59%, and an average floor heat flux of 60 W m^{-2}. A study of radiant floor heating was conducted by Hwang et al. [73]. The authors performed a simulation of a radiant heating floor assisted by air source heating systems in residential buildings, increasing the floor temperature by 1 °C (20–25 °C). The authors found that there are greater energy savings (59.2%) when only the air heat pump works and the floor temperature is lower (20 °C), whereas when only the radiant floor works, the temperature increases (25°C), thermal comfort decreases, and there is no energy saving. When the radiant floor is assisted by the heat pump, energy savings of 19.6% to 37.6% are obtained. In the case of Sharifi et al. [74], an algorithm was developed to determine the optimal load split of a hybrid TABS. The floor TABS was coupled with a ground source heat pump. They used a design methodology that guaranteed thermal comfort and a reduction in energy use. The authors analyzed the behavior of the system and found that the proposed algorithm reduced the cooling demand by 45% and the heating energy demand by 38%. Authors such as Zhu et al. [75] experimentally analyzed the coupling of a radiant floor and a fan coil cooling (RFCAFC) system in a place with a humid climate. The tubes were made of PE-RT and embedded in a concrete floor with polystyrene insulation and mortar. The system had the ability to automatically change the temperature of the fluid depending on weather conditions. The authors analyzed three different climatic conditions: (1) low temperature and high humidity; (2) high temperature and low humidity; and (3) average temperature and humidity. Zhu et al. found that the fan coil can reach 73.8% of the proportion of the cooling load. The case study with medium temperature and humidity consumed 11.36 kWh^{-1}, 15.4% less energy than the case with low temperature and high humidity. With this coupling the authors were able to maintain the soil surface temperature at 23 °C. Ren et al. [76] carried out an experimental analysis of radiant floor cooling (RFC) with a floor cooling source and displaced ventilation. The authors implemented control strategies with intermittent operation of the system and varying the speed and supply of water flow in the tubes, according to climatic conditions. The authors suggested that there should be a pre-cooling time to reduce the temperature inside the construction. With the proposed control strategies, the surface temperature of the floor decreased to 23.6 °C, with a difference of 4 °C between the radiant temperature and indoor air.

The floor studies presented in this section are summarized in Table 10. The table indicates that floor TABS have also been coupled with other systems in order to save energy and reduce the indoor temperature of buildings.

Table 10. Studies that analyzed the combination of floor TABS with other technologies.

Reference	Mode *	TABS Features *	Model	Findings
[69]	H, C	F = Water, DPB = 0.2 m, ϕ = 0.02 m	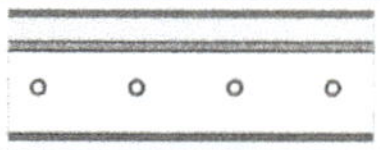	The indoor air temperature remained at 26 °C with a TABS and an air conditioning system.
[70]	C	F = Water		With a height of 5 m in a room, the radiant floor helps to improve thermal comfort, maintaining a temperature of 20.36 °C.
[71]	C	F = Water, DBP = 0.01 m, ϕ = 0.01 m	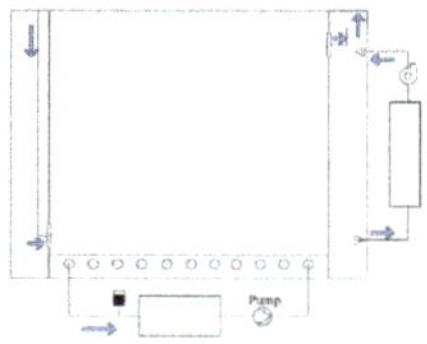	With the model predictive control, there is an energy saving of up to 17.5%, with an operating temperature of up to 24.5 °C.
[72]	C	F = Water, DBP = 0.05 m, ϕ = 0.01 m	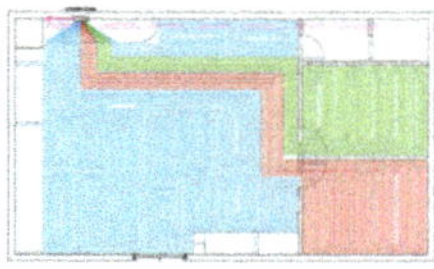	With the coupling of the radiant floor + fan coil + air source heat pump, a comfort temperature of 24.6–26.4 °C can be achieved indoors.
[73]	H	F = Water	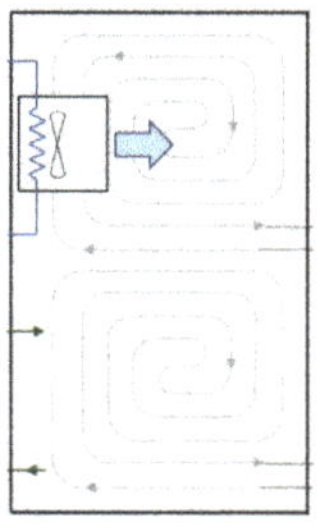	The radiant floor heating has a better performance when the floor temperature is 22 °C, providing an energy saving of 37.6%.
[74]	H, C	F = Water		TABS maintained the indoor air temperature between 21 and 26 °C.
[75]	C	F = Water, DBP = 0.06 m, ϕ = 0.012 m		With the RFCAFC, the indoor air temperature remained in the range of 25.4–26.6 °C.

Table 10. *Cont.*

Reference	Mode *	TABS Features *	Model	Findings
[76]	C	F = Water	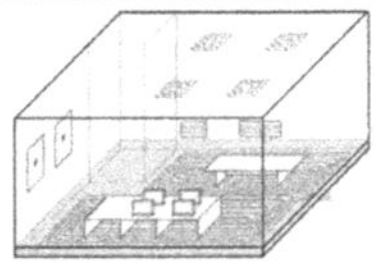	The interior temperature was maintained between 26 °C and 27 °C with the proposed system.

* H = heating, C = cooling, F = fluid, DBP = distance between the pipes, and ϕ = diameter of the pipes.

5. TABS Installed in Several Building Components

In this section are presented the TABS studied to decrease or increase the indoor temperature of buildings when they are installed in more than one building component, and in some cases, TABS was coupled with other technologies. Authors around the world have analyzed different parameters and scenarios with TABS and some have evaluated the utilization of TABS in the whole building envelope.

Indoor Temperature Behavior of Buildings with TABS Installed in Several Building Components

TABS has been analyzed to decrease or increase the temperature of indoor buildings. These systems can be embedded in one or several building envelope components and can be coupled with other technologies. Khan et al. [77] used a TABS embedded in the roof and floor working in cooling mode. The authors performed simulations using MATLAB and EnergyPlus to evaluate the thermal behavior and the energy-saving potential of TABS. The models were calibrated and validated with experimental data. The authors proposed two cases: one with a conventional air-cooling system and one with the proposed TABS. The authors found that the TABS provided up to 30% of energy savings compared to the traditional system. Leo Samuel et al. [78] studied a hybrid passive cooling system, which consisted of a cooling tower connected to a TABS. The system was proposed for five different climatic regions in twelve cities in India. The authors used COMSOL Multiphysics software to perform simulations of the hybrid cooling system. They compared different scenarios, such as floor and roof cooled TABS (RF) and all-surface cooled TABS (AS), in terms of cooling performance for various climatic zones. They concluded that TABS (RF) configuration in arid climates reduced the indoor air temperature up to 9.5 °C and the TABS (AS) configuration up to 14.4 °C. In contrast, in humid tropical climates, the reductions reached up to 4.4 °C and 6.6 °C, respectively. Later, Leo Samuel et al. [79] carried out a study using CFD and analyzed a TABS with embedded pipes in the roof and floor. In these pipes, they circulated water with outlet and return from a cooling tower. To validate the model, they built a prototype of dimensions 3.46 × 3.46 × 3.15 m, with a roof and floor thickness of 0.15 m. The authors found that the TABS maintained an indoor air temperature between 23.5 and 28 °C.

Some parameters were varied by Leo Samuel et al. [80] to analyze the thermal behavior of TABS. The authors numerically and experimentally analyzed the influence of three parameters: spacing, vertical position, and the arrangement of pipes embedded in the roof and floor. They found that by reducing the separation between pipes from 0.3 to 0.1 m and moving the pipes to the direction of the interior surface from 0.135 to 0.015 m reduced the indoor air temperature by between 1.6 and 2.7 °C, respectively. Meanwhile, changing the arrangement of the pipes from coil to parallel reduced the indoor air to 32.1 °C. The authors reached such reductions with a separation of 0.1 m, and a vertical position of 0.015 m, and a parallel arrangement of the pipes reduced the indoor air temperature by up to 6.8 °C, reaching a comfort temperature of 29°C in semi-arid weather. In the same year, Leo Samuel et al. [81] simulated the performance of TABS under a warm weather scenario. The authors used COMSOL Multiphysics to analyze the influence of the temperature and inlet velocity of water and the number of components with TABS for cooling. The CFD model was validated using experimental data from a previous study of the authors. The researchers found that the parameter that had the most significant effect on thermal comfort was the number of cooling surfaces. They showed that if all the room surfaces are cooled, with a flow of 19 L h^{-1} of water, it reduced the average indoor temperature by up to 5.7 °C. The same authors, Leo Samuel et al. [82], performed experimental tests of a scale room with a thermally activated construction system, using water pipes embedded in concrete in the roof, floor, and walls, with separate water flow controls. The experimental prototype measured 3.5 × 3.5 × 3.15 m with a 15 cm-thick reinforced concrete slab, surrounded by trees and structures that provided partial shade. They used ½″ schedule 40 PVC pipes, with a 10 cm separation between pipes. They studied temperature, relative humidity, air speed, and water flow through the pipes. The authors found that if only the cooling is activated on the roof, the indoor temperature remained around 33.1 °C. However, when the cooling is activated on the walls, floor, and roof, the temperature decreases to 29.2 °C. The authors concluded that this system, coupled with a passive ventilation system, increases its feasibility in climates with unfavorable conditions and works with a fluid at relatively high temperatures. To study the internal diameter of the heat exchanger pipes, the thermal conductivity of the pipes, and the thickness of the roof slab and floor, Leo Samuel et al. [83] analyzed the influence of those parameters of TABS on thermal comfort. They used a model built into COMSOL Multiphysics that was validated using experimental data obtained in the authors' previous work. They concluded that increasing the thermal conductivity of the pipes from 0.14 to 1.4 W m^{-1} K^{-1} considerably improves the cooling performance of the system. They found that the best combination of features studied was an internal diameter of the pipe of 0.0017 m, a thermal conductivity of the pipe of 0.14 W m^{-1} K^{-1} and a thickness in the roof and floor of 0.2 m. This combination reduced the indoor operating temperature by 4.7 °C. Michalak [18] carried out measurements and analyzed a TABS implemented in a building used as the primary heating and cooling source. The TABS was coupled with additional heating and cooling units such as fan coils, floor heating and air handling units (AHUs). The measurements were carried out during four months in an office with periods of occupation. The measurements were focused on variables such as indoor air temperature and the temperature of the floor. The authors also calculated predicted percentage of dissatisfied (PPD) and predicted mean vote (PMV). The average soil surface temperature was between 20.6 and 26.2 °C, while the average vertical air temperature was from 22.5 to 23.1 °C, the PMV ranged from 0.52 to 1.50, and less than 30% of the people expressed thermal dissatisfaction. The system analyzed by Michalak had 1275 kWh of exchange energy for cooling and 2500 kWh for heating. The author concluded that implementing a TABS with mechanical ventilation systems improves the thermal comfort conditions of an office. Other authors that conducted an experimental study of a hybrid TABS were Dharmasastha et al. [84]. They carried out an analysis of a TABS coupled with a gypsum roof reinforced with fiberglass (TAGFRG), where they varied the supply temperature of the water that passed through the pipes embedded in the roof, walls, and floors. The authors found that by decreasing the supply water temperature from 26 °C to 18 °C, the interior surface temperature of the roof can decrease by as much as 5.8 °C. However, the authors found that cooling just the roof of the building only lessened the effects that the outside ambient temperature had on indoor air temperature fluctuations. On the other hand, they found that if water was recirculated throughout the whole envelope of the test chamber, the thermal comfort increased, with a satisfaction percentage of 90%. Montenegro and Hongn [85] carried out a parametric study of TABS using a numerical model. The authors used experimental data from previous works to validate the model and subsequently compared the thermal behavior of two horizontal TABS configurations: floor and roof. The authors varied the separation between pipes (from 0.1 to 0.3 m), the volumetric flow, and the supply water temperature, as well as the distance between the pipes and the surface in contact with the interior environment of an enclosure. The authors concluded that the variables with the greatest influence on the thermal behavior of the TABS design are the separation between pipes and the water supply temperature, considered as the key parameters for increasing heat transfer between the construction element and the indoor environment. They proposed to reduce the separation between the tubes and the depth where they are installed, since this maximizes the removal of heat from the room to be cooled. The authors concluded that the potential for heat removal from a roof with TABS is 20–30% greater than the TABS on the floor. Oravec et al. [86] compared six radiant heating systems to make a guide that allows choosing a system according to its application. The authors compared six heating systems with PE-Xa pipes with different diameters, embedded in the floor, in the floor with metal fins, and in the wall (TABS or air gap). The authors analyzed the thermal performance, necessary heating area, thermal storage, and construction costs and the application of TABS in retrofitted buildings. They demonstrated that the behavior of TABS depends on the location of the tubes. The best performance was obtained by the Wall system (TABS) with a heating flux of 96 W m^{-2}. The authors suggest that floor heating shows an acceptable thermal performance, controllability, and storage capacity.

Table 11 summarizes the works that analyzed the installation of TABS in different building envelope components at the same time. In addition, TABS with an insulation system in the roof and the influence on the thermal comfort of the occupants were studied.

Table 11. Improvements in thermal comfort when TABS are used in several building components.

Reference	Weather	Mode *	TABS Features *	Model	Findings
[77]	-	C	F = Water, DPB = 0.10 and 0.15 m, ϕ = 0.015 m	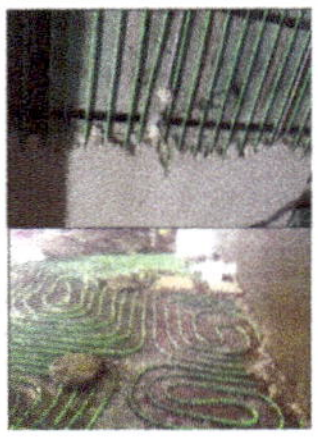	The radiant cooling systems provided energy savings of up to 30% compared to a traditional system.
[78]	Semi-arid, arid, humid subtropical, tropical wet and dry, tropical wet	C	F = Water	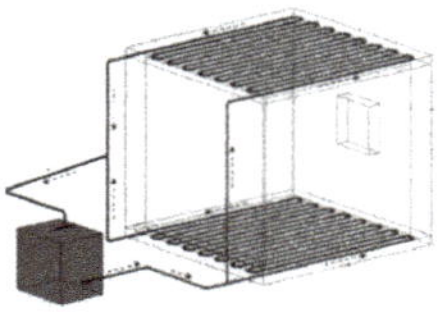	The TABS in the roof reduced the operative temperature by 9.5 °C, while the TABS in all surfaces reduced it by 14.4 °C.
[79]	Hot semi-arid	C	F = Water, DPB = 0.20 m, ϕ = 0.024 m	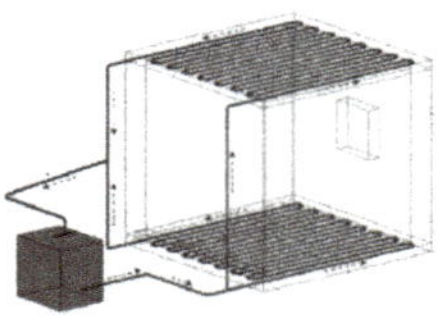	The system maintained indoor air temperature between 23.5 °C and 28 °C.
[80]	Hot semi-arid	C	F = Water, DPB = 0.02 m, ϕ = 0.00128 m		The indoor air temperature was reduced by 1.6 °C when the separation between pipes was increased; 2.7 °C by moving the pipes to the interior surface direction; and 32.1 °C by changing the arrangement of the tubes from coil to parallel.

Table 11. *Cont.*

Reference	Weather	Mode *	TABS Features *	Model	Findings
[81]	Hot and dry summer	C	F = Water, DPB = 0.2 m, ϕ = 0.013 m		The number of cooling surfaces was the parameter that had the most significant effect on thermal comfort. If all the room surfaces are cooled, the average indoor temperature is reduced by up to 5.7 °C.
[82]	Tropical wet, Dry climate	C	F = Water, DPB = 0.2 m, ϕ = 0.016 m	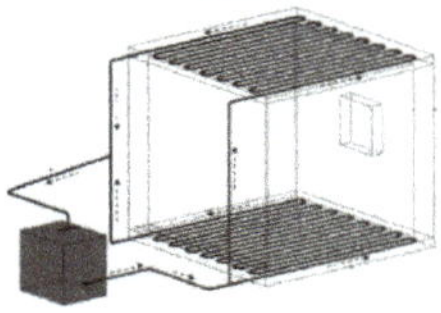	When TABS cooling was activated only on the roof, the indoor temperature remained at 33.1 °C. Meanwhile, when the TABS cooling was activated on the entire envelope, the temperature decreased to 29.2 °C.
[83]	-	C	F = Water, DPB = 0.2 m, ϕ = 0.01 m, v = 0.4 m s^{-1}	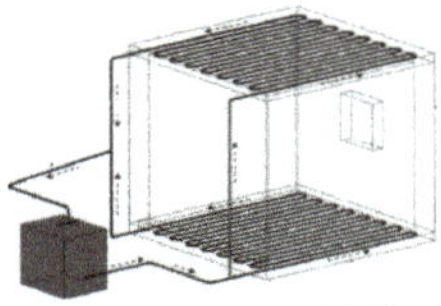	Increasing the thermal conductivity of the pipes improves the system's cooling performance. The system can reduced the indoor operating temperature by 4.7 °C.
[18]	-	H, C	F = Water	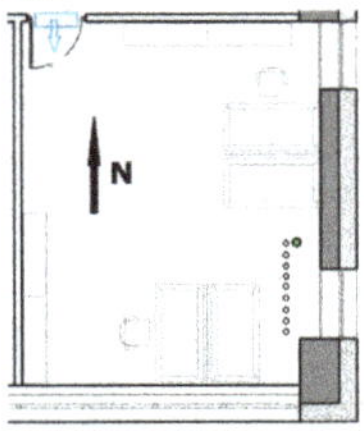	Implementing a TABS with mechanical ventilation systems improves the thermal comfort conditions in an enclosure.
[84]	Warm and humid	C	F = Water, DBP = 0.054 m, ϕ = 0.01 and 0.015 m		The TABS reduced the indoor air temperature by 2.1 °C when the temperature of the cooling water was reduced from 26 °C to 18 °C.
[85]	-	C	F = Water, $\dot{V}$ = 8 L min^{-1}	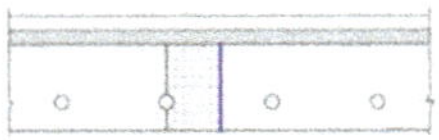	Heat removal in an enclosure increases when tube spacing and tube depth are decreased. The potential of a roof is higher (20–30%) compared to a floor TABS, with the same characteristics.
[86]	-	H	F = Water, DBP = 0.8–0.30 m, ϕ = 0.009–0.020 m	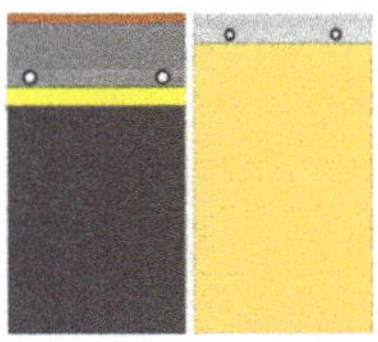	The thermal performance depends on the location of the tubes with respect to the indoor environment.

* H = heating, C = cooling, F = fluid, DBP = distance between the pipes, ϕ = diameter of the pipes, v = fluid velocity, and $\dot{V}$ = volumetric flow rate.

6. Results and Discussion

The objective of this study was to review the state of the art of TABS. In this study, the thermal behavior of TABS in roofs, walls, and floors was analyzed. TABS can be implemented both in a building component and in the entire envelope, helping to maximize its efficiency. In the review of the literature, it was found that the TABS can be named differently on some occasions depending on their location in the envelope and on their mode of operation: thermally activated building constructions, radiant cooling/heating systems, and active building storage systems, among others. The development of this study helped us to determine the main variables that were studied by the authors and conceptualize it as a summary in tables. Most of the reported works analyzed the behavior of the indoor ambient temperature in order to reach thermal comfort. Other aspects analyzed by the authors were the effect of changing the characteristics of the fluid on the indoor ambient temperature, the energy-saving capacity and the capacity of the cooling/heating load. The results of the manuscripts analyzed in this work are presented below.

Regarding the improvements in thermal comfort provided by TABS when installed in building roofs, the results are reported in terms of the reductions in indoor air temperature (6.7 °C) [35], the range in which the indoor air temperature remains (21–28 °C) [31,33] and the percentage of time in which the indoor temperature is the satisfaction zone (within 80–90%) [32]. On the other hand, the energy savings provided by TABS when embedded in building roofs were reported in a few works [34,43]. It was shown that TABS can provide energy savings between 13 and 50%.

The research on TABS embedded in building walls has shown that this technology can provide energy savings for heating by a factor ranging between 40 and 75% [56,58,59]. Several studies developed theoretical models validated with experimental results. These models were used to find the adequate values for pipe separation and pipe depth within the walls [45,48,50,51], water inlet temperature for cooling or heating [45,47], and water velocity and volumetric flow rate [47,48,50]. Modeling studies are relevant for the design of TABS because they allow researchers to find suitable values for the parameters mentioned above. Studies on TABS embedded in floors reveal that they are mostly made with tubes with diameters smaller than 0.012 m. The values reported in floor TABS in terms of indoor ambient temperature range from 21 °C to 27 °C [66,69,74,75]. The energy saving in floor TABS can reach 17.5%.

Other studies show that when TABS are installed in more than one building envelope component, they provide an essential contribution to the improvement in thermal comfort. The results are reported in terms of the indoor air temperature reductions or in terms of the interval in which the indoor air temperature remains. When the roof and floor had embedded TABS, and were used for cooling, it was shown that the indoor air temperature was reduced between 4.4 and 9.5 °C. When all the building envelope components (roofs, walls, and floor) have embedded TABS and are used for cooling, the indoor air temperature reductions range between 6.6 and 14.4 °C depending on the type of weather of the zone [78]. Other research shows that when TABS was activated in the whole envelope, it maintained the indoor air temperature at around 29 °C. When only the roof was activated, the indoor air temperature remained about 33 °C [82]. Some researchers showed that when TABS are installed in the building roof and the floor, the indoor air temperature is maintained between 23.5 and 28 °C [79].

Figure 3 classifies the research works considered in the current review according to the results presented by each work. Four main groups were formed: (1) research works that studied the influence of TABS on the thermal comfort conditions; (2) research works that studied TABS for heating; (3) research works that studied TABS for cooling; and (4) research works that studied TABS for heating and cooling. Regarding the first group, most of the existing studies were developed for buildings with TABS embedded in floor. Few studies on thermal comfort were developed for TABS embedded in the whole envelope. Regarding the second group, most of the existing studies for heating were developed for TABS embedded in walls; only few studies were developed for roofs. Regarding the third group, most of the studies on TABS were developed for roofs and the roof–floor, and a few studies for TABS embedded in walls. Finally, most studies were developed for TABS embedded in walls in the fourth group, and few were developed on the roof–floor.

The studies analyzed indicated that most TABS were developed for TABS embedded in roofs and walls. Embedded TABS have been combined with phase change materials (PCM) [40,57,58], with a reduction in the indoor temperature from 0.4 to 4.7 °C. TABS are mostly applied to cooling and are embedded in roofs, with an indoor temperature from 21 to 29.6 °C. Meanwhile, TABS embedded in walls are developed for heating and cooling/heating depending on the outdoor environment.

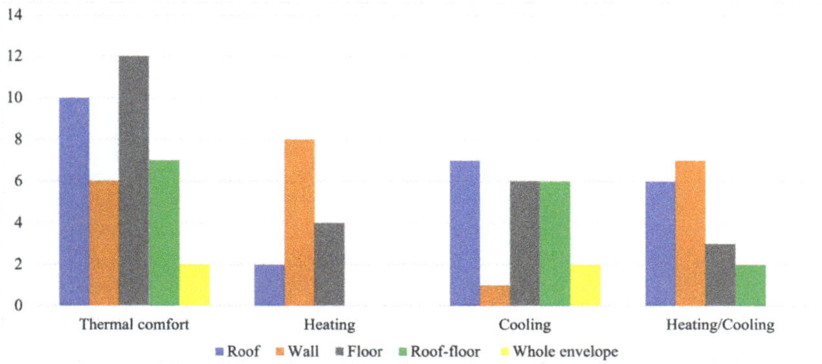

Figure 3. Studies developed for TABS in different building components.

From the analysis of the available literature, it is possible to identify alternatives that can contribute to thermal comfort in buildings. TABS is one of these alternatives that offers many benefits, but it has some limitations. Among the benefits that the authors report is the integration of TABS with systems that use renewable energy in the heating mode, and the recirculation of the working fluid of TABS for the needs of the users of the same building. Its energy storage capacity has been reported as a benefit, which can be increased by integrating a layer of PCM [20,26,57]. As part of the limitations of the TABS are the control strategies, because when there are several sections, it is necessary to activate them independently. According to some authors, it was found that appropriate control and operation strategies are required to reduce energy waste when changing the mode from cooling to heating (or vice versa) [32]. Furthermore, it has been reported that TABS cannot remove latent heat loads, which occur when the temperature of the building's interior environment drops below the dew point. This causes condensation to occur inside the building, which affects its hygrothermal behavior. To contribute to this, some authors propose the integration of systems with the ability to dehumidify the interior environment of the building and prevent the accumulation of condensate, especially in places with a humid climate [35]. On the other hand, another reported limitation of TABS is the complexity and costs of its implementation compared with the use of [20] air conditioning systems.

7. Conclusions

This study presents a review of the state of the art of TABS, where its different configurations and its implementation in the different building components (roof, wall, and floor) or the whole envelope were analyzed. Furthermore, their coupling with other systems was analyzed. Relevant results from the literature related to the thermal behavior and the critical parameters of these systems were discussed. TABS are becoming an attractive branch of research for those that analyze measures for improving the indoor environment of buildings. Several gaps were identified in the literature, and the following can be concluded:

- TABS have not been analyzed from a structural mechanics point of view. From the knowledge of the authors, there are not yet studies that have considered the effect of the embedded pipes on the mechanical behavior of building components such as roofs and walls. This fact is crucial in roofs because of its role in a building; researchers should find the maximum diameter of the pipes and the optimal separation between them that does not affect the structural behavior of the roof.

- The thermal behavior of building components with TABS depends on many parameters; some of these parameters are: (a) type of building component, (b) orientation of the building wall; (b) type of arrangement of the pipes; (c) separation between the pipes or pipe spacing; (d) diameter of the pipes; (e) material of the pipes; (f) thermophysical properties of the fluid that circulates within the pipes; and (g) volumetric or mass flow rate of the fluid. Thus, optimization methods such as genetic algorithms or other artificial intelligence techniques should be used to find the optimum value for the parameters involved in a good design of TABS embedded in building components.
- Regarding the type of arrangement of the pipes, TABS in series or in a serpentine-type arrangement have been extensively studied. However, other types of pipe arrangement, such as parallel, mixed or even tree-shaped, should be explored to find the best arrangement that benefits the thermal performance of TABS for each application.
- The effect of fins on the thermal performance of TABS embedded in building components needs further development. Few studies have analyzed this measure when TABS are installed on building floors; the results show that the system with fins improves the thermal storage capacity compared with the traditional system.
- A building component with embedded TABS designed for heating is expected to have a material in the exterior layer with a high solar absorptance. On the contrary, a component with embedded TABS designed for cooling is expected to have a material in the exterior layer with a low solar absorptance. However, when the building component performs both heating and cooling, a layer with a suitable value of solar absorptance should be selected. Future studies on finding the optimal value of solar absorptance should be performed to improve the efficiency of TABS.
- The coupling of TABS with other systems that contribute to improve the thermal behavior of a building, such as green roofs and walls, and ventilated roofs and walls, has not been explored. These passive techniques could help to improve the thermal behavior of TABS.

In accordance with that mentioned above, it can be said that TABS are systems with limitations and opportunities. Within the main limitations are the costs of installation and implementation. However, any new development that changes the paradigm of how it is built in the traditional way has implications that are reflected in the cost of installation, operation, and maintenance. However, the opportunities offered by TABS, according to the studies analyzed, can be said to far outweigh the limitations. This is due to the fact that TABS present the versatility to adapt to different constructions, climates, and types of materials, among others. Therefore, based on what was analyzed in this study, it can be said that TABS contribute to lowering the temperature inside a building, which is reflected in the reduction of up to 50% in energy consumption due to the use of of air conditioning systems. Therefore, the trend of the use of TABS is expected to increase as a strategy to contribute to the reduction in thermal loads in buildings. However, experimental studies are required under real conditions of use and structural criteria must be taken into account in order to implement TABS as a strategy that not only offers benefits from a thermal point of view, but also offers safety for building occupants.

Author Contributions: Conceptualization, I.H.-P. and M.M.V.-R.; writing—original draft preparation, M.M.V.-R., I.H.-P., K.M.A.-C., I.Z.-G., E.V.M.-M., I.H.-L. and J.S.-A.; writing—review and editing, I.H.-P. and M.M.V.-R.; supervision, I.H.-P. and K.M.A.-C. All authors have read and agreed to the published version of the manuscript.

Funding: This research received no external funding.

Institutional Review Board Statement: Not applicable.

Informed Consent Statement: Not applicable.

Data Availability Statement: Not applicable.

Acknowledgments: The first author acknowledges the grant provided by the National Council of Science and Technology (CONACYT-México).

Conflicts of Interest: The authors declare no conflict of interest.

Nomenclature

C	Cooling
CO_2	Carbon dioxide
DPB	Distance between pipes (m)
F	Fluid
H	Heating
$\dot{m}$	Mass flow rate (kg s^{-1})
$\dot{V}$	Volumetric flow rate (L h^{-1})
v	Fluid velocity (m s^{-1})
Abbreviations	
AHUs	Air handling units
AS	All-surface cooled TABS
Bio-PCM	Bio-based phase change material
CFD	Computational fluid dynamics
CLPHP	Closed-loop pulsating heat pipe
FD	Finite difference
FDFD	Frequency-domain finite difference
FEA	Finite element analysis
FVM	Finite volume method
GSHP	Ground source heat pump
HVAC	Heating, ventilation and air conditioning
IPCC	Intergovernmental Panel on Climate Change
MPC	Model predictive control
NTU	Numbers of transfer units method
PAC	Air conditioning system package
PCM	Phase change material
PE-RT	Polyethylene of raised temperature
PE-Xa	Cross-linked polyethylene
PMV	Predicted mean vote
PPD	Predicted percentage of dissatisfied
ppm	Parts per million
PPW	Parallel pipe-embedded wall
RC	Resistance–capacitance method
RF	Floor and roof cooled TABS
RFC	Radiant floor cooling
RFCAFC	Radiant floor and fan coil cooling
RFCUV	Radiant floor cooling with underfloor ventilation system
RFHS	Radiant floor heating system
SPW	Serial pipe-embedded wall
TABS	Thermally activated building system
TAGFRG	Gypsum roof reinforced with fiberglass
TAPCW	Thermo-activated PCM composite wall
TAW	Thermally activated wall system
TPTL	Two-phase thermosyphon loop
VRSP	Ventilated roof model with embedded pipes and a stabilized layer of PCM
WIPH	Wall implanted with heat pipes
XPAP	Aluminum–plastic pipe
Greek Symbols	
ϕ	Diameter of the pipes (m)

References

1. IPCC. *Climate Change 2014: Synthesis Report. Contribution of Working Groups I, II and III to the Fifth Assessment Report of the Intergovernmental Panel on Climate Change*; IPCC: Geneva, Switzerland, 2014; 151p. Available online: https://www.ipcc.ch/report/ar5/syr/ (accessed on 30 July 2022).
2. IEA. *Global Status Report for Buildings and Construction: Towards a Zero-Emissions, Efficient and Resilient Buildings and Construction Sector*; IEA: Paris, France, 2019. Available online: https://www.iea.org/reports/global-status-report-for-buildings-and-construction-2019 (accessed on 30 July 2022).
3. Tong, S.; Li, H. An efficient model development and experimental study for the heat transfer in naturally ventilated inclined roofs. *Build. Environ.* **2014**, *81*, 296–308. [CrossRef]
4. Ozarisoy, B. Energy effectiveness of passive cooling design strategies to reduce the impact of long-term heatwaves on occupants' thermal comfort in Europe: Climate change and mitigation. *J. Clean. Prod.* **2022**, *330*, 129675. [CrossRef]
5. Haghighi, A.; Maerefat, M. Design guideline for application of earth-to-air heat exchanger coupled with solar chimney as a natural heating system. *Int. J. Low-Carbon Technol.* **2015**, *10*, 294–304. [CrossRef]
6. Peñaloza Peña, S.A.; Jaramillo Ibarra, J.E. Potential Applicability of Earth to Air Heat Exchanger for Cooling in a Colombian Tropical Weather. *Buildings* **2021**, *11*, 219. [CrossRef]
7. Michalak, P. Impact of Air Density Variation on a Simulated Earth-to-Air Heat Exchanger's Performance. *Energies* **2022**, *15*, 3215. [CrossRef]
8. Kharrufa, S.N.; Adil, Y. Upgrading the building envelope to reduce cooling loads. *Energy Build.* **2012**, *55*, 389–396. [CrossRef]
9. Bunkholt, N.S.; Säwén, T.; Stockhaus, M.; Kvande, T.; Gullbrekken, L.; Wahlgren, P.; Lohne, J. Experimental Study of Thermal Buoyancy in the Cavity of Ventilated Roofs. *Buildings* **2020**, *10*, 8. [CrossRef]
10. Lee, H.; Ozaki, A.; Choi, Y.; Iqbal, M. Performance Improvement Plan of Air Circulation-Type Solar Heat-Storage System Using Ventilated Cavity of Roof. *Energies* **2021**, *14*, 1606. [CrossRef]
11. Silva, C.M.; Gomes, M.G.; Silva, M. Green roofs energy performance in Mediterranean climate. *Energy Build.* **2016**, *116*, 318–325. [CrossRef]
12. Zingre, K.T.; Kumar Devs, K.; Wan, M.P. Analysing the Effect of Substrate Properties on Building Envelope Thermal Performance in Various Climates. *Energies* **2020**, *13*, 5119. [CrossRef]
13. Barozzi, B.; Pollastro, M.C. Assessment of the Impact of Cool Roofs in Temperate Climates through a Comparative Experimental Campaign in Outdoor Test Cells. *Buildings* **2016**, *6*, 52. [CrossRef]
14. Ozarisoy, B.; Altan, H. A novel methodological framework for the optimisation of post-war social housing developments in the South-eastern Mediterranean climate: Policy design and life-cycle cost impact analysis of retrofitting strategies. *Sol. Energy* **2021**, *225*, 517–560. [CrossRef]
15. Suárez, R.; Escandón, R.; López-Pérez, R.; León-Rodríguez, L.; Klein, T.; Silvester, S. Impact of Climate Change: Environmental Assessment of Passive Solutions in a Single-Family Home in Southern Spain. *Sustainability* **2018**, *10*, 2914. [CrossRef]
16. Bugenings, L.A.; Kamari, A. Bioclimatic Architecture Strategies in Denmark: A Review of Current and Future Directions. *Buildings* **2022**, *12*, 224. [CrossRef]
17. Romaní Picas, J.; Pérez Luque, G.; Gracia Cuesta, A.D. Experimental evaluation of a heating radiant wall coupled to a ground source heat pump. *Renew. Energy* **2017**, *105*, 520–529. [CrossRef]
18. Michalak, P. Selected Aspects of Indoor Climate in a Passive Office Building with a Thermally Activated Building System: A Case Study from Poland. *Energies* **2021**, *14*, 860. [CrossRef]
19. Almeida, R.M.S.F.; Vicente, R.d.S.; Ventura-Gouveia, A.; Figueiredo, A.; Rebelo, F.; Roque, E.; Ferreira, V.M. Experimental and Numerical Simulation of a Radiant Floor System: The Impact of Different Screed Mortars and Floor Finishings. *Materials* **2022**, *15*, 1015. [CrossRef]
20. Romaní Picas, J.; Cabeza, L.F.; Pérez Luque, G.; Pisello, A.L.; Gracia Cuesta, A.D. Experimental testing of cooling internal loads with a radiant wall. *Renew. Energy* **2018**, *116*, 1–8. [CrossRef]
21. Resources, E.D. *Radiant Cooling, Energy Design Resources: Building Envelope Design*; Financial Times Energy, Inc.: Boulder, CO, USA, 2003.
22. Ma, P.; Wang, L.S.; Guo, N. Modeling of TABS-based thermally manageable buildings in Simulink. *Appl. Energy* **2013**, *104*, 791–800. [CrossRef]
23. Rhee, K.N.; Kim, K.W. A 50 year review of basic and applied research in radiant heating and cooling systems for the built environment. *Build. Environ.* **2015**, *91*, 166–190. [CrossRef]
24. Romaní Picas, J.; Gracia Cuesta, A.D.; Cabeza, L.F. Simulation and control of thermally activated building systems (TABS). *Energy Build.* **2016**, *127*, 22–42. [CrossRef]
25. Ma, P.; Wang, L.S.; Guo, N. Energy storage and heat extraction—From thermally activated building systems (TABS) to thermally homeostatic buildings. *Renew. Sustain. Energy Rev.* **2015**, *45*, 677–685. [CrossRef]
26. Krajčík, M.; Šikula, O. The possibilities and limitations of using radiant wall cooling in new and retrofitted existing buildings. *Appl. Therm. Eng.* **2020**, *164*, 114490. [CrossRef]
27. Krajčík, M.; Arıcı, M.; Šikula, O.; Šimko, M. Review of water-based wall systems: Heating, cooling, and thermal barriers. *Energy Build.* **2021**, *253*, 111476. [CrossRef]
28. Cai, R.; Sun, Z.; Yu, H.; Meng, E.; Wang, J.; Dai, M. Review on optimization of phase change parameters in phase change material building envelopes. *J. Build. Eng.* **2021**, *35*, 101979. [CrossRef]

29. Dong, W.; Chen, Y.; Bao, Y.; Fang, A. A validation of dynamic hygrothermal model with coupled heat and moisture transfer in porous building materials and envelopes. *J. Build. Eng.* **2020**, *32*, 101484. [CrossRef]
30. Wei, M.; Wang, B.; Liu, S. Numerical Simulation of Heat and Moisture Transfer of Wall with Insulation. *J. Phys. Conf. Ser.* **2019**, *1300*, 012029. [CrossRef]
31. Gwerder, M.; Lehmann, B.; Tödtli, J.; Dorer, V.; Renggli, F. Control of thermally-activated building systems (TABS). *Appl. Energy* **2008**, *85*, 565–581. [CrossRef]
32. De Wit, A.; Wisse, C. Hydronic circuit topologies for thermally activated building systems-design questions and case study. *Energy Build.* **2012**, *52*, 56–67. [CrossRef]
33. Rey Martínez, F.J.; Chicote, M.A.; Peñalver, A.V.; Gónzalez, A.T.; Gómez, E.V. Indoor air quality and thermal comfort evaluation in a Spanish modern low-energy office with thermally activated building systems. *Sci. Technol. Built Environ.* **2015**, *21*, 1091–1099. [CrossRef]
34. Chung, W.J.; Park, S.H.; Yeo, M.S.; Kim, K.W. Control of thermally activated building system considering zone load characteristics. *Sustainability* **2017**, *9*, 586. [CrossRef]
35. Dharmasastha, K.; Samuel, D.L.; Nagendra, S.S.; Maiya, M. Experimental investigation of thermally activated glass fibre reinforced gypsum roof. *Energy Build.* **2020**, *228*, 110424. [CrossRef]
36. Saw, L.; Yew, M.; Yew, M.; Chong, W.; Poon, H.; Liew, W.; Yeo, W. Development of the closed loop pulsating heat pipe cool roof system for residential buildings. *Case Stud. Therm. Eng.* **2021**, *28*, 101487. [CrossRef]
37. Wu, M.; Liu, X.; Tang, H. Simulation Analysis on the Solar Heating System Combined with Tabs in Lhasa, China of Annex 59. *Energy Proc.* **2015**, *78*, 2439–2444. [CrossRef]
38. Chung, W.J.; Park, S.H. Utilization of Thermally Activated Building System with Horizontal Ground Heat Exchanger Considering the Weather Conditions. *Energies* **2021**, *14*, 7927. [CrossRef]
39. Yu, J.; Yang, Q.; Ye, H.; Huang, J.; Liu, Y.; Tao, J. The optimum phase transition temperature for building roof with outer layer PCM in different climate regions of China. *Energy Procedia* **2019**, *158*, 3045–3051. [CrossRef]
40. Yu, J.; Leng, K.; Ye, H.; Xu, X.; Luo, Y.; Wang, J.; Yang, X.; Yang, Q.; Gang, W. Study on thermal insulation characteristics and optimized design of pipe-embedded ventilation roof with outer-layer shape-stabilized PCM in different climate zones. *Renew. Energy* **2020**, *147*, 1609–1622. [CrossRef]
41. Heidenthaler, D.; Leeb, M.; Schnabel, T.; Huber, H. Comparative analysis of thermally activated building systems in wooden and concrete structures regarding functionality and energy storage on a simulation-based approach. *Energy* **2021**, *233*, 121138. [CrossRef]
42. Hu, R.; Li, X.; Liang, J.; Wang, H.; Liu, G. Field study on cooling performance of a heat recovery ground source heat pump system coupled with thermally activated building systems (TABS). *Energy Convers. Manag.* **2022**, *262*, 115678. [CrossRef]
43. Lehmann, B.; Dorer, V.; Koschenz, M. Application range of thermally activated building systems tabs. *Energy Build.* **2007**, *39*, 593–598. [CrossRef]
44. Zhu, L.; Yang, Y.; Chen, S.; Sun, Y. Thermal performances study on a façade-built-in two-phase thermosyphon loop for passive thermo-activated building system. *Energy Convers. Manag.* **2019**, *199*, 112059. [CrossRef]
45. Romaní, J.; Cabeza, L.F.; de Gracía, A. Development and experimental validation of a transient 2D numeric model for radiant walls. *Renew. Energy* **2018**, *115*, 859–870. [CrossRef]
46. Todorović, R.I.; Banjac, M.J.; Vasiljević, B.M. Analytical and experimental determination of the temperature field on the surface of wall heating panels. *Therm. Sci.* **2015**, *19*, 497–507. [CrossRef]
47. Xie, J.; Xu, X.; Li, A.; Zhu, Q. Experimental validation of frequency-domain finite-difference model of active pipe-embedded building envelope in time domain by using Fourier series analysis. *Energy Build.* **2015**, *99*, 177–188. [CrossRef]
48. Zhu, Q.; Xu, X.; Gao, J.; Xiao, F. A semi-Dynamic simplified therm model of active pipe embedded building envelope based on frequency finite difference method. *Int. J. Therm. Sci.* **2015**, *88*, 170–179. [CrossRef]
49. Zhu, Q.; Li, A.; Xie, J.; Li, W.; Xu, X. Experimental validation of a semi-dynamic simplified model of active pipe-embedded building envelope. *Int. J. Therm. Sci.* **2016**, *108*, 70–80. [CrossRef]
50. Ibrahim, M.; Wurtz, E.; Anger, J.; Ibrahim, O. Experimental and numerical study on a novel low temperature façade solar thermal collector to decrease the heating demands: A south–north pipe-embedded closed-water-loop system. *Sol. Energy* **2017**, *147*, 22–36. [CrossRef]
51. Qu, S.; Hu, W.; Yuan, S.; Yin, R.; Ji, R. Optimal design and operation of thermally activated wall in the ultra-low energy buildings in China. *Buil. Simul.* **2020**, *13*, 961–975. [CrossRef]
52. Jiang, S.; Li, X.; Lyu, W.; Wang, B.; Shi, W. Numerical investigation of the energy efficiency of a serial pipe-embedded external wall system considering water temperature changes in the pipeline. *J. Build. Eng.* **2020**, *31*, 101435. [CrossRef]
53. Li, Z.; Zhang, Z. Dynamic heat transfer characteristics of wall implanted with heat pipes in summer. *Energy Build.* **2018**, *170*, 40–46. [CrossRef]
54. Iffa, E.; Hun, D.; Salonvaara, M.; Shrestha, S.; Lapsa, M. Performance evaluation of a dynamic wall integrated with active insulation and thermal energy storage systems. *J. Energy Storage* **2022**, *46*, 103815. [CrossRef]
55. Krzaczek, M.; Florczuk, J.; Tejchman, J. Improved energy management technique in pipe-embedded wall heating/cooling system in residential buildings. *Appl. Energy* **2019**, *254*, 113711. [CrossRef]

56. Stojanović, B.V.; Janevski, J.N.; Mitković, P.B.; Stojanović, M.B.; Ignjatović, M.G. Thermally activated building systems in context of increasing building energy efficiency. *Therm. Sci.* **2014**, *18*, 1011–1018. [CrossRef]
57. Guerrero Delgado, M.C.; Sánchez, J.; Álvarez, S.; Tenorio Ríos, J.A.; Cabeza, L.F.; Bartolomé, C.; Pavón Moreno, M.C. Evaluation of the behavior of an innovative thermally activated building system (TABS) with PCM for an efficient design. *E3S Web Conf.* **2019**, *111*, 8.
58. Chen, S.; Yang, Y.; Olomi, C.; Zhu, L. Numerical study on the winter thermal performance and energy saving potential of thermo-activated PCM composite wall in existing buildings. *Buil. Simul.* **2020**, *13*, 237–256. [CrossRef]
59. Delgado, M.G.; Ramos, J.S.; Domínguez, S.; Ríos, J.A.T.; Cabeza, L.F. Building thermal storage technology: Compensating renewable energy fluctuations. *J. Energy Storage* **2020**, *27*, 101147. [CrossRef]
60. Kisilewicz, T.; Fedorczak-Cisak, M.; Barkanyi, T. Active thermal insulation as an element limiting heat loss through external walls. *Energy Build.* **2019**, *205*, 109541. [CrossRef]
61. Qu, S.; Su, S.; Li, H.; Hu, W. Optimized control of the supply water temperature in the thermally activated building system for cold climate in China. *Sustain. Cities Soc.* **2019**, *51*, 101742. [CrossRef]
62. Kalús, D.; Gašparík, J.; Janík, P.; Kubica, M.; Šťastný, P. Innovative Building Technology Implemented into Facades with Active Thermal Protection. *Sustainability* **2021**, *13*, 4438. [CrossRef]
63. Arroyo, J.; Spiessens, F.; Helsen, L. Comparison of Model Complexities in Optimal Control Tested in a Real Thermally Activated Building System. *Buildings* **2022**, *12*, 539. [CrossRef]
64. Joe, J.; Karava, P. A model predictive control strategy to optimize the performance of radiant floor heating and cooling systems in office buildings. *Appl. Energy* **2019**, *245*, 65–77. [CrossRef]
65. Feng, J.D.; Schiavon, S.; Bauman, F. New method for the design of radiant floor cooling systems with solar radiation. *Energy Build.* **2016**, *125*, 9–18. [CrossRef]
66. Yang, X.; Pan, L.; Guan, W.; Tian, Z.; Wang, J.; Zhang, C. Optimization of the configuration and flexible operation of the pipe-embedded floor heating with low-temperature district heating. *Energy Build.* **2022**, *269*, 112245. [CrossRef]
67. Ma, J.; Yang, Y.; Zheng, X.; Dai, B.; Zhu, D.; Liu, Q. Impact on heat storage performance of concrete radiant floor with finned water supply pipes. *J. Build. Eng.* **2021**, *44*, 103351. [CrossRef]
68. Ruiz-Pardo, Á.; Rodríguez Jara, E.Á.; Conde García, M.; Ríos, J.A.T. Influence of Wood Properties and Building Construction on Energy Demand, Thermal Comfort and Start-Up Lag Time of Radiant Floor Heating Systems. *Appl. Sci.* **2022**, *12*, 2335. [CrossRef]
69. Park, S.H.; Chung, W.J.; Yeo, M.S.; Kim, K.W. Evaluation of the thermal performance of a Thermally Activated Building System (TABS) according to the thermal load in a residential building. *Energy Build.* **2014**, *73*, 69–82. [CrossRef]
70. Cen, C.; Jia, Y.; Liu, K.; Geng, R. Experimental comparison of thermal comfort during cooling with a fan coil system and radiant floor system at varying space heights. *Build. Environ.* **2018**, *141*, 71–79. [CrossRef]
71. Zhang, D.; Cai, N.; Cui, X.; Xia, X.; Shi, J.; Huang, X. Experimental investigation on model predictive control of radiant floor cooling combined with underfloor ventilation system. *Energy* **2019**, *176*, 23–33. [CrossRef]
72. Gu, X.; Cheng, M.; Zhang, X.; Qi, Z.; Liu, J.; Li, Z. Performance analysis of a hybrid non-centralized radiant floor cooling system in hot and humid regions. *Case Stud. Therm. Eng.* **2021**, *28*, 101645. [CrossRef]
73. Hwang, Y.J.; Jeong, J.W. Energy Saving Potential of Radiant Floor Heating Assisted by an Air Source Heat Pump in Residential Buildings. *Energies* **2021**, *14*, 1321. [CrossRef]
74. Sharifi, M.; Mahmoud, R.; Himpe, E.; Laverge, J. A heuristic algorithm for optimal load splitting in hybrid thermally activated building systems. *J. Build. Eng.* **2022**, *50*, 104160. [CrossRef]
75. Zhu, X.; Liu, J.; Zhu, X.; Wang, X.; Du, Y.; Miao, J. Experimental Study on Operating Characteristic of a Combined Radiant Floor and Fan Coil Cooling System in a High Humidity Environment. *Buildings* **2022**, *12*, 499. [CrossRef]
76. Ren, J.; Liu, J.; Zhou, S.; Kim, M.K.; Song, S. Experimental study on control strategies of radiant floor cooling system with direct-ground cooling source and displacement ventilation system: A case study in an office building. *Energy* **2022**, *239*, 122410. [CrossRef]
77. Khan, Y.; Khare, V.R.; Mathur, J.; Bhandari, M. Performance evaluation of radiant cooling system integrated with air system under different operational strategies. *Energy Build.* **2015**, *97*, 118–128. [CrossRef]
78. DG, L.S.; Nagendra, S.S.; Maiya, M. Feasibility analysis of passive thermally activated building system for various climatic regions in India. *Energy Build.* **2017**, *155*, 352–363.
79. Leo Samuel, D.; Nagendra, S.S.; Maiya, M. Simulation of indoor comfort level in a building cooled by a cooling tower–concrete core cooling system under hot–semiarid climatic conditions. *Indoor Built Environ.* **2017**, *26*, 680–693. [CrossRef]
80. Samuel, D.L.; Nagendra, S.S.; Maiya, M.P. Parametric analysis on the thermal comfort of a cooling tower based thermally activated building system in tropical climate–An experimental study. *Appl. Therm. Eng.* **2018**, *138*, 325–335. [CrossRef]
81. Leo Samuel, D.; Shiva Nagendra, S.; Maiya, M. An analysis of operating parameters in the cooling tower-based thermally activated building system. *Indoor Built Environ.* **2018**, *27*, 1175–1186. [CrossRef]
82. Leo Samuel, D.; Shiva Nagendra, S.; Maiya, M. A study of pipe parameters on the performance of cooling tower-based thermally activated building system. *Indoor Built Environ.* **2018**, *27*, 219–232. [CrossRef]
83. Samuel, D.L.; Nagendra, S.S.; Maiya, M. A sensitivity analysis of the design parameters for thermal comfort of thermally activated building system. *Sādhanā* **2019**, *44*, 48. [CrossRef]

84. Dharmasastha, K.; Leo Samuel, D.; Shiva Nagendra, S.; Maiya, M. Thermal comfort of a radiant cooling system in glass fiber reinforced gypsum roof-an experimental study. *Appl. Therm. Eng.* **2022**, *214*, 118842. [CrossRef]
85. Montenegro López, F.J.; Hongn, M.E. Estudio computacional de un sistema de acumulación estructural orientado a refrescamiento bioclimático: Primeros resultados. *Av. En Energías Renov. Y Medio Ambiente* **2020**, *24*, 51–61.
86. Oravec, J.; Šikula, O.; Krajčík, M.; Arıcı, M.; Mohapl, M. A comparative study on the applicability of six radiant floor, wall, and ceiling heating systems based on thermal performance analysis. *J. Build. Eng.* **2021**, *36*, 102133. [CrossRef]

buildings

MDPI

Article

Numerical Optimization for the Design of Geopolymer Concrete Walls with Phase Change Materials in the Mediterranean Climate

Fabrizio Ascione [1,*], Rosa Francesca De Masi [2], Nicoletta Del Regno [3], Filippo De Rossi [4], Antonio Gigante [2] and Silvia Ruggiero [2]

1 Department of Industrial Engineering, DII, Università degli Studi di Napoli Federico II, Piazzale Tecchio, 80, 80125 Napoli, Italy
2 Department of Engineering, DING, Università degli Studi del Sannio, Piazza Roma, 21, 82100 Benevento, Italy
3 Department of Medicine and Health Sciences, Vincenzo Tiberio, Università degli Studi del Molise, Via F. de Sanctis, 1, 86100 Campobasso, Italy
4 Department of Architecture, DIARC, Università degli Studi di Napoli Federico II, Via Monteoliveto, 3, 80134 Napoli, Italy
* Correspondence: fabrizio.ascione@unina.it; Tel.: +39-081-7682292

Abstract: The adoption of phase change materials (PCMs) is a promising solution for the improvement of building energy performances and indoor comfort, and the integration of geopolymer concrete (GPC) allows recycling materials and reducing the demand for raw materials in concrete production. Both materials contribute to reducing the carbon dioxide emission in the building lifecycle. In this frame, this paper proposes a complete numerical approach for selecting the optimal wall package made of GPC and PCMs in a Mediterranean climate. The first step of the method consists of a parametric analysis for evaluating the incidence on energy performance and thermal comfort of the main designing variables: insulation thickness, air cavity type and its thickness, and PCMs type. Then, assuming the discomfort hours as a limiting constraint, a multi-objective optimization is applied to a subset of solutions for determining the Pareto front solutions. The advantage of the proposed methodology is the combined evaluations of multiple variables with a simplicity in execution; for this reason, it is useful for other researchers aimed at studying innovative solutions. According to obtained results, the better exposure for the proposed wall package is the north or northeast one. The minimization of the cooling energy demand requires the adoption of two PCMs, on internal and external sides, with melting temperature of 26 °C. The optimization of yearly performance requires the adoption of the maximum insulation level on both sides and a not-ventilated air gap between the modules. The cooling and heating energy need can be reduced, respectively, by around −29% and −57%, compared to a reference configuration with vacuum insulation panels and thermal transmittance of 0.4 W/m^2 K.

Keywords: geopolymer concrete; phase change materials; simulations; multi-objectives optimization; Mediterranean areas

Citation: Ascione, F.; De Masi, R.F.; Del Regno, N.; De Rossi, F.; Gigante, A.; Ruggiero, S. Numerical Optimization for the Design of Geopolymer Concrete Walls with Phase Change Materials in the Mediterranean Climate. *Buildings* **2022**, *12*, 1353. https://doi.org/10.3390/buildings12091353

Academic Editor: Jan Fořt

Received: 5 August 2022
Accepted: 30 August 2022
Published: 1 September 2022

1. Introduction

Research on innovative materials for improving the building performance is of great interest in relation to the targets of energy saving and environmental impact reduction planned by the European strategy. On 14 July 2021, the European Commission adopted several important measures for achieving climate neutrality in the EU by 2050, including the intermediate target of 55% net reduction in greenhouse gas emissions by 2030 [1]. Construction material technologies are the driving force for enhancing and improving building energy efficiency and infrastructure functionality [2]. Among the most interesting materials, there are the geopolymeric concretes and the phase change materials (PCMs).

Using recycled materials as substitution for natural aggregate to produce geopolymer concrete is becoming a possible alternative for reusing construction and demolition waste with interesting eco-friendly benefits [3] and for supporting the sustainable development of the construction sector [4]. Several types of applications were proposed, such as the development of rubberized brick by utilizing crumb rubber as the sole fine aggregate in the production of geopolymer interlocking bricks [5]. Instead, Gerges et al. [6] underlined that the rubberized concrete mixture generally has a reduced compressive strength but also lower density, higher toughness, and higher impact resistance compared to conventional concrete. Cui et al. [7] introduced two new types of geopolymer composite lightweight sandwich panels; one with fiber-reinforced geopolymer composite skin layers and polyurethane foam core, and another one with basalt-fiber-reinforced polymer sheet. Longo et al. [8] worked on rheological behavior, as well as mechanical and thermal properties, of a geopolymer mortar incorporating fly ash and expanded glass aggregate.

Several papers have investigated the thermal characteristics of different compositions, and few papers give the indications on the reliable energy savings during the building use. Colangelo et al. [9] proposed to substitute part of the aggregates with plastic waste and to use a fly-ash-based geopolymeric binder for the production of low conductivity concrete. Wu et al. [10] used the basic oxygen furnace slag as a radiative cooling material in geopolymeric coating by reaching an emissivity of 0.95 within the range of 8–13 μm, and it also has high conductivity.

Foamed fly ash geopolymer was synthesized by Su et al. [11] to produce geopolymeric lightweight concrete. For densities from 1200 to 600 kg/m^3, the thermal conductivity diminished from 0.70 to 0.22 W/mK, which is much better than that the ordinary Portland cement. Parcesepe et al. [12] found that an alkali-activated concrete can achieve mechanical characteristics higher than those of ordinary Portland concrete but also an improvement of the thermal insulation capacity. Dhasindrakrishna et al. [13] underlined the interest for geopolymer foam concrete, thanks to their porous structure which brings the inherent merits of lightweight, acoustic, and thermal insulation, and fire resistance. For instance, Zang et al., found that the thermal conductivity is in a range of 0.15–0.48 W/m K [14]. Wang et al. [15] found that for the fly ash-based lightweight geopolymer concrete, the adoption of 0–1% of polypropylene fibers can increase the thermal conductivity by varying the moisture absorption. According to Henon et al. [16], by varying the pore volume fraction in a geopolymer foam between 65 and 85%, the thermal conductivity changes between 0.35 and 0.12 W/m K.

The heat capacity performance of geopolymer concretes can be also improved by means of microencapsulated phase change materials (PCMs). A review of the latest research results related to the use of PCMs in geopolymer materials was presented by Lach et al. [17]. The latent heat and melting temperature of investigated products are usually in the range 96.1–230 J/g and 21.9–33.8 °C, respectively [18].

There are several studies about the in-lab characterization of new composites and products; for instance, Hassan et al. [19] investigate the thermal and structural performance of geopolymer-coated polyurethane foam–phase change material capsules/geopolymer concrete composites with interesting results. Ramakrishnan et al. [20] modified an aerated/foamed geopolymer concrete with a paraffin/hydrophobic expanded perlite, and they found that the incorporation of 15% and 30% of PCM can reduce the peak indoor temperature by 1.85 °C and 3.76 °C, respectively, while enhancing the thermal storage capacity by 105% and 181%. Hassan et al. [21] tested novel geopolymer-coated expanded clay–phase change material macrocapsules added to geopolymer concrete. They observed a decrement of maximum surface temperatures of 8.0 °C compared to a mixture without PCM, and the thermal transmittance passed from 2.0 to 0.9 W/m^2 K. by means of a numerical model. Cao et al. [22] utilized multilayer walls integrating microencapsulated phase change materials into geopolymer concrete, and found annual energy reduction of 28–30%. Pilehvar et al. [23] underlined that the addition of PCM slows down the reaction rate of both geopolymer and Portland cement paste. The setting times were faster when the temperature was increased. However, the major disadvantage of adding microencapsulated phase change materials is the significant decrease in compressive strength [24,25].

As reported in the previous analysis, the recent literature is mainly focused on the characterization of the thermal and structural properties of the mixture for improving the thermal properties of geopolymer concrete with phase change materials. Instead, the proposed paper analyzes the behavior of the combination between macro-encapsulated PCM and geopolymer concrete block with the material mounted in series as two different layers. This solution could solve the problem of deterioration of structural performance.

Moreover, the estimation of the potential energy saving and the improving of the thermal comport is usually performed with a simplified model on short periods and there are few indications about the global energetic behavior. Regarding this, the paper proposes a complete approach for the evaluation of heating and energy need as well as for the yearly primary energy, because in some climates it can happen that the PCM improves the building only on one season, for instance, the summer, but the reduction of solar gains could worsen the winter performance. This is an optimization problem that requires the consideration of the main variables, such as the melting point, the thickness, and the position, and the paper suggests this approach to researchers with a discussion of a real case study that will be supported by experimental data in next months.

More in detail, a numerical investigation is proposed for selecting the optimal configuration of an innovative multilayered package to be installed in Mediterranean climates. This work was financially supported by Horizon 2020 with the project Green INSTRUCT (Green Integrated Structural Elements for Retrofitting and New Construction of Buildings). This project is focused on the development of a prefabricated modular structural building block that is superior to conventional precast reinforced concrete panels by virtue of its reduced weight, improved acoustic and thermal performance, and multiple functionalities. This study describes the design approach and the evaluations carried out for selecting the types and thickness of materials with the aim to improve energy performance and indoor microclimate of living space in a typical climate of South Italy. The obtained indications and the introduced methodological approach can be useful for designers and for researchers for identifying the best overall layout of innovative experimental installations under real external conditions.

Moreover, the discussed results will be used in the following months for comparing the expected performance with the experimental one. Indeed, a monitoring campaign is ongoing for the described wall package.

2. Research Method and Case Study

A detailed research method is proposed for the optimized design of a new passive solution considering both energy and indoor comfort aspects. The approach is based on two main steps: a preliminary parametric investigation followed by a constrained multi-objective optimization process, in order to provide the best compromise of configurations.

A case study is also developed because the results can be useful for other research in the field of innovative solution or when designers must select the optimal values for the design variables of a wall package with PCM materials to be installed in the Mediterranean areas.

2.1. Method for the Optimal Design of a Multilayered Sustainable Wall

The first step is a parametric study with which the influence of the main designing variables is investigated. In detail, we varied the insulation thickness, the air cavity type (ventilated or not) and its thickness, and the PCM type, also considering different integration methods and the position (internal side, external side, and both sides). The comparison of the possible design configurations is based on the evaluation of the cooling (E_C) and heating (E_H) energy need, the annual primary energy request (EP_{yearly}), and the number of discomfort hours (dH). E_C and E_H are the energy required to assure the comfort operative temperature, respectively, during the summer and winter. The comfort conditions are expressed according to the standard ASHRAE 55-2004 [26]. For the cooling and heating seasons, the resistance of clothing has to be set to 0.5 Clo and 1.0 Clo, respectively. The effectiveness of designed configurations was assessed throughout the evaluation of the percentage annual primary energy savings (ΔEP) and the percentage reduction of discomfort hours (ΔdH).

The multi-objective optimization is the second step for evaluating the best compromise design configuration between the most interesting solutions founded with the parametric analysis. The optimization tool has been usefully applied by the authors also in other study for selecting a living wall configuration in the Mediterranean climate [27]. The optimization problem needs the definition of the objective functions that in this case are the minimization of the heating and cooling energy needs; indeed, the adoption of passive solutions can vary the heat gains and losses throughout the building envelope and, thus, an appropriate choice can reduce the primary energy demand.

Often, the optimization problems for the building envelope are organized without considering the aspect of the thermal comfort [28]. Instead, in this study, the hours of discomfort are considered as a limiting constraint for determining the Pareto front solutions. For the optimization problem, the most interesting configurations obtained in the parametric analysis are evaluated by varying the exposure of the installation. The final outcome is the Pareto front [29], which is the set of the non-dominated solutions.

2.2. Presentation of the Case Study

The test room MATRIX (Multi Activity Test-Room for Innovating^{X}) of the Department of Engineering of University of Sannio is used as the case study because it is the site of installation for the experimental campaign. MATRIX is located in Benevento (130 m above sea level), a city of Campania in the south Italy.

MATRIX (Figure 1a) is a large-scale test room (36 m^2) with gross sizes of 6.00 × 6.00 × 5.50 m. Wooden basement and roofs and a vertical steel frame comprise the building structure. Plywood panels, without interruption, enclose all edges of the cube, in order to avoid the thermal bridges. MATRIX allows to test facades or components up to 3.0 m height and 5.0 m width because there is a mechanical system that allows to change three of the four vertical walls. Aerated cellular concrete and vacuum insulation panels are the basic elements of the existing configuration with thickness of 14 cm and an overall thermal transmittance (U) of 0.40 W/(m^2 K). There is one window with wood frame, double layer of elettrotropic glass; U is 2.2 W/(m^2 K).

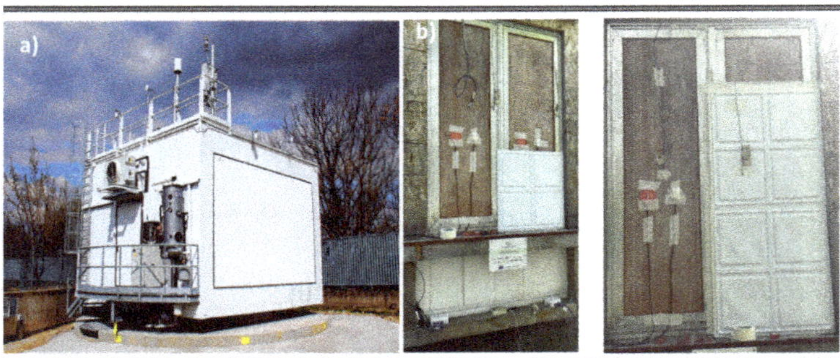

Figure 1. (**a**) MATRIX layout. (**b**) Installation of system under investigation for further studies.

A detailed description of the test room was already presented by Ascione et al. [30], who also built a numerical model of MATRIX by means of EnergyPlus [31], through an interface program, i.e., DesignBuilder [32]. The numerical model was simulated under real conditions, both internal and external, and it was calibrated according to the ASHRAE Guideline [33].

In the present study, the calibrated simulation model was modified to be an office with one occupant. The air-conditioning system consists of a hydraulic plant with in-room fan coils supplied by an air-cooled chiller (nominal energy efficiency ratio of 3.5 W_{TH}/W_{EL}) and a gas boiler (nominal efficiency of 92%). The adopted boundary conditions are as follows:

- The design set-point air temperature is 20 °C in winter and 26 °C in summer according to comfort requirements [34] and conditions imposed by technical standard [35].
- The heating period runs from 15 November to 31 March and the cooling period from 15 May to 30 September, and in both cases only during weekdays, from 8:00 a.m. to 8:00 p.m.
- The air change rate is equal to 0.3 h^{-1}.

For the simulations, the hourly weather file was created with data monitored from April 2014 to December 2021 by means of the climatic station on the roof of the laboratory. According to monitored values, Benevento has a typical Mediterranean climate and it is inside the Csa zone according to Köppen classification [36].

2.3. Description of Investigated Configurations

The analyzed multilayered system was designed in the frame of the Green INSTRUCT project. It is made of three main elements: internal panel, middle foam insulating layer, and the external panel. These panels are held together by an aluminum frame made of construction and demolition waste (CDW) for extrusion. The internal panel consists of an aluminum encasing with a magnesium-based concrete layer (MOC) followed by a geopolymer layer, a recycled polyurethane foam layer, an air gap, another PU foam layer, and a final geopolymer layer. The results of experiments on mechanical property, such as the compressive strength of the geopolymer synthesis, are described in the technical report [37]. The middle layer consists of 10 cm of insulation and it is connected to the internal panel by means of threaded bars that allow the creation of another air gap.

Table 1 shows the characteristics of all materials involved in the base design configuration: thickness (s), thermal conductivity (λ) or thermal resistance (R), density (ρ), and specific heat (cp).

Table 1. Properties of the basic design configuration.

Materials	s (m)	λ (W/m K)	ρ (kg/M3)	cp (kJ/kg K)
MOC	0.01	0.7	1100	1000
Geopolymer	0.015	1.0	1200	750
Polyurethane	0.04	0.025	100	1600
Air gap	0.06		R = 0.18 (m^2 K/W)	
Polyurethane	0.04	0.025	100	1600
Geopolymer	0.015	1.0	1200	750
Air gap	0.02		R = 0.088 (m^2 K/W)	
Polyurethane	0.10	0.026	100	1600

The conductivity and the density are furnished by producers during the project; the not-available properties are taken from international standard [38–40].

The thermal transmittance of this configuration is 0.13 W/m^2 K, the thermal capacity is 66.8 kJ/m^2 K, and the surface mass is 65.0 kg/m^2. According to this value, it can be stated that the insulation level is very high compared to the building stock in the Mediterranean zone but also with threshold values considered by the Italian Ministerial Decree [41]. Instead, the summer performance could be not good because the package is characterized by low thermal mass (<230 kg/m^2). In the following analysis, this configuration is considered the basic design configuration. The external layer is a green box for which a sensitivity analysis regarding the green layer was already presented by the authors [27].

Starting from this design proposal, the study is focused on the selection of the PCM to be coupled with the geopolymer blocks by varying the thicknesses of the main layers of the base design configuration. More in detail, the alternative configurations are defined by varying:

- Thickness of the polyurethane layers and of the air gap for the internal panel;
- Thickness and type of ventilation of air gap in the middle layer;
- The position and the type of PCM.

The variation of the air-gap thickness is also length to a variation of the thickness of the external polyurethane because the total thickness is kept constant (30 cm) as well as the dimension for the internal module (18 cm). The insertion of the layer with PCM was evaluated on the internal side of the described wall package (Figure 2a), on the external side (Figure 2b), and on both sides. The wallboards or plasters with microencapsulated PCMs were considered for the internal application in place of the MOC.

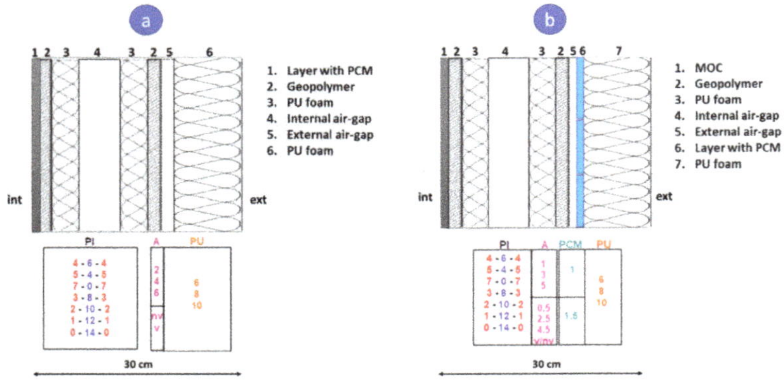

Figure 2. Scheme of simulated configurations: (**a**) PCM on the internal side; (**b**) PCM on the external side.

Macro-encapsulated PCMs were considered for external side insertions, and in this case, the MOC was the internal layer. To facilitate the solidification process, the PCM was inserted between the air gap and the external polyurethane. Consequently, with the objective to always have the same total thickness, in the external side applications the thickness of the air gap was lower than the internal side. Two thicknesses were evaluated for the external application of PCMs, namely, 1.0 cm and 1.5 cm. These are the most common commercial values and are also the available size for the selected product.

Finally, 210 configurations with PCM on the internal side, 504 configurations with PCM on the external side, and 60 configurations with PCM on both sides were evaluated. For the simulation, the proposed multilayered wall is applied on the south exposure because the most interesting researchers [42,43] indicate that the experimental assessment of PCMs is usually performed for this side, which is considered the best one.

It must be underlined that the adopted simulation model was validated with experimental data for the present configuration with vacuum insulation panel. Instead, the proposed configurations in Figure 2 were not verified by means of experimental data. Indeed, this study was performed with the aim to select the type of PCM for the experimental campaign to be conducted in the frame of the HORIZON project. In future months, with the data recorded, a comparison will be made between simulated and monitored variables for discussing the problem of the deviation of the in-field performance of the phase change materials.

Moreover, the discussed results will be used in the following months for comparing the expected performance with the experimental one. Indeed, a monitoring campaign is ongoing for the described wall package.

In the proposed analysis the reference nomenclature is

$$PI_{xyx}A_{zv}PU_t$$

where

- *PI* is the inner panel, "*x*" stands for the thickness (cm) of polyurethane, and "*y*" stands for the thickness (cm) of the air gap of the *PI*;
- *A* is the air gap of middle panel with "*z*" that indicates its thickness (cm), "*v*" means ventilated cavity (slightly) and alternatively "*nv*" not ventilated;
- "*t*" is the thickness (cm) of the polyurethane of the middle layer (*PU*).

All the solutions with phase change material on the internal side were identified by the name of the material before the described nomenclature. Instead, macro-encapsulated PCM were identified with the name and thickness of the phase change material between the term A_{zv} and the term PU_t. Finally, the solutions with PCMs on both sides were connoted by placing "in + out" before the previously described identification criteria.

2.4. Characteristics of Selected PCMs

PCMs were selected among available commercial products for which the melting and solidification curve is available. With regard to building materials containing microencapsulated PCMs, the following products were considered:

- Comfortboard® (wallboard);
- Weber.mur clima® (internal plaster);
- Dry Clay Plaster Trocken-Lehmputz-PCM® (internal plaster);
- SmartBoard™ (plasterboard);
- ENERCIEL® (finishing material).

The first three materials contain the same microencapsulated PCM (Micronal®), with a melting temperature (T_f) of 23 °C [44]. It can be integrated with different systems into building materials. The first examined wallboard is the Knauf Comfortboard® product that contains 80% gypsum and 20% paraffin. It has a thickness of 12.5 mm, melting point equal to 23 °C, and a heat storage capacity (LHS) of 207 kJ/m^2. Weber.mur clima and Dry Clay Plaster Trocken-Lehmputz-PCM® are two plasters for indoor applications, manufactured by Weber Saint-Gobain and Pro Lehm Frauwallner KG, respectively. In particular, the first one is a dry mineral mortar based on gypsum, in which Micronal® microcapsules are added with light mineral additives for better processing and to strengthen the bond. For the purposes of this study, a plaster thickness of 15 mm was considered, and thus the PCM is 20 wt%. The second product is a clay-based dry plaster with 30 wt% of Micronal® PCM; a thickness of 10 mm was considered in the following analysis. Briefly, all the information necessary for modeling the PCM in EnergyPlus was obtained from the web app made available by the manufacturer [45]. Thanks to this application, it is possible to evaluate the density, the thermal conductivity, the latent heat capacity (h_f), the thermal capacity (Q_f), and the melting curve reported in Figure 3.

The SmartBoard™ has thickness of 15 mm and contains 26% mass fraction of Micronal® PCM with a melting point of 26 °C. The heat storage capacity of the SmartBoard ™ is 330 kJ/m^2. In this case, the cumulative mean temperature–enthalpy curve (Figure 3d) was obtained thanks to the study of Ozdenefe et al. [46].

Finally, ENERCIEL® is a surface coating with biological microencapsulated PCM (INERTEK 23 in slurry form), 50 wt%, gypsum (10–15 wt%), and a titanium binder. The melting point can range from 23 °C to 26 °C. The product can have a maximum thickness of 3.0 mm and it is suitable for interior applications on walls and ceilings. In this study, a melting temperature of 26 °C and a latent storage capacity of 186.3 kJ/m^2 are considered. The cumulative mean enthalpy–temperature curve is already available in the EnergyPlus library.

Table 2 summarizes main characteristics of the construction materials with microencapsulated PCM.

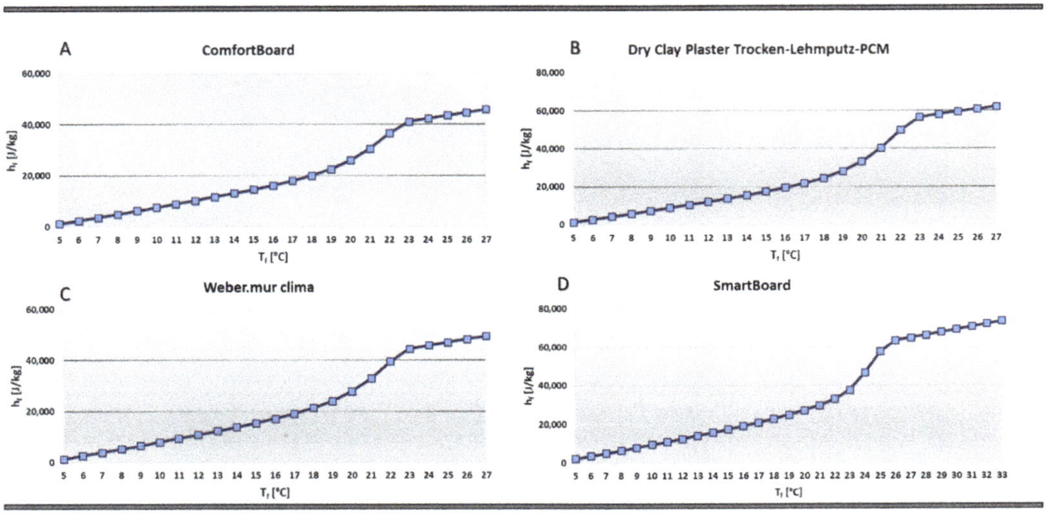

Figure 3. Enthalpy–temperature function: (**A**) Comfortboard®; (**B**) Weber.mur clima®; (**C**) Dry Clay Plaster; (**D**) SmartBoard™.

Table 2. Properties of building materials with microencapsulated PCM.

	T_f (°C)	s (mm)	ρ (kg/m³)	λ (W/m K)	c_p (J/kg K)	h_f (kJ/kg)	LHS (kJ/m²)	Q_f (kJ/m² K)
Comfortboard®	23	12.5	880	0.230	1170	18.90	207.9	12.87
Weber.mur clima®	23	15	1000	0.200	1380	20.70	205.3	20.70
Dry Clay Plaster PCM®	23	10	1100	0.400	2000	31.10	225.9	22.00
ENERCIEL®	23–26	3	832	0.148	2500	73.56	183.6	6.24
SmartBoard™	26.5	15	770	0.134	1200	28.57	330.0	13.86

The materials selected for external application are CSMs (compact storage modules) filled with macroencapsulated PCM [47]. The filling materials can be organic, indicated as RT®, or inorganic materials, indicated as SP®, and these assure a wide range of temperatures depending on the applications. The CSMs are available in the dimensions of 450 × 300 mm², and for different thicknesses. Table 3 summarizes the main characteristics of selected products and Figure 4 shows the melting curves; herein, the range of curves is different because these were built with the data available from manufacturers.

Table 3. Properties of macroencapsulated PCMs.

CSMs 450 × 300 mm²	T_f (°C)	ρ (kg/m³)	λ (W/m K)	c_p (J/kg K)	h_f (kJ/kg)	Q_f (kJ/m² K)
RT26®	26				180 (19–34 °C)	
RT28HC®	28	880	0.2		250 (21–36 °C)	17.6
RT31®	31			2000	165 (23–28 °C)	
SP26E®	26	1500			180 (17–32 °C)	30
SP29Eu®	29	1550	0.6		200 (22–37 °C)	31
SP31®	32	1350			210 (23–38 °C)	27

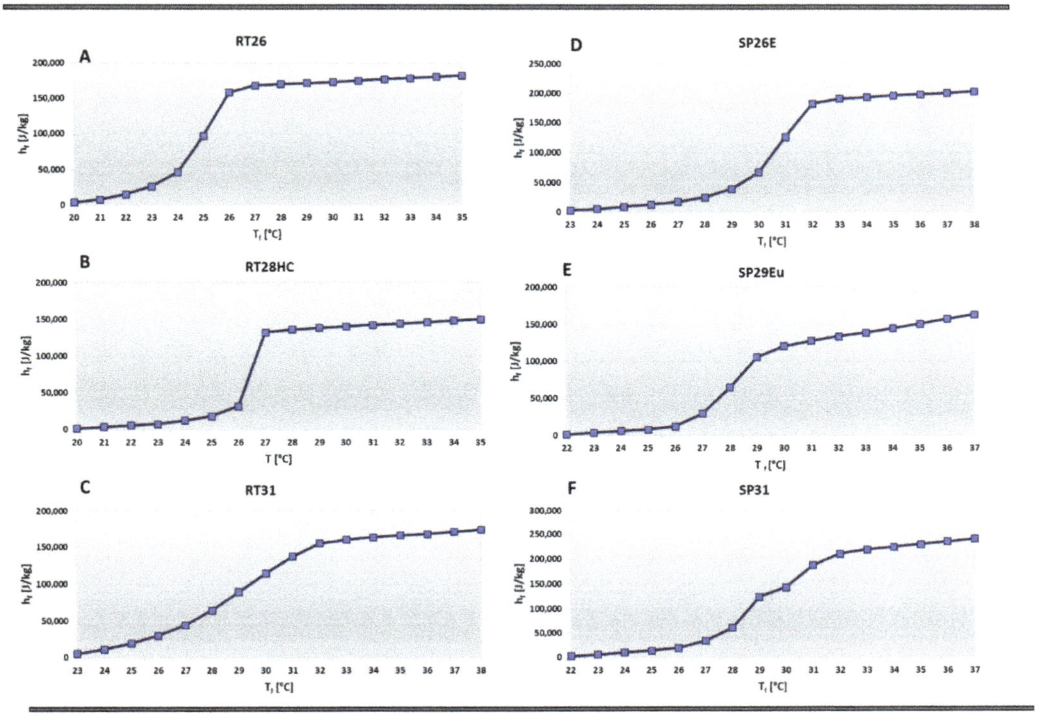

Figure 4. Enthalpy–temperature function of: (**A**) RT26®; (**B**) RT28HC®; (**C**) RT31®; (**D**) SP26E®; (**E**) SP29Eu®; (**F**) SP31®.

In the configuration with the MOC layer containing microencapsulated PCM, the melting curve was obtained by means of Microtek app [41] and, therefore, assuming the integration of Micronal PCM with a melting temperature of 23 °C. The MOC has cementitious matrix with aggregates such as wood chips recycled from CDW and reinforced by wood fibers and thermoplastic polymers. On the basis of this information, the percentage composition reported in [48] was considered suitable for the case under examination, with 60% by weight of cementitious compounds and 40% by weight of aggregates. Finally, a PCM integration percentage of 9% was calculated by making a weighted average of the microencapsulated PCM integration percentages recommended by the Microtek tool for cementitious materials and medium-density fiberboard derived from wood. The calculated properties are shown in Table 4 and the cumulative enthalpy–temperature curve is shown in Figure 5.

Table 4. Enthalpy–temperature function of MOC with Micronal PCM.

	s (mm)	ρ (kg/m3)	λ (W/m K)	c_p (J/kg K)	LHS (kJ/m^2)	Q_f (kJ/m^2 K)
MOC + Micronal®	10	1089	0.65	1182	9.3	12.9

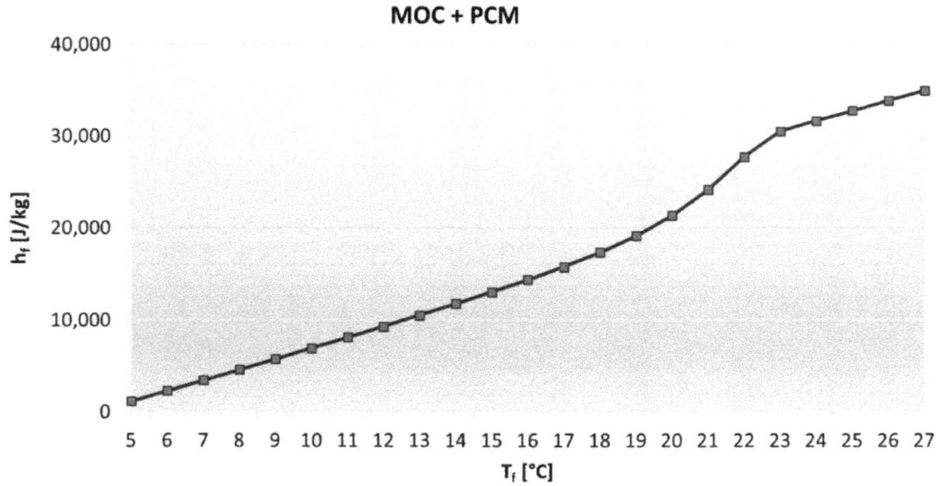

Figure 5. Enthalpy–temperature function of MOC with integrated PCM.

2.5. Numerical Model

EnergyPlus through the interface DesignBuilder is the tool used for the simulations. A one-dimensional conduction finite difference algorithm (CondFD) can be used for the simulation of PCM [49]. The CondFD algorithm is based on the implicit finite difference scheme, and it is possible to choose between Crank–Nicholson or fully implicit. The equation on which the calculation method is based for the fully implicit scheme within a homogeneous material is

$$c_p\, \rho \Delta X \frac{T_i^{j+1} - T_i^j}{\Delta t} = \lambda_w \frac{T_{i+1}^{j+1} - T_i^{j+1}}{\Delta X} + \lambda_E \frac{T_{i-1}^{j+1} - T_i^{j+1}}{\Delta X} \tag{1}$$

where

- T = temperature;
- i = node being modeled, $i + 1$ = adjacent node to interior of construction, $i - 1$ = adjacent node to exterior of construction;
- $j + 1$= new time step, j = previous time step;
- Δt = time step;
- Δx = finite difference layer thickness;
- c_p = specific heat of material;
- ρ = density of material;
- λ_W and λ_E are the thermal conductivity for interface between i node and $i + 1$ node, and between i node and $i - 1$ node.

If thermal conductivity is variable, it can be defined according to

$$\lambda_w = \frac{\lambda_{i+1}^{j+1} - \lambda_i^{j+1}}{2} \tag{2}$$

$$\lambda_E = \frac{\lambda_{i-1}^{j+1} - \lambda_i^{j+1}}{2} \tag{3}$$

In the CondFD algorithm, all elements are divided or discretized automatically using Equation (4), which depends on a space discretization constant (c), the thermal diffusivity of the material (α), and the time step. The default space discretization value of 3 (equivalent to a Fourier number (*Fo*) of 1/3) or input other values can be selected.

$$\Delta_x = \sqrt{c \cdot \alpha \cdot \Delta t} = \sqrt{\frac{\alpha \cdot \Delta t}{Fo}} \quad (4)$$

The CondFD method is coupled with an enthalpy–temperature function that must be defined for updating an equivalent specific heat at each time step. EnergyPlus offers two modeling methods, to associate the enthalpy–temperature curve:

- Basic: where the same temperature/enthalpy curve is used for both melting and freezing.
- Hysteresis: where the melting and freezing processes follow different temperature/enthalpy curves.

For the simulation of the case study, the basic method is used. The enthalpy–temperature function was defined through a set by means of a tabular form starting from the previous proposed figures. The tabular function usually considers the entire temperature range, from −20 °C to 60 °C. The limit to consider commercial material is the unavailability of data for defining the hysteresis phenomenon. The adopted model requires also a "temperature coefficient for thermal conductivity". This is used in order to introduce the temperature-dependent coefficient for thermal conductivity of the material. More in detail, this is the thermal conductivity change per unit of temperature excursion from 20 °C. Moreover, the time steps per hour equal to 12 are used, and the space discretization constant is set to 3.

More than 1000 simulations were carried out. Among the capabilities of the program, there is the use of a non-dominated sorting genetic algorithm based on the NSGA-II method [50], which is widely used as a "fast and elitist multi-objective" method providing a suitable trade-off between a well-converged and a well-distributed solution set. This tool was used for solving the optimization problem. For our scope, the maximum number of generations was set to 200 and it was typically in the range of 50–500. This value reflects the complexity of the analysis. The computational domain is the whole year; thus, simulations run for all months and both daily or monthly results are available; moreover, the time interval between two consecutive energy balances was fixed equal to 6 per hour.

3. Analysis of Simulation Results

The results of the parametric analysis and the multi-objective optimization are provided in the following sections.

3.1. Parametric Analysis with PCM on the Inner Side

3.1.1. Analysis of Heating and Cooling Energy Needs

Figure 6 shows the results in terms of heating and cooling energy needs for the configurations with the application of selected PCMs on the inner side. The yellow point is representative of the present configuration of MATRIX with vacuum insulation panels, and it was named base case (BC) The other point for the comparison is named $PI_{464}A_{2v}PU_{10}$; it is represented by a blue square and it is the configuration described in Table 1 with the default thickness and without PCM.

Considering the BC, all proposed solutions, also if applied only on one wall, allow a reduction of both heating and cooling energy demand without significant differences when the air-gap ventilation is compared with the same solution but with closed gap. Indeed, in the figure it can be noted that there are always two quite coincident points for each solution. The last branch of the graph where $PI_{464}A_{2v}PU_{10}$ is located collects all simulation results of the configurations without the PCM. The solution that minimizes the heating request is highlighted; it is $PI_{707}A_{2nv}PU_{10}$ for which E_C is 15.0 kWh/m^2 y and E_H is 21.4 kWh/m^2 y. This wall package, compared with the base design solution ($PI_{464}A_{2v}PU_{10}$), allows a slight improving of the performance because the cooling energy saving (ΔE_C) is −0.43% and the heating saving (ΔE_H) is −1.63%. The worst solutions, the highest points in the curve, are $PI_{0140}A_{6nv}PU_6$ and $PI_{0140}A_{6v}PU_6$, for which E_C is, respectively, 16.5 and 16.6 kWh/m^2 y, and E_H is, respectively, 23.5 and 23.6 kWh/m^2 y.

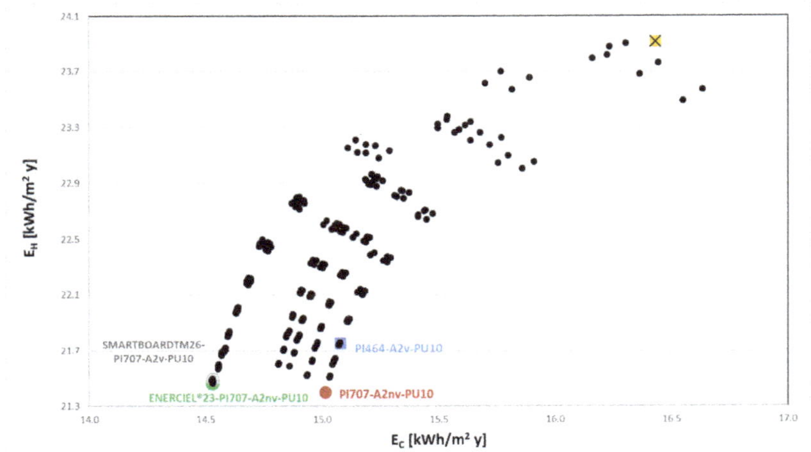

Figure 6. PCMs on inner side: comparison based on heating (E_H) and cooling load (E_C).

Instead, starting from the left, there is a first sequence of points starting from the solution *ENERCIEL®23-*$PI_{707}A_{2nv}PU_{10}$, which is characterized by the minimum of the cooling demand (14.5 kWh/m^2 y) and by a heating load of 21.5 kWh/m^2 y. This wall package, compared with BC, allows ΔE_C of −11.6% and ΔE_H of −10.2% and it can be considered a profitable solution for improving the performance achievable with an insulated system with low thermal mass. Considering the contribute of the phase change material, the evaluation of the heating and cooling loads reduction compared with the $PI_{464}A_{2v}PU_{10}$ indicates that the selected melting temperature contributes to slightly improving the thermal performance with ΔE_C equal to −3.6%. However, during the winter, the panel allows to increase the utilization of solar gains because it accumulates the energy and release during the day, allowing ΔE_H of −1.4%. The other points are representative of the same solution but with lower level of insulation until the extreme point *ENERCIEL®23-*$PI_{0140}A_{6v}PU_6$, for which E_C is 15.9 kWh/m^2 y and E_H is 23.7 kWh/m^2 y. This solution is not better than $PI_{464}A_{2v}PU_{10}$ because in both seasons the required energy increases; meanwhile, compared with BC, ΔE_C is −3.3% and ΔE_H is −1.1%. Thus, it can be concluded that the lower insulation level is not suitable for the considered climate and the PCM is not effective when included in this configuration. Along the same line, there are also the solutions with SmartBoard™. In this case, the package that minimizes the cooling energy demand is SMARTBOARD™26-$PI_{707}A_{2v}PU_{10}$; E_C is 14.5 kWh/m^2 y and E_H is 21.5 kWh/m^2 y. With higher melting temperature, the same energy saving compared with BC is achievable (ΔE_C −11.6%, ΔE_H −10.1%) for *ENERCIEL®23-*$PI_{707}A_{2nv}PU_{10}$. This happens because the ENERCIEL® product has the highest h_f of all the products examined (73.56 kJ/kg), thanks to the higher percentage of PCM; meanwhile, SmartBoard™ has the higher melting point, 26 °C, which is particularly suitable for combating indoor overheating during the summer months. The other products provide less benefit, and the simulation results occupy the other branches of the graph. For instance, with the same insulation level, the solution *WEBER.MUR CLIMA®23-*$PI_{707}A_{2nv}PU_{10}$ is characterized by E_C of 15.0 kWh/m^2 y and E_H of 21.5 kWh/m^2 y.

3.1.2. Analysis of Primary Energy Need and Discomfort Hours

Figure 7 shows the distribution of the simulation results considering the annual primary energy request (EP_{yearly}) and the number of discomfort hours (dH).

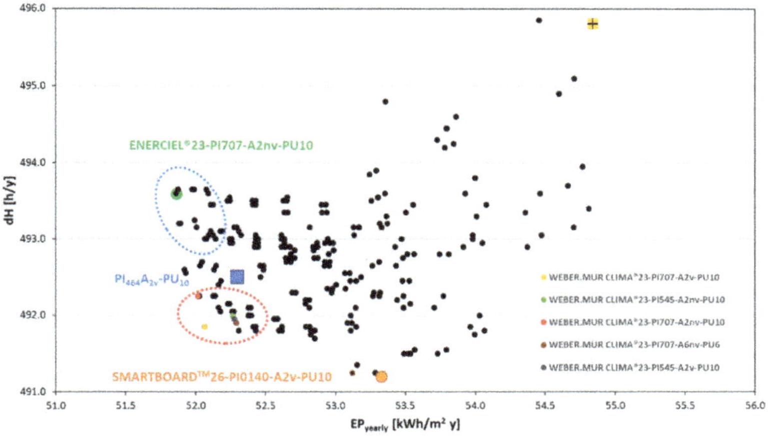

Figure 7. PCMs on internal side: comparison based on primary energy need and discomfort hours.

Additionally in this case there is no appreciable influence of the ventilation regime of the air gap. For the basic design configuration ($PI_{464}A_{2v}PU_{10}$), EP_{yearly} is 52.3 kWh/(m^2 y) and dH is equal to 492.5 h/y, and it is a good solution if compared with BC since ΔEP is -4.6% and the percentage reduction of discomfort hours is -0.7%.

The configuration that minimizes the number discomfort hours is $SMARTBOARD^{TM}26\text{-}PI_{0140}A_{2v}PU_{10}$ (dH = 491.2 h/y), but a very close result is obtained with $WEBER.MUR\ CLIMA^{®}\text{-}PI_{1121}A_{2nv}PU_{10}$ (dH = 491.3 h/y). At the same time, these configurations do not coincide with the one with the lowest primary energy requirement; in both cases the lower level of insulation, on the internal side, negatively affects the heating energy needs. Compared to the basic design, ΔEP is, respectively, +2.0% and +1.6%; instead, compared with BC, the achievable energy saving is, respectively, -2.8% and -3.2%.

In Figure 7, the points inside the red circle are characterized by the adoption of Weber.mur clima® plaster or the Comfortboard® panel on the internal side. Among these points, the $WEBER.MUR\ CLIMA^{®}\text{-}PI_{707}A_{2nv}PU_{10}$ solution has the lowest total primary energy need (EP_{yearly} = 1524.3 kWh/m^2 y), with ΔEP equal to -5.4% compared with BC. In the center of the figure (black points), the solutions without PCM are distributed; herein, the solution $PI_{464}A_{2v}PU_{10}$ is highlighted. In this area, the outermost points are related to the configurations with a maximum thickness of the external and internal polyurethane (therefore with a lower energy requirement for heating). Finally, the solutions with Dry Clay Plaster, the SmartBoard^{TM} panel, and the ENERCIEL® finish with the maximum level of thermal insulation are collected in the blue area. The case with minimum value of total primary energy need is $ENERCIEL^{®}23\text{-}PI_{707}A2_{nv}PU_{10}$, with ΔEP equal to -5.4% compared with BC, but only -0.8% compared with $PI_{464}A_{2v}PU_{10}$. This solution benefits from the high insulation level, meanwhile the PCM allows to reduce the cooling request compared with BC; however, considering the reduced difference with $PI_{464}A_{2v}PU_{10}$, it can be stated that for the considered climate and selected melting temperature, the PCM is not highly effective in reducing the whole energy demand.

In terms of EP_{yearly}, the worst cases with PCM are characterized by the lower insulation thickness on internal and external sides, and these are $COMFORTBOARD^{®}23\text{-}PI_{0140}A_{6v}PU_{6}$, $WEBER.MUR\ CLIMA^{®}23\text{-}PI_{0140}A_{6v}PU_{6}$, and $DRY\ CLAY\ PLASTER\ PCM^{®}23\text{-}PI_{0140}A_{6v}PU_{6}$; these have the same energy need of BC (around 54.8 kWh/m2 y) but lower discomfort hours thanks to the effect of PCM in the hottest period. Instead, considering the basic design solution, ΔEP *is* +4.6%.

3.2. Parametric Analysis with PCM on the External Side

3.2.1. Analysis of Heating and Cooling Energy Needs

With regard to the application of the CSMs on the external side, Figure 8 shows the distribution of the results in terms of energy need for heating and cooling. Herein, with different markers, the solutions without PCMs are represented, as well as the panels with different thickness of phase change material; it must be remarked that among these, six different macroencapsulated materials are considered (six different melting points). For the simulations, the PCMs are positioned in the external air gap for favoring the discharge process during the night. However, negligible differences are observed considering the ventilation type of the air gap. The points with the lower E_H are characterized by the maximum value (14 cm) of the thermal insulation on the internal side.

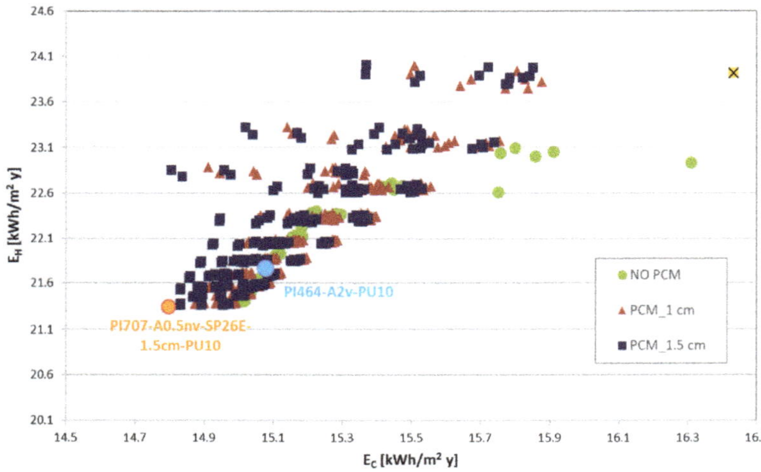

Figure 8. PCMs on external side: comparison based on heating and cooling load.

As expected, the cases with PCM thickness of 1.5 cm offer the greatest reductions in cooling requirements. For any configuration, the best phase change material to be placed on the external side is the one indicated by SP and a melting temperature of 26 °C with a thickness of 1.5 cm. Probably, for the considered climate, higher melting temperatures, if also useful to front extreme weather conditions, are not able to ensure a continued charge and discharge cycle of the material, and the transition is activated fewer times.

By adding to the aforementioned characteristics the maximum level of thermal insulation provided inside and outside, the best case is reached both in terms of E_C and E_H. In Figure 8 this is indicated with $PI_{707}A_{0.5nv}SP26E\text{-}1.5cm\text{-}PU_{10}$, with E_C = 14.8 kWh/(m^2 y) and E_H = 21.3 kWh/(m^2 y). Compared to the base case, ΔE_C is –9.95% while ΔE_H is −10.8%. Furthermore, differently from what was expected, the best configuration is characterized by an unventilated air cavity with the minimum thickness; this probably happens because the high value of insulation on the external side induced a delay in the phase transition, and the ventilation of the air cavity does not provide significant variations compared to the non-ventilated case. However, it can be observed that for the proposed wall package, when two PCMs have the same melting temperature and latent heat capacity also with a different thermal capacity, the performances are comparable. Indeed, in the proposed case study, $PI_{707}A_{0.5nv}RT26E\text{-}1.5cm\text{-}PU_{10}$, E_C = 14.9 kWh/(m^2 y) and E_H = 21.4 kWh/(m^2 y), and, thus, compared to the base case, ΔE_C is –9.4% while ΔE_H is −10.6%.

Considering the same configuration, with the same PCM, the increment of melting temperature does not guarantee higher energy saving. For instance, in the cases of $PI_{707}A_{0.5nv}SP26E\text{-}1cm\text{-}PU_6$ and $PI_{707}A_{0.5nv}SP31\text{-}1cm\text{-}PU_6$, with reference to the base case, ΔE_C is, respectively, -8.68% and −8.28%; with $PI_{707}A_{0.5nv}RT26E\text{-}1cm\text{-}PU_6$ and $PI_{707}A_{0.5nv}RT31\text{-}1cm\text{-}PU_6$, ΔE_C is, respectively, -8.49% and −8.18%. Comparable differences are obtained in all other cases. It can be also remarked that the products with higher melting temperature are also characterized by higher latent heat capacity; however, this characteristic seems to not improve the energy performance for the proposed wall package. Probably, the choice of an activation temperature of 26 °C is more advantageous given the external forcing of the considered location. The higher level of insulation does not allow the complete melting of the material, or the choice of air-gap thickness is not enough to assure a regular evacuation of the accumulated heat.

3.2.2. Analysis of Primary Energy Need and Discomfort Hours

For the application of the CSMs on the external side, Figure 9 shows the comparison between EP_{yearly} and dH.

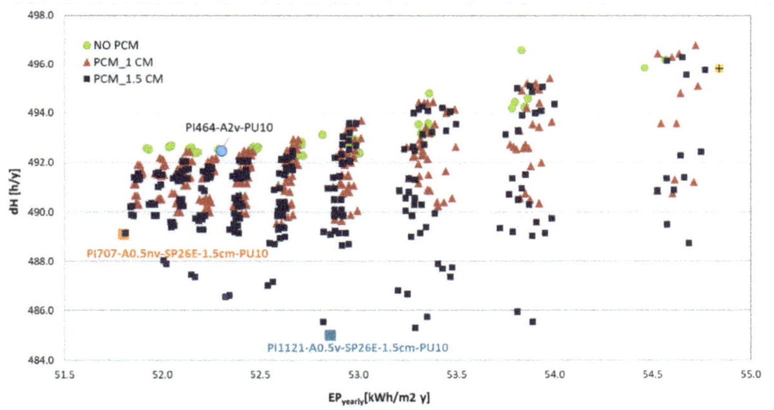

Figure 9. PCMs on external side: comparison based on primary energy need and discomfort hours.

The solutions are organized in several vertical distributions; each one is characterized by the same level of insulation but the configurations with the higher thickness of PCM are usually on the bottom of these lines. For instance, the first distribution on the right, where the base case is also highlighted, is composed of all solutions with the lower insulation level on the internal and external side. Herein, there is $PI_{0140}\text{-}A_{5nv}\text{-}RT26\text{-}1cm\text{-}PU6$ characterized by the lower energy demand (EP_{yearly} = 54.5 kWh/m^2 y) and dH = 490.8 h/y and $PI_{0140}\text{-}A_{4.5nv}\text{-}SP26E\text{-}1.5cm\text{-}PU6$ with the lower discomfort hours (dH = 489.5 h/y) and EP_{yearly} = 54.6 kWh/m^2 y. In both cases, when the melting temperature increases, the discomfort hours increase and dH becomes, respectively, 496.5 h/y and 495.8 h/y. This confirms that the most suitable melting point for the configuration under evaluation is 26 °C.

In the first line on the left, there are the solutions with the high levels of insulation (both on inner and outer sides) and the lowest energy needs. Different PCMs are coupled with the considered package: in the case of 1 cm there is an air gap of 1 cm, and the best solutions are RT26, RT28HC, and SP29Eu; with 1.5 cm and an air gap of 0.5 cm, the best solutions are RT26, SP26E, and SP29Eu. Among these, the minimum of EP_{yearly} is obtained with the $PI_{707}A_{0.5nv}SP26E\text{-}1.5cm\text{-}PU_{10}$ configuration, with a decreasing of ΔEP equal to −5.54% compared to the base case. Instead, if $PI_{464}\text{-}A_{2v}\text{-}PU_{10}$ is considered as reference, ΔEP is −0.95%. For the same configuration but adopting RT26, ΔEP is −5.44% and with SP31 it is −5.42%; more in general, there is a very slight difference with the other considered phase change materials and this confirms that the increment of melting temperature does not guarantee better performance.

At the bottom of the figure, there are the configurations with the lower number of discomfort hours. In this region it is possible to identify the configuration with the lowest number of hours of discomfort, namely, the one identified as $PI_{1121}A_{0.5v}SP26E\text{-}1.5cmPU_{10}$ is characterized by $\Delta dH_{\%}$ equal to −2.18% and $\Delta EP_{\%}$ of −3.62% compared to the base case. It has a thermal capacity equal to 102.2 kJ/m^2 K, i.e., increased by 78.6% compared to the same case without PCM and by 53% compared to $PI_{464}\text{-}A_{2v}\text{-}PU_{10}$. Compared with this basic design configuration, ΔEP is + 1.06% due to the low level of thermal insulation on the internal side of the wall.

3.3. Application of PCM Materials on Both Sides

3.3.1. Analysis of Heating and Cooling Energy Needs

The better configurations described in the previous section were combined for a further analysis. Table 5 summarizes the characteristics of these configurations and Figure 10 shows the simulation results, considering both the BC and the basic design configuration.

Table 5. Solutions for parametric analysis with PCM on both sides.

PCM on Internal Side	PU-Air Gap-PU	Air Gap-CSM-PU
Weber.mur clima® ENERCIEL® SmartBoard™ MOC with Micronal PCM	7-0-7 cm 1-12-1 cm 3-8-3 cm 5-4-5 cm 4-6-4 cm	A_{5v}RT31-1 cm-PU_6 $A_{0.5nv}$SP26E-1.5 cm-PU_{10} $A_{0.5v}$SP26E-1.5 cm-PU_{10}

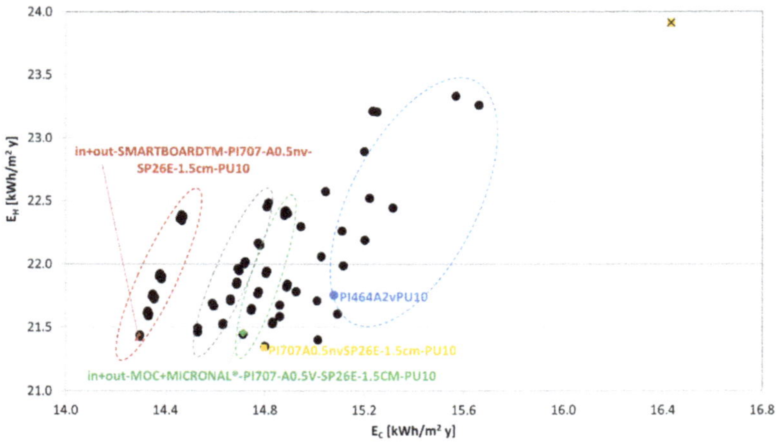

Figure 10. PCMs on both sides: comparison based on heating and cooling load.

Since fewer cases were evaluated, there is a clearer distinction between the results obtained for the examined configurations. Starting from the left, the first branch (red area) includes configurations with the SmartBoard™ or ENERCIEL® combined with 1.5 cm SP26E and 10 cm of external polyurethane. In this area, there is the configuration that minimizes E_C, that is, the one identified as *in + out-SMARTBOARD™*-$PI_{707}\text{-}A_{0.5nv}\text{-}SP26E\text{-}1.5cm\text{-}PU_{10}$. Compared to the base case, ΔE_C is −13% and it is −5.2% compared to $PI_{464}\text{-}A_{2v}\text{-}PU_{10}$. Its thermal capacity is equal to 124.26 kJ/m^2 and the heating energy need is reduced by −10.4% compared to the base case. Negligible differences are observed between this solution and the ones identified as *in + out-ENERCIEL®*-$PI_{707}A_{0.5v}SP26E\text{-}1.5cm\text{-}PU_{10}$ and *in + out-ENERCIEL®*-$PI_{707}A_{0.5nv}SP26E\text{-}1.5cm\text{-}PU_{10}$ that are included in the same branch.

The solution that minimizes E_H is $PI_{707}A_{0.5nv}SP26E\text{-}1.5cm\text{-}PU_{10}$ (−10.8% compared to the base case), but in this case the cooling energy need is higher, by +3.4% compared with *in + out-SMARTBOARD*^{TM}-PI_{707}-$A_{0.5nv}$-$SP26E$-$1.5cm$-PU_{10}; meanwhile, the heating demand is lower only by 0.4%.

In the second branch (gray area), the points with the lowest energy need for heating are characterized by SmartBoard™ or ENERCIEL® and the maximum thickness of the external polyurethane. In the upper part, there are the configurations with Weber.mur clima® combined with 1.5 cm SP26E and 10 cm external polyurethane. In the third branch (green area), there is the best solution with MOC + Micronal PCM, indicated with *in + out + MOC + MICRONAL*-$PI_{707}A_{5v}SP26E\text{-}1.5cm\text{-}PU_{10}$. Compared to the base case, $\Delta E_C = -10.5\%$ and $\Delta E_H = -10.3\%$.

Finally, in the fifth branch (blue area), there are the basic design configuration and configurations with 1 cm of RT31 combined with 6 cm external polyurethane on the external side and MOC + MICRONAL or Weber.mur clima® on the internal side.

More in general, it can be observed that the MOC with microencapsulated PCM performs worse in the cooling season than the other components evaluated for the internal application of phase change materials. This is mainly due to the low latent heat of fusion (9.2 kJ/kg). However, the other characteristics and the melting temperature make the performance of the MOC with PCM comparable to the plaster with Weber.mur clima®. For instance, considering the combinations *in + out + MOC + MICRONAL*-PI_{545}-A_{5v}-$RT31$-$1cm$-PU_6 and *in + out + WEBER.MUR CLIMA®*-PI_{545}-A_{5v}-$RT31$-$1cm$-PU_6, the heating demand is, respectively, 15.1 kWh/(m^2 ye) and 15.0 kWh/(m^2 year), and E_C is 22.0 kWh/(m^2 year) and 22.1 kWh/(m^2 year).

Furthermore, this analysis also confirms that the combination of SmartBoard™ and SP26 allows minimizing the energy need for cooling.

3.3.2. Analysis of Primary Energy Need and Discomfort Hours

The comparison in terms of the primary energy need and the hours of indoor discomfort is reported in Figure 11.

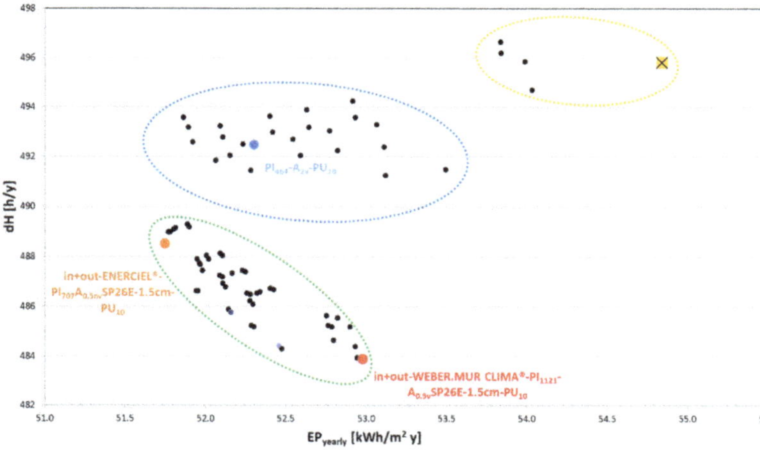

Figure 11. PCMs on both sides: comparison based on heating and cooling load.

Three groups of points were identified. The yellow one includes the base case and the solutions with the higher request of primary energy. Among these, the *in + out-ENERCIEL®*-$PI_{1121}A_{5v}$-$RT31$-$1cm$-PU_6 solution has the highest number of hours of discomfort ($dH \approx 497$ h/y), while the solution indicated by *in + out-MOC + MICRONAL*-$PI_{1121}A_{5v}RT31\text{-}1cm\text{-}PU_6$ implies the highest EP_{yearly}, equal to 54 kWh/(m^2 year).

In the central region there is the basic design configuration case (blue area). All other configurations are characterized by adoption of 1 cm of RT31 and 6 cm of insulation on the external side. The most interesting combinations, with ΔEP of 5.0%, are characterized on the internal side by the adoption of MOC + MICRONAL or Weber.mur clima® with a small variation of discomfort hours (−0.7%).

The most interesting region is evidenced with a green circle. Herein, there are solutions with SP26E and the maximum thickness of the external polyurethane, and in most cases with Weber.mur clima® plaster on the inner side.

However, the *in + out-ENERCIEL®-*$PI_{707}A_{0.5nv}SP26E\text{-}1.5cm\text{-}PU_{10}$ solution is the one that minimizes the total primary energy requirement. For this solution, $\Delta EP \approx -5.6\%$ and dH is 488 h/year. The solution with the lower number of discomfort hours ($dH \approx 484$ h/year) is *in + out-WEBER.MUR CLIMA®-*$PI_{1121}A_{0.5v}SP26E\text{-}1.5cm\text{-}PU_{10}$. In this case, the thermal transmittance is 0.188 W/m^2 K and thermal capacity is equal to 112 kJ/m^2 K. The energy consumption increases and ΔEP is +3.4% due to the low level of thermal insulation on the internal side.

3.4. Solution of the Multi-Objective Optimization

The second step of the proposed approach consists of the solution of a multi-objective optimization problem in order to minimize the energy need for heating and cooling by varying the orientation of the wall on which the wall package is installed. Based on the results obtained in the previous sections, the best solutions to be considered are *in + out-ENERCIEL®-*$PI_{707}A_{0.5nv}SP26E\text{-}1.5cm\text{-}PU_{10}$, $PI_{707}A_{0.5nv}SP26E\text{-}1.5cm\text{-}PU_{10}$, and *in + out-SMARTBOARD™-*$PI_{707}A_{0.5nv}SP26E\text{-}1.5cm\text{-}PU_{10}$. The base case and the basic configuration were also added to these solutions. The constraint is dH, equal to 491.3 h/year (minimum value according to the parametric analysis). The results are shown in Figure 12; here, the red dots represent the Pareto solutions. In detail, three optimal solutions were observed, as shown in Table 6, where it is clear that there are negligible differences both in term of heating and cooling request.

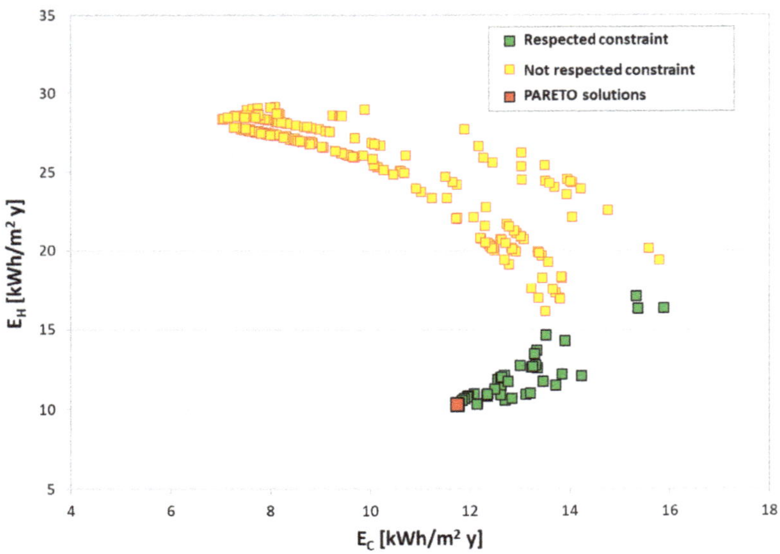

Figure 12. PCMs optimization: comparison based on heating and cooling load.

Table 6. Results of the optimal configurations.

Configuration	Orientation (°)	Exposure	E_C (kWh/m^2 y)	E_H (kWh/m^2 y)	dH (h/y)
in + out-ENERCIEL®- $PI_{707}A_{0.5nv}SP26E\text{-}1.5cm\text{-}PU_{10}$	336	north-east	11.73	10.31	415
$PI_{707}A_{0.5nv}SP26E\text{-}1.5cm\text{-}PU_{10}$	273	north	11.74	10.30	414
in + out-SMARTBOARD™- $PI_{707}\text{-}A0.5_{nv}\text{-}SP26E\text{-}1.5cm\text{-}PU_{10}$	332	north-east	11.74	10.30	415

The three cases are characterized by the maximum thermal insulation on the external and internal side of the wall, which allows minimizing the heating losses but it can cause an increment of the summer overheating; for this reason, the best orientation was sought. All configurations on the south exposure are characterized by the higher heating demand and the lower cooling load. This happens because the PCM reduces the utilization of solar gains during the wintertime.

Among the simulated ones, the configurations with internal and external application have the greatest potential mainly when the products ENERCIEL® and SmartBoard™ are applied on the internal side. These allow increasing the thermal capacity of the wall package, mitigating internal thermal fluctuations, reducing overheating, and ensuring greater microclimatic comfort. On the outside, the module with SP26E and thickness of 1.5 cm allows better results. Indeed, the nature of the incorporated PCMs (inorganic compounds with higher latent heat of fusion compared to organic compounds) and the melting point compatible with the average external temperature help to counteract the incoming heat flow in the cooling season. This effect limits indoor temperature peaks and provides greater microclimatic comfort. The northeast orientation is the best, probably because it requires greater isolation; meanwhile, in summer months, the sun's rays strike the wall perpendicularly in the morning and in the afternoon and the wall requires a material with a higher latent accumulation capacity. Indeed, for this exposure, the glass component of the test room is oriented to the southwest and is therefore exposed to intense solar radiation which would increase the solar gains. For this reason, it is required to contain the overheating in the environment with the material able to accumulate more latent heat.

Another interesting conclusion is that when the north exposure is selected, the adoption of one PCM as SP26E on the external side is enough to reach the proposed objectives.

In particular, the solution $PI_{707}A_{0.5nv}SP_{26E}\text{-}1.5cm\text{-}PU_{10}$ provides the lowest number of discomfort hours compared to the other solutions. This happens because the northern exposure in the cooling season is the side with the lowest surface temperature on the wall.

In all three cases, the heating and cooling loads are comparable; in particular, ΔEc is around −57% and ΔE_H is near −29%, compared to the base case, and these indices are, respectively, −22% and −53% compared to the basic design configuration.

4. Conclusions

In order to establish a wall package made of geopolymer concrete and phase change material that can provide sustainable solutions for building performance in Mediterranean climates, and to find the main outcomes of the overall design, a double step analysis is proposed for determining an optimal configuration of a real case study.

A numerical analysis is proposed by means of a dynamic simulation tool (EnergyPlus) for a simple building consisting of a large-scale test room (36 m^2), available at the Department of Engineering of University of Sannio. This was selected as the case study because in these next months, some experimental activities will be performed in the frame of the European project "Green Integrated Structural Elements for Retrofitting and New Construction of Buildings". The proposed dissertation is the starting point for investigating the effects of types and thickness of insulation, air gap, and PCM types for a base package under evaluation in the project. More in detail, the base wall of the test room with vacuum insulation panels and the base configuration of the project with internal panel with 4 cm polyurethane and 6 cm air gap (not ventilated), 2 cm of non-ventilated external air gap, and 10 cm of polyurethane on external side were compared with other packages in term of minimization of heating and cooling loads, minimization of discomfort hours, and yearly primary energy need.

Tree solutions were placed on the Pareto front; the one with north exposure requires the application only on the internal side of a PCM with melting temperature of 26 °C, and compared with the base case, ΔE_H is −57% and ΔE_C is −22%. This solution also minimizes the discomfort hours (−17%). Comparable energy saving can be reached with the application of two phase change materials (on the inner and outer sides) with melting temperature of 26 °C. All three solutions require the maximum insulation on internal and external sides, a not-ventilated air gap, and 1.5 cm of PCM thickness.

Author Contributions: Conceptualization, F.A., R.F.D.M., F.D.R. and S.R.; methodology, F.D.R. and S.R.; formal analysis, R.F.D.M., N.D.R. and A.G.; investigation, R.F.D.M., N.D.R. and A.G.; data curation, N.D.R., A.G. and S.R.; writing—original draft preparation, F.A. and R.F.D.M.; supervision, F.A. and F.D.R. All authors have read and agreed to the published version of the manuscript.

Funding: Horizon 2020 "Green INSTRUCT", grant number 723825.

Institutional Review Board Statement: Not applicable.

Informed Consent Statement: Not applicable.

Data Availability Statement: Data can be required to the corresponding authors.

Conflicts of Interest: The authors declare no conflict of interest.

References

1. Available online: https://ec.europa.eu/clima/eu-action/european-green-deal/delivering-european-green-deal_en (accessed on 15 June 2022).
2. Soliman, A.; Hafeez, G.; Erkmen, E.; Ganesan, R.; Ouf, M.; Hammad, A.; Eicker, U.; Moselhi, O. Innovative construction material technologies for sustainable and resilient civil infrastructure. *Mater. Today Proc.* **2022**, *60*, 365–372. [CrossRef]
3. Xie, J.; Zhao, J.; Wang, J.; Fang, C.; Yuan, B.; Wu, Y. Impact behaviour of fly ash and slag-based geopolymeric concrete: The effects of recycled aggregate content, water-binder ratio and curing age. *Constr. Build. Mater.* **2022**, *331*, 127359. [CrossRef]
4. Verma, M.; Dev, N.; Rahman, I.; Nigam, M.; Ahmed, M.; Mallick, J. Geopolymer Concrete: A Material for Sustainable Development in Indian Construction Industries. *Crystals* **2022**, *12*, 514. [CrossRef]
5. Mohammed, B.S.; Liew, M.S.; Alaloul, W.S.; Al-Fakih, A.; Ibrahim, W.; Adamu, M. Development of rubberized geopolymer interlocking bricks. *Case Stud. Constr. Mater.* **2018**, *8*, 401–408. [CrossRef]
6. Gerges, N.N.; Issa, C.A.; Fawaz, S.A. Rubber concrete: Mechanical and dynamical properties. *Case Stud. Constr. Mater.* **2018**, *9*, e00184. [CrossRef]
7. Cui, Y.; Hao, H.; Li, J.; Chen, W.; Zhang, X. Structural behavior and vibration characteristics of geopolymer composite lightweight sandwich panels for prefabricated building. *J. Build. Eng.* **2022**, *57*, 104872. [CrossRef]
8. Longo, F.; Lassandro, P.; Moshiri, A.; Phatak, T.; Aiello, M.A.; Krakowiak, K.J. Lightweight geopolymer-based mortars for the structural and energy retrofit of buildings. *Energy Build.* **2020**, *225*, 110352. [CrossRef]
9. Colangelo, F.; De Luca, G.; Ferone, C.; Mauro, A. Experimental and Numerical Analysis of Thermal and Hygrometric Characteristics of Building Structures Employing Recycled Plastic Aggregates and Geopolymer Concrete. *Energies* **2013**, *6*, 6077–6101. [CrossRef]
10. Wu, C.H.; Huang, C.H.; Li, Y.F.; Lee, W.H.; Cheng, T.W. Utilization of Basic Oxygen Furnace Slag in Geopolymeric Coating for Passive Radiative Cooling Application. *Sustainability* **2020**, *12*, 3967. [CrossRef]
11. Su, Z.; Hou, W.; Sun, Z.; Lvnd, W. Study of In Situ Foamed Fly Ash Geopolymer. *Materials* **2020**, *13*, 4059. [CrossRef]
12. Parcesepe, E.; De Masi, R.F.; Lima, C.; Mauro, G.M.; Maddaloni, G.; Pecce, M.R. Experimental Evaluation of the Mechanical Strengths and the Thermal Conductivity of GGBFS and Silica Fume Based Alkali-Activated Concrete. *Materials* **2021**, *14*, 7717. [CrossRef]
13. Dhasindrakrishna, K.; Pasupathy, K.; Ramakrishnan, S.; Sanjayan, H. Progress, current thinking and challenges in geopolymer foam concrete technology. *Cem. Concr. Compos.* **2021**, *116*, 103886. [CrossRef]
14. Zhang, Z.; Provis, J.L.; Reid, A.; Wang, H. Mechanical, thermal insulation, thermal resistance and acoustic absorption properties of geopolymer foam concrete. *Cem. Concr. Compos.* **2015**, *62*, 97–105. [CrossRef]
15. Wang, Y.; Zheng, T.; Zheng, X.; Liu, Y.; Darkw, J.; Zhou, G. Thermo-mechanical and moisture absorption properties of fly ash-based lightweight geopolymer concrete reinforced by polypropylene fibers. *Constr. Build. Mater.* **2020**, *251*, 118960. [CrossRef]

16. Henon, J.; Alzina, A.; Absi, J.; Smith, D.S.; Rossignol, S. Analytical estimation of skeleton thermal conductivity of a geopolymer foam from thermal conductivity measurements. *Eur. Phys. J. Spec. Top.* **2015**, *224*, 1715–1723. [CrossRef]
17. Łach, M.; Pławecka, K.; Bąk, A.; Adamczyk, M.; Bazan, P.; Kozub, B.; Korniejenko, K.; Lin, W. Review of Solutions for the Use of Phase Change Materials in Geopolymers. *Materials* **2021**, *14*, 6044. [CrossRef]
18. Asadi, I.; Baghban, H.M.; Hashemi, M.; Izadyar, N.; Sajadi, B. Phase change materials incorporated into geopolymer concrete for enhancing energy efficiency and sustainability of buildings: A review. *Case Stud. Constr. Mater.* **2022**, *17*, 01162. [CrossRef]
19. Hassan, A.; Rashid, Y.; Mourad, A.H.I.; Ismail, N.; Laghari, M.S. Thermal and Structural Characterization of Geopolymer-Coated Polyurethane Foam—Phase Change Material Capsules/Geopolymer Concrete Composites. *Materials* **2019**, *12*, 796. [CrossRef]
20. Ramakrishnan, S.; Pasupathy, K.; Sanjayan, J. Synthesis and properties of thermally enhanced aerated geopolymer concrete using form-stable phase change composite. *J. Build. Eng.* **2021**, *40*, 102756. [CrossRef]
21. Hassan, A.; Mourad, I.A.H.; Rashid, Y.; Ismail, N.; Laghari, M.S. Thermal and structural performance of geopolymer concrete containing phase change material encapsulated in expanded clay. *Energy Build.* **2019**, *191*, 72–81. [CrossRef]
22. Cao, V.D.; Bui, T.Q.; Kjøniksen, A.L. Thermal analysis of multi-layer walls containing geopolymer concrete and phase change materials for building applications. *Energy* **2019**, *186*, 115792. [CrossRef]
23. Pilehvar, S.; Sanfelix, S.G.; Szczotok, A.M.; Rodríguez, J.F.; Valentini, L.; Lanzón, M.; Pamies, R.; Kjøniksena, A.L. Effect of temperature on geopolymer and Portland cement composites modified with Micro-encapsulated Phase Change materials. *Constr. Build. Mater.* **2020**, *252*, 119055. [CrossRef]
24. Cao, V.D.; Pilehvar, S.; Salas-Bringas, C.; Szczotok, A.M.; Rodriguez, J.F.; Carmona, M.; Al-Manasir, N.; Kjøniksen, A.-L. Microencapsulated phase change materials for enhancing the thermal performance of Portland cement concrete and geopolymer concrete for passive building applications. *Energy Convers. Manag.* **2017**, *133*, 56–66. [CrossRef]
25. Cao, V.D.; Pilehvar, S.; Salas-Bringas, C.; Szczotok, A.M.; Valentini, L.; Carmona, M.; Al-Manasir, N.; Kjøniksen, A.-L. Influence of microcapsule size and shell polarity on thermal and mechanical properties of thermoregulating geopolymer concrete for passive building applications. *Energy Convers. Manag.* **2018**, *164*, 198–209. [CrossRef]
26. *Standard 55*; Thermal Environmental Conditions for Human Occupancy. ASHRAE: Peachtree Corners, GA, USA, 2004.
27. De Masi, R.F.; de Rossi, F.; Ruggiero, S.; Vanoli, G.P. Numerical optimization for the design of living walls in the Mediterranean climate. *Energy Convers. Manag.* **2019**, *195*, 573–586. [CrossRef]
28. Carlucci, S.; Cattarin, G.; Causone, F.; Pagliano, L. Multi-objective optimization of a nearly zero-energy building based on thermal and visual discomfort minimization using anon-dominated sorting genetic algorithm (NSGA-II). *Energy Build.* **2015**, *104*, 378–394. [CrossRef]
29. Nguyen, A.T.; Reiter, S.; Rigo, P. A review on simulation based optimization methods applied to building performance analysis. *Appl. Energy* **2014**, *113*, 1043–1058. [CrossRef]
30. Ascione, F.; De Masi, R.F.; de Rossi, F.; Ruggiero, S.; Vanoli, G.P. MATRIX, a multi activity test-room for evaluating the energy performances of 'building/HVAC' systems in Mediterranean climate: Experimental set-up and CFD/BPS numerical modeling. *Energy Build.* **2016**, *126*, 424–446. [CrossRef]
31. U.S. Department of Energy. *Energy Plus Simulation Software*, Version 8.9; U.S. Department of Energy: Washington, DC, USA, 2018.
32. DesignBuilder Software. Available online: https://designbuilder.co.uk/software/for-engineers (accessed on 20 June 2022).
33. ASHRAE—American Society of Heating, Refrigerating and Air-Conditioning Engineers. *Guideline 14: Measurement of Energy and Demand and Savings*; 1791 Tullie Circle: NE; ASHRAE: Atlanta, GA, USA, 2002; ISSN 1049-894X.
34. *UNI EN ISO 7730: 2006*; Ergonomics of the Thermal Environment—Analytical Determination and Interpretation of Thermal Comfort using Calculation of the PMV and PPD Indices and Local thermal Comfort Criteria. Italian Organisation for Stardardisation (UNI): Milan, Italy, 2006.
35. *UNI TS 11300-1:2014*; Energy Performance of Buildings—Part 1—Evaluation of Energy Need for Space Heating and Cooling. Italian Organisation for Stardardisation (UNI): Milan, Italy, 2014.
36. Rubel, F.; Kottek, M. Observed and projected climate shifts 1901-2100 depicted by world maps of the Köppen-Geiger climate classification. *Meteorol. Z.* **2010**, *19*, 135–141. [CrossRef]
37. Report D6.1. Available online: https://www.greeninstruct.eu/wp-content/uploads/2018/07/D-6_1_GEOPOLYMER-PRECURSORS.pdf (accessed on 16 May 2022).
38. *UNI EN ISO 6946:2008*; Building Components and Building Elements—Thermal Resistance and Thermal Transmittance—Calculation Method. Italian Organisation for Stardardisation (UNI): Milan, Italy, 2008.
39. *UNI 10351:2015*; Building Materials and Products—Hygrothermal Proprieties-Procedure for Determining the Design Values. Italian Organisation for Stardardisation (UNI): Milan, Italy, 2015.
40. *UNI EN ISO 10456:2008*; Building Materials and Products—Hygrothermal Properties-Tabulated Design Values and Procedures for Determining Declared and Design Thermal Values. Italian Organisation for Stardardisation (UNI): Milan, Italy, 2008.
41. Ministerial Decree 26.06.2015, Annex B: Minimum Requirements for Existing Buildings Subject to Energy Refurbishment. 2015. Available online: https://www.gazzettaufficiale.it/eli/id/2015/07/15/15A05198/sg (accessed on 12 September 2021). (In Italian).
42. Manso, M.; Castro-Gomes, J.P. Thermal analysis of a new modular system for green walls. *J. Build. Eng.* **2016**, *7*, 53–62. [CrossRef]
43. De Jesus, M.P.; Lourenço, J.M.; Arce, R.M.; Macias, M. Green facades and in situ measurements of outdoor building thermal behavior. *Build. Environ.* **2017**, *119*, 11–19. [CrossRef]

44. Microtek Laboratories. Available online: https://www.microteklabs.com/ (accessed on 1 September 2021).
45. Micronal-App. Available online: http://micronal-app.microteklabs.com/ (accessed on 10 June 2021).
46. Ozdenefe, M.; Dewsbury, J. T+hermal performance of a typical residential Cyprus building with phase change materials. *Build. Serv. Eng. Res. Technol.* **2015**, *37*, 85–102. [CrossRef]
47. Rubitherm. Available online: https://www.rubitherm.eu/en/productCategories.html (accessed on 4 July 2021).
48. Karthikeyan, N.; Sathishkumar, A.; Dennis Joseph Raj, W. Effects on setting, strength and water resistance of sorel cement on mixing fly ash as an additive. *Int. J. Mech. Eng. Robot. Res.* **2014**, *3*, 252–256.
49. Tabares-Velasco, P.C.; Christensen, C.; Bianchi, M. Verification and validation of EnergyPlus phase change material model for opaque wall assemblies. *Build. Environ.* **2012**, *54*, 186–196. [CrossRef]
50. Deb, K.; Pratap, A.; Agarwal, S.; Meyarivan, T. A fast and elitist, Multiobjective Genetic Algorithm: NSGA-II. *IEEE Trans. Evol. Comput.* **2002**, *6*, 182–197. [CrossRef]

buildings

MDPI

Article

A Numerical Method for Solving Global Irradiance on the Facades of Building Stocks

Haowei Xing [1,2], Yi Yang [2,3,*] and Shuqin Chen [1]

1 College of Civil Engineering and Architecture, Zhejiang University, Hangzhou 310027, China
2 The Architectural Design and Research Institute of Zhejiang University Co., Ltd., Hangzhou 310027, China
3 Center for Balance Architecture, Zhejiang University, Hangzhou 310027, China
* Correspondence: uadresearch@163.com

Abstract: Due to the influence of surrounding buildings on the radiation transfer process, the irradiance of individual buildings in building stocks is more uneven and different than that of individual buildings in open spaces. In view of the defect of the existing building surface irradiance calculation model in the sky radiation energy balance calculation, the complex surface reflection radiative transfer in diffuse irradiance, and complex processes, this paper combined the calculation of the complex surface narrow sky view, multiple reflections, and radiation characteristics of nonuniformity, and finally established the model for irradiance on the facade of a building stock (IFBS model) in a sheltered environment. The simulation results show that the IFBS model is superior to the traditional model in the calculation of sky diffuse irradiance and reflection irradiance of building stocks and is more suitable for the numerical calculation of the radiation transfer process of complex buildings.

Keywords: irradiance; building stock; building facade

Citation: Xing, H.; Yang, Y.; Chen, S. A Numerical Method for Solving Global Irradiance on the Facades of Building Stocks. *Buildings* **2022**, *12*, 1914. https://doi.org/10.3390/buildings12111914

Academic Editors: Shi-Jie Cao and Wei Feng

Received: 20 September 2022
Accepted: 2 November 2022
Published: 7 November 2022

1. Introduction

Radiation environment simulation is an important part of the performance simulation of the building complex. A building group performance simulation is generally attached to the third party building performance simulation module (e.g., BREDEM, EnergyPlus, DOE2). In the early days, due to the complexity of modeling and the calculation of single buildings, in order to reduce the demand for computing resources, scholars would first classify buildings into groups and then complete the calculation from single buildings to building groups by simply multiplying the number of similar buildings [1,2]. With the improvement of classification methods, scholars further classified similar buildings and made batch calculations, based on building shape [3,4], thermal parameters [5], building category [6,7] and other parameters. Although the classification processing can retain the thermal characteristics and other characteristic information of individual buildings, to a certain extent, the classification process itself is still the average processing of individual building information, so there is bound to be potential error when falling into the parameters of specific individual buildings. Along with the development of computer hardware and software, some scholars no longer categorize to simplify, but directly to each monomer building to simulate the whole compound, as per the obtained information. However, as a result of the third party building performance simulation module that is based on the monomer building performance simulation method, the impact of the surrounding buildings on the simulation of the target monomer building's outdoor thermal environment, is relatively lacking. Therefore, there are problems, such as data distortion and data interface accuracy mismatch in the input of boundary conditions (meteorological files, etc.). For example, in EnergyPlus, only one weather file can be input to a wall. The EnergyPlus framework cannot handle different boundary conditions on the same wall, which is particularly common in buildings. Due to the limitations of tools and methods, traditional building complex simulation methods will inevitably simplify or ignore the influence of surrounding building complexes on the outdoor thermal environment of individual buildings (such as the differences in radiant heat boundary conditions, local wind environment, etc.), thus losing some details of the radiant heat process description of building complexes.

Table 1 presents the information of the existing major simulation methods, in terms of the reflected radiation model and sky diffuse radiation models. By comparison, it can be concluded that the existing simulation methods mostly calculate the reflected radiation once. Further, the isotropic sky diffuse radiation distribution model is applied in most simulations, in order to simplify the calculation of the sky diffuse irradiance [8–10]. However, the accuracy of the isotropic model is acceptable under the assumption of a cloudy day or large sky view factor (SVF), which is not common in the building complex.

Table 1. Comparison of the methodology of existing buildings simulation models or software.

Model or Software *	Reflection Irradiance	Sky Diffuse Irradiance
Urban Canyon Model [11]	1 time reflection	isotropic
ENVI-met *	1 time reflection	isotropic
Town Energy Balance (TEB) [12]	Infinite reflection	isotropic
Temperature of Urban Facets in 3D (TUF-3D) [8]	Simplified multiple reflection	isotropic
Model for Urban Surface Temperature (MUST) [13]	Simplified multiple reflection	isotropic
DeST *	1 time reflection	isotropic
Fluent * + Solene *	1 time reflection	Perez model
TEB + EnergyPlus *	Infinite reflection	isotropic
Citysim *	1 time reflection	Perez model
INSEL * + ISO model	1 time reflection	Direct dispersion separation model
UMI *	Simplified multiple reflection	Perez model
Urban Energy Performance Calculator [14]	1 time reflection	isotropic

* Stands for software.

In the complex environment, the surrounding buildings will block the sky view field of the target building, which leads to a narrower view of the sky at the point on the surface of the target building. The narrow sky view will have an impact on the radiation transfer process of the sky diffuse radiation on the building surface, which will then have an impact on the numerical solution of the sky diffuse irradiance [15]. Sky diffuse radiation models are developed to interpret the distribution of sky diffuse radiation in the sky dome. Existing mainstream algorithms, based on the isotropic sky diffuse radiation model [16] or the Perez anisotropic sky diffuse radiation model [17,18] are under development. These two kinds of models use too many simplified assumptions, and their applicability in scenarios with a small sky view factor is limited. At the same time, it is also necessary to discuss whether the precision of the sky dome discretization needs to be improved in the current numerical calculation [19,20]. As for the reflection radiation, the current numerical method is based on Lambertian assumption and an applied finite reflection (mostly one time reflection) calculation algorithm, which leads to the heat gain of the reflected radiation being smaller [21], and the error will be significantly amplified with the increase of the building density [22].

To sum up, the study of building radiant heat processes should be based on a high-precision radiation model. The existing radiation simulation algorithms mostly follow the assumption of a single building. Therefore, there are many simplifications in the calculation of the reflection radiation and the sky diffuse radiation transfer. However, these simplifications can be adversely affected under certain conditions, especially in complex buildings. As a result, the calculation process cannot fully express the influence of building stocks on radiation transfer. The errors caused by this simplified algorithm, in calculating complex buildings and how to improve the algorithm to improve the simulation accuracy of radiation transfer processes under the influence of complex buildings remain to be studied.

In this work, a new numerical method will be constructed, based on the radiation transfer theory, and the algorithm in the solution process will be established to realize the accuracy improvement in the global irradiance simulation on the facades of building stocks. In Section 2, we analyze the radiant energy balance on the building surface, based on the net radiation analysis method. Further, we analyze the most appropriate theoretical algorithm for direct solar irradiance, sky diffuse irradiance, and reflection irradiance on the facades of building stocks. In Section 3, we develop the numerical solution of the radiation scheme. We mainly focus on the discretization of surfaces, constructing the matrix of each component of global irradiance, and obtaining a numerical equation for the global irradiance value of the building facades. In Section 4, we compare the new sky diffuse and reflection irradiance algorithm with the existing algorithm from the traditional method. The comparative analysis is based on the existing measurement study and the theoretical simulation in this work.

2. Radiation Scheme

2.1. Analysis of the Radiant Energy Balance on the Building Surface

Net radiation analysis was used to analyze the radiation energy on the surface i of building facades, as shown in Figure 1.

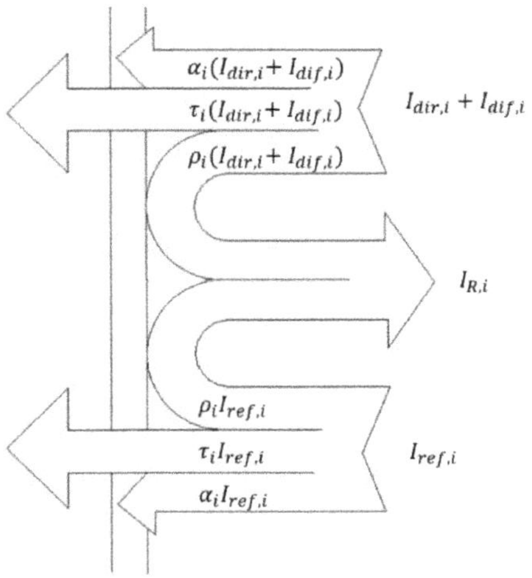

Figure 1. Schematic diagram of the analysis of the radiation energy balance on the element surface.

In the figure, α_i, τ_i, ρ_i represent the absorption rate, transmittance and reflectance of the element surface i, respectively, without dimensionality.

For a single element surface i, it can be seen from the schematic diagram of the radiation energy balance analysis, that the total irradiance received is:

$$I_{g,i} = I_{dir,i} + I_{dif,i} + I_{ref,i} \quad (1)$$

where $I_{dir,i}$, $I_{dif,i}$, $I_{ref,i}$, respectively, represent the direct solar irradiance value, sky diffuse irradiance value and the reflected irradiance from other microscopic surfaces received by surface i, W/m^2.

Meanwhile, the effective radiation (reflected radiation) intensity outward from element surface i is:

$$I_{R,i} = \rho_i I_{g,i} = \rho_i \left(I_{dir,i} + I_{dif,i} + I_{ref,i} \right) \tag{2}$$

The reflected irradiance received by a cell surface is the cumulative value of the reflected radiation intensity from other cells:

$$I_{ref,i} = \sum_{j=1}^{N} I_{R,j} F_{ij} \tag{3}$$

where F_{ij} is the shape factor of the surface i to the surface j, dimensionless. The shape factor F_{ij} is defined as the fraction of energy leaving a black surface element i that arrives at another black surface element j [23].

To sum up, the total radiation received by a single micro-element surface i can be calculated as:

$$I_{g,i} = I_{dir,i} + I_{dif,i} + \sum_{j=1}^{N} \rho_j I_{g,j} F_{ij} \tag{4}$$

2.2. Solar Direct Irradiance on the Facades of Building Stocks

2.2.1. Calculation of the Solar Orientation

In the calculation of the surface direct solar irradiance, the traditional method mainly applies the solar azimuth and solar altitude angles, which are calculated, based on latitude, declination angle, and solar hour angle, to locate the sun's position in the sky [13,24]. In the IFBS model, the same calculation equation is used to calculate the basic parameters related to the solar azimuth, such as the solar declination angle, solar angle, altitude angle, and azimuth angle.

The calculation equation of the solar altitude angle is as follows:

$$sinh = sin\delta sin\phi + cos\delta cos\phi cos\omega \tag{5}$$

where h is the solar altitude angle, °; ϕ is the local latitude, °; ω is solar hour angle, °; δ is declination angle, °.

The calculation equation of the solar azimuth is:

$$cosA = (sinhsin\phi - sin\delta)/coshcos\phi \tag{6}$$

2.2.2. Judgment of Occlusion

The solar radiation obtained on any plane is related to the incidence angle of sunlight on the plane. The intensity of the direct solar irradiance, received by the surface i at a fixed incidence angle can be calculated as [25]:

$$I_{dir,i} = I_0 cos\theta_i \tag{7}$$

where I_0 is the direct solar radiation intensity on the surface i without occlusion, W/m^2; θ_i is the solar incident angle of the micro-surface i, °.

In fact, the occlusion phenomenon is quite common in building stocks. Therefore, whether the plane is occluded, needs to be calculated. Due to the parallel characteristics of sunlight, the concept of "backward ray tracing" [26] is designed to make ray tracing more efficient. In backward ray tracing, an eye ray is created at one point; the vector direction of the ray is in the opposite direction of the sunlight. The sun will be visible from that point if the eye ray hits no object. In the backward ray tracing method, the calculation process will trace rays of interest and not consume resources tracing rays that do not contribute to the solution of the engineering/design problem. Therefore, the backward ray tracing method is wildly adopted in solar energy calculation [27–29] and will be applied in this paper.

Based on the above calculation of the sun's altitude angle and azimuth, the X-axis represents the east-west direction (positive east direction), the Y-axis represents the north-south direction (positive north direction), and the Z-axis represents the altitude direction (positive upward direction). The vector of the backward ray $\overrightarrow{\boldsymbol{I_{rev,i}}} = \left(I_{i,x}, I_{i,y}, I_{i,z}\right)$ from surface i at a given time

$$\begin{cases} I_{i,x} = coshsinA \text{ if solar time is less than 12 else } - coshsinA \\ I_{i,y} = -coshcosA \\ I_{i,z} = sinh \end{cases} \tag{8}$$

Figure 2 presents the spatial geometric relation between the backward ray and the building interface. Taking the roof's occlusion of light as an example, let $\alpha_j \in V_{roof}$, where Δd_j and Δl_j represent the size of the roof interface α_j on Y-axis and X-axis, respectively, and V_{roof} is the set of all roof interfaces. $\overrightarrow{\boldsymbol{I_{rev,i}}}$ represents the reverse beam vector and intersects with the plane of the roof interface α_j at the point A_{ij}. The geometric center of the roof interface α_j is A_0. Then, the judgment of the occlusion relation between the reverse beam and the roof interface α_j can be transformed into the judgment of A_{ij} and A_0 distance. If the distance between A_{ij} and A_0 is greater than $\Delta l_j/2$ on the X-axis or greater than $\Delta d_j/2$ on the Y-axis, it indicates that the reverse beam is not obscured by the roof interface α_j.

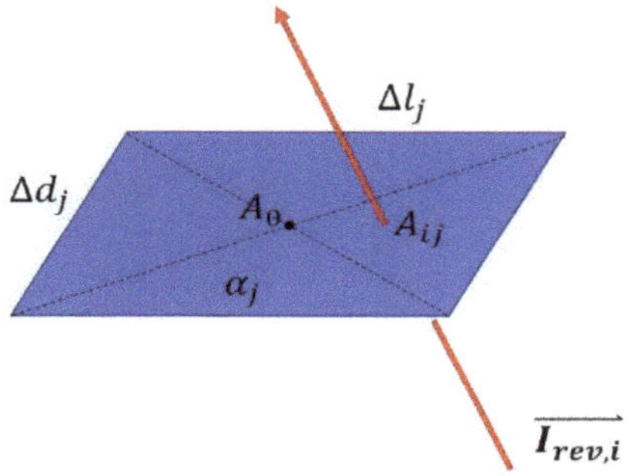

Figure 2. Light beam occlusion judgment, based on the backward ray tracing method.

The three-dimensional coordinates of A_{ij} and A_0 are (x_{ij}, y_{ij}, z_{ij}) and (x_0, y_0, z_0), respectively. Then the Boolean value of the occlusion judgment between element surface i and roof interface α_j, can be calculated as follows:

$$Flag_{ij} = |x_{ij} - x_0| > \Delta l_j/2 \mid |y_{ij} - y_0| > \Delta d_j/2 \tag{9}$$

where $Flag_{ij} = 1$ represents no occlusion. For all roof interfaces $\alpha_j \in V_{roof}$, after traversing the judgment process, it can be obtained:

$$\boldsymbol{Flag_{i-roof}} = (Flag_{i1}, Flag_{i2}, \ldots, Flag_{iN})^T \tag{10}$$

Similarly, for the set of other facades of the building, the value of $\boldsymbol{Flag_{i-wall}}$ can be obtained, and the calculation equation of the Boolean value of the final judgment of whether the element surface i is covered is as follows:

$$Flag_i = \boldsymbol{Flag_{i-roof}} \,\&\, \boldsymbol{Flag_{i-wall}} \tag{11}$$

where $Flag_i = 1$ means that the element surface i is not obtained.

Finally, the equation for calculating the direct solar irradiance value on the element surface i at a given time is:

$$I_{dir,i} = Flag_i I_{dir0} cos\theta_i \tag{12}$$

where I_{dir0} is the direct solar radiation intensity at this moment, W/m^2; θ_i is the solar incident angle of the micro surface i, °.

2.3. *Sky Diffuse Irradiance on the Facades of Building Stocks*

2.3.1. Determination of the Discrete Precision of the Sky Vault

The surface sky diffuse irradiance is obtained by integrating the product of the sky radiation intensity with the cosine of the incident angle of the sky points over all solid angles of the sky. Among them, the solid angle [30] (Figure 3) is defined as:

$$d\Omega = \frac{df}{R^2} = \frac{(Rd\theta)(rd\phi)}{R^2} \quad (13)$$

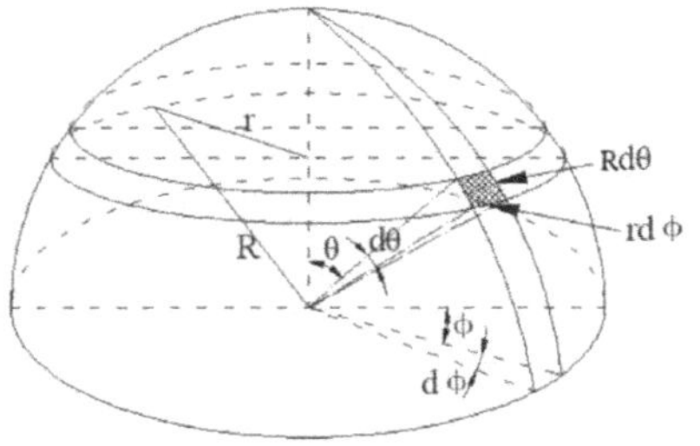

Figure 3. Definition of the solid angle in the spherical coordinate system.

Assuming that the radiation intensity along the θ direction is I_θ, the equation for calculating the sky diffuse irradiance on one surface as follows:

$$I_{dif} = \int_0^{2\pi} I_\theta \cos\theta d\Omega = \int_{\phi=0}^{\phi=2\pi} \int_{\theta=0}^{\theta=\frac{\pi}{2}} I_\theta \cos\theta \sin\theta d\phi d\theta \quad (14)$$

In a non-sheltered environment, the analytical solution of the sky diffuse irradiance on the surface of the building walls can be obtained by the above equation. This method is widely used in the calculation of the surface irradiance of traditional single buildings. However, in building complexes, the shape of the visible sky area is irregular (see Figure 4), due to the narrow perspective of the sky on the surface of the buildings. Therefore, the finite element numerical method is more suitable for the calculation of the sky diffuse radiation in building complexes. The numerical method first discretizes the sky into the set of sky lattices, in which the sky diffuse radiation intensity of the micro surface j along the θ direction is $I_{\theta,sky,j}$, and then the surface sky diffuse irradiance of the plane i is calculated as follows:

$$I_{dif,i} = \int_0^{2\pi} IF_\theta \; I_{\theta,sky,j} \cos\theta_{ij} d\Omega = \sum_{j=0}^{n} IF_\theta \; I_{\theta,sky,j} \cos\theta_{ij} Area_j \quad (15)$$

where, θ_{ij} is the incident angle from the micro-element surface to the plane, °; IF_θ is the Boolean value of the sky visible judgment along the θ direction (1 if the sky micro point is visible, 0 if it is not visible), it is dimensionless; $Area_j$ is the Area of the sky element surface j, m^2; $Area_j = sin\theta d\phi d\theta$, m^2; $I_{\theta,sky,j}$ is the diffuse radiation intensity at the sky micro surface j, $W/(sr{\cdot}m^2)$.

Figure 4. Fish eye photo in the building stocks environment.

Among the variables in Equation (15), the discrete accuracy of the sky lattice model is particularly important for the expression of the sky diffuse radiation. Existing scholars have constructed several types of sky dome partition methods, such as the Tregenza145 sky lattice model [31], which is the most widely used [32,33]. However, it is easy to have calculation errors in the calculation scenes of a narrow sky view in a building complex, due to its coarse discrete accuracy. For example, the yellow dot in Figure 5a marks a certain area A in the real sky. Due to the intersection of area A and three sky pieces B_j $(j = 1,2,3)$ in the Tregenza145 lattice model, area A will be rendered as the area B, marked yellow in Figure 5b. Furthermore, the average sky diffuse radiation intensity of the element surface B_j will be set to be equal to the area A, so the calculated value of the diffuse irradiance will be larger, due to the large rendering area in the calculation process.

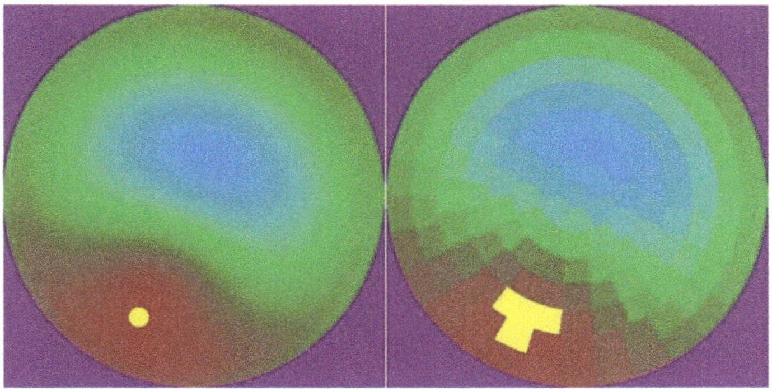

(a) (b)

Figure 5. Influence of the sky lattice model on the expression of the sky diffuse radiation intensity. (**a**) Rendering of a real sky diffuse radiation intensity distribution. (**b**) Rendering, based on the Tregenza145 lattice model.

The sky lattice model not only has an impact on the rendering expression of the sky diffuse radiation intensity, but also directly determines the accuracy of the occlusion judgment between the sky dome element and the building surface element. Figure 6a shows a real sky occlusion rendering in the form of a small sky view (the pure white part represents the sky), and Figure 6b shows the occlusion rendering by the Tregenza145 model. By comparison, it is observed that the occlusion judgment result obtained by the Tregenza145 lattice model is incorrect in both buildings and the sky (i.e., the outline of the building is not clear, and the boundaries of the visible part of the sky are blurred). Especially for the rendering of the visible part of the sky, the Tregenza145 lattice model results in a larger sky $Area_j$ (the pure white part), in Figure 6b, compared with the real visible sky part in Figure 6a, which will further lead to a larger sky diffuse irradiance calculation result, based on Equation (15). Therefore, the sky lattice model with a finer granularity should be applied to a smaller sky view and a more complex surrounding environment. By the accuracy improvement trial calculation, the area difference between the real sky and the sky pieces discretized by the sky lattice model, is less than 2° (Tregenza145's discretization accuracy is 12° at the height angle). Thus, the IFBS model divides the sky lattice every 2° along the azimuth and height angles. The hemispherical sphere of the whole sky is divided into 8100 micro-elements (360°/2 × 90°/2). The spatial coordinate distribution of the center point of the sky lattice after the dispersion is shown in Figure 7.

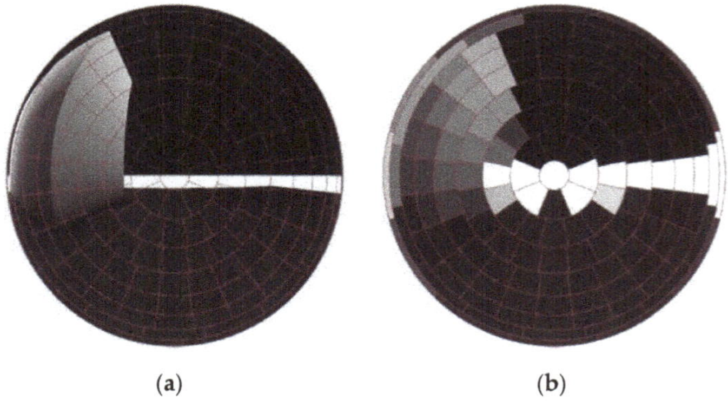

Figure 6. Influence of the sky lattice model on the occlusion rendering. (**a**) Real sky occlusion rendering. (**b**) Tregenza145 model occlusion rendering.

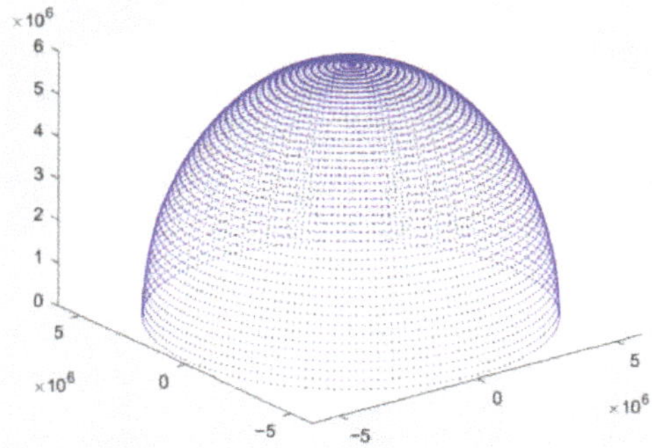

Figure 7. Schematic diagram of the sky lattice.

2.3.2. Determination of the Sky Diffuse Radiation Distribution Algorithm

As for the sky diffuse radiation model, the measurement of the existing sky diffuse radiation model is mostly carried out in the open scene, which is seriously inconsistent with the fact that the building surface in the building complex has a narrow sky perspective [34]. Therefore, the measurement and comparison of the sky diffuse radiation model should be carried out, based on the characteristics of the physical environment of the building surface in the building complex. The sky diffuse radiation model that is best suited for simulating the surface irradiance in the building complexes is chosen, and the calculation equation of $I_{\theta,sky,j}$ is determined. In this paper, it is concluded that, compared with the isotropic model and the Perez model, the Igawa model is more suitable for the simulation of the surface sky diffuse irradiance in building complexes. Therefore, this paper uses the Igawa model to calculate the sky diffuse irradiance of the building surface.

The Igawa model revises the diffuse radiation intensity in the azimuth angle and the height angle, respectively, to approximate the true anisotropic sky diffuse radiation intensity distribution, and its calculation equation is as follows:

$$I_{\theta,sky,j} = I_{dif,z}\varphi(\gamma)f(\zeta)/(\varphi(\pi/2)f(z)) \quad (16)$$

where $I_{dif,z}$ is the zenith radiation, $W/(m^2 \cdot sr)$; $\varphi(x)$ is a gradient function, as shown in Equation (17); $f(x)$ is a discrete function, as shown in Equation (18).

$$\varphi(x) = 1 + a \cdot exp(b/sinx) \quad (17)$$

$$\varphi f(x) = 1 + c \cdot [exp(d \cdot x) - exp(d \cdot \pi/2)] + e \cdot cos^2 x \quad (18)$$

where, a, b, c, d, e are the constants involved in the calculation equation of the gradient function $\varphi(x)$ and the discrete function $f(x)$.

Since the zenith radiation of $I_{dif,z}$ is still unknown in Equations (16)–(18), it will be solved by the simultaneous equation below.

Firstly, the diffuse irradiance value of an unshielded horizontal plane can be obtained by the numerical solution:

$$I_{dif,h} = \sum_{j=0}^{n} I_{dif,z} \varphi(\gamma) f(\zeta) / (\varphi(\pi/2) f(z)) \cos\theta_{ij} Area_j \quad (19)$$

The horizontal diffuse irradiance without shielding can be obtained from the total irradiance and direct solar radiation intensity:

$$I_{dif,h} = I_{g,h} - I_{dir0} cos\theta \quad (20)$$

The total horizontal irradiance $I_{g,h}$ and the direct solar radiation intensity I_{dir0} are given by the meteorological documents (usually taken from meteorological stations or typical year meteorological documents), so $I_{dif,z}$ can be theoretically obtained by Equations (19) and (20) simultaneously.

2.4. Reflected Irradiance on the Facades of Building Stocks

The existing building surface irradiance model adopts the simplified treatment of primary reflection in the calculation of the reflected irradiance, so the simplified treatment is made in the calculation of the energy balance analysis. That is, when calculating the reflected radiation intensity outward from the cell surface i, equation is not $I_{R,i} = \rho_i \left(I_{dir,i} + I_{dif,i} + I_{ref,i} \right)$, but $I_{R,i} = \rho_i \left(I_{dir,i} + I_{dif,i} \right)$, because of the simplified process, the calculated results of the surface reflection irradiance are small. Since the reflection between the interfaces of the building complexes is actually an infinite reflection, and the proportion of the reflected radiation in building complexes is higher than that in individual buildings, the surface irradiance value in the environment of the building complexes should be calculated, based on the net radiation analysis method for the simultaneous equations of the discrete grids, and then the actual situation under the infinite reflection can be solved.

According to the net radiation analysis method, the effective irradiance value of the cell surface I received from all other cells can be calculated by the equation:

$$A_i I_{ref,i} = \sum_{j=1}^{N} I_{R,j} F_{ji} A_j \quad (21)$$

where A_i and A_j are the areas of the micro surface i and j, respectively, m^2; F_{ji} is the dimensionless shape factor of the surface j to the surface i.

3. Numerical Solution of the Radiation Scheme

3.1. Surface Discretization

For a single building with no shelter around, the irradiance distribution on its surface is relatively uniform; thus, it is reasonable to calculate the wall as a whole. However, the irradiance distribution of the facades in a building complex is obviously different, due to the characteristics of mutual occlusion and a narrow sky view [21,35]. Therefore, it is not advisable to calculate the facades as a whole. The IFBS model first discretizes the facades of the building complex and the ground before the irradiance simulation, as shown in Figure 8.

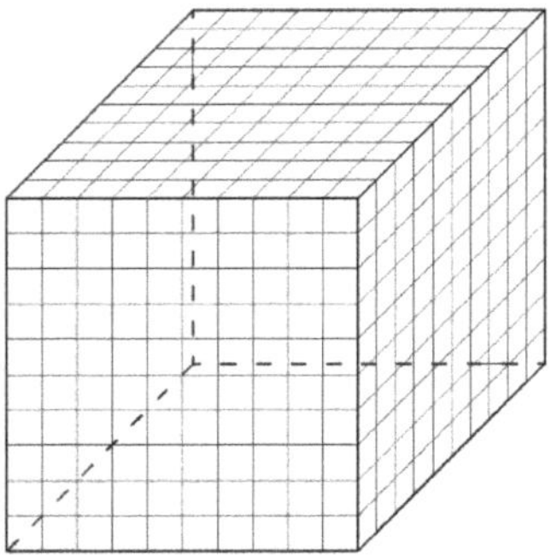

Figure 8. Schematic diagram of the surface discretization.

For the discretized interface, the total surface irradiance in this paper can be expressed as matrix:

$$\boldsymbol{I_g} = \left(I_{g,1}, I_{g,2}, \ldots, I_{g,i}, \ldots, I_{g,N}\right)^T \tag{22}$$

where $\boldsymbol{I_g}$ is the total irradiance value matrix; $I_{g,i}$ is the total surface irradiance of the element surface i, W/m^2.

The total surface irradiance value is expressed in the matrix form, which can not only directly express the storage form of the discretized data but also facilitate the numerical solution of the surface irradiance by the matrix operation. Further, each component of the global irradiance is expressed in the matrix form:

$$\boldsymbol{I_{dir}} = \left(I_{dir,1}, I_{dir,2}, \ldots, I_{dir,i}, \ldots, I_{dir,N}\right)^T \tag{23}$$

$$\boldsymbol{I_{dif}} = \left(I_{dif,1}, I_{dif,2}, \ldots, I_{dif,i}, \ldots, I_{dif,N}\right)^T \tag{24}$$

$$\boldsymbol{I_{ref}} = \left(I_{ref,1}, I_{ref,2}, \ldots, I_{ref,i}, \ldots, I_{ref,N}\right)^T \tag{25}$$

where $\boldsymbol{I_{dir}}$ is the direct solar irradiance value matrix; $\boldsymbol{I_{dif}}$ is the sky diffuse irradiance value matrix; $\boldsymbol{I_{ref}}$ is the reflection irradiance value matrix; $I_{dir,i}$ is the surface direct solar irradiance value of the micro-surface i, W/m^2; $I_{dif,i}$ is the surface sky diffuse irradiance of the micro-surface i, W/m^2; $I_{ref,i}$ is the surface reflection irradiance value of the micro-surface i, W/m^2;

Then, the solution equation of the total surface irradiance value matrix $\boldsymbol{I_g}$ of the discrete element surface at a certain time is:

$$\boldsymbol{I_g} = \boldsymbol{I_{dir}} + \boldsymbol{I_{dif}} + \boldsymbol{I_{ref}} \tag{26}$$

3.2. Matrix of Each Component of the Global Irradiance

For the direct solar irradiance, the direct solar irradiance value matrix $\boldsymbol{I_{dir}}$ is obtained, based on Equations (12) and (23):

$$\begin{aligned}\boldsymbol{I_{dir}} &= I_{dir0}(Flag_1 cos\theta_1, Flag_2 cos\theta_2, \ldots, Flag_i cos\theta_i, \ldots, Flag_N cos\theta_N)^T \\ &= I_{dir0}\, \boldsymbol{Flag.} * \boldsymbol{cos\theta}\end{aligned} \tag{27}$$

For the sky diffuse irradiance, the sky diffuse irradiance value matrix can be obtained by Equations (15) and (24). However, this method needs to calculate $I_{dif,z}$ before rendering the calculation of the sky diffuse radiation distribution, which leads to complicated solving steps. Therefore, the IFBS model builds a new solution method, based on the matrix operation. Firstly, it takes the diffuse irradiance of the roof of the tallest building in the building complex as the diffuse irradiance of the unsheltered horizontal plane (there is no occlusion in the horizon of the roof of the tallest building, which can be equivalent to the unsheltered horizontal plane. in theory), and sets it as $I_{dif,j}$, and sets $I_{dif,z} = 1$.

Secondly, the matrix of the relative diffuse irradiance value of the micro-surface of all building complexes can be calculated:

$$\mathbf{I'_{dif}} = \left(I'_{dif,1}, I'_{dif,2}, \ldots I'_{dif,j}, \ldots I'_{dif,N}\right)^{T} \tag{28}$$

At the same time, the true diffuse irradiance of an unsheltered horizontal plane can be calculated from the total irradiance of the horizontal plane and the direct solar radiation:

$$I_{dif,j} = I_{g,h} - I_{dir0}cos\theta \tag{29}$$

Thirdly, the correction coefficient $Index_{dif}$ can be obtained from the combination of Equations (28) and (29):

$$Index_{dif} = I_{dif,j} / I'_{dif,j} \tag{30}$$

Finally, the true diffuse irradiance matrix of a building complex $\mathbf{I_{dif}}$ can be calculated from the relative irradiance matrix $\mathbf{I'_{dif}}$ and the correction coefficient $Index_{dif}$.

$$\mathbf{I_{dif}} = Index_{dif}\mathbf{I'_{dif}} \tag{31}$$

For the reflection irradiance, the deduction of the numerical calculation is shown in the Appendix A and the result can be expressed as:

$$\mathbf{I_{ref}} = \left(\mathbf{E} - \mathbf{F_{ij}}. * \boldsymbol{\rho_{N\times N}}\right)^{-1}\mathbf{F_{ij}}\left(\boldsymbol{\rho_{N\times 1}}. * \left(\mathbf{I_{dir}} + \mathbf{I_{dif}}\right)\right) \tag{32}$$

where $\mathbf{I_{ref}}$ is the reflection irradiance value matrix; $\mathbf{E}$ is the identity matrix; $\mathbf{F_{ij}}$ is the angular coefficient matrix; $\boldsymbol{\rho_{N\times N}}$ and $\boldsymbol{\rho_{N\times 1}}$ are the reflectance matrices. $\mathbf{I_{dir}}$ is the direct solar irradiance value matrix; $\mathbf{I_{dif}}$ is the sky diffuse irradiance value matrix. The matrix of $\mathbf{F_{ij}}$, $\boldsymbol{\rho_{N\times N}}$ and $\boldsymbol{\rho_{N\times 1}}$ is shown in the Appendix A.

3.3. Numerical Equation for the Global Irradiance Value of the Building Facades

The numerical solution equation of the total surface irradiance value matrix $\mathbf{I_g}$ of the building complex at a certain time is:

$$\mathbf{I_g} = \mathbf{I_{dir}} + \mathbf{I_{dif}} + \mathbf{I_{ref}} \tag{33}$$

where, the matrix of the direct solar irradiance value of the building surface in the building complex obtained, based on the backward ray tracing method is:

$$\mathbf{I_{dir}} = \left(I_{dir,1}, I_{dir,2}, \ldots, I_{dir,i}, \ldots, I_{dir,N}\right)^{T} = I_{dir0}\,\mathbf{Flag}. * \boldsymbol{cos\theta} \tag{34}$$

The matrix of the sky diffuse irradiance value of the building surface in the building complex, based on the Igawa sky diffuse radiation model is:

$$\mathbf{I_{dif}} = \left(I_{dif,1}, I_{dif,2}, \ldots, I_{dif,i}, \ldots, I_{dif,N}\right)^{T} = Index_{dif}\mathbf{I'_{dif}} \tag{35}$$

The surface reflection irradiance value matrix of the building complex obtained, based on the infinite reflection radiation process is:

$$\mathbf{I_{ref}} = \left(I_{ref,1}, I_{ref,2}, \ldots, I_{ref,i}, \ldots, I_{ref,N}\right)^{T} = \left(\mathbf{E} - \mathbf{F_{ij}}. * \boldsymbol{\rho_{N\times N}}\right)^{-1}\mathbf{F_{ij}}\left(\boldsymbol{\rho_{N\times 1}}. * \left(\mathbf{I_{dir}} + \mathbf{I_{dif}}\right)\right) \tag{36}$$

To sum up, the expression equation of the numerical solution model of the building surface irradiance (IFBS model) in the building complex, namely, the solution equation of the total surface irradiance value matrix, at a certain time, after the discrete element surface is:

$$\mathbf{I_g} = I_{dir0}\,\mathbf{Flag}. * \boldsymbol{cos\theta} + Index_{dif}\mathbf{I'_{dif}} + \left(\mathbf{E} - \mathbf{F_{ij}}. * \boldsymbol{\rho_{N\times N}}\right)^{-1}\mathbf{F_{ij}}\left(\boldsymbol{\rho_{N\times 1}}. * \left(I_{dir0}\,\mathbf{Flag}. * \boldsymbol{cos\theta} + Index_{dif}\mathbf{I'_{dif}}\right)\right) \tag{37}$$

where ***Flag*** is the Boolean value matrix of the micro-element surface occlusion judgment (1 represents visibility, 0 represents occlusion), dimensionless;I_{dir0} is the intensity of the direct solar radiation, W/m^2; θ is the incidence angle of the sun, °. $Index_{dif}$ is the surface sky diffuse irradiance correction coefficient, which is dimensionless; $\boldsymbol{I}'_{dif}$ is the matrix of the diffuse irradiance value of the micro-plane, relative to the sky, W/m^2; ***E*** is the identity matrix, dimensionless; $\boldsymbol{F}_{ij}$ is the angular coefficient matrix, dimensionless; $\rho_{N\times N}$ and $\rho_{N\times 1}$ are the reflectivity matrices.

4. Results and Discussion

4.1. Comparison between the New Sky Diffuse Irradiance Algorithm and the Existing Algorithm

This paper established the algorithm of the building surface sky diffuse irradiance in a building complex, based on the characteristics of the surrounding environment of the building envelope and the shortcomings of the existing surface sky diffuse irradiance algorithm, applied to the building complex. The sky diffuse irradiance algorithm of the building surface in this paper (referred to as the IFBS model algorithm in this paper), is compared with the existing algorithm (based on the Perez sky radiation model + Tregenza145 sky lattice).

Figures 9 and 10 show the distribution of the sky radiation intensity, based on the sky lattice of the Tregenza145 sky lattice model and the IFBS model, respectively. By comparison of Figure 9a,b and Figure 10a,b in the two figures, it can be seen that the sky lattice model in the IFBS model algorithm has a higher rendering accuracy than the traditional Tregenza145 model, in both the Perez model and the Igawa model. The reason is that the IFBS model algorithm has a higher dispersion of the sky dome.

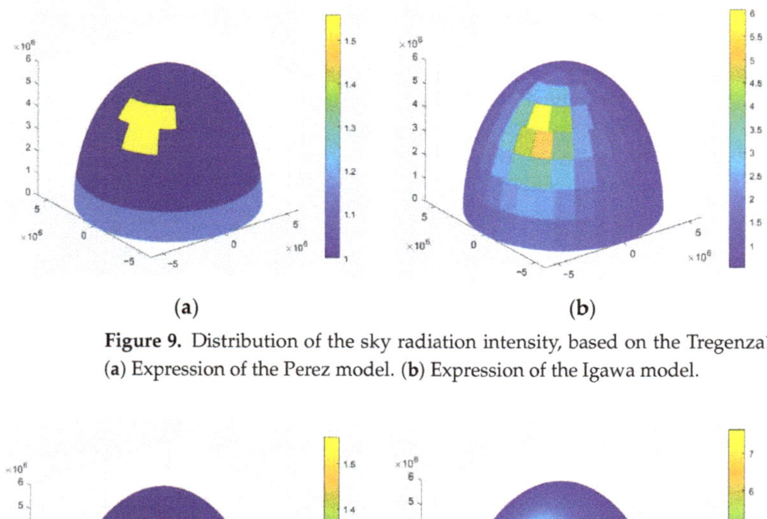

(a) (b)

Figure 9. Distribution of the sky radiation intensity, based on the Tregenza145 sky lattice model. (**a**) Expression of the Perez model. (**b**) Expression of the Igawa model.

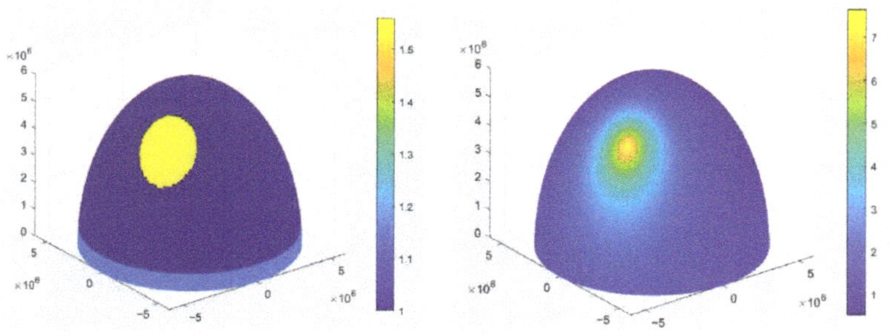

(a) (b)

Figure 10. Distribution of the sky radiation intensity, based on the sky lattice of the IFBS model algorithm. (**a**) Expression of the Perez model. (**b**) Expression of the Igawa model.

In terms of the calculation of the intensity distribution of the sky diffuse radiation, some actual measurements of the intensity distribution of the sky diffuse radiation have been made by the existing research (e.g., the diffuse radiation intensity distribution in the sky when the solar altitude angle measured, by Hay and Davies [36], is 67.6°; the distribution of the sky diffuse radiation intensity measured, by Temps and Coulson [37], by shielding the direct solar radiation). These measured studies give sufficient reference materials for scholars to study the distribution of the sky diffuse radiation intensity in the sky dome, and provide data support for the comparison of the models in this paper. Figures 9 and 10 are the renderings of the calculation results of the distribution of the sky diffuse radiation by the existing algorithm and the IFBS model algorithm, respectively. Based on the measurements mentioned above, it can be seen that the IFBS model algorithm is closer to the measured results of Hay, Temps and other scholars, when simulating the distribution of the sky diffuse radiation intensity, and the particle accuracy of rendering the distribution of the sky diffuse radiation intensity is significantly improved. Therefore, it is more suitable for calculating the case of the sky diffuse radiation under the narrow perspective of the sky in building complexes.

4.2. Comparison between the New Reflection Irradiance Algorithm and the Existing Algorithm

The numerical solution in this paper is to calculate the reflected irradiance of the surface under infinite reflection, derived from net radiation analysis. Moreover, the matrix equation for solving the reflected irradiance of the surface under finite reflection in the existing traditional irradiance model is also given, as follows.

If the number of reflections is 1, then:

$$I^{1}_{ref} = F_{ij}\left(\rho_{N\times1}.*\left(I_{dir}+I_{dif}\right)\right) \tag{38}$$

If the number of reflections is greater than 1, the calculated result of the reflection irradiance value of the micro-surface after N reflections is:

$$I^{N}_{ref} = I^{1}_{ref} + F_{ij}\left(\rho_{N\times1}.*I^{N-1}_{ref}\right) \tag{39}$$

Taking the dot-matrix building complex with the ratio of length to width to height of 1:1:5 as an example (the building spacing is the same width as the building, and all the interface reflectance is set to 0.4). If the building spacing is the same width as the building, 1–10 finite times of the reflection will be used. The time to process calculations (CPU: Intel I7-10710U, RAM: 16 GB), based on the infinite times of the reflection will be recorded, as shown in Table 2. The comparison shows that the calculation time of the finite reflection algorithm is positively correlated with the number of reflections, while the calculation time of the infinite reflection radiation algorithm is in the same order of magnitude as the finite reflection algorithm, and the calculation time of the infinite reflection radiation algorithm is less than four or more reflections.

Table 2. Comparison of the calculation time(s) of the different algorithms.

1 Time of Reflection	2 X [1]	3 X [1]	4 X [1]	5 X [1]	6 X [1]	7 X [1]	8 X [1]	9 X [1]	10 X [1]	Infinite Reflections
16.4	28.4	44.0	64.5	73.8	91.7	110.9	128.4	154.9	167.1	54.2

[1] X is the shortform for the times of reflection.

In the performance calculation of a building complex, the algorithm itself needs to consider not only the calculation time but also the calculation accuracy. Although the mainstream algorithm (one-time reflection algorithm) performs best in computing time, the potential error caused by the reflection time simplification is rarely mentioned in existing studies. In this paper, the calculated result of the surface reflection irradiance under infinite reflection, is assumed to be true. Taking the dot-matrix building complex with the ratio of length to width to height of 1:1:5 as an example, the calculation errors of the 1–10 time reflection algorithm and the infinite reflection algorithm are shown in Table 3. Meanwhile, the maximum error and average error were plotted, as shown in Figure 11.

Table 3. Comparison of the calculation errors of the finite reflection radiation algorithms.

Number of Reflections	1	2	3	4	5	6	7	8	9	10
average error (%)	−27.63	−3.47	−1.00	−0.25	−0.07	−0.02	−0.01	0.00	0.00	0.00
standard deviation (%)	16.91	3.09	1.07	0.32	0.10	0.03	0.01	0.00	0.00	0.00
maximum error (%)	−60.21	−23.47	−8.52	−2.82	−0.90	−0.28	−0.08	−0.03	−0.01	0.00
25% quantile	−42.11	−4.06	−1.37	−0.29	−0.09	−0.02	−0.01	0.00	0.00	0.00
50% quantile	−23.40	−2.36	−0.62	−0.14	−0.04	−0.01	0.00	0.00	0.00	0.00
75% quantile	−13.54	−1.68	−0.33	−0.08	−0.02	−0.01	0.00	0.00	0.00	0.00

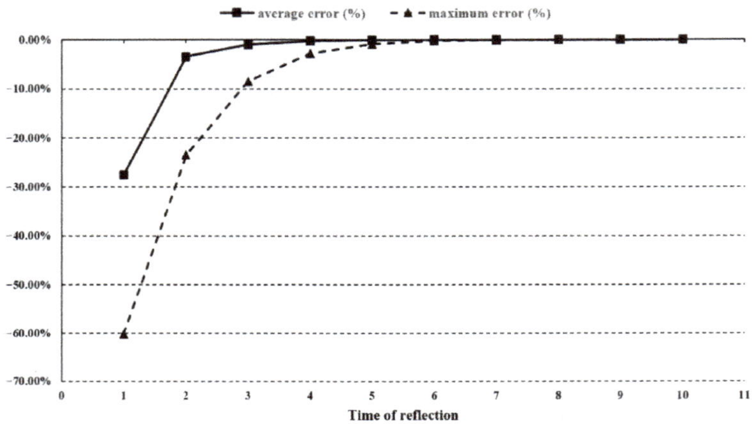

Figure 11. Comparison of the calculation errors of the finite reflection radiation algorithms.

The calculation results show that the one-time reflection algorithm has a significant error in the calculation of the building reflection irradiance, and its average error is −27.63%, indicating that the one-time reflection algorithm is underestimated in the simulation of the reflection irradiance. The maximum error of the one-time reflection algorithm, in this case, is −60.21%. Figure 12 shows the relative deviation caused by the limited time reflection algorithm (example: −0.6 means the error is negative and the absolute value is equal to 60% of the true value). The figure shows the maximum error is in the backlight surface of the buildings, where the reflection radiation transfers by multiple reflections. The calculation error of the traditional algorithm is quite significant because of the simplification of the infinite reflections.

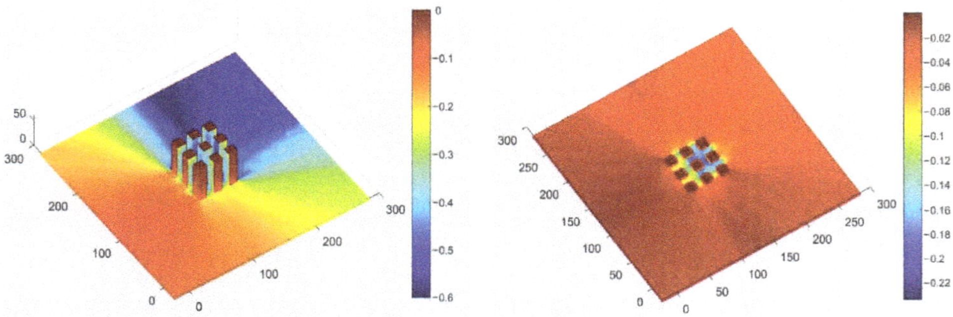

(a) (b)

Figure 12. Relative deviation of the reflection irradiance calculated by the finite time reflection algorithm for the building complexes. (**a**) one-time reflection algorithm. (**b**) two-time reflection algorithm.

Compared with the one-time reflection algorithm, the accuracy of the two-time reflection algorithm is significantly improved. The average error decreases from −27.63% to −3.47%, but the maximum error is still as high as −23.47%, indicating that the two-time reflection algorithm still has a large error in the simulation of the reflection irradiance of the buildings in a particular position. It is not difficult to find that the maximum error is located near the ground in the center of the building complex or on the sunny side of the building. In such areas, the radiation usually reaches the surface after multiple reflections. Therefore, the limitations of the finite reflection algorithm in simulating the real radiation transfer process can be seen from the heatmap of the simulation errors.

If the limit of acceptable error threshold is 10%, the mean error and maximum error are both less than the threshold for the three-time reflection, in this case. If the threshold of acceptable error is limited to 5%, the number of reflections can be increased to four.

In this case, the building density is only 25%, and the floor area ratio is 1.25. In order to avoid the limitation of individual cases, this paper also builds buildings with different floor area ratios for the simulation analysis (floor area ratio is 0.25–2.75, building length, width, and spacing are equal, and all interface reflectance is set to 0.3). The simulation results are shown in Tables 4 and 5.

Table 4. Comparison of the average deviation calculated by the finite reflection radiation algorithm (%).

Plot Ratio	1 X [1]	2 X [1]	3 X [1]	4 X [1]	5 X [1]	6 X [1]	7 X [1]	8 X [1]	9 X [1]	10 X [1]
0.25	−16.74	−1.52	−0.25	−0.04	−0.01	0.00	0.00	0.00	0.00	0.00
0.75	−20.49	−1.95	−0.41	−0.07	−0.02	0.00	0.00	0.00	0.00	0.00
1.25	−22.30	−1.98	−0.44	−0.08	−0.02	0.00	0.00	0.00	0.00	0.00
1.75	−23.33	−1.94	−0.44	−0.08	−0.02	0.00	0.00	0.00	0.00	0.00
2.25	−24.00	−1.89	−0.43	−0.08	−0.02	0.00	0.00	0.00	0.00	0.00
2.75	−24.28	−1.88	−0.43	−0.08	−0.02	0.00	0.00	0.00	0.00	0.00

[1] X is the shortform for the times of reflection.

Table 5. Comparison of the maximum deviation calculated by the finite reflection radiation algorithm (%).

Plot Ratio	1 X [1]	2 X [1]	3 X [1]	4 X [1]	5 X [1]	6 X [1]	7 X [1]	8 X [1]	9 X [1]	10 X [1]
0.25	−46.39	−5.05	−1.01	−0.16	−0.03	0.00	0.00	0.00	0.00	0.00
0.75	−50.00	−11.00	−3.03	−0.64	−0.15	−0.03	−0.01	0.00	0.00	0.00
1.25	−52.36	−14.72	−4.05	−1.01	−0.24	−0.06	−0.01	0.00	0.00	0.00
1.75	−55.41	−16.48	−4.62	−1.20	−0.30	−0.07	−0.02	0.00	0.00	0.00
2.25	−57.58	−17.25	−4.88	−1.29	−0.32	−0.08	−0.02	0.00	0.00	0.00
2.75	−58.48	−17.81	−5.02	−1.34	−0.33	−0.08	−0.02	0.00	0.00	0.00

[1] X is the shortform for the times of reflection.

The floor area ratio parameter is set to 0.25–2.75 in Tables 4 and 5. The simulation results show that as the building floor area ratio and density increase, the finite reflection algorithm produces more errors, with the maximum error increasing faster than the average error. For the impact of reflectance, Table 3 (reflectance is set to 0.4) and Tables 4 and 5 (reflectance is set to 0.3) compare reflectance's effect on the simulation accuracy. It can be concluded that with the decrease in the reflectance of the building interface, the error of the finite (one-time) reflection algorithm decreases in the same case as the building stocks model (plot ratio of 1.25). The average error decreases from −27.63% to −22.30%, while the maximum error decreases from −60.21% to −52.36%.

To sum up, the finite reflection algorithm will introduce errors in the simulation of the reflection irradiance of the building. Such errors are positively correlated with the reflectivity and floor area ratio of building facades, and the maximum error is more sensitive to parameter changes than the average error. In high-rise, high-density, or high-reflectivity building stocks, in this paper, the times of reflection should be higher than three with a 5% maximum error threshold, based on the error analysis in Table 5. However, it can be seen from the time comparison of the different algorithms in Table 3, that the four-time reflection algorithm takes longer than the optimized algorithm, so it is inferior to the optimized algorithm, in both accuracy and calculation time. Therefore, it is recommended to use the infinite reflection algorithm optimized in this paper in the numerical calculation of the reflective radiation transfer of high-rise, high-density, or high-reflectivity buildings. In the case of low-rise buildings with a low density and low reflectivity, the optimized infinite reflection algorithm is also recommended to minimize the calculation error, as there is no significant difference between the infinite reflection algorithm and the multiple reflection algorithm, in the calculation time. If computing resources are limited or equations cannot be solved by the matrix method, appropriate reflection times can be selected for calculation, based on the actual accuracy requirements and computing resources. If the acceptable error threshold is 5%, the four-time reflection algorithm is recommended.

5. Conclusions

This paper first analyzes the radiation energy balance on the surface of the envelope structure, based on the influence of the building complex on the radiation transfer process. The framework of irradiance on the facade of the building stock (IFBS Model) and a radiation scheme for each component of the global irradiance are proposed. The main focus is on the sky diffuse irradiance and reflection irradiance under the narrow sky view of the building complex. Secondly, the surface direct solar irradiance value matrix, the surface sky diffuse irradiance value matrix, and the surface reflection irradiance value matrix in the IFBS model were solved by the numerical solution, based on the surface discretization and deduction of the matrix. Finally, based on the simulated and measured results, the difference between the IFBS Model and the existing model is analyzed, and the error distribution of the finite reflection algorithm in calculating the reflection irradiance of buildings is illustrated by using the error heatmap, and the algorithm selection suggestions are given. The main conclusions of this paper are as follows:

(1) A model for irradiance on the facade of building stocks (IFBS Model) was constructed, based on the characteristics of the uneven surface radiation, narrow surface sky view, and the multiple reflected radiation processes of the building groups. The equation of the global irradiance value matrix is obtained. The model is based on the analysis of the radiation energy balance of the building surface after discretization, and the numerical expression of the influence of the building complex on the radiation transfer process is perfected.

(2) In calculating the sky diffuse irradiance in a narrow surface sky view, the traditional solution of the sky diffuse irradiance can be improved by ascending the sky lattice discretization precision and applying a more accurate sky diffuse radiation model. It is suggested to replace the traditional Perez model with the Igawa model for the radiation intensity distribution rendering.

(3) Compared with the infinite reflection algorithm, the existing mainstream one-time reflection algorithm has a significant error, especially on the surface where the radiation can reach only after multiple reflections. Such an error increases with the increase of the floor area ratio and the reflectivity of the building complex. It is recommended to use the infinite reflection algorithm, based on the net radiation analysis method in simulating the reflection irradiance of building facades. When the calculation resources are limited, the maximum error can be maintained within 5% by applying the four-time reflection algorithm.

Potential limitations of this work include the relatively limited measurements carried out by other scholars, which are related to the validity of the results by using the IFBS algorithms. As theoretically illustrated in this paper, the IFBS model will achieve a higher level of accuracy than traditional models, by applying more advanced models and algorithms. However, getting as many measurement studies as possible is necessary to verify the improvement in the accuracy of the IFBS model, compared with the traditional model. Unfortunately, few measurement studies have been carried out under the narrow sky view, and the corresponding methods and instruments are difficult to obtain. Therefore, future work on this paper will focus on the above problem.

Author Contributions: Conceptualization, H.X. and Y.Y.; methodology, H.X.; software, H.X.; formal analysis, H.X. and S.C.; writing—original draft preparation, H.X.; writing—review and editing, H.X. and S.C.; funding acquisition, Y.Y. All authors have read and agreed to the published version of the manuscript.

Funding: This research was funded by [Postdoctoral Foundation of Zhejiang Province] grant number [ZJ2021018]; [International Science and Technology Cooperation Program of the Ministry of Housing and Urban Rural Development] grant number [H20200014]; [Zhejiang Provincial Construction Department] grant number [2021K073].

Acknowledgments: The authors acknowledge the financial support from the Centre for Balance Architecture, Zhejiang University. This work is also funded by the Postdoctoral Foundation of Zhejiang Province (No. ZJ2021018), the International Science and Technology Cooperation Program of the Ministry of Housing and Urban Rural Development (No. H20200014), and the Zhejiang Provincial Construction Department (No. 2021K073).

Conflicts of Interest: The authors declare no conflict of interest.

Appendix A Numerical Solution of the Reflection Irradiance Matrix

Based on the net radiation analysis method in Section 2, the effective irradiance value of the cell surface I received from all other cells can be calculated by the equation:

$$A_i I_{ref,i} = \sum_{j=1}^{N} I_{R,j} F_{ji} A_j \tag{A1}$$

where A_i and A_j are the areas of the micro surface i and j, respectively, m^2; F_{ji} is the dimensionless shape factor of the surface j to the surface i.

According to the principle of the shape factor interchangeability,

$$A_i F_{ij} = A_j F_{ji} \tag{A2}$$

The equation for calculating the reflection irradiance value of the micro-element surface can be obtained in conjunction with the above equations, as:

$$I_{ref,i} = \sum_{j=1}^{N} I_{R,j} F_{ij} = \sum_{j=1}^{N} F_{ij} \rho_j \left(I_{dir,j} + I_{dif,j} + I_{ref,j} \right) \tag{A3}$$

Equations of order N can be obtained by expanding Equation (A3):

$$I_{ref,1} - F_{11}\rho_1 I_{ref,1} - F_{12}\rho_2 I_{ref,2} - \ldots - F_{1N}\rho_N I_{ref,N} = F_{11}\rho_1\left(I_{dir,1} + I_{dif,1}\right) + F_{12}\rho_2\left(I_{dir,2} + I_{dif,2}\right) + \ldots + F_{1N}\rho_N\left(I_{dir,N} + I_{dif,N}\right) \tag{A4}$$

$$I_{ref,2} - F_{21}\rho_1 I_{ref,1} - F_{22}\rho_2 I_{ref,2} - \ldots - F_{2N}\rho_N I_{ref,N} = F_{21}\rho_1\left(I_{dir,1} + I_{dif,1}\right) + F_{22}\rho_2\left(I_{dir,2} + I_{dif,2}\right) + \ldots + F_{2N}\rho_N\left(I_{dir,N} + I_{dif,N}\right) \tag{A5}$$

$$I_{ref,N} - F_{N1}\rho_1 I_{ref,1} - F_{N2}\rho_2 I_{ref,2} - \ldots - F_{NN}\rho_N I_{ref,N} = F_{N1}\rho_1\left(I_{dir,1} + I_{dif,1}\right) + F_{N2}\rho_2\left(I_{dir,2} + I_{dif,2}\right) + \ldots + F_{NN}\rho_N\left(I_{dir,N} + I_{dif,N}\right) \tag{A6}$$

Then, the equations given above can be arranged into the matrix form, as follows:

$$\left(\boldsymbol{E} - \boldsymbol{F_{ij}} .* \boldsymbol{\rho_{N\times N}}\right)\boldsymbol{I_{ref}} = \boldsymbol{F_{ij}}\left(\boldsymbol{\rho_{N\times 1}} .* \left(\boldsymbol{I_{dir}} + \boldsymbol{I_{dif}}\right)\right) \tag{A7}$$

where $\boldsymbol{I_{ref}}$ is the reflection irradiance value matrix; $\boldsymbol{E}$ is the identity matrix; $\boldsymbol{F_{ij}}$ is the angular coefficient matrix; $\boldsymbol{\rho_{N\times N}}$ and $\boldsymbol{\rho_{N\times 1}}$ are the reflectance matrix. $\boldsymbol{I_{dir}}$ is the direct solar irradiance value matrix; $\boldsymbol{I_{dif}}$ is the sky diffuse irradiance value matrix. For $\boldsymbol{I_{ref}}$, $\boldsymbol{F_{ij}}$, $\boldsymbol{\rho_{N\times N}}$ and $\boldsymbol{\rho_{N\times 1}}$, see Equations (A8)–(A11).

$$\boldsymbol{I_{ref}} = \left(I_{ref,1}, I_{ref,2}, \ldots, I_{ref,N}\right)^T \tag{A8}$$

$$F_{ij} = \begin{pmatrix} F_{11} & \cdots & F_{1N} \\ \vdots & \ddots & \vdots \\ F_{N1} & \cdots & F_{NN} \end{pmatrix} \quad (A9)$$

$$\rho_{N\times N} = \begin{pmatrix} \rho_1 & \cdots & \rho_N \\ \vdots & \ddots & \vdots \\ \rho_1 & \cdots & \rho_N \end{pmatrix} \quad (A10)$$

$$\rho_{N\times 1} = (\rho_1, \rho_2, \ldots, \rho_N)^T \quad (A11)$$

Based on the Equations (A7)–(A11), I_{ref}, the diffuse irradiance value matrix of the micro-element surface can be obtained by:

$$I_{ref} = \left(E - F_{ij}.*\rho_{N\times N}\right)^{-1} F_{ij}\left(\rho_{N\times 1}.*\left(I_{dir} + I_{dif}\right)\right) \quad (A12)$$

References

1. Firth, S.K.; Lomas, K.J.; Wright, A. Investigating CO_2 emission reductions in existing urban housing using a community domestic energy model. *Proc. Build. Simul.* **2009**, *9*, 2098–2105.
2. Heiple, S.; Sailor, D.J. Using building energy simulation and geospatial modeling techniques to determine high resolution building sector energy consumption profiles. *Energy Build.* **2008**, *40*, 1426–1436. [CrossRef]
3. Shimoda, Y.; Fujii, T.; Morikawa, T.; Mizuno, M. Residential end-use energy simulation at city scale. *Build. Environ.* **2004**, *39*, 959–967. [CrossRef]
4. Dascalaki, E.G.; Droutsa, K.G.; Balaras, C.A.; Kontoyiannidis, S. Building typologies as a tool for assessing the energy performance of residential buildings—A case study for the Hellenic building stock. *Energy Build.* **2011**, *43*, 3400–3409. [CrossRef]
5. Famuyibo, A.A.; Duffy, A.; Strachan, P. Developing archetypes for domestic dwellings—An Irish case study. *Energy Build.* **2012**, *50*, 150–157. [CrossRef]
6. Caputo, P.; Costa, G.; Ferrari, S. A supporting method for defining energy strategies in the building sector at urban scale. *Energy Policy* **2013**, *55*, 261–270. [CrossRef]
7. Aksoezen, M.; Daniel, M.; Hassler, U.; Kohler, N. Building age as an indicator for energy consumption. *Energy Build.* **2015**, *87*, 74–86. [CrossRef]
8. Krayenhoff, E.S.; Voogt, J.A. A microscale three-dimensional urban energy balance model for studying surface temperatures. *Bound.-Layer Meteorol.* **2007**, *123*, 433–461. [CrossRef]
9. Bruse, M.; Fleer, H. Simulating surface–plant–air interactions inside urban environments with a three dimensional numerical model. *Environ. Modell. Softw.* **1998**, *13*, 373–384. [CrossRef]
10. Hong, T.; Jiang, Y. A new multizone model for the simulation of building thermal performance. *Build. Environ.* **1997**, *32*, 123–128.
11. Nunez, M.; Oke, T.R. The energy balance of an urban canyon. *J. Appl. Meteorol.* **1977**, *16*, 11–19. [CrossRef]
12. Masson, V. A physically-based scheme for the urban energy budget in atmospheric models. *Bound.-Layer Meteorol.* **2000**, *94*, 357–397. [CrossRef]
13. Yang, X.; Li, Y. Development of a Three-Dimensional Urban Energy Model for Predicting and Understanding Surface Temperature Distribution. *Bound.-Layer Meteorol.* **2013**, *149*, 303–321. [CrossRef]
14. Quan, S.J.; Li, Q.; Augenbroe, G.; Brown, J.; Yang, P.P.-J. Urban data and building energy modeling: A GIS-based urban building energy modeling system using the urban-EPC engine. In *Planning Support Systems and Smart Cities*; Springer: Berlin, Germany, 2015; pp. 447–469.
15. Li, Z.; Xing, H.; Augenbroe, G. Criterion based selection of sky diffuse radiation models. *Sustain. Cities Soc.* **2019**, *50*, 101692. [CrossRef]
16. Liu, B.Y.H.; Jordan, R.C. The long-term average performance of flat-plate solar-energy collectors. *Sol. Energy* **1963**, *7*, 53–74. [CrossRef]
17. Perez, R.; Stewart, R.; Arbogast, C.; Seals, R.; Scott, J. An anisotropic hourly diffuse radiation model for sloping surfaces: Description, performance validation, site dependency evaluation. *Sol. Energy* **1986**, *36*, 481–497. [CrossRef]
18. Perez, R.; Seals, R.; Ineichen, P.; Stewart, R.; Menicucci, D. A new simplified version of the perez diffuse irradiance model for tilted surfaces. *Sol. Energy* **1987**, *39*, 221–231. [CrossRef]
19. Reinhart, C. Daysim, Advanced Daylight Simulation Software. 2013. Available online: http://daysim.ning.com (accessed on 27 June 2015).
20. Ward, G.; Shakespeare, R. *Rendering with Radiance: The art and science of lighting visualization*; Morgan Kaufmann: San Francisco, CA, USA, 1998.
21. Arnfield, A.J. Two decades of urban climate research: A review of turbulence, exchanges of energy and water, and the urban heat island. *Int. J. Climatol. J. R. Meteorol. Soc.* **2003**, *23*, 1–26. [CrossRef]
22. Terjung, W.H.; Louie, S.S.F. A Climatic Model of Urban Energy Budgets. *Geogr. Anal.* **1974**, *6*, 341–367. [CrossRef]

23. Siegel, R.; Howell, J.R. *Thermal Radiation Heat Transfer*; McGraw Hill: New York, NY, USA, 1972.
24. Wang, B.; Liu, G. Recalculation of common astronomical parameters in solar radiation observation. *J. Sol. Energy* **1991**, *1*, 27–32.
25. Yan, Q.; Zhao, Q. *Heat Process in Buildings*; Building Industry Press of China: Beijing, China, 1986.
26. Jensen, H.W. *Realistic Image Synthesis Using Photon Mapping*; Ak Peters: Natick, MA, USA, 2001; Volume 364.
27. Chen, C.W.; Hopkins, G.W. Ray tracing through funnel concentrator optics. *Appl. Opt.* **1978**, *17*, 1466–1467. [CrossRef]
28. Daly, J.C. Solar concentrator flux distributions using backward ray tracing. *Appl. Opt.* **1979**, *18*, 2696–2699. [CrossRef]
29. Leutz, R.; Annen, H.P. Reverse ray-tracing model for the performance evaluation of stationary solar concentrators. *Sol. Energy* **2007**, *81*, 761–767. [CrossRef]
30. Jaffey, A.H. Solid angle subtended by a circular aperture at point and spread sources: Formulas and some tables. *Rev. Sci. Instrum.* **1954**, *25*, 349–354. [CrossRef]
31. Tregenza, P.R.; Waters, I. Daylight coefficients. *Light. Res. Technol.* **1983**, *15*, 65–71. [CrossRef]
32. Ng, E.; Cheng, V.; Gadi, A.; Mu, J.; Lee, M.; Gadi, A. Defining standard skies for Hong Kong. *Build. Environ.* **2007**, *42*, 866–876. [CrossRef]
33. Li, D.H. A review of daylight illuminance determinations and energy implications. *Appl. Energy* **2010**, *87*, 2109–2118. [CrossRef]
34. Oke, T.R. Canyon geometry and the nocturnal urban heat island: Comparison of scale model and field observations. *J. Climatol.* **1981**, *1*, 237–254. [CrossRef]
35. van Esch, M.M.E.; Looman, R.H.J.; de Bruin-Hordijk, G.J. The effects of urban and building design parameters on solar access to the urban canyon and the potential for direct passive solar heating strategies. *Energy Build.* **2012**, *47*, 189–200. [CrossRef]
36. Davies, J.; Hay, J. Calculation of the Solar Radiation Incident on an Inclined Surface. In Proceedings of the First Canadian Solar Radiation Data Workshop, Toronto, ON, Canada, 17–19 April 1978; Hay, J.E., Won, T.K., Eds.; 1980; pp. 32–58.
37. Temps, R.C.; Coulson, K.L. Solar radiation incident upon slopes of different orientations. *Sol. Energy* **1977**, *19*, 179–184. [CrossRef]

designs

MDPI

Article

Simulation and Experimental Study of CO_2 Transcritical Heat Pump System with Thermoelectric Subcooling

Junlan Yang *, Linxiu Wang, Yifei Han, Xin Zhang and Yufan Du

School of Energy and Security Engineering, Tianjin Chengjian University, Tianjin 300384, China
* Correspondence: yjlfg@163.com; Tel.: +86-13820491855

Abstract: In order to improve the efficiency of the system and promote its application in other industries, the performance of a thermoelectric subcooled CO_2 transcritical heat pump system was studied. A simulation model of the system was established using steady-state lumped parameter technology, and the experimental data were compared with the simulation results. The effects of cooling and chilled water flow rate and temperature, subcooling degree, compressor discharge pressure on the coefficient of performance (COP), and heating coefficient of performance (COPh) were analyzed. The results showed that COP/COPh increased with the increase in cooling and chilled water flow rate and chilled water temperature and decreased with the increase in cooling water temperature. The experimental COPh and COP of the system with a thermoelectric subcooler increased by 4.19% and 4.62%, respectively, compared to the system without it. The simulated data was in good agreement with the experimental data, and the error was within 10%, thus verifying the correctness of the model. When the subcooling degree increased to 11 °C, the system simulation results showed that COP/COPh increased by about 40% and 13.3%, respectively. The optimal high pressure was about 8.0 MPa, which corresponded to the maximum COP and COPh of the system of 3.25 and 4.25, respectively. The research results can provide a theoretical basis for future system optimization.

Keywords: thermoelectric subcooler; CO_2 transcritical cycle; simulation; experimental measurement

Citation: Yang, J.; Wang, L.; Han, Y.; Zhang, X.; Du, Y. Simulation and Experimental Study of CO_2 Transcritical Heat Pump System with Thermoelectric Subcooling. *Designs* **2022**, *6*, 115. https://doi.org/10.3390/designs6060115

Academic Editors: Luis Le Moyne and Danial Karimi

Received: 31 August 2022
Accepted: 11 November 2022
Published: 16 November 2022

1. Introduction

The world is paying more attention to environmental concerns due to fast economic expansion. At the 75th session of the United Nations General Assembly, China proposed the targets of "carbon peak" by 2030 and "carbon neutral" by 2060 to address the issue of global warming [1]. The main solution to the problem of carbon dioxide emission is to reduce the use of fossil fuels [2]. Heat pump technology has the potential to minimize the usage of fossil fuels while enhancing energy efficiency. The use of artificial (unnatural) working medium in heat pump units will cause environmental issues. For a long time, the widespread use of refrigerants, such as HFC, has intensified the global greenhouse effect [3,4]. To address this issue, China ratified the Kigali Amendment to the Montreal Protocol, which came into effect on 1 June 2022. The use of HFCs is also anticipated to be significantly reduced in the Chinese market [5,6]. The use of natural ingredients as refrigerants has gained widespread attention in recent years. Due to its advantages of large volume cooling capacity, good compatibility, low price, low viscosity, and low pressure ratio, carbon dioxide has begun to be used as a refrigerant worldwide [7]. However, CO_2 transcritical heat pump cycles also have several drawbacks, such as high operating pressure and large throttling losses, which result in low circulation efficiency [8,9].

In order to improve the system efficiency and reduce throttling loss, Dai et al. [10] proposed a new type of thermoelectric subcooler–expander coupled CO_2 transcritical refrigeration cycle and analyzed the energy losses and efficiencies in detail. Rigola et al. [11] used theoretical and experimental results to show that the CO_2 transcritical cycle with an internal heat exchanger could increase the cooling capacity and COP.

A thermoelectric subooler (TESC) is composed of multiple thermoelectric elements in series or in parallel. Thermoelectric modules are able to pump heat from a cold surface to a hot surface through the Paltier effect. Its advantages include small size, light weight, reliable performance, and ease of use. It has been documented that the performance of a thermoelectric cooler in a heat exchanger is related to the influence of heat transfer area, thermal conductivity, and heat transfer mechanism [12,13]. In addition, the performance and operational reliability of TESC are significantly affected by the joule heat generated by the input current inside the module [14].

To improve the effectiveness of cooling facilities, many authors have proposed thermoelectric subcooling in transcritical CO_2 refrigeration systems. Koeln et al. [15] found that subcooling the outlet of the gas cooler of a CO_2 transcritical refrigeration system could significantly improve the efficiency of the system. In a heat pump experiment, Wang [16] discovered that including a subcooler might raise the product's energy efficiency ratio. Yang et al. [17] found that the application of a thermoelectric subcooler at the outlet of the air cooler in the transcritical CO_2 cycle could effectively improve the efficiency of the whole system. Li et al. [18] designed a subcooling device based on the principle of thermoelectricity and found that the cooling effect of the thermoelectric subcooling device was the best at 12 V working voltage. Astrain et al. [19] compared a CO_2 transcritical refrigeration system with a thermoelectric module, an air-cooled CO_2 transcritical system, and a system with internal heat exchange and found that the cooling efficiency of the system with a thermoelectric module was higher than the other two systems. Sánchez et al. [20] proposed a thermoelectric subcooling system and tested it in a CO_2 transcritical refrigeration unit. The results showed that under optimal operating conditions, the COP and cooling capacity of the refrigeration unit could be increased by 9.9% and 16.0%, respectively. Aranguren [21] conducted an experimental study on a transcritical CO_2 compression cycle with a thermoelectric subcooler, and the results showed that the experimental COP increased by 11.3% and the cooling capacity improved by 15.3%.

Most scholars have established models to analyze the performance of CO_2 transcritical refrigeration cycles with thermoelectric subcoolers. In addition, some scholars have conducted research only through experiments. In this study, the performance of a transcritical CO_2 refrigeration cycle with a thermoelectric subcooler was investigated by experiments and simulation models. The system model was simulated by MATLAB software. In addition, the influence of chilled water flow rate and temperature, cooling water flow rate and temperature, compressor discharge pressure, and subcooling degree on the performance of the system was also analyzed. The purpose of this study was to provide theoretical suggestions for further improving the performance and optimization of such systems.

2. Experiment Test

2.1. Refrigeration System

This section describes the configuration of a single-stage vapor compression system, including a thermoelectric subcooler (TESC). Figure 1 shows the schematic diagram of a refrigeration system and the system's P–H diagram.

The four main components of the experimental system were the CO_2 heat pump system, the water system, the data collecting system, and the control system. Figure 2 provides a flow chart of the system.

The main components and technical parameters of the heat pump system are shown in Table 1.

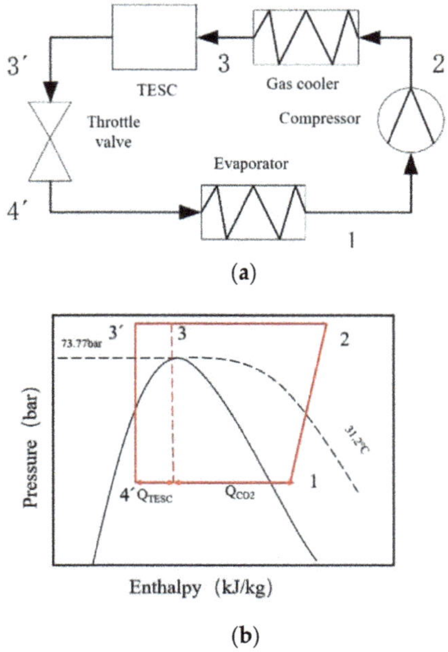

Figure 1. (**a**) Schematic diagram, (**b**) P–H diagram of the system.

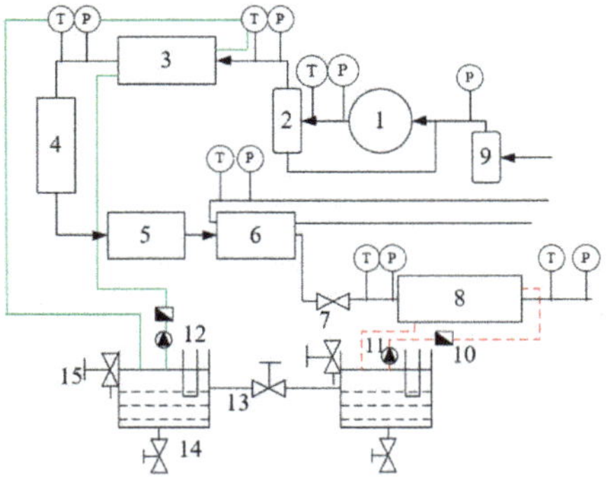

Figure 2. CO_2 water–water heat pump system with a thermoelectric subcooler. 1—compressor, 2—oil separator, 3—gas cooler, 4—thermoelectric subcooler, 5—mass flow meter, 6—regenerator, 7—throttle valve, 8—evaporator, 9—gas–liquid separator, 10—water flow meter, 11—water pump, 12—electric heater, 13—water tank, 14—drain valve, 15—inlet valve, T—thermocouple, P—pressure transmitter.

Table 1. The main components and technical parameters of the heat pump system.

Equipment	Details
Compressor	The CO_2 special compressor produced by Dorin (Torin) in Italy, model CD380H, speed 1450 rpm, rated input power 3.3 kW, oil injection capacity 1.3 kg, net weight 77 kg.
Gas cooler	Self-made casing heat exchanger, Φ22.2 seamless steel pipe as outer pipe, Φ12 nickel white copper threaded pipe as inner pipe, water pipe layer, CO_2 shell layer, pipe length 16 m, total heat exchange area of 1.2 m^2, countercurrent form adopted to enhance heat transfer, maximum pressure 14 MPa.

Table 1. *Cont.*

Equipment	Details
Evaporator	Self-made casing heat exchanger, Φ22.2 seamless steel pipe as outer pipe, Φ12 nickel white copper threaded pipe as inner pipe, water pipe layer, CO_2 shell layer, pipe length 16 m, total heat exchange area of 1.2 m^2, countercurrent form adopted to enhance heat transfer, maximum pressure 14 MPa.
Thermoelectric subcooler	Composed of thermoelectric refrigeration sheet, water cooler, and cold end fin of model TEC1-12710, with a size of 16 × 1 × 6 cm.
Regenerator	Self-made casing-type internal heat exchanger, stainless steel material, copper tube with inner tube Φ12, copper tube with outer tube Φ19, heat exchange area 0.3 m^2, high temperature CO_2 pipe from gas cooler, low temperature CO_2 from evaporator flow outside the tube, countercurrent heat exchange.
CO_2 expansion control valve	Inner diameter of the connecting pipe of 2.4 mm, cooling capacity of 8.6 kW, maximum pressure of 15 MPa, and working pressure difference of 0–10 MPa.
Gas–liquid separator	Self-made stainless steel gas–liquid separator, outer tube Φ50, inner tube Φ6, height of 0.5 m.
Oil separator	Homemade stainless steel oil separator, outer tube Φ100, height 0.5 m, interface size 1–1/8 (28 mm).
CO_2 high-pressure reservoir	Outer tube Φ100, height 0.5 m.

2.2. Thermoelectric Subcooler (TESC)

The thermoelectric subcooler is composed of thermoelectric refrigerating sheets, a cold plate, and a radiator. The cold plate is mounted at the cold end of the stack. The thermoelectric subcooler uses the Peltier principle. Semiconductors are divided into N-type and P-type according to the different charge carriers. When the power is turned on, an electron transition occurs at the contact of these two semiconductor materials, which generates or absorbs energy, forming a cold and hot junction.

For a thermoelectric refrigerating sheet, the theoretical cold end cooling capacity (Q_c) and power consumption (W_e) can be calculated using Equations (1) and (2), respectively [10].

$$Q_c = (\alpha_p - \alpha_n)AT_c - 0.5A^2R - K(T_H - T_C) \quad (1)$$

$$W_e = (\alpha_p - \alpha_n)AT_c + A^2R \quad (2)$$

where α refers to the Seebeck coefficient, V/K; P and N refer to the subscripts; A refers to the current; T_C refers to the cold end temperature in K; R refers to the resistance in Ω; K refers to the thermocouple thermal conductivity, W/K; and T_h refers to the hot end temperature in K.

In order to better measure the pros and cons of the thermoelectric subcooler, the ratio of the cooling capacity to the power consumption of the thermoelectric subcooler, namely, the efficiency *COPsc*, can be calculated as follows:

$$COP_{sc} = \frac{Q_c}{W_e} \quad (3)$$

Due to the hot end of the thermoelectric subcooler constantly emitting heat during operation, the water cooled method is utilized to quickly disperse heat and prevent overheating damage.

Fins are added to the thermoelectric subcooler's cold end in order to expand the heat exchange area to cool the refrigerant in the pipeline. Here, the thermal contact resistance between the thermoelectric tube and the pipe is reduced by the thermal paste. The size of the thermoelectric subcooler is marked in Figure 3. Figure 4 is a physical diagram of the thermoelectric subcooler.

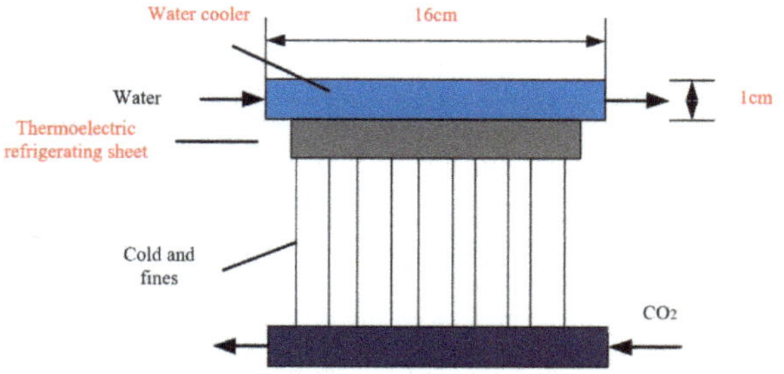

Figure 3. Structure of the thermoelectric subcooler.

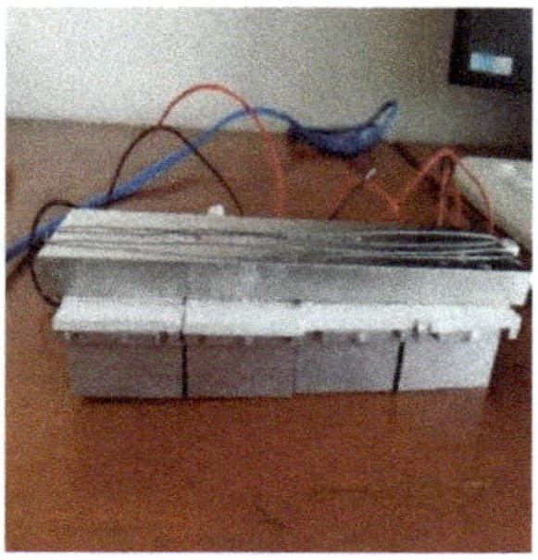

Figure 4. Physical diagram of the thermoelectric subcooler.

In order to achieve the subcooling degree of 5 °C, the cooling capacity of the thermoelectric subcooler should be 1.5 kW according to calculation using Equation (4) assuming the evaporation temperature is 0 °C, the discharge pressure of compressor is 8.5 MPa, the outlet temperature of the gas cooler is 35 °C, and the mass flow of the CO_2 is 180 kg/h.

Thus, 22 thermoelectric refrigerating sheets of the model TEC1-12710 were chosen with the size of 40 × 40 × 3.4 mm. The specific parameters are shown in Table 2.

$$Q = Gr(h_3 - h_3') \quad (4)$$

where h_3 refers to the enthalpy value of the gas cooler outlet; h_3' refers to the enthalpy value of the thermoelectric subcooler outlet; Gr refers to the refrigerant mass flow; and Q refers to the refrigeration capacity of the thermoelectric subcooler.

Table 2. Performance parameters of thermoelectric refrigerating sheet.

Model	Maximum Operating Temperature/ °C	Maximum Cooling Capacity/ W	Maximum Temperature Difference/ °C	Input Voltage/ V	Maximum Current/ A
TEC1-12710	80	89	65	12	10

2.3. Experimental Condition

The performance of the CO_2 transcritical water–water heat pump system was evaluated under various operating conditions, including with and without a thermoelectric subcooler.

The experiment's rated working conditions were as follows: CO_2 mass flow rate of 180 kg/h, cooling water flow rate of 0.5 m^3/h, chilled water flow rate of 1.2 m^3/h, inlet temperature of cooling water of 20 °C, and inlet temperature of chilled water of 12 °C.

The variable working conditions of the experiment were as follows: (1) variation of the mass flow rate of CO_2 from 160 to 200 kg/h, (2) variation of cooling water flow rate from 0.8 to 2 m^3/h and cooling water temperature from 20 to 30 °C, (3) variation of chilled water flow rate from 0.4 to 1 m^3/h and chilled water temperature from 10 to 20 °C.

2.4. *Experimental Data Processing*

2.4.1. System Cooling Capacity

Calculations were made based on the exothermic heat dissipation on the chilled water side of the evaporator:

$$Q_1 = c_{p1}\frac{g_{w1} \cdot \rho_{w1}}{3600}(t_{win} - t_{wout}) \tag{5}$$

where Q_1 refers to the refrigeration capacity; c_{p1} refers to the constant pressure specific heat of the chilled water; g_{w1} refers to the volume flow of the chilled water; ρ_{w1} refers to the density of the chilled water; and t_{win} and t_{wout} refer to the inlet and outlet water temperature of the chilled water, respectively.

2.4.2. The Heat Dissipation of the Gas Cooler

The heat absorption on the cooling water side of the gas cooler was calculated as follows:

$$Q_2 = c_{p2}\frac{g_{w2} \cdot \rho_{w2}}{3600}(t_{w,out} - t_{w,in}) \tag{6}$$

where Q_2 refers to the heat absorption; c_{p2} refers to the specific heat of the cooling water; g_{w2} refers to the volume flow of the cooling water; ρ_{w2} refers to the density of the cooling water; and $t_{w,in}$ and $t_{w,out}$ refer to the inlet and outlet water temperature of the cooling water, respectively.

2.4.3. Coefficient of Performance

The COP and COPh of the entire refrigeration system were calculated using the following formulas. The system's total power consumption included the power consumed by the compressor and the TESC (Equation (9)).

$$COP = \frac{Q_1}{W_{com} + W_{TESC}} \tag{7}$$

$$COPh = \frac{Q_2}{W_{com} + W_{TESC}} \tag{8}$$

$$W_{TESC} = V_{TEM} I_{TEM} \tag{9}$$

2.5. *Experimental Error Analysis*

This section analyzes the possible errors in the experiment resulting from many uncertain factors in the operational process.

2.5.1. Data Acquisition System

The data acquisition equipment included a platinum resistance temperature sensor, pressure sensor, electric power transmitter, turbine water flow meter, and electromagnetic CO_2 mass flow meter. The parameters of each data acquisition device are shown in Table 3 below.

Table 3. Technical parameters of each data acquisition equipment.

	Equipment	Measuring Range	Precision or Grade of Precision	Conditions of Use	Instructions
Temperature	Platinum resistance temperature sensor	−50 to 400 °C	A grade 0.1%	—	Siemens 7MC1006-1DA16-Z T10
Pressure	Pressure transducer	1 kPa to 40 MPa	±0.25%	—	Siemens 7MF1567-3DE00-3AA1
Power	Electric power transmitter	0–866 W	0.2%	Operating temperature: 0–45 °C	Suzhou honow FPW-201

Table 3. *Cont.*

	Equipment	Measuring Range	Precision or Grade of Precision	Conditions of Use	Instructions
Water flow	Turbine flow meter	0.2–1.2 m^3/h	Level 0.5 (water calibration)	Temperature of measured medium: −20 to 120 °C; pressure: ≤25 Mpa	Dayt and LWGAYA-15 The connection mode is threaded connection
CO_2 mass flow rate	Coriolis mass flow meter	0–250 kg/h	0.5 grade	Standard temperature: −50 to 150 °C, fluid Pressure measurement tube: 23 MPa	Siemens 7ME4100-1CL10-1DA1

2.5.2. Uncertainty of Chilled Water Flow

The flowmeter used to measure the flow rate of chilled water was 1.6 m^3/h, the measurement accuracy of the flowmeter was 0.5 level, and the uncertainty was δv_w = 0.008 m^3/h. The smallest chilled water flow in the measurable range was 0.8 m^3/h, and the maximum relative uncertainty of chilled water flow was 1%.

2.5.3. Uncertainty of Refrigerant Mass Flow Rate

The mass flow meter used to measure the mass flow of the refrigerant had a range of 0–250 kg/h, and the uncertainty of the mass flowmeter was δq = 0.1 kg/h. The maximum relative uncertainty of mass flow was 0.063%.

2.5.4. The Uncertainty of Cooling Capacity and COP

Because the cooling capacity and COP were calculated indirectly from other data collected, their errors can be analyzed using the power of second method, that is, if Y is a function of n independent variables, x_ζ is the independent variable affecting the function Y, and the error of Y can be determined by Equation (10):

$$\frac{\delta Y}{Y} = \sum_{\varsigma=1}^{n}\left[\left(\frac{\delta x_\varsigma}{x_\varsigma}\right)^2\right]^{\frac{1}{2}} \tag{10}$$

Due to $Q = f(m_w, t_{wi}, t_{wo})$, the uncertainty of the cooling capacity Q can be calculated as follows:

$$\begin{aligned}\tfrac{\delta Q}{Q} &= \left[\left(\tfrac{\delta m_w}{m_w}\right)^2 + \left(\tfrac{\delta t_{win}}{t_{win}}\right)^2 + \left(\tfrac{\delta t_{wout}}{t_{wout}}\right)^2\right]^{\frac{1}{2}} \\ &= \left[(0.5\%)^2 + (0.1\%)^2 + (0.1\%)^2\right]^{\frac{1}{2}} \\ &= 0.52\%\end{aligned} \tag{11}$$

Due to $W = f(Gr, t_{in,com}, t_{out,com}, P_{in,com}, P_{out,com})$, the uncertainly of the compressor power consumption W can be calculated as follows:

$$\begin{aligned}\tfrac{\delta W}{W} &= \left[\left(\tfrac{\delta Gr}{Gr}\right)^2 + \left(\tfrac{\delta t_{in,com}}{t_{in,com}}\right)^2 + \left(\tfrac{\delta t_{out,com}}{t_{out,com}}\right)^2 + \left(\tfrac{\delta P_{in,com}}{P_{in,com}}\right)^2 + \left(\tfrac{\delta P_{out,com}}{P_{out,com}}\right)^2\right]^{\frac{1}{2}} \\ &= \left[(0.5\%)^2 + (0.1\%)^2 + (0.1\%)^2 + (0.25\%)^2 + (0.25\%)^2\right] \\ &= 0.62\%\end{aligned}$$

Due to $COP = f(Q, W)$, the uncertainty of COP can be calculated as follows:

$$\begin{aligned}\tfrac{\delta COP}{COP} &= \left[\left(\tfrac{\delta Q}{Q}\right)^2 + \left(\tfrac{\delta W}{W}\right)^2\right]^{\frac{1}{2}} \\ &= \left[(0.52\%)^2 + (0.62\%)^2\right]^{\frac{1}{2}} \\ &= 0.81\%\end{aligned} \tag{12}$$

where m_{w1} refers to the flow rate of chilled water in kg/s; t_{win} and t_{wout} refer to the inlet and outlet water temperatures of the chilled water of the evaporator, respectively, respectively, in °C; P refers to the measured pressure in MPa; and W refers to the compressor consumption power in kW.

3. Simulation Model Establishment

In this section, the CO_2 transcritical water–water heat pump model is discussed in detail by establishing a mathematical model and using MATLAB to call physical parameters in Refprop software. The system is mainly composed of a compressor, gas cooler, thermoelectric subcooler, throttle valve, and evaporator. The use of energy conservation and related principles to establish the model can effectively supplement the problem of incomplete data caused by the limitation of test conditions. The model can be used to comprehensively analyze the impact of different parameters on the performance of the system and provide theoretical guidance to further understand the performance of and investment required for a heat pump system.

3.1. Compressor Model

Mass flow rate of CO_2 refrigerant

$$G_r = \frac{V_{th}\eta_v}{3600 v_s} \tag{13}$$

Volume efficiency [22]:

$$\eta_v = 0.976728 - 0.0921418\left(\frac{P_2}{P_1}\right)^{0.714} \tag{14}$$

where V_{th} refers to the calculated exhaust volume in m^3/h; v_s refers to the compressor refrigerant specific capacity; P_2 refers to the compressor exhaust pressure in MPa; and P_1 refers to the compressor suction pressure in MPa.

The compressor power consumption can be calculated by Equation (15):

$$W_{com} = \frac{G_r(h_2' - h_1)}{\eta_{is}\eta_m} \tag{15}$$

where $\eta_{\rm is}$ and $\eta_{\rm m}$ are calculated using empirical formula [22,23]:

$$\eta_m = 0.26 + 0.7952\left(\frac{P_2}{P_1}\right) - 0.2803\left(\frac{P_2}{P_1}\right)^2 + 0.414\left(\frac{P_2}{P_1}\right)^3 - 0.0022\left(\frac{P_2}{P_1}\right)^4 \tag{16}$$

$$\eta_{is} = 0.995541 - 0.107987\left(\frac{P_2}{P_1}\right)^{0.714} \tag{17}$$

In Equations (16) and (17), h_2' refers to the isentropic enthalpy value of compressor outlet state point; h_1 refers to the enthalpy value when the machinery is inhaled; $\eta_{\rm m}$ refers to the mechanical efficiency; and $\eta_{\rm is}$ refers to the isentropic efficiency.

3.2. Gas Cooler

The model of the gas cooler was constructed by the centralized parameter method, and the following assumptions were made:

1. When the refrigerant and water are exchanged heat, it is a one-dimensional steady-state model, and the temperature and flow rate of the refrigerant and the water are evenly distributed in the corresponding cross section.
2. All the heat losses of the gas cooler are ignored, and the outer pipe wall is considered to be adiabatic.
3. The pressure drop of the water in the tube is ignored.

4. The thermal conduction process only occurs in the horizontal direction of fluid flow.
5. The system operation state is steady.
6. The refrigerant flows along the tube and is evenly distributed.

According to the energy conservation law, the heat released by the refrigerant is the same as that absorbed by cooling water. Thus, the following equation can be obtained:

$$Q_{CO_2} = G_r(h_2 - h_3) \tag{18}$$

Cooling water side heat absorption equation:

$$Q_w = m_{w2}cp_2(t_{w,out} - t_{w,in}) \tag{19}$$

Total heat transfer equation:

$$Q_{CO_2} = Q_w = KA_2 \triangle t \tag{20}$$

where m_{w2} refers to the cooling water flow in kg/s; c_{pw} refers to the specific heat capacity of the cooling water at constant pressure in kJ/(kg·°C); A_2 refers to the heat exchange area of the gas cooler in m^2; and $\triangle t$ refers to the logarithmic average temperature difference in °C.

The parameters involved can be calculated as follows:

(1) Using the outer surface of the inner tube as a reference, the total heat transfer coefficient solution equation is established.

$$K = \frac{1}{\left(\frac{1}{h_{CO_2}+r_1}\right) + \frac{\delta}{\lambda}\ln\frac{d_{w,o}}{d_m} + \left(\frac{1}{h_w} + r_2\right)\frac{d_{w,o}}{d_{w,i}}} \tag{21}$$

where r_1 and r_2 refer to the fouling coefficient of the inner and outer tubes, respectively; $d_{w,o}$ refers to the inner tube outside diameter in mm; and $d_{w,i}$ refers to the inner diameter of the inner tube in mm.

(2) Logarithmic mean temperature difference:

$$\Delta t = \frac{\left(t_{w,in} - t_{CO_2,out}\right) - \left(t_{w,out} - t_{CO_2,in}\right)}{\ln\left(\frac{t_{w,in} - t_{CO_2,out}}{t_{w,out} - t_{CO_2,in}}\right)} \tag{22}$$

(3) Heat transfer area:

$$A = \pi d_{w,o} l \tag{23}$$

where l refers to the tube length in m.

In the cycle process, the gas cooler exothermic heat in the transcritical and the conventional cycle in the subcritical exothermic heat release are very different, which is caused by the special thermal properties of CO_2. At present, more and more researchers have started studying the heat exchange correlation type of the air cooler in depth. According to the literature, the heat exchange working conditions of the heat exchange correlation type established by Yoon et al. were similar to this paper; thus, we selected the heat exchange correlation type of Yoon [24]:

$$Nu_{CO2} = a\mathrm{Re}_{CO_2}{}^{b}\mathrm{Pr}_{CO_2}{}^{c}\left(\frac{\rho_{PC}}{\rho_f}\right)^{n} \tag{24}$$

where ρ_{pc} refers to the critical density of fluids, and ρ_f refers to the fluid density.

Heat transfer coefficient on the cooling water side:

$$h_w = \frac{Nu_w \cdot \lambda_w}{l} \tag{25}$$

3.3. Thermoelectric Subcooler

When the model was established, the input parameters included the cooling capacity and the number of thermoelectric refrigerating sheets. The output parameter was the degree of subcooling.

By calculation, 22 refrigerants sheets with a cooling capacity of 70 W and type TEC1-12710 constituted a thermoelectric subcooler, and the total cooling capacity of the thermoelectric subcooler was 1.5 kW.

1. The CO_2 side cooling capacity:

$$Q = Gr(h_3 - h_3') \tag{26}$$

2. The cooling capacity of the thermoelectric subcooler [10]:

$$Q_c = n\left[(\alpha_p - \alpha_n)AT_c - 0.5A^2R - K(T_H - T_C)\right] \tag{27}$$

3. Equation for conservation of energy:

$$Q = Q_c \tag{28}$$

3.4. Throttle Valve

The throttle process in the throttle valve assumes that the enthalpy values before and after the throttling are equal:

$$h_{3'} = h_{4'} \tag{29}$$

3.5. Evaporator

The simulation model using a centralized parametric method was built on a lab jacketed evaporator based on the following assumptions:

1. The casing used is uniform and regularly round.
2. The chilled water and refrigerant both flow in a certain dimensional direction.
3. The chilled water and refrigerant are evenly distributed in the tube.
4. The heat transfer loss of the evaporator is not considered.
5. The interference caused by the lubricating oil and other similar factors on the heat exchange is ignored.

Heat absorption of refrigerant:

$$Q_r = Gr(h_1 - h_{4'}) \tag{30}$$

Heat release on the side of chilled water:

$$Q_{ld} = cp_1 m_{w1}(t_{win} - t_{wout}) \tag{31}$$

Total heat exchange:

$$Q_r = Q_{ld} = KA_1 \triangle t_1 \tag{32}$$

where m_{w1} refers to the flow rate per second of chilled water in kg/s; t_{win} refers to the temperature of the chilled water inlet in °C; t_{wout} refers to the outlet temperature of chilled water in °C; A_1 refers to the heat transfer area of chilled water in m^2; K refers to the heat transfer rate of the evaporator, $W/(m^2{\cdot}K)$; and $\triangle t_1$ refers to the logarithmic average temperature difference in °C.

The parameters involved can be calculated as follows:

1. Using the outer surface of the inner tube as a reference, the total heat transfer coefficient solution equation is established as shown in Equation (33):

$$K = \frac{1}{\left(\frac{1}{h_r} + r_1\right) + \frac{\delta}{\lambda}\ln\frac{d_{ld,o}}{d_m} + \left(\frac{1}{h_{ld}} + r_2\right)\frac{d_{ld,o}}{d_{ld,i}}} \tag{33}$$

where r_1 and r_2 refer to the fouling coefficient on the CO_2 side and the chilled water side, respectively, and d_m refers to the average diameter of the tube.

2. The heat transfer coefficient on the chilled water side is calculated using the Dittus–Boelter correlation [25]:

$$Nu_{ld} = 0.023\mathrm{Re}^{0.8}\mathrm{Pr}^{n} \tag{34}$$

where n = 0.4 when the fluid is heated, and n = 0.3 when the fluid is cooled.

Compared to [26,27], Kew and Cornwell [28] heat transfer related formulas were selected as the correlation relationship of CO_2 boiling heat transfer coefficient in the evaporator due to the similar dimensions and other relevant parameters with the laboratory evaporator model. The details are as follows [28]:

$$h_r = 30{\cdot}\mathrm{Re}_r{}^{0.857}{\cdot}Bo^{0.714}{\cdot}(1-x)^{-0.143}{\cdot}\frac{\lambda_r}{n \times d_r} \tag{35}$$

where h_r refers to the heat transfer coefficient on the refrigerant side, $W/(m^2{\cdot}K)$; λ_r refers to the thermal conductivity coefficient for the refrigerant side, $W/(m{\cdot}K)$; x refers to dryness; Re_r refers to Reynolds number; and Bo refers to boiling number.

3.6. Solving the System Model

The matching module was developed in MATLAB and solved in accordance with the mathematical model of each component. Characteristics such as cooling/heating capacity and COP/COPh were determined by inputting the compressor discharge pressure, the tube diameter of the evaporator and gas cooler, and the temperature and flow rate of the chilled water and the cooling water. A compressor module, gas cooler module, thermoelectric subcooler module, throttle valve module, and evaporator module made up the overall system. Each component was meticulously simulated using the defined model, and data on endothermic and exothermic heat were calculated. The absolute value of the relative error of cooling capacity and heat absorption was taken as the convergence condition. If the error was less than 5%, the program continued calculation; otherwise, the parameters were reassumed. Figure 5 is the flow chart of system model calculation.

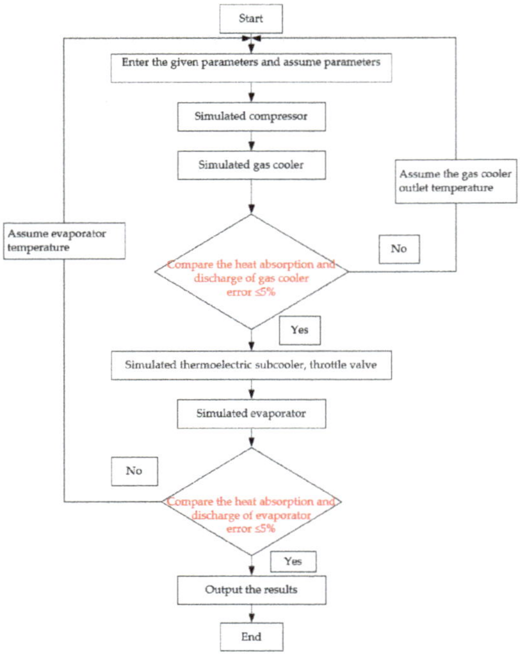

Figure 5. The flow chart of system model calculation.

4. Results and Discussion

4.1. Analysis of Experimental Results

4.1.1. The Variation of the Cooling Water Flow Rate

Figure 6 shows the relationship between COPh and cooling water flow. With continuous increase in the cooling water flow, the heating coefficient of performance and the variation trend were similar for the systems with and without a subcooler. At the same time, the heating coefficient of the system with a subcooler increased by 3.14% compared to that without it under the same conditions.

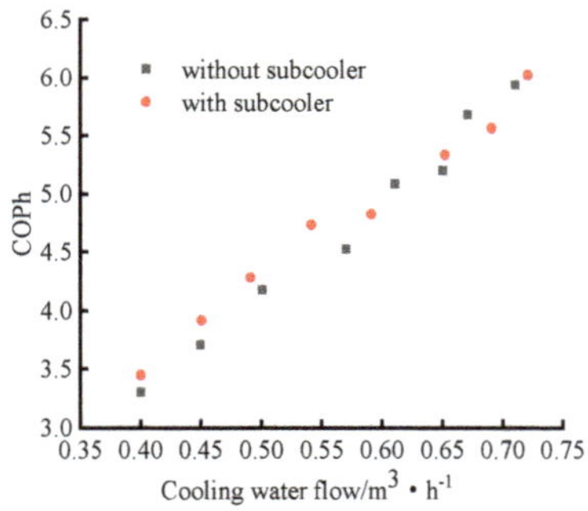

Figure 6. Influence of cooling water flow rate on COPh.

4.1.2. Variation of Cooling Water Temperature

As can be seen from Figure 7, regardless of whether the system was equipped with a subcooler, the COPh decreased as the cooling water temperature increased, which was similar to the trend of heating coefficient of performance of the system without a subcooler.

Under the same conditions, the COPh efficiency of the system with a subcooler increased by 2.63% compared to the system without a subcooler.

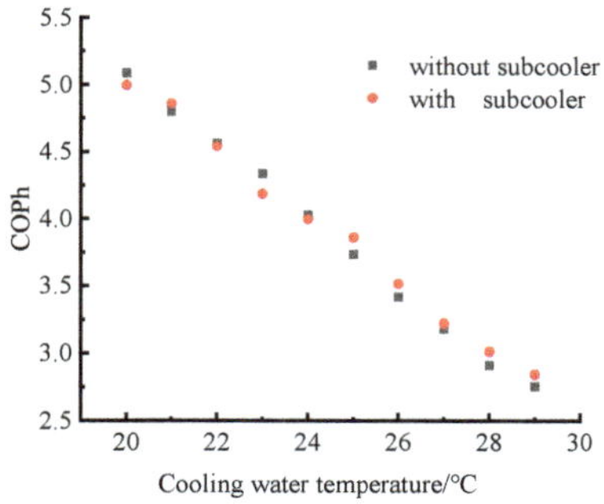

Figure 7. Influence of cooling water temperature on COPh.

4.1.3. Variation of Chilled Water Flow

Figure 8 shows the variation trend of coefficient of performance with increasing chilled water flow rate with and without a subcooler. From Figure 8, it can be seen that the COP of the system increased with the increase in chilled water flow rate with or without a subcooler, and the coefficient of performance of the system with a subcooler was 1.62% higher than that of the system without a subcooler.

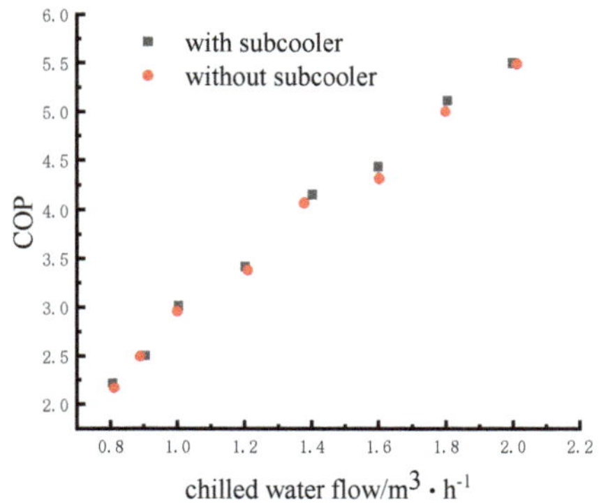

Figure 8. Influence of chilled water flow on COP.

4.1.4. Variation of Chilled Water Temperature

Figure 9 shows the trend of coefficient of performance with and without a subcooler. As can be seen from Figure 9, the system cooling efficiency COP increased with the increase in chilled water temperature regardless of whether the system was equipped with a thermoelectric subcooler. The COP of the system with a subcooler was significantly higher than that of the system without a subcooler by 3.14%.

4.2. System Model Validation

When the experimental and simulated working conditions of the transcritical CO_2 heat pump system with a thermoelectric subcooler were the same, the results obtained by the two methods were compared and analyzed, and the relative error was used in the analysis process:

$$\text{relative error} = \frac{\text{simulation value} - \text{experimental value}}{\text{experimental value}} \times 100\% \quad (36)$$

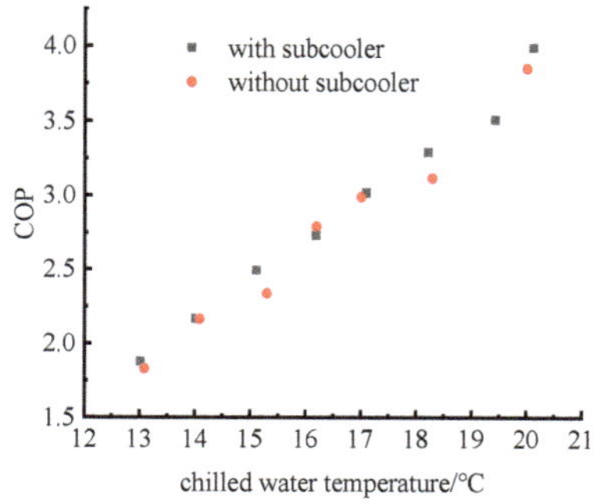

Figure 9. Influence of chilled water temperature on COP.

Figures 10 and 11 show the experimental and simulated values of COPh when the flow rate and temperature of cooling water were changed. It can be seen that when the cooling water flow rate increased, the experimental data and simulation data showed an upward trend, and the consistency was higher at 0.4–0.55 m^3/h. When the cooling water temperature gradually increased, COPh continued to decrease, and the analog value was generally slightly higher than the experimental results with an error margin of about 8.6%.

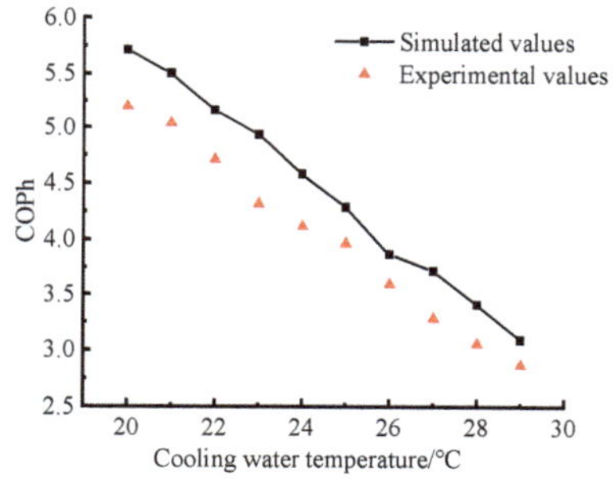

Figure 10. Influence of cooling water temperature on COPh.

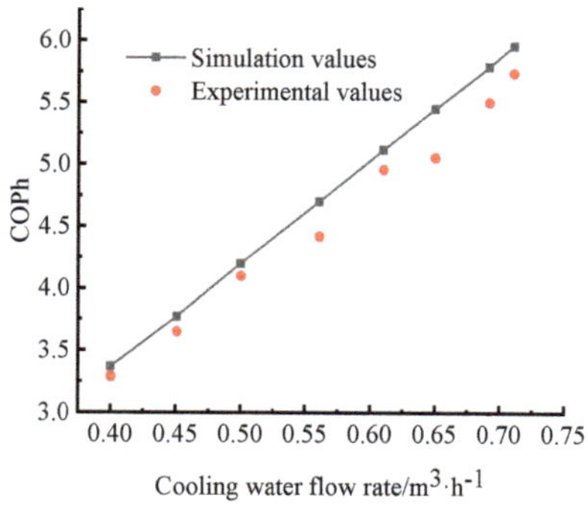

Figure 11. Influence of cooling water flow rate on COPh.

Figure 12 compares the refrigeration coefficient of performance of the experimental data and simulated data for different chilled water temperatures. As the temperature of chilled water gradually increased, the experimental value and the simulated value of COP gradually increased. The trend of the two was similar, and the simulation results were slightly higher than the experimental results.

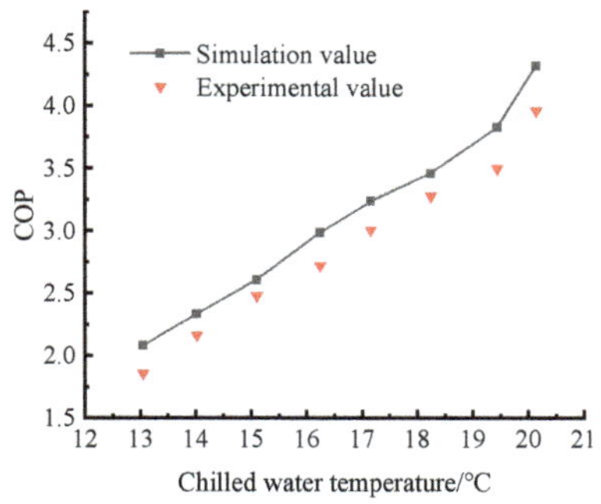

Figure 12. Influence of chilled water temperature on COP.

4.3. Simulation Result Analysis

4.3.1. Influence of the Subcooling Degree

With the increase in thermoelectric subcooling sheets, the degree of subcooling increases. As can be seen from Figures 13 and 14, the cooling capacity/heating capacity was positively correlated with COP/COPh and the degree of subcooling. When the subcooling increased from 1 to 11 °C, the cooling capacity increased from 1 to 7 kW, the heating capacity increased from 5.75 to 11.75 kW, COP increased by 40%, and COPh increased by 13.3%. This was due to the increase in thermoelectric cooling sheets, which led to an increase in the degree of subcooling.

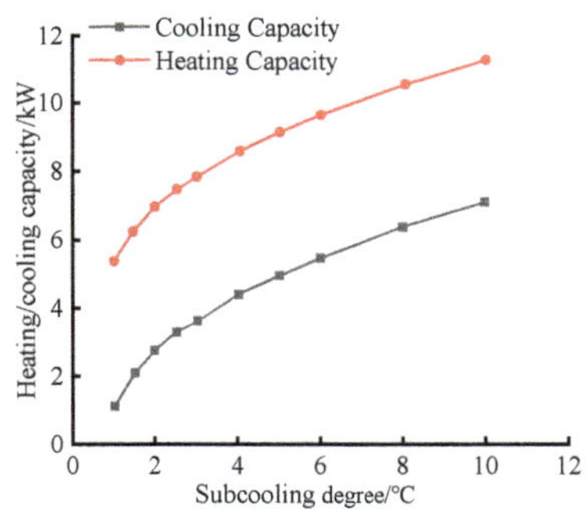

Figure 13. Influence of subcooling degree on heating/cooling capacity.

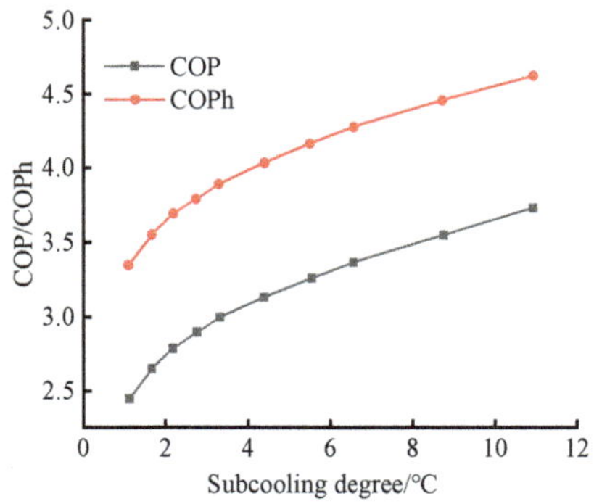

Figure 14. Influence of subcooling degree on COP/COPh.

4.3.2. Influence of Cooling Water Flow Rate and Temperature

At rated conditions, COP/COPh is shown against cooling water flow rate in Figure 15. The chart shows a considerable positive correlation between COP/COPh and cooling water flow. According to the calculation results, increasing the cooling water flow would cause a heat exchange between the refrigerant and the cooling water.

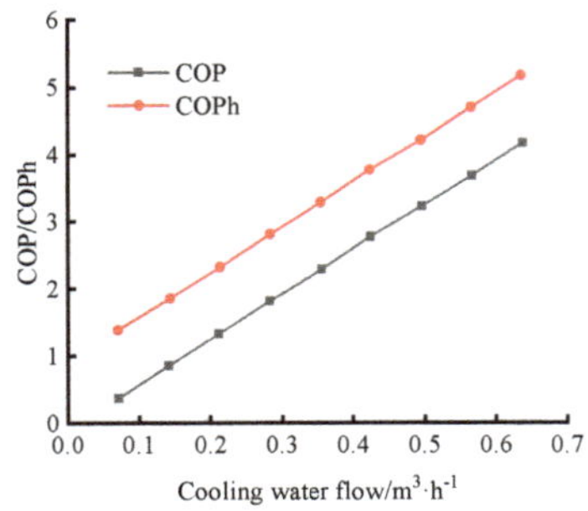

Figure 15. Influence of cooling water flow rate on COP/COPh.

As can be seen in Figure 16, there was a slight inverse relationship between cooling water temperature and COP/COPh. The COPh was around 2.5 and the COP was approximately 1.5 when the cooling water temperature was 30 °C.

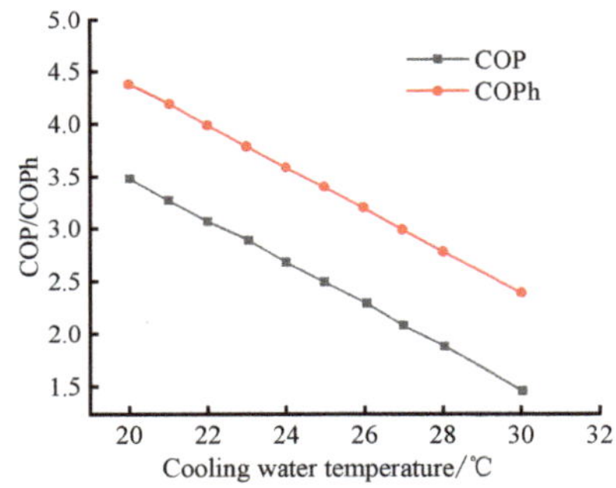

Figure 16. Influence of cooling water temperature on COP/COPh.

4.3.3. Influence of Chilled Water Flow Rate and Temperature

From Figures 17 and 18, it can be seen that COP/COPh had a positive correlation with chilled water flow rate and temperature. As the chilled water flow increased, COP increased from 1.2 to 3.2 and COPh increased from 2 to 4.5. It can be seen that the heat exchange between the chilled water and refrigerant in the evaporator was strengthened due to increased chilled water flow rate. The evaporation process was endothermic. With the increase in chilled water temperature, the heat exchange between the refrigerant and chilled water in the evaporator was strengthened, so the system efficiency increased.

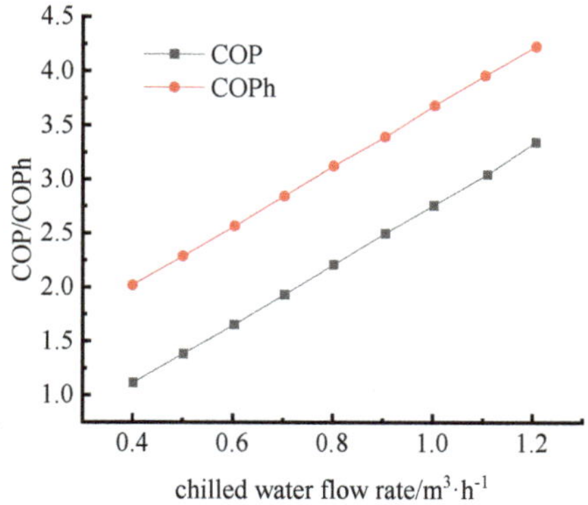

Figure 17. Influence of chilled water flow rate on COP/COPh.

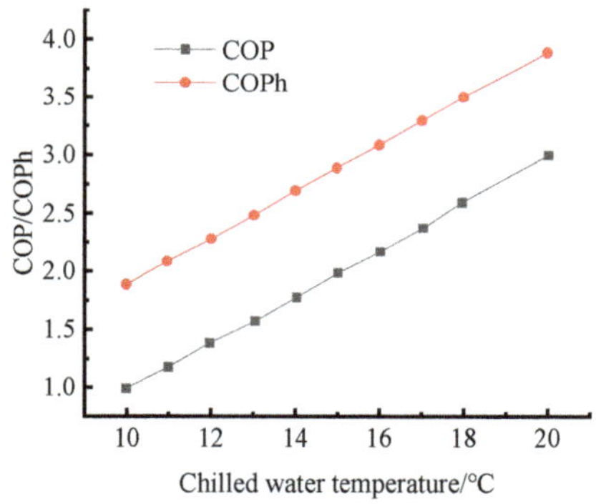

Figure 18. Influence of chilled water temperature on COP/COPh.

4.3.4. Influence of Compressor Discharge Pressure

As can be seen from Figure 19, the system's COP and COPh increased as the discharge pressure increased, and the variation trend gradually decreased, with the optimal high pressure existing. The highest values of COP and COPh of the system were 3.25 and 4.25, respectively, when the compressor discharge pressure was about 8.0 MPa.

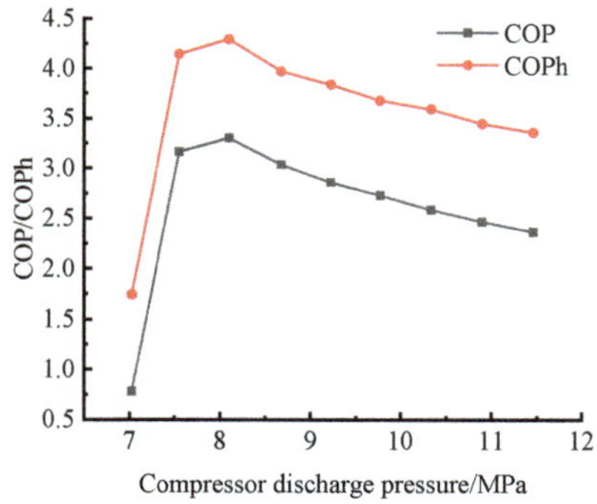

Figure 19. Influence of compressor discharge pressure on COP/COPh.

5. Conclusions

Based on the existing experimental bench, the corresponding model of a CO_2 transcritical water–water heat pump system with a thermoelectric subcooler was established by MATLAB. The compressor, gas cooler, subcooler, throttle valve, and evaporator were simulated and tested, and the simulation results were compared with the experimental results. The results are as follows:

1. Through calculation, it was found that the uncertainty of the experiment was less than 1%, indicating that the accuracy of the experiment was high. When the cooling water flow increased, COPh continued to rise, regardless of whether the system was equipped with a thermoelectric subcooler. COP increased with increased chilled water flow and temperature.
2. The simulation results of the system were compared with the experimental results, and the error was generally less than 10%, thus verifying the high accuracy of the established simulation model.
3. Through simulation calculation, it was found that with the increase in chilled water flow and temperature, COP and COP showed a gradual upward trend.
4. When the discharge pressure of the compressor changed, COP and COPh corresponded to an optimal high pressure of about 8 MPa.

Author Contributions: Data curation, J.Y.; writing—original draft preparation, L.W.; material collection, Y.H.; literature collection, X.Z.; data reduction, Y.D. All authors have read and agreed to the published version of the manuscript.

Funding: This research was funded by the Tianjin Natural Science Foundation Project, grant number 17JCZDJC31400.

Data Availability Statement: Data available on request due to restrictions, e.g., privacy or ethical. The data presented in this study are available on request from the corresponding author.

Conflicts of Interest: The authors declare no conflict of interest.

Nomenclature

A	current, A
Bo	boiling number
COP	coefficient of performance
COP_h	heating coefficient of performance
c_p	specific heat at constant pressure, kJ/(kg·K)
$d_{w,o}$	Outside diameter of the inner tube, mm
$d_{w,i}$	inner diameter of the inner tube, mm
Gr	refrigerant mass flow, kg/s
g_{w1}	volume flow of the chilled water, m^3/h
g_{w2}	volume flow of the cooling water, m^3/h
h	specific enthalpy, kJ/kg
K	thermocouple thermal conductivity, W/K
l	tube length, m
m_{w1}	cooling water mass flow rate, kg/s
m_{w2}	chilled water mass flow rate, kg/s
P	pressure, MPa
Q	refrigeration capacity of thermoelectric subcooler, kW
Q_1	refrigeration capacity, kW
Q_2	heating capacity, kW
R	resistance, Ω;
Re_r	Reynolds number
r	fouling coefficient, $m^2\cdot°C/W$
t	temperature, °C

T	thermoelectric subcooler temperature, K
v_s	compressor refrigerant specific capacity
W	power consumption, kW
x	dryness
Greek symbols	
α	Seebeck coefficient, V/K
η_{m}	mechanical efficiency
η_{is}	isentropic efficiency
η_v	volumetric efficiency
ρ	density, kg/m^3
Subscript	
c	cold end
com	compressor
h	hot end
n	N-type
p	P-type
TESC	thermoelectric subcooler
w_1	chilled water
w_2	cooling water
w, in	inlet water of the cooling water
w, out	outlet water of the cooling water
win	inlet of the chilled water
$wout$	outlet of the chilled water

References

1. Li, Y.-F. Do a good job of carbon peaking and carbon neutrality, and usher in a new ea of low-emission development. *World Environ.* **2021**, *1*, 16–19.
2. Zhang, B.-T. How to fulfill the commitment to reduce emissions: The inspiration for China's first carbon peak. *Hydropower New Energy* **2021**, *3*, 1–6.
3. Jiang, H. Analysis of ozone and ozone layer destruction and its protection mechanism. *Low-Carbon World* **2017**, *3*, 27–28.
4. Wang, G.; Xie, G.-Z.; Zhang, L.-R. Analysis of HCFCs working fluid substitution in the refrigeration and air conditioning industry. *Refrig. Air Cond.* **2012**, *26*, 18–23+43.
5. Wang, X.-C. Interpretation, review and initiatives on the Montreal Protocol amendments to reduce hydrofluorocarbons (HFCs). *HVAC* **2017**, *5*, 72–76.
6. Zhang, Z.-Y.; Fang, X.-K.; Bie, P.-J. Analysis of the contribution of China's control of HFCs emissions to climate change mitigation. *Environ. Prot.* **2017**, *45*, 65–67.
7. Liu, Y.-H.; Feng, L.-W.; Yang, Z.-F. Theoretical analysis and development of air source heat pump water heater based on CO_2 refrigerant. *Refrigeration* **2022**, *1*, 36–40.
8. Wang, G.-B.; Zhang, X.-R. Thermoeconomic Optimization and Comparison of the Simple Single-Stage Transcritical Carbon Dioxide Vapor Compression Cycle with Different Subcooling Methods for District Heating and Cooling. *Energy Convers. Manag.* **2019**, *185*, 740–757. [CrossRef]
9. Ma, Y.-T.; Yuan, Q.-X.; Li, M.-X. Performance comparison of transcritical CO_2 with expander and back-cycle with injector. *Low Temp. Subconductivity* **2011**, *39*, 36–41.
10. Dai, B.-M.; Liu, S.-C.; Pan, H.-R.; Sun, Z.-L.; Yang, Q.-R.; Ma, Y.-T. Thermoelectric subcooler-expander coupled CO_2 transcritical refrigeration cycle analysis. *Fluid Mach.* **2019**, *47*, 76–82+46.
11. Rigola, J.; Ablanque, N.; Pérez-Segarra, C.D.; Oliva, A. Numerical Simulation and Experimental Validation of Internal Heat Exchanger Influence on CO_2 Trans-Critical Cycle Performance. *Int. J. Refrig.* **2010**, *33*, 664–674. [CrossRef]
12. Lin, X.; Mo, S.; Mo, B.; Jia, L.; Chen, Y.; Cheng, Z. Thermal Management of High-Power LED Based on Thermoelectric Cooler and Nanofluid-Cooled Microchannel Heat Sink. *Appl. Therm. Eng.* **2020**, *172*, 115165. [CrossRef]
13. Astrain, D.; Aranguren, P.; Martínez, A.; Rodríguez, A.; Pérez, M.G. A Comparative Study of Different Heat Exchange Systems in a Thermoelectric Refrigerator and Their Influence on the Efficiency. *Appl. Therm. Eng.* **2016**, *103*, 1289–1298. [CrossRef]
14. Gong, T.; Gao, L.; Wu, Y.; Zhang, L.; Yin, S.; Li, J.; Ming, T. Numerical Simulation on a Compact Thermoelectric Cooler for the Optimized Design. *Appl. Therm. Eng.* **2019**, *146*, 815–825. [CrossRef]
15. Koeln, J.P.; Alleyne, A.G. Optimal Subcooling in Vapor Compression Systems via Extremum Seeking Control: Theory and Experiments. *Int. J. Refrig.* **2014**, *43*, 14–25. [CrossRef]
16. Wang, Q.-Z.; Liu, S.-F. Test and practice of installing subcoolers in heat pumps. *Jiangxi Coal Technol.* **2005**, 29–31. [CrossRef]

17. Yang, B.; Ahuja, H.; Tran, T. Review Article: Thermoelectric Technology Assessment: Application to Air Conditioning and Refrigeration. *HVACR Res.* **2008**, *14*, 635–653. [CrossRef]
18. Li, G.-L.; Lv, Z.-J.; Jin, T.-X. Experimental study on the optimization of the working voltage of thermoelectric subcooler. *Low Temp. Supercond.* **2018**, *46*, 86–90.
19. Astrain, D.; Merino, A.; Catalán, L.; Aranguren, P.; Araiz, M.; Sánchez, D.; Cabello, R.; Llopis, R. Improvements in the Cooling Capacity and the COP of a Transcritical CO_2 Refrigeration Plant Operating with a Thermoelectric Subcooling System. *Appl. Therm. Eng.* **2019**, *155*, 110–122. [CrossRef]
20. Sánchez, D.; Aranguren, P.; Casi, A.; Llopis, R.; Cabello, R.; Astrain, D. Experimental Enhancement of a CO_2 Transcritical Refrigerating Plant Including Thermoelectric Subcooling. *Int. J. Refrig.* **2020**, *120*, 178–187. [CrossRef]
21. Aranguren, P.; Sánchez, D.; Casi, A.; Cabello, R.; Astrain, D. Experimental Assessment of a Thermoelectric Subcooler Included in a Transcritical CO_2 Refrigeration Plant. *Appl. Therm. Eng.* **2021**, *190*, 116826. [CrossRef]
22. Liu, Z.-Y. Numerical Simulation and Experimental Study of CO_2 Transcritical Cycle Water-Water Heat Pump. Ph.D. Thesis, Tianjin University, Tianjin, China, 2011.
23. Austin, B.T.; Sumathy, K. Parametric Study on the Performance of a Direct-Expansion Geothermal Heat Pump Using Carbon Dioxide. *Appl. Therm. Eng.* **2011**, *31*, 3774–3782. [CrossRef]
24. Yoon, S.H.; Kim, J.H.; Hwang, Y.W.; Kim, M.S. Heat Transfer and Pressure Drop Characteristics during the In-Tube Cooling Process of Carbon Dioxide in the Subcritical Region. *Int. J. Refrig.* **2003**, *26*, 857–864. [CrossRef]
25. Searle, M.; Black, J.; Straub, D.; Robey, E.; Yip, J.; Ramesh, S.; Roy, A.; Sabau, A.S.; Mollot, D. Heat Transfer Coefficients of Additively Manufactured Tubes with Internal Pin Fins for Supercritical Carbon Dioxide Cycle Recuperators. *Appl. Therm. Eng.* **2020**, *181*, 116030. [CrossRef]
26. Chen, D.-S.; Shi, Y.-M.; Gao, Z.-Q. Some general flow boiling estimates the applicability of correlations to several elements. *J. Refrig.* **2013**, *34*, 8–14.
27. Yang, J.-L.; Tang, J.-B.; Miao, Y.-F. Optimal design calculation of cased CO_2 evaporator. *Fluid Mech.* **2016**, *44*, 83–86.
28. Kew, P.A.; Cornwell, K. Correlations for the Prediction of Boiling Heat Transfer in Small-Diameter Channels. *Appl. Therm. Eng.* **1997**, *17*, 705–715. [CrossRef]

buildings

MDPI

Article

Experimental Study of LiCl/LiBr-Zeolite Composite Adsorbent for Thermochemical Heat Storage

Depeng Chen [1], Xin Chen [1], Zhiwei Ma [2], Yaodong Wang [2], Anthony Paul Roskilly [2] and Jian Zhou [2,*]

[1] School of Architecture and Engineering, Anhui University of Technology, Maanshan 243000, China
[2] Department of Engineering, Durham University, Durham DH1 3LE, UK
* Correspondence: jian.zhou@durham.ac.uk; Tel.: +44-(0)-7533095659

Abstract: Adsorption-based thermochemical heat storage is a promising long-term energy storage technology that can be used for seasonal space heating, which has received significant amount of efforts on the research and development. In this paper, the heat storage capacity of composite adsorbents made by LiCl + LiBr salt and 3A zeolite was investigated. The basic characteristics of composite material groups were experimentally tested, and it was found that the adsorption composite with 15 wt% salt solution had excellent adsorption rate and adsorption capacity, which was considered as the optimal composite material. Furthermore, the heat storage density of the composite material could be as high as 585.3 J/g, which was 30.9% higher than that of pure zeolite. Using 3 kg of the composite material, the adsorption heat storage experiment was carried out using a lab-scale reactor. The effects of air velocity and relative humidity on the adsorption performance were investigated. It was found that a flow rate of 15 m^3/h and a relative humidity of 70% led to the most released adsorption heat from the composite material, and 74.3% of energy discharge efficiency. Furthermore, an adsorption heat storage system and a residential model were built in the TRNSYS software to evaluate the building heating effect of such heat storage system. It is found that the ambient temperature will affect the heating effect of the adsorption heat storage system. The coefficient of performance (*COP*) of this model is as high as 6.67. Compared with the gas boiler heating system, the adsorption heat storage energy can replace part of the gas consumption to achieve energy savings.

Keywords: thermochemical heat storage; hygroscopic salt; zeolite; water vapour sorption; energy discharge efficiency; TRNSYS; building model

Citation: Chen, D.; Chen, X.; Ma, Z.; Wang, Y.; Roskilly, A.P.; Zhou, J. Experimental Study of LiCl/LiBr-Zeolite Composite Adsorbent for Thermochemical Heat Storage. *Buildings* **2022**, *12*, 2001. https://doi.org/10.3390/buildings12112001

Academic Editors: Shi-Jie Cao and Wei Feng

Received: 10 October 2022
Accepted: 11 November 2022
Published: 17 November 2022

1. Introduction

Energy consumption in buildings has steadily increased in recent years. Fossil fuels are increasingly becoming unviable as an energy source due to the rapidly increasing global population and growing demands of urban construction. Instead, a range of ecologically sound renewable energy sources, such as solar, wind, and biomass energy, are increasingly being developed to supply energy demands [1–4]. This transition to renewable energy is important to slow climate change and reduce the usage of fossil fuel resources. However, at present, there is a mismatch between user demand and the supply of most renewable energy sources. For example, solar energy is an effective alternative to fossil fuels for heating buildings, which is abundant, however it is intermittent and unstable because the solar radiation varies with weather conditions and time of a day [5]. Most of the solar energy supply is in the summer, heating is most needed in the winter when solar energy is less abundant [6]. Therefore, the reasonable utilization of solar energy requires new technologies to realise short to long term heat storage.

Thermal energy storage (TES) can be used to resolve the timing mismatch between solar energy supply and energy demand [7,8]. There is increasing interest in technically advanced and economical TES systems for applications such as hot water supply and space heating. Among them, thermochemical heat storage (TCHS) systems have many advantages over other types of TES [9,10]. This method involves storing energy through reversible adsorption processes or chemical reaction, thus, high reaction enthalpies can be used to achieve greater energy storage density, making the system appropriate for large-scale applications. Using this method, the heat loss during the energy storage process is almost zero [11,12]; in addition, due to controllable gas partial pressure, this technology is capable of heating, cooling and thermal energy storage functions, allowing the conversion of heat within a specific temperature range [13]. Such feature provides more flexibility than traditional heat storage systems. Moreover, long-term seasonal storage for buildings can be achieved by using TCHS because the heat is stored as chemical potential, which does not degrade over time [14–17].

Hygroscopic salts have been extensively investigated as one of the most promising materials for TCHS applications. These salts have high water absorption capability and heat storage density [18,19] and have been identified as suitable thermal storage candidates in the temperature range of 20 to 200 °C [20,21]. Typical hygroscopic salts for TCHS are $CaCl_2$ and $MgCl_2$ [22], LiCl [23,24] and $MgSO_4$ [25,26]. In [27], LiCl composites showed higher adsorption capacity, closely followed, in order, by $CaCl_2$ and LiBr composites. By mixing other salts, LiI, $LiNO_3$, LiCl, etc., into the LiBr solution, mixed materials have been developed aiming to improve solubility, corrosion and stability [28,29], which inspired the development of composite material using mixed salts. Gordeeva et al. [30] found that dual salt composite adsorbents could reduce adsorption hysteresis while having the potential to significantly improve flexibility and operational reliability, and they designed binary LiCl-LiBr systems confined in silica pores as effective materials with predetermined adsorption properties. Entezari et al. [31] discovered that adding a small amount of LiBr to LiCl can improve the adsorption capacity of the binary salt composites by up to 5.5%, and he used simulations to evaluate the dehumidification performance of the adsorbent.

In practical system applications, deliquescence of hygroscopic salts leads to the leakage of the liquid salty solution and loss of salt, thus reducing the cyclic stability [32]. To improve the usage of hygroscopic salts, researchers have investigated embedding the salts in porous structures, which are used as carriers to immobilize inorganic salts and distribute them uniformly. Furthermore, the porous properties of the carrier increase the surface area of the composite adsorbent—these materials enhance the heat and mass transfer efficiency with increasing surface area [33]. Commonly used porous matrix (CSPM) include silica gel [34,35], zeolite [36,37] and expanded graphite [38]. Zeolites are porous materials with good water absorption capacity and the ability to function at high temperatures, thus forming excellent matrices for composite materials [39]. Thomas et al. [40] developed composite adsorbent by impregnating zeolites with salt solutions and concluded that the impregnation of these salts by zeolite allowed salt deliquescence and water absorption above the deliquescence humidity, without the problem of leakage and loss of salt. With high salt loading ratio and at high air humidity levels, the heat storage density of composite material can be up to 153% of that of pure zeolite. Gareth et al. [6] developed four zeolite composites impregnated with different mass fractions of $MgSO_4$; their results showed a strong correlation between the microstructure of zeolite and the heat generated by the composite. To avoid deliquescence, the mass concentration of hygroscopic salts in the composites was usually limited to below 35%.

Previous studies on composite adsorbents have focused on characterising small amount of materials [41–43]. These studies have been less further investigated by applying the materials to real practical systems. System design is the key to developing architectural applications for TCHS. In the current study, the deployment of conventional filled beds in energy storage systems suffers from low energy storage density and large volume. Effective ways to solve such problems and reduce potential costs are worth being explored [44]. Addressing this research goal, Calabrese [45] evaluated the performance of foam-based adsorbent materials in open and closed adsorption heat storage applications and pointed out that matrix maintenance of cyclic stability in practical applications is an important issue to be addressed. Xu et al. [46] developed a numerical model to investigate the thermochemical reaction process in a reactor filled with composite adsorbent and calculated the heat transfer between the composite material and water vapour. Li et al. [47] numerically developed a novel multi-layered sieve reactor focusing on the interaction between the composite material and the air in an open system; in addition, they performed a parametric analysis of the inlet temperature, humidity and airflow rate. Future more, it is expected that the TCHS system will be applied to actual residential heating. A laboratory prototype of a fixed-bed open TCHS reactor developed by Zondag et al. [22] can provide a certain amount of thermal energy for home heating. The system contains 17 dm^3 of adsorbent material, which can generate 150 W of thermal energy and achieve an effective energy storage density of about 0.5 GJ/m^3. Consider time period and material cost constraints when using energy from TCHS systems for home construction. Researchers can use the TRNSYS program to simulate and analyze transient systems. Safa et al. [48] simulated the heating performance of the heat pump system in the TRNSYS environment and found that the coefficient of performance (*COP*) under different heat source temperatures was 3.05–3.44 during heating. In TRNSYS-EES, Sakellari et al. [49] analyzed the performance of a heating system based on an exhaust air heat pump using simulation. And several strategies have been proposed to keep comfort within a reasonable range.

The purpose of the present study is to develop suitable LiCl/LiBr–zeolite composite adsorbents for TCHS applications and to perform test of the thermal storage performance in a lab-scale adsorption heat storage reactor. Finally, TRNSYS was used to evaluate the heating effect of this system. The studied pure zeolite and composite adsorbents were subjected to electron microscopy measurement, kinetic adsorption performance and DSC tests to investigate their pore characteristics, water adsorption, heat transfer and mass transfer properties. Based on the identified optimal LiCl/LiBr–zeolite composite adsorbent, a set of adsorption heat storage test rig was constructed. Considering the effects of both air flow rate and air humidity, the water adsorption performance of the developed composite adsorbent under different operating conditions was explored, and the energy storage density and cycle stability of the material were also obtained and analysed. Although the heat storage performance of the open system TCHS has received more attention, the performance of the combination of solar thermal collection and TCHS system heating to meet the space heating needs is still less researched. In this study, a simulation model of the dwelling was developed in Trnsys software. In the winter climate environment, the reactor is charging by solar panels during the day, and discharging by hydration reaction of the adsorbent in the reactor during the night. The heat released is used for space heating of the house to maintain the room temperature in the range of 21 $\pm$ 1 °C. When the material reaction completely stops exothermic, start gas as auxiliary energy supplement, continue to heat the house. This paper also simulates a reference system with gas as the heating energy source. The heating effect of the TCHS system with a time period of 7 months was studied. Analyze and calculate gas energy savings to assess the benefits of the system.

2. Materials

2.1. Preparation of Composite Adsorbent

Anhydrous LiCl and anhydrous LiBr were respectively purchased from Sinopharm Chemical Reagent Co., Ltd., (Shanghai, China). and Hefei BASF Biotechnology Co., Ltd., (Hefei, China). 3A zeolite was purchased from Clear Spring Technology Co., Ltd., (Singapore).

Figure 1 shows the preparation process of the composite adsorbent. First, anhydrous LiCl and anhydrous LiBr solids with pre-defined mass ratio (1:1 to 1:5) were separately added to distilled water and continuously mixed to allow the salt powder to completely dissolve in the water to ensure uniform mixing with the 3A zeolite. To determine the optimal ratio for a composite adsorbent, five salt solution groups with various mass fractions (5 wt%, 10 wt%, 15 wt%, 20 wt% and 25 wt%) were prepared. Considering the reasonably cost effective and easy in process, this study used the water impregnation method to manufacture the composite adsorbent, which can achieve the desired effect of zeolite loading salt. In the first step, zeolite spheres of 3–5 mm size were soaked in deionised water for 12 h for pre-treatment; after soaking, the zeolite was heated and dried at 200 °C for six hours before being colded. The second step involved impregnating the previously treated dry zeolite with salt solution. 30 g of zeolite was soaked in 50 g of salt solution. To ensure that liquid solution was evenly filled in the pores of zeolite, this impregnation process was conducted with continuous stirring using a glass rod. Finally, the salt solution-carrying zeolite was dried again in an oven at 200 °C to form a dehydrated composite adsorbent. The samples were numbered Z0 (pure zeolite) and Z5, Z10, Z15, Z20 and Z25 for the samples using 5–25 wt% salt solution (e.g., Z5 corresponds to the composite adsorbent manufactured using 5 wt% salt solution). The ingredient with various ratios of the manufactured composites are showen in Table 1. Notably, it was found during the preparation that salt concentrations above 25 wt% were not easily made into composite adsorbents. The structure of zeolite particles is destroyed during the impregnation process, and some of the zeolite particles will be completely broken into powder form. Their performance was tested to be inferior to that of the intact sample set, so the maximum concentration of salt solution used in this experiment was 25 wt%.

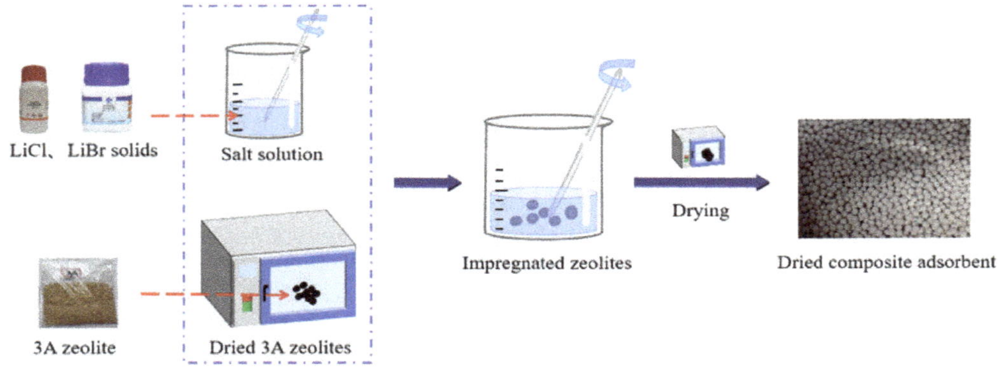

Figure 1. Preparation process of the composite adsorbent.

Table 1. The ingredient with various ratios of the manufactured composites.

No.	Z0	Z5	Z10	Z15	Z20	Z25
Salt solution concentration	0 wt%	5 wt%	10 wt%	15 wt%	20 wt%	25 wt%
LiCl:LiBr mass ratio	-	1:1	1:2	1:3	1:4	1:5

2.2. *Properties of the Composite Adsorbent*

2.2.1. Morphologies

The samples were imaged using a Sigma 300 model scanning electron microscope (SEM). Figure 2 shows the microscopic surface morphology of the samples at a magnification of 20,000×. As shown in the Figure 2a–f, the pure zeolite (sample Z0) exhibits clear pores, while the remaining samples demonstrate that the salt and 3A zeolite are well mixed, with salt tending to collect in the uneven areas of the zeolite surface. As the salt content of the composite increases, they gradually adhere to the surface, thereby increasing the surface roughness. The surface of the adsorbed sample displayed clear salt crystallisation, with increased crystallisation observed with increasing salt concentration. Such a phenomenon is consistent with previous descriptions of the microstructure of adsorbents loaded with salts. The outer surface of sample Z25 was almost completely covered with salt crystals, however, the channels leading from the surface of the zeolite to its core were unobstructed. More salt loading can increase the water adsorption capacity because salt embedded in the sample creates numerous micropores; this process significantly increases the surface area of the composite material and the extent of its contact area with water vapour. However, high-concentration salt solutions are also susceptible to crystallisation, which may limit the material's mass transfer capacity. It is therefore essential to identify the optimum concentration of salt solution used for manufacturing the composite adsorption material.

Figure 2. SEM photos of the composites with different solution concentrations at the magnification of 2000 times: (**a**) Z0; (**b**) Z5; (**c**) Z10; (**d**) Z15; (**e**) Z20; (**f**) Z25.

2.2.2. Pore Properties

The pore volume and pore size distribution of the materials were examined by a fully automatic surface and pore size analyzer (BET, Micromeritics APSP2460) using nitrogen adsorption-desorption method. The pore size distributions and specific surface areas of porous materials were calculated using the Barrett-Joyner-Halenda (BJH) theory and the Brunauer-Emmet-Teller (BET) theory. The porous properties of the samples, such as pore size distribution, pore volume and surface area, are listed in Table 2. Figure 3a,b show the pore volume versus different pore widths. A sufficiently large pore volume can maintain the passage of water vapor in and out of the adsorption process. Otherwise, the mass transfer of water adsorption kinetics would be limited. All samples showed a similar trend, increasing rapidly at the beginning and stabilizing at the end. They all had a peak in pore volume at about 5–7 nm pore width, which means that most of the pore sizes were distributed in this range. Moreover, the pore volume or average pore size may decrease as the concentration of the solution increases. Although a high salt concentration theoretically provides more reactants for water adsorption, it also limits water adsorption by sacrificing some mesoporous mass transport channels in the support structure. This suggests that the balance between LiCl/LiBr loading and pore volume in the composite should be carefully maintained to accomplish an optimal thermochemical thermal storage material.

Table 2. The essential porous characteristics of the manufactured composite adsorbents.

Group	Surface Area (m^2/g)	Pore Volume (cm^3/g)	Adsorption/Desorption Average Pore Diameter (nm)
Z0	27.73	6.36×10^{-2}	9.02/10.71
Z5	28.11	6.22×10^{-2}	8.74/10.41
Z10	28.76	6.35×10^{-2}	8.65/10.21
Z15	26.90	6.07×10^{-2}	8.89/10.53
Z20	25.67	5.88×10^{-2}	9.00/10.84
Z25	25.83	5.83×10^{-2}	8.86/10.63

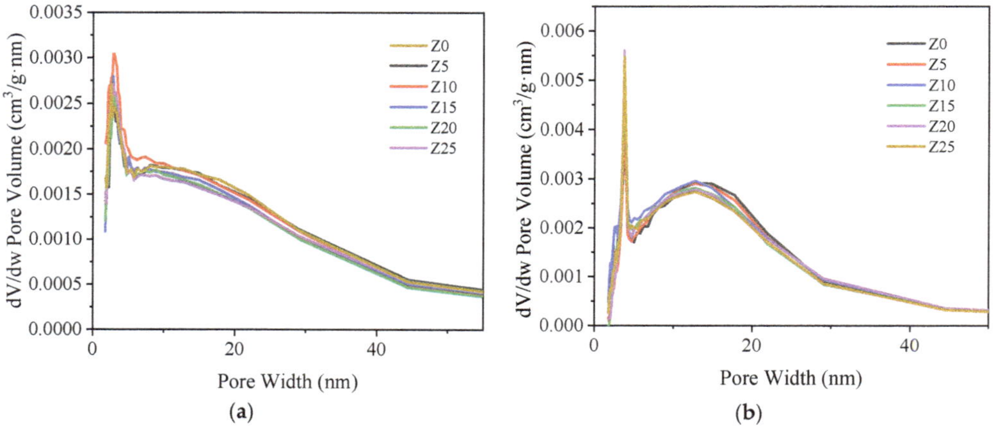

Figure 3. BET results of the prepared composite adsorbents. (**a**) Adsorption pore volume vs. pore width, (**b**) Desorption pore volume vs. pore width.

2.2.3. Adsorption Kinetic Performance

The adsorption rate and capacity of water vapour have significantly impact on the energy storage capacity of LiCl-LiBr-zeolite adsorption systems. The adsorption tests were conducted in a constant temperature and humidity chamber (climatic chamber) under a temperature of 25 °C and relative humidity (RH) levels of 60%, 70%, 80%, and 90% to analyse the adsorption kinetics. The mass change of the adsorbent samples over 24 h was measured by using an electronic balance. The water uptake kinetics of samples are depicted in Figure 4. When the weight difference is below 5% in two consecutive 60 min intervals, adsorption equilibrium was deemed to have been achieved.

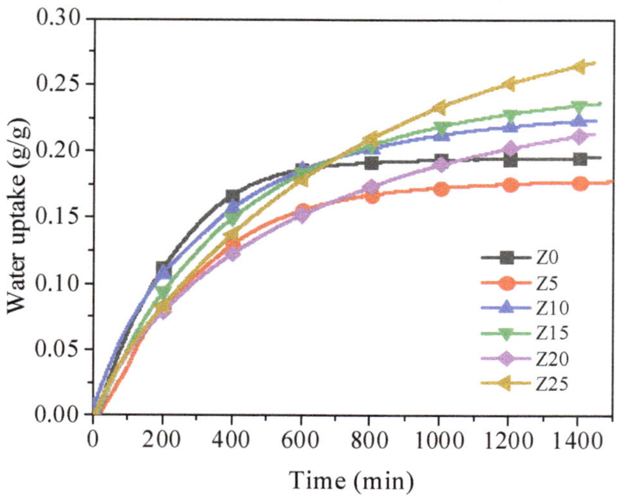

Figure 4. Water adsorption kinetics of samples with different composites.

The pure zeolite, Z0, exhibits high adsorption rate at the early adsorption stage, reaching a maximum value of 0.18 g/g after 600 min of adsorption at 25 °C and 80% RH, as shown in Figure 4. Comparing to Z0, Z5 performs more poorly as an adsorbent with lower adsorption rate and capacity, which is due to its low salt content and partially blocked water transfer channels. Higher salt content adsorbents take longer to achieve equilibrium, suggesting that the water transfer channels were blocked more significantly with increasing salt content; however, the adsorption rate are larger at late stage of adsorption and the final adsorption capacity is also promoted. In contrast, samples Z10, Z15 and Z25 demonstrate reasonable adsorption rates and capacities. However, the salt swells upon hydration, leading to the blockage of the zeolite pore channels, especially in Z20 as shown in BET result (Table 2). Both the active adsorption area (specific surface area) and the mass transfer channel (pore volume) were reduced. This self-clogging effect is detrimental to the water adsorption behavior of the material. Thus, Z20 exhibited a refrained adsorption rate than other candidates until 700 min due to the corresponding significant mass transfer barrier; thereafter due to the enhanced water adsorption capacity, i.e., increased salt loading, the overall water adsorption capacity is higher than Z0 and Z5. Finally, the adsorption capacity of Z10, Z15 and Z25 are 0.22 g/g, 0.24 g/g and 0.27 g/g, respectively. Z15 always maintains a high adsorption rate and has a high adsorption capacity when the adsorption equilibrium is reached.

The RH significantly impacts the equilibrium adsorption capacity of every sample. Figure 5 shows the adsorption capacity of the samples after reaching equilibrium at 40%, 60% and 80% RH. The overall pattern indicates that high RH encourages wate vapour adsorption, thus increasing the adsorption capacity of all samples. This occurs because the increased RH elevates the water vapour pressure difference between the water vapour in the air and the equilibrium water vapour pressure of the adsorbent material, which facilitates the effective diffusion of water vapour into the adsorbent. Another significant factor is that salt deliquescence in a high-humidity environment reduces zeolite micropore blockage, which is advantageous for increasing water adsorption. Nevertheless, the swelling of salt hydration occurs at the same relative humidity conditions, which may lead to the clogging of zeolite pores. The blocking extent referring to the solution concentration was partly revealed in the BET result, which has been discussed in last section.

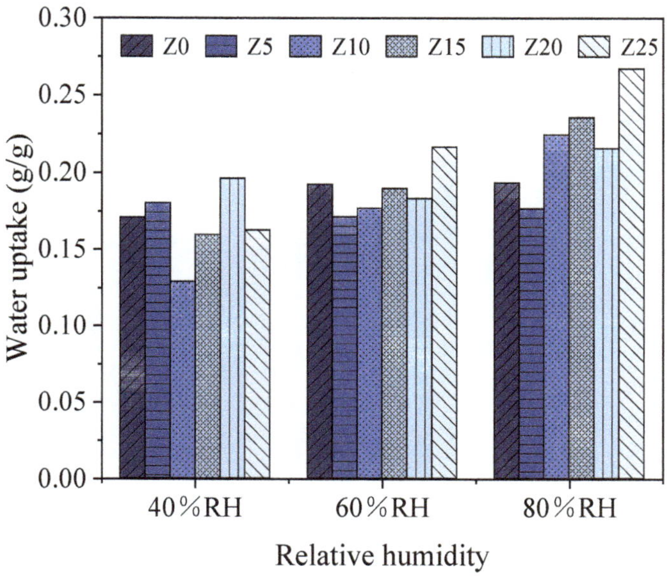

Figure 5. Water adsorption capacity with different relative humidifies.

2.2.4. Adsorption Heat

Differential scanning calorimetry (DSC) were performed on Z0 to Z25. Using a simultaneous thermal analyser, with N_2 selected as the test atmosphere. The experimental temperature conditions were set to increase from 30 °C to 300 °C, the heating rate was 5 °C/min, and the experimental time was 1 h. The heat flow was recorded throughout the process.

The DSC curves of the endothermic dehydration process of each group of samples are shown in Figure 6a. The theoretical heat storage density of the composite heat storage material is obtained by integrating the curve inland in a certain range, and plotted in Figure 6b. Sample Z0 exhibites a heat storage density of 447.1 J/g; however, in all instances, the heat storage density of the composites are higher than that of the pure zeolite. This effect is due to the contribution of salt to the water absorption capacity, thereby increasing its thermal storage density. The most significant thermal storage density is recorded in Z5 (32.6% higher than Z0). Although no positive correlation was identified between the heat storage density of the composite adsorbent and the salt solution concentration, a higher solution concentration may nonetheless hinder mass transfer during water adsorption, thus resulting in a lower heat storage density. Therefore, samples Z5, Z10 and Z15 can be used as the three 'ideal' samples for the analyses in this study because their heat storage densities do not markedly differ.

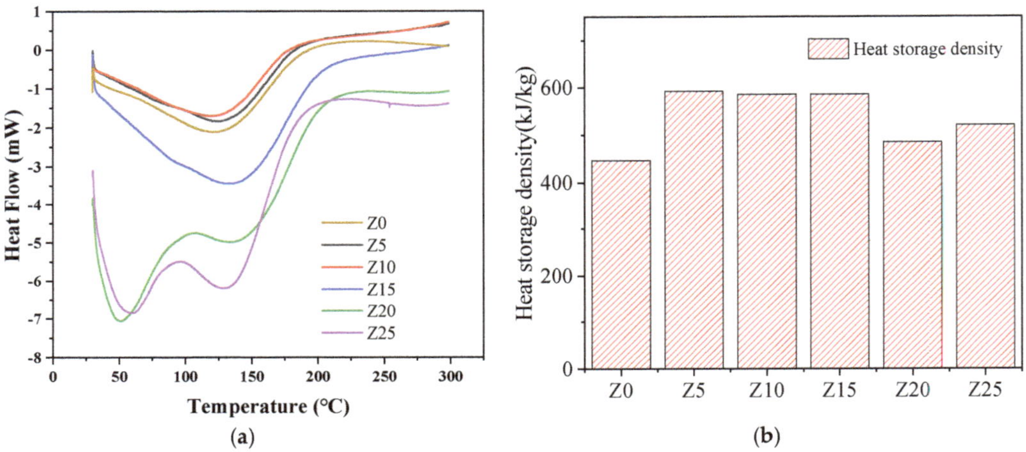

Figure 6. DSC results of the prepared composite adsorbents. (**a**) Heat flow curve, (**b**) Heat storage density.

2.3. *Selection of the Optimal Adsorbent*

Based on the analyses above, the adsorption performance of composite materials are found to be variably impacted by the salt concentration with no clear linear relationship. As confirmed by the SEM experiments, changing the salt content influenced the microscopic scale size characteristics of the composite adsorbent. The investigation of the adsorption kinetics of the samples revealed that Z15 had both a high water adsorption rate and capacity. Further DSC studies show the heat storage density of Z15 is up to 585.3 J/g, which is 30.9% higher compared to Z0. On this basis, sample Z15 was selected as the optimal composite adsorbent for lab-scale open system experiment.

3. Experiment Overview

3.1. *Test Rig*

Figure 7a shows a schematic diagram of a reaction system. Figure 7b is the photo of the thermochemical heat storage reaction system, which comprises a pipe with a valve, a fan, a humidifier and a reactor. The reactor is placed inside the hollow cylindrical pipe that houses the entire apparatus. The rightmost side of the pipe is connected to an inclined flow booster pipe fan. In addition, the humidifier is connected to the inlet air duct. The layout of the reactor is shown in Figure 7c. The wire mesh container containing the composite adsorbent Z15 is covered with a metal disc-shaped covering. To achieve a more thorough and uniform adsorbent reaction, the reactor is fitted with a thin tube with perforations in the centre shaft. In total, 3 kg of Z15 adsorbent was placed inside the reactor. During the experiment, the pipes were covered in aluminium foil insulation foam to minimize heat loss to the exterior. Temperature and humidity sensor (A-Pt100) were installed at the inlet and outlet of the reactor to measure the temperature and humidity of the air. To assess the volume of air passing through the pipe, a gas flow meter was installed at the end of the pipe.

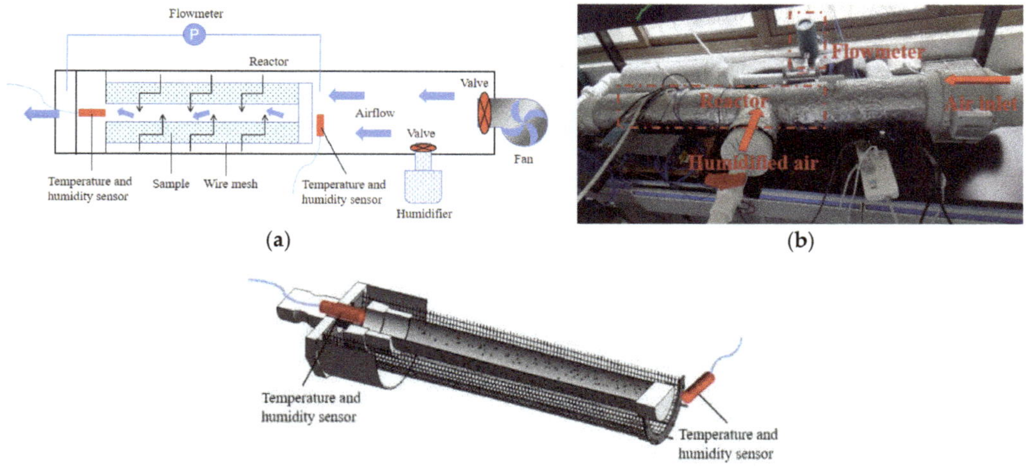

Figure 7. Adsorption experimental pipeline diagram: (**a**) piping schematic diagram; (**b**) reaction system assembly diagram; (**c**) reactor cross-sectional view.

3.2. Experimental Procedures

In this study, the air flow rate and relative humidity were chosen as the main variables for investigation; thus, to examine the impacts of both on the thermal storage performance of the composite salt, orthogonal experiments were configured with airflow rates of 5 m^3/h, 10 m^3/h, 15 m^3/h and 20 m^3/h and humidity levels of 60%, 70%, 80% and 90% RH.

The 3 kg of LiCl/LiBr-zeolite composite thermal storage material was placed in the reactor, and the fan and humidifier were adjusted to ensure that the experimental conditions were within the predetermined ranges. Overall, the adsorption reaction procedure for the composite material takes 12 to 20 h to complete. In the initial stage of the reaction, the outlet temperature rises at a certain rate and then begins to fall after reaching its peak. The temperature curve subsequently tends to slowly level off and return to the inlet temperature, indicating the reaction's completion. The adsorption material in the reactor was heated in an oven following each experiment, totally dehydrated, and then cooled in preparation for the following experiment.

3.3. Energy Density Calculate

For the thermochemical heat storage system, the energy density is a crucial indicator to evaluate the performance of the composite salt heat storage, which can be calculated by the following equation. Since the humidity data measured by the temperature and humidity sensor is 0–100% RH, the water vapor partial pressure P_v in the wet air can be obtained by the following Equation (1) [50]:

$$P_v = \varphi P_s \tag{1}$$

P_s is the saturated water vapor partial pressure [50], and the absolute humidity d of wet air can be expressed as:

$$d = 0.622\frac{P_V}{P - P_V} \quad (2)$$

The enthalpy H of wet air can be calculated from Equation (3) [39]:

$$H = 1.005t + d(2500 + 1.84t) \quad (3)$$

The energy storage density Q_E can be calculated by Equation (4) [51]:

$$Q_E = \frac{\int_0^{t_d} \rho_f q_v (H_{exit} - H_{inlet}) dt}{M} \quad (4)$$

ρ_f is the density of the dry air, q_V is the volume flow rate of the gas, M is the total mass of the adsorbent material in the reactor, this experiment uses M = 3 kg.

The energy discharge efficiency of this reactor can be estimated using Equation (5) [39]:

$$\eta = \frac{Q_E}{Q_D} \times 100\% \quad (5)$$

Q_D is the theoretical heat storage density of the composite adsorbent, which is derived from the DSC experiment in Section 2.2.3.

4. Experimental Results and Discussion

4.1. Effect of Air Flow Rate

For each of the four relative humidity levels, Figure 8 illustrates the impact of various air flow rates on the exit air temperature. All the temperature differential variations in Figure 8 initially climb rapidly and then slowly decline until the temperature rise becomes close to zero. The final temperature increase reduces to zero at the point of which the adsorbent is saturated, the stored heat is completely released and the reaction is over. The rate of temperature rise increases with increasing air flow rate, with the recorded temperature rise rate at a flow rate of 5 m^3/h noticeably lower than the other measured flow rates. Since the air flow rate affects the rate of moisture transfer and the low air flow rate limits the process of mass and heat transfer between the wet air and the adsorbent [52]. However, the effective reaction time of the material becomes longer at low flow rate conditions. The reaction time decreases consistently with increasing flow rate under conditions of 60% RH. The adsorbent still tends to have a shorter reaction time at high flow rates when the air has an RH of 70% or below, as shown in Figure 8a,b, however, as the relative humidity rises, this difference becomes less noticeable at flow rates of 10 m^3/h, 15 m^3/h, and 20 m^3/h, as shown in Figure 8c,d.

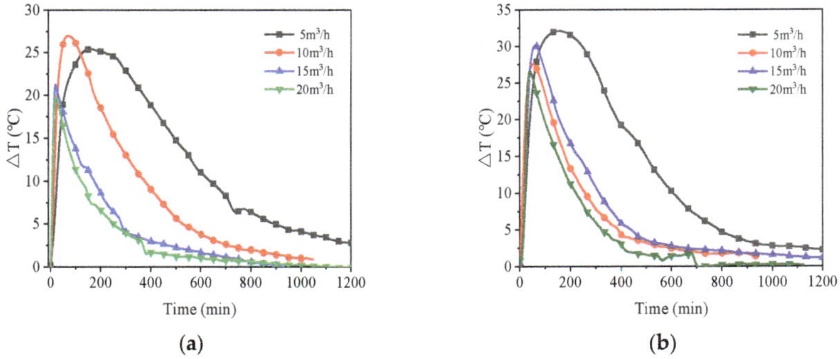

Figure 8. *Cont.*

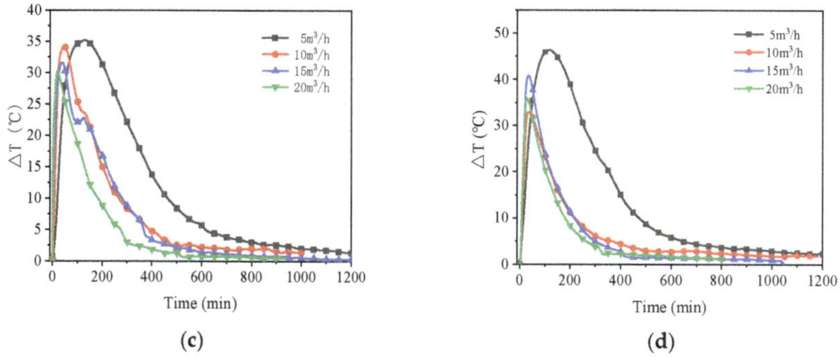

Figure 8. Air temperature rises with time for different air flow rates at four air relative humidity: (**a**) 60%RH; (**b**) 70%RH; (**c**) 80%RH; (**d**) 90%RH.

4.2. Effect of Air Relative Humidity

As the results from Figure 9, the maximum temperature difference increases as the relative humidity increases. The air temperature increased by 84.1%, 28.1%, 93.3%, and 79.8%, respectively, when the relative humidity was elevated from 60% to 90%. Similarly, when more humid air flowed through the reaction pipeline, the rate of temperature rise of the air at the initial stage was higher. This behaviour was especially noticeable at a flow rate of 5 m^3/h. It is because that the high relative humidity provides higher pressure difference to drive the adsorption reaction, which accelerates the hydration reaction of the material driving the heat generation. This causes the increase in temperature rise value and rate.

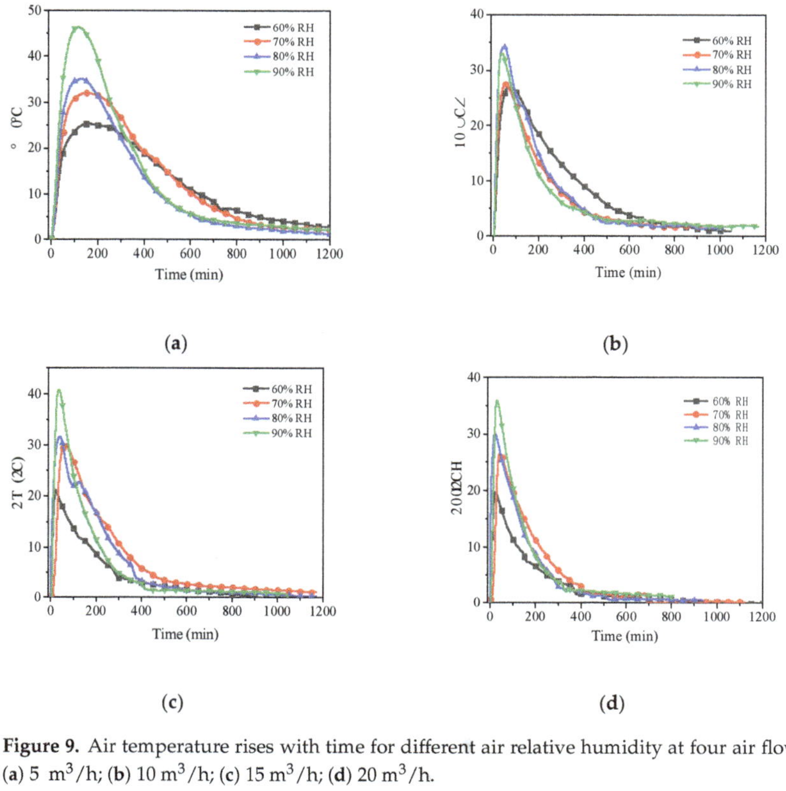

Figure 9. Air temperature rises with time for different air relative humidity at four air flow rates: (**a**) 5 m^3/h; (**b**) 10 m^3/h; (**c**) 15 m^3/h; (**d**) 20 m^3/h.

4.3. Energy Storage Density

For each group of experiments, the energy storage density was calculated and plotted in Figure 10. It can be seen from the experimental results that the maximum temperature difference and the temperature rise rate have a positive correlation trend with the gas flow rate and relative humidity. However, energy storage density does not have such a law. From the formula, it shows that the energy storage density is not only related to the maximum temperature difference, but also to the effective reaction time. In addition, it cannot be ignored that in the same room temperature environment, the heat loss under different experimental conditions is different. The higher the flow rate, the more heat is lost through the pipe, metal reactor. In this experiment, the composite material has the highest heat release value of 434.4 J/g when the gas flow rate is 15 m^3/h and the relative humidity is 70%, and its energy discharge efficiency is 74.3%.

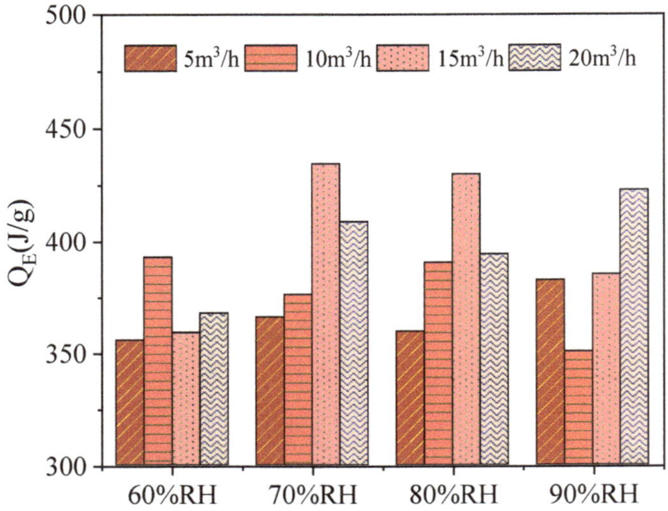

Figure 10. Energy storage density with different experimental conditions.

4.4. Cyclic Experiments

To get the cyclic performance, the adsorbent material was subjected to ten cycles of adsorption/desorption under experimental conditions with an air flow rate of 15 m^3/h and a relative humidity of 70%. Figure 11 shows the variation of the outlet temperature with time. It can be seen that the rate of temperature increase decreases with the number of cycles. Figure 12 shows the variation of the energy storage density for the ten cycling experiments. The average error of the cycling result was 0.41%, which proved the experimental result to be convincing. Although the adsorption heat decreased significantly after the first three trials (by 0.53%), the decrease slowed down in the subsequent studies, with a final rate of change of only 0.17%. This indicates that while certain microstructure changes do occur in the composite adsorption material during cycling, these changes become smaller and smaller as the number of cycles increases. Cycling experiments show that this material can be recycled for a long time in cross-season heat storage.

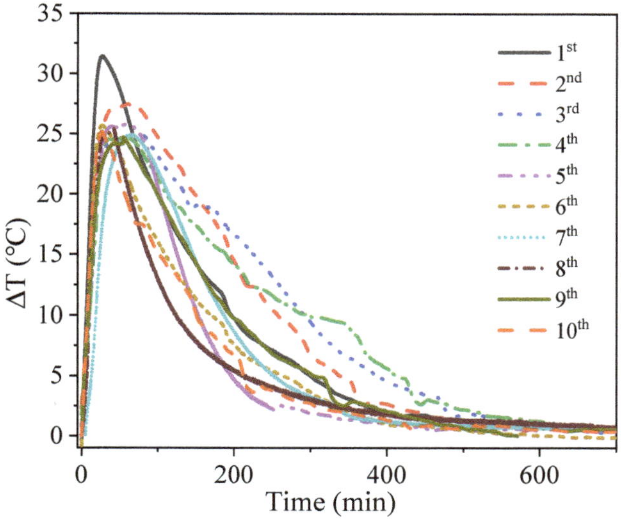

Figure 11. Air temperature rises with time for cyclic experiments.

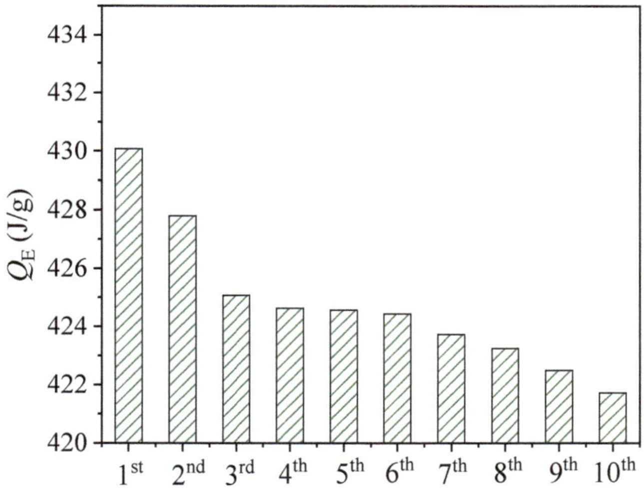

Figure 12. Energy storage density for cyclic experiments.

5. Numerical Simulation

5.1. System Description

In this study, the TRNSYS software was used to dynamically simulate the charging mode and the discharging mode. The system mainly consists of three parts. The first part is solar energy collection and energy storage; the second part is the reactor heat release as well as gas heating; the last part is the residential model as the heated unit. Figure 13 shows the charge mode and discharge mode for the entire operation. During the day, the sorbent is heated and dehydrated by a solar-triggered charging mode. All heat is stored, and there is no heat loss in this process (the heat conversion efficiency is 74.3%). The input heat from the solar panels regenerated the adsorbent in the reactor. At night, the discharge mode is triggered by blowing air at a certain flow rate and humidity into the reactor. All the heat of adsorption released by the reactor heats the air, which is mixed with ambient air and passed into the dwelling to heat the space. The specific content of the heat release analysis module is described in Section 3. The optimal heat extraction efficiency based on experimental data converts the energy stored during the day. In addition, gas is used as an auxiliary energy to supplement the insufficient power required by the house. Considering real life, assume that a simple thermostatic controller is used to maintain the indoor temperature at 22 ± 1 °C. When the indoor average area temperature is lower than 21 °C, the fan starts to work, and the reactor provides heat; when it is higher than 23 °C, the fan stops running. In addition, when the reaction of the adsorbent in the reactor is no longer enough to supply the indoor temperature to 20 °C, no heat is provided, and the gas is started for heating work.

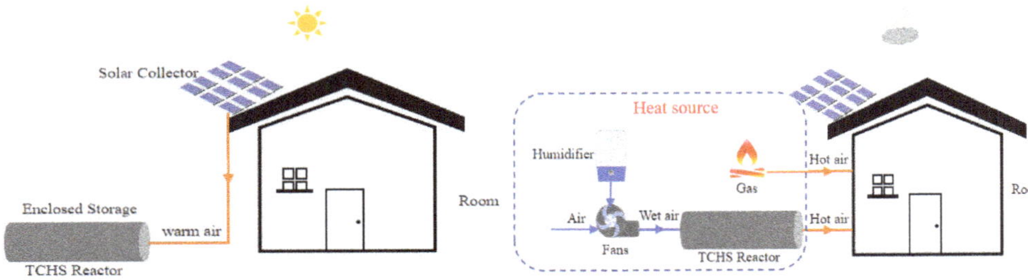

Figure 13. Charge and discharge modes of an open TCHS system.

To realize the residential model of the study, the 3D architectural design software SketchUp was used. A typical two-story residential model (actual size is 6150 mm × 8350 mm × 4550 mm). As shown in Figure 14, the house model is then imported into TRNSYS through the TRNSYS 3d plugin. Next, transfer the module in TRNSYS Simulation Studio to build the TCHS-building simulation platform. Figure 15 shows the dynamic simulation flow, with controllers for each zone. Table 3 gives the structural information of the building used to determine the parameters of the walls, roof, floor and windows. The TRNSYS program generally creates a flexible environment for modelling buildings, where different building descriptions can be obtained by changing design parameters. However, this study mainly focuses on the performance research of the TCHS system. It explores the degree of its effect applied to the building, so that the parameter setting of the building itself is not changed. The system lays half-roof synchronous monocrystalline silicon solar modules and uses Type103 modules to simulate the photovoltaic power generation process. The conversion efficiency of this component can reach 21.3% under standard test condition. The climatic conditions entered refer to the climatic data of a typical weather year in Newcastle, including ambient temperature, instantaneous solar radiation on the horizontal plane, wind speed, and ambient air relative humidity. In the control module of the software, temperature control, time control and heat control are adopted. At the same time, the indoor temperature is within the range of 22 ± 1 °C, and the TCHS heating working time is from 20:00 at night to 9:00 in the morning of the next day. During this time period, when the room needs heating and the energy storage of the TCHS system is not exhausted, the system heats the room. The hot air leading to the room is distributed according to the area ratio of each room. In the analysis process of building energy consumption, the Type56 analysis module is used. The terminal calls the multi-region building model TRNBUILD to simulate the thermal process of the room. By simulating, the domestic temperature, the data of photoelectric conversion and energy consumption are calculated. The above data are monitored by the Type65 online plotter and monthly statistical output by Type46. The simulation time step is 0.05 h, and the simulation period is from October to April of the following year when heating is required.

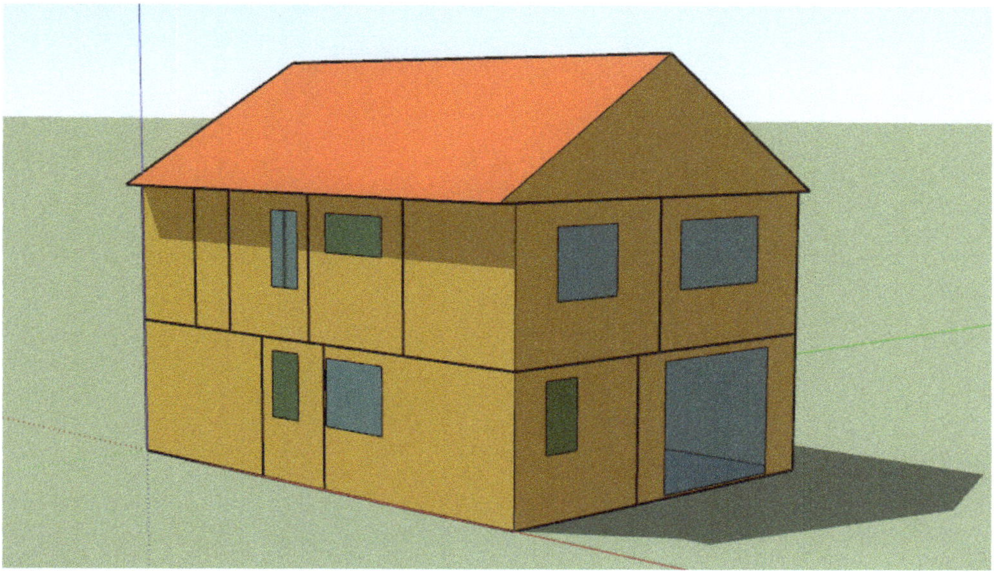

Figure 14. 3D model of the building.

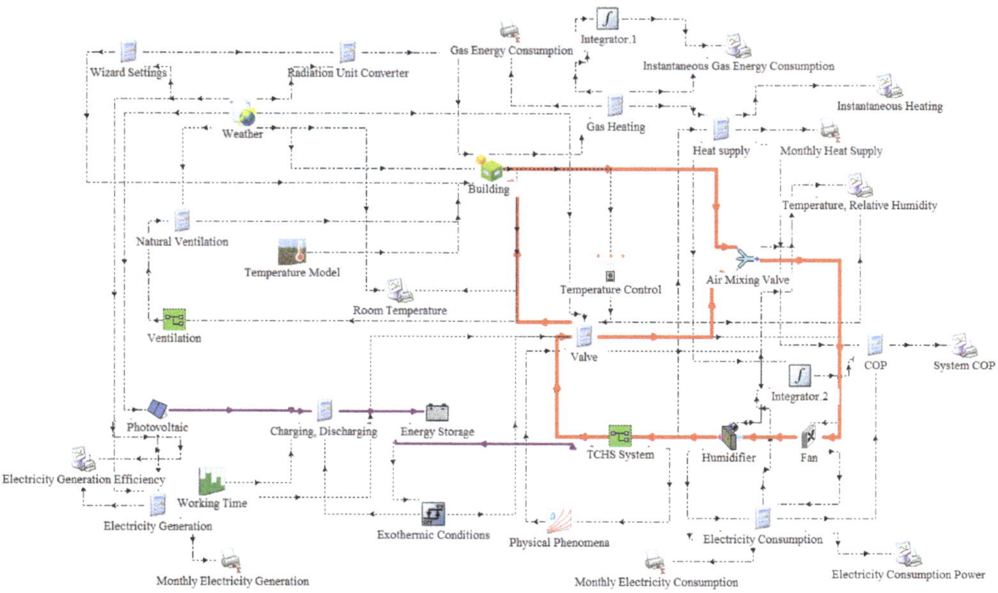

Figure 15. TRNSYS model for building with TCHS system.

Table 3. Physical and thermal properties of building materials used in modelling using TRNSYS.

Surfaces	Thermal Properties
Doors	U = 2.82 W/m^2·K Longwave emission coefficient = 0.9 Convection coefficient = 11 kJ/h·m^2·k (inside), 64 kJ/h·m^2·k (outside)
Windows	U = 1.27 W/m^2·K g-value = 0.74 coefficient = 11 kJ/h·m^2·k (inside), 64 kJ/h·m^2·k (outside)
Floors	U = 0.25 W/m^2·K Solar absorptance = 0.8 Longwave emission coefficient = 0.9 Convection coefficient = 11 kJ/h·m^2·k
Walls	U = 0.54 W/m^2·K Solar absorptance = 0.7 Longwave emission coefficient = 0.9 Convection coefficient = 11 kJ/h·m^2·k
Roof	U = 2.69 W/m^2·K Solar absorptance = 0.8 Longwave emission coefficient = 0.9 Convection coefficient = 11 kJ/h·m^2·k (inside), 64 kJ/h·m^2·k (outside)

5.2. Performance Analysis

5.2.1. Evaluation Indicators

Compared with the energy consumption Q_1 of the traditional (gas) system, the energy consumption Q_2 achieved by the retrofitted system (TCHS and gas) is calculated according to the following Equation (6) [53].

$$Q_2 = 2.42 \times (Q_{hum} + Q_{fan}) + Q_{gas} \tag{6}$$

The heating of TCHS requires raw power from fans and humidifiers. Q_{hum} and Q_{fan} represent the power consumption of the humidifier and fan, respectively, and Q_{gas} is the gas consumption. Among them, 2.42 is the energy conversion factor of electricity equivalent to primary energy. For TCHS systems, the average heating situation is expressed using the coefficient of performance *COP*, which is defined as [54]:

$$COP = \frac{Q_{sor}}{Q_{hum} + Q_{fan}} \tag{7}$$

Among them, Q_{sor} provides heat for the TCHS system. It can be calculated from the heat capacity obtained by air flowing through the adsorbent material.

5.2.2. Analysis of Residential Heating Performance Based on TCHS System

The dynamic simulation of the charging and discharging process of the established system is carried out for a period of 7 months. The results show the operating state and performance of the system under various climatic conditions. Figure 16 shows the changes in ambient temperature and room temperature during the simulation cycle. Although there is temperature control to set the temperature in the range of 21 ± 1 °C, it can be seen that the overall trend is still affected by the ambient temperature.

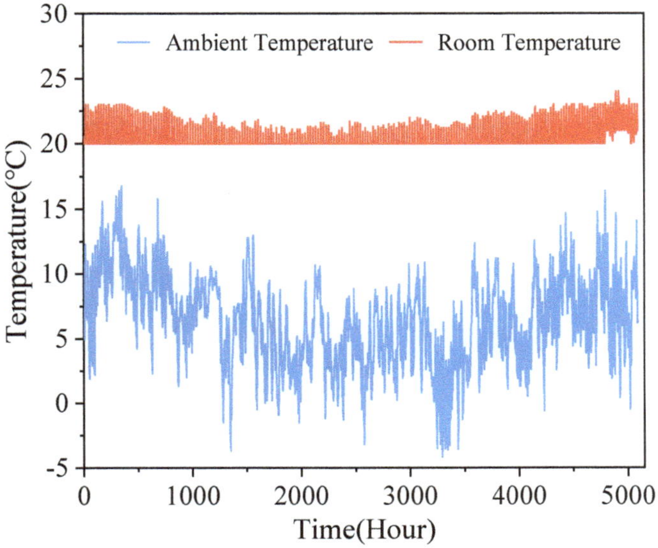

Figure 16. Variation of ambient temperature and room temperature under simulation cycle.

Figure 17 shows the solar power generation, the heat of desorption Q_{des} of the adsorbent and the heat of adsorption Q_{sor} of the adsorption system for each month in charge mode and discharge mode. As the climate temperature rises, the better the solar heat collection effect is, the more energy can be provided to the adsorption heat storage system, and the better the effect of the adsorption heat supply. The conversion efficiency between the two follows the experimental results in Section 4.3, which is close to the optimal energy conversion efficiency of 74.3%.

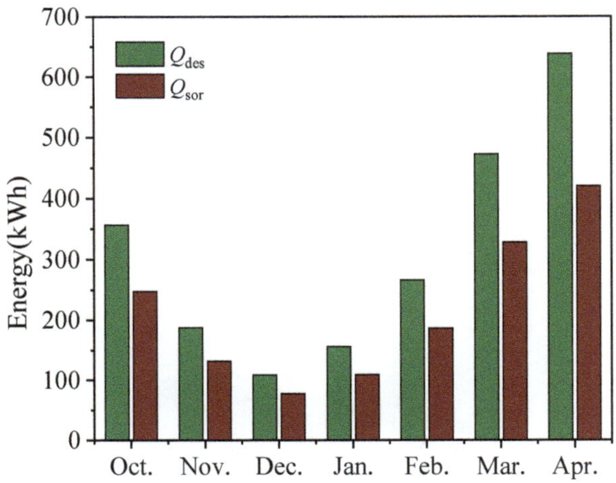

Figure 17. Monthly change in thermal capacity.

To evaluate the energy consumption reduction of the heating system by the heat release of TCHS. Figure 18 plots the energy consumption Q_1 of gas heating alone, and the energy consumption Q_2 of the adsorption system and gas co-heating. It can be seen from the figure that the TCHS system can reduce the consumption of gas, and the total energy consumption in seven months can reduce the gas consumption by 8.8%. However, natural gas still accounts for most energy consumption, as TCHS is limited by solar energy harvesting. After the TCHS releases all the heat of adsorption, the fans and humidifiers stop working. If the ability to increase the TCHS reserve solar energy is improved, the heat release from TCHS can be increased. The power consumption of fans and humidifiers accounts for the central part, which is expected to save most of the gas.

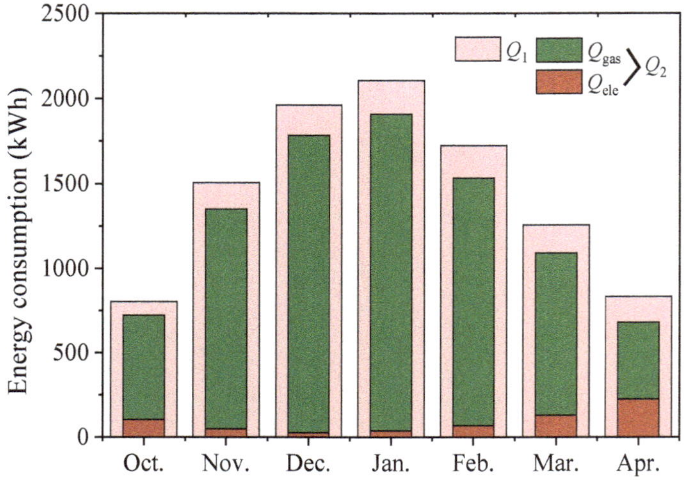

Figure 18. Monthly energy consumption of the system.

Figure 19 plots the *COP* values for TCHS for each month. The maximum was in December at 6.67. This is because the TCHS system relies on the hydration reaction of the adsorbent, which is affected by the climatic environment. Because the air humidity in Newcastle in December is relatively high, it can provide natural moisture for the reaction of the TCHS system, reducing the burden of electricity consumption of the humidifier. Similarly, the minimum value of 4.48 appeared in April. Although solar energy was abundant in April, the charging capacity reached the maximum (Figure 17); but the natural air was dry, which significantly increased the power consumption of the humidifier (Figure 18). The average monthly *COP* was calculated to be 5.56. If the investment in the TCHS system is increased, it is likely to replace most of the gas, and the heating demand of the house can be met with less power consumption.

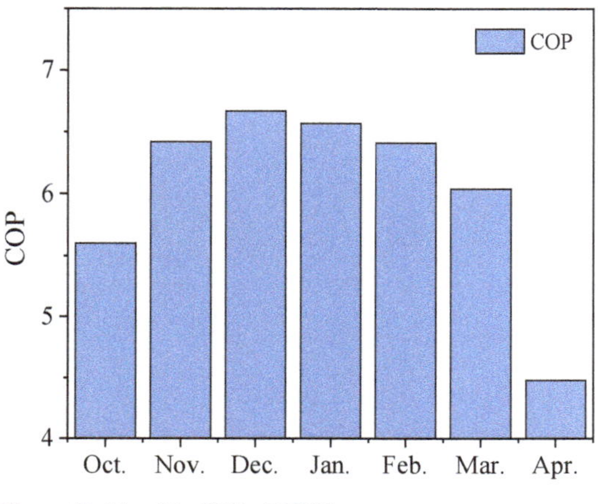

Figure 19. Monthly COP of TCHS system.

6. Conclusions

In this study, a LiCl/LiBr–zeolite composite adsorbent was developed. Subsequently, it was applied to a lab-scale reaction unit for experimental study. For the further evaluation of the practical building heating performance of TCHS system, it was applied to the domestic house. The TCHS-building model was built in the TRNSYS program. The main conclusions are as follows:

- The different salt concentrations resulted in different adsorption capacities and thermal storage properties. Under the same experimental conditions, composite Z15 shows the best adsorption performance.
- In the adsorption heat storage experiment using a lab-scale reactor, the temperature rise rate at the outlet of the reactor is influenced by the air flow rate and humidity. The studied composite adsorbent exhibited its highest heat density of 434.4 J/g at an air flow rate of 15 m^3/h and an RH value of 70%, achieving an energy discharge efficiency value of 74.3%.
- The average error of 10 cycles repeating adsorption experiment was 0.41%. The decrease in temperature lift and adsorption heat showed a slowing down trend with the number of cycling experiments, and the heat density of the last cycle test decreased by only 0.17% which indicated a good periodicity and stability of composite adsorbent.
- The TRNSYS simulation found that the heating effectiveness of the TCHS system is dependent on the ambient temperature and humidity, and that the power consumption (i.e., fans and humidifiers) is lower in December, reaching a maximum COP of 6.67.

Author Contributions: Conceptualisation, J.Z. and Z.M.; methodology, J.Z. and Z.M.; software, X.C. and J.Z.; validation, J.Z. and X.C.; formal analysis, J.Z. and X.C.; investigation, J.Z. and X.C.; resources, Y.W.; data curation, X.C., J.Z. and Z.M.; writing—original draft preparation, J.Z.; writing—review and editing, Z.M. and J.Z.; visualisation, X.C. and J.Z.; supervision, D.C., Z.M., Y.W. and A.P.R.; project administration, Z.M. and J.Z.; funding acquisition, Z.M. and J.Z. All authors have read and agreed to the published version of the manuscript.

Funding: This research was funded by Engineering and Physical Sciences Research Council, grant number EP/V042564/1.

Data Availability Statement: Not applicable.

Acknowledgments: I would like to express my deep gratitude to Ke Tang, for her patient guidance, enthusiastic encouragement, and useful critiques of this research work. I would also like to thank Pan Wang, for her advice and assistance in keeping my progress on schedule. I would also like to extend my thanks to the technicians of the laboratory for their help in offering me the resources in running the program.

Conflicts of Interest: The authors declare no conflict of interest. The funders had no role in the design of the study; in the collection, analyses, or interpretation of data; in the writing of the manuscript; or in the decision to publish the results.

References

1. Kuravi, S.; Trahan, J.; Goswami, Y.D.; Rahman, M.M.; Stefanakos, E.K. Thermal energstorage technologies and systems for concentrating solar power plants. *Prog. Energy Combust. Sci.* **2013**, *39*, 285–319. [CrossRef]
2. Mahlia, T.M.I.; Saktisahdan, T.J.; Jannifar, A.; Hasan, M.H.; Matseelar, H.S.C. A review of available methods and development on energy storage; technology update. *Renew. Sustain. Energy Rev.* **2014**, *33*, 532–545. [CrossRef]
3. Pardo, P.; Deydier, A.; Anxionnaz-Minvielle, Z.; Rougé, S.; Cabassud, M.; Cognet, P. A review on high temperature thermochemical heat energy storage. *Renew. Sustain. Energy Rev.* **2014**, *32*, 591–610. [CrossRef]
4. Yan, T.; Wang, R.Z.; Li, T.X.; Wang, L.W.; Fred, I.T. A review of promising candidate reactions for chemical heat storage. *Renew. Sustain. Energy Rev.* **2015**, *43*, 13–31. [CrossRef]
5. Li, T.X.; Wu, S.; Yan, T.; Wang, R.Z.; Zhu, J. Experimental investigation on a dual-mode thermochemical sorption energy storage system. *Energy* **2017**, *140*, 383–394. [CrossRef]
6. Whiting, G.; Grondin, D.; Bennici, S.; Auroux, A. Heats of water sorption studies on zeolite-MgSO4 composites as potential thermochemical heat storage materials. *Sol. Energy Mater. Sol. Cells* **2013**, *112*, 112–119. [CrossRef]

7. Liu, M.; Tay, N.H.S.; Bell, S.; Belusko, M.; Jacob, R.; Will, G.; Saman, W.; Bruno, F. Review on concentrating solar power plants and new developments in high temperature thermal energy storage technologies. *Renew. Sustain. Energy Rev.* **2016**, *53*, 1411–1432. [CrossRef]
8. Mazzoni, S.; Ooi, S.; Nastasi, B.; Romagnoli, A. Energy storage technologies as techno-economic parameters for master-planning and optimal dispatch in smart multi energy systems. *Appl. Energy* **2019**, *254*, 113682. [CrossRef]
9. Ryu, J.-Y.; Alford, A.; Lewis, G.; Ding, Y.; Li, Y.; Ahmad, A.; Kim, H.; Park, S.-H.; Park, J.-P.; Branch, S.; et al. A novel liquid air energy storage system using a combination of sensible and latent heat storage. *Appl. Therm. Eng.* **2021**, *203*, 117890. [CrossRef]
10. Bellan, S.; Gonzalez-Aguilar, J.; Romero, M.; Rahman, M.M.; Goswami, D.Y.; Stefanakos, E.K.; Couling, D. Numerical analysis of charging and discharging performance of a thermal energy storage system with encapsulated phase change material. *Appl. Therm. Eng. Des. Process. Equip. Econ.* **2014**, *71*, 481–500. [CrossRef]
11. Abedin, A.H.; Rosen, M.A. A critical review of thermochemical energy storage systems. *Open Renew. Energy J.* **2011**, *4*, 42–46. [CrossRef]
12. Aydin, D.; Casey, S.P.; Riffat, S. The latest advancements on thermochemical heat storage systems. *Renew. Sustain. Energy Rev.* **2015**, *41*, 356–367. [CrossRef]
13. Schmidt, M.; Szczukowski, C.; Roßkopf, C.; Linder, M.; Wörner, A. Experimental results of a 10 kW high temperature thermochemical storage reactor based on calcium hydroxide. *Appl. Therm. Eng.* **2014**, *62*, 553–559. [CrossRef]
14. Li, W.; Luo, X.; Yang, P.; Wang, Q.; Zeng, M.; Markides, C.N. Solar-thermal energy conversion prediction of building envelope using thermochemical sorbent based on established reaction kinetics. *Energy Convers. Manag.* **2021**, *252*, 115117. [CrossRef]
15. Scapino, L.; Zondag, H.A.; Van Bael, J.; Diriken, J.; Rindt, C.C.M. Sorption heat storage for long-term low-temperature applications: A review on the advancements at material and prototype scale. *Appl. Energy* **2017**, *190*, 920–948. [CrossRef]
16. Lizana, J.; Chacartegui, R.; Barrios-Padura, A.; Valverde, J.M. Advances in thermal energy storage materials and their applications towards zero energy buildings: A critical review. *Appl. Energy* **2017**, *203*, 219–239. [CrossRef]
17. Cabeza, L.F.; Solé, A.; Barreneche, C. Review on sorption materials and technologies for heat pumps and thermal energy storage. *Renew. Energy* **2017**, *110*, 3–39. [CrossRef]
18. Gaeini, M.; Zondag, H.; Rindt, C. Effect of kinetics on the thermal performance of a sorption heat storage reactor. *Appl. Therm. Eng.* **2016**, *102*, 520–531. [CrossRef]
19. Donkers, P.A.J.; Sogutoglu, L.C.; Huinink, H.P.; Fischer, H.R.; Adan, O.C.G. A review of salt hydrates for seasonal heat storage in domestic applications. *Appl. Energy* **2017**, *199*, 45–68. [CrossRef]
20. Bertsch, F.; Mette, B.; Asenbeck, S.; Kerskes, H.; Müller-Steinhagen, H. Low temperature chemical heat storage—An investigation of hydration reactions. In Proceedings of the 11th Int. Conference on Thermal Energy Storages (Effstock), Stockholm, Sweden, 14–17 June 2009.
21. Molenda, M.; Bouché, M.; Linder, M.; Blug, M.; Busse, J.; Wörner, A. Thermochemical energy storage for low temperature applications: Materials and first studies in a gas-solid reactor. In Proceedings of the 12th Int. Conference on Energy Storage (Innostock), Lleida, Spain, 16–18 May 2012.
22. Zondag, H.; Kikkert, B.; Smeding, S.; de Boer, R.; Bakker, M. Prototype thermochemical heat storage with open reactor system. *Appl. Energy* **2013**, *109*, 360–365. [CrossRef]
23. Yu, N.; Wang, R.Z.; Lu, Z.S.; Wang, L.W. Development and characterisation of silica gel-LiCl composite sorbents for thermal energy storage. *Chem. Eng. Sci.* **2014**, *111*, 73–84. [CrossRef]
24. Yu, N.; Wang, R.Z.; Lu, Z.S.; Wang, L.W. Study on consolidated composite sorbents impregnated with LiCl for thermal energy storage. *Int. J. Heat Mass Trans.* **2015**, *84*, 660–670. [CrossRef]
25. Brancato, V.; Calabrese, L.; Palomba, V.; Frazzica, A.; Fullana-Puig, M.; Solé, A.; Cabeza, L.F. $MgSO_4 \cdot 7H_2O$ filled macro cellular foams: An innovative composite sorbent for thermo-chemical energy storage applications for solar buildings. *Sol. Energy* **2018**, *173*, 1278–1286. [CrossRef]
26. Piperopoulos, E.; Calabrese, L.; Bruzzaniti, P.; Brancato, V.; Palomba, V.; Caprì, A.; Frazzica, A.; Cabeza, L.F.; Proverbio, E.; Milone, C. Morphological and Structural Evaluation of Hydration/Dehydration Stages of $MgSO_4$ Filled Composite Silicone Foam for Thermal Energy Storage Applications. *Appl. Sci.* **2020**, *10*, 453. [CrossRef]
27. Zhao, H.Z.; Wang, Z.Y.; Li, Q.W.; Wu, T.H.; Zhang, M.; Shi, Q.Q. Water sorption on composite material "zeolite 13X modified by LiCl and $CaCl_2$". *Microporous Mesoporous Mater.* **2020**, *299*, 110109. [CrossRef]
28. Macriss, R.A.; Gutraj, J.; Zawacki, T.S. *Absorption Fluid Data Survey: Final Report on Worldwide Data*; Oak Ridge National Lab.: Oak Ridge, TN, USA; Institute of Gas Technology: Chicago, IL, USA, 1988.
29. Posern, K.; Kaps, C. Calorimetric studies of thermochemical heat storage materials based on mixtures of $MgSO_4$ and $MgCl_2$. *Thermochim. Acta* **2010**, *502*, 73–76. [CrossRef]
30. Gordeeva, L.G.; Grekova, A.D.; Krieger, T.A.; Aristov, Y.I. Adsorption properties of composite materials (LiCl+LiBr)/silica. *Microporous Mesoporous Mater.* **2009**, *126*, 262–267. [CrossRef]
31. Entezari, A.; Ge, T.S.; Wang, R.Z. Water adsorption on the coated aluminum sheets by composite materials (LiCl + LiBr)/silica gel. *Energy* **2018**, *160*, 64–71. [CrossRef]
32. Druske, M.; Fopah-lele, A.; Korhammer, K.; Urs, H. Developed materials for thermal energy storage: Synthesis and characterization. *Energy Procedia* **2014**, *61*, 96–99. [CrossRef]
33. Mohapatra, D.; Nandanavanam, J. Salt in matrix for thermochemical energy storage—A review. *Mater. Today Proc.* **2022**. [CrossRef]

34. Lim, K.; Che, J.; Lee, J. Experimental study on adsorption characteristics of a water and silica-gel based thermal energy storage (TES) system. *Appl. Therm. Eng.* **2017**, *110*, 80–88. [CrossRef]
35. Erlund, R.; Zevenhoven, R. Thermal energy storage (TES) capacity of a lab scale magnesium hydro carbonates/silica gel system. *J. Energy Storage* **2019**, *25*, 100907. [CrossRef]
36. Mahon, D.; Claudio, G.; Eames, P. A study of novel high performance and energy dense zeolite composite materials for domestic interseasonal thermochemical energy storage. *Energy Procedia* **2019**, *158*, 4489–4494. [CrossRef]
37. Whiting, G.; Grondin, D.; Bennici, S.; Auroux, A.; Boer, D.R.; Smeding, S.F.; Zondag, H.A.; Krol, G. *Development of a Prototype System for Seasonal Solar Heat Storage Using an Open Sorption Process*; ECN: Petten, The Netherlands, 2014; Volume 99, pp. 1–9.
38. Singh, A.P.; Khanna, S.; Paneliya, S.; Hinsu, H.; Patel, Y.; Mehta, B. Preparation and characterisation of solid-state neopentyl glycol/expanded graphite micro composite for thermal energy storage applications. *Mater. Today Proc.* **2021**, *47*, 621–625. [CrossRef]
39. Xu, C.; Yu, Z.B.; Xie, Y.Y.; Ren, Y.X.; Ye, F.; Ju, X. Study of the hydration behavior of zeolite-$MgSO_4$ composites for long-term heat storage. *Appl. Therm. Eng.* **2018**, *129*, 250–259. [CrossRef]
40. Nonnen, T.; Preißler, H.; Kött, S.; Beckert, S.; Gläser, R. Salt inclusion and deliquescence in salt/zeolite X composites for thermochemical heat storage. *Microporous Mesoporous Mater.* **2020**, *303*, 110239. [CrossRef]
41. Zhang, Y.N.; Wang, R.Z.; Zhao, Y.J.; Li, T.X.; Riffat, S.B.; Wajid, N.M. Development and thermochemical characterisations of vermiculite/SrBr2 composite sorbents for low-temperature heat storage. *Energy* **2016**, *115*, 120–128. [CrossRef]
42. Wang, Q.; Xie, Y.Y.; Ding, B.; Yu, G.L.; Ye, F.; Xu, C. Structure and hydration state characterisations of $MgSO_4$-zeolite 13x composite materials for long-term thermochemical heat storage. *Sol. Energy Mater. Sol. Cells* **2019**, *200*, 110047. [CrossRef]
43. Wei, S.; Zhou, W.; Han, R.; Gao, J.; Zhao, G.; Qin, Y.; Wang, C. Influence of minerals with different porous structures on thermochemical heat storage performance of CaCl2-based composite sorbents. *Sol. Energy Mater. Sol. Cells* **2022**, *243*, 111769. [CrossRef]
44. Ge, Y.Q.; Zhao, Y.; Zhao, C.Y. Transient simulation and thermodynamic analysis of pumped thermal electricity storage based on packed-bed latent heat/cold stores. *Renew. Energy* **2021**, *174*, 939–951. [CrossRef]
45. Calabrese, L.; Antonellis, D.S.; Vasta, S.; Brancato, V.; Freni, A. Modified Silicone-SAPO34 Composite Materials for Adsorption Thermal Energy Storage Systems. *Appl. Sci.* **2020**, *10*, 8715. [CrossRef]
46. Xu, C.; Xie, Y.Y.; Liao, Z.R.; Ren, Y.; Ye, F. Numerical study on the desorption process of a thermochemical reactor filled with $MgCl_2 \cdot 6H_2O$ for seasonal heat storage. *Appl. Therm. Eng.* **2019**, *146*, 785–794. [CrossRef]
47. Li, W.; Klemeš, J.J.; Wang, Q.W.; Zeng, M. Development and characteristics analysis of salt-hydrate based composite sorbent for low-grade thermochemical energy storage. *Renew. Energy* **2020**, *157*, 920–940. [CrossRef]
48. Safa, A.A.; Fung, A.S.; Kumar, R. Heating and cooling performance characterisation of ground source heat pump system by testing and TRNSYS simulation. *Renew. Energy* **2015**, *83*, 565–575. [CrossRef]
49. Sakellari, D.; Forsén, M.; Lundqvist, P. Investigating control strategies for a domestic low-temperature heat pump heating system. *Int. J. Refrig.* **2006**, *29*, 547–555. [CrossRef]
50. Voottipruex, K.; Sangswang, A.; Naetiladdanon, S.; Mujjalinvimut, E.; Wongyoa, N. PEM fuel cell emulator based on dynamic model with relative humidity calculation. In Proceedings of the 2017 14th International Conference on Electrical Engineering/Electronics, Computer, Telecommunications and Information Technology (ECTI-CON), Phuket, Thailand, 27–30 June 2017; pp. 529–532.
51. Zhao, Y.J.; Wang, R.Z.; Zhang, Y.N.; Yu, N. Development of $SrBr_2$ composite sorbents for a sorption thermal energy storage system to store low-temperature heat. *Energy* **2016**, *115*, 129–139. [CrossRef]
52. Zhang, Y.N.; Wang, R.Z.; Li, T.X. Experimental investigation on an open sorption thermal storage system for space heating. *Energy* **2017**, *141*, 2421–2433. [CrossRef]
53. Jarre, M.; Noussan, M.; Simonetti, M. Primary energy consumption of heat pumps in high renewable share electricity mixes. *Energy Convers. Manag.* **2018**, *171*, 1339–1351. [CrossRef]
54. Martinek, J.; Jorgenson, J.; McTigue, J.D. On the operational characteristics and economic value of pumped thermal energy storage. *J. Energy Storage* **2022**, *52*, 105005. [CrossRef]

energies

Article

Sugarcane Bagasse as Aggregate in Composites for Building Blocks

Patrícia P. L. de Souza [1], Rute Eires [2] and Raphaele Malheiro [2,*]

[1] Department of Civil Engineering, University of Minho, 4800-058 Guimarães, Portugal
[2] Centre for Territory, Environment and Construction (CTAC), Department of Civil Engineering, University of Minho, 4800-058 Guimarães, Portugal
* Correspondence: raphaelemalheiro@civil.uminho.pt

Abstract: Each year, hundreds of millions of tons of processed sugarcane generate, by weight, 25 to 30% of bagasse as waste, whose destination is combustion for energy cogeneration. This research proposes an alternative and more sustainable use for this waste. The use of sugarcane bagasse (SCB) as the single aggregate in composites for building blocks was studied. The raw bagasse was used without any treatment. As the binder, aerial lime and/or soil were used. Both provided enough mechanical strength for non-load-bearing walls. The composite of SCB with soil achieved the best performance in terms of mechanical resistance: 2.6 MPa in compressive strength and 2.1 MPa in bending strength, while the composite of SCB with lime achieved 1.76 MPa and 1.7 MPa, respectively. The higher number of fibers in the SCB/lime mixture provides better thermal insulation than clay brick or conventional concrete, such as "hempcrete". The lime composites obtained greater water resistance and less loss of mechanical strength when saturated. However, the higher water absorption coefficient makes it necessary to apply a waterproof mortar on surfaces exposed to the weather. The replacement of supplied blocks by SCB blocks can offer a better and more economical solution that improves the quality of the built environment and is more ecofriendly.

Keywords: building blocks; sugarcane bagasse; lime; soil; agro-industrial waste

Citation: Souza, P.P.L.d.; Eires, R.; Malheiro, R. Sugarcane Bagasse as Aggregate in Composites for Building Blocks. *Energies* **2023**, *16*, 398. https://doi.org/10.3390/en16010398

Academic Editor: Chi-Ming Lai

Received: 1 December 2022
Revised: 21 December 2022
Accepted: 25 December 2022
Published: 29 December 2022

1. Introduction

As the goals established for sustainable global development, among other measures, the civil construction sector must strive to reduce the consumption of non-renewable materials, the consumption of fossil fuel energy, and the emissions of greenhouse gases.

Finding a destination for the unquantifiable wastes has been a constant task. The sugar and alcohol industry by sugarcane produces a significant amount of natural waste, sugarcane bagasse (SCB). This is the first by-product of this production and has great potential for use as a raw material for the production of other materials.

The sugar and alcohol industries have autonomy, prominence, and influence in several global segments (economic, social, environmental, and agricultural). Brazil is the world's largest producer of sugarcane. The harvest forecast for the period 2022/2023 is 572.9 million tons [1]. Processed sugarcane generates, by weight, 25 to 30% of bagasse [1]. In Brazilian mills, SCB is used for the cogeneration of electrical energy and the surplus, as animal nutritional supplementation and fertilizer for agriculture. Due to the inconsistency of nutrients, this process serves more to discard than to use the material [2].

Thus, this research work aims to contribute to the recovery of this residue (SCB) in a more positive way than combustion for energy cogeneration. In this regard, SCB-based composites and environmentally friendly binders (soil and/or aerial lime) have been developed, which can be used as lightweight and insulating concretes for construction block manufacture, which may be applied in non-load-bearing masonry walls.

As with hemp and under the same arguments, sugarcane is a renewable resource that can be cultivated in annual cycles. In addition, as with other plants, during its development, it extracts carbon dioxide from the atmosphere. Its application in construction materials promotes carbon sequestration during the useful life of the building. Sugarcane bagasse concrete captures more CO_2 in construction than is emitted during its production and any process involved in its disposal at the end of its life. For this reason, it can be considered a "carbon-negative" material [3]; these composites can also be recyclable at the end of the building's life cycle and can be reused, crushed, and mixed with lime binder to make new blocks or recycled in the preparation of plastering mortar. In another hypothesis, the compost crushed and spread over agricultural land can be used to correct soil acidity [4]. As the material is naturally biodegradable, landfilling would also have a minimal environmental impact.

The knowledge presented in the below review focuses on the applications of SCB and the studies that inspired the development of these new light and insolating compositions, especially hemp concrete and other light concrete with agro-industrial residues with lime-based binders and as the only aggregate.

1.1. *Applications of SCB*

The application of SCB in civil construction has been studied: (i) in addition to cementitious compositions [5]; (ii) with cement and SCB ash (the second by-product of this industry, by the SCB burning to produce energy) [6]; (iii) in composites of SCB with cement and polymers [7]; (iv) in addition to plastic waste [8]; (v) in polymeric tiles of hybrid composites of fiber glass and natural fibers from SCB [9]. In other industries, such as the automotive, a composite of SCB with recycling PET was studied [10] and investigations such as a composite of EPS with SCB [11] seek to prospect innovative applications, based on the study of resistance and new physical characteristics of the resulting materials.

However, in composites of SCB with earth, as in the studies of Bock-Hyeng et al. [12], investigations are scarce and the amount of SCB fiber is very reduced.

The present study sought to develop a material capable of responding in a complementary way to the various contemporary demands in a sustainable context. It is intended to rescue ancient technologies, such as adobe, the use of natural fibers, and lime. The innovation in this work is characterized by the use of a relevant amount of SCB fibers, in the order of 30 to 35%, in the studied composites. Moreover, compositions of SCB fibers with lime as the main binder do not seem to exist at this moment.

1.2. *The "Sugar Cane Concrete"*

The proposal for the sugarcane bagasse (SCB) and lime mixtures for building blocks is inspired by the existing similar composite based on hemp and lime, briefly presented below. The composite obtained by replacing lime with soil as a natural binder, such as adobe, but with a greater number of fibers as usual was also studied in the search of a more sustainable solution, as it does not use a large quantity of calcined materials in the formulation. Depending on the characteristics of the materials, in a composite with fibers and soil, it is convenient to add a binder capable of improving resistance to water action. Lime has this advantage because reduces water absorption and favors vapor permeability, forming a limestone barrier around the vegetable fiber [13].

In the composite without soil in the matrix, lime is capable of adding and maintaining essential properties for the resistance and durability of the material. As a disadvantage, the long carbonation time impacts the curing time, as verified in studies with hemp [13].

SCB ash has proven a good performance as an efficient pozzolanic material in composites prepared with cement [14]. As long as it can improve or preserve the resistance and durability of the material, it is possible that its addition as a pozzolanic material can bring another environmental benefit, namely through the use of a final residue.

In this work, three composites with sugarcane bagasse were developed, with a view to making building blocks. These are composites with relevant fiber additions, in the order of 30% to 35% of fibers, in formulations with hydrated aerial lime, with lime and soil, and only with soil. Additionally, the addition of SCB ash was also tested in order to compare the pozzolanic effect of this material with the effect of using metakaolin.

The specimens prepared during the study were subjected to laboratory tests with the objective of measuring the mechanical, hygroscopic, and thermal performance of the composites and, thus, verifying the feasibility of applying the product, and also prospecting the potential to act as a structural complement.

1.3. The "Hemp Concrete" and Other Lightweight Concretes with Agroindustry Wastes

The use of hemp fibers in construction goes back a long time. There are hemp/soil mortars in India that are about 1500 years old [15]. The first "hemp/lime concrete" was developed by Charles Rasetti in 1987 in France [16]. The woody core, with a high silica content, interacts with the lime and promotes the hardening of the mixture [16]. Used for the production of hemp concrete (or hempcrete), the construction material is much studied and used in the European Union.

The hemp block walls function as thermal and acoustic insulation and a good compromise between thermal conductivity and thermal inertia [17]. The material is easy to adapt to the climate and easy to produce locally. It prevents the occurrence of condensation due to the wall's breathability and absorption capacity and resistance to water, which impacts the quality of the environment and the health of the inhabitants [16].

In the life cycle assessment, in addition to being biodegradable, it captures CO_2 from the atmosphere, reduces the use of toxic materials and waste production, uses renewable resources, and can be recycled at the end of the useful life of the building, thus reducing the environmental impact [13–16]. Its ductility and ability to adjust to building movements prevent the appearance of cracks. It is a non-flammable material, does not release toxic fumes, and is resistant to insects, fungi, and bacteria [17].

In addition to the hempcrete research, other agroindustry or agri-food wastes have been studied in lightweight concretes, mainly with lime-based binders and replacing the aggregates in full. For example, the study of Chabannes et al. showed the use of rice husk and hemp as aggregates (without and with previously treated aggregates with $Ca(OH)_2$) using a lime-based binder for lightweight concretes, Lime and Hemp Concrete (LHC), and Lime and Rice husk Concrete (LRC) [18]. Chabannes et al. also studied sunflower stem aggregates with eco-friendly binders and their multi-physical properties as insulating concrete [19]. However, there are also studies with cementitious binders and only with a partial replacement of aggregates such as the study of Gradinaru et al. with 50% of sunflower aggregates (treated with sodium silicate solution), with 50% of sand and cement as a binder, and with the addition of superplasticizer to reduce the amount of water and obtain greater resistances [20].

As no lightweight lime-based concrete study has yet been conducted with total aggregate substitution by sugarcane bagasse, the present study was intended to study simple mixtures, without any treatment of the vegetable aggregate and without additives that change its rheology or its binder adhesion. Only in this way will it be possible to determine the material behavior. In addition, it is also intended to obtain an economically accessible construction product; therefore, a minimum of additions and processes is necessary.

2. Materials and Methods

The materials used in the experimental research were sugarcane bagasse (SCB) as an aggregate; aerial lime, SCB ash, Alentejo and Labruge soil, and metakaolin as binders; as an additive, sodium borate or borax.

2.1. Materials

2.1.1. Sugarcane Bagasse—SCB

The SCB used came from Madeira Island, where sugar cane is cultivated for the production of national rum. The preparation of the material used in the mixtures required spreading the wet bagasse over a plastic sheet, where it was turned over twice a week to dry naturally and homogeneously, for 4 weeks, until it was dry. This drying was necessary because, otherwise, fungi would appear, just as it was already appearing in the most humid parts when we received the material. In morphological analysis (Figure 1a,b) of SCB samples, it was found that the SCB length varies between 10 and 30 mm, but most are 15 mm. The pieces of SCB are made up of long fibers [21] with diameters ranging from 0.2 to 0.5 mm below 100 μm and scaly surfaces (Figure 1c).

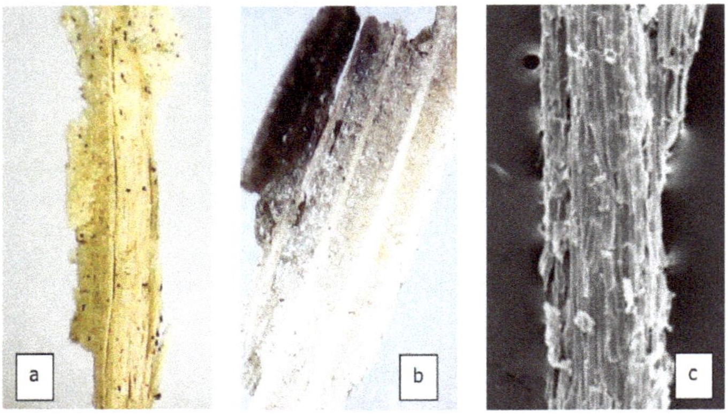

Figure 1. (**a**,**b**) Dried SCB samples were analyzed under an optical microscope (20× magnification) (figures from the authors). (**c**) Bundle of dried fibers analyzed under the microscope (adapted from Oliveira [2]).

SCB is generically composed of cellulose (50%), hemicellulose (25%), and lignin (20%) [2].

2.1.2. Used Soils

Two types of soils were used separately, and Figure 2 shows their particle size distribution curves. The soil used in the first and second trial mixes came from Alentejo, a southern region of Portugal, designated here as Soil 1. It has a good granulometric distribution, with 15.9% of gravel, 47.2% of sand, 17.6% of silt, and 19.4% of clay. The percentage of clay in this soil is considered sufficient for construction, as it reaches just over 20% of the analyzed volume. Soil sifting is necessary to adjust the amount of gravel and obtain 4 mm as the maximum dimension. In the third trial mix, soil from Labruge, Vila do Conde, designated here as Soil 2, was used. This soil presented 65% of fine material (silt + clay), with the clay fraction equivalent to 15%. The type of aggregate present is fine sand, with a maximum diameter of 2 mm and with a well-graded granulometry distribution. The percentage of sand present in this soil is small.

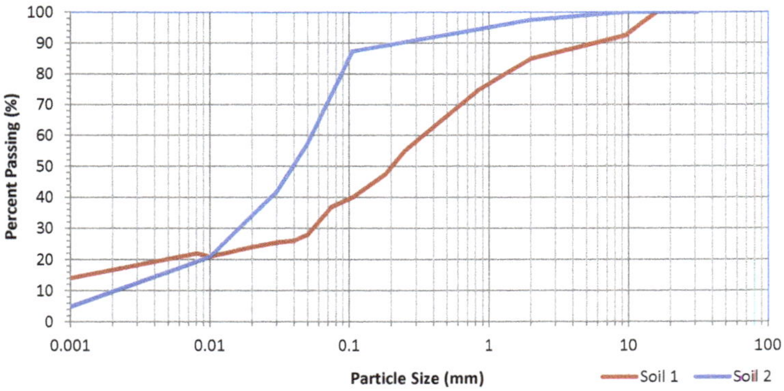

Figure 2. Particle size distribution of the soils used.

2.1.3. Hydrated Lime

Hydrated aerial lime (calcium hydroxide), manufactured in Portugal, was used. It is calcium lime, classified as CL90-S according to the EN 459-1:2015 [22] standard, which has an apparent density of 0.46 g/cm^3.

2.1.4. Metakaolin

Metakaolin is obtained from the calcination (at temperatures between 700 and 800 °C) of kaolinitic materials, which is a mineral clay with a high content of silicon dioxide. The reaction of metakaolin with lime produces hydrated calcium silicate (CSH) and aluminum hydrates. Processed with less energy than cement, metakaolin is used as a pozzolana to improve the mechanical strength of mixtures with lime or cement. The metakaolin used in this study was produced in Portugal and is characterized by a light orange color, which influences the final composite color. The main components of its chemical composition are the silica (60.7%) and alumina (34.3%) according to the producer's datasheet, mentioned by Kropidłowska [13].

2.1.5. Ashes of SCB (ASCB)

For the use of SCB ashes in this research, a portion of SCB, 800 g, was placed into the muffle furnace at 600 °C. The process was carried out in 3 cycles: heating (1 h), burning (4 h), and cooling slowly, obtaining 106 g of ash. It was observed that this process did not produce calcination identical to that of the industrially obtained process, as the reference studies indicate that the ASCB corresponds to 0.6% of the initial weight of the SCB [23], and was obtained here in the proportion of 13.25% of the initial weight of the SCB.

2.2. Methods

2.2.1. Mixtures

An outline of all the steps carried out in the methodology can be seen in Figure 3.

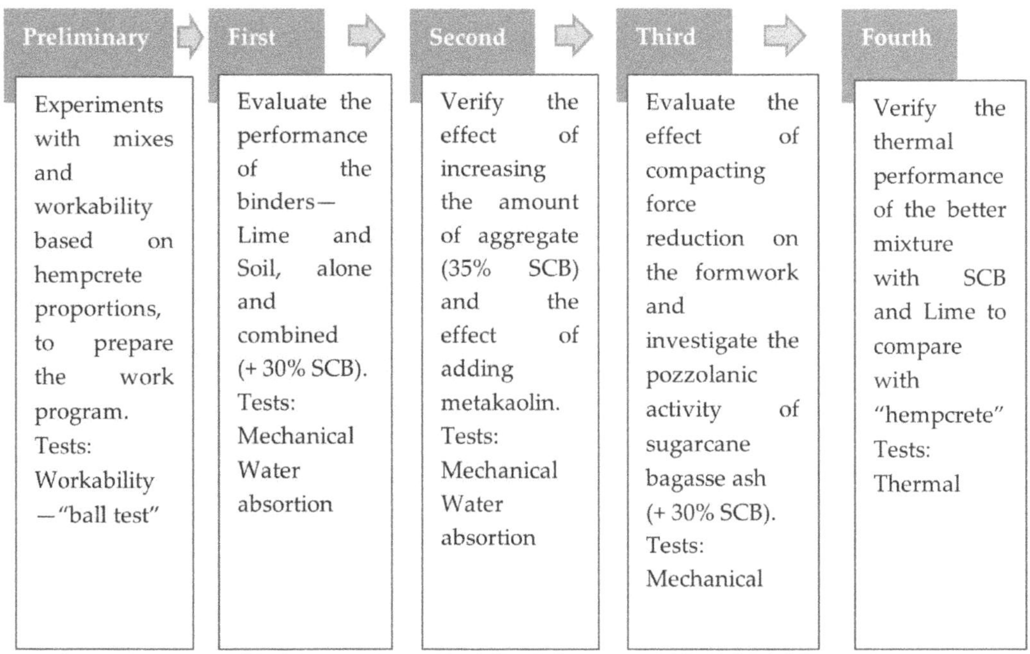

Figure 3. Stages of the study and the purpose of each.

To prospect the material, potential preliminary mixes were prepared to evaluate the binder amount of water necessary to form a composite with adequate workability and minimum resistance for the manufacture of the blocks, to establish a plan in line with the intended results. The results obtained in this preliminary stage were guidelines for the planning of the 1st trial mix of the work. Three different trial mixes were then carried out, as shown in Table 1:

- The first is to evaluate the performance of the main binders (lime and soil) and the combination of both, with 30% of SCB.
- The second is to seek the maximum sustainability potential by increasing the fiber content and further reducing the weight of the material to improve the thermal behavior. For this, we tried to verify the effect of increasing the amount of SCB fibers in the mixtures to 35%. Furthermore, to reduce the amount of lime, it was decided to add metakaolin and evaluate its pozzolanic effect, using the proportions tested by Kropidłowska [13] with hemp concrete mixes, including the use of borax as an additive.
- The third is to evaluate the addition of SCB ash as a pozzolanic material, readjusting the number of fibers to the content of the first trial, but slightly reducing the compaction force in the formwork, to not interfere with the thermal behavior. The MR3 reference mixture is an adobe without the addition of fibers and was prepared with the purpose of comparing the results and evaluating the resistance of the soil used.

Table 1. Studied compositions and used symbols.

Studied Compositions										
		Solid Materials								
	Symbol	Aggregate %	Binder %						Binder %	Water %
		SCB	Lime	Soil	MKL	Ash	BX			
1st trial mix	S.L-1	30	70	-	-	-	-		46	54
	S.So1-1	30	-	70 *	-	-	-		70	30
	S.L.So1-1	30	10	60 *	-	-	-		48	52
2nd trial mix	S.L-2	35	65	-	-	-	-		36	64
	S.So1-2	35	-	65 *	-	-	-		41	59
	S.L.Mk.B-2	32	40.6	-	27	-	0.4		44	56
3rd trial mix	S.L-3	30	70	-	-	-	-		36	64
	S.So2-3	22	-	78 **	-	-	-		49	51
	S.L.So2-3	20	60	20 **	-	-	-		47	53
	S.L.A-3	26	68	-	-	6	-		38	62
	So-2	-	-	100 **	-	-	-		77	23

* Soil 1, ** Soil 2.

To verify the workability of the composites obtained, an empirical test known as the "ball method" was carried out, which consists of taking a sample of the mixture and making a ball by hand. If the ball remains well formed and with good consistency, it means that the workability of the material is suitable to produce specimens, as was used by Martins in soil mixtures for compressed earth [24], and, in this case, its procedure helps to verify the adhesion of the materials.

2.2.2. Manufacture and Curing Conditions of Samples

Specimens measuring 40 × 40 × 160 mm (Figure 4a) were used for mechanical resistances and water absorption by immersion test, while specimens measuring 150 × 150 × 59 mm (Figure 4b), dimensions compatible with the hot box, were only used for the thermal conductivity test.

Figure 4. (**a**,**b**). Specimens for general tests (**a**) and sample for thermal conductivity (**b**).

The mixing process was carried out in a pan mixer with a rotating drum. The procedure is based on the methods used to produce hemp concrete, such as those used by Kropidłowska [13]. Water was added partially before and after the binder ($^1/_3$ of the amount before and the rest after). Each sample was manually compacted in three layers, using a stone block and a rubber mallet to compact evenly. The formwork took place after 2 h of rest and the specimens were stored in a dry, semi-closed cupboard, at room temperature and humidity for the curing process (around 20 °C and relative humidity of approximately 50%).

2.2.3. Performed Tests

The mixtures made were tested for their mechanical performance: concerning resistance to compression and bending, at 28, 60, and 90 days, with the exception of the mixtures from the second trial mix, which, due to the COVID-19 pandemic, were only tested at 28 and 90 days.

The flexural strength tests were carried out in accordance with EN 1015-11:1999 [25]. Tests were conducted on Lloyds Instruments (universal test hydraulic press with a maximum capacity of 50 kN), with an applied load of 10 N/s. Three to six specimens were used per lot.

The compressive strength tests were carried out in accordance with the EN 1015-11:1999 [25] standard, using the parts left after the rupture of the specimens used in the bending test. A force of 10 N/s was applied in the most unfavorable direction, perpendicular to the compaction of the fibrous material.

With regard to the performance against the water action, tests of water absorption by capillarity and by immersion were carried out. The fibrous nature of the SCB and the porous nature of the raw soil demanded an adaptation of the measurement process of the capillary water absorption test, commonly presented by EN 1015-18:2002 [26]. The method used was described by Hall and Djerbib [27] and this was applied to obtain a good indication of the performance of compacted earth blocks, which is an adaptation of the British Standard BS3921 (IRS) test [28]. Called IRS Oasis, the method was very useful, due to the special vulnerability of earth and fiber when in contact with water. For the test, specimens of 40 × 40 × 160 mm were used, one of each mixture from the first and second trials (213 and 107 days, respectively), cured at room temperature and humidity. The specimens were cut in half, measured, weighed, and arranged on a spongy block (known as Oasis) immersed in a box with water until its maximum absorption capacity stabilized. The Oasis was kept saturated, with 2 cm out of the water, as seen in Figure 5. The contact times began to be measured and the samples were weighed after 5, 10, 15, and 30 min and then 1, 2, 3, 4, 5, 7, 9, and 24 h.

Figure 5. Specimens cut and accommodated over the saturated Oasis.

Regarding water absorption by immersion, specimens measuring 40 × 40 × 160 mm were used. One from the first trial mix at 210 days of curing and two from the second, at 105 days, always under the same conditions of ambient temperature and humidity. After drying in an oven at 60 °C, they were immersed in water at room temperature. After the first hour, the specimens were removed from the water and excess water was removed with absorbent paper. Then, they were weighed and returned to the water, where they remained for 24 h, calculating the water absorption after this time. At the end of the absorption period, the specimens were subjected to the compressive strength test, according to the procedure described above, to measure the loss of strength due to exposure to water. However, only the SCB with lime and SCB with lime and soil specimens were subjected to the test, as the specimens without hydraulic binders were very fragile and would not withstand the immersion test.

As it is possible that the SCB composite has a thermal performance similar to that of hempcrete, a specimen of the BC1 mixture (150 × 150 × 59 mm) was prepared to be evaluated. The thermal performance of SCB composites was evaluated by considering their thermal resistance (Re) and thermal conductivity (λ). These parameters were determined using a calibrated hot box designed and built at the Department of Civil Engineering of the University of Minho [29], based on ASTM specifications ASTM C1363-19 [30].

The hot box is composed of two chambers, the cold and the hot one, and one mounting ring is placed between the two chambers (Figure 6). The SCB composites were placed in the center of the mounting ring.

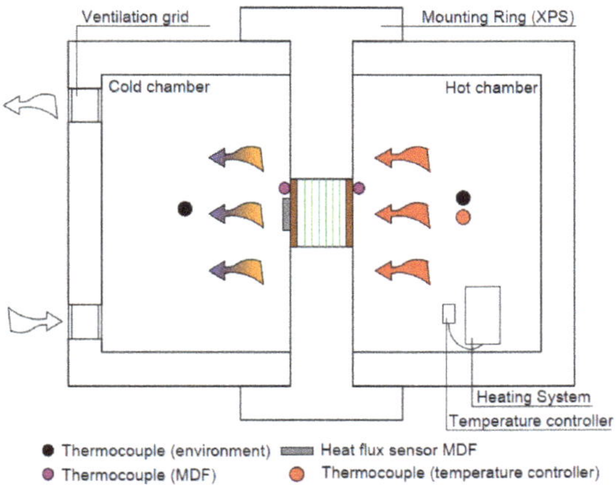

Figure 6. Schematic representation of the hot box used [31].

The tests were carried out considering the heat flow meter method, defined in ISO 9869-1 [32]. The heat flux was measured through a heat flux sensor installed in the SCB composites' central part, and thermocouples measured the temperatures. With the values of the heat flux (q) and the surface temperatures (T), it was possible to determine the thermal resistance (Re) of the material, using Equation (1). ΔT is the difference between the surface temperature of the SCB composites in the hot and cold chambers. The thermal resistance of the SCB composites was determined using Equation (2). The SCB composites' thermal conductivity (λ) was assessed using Equation (2), where e is the SCB composites' thickness.

$$\text{Re } [(\text{m2.}^{\circ}\text{C})/\text{W}] = \Delta T/q \tag{1}$$

$$\lambda \text{ } [\text{W/m. }^{\circ}\text{C}] = e/\text{Re} \tag{2}$$

3. Results and Discussions

Considering that there are still no standards for this type of lightweight composite material with fibers and lime-based binders, considerations are made considering the aforementioned studies on hempcrete and the technical guides about this construction product.

3.1. Mechanical Resistance Performance

As seen in Figures 7 and 8, the adobe without fibers So-2 had better compressive strength than the mixtures with SCB, which was expected considering the large volume of fibers added. Among the mixtures with SCB, the S.So1-1 (30% SCB and soil) obtained the best values, followed by the S.L.So1-1 (30% SCB with soil and lime) and the S.L-1 (30% SCB with lime). This means that the addition of lime seemed to reduce the resistance, alone or combined with soil. Other authors noted a reduction in resistance in mixes of soil and lime as binder material [33]. However, what happens with the isolated use of lime as a binder with SCB should be tested further. Microscopy tests would help to understand better the adhesion between these fibers and the aerial calcic lime. In terms of flexural strength, a similar behavior can be seen, and in general, the following considerations can be made:

- The increase in the proportion of fibers from 30% to 35% of SCB reduced the mechanical performance of mixtures (comparing the better results in the first trial mix with the other trials);
- In addition, the compaction force positively affected the mixtures, as the best performances were registered in the first trial, in which all specimens presented higher density;
- The addition of SCB ash did not show efficacy in the composite's behavior, S.L.A-3;
- Comparing the mixtures of soil/lime with the mixtures only with soil, it was observed that mixtures only with soil presented a greater magnitude of resistance, both to compression and flexion, seeming to have no advantages in terms of mechanical resistance in lime addition. It should be noted that the results were inconsistent at some ages, but this can be considered normal as the study was carried out with natural materials and subject to a greater dispersion of results.

Comparing the obtained results with standards or guidelines, the following may be considered:

- Considering the minimum compressive strength performance values required by a French hempcrete technical guide (>0.2 MPa for walls) [34] and the results obtained in the sample tests, the composites fell within the range of materials with sufficient strength for non-structural walls.
- The strengths of the S.So1-1 mixture were higher than the minimum 1.5 MPa, established for adobe blocks, by NBR 16814:2020 [35], in compression. However, its density, very close to that of adobe, eliminated the virtue of lightness, compared with the mixtures without soil. Among SCB composites with lime and soil, only S.L.So1-1 showed recommendable mechanical efficiency.
- The mechanical performance tests of the SCB blocks, in the various mixtures studied in this investigation, showed resistances lower than the 3 MPa established by

NBR 15.270-2 [36] for ceramic blocks and conventional hydraulic concrete blocks, established by NBR 6136: 2016 [37].

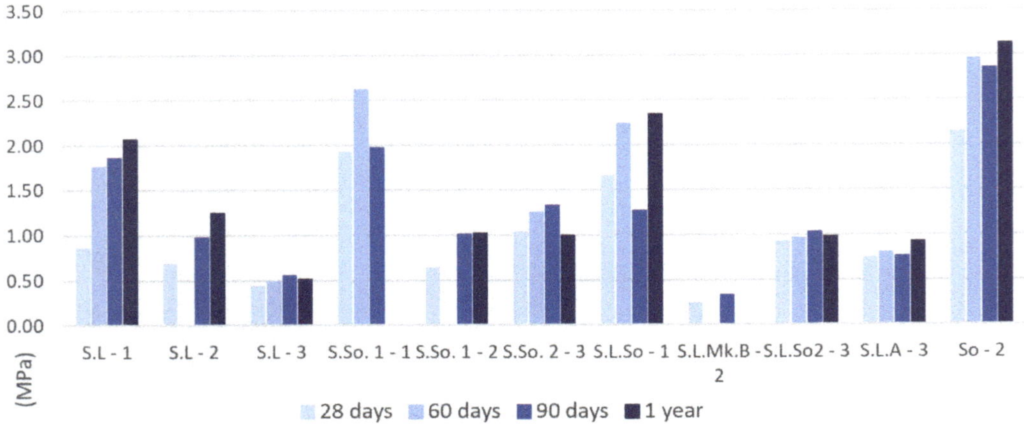

Figure 7. Compressive strength performance of all mixes.

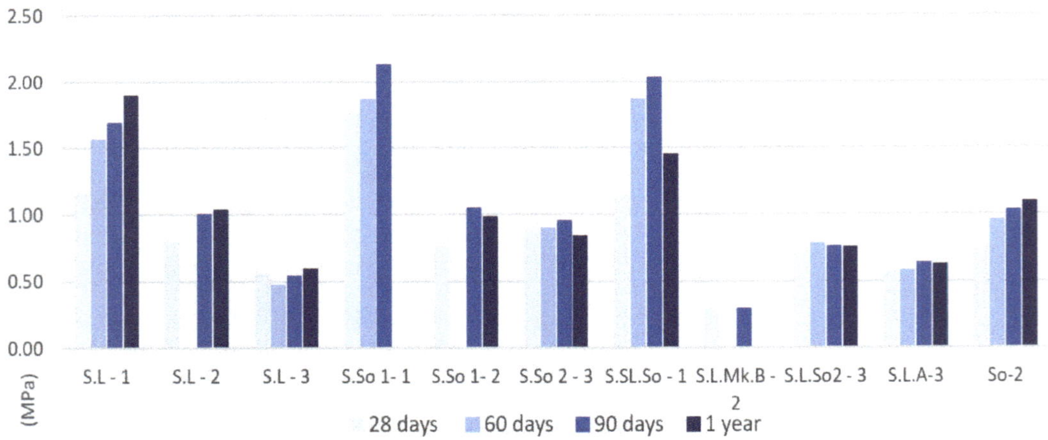

Figure 8. Bending performance of all mixtures.

In this way, the results seem to be good, even considering the lower obtained values by the increase in the number of fibers and reduced compaction. However, for better conclusions, it should be interesting to develop more research, with more variation in the number of fibers and greater control of compaction through sample density.

For a better analysis of the results obtained, it is important to remember that the sugarcane bagasse is a unique aggregate, it is not a mineral and rigid material, and no cement is used, so the resistances are very low when compared to conventional concrete or even with concrete reinforced with vegetable fibers for hollow building blocks. Some studies noted a compressive strength between 13 and 33 MPa in cement-based composites to produce hollow blocks [38,39].

However, it can be compared with identical composites, such as adobe (for soil-based mixtures) and lightweight concretes of industrial hemp and lime binders, also without sand or cement.

Comparison of Compressive Strength Performance of SCB Composites with "Hempcrete" and Other Agro-Wastes Lime-Based Composites

For the reasons mentioned above, a comparison of the obtained results (considering the average values of the three phases for each mixture) and the results of other researchers was made. Table 2 shows the main methods used in the research, as mixtures, specimens, and curing conditions, and Figure 9 shows the results of the compressive strength and density for a better interpretation of the results.

Table 2. Comparison with other authors—mixtures and their methods used.

Mixtures	Binders	Aggregate % (Mass)	Specimens (Dimensions-mm^3)	Curing Conditions	References
Light Rice Lime Concrete	NHL3.5 and CL90-S at 50/50 wt.%	33.33%	Ø110 × 220 cylindrical	climate-controlled room at 20 °C and 50% RH	Chabannes et al. [18]
Light Hemp Lime Concrete					
Light Hemp Lime Concrete	CL90	30%	100 × 100 × 100	room temperature and humidity (18 to 22 °C) in formwork 5 days	Araújo [40]
Light Sunflower Lime 10% Poz Concrete	75% CL, 15% NHL 10% pozzolanic binder	33.33%	Ø110 × 220 cylindrical	room temperature 20 °C and35 ± 5% RH	Chabannes et al. [19]
Light Hemp Lime 27% Mk Concrete	77% CL90 27% Metakaolin	32–35%	100 × 100 × 100	humidity chamber 7days (87%RH) room temperature (21 °C) and humidity (65–75%)	Kropidłowska [13]
SCB Lime Concrete	CL90	30–35%	40 × 40 × 160 Load area-40 × 40	semi-closed cupboard, room temperature 20 °C and 50 ± 5% RH	Present study
SCB Soil	Soil	22–35%			
SCB Soil Lime Concrete	Soil CL90	20–30%			
SCB Lime 27% Mk Concrete	CL90 Metakaolin	32%			
SCB Lime 6% SCBA Concrete	CL90	26%			

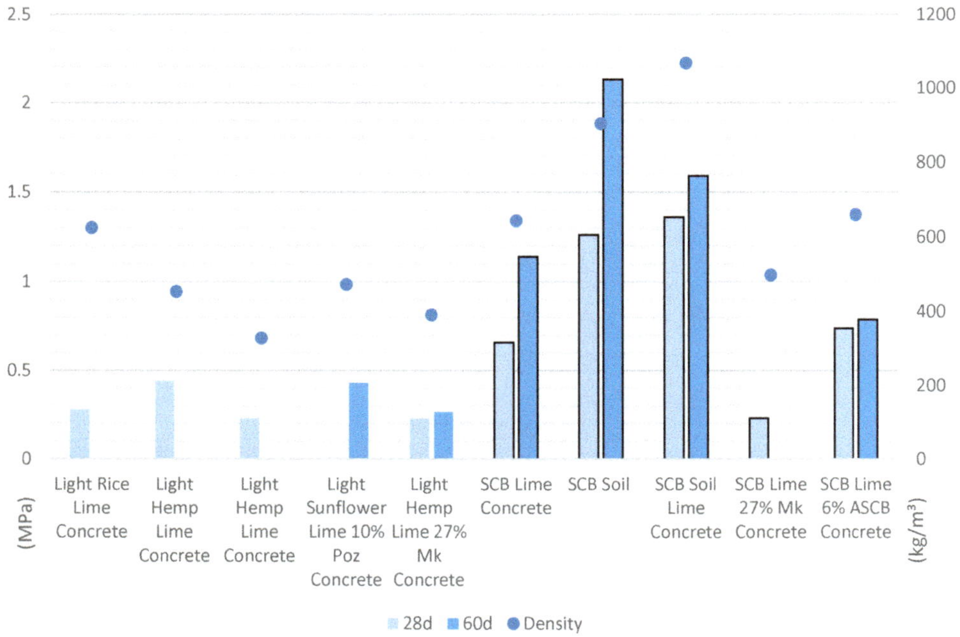

Figure 9. Comparison of compressive strength results with other authors.

Considering the methods presented in Table 2, it was verified that the percentage of plant aggregates varied between 20 and 35%. The curing conditions were somewhat similar to the air, at an approximate temperature of 20 °C, and the difference in humidity in most of the curing time was not very significant. The shape and size of the samples in the different studies presented differences compared to the size used in this study. In this case, the size used was smaller.

Comparing the values, the following may be considered:

- Regarding density versus resistance, there is a general trend of increased resistance as the density increases, as expected. However, there is the exception of Light Rice Concrete, which, even with reduced resistance, achieves values above the average of concrete with hemp. This value is most likely justified by the use of hydraulic lime together with the aerial lime (50% of each) as this type of lime usually leads to greater resistance.
- Regarding the values obtained from resistance, considering a smaller sample size used in this study, it can be considered that the size of the specimens does not seem to have much influence on the resistance increase, because the mixture SCB Lime 27% Mk Concrete, performed with the same proportion of aggregate and binders as Kropidłowska [13], had an identical resistance. Moreover, in the literature, there also does not seem to be certainty in the true influence of the size and shape of the specimens, and there are many contradictory results [40];
- The preformed mixtures with sugarcane bagasse (SCB) generally have higher resistances, except for the mixture with referred metakaolin. These values seem to be related to the higher density of mixtures with SCB. In reverse, the mixture with metakaolin has a lower density and also lower resistance compared to the other mixtures with SCB.
- For the use of pozzolanic material, there tends to be better results with a lower percentage added to the aerial lime according to the results of mixing sunflower concrete with 10% of pozzolanic material and the mixture of SCB with 6% Ash of SCB (or ASCB). However, this difference may also be related to the type of pozzolans added. In addition, concerning the use of the Ash, it is possible to make some relationships with another study of [41] where this Ash was used with lime in soil stabilization for other uses. These authors obtained, with a lower amount of lime, higher resistance values with the increase in Ash amount, between 8% and 16% of Ash. This indicates that the results are consistent and probably, with a higher percentage of gray, the result would be even better.
- It is also observed that the mixture of SCB with soil, without lime, also presents better results than "hemp concrete". This higher strength is justified by the higher density of the mixtures once the soil is used.

In general, is important to note that, for a precise conclusion, the use of a similar density is usually obtained with hemp. However, as the SCB aggregate has a smaller dimension than hemp hurds, for the cohesion of the mixture, it requires a greater amount of binder material in volume, and this leads to a higher density.

3.2. Thermal Performance

Figure 10 represents the temperatures and heat flow reached for the 72 h test [32] in the hot box for the SCB composite studied. From Figure 10, it is possible to see that the temperature in the chambers (hot and cold) and heat flow remained very stable during the test period.

The outputs from the hot box were used to calculate the thermal parameters considering Equations (1) and (2).

Table 3 shows the value of the thermal conductivity coefficient (λ) obtained by the SCB/lime composite (S.L-1 mixture) in comparison with the typical values of several conventional construction materials and some studies of hemp concrete.

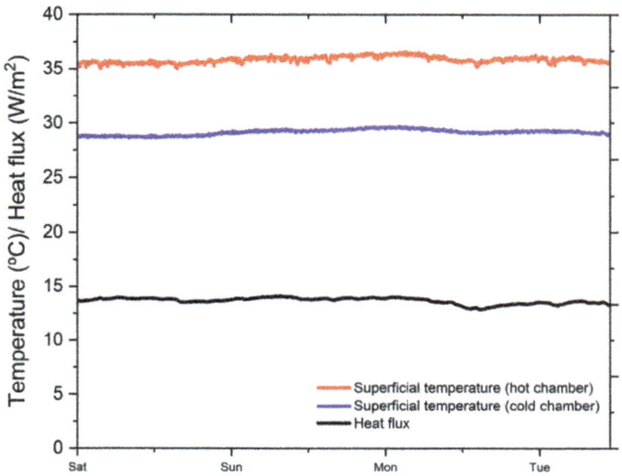

Figure 10. S.L-1 mixture: temperatures and heat flux.

Table 3. Thermal conductivity coefficient of SCB composite and other building materials.

Building Materials	Thermal Conductivity Coefficient (λ) (W/m. °C)	References
Hard limestone	1.7	Nunes [42]
Adobe	1.1	Nunes [42]
Brick	0.41	Carvalho [43]
S.L-1 mixture (70% lime, 30% SCB; 842 kg/m^3)	0.12	This study
"Hempcrete" (65% de hydrated lime, 30% de hemp)	0.11	Araújo [40]
Hemp concrete (220–627 kg/m^3)	0.06–0.14	Abdellatef and Kavgic [17]

The results show a satisfactory thermal performance for the SCB composite. Its thermal conductivity, 0.12 W/m. °C, is less than those of hard limestone, adobe, and brick. Considering the hempcrete, the SCB composite studied shows a very similar performance to hempcrete studied by Araújo [40], 0.11 W/m. °C. Furthermore, the value achieved is in between the values shown by Abdellatef and Kavgic [17] for different hemp concrete compositions. However, for a more assertive conclusion of the results, a more detailed investigation would be important, such as testing different variations of binder material and mixtures with different densities.

3.3. *Water Absorption and Resistance*

3.3.1. Water Absorption by Capillarity and Immersion

As vegetable materials and soil binder, the water absorptions values are, in general, very high. As can be seen in Figure 11, the study of water absorption by capillarity showed reduced tolerance to water, requiring the application of waterproof mortar and the preparation of a base with conventional masonry or stonework, to keep the material away from contact with water, which is essential to guarantee the durability of the building.

The results obtained in capillary absorption, considering the values in kg/m^2. min$^{0.5}$, also reveal that:

- Mixtures only with SCB and lime have a smaller water absorption and the replacement of lime by metakaolin does not present an advantage at this level, on the contrary;
- In general, mixtures with soil have higher water absorption, except the mixture with lime and metakaolin, probably due to its reduced compaction and mass;
- The addition of lime to the soil does not seem to have an advantage; on the contrary, water absorption is lower than expected.

During the capillary test, fungal growth was also observed in the mixtures with soil and without lime. Thus, it can be considered that lime offers the advantage of not facilitating the development of living matter. In those where lime was mixed with soil, contamination was delayed and greatly reduced.

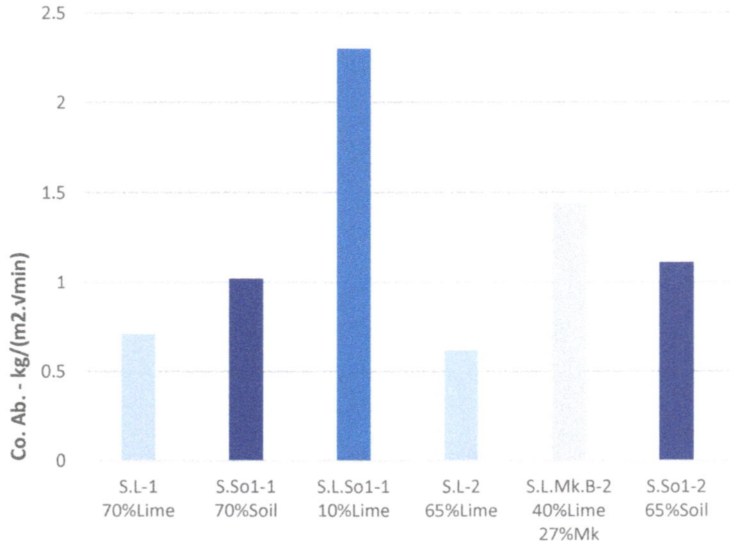

Figure 11. Coefficient of absorption by capillarity.

The water immersion test results presented some differences in relation to the capillarity test. In this case, the addition of lime seemed to favor the reduction in water absorption, as shown in Table 4, comparing S.L-1 with S.L-2 (with less than 5% of lime) and S.So1-1 with S.L.So1-1 (10% lime). In addition, a lower absorption in two mixtures with soil (S.So1-1 and S.L.So1-1 (with 10% lime)) can be seen, contrary to what would be expected for soil mixtures. This lower absorption may be justified by the greater mass of the specimens compared to the others, which indicates that the higher compaction of the mixtures will have fewer voids and reduce the absorption.

Table 4. Percentage of water absorption by immersion after 24 h and nominal value per m^3.

Mixes	Dry Weight (g)	24 h (g)	Absorption (%)	Water (kg)/m^3
S.L-1 (70% Lime)	155	268.9	73.5	445
S.So1-1	255.7	393.9	54	539
S.L.So1-1 (10% Lime)	254	362.1	42.5	421
S.L-2 (65% Lime)	125	229.7	83.8	409
S.So1-2	119	283	137.81	640

Figure 12 presents a comparison of the results obtained in this study with the values of research on hempcrete [13], ceramic masonry blocks, and concrete blocks [44]. The relationship between the density and the absorption coefficient of the tested compositions was examined.

As expected, there was less absorption in brick and ordinary concrete. It was observed that denser materials seemed to have less absorption in general, but in the SCB's studied mixtures or in hempcrete, this was not always true. The obtained results showed that the hempcrete based on hydrated lime [13–17] presented a lower density and a lower absorption than the composites with SCB and also the mixture S.L-2.

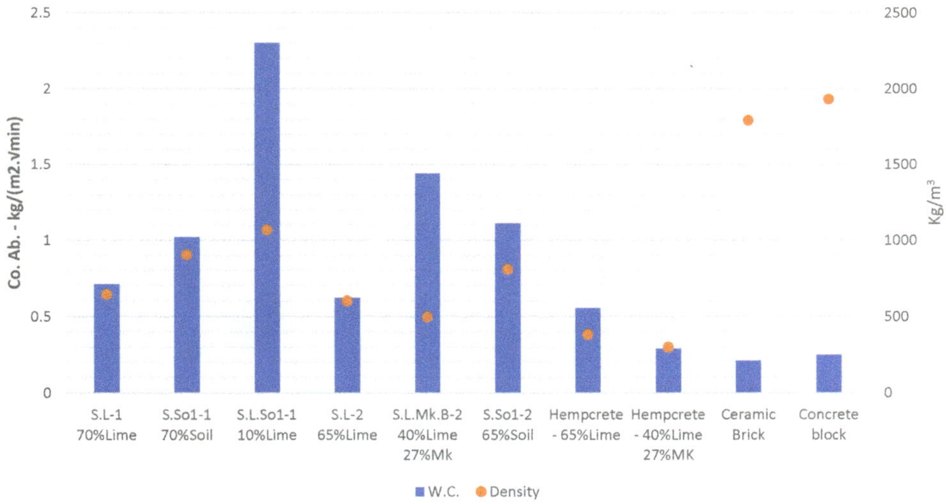

Figure 12. Comparison between densities and absorption coefficients of studied compositions and other materials.

3.3.2. Compressive Strength in Saturated Specimens

The results of the compression test on specimens subjected to the absorption test by immersion are presented in Table 5, where it can be seen that the water absorption did not affect the compressive strength of S.L-1 and S.L.So1-1. On the contrary, there was a slight increase in resistance. However, there was a small loss of resistance in S.L-2. As stated by Pinto [45], when saturated, the tensile strength of lignocellulosic fibers is also slightly higher, which could justify this increase. It was also registered that the composite with only soil could not be tested, due to the lack of cohesion of the material after absorption. As such, it is not water-resistant.

Table 5. Compressive strength in specimens subjected to the immersion absorption test.

Mixes	Compressive Strength (MPa) 90 d, Dry	Compressive Strength (MPa), Saturated
S.L-1	1.60	1.65
S.L.So1-1	1.26	1.38
S.L-2	0.98	0.82

4. Conclusions

The main conclusions that can be reached about this study are the following.

Considering the minimum performance values required by the hempcrete technical guide and the results obtained in the sample tests, the composites fall within the range of materials with sufficient strength for non-structural walls.

Aerial lime and/or soil used as binder provide enough mechanical strength for this application. The composite of SCB with soil achieves the best performance in terms of mechanical resistance: 2.6 MPa in compressive strength and 2.1 MPa in bending strength, while the composite of SCB with lime achieves 1.76 MPa and 1.7 MPa, respectively.

The lime composites obtain greater water resistance and less loss of mechanical strength when saturated. However, the higher water absorption coefficient makes it necessary to apply a waterproof mortar on surfaces exposed to the weather.

Concerning the thermal properties, it is verified that the composition reaches low values of thermal conductivity; therefore, it presents good behavior, far superior to conventional materials used in masonry construction.

Alongside the results, it is concluded that the researched composites can have several applications as a non-structural material (Figure 13a,b).

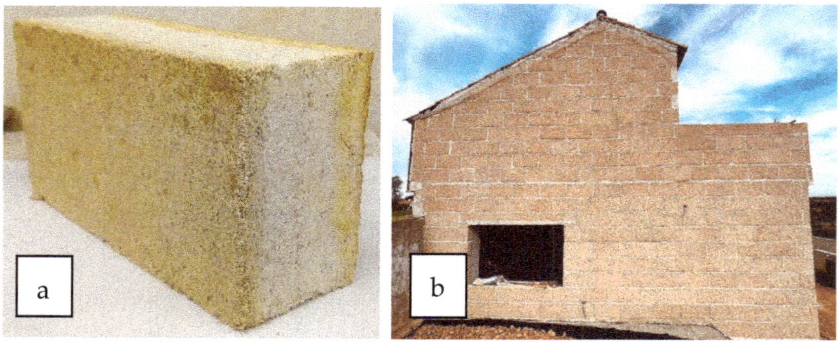

Figure 13. (**a**) Block with SCB and lime (credits: Souza [46]); (**b**) house with hemp concrete blocks (credits: Cânhamor [47]).

Regarding the sustainability of the studied composites, it can be considered that the mixtures with only SCB and soil will be the most sustainable, as they do not contain a binder and, as such, will have a lower energy expenditure. For the same reason, the compositions of SCB with soil and lime will follow, and finally, the composite of SCB with lime. However, due to the lower density, the mixture with lime may have better thermal behavior, which will lead to benefits in the sustainability of construction in general. In addition, due to its lightness, the composite with SCB and lime contributes significantly to lightening the structure.

Furthermore, the products in this study show a potential for a differentiated finish, with comfort, durability, and economy that compensate for the difficulties. The fact that this can be manufactured in a simple way means that it would be constructed at an affordable cost. As such, the intensive sugarcane industry could easily sponsor the construction of houses for local communities to partially compensate them for the inconvenience caused.

All waste produces discontent. Far from agreeing with the monoculture of sugar cane, while the agroindustry does not find solutions with lesser environmental, social, and economic impact, this type of composite for building blocks can be a viable solution for increasing the value of sugarcane bagasse.

Author Contributions: P.P.L.d.S. undertook the main part of the research that was the base of this paper. She developed the research method and analysed the results with the contribution of R.M. and R.E., P.P.L.d.S., R.M. and R.E. wrote the document. R.M. and R.E. helped to develop the discussion sections of the paper and provided critical judgment on the undertaken research. They also revised the document. Adicionally, R.E. supervised all the work. All authors have read and agreed to the published version of the manuscript.

Funding: The authors would also like to acknowledge FCT/MCTES—Foundation for Science and Technology/Ministry of Science, Technology and Higher Education, within the scope of CTAC—Center for Territory, Environment and Construction—UIDB/0407/2020 e UIDP/04047/2020.

Acknowledgments: The authors would like to acknowledge J. Faria & Filhos, Lda., located in Madeira Island, for kindly sending us a generous shipment of sugarcane bagasse, without which it would not have been possible to carry out this study.

Conflicts of Interest: The authors declare no conflict of interest. The funders had no role in the design of the study; in the collection, analyses, or interpretation of data; in the writing of the manuscript; or in the decision to publish the results.

References

1. Companhia Nacional de Abastecimento (CONAB). Acompanhamento Da Safra Brasileira de Cana-de-Açúcar—V. 6—SAFRA 2019/20 N.3—Terceiro Levantamento. CONAB: Brasília, Brazil, 2019; ISSN 2318-7921.
2. Oliveira, O.C. Avaliação de Fibras de Bagaço de Cana-de-Açúcar in Natura e Modificada Para Aplicação Em Compósitos. Master's Thesis, Universidade Estadual do Norte Fluminense, Rio de Janeiro, Brazil, 2018.
3. Stanwix, W.; Sparrow, A. *The Hempcrete Book: Designing and Building with Hemp-Lime*; UIT Cambridge Ltd.: Cambridge, UK, 2014.
4. Daly, P.; Ronchetti, P.; Woolley, T. *Hemp Lime Bio-Composite as a Building Material in Irish Construction*; ENVIRONMENTAL PROTECTION AGENCY: Wexford, Ireland, 2012.
5. Rossi, F.G.; Martins, J.; Denari, M.; Hassaine, H.; Gregório, T.; Reis, C.E.N. Incorporação Da Fibra Do Bagaço de Cana-de-Açúcar No Cimento. In Proceedings of the XX Congresso Brasileiro de Engenharia Química, Florianópolis, Brazil, 19–22 November 2014.
6. Paula, M.O.; Tinôco, I.F.F.; Rodrigues, C.S.; Silva, E.N.; Souza, C.F. Potencial Da Cinza Do Bagaço Da Cana-de-Açúcar Como Material de Substituição Parcial de Cimento Portland. *Rev. Bras. Eng. Agrícola E Ambient.* **2009**, *13*, 353–357. [CrossRef]
7. Ligowski, E.; Santos, B.C.; Fujiwara, S.T. Materiais Compósitos a Base de Fibras Da Cana-de-Açúcar e Polímeros Reciclados Obtidos Através Da Técnica de Extrusão. *Rev. Polímeros Ciência E Tecnol.* **2015**, *25*, 70–75. [CrossRef]
8. Coelho, K.V.S. *Desenvolvimento de Compósitos de Polietileno de Alta Densidade Pós Consumo (PEADpc) Com Fibras de Bagaço de Cana-de-Açúcar (SCB): Influência Do Teor de Fibras e Do Compatibilizante*; Final Year Project; Centro Universitário Zona Oeste: Rio de Janeiro, Brazil, 2017.
9. Araújo, A.O.; Martins, K.P.; Junior, V.D.L. *Obtenção de Compósitos Híbridos de Fibras de Vidro e Fibras Do Bagaço de Cana-de-Açúcar Com Foco Na Aplicação Em Telhas Poliméricas*; Final Year Project; Faculdade Evangélica de Goianésia—FACEG: Goianésia, Brazil, 2022.
10. Gandara, M.; Gonçalves, A.R.; Saron, C. Compósitos de PET Reciclado Com Fibra de Cana-de-Açúcar Tratada Por Explosão a Vapor. *Rev. Matéria* **2017**, *22*, 1–10. [CrossRef]
11. Jesus, L.C.C. Obtenção e Caracterização de Compósitos de Poliestireno Expandido Pós-Consumo Reforçados Com Celulose de Bagaço de Cana-de-Açúcar. Master's Thesis, Universidade de Brasília, Brasília, Brazil, 2014.
12. Bock-Hyeng, C.; Ofori-Boadu, A.N.; Yamb-Bell, E.; Shofoluwe, M.A. Mechanical Properties of Sustainable Adobe Bricks Stabilized with Recycled Sugarcane Fiber Waste. *Int. J. Eng. Res. Appl.* **2016**, *6*, 50–59.
13. Kropidłowska, A.M. A Investigação Experimental Das Propriedades Mecânicas e Higroscópicas Do Hempcrete Como Um Material de Isolamento Para Painéis Pré-Fabricados. Master's Thesis, Universidade do Minho, Guimarães, Portugal, 2019.
14. Valenciano, M.D.C.N.; Freire, W.J. Características Físicas e Mecânicas de Misturas de Solo, Cimento e Cinzas Do Bagaço de Cana-de-Açúcar. *Construções Rurais E Ambiência* **2004**, *24*, 484–492.
15. Singh, M.; Mamania, D.; Shinde, V. The Scope of Hemp (*Cannabis sativa* L.) Use in Historical Conservation in India. *Indian J. Tradit. Knowl.* **2018**, *17*, 314–321.
16. Prieu, F. Comment Le Chanvre Est-Il Utilisé Dans La Construction? Available online: https://greentropics.co/chanvre/chanvre-construction/ (accessed on 20 September 2022).
17. Abdellatef, Y.; Kavgic, M. Thermal, Microstructural and Numerical Analysis of Hempcrete-Microencapsulated Phase Change Material Composites. *Appl. Therm. Eng.* **2020**, *178*, 115520. [CrossRef]
18. Chabannes, M.; Garcia-Diaz, E.; Clerc, L.; Bénézet, J.C. Effect of Curing Conditions and $Ca(OH)_2$-Treated Aggregates on Mechanical Properties of Rice Husk and Hemp Concretes Using a Lime-Based Binder. *Constr. Build. Mater.* **2016**, *102*, 821–833. [CrossRef]
19. Chabannes, M.; Nozahic, V.; Amziane, S. Design and Multi-Physical Properties of a New Insulating Concrete Using Sunflower Stem Aggregates and Eco-Friendly Binders. *Mater. Struct.* **2014**, *48*, 1815–1829. [CrossRef]
20. Helepciuc, C.M.; Barbuta, M.; Serbanoiu, A.A. Characterization of a Lightweight Concrete with Sunflower Aggregates. *Procedia Manuf.* **2018**, *22*, 154–159. [CrossRef]
21. Lenço, P.C. Caracterização Do Bagaço de Cana-de-Açúcar Para Geração de Energia. Ph.D. Thesis, Universidade Estadual de Campinas, Campinas, Brazil, 2010.
22. EN 459-1. Building Lime—Part 1: Definitions, Specifications and Conformity Criteria. British Standards Institution: London, UK, 2015. Available online: https://standards.iteh.ai/catalog/standards/cen/588081bb-ff4e-4421-997c-2d7dcab1b6ac/en-459-1-2015 (accessed on 30 November 2022).
23. Vasconcelos, M.C.A. Avaliação Da Atividade Pozolânica Da Cinza Do Bagaço de Cana-de-Açúcar Utilizando Métodos Físicos. Master's Thesis, Universidade Federal de Pernambuco, Recife, Brazil, 2013.
24. Martins, T.J.R. Incorporação de Materiais de Mudança de Fase Em Blocos de Terra Comprimida. Master's Thesis, Universidade do Minho, Guimarães, Brazil, 2018.
25. EN 1015-11. Methods of Test for Mortar for Masonry—Determination of Flexural and Compressive Strength of Hardened Mortar. British Standards Institution: London, UK, 1999. Available online: https://standards.iteh.ai/catalog/standards/cen/14596d4c-119b-4a78-94e1-3fe481a29bde/en-1015-11-2019 (accessed on 30 November 2022).
26. EN 1015-18. Methods of Test for Mortar for Masonry—Part 18: Determination of Water Absorption Coefficient Due to Capillary Action of Hardened Mortar. British Standards Institution: London, UK, 2002. Available online: https://standards.iteh.ai/catalog/standards/cen/d9e520b3-5f88-4368-ae91-e568d18aef5a/en-1015-18-2002 (accessed on 30 November 2022).
27. Hall, M.; Djerbib, Y. Moisture Ingress in Rammed Earth: Part 1—The Effect of Soil Particle-Size Distribution on the Rate of Capillary Suction. *Constr. Build. Mater.* **2004**, *18*, 269–280. [CrossRef]

28. EN 772-7. Methods of Test for Masonry Units Determination of Water Absorption of Clay Masonry Damp Proof Course Units by Boiling in Water. British Standards Institution: London, UK, 1998.
29. Malheiro, R.; Ansolin, A.; Guarnier, C.; Fernandes, J.; Amorim, M.; Silva, S.M.; Mateus, R. The Potential of the Reed as a Regenerative Building Material—Characterisation of Its Durability, Physical, and Thermal Performances. *Energies* **2021**, *14*, 4276. [CrossRef]
30. ASTM C1363-19. Test Method for Thermal Performance of Building Materials and Envelope Assemblies by Means of a Hot Box Apparatus. ASTM International: West Conshohocken, PA, USA, 2019. Available online: https://www.astm.org/c1363-19.html (accessed on 30 November 2022).
31. Ansolin, A. Characterization of the Reed Arundo Donax L. and Evaluation of Its Potential as a Thermal Insulation Material in Portugal (in Portuguese). Master's Thesis, Universidade do Minho, Guimarães, Portugal, 2021.
32. ISO 9869-1. Thermal Insulation-Building Elements-In Situ Measurement of Thermal Resistance and Thermal Transmittance-Calculation Methods. ISO: Geneva, Switzerland, 2014. Available online: https://www.iso.org/obp/ui/#iso:std:iso:9869:-1:ed-1:v1:en (accessed on 30 November 2022).
33. Gomes, M.I.; Lima, J.; Santos, T.; Gomes, J.; Faria, P. The Benefits of Eco-Efficient Plasters for Occupant's Health—A Case Study. In *Ecological and Health Effects of Building Materials*; Malik, J.A., Marathe, S., Eds.; Springer: Cham, Switzerland, 2022. [CrossRef]
34. Nieuwenhuyze, L.V. Chanvre: Tome 2: Technologie de Chantier; Construire en Chanvre. 2021. Available online: https://www.ressources-caue.fr/GED_K/119703593798/CenC_Bonnes_Pratiques_Tome_2_com.pdf (accessed on 30 November 2022).
35. NBR 16814. Adobe—Requisitos e Métodos de Ensaio. Associação Brasileira de Normas Técnicas: Rio de Janeiro, Brazil, 2020. Available online: https://edisciplinas.usp.br/pluginfile.php/5711732/mod_resource/content/2/NBR16814%20norma%20adobe%20Arquivo%20para%20impress%C3%A3o.pdf (accessed on 30 November 2022).
36. NBR 15270-2. Componentes Cerâmicos Parte 2: Blocos Cerâmicos Para Alvenaria Estrutural—Terminologia e Requisitos. Associação Brasileira de Normas Técnicas: Rio de Janeiro, Brazil, 2005. Available online: https://www.studocu.com/pt-br/document/universidade-estadual-de-campinas/sistemas-prediais-hidraulicos-e-sanitarios/abnt-nbr-15270-2-2005-blocos-ceramicoas/14302660 (accessed on 30 November 2022).
37. NBR 6136. Blocos Vazados de Concreto Simples Para Alvenaria—Requisitos. Associação Brasileira de Normas Técnicas: Rio de Janeiro, Brazil, 2016. Available online: https://www.academia.edu/40801258/ABNT_NBR_Blocos_vazados_de_concreto_simples_para_alvenaria_Requisitos_Hollow_concrete_blocks_for_concrete_masonry_Requirements (accessed on 30 November 2022).
38. Hanuseac, L.; Barbuta, M.; Bejan, L.; Rosu, R.; Timu, A. Experimental Study on Hollow Blocks with Wastes. *Proceedings* **2021**, *63*, 79. [CrossRef]
39. Jonaitis, B.; Zavalis, R. Experimental Research of Hollow Concrete Block Masonry Stress Deformations. *Procedia Eng.* **2013**, *57*, 473–478. [CrossRef]
40. Araújo, E.F. Materiais Compósitos Com Incorporação de Cânhamo Industrial. Master's Thesis, Universidade do Minho, Guimarães, Portugal, 2015.
41. Sarosh, H.Z.A.; Suhail, K.A.; Mariyam, R.; Tishana, K.N.; Rashida, P.K. Soil Stabilization Using Lime and Sugarcane Bagasse Ash. *Int. J. Eng. Res. Technol.* **2021**, *10*, 1014–1019.
42. Nunes, P. Comportamento Térmico de Diferentes Soluções de Envolvente Opaca Vertical Em Edifícios. Master's Thesis, Instituto Superior Técnico, Lisboa, Portugal, 2014.
43. Carvalho, P.D. Caracterização Do Desempenho Térmico de Soluções Construtivas Sujeitas a Um Regime de Transferência de Calor Dinâmico. Master's Thesis, Universidade de Coimbra, Coimbra, Portugal, 2017.
44. Eires, R.; Jalali, S.; Camões, A. Novos Compósitos Eco-Eficientes Para Aplicações Não Estruturais Na Construção. *Rev. Int. Construlink* **2010**, *8*, 45–55.
45. Pinto, E.S. Solo-Cimento Compactado: Proposta de Métodos de Ensaio Para Dosagem e Caracterização Física e Mecânica. Master's Thesis, Universidade Estadual Paulista, Bauru, Brazil, 2016.
46. Souza, P.P.L. de; Compósitos de Bagaço de Cana-de-Açúcar Para Blocos Construtivos. Master's Thesis, University of Minho, Guimarães, Portugal, 2021.
47. Cânhamor—Os Nossos Blocos. Available online: https://canhamorhemp.com/os-nossos-blocos/ (accessed on 20 December 2022).

designs

MDPI

Article

CFD Investigation of Ventilation Strategies to Remove Contaminants from a Hospital Room

Mustafa Alkhalaf [1,*], Adrian Ilinca [2] and Mohamed Yasser Hayyani [1]

1 Wind Energy Research Laboratory, The University of Quebec at Rimouski, Rimouski, QC G5L 3A1, Canada
2 École de Technologie Supérieure, Montréal, QC H3C 1K3, Canada
* Correspondence: mustafa.alkhalaf@uqar.ca

Abstract: The primary requirement in designing air conditioning systems in healthcare facilities is eliminating contaminants. It is considered one of the crucial health elements in building design, particularly in the presence of many airborne diseases such as COVID-19. The purpose of this numerical research is to simulate various ventilation designs for a hospital room model by taking into account results obtained by previous researchers. Four designs with three airflows, 9, 12, and 15 ACH (Air Change per Hour), are applied to explore the capacity of the ventilation system to remove contaminants. The objective is to determine the influence of airflow and the diffuser location distribution on the pollutants elimination represented by carbon dioxide. The Reynold Averaged Navier–Stokes (RANS) equations and the k-ε turbulence model were used as the underlying mathematical model for the airflow. In addition, boundary conditions were extracted from ASHRAE (American Society of Heating, Refrigeration, and Air-Conditioning Engineers Society) ventilation publications and relevant literature. Contrary to what was expected, this study's results demonstrated that increased ventilation alone does not always improve air distribution or remove more contaminants. In addition, pollutant removal was significantly affected by the outlet's location.

Keywords: air quality; CFD simulation; trace study; contaminant removal

Citation: Alkhalaf, M.; Ilinca, A.; Hayyani, M.Y. CFD Investigation of Ventilation Strategies to Remove Contaminants from a Hospital Room. *Designs* **2023**, *7*, 5. https://doi.org/10.3390/designs7010005

Academic Editors: Shi-Jie Cao and Wei Feng

Received: 30 November 2022
Revised: 26 December 2022
Accepted: 29 December 2022
Published: 4 January 2023

1. Introduction

The coronavirus pandemic has resulted in more than 15 million infections and over 619,000 deaths worldwide in 2020. The regions most affected by the pandemic are Asia, particularly China, Europe, the United States, South America, and Mexico. Severe acute respiratory syndrome (SARS-CoV-2) can spread over long distances in the air. Therefore, airborne transmission played a significant role in the rapid propagation of the epidemic [1]. Given the rapid spread of the disease, hospitals are compelled to treat patients in isolation. When an infected individual sneezes or coughs, microscopic particles are disseminated throughout the environment. If another person inhales these particles, they may become sick. Therefore, airborne infectious diseases can rapidly spread in an inadequately ventilated isolation room. Rooms with a proper ventilation system will be free of infectious airborne particles, such as viruses, bacteria, or microorganisms [2,3].

During the ventilation system operation, the occupants' attention is focused on thermal comfort because the sensation of thermal comfort is immediate and thermal discomfort cannot be tolerated. However, poor air quality is hard to notice, so the occupants' response takes longer [4]. Therefore, efficient ventilation and adequate indoor air quality (IAQ) are essential for human health, well-being, and productivity. For instance, a good IAQ effectively removes indoor pollutants and introduces an adequate quantity of fresh outdoor air for occupants [5]. Several indicators assess ventilation efficiency.

Contaminant removal efficiency (CRE) and air exchange efficiency (ACE) are indicators regularly used as they can be easily measured both in a laboratory and in the field and can be applied to all ventilation methods. In addition, they are generic, and almost all other indicators are extensions of them. For example, CRE is an indicator of the pollution level in a room that depends not only on the airflow pattern but also on the characteristics of pollutant sources, such as density, area, and position [4].

IAQ can be assessed in terms of the concentration of gaseous ingredients. Many gaseous ingredients such as volatile organic compounds (VOCs), formaldehyde, nitrogen oxides, sulfur oxide, carbon dioxide, carbon monoxide, particulates, and infectious pollutants often degrade indoor air quality. The high concentration of these pollutants leads to serious health effects and inconvenience for patients. However, pollutants other than carbon dioxide are usually recorded as sufficiently below the standard limit [6]. In addition, the carbon dioxide concentration depends on the number of humans inside an occupied space. Therefore, carbon dioxide concentration is widely used as an indicator of IAQ.

No sole ventilation design solution can solve all the airborne transferable particle concentration problems and consistently be cost-effective. Therefore, ventilation designers must always consider the cost of installing and operating their systems and need effective control strategies for air systems to make a feasible design [7]. Therefore, various healthcare facility ventilation designs provide a minimum ventilation equivalent to 12 ACH (Air Change per Hour) for isolation rooms [8]. According to [9], the transmission of air pollutants can spread disease to healthcare workers and patients.

The pressure value of the air inside a hospital room should be checked and adjusted correctly accordingly to its usage. For example, if the room is host to a transmissible disease, the room's pressure is lowered to prevent particles from leaking out of the room; if the room is used for an operation, the room's pressure is increased to prevent particles from entering so the space can be kept sterile [10]. Therefore, for the systems serving Airborne Infection Isolation Rooms (AIIR), negative air pressure must be designed relative to adjacent rooms or hallways. However, an AIIR plan with negative pressure involves a complex decision-making process [11].

Several studies have noted an interdependence between ventilation and health, but the actual relationship and attributable mechanisms remain unclear. In addition, the data was insufficient to define minimum ventilation standards to control the prevalence of airborne disease in any setting. Consequently, it revealed the uncertainty about the connection between ventilation and health [12].

Researchers have examined how different vent placements affect the flow and heat transmission characteristics inside an enclosed system with a volumetric heat source. Three aspect ratios H/W = 1, 2, 3, and three Ra_h = 103, 104, 105, have been examined together with two global factors, the mass flow rate and Nu. The conclusion is that the position of the vent should minimize the dead zone size. In addition, the transfer properties of the system would improve if the dead zones were reduced [13].

The airflow direction depends on the room pressure and the inlet/outlet distribution through the space. Accordingly, a cautious and intentional ventilation device can be more effective in containing and removing airborne contaminants [12]. Unfortunately, the literature does not provide enough indication of the location of outlets and inlets. For example, it does not limit the maximum and minimum distances between the two openings [14]. As per ASHRAE standards, space air diffusion has defined some particular rules, trying to guide the designers toward the best solution. For example, it mentions that the outlet diffuser should be located on the side of the room away from the supply diffuser to reduce short-circuiting of supplied air [14].

Using computational fluid dynamics modeling and field measurement, a comparison of the efficiency of three ventilation systems has been made in a study by Cho [11]. The research aimed to protect medical personnel from inhaling patients' respiratory droplets of sputum in an isolation room provided with negative pressure. A new ventilation approach for isolation rooms is proposed based on empirical data and simulated findings from three ventilation systems that have been effective in removing pollutants. The results show that ventilation systems using the "low-level extraction" method are superior for removing contaminants from the breathing zone.

The air change rate and pollutant removal efficiency were used to assess AIIR ventilation performance in three Finnish hospitals [14,15]. The results showed that the AIIR's and anteroom's high ventilation rates (4–24) ACH were insufficient to stop the spread of infectious microorganisms due to improper airflow.

Researchers at the University of Cordoba [16] conducted an experimental and numerical study to assess ventilation at three different rates of air change of 6, 9, and 12 ACH in the transport of pollutants to a patient (P) lying in a hospital bed to a healthcare worker (HCW) standing beside the bed. They found that increasing ACH cannot reduce exposure and, in some circumstances, may increase it.

Another numerical study [17] aimed to recognize the role of ventilation in preventing and controlling infection in general hospital wards and obtain a simple design, cost-effective ventilation system to reduce infection. The study's results revealed that rearranging the air return diffuser position and increasing the aeration rate to 12 ACH kept the ventilation under control and enhanced its ability to reduce the risk of transmitting diseases to public wards.

Experimental and numerical tests have been performed [18] in a hospital to monitor the circulation and removal of inhalable aerosols (0.5–10 μm) based on the ventilation rate in the isolation room. It was found that increased ventilation from 2.5 to 5.5 ACH resulted in reducing aerosol concentrations by only 30%. Higher ventilation rates were not relatively effective in reducing the concentration of pollutants.

A numerical computational fluid dynamics (CFD) approach is applied by Lu et al. [19] to investigate if the representation of CO_2 could be achieved and explore contaminant distribution in a two-bed hospital ward with two patients and one healthcare worker under different types of ventilation. For simulating the exhaled and coughed contaminants by patients with different postures, a tracer gas (CO_2) is applied. The results demonstrate that stratum ventilation minimizes the exposure risk of healthcare workers in hospital wards. Furthermore, under stratum ventilation, the contaminant concentration in the breathing zone at 1.3–1.7 m above the floor is lower, and the contaminant removal effectiveness is comparably higher.

The primary limitations and solutions in the epidemic era have been assessed by Fan et al. [20]. They reviewed the most up-to-date scientific literature on indoor ventilation modes and manuals from different countries, identifying characteristics of different ventilation modes and evaluating effects in different application occasions. In addition, they studied which virus spread regulations and operating modes, including non-uniform and unstable ones, demonstrated the best performance for air quality.

To determine the importance of the air outlet near the patient, Borro et al. [1] investigated the role of HVAC systems in spreading infection through CFD simulation cough at Bambino Gesu Children's Hospital in the Vatican State. In addition, the potential role of exhaust ventilation systems placed over the mouth of a coughing patient was also assessed. Despite doubling the airflow in the HVAC system providing a significant reduction in the concentration of airborne pollutants, it also results in a significant increase in turbulent air movement, which gives the droplets and air pollutants a spread of increased and faster long-range in the room. However, the presence of the local exhaust ventilation (LEV) unit above the patient's face illustrates a very high capacity to reduce droplets and polluted air in the room, guaranteeing the total absence of exposure to infection risks for the patient.

Another study [21] examined the ventilation system's performance in a standard hospital room using SolidWorks flow simulation software. Achieving indoor air quality (IAQ) depends on many factors, including removing pollutants. Therefore, CRE greater than one is strongly recommended for the optimal ventilation system. The simulation gives a CRE of 1.23 for exhaled air, which means that the ventilation system is reasonably efficient in removing polluted air, although the study result was satisfactory. However, the results obtained through this study are not blindly applicable to hot and humid countries.

The primary goal of this research is to conduct a numerical study of several ventilation system strategies for a hospital isolation room. This study focuses on determining the optimum strategy to protect healthcare workers from being inoculated, eliminate infectious sources, or reduce their spread. The findings were explored in detail via the Local Air Quality Index (LAQI) by demonstrating the efficiency of the ventilation to eliminate contaminants at each point in the space. In addition, we present the trace study to illustrate the steady-state diffusion of the contaminants and CRE to demonstrate the efficiency with which the ventilation system eliminates them.

2. Methodology

CFD is the most advanced building simulation method and uses Navier–Stokes equations to solve the flow field in the fluid domain inside the building [22,23].

2.1. Governing Equations

2.1.1. Turbulence Model

Reynolds number is the product of representative velocity and length scales divided by kinematic viscosity and characterizes laminar and turbulent flows. Most of the fluid flows encountered in engineering practice are considered turbulent, so flow simulations were mainly developed to simulate and study turbulent flows. Average Favre-Navier–Stokes equations are used to predict turbulent flows, in which the effects of time-averaged flow turbulence on flow parameters are considered.

Previous publications have established that the Reynolds-averaged-Navier–Stokes approach is adequate for modeling airflow in closed spaces. Two additional partial differential equations associated with the k-ε turbulence model complete mathematical closure [12,23].

The general laws of mass, angular momentum, and energy conservation can be written in a cartesian frame rotating at an angular speed Ω around an axis passing through the origin of the frame in the following conservation form (Solidworks Flow Simulation). The subscripts, unless stated otherwise, are used to denote summation over the three coordinate directions, x, y, and z, associated with $i = 1$, $i = 2$, and $i = 3$, respectively [24,25]:

$$\frac{\partial \rho}{\partial t} + \frac{\partial}{\partial x_i}(\rho u_i) = 0 \tag{1}$$

$$\frac{\partial \rho u_i}{\partial t} + \frac{\partial}{\partial x_i}(\rho u_i u_j) + \frac{\partial p}{\partial x_i} = \frac{\partial}{\partial x_i}\left(\tau_{ij} + \tau_{ij}^R\right) + S_i; i = 1,2,3 \tag{2}$$

$$\frac{\partial \rho H}{\partial t} + \frac{\partial \rho u_i H}{\partial x_i} = \frac{\partial}{\partial x_i}\left(u_j\left(\tau_{ij} + \tau_{ij}^R\right) + q_i\right) + \frac{\partial p}{\partial t} - \tau_{ij}^R \frac{\partial u_i}{\partial x_j} + \rho\varepsilon + S_i u_i + Q_H, \tag{3}$$

$$H = h + \frac{u^2}{2} + \frac{5}{3}k - \frac{\Omega^2 r^2}{2} - \sum_m h_m^0 y_m, \tag{4}$$

Here, u is the fluid velocity, ρ is the fluid density, S_i is a mass-distributed external force per unit mass due to a porous media resistance $\left(S_i^{porous}\right)$, a buoyancy $\left(S_i^{gravity} = -\rho g_i\right)$, where g_i is the gravitational acceleration component along the i-th coordinate direction, and the coordinate system's rotation $(S_i^{rotation})$, i.e., $S_i = (S_i^{porous} + S_i^{gravity} + S_i^{rotation})$. h is the thermal enthalpy, Q_H is a heat source or sink per unit volume, τ_{ij} is the viscous shear stress tensor, q_i is the diffusive heat flux, Ω is the angular velocity of the coordinate system in rotation, r is the distance between a point and the axis of rotation in the frame of rotation, k is the kinetic energy of the turbulence, h_m^0 is an individual thermal enthalpy of the m-th component of the mixture, y_m is a concentration of the m-th component of the mixture. Subscripts are used to indicate grouping across the three coordinate directions. In our study, there is no rotation of the flow domain and no porous media resistance. Therefore, those terms are set to zero in the simulations.

The viscous shear tensor for Newtonian fluids is defined as:

$$\tau_{ij} = \mu\left(\frac{\partial u_i}{\partial x_j} + \frac{\partial u_j}{\partial x_i} - \frac{2}{3}\delta_{ij}\frac{\partial u_k}{\partial x_k}\right) \tag{5}$$

The Reynolds-stress tensor as the following Boussinesq assumption has the following form:

$$\tau_{ij}^{R} = \mu_t\left(\frac{\partial u_i}{\partial x_j} + \frac{\partial u_j}{\partial x_i} - \frac{2}{3}\delta_{ij}\frac{\partial u_k}{\partial x_k}\right) - \frac{2}{3}\rho k\delta_{ij} \quad (6)$$

Here, is δ_{ij} the Kronecker delta function (it is equal to unity when $i = j$, and zero otherwise), μ is the dynamic viscosity parameter, μ_t is the turbulent eddy viscosity coefficient and k is the turbulent kinetic energy. Note that μ_t and k are zero for laminar flows. Within k-ε turbulence model, μ_t is computed using the turbulent dissipation ε and the turbulent kinetic energy k.

$$\mu_t = f_\mu \frac{C_\mu \rho k^2}{\varepsilon} \quad (7)$$

Here f_μ is a turbulent viscosity factor. It is defined by the expression

$$f_\mu = \left[1 - \exp(-0.0165R_y)\right]^2 \times \left(1 + \frac{20.5}{R_T}\right), \quad (8)$$

where: $R_T = \frac{\rho k^2}{\mu\varepsilon}$, $R_y = \frac{\rho\sqrt{k}y}{\mu}$. with y the distance to the wall. This function makes it possible to take into account the laminar-turbulent transition. Two additional transport equations are used to describe turbulent kinetic energy and dissipation,

$$\frac{\partial \rho k}{\partial t} + \frac{\partial}{\partial x_i}(\rho u_i k) = \frac{\partial}{\partial x_i}\left(\left(\mu + \frac{\mu_t}{\sigma_k}\right)\frac{\partial k}{\partial x_i}\right) + S_k, \quad (9)$$

$$\frac{\partial \rho \varepsilon}{\partial t} + \frac{\partial}{\partial x_i}(\rho u_i \varepsilon) = \frac{\partial}{\partial x_i}\left(\left(\mu + \frac{\mu_t}{\sigma_\varepsilon}\right)\frac{\partial \varepsilon}{\partial x_i}\right) + S_\varepsilon, \quad (10)$$

Here, the source terms S_k and S_ε are given by:

$$S_k = \tau_{ij}^{R}\frac{\partial u_i}{\partial x_j} - \rho\varepsilon + \mu_t P_B \quad (11)$$

$$S_\varepsilon = C_{\varepsilon 1}\frac{\varepsilon}{k}\left(f_1 \tau_{ij}^{R}\frac{\partial u_i}{\partial x_j} + \mu_t C_B P_B\right) - C_{\varepsilon 2} f_2 \frac{\rho\varepsilon^2}{k} \quad (12)$$

Here, P_B represents the turbulent generation due to buoyancy forces and can be written as: $P_B = -\frac{g_i}{\sigma_B}\frac{1}{\rho}\frac{\partial \rho}{\partial x_i}$, where g_i is the component of gravitational acceleration in the direction x_i, the constant σ_B = 0.9, and constants C_B is defined as: C_B = 1 when $P_B > 0$, and 0 otherwise;

$$f_1 = 1 + \left(\frac{0.05}{f_\mu}\right)^3, \; f_2 = 1 - \exp\left(-R_T^2\right) \quad (13)$$

The constants C_μ, $C_{\varepsilon 1}$, $C_{\varepsilon 2}$, σ_k, σ_ε are defined empirically. In Solidworks Flow Simulation, the following typical values are used [24]:

$$C_\mu = 0.09,\; C_{\varepsilon 1} = 1.44,\; C_{\varepsilon 2} = 1.92,\; \sigma_k = 1.3, \sigma_\varepsilon = 1$$

With the Lewis number $Le = 1$, the diffusive heat flux is defined as:

$$q_i = \left(\frac{\mu}{Pr} + \frac{\mu_t}{\sigma_c}\right)\frac{\partial h}{\partial x_i}; i = 1,2,3. \quad (14)$$

The constant is fixed at σ_c = 0.9, Pr is the Prandtl number, and h is the enthalpy. These equations apply to laminar and turbulent flows. In addition, it is possible to switch from one state to another and vice versa. Parameters k and μ_t are zero for purely laminar flows.

2.1.2. Contaminant Substance

The contaminant substance in the room diffuses in a gaseous (or liquid) form in the ambient air, called the carrier fluid. As the contaminant substance mass fraction y is minimal ($y << 1$), it will not impact the carrier fluid flow's properties. The contaminant substance distribution in the room is modeled using the "Tracer Study Option" of the software SolidWorks Flow Simulation. The following equation integrates the substance's non-uniform concentration and the carrier fluid's pressure gradient [11,24].

$$\begin{aligned} \frac{\partial \rho y}{\partial t} + \frac{\partial}{\partial x_i}\left[\rho y u_i - \frac{\rho RT}{pm}\left(\frac{\mu}{p_r \cdot L_e} + \frac{\mu_t}{p_{r_t} \cdot L_{e_t}}\right)\frac{\partial y}{\partial x_i}\right] \\ = \frac{\partial}{\partial x_i}\frac{m_1 m_2}{m^2}\left[-\frac{\rho y v_1 - y}{p}\left(\frac{\mu}{p_r \cdot L_e} + \frac{\mu_t}{p_{r_t} \cdot L_{e_t}}\right)\frac{\partial p}{\partial x_i}\right] \end{aligned} \quad (15)$$

Here, ρ is the density of both carrier fluid and substance's mixture (as $y << 1$, ρ can be considered similar for both), t is time, x_i is the i-th component of the coordinate system used, u_i is the i-th component of the velocity of the carrier fluid (the substance has the same velocity), p is the static pressure of the carrier fluid, R is the universal constant of gas. In addition, m is the molar mass (for both the carrier fluid and the substance's mixture), m_1 is the substance's molar mass, m_2 is the molar mass of the carrier fluid, is the specific volume of the substance. The other values are μ-the laminar viscosity, μ_t-the turbulent viscosity, Pr, Pr_t-the laminar and turbulent Prandtl numbers, Le, Le_t-the laminar and turbulent Lewis numbers, all of the carrier fluid.

2.1.3. Removal Effectiveness

(a) *Contaminant Removal Efficiency (CRE)*

CRE measures how well the ventilation system works to clear a room of contaminants. It is defined, when more than one fluid is present in the control space [26], as:

$$CRE = \frac{C_e}{< C >} \quad (16)$$

Here, C_e is the average contaminant's mass fraction flowing outside the computational domain, and $< C >$ is the average contaminant's mass fraction inside the computational domain.

A value of $CRE = 1$ means an equilibrium, an uniformly mixed system. *CRE* values greater than 1 mean that contaminant is removed from space, while a value less than 1 refers to an increasing contaminant concentration.

(b) *The Local Air Quality Index (LAQI)*

The *LAQI* indicates the ventilation system's efficiency in removing polluted air from a specific point in the computational domain. It can be defined, when there is more than one fluid in the control space, [19] as:

$$LAQI = \frac{C_e}{C} \quad (17)$$

where C_e is the average contaminant's mass fraction flowing outside the computational domain, and C is the mass fraction of the contaminant at a specific point.

A value of *LAQI* = 1 characterizes a perfectly mixed system. Otherwise, a higher *LAQI* characterizes a better capacity of the ventilation system to exhaust polluted air from that specific point.

2.2. Model Description

A sample of a hospital isolation room (Tables 1 and 2) was developed to conduct the present research. Inside the room, three models have been added. The first is the bed, then the health care worker (HCW) standing beside the bed and the patient lying on the bed. The air enters the room conditioned and refreshed at the inlet, while the contaminated air is eliminated through the outlet. Three light bulbs are mounted overhead, as shown in Figure 1. Previous studies suggested some designs that achieved optimum performance for either removing contaminants or enhancing thermal comfort. Based on that, we investigated how these designs influence the environment inside the space. As a result, the designs of (Figure 2a,b) were recommended by Çuhadaroğlu and Sungurlu [27], and the design of (Figure 2c) was selected by Thatiparti et al. [28]. Lastly, the design shown in (Figure 2d) was chosen by Cho [11] for optimal results.

Table 1. Dimensions Details of The Room's Computational Domain and The Models [29,30].

	Length (m)	Width (m)	Height (m)
Room	5.00	4.00	2.8
Diffusers	0.4	0.4	-
Bed	2.2	0.9	0.4
Mannequin	1.75	0.6	-

Table 2. Locations of Air Diffusers and Ventilation Volume for Each Computational Case.

Case No.	Air Inlet (AI)	Air Outlet (AO)	Distance (m)	Ventilation Flow Rate (m^3/s)
1, 2, 3	Sidewall	Sidewall	AI = 0.5 roof, AO = 1 floor	0.14, 0.18, 0.23 for ACH 9, 12 and 15
4, 5, 6	Sidewall	Sidewall	AI = 0.5 roof, AO = 0.5 floor	0.14, 0.18, 0.23 for ACH 9, 12 and 15
7, 8, 9	Behind the HCW	In front of HCW	AI = AO = 0.5 roof & floor	0.14, 0.18, 0.23 for ACH 9, 12 and 15
10, 11, 12	Roof	Roof	AI = 1 SW AO = 0.5 SW	0.14, 0.18, 0.23 for ACH 9, 12 and 15

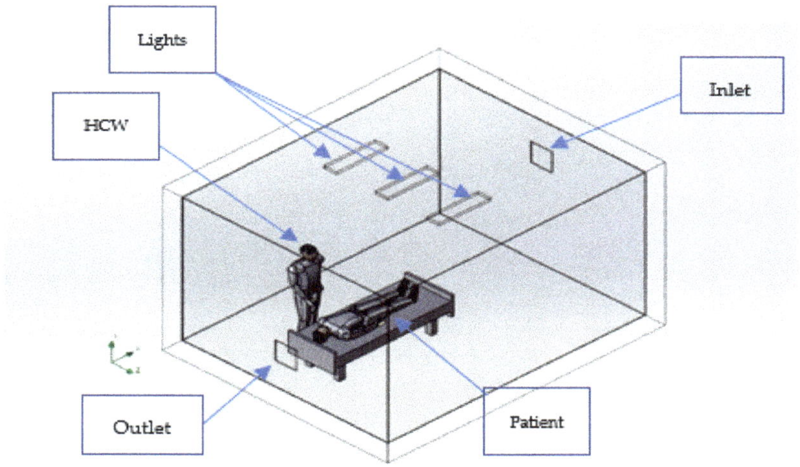

Figure 1. Isolation room and its models.

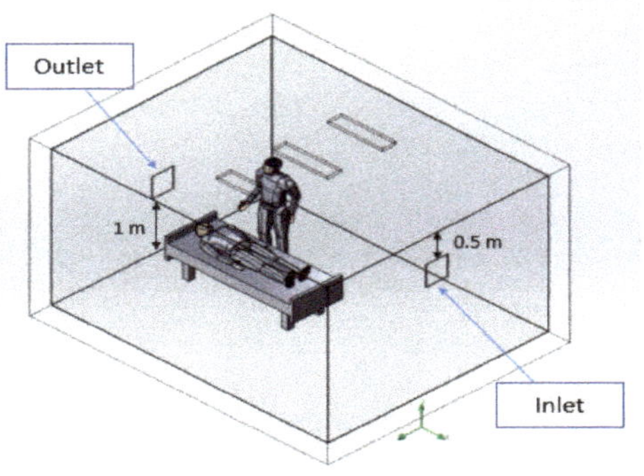

(**a**) Geometry for cases 1, 2 and 3

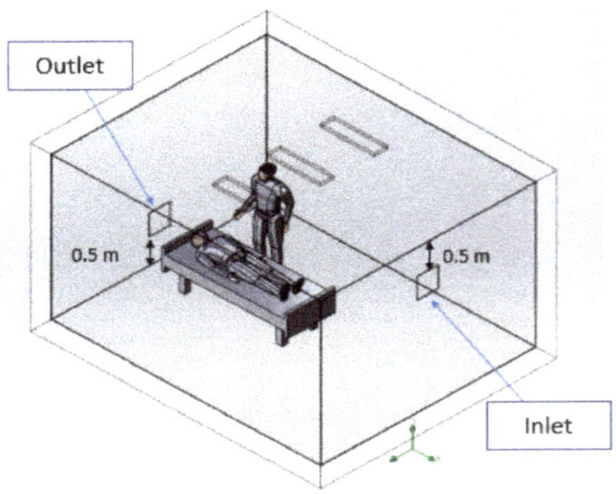

(**b**) Geometry for cases 4, 5 and 6

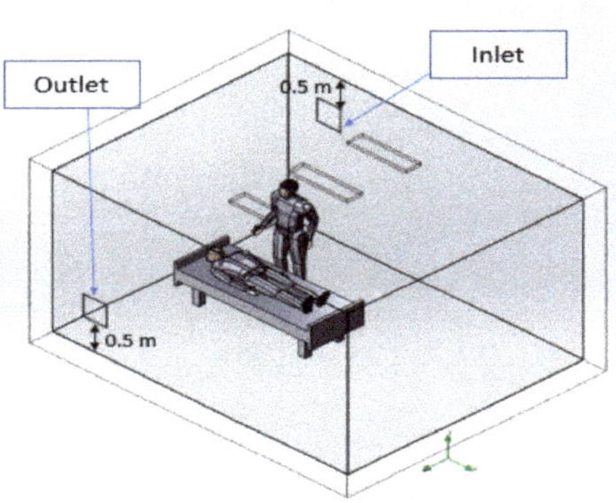

(**c**) Geometry for cases 7, 8 and 9

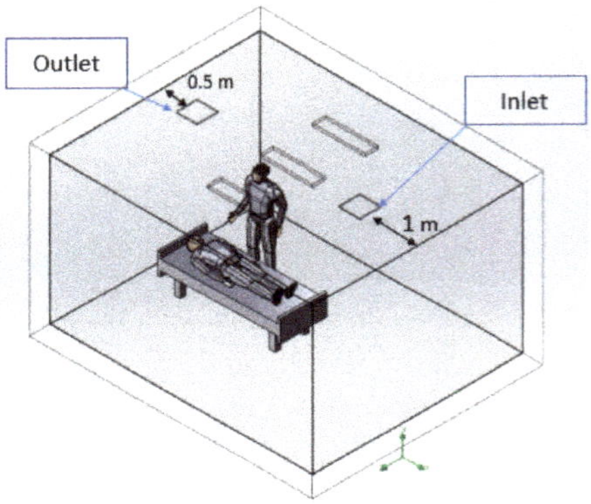

(**d**) Geometry for cases 10, 11 and 12

Figure 2. The Layout of The Isolation Room With Different Diffusers Locations.

2.3. *Boundary Conditions*

Steady-state simulations are performed to verify pollutant concentration and thermal comfort through different ventilation designs. Initially, the airflow supply was maintained at 9 ACH with no return air [19]. Afterward, the 12 and 15 ACH were applied. For the system airborne infectious isolation rooms (AIIR), according to the American standard ASHRAE 170 [31], the pressure difference is essential to be maintained, and it was set at 2.5 Pa at the outlet. Heat sources for both the health care worker and patient, as well as the ceiling light, have been chosen to be 144 W/m^2, 81 W/m^2, and 11 W/m^2, respectively [21,31]. In addition, it was assumed that the walls are adiabatic, with no heat transfer or storage within, and the air inlet temperature is 25 °C, while the initial room temperature is 16 °C, the pressure is 101,325 Pa and the relative humidity is 50%. Moreover, pollutants are represented by carbon dioxide, which is excreted through the patient's mouth with a mass flow of 0.00014 kg/s, at a temperature of 34 °C with 100% humidity [19,32]. The turbulence intensity in the inlet was set at 20% within the recommended limits (10–30%) [33]. For the low-velocity value, 2% has been chosen in the range of 1–5% [34]. The length scale variation does not influence the flow pattern at a low inlet turbulence intensity value. However, its effect starts to appear from the 40% value of turbulence intensity. The turbulence length was chosen for both the inlet and the patient's mouth based on the recommendations in reference [35].

2.4. Mesh Independency Study and Solution Convergence Criteria

The domain has been split through a computational mesh along the coordinate system's axes as rectangular parallelepipeds cells using a set of orthogonal planes. First, the original parallelepiped cells containing boundaries are split into several parts. The basic mesh is coarse. It is constructed for the whole domain at the beginning of the process and formed by dividing the domain by parallel planes into slices orthogonal to the Global Coordinate System's axes. Next, the mesh is refined locally by splitting a cell into eight through three orthogonal planes that divide the cell's edges into halves. The level of the initial mesh (L_{ini}) is specified between (3–7) for each mesh, respectively, and the minimum gap size (h_{gap}) is 0.01 m. This ensures that the flow passage through the gap is a width larger than the specified minimum gap size.

The local mesh was applied as a cube surrounding the occupants (the area of interest). The mesh is refined near the fluid/solid boundary to have the first grid point inside the viscous sublayer ($y^+ \leq 5$). We encountered convergence issues for higher values of y^+ for the first grid point, although the study is steady state. Therefore, the log-law for the mean velocity near the walls is applied when $5 < y^+ < 11.225$, and the laminar stress–strain relationship is applied for lower values.

Furthermore, the enhanced wall treatment is applied for the standard k-epsilon with the low Reynolds model. This near-wall modeling method combines the two-layer model with enhanced wall functions [2,12,36]. Different grid sizes have been tested for accuracy. Five simulations were carried out from 125,843 to 3,680,531 cells to increase the precision of the results. An internal point was chosen in the domain to determine the temperature error throughout the refined meshes to assess mesh convergence. Figure 3 illustrates LAQI error fluctuation at the outlet for different mesh sizes. The convergence is achieved with a 2 million cells mesh.

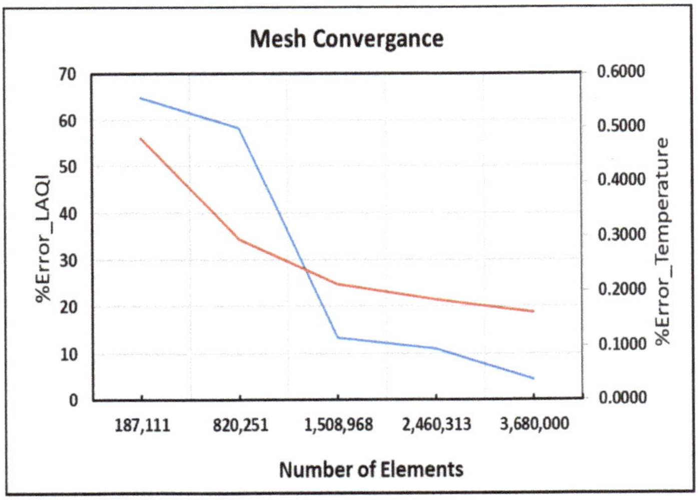

Figure 3. Mesh Independence (%Error_LAQI and Temperature) vs. Number of Elements.

Regarding the last two meshes (until 3.68 million cells), the convergence error is reduced to 4.43% for LAQI and less than 0.2% for the temperature, as shown in Figure 3. According to [2], a convergence error of less than 5% for the LAQI is acceptable. This accuracy is sufficient for doing the simulations, considering the power of the CPU, the domain magnitude, and the complicated curvy surfaces [37].

Convergence occurs when the solution no longer changes with successive iterations [24,25]. The convergence takes around 1100 iterations with a computation time of around 28 hours for each case on a single-processor desktop computer.

3. Results and Discussion

The results allow an understanding of how the mechanism of pollution transmission would be affected by adjusting the flow and the location of the air diffusers to find the optimum case through the results.

3.1. Cases 1, 2 and 3

As mentioned, each design has been tested using three cases with different airflows. In this design, the inlet is close to the ceiling at 0.5 m, and the outlet is located on the other side behind the patient at level 1 m from the floor, as shown in Figure 2a.

3.1.1. Local Air Quality Index (LAQI)

Using carbon dioxide as a proxy for the exhaled contaminant is one of the ways to study the air quality. Results were taken at the height of 1 m from the ground. Pollutant concentrations are lower when the value of LAQI is high. Figure 4a at 9 ACH indicates that the area near the exhaust is where most pollutants are removed, especially at the center of the outlet. LAQI gradually decreases away from the outlet.

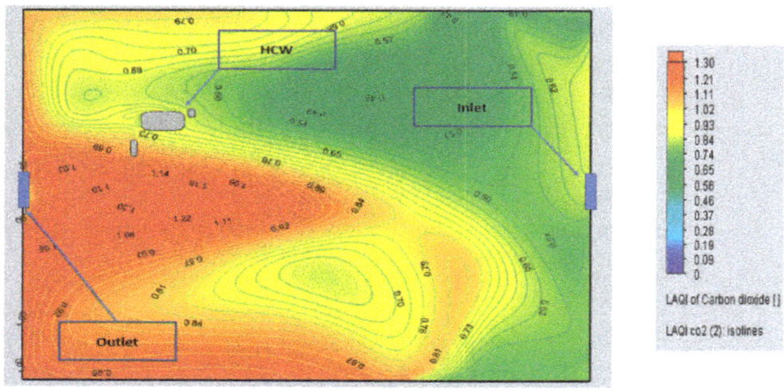

(**a**) 9 ACH

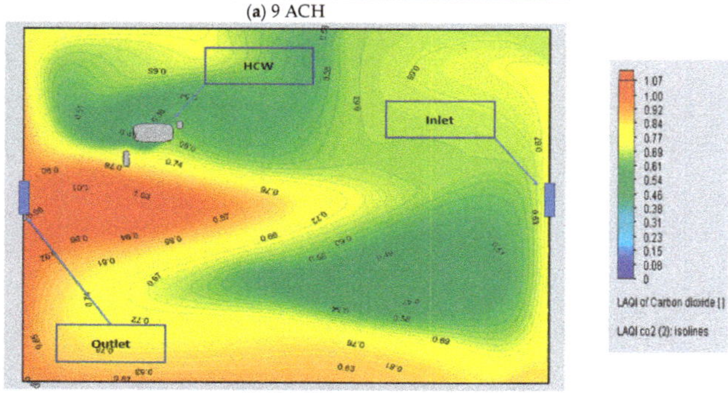

(**b**) 12 ACH

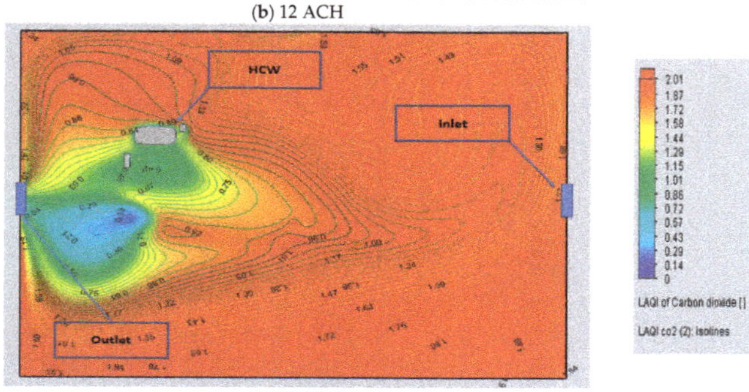

(**c**) 15 ACH

Figure 4. Top View of LAQI of CO_2 (**a**) Case 1 at 9 ACH (**b**) Case 2 at 12 ACH (**c**) Case 3 at 15 ACH.

In contrast, the HCW area behind the bed and the area near the inlet is characterized by a LAQI of approximately 0.6. It is less effective in removing contaminants as the force of the air is not strong enough to push the pollutants toward the outlet. This influence of ACH number on the airflow and the mixing process has also been illustrated by Berlanga, F. A. et al. [30].

At 12 ACH (Figure 4b), it can be observed that the removal is less effective at the lower part of the outlet level, while it is higher in the upper part, up to the inlet level. The removal effectiveness in the upper part will also be apparent in the Trace Study of CO_2 (Flow Trajectories) section of each case.

3.1.2. Tracer Study of CO_2 (Flow Trajectories) Section of Each Case

When the ventilation increases to 15 ACH (Figure 4c), the pollutants are concentrated near the outlet. LAQI value is about 0.15 near the source (patient's mouth), and 0.54 near the outlet as the particles are forced out of the outlet more quickly. However, the contaminants are extended over a larger area. This result agrees with Ameer et al. [26]. They found that a higher airflow velocity results in a larger polluted region with a lower concentration of pollutants, as the particles are forced out of the outlet more quickly. This finding corresponds with an experimental study concluding that an increased airflow decreases exposure to contaminants [30].

3.1.3. Tracer Study of CO_2 (Flow Trajectories)

As mentioned, the pollutant is represented by the carbon dioxide emitted by the patient's exhale, and its movement and concentration are explored over the room. The concentration is given as parts per million, ppm. The flow from the patient's mouth has been considered 0.00014 kg/s. At a level of 1 m, there is noticeably less concentration at the outlet, around 0.0008 ppm, and about 0.001 ppm inside the air stream at 9 ACH (Figure 5a). However, at 12 ACH (Figure 5b), it is found that the mixing is more significant, and the pollutant concentration is approximately 0.00065 ppm lower on the opposite side of HCW. However, the area of HCW still has a high pollution concentration of about 0.0008 ppm. On the contrary, at 15 ACH (Figure 5c) demonstrates that a more extensive air volume circulation improves pollutant removal. The concentration is about 0.0004 ppm. These results correspond to a similar experimental study, which shows that a higher ACH increases removal [30] while some contaminants are still contained under the inlet at 0.0007 ppm.

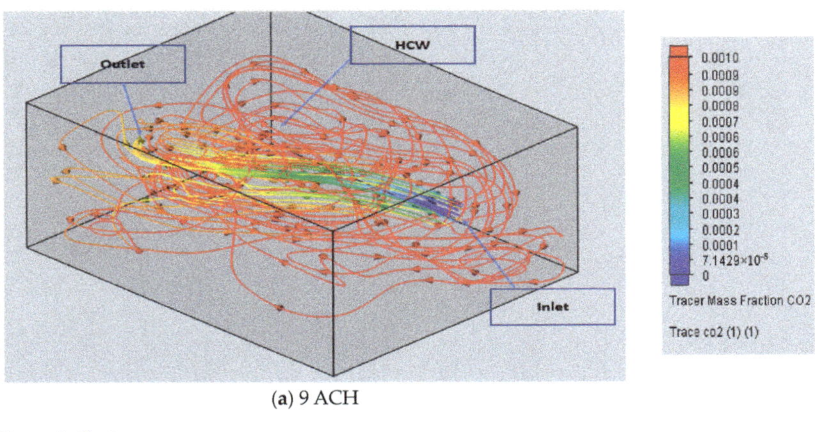

(**a**) 9 ACH

Figure 5. *Cont.*

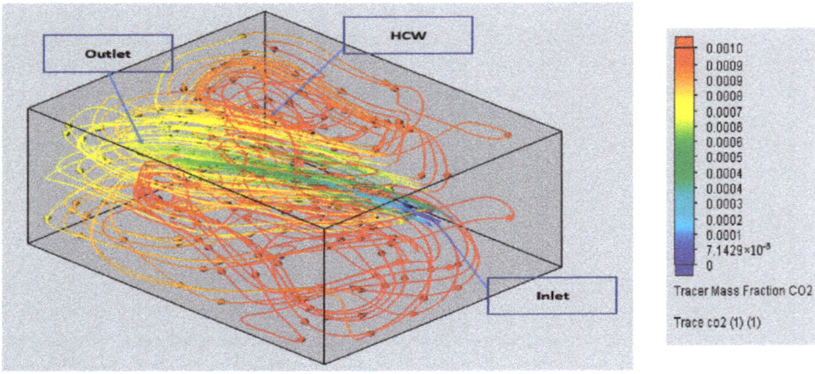

(**b**) 12 ACH

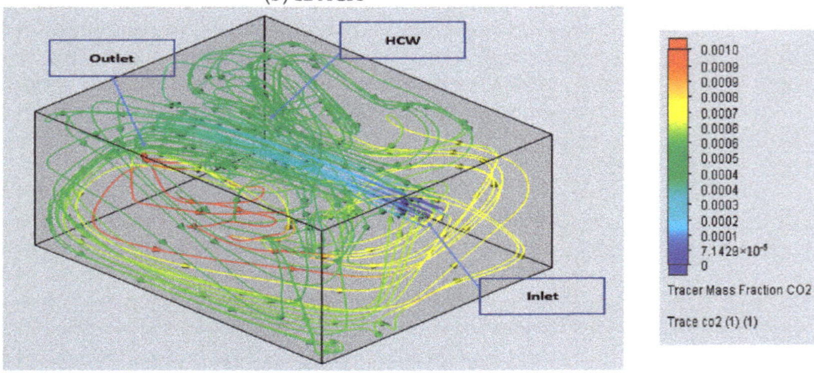

(**c**) 15 ACH

Figure 5. Flow Trajectories of Trace Study of CO_2 (**a**) Case 1 at 9 ACH (**b**) Case 2 at 12 ACH (**c**) Case 3 at 15 ACH.

3.2. *Cases 4, 5 and 6*

In the following cases, the design is similar to the previous one, with a different outlet location. The new position is 0.5 m from the floor instead of 1 m, as shown in Figure 2b.

3.2.1. Local Air Quality Index (LAQI)

The elimination of pollutants after adjusting the outlet place at the level of the head is seen as supplementary and practical. This is due to removing pollutants from the patient's mouth directly towards the outlet. This behavior corresponds to an experimental study showing that the outlet position affects the removal effectiveness [30].

The LAQI value in Figure 6a, at 9 ACH, is around 5 in the HCW area, while it is 0 in the area behind HCW, close to the corner. This area has a lot of contaminants due to the shortness of mixing.

In Figure 6b, for 12 ACH, the LAQI values are, in general, close to 10, which is excellent. However, the HCW area is still exposed to contamination, with LAQI values between 3.5–7.

For 15 ACH, in Figure 6c, the strong airflow and strategic placement of the outlet, as indicated by Ameer et al. [26], keep the whole room relatively clean. The most contaminated area is near the outlet and has a LAQI value of between 1.45 and 3.3.

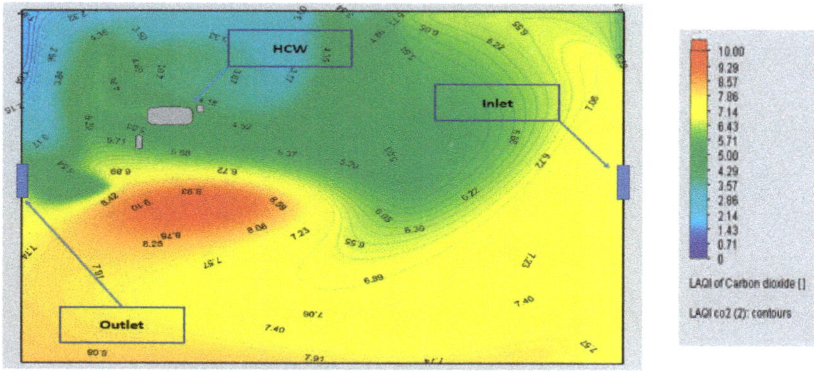

(**a**) 9 ACH

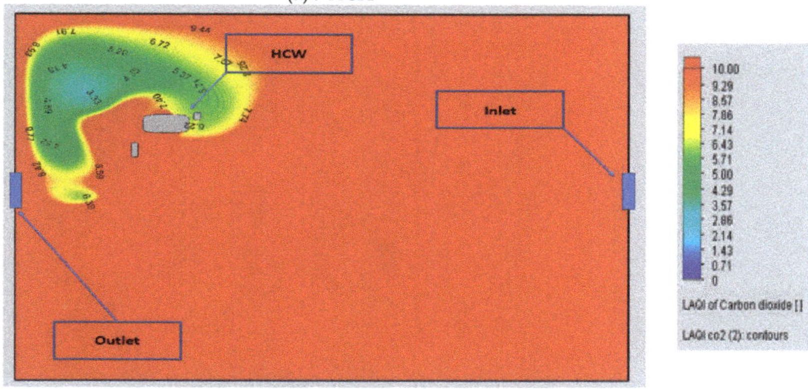

(**b**) 12 ACH

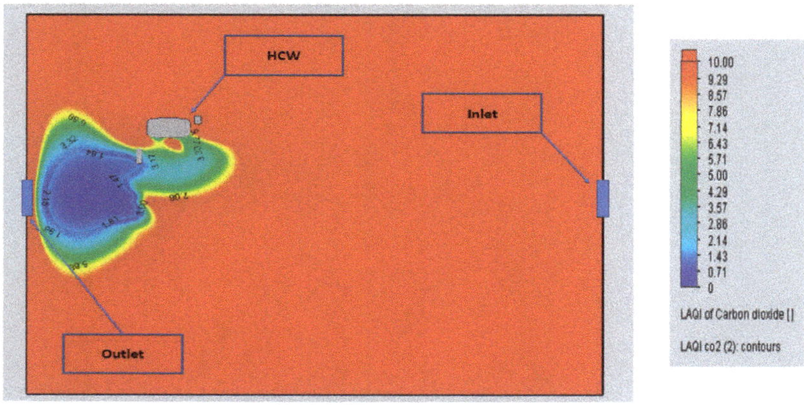

(**c**) 15 ACH

Figure 6. Top view for LAQI of CO_2 (**a**) Case 4 at 9 ACH (**b**) Case 5 at 12 ACH (**c**) Case 6 at 15 ACH.

3.2.2. Tracer Study of CO_2 (Flow Trajectories)

Pollutant removal must also consider the pollutant's particle circulation inside the room. The simulation results were shown for 9 ACH in (Figure 7a)). The presence of contaminants is at 0.0005 ppm behind the HCW, which is considered a high value. The majority of the area of the HCW side has contaminants around 0.0004 ppm. In Figure 7b, for 12 ACH, the concentration decreases in the same area to 0.0003 ppm.

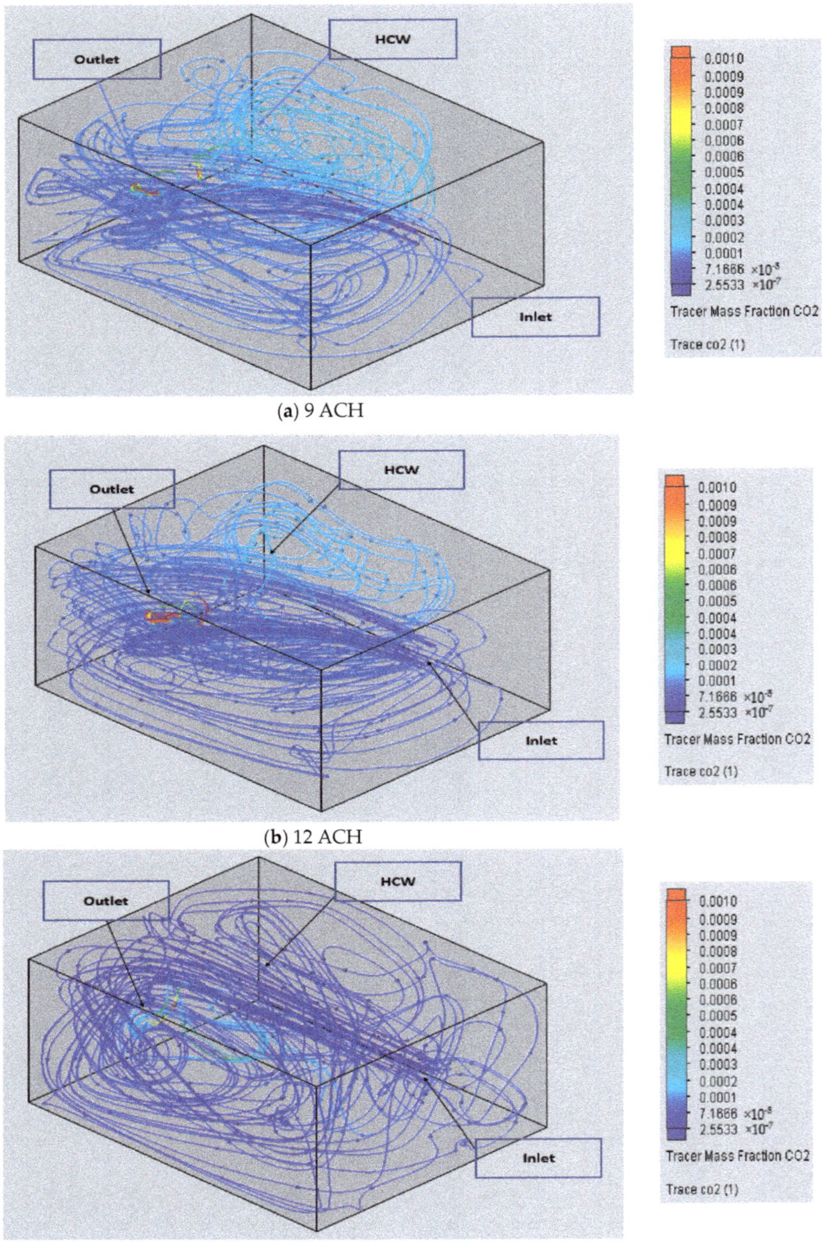

(**c**) 15 ACH

Figure 7. Flow trajectories for Trace study of CO_2 (**a**) Case 4 at 9 ACH (**b**) Case 5 at 12 ACH (**c**) Case 6 at 15 ACH.

On the other hand, near the outlet, the value exceeds 0.001 ppm due to the outlet being near the source of CO_2 (patient's mouth). Moreover, as shown in Figure 7c, for 15 ACH, almost all concentrations within the room are less than 0.0001 ppm except for the area above the bed. This configuration, associated with the ACH value and the outlet position, has been predicted and recommended by Berlanga, F. A. et al. [30].

3.3. *Cases 7, 8 and 9*

In this design, the inlet is at the top of the wall behind the HCW at a distance of 0.5 m from the roof, and the outlet is located on the bottom, close to the floor on the other wall's side at a distance of 0.5 m as shown in Figure 2c.

3.3.1. Local Air Quality Index (LAQI)

The placement of the inlet and outlet is a significant factor in determining how well pollutants are eliminated [26]. Figure 8a for 9 ACH shows that the area of minimum contamination appears on the right side of the HCW with LAQI values between 0.8 and 1. However, the region, as a whole, is still exposed to pollution due to the air pushing pollutants and not pulling them directly.

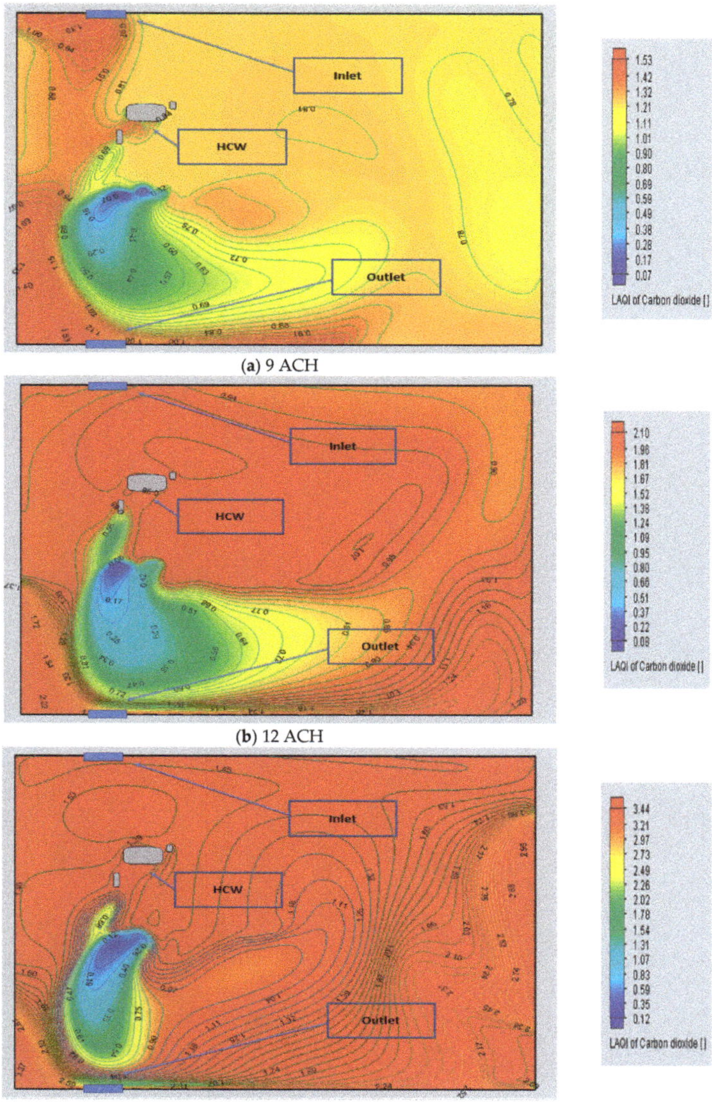

(**a**) 9 ACH

(**b**) 12 ACH

(**c**) 15 ACH

Figure 8. Top view for LAQI OF CO_2 (**a**) Case 7 at 9 ACH (**b**) Case 8 at 12 ACH (**c**) Case 9 at 15 ACH.

An improved contamination removal is observed at 12 ACH (Figure 8b), the area surrounding the HCW has a LAQI value near 0.95. However, there is still some spread toward the unoccupied region because of the air path, which makes the other side (close to the outlet) full of pollutants. It seems that the air velocity was not sufficient to remove the pollutants effectively.

The 15 ACH (Figure 8c) appears as the best flow rate for this design in eliminating pollutants based solely on the force of the air. As a result, most of the space has an optimum value of LAQI of more than 1, except the region between the pollutants source and the outlet. This result corresponds with the experimental study by Berlanga, F. A. et al. Furthermore, higher airflow enhances the elimination of the existing contaminants inside the space [30].

3.3.2. Tracer Study of CO_2 (Flow Trajectories)

Figure 9a at 9 ACH shows that the pollutant concentration is reduced in the air path between the inlet and the outlet, which helps to ensure a clean area. The concentration near the patient is 0.0005 ppm. Nevertheless, the rest of the room has a concentration of approximately 0.0018 ppm, which corresponds with the outcomes from the LAQI analysis. For 12 ACH (Figure 9b), the results show the area far from the ventilation pathway has a slightly lower concentration, about 0.0011 ppm. As for 15 ACH (Figure 9c), the efficiency improves, where the values are approximately 0.0007 ppm for the unoccupied region. Kong X. et al. observed a similar impact of various ventilation systems on pollutant emissions through an experimental study [38].

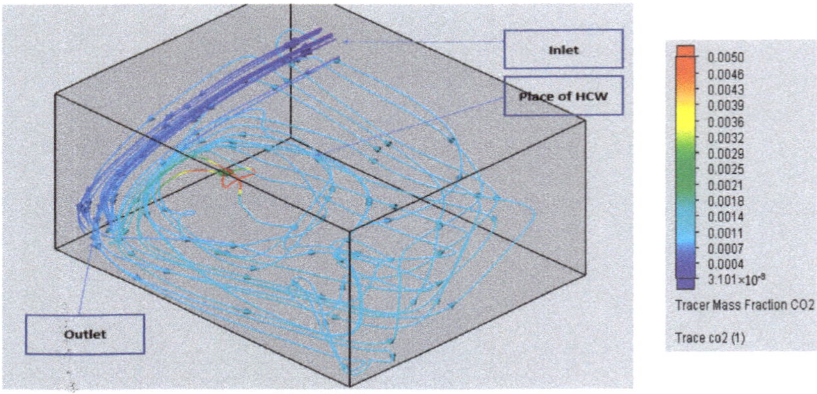

(**a**) 9 ACH

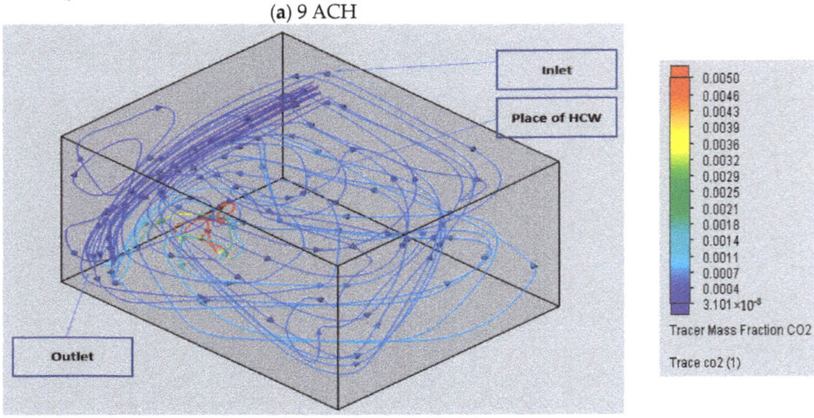

(**b**) 12 ACH

Figure 9. *Cont.*

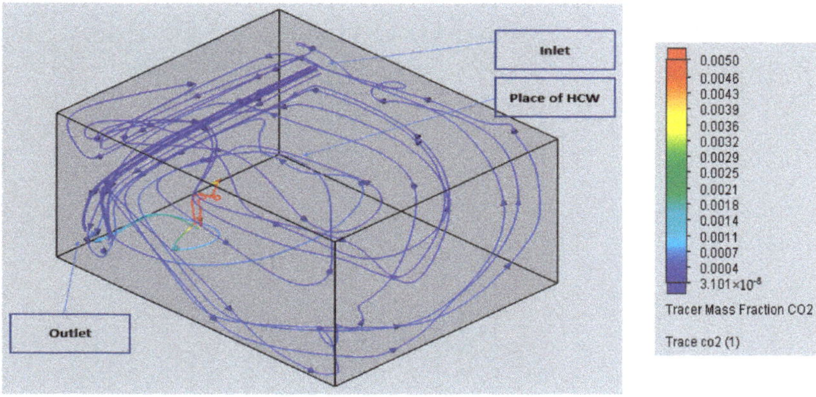

(**c**) 15 ACH

Figure 9. Flow Trajectories for Trace study of CO_2 (**a**) Case 7 at 9 ACH (**b**) Case 8 at 12 ACH (**c**) Case 9 at 15 ACH.

3.4. *Cases 10, 11 and 12*

Both inlet and outlet are roof-mounted in this design. The distance from the side wall is 1 m for the inlet and 0.5 m for the outlet, as shown in Figure 2d.

3.4.1. Local Air Quality Index (LAQI)

As illustrated in Figure 10a for 9 ACH, the outlet absorbs contaminants emitted from the patient's mouth, but some fall due to gravity and are stuck in the lower level. The value of LAQI starts around 0.1 close to the source, and it goes up gradually as the concentration of the contaminants reduces towards the outlet. At the same time, Figure 10b at 12 ACH results show that the increase in the airflow leads to more polluted regions. The mixing increases due to the higher airflow as emitting the pollutants here depends on the buoyancy. Low LAQI values are observed near the patient. At 15 ACH (Figure 10c), the contaminated area is confined near the inlet region towards the working area. Therefore, a correlation between airflow and pollution levels is noticed.

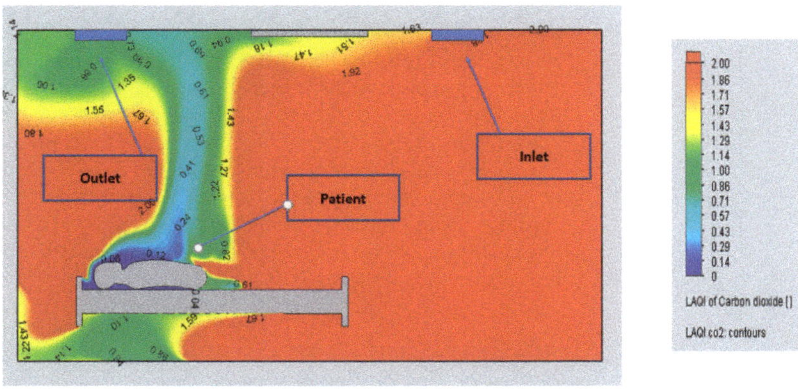

(**a**) 9 ACH

Figure 10. *Cont.*

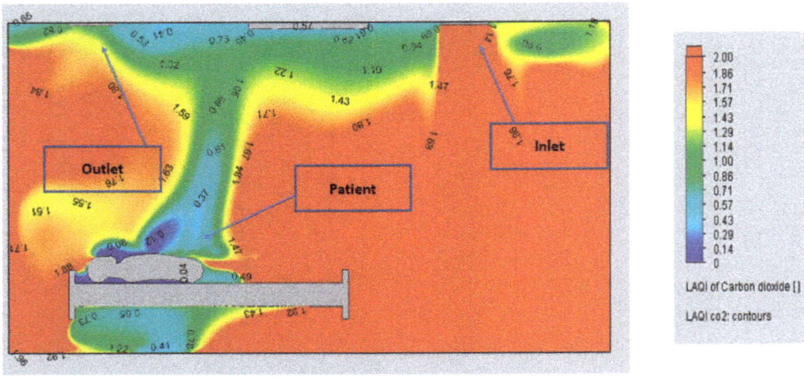

(**b**) 12 ACH

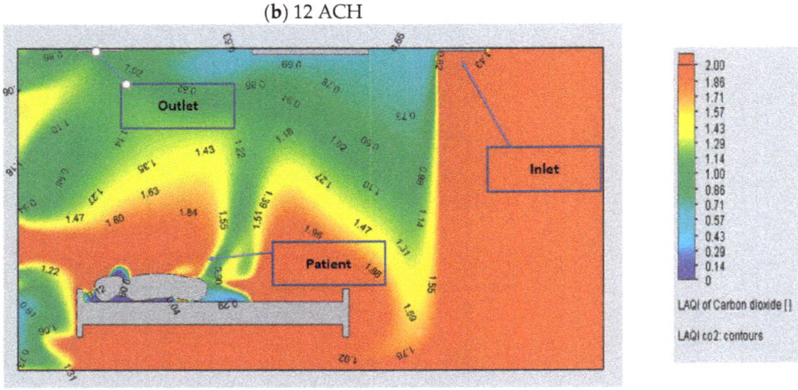

(**c**) 15 ACH

Figure 10. Front view for LAQI OF CO_2 (**a**) Case 10 at 9 ACH (**b**) Case 11 at 12 ACH (**c**) Case 12 at 15 ACH.

3.4.2. Tracer Study of CO_2 (Flow Trajectories)

At 9 ACH, it can be seen in Figure 11a that the pollutants are present at a high concentration of about 0.004 PPM at the source location (patient's mouth). It gradually decreases towards the outlet, while the unoccupied area has a minor concentration of around 0.0004 ppm. As shown in Figure 11b at 12 ACH, the mixing of pollutants increases, which helps spread their concentration over a larger volume, especially the HCW side, where the result is 0.0035 ppm. The airflow increase affects the distribution of these pollutants around the space. At 15 ACH, as seen in (Figure 11c), there is a similar effect with case 11 at 12 ACH, the region where the inlet exists has lower contaminants, with a value of 0.0004 ppm. Referring to the inlet location in this design, removing the contaminants is considered inappropriate. This confirms the results of the experimental study [30], which found that the ventilation performance of the side inlet/ outlet design was significantly higher than that of the top inlet/outlet design.

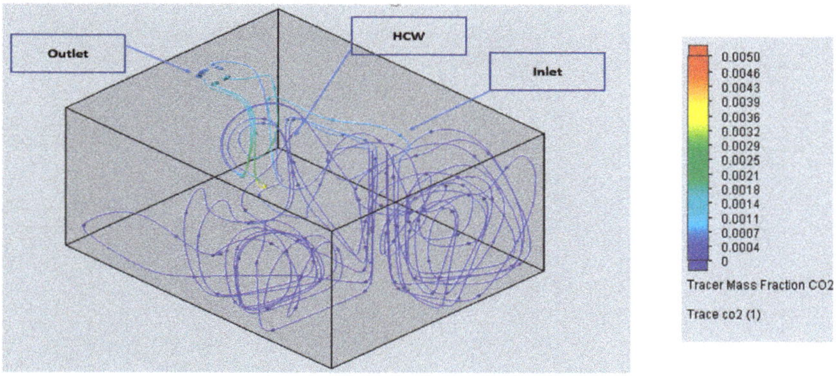

(**a**) 9 ACH

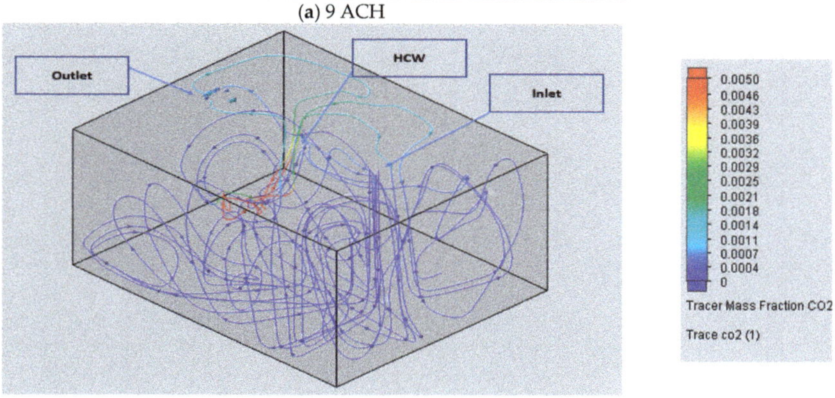

(**b**) 12 ACH

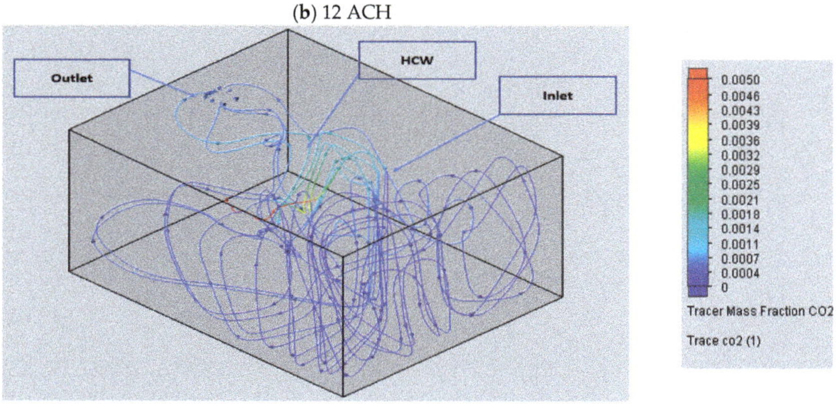

(**c**) 15 ACH

Figure 11. Flow Trajectories for Trace study of CO_2 (**a**) Case 10 at 9 ACH (**b**) Case 11 at 12 ACH (**c**) Case 12 at 15 ACH.

3.5. Contaminant Removal Effectiveness (CRE)

In the diagram in Figure 12, it can be noticed from the model (Figure 2b) that the best CRE values occur in cases 4, 5, and 6, where the outlet is situated close to the head at a low elevation. Compared to other designs, this one effectively removed contaminants from the space. At the same time, a similar design (Figure 2a) was used for cases 1, 2, and 3, where the outlet was situated above the patient's head, resulting in an undesirable outcome. Meanwhile, for the third design, case 9 at 15 ACH (Figure 2c) has the optimum

CRE with a value of more than 1.5. Finally, in the fourth design case (10, 11, and 12) illustrated in Figure 2d, all of these airflows achieved the minimum required contaminant removal performance.

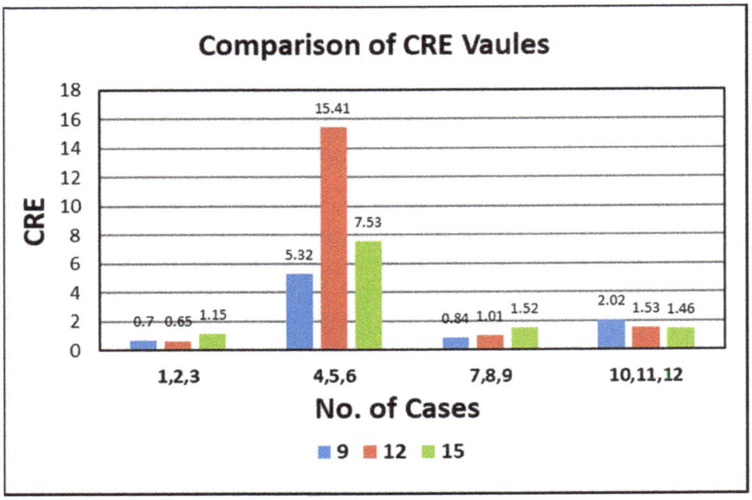

Figure 12. Comparison of room's CRE values for the different test cases.

The following charts are drawn along a 2 m horizontal line located 1 m from the ground, above and parallel to the patient. We compare LAQI through aeration value. Results in Figures 13 and 14 illustrate the excellent values for pollutant removal efficiency in cases 4 and 5. However, case 6 at 15 ACH, as shown in Figure 15, does not show adequate results at the beginning of the bed area where the contaminants' source exists. Away from the patient's mouth, LAQI values rise dramatically. The other cases, even with the proximity from the outlet, still have low and fluctuating values, which could not be recommended.

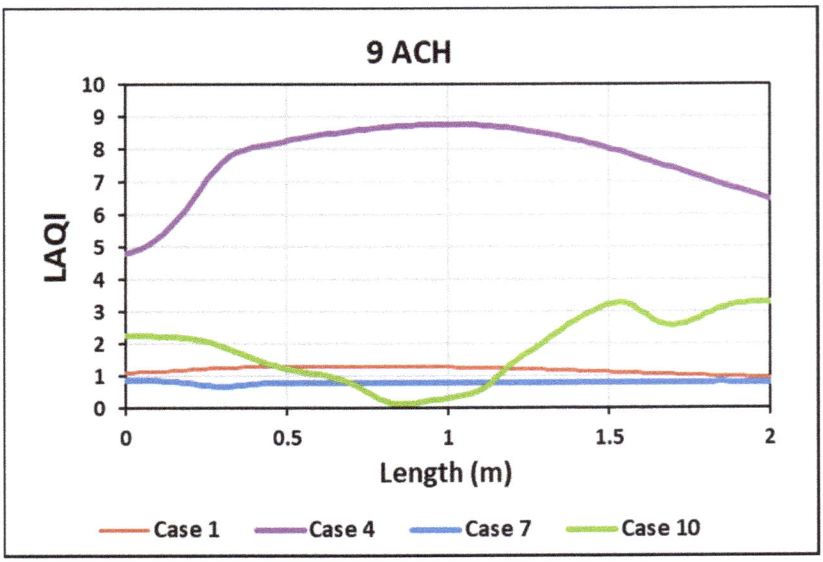

Figure 13. Comparison of LAQI at 9 ACH.

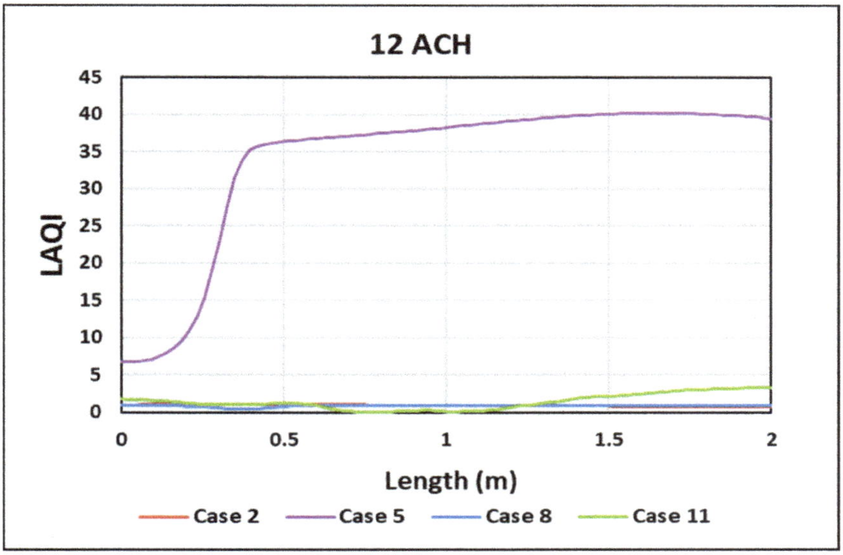

Figure 14. Comparison of LAQI at 12 ACH.

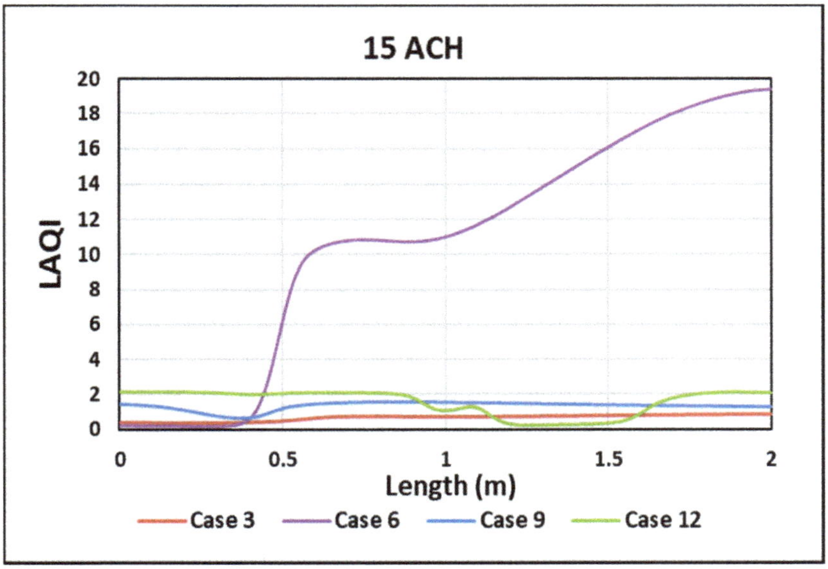

Figure 15. Comparison of LAQI at 15 ACH.

4. Conclusions

Ventilation dramatically impacts human health because of its effect on the dispersal of contaminants like SRAS-COVID-19. Therefore, it is crucial to have a ventilation system that can prevent the spread and decrease the transmission of contaminants in buildings, especially in healthcare facilities. Cases 4, 5, and 6 of the second design yielded optimum results for pollution removal. All the cases achieved a CRE value higher than 5. The contaminants were removed quickly due to the proximity of the outlet to the source of CO_2. The outlet is located at 0.5 m from the floor in these cases. These cases correspond to the conclusions of Cho [11].

On the other hand, the worst position for the outlet was in the first design, where the outlet was above the level of the patient in cases 1, 2, and 3. In this configuration, case 3 is the only one with a CRE value higher than 1. However, in cases 3 and 6, at 15 ACH, the airflow becomes horizontal due to the force of air which overcomes gravity.

The analysis shows that LAQI values are of similar magnitude, except for cases 4, 5, and 6, which show significantly better performances. These results correspond with several experimental studies, showing similar performances for different airflow intensities and inlet/outlet positions [30]. Additionally, dead zones should be avoided; the airflow cannot reach these regions to improve the removal of the contaminants [13].

Despite abundant studies in this field, considerable effort is still required to identify the best solutions. In addition, many difficulties remain during the modeling process, such as the magnitude of the model and the human body's unique design, making meshing challenging. Furthermore, the experimental work will play a vital part in confirming the outcomes alongside these simulations.

Author Contributions: Conceptualization, M.A.; methodology, M.A.; software, M.A.; validation, M.A.; formal analysis, M.A.; investigation, M.A.; resources, M.A.; data curation, M.A.; writing—original draft preparation, M.A., A.I.; writing—review and editing, M.A., A.I. and M.Y.H.; visualization, M.A., A.I.; supervision, A.I. and M.Y.H.; project administration, A.I. and M.Y.H.; funding acquisition, A.I. All authors have read and agreed to the published version of the manuscript.

Funding: The authors acknowledge the financial contribution from the Natural Sciences and Engineering Research Council (NSERC) Canada through a Discovery Grant.

Data Availability Statement: Not applicable.

Conflicts of Interest: The authors declare no conflict of interest.

References

1. Borro, L.; Mazzei, L.; Raponi, M.; Piscitelli, P.; Miani, A.; Secinaro, A. The role of air conditioning in the diffusion of Sars-CoV-2 in indoor environments: A first computational fluid dynamic model, based on investigations performed at the Vatican State Children's hospital. *Environ. Res.* **2021**, *193*, 110343. [CrossRef] [PubMed]
2. Hallé. Prediction of Bioparticles Dispersion and Distribution in a Hospital Isolation Room. Ph.D. Thesis, École De Technologie Supérieure, Montreal, QC, Canada, 2016.
3. Anuraghava, C.; Abhiram, K.; Reddy, V.N.; Rajan, H. CFD modelling of airborne virus diffusion characteristics in a negative pressure room with mixed mode ventilation. *Int. J. Simul. Multidiscip. Des. Optim.* **2021**, *12*, 1. [CrossRef]
4. Novoselac, A.; Srebric, J. Comparison of Air Exchange Efficiency and Contaminant Removal Effectiveness as IAQ Indices. *Trans.-Am. Soc. Heat. Refrig. Air Cond. Eng.* **2003**, *109*, 339–349. Available online: https://www.caee.utexas.edu/prof/novoselac/Publications/Novoselac_ASHRAE_Transactions_2003.pdf (accessed on 11 July 2003).
5. Tian, X.; Zhang, S.; Awbi, H.B.; Liao, C.; Cheng, Y.; Lin, Z. Multi-indicator evaluation on ventilation effectiveness of three ventilation methods: An experimental study. *Build Environ.* **2020**, *1*, 180. [CrossRef]
6. Daisey, J.M.; Angell, W.J.; Apte, M.G. Indoor Air Quality, Ventilation And Health Symptoms In Schools: An Analysis of Existing Information. *Indoor Air Berkeley USA* **2003**, *13*, LBNL-48287. [CrossRef]
7. American Society of Heating R and ACEngineers. HVAC Design Manual for Hospitals and Clinics. 2012. 301p. Available online: https://www.ashrae.org/technical-resources/bookstore/hvac-design-manual-for-hospitals-and-clinics (accessed on 11 July 2003).
8. Bolashikov, Z.D.; Melikov, A.K.; Kierat, W.; Popioek, Z.; Brand, M. Exposure of health care workers and occupants to coughed airborne pathogens in a double-bed hospital patient room with overhead mixing ventilation. *HVAC R Res.* **2012**, *18*, 602–615.
9. Chow, T.T.; Kwan, A.; Lin, Z.; Bai, W. Conversion of operating theatre from positive to negative pressure environment. *J. Hosp. Infect.* **2006**, *64*, 371–378. [CrossRef]
10. Alhamid, M.I.; Budihardjo Raymond, A. Design of the ventilation system and the simulation of air flow in the negative isolation room using FloVent 8.2. *AIP Conf. Proc.* **2018**, *1984*, 020016.
11. Cho, J. Investigation on the contaminant distribution with improved ventilation system in hospital isolation rooms: Effect of supply and exhaust air diffuser configurations. *Appl. Therm. Eng.* **2019**, *148*, 208–218. [CrossRef]
12. Mousavi. Airborne Infection in Healthcare Environments: Implications to Hospital Corridor Design [Internet]. 2015. Available online: http://digitalcommons.unl.edu/constructiondisshttp://digitalcommons.unl.edu/constructiondiss/21 (accessed on 17 July 2015).
13. Abhinav, R.; Sunder, P.B.S.; Gowrishankar, A.; Vignesh, S.; Vivek, M.; Kishore, V.R. Numerical study on effect of vent locations on natural convection in an enclosure with an internal heat source. *Int. Commun. Heat Mass Transf.* **2013**, *49*, 69–77. [CrossRef]

14. Abdel, K.; Saadeddin, R. The Effects of Diffuser Exit Velocity and Distance Between Supply The Effects of Diffuser Exit Velocity and Distance Between Supply and Return Apertures on the Efficiency of an Air Distribution and Return Apertures on the Efficiency of an Air Distribution System in an Office Space System in an Office Space [Internet]. 2016. Available online: https://openprairie.sdstate.edu/etd (accessed on 5 February 2016).
15. Kokkonen, A.; Hyttinen, M.; Holopainen, R.; Salmi, K.; Pasanen, P. Performance testing of engineering controls of airborne infection isolation rooms by tracer gas techniques—Enhanced Reader. *Indoor Built Environ.* **2014**, *23*, 994–1001. [CrossRef]
16. Villafruela, J.M.; Olmedo, I.; Berlanga, F.A.; Ruiz de Adana, M. Assessment of displacement ventilation systems in airborne infection risk in hospital rooms. *PLoS ONE* **2019**, *1*, 14. [CrossRef] [PubMed]
17. Yam, R.; Yuen, P.L.; Yung, R.; Choy, T. Rethinking hospital general ward ventilation design using computational fluid dynamics. *J. Hosp. Infect.* **2011**, *77*, 31–36. [CrossRef] [PubMed]
18. Mousavi, E.S.; Grosskopf, K.R. Ventilation Rates and Airflow Pathways in Patient Rooms: A Case Study of Bioaerosol Containment and Removal. *Ann. Occup. Hyg.* **2014**, *59*, 1190–1199. [CrossRef]
19. Lu, Y.; Oladokun, M.; Lin, Z. Reducing the exposure risk in hospital wards by applying stratum ventilation system. *Build Environ.* **2020**, *183*, 107204. [CrossRef]
20. Fan, M.; Fu, Z.; Wang, J.; Wang, Z.; Suo, H.; Kong, X.; Li, H. A review of different ventilation modes on thermal comfort, air quality and virus spread control. *Build. Environ.* **2022**, *212*, 108831. [CrossRef] [PubMed]
21. Ahmed, T. Performance Investigation of Building Ventilation System by Calculating Comfort Criteria through HVAC Simulation. *IOSR J. Mech. Civ. Eng.* **2012**, *3*, 7–12. [CrossRef]
22. Risberg, D. Daniel Risberg Analysis of the Thermal Indoor Climate with Computational Fluid Dynamics for Buildings in Sub-arctic Regions. Ph.D. Thesis, Luleå University of Technology, Luleå, Sweden, 2018.
23. Ghanta, N. Meta-Modeling and Optimization of Computational Fluid Dynamics (CFD) Analysis in Thermal Comfort for Energy-Efficient Chilled Beams-Based Heating, Ventilation and Air-Conditioning (HVAC) Systems. Ph.D. Thesis, Massachusetts Institute of Technology, Cambridge, MA, USA, May 2020.
24. Solidworks Flow Simulation. Technical Reference Solidworks Flow Simulation. 2021. Available online: https://www.solidworks.com/product/solidworks-flow-simulation (accessed on 5 February 2021).
25. Mekbib Kifle, P. CFD Simulation of Heavily Insulated Office Cubicle Heated by Ventilation Air Done in Collaboration with [Internet]. 2018. Available online: https://www.oslomet.no (accessed on 23 May 2018).
26. Azmi, M.A.; Hassan, N.N.; Salleh, Z.M. Investigation of Airflow in A Restaurant to Prevent COVID-19 Transmission Using CFD Software. *Prog. Eng. Appl. Technol.* **2021**, *3*, 977–991.
27. Çuhadaroğlu, B.; Sungurlu, C. ID 8-A CFD Analysis of Air Distributing Performance of a New Type HVAC Diffuser. 2015. Available online: https://www.researchgate.net/publication/288324226_ID_8_-A_CFD_analysis_of_air_distributing_performance_of_a_New_Type_HVAC_Diffuser. (accessed on 27 December 2015).
28. Thatiparti, D.S.; Ghia, U.; Mead, K.R. Computational fluid dynamics study on the influence of an alternate ventilation configuration on the possible flow path of infectious cough aerosols in a mock airborne infection isolation room. *Sci. Technol. Built. Environ.* **2017**, *23*, 355–366. [CrossRef]
29. Solidworks Flow Simulation 201. 2021. Available online: https://www.solidworks.com/media/solidworks-2021-flow-simulation (accessed on 5 February 2021).
30. Berlanga, F.A.; Olmedo, I.; de Adana, M.R.; Villafruela, J.M.; José, J.F.S.; Castro, F. Experimental assessment of different mixing air ventilation systems on ventilation performance and exposure to exhaled contaminants in hospital rooms. *Energy Build.* **2018**, *177*, 207–219. [CrossRef]
31. Ventilation of Health Care Facilities [Internet]. 2020. Available online: https://www.ashrae.org (accessed on 5 February 2020).
32. Yoon, S.H.; Ahn, H.S.; Choi, Y.H. Numerical Study To Evaluate The Characteristics Of Hvac-Related Parameters To Reduce CO_2 Concentrations In Cars. *Int. J. Automot. Technol.* **2016**, *17*, 959–966. [CrossRef]
33. Cehlin, M.; Moshfegh, B. Numerical modeling of a complex diffuser in a room with displacement ventilation. *Build. Environ.* **2010**, *45*, 2240–2252. [CrossRef]
34. Cao, G.; Ruponen, M.; Paavilainen, R.; Kurnitski, J. Modelling and simulation of the near-wall velocity of a turbulent ceiling attached plane jet after its impingement with the corner. *Build. Environ.* **2011**, *46*, 489–500. [CrossRef]
35. Gao, N.; Niu, J. Transient CFD simulation of the respiration process and inter-person exposure assessment. *Build. Environ.* **2006**, *41*, 1214–1222. [CrossRef]
36. Vasilopoulos, K.; Sarris, I.E.; Tsoutsanis, P. Assessment of air flow distribution and hazardous release dispersion around a single obstacle using Reynolds-averaged Navier-Stokes equations. *Heliyon* **2019**, *5*, e01482. [CrossRef]
37. Georges, L. CFD Simulation of Active Displacement Ventilation Tollef Hjermann. 2017. Available online: http://hdl.handle.net/11250/2454912org (accessed on 15 June 2017).
38. Kong, X.; Guo, C.; Lin, Z.; Duan, S.; He, J.; Ren, Y.; Ren, J. Experimental study on the control effect of different ventilation systems on fine particles in a simulated hospital ward. *Sustain. Cities Soc.* **2021**, *73*, 103102. [CrossRef]

buildings

MDPI

Article

Assessing Efficiency and Environmental Performance of a Nearly Zero-Energy University Building's Energy System in Norway

Fredrik Ege Abrahamsen [1,*,†], Sturla Grina Ruud [2,†] and Alemayehu Gebremedhin [2]

1 Department of Electronic Systems, Faculty of Information Technology and Electrical Engineering, NTNU—Norwegian University of Science and Technology, 2815 Gjøvik, Norway
2 Department of Manufacturing and Civil Engineering, Faculty of Engineering, NTNU—Norwegian University of Science and Technology, 2815 Gjøvik, Norway
* Correspondence: fredrik.e.abrahamsen@ntnu.no
† These authors contributed equally to this work.

Abstract: Increasing awareness of climate issues in recent decades has led to new policies on buildings' energy consumption and energy performance. The European Union (EU) directive 2010/31/EC, i.e., the energy performance of buildings directive (EPBD), is one of the measures initiated to achieve climate and energy goals by reducing energy use and greenhouse gas emissions in the building sector. The EPBD required all new buildings to be nearly zero-energy buildings (nZEBs) by 2021. Nearly zero-energy buildings (nZEBs) are buildings with a very-high-energy performance and nearly zero or low-energy requirements covered to a very significant extent by energy from renewable sources produced on-site or nearby. The utilisation of solar photovoltaic (PV) panels is a common approach for achieving the nZEB standard. The carbon footprint of PV panels is often not discussed as a parameter. This paper aimed to analyse the environmental performance of an existing nearly zero-energy university building in a Norwegian use case scenario. This analysis is performed by assessing annual electricity and heat consumption from both energetic and environmental perspectives. The energy required for the building during the studied period is then used to analyse the environmental and energy performance of the building. When it comes to the environmental assessment, the commercial software SimaPro was used. The proposed revision EPBD and nZEB definition from 2021 suggests that nZEBs should also consider operational greenhouse gas emissions and life-cycle global warming potential from 2027. The life cycle assessment (LCA) of the building's energy sources looks at the global warming potential (GWP) and greenhouse gas (GHG) emissions, and how they compare to Norwegian grid electricity. The results of the analysis highlights potential challenges to justifying the use of alternative energy sources to fulfil the criteria of nZEBs. When installing solar PV, it is important to consider the energy mix of the country where the solar PVs are produced. To solely consider the energy performance of the building, the installation of solar PV panels in countries with a high share of renewable energy may result in a reduced impact in terms of emission reduction from a life cycle perspective.

Keywords: building performance; energy accounting; zero-energy buildings; nearly zero-energy buildings; renewable energy; photovoltaic electricity generation; district heating; life cycle assessment

Citation: Abrahamsen, F.E.; Ruud, S.G.; Gebremedhin, A. Assessing Efficiency and Environmental Performance of a Nearly Zero-Energy University Building's Energy System in Norway. *Buildings* **2023**, *13*, 169. https://doi.org/10.3390/buildings13010169

Academic Editors: Shi-Jie Cao and Wei Feng

Received: 26 October 2022
Revised: 14 December 2022
Accepted: 26 December 2022
Published: 6 January 2023

1. Introduction

The release of the 2021 report from the Intergovernmental Panel on Climate Change (IPCC) [1] states that human-induced activity is responsible for the warming of the atmosphere, ocean and land. Globally, the building sector accounts for 37% of the total final energy consumption, and 55% of the electricity consumption [2,3]. At the United Nations (UN) climate change conference (COP21) in Paris in 2015, 196 countries committed to complying with the Paris agreement. The agreement establishes a framework for keeping the global average temperature well below 2.0 °C, preferably to 1.5 °C compared to pre-industrial levels to mitigate the impacts of climate change and achieve a climate neutral world by 2050 [4]. In 2015, the Norwegian government set out to reduce emissions by 40% in 2030 compared to 1990 emission levels. Today, the goal is expanded to at least a 50% reduction by the year 2030, and a 90–95% emission reduction by the year 2050 [5–7]. The building and household sectors are the largest contributors to the total energy use in the European Union (EU) at approximately 40%, and accounts for 36% of the associated GHG emissions [2,8–10]. The EU has set forth a ambitious program to achieving climate neutrality by 2050 [11]. Part of this is the Energy Performance of Buildings Directive (EPBD) aiming for all new buildings to meet nearly zero-energy building (nZEB) standard [12]. In December 2021, a recast of EPBD was proposed, wherein one of the changes was to move from zero-energy buildings to zero-emission buildings. Improving energy efficiency in buildings is one of the paths to achieving climate and energy goals [9,12].

The environmental assessment of local and on-site renewable energy conversion systems and the GWP and GHG emission reduction potential of nZEBs with various different energy systems were previously assessed in the literature [13–21]. The authors of [16] investigated the contribution of nZEB standards to China's CO_2 emission reduction targets by 2060. The authors suggested that nZEBs should be promoted for both new and retrofit buildings in order to achieve emission targets. The authors of [17] performed the LCA of an nZEB with integrated solar PV systems with and without energy storage for different climates. The authors highlighted the importance of achieving a balance between energy savings and renewable energy integration. The author of [18] investigated the potential environmental impact of residential rooftop solar PV systems at 76 different locations in Europe. The author investigated how the location and local electricity mix affects the environmental impact of the analysed PV system. The author of [19] considered a pre-existing detached single-family house which was converted into net zero-energy building, and performed an LCA of the solar PV system used for the conversion. The authors concluded that the impact of the analysed PV system is considerably less than that of the substituted conventional energy systems.

The EPBD set goals on the nZEB standard for new buildings in the EU. The requirements applies from 31 December 2018 for public buildings, and all other buildings from 31 December 2020 [12,21]. When performing the major renovation of older buildings, the same requirements for energy efficiency apply. According to EPBD, all new buildings should be nZEBs [22]. Member states are required to define numerical threshold values to achieve a "very high energy performance" and define local nZEB requirements based on the local situation [23]. Additionally, the EU directive 2018/844, amending 2010/31/EC states that, due to the high energy consumption in the building sector, each member state should seek to renovate buildings, to promote energy efficiency as well as considering the integration of renewable energy sources to reach the goals put forth by the Paris agreement. Furthermore, the implementation of nZEBs is listed as important for the transition to a more sustainable energy supply system [24]. Renewable energy will play a key role in low-emission buildings and in the transition to a sustainable energy system.

The purpose of this paper is to conduct an analysis of the emissions associated with the energy systems of an existing nearly zero-energy university building in a Norwegian use case scenario. The aim of this was to identify environmental impacts and potential challenges to utilising alternative energy resources as a means to achieve nZEB status in a public building. This analysis was performed by assessing annual electricity demand and heat consumption from both energetic and environmental perspectives. In order to evaluate the GWP and GHG emissions of the studied system, the software SimaPro was used. Lastly, the finding will be used to try to highlight the potential challenges associated with justifying the nZEB status by the integration of alternative energy sources, which has become a trend. In Norway, it has become common to install solar PV on public buildings in order to present the buildings as nZEBs. The building operators feel the pressure to present their buildings as green. The studied building in this case was built as a passive house, and the installation of solar PV was performed with the motive to achieve nZEB status. This has been the main cause which triggered the motivation for this study. To the best of our knowledge, there is no study that highlights these challenges related to Norwegian conditions. It is by no means not the intention of the authors to oppose the installation of solar PV, but rather to highlight the issues of uncritically justifying the installation of solar PV without considering the environmental aspects and emissions from a life cycle perspective. We believe that solar energy holds great potential in combatting the energy crisis and global warming.

The remainder of this paper is organised as follows: Section 1 provides an introduction by presenting relevant background information underlining the motivation and main objective of this study. Section 2 describes the different definitions of zero-emission buildings and nearly zero-emission buildings relevant to this study as well as renewable energy in nearly zero-energy buildings and the Norwegian energy mix. Section 3 gives information on the building of which the study is focused and describes the energy system in the building. Section 4 describes the methodology for conducting the study, and the collection and analysis of building and energy data. The life cycle assessment of solar PV, the assessment of district heating (DH) and imported electricity. Section 5 gives an overview of the results of the energy analysis, and the environmental performance of solar PV, DH, and imported electricity. Section 6 discusses the results, and Section 7 provides a conclusion.

2. Definitions of ZEB and nZEBs

The European Union encourages energy efficiency and renewable energy production to achieve carbon neutrality by 2050. Improving the energy efficiency and utilising renewable energy in buildings is one of the paths to achieving climate and energy goals [25]. In Europe, the EPBD 2010/31/EU states that and after 31 December 2018, all new buildings owned by public authorities are nZEBs, and by 31 December 2020 all new buildings are nZEBs [12]. A uniform strategy for implementing nZEBs is not established in EPBD, and the member states must develop their own nZEB definition based on variations in primary energy sources utilised for electricity generation, national, regional or local conditions [12,25]. In a report from the European Commission in 2020, it was stated that 23 of the member countries defined the requirements for nZEBs, while the remaining four were in the process of developing or revising the definitions [26,27]. Most definitions include the requirements for primary energy use, or the share of energy covered by renewable sources. Some definitions also include requirements for the thermal transmittance (u-value) for walls, roofs, floors, windows and doors [26]. Due to variations in local climate throughout the European member states, there is a need for individual definitions of energy efficiency. The indicators for determining the local nZEB definitions are not comparable due to variations in energy performance calculations and variations in the inclusion of different non-mandatory energy sources, e.g., lighting and appliances [28].

In December 2021, the European Commission proposed a recast of EPBD with new standards for energy performance to decarbonise the building sector by reducing GHG emissions and energy consumption, and new nZEB definitions. The new nZEB definition refers to the term "nearly zero-emission building" [29]. Zero-emission buildings will replace nearly zero-energy buildings as the standard for new buildings from 1 January 2030, and as of 1 January 2027, for all renovated buildings [30]. According to the redefined directive proposal, a zero-emission building is defined as a building with very-high-energy performance, with a very low amount of energy still required which is fully covered by energy from renewable sources and without on-site carbon emissions from fossil fuels [29]. Although Norway is not an EU member state, as part of the European Economic Area (EEA) agreement, Norway is obliged to comply with EU directives and participate in the adoption of the directive on energy performance for buildings (EPBDs) [23]. The Norwegian building code "Regulations on technical requirements for construction works" (TEK17) [31] was last updated in 2017. The inclusion and definition of nZEBs in Norway are expected in the next version of the building code.

2.1. Zero-Energy Buildings (ZEBs)

Zero-energy buildings (ZEBs) are grid-connected energy-efficient buildings that balance their total annual energy consumption by utilising the on-site generation of electricity [32]. The definition of on-site or locally produced renewable energy includes, among others, wind, solar, hydropower, biomass, etc. [26]. A summary of relevant ZEB terminology is given in Table 1.

2.2. Net Zero-Energy Buildings (Net ZEBs)

The term "net" refers to the annual balance of primary energy based on the supplied and exported thermal and electric energy. Net zero-energy buildings are defined as grid-connected buildings with a very high energy performance, meaning that the primary energy usage is less than or equal to zero. In other words, this means that a net ZEB produces the same amount of energy from renewable sources as the energy required for its operation [32,33].

2.3. Nearly Zero-Energy Buildings (nZEBs)

The EPBD defines a "nearly zero-energy building" as a building that has very high energy performance. It also states that the nearly zero or very low amount of energy required should be covered to a very significant extent by energy from renewable sources, including energy from renewable sources produced on-site or nearby [12]. A definition of "very high energy performance" and "a very significant extent by energy from renewable sources" as well as "nearby" must be defined by each member state. The energy performance is the required energy needed to meet the energy demand associated with the typical use of the building, including energy for heating, cooling, ventilation, hot water and lighting [28]. It is up to every member state to define its requirements for nZEBs based on national, regional, or local conditions, including a numerical indicator of primary energy use expressed in kWh/m^2 per year [34].

2.4. Zero-Emission Buildings (ZEBs) (Proposed)

As part of the Fit for 55 packages, the European Commission proposed a recast of the EPBD in 2021 moving from assessing the energy performance in nearly zero-energy buildings (nZEBs) to also including the emissions. The proposed recast introduced the new term "zero-emission building". By aligning the energy performance requirement for new buildings to the longer-term climate neutrality goal and "energy efficiency first principle", the main goal of the revised EPBD is the reduction in operational greenhouse gas emissions and final energy consumption. Zero-emission buildings include the calculated life-cycle GWP and its disclosure through the energy performance certificate of the building. The new definition and requirements apply from 1 January 2030 for all new buildings, and as of 1 January 2027 for new public buildings [10,30].

2.5. Norwegian ZEB and nZEB Definitions

Norway has defined specific criteria for passive house and low-energy buildings; however, a definition of the nZEB requirements are yet to be determined. The current building code (TEK17) states that the net annual energy demand should not exceed the requirements for passive houses. The minimum requirements for the total net energy demand in the current building code are 125 kWh/m^2 usable heated floor space area per year for Norwegian university buildings [31].

A passive house in a Norwegian context is often described as a building with high quality, a good indoor climate, and low energy demand. It is claimed that passive houses consume up to 80% less energy than traditional buildings in Europe [35]. The basic components of any passive house are excellent thermal insulation, the avoidance of thermal bridges, and low heat loss from windows. Airtight building envelope, ventilation system with heat recovery and the passive use of solar energy [36]. There are also specific requirements for different types of buildings related to the heat loss from transmission and infiltration, heating and cooling demand-based, and u-value requirements for building components [31,37,38]. In comparison, ZEBs and nZEBs will require even less energy than low-energy buildings and passive houses. Due to the lack of formal nZEB requirements in the current Norwegian building code, unofficial definitions have been conceived by building developers, among others. A proposal for Norwegian nZEB definition and target values for energy use was presented in [39], which was carried out by Rambøll and LINK on behalf of the Norwegian Building Authority. The proposed target for nZEB states that a Norwegian nZEB should have a reduction of 70% of the energy use compared to the previous Norwegian building code TEK10 at 160 kWh/m^2. The Research Centre on Zero Emission Buildings (ZEB Lab) in Norway defines ZEBs as zero-emission buildings [40,41]. In this definition, the energy balance is measured in terms of GHG emissions in CO_2-eq during the buildings' life cycle rather than direct energy demand. In the following report [42], the authors based their definition of nZEBs on the work of [39], and used the current building code as a reference for determining the energy requirement for different categories of buildings. For university buildings, the energy requirement is determined to be 40 kWh/m^2 per year.

Some examples of Norwegian ZEB pilot projects in Norway include Powerhouse Kjørbo in Sandvika, the ZEB lab in Trondheim, Ydalir Living Lab in Elverum, and Zero Village in Bergen [43–47].

Table 1. Summary of ZEB terminology.

Case	Description	Reference
ZEB in EU	Used to describe a building which uses little to no energy	[26]
ZEB in Norway	No formal definition. Unofficial definitions have proposed using emissions as unit.	[32]
Net ZEB	A building that generates the same amount of energy from on-site renewable energy sources in order to achieve net zero-energy balance.	[32,33]
nZEB in EU	A building that uses little energy, and where the energy that is used are being produced on-site or locally from renewable sources. Member countries specify its own requirements based on national, regional or local conditions.	[12,28,34]
nZEB in Norway	No formal definition. Different definitions and targets have been proposed using current legislation as a reference. Highly energy-efficient building, using on-site or local energy generation.	[39–42]
New proposed ZEB in EU	Zero-emission building. A building with very high energy performance. The very low amount of energy required shall be fully covered by local or on-site renewable energy sources. Consider emissions and global warming potential (GWP)	[10,30]

2.6. Renewable Energy Sources in nZEBs

As stated in Table 1, a part of the European Commission's recommendations for achieving nZEB criteria is implementing the on-site production of renewable energy. This was also promoted in proposals for a Norwegian nZEB definition by the Norwegian Building Authority [48]. Solar PV panels are among the options for on-site energy generation and can be integrated into pre-existing buildings. Solar thermal systems can be utilised for water and space heating. Depending on the PV systems and energy demand, the building may export electricity during periods where the PV production exceeds the electricity demand and import electricity when the demand is higher than production. In Norway, electricity demand is typically at its highest during winter when the solar conditions are most challenging. However, this paper only considers the environmental assessment of the energy system of the building and not grid-technical aspects such as these. The authors of [49] reviewed a representative sample size of nZEBs across Europe, where 60% of the analysed nZEBs used a single system for heating (ventilation, hot water, or space heating). The authors found that, for cold climate conditions, heat pumps (31%), boilers (21%), and DH (25%) were the most common means of meeting the heating demand. Furthermore, the authors found that, for the implementation of renewable energy sources, 15% had photovoltaic (PV) panels, 18% had solar thermal, 18% had PV and solar thermal, and 41% had no PV or solar thermal systems for cold-climate nZEBs. A Norwegian study on the impact of solar PV in residential ZEB pilots found that PV contributes to a relatively large share of material emissions. Furthermore, when the emissions from the PV system were compared with the emissions from the Norwegian hydropower, it was found that the PV system does not compensate for these emissions within 30 years [50]. This result is supported by the findings of [51], wherein the author found that PV technology installed in Norway had the least potential of the 31 studied locations. These findings are also supported by an International Energy Agency (IEA) report [52], which stated that Norway has the lowest potential for CO_2 mitigation among OECD countries.

2.7. Norwegian Energy Mix

Norway has a high share of renewable energy incorporated in its energy system. Historically, electricity production has mainly been based on hydropower. In recent years, the trend has shifted, and an increase in wind power has been developing [53]. In the year 2020, a total of 154,197 GWh of electric energy was produced. In which hydro-power production made up 91.8%, wind power made up 6.4% and thermal-power production made up 1.7% [54]. The total hydropower production was 143,699 GWh in 2021, the total production from wind power was 11,768 GWh, and the total electricity production from thermal power plants was 1646 GWh in 2021 [54]. In Europe, France and Germany are the main exporters of energy and are central participants within the European power market. Within Europe, most energy is transported over the shortest distances. Norway has predominantly imported energy from Sweden, Germany, and other smaller exporters [55]. As such, power exchange with Europe is an integral part of the Norwegian central grid. Cables are between Norway and Germany, Denmark, England, the Netherlands, Finland, and Russia. In 2020, the net exchange was 20,472 GWh, the total exported energy was 24,968 GWh, and the total imported energy was 4496 GWh. For the year 2021, the exchange was 17,584 GWh, the amount imported 8235 MWh, and the exported amount was 25,819 MWh [54]. Norway has a cold climate and the annual energy use in households and commercial sectors are approximately 80 TWh, around half of which is used for space heating [56]. In Norway, the electricity production mix is based mainly on renewable sources. The authors of [46] explained how Ecoinvent processes may be representative of Norwegian material production lines, and overestimate the total associated embodied GHG emissions due to a European or global electricity mix based on fossil fuels [46,57].

3. Description of the Smaragd Building and Its Energy System

The Smaragd-building opened on campus Gjøvik in 2017. The building is 4980 m^2 divided over five floors and includes offices, laboratories, auditoriums, and a coffee shop. The total heated area is of 4775 m^2 [58]. The building is considered to be an nZEB building according to the building operator's own definition. Even though there is no formal definition from the Norwegian government to date, this label is justified by using a modified version of the original nZEB proposal, described in Section 2.5. In the case of Smaragd, the criteria to establish nZEB conditions were set to an energy demand which was 25% lower than the passive house standard. The building was therefore planned as a passive house, but through the installation of solar PV the building was considered to be nZEBs. An overview of the established criteria from the builder can be seen in Table 2. The negative sign of PV indicates that it is to be subtracted from the energy budget. To achieve the passive house criteria, a specific energy target was established, which is equal to 78.1 kWh/m^2. This is the reason that solar PV was added to the building. The production was estimated to be approximately 70 MWh per year.

Table 2. Energy budget for Smaragd.

Classification	Specific Value [kWh/m^2]
Imported Electricity	44.4
Heating	38.7
PV	−11
SUM	58.2

3.1. Energy System in the Smaragd-Building

The following energy sources are used in the building: electricity from the grid, on-site production of electricity from the solar PV system, bio-based DH, and air-to-water cooling system. The cooling system is not considered in this work, due to data limitations.

3.1.1. District Heating

In the building, DH is used for space heating, tap water heating, and snow melting at the entrances [59]. The heat production plant is located approximately 1.5 km from the university campus. The fuel distribution of heat conversion is illustrated in Figure 1 [60].

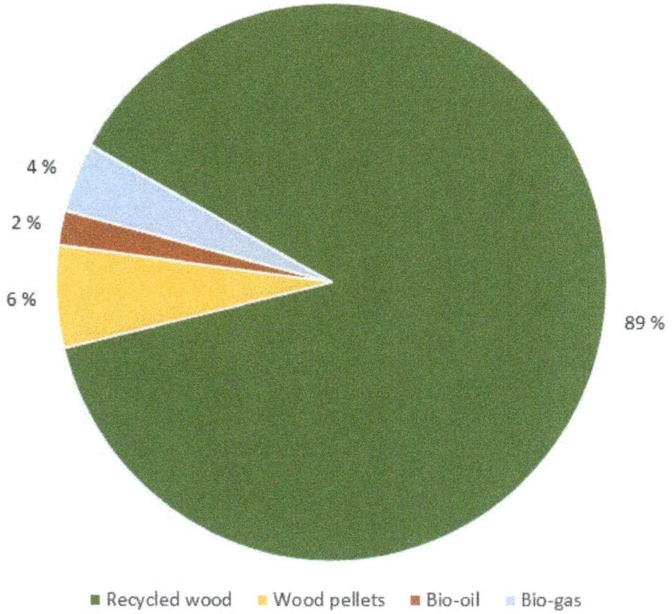

Figure 1. District heating fuel mass distribution.

With respect to the annual heating demand from the DH to the building (0.523% of total energy production), the calculated annual fuel distribution is as follows: 103.368 tonnes of recycled wood, 5.413 tonnes of wood pellets, 0.527 tonnes of bio-oil, and 2.244 tonnes of biogas. The values are derived with system efficiency and fuel heating values with a mean moisture content. The recycled wood used for the production of DH comes from local recycling stations. Based on data obtained from Gjøvik DH central, the plant has an estimated total efficiency of 88%. The reported annual distribution net losses were given as 13%, and the recycled wood with an average moisture content of approximately 35%.

3.1.2. Solar PV

A PV system is installed to reduce the grid energy demand as a measure to achieve the nZEB energy requirement [61]. The panels used on the system are of type SunPower X20-327-COM and can be seen in Figure 2. Each panel consists of 96 mono-crystalline cells. The PV system consists of 280 modules divided into six arrays, mounted on a flat roof. A total of 258 modules are located on the fifth-floor roof, and 22 modules are on the lower level. The inclination is 10°, and the azimuth angles are 50° and −130°. The total peak power of the system is estimated to be 91.56 kWp. The orientation of each array can be seen in Table 3. The area of each panel is 1.63 m^2, and it has a mass of 18.6 kg. In total, the total PV is area 456.6 m^2 and the total is mass 5200 kg.

Table 3. Solar PV array description.

Array	No. of PV Modules	In Series	In Parallel	Tilt/Azimuth
Sub-array No 1	48	12 modules	4 strings	10°/50°
Sub-array No 2	44	11 modules	4 strings	10°/50°
Sub-array No 3	48	12 modules	4 strings	10°/50°
Sub-array No 4	48	12 modules	4 strings	10°/−130°
Sub-array No 5	44	11 modules	4 strings	10°/−130°
Sub-array No 6	48	12 modules	4 strings	10°/−130°

(a) (b)

Figure 2. The Smaragd building PV system. (**a**) Fifth-floor PV system. (**b**) Second-floor PV system.

4. Methodology and Data

A combination of a cradle-to-gate LCA assessment on solar PV used for on-site generated power production and data analysis for electricity and heating demand was conducted. This study was conducted in two parts. The first part is energy mapping, and the second part consists of a cradle-to-gate LCA of the PV system. A shorter assessment of the use phase of the DH plant was also conducted. Due to limitations in the available data, the fuel distribution of the DH was based on approximations. In order to establish the energy flow used in the environmental assessment, hourly data were organised to represent the imported electricity and heating demand. In the following paragraphs, the methodology applied during this study was presented.

4.1. Collecting Building Data

Energy use in the building was registered and logged since its opening in 2017. Measurements are registered every hour and include the DH inlet and outlet temperature at the substation, solar PV electricity generation, electricity consumption as well as outdoor temperature. As the first task, relevant building features were compiled, such as the heating system, energy generation system and gross floor area.

4.2. Collection and Analysis of Energy-Use and Generation Data

Electricity and DH consumption were compiled with an hourly resolution for the year 2020. The measurement data are a combination of the instantaneous values and cumulative values that depend on the type of measurement. The energy data used in the mapping were collected as raw data from the sensors of the building's internal energy monitoring system. Data for on-site electricity generation were collected as hourly raw data and were confirmed by the building manager, who is responsible for building maintenance. The amount of electricity fed to the grid was requested but was unavailable. Therefore, this was calculated by determining the surplus with respect to the production and demand for each hour of the year. This made it possible to calculate the amount of electricity from the grid for 2020. The building has its own separate DH substation where the heat from the network is further distributed to the building.

4.3. Environmental Assessment of Energy System

To conduct the environmental assessment, the Simapro version 9.1.1 and Ecoinvent 3 database were chosen for the inventory and the results database of the impact assessment. SimaPro is a tool that is used to perform LCA studies on different systems which was developed by Pré Sustainability [62]. Simapro offers a selection of different modelling approaches for different parameters and is a part of the results database. In this paper, the ReCiPe-a hierarchical (H) mid- and endpoint model was used, in addition to the IPCC 2013 GWP over a period of 100 years for the modelling of global warming potential. More information about these modelling approaches can be found in the literature. The cradle-to-gate analysis used ISO 14040 and ISO 14044 as the bases for the methodology [63,64] together with the methodology presented by the authors of [65] for solar PV. Further PV-technical parameters were based on [14,66,67]. The analysis of the DH demand and imported electricity was calculated based on the fuel consumption and local data. The entire energy system was viewed over a period of 30 years, and the PV system was considered from cradle to gate, whereas the DH and imported electricity were analysed from the use phase only. The results of the PV system LCA are related to the specific electricity mix of the country. The authors of [18] assessed the overall impact of the cradle-to-grave LCA of a PV system in 76 European locations over 100 years. The results indicate that, in countries with a high penetration of renewable energy sources in the energy mix, the net environmental impact of PV systems applied to buildings might be negative if the production is not substantial. In countries where electricity production is based on fossil fuels, PV systems can be considered environmentally friendly [17,18].

4.4. System Boundaries

Figures 3 and 4 give an illustration of the system boundaries of the PV and the DH system. The figures are colour-coded to show that the life cycle inventory is organised. Included in the system boundaries of the PV system are the extraction and processing of raw materials, the manufacturing of panels, transportation, and installation. The outputs from the system are emissions to air and water and energy. The DH system boundaries include the production of wood pellets and biogas, transportation of fuels to DH plant, incineration of fuels and transportation of waste. The associated outputs are emissions to air and water and heat. DH from biomass is commonly viewed as climate neutral due to the carbon storage in organic material. Therefore, the emission of CO_2 was considered to be net zero in this paper. The PV cells are set to manufacturing in Malaysia and module assembly in Mexico, and are stored in Germany and Norway before being shipped to the installation location. DH includes the delivery of fuel and transportation of fly ash.

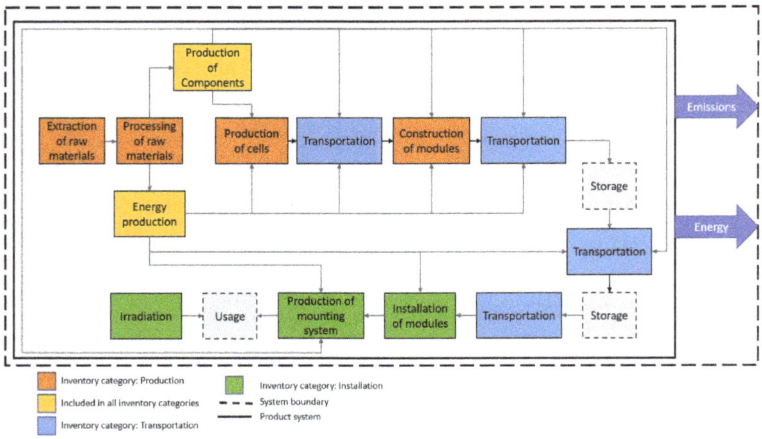

Figure 3. Photovoltaic system boundary.

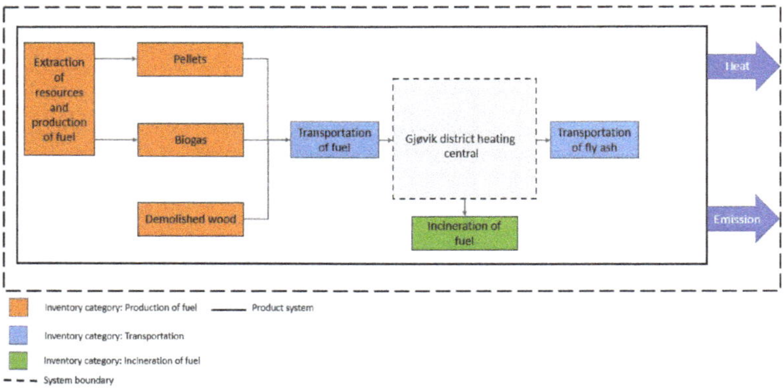

Figure 4. District heating system boundary.

5. Results

This section of the paper presents the results of the energy mapping of the buildings' energy system.

5.1. Electricity Consumption and Generation

Figure 5 illustrates a monthly overview of the following: the total electricity consumption, energy production from solar PV, the total electricity from the grid, and the total electricity to the grid. It can be seen that the total consumption in 2020 was 311.2 MWh, the production from PV was 69.5 MWh, import of 271.8 MWh, and exported electricity was 30.2 MWh. The total consumption and import of electric energy was highest during the months of January and December. Solar PV production and export of electric energy peaked in June.

5.2. District Heating

The heating demand shown in Table 4 shows the DH consumption in kWh and kWh/m^2 for each month of the year. Figure 6 illustrates the heating demand in relation to the average temperature for each month registered by a temperature sensor on the rooftop of the studied building. It can be seen that the peak demand is in the winter months, more specifically in February and December. The total DH demand for the year 2020 was 259.073 MWh.

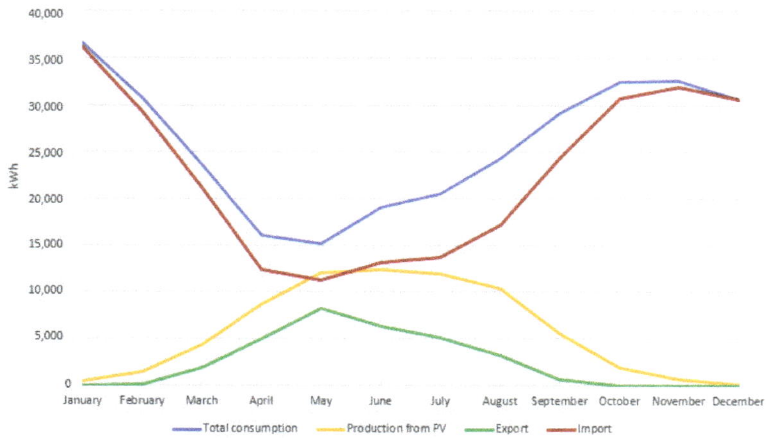

Figure 5. Electricity consumption, production, import and export in 2020.

Table 4. District heating demand. Total and per square meter per month.

Month	Heating Demand (kWh)	Heating Demand (kWh/m^2)
January	40,278	8.44
February	41,008	8.59
March	30,142	6.31
April	14,240	2.98
May	10,241	2.14
June	4411	0.92
July	6799	1.42
August	5444	1.14
September	10,681	2.24
October	22,049	4.62
November	30,026	6.29
December	43,754	9.16
Total	259,073	-
Average	21,589 kWh/month	4.52 kWh/m^2/month

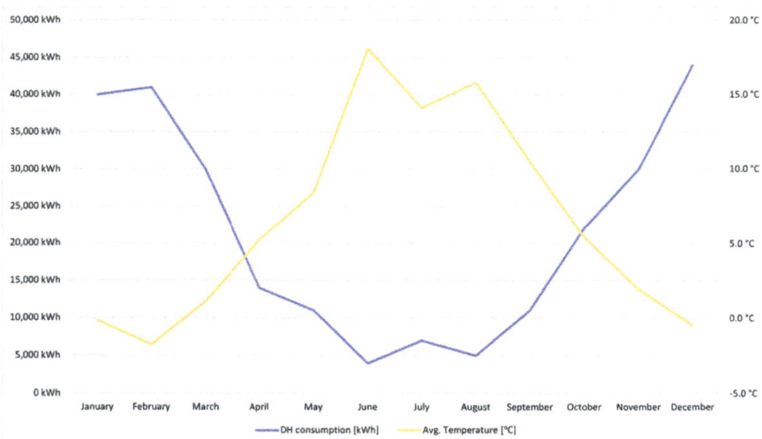

Figure 6. District heating delivered with the correlating mean temperature.

5.3. Environmental Performance

This section of the paper will present the environmental assessment results. The results will be presented as GWP, damage assessment, and at the end, a comparison between the energy systems.

5.3.1. Global Warming Potential

Figure 7 portrays a percentage-wise distribution of the global warming potential from each of the energy sources—solar PV, DH, and imported electric energy—in the following climate impact categories: fossil, biogenic, CO_2 uptake, and land use and transformation. A positive value in percent represents the release of GHG gases, whilst a negative value represents the uptake of CO_2. The values are the output results of the IPCC 2013 GWP modelling.

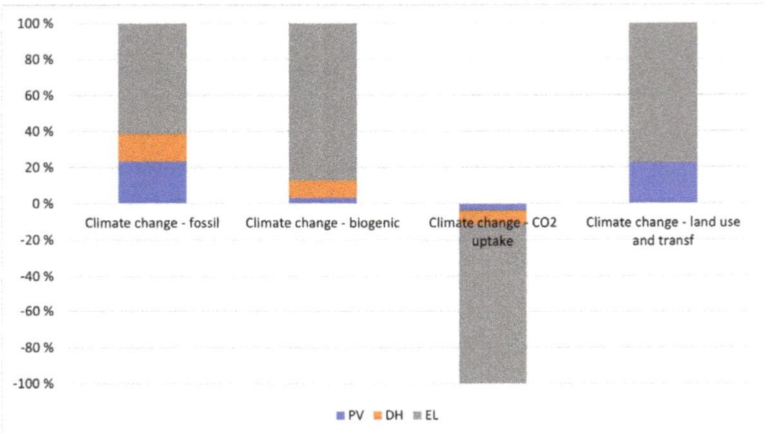

Figure 7. Global warming potential from the Smaragd buildings energy system.

It can be seen from Figure 7 that electricity from the grid will contribute the most to emissions, except biogenic emissions during the energy systems considering a time horizon. The biogenic category mostly consists of DH. The total emissions for the entire energy system are 386.867 tonnes of CO_2-eq. From the total emissions, the PV systems contribute to 22.831% of the emissions, with the majority from the inventory category "Manufacturing of photovoltaic panel". The DH system stands for 15.320% of the total emissions where the majority comes from "incineration of fuel" . This has, however, not been included in the study. In the final part of the energy system, the importation of electricity is responsible for 61.849% of the total emissions.

5.3.2. Damage Assessment

An overview of different emissions indicating the percentage-wise distribution from each component of the energy system can be seen in Figure 8. The result values are the output of the ReCiPe(H) modelling.

Figure 8 illustrates the percentage-wise distribution of different emissions to land and water. Imported electrical energy is responsible for most emissions, as represented in grey. The DH system is percentage-wise the second largest contributor of emissions. However, the percentage-wise distribution does not indicate the damage done to the environment. An overview of the damage assessment from each system can be seen in Table 5. The results are the output of the ReCiPe(H) endpoint modelling.

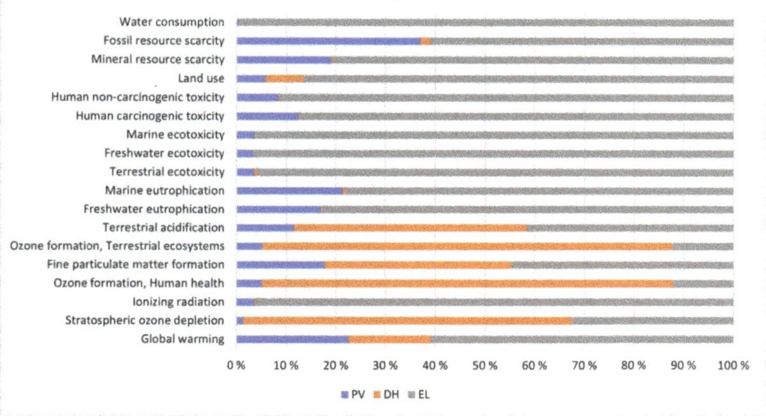

Figure 8. Midpoint overview from the Smaragd buildings energy system.

Table 5. Damage assessment.

Damage Category	Unit	PV	DH	Imported Electricity	Sum
Human health	DALY	0.2764082	0.34410322	1.4552339	2.07574532
Ecosystems	species.yr	0.000424017	0.000980697	0.001943737	0.00334845
Resources	USD2013	5238.0298	532.32473	9571.3607	15,341.7152

Table 5 consists of the following damage categories, namely human health, ecosystems, and resources. Disability-Adjusted Life Years (DALY) is a measurement of overall disease burden, represented by the damage done to human health measured in lives lost due to premature mortality and years lost due to disability. Ecosystems are a representative number of how many species go extinct (species.year). Resources are a value representing the increased cost of extracting resources in USD (USD2013). It can be seen in the table that imported electricity will in sum contribute more to environmental damage than both the PV system and the DH system. From Table 5, it can be seen that over the modelled period, the simulated energy system will cause the loss of 2.0757 years of human life, the extinction of 33.4845×10^{-4} species, and an increased cost for extraction of materials to USD 15,341.71.

5.3.3. Comparison of Energy Supply Systems

Until this point, the results are presented in net values. In this section, values per produced kWh are presented. Figure 9 shows the different emission levels for each of the different energy systems. The results are based on the average production over 30 years, and the associated emissions. It can be seen that the PV system has the highest relative level of emission of 0.045 kg CO_2-eq/kWh. Grid electricity comes in at a specific emission factor of 0.0237 kg CO_2-eq/kWh. DH has the lowest specific emission levels of 0.0076 kg CO_2-eq/kWh.

Figure 10 illustrates the CO_2-eq mitigation potential if the electric demand covered by solar PV were replaced with grid-imported electricity. The figure further shows that the mitigation potential would be equivalent to 41 tonnes of CO_2-eq over the panel's lifetime. The results are calculated with emission factors from Figure 9.

Figure 9. Comparison of emission factors between the energy sources of the Smaragd building.

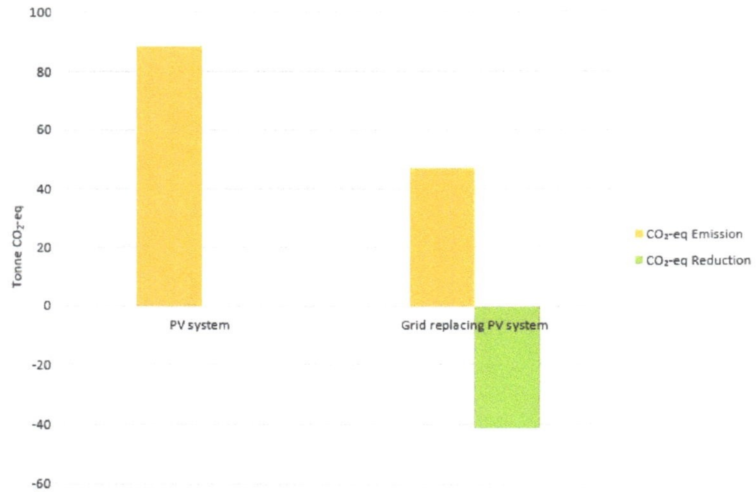

Figure 10. Impact mitigation potential of the photovoltaic system.

6. Discussion

This section discusses the obtained findings from the energy and LCA analysis of the building's energy system. The energy system under consideration for this building consists of solar PV, grid supply, and DH. As mentioned earlier, the main reason for the installation of solar PV on the roof of the studied building is to balance the building's energy budget and achieve a nZEB status, as defined in Section 3. The premise of local and on-site energy production, which was also included in a Norwegian nZEB proposal as described in Section 2.5, is a commonly accepted approach to nZEBs. Although locally generated electricity can reduce the amount of imported/purchased electricity, this should be seen in comparison with a wider system perspective. The purpose of nZEB regulation is to reduce emissions from the building sector. A common denominator of nZEBs is a building with low energy demand and local energy production. When using energy as a measurement, real emissions can be misinterpreted. Using local energy production such as solar PV has the potential to reduce grid-imported electricity. The problem arises when local energy production has a higher emission factor from a life cycle perspective than the alternative, in this case, grid-imported electricity. When looking at a net "reduction" in energy, the environmental impact of the case study is low compared to conventional buildings. However, as it is illustrated in this paper, the reduction in imported or purchased energy does not necessarily mean the reduction in emissions from a life cycle perspective.

The results from Figure 9 show that solar PV panels, which in this case are produced in Malaysia (with a predominantly fossil fuel-based power production mix), have higher emission levels per kWh produced than the electricity bought from the national power grid. If the emission factor of an energy conversion technology such as solar PV exceeds the current country's electricity production mix, the net contribution from the PV can be negative. In general terms, the cleaner the original energy mix becomes, the more important is it to thoroughly assess the overall climate impact. This is the reason that, if the criteria for nZEBs is solely dependent on energy requirements, it can be misleading in terms of actual environmental impact. This exact scenario can be argued in this paper. Using imported grid electricity in contrast to the installed solar PV would lead to a GHG emission reduction, given the reference frame used in this work. It is therefore also important to assess emissions in connection with the building, which will be part of the revised EPBD. To assume that local energy production is always the better alternative can be misleading. However in this case, if the manufacturing of the solar PV panels was placed in a country with a cleaner energy supply mix, the emission levels would be decreased. This was also the conclusion from the study [18], where the author concluded that even though solar PV is considered "clean", the environmental impact is dependent on the amount of renewable energy in the electricity production mix as well as the PV systems' annual production. Therefore, environmental performance should be also considered as a criteria in nZEB legislation. This is also supported by the proposed recast of the EPBD from 2021. Since Norway is considered to have a "clean" energy mix, it is perhaps more appropriate to assess emission factors instead of energy needs. As illustrated in this paper, a so-called low energy demand does not necessarily translate into lower emissions.

Furthermore, the total energy demand for the building for the year 2020 amounts to 570 MWh, of which the electric energy demand accounts for about 55%, while the remaining 45% is heating. In this case, solar power production of 70 MWh is also included. As characterised in Figure 5 and Table 4, it appears that the heating demand exceeds the energy demand for a passive house (although this is unknown due to insufficient data), and therefore the proposed nZEB definition. However, it is difficult to make any conclusion regarding the source of high heating demand in 2020. It should also be mentioned that the campus grounds were closed for periods of time during the year 2020, starting in March due to the COVID-19 pandemic. It is due to this that the energy demand in previous and subsequent years can be higher than what was investigated in this paper.

7. Conclusions

Energy use and environmental emissions are interlinked issues. With energy usage, there is a corresponding emission. To have a low energy demand is perhaps not the right viewpoint if it leads to a higher impact in terms of environmental emissions. It was found that the installed PV modules had a higher specific emission factor than imported electric energy from the power grid. The major cause of the emission factor was the energy mix in the PV cell manufacturing country. It is therefore considered to be important that future legislation regarding nZEBs should take into account the environmental parameters as well as the energy accounting. This paper is meant to provide a preliminary indication of the possible consequences of a lack of focus on the climate impact. It is therefore crucial that more work is performed on this subject. The results in this paper should by no means be interpreted as discouraging the use of solar PV systems as harmful for the environment, but rather that it is important to assess the climate impact factors of energy systems.

Author Contributions: Conceptualisation, F.E.A. and S.G.R.; methodology, F.E.A. and S.G.R.; software, S.G.R.; validation, F.E.A. and S.G.R.; formal analysis, F.E.A. and S.G.R.; data curation, F.E.A. and S.G.R.; writing—original draft preparation, F.E.A., S.G.R. and A.G.; writing—review and editing, F.E.A., S.G.R. and A.G.; visualisation, F.E.A. and S.G.R.; supervision, A.G. All authors have read and agreed to the published version of the manuscript.

Funding: The APC was funded by NTNU—Norwegian University of Science and Technology.

Data Availability Statement: Not applicable.

Conflicts of Interest: The authors declare no conflict of interest.

Abbreviations

The following abbreviations are used in this manuscript:

CO_2	Carbon dioxide
CO_2-eq	Carbon dioxide equivalent
DALY	Disability-Adjusted Life Years
DH	District heating
EEA	European Economic Area
EPBD	Energy performance of buildings directive
EU	European union
GHG	Greenhouse gas
GWP	Global warming potential
IEA	International Energy Agency
IPCC	Intergovernmental Panel on Climate Change
LCA	Life Cycle Assessment
net ZEB	Net zero-energy building
nZEB	Nearly zero-energy building
OECD	Organisation for Economic Co-operation and Development
PV	Photovoltaic
UN	United Nations
ZEB	Zero-energy building

References

1. Masson-Delmotte, V.; Zhai, P.; Pirani, A.; Connors, S.L.; Péan, C.; Berger, S.; Caud, N.; Chen, Y.; Goldfarb, L.; Gomis, M.I.; et al. IPCC, 2021: Summary for Policymakers. In *Climate Change 2021: The Physical Science Basis. Contribution of Working Group I to the Sixth Assessment Report of the Intergovernmental Panel on Climate Change*; Cambridge University Press: Cambridge, UK, 2021.
2. Hamilton, I.; Kennard, H.; Rapf, O.; Kockat, D.J.; Zuhaib, D.S.; Abergel, T.; Oppermann, M.; Otto, M.; Loran, S.; Fagotto, I.; et al. *2020 Global Status Report for Buildings and Construction: Towards a Zero-emission, Efficient and Resilient Buildings and Construction Sector*; UNEP: Nairobi, Kenya, 2020.
3. European Commission. Commission recommendation (EU) 2019/786 of 8 May 2019 on building renovation. *Off. J. Eur. Union* **2019**, *L 127*, 34–79.
4. United Nations Framework Convention on Climate Change. The Paris Agreement. Available online: https://unfccc.int/process-and-meetings/the-paris-agreement/the-paris-agreement (accessed on 25 March 2021).
5. United Nations Framework Convention on Climate Change. Update of Norway's Nationally Determined Contribution. Available online: https://www4.unfccc.int/sites/ndcstaging/PublishedDocuments/Norway%20First/Norway_updatedNDC_2020%20(Updated%20submission).pdf (accessed on 25 March 2021).
6. Klima- og Miljødepartementet. Norge Forsterker Klimamålet for 2030 til Minst 50 Prosent og Opp Mot 55 Prosent. Available online: https://www.regjeringen.no/no/aktuelt/norge-forsterker-klimamalet-for-2030-til-minst-50-prosent-og-opp-mot-55-prosent/id2689679/ (accessed on 25 March 2021).
7. Statsministerens Kontor. Granavolden-Plattformen. Available online: https://www.regjeringen.no/no/dokumenter/politisk-plattform/id2626036/ (accessed on 25 March 2021).
8. Hermelink, A.; Schimschar, S.; Boermans, T.; Pagliano, L.; Zangheri, P.; Armani, R.; Voss, K.; Musall, E. *Towards Nearly Zero-Energy Buildings. Definition of Common Principles under the EPBD*; Final Report; Ecofys: Koeln, Germany, 2013.
9. European Commission. Energy Performance of Buildings Directive. Available online: https://ec.europa.eu/energy/topics/energy-efficiency/energy-efficient-buildings/energy-performance-buildings-directive_en (accessed on 12 April 2021).
10. European Commission. Proposal for a Directive of the European Parliament and of the Council on the Energy Performance of Buildings (Recast) COM/2021/802 Final. Available online: https://eur-lex.europa.eu/legal-content/EN/TXT/?uri=celex%3A52021PC0802 (accessed on 22 August 2021).

11. European Commission. *Communication from the Commission to the European Parliament, the Council, the European Economic and Social Committee and the Committee of the Regions Youth Opportunities Initiative*; European Commission: Luxembourg, 2011.
12. Directive 2010/31/EU of the European Parliament and of the Council of 19 May 2010 on the Energy Performance of Buildings. Available online: https://eur-lex.europa.eu/legal-content/EN/TXT/?uri=celex:32010L0031 (accessed on 12 April 2021).
13. Fthenakis, V.; Kim, H.; Frischknecht, R.; Raugei, M.; Sinha, P.; Stucki, M. *Life Cycle Inventories and Life Cycle Assessment of Photovoltaic Systems*; IEA-PVPS-TASK 12; IEA: Paris, France, 2011.
14. Fthenakis, V.; Betita, R.; Shields, M.; Vinje, R.; Blunden, J.; Shields, M. Life cycle analysis of high-performance monocrystalline silicon photovoltaic systems: Energy payback times and net energy production value. In Proceedings of the 27th European Photovoltaic Solar Energy Conference and Exhibition, Frankfurt, Germany, 24–28 September 2012; pp. 4667–4672.
15. Deng, S.; Wang, R.; Dai, Y. How to evaluate performance of net zero energy building–A literature research. *Energy* **2014**, *71*, 1–16. [CrossRef]
16. Zhang, S.C.; Yang, X.Y.; Xu, W.; Fu, Y.J. Contribution of nearly-zero energy buildings standards enforcement to achieve carbon neutral in urban area by 2060. *Adv. Clim. Chang. Res.* **2021**, *12*, 734–743. [CrossRef]
17. De Masi, R.F.; Gigante, A.; Vanoli, G.P. Are nZEB design solutions environmental sustainable? Sensitive analysis for building envelope configurations and photovoltaic integration in different climates. *J. Build. Eng.* **2021**, *39*, 102292. [CrossRef]
18. Martinopoulos, G. Are rooftop photovoltaic systems a sustainable solution for Europe? A life cycle impact assessment and cost analysis. *Appl. Energy* **2020**, *257*, 114035. [CrossRef]
19. Martinopoulos, G. Life Cycle Assessment of solar energy conversion systems in energetic retrofitted buildings. *J. Build. Eng.* **2018**, *20*, 256–263. [CrossRef]
20. Sherwani, A.; Usmani, J.; Varun. Life cycle assessment of solar PV based electricity generation systems: A review. *Renew. Sustain. Energy Rev.* **2010**, *14*, 540–544. [CrossRef]
21. D'Agostino, D.; Zangheri, P.; Castellazzi, L. Towards Nearly Zero Energy Buildings in Europe: A Focus on Retrofit in Non-Residential Buildings. *Energies* **2017**, *10*, 117. [CrossRef]
22. Buildings Performance Institute Europe. Nearly Zero: A Review of EU Member State Implementation of New Build Requirements. Available online: https://www.bpie.eu/wp-content/uploads/2021/06/Nearly-zero_EU-Member-State-Review-062021_Final.pdf.pdf (accessed on 20 October 2021).
23. Allard, I.; Nair, G.; Olofsson, T. Energy performance criteria for residential buildings: A comparison of Finnish, Norwegian, Swedish, and Russian building codes. *Energy Build.* **2021**, *250*, 111276. [CrossRef]
24. European Commission. European Commission Directive (EU) 2018/844 of the European Parliament and of the Council of 30 May 2018 Amending Directive 2010/31/EU on the Energy Performance of Buildings and Directive 2012/27/EU on Energy Efficiency (Text with EEA Relevance). *Off. J. Eur. Union* **2018**, *L 156*, 75–91.
25. D'Agostino, D.; Tzeiranaki, S.T.; Zangheri, P.; Bertoldi, P. Assessing nearly zero energy buildings (NZEBs) development in Europe. *Energy Strategy Rev.* **2021**, *36*, 100680. [CrossRef]
26. European Commission. Report from the Commission to the European Parliament and the Council 2020 Assessment of the Progress Made by Member States towards the Implementation of the Energy Efficiency Directive 2012/27/EU and towards the Deployment of Nearly Zero-Energy Buildings and Cost-Optimal Minimum Energy Performance Requirements in the EU in Accordance with the Energy Performance of Buildings Directive 2010/31/EU. Available online: https://eur-lex.europa.eu/legal-content/EN/TXT/?uri=CELEX:52020DC0954 (accessed on 12 April 2021).
27. Tenhunen, S. *Energy Performance of Buildings Directive 2010/31/EU: Fit for 55 Revision—Implementation in Action*; EPRS: European Parliamentary Research Service: Bruxelles, Belgium, 2021.
28. European Commission. Commission Recommendation (EU) 2016/1318 of 29 July 2016 on guidelines for the promotion of nearly zero-energy buildings and best practices to ensure that, by 2020, all new buildings are nearly zero-energy buildings. *Off. J. Eur. Union* **2016**, *L 208*, 46–57.
29. Wilson, A. *Revision of the Energy Performance of Buildings Directive: Fit for 55 Package*; Technical Report; European Parliamentary Research Service: Bruxelles, Belgium, 2022.
30. European Commission. Nearly Zero-Energy Buildings. Available online: https://energy.ec.europa.eu/topics/energy-efficiency/energy-efficient-buildings/nearly-zero-energy-buildings_en (accessed on 16 August 2021).
31. Direktoratet for byggkvalitet. Byggteknisk Forskrift (TEK17) med Veiledning. Available online: https://dibk.no/regelverk/byggteknisk-forskrift-tek17/ (accessed on 24 February 2021).
32. Stene, J.; Alonso, M.J.; Rønneseth, Ø.; Georges, L. State-of-the-Art Analysis of Nearly Zero Energy Buildings; Country Report IEA HPT Annex 49 Task 1–Norway; SINTEF Academic Press: Blindern, Norway, 2018.
33. Sartori, I.; Graabak, I.; Dokka, T. Proposal of a Norwegian ZEB definition: Storylines and Criteria. In Proceedings of the Renewable Energy Beyond 2010, Trondheim, Norway, 7–8 June 2010; pp. 179–190.
34. D'Agostino, D.; Mazzarella, L. What is a Nearly zero energy building? Overview, implementation and comparison of definitions. *J. Build. Eng.* **2019**, *21*, 200–212. [CrossRef]
35. Colclough, S.; Kinnane, O.; Hewitt, N.; Griffiths, P. Investigation of nZEB social housing built to the Passive House standard. *Energy Build.* **2018**, *179*, 344–359. [CrossRef]
36. Feist, W.; Schnieders, J. Energy efficiency—A key to sustainable housing. *Eur. Phys. J. Spec. Top.* **2009**, *176*, 141–153. [CrossRef]

37. *NS 3701*; Kriterier for Passivhus og Lavenergibygninger: Yrkesbygninger = Criteria for Passive Houses and Low Energy Buildings: Non-Residential Buildings. Standard Norge: Lysaker, Norway, 2012.
38. *NS 3700*; Kriterier for Passivhus og Lavenergibygninger: Boligbygninger = Criteria for Passive Houses and Low Energy Buildings: Residential Buildings. Standard Norge: Lysaker, Norway, 2013.
39. Killingland, M.; Lånke, A.F.; Ragnøy, M.M.; Aga, P.; Smits, F.; Andresen, I.; Elvebakk, K.; Holthe, F. Utredning Nesten Nullenergibygg for Norge. 2013. Available online: https://dibk.no/globalassets/energi/nesten_nullenergibygg_for_norge_ramboll_og_link.pdf (accessed on 23 November 2021).
40. Fufa, S.M.; Schlanbusch, R.D.; Sørnes, K.; Inman, M.R.; Andresen, I. *A Norwegian ZEB Definition Guideline*; SINTEF Academic Press: Blindern, Norway, 2016.
41. Kristjansdottir, T.; Fjeldheim, H.; Selvig, E.; Risholt, B.D.; Time, B.; Georges, L.; Dokka, T.H.; Bourelle, J.; Bohne, R.A.; Cervenka, Z. *A Norwegian ZEB-Definition Embodied Emission*; SINTEF Academic Press: Blindern, Norway, 2014.
42. Andresen, I.; Dokka, T.H.; Lassen, N. Kriterier for NZEB for FutureBuilt-Prosjekter. Revisjon August 2022. Available online: https://www.futurebuilt.no/content/download/28111/157863 (accessed on 23 November 2021).
43. Hestnes, A.; Eik-Nes, N. *Zero Emission Buildings*; Fagbokforl: Bergen, Norway, 2017.
44. Andresen, I.; Wiik, M.K.; Fufa, S.M.; Gustavsen, A. The Norwegian ZEB definition and lessons learnt from nine pilot zero emission building projects. *IOP Conf. Ser. Earth Environ. Sci.* **2019**, *352*, 012026. [CrossRef]
45. Yttersian, V.L.; Fuglseth, M.; Lausselet, C.; Brattebø, H. OmrådeLCA, assessment of area development: Case study of the Zero-Emission Neighbourhood Ydalir. *IOP Conf. Ser. Earth Environ. Sci.* **2019**, *352*, 012041. [CrossRef]
46. Lausselet, C.; Crawford, R.H.; Brattebø, H. Hybrid life cycle assessment at the neighbourhood scale: The case of Ydalir, Norway. *Clean. Eng. Technol.* **2022**, *8*, 100503. [CrossRef]
47. Wiik, M.R.K.; Fjellheim, K.; Vandervaeren, C.M.L.; Lien, S.K.; Meland, S.; Nordström, T.; Baer, D.; Cheng, C.Y.; Truloff, S.; Brattebø, H.; et al. *Nullutslippsnabolag i Smarte Byer: Definisjon, Vurderingskriterier og Nøkkelindikatorer*, Versjon 3.0.; SINTEF Academic Press: Blindern, Norway, 2022.
48. Norwegian Building Authority. Hva er et Nesten Nullenergibygg? Available online: https://dibk.no/om-oss/Nyhetsarkiv/hva-er-et-nesten-nullenergibygg/ (accessed on 13 September 2021).
49. Paoletti, G.; Pascual Pascuas, R.; Pernetti, R.; Lollini, R. Nearly zero energy buildings: An overview of the main construction features across Europe. *Buildings* **2017**, *7*, 43. [CrossRef]
50. Kristjansdottir, T.F.; Good, C.S.; Inman, M.R.; Schlanbusch, R.D.; Andresen, I. Embodied greenhouse gas emissions from PV systems in Norwegian residential Zero Emission Pilot Buildings. *Sol. Energy* **2016**, *133*, 155–171. [CrossRef]
51. Stoppato, A. Life cycle assessment of photovoltaic electricity generation. *Energy* **2008**, *33*, 224–232. [CrossRef]
52. Gaiddon, B.; Jedliczka, M. Compared assessment of selected environmental indicators of photovoltaic electricity in OECD cities. *Int. Energy Agency Photovolt. Power Syst. Programme* **2006**, *11*, 363–385.
53. Aanensen, T. Wind Power Generation Continues to Rise. Available online: https://www.ssb.no/en/energi-og-industri/artikler-og-publikasjoner/wind-power-generation-continues-to-rise (accessed on 24 June 2022).
54. Statistics Norway. 08307: Production, Imports, Exports and Consumption of Electric Energy (GWh) 1950–2021. Available online: https://www.ssb.no/en/statbank/table/08307 (accessed on 24 June 2022).
55. Schäfer, M.; Hofmann, F.; Abdel-Khalek, H.; Weidlich, A. Principal Cross-Border Flow Patterns in the European Electricity Markets. In Proceedings of the 2019 16th International Conference on the European Energy Markets, Ljubljana, Slovenia, 18–20 September 2019; pp. 1–6. [CrossRef]
56. Rosenberg, E.; Lind, A.; Espegren, K.A. The impact of future energy demand on renewable energy production–Case of Norway. *Energy* **2013**, *61*, 419–431. [CrossRef]
57. Treyer, K.; Bauer, C. Life cycle inventories of electricity generation and power supply in version 3 of the ecoinvent database—Part I: Electricity generation. *Int. J. Life Cycle Assess* **2016**, *21*, 1236–1254. [CrossRef]
58. Norges Teknisk-Naturvitenskapelige Universitet. Teknologiveien 22-Smaragdbygget . Available online: https://www.ntnu.no/campusutvikling/teknologiveien22 (accessed on 15 January 2022).
59. Statsbygg. 1021001 Høgskolen i Gjøvik. Available online: https://www.ntnu.no/documents/1268425101/1276521140/4.+Byggeprogram+med+vedlegg.pdf/187dc23b-beca-416a-b40f-3435ce027a06 (accessed on 15 January 2022).
60. Norsk Fjernvarme. Fjernkontrollen. Available online: https://www.fjernkontrollen.no/ (accessed on 30 November 2022).
61. Statsbygg. Kjempepotensial for Solceller. Available online: https://www.statsbygg.no/nyheter/kjempepotensial-for-solceller (accessed on 6 May 2022).
62. PRé Sustainability. SimaPro | The World Leading LCA Software. Available online: https://simapro.com/ (accessed on 14 April 2022).
63. *ISO 14040:2006*; Environmental Management: Life Cycle Assessment; Principles and Framework. International Organization for Standardization: Geneva, Switzerland, 2006.
64. *ISO 14044:2006*; Environmental Management: Life Cycle Assessment; Requirements and Guidelines. International Organization for Standardization: Geneva, Switzerland, 2006.
65. Frischknecht, R.; Heath, G.; Raugei, M.; Sinha, P.; de Wild-Scholten, M. *Methodology Guidelines on lIfe Cycle Assessment of Photovoltaic Electricity*; NREL: Golden, CO, USA, 2016.

66. Dones, R.; Frischknecht, R. Life-cycle assessment of photovoltaic systems: Results of Swiss studies on energy chains. *Prog. Photovolt. Res. Appl.* **1998**, *6*, 117–125. [CrossRef]
67. Jungbluth, N. Life cycle assessment of crystalline photovoltaics in the Swiss ecoinvent database. *Prog. Photovolt. Res. Appl.* **2005**, *13*, 429–446. [CrossRef]

energies

MDPI

Article

Energy Conservation in a Livestock Building Combined with a Renewable Energy Heating System towards CO_2 Emission Reduction: The Case Study of a Sheep Barn in North Greece

Antonios A. Lithourgidis [1], Vasileios K. Firfiris [1], Sotirios D. Kalamaras [1], Christos A. Tzenos [1], Christos N. Brozos [2] and Thomas A. Kotsopoulos [1,*]

[1] Department of Hydraulics, Soil Science and Agricultural Engineering, School of Agriculture, Aristotle University of Thessaloniki, GR-54124 Thessaloniki, Greece
[2] Clinic of Farm Animals, Faculty of Veterinary Medicine, Aristotle University of Thessaloniki, 11 Stavrou Voutyra Str, GR-54627 Thessaloniki, Greece
* Correspondence: mkotsop@agro.auth.gr

Abstract: Cold stress in sheep is usually overlooked, even though the animals' welfare and productivity are affected by low temperatures. The aim of this research was to find out if and to what extent the temperature inside a sheep barn could be maintained within the range of the thermoneutral zone during winter, primarily to increase feed conversion and to reduce GHG emissions. For this reason, an automation system was installed at a sheep barn in northern Greece, and heat losses from the building were calculated. The biogas potential of the sheep barn waste was examined in the laboratory via the BMP method. The results showed that the installation of an automation system together with a hypothetical biogas heating system could maintain the barn's temperature in the range of a sheep's thermoneutral zone during winter for the 94% of the scenarios examined if the total energy of the biogas was utilized, while heating energy that was instantly and continuously used succeeded in 48% of the investigated cases. The surplus of energy produced by biogas could potentially raise the water temperature that animals drink up to 2.9 °C. The absence of cold stress decreases the dry matter intake and the CH_4 produced by ruminal fermentation. Moreover, lower GHG emissions are achieved as waste is treated through anaerobic digestion, which would likely be released into the environment if left untreated.

Keywords: greenhouse gas emissions; climate control; anaerobic digestion; biogas; cold stress; sheep barn

Citation: Lithourgidis, A.A.; Firfiris, V.K.; Kalamaras, S.D.; Tzenos, C.A.; Brozos, C.N.; Kotsopoulos, T.A. Energy Conservation in a Livestock Building Combined with a Renewable Energy Heating System towards CO_2 Emission Reduction: The Case Study of a Sheep Barn in North Greece. *Energies* **2023**, *16*, 1087. https://doi.org/10.3390/en16031087

Academic Editor: Andres Siirde

Received: 23 December 2022
Revised: 5 January 2023
Accepted: 14 January 2023
Published: 18 January 2023

1. Introduction

Livestock and climate change are strongly associated. Livestock is contributing to the increase of greenhouse gas (GHG) emissions with the direct (waste) or indirect production (feed production and feed waste) of gases such as methane (CH_4), nitrous oxide (N_2O) and carbon dioxide (CO_2) [1]. According to the literature, more than 30 billion livestock contribute annually 14–16% of the total global GHG emissions [2]. Livestock production has intensified throughout the years and is expected to increase in order to meet the increasing population demands [3]. As a result, GHG emissions are also expected to rise. On the other hand, climate change affects livestock as the parameters that define the optimum growth conditions of animals, such as temperature and relative humidity, are subjected to an ongoing change in many regions globally. In addition, many sudden variations and extreme weather phenomena are more frequently observed [4]. This predicament affects the productivity of animals (milk and meat production, reproduction), as well as other factors that are connected with livestock production such as crops, land, feed quality, etc. [5].

All animals have a thermoneutral zone (TNZ), which is a range of ambient temperatures where normal metabolism offers enough heat to keep an optimum constant temperature. The temperature boundaries of the TNZ are the upper critical temperature (UCT) and the lower critical temperature (LCT). Much research has been conducted regarding the heat stress problem. However, cold stress also affects the animals' welfare. At temperatures lower than LCT, animals need to increase their metabolism to maintain normal body temperature; at higher temperatures than UCT, the animal's body temperature increases above normal because of inadequate evaporative heat loss. Animals consume feed to sustain their body temperature inside the boundaries of the TNZ [6]. Thus, the feed intake is converted to body heat instead of milk weight gain and other productivity factors. Temperatures beyond these limits increase the animal GHG emissions as the efficiency of feed conversion is decreased and their respiration is increased. Apart from the inefficient conversion of feed, when animals are below the LCT they consume larger amounts of feed [6] in order to retain their body temperature at the proper level and not to increase productivity. The control of the environmental conditions inside a livestock building is performed with the installation of automation and climate control systems [7–12]. These systems consume energy, and they are contributing to the increase in GHG emissions as they are mainly powered by electricity or fossil fuels. This multidimensional issue, enhanced by climate change, concerns all livestock, both in colder and warmer periods, according to the region and type of livestock.

Small ruminants contribute to a percentage between 7–10% of the total GHG emissions from livestock production [13]. Sheep are responsible for near half of these emissions [14,15]. As in other livestock types, ruminal fermentation and feed production are the main sources of GHG emissions from sheep [15]. Greece has a significant number of sheep farms; sheep are usually housed in simple constructions without the support of equipment for climate regulation. Most sheep barns in Greece are operating without an artificial ventilation system. The ventilation is performed via the natural movement of air through the side openings and ridge windows. The most common practice for the producers is to keep the windows open all year long. This practice tackles the heat stress problem during warmer periods. During colder periods, though, the temperature inside the sheep barn is lower than the lower critical temperature of the animals' TNZ. As a result, the animals consume more feed to retain their body temperature within their TNZ, and they face a change in their metabolic reactions [6,16], which leads to an increase in feed intake during the winter months [17–19] without leading to a weight increase [20]. As reported in the literature, an increase in feed intake leads to an increase in GHG emissions [21–23]. Actually, the relationship between dry matter intake (DMI) and GHGs emissions is directly proportional, and it was reported previously in several research works [22,24,25]. So, an efficient and environmentally friendly method that aims to retain proper conditions inside a sheep barn all year long, in combination with a reduction of sheep waste and GHG emissions, is of great importance.

Drinking water temperatures, especially in the winter months, are also an important aspect that affects animal performance. According to the literature, water should be kept in a specific temperature range (2–20 °C) in order to enhance the productivity of sheep [26–28]. It is strongly related to the animal's digestion performance, and under cold environmental conditions, heating of the drinking water may be necessary. Especially, when it is usually stored in open tanks inside non-heated storage rooms or outside the barn.

Up until now, sheep barns in Greece have operated during the winter without considering any of the above-mentioned issues (cold stress, water temperature and feed conversion rates), resulting in low productivity rates while causing an environmental burden. The aggregation of heat produced by each animal's biological functions could potentially contribute to mitigating the effect of low temperatures in the external environment on the internal climate of the sheep barn. In the case of sheep, the heat produced depends on the body weight and species [29,30]. This heat could be used to increase the temperature inside the barn during colder periods. On the other hand, waste from sheep could also be a source of heat if it was converted to biogas through anaerobic digestion and used as fuel for a heating system.

Anaerobic digestion is the microbiological process of decomposing organic matter in the absence of oxygen. Biogas is the main product of anaerobic digestion, and it primarily consists of CH_4 and CO_2. It can be used directly in specified internal combustion engines to generate electricity and heat, but it can also be used only for heat generation. It has been reported that utilization of biogas through a boiler reached an energy efficiency of 83% for heat generation [31]. The combination of retaining the heat generated by animal bodies inside the barn and utilizing the sheep waste through anaerobic digestion for biogas heat generation may be the solution for reducing GHG emissions and increasing productivity of livestock during the winter. This sustainable solution for the regulation of the microclimate conditions inside a sheep barn has never been examined extensively.

The aim of this research was to find out if and to what extent the temperature inside a sheep barn could be maintained within the range of TNZ during winter, primarily to increase feed conversion and reduce GHG emissions. Thus, a hypothesis was made that the energy needs of a sheep barn in northern Greece will be covered from the animals' body heat and the biogas produced from the utilization of the sheep waste. For this reason, an automation system was installed, and the biogas potential of the sheep waste was examined in the laboratory. Specifically, an automated system for opening and closing the window openings of the sheep barn was installed and programmed to operate based on the sheep TNZ. This system also monitored the air quality inside the barn in terms of relative humidity, CH_4, ammonia and CO_2 levels. Sheep barns' waste biogas potential was examined to calculate the heat that could be generated by directly burning the biogas and to estimate the GHG emissions of the sheep waste if it was not subjected to anaerobic digestion. The outcome of this research is important as it proposes an agricultural practice that could lead to the sustainable operation of simple structures, such as sheep barns. This practice is accompanied by low energy consumption, can be applied locally, has a relatively low cost and is simple to use. In addition, it is in line with the targets set by the EU for the mitigation of GHG emissions by all economic sectors, including agriculture [32].

2. Materials and Methods

2.1. Experimental Sheep Barn

The sheep barn was near Galatista town of the Chalkidiki regional unit in northern Greece. The exact location coordinates were 40°26′21.6″ N/23°16′16.4″E, and the elevation was 496 m [30]. An outside view of the sheep barn and a schematic view of the building, as well as the building materials and structure dimensions, are presented in Figure 1. The materials of which the building was constructed and their geometrical characteristics as well as the thermal transmittance (U value) are presented in Table 1.

For the scope of the experiment, specific parts of the barn were modified, and new equipment was installed. Hand operation for the side windows was removed, and electric motors were installed for automated operation. Also, the mechanical part of the automated electrical roof window was repaired, and a programmable logic controller (PLC) with a touch screen was installed with appropriate sensors (see Section 2.2—Automation system, data logging and sensors) for monitoring the air quality inside the sheep barn. Installation works and equipment are highlighted in photos provided as Supplementary Material (Section S5).

The Assaf sheep was the housed animal of the barn. The number of housed sheep varied from 150–250, depending on the birth replacement animal rates. The animals were growing lambs (yearlings), and their weight was about 30 kg. During the experimental period, the number of animals was 180. The type of housing is freestall.

The modifications to the barn and the installation of the automatic control system aim to enhance the microclimatic conditions inside the barn, which strongly relate to animal welfare. With the proposed operation rationale, the ventilation of the barn is improved, as it was performed only in cases where the gas concentrations or relative humidity were not at acceptable levels. In this automatic system operation, the temperature parameter was also included, which is also a crucial factor for animal welfare. Thus, ventilation was also performed, considering the optimum temperature for the sheep and reducing the heat losses, when possible, during the winter. All the parameters listed above (gas concentration, relative humidity and temperature) are microclimatic parameters that need to be maintained at an appropriate level for animal welfare. Through the implementation of the automation system, this is achieved with low energy consumption and operation costs.

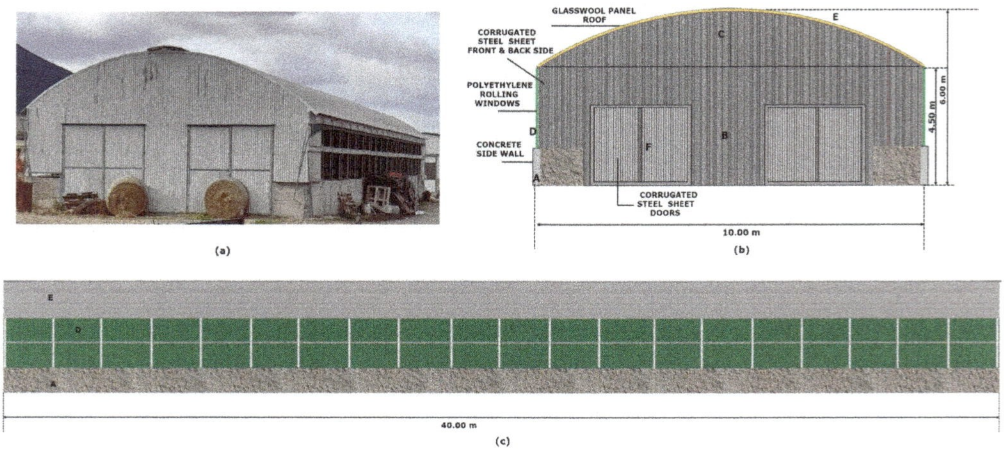

Figure 1. (**a**) Outsidze view of the sheep barn near Galatista town; schematic views with dimensions of (**b**) the front and (**c**) the side of the building.

Table 1. Sheep barn material properties, geometrical characteristics and the U value [33–38].

Symbol	Description	Material	Dimensions	U Value ($W{\cdot}m^{-2}{\cdot}K^{-1}$)	Heat Exchange Surface (m^2)
A	Side front and back side part	Concrete	Height: 1.50 m Thickness: 0.15 m Length: Lengthwise the right and left side (40 m) / 1 m in front and back side on the left and right (Figure 1)	3.89	126
B	Front and back side rectangular part	Corrugated metal sheet (steel)	Thickness: 0.01 m Rest dimensions (Figure 1)	7.26	84
C	Front and back side pediment:	Corrugated metal sheet (steel)	Thickness: 0.01 m Rest dimensions (Figure 1)	7.26	23.5
D	Side double rolling windows	PVC sheets	Height 3.00 m Thickness: 0.01 m Length: 40.00 m	4.99	240
E	Roof	Curved insulation panel (glass wool)	Thickness: 0.1 m Rest dimensions (Figure 1)	0.43	438
F	Doors	Corrugated metal sheet (steel)	Height 3.00 m Width: 3.00 m	Included in front and back sides	Included in front and back sides

2.2. Automation System, Data Logging and Sensors

An automatic operation and monitoring system was installed in the sheep barn. The system consisted of a PLC unit, electrical installations (board, cables and window motors) and sensors. Three combined temperature and relative humidity sensors (Autonics, THD-DD1-C, Mundelein-IL-USA) were used for monitoring the internal and the external environments of the sheep barn. Two of them were positioned in the sheep barn at a height of about 1.5 m from the ground, while the second was positioned close to the roof window at about 5.0 m from the ground. The third one was installed on the roof of the sheep barn to monitor the ambient outdoor temperature and relative humidity. Air quality inside the barn was monitored with a CH_4 sensor (Bacharach MGS 550 gas detector, Pittsburgh, PA, USA), a CO_2 sensor (Bacharach MGS 450 gas detector, Pittsburgh, PA, USA) and an NH_3 sensor (Bacharach MGS 450 gas detector, Pittsburgh, PA, USA), which were positioned at a height as close as possible to the sheep's height. All the above-mentioned parameters were measured and stored every 15 min by the PLC unit. All the sensors in the PLC unit were new, and they were already factory calibrated. The experimental measurements occurred from 23 January to 27 January of 2022 from 6:00 am to 16:00 pm every day. The specific period was chosen because the conditions were favorable for the scope of the experiment.

2.3. Operation of the Sheep Barn Automation System

The system was programmed to open and close the windows based on specific ranges of parameters that are presented in Table 2. The rationale behind this operation design was to keep the temperature inside the TNZ without endangering the air quality of the internal environment. The range of the upper and lower limits of the parameters were reported in previous research [6,39–41].

Table 2. Parameters range for opening and closing of windows.

Parameter	Lower Value (Windows Close)	Higher Value (Windows Open)	Priority
Average temperature (2 sensors)	8 °C	12 °C	3
Average relative humidity (2 sensors)	65%	70%	2
NH_3	5 ppm	20 ppm	1
CH_4	-	-	5
CO_2	700 ppm	2500 ppm	4

2.4. Biogas Potential

Biogas potential tests of sheep manure, wheat straw and different mixtures of them were performed in the laboratory. In total, nine substrates were examined, and they are presented in Table 3.

Table 3. Substrates for biogas potential and their attributed acronyms.

Substrates	Acronyms
Wheat Straw	WS
Sheep manure—fresh	BL—0d
Sheep manure—15 days	BL—15d
Sheep manure—30 days	BL—30d
75% SM: 25% WS—fresh	75:25 0d
75% SM: 25% WS—15 days	75:25 15d
75% SM: 25% WS—30 days	75:25 30d
60% SM: 40% WS—fresh	60:40 0d
90% SM: 10% WS—fresh	90:10 0d

The manure produced per sheep (SM) was estimated bibliographically at 0.04 $kg{\cdot}kg_{sheep}{}^{-1}$ [42], 50% of which is considered feces and 50% urine [43], while the quantity of wheat straw (WS) used in the sheep barn as bedding material was provided by the owner and was 0.2 $kg_{WS}{\cdot}sheep^{-1}{\cdot}day^{-1}$. The barn's average sheep weight was about 30 kg, which results in an estimated manure generation of 0.6 $kg{\cdot}sheep^{-1}{\cdot}d^{-1}$. From the selected mixtures for biogas potential, the mixture of 75% SM with 25% WS (*w/w*) was the ratio closest to the real conditions inside the experimental sheep barn.

Sheep manure and bedding (wheat) straw were collected at certain time periods (fresh, 15 days old and 30 days). Prior to anaerobic digestion, manure samples were homogenized, and wheat straw was cut into a length of approximately 2.5 cm. The inoculum used was obtained from previous anaerobic digestion experiments and pre-incubated at 37 ± 0.5 °C until no significant methane production was observed [44]. Blank biogas potential tests with only inoculum were performed to monitor residual biogas production. All the biogas potential tests were performed in triplicate.

Glass bottles of 0.3 L were used as batch reactors for anaerobic digestion. The batch reactors were filled with the different mixtures of substrate and inoculum, with a feed to inoculum ratio (F/I) of 1 in terms of volatile solids (VS). The final operating volume was adjusted to 0.17 L by adding deionized water, and the headspace of each batch reactor was purged with nitrogen gas (99.99% purity) for approximately 2 min to ensure anaerobic conditions before being sealed and placed into an automated thermal chamber that retained the temperature at 37 ± 0.5 °C. Once per day, the batch reactors were manually mixed vigorously. Biogas production was converted to standard temperature and pressure conditions (0 °C and 1 atm). Procedures and analysis were performed according to a proposed protocol for batch essays that has been previously established [44].

2.5. Calculations

2.5.1. Energy Conservation

Energy conservation is achieved based on the hypothesis that to retain the air temperature inside the barn and the drinking water temperature within an acceptable range, an artificial heating system should be used. Thus, with the proposed combined operation, the heat is produced by the sheep's metabolism and by a heating system that burns biogas, which is produced by the anaerobic digestion of sheep waste. Therefore, the use of artificial heating systems (commonly powered with conventional fuels) for tackling cold stress in sheep is mitigated.

The barn is handled as a closed system, so based on the 1st law of thermodynamics, if a certain temperature value needs to be retained inside the structure, a balance between the heat losses and heat gains should be achieved (Equation (1)). It should be noted that in this case the loss from ventilation is considered negligible since the windows are closed.

$$\dot{Q}_{an} + \dot{Q}_{bg} = \dot{Q}_{ch} + \dot{Q}_{inf} \tag{1}$$

$\dot{Q}_{an}$: Total heat produced by animals in the sheep barn (W)

$\dot{Q}_{bg}$: Potential heat produced by a heating system utilizing biogas (W)

$\dot{Q}_{ch}$: Heat losses by combined heat transfer mechanism (W)

$\dot{Q}_{inf}$: Heat losses by infiltration (W)

The value of $\dot{Q}_{an}$ is calculated by the number of animals and the heat produced by each animal individually. According to literature, sheep produce about 2.6 $W \cdot kg_{bodyweight}{}^{-1}$ [29].

The $\dot{Q}_{bg}$ will be calculated by using a typical heating system performance rate equation, based on the hypothesis that part of the energy of the produced biogas will be transformed into effective heat (Equation (2)).

$$\eta_{th} = \frac{\dot{Q}_{bg}}{\dot{Q}_{bgp}} \tag{2}$$

Biogas exploitation for the production of energy has been studied previously and it was found that it can attribute from 5.0 $kWh \cdot m^{-3}$ up to 7.5 $kWh \cdot m^{-3}$ [45]. An average value of 6.25 $kWh \cdot m^{-3}$ was used for the energy calculations regarding biogas utilization.

η_{th}: Performance rate of the heating system

$\dot{Q}_{bg}$: Potential heat produced by a heating system utilizing biogas (W)

$\dot{Q}_{bgp}$: Input energy of the system from the produced biogas (W)

The heat losses by combined heat transfer phenomena are calculated by Equation (3) [33].

$$\dot{Q}_{ch} = U \cdot A \cdot (\Delta T) \tag{3}$$

U: U value ($W \cdot m^2 \cdot K^{-1}$)

A: Heat exchange surface (m^2)

ΔT: Temperature difference between the prevailing temperature inside the barn and the ambient outdoor temperature (°C)

The U value depends on the constructive element's thermal properties and thickness, and it was calculated or referred to according to the literature [37,39]. The heat exchange surface depends on the dimensions and shape of each constructive element. ΔT is usually taken as the temperature difference between the temperature that is required to prevail inside the structure (optimum growth within the thermoneutral zone or lower acceptable, etc.) and the average lower temperature of the region. In the current experiment, the values of the ambient outdoor temperature will be provided by the monitoring system (PLC). The temperature inside the barn was set at 10 °C based on the literature for optimum performance of sheep [6,41].

The heat losses by infiltration are calculated by Equation (4) [46].

$$\dot{Q}_{inf} = \rho \cdot c_p \cdot n \cdot \frac{V}{3600} \cdot (\Delta T) \quad (4)$$

ρ: Air density ($kg \cdot m^{-3}$)
c_p: Water specific heat ($kJ \cdot kg^{-1} \cdot K^{-1}$)
n: Air exchanges (h^{-1})
V: Air volume of the structure (m^3)
ΔT: Temperature difference between the prevailing temperature inside the barn and the ambient outdoor temperature (°C)

Based on the bibliography, the values for air density, air specific heat and air exchanges (considering the barn as an old building) were 1 $kg \cdot m^{-3}$, 1 $kJ \cdot kg^{-1} \cdot K^{-1}$ [47], and 1 h^{-1} [46], respectively. The volume of the structure was calculated to be equal to 2742 m^3, while the ΔT was calculated by Equation (3).

All the above-mentioned energy flow parameters can be converted to energy values if multiplied with time to work, with energy values expressed in kWh.

Keeping the temperature of the drinking water at proper levels also requires the consumption of energy. Heat produced from biogas will be examined to see if it could contribute to retaining the water at a proper temperature. For this reason, the 1st law of thermodynamics was used as presented in Equation (5) [47].

$$Q_w = m \cdot c_p \cdot (\Delta T_{water}) \cdot 0.00027 \quad (5)$$

Q_w: Heat required for water heating (kWh)
m: Total water mass (kg)
c_p: Water specific heat ($kJ \cdot kg^{-1} \cdot K^{-1}$)
ΔT_{water}: Water temperature difference (K)

The water mass required for a sheep weight between 27–50 kg is about 3.8–5.7 $L \cdot d^{-1}$ [48]. The average value of 4.75 $L \cdot d^{-1}$ was used. The water specific heat is 4.2 $kJ \cdot kg^{-1} \cdot K^{-1}$ [47]. The lower temperature of drinking water was chosen at 2 °C, while the highest will be calculated based on the production of biogas by anaerobic digestion.

2.5.2. Reduction of GHG Emissions

An approximate estimation was performed about the GHG emissions and their reduction in the case of utilizing organic carbon due to anaerobic processes to produce biogas for tested substrates. To calculate the $g_{VS} \cdot sheep^{-1} \cdot d^{-1}$, if stored fresh and used at 15 days and 30 days' time periods, the fresh quantity of 0.6 $kg \cdot sheep^{-1} \cdot d^{-1}$ was multiplied by the weight loss that occurred at those periods based on the total solids (TS) measurements of the samples.

The volume of biogas potential and GHG emissions were calculated, in $L_{Biogas\text{-}GHGs} \cdot sheep^{-1} \cdot d^{-1}$, using the Equation (6).

$$Volume_{Biogas-GHGs} = \left(V_{CH_4} + V_{CO_2}\right) \cdot \frac{600 \cdot (100 - WL_{t=n})}{MW_{vs=1} \cdot 100} \quad (6)$$

where:
V_{CH_4}: total methane production volume of sample (L)
V_{CO_2}: total carbon dioxide production volume of sample (L)
$WL_{t=n}$: percentage of weight loss of manure (%), where t = n refers to the collection time period of the manure sample (n = 0 days, 15 days, 30 days)

$MW_{vs=1}$: Weight of manure to obtain 1 g_{VS} for that substrate (g)

The daily production volumes of CH_4 and CO_2 (mL) were calculated using Equation (7) [44].

$$Volume_{dailyGas} = \frac{C_n - C_{n-1}}{100} \cdot V_{hs} \tag{7}$$

where:

C_n: daily gas concentration measurement (%)

C_{n-1}: previous-day gas concentration measurement (%)

V_{hs}: batch reactor headspace volume (mL)

2.6. Analytical Methods

Total solids (TS) and volatile solids (VS) of the sheep manure, wheat straw and inoculum were determined according to Standard Methods [49]. Headspace samples from the reactors were taken for biogas composition analysis (for the first 11 days daily and every few days after). They were collected with a gas-tight syringe and immediately injected in a gas chromatograph, as described by Kalamaras et al. (2020) [50]. The GC was calibrated by using a certified gas of known composition (60% methane- 40% carbon dioxide) and helium gas was used as a carrier gas. The measured values were adjusted to the volumes at standard temperature (0 °C) and pressure (1 atm). The termination of the experiment occurred when methane production was <1% of the accumulated volume of methane for three consecutive days, as established by Holliger et al. (2016) [51].

At the end of the experiment, samples were extracted from each batch reactor to prepare them for VFA analysis. Samples were centrifuged twice for 10 min at 12,000 rpm, and pH was measured. Analysis was performed (1 μL injection sample volume) in a gas chromatograph, and helium was used as a carrier gas, as described in previous research studies [52,53].

2.7. Statistical Analysis

Statistical data analysis was performed with the software IBM SPSS Statistics, version 28. Analysis included descriptive statistics and mean values, standard errors and standard deviations were calculated. Comparisons of the means were performed using one-way analysis of variance, and they were evaluated using the least-significant-difference (Tukey) test. Readings were considered significant when the p value was < 0.05.

2.8. Ethical Statement

In the present study, no handling, and no harm was caused, in any case, or disruption to the behavior and well-being of the animals. Samples were collected from the livestock building's floor only after the farmer provided his consent. The study was carried out following Directives 2010/63/EU [54] and 86/609/EEC [55] regarding the protection of animals used for experimental and other scientific purposes, and activities were performed in compliance with the regulations.

3. Results

3.1. Biogas Potential of Sheep Barn Waste

Sheep manure collected at certain time periods (0 days, 15 days and 30 days), wheat straw and mixtures of them were used as substrates for anaerobic digestion and were compared regarding methane production. The results of the total methane yield and the cumulative methane production as a function of time for all substrates are presented in Figure 2.

Methane production started immediately in all the reactors. The wheat straw was used both as feed and bedding material in the study's experimental sheep barn. Its high nutritional value is clearly reflected in the results below, as it had the highest methane production. This explains the fact that all three substrates with only sheep manure had less production than their corresponding substrate, which contained 25% (weight) of wheat straw. The highest methane production of manure-based substrates was observed with fresh manure. Methane production was lower in all substrates where fresh manure was not used (15 days and 30 days manure), regardless of the addition of wheat straw. The 75:25 30d had a higher yield than the 75:25 15d. In all batch reactors, at least 80% of their total CH_4 volume was achieved by the 27th day and at least 90% by the 35th. The 75% SM: 25% WS ratio represented the real condition in the study's barn case. However, different ratios (*w/w*) of fresh sheep manure (0 days) and wheat straw (60% SM: 40% WS and 90% SM: 10% WS) were also tested to evaluate their CH_4 potential. In terms of CH_4, they did not show any significant statistical difference with the sheep barn's ratio (75% SM: 25% WS), regarding yield and production rate (Figure 2). Therefore, it can be concluded that altering the weight ratio by a step of 15% of sheep manure to wheat straw did not significantly affect the CH_4 production. The above results indicate the higher efficiency of anaerobic co-digestion regarding the CH_4 production of sheep manure and bedding straw, as well as the importance of the freshness of the manure collected.

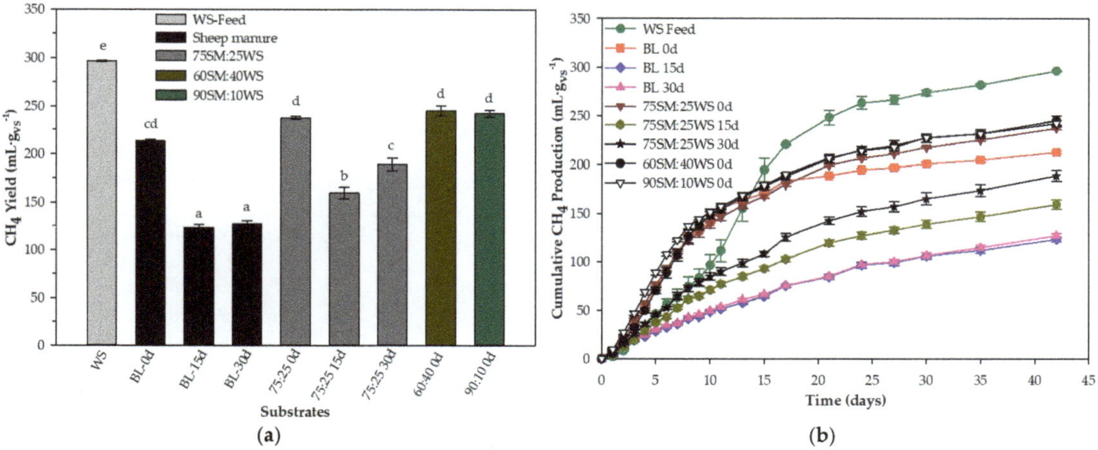

Figure 2. Methane yields (**a**) and cumulative methane productions (**b**) from the anaerobic mono- and co-digestion of each substrate. Different letters above the bars signify distinct statistical groups ($p < 0.05$) between the different tests. The bars designate the standard deviation.

CO_2 production was also measured during anaerobic mono- and co-digestion of all substrates to evaluate the organic carbon in waste that will transform over time into CO_2 emissions due to waste biological activity inside the sheep barn. The results of the CO_2 yield and cumulative production for all substrates are presented in Figure 3.

The CO_2 produced by anaerobic digestion followed, mostly shows the same trend as CH_4 productions mentioned previously. The highest CO_2 yield was produced from the wheat straw (WS-Feed) and had a significant statistical difference with the rest of the samples. In accordance with this, all three mixed substrates had a higher production in comparison with the mono-digestion of manure. Between the mixed substrates, the ratio of 60:40 0d had the highest CO_2 production. Thus, it can be concluded that by increasing the addition of wheat straw to the mixture, the CO_2 production increases accordingly. The lowest production rate and total volume were observed from BL—15d and BL—30d, which indicates that time was an important factor regarding CO_2 yield.

Total VFA concentrations were less than 0.06 g·L^{-1} in all the tested substrates (see Supplementary Section S2). Therefore, there was no inhibition on the biogas production, and a stable anerobic digestion process was performed. The total volume of CH_4 and CO_2 produced at the end of the experiments for all substrates is summarized below in Table 4.

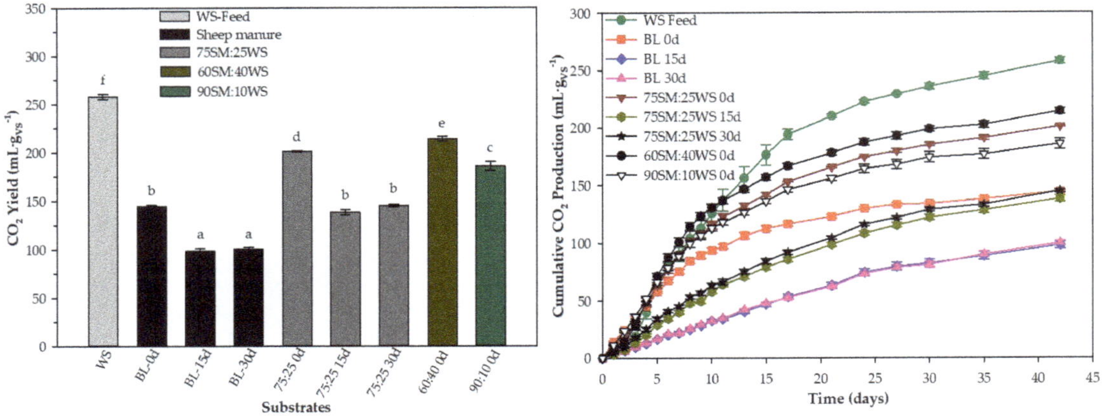

(a) (b)

Figure 3. Carbon dioxide yields (**a**) and cumulative carbon dioxide productions (**b**) from the anaerobic mono- and co-digestion of each substrate. Different letters above the bars signify distinct statistical groups ($p < 0.05$) between the different tests. The bars designate the standard deviation.

Table 4. Total volume of methane and carbon dioxide production from the tested substrates.

Substrates	Total CH_4 Production, $mL \cdot g_{VS}{}^{-1}$	Total CO_2 Production, $mL \cdot g_{VS}{}^{-1}$
Wheat Straw	296.71 ± 0.86 *	258.14 ± 2.72
Sheep manure—fresh	212.95 ± 1.89	144.45 ± 0.97
Sheep manure—15 days	123.25 ± 3.11	98.25 ± 2.96
Sheep manure—30 days	127.12 ± 3.00	100.13 ± 2.02
75% SM: 25% WS—fresh	237.42 ± 1.76	201.07 ± 0.69
75% SM: 25% WS—15 days	159.65 ± 6.02	138.31 ± 2.75
75% SM: 25% WS—30 days	189.05 ± 6.84	144.90 ± 1.38
60% SM: 40% WS—fresh	245.39 ± 5.41	214.15 ± 2.58
90% SM: 10% WS—fresh	242.32 ± 3.57	185.83 ± 4.43

* Standard deviation.

3.2. GHGs Emissions Based on Anaerobic Digestion

A rough estimation of the GHG emissions was performed based on the anaerobic digestion tests. The sum of CH_4 and CO_2 production for each tested substrate represents the organic carbon that could have been released due to biological activities. Specifically, composting processes are intensified due to aerobic conditions, and CO_2 is released by aerobic microorganisms; CH_4 could also be produced when anaerobic conditions occur. The total volume of CO_2 emissions per g of VS inside the sheep barn could be equal to the sum of CO_2 and CH_4 volumes per g of VS added in the batch experiments, as each mole of CH_4 produced anaerobically could correspond to 1 mole of CO_2 produced aerobically. The GHG estimation was performed by multiplying the amount of manure (g_{VS}) produced per sheep per day with the biogas potential (Equation (6)). The GHG emissions expressed as biogas potential based on the substrates used for anaerobic digestion are presented in Table 5.

The substrate with a ratio of 60SM:40WS obtained the highest biogas potential. Nevertheless, it should be highlighted that this ratio does not represent a realistic scenario due to being impractical and economically unviable, and it was tested in terms of investigating a possible higher methane production from the sheep barn wastes. This fact applies also to the 100% wheat straw ratio, which is not presented in Table 5.

Table 5. Estimation of GHG emissions expressed as biogas potential based on the availability of organic carbon during anaerobic digestion of substrates.

Substrates	Period (d)	Biogas-GHGs ($mL \cdot g_{VS}^{-1}$)	Available $g_{VS} \cdot Sheep^{-1} \cdot d^{-1}$ for Each Substrate	$L_{Biogas\text{-}GHGs} \cdot Sheep^{-1} \cdot d^{-1}$
Sheep manure	0 15 30	357.40 ± 2.20 * 221.50 ± 5.49 227.25 ± 4.45	128.21 54.57 52.18	45.82 ± 0.28 12.09 ± 0.30 11.86 ± 0.23
Sheep manure 75% + Wheat straw 25%	0 15 30	438.49 ± 1.24 297.97 ± 7.62 333.95 ± 6.84	300 74.72 72.27	131.55 ± 0.37 22.26 ± 0.57 24.13 ± 0.49
Sheep manure 90% + Wheat straw 10%	0	428.15 ± 6.57	185.76	79.53 ± 1.22
Sheep manure 60% + Wheat straw 40%	0	459.54 ± 6.60	472.44	217.11 ± 3.12

* Standard deviation.

3.3. *Sheep Barn Energy Analysis*

All the substrates presented in Table 5 were used, except for the ratio of 60% SM: 40% WS, mainly for economic viability reasons. The potential heat gained by burning the biogas for energy analysis was named $\dot{Q}_{bgX}$ (where X is the numerical order of the first seven substrates in Table 5. The automation system operated for 10 h daily without technical problems. The time durations for which the windows remained closed for each day are presented in Table 6. Temperature and relative humidity graphs of the climate inside and outside the sheep barn as well as gas measurements during the experimental period are presented in detail in the Supplementary Materials (Sections S3 and S4) .

Table 6. Time duration of closed windows during the operation of the automation system.

Day	23 January	24 January	25 January	26 January	27 January
Time (min) duration of closed windows	240	345	300	180	90
Percentage of time (%) with closed windows during operation	40%	57.5%	50%	30%	15%

The potential biogas production and the relevant energy exchanges in the barn were calculated by Equations (1)–(4) and (6), and they are presented in Table 7. Energy exchanges in the sheep barn for all of the scenarios are presented in detail in the Supplementary Materials (Section S1). Heat losses from the sheep barn and body heat generated from sheep were valued for the energy analysis only when the windows were closed during the operation period of 10 h. Two different settings for energy analysis were examined. The first setting considered all the heating energy produced from biogas by the anaerobic digestion of the substrates. So, additional equipment, such as heat or biogas storage, was utilized for the exploitation of heat energy when it was required to raise the temperature inside the barn and when the windows were closed. In the second setting, the storage equipment was not present, and the biogas was used continuously by the heating system. Biogas heat energy in this setting was calculated based on the volume of biogas produced via anaerobic digestion of substrates generated inside the barn and only when the windows were closed during the 10 h of the automated system operation. The exact hourly heat exchange for each category is illustrated in the Supplementary Materials.

Table 7. Sheep barn energy analysis for the two settings when the windows were closed during the operation of the automated system.

Day	23 January	24 January	25 January	26 January	27 January
Total energy amounts with a biogas or a heat storage (kWh)					
Heat losses ($\dot{Q}_{ch} + \dot{Q}_{inf}$)	160.4	149.5	229.3	124.5	28.8
Total body heat from sheep ($\dot{Q}_{an}$)	59.7	77.2	63.2	31.6	21.1

Table 7. *Cont.*

Day		23 January	24 January	25 January	26 January	27 January
Qbiogas ($\dot{Q}_{bg}$)	$\dot{Q}_{bg1}$	* (+) 417.5	(+) 417.5	(+) 417.5	(+) 417.5	(+) 417.5
	$\dot{Q}_{bg2}$	(+) 110.2	(+) 110.2	** (−) 110.2	(+) 110.2	(+) 110.2
	$\dot{Q}_{bg3}$	(+) 108.1	(+) 108.1	(−) 108.1	(+) 108.1	(+) 108.1
	$\dot{Q}_{bg4}$	(+) 1198.7	(+) 1198.7	(+) 1198.7	(+) 1198.7	(+) 1198.7
	$\dot{Q}_{bg5}$	(+) 202.8	(+) 202.8	(+) 202.8	(+) 202.8	(+) 202.8
	$\dot{Q}_{bg6}$	(+) 211.9	(+) 211.9	(+) 211.9	(+) 211.9	(+) 211.9
	$\dot{Q}_{bg7}$	(+) 724.7	(+) 724.7	(+) 724.7	(+) 724.7	(+) 724.7
Total energy amounts with continuous heating from biogas (kWh)						
Heat losses ($\dot{Q}_{ch} + \dot{Q}_{inf}$)		160.4	149.5	229.3	124.5	28.47
Total body heat from sheep ($\dot{Q}_{an}$)		59.7	77.2	63.2	31.6	21.1
Qbiogas ($\dot{Q}_{bg}$)	$\dot{Q}_{bg1}$	(−) 56.2	(+) 80.7	(−) 70.2	(−) 42.1	(+) 21.1
	$\dot{Q}_{bg2}$	(−) 44.1	(−) 63.4	(−) 55.1	(−) 33.1	(+) 16.5
	$\dot{Q}_{bg3}$	(−) 43.2	(−) 62.2	(−) 54.1	(−) 32.4	(+) 16.2
	$\dot{Q}_{bg4}$	(+) 479.5	(+) 689.3	(+) 599.4	(+) 359.6	(+) 179.8
	$\dot{Q}_{bg5}$	(−) 81.1	(−) 116.6	(−) 101.4	(−) 60.8	(+) 30.4
	$\dot{Q}_{bg6}$	(−) 84.8	(+) 121.8	(−) 106.0	(−) 63.6	(+) 31.8
	$\dot{Q}_{bg7}$	(+) 289.9	(+) 416.7	(+) 362.4	(+) 217.4	(+) 108.7

*: (+) when heat gain ($\dot{Q}_{an} + \dot{Q}_{bg}$X) is higher than heat losses, **: (−) when heat gain ($\dot{Q}_{an} + \dot{Q}_{bg}$X) is lower than heat losses.

3.4. Potential Energy for Heating of the Drinking Water

As shown in Table 7, in many cases, the $Q_{\cdot bg}$ potential was greater than required to cover the sheep barn heating needs. This surplus of heating energy could be used to increase the temperature of the water according to Equation (5). In the case of using the heat surplus to raise the temperature of the water, the biogas heating system should have a hot water distribution system (not a hot air distribution system) to have the capability to direct this heat to the drinking water tank each time it is not operating to cover the heating needs of the building. In Table 8, the remaining heat for each day with the $Q_{\cdot bg}$ scenario as well as the water temperature potential rise are presented.

Table 8. Residual heating energy used for the heating of the drinking water after being used for sheep barn heating and the corresponding rise in the drinking water temperature.

Day	23 January	24 January	25 January	26 January	27 January
Q_{w1} (kWh)	202.0	324.3	258.4	328.1	409.7
ΔT_{water1} (°C)	0.6	0.9	0.7	0.9	1.1
Q_{w2} (kWh)	0.0	32.4	0.0	0.0	102.4
ΔT_{water2} (°C)	0.0	0.1	0.0	0.0	0.3
Q_{w3} (kWh)	0.0	30.4	0.0	0.0	100.3
ΔT_{water3} (°C)	0.0	0.1	0.0	0.0	0.3
Q_{w4} (kWh)	768.4	1066.5	1039.6	1109.3	1190.9
ΔT_{water4} (°C)	2.1	3.0	2.9	3.1	3.3
Q_{w5} (kWh)	46.3	120.3	0.0	0.0	195.0
ΔT_{water5} (°C))	0.1	0.3	0.0	0.0	0.5
Q_{w6} (kWh)	52.9	129.0	0.0	0.0	204.1
ΔT_{water6} (°C)	0.1	0.4	0.0	0.0	0.6
Q_{w7} (kWh)	424.7	616.2	565.6	635.3	716.9
ΔT_{water7} (°C)	1.2	1.7	1.6	1.8	2.0

4. Discussion

Renewable energy heating systems in livestock buildings represent an efficient and environmentally friendly technology that contributes to the implementation of carbon neutral farming practices. The proposed operation for the experimental sheep barn in this work has a low initial investment and operating cost and promotes an efficient waste management approach. Previous studies have examined biogas production from animal waste and its utilization for generating electricity and heat [56–58], but not specifically for covering the heating needs of a sheep barn while achieving optimum welfare conditions for the animals.

The total methane production of all substrates varied from 296.71 $mL \cdot g^{-1}{}_{VS}$ to 123.25 $mL \cdot g^{-1}{}_{VS}$. The results showed that as the time from generation to collection of waste increased, biogas production decreased. An interesting finding was that samples collected after 30 days obtained a slightly higher production than samples collected after 15 days. A possible explanation is that lignocellulosic substances were possibly degraded by aerobic microorganisms and became more accessible for biogas production during anaerobic digestion. Moreover, the addition of nitrogen can significantly increase the degradation rate of lignin. The application of even 0.12% nitrogen could increase the degradability of lignin up to 29.8%, as studied by Vázquez et al. [59]. Thus, the presence of sheep urine inside the barn, which contains N [60], could probably have had an impact on the lignocellulosic substances. The 75 SM: 25 WS ratio that was already used by the farmer in the experimental sheep barn was lower in terms of methane production by 2.02% and 3.25% compared with the ratios of 90SM:10WS and 60SM:40WS, respectively. These ratios were investigated to obtain higher methane productions in comparison with the current practice. However, the higher methane production obtained by these substrates was not enough in comparison with the ratio already used in the sheep barn. Therefore, the best of the examined ratios for producing methane for heating purposes in the sheep barn was the farmer's current practice (75 SM: 25 WS). Similar wheat straw methane yields were reported in previous studies [61]. Regarding fresh sheep manure mono-digestion, the 212.95 $mL \cdot g_{VS}{}^{-1}$ obtained by anaerobic digestion was also comparable with previous research [62].

An estimation of the biogas potential per sheep per day was performed based on the conducted anaerobic digestion and co-digestion tests. In all cases, substrates that contained fresh sheep manure had a significantly higher biogas potential than their equivalents, containing 15 days and 30 days old samples. The current sheep barn's ratio (75 SM: 25 WS) could provide the highest potential of up to 131.55 $L_{biogas} \cdot sheep^{-1} \cdot d^{-1}$. Time seemed to have a significant effect on the reduction of biogas potential. Specifically, for the sheep manure mono-digestion, at least 73.61% of the biogas was lost when fresh manure was not used, whereas for the mixed 75 SM: 25 WS tests, at least 81.66% was observed. It could be indicated that these reductions would possibly result in lost emissions if the substrates were obtained at those later time periods. By performing the anaerobic digestion process, organic carbon can be utilized that otherwise could have been emissions, which results in GHG savings.

The installation of the automation system in combination with heat from biogas had a positive effect on the microclimate of the structure. In particular, the heat gained from the sheep contributed to a rise in the temperature inside the structure, providing an environment within the TNZ range. When the windows were closed, the trapped heat produced by the sheep was sometimes enough to fully cover the structure's heat losses. When this was not possible, heat losses ($\dot{Q}_{ch} + \dot{Q}_{inf}$) were covered in a percentage between 27–60%, 30–56%, 25–52%, 25–30% and 55–58% during 23, 24, 25, 26 and 27 January, respectively. In the morning hours, when the ambient environment temperature was sometimes below zero, the energy covered was the lower threshold of the ranges mentioned above for each day. The higher threshold occurred during hours when the ambient environment temperature was close to 10 °C.

The examined setting of the hypothetical biogas heating system that was equipped also with a biogas or heat storage system covered all of the heat losses from the sheep barn (Table 7, $\dot{Q}_{bg1}$) when fresh manure and ratios of manure and wheat straw were used. However, manure that was left in the sheep barn for the examined two periods covered most of the heating needs of the barn, except for one day. At the second setting, where the biogas heating system did not have storage equipment, only the weight ratios (%) of 75:25 and 90:10 of fresh sheep manure and wheat straw succeeded in covering the heat loss from the sheep barn.

The percentage of the losses covered by the hypothetical biogas heating system is strongly dependent on the heat distribution system. In the first setting, in 94% of the examined cases (Table 7), the heat losses were covered by the combined heat gain from the sheep's body metabolism and from the biogas heating unit. This could be done by storing the energy in a biogas or a heat-insulated buffer tank and using it only when the temperature of the barn is lower than the LCT of sheep. In the option of a heated buffer tank, the heating medium, possibly water, will help to dissipate the heat through a distribution system. Therefore, this setting has a more complex installation and a higher initial investment cost than the second setting, where no energy storage is present. In the second setting, the produced biogas was continuously burned, and the heat gain was distributed inside the barn by an air heater, and it was not possible to store the heating energy. For the second setting, the heat energy losses were fully covered at a percentage of 48%, but the investment and operation costs of this system are lower. Up until now, low temperatures in sheep barns were not systematically addressed in most cases, both in research and in real-world conditions. Some passive methods are mentioned in the literature [63,64] for preventing wind speed (natural or artificial shelter belts), which leads to low temperatures inside the barn. The main problem with non-controlled ventilation systems in sheep barns is that windows or doors remain open most of the time to avoid inadequate ventilation (high relative humidity or NH_3 concentrations). It has been found that this approach leads to an indoor temperature that is a little higher than the ambient outdoor temperature (1–2 °C) [64]. This temperature is outside of the thermoneutral zone of the sheep, with all the negative consequences that have already been mentioned. Thus, the proposed system provides an overall management of the microclimatic conditions in the barn without affecting any aspect of the animals' welfare or health compared to traditional or common practices. Some suggestions for maintaining the proper temperature for newborn sheep are also mentioned, such as the use of heating lamps [65], but this system is accompanied by energy consumption and requires careful installation to avoid possible accidents.

The heating of the drinking water could be provided by the biogas heating system (1st setting with biogas or heat storage). Installation of an insulated drinking water storage tank is crucial to delivering the surplus of energy (Table 7) and achieving the corresponding drinking water temperature. From a technical point of view, heat surplus will be driven to the water storage tank and can be easily implemented as an extension of the aforementioned heating system. The temperature rise was significant in the co-digestion of fresh manure and wheat straw mixtures of 75:25 (Q_{w4}, Table 8) and 90:10 (Q_{w7}, Table 8), where the temperature could rise, on average, for all the examined days by 2.9 °C and 1.7 °C, respectively. In the rest of the cases, the temperature rise was not significant. It should also be noticed that the use of surplus energy for water heating may contrast with the use of energy to cover heating needs inside the building at a later time, as described above. It must be decided by the producer which option is more beneficial for the sheep and where to utilize this energy amount for heating the building or for drinking water. In practice the artificial heating of water [65,66] or continuous replacement are suggested [65]. These practices result in increased energy consumption and labor costs compared to the proposed system, which utilizes a renewable energy source.

The temperature inside the barn was higher in comparison with the environment outside the sheep barn, so the sheep's dry matter intake (DMI) was reduced. This is portrayed by the Equation (8), as DMI depends on the temperature of the environment [6].

$$DMI = 111.3 - 0.52 \cdot Ti \quad (8)$$

The DMI reduction is higher when the temperature difference between indoor and ambient environment temperatures is wider. The decrease in DMI also reduces the feed intake by the sheep, which probably results in lower waste generation and, consequently, lower GHG emissions. Moreover, it leads to a further reduction of CH_4 emissions, which originate by ruminal fermentation. The reduction of CH_4 emissions ($kg_{ch4} \cdot sheep^{-1}\ d^{-1}$) based on the DMI ($kg_{feed} \cdot sheep^{-1} \cdot d^{-1}$) decrease is estimated by Equation (9) [19].

$$(CH_{4w}) = 5.6488 \cdot (DMI_w) + 0.6518 \quad (9)$$

The reduction of methane from ruminal fermentation is an added environmental benefit of the proposed operation rationale, but it could also lead to energy conservation since less energy is consumed for feed production and transportation to the sheep barn.

The extent of the internal temperature of the sheep barn that can be achieved, aiming to achieve increased feed conversion and GHG emissions reduction.

5. Conclusions

The sustainability and benefits of an automation system in a sheep barn combined with a hypothetical biogas heating system utilizing the barn's waste were examined in this work. The results indicated that this method has a positive effect on the microclimate of the livestock building and the animals' welfare by maintaining the temperature within the range of the thermoneutral zone. The proposed solution covered the heat losses of the sheep barn in 94% of the examined cases in the first setting where energy storage was used. Thus, the aim of this study was achieved, and the initial hypothesis was confirmed. When the sheep barn heat losses were fully covered by the biogas heating system, surplus heat could also be used to increase the temperature of the drinking water from 1.7 °C to 2.9 °C. The outcome of this research highlights an agricultural practice for livestock buildings that is sustainable and reduces GHG emissions. Further research may be conducted to evaluate the addition of a solar system to increase the capability and capacity of the proposed practice.

Supplementary Materials: The following supporting information can be downloaded at: https://www.mdpi.com/article/10.3390/en16031087/s1, Section S1: Illustration of dynamic heat exchanges in the sheep barn, Section S2: Volatile Fatty Acids graphic illustration, Section S3: Temperature and relative humidity graphs, Section S4: Internal air quality of the sheep barn, Section S5: Repairs and installation of equipment for the automation system for opening and closing side and the roof windows, as also the location of sensors for monitoring the internal air quality of the sheep barn, Section S6: U—Value calculation.

Author Contributions: Conceptualization, A.A.L., V.K.F. and T.A.K.; methodology, A.A.L., V.K.F., S.D.K. and T.A.K.; investigation, A.A.L. and V.K.F.; writing—original draft preparation, A.A.L.; writing—review and editing, V.K.F., S.D.K., C.A.T., C.N.B. and T.A.K.; supervision, T.A.K.; funding acquisition, T.A.K. All authors have read and agreed to the published version of the manuscript.

Funding: This research was funded by the European Union's Horizon 2020 under the reSilienT fARminG by Adaptive microclimaTe managEment (STARGATE) project (grant number no. 818187).

Data Availability Statement: Not applicable.

Acknowledgments: The authors would like to acknowledge the European Union's Horizon 2020 and STARGATE project for funding the work and Manolis Stroikos for allowing us to use his sheep barn for experimental procedures.

Conflicts of Interest: The authors declare no conflict of interest.

Abbreviations

GHGs	Greenhouse gases	c_p	Water specific heat ($kJ \cdot kg^{-1} \cdot K^{-1}$)
BMP	Biomethane potential	n	Air exchanges (h^{-1})
TNZ	Thermoneutral zone	V	Air volume of the structure (m^3)
UCT	Upper Critical Temperature	Q_w	Heat required for water heating (kWh)
LCT	Lower Critical Temperature	m	Total water mass (kg)
PLC	Programmable logic controller	ΔT_{water}	Water temperature difference (K)
SM	Sheep manure	$Volume_{Biogas-GHGs}$	Volume of biogas potential—GHGs emissions
WS	Wheat straw	V_{CH_4}	Total methane production volume of sample (L)
w/w	Weight to weight	V_{CO_2}	Total carbon dioxide production volume of sample (L)
$\dot{Q}_{an}$	Total heat produced by animals in the sheep barn (W)	$WL_{t=n}$	Percentage of weight loss of manure (%), where t = n refers to the collection time period of the manure
$\dot{Q}_{bg}$	Potential heat produced by a heating system utilizing biogas (W)	$MW_{vs=1}$	Weight of manure to obtain 1 g_{VS} for that substrate (g)
$\dot{Q}_{ch}$	Heat losses by combined heat transfer mechanism (W)	$Volume_{dailyGas}$	The daily production volume of measured gas (mL)
$\dot{Q}_{inf}$	Heat losses by infiltration (W)	C_n	Daily gas concentration measurement (%)
	Performance rate of the heating system	C_{n-1}	Previous day gas concentration measurement (%)
$\dot{Q}_{bgp}$	Input energy of the system by the produced biogas (W)	V_{hs}	Batch reactor headspace volume (mL)
U	U value ($W \cdot m^2 \cdot K^{-1}$)	VS	Volatile solids
A	Heat exchange surface (m^2)	TS	Total solids
ΔT	Temperature difference of the prevailing temperature in the barn and the ambient outdoor temperature (°C)	GC	Gas chromatography
ρ	Air density ($kg \cdot m^{-3}$)	VFA	Volatile fatty acids

References

1. Cheng, M.; McCarl, B.; Fei, C. Climate Change and Livestock Production: A Literature Review. *Atmosphere* **2022**, *13*, 140. [CrossRef]
2. Barthelmie, R.J. Impact of Dietary Meat and Animal Products on GHG Footprints: The UK and the US. *Climate* **2022**, *10*, 43. [CrossRef]
3. Gordon, I.J. Review: Livestock production increasingly influences wildlife across the globe. *Animal* **2018**, *12*, s372–s382. [CrossRef]
4. Clarke, B.; Otto, F.; Stuart-Smith, R.; Harrington, L. Extreme weather impacts of climate change: An attribution perspective. *Environ. Res. Clim.* **2022**, *1*, 012001. [CrossRef]
5. Collier, R.J.; Baumgard, L.H.; Zimbelman, R.B.; Xiao, Y. Heat stress: Physiology of acclimation and adaptation. *Anim. Front.* **2018**, *9*, 12–19. [CrossRef] [PubMed]
6. Subcommittee on Environmental Stress and National Research Council. *Effect of Environment on Nutrient Requirements of Domestic Animals*; Subcommittee on Environmental Stress & National Research Council: Wahsington, DC, USA, 1981. [CrossRef]
7. Cui, Y.; Theo, E.; Gurler, T.; Su, Y.; Saffa, R. A comprehensive review on renewable and sustainable heating systems for poultry farming. *Int. J. Low-Carbon Technol.* **2019**, *15*, 121–142. [CrossRef]
8. Boltyanska, N. Justification of Choice of Heating System for Pigsty. *TEKA Int. Q. J. Mot. Veh. Oper. Energy Effic. Mech. Eng.* **2018**, *18*, 57–62. Available online: https://core.ac.uk/download/pdf/185694988.pdf (accessed on 22 December 2022).
9. Chernoivanov, V.; Katkov, A.; Gabitov, I.; Yukhin, G.; Martynov, V.; Khasanov, E.; Mudarisov, S.; Baltikov, D.; Khammatov, R.; Kovalev, P. Technical equipment of farms for comfortable cow keeping in winter conditions. *Bulg. J. Agric. Sci.* **2019**, *25*, 45. Available online: https://www.agrojournal.org/25/02s-06.pdf (accessed on 22 December 2022).
10. Fournel, S.; Ouellet, V.; Charbonneau, É. Practices for Alleviating Heat Stress of Dairy Cows in Humid Continental Climates: A Literature Review. *Animals* **2017**, *7*, 37. [CrossRef]
11. Burgos, R.; Odens, L.; Collier, R.; Baumgard, L.; VanBaale, M. Evaluation of Different Cooling Systems in Lactating Heat Stressed Dairy Cows in a Semi-Arid Environment. In *Arizona And New Mexico dairy newsletter*; N.M.S.U. Cooperative Extension-The University of Arizona, Ed.; Cooperative Extension-The University of Arizona: Tucson, AZ, USA; New Mexico State University: Tucson, AZ, USA, 2007; Available online: https://cals.arizona.edu/extension/dairy/az_nm_newsletter/2007/april.pdf (accessed on 22 December 2022).
12. Anarbaev, A.; Zakhidov, R.; Tursunov, O.; Kodirov, D.; Vakhidov, U.; Khaliknazarov, U.; Yusupov, Z.; Kushev, A. Using of evaporative cooling systems in poultry farms. *IOP Conf. Series: Earth Environ. Sci.* **2020**, *614*, 012015. [CrossRef]
13. Gerber, P.J.; Steinfeld, H.; Henderson, B.; Mottet, A.; Opio, C.; Dijkman, J.; Falcucci, A.; Tempio, G. *Tackling Climate Change through Livestock—A Global Assessment of Emissions and Mitigation Opportunities*; Food and Agriculture Organization of the United Nations (FAO), Ed.; Food and Agriculture Organization of the United Nations (FAO): Rome, Italy, 2013; Available online: http://www.fao.org/docrep/018/i3437e/i3437e.pdf (accessed on 22 December 2022).
14. Food and Agriculture Organization of the United Nations (FAO). *Results | Global Livestock Environmental Assessment Model (GLEAM) | Food and Agriculture Organization of the United Nations*; Food and Agriculture Organization of the United Nations (FAO): Rome, Italy, 2019; Available online: http://www.fao.org/gleam/results/en/ (accessed on 22 December 2022).
15. Bhatt, A.; Abbassi, B. Review of environmental performance of sheep farming using life cycle assessment. *J. Clean. Prod.* **2021**, *293*, 126192. [CrossRef]

16. Ekpe, E.D.; Christopherson, R.J. Metabolic and endocrine responses to cold and feed restriction in ruminants. *Can. J. Anim. Sci.* **2000**, *80*, 87–95. [CrossRef]
17. Mullender, S.; Zaralis, K.; Pardo, G.; del Prado, A.; Dellar, M.; Ruiz, D.Y.; Carabaño, M.J. D.3.1: Report on review of information on FP7 projects and literature on climate change and small ruminants. In *ISAGE (Innovation for Sustainable Sheep and Goat Production in Europe)/Grant Agreement Number: 679302*. 2018. Available online: https://ec.europa.eu/research/participants/documents/downloadPublic?documentIds=080166e5b7c851c9&appId=PPGMS (accessed on 22 December 2022).
18. Pardo, G.; del Prado, A.; Batalla, I.; Carabaño, M.J.; Ramón, M.; Belanche, A. D.3.2: Report on development of meta-models on the effect of climate parameters on animal productivity and welfare. In *ISAGE (Innovation for Sustainable Sheep and Goat Production in Europe)/Grant Agreement Number: 679302*. 2019. Available online: https://www.isage.eu/wp-content/uploads/D3.2_Report-on-development-of-meta-models-on-the-effect-of-climate-parameters-on-animal-productivity-and-welfare.pdf (accessed on 22 December 2022).
19. del Prado, A.; Batalla, I.; Pardo, G.; Jebari, A.; Ragkos, A.; Theodoridis, A.; Arsenos, G. D.4.3: New holistic model that can be used to redesign terrestrial small ruminant's livestock systems. In *ISAGE (Innovation for Sustainable Sheep and Goat Production in Europe)/Grant Agreement Number: 679302*. 2019. Available online: https://www.isage.eu/wp-content/uploads/D4.3_Holistic-farm-model.pdf (accessed on 22 December 2022).
20. Shi, L.; Xu, Y.; Jin, X.; Wang, Z.; Mao, C.; Guo, S.; Yan, S.; Shi, B. Influence of Cold Environments on Growth, Antioxidant Status, Immunity and Expression of Related Genes in Lambs. *Animals* **2022**, *12*, 2535. [CrossRef] [PubMed]
21. Opio, C.; Gerber, P.; Mottet, A.; Falcucci, A.; Tempio, G.; MacLeod, M.; Vellinga, T.; Henderson, B.; Steinfeld, H. *Greenhouse Gas Emissions from Ruminant Supply Chains—A Global Life Cycle Assessment*; Food and Agriculture Organization of the United Nations (FAO): Rome, Italy, 2013; Available online: https://www.fao.org/3/i3461e/i3461e.pdf (accessed on 22 December 2022).
22. Paganoni, B.; Rose, G.; MacLeay, C.; Jones, C.; Brown, D.J.; Kearney, G.; Ferguson, M.; Thompson, A.N. More feed efficient sheep produce less methane and carbon dioxide when eating high-quality pellets. *J. Anim. Sci.* **2017**, *95*, 3839–3850. [CrossRef] [PubMed]
23. Robinson, D.L.; Goopy, J.P.; Donaldson, A.; Woodgate, R.; Oddy, V.; Hegarty, R. Sire and liveweight affect feed intake and methane emissions of sheep confined in respiration chambers. *Animal* **2014**, *8*, 1935–1944. [CrossRef]
24. Hammond, K.; Burke, J.; Koolaard, J.; Muetzel, S.; Pinares-Patiño, C.; Waghorn, G. Effects of feed intake on enteric methane emissions from sheep fed fresh white clover (Trifolium repens) and perennial ryegrass (Lolium perenne) forages. *Anim. Feed. Sci. Technol.* **2013**, *179*, 121–132. [CrossRef]
25. Muir, S.; Linden, N.; Kennedy, A.; Knight, M.; Paganoni, B.; Kearney, G.; Thompson, A.; Behrendt, R. Correlations between feed intake, residual feed intake and methane emissions in Maternal Composite ewes at post weaning, hogget and adult ages. *Small Rumin. Res.* **2020**, *192*, 106241. [CrossRef]
26. Da-lu, C.; Yu-hong, G.; Wei-ting, G.; Xin-sheng, S.; Chao, W.; Wei-tao, Z.; Juan-juan, Z.; Shou-pei, Z. Effect of drinking water temperature on production performance, blood biochemical parameters and rumen microstructure of fattening sheep in winter. *Acta Prataculturae Sin.* **2020**, *29*, 57–66. [CrossRef]
27. Wiseman, T. Livestock Water Is Essential, Even in Winter. 2018. Available online: https://u.osu.edu/sheep/2018/02/13/livestock-water-is-essential-even-in-winter/ (accessed on 22 December 2022).
28. Golher, D.M.; Thirumurugan, P.; Patel, B.H.M.; Upadhyay, V.K. Effect of drinking water temperature on water intake, feed intake and milk production of crossbred dairy cattle at high altitude temperate Himalayas. *Indian J. Anim. Sci.* **2014**, *84*, 1101–1104.
29. McArthur, A.J. Air movement and heat loss from sheep. III. Components of insulation in a controlled environment. In *Proceedings of the Royal Society of London. Series B, Biological Sciences*; Royal Society: London, UK, 1980; pp. 219–237. [CrossRef]
30. Stockman, C. The Physiological and Behavioural Responses of Sheep Exposed to Heat Load within Intensive Sheep Industries. In *School of Veterinary and Biomedical Sciences*; Murdoch University, Western Australia: Murdoch, WA, Australia, 2006; Available online: http://researchrepository.murdoch.edu.au/id/eprint/437 (accessed on 22 December 2022).
31. Hakawati, R.; Smyth, B.M.; McCullough, G.; De Rosa, F.; Rooney, D. What is the most energy efficient route for biogas utilization: Heat, electricity or transport? *Appl. Energy* **2017**, *206*, 1076–1087. [CrossRef]
32. European Comission. Communication from the commission to the european parliament, the council, the european economic and social committee and the committee of the regions: 'Fit for 55': Delivering the EU's 2030 Climate Target on the way to climate neutrality. In *52021DC0550*; European Comission: Brussels, Belgium, 2021; Available online: https://eur-lex.europa.eu/legal-content/EN/TXT/?uri=CELEX%3A52021DC0550 (accessed on 22 December 2022).
33. Albright, L.D. *Environment Control for Animals and Plants*; De Vore-Hansen, P., Ed.; American Society of Agricultural Engineers: St. Joseph, MI, USA, 1990.
34. Nikita-Martzopoulou, C. *Livestock structures*; Giahoudis Publications: Thessaloniki, Greece, 2006; Available online: http://libsearch.teiep.gr/Record/00014445 (accessed on 22 December 2022)ISBN 9607425952.
35. The engineering toolbox official website, Solids, Liquids and Gases - Thermal Conductivities. Available online: https://www.engineeringtoolbox.com/thermal-conductivity-d_429.html (accessed on 22 December 2022).
36. The engineering toolbox official website. Metals, Metallic Elements and Alloys - Thermal Conductivities. Available online: https://www.engineeringtoolbox.com/thermal-conductivity-metals-d_858.html (accessed on 22 December 2022).
37. Knauf offisial Website. Glass Mineral Wool Insulation. 2022. Available online: https://www.knaufinsulation.co.uk/glass-mineral-wool-insulation (accessed on 22 December 2022).

38. Strip curtains direct official website, The benefits of pvc as a material. Available online: https://stripcurtainsdirect.co.uk/blogs/news/the-benefits-of-pvc-as-a-material (accessed on 22 December 2022).
39. Seyfi, S.U.; Ilhan, F. Effects of hourly, daily and seasonal variation of hazardous gases and climatic factors on the welfare of sheep housed in solid-floor confinement barns. *South Afr. J. Anim. Sci.* **2019**, *49*, 441. [CrossRef]
40. KiliÇ, İ.; Onuk, A.; Şimşek, E.; Yaslıoğlu, E. *Bir Koyun Ağılında Amonyak ve Karbondioksit Konsantrasyonları*; Kahramanmaraş Sütçü İmam Üniversitesi Doğa Bilimleri Dergisi: Kahramanmaraş, Türkiye, 2017. [CrossRef]
41. da Silva, R.G. Chapter 12 Weather and Climate and Animal Production. Guide to Agricultural Meteorological Practices. *Switzerland: World AgroMeteorological Information Service.* 2010. Available online: http://www.wamis.org/agm/gamp/GAMP_Chap12.pdf (accessed on 22 December 2022).
42. U.S. Department of Agriculture. *Soil Conservation Service, Agricultural Waste Management Field Handbook, Part 651, Chapter 4. USDA*; 1992. Available online: https://directives.sc.egov.usda.gov/OpenNonWebContent.aspx?content=31475.wba (accessed on 22 December 2022).
43. Greek Ministry, Code of Good Agricultural Practice No. 85167/820, Ministerial Decision, (2000), Annex, Table 2, Official Gazette (477 B′). Available online: https://www.elinyae.gr/sites/default/files/2019-07/477b_2000.1438762549093.pdf (accessed on 22 December 2022).
44. Angelidaki, I.; Alves, M.M.; Bolzonella, D.; Borzacconi, L.; Campos, J.L.; Guwy, A.J.; Kalyuzhnyi, S.; Jenicek, P.; Van Lier, J.B. Defining the biomethane potential (BMP) of solid organic wastes and energy crops: A proposed protocol for batch assays. *Water Sci. Technol.* **2009**, *59*, 927–934. [CrossRef] [PubMed]
45. Balat, M.; Balat, H. Biogas as a Renewable Energy Source—A Review. *Energy Sources Part A Recover. Util. Environ. Eff.* **2009**, *31*, 1280–1293. [CrossRef]
46. ISSO (Institution for the Study and Promotion of Research in the Field of Building Services). *Method for the Calculation of the Design Heat Loss for High Spaces—Calculation of the Design Heat Loss of Large Enclosures and Rooms with a Height that Exceeds 5 Metres*; ISSO: Rotterdam, The Netherlands, 2006; Available online: https://euro-air.com/wp-content/uploads/2021/08/ISSO-publication-57E.pdf (accessed on 22 December 2022).
47. Cengel, Y.; Boles, C. *Thermodynamics: An Engineering Approach*, 8th ed.; McGraw-Hill Education: New York, NY, USA, 2014; Available online: https://studylib.net/doc/25885968/thermodynamics-an-engineering-approach-8th-edition-by-yun (accessed on 22 December 2022).
48. Meehan, M.A.; Stokka, G.; Mostrom, M. *Livestock Water Requirements*; North Dakota State University: Fargo, NA, USA, 2021; Available online: https://www.ag.ndsu.edu/publications/livestock/livestock-water-requirements (accessed on 22 December 2022).
49. Rice, E.W.; Baird, R.; Eaton, A. *Standard Methods for the Examination of Water and Wastewater*, 23rd ed.; American Public Health Association: Washington, DC, USA; American Water Works Association: Austin, TX, USA; Water Environment Federation: Alexandria, VA, USA, 2017; Volume 1, p. 1546.
50. Kalamaras, S.D.; Vasileiadis, S.; Karas, P.; Angelidaki, I.; Kotsopoulos, T.A. Microbial adaptation to high ammonia concentrations during anaerobic digestion of manure-based feedstock: Biomethanation and 16S rRNA gene sequencing. *J. Chem. Technol. Biotechnol.* **2020**, *95*, 1970–1979. [CrossRef]
51. Holliger, C.; Alves, M.; Andrade, D.; Angelidaki, I.; Astals, S.; Baier, U.; Bougrier, C.; Buffière, P.; Carballa, M.; De Wilde, V.; et al. Towards a standardization of biomethane potential tests. *Water Sci. Technol.* **2016**, *74*, 2515–2522. [CrossRef]
52. Christou, M.; Vasileiadis, S.; Kalamaras, S.; Karpouzas, D.; Angelidaki, I.; Kotsopoulos, T. Ammonia-induced inhibition of manure-based continuous biomethanation process under different organic loading rates and associated microbial community dynamics. *Bioresour. Technol.* **2020**, *320*, 124323. [CrossRef]
53. Kalamaras, S.D.; Vitoulis, G.; Christou, M.L.; Sfetsas, T.; Tziakas, S.; Fragos, V.; Samaras, P.; Kotsopoulos, T.A. The Effect of Ammonia Toxicity on Methane Production of a Full-Scale Biogas Plant—An Estimation Method. *Energies* **2021**, *14*, 5031. [CrossRef]
54. European Comission. *Directive 2010/63/EU of the European Parliament and of the Council of 22 September 2010 on the Protection of Animals Used for Scientific Purposes*; European Comission: Brussels, Belgium, 2019; Available online: https://eur-lex.europa.eu/eli/dir/2010/63/2019-06-26 (accessed on 22 December 2022).
55. European Comission. *Council Directive of 24 November 1986 on the Approximation of Laws, Regulations and Administrative Provisions of the Member States Regarding the Protection of Animals Used for Experimental and Other Scientific Purposes (86/609/EEC)*; European Comission: Brussels, Belgium, 2010; Available online: https://eur-lex.europa.eu/eli/dir/1986/609/2013-05-10 (accessed on 22 December 2022).
56. Abdeshahian, P.; Lim, J.S.; Ho, W.S.; Hashim, H.; Lee, C.T. Potential of biogas production from farm animal waste in Malaysia. *Renew. Sustain. Energy Rev.* **2016**, *60*, 714–723. [CrossRef]
57. Ardebili, S.M.S. Green electricity generation potential from biogas produced by anaerobic digestion of farm animal waste and agriculture residues in Iran. *Renew. Energy* **2020**, *154*, 29–37. [CrossRef]
58. Akbulut, A.; Kose, R.; Akbulut, A. Technical and Economic Assessments of Biogas Production in a Family Size Digester Utilizing Different Feedstock Rotations: Döğer Case Study. *Int. J. Green Energy* **2013**, *11*, 113–128. [CrossRef]
59. Vázquez, G.; Antorrena, G.; González-Álvarez, J.; Freire, M.S.; López, S. Acetosolv pulping of pine wood. Kinetic modelling of lignin solubilization and condensation. *Bioresour. Technol.* **1997**, *59*, 121–127. [CrossRef]

60. Bristow, A.W.; Whitehead, D.C.; Cockburn, J.E. Nitrogenous constituents in the urine of cattle, sheep and goats. *J. Sci. Food Agric.* **1992**, *59*, 387–394. [CrossRef]
61. Raposo, F.; De la Rubia, M.A.; Fernández-Cegrí, V.; Borja, R. Anaerobic digestion of solid organic substrates in batch mode: An overview relating to methane yields and experimental procedures. *Renew. Sustain. Energy Rev.* **2012**, *16*, 861–877. [CrossRef]
62. González, R.; Blanco, D.; Cascallana, J.G.; Carrillo-Peña, D.; Gómez, X. Anaerobic Co-Digestion of Sheep Manure and Waste from a Potato Processing Factory: Techno-Economic Analysis. *Fermentation* **2021**, *7*, 235. [CrossRef]
63. National Farm Animal Care Council (NFACC). *Code of Practice for the Care and Handling of Sheep*; Canadian Sheep Federation and the National Farm Animal Care Council: Lacombe, AB, Canada, 2013; Available online: https://www.nfacc.ca/pdfs/codes/sheep_code_of_practice.pdf (accessed on 22 December 2022)ISBN 978-0-9919585-8-0.
64. Şimşek, E.; Yaslıoğlu, E.; Arıcı, I. A Research on Applicabilitiy of Greenhouse Sheep Barns in Bursa Contidions. *Atatürk Univ. J. Agric. Fac.* **2003**, *34*, 167–172. Available online: https://dergipark.org.tr/en/pub/ataunizfd/issue/2945/40809 (accessed on 22 December 2022).
65. Metzger, M. *Winter Management Tips for Sheep*; Michigan State University Externtion Program- Michigan State Univeristy: East Lansing, MI, USA, 2018.
66. Coffey, L.; Hale, M.; Scott, D. *Providing Water to Livestock in Frigid Weather*; ATTRA Sustainable Agriculture- The National Center for Appropriate Technology: Butte, MT, USA, 2021.

buildings

MDPI

Article

A Case Study of Empirical Validation of EnergyPlus Infiltration Models Based on Different Wind Data

Gabriela Bastos Porsani and Carlos Fernández Bandera *

School of Architecture, University of Navarra, 31009 Pamplona, Spain
* Correspondence: cfbandera@unav.es

Abstract: Building retrofitting is an efficient means of reducing greenhouse gas emissions. Its first focus is on building façade, as transmission and air leakage are the main sources of energy loss in buildings. Nowadays, building modellers cannot easily implement envelope air leakage and assume constant values, which results in erroneous energy estimates. Additionally, in energy simulations, a weather file is usually inserted with measurements provided by a weather station. In this study, we revealed the use of wind data from the weather file (herein as global wind) to calculate the infiltration of a test case in Spain, using the three algebraic equations of EnergyPlus. Furthermore, four other wind data were applied: eastbound and westbound winds from the weather file and two from in situ measurements (on the southeast and on the northwest façades). The fifteen combinations of the three infiltration models and the five wind data were empirically evaluated, using the tracer gas results performed during three different periods. The combinations were validated according to the American Society for Testing Materials D5157 standard criteria, and the best and the only ones that complied with the standard were those using the wind data from the southeast in situ sensor and the west wind from the weather station. The global wind was not able to generate accurate infiltration models, which raises doubts about its use in the highly-time calibration of energy models. However, its disaggregation was a cost-effective strategy to estimate the infiltration of this case study.

Keywords: wind data; tracer gas test; decay method; EnergyPlus; infiltration modelling; building retrofitting; ASTM D5157

Citation: Bastos Porsani, G.; Fernández Bandera, C. A Case Study of Empirical Validation of EnergyPlus Infiltration Models Based on Different Wind Data. *Buildings* **2023**, *13*, 511. https://doi.org/10.3390/buildings13020511

Academic Editors: Shi-Jie Cao, Wei Feng, Alessandro Cannavale and Eusébio Z.E. Conceição

Received: 14 November 2022
Revised: 12 December 2022
Accepted: 7 February 2023
Published: 13 February 2023

1. Introduction

In 2020, the European Union (EU) provided an unprecedented response to the coronavirus crisis that hit Europe and the world through hlNext Generation EU (NGEU): a temporary instrument called Recovery and Resilience Facility (RRF). The RRF makes EUR 750 billion (in 2018 prices) in grants to ease the recovery actions and investments carried out by Member States [1]. The main goal of the RRF is to reduce the social and economic footprint created by the pandemic and to make European societies and economies more resilient, sustainable, and ready for a new paradigm based on green and digital transitions. A minimum of 30% of expenditure should be dedicated to climate investment and reforms [2]. Among the flagship areas, buildings renovation appears as the most crucial, because building retrofitting is probably the most cost-effective way of cutting down greenhouse gas emissions [3].

Furthermore, the European Union proposed a set of directives to eradicate inefficient buildings, by enforcing the Energy Performance of Buildings Directive (EPBD). In this framework, building energy retrofit projects will rely on using digital twins, which can be created by using building information modelling (BIM) technology, followed by a building energy model (BEM) to quantify energy savings. For this purpose, the interoperability between BIM and BEM should be considered, in order to guarantee the confidence of investors in the energy efficiency sector [4]. However, BEM requires adjustments of input parameters that are unknown and difficult to measure, leading to high unpredictability in energy savings. Therefore, the limitations in the analysis of building energy retrofit are mainly due to the lack of accuracy of the model.

1.1. *Background and Motivation*

Retrofit projects typically use a calibrated BEM to ensure that building systems are properly modeled. There are challenges in the calibration process for the measurement and verification of energy savings, which can be based on mathematical algorithms and physical-based models and are evaluated according to uncertainty analysis [5–7]. As stated by the *ASHRAE Handbook: Fundamentals: SI edition* (2017) [8], several difficulties prevent achieving a calibrated simulation [9–12]. One of them is the method used to measure the input parameters needed for the simulation, i.e., infiltration values [13–16].

Infiltration is also known as the flow of outdoor air into a building through unintentional openings. Similar to natural ventilation, infiltration is driven by the pressure differences across the envelope caused by air and wind density variability generated by the temperature differences between indoor and outdoor air. For that reason, infiltration has two components: stack pressure and wind pressure. Stack pressure is the hydrostatic pressure produced by the mass of a column of air inside or outside a building [17]. When wind impinges on a façade, it creates a distribution of static pressures that depends on the wind speed, wind direction, surface orientation, air density, and surrounding conditions [18]. Moreover, if there is infiltration in a building envelope, it could affect the heating and cooling loads. If the outside air entering the building is cold, the heating load could increase by 13% to 30%. On the other hand, if it is warm, the cooling load could increase by 4% to 14% [19,20]. Furthermore, some studies estimated that air leakage could be responsible for 50% of energy loss [21,22]. In the current context, where regulations are being created to reduce carbon emissions [23], it is relevant to control any cause of increasing building loads, one of the most important of which is leakage airflow, and, therefore, its precise measurement and correct input in BEM software should be carried out.

1.2. *Infiltration Modelling*

Detailed models for air leakage can be produced by using multizone airflow or CFD software. In EnergyPlus, the AirFlowNetwork (AFN) model can be used to determine model infiltration and mixing airflow between zones with or without HVAC operation. It presents three empirical equations to calculate infiltration: ZoneInfiltration:DesignFlowRate, ZoneInfiltration:FlowCoefficient, and ZoneInfiltration:EffectiveLeakageArea. In addition to facilitating more accurate calculations of wind-driven infiltration, EnergyPlus calculates the wind speed as a function of height by using the input or default wind speed profile coefficients [24].

CONTAM is another multizone simulation software, which was developed at the National Institute of Standards and Technology (NIST) and is widely accepted to estimate infiltration [25]. In CONTAM models, it is possible to implement wind direction and wind speed, ideally for each thermal zone. Therefore, another possibility is coupling CONTAM with EnergyPlus (EP), but it can be a cumbersome process, and the main limitation of this co-simulation technique is related to the synchronisation time-step size of the quasi-dynamic method [26]. Additionally, when translating these results to EnergyPlus, infiltration rates are averaged over the entire exterior surface area, as was explained by Ng et al. [25,27]. Empirical approaches simplify multizone building airflow models and represent cost-effective solutions for non-expert users.

The wind is a key factor in the generation of air leakage, and since the EP empirical equations consider the wind speed data to quantify infiltration, it is important to analyse which wind speed values should be applied. In some studies, authors installed a dedicated weather station on the roof of their test case to use actual weather data in modelling, as is the case in Shrestha et al. and Bae et al. [28,29]. In contrast, Taddeo et al. [16] collected the wind speed data from a weather station 1 km far from their test space. Their wind speed values were corrected according to the height of the building. Winkler et al. [30] evaluated

EnergyPlus AFN models for residential infiltration. They determined five cases, and in each, they changed the test conditions. In relation to the wind speed, in the second and fifth cases, they used a typical meteorological year (TMY) wind speed data, while in the first and third cases, they applied a wind speed value of 5 m/s, and in the fourth case, 0 m/s. The aim of their study was to compare EP AFN models with CONTAM and building energy optimisation (BEopt) [31] models, but they did not use in situ wind speed data, and it was not in their scope to empirically assess the models.

1.3. Tracer Gas Technique

The most accurate way to determine a building's infiltration rate is to measure it. According to ASHRAE [8], tracer gas measurement is the reference technique. There are procedures (e.g., the Standard ASTM E741 test method [32]) that use gas to label indoor air, as stated by Sherman in different works [33–35]. There are three methods for tracer gas test: constant injection, constant concentration, and concentration decay; the latter is the easiest to implement [36].

As described by Cui et al. [37], the concentration decay method consists of injecting a dose of CO_2 and mixing it into the room. The decay method is based on the assumptions that (1) the background concentration is known, (2) infiltration out of the building is the main way of removing the CO_2 from the room, and (3) the CO_2 concentration within the room should be uniform. The decrease in CO_2 is recorded during a given period.

1.4. Contribution and Originality of the Research

This preliminary study aims to reduce the uncertainty of infiltration in the building energy model calibration, using only the EnergyPlus infiltration objects and different wind data. Although many studies evaluate wind-driven infiltration modelling, as far as the authors know, none of the published validation reports on air leakage empirically compare different wind data applied with the EnergyPlus infiltration models, with one of the wind data being measured in situ. In this study, we revealed the use of the wind data from the weather file (hereafter, the global wind) against four other wind data to estimate the infiltration of a test room in an apartment in Spain. The global wind is usually applied in energy simulations. There is a general consensus that EP only accounts for wind speed and does not have a wind direction component in the infiltration objects [38]. Therefore, we focused on the global wind and disaggregated it into two types of data: the eastbound wind and the westbound wind. In addition, we used the wind data recorded with sensors in situ: one on the southeast façade and the other on the northwest façade. We applied the five wind data to the three infiltration models of EnergyPlus, resulting in fifteen combinations, to verify which of these most accurately represented the infiltration of the test space.

We performed a tracer gas test based on the concentration decay method [37], to measure infiltration and empirically validate the results. CO_2 was chosen as tracer gas because it complies with the desirable qualities such as detectability, non-reactivity, and non-toxicity at low concentrations, and it is well stirred with air (similar density), so it should be differentiated from other components of air [39]. This in situ experiment was performed over 31 days in three different periods (summer, winter, and spring), and a total of 48.439 time-steps of one-minute data of CO_2, wind speed, and temperature were recorded.

The results were statistically verified. We calculated the standard deviation values for the analysis of the measurements. In addition, we assessed the accuracy of the 15 combinations with the mean bias error (MBE) values [40] and presented in Equation (A4). Then, we evaluated their correlation between the measured and predicted values according to the Standard Guide for Statistical Evaluation of Indoor Air Quality Models (ASTM D5157) requirements [41], which is suggested by ASHRAE [8] for the empirical validation of experimental evidence of indoor environment modelling. To the authors' knowledge, no previous study has ever used all ASTM D5157 requirements to evaluate EP infiltration models.

The remainder of this paper is organised as follows: Section 2 describes the test room, the monitoring system, and the tracer gas experiment; Section 3 explains the method to calculate the air leakage of the test space, and Section 4 presents the MBE and the ASTM D5157 Standard used to evaluate the models. Finally, in Section 5, we analyse the recorded data, and Section 6 shows the results of the fifteen combinations. Section 7 concludes this research.

2. Experimental Procedure

2.1. Test Room and Instrumentation

A tracer gas test of the concentration decay method was conducted in the living room of an attic of a seven-story apartment building in Pamplona, Spain (see Figure 1). The room with an area of 29.50 m² has two main façades (southeast and southwest) made of perforated brick of 115 mm, with an air cavity of 30 mm, as well as 50 mm EPS foam, 70 mm hollow brick, and a last layer of gypsum plaster of 15 mm. The interior walls are constructed with gypsum plaster (20 mm), hollow brick (75 mm), and gypsum plaster (20 mm). Figure 2 shows the dimensions of the openings and their position in the room.

The in situ monitoring system consisted of two types of sensors: DELTA (model OHM HD37VBTV.1) and EXTECH (model CO210). Besides the two CO_2 DELTA sensors (ppm) installed in the living room, three CO_2 EXTECH sensors (ppm) were also implemented, in order to verify the homogeneity of the injected gas (see Figures 3 and 4). Both had the same accuracy of $\pm$ 50 ppm, but the DELTA sensors were connected to the HOBO management system of the room, making it easier to manage and download their data.

Figure 1. External view of the apartment is indicated in yellow.

Moreover, two sensors of indoor ambient temperature (°C), model HOBO ZW-006 (with $\pm$2% precision), were installed at different heights (0.80 m and 1.75 m above the ground), and their average was used in the equations.

In relation to weather conditions, a total of five sensors were placed outside the apartment. Two wind speed sensors (m/s) were installed, model AHLBORN FVA 615-2 with $\pm$0.5 m/s accuracy: one at 1.60 m above the ground on the northwest terrace and the other at 1.90 m above the ground on the southeast terrace. The three other sensors were placed 2.32 m above the ground on the southeast façade: one for CO_2 (ppm), model Delta OHM HD37VBTV.1, and two for outdoor ambient temperature (°C), model HOBO ZW-006.

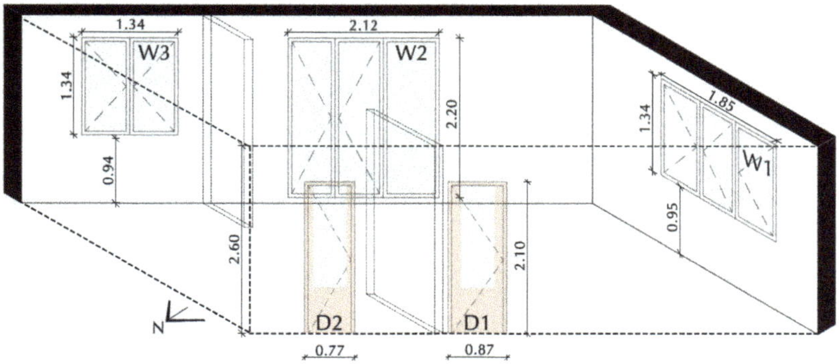

Figure 2. Isometric representation of the five openings of the room. The panels in two interior wooden doors are divided into two parts: The first is glazed with wooden mullions, and the lower part is only made of wood. The main door (D1) is 1.72 m², and the secondary door (D2) is 1.61 m². The southwest façade has a tilt-and-turn window (W1) (2.47 m²), the southeast façade has a door window (W2) with an area of 4.66 m², and a tilt-and-turn window (W3) with an area of 1.79 m². All windows are made of aluminium, double-clear glass of 3 mm each, and an air cavity of 13 mm. See Figure 3 for the plan view.

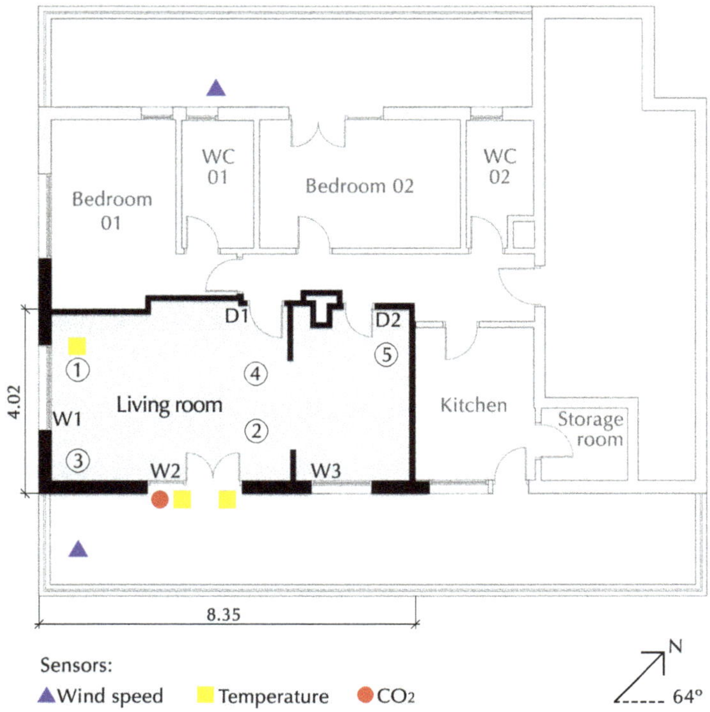

Figure 3. Plan of the apartment. W, window; D, door. Numbers are CO_2 sensors. Sensor 1 was installed at 0.40 m above the ground; sensor 2 at 0.74 m; sensor 3 at 1.19 m; sensor 4 at 0.74 m, and sensor 5 at 1.52 m.

Figure 4. Northeast side view of the room. D, door; numbers are CO_2 sensors.

In order to capture the variation in the data as completely as possible, both indoor and outdoor data were recorded at a time-step of one minute.

In addition to the in situ data, the wind speeds collected at a weather station were also used. The station was installed on the roof of a commercial building located 2 km away from the test house. Therefore, five different wind data were used to calculate infiltration, and they were organised as explained in Figure 5.

2.2. *Tracer Gas Test*

The tracer gas concentration decay test was carried out during three different seasons, and its data were organised into three periods, where T represents training, and C represents checking. The first period (P_1_T) refers to 9 days of summer, from 20 June to 2 July 2021 (10.545 time-steps of data); P_1_C represents 11 days of winter from 10 December 2021 to 9 January 2022 (24.869 time-steps); and P_2_C is the last period with 11 days of experiment in spring from 24 March to 24 April 2022 (13.025 time-steps). Normally, the apartment is occupied, but in all periods, it was maintained unoccupied, in order to avoid occupancy contamination in the data.

It is important to have in mind that the interior doors of the living room were closed and sealed from the other rooms, which physically constitute the thermal zones in a BEM. If the zones are not defined and analysed in a separate manner, modellers should use ZoneMixing EnergyPlus object. Furthermore, the openings of the test room were kept closed during the experiment. Under these conditions, the procedure consisted of an injection of CO_2 twice into the room, once to the east and once to the west. Before spraying the CO_2, the windows were opened with the aim of not over-pressurising the test room.

This experiment was applied as a tool to empirically evaluate the estimated infiltration values. Therefore, the analysis of the results presented in Section 6 was carried out by comparing the observed CO_2 versus predicted CO_2 .

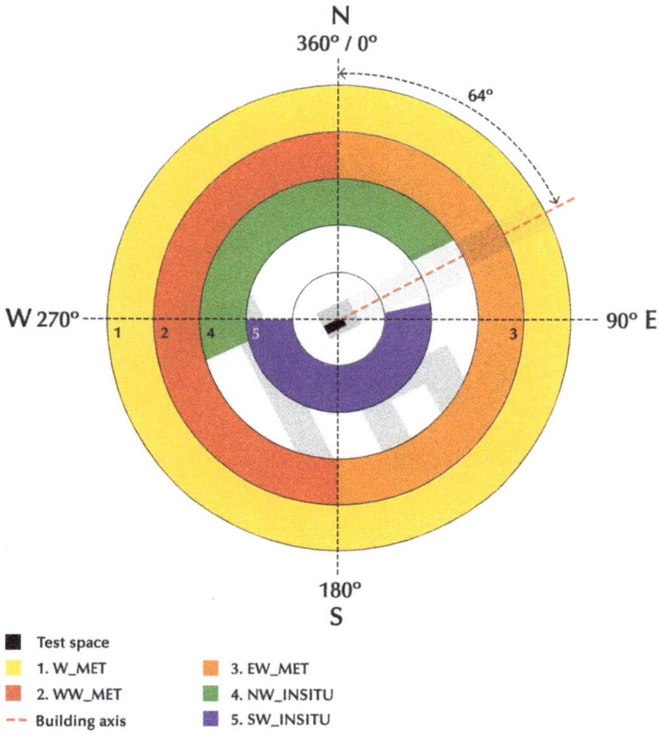

Figure 5. Graphical representation of selected wind speed data. W_MET is the weather station winds (herein as global wind) from 0° to 360°; WW_MET and EW_MET are the weather station west and east winds from 181° to 360° and 0° to 180°, respectively; NW_INSITU is the wind data recorded by the wind speed sensor on the northwest terrace; SW_INSITU is the wind data collected on the southeast terrace.

3. Method of Calculating Air Leakage

To accurately calculate infiltration, the analysis of the fifteen combinations was performed in three main steps, which are explained in the following subsections.

3.1. First Step: State of the Art of Infiltration in EnergyPlus Software

The first step was the air leakage calculation using the three airflow objects provided by EnergyPlus: ZoneInfiltration: DesignFlowRate (DFR), ZoneInfiltration: FlowCoefficient (IFC), and ZoneInfiltration: EffectiveLeakageArea (ELA). All equations are detailed in Appendix A: Equations (A1)–(A3).

Each equation requires coefficient values that are often debated. For DFR, the EnergyPlus Input–Output manual [24] defaults to constant infiltration, A = 1, B = C = D = 0.0. DOE-2, a predecessor of EnergyPlus, uses a base wind speed of 4.47 m/s to calculate I_{design} with C = 0.224, A = B = D = 0.0. BLAST, another predecessor of EnergyPlus, uses a base wind speed of 3.35 m/s to calculate I_{design} with A = 0.606, B = 0.03636, C = 0.1177, D = 0.0. Other methods have been developed and published to calculate the coefficients and I_{design}, such as those by Ng et al. [42]. For IFC and ELA, the coefficients were determined by EnergyPlus for a three-story building.

The Input–Output document [24] of EnergyPlus recommends using ad hoc coefficients for a specific site. For this reason, in this work, there was no limitation in the range of the coefficients during model fitting, except I_{design}, which was set to 1 in order to easily compare the results of the other coefficients, and n value for IFC was limited between 0.60 and 0.70, as EnergyPlus determines in its document. The calculation of infiltration was carried out for each day and with each of the five wind data. To initiate the calculations, random coefficients were implemented in these objects, resulting in inaccurate infiltration values.

3.2. Second Step: CO_2 Decay Method

The second step was the generation of the CO_2 predicted curve. In this regard, Equation (1) is fed by the infiltration values generated by the three EnergyPlus models Equations (A1)–(A3). For a better estimation of the CO_2 concentration, the first 40 min, which refer to the concentration peaks, were removed, so that only the uniform mixture of CO_2 was used in Equation (1). Although there were five CO_2 sensors in the room, only the mean data of the two CO_2 DELTA sensors were used in the calculations, to facilitate the data management of the HOBO system.

In this study, the multi-point decay method was implemented, which yields more accuracy and fewer uncertainties than the two-point method [43]. The first value of C_p is calculated with $C_o - C_{bg}$ equal to the observed CO_2 concentration minus the daily average of outdoor CO_2 concentration at t = 0. Then, to estimate the second value of C_p, $C_o - C_{bg}$ is equal to the C_p value of the time-step before. This process was repeated for each time-step to generate the CO_2 predicted curve.

The decay method is described by [44]:

$$C_p = \left(C_o - C_{bg}\right)e^{-It} \quad (1)$$

where
C_p = predicted CO_2 concentration at time, t;
C_o = average of observed indoor CO_2 concentration in the space;
C_{bg} = daily average of measured outdoor CO_2 concentration in the air;
t = time, s;
I = infiltration of each time-step calculated by EP models.

As for infiltration, the simulation of CO_2 was performed for each decay day and with each wind data. As random coefficients of EP models were selected to start this process, a third step was necessary to increase the accuracy of the calculations.

3.3. Third Step: Model Fitting

The last step was to perform multivariate regression to find the suitable coefficients for Equations (A1)–(A3) based on each wind data. Therefore, the regression model was based on the objective function of minimising the sum of mean absolute error (MAE) between the observed and predicted CO_2 concentration, C_o, and C_p, and to this end, the model searched for the most accurate coefficients for the period and wind data. In this study, the coefficients found for P_1_T were applied to the winter and spring data as checking periods.

4. Statistical Evaluation

ASHRAE Handbook: Fundamentals: SI edition (2017) declares that it is crucial to apply valid statistical tools in order to compare predictions and measurements and suggests the use of the "American Society for Testing Material (ASTM) D5157: Standard Guide for Statistical Evaluation of Indoor Air Quality Models" for evaluating empirical models [8]. It focuses on the accuracy of indoor concentrations predicted by a model, instead of operational details (for example, the ease of model implementation) [41]. Additionally, it provides details on setting evaluation objectives, statistical instruments for assessing IAQ model performance, choosing datasets for evaluation, and considerations in applying these instruments. Moreover, the standard highlights the idea of using two independent datasets, one for the training process and the other for the validation of the trained model; in this way, model overfitting is avoided. In this study, each model was checked twice in two different periods, thus going beyond the standard criteria. Both types of periods must reach the standard requirements to be classified as an accurate model. This study follows other similar works for the validation of multizone airflow and contaminants [8,45].

The ASTM D5157 Standard provides three statistical instruments for assessing the agreement between predictions and measurements and two others for assessing bias [41]. R^2, NMSE, and the line of regression (m) must be $\geq$0.90, $\leq$0.25, and between 0.75 and 1.25, respectively. The intercept of the average measured concentration, $b/\overline{C}_o$, should be $\leq$25%. The indices to assess bias, FB, and FS have limits equal to $\leq$0.25 and $\leq$0.50, in this order.

In addition to these statistical indices proposed by ASTM D5157, we calculated the mean bias error (MBE) of each combination to understand their accuracy. MBE (Appendix B) indicates how much bias there is in a model and what is its direction. If an MBE value is positive, it means the model overestimates the values in comparison to the observed values. If it is negative, it underestimates the values.

Moreover, in order to understand the degree of errors in the results, we calculated the standard deviation of the measurements (σ) according to the International Performance Measurement and Verification Protocol (IPMVP) [46]. The σ is calculated from the mean value of each measured data for each period (see Appendix C). It represents the precision of the measurement, i.e., how close the measured values are to each other.

5. Measurements Analysis

5.1. Weather Conditions of Each Period

The precision of the weather conditions data is shown in Table 1. P_1_T has 70% of the highest σ values, which confirms it is a valuable period to train the models. Additionally, in this period, EW_MET has the highest dispersion of the data spread around the mean, mean = 4.17, and σ = 2.12. On the other hand, the lowest one in the same period is SW_INSITU with a mean = 0.18, and σ = 0.33, which means it is the most precise measurement. Figures 6–8 demonstrate the wind speed and ΔT curves.

Table 1. Mean (μ) and standard deviation (σ) values of each measured data for each period.

Parameter	Index	P_1_T	P_2_C	P_3_C
CO_2	μ (ppm)	613.75	561.14	629.78
	σ (ppm)	316.80	378.57	278.47
ΔT	μ (°C)	4.80	13.08	11.26
	σ (°C)	10.29	3.17	3.85
W_MET	μ (m/s)	4.54	0.13	0.14
	σ (m/s)	1.99	0	0
WW_MET	μ (m/s)	0.56	0	0
	σ (m/s)	1.78	0	0
EW_MET	μ (m/s)	4.17	0.13	0.14
	σ (m/s)	2.12	0	0
NW_INSITU	μ (m/s)	0.33	0.06	0.34
	σ (m/s)	0.59	0.23	0.64
SW_INSITU	μ (m/s)	0.18	0.13	0.13
	σ (m/s)	0.33	0.11	0.16

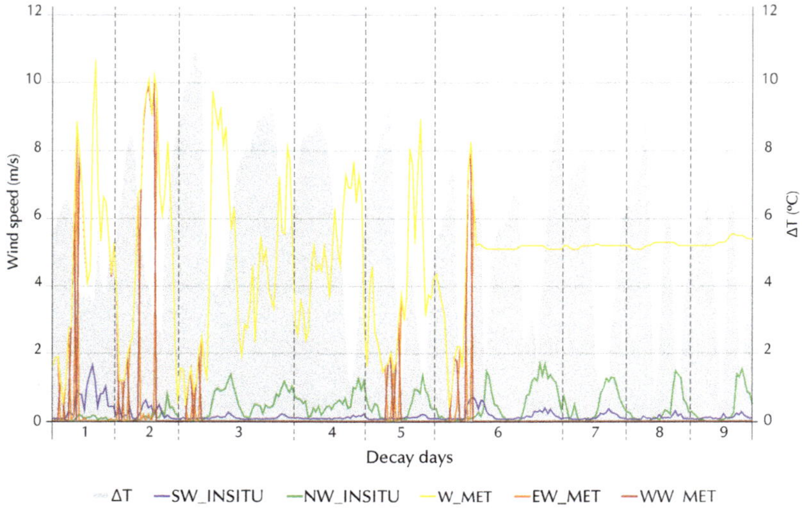

Figure 6. Wind speed and ΔT averages of summer (P_1_T) at ten-minute time-step. This period presents higher wind and ΔT variability, especially in the weather station's wind data, according to which the wind speed sometimes exceeds 10 m/s.

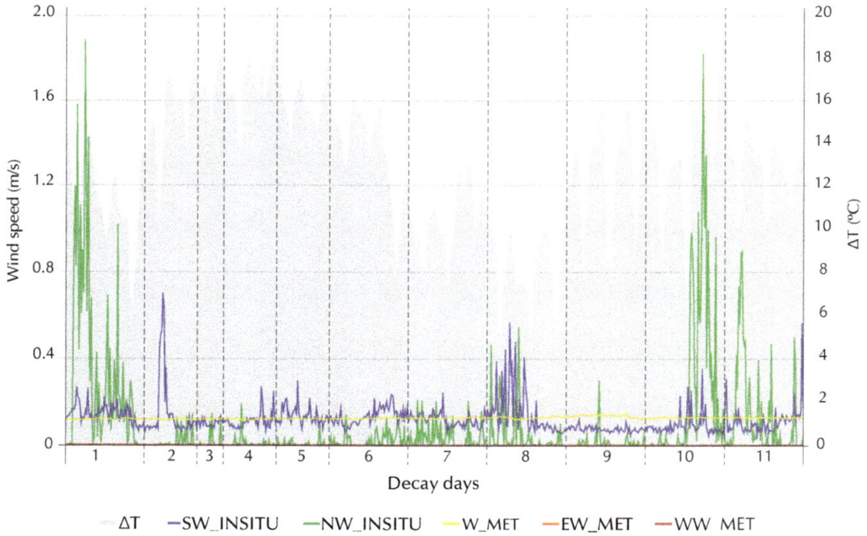

Figure 7. Wind speed and ΔT averages of winter (P_2_C) at ten-minute time-step. Most days have practically the same wind rhythm pattern. In contrast to P_1_T, the weather station data are equal to zero or maintain the same value almost every day. In this figure, it is difficult to see the EW_MET because its curve is behind the W_MET.

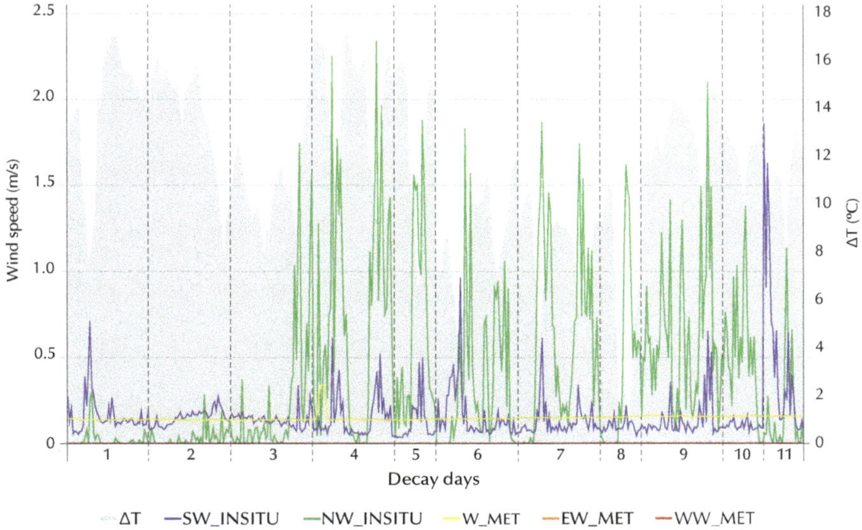

Figure 8. Wind speed and ΔT averages of spring (P_3_C) at ten-minute time-step. As in the second period, there is a rise and fall pattern, which is more evident for W_MET and WW_MET. In this figure, it is difficult to see the EW_MET because its curve is behind the W_MET.

5.2. CO_2 *Uniformity*

The tracer gas was homogeneous in the whole zone. We based the CO_2 uniformity analysis on the method and requirements of ASTM E741 Standard for Air Change Measurements. This standard states that gas concentrations at representative locations throughout the zone should differ by less than 10% of the average concentration for the zone. Table 2 shows the standard deviation (σ) in % of each sensor with respect to the mean value. This confirms that the calculations are according to the standard criteria.

Table 2. Standard deviation (σ) values in % of each sensor from the average CO_2 concentration.

Number	Model	σ
1	OHM HD37VBTV.1	6.78
2	OHM HD37VBTV.1	4.87
3	CO210	4.98
4	CO210	9.15
5	CO210	6.52

5.3. *Daily* CO_2 *Measurements*

Table 3 shows CO_2 concentrations after injection, at the initial point t = 0. The color scale of the standard deviation is according to the values of each period. Although the correlation between the CO_2 concentration and the wind is not linear, and infiltration also depends on the stack effect, it is possible to see some matching between them: When the SW_INSITU and WW_MET speeds are high, the standard deviation of that period is low, as happens in the second day of P_1_T and in the eleventh day of P_3_C. As illustrated in this table, the P_2_C has a high standard deviation on all days and a narrow range of σ values. In contrast, the training period has a wide range of 81 ppm to 632 ppm of standard deviation, which is logical, because one of the main causes of air leakage is the wind speed, and the first nine days present more variable wind speed data than the other periods.

Table 3. CO_2 daily measurements at t = 0 in ppm. C_{bg} means the daily average of the measured outdoor CO_2 concentration in the air; C_o is the average of the observed indoor CO_2 concentration in the space, and σ is the standard deviation of the day in ppm. Red is the highest value and green the lowest one.

Period	Days	C_{bg}	C_0	σ
P_1_T	1	379.19	1478.75	235.83
	2	390.02	1424.80	151.67
	3	378.06	3042.00	632.15
	4	380.58	1407.90	199.49
	5	394.63	1621.75	125.36
	6	382.85	1734.20	197.68
	7	389.84	2510.30	190.40
	8	394.37	1340.95	81.94
	9	389.67	2029.95	342.26
P_2_C	1	387.14	2001.00	384.65
	2	407.09	1992.00	429.72
	3	430.93	2010.00	326.85
	4	423.53	2002.2	389.57
	5	396.32	1996.80	390.57
	6	396.32	1831.20	300.96
	7	396.32	1898.40	344.78
	8	396.32	1509.00	260.68
	9	396.32	1798.80	343.96
	10	396.32	1898.40	389.03
	11	396.32	2002.20	396.81
P_3_C	1	393.89	1995.00	359.09
	2	435.66	1831.20	273.82
	3	408.42	1898.40	316.03
	4	377.11	1264.20	134.08
	5	381.34	1987.80	369.05
	6	385.24	1665.60	209.39
	7	386.35	1383.00	178.41
	8	399.52	1483.80	206.08
	9	373.74	1123.80	106.45
	10	371.24	1522.20	166.58
	11	384.37	1097.40	104.73

6. Results and Discussion

Before introducing the statistical indices proposed by ASTM D5157 for each of the fifteen combinations, it is important to look at the MBE values of the combinations. As shown in Table 4, the models with WW_MET and SW_INSITU present values very close to zero, having the lowest bias to represent the observed CO_2 measurements: DFR with WW_MET, IFC with SW_INSITU, and ELA with WW_MET. On the other hand, the EP models with W_MET, EW_MET, and NW_INSITU data are the most inaccurate combinations, with the lowest MBE values, reaching as low as −81% underestimation, e.g., ELA with NW_INSITU. MBE was calculated only for P_1_T because the coefficients of the models were calculated for this period.

Table 4. MBE values for each combination in P_1_T. The colors of the scale refer to each row, from the lowest (red) to the highest (green) values.

Model	Wind Speed Data				
	W_MET	WW_MET	EW_MET	NW_INSITU	SW_INSITU
DFR	−58.42	0.56	−73.58	−69.20	−4.48
IFC	−65.44	−2.67	−65.43	−80.97	1.42
ELA	−58.99	1.14	−78.68	−81.88	−6.53

All results are presented in Table 5, followed by analysis and dispersion graphs, to facilitate a comparison between the three EnergyPlus models and the five wind inputs. As mentioned before, they were evaluated according to the criteria of the ASTM D5157 Standard, which requires compliance in all periods.

Table 5. Results of the fifteen combinations according to ASTM D5157. The colors are the same as those defined in Figure 5. (Text in red color the values that do not meet the standard.)

Model	Wind	Period	$\overline{C}_o$(ppm)	$\overline{C}_p$(ppm)	R^2	m	b	b/$\overline{C}_o$ (%)	NMSE	FB	FS
DFR	W_MET	P_1_T	613.75	672.17	0.77	0.74	218.75	35.64	0.066	0.091	−0.335
		P_2_C	561.14	561.08	0.99	1.01	−4.11	−0.73	0.025	0.000	0.029
		P_3_C	629.78	633.54	0.93	1.06	−34.09	−5.41	0.017	0.006	0.095
	WW_MET	P_1_T	613.75	613.19	0.98	1.02	−14.32	−2.33	0.005	−0.001	0.062
		P_2_C	561.14	513.62	0.98	1.02	−58.20	−10.37	0.061	−0.088	0.053
		P_3_C	629.78	654.50	0.92	1.05	−5.58	−0.89	0.019	0.038	0.086
	EW_MET	P_1_T	613.75	689.16	0.71	0.71	254.43	41.45	0.083	0.116	−0.340
		P_2_C	561.14	453.47	0.97	1.01	−115.00	−20.49	0.237	−0.212	0.054
		P_3_C	629.78	608.74	0.93	1.08	−69.38	−11.02	0.021	−0.034	0.112
	NW_INSITU	P_1_T	613.75	682.95	0.73	0.71	249.37	40.63	0.077	0.107	−0.369
		P_2_C	561.14	546.98	0.99	1.02	−23.54	−4.19	0.026	−0.026	0.046
		P_3_C	629.78	667.60	0.99	1.02	−21.02	−3.34	0.021	0.058	0.076
	SW_INSITU	P_1_T	613.75	618.23	0.99	1.00	4.34	0.71	0.003	0.007	0.011
		P_2_C	561.14	478.58	0.99	1.03	−97.75	−17.42	0.034	−0.159	0.034
		P_3_C	629.78	617.04	0.99	1.03	−104.72	−16.63	0.014	−0.020	0.096
IFC	W_MET	P_1_T	613.75	679.19	0.68	0.73	230.51	37.56	0.087	0.101	−0.246
		P_2_C	561.14	402.32	0.96	1.00	−158.63	−28.27	0.533	-0.330	0.043
		P_3_C	629.78	625.57	0.93	1.07	−48.55	−7.71	0.019	-0.007	0.105
	WW_MET	P_1_T	613.75	616.42	0.97	1.05	−27.43	−4.47	0.011	0.004	0.130
		P_2_C	561.14	436.66	0.97	1.01	−130.01	−23.17	0.317	−0.250	0.052
		P_3_C	629.78	594.10	0.92	1.08	−86.08	−13.67	0.025	−0.058	0.118
	EW_MET	P_1_T	613.75	679.18	0.68	0.73	230.50	37.56	0.087	0.101	−0.246
		P_2_C	561.14	402.31	0.96	1.00	−158.63	−28.27	0.533	−0.330	0.043
		P_3_C	629.78	560.19	0.91	1.09	−128.74	−20.44	0.041	−0.117	0.136
	NW_INSITU	P_1_T	613.75	694.72	0.66	0.69	271.81	44.29	0.094	0.124	−0.333
		P_2_C	561.14	478.56	0.98	1.01	−90.21	−16.08	0.148	−0.159	0.051
		P_3_C	629.78	600.35	0.98	1.02	−103.60	−16.45	0.017	−0.048	0.092
	SW_INSITU	P_1_T	613.75	612.33	0.99	1.01	−9.64	−1.57	0.004	−0.002	0.041
		P_2_C	561.14	469.69	0.99	1.04	−82.75	−14.75	0.040	−0.177	0.037
		P_3_C	629.78	603.99	0.99	1.03	−111.95	−17.78	0.017	−0.042	0.097
ELA	W_MET	P_1_T	613.75	672.74	0.76	0.73	222.24	36.21	0.068	0.092	−0.336
		P_2_C	561.14	567.94	0.99	1.01	2.83	0.50	0.024	0.012	0.027
		P_3_C	629.78	642.73	0.92	1.05	−19.61	−3.11	0.019	0.020	0.092
	WW_MET	P_1_T	613.75	612.61	0.97	1.04	−25.45	−4.15	0.009	−0.002	0.107
		P_2_C	561.14	388.64	0.98	1.02	−112.34	−20.02	0.214	−0.202	0.058
		P_3_C	629.78	609.01	0.92	1.07	−65.59	−10.41	0.022	−0.034	0.110
	EW_MET	P_1_T	613.75	692.43	0.67	0.68	272.34	44.37	0.093	0.120	−0.353
		P_2_C	561.14	511.58	0.98	1.02	−59.28	−10.56	0.067	−0.092	0.052
		P_3_C	629.78	657.42	0.93	1.04	0.53	0.08	0.018	0.043	0.079
	NW_INSITU	P_1_T	613.75	695.63	0.68	0.69	270.87	44.13	0.091	0.125	−0.345
		P_2_C	561.14	508.33	0.98	1.02	−61.83	−11.02	0.073	−0.099	0.050
		P_3_C	629.78	624.93	0.99	1.02	−70.62	−11.21	0.015	−0.008	0.080
	SW_INSITU	P_1_T	613.75	620.28	0.99	0.98	19.65	3.20	0.004	0.011	−0.029
		P_2_C	561.14	498.13	0.99	1.03	−108.97	−19.42	0.020	−0.119	0.040
		P_3_C	629.78	623.77	0.99	1.04	−82.69	−13.13	0.014	−0.010	0.086

The SW_INSITU with the three infiltration models is the wind data that delivers the best results in all periods. This was expected because these data refer to the wind that impinges directly on the main façade of the test room. The WW_MET was used as the second wind data, and it performs the best in the three seasons and with the three models. These two wind data that meet the ASTM D5157 criteria are related to the room orientations, southeast and southwest, which could be the reason for their good agreement. In Appendix D, it can be seen that these combinations are the only ones that present a quadratic relation of the wind (values of D coefficient), which also could be the reason why these data resulted in the best CO_2 prediction. Although P_1_T presents a lower standard deviation (316.80 ppm) of CO_2 than P_2_C (378.57 ppm), EP models with SW_INSITU and WW_MET are still capable of predicting the measured CO_2 concentration.

On the other hand, W_MET, EW_MET, and NW_INSITU do not approve the standard requirements in P_1_T with any air leakage object, presenting values of R^2 from 23% to 29%, worse than SW_INSITU in the same period. As aforementioned, these wind data are also the most inaccurate ones. The data variability of P_1_T requires the right wind data to produce high-quality models. Dispersion graphs in Figure 9 show the good and the bad correlation between the predicted and measured CO_2 concentration in the first period provided by DFR using SW_INSITU and EW_MET, respectively.

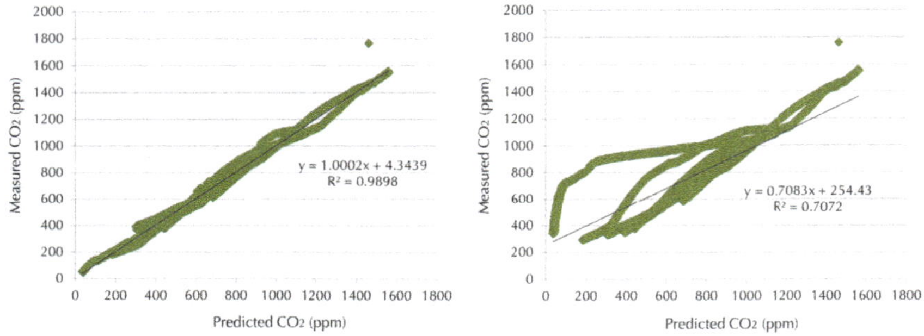

(**a**) SW_INSITU. (**b**) EW_MET.

Figure 9. Correlation between predicted and measured CO_2 concentration by DFR models with SW_INSITU (**a**) and EW_MET (**b**) in P_1_T.

Similar results are found when combining the IFC and ELA objects with the five wind data. W_MET, EW_MET, and NW_INSITU again do not meet the ASTM D5157 criteria, as many parameters such as R^2, m, and $b/\overline{C}_o(\%)$ were outside the requirements in P_1_T. Furthermore, in P_2_C the IFC model with the wind data from the weather station (W_MET, EW_MET, and WW_MET) have NMSE values higher than those demanded by the standard. Despite this, IFC with WW_MET has the best NMSE value with a difference of 21% with respect to the limit (0.25), while the others have more than twice the difference. Dispersion graphs in Figures 10 and 11 clearly illustrate the distinction between the best combination (IFC with WW_MET and ELA with SW_INSITU), and the worst (IFC with NW_INSITU, and ELA with W_MET).

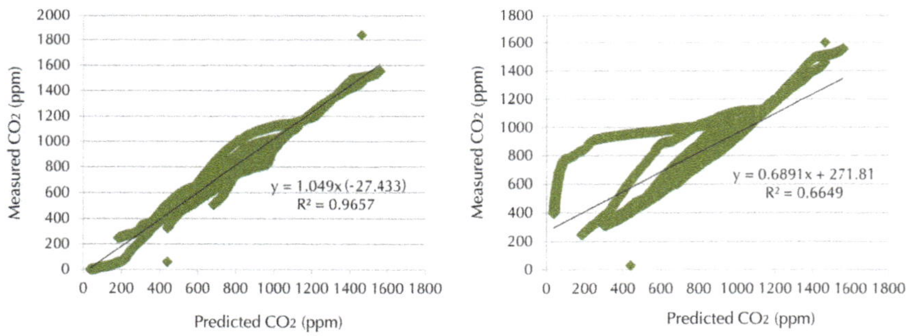

(**a**) WW_MET. (**b**) NW_INSITU.

Figure 10. Correlation between predicted and measured CO_2 concentration by IFC models with WW_MET and NW_INSITU in P_1_T.

In summary, the wind data W_MET, EW_MET, and NW_INSITU are inadequate to represent the actual infiltration of this case study. This is especially noteworthy in the case of W_MET because it is the global wind used by any BEM software and many energy modellers. There is useful information inside W_MET, but when applied without wind disaggregation, it produces a misleading effect. This raises doubts about the direct use of this type of wind in energy estimations and calibration of energy models without wind disaggregation. Therefore, for this study, the WW_MET ASTM D5157 approval in the training period represents an alternative and cost-effective option to select the best wind data to calculate air leakage, in case it is not possible to install in situ sensors. Figure 12 summarises the results of this research.

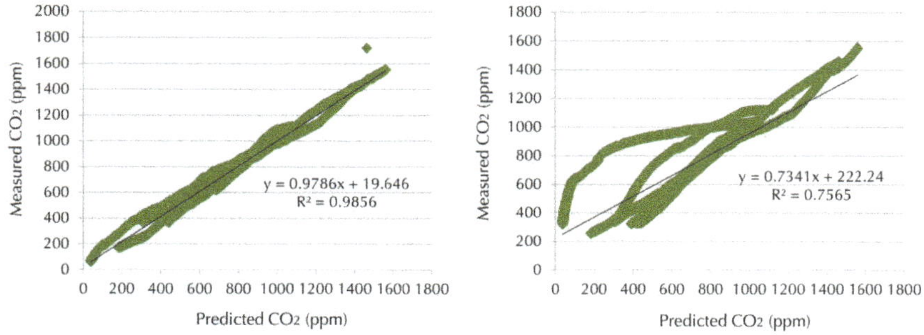

(**a**) SW_INSITU. (**b**) W_MET.

Figure 11. Correlation between predicted and measured CO_2 concentration by ELA models with SW_INSITU and W_MET in P_1_T.

It is noteworthy that in this study, we did not analyse whether the wind data and the infiltration model were wrong or right. We only highlighted that since our main purpose was to calibrate building energy models, we needed real-time wind data; otherwise, the infiltration values would not be accurate in high-time estimations. On the other hand, for annual energy simulations, the application of the TMY format might be suitable.

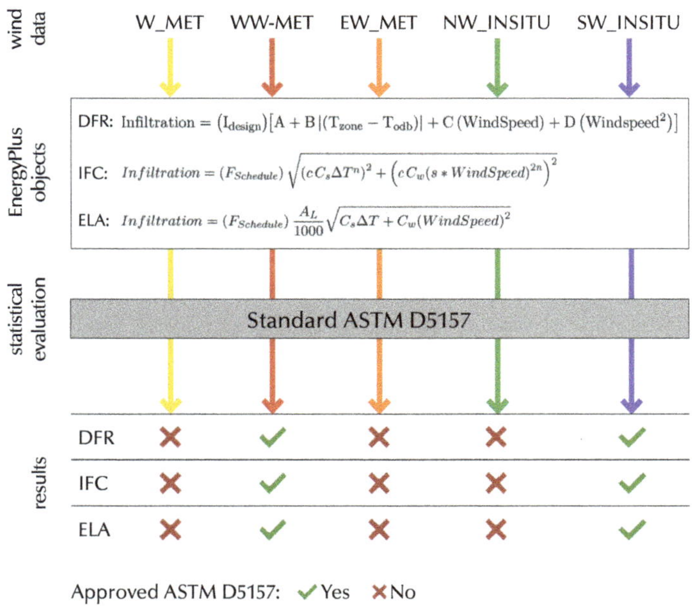

Figure 12. Summary of the paper results.

7. Conclusions

In this paper, the application of the wind data from a weather file (herein as global wind) to estimate the infiltration of a test space was evaluated. To calculate air leakage, three infiltration models of EnergyPlus with algebraic equations were used: DesignFlowRate (DFR), FlowCoefficient (IFC), and EffectiveLeakageArea (ELA). These models were applied with four other wind data: from the weather file, eastbound and westbound wind data were used, as well as the wind data from in situ sensors (on southeast and northwest façades), to understand which combination delivered the best result. All fifteen combinations were empirically evaluated according to an experiment of CO_2 concentration decay carried out in a test room. The empirical validation of the results was carried out taking into account the requirements of ASTM D5157 to evaluate IAQ in three periods: The summer data were used as training, and winter and spring data were used for validation.

As far as the authors know, the empirical comparison of different wind data with EnergyPlus infiltration models, as well as the application of ASTM D5157 to assess these models, have never been carried out before. The results are specific to this test room and the main conclusions are as follows:

- The best combinations and the only ones that meet the ASTM D5157 criteria use the wind speed measured in situ on the southeast façade and the west wind from the weather file. Both wind data are related to the unsealed orientations of the test space.
- The use of the global wind from the weather file is not the most accurate option to estimate infiltration with any EP model. This raises doubts about the use of this wind in the calibration of energy models without wind disaggregation.
- Global wind disaggregation is a good and cost-effective strategy to apply with EnergyPlus air leakage models and results in accurate infiltration values.

Further research should be carried out to verify whether these results occur in other rooms of the same apartment, and other buildings as well. In addition, the application of disaggregated wind data should be implemented in the calibration process of BEMs, by using a schedule activation in EnergyPlus to accurately account for infiltration and achieve better energy predictions. As a recommendation, EnergyPlus should take wind direction into account in air leakage estimates, just as it does in calculating façade convection coefficients.

Author Contributions: Conceptualisation, G.B.P. and C.F.B.; methodology, G.B.P. and C.F.B.; software, G.B.P. and C.F.B.; validation, G.B.P. and C.F.B.; investigation, G.B.P. and C.F.B.; resources, G.B.P. and C.F.B; writing—original draft preparation, G.B.P. and C.F.B.; writing—review and editing, G.B.P. and C.F.B.; supervision, C.F.B.; project administration, C.F.B.; funding acquisition, C.F.B. All authors have read and agreed to the published version of the manuscript.

Funding: The author, Gabriela Bastos Porsani, has received funding from a PhD scholarship programme called "Asociación de Amigos de la Universidad de Navarra" of the University of Navarra, Spain.

Data Availability Statement: Data sharing is not applicable to this article.

Conflicts of Interest: The authors declare no conflicts of interest.

Abbreviations

The following abbreviations are used in this manuscript:

EU	European Union
NGEU	Next-Generation EU
RRF	Recovery and Resilience Facility
EPBD	Energy Performance of Buildings Directive
BIM	Building Information Modelling
BEM	Building Energy Model
ASHRAE	American Society of Heating, Refrigerating and Air-Conditioning Engineers
CFD	Computational Fluid Dynamics
AFN	AirFlowNetwork
HVAC	Heating, Ventilation, and Air-conditioning
NIST	National Institute of Standards and Technology
EP	EnergyPlus
TMY	Typical Meteorological Year
BEopt	Building Energy Optimisation

ASTM	American Society for Testing Material
MBE	Mean Bias Error
ppm	Parts Per Million
°	Degrees
°C	Celsius Degrees
m	Metre
T	Temperature
m/s	Metres per Second
%	Percentage
DFR	ZoneInfiltration: DesignFlowRate
IFC	ZoneInfiltration: FlowCoefficient
ELA	ZoneInfiltration: EffectiveLeakageArea
MAE	Mean Absolute Error
IAQ	Indoor Air Quality
IPMVP	International Performance Measurement and Verification Protocol

Appendix A. EnergyPlus Infiltration Models

- DFR
  The most commonly used infiltration model is the DFR based on the work developed by Coblenz and Achenbach [47]. The general equation is as follows:

$$I = \left(I_{design}\right)(F_{sch})\left[A + B|(T_{zone} - T_{odb})| + C(WS) + D\left(WS^2\right)\right] \tag{A1}$$

  where
  I_{design} = is the design infiltration rate in air changes/hour;
  F_{sch} = is the infiltration schedule;
  T_{zone} and T_{odb} = are temperatures °C, the absolute difference in temperature between the average dry bulb of the zone and the average outdoor dry bulb;
  WS = is the wind speed in m/s.

- IFC
  Another air leakage implementation in EnergyPlus and some other programs is based on the AIM-2 model by Walker and Wilson [48]. It is presented in the ZoneInfiltration: FlowCoefficient object of EnergyPlus and can be expressed as follows:

$$I = (F_{schedule})\sqrt{(cC_s\Delta T^n)^2 + \left(cC_w(s \times WS)^{2n}\right)^2} \tag{A2}$$

  where
  $F_{Schedule}$ is a value from a user-defined schedule;
  c is the flow coefficient in $m^3/(sPa^n)$;
  C_s is the coefficient for stack-induced infiltration in $(Pa/K)^n$;
  ΔT is the absolute difference in temperature between the average dry bulb of the zone and the average outdoor dry bulb;
  n is the pressure exponent;
  C_w is the coefficient for wind-induced infiltration in $(Pas^2/m^2)^n$;
  s is the shelter factor;
  WS is the local wind speed.

- ELA
  Furthermore, EnergyPlus and other whole-building energy software programmes implement infiltration based on the effective leakage area calculation in the ASTM Standard E779 [49]. Sherman and Grimsrud developed correlations for small detached residential buildings [50]. The ELA equation is as follows:

$$I = (F_{schedule})\frac{A_L}{1000}\sqrt{C_s\Delta T + C_w(WS)^2} \tag{A3}$$

where
$F_{Schedule}$ is a value from a user-defined schedule;
A_L is the effective air leakage area in cm^2 that corresponds to a 4 Pa pressure differential;
C_s is the coefficient for stack-induced infiltration in $(L/s)^2/(cm^4K)$;
ΔT is the absolute difference in temperature between the average dry bulb of the zone and the average outdoor dry bulb;
C_w is the coefficient for wind-induced infiltration in $(L/s)^2/(cm^4(m/s)^2)$;
WS is the local wind speed.

Appendix B. Mean Bias Error (MBE)

$$MBE = \frac{1}{n}\sum_{i=1}^{n}(C_{pi} - C_{oi}) \quad \text{(A4)}$$

where
n is the number of time-steps for P_1_T;
C_pi is the predicted CO_2 concentration in ppm;
C_oi is the observed CO_2 concentration in ppm.

Appendix C. Standard Deviation (σ)

$$\sigma = \sqrt{\frac{\sum(C_{oi} - \overline{C}_o)^2}{n-1}} \quad \text{(A5)}$$

where
n is the number of time-steps of each period;
C_oi is the observed CO_2 concentration in ppm;
$\overline{C}_o$ is the mean observed CO_2 concentration in ppm.

Appendix D. Coefficients of the Infiltration Models

The following tables present the coefficients of DFR, IFC, and ELA by wind data for the training period.

Table A1. Coefficients of DFR Equation (A1) by wind data.

Model	Wind	I	A	B	C	D
DFR	W_MET	1	0.000541	0.000061	0.000002	0.000012
	WW_MET	1	0.000800	0.000054	0	0.000172
	EW_MET	1	0.000656	0.000083	0	0.000001
	NW_INSITU	1	0.000929	0.000035	0.000098	0
	SW_INSITU	1	0.000583	7.52×10^{-5}	0	0.002486

Table A2. Coefficients of IFC Equation (A2) by wind data.

Model	Wind	c	s	C_s	C_w	n
IFC	W_MET	6.0128×10^{-5}	0	7.178857	0.521321	0.600000
	WW_MET	0.06115258	0.101024	0.006429	0.267782	0.600000
	EW_MET	6.0128×10^{-5}	0	7.178973	0.521321	0.600000
	NW_INSITU	0.00198763	0.663317	0.175831	0.726885	0.600000
	SW_INSITU	9.0681×10^{-5}	7.677816	3.746957	4.536288	0.600000

Table A3. Coefficients of ELA Equation (A3) by wind data.

Model	Wind	A_L	C_s	C_w
ELA	W_MET	0.61379371	0.358482	0.070889
	WW_MET	1.06673691	0.203145	1.767608
	EW_MET	1.58276916	0.070587	0.005655
	NW_INSITU	2.19812103	0.036404	0.176512
	SW_INSITU	2.10798779	0.035641	4.061227

References

1. Bekker, S. The EU's recovery and resilience facility: A next phase in EU socioeconomic governance? *Politics Gov.* **2021**, *9*, 175–185.
2. Unión Europea Reglamento (UE) 2021/241 del Parlamento Europeo y del Consejo de 12 de febrero de 2021 por el que se establece el Mecanismo de Recuperación y Resiliencia. *Diario Oficial de la Unión Europea del*, 30 September 2010, pp. 1–59.
3. De la Porte, C.; Jensen, M.D. The next generation EU: An analysis of the dimensions of conflict behind the deal. *Soc. Policy Adm.* **2021**, *55*, 388–402. https://doi.org/https://doi.org/10.1111/spol.12709.
4. Bastos Porsani, G.; Del Valle de Lersundi, K.; Sánchez-Ostiz Gutiérrez, A.; Fernández Bandera, C. Interoperability between Building Information Modelling (BIM) and Building Energy Model (BEM). *Appl. Sci.* **2021**, *11*, 2167.
5. Ruiz, G.R.; Bandera, C.F.; Temes, T.G.A.; Gutierrez, A.S.O. Genetic algorithm for building envelope calibration. *Appl. Energy* **2016**, *168*, 691–705.
6. Fernández Bandera, C.; Ramos Ruiz, G. Towards a new generation of building envelope calibration. *Energies* **2017**, *10*, 2102.
7. Manfren, M.; Aste, N.; Moshksar, R. Calibration and uncertainty analysis for computer models–a meta-model based approach for integrated building energy simulation. *Appl. Energy* **2013**, *103*, 627–641.
8. The American Society of Heating, Refrigerating and Air-Conditioning Engineers. *The 2017 ASHRAE Handbook—Fundamentals*; ASHRAE: Peachtree Corners, GA, USA, 2017.
9. Du, H.; Jones, P.; Segarra, E.L.; Bandera, C.F. Development of a REST API for obtaining site-specific historical and near-future weather data in EPW format. In Proceedings of the 4th IBPSA-England Conference on Building Simulation and Optimization, Cambridge, UK, 11–12 September 2018.
10. Bhandari, M.; Shrestha, S.; New, J. Evaluation of weather datasets for building energy simulation. *Energy Build.* **2012**, *49*, 109–118.
11. Segarra, E.L.; Ruiz, G.R.; González, V.G.; Peppas, A.; Bandera, C.F. Impact Assessment for Building Energy Models Using Observed vs. Third-Party Weather Data Sets. *Sustainability* **2020**, *12*, 6788.
12. Gutiérrez, V.; Ramos Ruiz, G.; Fernández Bandera, C. Impact of Actual Weather Datasets for Calibrating White-Box Building Energy Models Base on Monitored Data. *Energies* **2021**, *14*, 1187.
13. González, V.G.; Ruiz, G.R.; Segarra, E.L.; Gordillo, G.C.; Bandera, C.F. Characterization of Building Foundation in Building Energy Models. In Proceedings of the Building Simulation, Rome, Italy, 2–4 September 2019.
14. Lee, S.H.; Hong, T. Validation of an inverse model of zone air heat balance. *Build. Environ.* **2019**, *161*, 106232.
15. Hong, T.; Lee, S.H. Integrating physics-based models with sensor data: An inverse modeling approach. *Build. Environ.* **2019**, *154*, 23–31.
16. Taddeo, P.; Ortiz, J.; Salom, J.; Segarra, E.L.; González, V.G.; Ruiz, G.R.; Bandera, C.F. Comparison of experimental methodologies to estimate the air infiltration rate in a residential case study for calibration purposes. In Proceedings of the 39th AIVC 2018-Smart Ventilation for Buildings, Antibes Juan-Les-Pins, France, 18–19 September 2018; p. 68.
17. Han, G.; Srebric, J.; Enache-Pommer, E. Different modeling strategies of infiltration rates for an office building to improve accuracy of building energy simulations. *Energy Build.* **2015**, *86*, 288–295.
18. Davenport, A.; Hui, H. *External and Internal Wind Pressures on Buildings*; BLWT820133; Boundary Layer Wind Tunnel Laboratory, University of Western Ontario: London, ON, Canada, 1982.
19. Raman, G.; Chelliah, K.; Prakash, M.; Muehleisen, R.T. Detection and quantification of building air infiltration using remote acoustic methods. In *INTER-NOISE and NOISE-CON Congress and Conference Proceedings*; Institute of Noise Control Engineering: Washington, DC, USA, 2014; Volume 249, pp. 3976–3985.
20. Persily, A.K.; Emmerich, S.J. Energy Impacts of Infiltration and Ventilation in US Office Buildings Using Multizone Airflow Simulation. *Proc. Iaq Energy* **1999**, *98*, 191–206.
21. Miszczuk, A.; Heim, D. Parametric study of air infiltration in residential buildings—The effect of local conditions on energy demand. *Energies* **2021**, *14*, 127.
22. Jokisalo, J.; Kurnitski, J.; Korpi, M.; Kalamees, T.; Vinha, J. Building leakage, infiltration, and energy performance analyses for Finnish detached houses. *Build. Environ.* **2009**, *44*, 377–387.
23. Wilkki, C.M.; Reeve, N. *Communication from the Commission to the European Parliament, the Council, the European Economic and Social Committee and the Committee of the Regions on European Missions European Commission Directorate-General for Research and Innovation Directorate G—Common Policy Centre*; European Commission: Brussels, Belgium, 2021.
24. DoE. *EnergyPlus Engineering Reference: The Reference to EnergyPlus Calculations*; Lawrence Berkeley National Laboratory: Berkeley, CA, USA, 2021.
25. Ng, L.C.; Quiles, N.O.; Dols, W.S.; Emmerich, S.J. Weather correlations to calculate infiltration rates for US commercial building energy models. *Build. Environ.* **2018**, *127*, 47–57.

26. Dols, W.S.; Emmerich, S.J.; Polidoro, B.J. Coupling the multizone airflow and contaminant transport software CONTAM with EnergyPlus using co-simulation. *Build. Simul.* **2016**, *9*, 469–479.
27. Ng, L.C.; Dols, W.S.; Emmerich, S.J. Evaluating potential benefits of air barriers in commercial buildings using NIST infiltration correlations in EnergyPlus. *Build. Environ.* **2021**, *196*, 107783.
28. Shrestha, S.; Hun, D.; Moss, C. Modeling Whole Building Air Leakage and Validation of Simulation Results against Field Measurements. In *Whole Building Air Leakage: Testing and Building Performance Impacts*; ASTM International: West Conshohocken, PA, USA, 2019.
29. Bae, Y.; Joe, J.; Lee, S.; Im, P.; Ng, L. *Evaluation of Existing Infiltration Models Used in Building Energy Simulation*; Technical report; Oak Ridge National Lab (ORNL): Oak Ridge, TN, USA.
30. Winkler, J.M.; Horowitz, S.G.; DeGraw, J.W.; Merket, N.D. *Evaluating EnergyPlus Airflow Network Model for Residential Ducts, Infiltration, and Interzonal Airflow*; Technical report; National Renewable Energy Lab (NREL): Golden, CO, USA, 2017.
31. BEopt: Building Energy Optimization Tool. Available online: https://www.nrel.gov/buildings/beopt.html (accessed on 2 December 2022).
32. *ASTM 11 (2017). Standard Test Method for Determining Air Change in a Single Zone by Means of a Tracer Gas Dilution. Standard E741*; Technical report; American Society for Testing and Materials: West Conshohocken, PA, USA, 2017.
33. Sherman, M.H. Tracer-gas techniques for measuring ventilation in a single zone. *Build. Environ.* **1990**, *25*, 365–374.
34. Sherman, M. Uncertainty in air flow calculations using tracer gas measurements. *Build. Environ.* **1989**, *24*, 347–354.
35. Sherman, M. On the estimation of multizone ventilation rates from tracer gas measurements. *Build. Environ.* **1989**, *24*, 355–362.
36. Chao, C.Y.; Wan, M.; Law, A.K. Ventilation performance measurement using constant concentration dosing strategy. *Build. Environ.* **2004**, *39*, 1277–1288.
37. Cui, S.; Cohen, M.; Stabat, P.; Marchio, D. CO_2 tracer gas concentration decay method for measuring air change rate. *Build. Environ.* **2015**, *84*, 162–169.
38. Gowri, K.; Winiarski, D.W.; Jarnagin, R.E. *Infiltration Modeling Guidelines for Commercial Building Energy Analysis*; Technical report; Pacific Northwest National Lab (PNNL): Richland, WA, USA, 2009.
39. Hunt, C. Air infiltration: A review of some existing measurement techniques and data. In *Building Air Change Rate and Infiltration Measurements*; American Society for Testing and Materials: West Conshohocken, PA, USA, 1980.
40. Duda, S. Common Evaluation Metrics for Regression Analysis, 2021. Available online: https://scottmduda.medium.com/common-evaluation-metrics-for-regression-analysis-4b62726f1aad (accessed on 2 December 2022).
41. *ASTM 2019. Standard Guide for Statistical Evaluation of Indoor Air Quality Models. Standard D5157*; Technical report; American Society for Testing and Materials: West Conshohocken, PA, USA, 2019.
42. Ng, L.C.; Persily, A.K.; Emmerich, S.J. Improving infiltration modeling in commercial building energy models. *Energy Build.* **2015**, *88*, 316–323.
43. Remion, G.; Moujalled, B.; El Mankibi, M. Review of tracer gas-based methods for the characterization of natural ventilation performance: Comparative analysis of their accuracy. *Build. Environ.* **2019**, *160*, 106180.
44. The American Society of Heating, Refrigerating and Air-Conditioning Engineers. *Handbook*; ASHRAE: Peachtree Corners, GA, USA, 2017.
45. Emmerich, S.; Howard-Reed, C.; Nabinger, S. Validation of multizone IAQ model predictions for tracer gas in a townhouse. *Build. Serv. Eng. Res. Technol.* **2004**, *25*, 305–316.
46. E.V.O. *Uncertainty Assessment for IPMVP*; E.V.O.: Washington, DC, USA, 2018, pp. 1–82.
47. Achenbach, P.R.; Coblenz, C. Field measurements of air infiltration in ten electrically heated houses. *Ashrae Trans.* **1963**, *69*, 358–365.
48. Walker, I.S.; Wilson, D.J. Field validation of algebraic equations for stack and wind driven air infiltration calculations. *HVAC&R Res.* **1998**, *4*, 119–139.
49. *ASTM (2019). Standard Test Method for Determining Air Leakage Rate by Fan Pressurization. Standard E779*; Technical report; American Society for Testing and Materials: West Conshohocken, PA, USA, 2019.
50. Sherman, M.H. *Infiltration-Pressurization Correlation: Simplified Physical Modeling*; Lawrence Berkeley National Laboratory: Berkeley, CA, USA, 1980.

MDPI
St. Alban-Anlage 66
4052 Basel
Switzerland
www.mdpi.com

MDPI Books Editorial Office
E-mail: books@mdpi.com
www.mdpi.com/books

www.ingramcontent.com/pod-product-compliance
Lightning Source LLC
Chambersburg PA
CBHW061444220926
51850CB00004B/9
9783036596242